WILLIAM F. MAAG LIBRARY
YOUNGSTOWN STATE UNIVERSITY

ANNUAL REVIEW OF PHYSICAL CHEMISTRY

EDITORIAL COMMITTEE (1982)

ANDREAS C. ALBRECHT
DONALD M. CROTHERS
GEORGE W. FLYNN
ROBERT GOMER
B. SEYMOUR RABINOVITCH
J. MICHAEL SCHURR
WILLIAM STEELE
HERBERT L. STRAUSS

Responsible for the organization of Volume 33
(Editorial Committee, 1980)

ANDREAS C. ALBRECHT
VICTOR A. BLOOMFIELD
GEORGE W. FLYNN
ROBERT GOMER
B. SEYMOUR RABINOVITCH
J. MICHAEL SCHURR
WILLIAM STEELE
HERBERT L. STRAUSS
HANS C. ANDERSON (Guest)
WILLIAM A. GODDARD III (Guest)

Production Editor	SUZANNE PRIES COPENHAGEN
Indexing Coordinator	MARY A. GLASS
Subject Indexer	STEVEN SORENSEN

ANNUAL REVIEW OF PHYSICAL CHEMISTRY

VOLUME 33, 1982

B. SEYMOUR RABINOVITCH, *Editor*
University of Washington

J. MICHAEL SCHURR, *Associate Editor*
University of Washington

HERBERT L. STRAUSS, *Associate Editor*
University of California, Berkeley

ANNUAL REVIEWS INC.   4139 EL CAMINO WAY   PALO ALTO, CALIFORNIA 94306 USA

ANNUAL REVIEWS INC.
Palo Alto, California, USA

COPYRIGHT © 1982 BY ANNUAL REVIEWS INC., PALO ALTO, CALIFORNIA, USA. ALL RIGHTS RESERVED. The appearance of the code at the bottom of the first page of an article in this serial indicates the copyright owner's consent that copies of the article may be made for personal or internal use, or for the personal or internal use of specific clients. This consent is given on the condition, however, that the copier pay the stated per-copy fee of $1.00 per article through the Copyright Clearance Center, Inc. (21 Congress Street, Salem, MA 01970) for copying beyond that permitted by Sections 107 or 108 of the US Copyright Law. The per-copy fee of $1.00 per article also applies to the copying, under the stated conditions, of articles published in any Annual Review serial before January 1, 1978. Individual readers, and nonprofit libraries acting for them, are permitted to make a single copy of an article without charge for use in research or teaching. This consent does not extend to other kinds of copying, such as copying for general distribution, for advertising or promotional purposes, for creating new collective works, or for resale. For such uses, written permission is required. Write to Permissions Dept., Annual Reviews Inc., 4139 El Camino Way, Palo Alto, CA 94306 USA.

International Standard Serial Number: 0066-426X
International Standard Book Number: 0-8243-1033-0
Library of Congress Catalog Card Number: A-51-1658

Annual Review and publication titles are registered trademarks of Annual Reviews Inc.

Annual Reviews Inc. and the Editors of its publications assume no responsibility for the statements expressed by the contributors to this Review.

PRINTED AND BOUND IN THE UNITED STATES OF AMERICA

PREFACE

The contents of the present volume illustrate anew the remarkable breadth of application of physical chemistry throughout the basic physical, chemical, and biological sciences. Progress in understanding diverse phenomena at almost startling levels of complexity, resolution, and molecular detail is chronicled in these chapters. The Editorial Committee gratefully acknowledges the unselfish efforts of the contributing authors in their difficult task of organizing and communicating so well these advances to the scientific community.

The Production Editor, Suzanne Pries Copenhagen, receives our special thanks for her indispensable contributions; also, we thank Steven Sorensen, who prepared the subject index, and others of the staff who have assisted us.

THE EDITORIAL COMMITTEE

SOME RELATED ARTICLES IN OTHER *ANNUAL REVIEWS*

From the *Annual Review of Biochemistry*, Volume 51 (1982)
  Proton Electrochemical Gradients and Energy-Transduction Processes, S. J. Ferguson and M. C. Sorgato
  The Three-Dimensional Structure of DNA, S. B. Zimmerman
  Molecular Mechanisms in Genetic Recombination, D. Dressler and H. Potter
  Magnetic Resonance Studies of Active Sites in Enzymic Complexes, M. Cohn and G. H. Reed
  The Conformation, Flexibility, and Dynamics of Polypeptide Hormones, T. Blundell and S. Wood
  Specific Intermediates in the Folding Reactions of Small Proteins and the Mechanism of Protein Folding, P. S. Kim and R. L. Baldwin

From the *Annual Review of Biophysics and Bioengineering*, Volume 11 (1982)
  High Pressure Effects on Proteins and Other Biomolecules, K. Heremans
  ^{18}O and ^{17}O Effects on ^{31}P NMR as Probes of Enzymatic Reactions of Phosphate Compounds, M. Cohn
  Light Scattering and Absorption as Methods of Studying Cell Population Parameters, P. Latimer
  Diffusion-Enhanced Fluorescence Energy Transfer, L. Stryer, D. D. Thomas, and C. F. Meares
  Optical Detection of Magnetic Resonance in Biologically Important Molecules, A. L. Kwiram and J. B. Alexander Ross

From the *Annual Review of Earth and Planetary Science*, Volume 10 (1982)
  Dynamical Constraints of the Formation and Evolution of Planetary Bodies, A. W. Harris and W. R. Ward
  Interiors of the Giant Planets, D. J. Stevenson
  Heat and Mass Circulation in Geothermal Systems, I. G. Donaldson
  Applications of the Ion Microprobe to Geochemistry and Cosmochemistry, N. Shimizu and S. R. Hart

From the *Annual Review of Materials Science*, Volume 12 (1982)
  Recent Developments in Rapidly Melt-Quenched Crystalline Alloys, R. W. Cahn
  Metalorganic Chemical Vapor Deposition, P. D. Dapkus
  Rutherford Backscattering and Channeling Analysis of Interfaces and Epitaxial Structures, L. C. Feldman and J. M. Poate
  Magnetic Susceptabilities of Highly Conducting One-Dimensional Materials, W. E. Hatfield and L. W. ter Haar
  Recombination-Enhanced Reactions in Semiconductors, D. V. Lang
  Transition Metal Oxide Gels and Colloids, L. Livage and J. Lemerle

From the *Annual Review of Nuclear and Particle Science*, Volume 32 (1982)
  Beam-Foil Spectroscopy, H. G. Berry and M. Hass
  Statistical Spectroscopy, J. B. French and V. K. B. Kota
  Electric-Dipole Moments of Particles, N. F. Ramsey

CONTENTS

THE WAY IT WAS, *Joseph E. Mayer*	1
HIGH PRESSURE STUDIES OF MOLECULAR LUMINESCENCE, *H. G. Drickamer*	25
DYNAMICS OF ENTANGLED POLYMER CHAINS, *P. G. de Gennes and L. Léger*	49
DYNAMICS OF MOLECULES IN CONDENSED PHASES: PICOSECOND HOLOGRAPHIC GRATING EXPERIMENTS, *M. D. Fayer*	63
MOLECULAR FEATURES OF METAL CLUSTER REACTIONS, *E. L. Muetterties, R. R. Burch, and A. M. Stolzenberg*	89
PHOTOFRAGMENT ALIGNMENT AND ORIENTATION, *Chris H. Greene and Richard N. Zare*	119
ELECTRONIC STRUCTURE CALCULATIONS USING THE $X\alpha$ METHOD, *David A. Case*	151
COLLISIONS OF RYDBERG ATOMS WITH MOLECULES, *F. B. Dunning and R. F. Stebbings*	173
POLYELECTROLYTE THEORIES AND THEIR APPLICATIONS TO DNA, *Charles F. Anderson and M. Thomas Record, Jr.*	191
COMPLEX COORDINATES IN THE THEORY OF ATOMIC AND MOLECULAR STRUCTURE AND DYNAMICS, *William P. Reinhardt*	223
LIGHT AND RADICAL IONS, *Terry A. Miller*	257
DYNAMICS OF PROTEINS, *P. G. Debrunner and H. Frauenfelder*	283
RECENT DEVELOPMENTS IN ELECTRON PARAMAGNETIC RESONANCE: TRANSIENT METHODS, *S. I. Weissman*	301
SOME PHYSICAL STUDIES OF THE CONTRACTILE MECHANISM IN MUSCLE, *Manuel F. Morales, Julian Borejdo, Jean Botts, Roger Cooke, Robert A. Mendelson, and Reiji Takashi*	319
RESONANCE RAMAN SCATTERING: THE MULTIMODE PROBLEM AND TRANSFORM METHODS, *P. M. Champion and A. C. Albrecht*	353
IONIZATION IN SOLUTION BY PHOTOACTIVATED ELECTRON TRANSFER, *D. Mauzerall and S. G. Ballard*	377
THEORIES OF THE DYNAMICS OF PHOTODISSOCIATION, *M. Shapiro and R. Bersohn*	409
POLYACETYLENE, $(CH)_x$: THE PROTOTYPE CONDUCTING POLYMER, *S. Etemad, A. J. Heeger, and A. G. MacDiarmid*	443
TIME-RESOLVED RESONANCE RAMAN STUDIES OF HEMOGLOBIN, *J. M. Friedman, D. L. Rousseau, and M. R. Ondrias*	471
HYDROCARBON BOND DISSOCIATION ENERGIES, *Donald F. McMillen and David M. Golden*	493
SOLID STATE NMR OF BIOLOGICAL SYSTEMS, *Stanley J. Opella*	533
INDEXES	
Author Index	563
Subject Index	580
Cumulative Index of Contributing Authors, Volumes 29–33	592
Cumulative Index of Chapter Titles, Volumes 29–33	594

ANNUAL REVIEWS INC. is a nonprofit scientific publisher established to promote the advancement of the sciences. Beginning in 1932 with the *Annual Review of Biochemistry,* the Company has pursued as its principal function the publication of high quality, reasonably priced *Annual Review* volumes. The volumes are organized by Editors and Editorial Committees who invite qualified authors to contribute critical articles reviewing significant developments within each major discipline. The Editor-in-Chief invites those interested in serving as future Editorial Committee members to communicate directly with him. Annual Reviews Inc. is administered by a Board of Directors, whose members serve without compensation.

1982 Board of Directors, Annual Reviews Inc.

Dr. J. Murray Luck, Founder and Director Emeritus of Annual Reviews Inc.
Professor Emeritus of Chemistry, Stanford University
Dr. Joshua Lederberg, President of Annual Reviews Inc.
President, The Rockefeller University
Dr. James E. Howell, Vice President of Annual Reviews Inc.
Professor of Economics, Stanford University
Dr. William O. Baker, *Retired Chairman of the Board, Bell Laboratories*
Dr. Robert W. Berliner, *Dean, Yale University School of Medicine*
Dr. Winslow R. Briggs, *Director, Carnegie Institution of Washington, Stanford*
Dr. Sidney D. Drell, *Deputy Director, Stanford Linear Accelerator Center*
Dr. Eugene Garfield, *President, Institute for Scientific Information*
Dr. Conyers Herring, *Professor of Applied Physics, Stanford University*
Dr. D. E. Koshland, Jr., *Professor of Biochemistry, University of California, Berkeley*
Dr. Gardner Lindzey, *Director, Center for Advanced Study in the Behavioral Sciences, Stanford*
Dr. William D. McElroy, *Professor of Biology, University of California, San Diego*
Dr. William F. Miller, *President, SRI International*
Dr. Esmond E. Snell, *Professor of Microbiology and Chemistry, University of Texas, Austin*
Dr. Harriet A. Zuckerman, *Professor of Sociology, Columbia University*

Management of Annual Reviews Inc.

John S. McNeil, Publisher and Secretary-Treasurer
Dr. Alister Brass, Editor-in-Chief
Mickey G. Hamilton, Promotion Manager
Donald S. Svedeman, Business Manager

ANNUAL REVIEWS OF		SPECIAL PUBLICATIONS
Anthropology	Medicine	Annual Reviews Reprints:
Astronomy and Astrophysics	Microbiology	Cell Membranes, 1975–1977
Biochemistry	Neuroscience	Cell Membranes, 1978–1980
Biophysics and Bioengineering	Nuclear and Particle Science	Immunology. 1977–1979
Earth and Planetary Sciences	Nutrition	Excitement and Fascination
Ecology and Systematics	Pharmacology and Toxicology	of Science, Vols. 1 and 2
Energy	Physical Chemistry	
Entomology	Physiology	History of Entomology
Fluid Mechanics	Phytopathology	
Genetics	Plant Physiology	Intelligence and Affectivity.
Immunology	Psychology	by Jean Piaget
Materials Science	Public Health	
	Sociology	Telescopes for the 1980s

For the convenience of readers, a detachable order form/envelope is bound into the back of this volume.

THE WAY IT WAS

Joseph E. Mayer

Department of Chemistry, University of California at San Diego, La Jolla, California 92093

Introduction

At one of Bill Libby's bacchanalian birthday parties in Chicago in the early 1950s, Bill made a solemn pronouncement, which he often did out of the blue sky: "Of all people who have ever lived, 15% of them are living now." The statement was received solemnly by the assembled graduate students and young postdoctorals and elicited very little comment. Sometimes these pronouncements of Bill's formed the theme of discussion for the rest of the party, with a great deal of conversation and argument as to whether the statement was reasonable or nonsense. This one did not, as I remember. It was only months or even years later after I happened to run into some discussion by an historian of census taking in antiquity that I remembered Bill's statement. I tried to see if I could fit some simple algebraic curve to the data given in the article and to a few other figures that I found or remembered. I succeeded; I do not remember the formula; but not to my surprise, the answer was close to Bill's. I concluded that a maximum of about 15% of all humans were still alive, and almost certainly as many as 7%.

Obviously, whatever date one selects as a starting date for such a calculation, one selects a generation that has ancestors for millions of years back to the first lungfish that climbed out of the primordial ocean. One has to start with some arbitrary date. I felt that one might start with ten thousand years ago, which presumably predated any of the large city states in China and also in the Near East and Egypt. If one asks what fraction of *literate* humans who have ever lived are now alive, the number must be greatly increased.

The same kind of difficulty arises when one tries to set numbers for the fraction of scientists living at any year. What does one have to do to be called a scientist? I decided that anyone who spent on science more than

10% of his waking, thinking time for a period of more than a year would be called a scientist, at least for that year. I suppose most young scientists of the present era believe they spend more, very much more indeed, than 10% of their time thinking about science. But I wonder if they do not find that most of that time is spent filling out forms and applying for more research money. In any case, one has to be able to include a man whose paid occupation was director of the British mint, and also one employed by the Swiss patent office.

If one asks what fraction of published science has appeared since I began proudly calling myself "scientist" after my 1927 PhD, the answer is almost all of it! A walk through the stacks of any research library shows much more than ten-fold linear feet of shelf space devoted to "archive" science journals per year now than then. An increase of more than one order of magntidue in any comparison of world science in 54 years from 1927 to 1981 is a completely quantitive change of the characteristics of the milieu.

Childhood

My father was one of 18 siblings, 16 of whom grew to maturity, born to his parents in the small town of Schruns in the Montafon Valley of Vorahlberg, the most westerly province of Austria. As far as I know he was the only one whose education went to collegiate level. He attended the "Technische Hochschule" in Innsbruck in the Tirol. After graduation he went to Paris to the Sorbonne, where he received the degree of "Docent" of applied mathematics, after which he emigrated to the United States and was employed there as a civil engineer by the Union Bridge Company.

My mother, Catherine Proescia, an American-born New York City school teacher, was twenty years my father's junior in age. When I was less than five we moved to Montreal. My father became assistant chief engineer employed by the Canadian government to be responsible for the design of the new Quebec Bridge to replace the first-designed structure which had collapsed.

I learned and understood that if you made a bridge just like the one that stood up, with a given span between piers, and then doubled the span as well as the linear dimensions of every steel member in the design, the strength of the units would be four-fold greater but the weight eight-fold more. Any five-year-old who could count could understand that, if he were sufficiently interested to pay attention to a clearly illustrated lecture by a patient teacher.

My fascination with mechanical construction was greatly stimulated by a fabulous toy brought to me by a young German engineer from Karlsruh: this was a big "Mechano" set. My father was dissatisfied with the mathematical training of American civil engineers. The young German, Hans Grether, became a constant visitor to our household in Montreal until early August 1914, when he avoided the British blockade by shipping as a crewman on a Greek freighter bound for Hamburg, where he jumped ship and assumed his German military duties.

I think my father always considered himself a scientist, an avid reader of Darwin, Huxley, Herbert Spencer, as well as of Freud and Oswald Veblen. Civil engineering was a practical source of income. His family in Schruns were practicing Roman Catholics, but Dad disliked catholicism. He and my mother regularly attended the Unitarian church on Sundays on Sherbrooke Street and I its sunday school. The minister, Frederick Griffin, lived down the street from us. He was a frequent guest at our house, tolerant but disapproving of my father's professed agnosticism.

I have adopted the same agnostic belief, modified by two events. One, through a scolding by a peer when I was a graduate student in Berkeley that the claim of agnostic was a cowardly dodge: Was I not really a convinced atheist? But second, after two round trips of the world, I have attained a theistic realization that all gods worshipped by man have existed or still exist and have been very important in human history. I have seen many! I have seen Shiva's most important organ three meters in circumference and five meters high worshipped by Nandi the bull from without the temple door. I have seen the enormous benign bronze Buddha at Kamakura and the sleeping Buddha in Bangkok. I have seen the great lions of the gods of the Phoenicians shattered by their fall from the lofty pillars of the greatest temple of antiquity dedicated originally to the mighty Baal, but later defeated by Rome who renamed him Jupiter. Beside the fallen lions of Baalbeck is the more modest, but more durable, still-roofed temple to a god at least surreptitiously worshipped throughout the world since the origin of man, whom the Romans called Bacchus.

I suppose that north of cancer 23°N and west of 70°E, for almost two millennia, worship on the Eurasian continent has been pretty much dominated by the various sects that originate from the Judaic tradition—Judaism, Christian, and Moslem—and for the past half millennium, through conquest this is also more or less true of both Americas.

The gods created by man require a priesthood to codify their worship and see to it that temples are erected to demonstrate their power: the might of the doctrine and of the priests of the church. Many are the bloody wars that have been fought about trivia in the theological phraseology of doctrine. It seems as though the teaching of the gentle Jesus has been used as the cause and excuse for many of the cruelest wars of history.

My Introduction to Chemistry

I had a fortunate experience in Hollywood High School, California, in having an excellent chemistry teacher, Mr. Gray, who interested me in the methods of chemistry and with whom I took a second course with a very few (four or five) other students on quantitative analysis. At that time the worldwide price of sugar had recently more than doubled. I spent two campaigns after high school graduation working in two sugar mills, one in Huntington Beach, California, and one in Hooper, Utah. But the price of sugar plummeted at the end of the second campaign and it was clear that chemists were not going to be employed at the bench level. One could guess how to run a sugar mill without paying slaves to titrate the out-put every half hour.

Cal Tech

One of the young high-school educated chemists, Lee Prentice, and I applied to Cal Tech in the midyear class of 1921, the last year that class was given. The name California Institute of Technology had just been adopted by Troop Institute that January with Mullikan's ascension to the presidency of the institution. The midyear class started in January, 1921, and went until the Friday preceding the normal sophomore year beginning on Monday. Without a vacation, we joined the regular sophomore class of the students who had entered the Troop Institute in fall of 1920.

Cal Tech at that time was, as it always has been since, an excellent school. A. A. Noyes was head of the Chemistry Department. He took a great interest in the students although he did not teach any courses. I was amazed to have him greet me by name once in the hall when I was not quite sure who he was. Physical chemistry at that time was pretty much limited to kinetics and thermodynamics.

Richard Tolman came in 1922 as professor of both physical chemistry and mathematical physics. In the fourth year that I was at Cal Tech, Paul Ehrenfest from Leiden, Netherlands, came and spent a semester. Linus Pauling and Paul Emmett came as graduate students in their first year from Oregon Aggie in Corvallis, as I was in my third year as an undergraduate. In my senior year, one of our courses taught by Tolman was based on Summerfeld's *Atomic Structure and Spectral Lines*. The English edition had just come out. The course was listed as graduate, but a few undergraduates took it. Since the undergraduate chemists had not had a thorough course in mechanics, we were pretty much floored. I failed the final examination completely and received a grade of F, which threatened my graduation. Somehow without getting the grade changed they permitted me to graduate.

I had a job as a slavey assistant in Roscoe Dickinson's laboratory, where Linus Pauling was starting his dissertation work on x-ray crystal structure. My first and most serious occupation was to put up a chicken wire grid to protect Linus. Roscoe Dickinson was quite short. The laboratory was in the cellar of Gates Hall and the ceiling was fairly low. The line from the transformer to the x-ray tube went under the ceiling most of the length of the laboratory. Chicken wire had to go under the line to keep Linus's hair from standing on end and shorting the line.

Between Linus's first and second graduate year, Linus purchased Roscoe Dickinson's old Ford—I think it was a 1911 model, in any case it was somewhat antique—and drove up to Oregon where he had done his undergraduate work at Oregon Agricultural. When he came back, he brought his bride, Ava Helen. I was in the lab when Linus walked in upon his return. Roscoe greeted him with obvious relief and the immediate question: How did the Ford behave? "Fine." "It stood up to Corvallis and back without breaking down!?" "Yes." "No troubles at all?" "Well,...we did tip over once on the way back." "My God, weren't either of you hurt?" "Only a few black and blue bruises, but when we got the Ford upright again she started fine."

I had had a very pleasant relationship of my own as a child in Montreal with Mary Bishop. Mary had since gone to Pratt Institute and graduated in fine arts there. Following graduation she had an assignment to interior decorate a new elegant hotel that was being built at the mouth of the Saginaw River where it flows into the Gulf of St. Lawrence, and

Roscoe used Noyes & Sherrill's *Chemical Principles*, which I think was a remarkable textbook. The book was new; the very beat-up edition that I still have was copyrighted in May 1922, but has references to preliminary editions of parts by Arthur A. Noyes alone, copyrighted in 1917 and 1920. The most used phrase in the text that I have is "an important principle is illustrated by the following problem." The student learns nothing unless he does the problems: there are many, and they are mostly hard. I took the course checking, as I now find in my copy, all but a very, very few of the problems. The next year I corrected the problem homework for the following junior class. Since most of the problems were sophisticated, there were several correct approaches as well as an infinity of wrong methods. I learned the material treated in that text very well.

In my senior year I undertook an experimental research problem under the direction of David F. Smith, a National Research Fellow (1). The work was published in the *Journal of the American Chemical Society* in

1924—my first publication! My most lasting lesson from that research was to distrust labels. The title of the paper was "The Free Energy of Aqueous Sulfuric Acid," which we determined by establishing equilibrium at 80°C for the reaction $H_2SO_4 + 6HI \rightleftarrows 3I_2 + 4H_2O + S_{rh}$, approaching equilibrium from both sides in 20 or more days. Our first attempts used "chemically pure HI" from the stockroom: a clear colorless liquid that reacted instantaneously with even dilute sulfuric acid, due to some phosphorous-containing compound producing H_2S but no I_2, probably phosphonium iodide at about 0.3 normal concentration, which seemed a mite high concentration of impurity in a bottle labeled "C. P." We made our own HI after that.

One of the very long-standing mysteries of science was the behavior of relatively dilute solutions of salts in water. The laws of perfect solutions were known, and relatively low molecular weight, nonpolar molecules were known to obey the laws well at the two ends of the composition mol-fraction diagram. Molecules of similar size and shape deviated but little from the perfect solution laws, even in the middle of the diagram. A. A. Noyes had recognized the quite different behavior of the highly ionized salts in water solution and even before the turn of the century had published some interesting papers calling attention to peculiarities in ionic solution behavior. G. N. Lewis in 1911 and 1912 introduced the concept of "ionic strength," μ, the sum over all ions (i) in a solution of one half the molar concentration times the charge z_i squared, $\Sigma_i \frac{1}{2} c_i z_i^2$. Thus, for a one-one salt like NaCl, the ionic strength is equal to the molar concentration but four-fold greater for the same molar concentration of magnesium sulfate. Lewis then observed that the activity coefficient of any solution of all salts with given valence type z_+, z_- depended on the ionic strength alone at reasonably low concentrations, and its deviation from unity at very low concentration was proportional to $\sqrt{\mu}$.

There was no theoretical explanation until in 1923 the Debye-Hückel treatment of the thermodynamics of ionic solutions was published in the Physikalische Zeitschrift. A. A. Noyes immediately recognized its importance and gave a special colloquium lecture on it. As far as I remember it was the only scientific lecture I ever heard from Noyes, which is a shame because he was so clear and precise that I left with the illusion that I understood it perfectly.

Interlude

After graduating from Cal Tech, my mother, father, and I took a trip to Hawaii. At that time the Moana Hotel was the only hotel in Waikiki except for some bungalow cottages, located where the Royal Hawaiian is now. We stayed at the seaside cottages but visited some of the other islands. It was the first time I had seen true tropical luxuriance. When we returned I stayed in Berkeley, and my mother and father returned to Pasadena. It was the last I saw of my father before his death, which was just before Christmas of that year.

My mother was terribly shocked at my father's death, and more so because she had been so anxious to go to Europe with him and to meet his brothers and sisters in Schruns, most of whom were still there. I agreed to go with her and we arranged to leave by train and boat at the end of the spring term.

I had had a very pleasant relationship of my own as a child in Montreal with Mary Bishop. Mary had since gone to Pratt Institute and graduated in fine arts there. Following graduation she had an assignment to interior decorate a new elegant hotel that was being built at the mouth of the Saginaw River where it flows into the Gulf of St. Lawrence, and she wished to spend her remuneration for that work on a trip to Europe. My mother suggested that she come with us. This was a very happy thing for me. Mary's knowledge of European art was of course exactly what I needed. Mary used the method on museums that I have since learned to adopt, namely to rush through the complete museum, glancing at everything, and then to go back and repeat, stopping only at the things found interesting. I fell in with Mary's tastes very quickly. It was a real revelation to me how much more one saw in a museum with a really good guide than going through by one's self. The present arrangement in many museums of renting a talking machine that tells you what you are looking at and where to go next of course did not exist then. The guides that took groups through then were something of a nuisance.

The three of us "did" Europe together, enjoying the usual tourist traps, but the high point was the visit to Schruns. There we stayed with Uncle Wilhelm and his wife Victoria. Uncle Wilhelm had the commodious second floor over the one bakery in town. I hadn't ever seen a really modern electric kitchen until Tanta Victoria's kitchen in Schruns. Of course another Mayer, a brother of my father's, owned the Electricitätes Werke that supplied the electrical railroad from Bludenz to Schruns as well as the city lighting. Another uncle owned the one inn in town. Schruns at that time was not a well-known tourist resort. Since then it has hosted the International Winter Olympics.

Wilhelm Mayer (Myer is the German pronunciation of the name) had purchased the flour mill in St. Poltern, which is a city much closer to Vienna. Since it was summer when we were there with Mary, the family moved to their summer cottage, away from the big city of Schruns, into the higher mountain valley of Gargellen. We traveled by horse and buggy four hours, from Schruns to Gargellen, a difference of about two thou-

WILLIAM F. MAAG LIBRARY
YOUNGSTOWN STATE UNIVERSITY

sand meters in altitude. We spent a few days in Gargellen where the cows were for their summer pasture and the cheese was made. We went home about Christmas time so that I would be in Berkeley to begin the spring semester, which began in January.

The last time I visited Schruns with my present wife, Peg, only a few years ago. There were practically no Mayers; the uncles and aunts of course were all dead.

Berkeley and Gilbert Lewis

My textbooks as an undergraduate often contained an introduction quoting Roger Bacon's thirteenth century admonition that no theory was valid if it contradicted observation, that is, experimental fact. Another caution known as Occam's Razor is an added lemma that was then seldom discussed, but states that no unnecessary verbiage should be added to the bare bones of the theory necessary to account for all pertinent facts. These two principles have dominated all good science for several centuries. In my student days there were rare but occasional polemics in the literature concerning the validity of a particular theory versus an apparently contradictory experiment. It was a hardy theoretician who dared enter such a polemic unless he knew the theory had been misapplied or that the experiment had an obvious error. James Franck often remarked that nothing looked more like an important new discovery than a poorly conducted experiment. The other explanation, that the experimentalist misused the theory, also occurs. In the early 1930s I had an evening's discussion with a Johns Hopkins faculty colleague who insisted that themodynamics did not apply to organic chemical reactions! He had a Harvard chemical PhD too!

In the five decades since my student days there are still disagreements between scientists, a negligible fraction of which concern pure science, but more the wisdom of social action in a technical matter. The few serious disagreements in science itself practically never involve identifiable conflict between pure theory and pure experiment. Experimental procedures have become so involved that when conflicts in interpretation occur, it is usually unclear which party is invoking theory and which simple demonstrable experimental fact. Both parties usually use a mixture of both.

The seven-century-old heretical philosophies of Roger Bacon and William of Occam are now so deeply ingrained in the thinking of all scientists, and even, although far less thoroughly, in the consciousness of all scholars, that their principles are seldom now discussed. The methods of science are now, as they always have been, the methods used by the individual practitioners of the disciplines. Since all human individuals are unique, and no two are identical, there is a wide diversity of approach. Nevertheless, there are some characteristics of the ratiocination of scientists, particularly in the physical sciences, that differ from those most commonly met in scholars of other fields.

One of these is both the cause and the effect of the diversity of fields of research. Science seldom, perhaps one should say never, discusses a vague question and never answers one. The vague question is broken down piecemeal into small well-defined parts, and answers to each is painfully sought. The essential feature is that the individual, smaller question is clearly formulated and defined.

One of my early experiences as a graduate student taught me this from a great master. After graduating with a B.S. in chemistry from Cal Tech, I applied to the University of California for a teaching fellowship, which was awarded to me. The next four years I spent in Berkeley. Gilbert Newton Lewis, known as the "Chief" or just as "G. N.," was head of the School of Chemistry. He soon became and remains to this day one of my greatest idols. He was a great and tolerant person, a very important scientist, and a fabulous teacher. The department was a happy one with a very strong leaning towards physical chemistry, which in those days was almost exclusively the application of thermodynamics. Even the organic chemistry faculty knew and used thermodynamics, which was unusual at that time. There were no regularly scheduled graduate courses. Most graduate students from other colleges were advised to take the senior thermo-course. We were advised to take graduate courses in mathematics and physics, and some took courses in biology.

I can imagine no milieu more beneficial to the development of a graduate student than that department at that time. The atmosphere was that of unravelling the intricacies of nature in one of its important aspects. Pure knowledge of an assortment of unconnected facts was seldom emphasized, but a deep understanding of principles and originality in interpretation were most admired. I was never aware of jealousy or friction between faculty members and in four years I grew to know most of them very well. All of them seemed to admire and love G. N. That the atmosphere was good for students has been evidenced by the relatively large fraction of them elected to the National Academy of Sciences some ten to fifteen or more years after their doctorates.

One of the nearest to a required graduate course at that time in the Berkeley Chemistry Department was the Monday evening seminar which all graduate students were expected to attend, and which I think most of us did gladly. Since travel from the East Coast, or even from the Midwest, was then a five-day train ride, we had few visiting scientists, and the notables who did come generally stayed a semester or more and gave a full series of lectures on a special topic. There were seldom more than one on hand. The Monday seminar was rarely attended by any but chemistry faculty, one or two post-docs, and graduate students. During my stay the routine did not vary. A faculty member or a graduate student presented a summary of a paper in the literature that some faculty member had suggested would be interesting. The faculty sat at a long table, students along the wall. The assigned time was twenty minutes. I do not remember slides ever being used, and certainly there was no vue-graph, but blackboard and eraser were in constant use.

After the presentation there was discussion, often for longer than the time taken by the speaker and sometimes involving heated argumentation. When the discussion began to lag G. N. would look around the room, select one of the attendees, student or faculty, and address him: "Mr. John Doe, tell us about your research," or some other equivalent request. If addressed to a beginning student in his first year, the nuance of the request was to describe the question to be answered, the importance of the problem, and the projected experimental approach. Completed research was delivered elsewhere, in special announced lectures by faculty, or in the public PhD examination of students.

I think that the emphasis, which we soon began to recognize, on problems of general importance, rather than on adding only one more example to many similar worked out and well understood cases, was probably the prime characteristic that we, as students, took away from our experience at Berkeley as our most important legacy. Of course, then as now, the research problem was almost always suggested by a faculty member, but the student was expected to be able to critically and dispassionately evaluate its significance and importance to scientific understanding. The answer, occasionally given by students now, "It is part of Professor X's project," by itself would have been regarded as unsatisfactory without an understanding of its place in the project and the project's place in science.

The answer to G. N.'s request seldom exceeded a quarter hour, and the faculty discussion followed another quarter hour with many helpful suggestions of procedure or experimental technique. I think that most of us enjoyed being called on. As I remember, usually two or three students were called upon each week, so that each of us reported several times on the progress of our incomplete doctoral dissertations. The seminar was open-ended in time and often lasted two hours, but the diversity of subject matter kept it from becoming tiresome.

One experience in that seminar I shall always remember. I was elected to give the twenty minute report on a paper, I think in the *Zeitschrift für*

Physik, on a subject I have now forgotten but which involved high vacuum technique as did my doctoral research. Towards the end of my carefully prepared and, I hoped, very clear presentation, I became quite aware of incipient troubles. Unusually dense clouds of aromatic smoke were being emitted from G. N. and his ever present "Fighting Bob" nickel cigar to which he had become addicted in the Philippines. Worse still, the cigar was being shortened from the chewed end rapidly and G. N.'s scowl looked ferocious. I finished and awaited the explosion. Smoke and scowl continued for some time. Finally the cigar came out and after that a question, long and involved and obviously concerned with the scientific conclusion of the paper. I found the question incomprehensible and after some hesitation said, "I'm sorry Professor Lewis, but I don't understand the question." More smoke, more scowl, and finally, "Damn it Mayer, of course you don't. If I understood it I'd probably know the answer." Laughter in the room.

I have often thought that that episode illustrates a rather frequent stage in the development of science. The most important part of any real advance is to formulate clearly the correct question.

It was typical of G. N.'s eclectic interest in all of science that he became interested in relativity, which is as far from physical chemistry as one can imagine in any physical theory. I remember a fascinating lecture on special relativity that he gave, at which I saw for the first time the now oft repeated time-space diagram.

I don't remember exactly when I actually began work on what became my dissertation. It was a fairly difficult experimental stunt, and I think that actually had we started it after we really understood quantum mechanics, we would have thought it not worth doing. The dissertation was finally published under the title, "The Disproof of the Radiation Theory of Unimolecular Reactions" (2). If chemical reagents require activation energy to react, this activation energy must either come from radiation or from collisions between molecules. If it is due to collisions between molecules, the rate should be proportional to at least the square of the pressure and the reaction would not be unimolecular, but bimolecular. This had been observed, of course, and Jean Perrin postulated that all unimolecular reactions were due to the absorption of radiation. I set up a very high temperature radiation field such that, if activation were by radiation, a beam of molecules passing through the field would react. The reaction we studied was the racemization of pinene which, from its known rate measured at room temperature and extrapolated to our field temperature, should show racemization on passing through a few centimeters if it were indeed activated by radiation. The radiation bath was simply a quartz cylinder about three centimeters long, through which the pinene coming through two holes in platinum foil was caught in liquid air at the other end and later transferred to a capillary tube and the optical rotation measured. The effect was nil. Of course this is no mystery now and I think it was recognized at that time or very soon afterwards that at sufficiently low pressure the racemization might indeed go with the square of the pressure. However, the negative result of the measurement was accepted for publication and I was granted a doctorate degree.

I do not know how it is now, but at that time teaching fellows in the chemistry department had the privilege of belonging to the faculty club. I ate there at lunchtime rather regularly, usually sitting at a table with other chemists, although quite often when the chemistry table was full I would sit with faculty members of other departments. The chemists, along with one mathematician, frequently played cards for a half hour after lunch.

I became pretty good friends with Wendell Latimer. Latimer at that time had lost his first wife and was, I guess, in a really unhappy period of his life. However, he met his second wife, Latha, and they were happily married before I got my degree. Before that time we often went to San Francisco. Several of the graduate students, often with Wendell, had dinner at a favorite second floor Italian restaurant, "Mimi's." We were always well fortified with drinks, particularly with good Italian red wine. Mimi used to regale us with stories of the opera singers he knew. He had been an impresario in Vienna and in Rome. We always took Mimi's tales with a grain of salt, but one day the New York Metropolitan Opera was playing in San Francisco. The two stars showed up after the opera at Mimi's while we were there having dinner. The two stars, if my memory is correct, were Martinelli and Galicurci. Galicurci embraced Mimi effusively as her old teacher. We were greatly impressed and we really believed all of Mimi's stories after that. The two opera singers had a postman's holiday and we enjoyed it thoroughly. They sang loud and lustily for several hours after a tiring opera.

Hildebrand, of course, had the freshman chemistry course at that time. I was employed as a teaching fellow; my memory is that every entering student was a teaching fellow. I was impressed by the system. The younger faculty members were in charge of laboratory sections. I think all laboratory sections were about 25 students. At that time, experienced teaching fellows were assigned with the beginning teaching fellows for their first year in a laboratory section which was officially in charge of a faculty member. The faculty member would not necessarily stay throughout the whole afternoon after the first few weeks, if he knew he could trust the teaching fellow.

There was a careful way of correcting for differences in the grading of papers so that the students were not penalized by getting a very strict teaching fellow doing the grading. At the end of the semester there was of course a big final examination, and the grading was done by the teaching fellows all at once in a big room, usually one teaching fellow taking one question. In following years, I did not have a faculty member in the laboratory section but had it all to myself.

As I mentioned above, there were no graduate courses in chemistry listed in the catalog and none given, except that most of the beginning graduate students were advised to take the undergraduate thermodynamics course, which was based on Lewis & Randall. I was excused from that since it was assumed that Cal Tech would have given me a thorough basis in thermodynamics. My final examination was during the 1927 summer session and was open to the public. At that time there was a certain amount of pressure on the high school teachers to know the subject matter that they were teaching rather than only to have had courses in education. There were quite a number of them attending my examination. I had a pleasant introduction to my committee by having lunch with them at the faculty club and playing cards afterwards until it was time to go across the street to Gilman Hall for the examination itself.

Gilbert Lewis at that time was known to have usually one tricky question and he certainly had one for me: Take two independent tungsten filaments, in high vacuum, one of them at a temperature approximately 2000 degrees or below, and the other one really white hot, and both filaments are on separate lines by which the current and voltage across can be controlled. A very small pressure of chlorine is allowed into the apparatus, something like 10^{-4} or 10^{-5} mm pressure. The observation is that the higher temperature filament's resistance decreases: it gets thicker, whereas that at a lower temperature gets thinner. After a few promptings to consider possible chemical reactions, I saw the light: The answer was that at the lower temperature the chlorine attacks tungsten, forming a volatile tungsten chloride, but at the very high temperature this compound dissociates, depositing tungsten on the hotter filament itself. I was very proud that I got the answer correctly.

After the examination, Lewis asked me if I would stay as a postdoc, that he would like to discuss a few problems with me. I do not remember what I was paid but I did get a little more than the teaching fellowship and I felt very flattered by the invitation. I actually stayed until the autumn of 1929 when I went on a National Research fellowship to Göttingen to work with James Franck.

With Gilbert Lewis I had a very stimulating two-year experience. I had no knowledge of statistical mechanics and Lewis had never worked in the field either. He had become interested in the discovery that had just been made of the difference between quantum mechanical statistical mechanics and the classical, and the Bose-Einstein versus Fermi-Dirac systems.

During the day I tried to learn statistical mechanics using Tolman's two books, the first of which I found clear and interesting, but the second seemed to me to be too talkative. I also tried Fowler & Guggenheim, which I did not really like, probably mostly because I was unacquainted with the mathematical methods used.

Gilbert and I spent the evenings together, usually at about eight o'clock, sometimes until about midnight. That was really an experience. It was most interesting to see how Gilbert Lewis thought. He was not infinitely brilliant, but he would go over and over a problem until he really understood. Eventually he decided that we ought to publish and the result was three papers that appeared in the *Proceedings* of the National Academy (3).

I still like the method that we evolved for deriving thermodynamics from statistical mechanics, that is, from the mechanical laws for the motion of molecules. Of course the black body radiation problem had been solved but hardly explained. I remember one remark of Gilbert Lewis at that time. He said that the first papers on black body radiation by Planck were clear and concise. The later papers got fuzzier. Actually, we were quite conscious of the fact that the quantum mechanical interpretation was so strange that it was scarcely believable.

The Michaelson-Morley experiment uses half-silvered mirrors which split a beam of photons into two paths which then are reflected back by two mirrors at right angles to each other, such that there are two beams at right angles going through the half-silvered mirror at 45 degrees, which come back and then half of both beams are coalesced again and interfere. There is a beam of light, half of it moving in one direction, half of it moving at 90 degrees in another direction; both beams are completely reflected approximately the same distance and one-half of each of the beams goes to a receiver after passing through the half-silvered mirrors. If one of the two totally reflecting mirrors is blacked out, the pattern is a simple one: one quarter of the beams arrive at the receiving station. If the other mirror is blacked out, the pattern looks almost identical to the naked eye, but the peaks in intensity are actually shifted. Of course when both of the totally reflecting mirrors operate, the two beams interfere with each other and give the interference pattern. There is no mystery with intense classical beams of electromagnetic waves. The mystery, of course, is that if you picture beams consisting of single photons which always have the energy $h\nu$, how do the photons know what the other photon does at a large distance, that is, whether or not it is reflected back? This, of course, is the essence of the difficulty that Niels Bohr and Einstein argued over years later and is responsible for Einstein's statement, "Rafiniert ist der lieber Gott aber Boshaft ist er nicht." (God is sophisticated but not mean.)

In 1929 I was the proud possessor of a National Research Fellowship paid by the Rockefeller Institute for postdoctoral work in Europe, and I had arranged to be able to study in Göttingen and work with James Franck. Franck had been in Berkeley. I gave him a letter of introduction from Hans Grether. After World War I, Hans Grether had gone to Peru and worked on the railroad that was to run over the Andes into the Amazon Valley. His brother had a ranch in Somis and also one in Salinas, California. The brother, Karl, had come to California from Germany while we were still in Montreal before World War I, and Hans was still in Montreal. Karl stayed with us there in Montreal for a few days. We made contact with Karl in California and I visited the ranch several times. Hans came up once from Peru to visit his brother, and when I told him that I was hoping to go to Germany to work with Franck he gave me a letter of introduction. He had worked with Franck during the war, actually, I think, on poison gas.

Göttingen at that time was known as the source of quantum mechanical knowledge. Heisenberg, working with Max Born, simultaneously developed, at the time that Schrödinger developed wave mechanics, an equally logical system of quantum mechanics. As you may remember, for quite a while there was some difficulty concerning this, as Schrödinger's wave mechanical approach seemed to do exactly the same things as the matrix approach of Heisenberg and Born. It was Schrödinger who finally reconciled the two systems as being really different mathematical formulations of the same theory. Max Born at the time was in Berkeley, and I heard later from Maria that he was rather annoyed that he was scooped by Schrödinger simply because he had too much to do in Berkeley.

Thorfin Hogness, who was a member of the Berkeley faculty, had been in Göttingen for a year previously. He told me that most Americans in Göttingen stayed at a pension on Nicholausberger Weg, the Kreuznacker House, but if I could possibly get a room in a private dwelling, it might be more pleasant than a pension. He mentioned that his pediatrician, Professor Göppert, who was a professor of pediatrics, had died and he understood that Frau Göppert might have a room.

When I went to see James Franck I asked him where I should stay and he gave me almost the identical advice—namely, that there was a pension on Nicholausberger Weg, the Kreuznacker House, and it was very satisfactory and most of the Americans stayed there. I'd probably be happier if I could possibly get a room somewhere alone, and he suggested that Frau Professor Göppert had rented a room to an American (it turned out that the American was Robert Mulliken) a year ago and I might try there.

The maid who answered the door said the Frau Professor Göppert was ill and wouldn't see anybody. Actually she had a nasty cold. However, the maid told me that she would ask the daughter to come down and speak to me. Well, the daughter came, smiled benignly at my frantic German, and then answered in beautiful Cambridge English, that her mother was sick, that it was just a cold, but she did not want to see anybody, that I should come back in a day or two, which I did. I was staying at a hotel close to the railroad station. I was much impressed with the daughter and particularly by her perfect English, which I later found she had acquired in one semester at Cambridge on a student fellowship from Germany, in Rutherford's laboratory. She had lived in Girton College while she was in Cambridge, which was the only girl's student house at that time.

Well, I was feeling relatively wealthy and I purchased on Opel. The Opel was not a General Motors car at that time; the firm was later purchased by General Motors. The Opel was a wonderful car. It was put together with picture-hanging wire and sealing wax. I think the existence of the Opel changed my future life. It was a beautiful machine and I had the only automobile of any of the students or of any of the young faculty.

Maria was the belle of Göttingen, as I soon found out. She and the two daughters of Marianna and Herr Professor Landau, along with Titi Stein, seemed to make up the acceptable female contingent of every student party.

On Wednesday afternoons the students of Göttingen frequented Maria Springs—nothing to do with my Maria. Maria Springs was a natural amphitheater in the woods about 10–12 miles north of Göttingen on the main road and on the railroad. The German band played there and there was dancing on the floored area at the bottom of the amphitheater. The Corps, that is the fraternities, all came in decorated coach-and-fours and made quite a show of their arrival. The girls of Göttingen were welcomed in the early afternoon. The last train from Maria Springs to Göttingen left at 6 o'clock in the evening, which in summer time, of course, was bright daylight still. Proper girls had to go home by the 6 o'clock train. Even with my car I could not get Maria to stay longer than the time the train left.

The German school system, like the American, is a state affair, that is, it is different in the different provinces, but essentially similar, just as the American states have very similar schooling arrangements. Until the students are 10 years old, schooling is uniform, simply a neighborhood school; after that there is a separation. The Volkschule does not prepare students to go into a university, and in general students who start in at the Volkschule never can get into a university. Now I say "in general" because I know at least one exception, and that was Hans Jensen, who was the son of a poor gardener and could not afford to go to an Oberrealschule or Gymnasium, the two schools that do prepare for the university. His instructor at the Volkschule recognized his ability and managed with great difficulty to obtain permission for him to take the abitur, which he passed perfectly. The Oberrealschule, the purely classical school preparation for the university, had both Latin and Greek as required subjects. The Gymnasium permitted one to take an abitur also but was more technical. I think both required calculus. The abitur was very much feared. It was generally regarded as disastrous to fail it and one could not get into a university without passing it. The American department idea with several professors in the university was nonexistent. On the contrary, each full professor had his own institute and apparently operated completely independently of any other full professor in the same field. For instance Göttingen had three physics institutes. The Erstes, the first institute, was under Herr Professor Pohl and was primarily solid state work. The Zweites was James Franck's institute to which I was assigned. The Drittes institute was that of Max Born and was purely theoretical. The only common feature was a weekly seminar that all members of all three institutes attended. The seminar had other attendees, for instance both Hilbert and Courant quite frequently came; less often the Professor of aerodynamics, Prandtl, came.

I know very little about the chemistry departments. The physical chemist was Professor Gustav Tamman, who was regarded as a fierce man to have on an examination. One student, who had Tamman on his PhD examination, came out trembling. Tamman had asked him: "Na Kerl was ist lambda?" "Lambda ist Wellenlänge." "Falsch! Lambda ist specifische Conductivität!" Many physics students had Tamman on their examination, but I don't know of any student who was ever flunked because of Tamman's questions, although a good many students were not passed on their first examination.

Only a few weeks after I got to Göttingen there was a meeting of the German Physical Society in one of the Hartz mountain towns. There I met quite a few German physicists whose names I had heard, but had not met before. Among them was Polanyi, the Hungarian physicist philosopher. His son is now professor at the University of Toronto and was a long time in Ottawa.

In Göttingen, Professor Franck came around on a certain afternoon each week with his entourage of assistants and visited everybody working in the laboratory. The group would always include Herta Sponer, who was Franck's assistant, and later after Franck's wife died he married Herta. Otto Oldenberg was the oldest of the people who came on these tours. He was Auserordentlich professor, which corresponds pretty much to an associate professor in America and implies tenure. Otto Oldenberg later emigrated and was in Harvard.

One of the important holidays in northern Germany is Pfingsten, which I think is the same as Whitsuntide in England. Dick Badger, who was in Göttingen from Cal Tech, and I took our first Pfingsten vacation together walking from Lyon to Marseille down the Rhone Valley. It was a long walk, rather too long, walking is slow; however, we did see the French countryside and enjoyed the Rhone wines enormously on the way. One afternoon a car stopped and the two Canadian occupants asked us if we wanted a ride. We did. Later in the afternoon they stopped at an inn and decided they were going to stay there for the night. Dick Badger and I also found it a good idea. The two Canadians asked for a room with bath. Dick Badger and I were satisfied with a room without a bath. At supper the Canadians suggested that we ought to come up and look at their room, that it was worthy of a visit. We did after supper and found the room gorgeously furnished—an enormous room with a platform in the middle of it and a big bathtub on the platform! Dick Badger and I parted in Marseille and both of us went back to Göttingen independently.

In Göttingen I found several letters from Johns Hopkins University offering me a position as associate. The first one had evidently come just about as I left on the Pfingsten holiday. I responded affirmatively and felt very happy that I had a position assured when I got back to the United States. In the meantime I was getting more and more interested in trying to induce Maria to come back as my wife to the United States. This was not completely trivial; the German immigration quota was filled for several years in advance. However, I found out from the consulate that my wife could get in on a special visa. Well, that worked out. I remember that Maria's favorite aunt, who was not very much older than Maria, the wife of the youngest brother of her father, said to her: "You are fortunate in going to America. My sons will be caught up in the next war." They were. One of them survived but was badly wounded.

Return to USA

Maria and I married, and Maria finished her exam shortly before we left. We traveled on the Nord Deutscher Loyd ship, Europa, on its maiden voyage. Arriving in New York on April Fools day, we were met by my cousin and his wife. We stayed with them a few days and then went on to

Baltimore where we found accommodations in a pension. However, the summer vacation broke out pretty soon and we found that there was going to be a special summer session for graduate work at Ann Arbor, Michigan. Enrico Fermi and Paul Ehrenfest were to be the two lecturers. We both knew Ehrenfest quite well. He had been a regular visitor in Göttingen and on his invitation we had once driven to Leiden and stayed with him and his wife. He locked Maria in his study and scolded her that she was not working on her dissertation. He let her out only after she produced n pages, and I forget the number n. In the meantime I was assigned the guest room; the guest room was on the third floor of the house or maybe it was the fourth. It was a whitewashed room lined with bookcases which contained paperback detective stories. The detective stories were in all languages. There were many in English, some in Russian, and of course various ones in Dutch and German. And there were signatures on the whitewashed walls of guests who had stayed there, signatures with dates attached. I signed under that of the last guest, Albert Einstein!

That particular Ann Arbor summer session was enormously successful. Both Enrico Fermi and Paul Ehrenfest were extremely good lecturers. Each sat in the front row when the other was lecturing and corrected the other's English, much to the amusement of the audience. But both were extremely clear. The audience included Robert Atkinson, an English astronomer and physicist, Lars Onsager, Serge Korff, Donald Andrews, Charles Squire, and of course Sam Goudsmit and George Uhlenbeck, both professors at Ann Arbor at the time.

We became particularly good friends of the Fermi's. Laura Fermi was always a delight and Enrico was always interesting and informative. He was a lot of fun too.

After the session was over I undertook to show America to Maria. We drove in our very conservative secondhand Buick, which we labeled "Connie" for conservative, and tried to see as much of the West as we could. We included the Black Hills, the Tetons, the Yellowstone and Glacier Parks, and then went further west to Seattle, stopping at Mt. Rainier, which I climbed. In Seattle, we visited Henry Frank who was at a summer resort on one of the islands in Puget Sound at that time. Actually we dug clams, much to the horror of the natives, who were not brought up in New England. On driving up to the village at the bottom of Mt. Rainier, as far as we could go in a car, we passed Nisqually Glacier, which was then down almost to the road. We chilled the clams on the glacier and enjoyed them.

We drove south from Seattle, stopped at Crater Lake, and then took the Redwood Highway from there to the San Francisco Bay area and

Berkeley. At Berkeley we visited old friends of mine, including Robert Oppenheimer and his wife Kitty. In the last years that I was at Berkeley, Oppenheimer had come back from Germany and gave a lecture on quantum mechanics. I don't think I understood anything of it but I was enormously impressed and felt that I was getting quite a bit from hearing his stories. I was amused recently at reading an article in the Cal Tech magazine by Carl Anderson who had listened to Oppenheimer's lectures on quantum mechanics in the same year at Cal Tech. According to Anderson's story, the class kept getting smaller and smaller until he was the only one left. Oppenheimer then came to him and said, "Please don't leave, I can't go on with nobody in the class. Let me have at least one student to the end of the quarter." At Berkeley there were several of us who went through the whole semester, or whatever the length of time that Oppenheimer was scheduled to lecture, and we enjoyed it, but I think we were pretty well snowed. Oppenheimer was not a good lecturer at that time. His great facility for making things clear came only later. I think actually that this is a common failing of very brilliant young fresh PhD's in not being able to talk down to students that are not as able as they are. Or even if the students are actually very good, they still tend to talk over their heads.

From Berkeley we went down the coast to Pasadena where we visited the Paulings. We camped in the Pauling yard. They had a house very close to Cal Tech, and we imposed ourselves on them by setting up our tent in their yard. It was a delightful visit for us. I was very sad to see in the newspaper recently that Ava Helen died. She was a delightful person.

The most exciting thing that happened on the trip back from Pasadena to Baltimore was simply a series of tire failures which depleted the ready cash that we had. At that time it was almost impossible to cash checks, although I think we had some money in the bank in Baltimore. If I remember correctly, we simply drove through the bridge at Harper's Ferry without paying any toll. When we got to Baltimore late at night, we had no key to our house, having left it with Frank and Kitty Rice, who lived very close to us on the second floor of an apartment building. We went there and after a while managed to awaken the Rices, went up to their apartment, where we were given a few drinks and then, with the key to our house, proceeded to our home. Frank and Kitty Rice were our most intimate friends at that time. Frank was professor at Johns Hopkins and Kitty was a student in the Johns Hopkins Hospital Medical School, where she later got her degree, Doctor of Medicine with a specialty in psychiatry. Frank left Johns Hopkins for Catholic University in Washington in 1938 shortly before I was fired and went to Columbia. The administration at Johns Hopkins was making it uncomfortable for people to stay. I believe there was a real financial difficulty.

In summer of 1931 we went to Göttingen. Maria was employed by Max Born to help write an article on crystal dynamics in the *Handbuch der Physik*. I went with her, or course, but in addition, my first real student at Hopkins, Lindsay Helmholz, came with us.

It was impractical to start any experimental work in the summer, at least trying to do the experimental work Lindsay and I were interested in, so I worked on crystal theory, that is, lattice energy theory. I consulted with Max Born, and Lindsay helped me. We published two papers, one of them under the names of Max Born and Joseph E. Mayer (4) and the second one Joseph E. Mayer and Lindsay Helmholz (5). This was really very much a copy of the methods used by Born and Haber years earlier, but went into much more detail, attempting to get really good semiimpirical values for the lattice energies of the alkali halide series using exponential repulsive potential. The results were extremely good, really, and I think that for many years they were the best theoretical calculations of the energies for any chemical reactions. Of course, the lattice energy, which is the energy difference between the vapor ions and the normal crystal, is very large, much larger than any directly observable chemical transformation of the ions. Actually there is very little difference in energy between the crystalline salts and the salts in water solution where the ions are dissociated. However, the differences in the chemically observable solubilities were actually fairly well given by our theoretical values at that time. Certainly rather better than almost any quantum mechanical calculation of the energy of the chemical reaction then current. Lattice energies are of the order of 100 kcal per mole or more and the results showed fairly good values to the order of 1 kcal. Later on we evolved a different method of getting at lattice energies. The only unknown at that time was always the electron affinity of the halogen, that is the reaction chlorine neutral atom as a perfect gas to chlorine minus, also a perfect gas. The new method was simply to observe the ratio of electrons to ions coming off a hot filament in an atmosphere of very low pressure of chlorine or any other halide. The electrons could be deflected by a relatively small magnetic field parallel to the length of the filament; they curled themselves up and did not go to a positively charged plate cylinder of a centimeter diameter. This was a far simpler and more satisfactory method than the awkward treatment of the salt crystals. It worked quite well on the halides but did not give good results when used to measure the electron affinity of oxygen.

In the meantime, a Wiley representative got Maria and I interested in trying to produce a book on statistical mechanics and it actually materialized in 1940, the first edition of Mayer & Mayer's *Statistical Mechanics* (6). The book was quite successful as books on statistical mechanics go—enough so that Wiley even tried to get us to produce a second edition very much later. It was in doing that that I got interested in attempting to improve the method of deriving equations for the virial coefficients of normal molecules. Philip Ackermann was working on an experimental problem of seeing whether we could use low energy electron beams for molecular structure studies. Ackermann had trouble getting a satisfactory position and he stayed and helped me with the statistical mechanics calculations. The first two papers on the theory of condensing systems (which was an unfortunate choice of titles) were done with Ackermann's help (7, 8).

In general, I had a most interesting and excellent group of students while I was at Hopkins. It was an unfortunate time, much worse than at present, to get positions. Science was not a popular subject and well supported as it still is now, in spite of our complaints.

Maria and I tried to build an electron microscope. It was a joke and we took it as a joke; we never even tried to publish it. The apparatus was essentially made of wood with window screening to produce the electric fields. I give this only as an example of the sort of tomfoolery that we often were forced to use. Of course, essentially we did not have the courage to think that an electron microscope would have a real use in the future. There are other scientists who have occasionally thought ahead of their time in trying to develop experimental methods, but given them up, wisely probably, because everything has a time to be successful. The only thing I object to is that in many cases these people think they have been cheated. We knew what we were doing was foolish, but it was fun.

The sad thing of that time was how extremely difficult it was to get good positions for our students when they graduated. Particularly the women. I was very fortunate in having several excellent women students at Hopkins: there were Sally Harrison, Sally Streeter, Irmgaard Holdner Wintner, who was the wife of a mathematician, a professor at Hopkins, Aurel Wintner. I also had Willard Bleick and Louis Roberts as well as J. J. Mitchel. Louis Roberts came with me to Columbia and actually took his PhD degree at Columbia instead of Hopkins. Willard Bleick was one of the more impressive students.

In 1938 I was informed that I would be discontinued at Hopkins after a year and a half. It was very fortunate for me, although I was happy at Hopkins and I would have stayed there had I not been fired. I wrote to various friends, including Harold Urey at Columbia and James Franck, who had left Hopkins to go to the University of Chicago. At both places I was given an offer to come as associate professor, presumably with tenure or at least tenure after one year; I felt very happy to go to Isaiah Bowman, the president of Hopkins, and resign long before I was obliged to leave. I finally decided after visiting both Columbia and the University of Chicago to go to Columbia. I think it was probably a very good choice at the time; in any case I have followed Harold Urey ever since.

Maria and I went often in summer to Germany. In 1937 she received a telegram that her mother had had a stroke. We managed to get her on the Europa the same evening, which was quite a feat and managed only because we knew the German consul in Baltimore who was also the representative of Nord Deutscher Loyd. Maria was received at disembarkation in Germany with the news that her mother had died. We managed to get a load of furniture out of her house but never succeeded in being paid for the sale of the house.

Of course, the trips to Germany were always sad after Hitler's arrival as Führer.

I learned many things at the time. Americans are too polite and not nearly as direct as the Germans or, particularly, as the Dutch. Several times Germans told me they had a position in the United States and showed me the letters they had received. Those letters did not offer a position at all. They were usually in the tone, "Of course we would love to have you here, all my colleagues would like it, they would all recommend to the administration that we create a position for you, but the financial situation is so bad that it is almost impossible that anything will happen..." and so on.

Literature Cited

1. Smith, D. F., Mayer, J. E. 1924. *J. Am. Chem. Soc.* 46:75–83
2. Lewis, G. N., Mayer, J. E. 1927. *Proc. Natl. Acad. Sci. USA* 13:623
3. Lewis, G. N., Mayer, J. E. 1928. *Proc. Natl. Acad. Sci. USA* 14:569; 14:575; 15:172; 15:208
4. Born, M., Mayer. J. E. 1932. *Z. Phys.* 75:1
5. Mayer, J. E., Helmholz, L. 1932. *Z. Phys.* 75:18
6. Mayer. J. E., Mayer, M. G. 1940. *Statistical Mechanics*, pp. xi, 495. New York: Wiley
7. Mayer, J. E. 1937. *J. Chem. Phys.* 5:67
8. Mayer, J. E., Ackermann, P. G. 1937. *J. Chem. Phys.* 5:74

HIGH PRESSURE STUDIES OF MOLECULAR LUMINESCENCE

H. G. Drickamer

School of Chemical Sciences and Materials Research Laboratory, University of Illinois, Urbana, Illinois 61801

Over the past three decades, high pressure has proved to be a powerful, versatile tool for investigating electronic phenomena in condensed systems. In this review I discuss high pressure luminescence of molecules. High pressure studies of purely inorganic crystals have been reviewed in a number of places (1–4), so I restrict coverage here to organic molecules and metal chelates.

Except for a single early paper (5), much of the pioneering work in this area has been by Offen, with very significant contributions by Nicol also. The first studies emphasized changes in peak location along with qualitative or semiquantitative measurements of intensity and some lifetime measurements. More recently, with modern instrumentation it has been possible to measure radiative and nonradiative rates for dissipation of the excitation (k_r and k_{nr}).

I present in outline the major features that have been observed in four categories: (a) fluorescence, (b) phosphorescence, (c) excimer formation, and (d) luminescence as a biological probe, referring the readers to the literature for details. I then discuss more completely two kinds of studies where pressure has resolved problems difficult to establish based on atmospheric pressure data.

FLUORESCENCE

Much of the early work on fluorescence of aromatic and related hydrocarbons in condensed phases centered on the peak shift with pressure (6–25). Several generalizations can be made. The emission (as well as the absorption) shifts to lower energy (red) with increasing pressure. The rate of shift with pressure is usually larger in the crystalline state than when the molecule is dissolved in polymer films or in liquids. The shift for states of ^1L_a symmetry is greater than that for states of ^1L_b symmetry. The conclusion is that the degree of red shift is dominated by the polarizability of the medium, and of the excited state relative to the ground state. This effect is illustrated in Figure 1 where pressure-induced changes in peak location for three diphenylpolyenes are plotted as a term $(n^2-1)/(n^2+2)$ (n = refractive index), which is proportional to the polarizability for a series of media as diverse as fluorocarbons and polystyrene (26). The dominant effect of the polarizability is clear.

Lifetime measurements (13, 15, 19) indicate that at ordinary temperatures a modest decrease with increasing pressure is observed; this is probably associated with an increase in the nonradiative rate of energy dispersion k_{nr} due to increased vibrational overlap as the excited state moves closer in energy to the ground state. This conclusion is fortified by measurements of luminescent efficiency vs peak energy. The situation is best illustrated by studies on azulene derivatives (27). These molecules can emit from either of two excited states, S_1 and S_2, one of which (S_2) is much more polarizable than the other. In Figures 2 and 3 we observe peak shift and relative emission efficiency plotted vs pressure for these two states for an azulene derivative dissolved in polymethylmethacrylate (PMMA). For S_2 we observe a red shift of ~ 1500 cm^{-1}, accompanied by

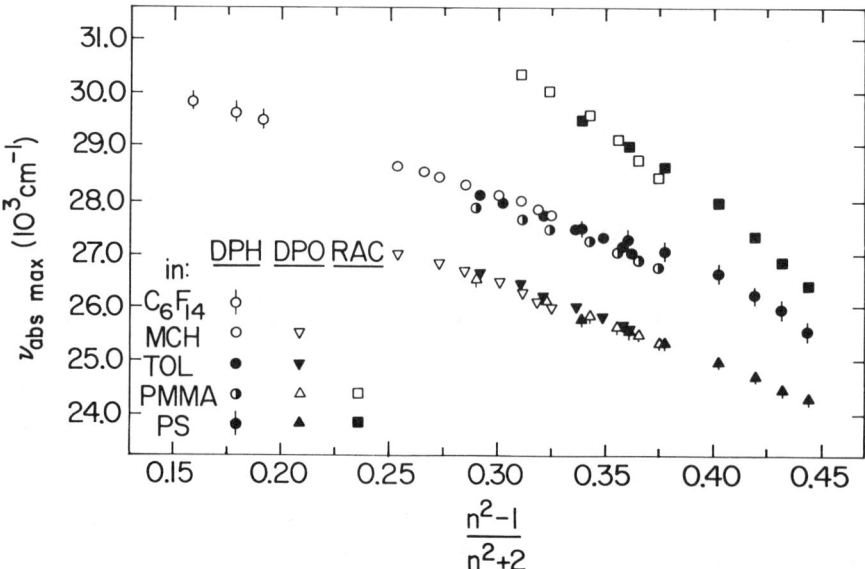

Figure 1 Peak location vs $(n^2-1)/(n^2+2)$ a measure of polarizability (n = refractive index).

a decrease by a factor of ~50 in emission efficiency, while for the other state (S_1) a blue shift of the same magnitude results in an increase in efficiency by a factor of ~45.

Also in a study of metalloporphyrins in PMMA and polystyrene (PS) (28), the relative nonradiative rates for fluorescence (and phosphorescence) in the two media scaled very closely with the relative peak shifts; the larger shifts in the more polarizable PS are accompanied by larger changes in k_{nr}.

High pressure fluorescence studies for heterocyclic molecules, while somewhat less extensive, have also yielded important results (29–37). Although it is difficult to generalize, red shifts with pressure are apparently somewhat smaller for $n-\pi^*$ than for $\pi-\pi^*$ emissions, due to smaller transition moments and smaller differences in polarizability in the ground and excited states; indeed, blue shifts may be observed.

Thus, increased interaction with the environment tends to lower the energy of the $\pi-\pi^*$ excitation vis-à-vis the $n-\pi^*$. This effect is illustrated by a study of fluorenone (34) in polyisobutylene (PIB), poly(4-methyl-l-pentene) (PMB), polymethylmethacrylate (PMMA), polystyrene (PS), and in the crystalline form (C). The molecule-environ-

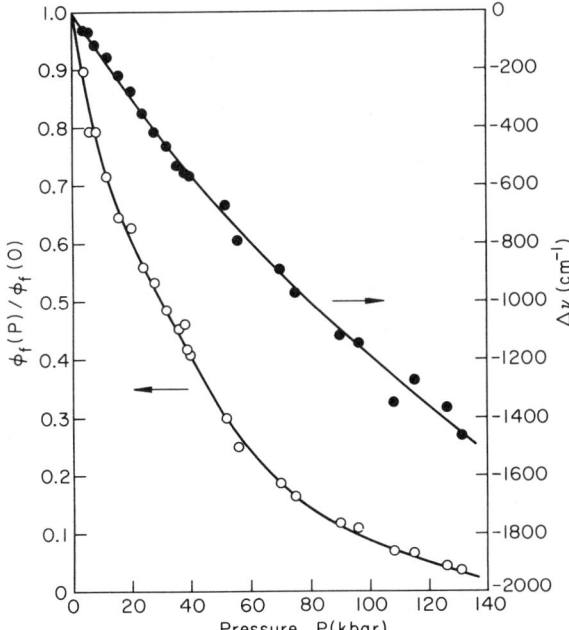

Figure 2 Change in peak location and emission intensity vs pressure for S_2-S_0 emission in azulene derivative.

ment interaction increases in the order PIB ≃ PMB < PMMA < PS < C. In PIB and PMB, the one atmosphere emission is weak, and almost entirely $n-\pi^*$ in character. The crystalline emission is a factor of 200 more intense and mostly $\pi-\pi^*$. The other solvents are intermediate in intensity and character. In PIB and PMB there is an increase in intensity by a factor of ~ 80 from one atmosphere to 120 kb as the $\pi-\pi^*$ emission becomes stabilized and takes over. At higher pressure, intersystem crossing and thermal dissipation to the ground state become important. In PMMA, the intensity starts at a higher level and increases less, with a maximum at lower pressure (60 kb) because of the increased $\pi-\pi^*$ character at one atmosphere; this effect is stronger in PS. In the crystal, the intensity decreases monotonically with increasing pressure as it is initially almost entirely $\pi-\pi^*$ in character. The large red shift enhances intersystem crossing and thermal dissipation of the energy.

Both the absorption and fluorescence peaks in electron donor-acceptor (EDA) complexes generally shift to lower energies with increasing pressure and the peaks increase significantly in intensity (38, 39). The

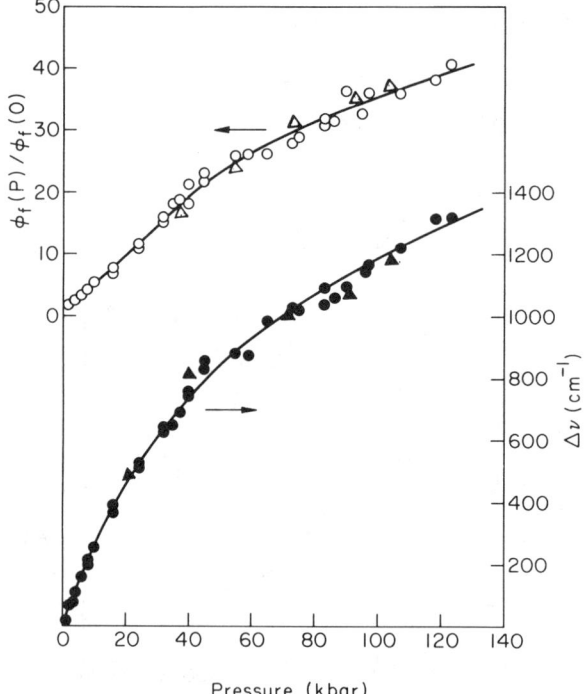

Figure 3 Change in peak location and emission intensity vs pressure for S_1-S_0 emission in same compounds as in Figure 2.

intensity effects can be explained in terms of increasing probability of occupation of the excited state with compression as predicted by Mulliken (40). The Stokes shift usually decreases with increasing pressure; this can be explained in terms of reasonable potential energy curves for the ground and excited states.

Recently there was an extensive high pressure study of four intramolecular charge transfer complexes under pressure, including fluorescence (and phosphorescence) quantum yields, lifetimes, and peak energies in a variety of liquid and polymeric solvents as well as in the crystalline state (41). The results are too complex to be summarized easily, but generally the effect of increasing pressure has many analogies to that of increasing solvent polarity for these materials.

PHOSPHORESCENCE

In relatively simple molecules, the shift of the phosphorescence peak is in the same direction as the fluorescence, but smaller by a factor of 4–5 for states of the same symmetry (42, 43). One can rationalize this result in a rather simple way: the polarizability of an excited triplet state should be less than that of the corresponding singlet, since the necessity for the two electrons of the same spin to stay apart places a restriction on the deformation of the electron cloud in its interaction with its environment.

A study of the intensity of fluorescence emission from anthraldehyde derivatives as a function of pressure (44) demonstrates this relative shift in a dramatic manner. In these molecules the fluorescence is strongly quenched at ambient pressure, since a triplet state lies not far below the excited singlet, so that intersystem crossing is rapid. As the singlet decreases in energy with respect to the triplet state, the barrier to intersystem crossing increases and the quantum efficiency of fluorescence emission increases by four orders of magnitude in 100 kb.

By far the most extensive studies of phosphorescence of organic compounds have been performed by Offen and his co-workers. These studies fall into several categories: (a) the spectra and lifetimes of aromatic hydrocarbons at room temperature and at 77 K (45–53), (b) the temperature coefficient of the lifetime at constant pressure (54, 55), and (c) studies of $n-\pi^*$ emissions in aromatic carbonyl compounds at room temperature and 77 K (56–60).

For the hydrocarbons, the lifetime shortens by 20–30% in 30 kb. The lifetime naturally increases with decreasing temperature at all pressures, but the temperature coefficient is less at high pressure. Similar effects were observed in halogenated hydrocarbons but the temperature coefficient is markedly smaller. Pressure strongly inhibits the quenching effect of oxygen, since diffusion rates are reduced in polymeric matrices. Effects of deuteration have also been studied (57, 59).

For the $n-\pi^*$ phosphorescence in aromatic carbonyls (56–58), small red or blue shifts were observed depending on the compound and the matrix. Lifetimes decreased with increasing pressure as for $\pi-\pi^*$ transitions.

EXCIMER AND EXCIPLEX EMISSION

In a variety of materials, a broad structureless emission is observed at an energy lower than the normal lowest singlet, but identifiable on a number of grounds from phosphorescence. This emission is assigned to molecules associated in the excited state, either with identical molecules (*excimers*) or in unlike pairs (*exciplexes*). In a situation in which these complexes can form, one might expect pressure to favor them, as they should be accompanied by a decrease in volume of the total system. Early observations by Jones & Nicol (61–63) indeed showed increased excimer formation with pressure for a number of aromatic hydrocarbons. In addition to a red shift and reversible increase in intensity, they observed an irreversible increase, which they associated with defects introduced in the crystal by quasihydrostatic pressure.

Johnson & Offen (64–66) studied excimers of pyrene and perylene in toluene, ethanol, and cyclohexane at 296 and 77 K. They discussed in some detail the competition between viscosity effects that inhibit excimer formation and the volume decrease that should favor excimers.

In polyvinyl carbazole (PVCA), there exist two excimer emissions, formed by traps on the chain, which capture hopping excitons. The emission was studied as a function of pressure in neat PVCA and for PVCA in dilute solution in polymethylmethacrylate (PMMA), polystyrene (PS), and polyisobutylene (PIB) (67). At low pressures, emission from the trap requiring chain rotation dominated, while at high pressure, the trap requiring no movement gave most of the emission. The pressure for changeover is in the order PIB > PMMA > PS > PVCA. The glass transition temperatures are in the inverse order; this demonstrates the effect of the rigidity of the matrix on the type of excimer formed as well as the usefulness of high pressure luminescence to trace local mobility in polymers.

LUMINESCENCE AS A BIOLOGICAL PROBE

The use of luminescence as a biological probe, especially of protein conformation and motion, is now over 25 years old. For the past 7–8 years, significant high pressure studies have been made. Since there has been a recent review of high pressure in biochemistry (68), and one specifically on high pressure luminescence is in preparation (69), I give only the briefest outline here.

Two types of luminescent probes have been used (70–78). The amino acid tryptophan emits because of the indole group it contains, with characteristics that vary drastically with its environment. (See the next section for indole studies.) Tyrosine can also be used. It is also possible to use ligands such as *l*-anilino-8-napthalene-sulphonate (ANS) or 6-propionyl-2-dimethylaminonaphthalene (PRODAN) to probe the availability of covalent sites for attachment. In this case it is necessary to separate the effects of pressure on site availability and equilibrium between attached and free ligand. For this purpose, a study of model compounds is useful (73).

The effect of pressure on the equilibrium between ligand and protein favors detachment of the ligand, but an increase in availability of sites may counteract this effect. Both reversible and irreversible changes of conformation have been observed, at times in the same protein at different pressures.

Depolarization of fluorescence has also proved to be a powerful tool (74), reinforcing the results of peak shift and intensity studies. Finally, a recent investigation of pressure effects on tRNA conformation has been made (75).

ENVIRONMENTAL EFFECTS

One of the significant problems in condensed phase science is to relate the electronic properties of a localized center to the bulk properties of the medium in which the center is contained. Pressure is a particularly useful parameter for this problem, in that one can vary these macroscopic properties in a continuous and controllable way over large ranges without changing the temperature or the chemical nature of the medium. In this section I consider the relationship of viscosity or dielectric constant to various luminescence properties of molecules or complexes.

For a number of diphenylmethane and triphenylmethane dyes it has been noted that the luminescence efficiency increases with increasing solvent viscosity. Förster & Hoffman (76) developed a theory of luminescence efficiency for these dyes which involves the rate of relaxation in position of the planar phenyl groups vis-à-vis the rate of emission. Assuming a quadratic coupling to the environment, they established that over a considerable range the efficiency should vary as the two-thirds power of the viscosity (η). At very high viscosities the emission efficiency should approach unity asymptotically and they assumed that at very low viscosities the efficiency would become independent of η.

Attempts to test this theory by using temperature or composition to control η gave equivocal results. With pressure one can cover a range of 10^6 in viscosity at constant temperature with molecules of very similar character. Figure 4 shows the result of such a study on crystal violet (77). The line has a slope of 2/3. Over 3.5 orders of magnitude in viscosity the theory holds. The leveling at high viscosity is as expected. There is an as yet unexplained anomaly at very low viscosity where the intensity changes more rapidly with viscosity rather than leveling as predicted. The results for a similar dye, Auramine-O, were identical except that the low viscosity anomaly was a little larger, as represented by the *dashed line* in Figure 4.

As indicated above, high pressure fluorescence has become a useful probe for understanding conformation of biological molecules and, in particular, protein conformation. One of the important probes is the luminescent amino acid tryptophan containing the chromophore indole. It is useful because its peak energy and radiative and nonradiative decay rates are sensitive to changes in conformation of the protein, with resulting changes in the environment of the tryptophan. To investigate what macroscopic property is dominant in bringing about these changes

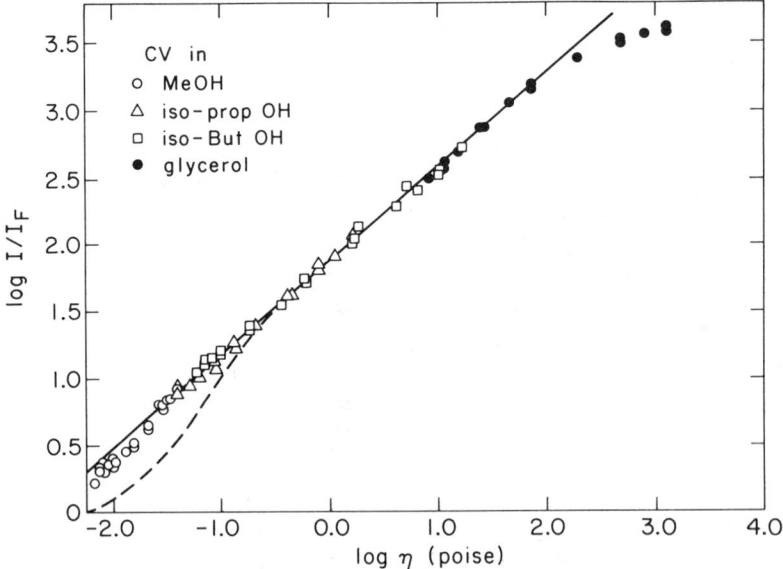

Figure 4 Log η (viscosity) vs log of the relative emission intensity for crystal violet (---Auramine O) in various alcoholic solvents.

in localized properties, experiments were performed on indole, tryptophan, and 5-methoxyindole (78). The first two of these have two closely spaced excited states so that the emitting state may be ^1L_a or ^1L_b, depending on external influences; their behavior as a function of pressure is very similar. The 5-methoxyindole emits always from the ^1L_b state. (The ^1L_a, ^1L_b nomenclature is not strictly applicable but is widely used.)

A variety of factors, including level inversion, exciplex formation, solvent reorientation, thermal isomerization, and emission from a solvated Rydberg state, have been proposed to account for the changes of peak location and other emission properties with solvent. These, together with some specific anomalies observed in water and methanol under pressure, are discussed in the original paper (78). Here we emphasize only the major common factors in establishing $h\nu$, k_r, and k_{nr} in a variety of solvents including hydrocarbons, alcohols, and water to a pressure of 10 kb.

In a liquid solution one would expect that the dispersion of thermal energy would be a collisional process. As discussed in the original paper, considerable care was exercised to eliminate impurity, oxygen, and concentration effects so that one should be primarily involved with solvent-solute collisions. The relevant macroscopic parameter is the viscosity. Figure 5 presents plot of log η vs k_{nr} for indoles in a series of solvents. Clearly the macroscopic viscosity controls the thermal dissipa-

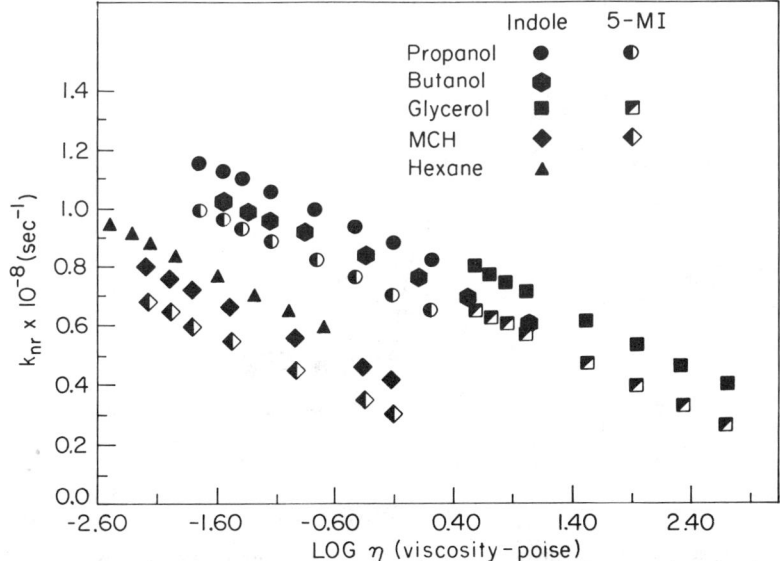

Figure 5 Log η vs k_{nr} for indoles in various solvents.

tion of energy. Incidentally, a similar correlation was observed for metalloporphyrins in various solvents (79).

For indole the emission in the hydrocarbons is at higher energy than in alcohols, which reduces the probability of a successful thermal transfer in a given collision. Figure 6 illustrates the effect of rigidity on thermal dissipation of excitation very dramatically for porphyrins in m-xylene and $CHCl_3$, which freeze at ~ 4.5 and 5.5 kb, respectively, at 25°C; k_{nr} decreases discontinuously by a factor of 5–7 when the solvent freezes.

What is the effect of the environment on the radiative rate k_r? Figure 7 presents the effect of the dielectric constant (ε) on k_r for several indoles in solvents from hexane and methylcyclohexane (MCH) through a series of alcohols to water. ε varies from 2 to 110. For the hydrocarbons and higher alcohols, ε varies little with pressure, but for water it increases from ~ 80 to ~ 110 in 10 kb. In spite of secondary variations in some alcohols, ε is clearly the controlling parameter over this very large range

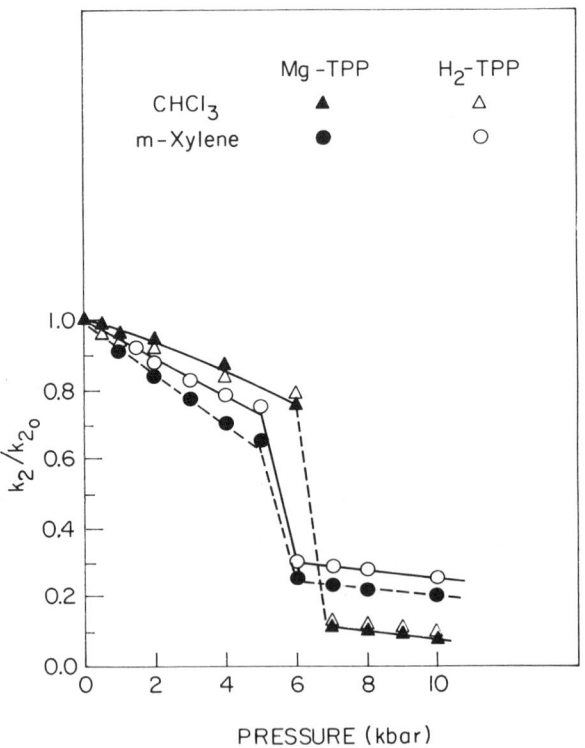

Figure 6 Change in nonradiative rate ($k_2 = k_{nr}$) vs pressure for porphyrins in m-xylene and $CHCL_3$ showing discontinuity at the freezing point.

HIGH PRESSURE MOLECULAR LUMINESCENCE   35

of solvents. A group of studies in which pH and ε were varied separately at 1 atm in aqueous solutions confirmed this conclusion.

Figure 8 shows the correlation of peak energy with dielectric constant for 5-methoxy-indole (5MI) and indole (IN). 5MI has an excited state of ^1L_b symmetry under all conditions. This is known to be a state of relatively low polarizability. There is considerable debate about the

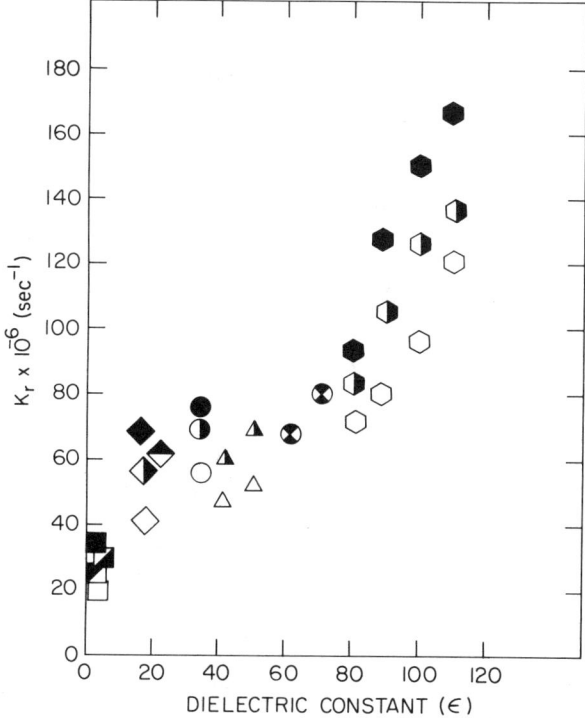

Figure 7 k_r vs dielectric constant (ε). The symbols have the following meaning:

	IND	5-MO	TRYPT.
WATER	◐	○	⬢
METHANOL	◐	○	●
PROPANOL	◆		
BUTANOL	◆	◇	◆
GLYCEROL	▲	△	
MCH	◪	☐	■
HEXANE	◪		
MeOH-H₂O	⊛		

excited state of indole. In alcohols, it probably has the relatively polarizable ^1L_a configuration, but it is frequently considered to be entirely different in water than in less polar solvents such as alcohols. The energy varies continuously with ε, however, so this macroscopic parameter can account for all changes of the emission energy. The second order effects

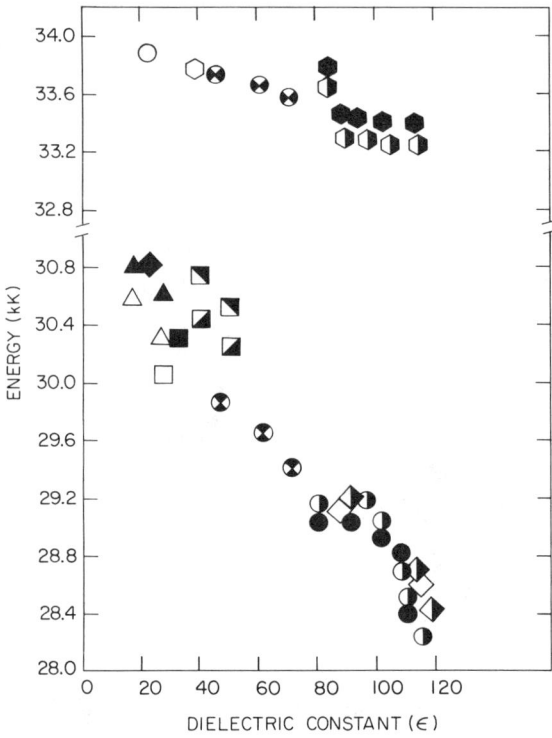

Figure 8 Peak energy vs dielectric constant (ε). The symbols have the following meaning:

	INDOLE	5-MO	TRYPT.
GLYCEROL	◣		◪
PROPANOL	◆		
BUTANOL	▲	○	△
METHANOL	■	◯	□
MeOH-H₂O	⊗	⊖	
IN WATER			
EXPERIMENTAL VARIABLE	INDOLE	5-MO	TRYPT.
PRESSURE	●	⬢	◇
DIELECTRIC CONSTANT	◐	⬢	◈

for k_r and $h\nu$ in water and methanol discussed in the original paper do not affect this overall view.

An area of considerable interest in recent years is the luminescence of molecules or complexes that can emit from two excited states which may differ in geometry or interaction with the surrounding solvent molecules. Two questions arise: 1. To what extent is the relative emission intensity governed by an equilibrium distribution of excitation between two excited states, or by kinetic processes either within the molecule or in the surroundings? 2. Is the kinetic governing process a matter of the height of an energy barrier to molecular rearrangement or a diffusional process among the surrounding molecules? High pressure has contributed to understanding both of these problems.

The first question has been studied through the luminescence of 3-hydroxyflavone (80). Sengupta & Kasha (81) studied flavones extensively as a function of solvent (81) and temperature and found that the rigidity of the medium may be an important factor in the distribution of excitation between excited states. The effect of pressure on the emission in isobutanol, glycerol, and two mixtures has been studied.

The fraction of tautomer emission as a function of pressure is shown in Figure 9 (80). In isobutanol, the fraction of tautomer increases with pressure throughout the range. Lifetimes measured from each of the two peaks were the same at all pressures, so the two excited states are in equilibrium. From the pressure dependence of the equilibrium constant one can extract the volume change of the system when the tautomerization occurs, which is a decrease of 2–3 cc/mole.

For glycerol and the mixed solvents, the initial increase in tautomer fraction with pressure is similar, but in glycerol the fraction of tautomer maximizes at 2.5–3 kb and decreases thereafter. In the 90-10 (glycerol/isobutanol) mixture, the maximum occurs at 5–6 kb and in the 50-50 mixture at about 7–7.5 kb. In each case the viscosity at the maximum is 45–50 P. It is reasonable to assume that at lower pressures (viscosities), equilibrium obtains, but that starting just before the maximum the distribution is kinetically limited. This assumption is confirmed by the lifetime measurements. Figure 10 (80) exhibits results from glycerol and for the 50-50 mixture. In each case, the lifetimes from the two states are identical at lower pressures, but at or near the maximum they start to diverge sharply. Results for the 90-10 mixture are consistent.

A variety of electron donor-acceptor (EDA) complexes exhibit emission from either of two states, depending on the rigidity of the medium. Based on the effects of temperature and solvent, workers have speculated that the effect of increasing viscosity is to increase the energy barrier to transformation along an electronic coordinate associated with the emission (81). One would expect that the rate of rearrangement should be strongly dependent on temperature at constant viscosity; in fact, the temperature coefficient should increase with increasing viscosity.

A series of EDA complexes of tetracyanobenzene (TCNB) with aromatic hydrocarbons, especially xylenes, have been studied (82, 83). These EDA complexes exhibit emission from a geometry like the ground state (the FC state) in rigid media and from quite a different geometric arrangement (the EQ state) in fluid media. A series of paraffin hydrocarbon solvents were used: methylcyclohexane (MCH), heptamethylnonane (HMN), tetramethyl pentadecane (TMPD), and a mixed solvent. These have in common about the same polarizability but have very different pressure coefficients of viscosity, so a range of $\sim 10^5$ in viscosity can be covered at constant temperature.

At both 25°C and 0°C, the distribution of emission at steady state between the two states and the time dependent emission (decay) have

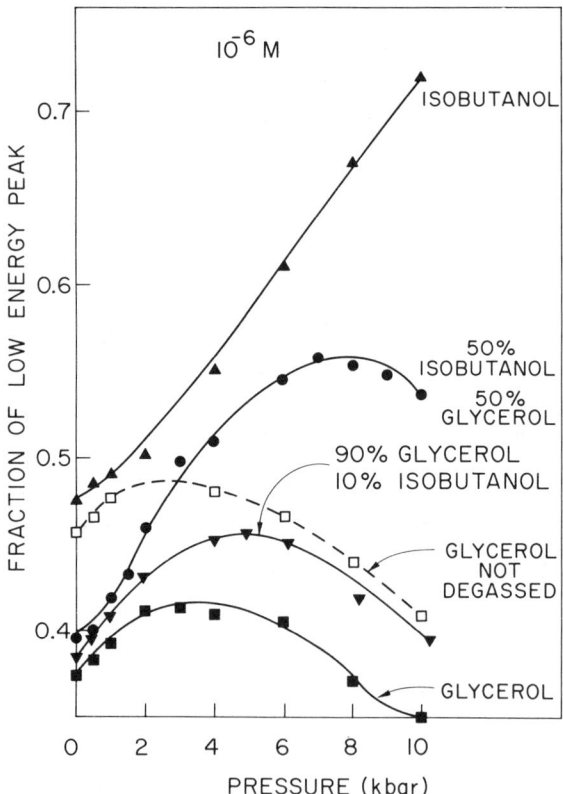

Figure 9 Fraction of low energy (tautomer) emission vs pressure—3-hydroxyflavone in various solvents.

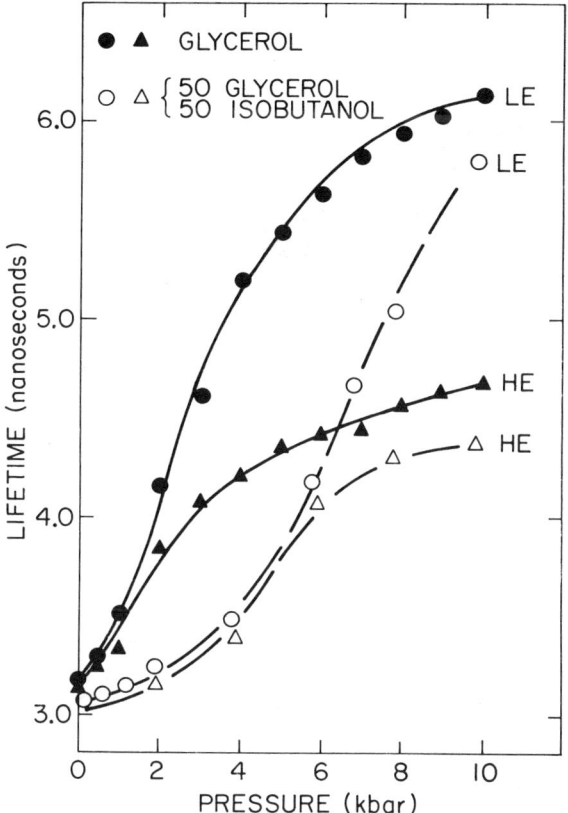

Figure 10 Radiative lifetime vs pressure for low energy (LE) and high energy (HE) peaks of 3-hydroxyflavone in two solvents.

been measured as a function of pressure. From these results one can extract the rate of rearrangement k_{RE} from the FC to EQ state. Figure 11 plots $\log k_{RE}$ vs $\log \eta$. Clearly k_{RE} is independent of solvent, donor, or temperature except insofar as these affect viscosity. The slope at quite low viscosity is ~ -1, but above 3 P it is constant at -0.39, independent of temperature. A more direct check of the slope is obtained from the ratio

$$x = \left[\frac{\log \dfrac{k_{RE}(25)}{k_{RE}(0)}}{\log \dfrac{\eta(25)}{\eta(0)}} \right]_{P, \text{solvent}}$$

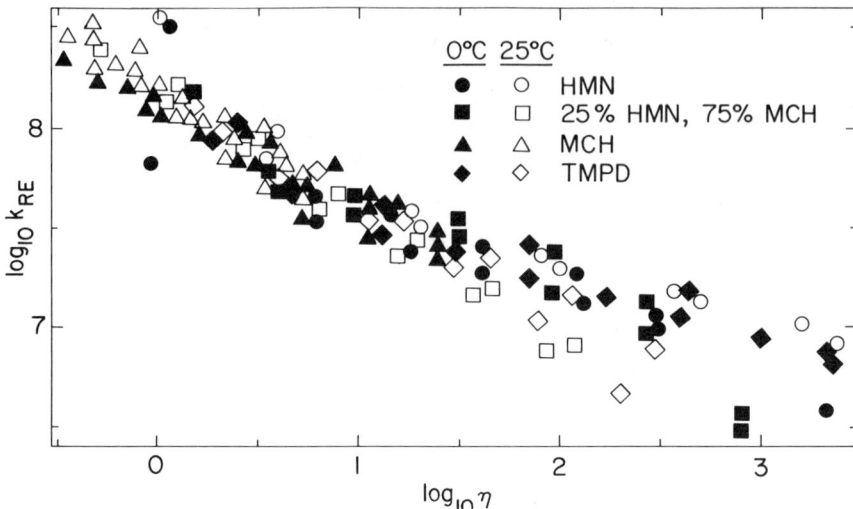

Figure 11 Log k_{RE} (rate of rearrangement of complex and surroundings) for TCNB-xylene EDA complexes vs log η (viscosity) at 0 K and 25 K.

x was found to vary from -1 at low viscosities to an average value of -0.42 above 2–3 P. The effect of viscosity is clearly a diffusional one, involving rearrangement of surrounding molecules to provide an appropriate environment for one state or the other.

From the Debye equation for molecular relaxation as a function of viscosity, one can write

$$k_{RE} = C\frac{T}{\eta}.$$

This predicts a relaxation rate inversely proportional to viscosity, with a small temperature dependence. Exactly this condition obtains in the low viscosity region. The Debye equation was derived for rigid spheres. MCH, the solvent used for the lower viscosity data, is a reasonable approximation to this condition. The other solvents are long flexible molecules and segmental motion or rotation around the long axis could permit rearrangement without contributing significantly to viscous flow.

The studies presented in this section demonstrate the powerful nature of pressure as a tool for understanding the relationship between localized electronic phenomena and the bulk properties of the medium in which they occur.

PRESSURE TUNING OF ENERGY LEVELS

From the electronic viewpoint, the most general effect of pressure is to increase overlap of electronic orbitals on adjacent atoms or molecules and thus perturb different types of orbitals by different amounts. This pressure tuning of energy levels has been used to characterize electronic states, to test theories, and to bring about electronic transitions to new ground states as shown by insulator-metal transitions, spin flip transitions, etc. Pressure tuning has been very effective in testing theories of luminescence in inorganic crystals, as reviewed in Refs. (1–4).

I present here a study of some rare earth chelates in which the strong perturbation of the ligand orbitals by pressure has dramatic effects on the energy transfer process. The original study (84) involved three rare earths and four sets of ligands, all in tris conformation; here I discuss only two chelates of europium (Eu) with acetylacetonate derivatives.

The rare earth ion is surrounded by six oxygens whose electronic character is modified by the substituents on the terminal carbons of the acetylacetonate. These modifications establish the energies of the excited singlet and triplet states of the ligands. I discuss primarily the ligand dibenzilmethide (DBM) with two phenyl substituents, but briefly mention results for a complex with thenoyltrifluoroacetylacetonate (TTF).

The usual excitation is to the first excited singlet state (S_1) of the ligand, while emission is observed from the lowest excited $4f$ level (labeled $^5D_0$) of the rare earth.

The question under review is regarding the nature of the path from S_1 to $^5D_0$ and the effect of various paths on the emission intensity. If the lowest triplet state of the ligand lies high in energy above the $^5D_1$ and $^5D_0$ states of the Eu^{+3} ion, it is not involved in the energy transfer of e.g. $^5D_1 \rightarrow {}^5D_0$. If it lies between the $^5D_1$ and $^5D_0$ levels or lower, it can be profoundly involved in the process.

The location of the ligand S_1 and T_1 levels varies with the type of substituents on the ligand and with the medium, crystalline or polymeric. With increasing pressure, the S_1 and T_1 levels shift to lower energy vis-à-vis the Eu^{+3} levels, so from pressure studies we can test various hypotheses concerning the effect of changing the character of the ligands or the medium on the luminescence efficiency.

Figure 12 shows the energy of the $^5D_0$ and $^5D_1$ levels of Eu^{+3} on the right. On the left are the S_1 and T_1 levels of DBM in the crystalline state and dissolved in PMMA (polymethylmethacrylate). The *solid lines* represent the energies at 1 atm pressure, while the *dashed lines* represent the energies at 40 kb. (By 90 kb the S_1 and T_1 levels shift again as far to lower energy.) As one can see, the ligand levels lie at higher energy and shift less in the PMMA than in the crystalline state. As a result, there should be no important change in the ligand-metal energy transfer with pressure in PMMA. In the crystalline state there are two possible paths for transfer of excitation from $^5D_1$ to $^5D_0$. The direct transfer occurs only slowly. There is, however, an alternative possibility. The path from $^5D_1$ to

T_1 and back to $^5D_0$ is allowed. As the energy difference between T_1 and $^5D_0$ decreases, the transfer rate will increase. However, when T_1 comes within a few k_BT of $^5D_0$, back transfer can become significant; as it comes below $^5D_0$, this quenching mechanism will dominate. In addition, as T_1 moves to lower energy, the rate of quenching via thermal crossing to the ground state will increase. If transfer via T_1 is the most important process, one would observe for DBM in PMMA a modest increase in intensity with pressure as $T_1 \rightarrow {}^5D_1$ transfer improves, and possibly a slight decrease at very high pressure due to back transfer.

In the crystal one would predict a distinct increase at lower pressures followed by a dramatic drop as quenching takes over. Superimposed on these effects is a modest increase in emission intensity (a factor of 1.5 to 2 in ~60 kb) because of increase in the radiative rate due to increased spin-orbit coupling and better d-f orbital mixing at high pressure. In Figure 13 we see the relative emission efficiency from $^5D_0$ vs pressure. The results are exactly as predicted above. Evidently, transfer via T_1 is dominant.

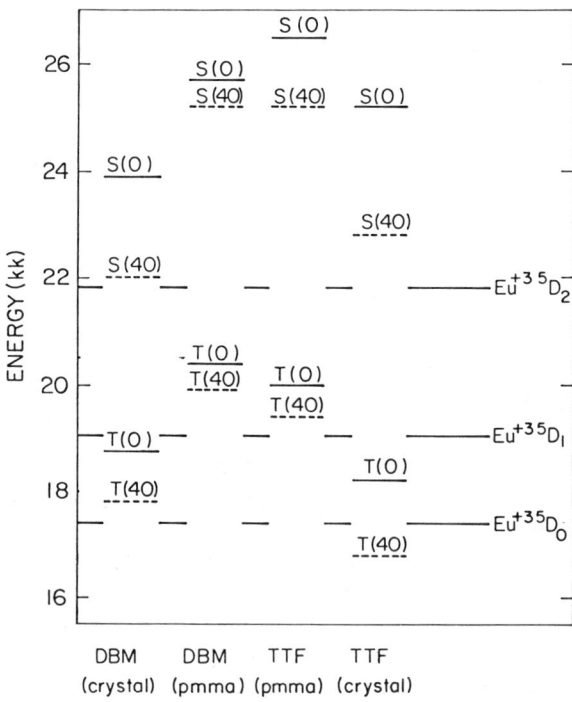

Figure 12 Excited state energy levels for Eu^{+3} and two ligands. For the ligands (0) represents one atmosphere energy while (40) represents energy at 40 kb.

There are a couple of ways to check this conclusion. From Figure 12 we see three possible methods to stimulate emission from $^5D_0$. If one excites directly to $^5D_0$, there should be only a modest change in intensity due to the change in radiative rate mentioned above. If one excites in the ligand, for the crystal, one would expect, relative to what one observes by exciting to $^5D_0$, a sharp increase at low pressure and a drop beyond 30 kb. When one excites to $^5D_1$, the relative effect should be the same as for excitation in the ligand. However, for the PMMA solution there should be no effect of the type of excitation. In Figure 14 we observe that the relative intensity effects are exactly as predicted.

Another check involves the rate of back donation from $^5D_0$ to T_1, which is limited by an energy barrier E. One can extract E from the temperature coefficient of either the emission intensity or lifetime. (Both have been used.) E should decrease as T_1 shifts to lower energy. The shift of T_1 can be established from the shift of phosphorescence from the

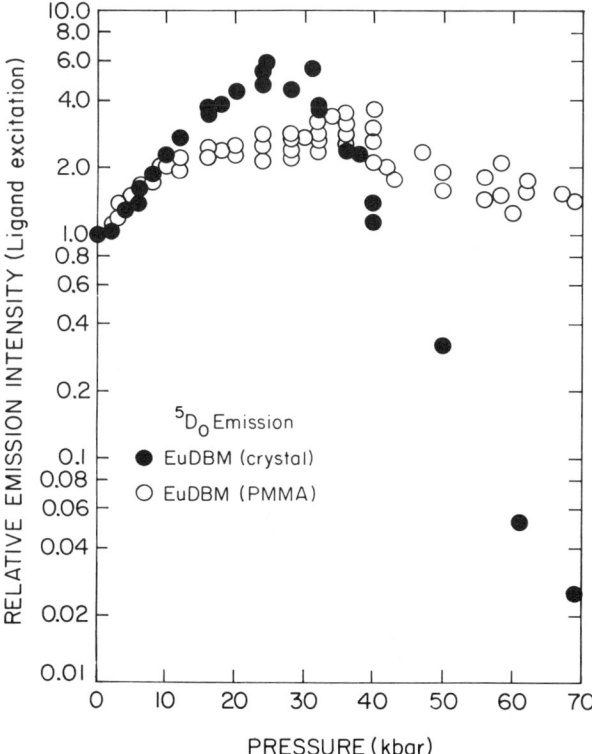

Figure 13 Relative emission efficiency vs pressure—$^5D_0$ emission of EuDBM in crystal and in polymethylmethacrylate (PMMA).

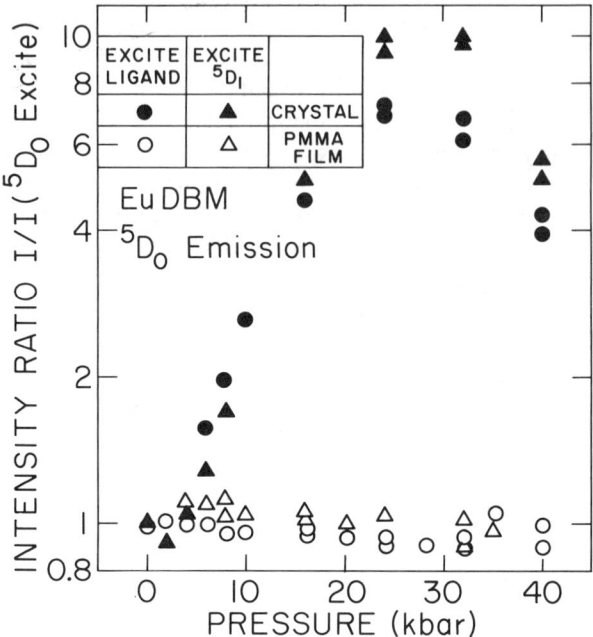

Figure 14 Emission efficiency vs pressure for EuDBM in crystal and PMMA when excited in ligand and to $^5D_1$ relative to excitations to $^5D_0$.

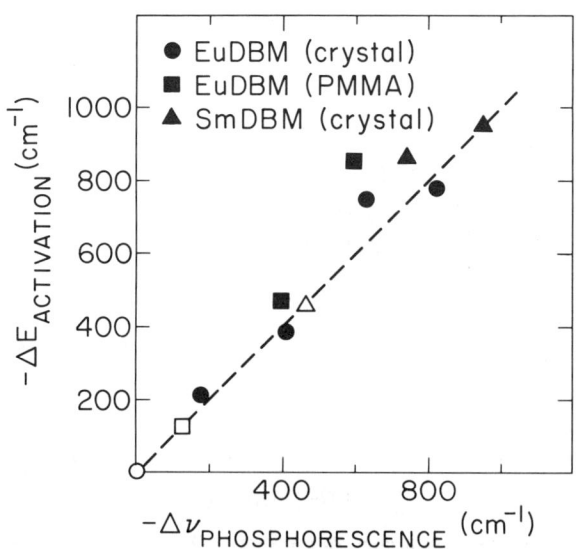

Figure 15 Change in energy barrier for $^5D_0$ to ligand T_1 energy transfer vs shift in T_1 phosphorescence peak for rare earth chelates.

gadolinum derivative. Figure 15 shows the change in E plotted against the shift of the phosphorescence. The *dotted line* has a slope of 45°, precisely confirming the hypothesis.

Finally, from Figure 12, we deduce the behavior of the TTF complex with Eu^{+3}. The ligand levels lie slightly lower than for DBM. In the plastic, this should have no important effect, but in the crystal, T_1 already lies near the optimum for ligand to $^5D_0$ transfer. Thus we can anticipate little if any increase at low pressure and a dramatic decrease at high pressure. These predictions are confirmed in Figure 16. One could thus, in principle, make first order estimates of emission efficiency of complexes knowing the location of T_1 and of the excited 4-f levels of the rare earth ion.

This is only one of a great many examples of how pressure tuning of energy levels can lead to a better understanding of atmospheric pressure phenomena.

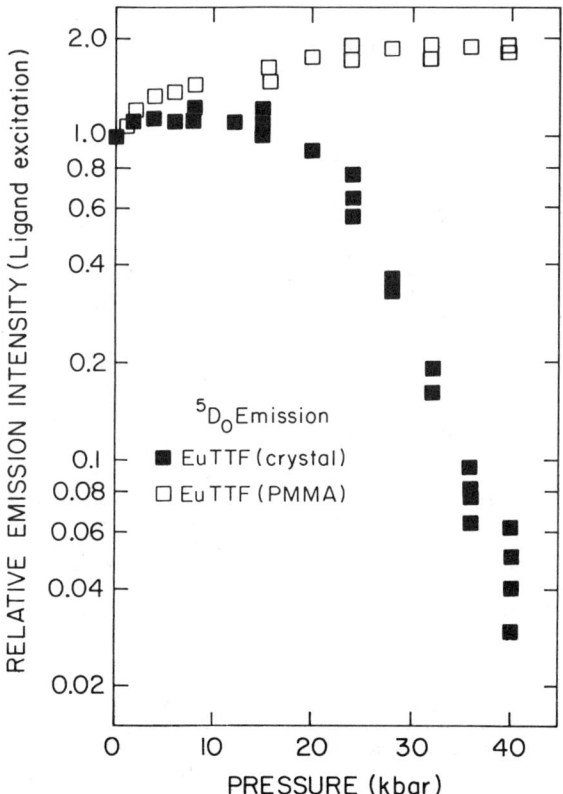

Figure 16 Relative emission efficiency vs pressure—$^5D_0$ emission of EuTTF in crystal and in PMMA.

SUMMARY

The studies of high pressure molecular luminescence reviewed here, together with results for inorganic systems discussed elsewhere, provide one more piece of evidence as to the versatility and power of high pressure as a tool for characterizing electronic states, testing theories concerning electronic phenomena, and, in general, obtaining a better understanding of electronic behavior in condensed systems.

ACKNOWLEDGMENT

The author thanks H. W. Offen and M. F. Nicol for sending copies of their very important papers on high pressure luminescence. Financial support from the Department of Energy under contract DE-AC02-76ER01198 is gratefully acknowledged.

Literature Cited

1. Drickamer, H. G. 1979. In *High Pressure Science and Technology*, ed. K. D. Timmerhaus, pp. 1–18. New York: Plenum
2. Drickamer, H. G., Mitchell, D. J., Schuster, G. B. 1980. In *Radiationless Transitions*, ed. S. H. Lined, pp. 289–316. New York: Academic
3. Drickamer, H. G., Klick, D. I. 1978. *J. Less Common Met.* 62:381–96
4. Drickamer, H. G. 1980. *Rev. Phys. Chem. Jpn.* 50:1–18
5. Drickamer, H. G., Gregg, D. W. 1961. *J. Chem. Phys.* 35:1780–88
6. Nicol, M. 1965. *J. Opt. Soc. Am.* 55:1176–78
7. Nicol, M. F. 1966. *J. Chem. Phys.* 45:4753–54
8. Nicol, M., Vernon, M., Woo, J. T. 1975. *J. Chem. Phys.* 63:1992–99
9. Ebisuzaki, Y., Taylor, T. J., Woo, J. T., Nicol, M. 1977. *J. Chem. Soc. Faraday Trans. 2* 73:253–64
10. Nicol, M. F., Hara, Y., Wiget, J. M., Anton, M. 1978. *J. Mol. Struct.* 47:371–78
11. Hara, Y., Nicol, M. F. 1978. *Bull. Chem. Soc. Jpn.* 51:1985–87
12. Offen, H. W., Eliason, R. R. 1965. *J. Chem. Phys.* 43:4096–4101
13. Offen, H. W., Phillips, D. T. 1968. *J. Chem. Phys.* 49:3995–97
14. Hein, D. E., Offen, H. W. 1969. *Mol. Cryst.* 5:217–24
15. Kim, J. J., Beardslee, R. A., Phillips, D. T., Offen, H. W. 1969. *J. Chem. Phys.* 51:2761–62
16. Offen, H. W. 1966. *J. Chem. Phys.* 44:699–703
17. Offen, H. W., Beardslee, R. A. 1968. *J. Chem. Phys.* 48:3584–87
18. Nakashima, T. T., Offen, H. W. 1968. *J. Chem. Phys.* 48:4817–21
19. Johnson, P. C., Offen, H. W. 1970. *Chem. Phys. Lett.* 6:505–7
20. Beardslee, R. A., Offen, H. W. 1971. *J. Chem. Phys.* 55:3516–19
21. Johnson, P. C., Offen, H. W. 1972. *J. Chem. Phys.* 57:336–38
22. Johnson, P. C., Offen, H. W. 1972. *J. Chem. Phys.* 57:1473–75
23. Okamoto, B. Y., Drickamer, H. G. 1974. *J. Chem. Phys.* 61:2878–83
24. Okamoto, B. Y., Drickamer, H. G. 1974. *Proc. Natl. Acad. Sci. USA* 71:4757–59
25. Brey, L., Schuster, G. B., Drickamer, H. G. 1979. *J. Am. Chem. Soc.* 101:129–34
26. Brey, L., Schuster, G. B., Drickamer, H. G. 1979. *J. Chem. Phys.* 71:2765–72
27. Mitchell, D. J., Schuster, G. B., Drickamer, H. G. 1977. *J. Am. Chem. Soc.* 99:7489–95
28. Politis, T. G., Drickamer, H. G. 1981. *J. Chem. Phys.* 74:263–72
29. Nicol, M., Wild, S. M., Yancey, J. 1973. *J. Chem. Phys.* 58:4350–57
30. Park, E. H., Kadhim, A. H., Offen, H. W. 1968. *Photochem. Photobiol.* 8:261–72
31. Johnson, P. C., Offen, H. W. 1971. *J. Chem. Phys.* 55:2945–51
32. Mastrangelo, C. J., Offen, H. W. 1975. *J. Solution Chem.* 4:965–67
33. Hara, K., Schuster, G. B., Drickamer, H. G. 1977. *Chem. Phys. Lett.* 47:462–65

34. Mitchell, D. J., Schuster, G. B. Drickamer, H. G. 1977. *J. Chem. Phys.* 67:4832–35
35. Hook, J. W. III, Drickamer, H. G. 1978. *J. Chem. Phys.* 69:811–15
36. Rollinson, A. M., Schuster, G. B., Drickamer, H. G. 1978. *Chem. Phys. Lett.* 59:559–61
37. Mitchell, D. J., Schuster, G. B., Drickamer, H. G. 1979. *J. Chem. Phys.* 70:2443–49
38. Offen, H. W., Studebaker, J. F. 1967. *J. Chem. Phys.* 47:253–55
39. Kadhim, A. H., Offen, H. W. 1968. *J. Chem. Phys.* 48:749–53
40. Mulliken, R. S. 1952. *J. Am. Chem. Soc.* 64:811–17
41. Rollinson, A. M., Drickamer, H. G. 1980. *J. Chem. Phys.* 73:5981–96
42. Nicol, M., Somekh, J. 1968. *J. Opt. Soc. Am.* 58:233–35
43. Shaw, R. W., Nicol, M. 1976. *Chem. Phys. Lett.* 39:108–12
44. Mitchell, D. J., Schuster, G. B., Drickamer, H. G. 1977. *J. Am. Chem. Soc.* 99:1145–48
45. Offen, H. W., Baldwin, B. A. 1966. *J. Chem. Phys.* 44:3642–43
46. Baldwin, B. A., Offen, H. W. 1967. *J. Chem. Phys.* 46:4509–13
47. Baldwin, B. A., Offen, H. W. 1968. *J. Chem Phys.* 48:5398–60
48. Baldwin, B. A., Offen, H. W. 1968. *J. Chem. Phys.* 49:2933–36
49. Baldwin, B. A., Offen, H. W. 1968. *J. Chem. Phys.* 49:2937–39
50. Simpson, J. D., Offen, H. W., Burr, J. G. 1968. *Chem. Phys. Lett.* 2:383–84
51. Offen, H. W., Hein, D. E. 1969. In *Molecular Luminescence*, ed. E. C. Lim. New York: Benjamin. 83 pp.
52. Offen, H. W., Hein, D. E. 1969. *J. Chem. Phys.* 50:5274–78
53. Beardslee, R. A., Offen, H. W. 1973. *J. Chem. Phys.* 59:4633–36
54. Rodriguez, S., Offen, H. W. 1970. *J. Chem. Phys.* 52:586–89
55. Beardslee, R. A., Offen, H. W., 1970. *J. Chem. Phys.* 52:6016–20
56. Kadhim, A. H., Offen, H. W. 1967. *J. Chem. Phys.* 47:1289–92
57. Simpson, J. D., Offen, H. W. 1970. *J. Chem. Phys.* 52:1467–71
58. Simpson, J. D., Offen, H. W. 1970. *Mol. Photochem.* 2:115–20
59. Simpson, J. D., Offen, H. W. 1971. *J. Chem. Phys.* 55:1323–26
60. Simpson, J. D., Offen, H. W. 1971. *J. Chem. Phys.* 55:4832–36
61. Jones, P. F., Nicol, M. 1965. *J. Chem. Phys.* 43:3759–60
62. Jones, P. F., Nicol, M. 1965. *J. Chem. Phys.* 48:5440–47
63. Jones, P. F., Nicol, M. 1968. *J. Chem. Phys.* 48:5457–64
64. Johnson, P. C., Offen, H. W. 1972. *J. Chem. Phys.* 56:1638–42
65. Johnson, P. C., Offen, H. W. 1973. *Chem. Phys. Lett.* 18:258–60
66. Johnson, P. C., Offen, H. W. 1973. *J. Chem. Phys.* 59:801–6
67. Chryssomallis, G., Drickamer, H. G. 1979. *J. Chem. Phys.* 71:4817–23
68. Heremans, K. 1980. *Rev. Phys. Chem. Jpn.* 50:259–73
69. Weber, G., Drickamer, H. G. 1982. *Q. Rev. Biophys.* In press
70. Li, T. M., Hook, J. W. III, Drickamer, H. G., Weber, G. 1976. *Biochemistry* 15:3205–11
71. Li, T. M., Hook, J. W. III, Drickamer, H. G., Weber, G. 1976. *Biochemistry* 15:5571–80
72. Visser, A. J. W. G., Li, T. M., Drickamer, H. G., Weber, G. 1977. *Biochemistry* 18:3079–83
73. Torgerson, P. M., Drickamer, H. G., Weber, G. 1979. *Biochemistry* 18:3079–83
74. Chryssomallis, G. S., Torgerson, P. M., Drickamer, H. G., Weber, G. 1981. *Biochemistry* 20:3955–59
75. Torgerson, P. M., Drickamer, H. G., Weber, G. 1980. *Biochemistry* 19:3957–60
76. Förster, T. H., Hoffmann, G. 1971. *Z. Phys. Chem.* NF75:63–69
77. Brey, L. A., Schuster, G. B., Drickamer, H. G. 1977. *J. Chem. Phys.* 67:2648–50
78. Politis, T. G., Drickamer, H. G. 1981. *J. Chem. Phys.* 75:3203–10
79. Politis, T. G., Drickamer, H. G. 1982. *J. Chem. Phys.* 76:285–91
80. Salman, O. A., Drickamer, H. G. 1981. *J. Chem. Phys.* 75:572–76
81. Sengupta, P. K., Kasha, M. 1979. *Chem. Phys. Lett.* 68:382–86
82. Thomas, M. M., Drickamer, H. G. 1981. *J. Chem. Phys.* 74:3198–3204
83. Thomas, M. M., Drickamer, H. G. 1981. *J. Chem. Phys.* 75:5246–49
84. Hayes, A. V., Drickamer, H. G. 1982. *J. Chem. Phys.* 76:114–25

DYNAMICS OF ENTANGLED POLYMER CHAINS

P. G. de Gennes and L. Léger

Collège de France, 75231 Paris Cedex 05, France

Polymer melts flow like conventional (viscous) liquids if they are subjected to slowly varying perturbations; but, at somewhat higher frequencies ω, they behave like an elastic rubber. These mechanical properties depend strongly on the polymer structure (linear/branched, flexible/rigid, etc). In this review, we concentrate on the best understood case: linear, flexible chains, with a large number (N) of monomers per chain. The crucial feature is that these chains are *entangled*. The effects of entanglements on the *mechanical* properties of polymers have been recognized and analyzed in some detail. More recently, entanglements have been probed at a more microscopic level: (*a*) self-diffusion, following the motions of one labeled chain; (*b*) measurements involving spatial effects at distances smaller than the coil size: polymer/polymer welding; spinodal decomposition of blends; neutron measurements on stretched samples; chemical reactions in melts.

We discuss all these phenomena within a single model, based on the "reptation" idea (1). This model is not final, and may well require future adjustments, but it gives a comparatively simple framework with which to unify these very different types of experiments.

The viscoelastic properties of polymers have been measured with great care, beginning with the pioneering work of J. Ferry (2). These properties display a certain characteristic time τ, which increases very rapidly with the molecular weight (or equivalently, with the number N of monomers per chain)

$$\tau = \tau_o N^a$$

where τ_o is a microscopic time (of order 10^{-10} sec in melts) and a is an exponent, of order 3.2–3.4. Since N can reach values as high as 10^4–10^5, the time τ may become extremely long (minutes).

The meaning of the long time τ was understood qualitatively many years ago (2, 3). At times $t < \tau$, the "knots" between chains cannot be undone, and the mechanical behavior of the melt is similar to the behavior of a permanently cross-linked network. At times $t > \tau$, the "knots" open up by Brownian motion; the chains can slip with respect to each other, and flow is observed.

The reptation model makes this picture slightly more quantitative, and leads to a definite prediction for the exponent $a = 3$. Edwards (4) visualized each chain as being confined in a "tube": it is therefore impossible for the chain under study (the "test chain") to intersect its neighbors. The test chain moves inside its own tube by snake-like motions (1). We can characterize this by a certain *mobility* μ_{tube} giving the velocity (along the tube), which the whole chain takes under an imposed pulling force. This μ_{tube} is inversely proportional to the chain length

$$\mu_{\text{tube}} = \mu_o N^{-1}.$$

To the mobility μ_{tube}, one can associate a tube diffusion coefficient D_{tube}, through the Einstein relation

$$D_{\text{tube}} = kT\mu_{\text{tube}}.$$

Knowing D_{tube}, we can picture the Brownian motion of the chain along its own tube: the curvilinear displacement s after a time t is ruled by

$$s^2(t) = 2D_{\text{tube}}t.$$

These very simple ideas are enough to determine the relaxation time τ, i.e. the time required for a displacement comparable to the total length of the tube L

$$\tau \sim L^2/D_{\text{tube}}.$$

Since L is proportional to N, while $D_{\text{tube}} \sim N^{-1}$, the reptation model predicts $\tau \sim N^3$. Experimentally, the exponent a in $\tau = \tau_o N^a$ is usually slightly larger than 3.

Various explanations have been proposed for this discrepancy:

1. Polydispersity effects (M. Adam and M. Delsanti, private communication): In view of the high exponent a, the mechanical data are very sensitive to the large N tail of the length distribution. This tail is probably stronger at large N (because of various complications either in synthesis or in extraction of a given N fraction). Thus, the deviation $a - 3$ may be caused by purely technical problems.

2. Cross-over from nonentangled to entangled regimes (5, 6) at relatively small N ($N = N_e \sim 300$ in typical polymers): For $N < N_e$ the relaxation time τ is shorter ($\tau \sim N^2$). When $N > N_e$, τ may increase sharply (reaching ultimately $\tau \sim N^3$ at $N \gg N_e$). In this region of sharp increase, the apparent exponent (a) could be larger than 3.
3. Fundamental revisions of the reptation concept: If some exceptionally tight knots had a lifetime much longer that the reptation time τ, the time decays would be deeply modified (J. Noolandi and H. Wendel, private communication). A precise model displaying these long-lived knots, however, is still lacking.

Doi & Edwards (7) have constructed a very elegant theory for the viscoelastic behavior of melts, using the reptation concept. One major component, the "stress memory function," gives the correlation between the stress at time t and the stress at time 0. The stress memory function turns out to be exactly proportional to the "tube memory," i.e. to the fraction $M(t)$ of the chain which at time t is still in the tube defined at time 0. $M(t)$ has been computed in Ref. (1) and is exponential at large times $M(t) \sim e^{-t/\tau}$.

The Doi-Edwards theory gives a good description for many viscoelastic properties of polymer melts, not only in the linear regime (weak perturbations) but also in the strong deformation regime. Certain detailed comparisons are limited by polydispersity effects, but, on the whole, the melts of linear, flexible polymers may well become the best model system of rheology.

In the present review, we do not dwell on these mechanical properties, which are now reasonably well understood; rather, we emphasize microscopic studies on entangled systems.

SELF-DIFFUSION

The self-diffusion coefficient of the polymer chain can provide an interesting way of testing the reptation model; however, it is not easy to measure.

We introduced above the diffusion coefficient of the chains along their own tube D_{tube}. Since the tube is contorted, the self-diffusion coefficient D_{self} measured in a real experiment will be different from D_{tube}. D_{self} characterizes a random walk with an elementary step of time τ and an elementary step of length equal to the radius of gyration of the chain $R_g \sim N^{1/2}$. The reptation prediction for D_{self} is then (b is monomer size)

$$D_{self} \cong \frac{R_g^2}{\tau} \cong \frac{b^2}{\tau_0} N^{-2}.$$

This is indeed a very slow process, and a typical order of magnitude, for $N \sim 10^3$, $b \sim 3.10^{-8}$ cm, and $\tau_o \sim 10^{-10}$ sec, is $D_{\text{self}} \sim 10^{-11}$ cm^2/sec. The time t required to perform one experiment is related to the distance l over which the diffusion will take place through $D_{\text{self}} t \sim l^2$. In usual tracer techniques, $l \gtrsim 1$ mm and $t \gtrsim 10^9$ sec. To perform such experiments one must either wait very long or try to decrease the diffusion length by several orders of magnitude.

Another difficulty in performing the experiment stems from the necessity to label the chains. Because mixing chemically different polymers is difficult (a small repulsion between monomers will result in N times larger repulsion between chains), only small chemical modifications of the polymer chain are usually allowed (deuteration, or change of only few monomers among N). This causes a problem regarding the sensitivity of detection of the labeled chains. Several sets of data, however, are now available, both in polymer melts and in entangled solutions.

Self-Diffusion in Polymer Melts

The simplest conceivable experiment to measure a self-diffusion coefficient is first to build a step-like concentration profile of labeled chains in the sample, then follow the broadening with time of that concentration profile. Early attempts along these lines were performed by Bueche et al (8) and Kumagai and co-workers (9). They used polymer chains labeled with radioactive tracers and started the experiment by putting into close contact two polymer samples, one containing a mixture of labeled and unlabeled chains, the other free of labels. They measured the decrease of the emitted radioactivity as a function of time due to absorption as the labeled chains penetrated the absorbing polymer matrix. The main difficulty of the experiment was that it did not provide any means of testing the quality of the contact initially established between the two specimens; hence, the penetration of the labeled chains in the unlabeled polymer matrix may be entirely dominated by interfacial resistance to the diffusion.

An improved version of that experiment was performed by Klein & Briscoe (10). They used deuterated polyethylene as labeled chains diffusing in a protonated polyethylene melt, and measured the local concentration of labeled monomers through the amplitude of the IR absorbtion of the C–D bond. They then had access to the total concentration profile, with a spatial resolution of about 100 μm, and could estimate the importance of interfacial resistance to diffusion by comparing their experimental profile to the predicted Fick form. Retaining only the experiments exhibiting no interfacial problems, they measured the self-diffusion coefficient as a function of molecular weight. Their samples had a large polydispersity ($M_w/M_n \sim 20$), but correcting for it they were able to find $D_{\text{self}} \sim M_w^{-2\pm0.1}$, a result very close to the reptation prediction.

A completely different technique, the pulsed-field gradient nuclear magnetic resonance, has been used by Tanner (11) to measure the self-diffusion coefficient in polydimethysiloxane melts. This method presents many advantages over the preceding ones: the labeling is performed through the characteristic Larmor frequencies of the macromolecular protons; it is not a chemical labeling; it is achieved in the bulk sample and automatically avoids interfacial problems. However, the method necessitates working with species having a long enough spin-spin relaxation time T_2, and cannot be applied for very long chains ($N \leqslant 10^4$). Tanner's data for the two largest molecular weights used are compatible with the reptation prediction (R. Ullman, private communication).

Another set of data on the molecular weight dependence of the *mutual* diffusion coefficient in a miscible binary polymer system has been reported by Laurence and co-workers (12). They used a combination of scanning electron microscopy and energy dispersive analysis of x-ray fluorescence to observe directly the interface between the two polymer samples; they measured the evolution of the concentration gradient of one material as a function of time. They took advantage of the good contrast for X-rays of their particular system [poly(vinyl chloride)-poly(ε-caprolactone)] and attained a spatial resolution of a few microns. They could then measure self-diffusion coefficients in the range 10^{-12} to 10^{-13} cm^2/sec. Their results yielded $D \sim M_w^{-1}$, which they concluded was incompatible with the reptation model. This is in fact not the case: because the polymers in the mixture were compatible, the Flory interaction parameter between the two species is negative (attraction) and large. There is then a chemical driving force for the mixing. The reptation under such driving force has been worked out (13), and corresponds to a diffusion coefficient inversely proportional to the molecular weight of each species.

Thus, all the available data for self-diffusion seem compatible in polymer melts with the reptation picture.

Self-Diffusion in Entangled Polymer Solutions

The description of chain motions in entangled polymer solutions seems much more difficult a priori than in a melt, as concentration effects have to be taken into account. The chains still cannot cross each other, but the size of the tube is now related to the average distance between entanglements ξ. From scaling arguments (14a,b), ξ is independent of the polymerization index N, and only depends on the monomer concentration c through a power law $\xi \sim c^{-\alpha}$ ($\alpha = 0.75$ in the case of a good solvent). ξ is the screening length for both excluded volume and hydrodynamic interactions (14a,b). The test chain can then be considered as made of a succession of independent subunits of size ξ.

If g is the number of monomers in each subunit, $g = c\xi^3$. The mobility of the chain along the tube is

$$\mu_t = \mu_1 (N/g)^{-1}$$

where $\mu_1 \sim 1/\eta_s \xi$ is the mobility of each subunit (η_s is the solvent viscosity). The diffusion coefficient along the tube D_t is related to μ_t by an Einstein relation

$$D_t = kT\mu_t = \frac{kT}{\eta_s \xi} \frac{g}{N}.$$

In a way similar to the melt case we define the reptation time τ as the time it takes the chain to completely renew its configuration, i.e. $D_t \tau \sim L_t^2 \sim (N/g)^2 \xi^2$, and the self-diffusion coefficient as $D_{\text{self}} \tau \sim R^2(c)$ where $R(c)$ is the radius of gyration of the chain $R(c) \sim (N/g)^{1/2} \xi$. The scaling plus reptation prediction for the self-diffusion of entangled chains in good solvent is then

$$D_{\text{self}} \sim kT/\eta_s \xi (g/N)^2 \sim N^{-2} c^{-1.75}.$$

Experimentally, as we are now dealing with liquid samples with a relatively small viscosity, all the techniques starting with a step-like concentration profile of labeled species are difficult to handle, and the labeling has to be performed in situ.

The pulse-field gradient NMR technique has been used very recently by Callaghan & Pinder (15a,b) to follow self-diffusion of polystyrene chains in carbon tetrachloride. They obtained a concentration dependence $D_{\text{self}} \sim c^{-1.8}$, but no clear answer for the molecular weight dependence. For the largest molecular weights $M(\sim 2.10^6)$, the spin echo signals become complex: the time interval Δ between two pulses becomes smaller than the reptation time τ, and macroscopic diffusion does not hold. The authors discussed this regime in terms of elastic fluctuations of the transient network. We prefer another approach, based on the semilocal motions (discussed below). The central parameter is the mean square displacement $\langle Z^2 \rangle$ of one nuclear spin during the time Δ. The semilocal prediction for $\Delta < \tau$ with entangled chains is $\langle Z^2 \rangle \cong R_g^2 (\Delta/\tau)^{1/2}$. This form of $\langle Z^2 \rangle$ agrees rather well with the raw data of Callaghan & Pinder for high M and $\tau > \Delta$. The M dependence of $\langle Z^2 \rangle$ at fixed Δ remains to be checked.

In the forced Rayleigh light scattering (FRS) technique (16, 17a,b), a photochromic molecule is covalently bonded at one extremity of a few chains in the solution. A periodic distribution of photoexcited molecules is then built in the sample by irradiating it with a pulsed interference pattern and is relaxed by diffusion after the flash excitation. This periodic distribution of photoexcited molecules acts as an absorption grating for a second laser beam, sensitive only to the photoexcited species. The decrease of the diffracted intensity directly reflects the kinetics of the diffusion of the chains. FRS is a tracer technique in which the diffusion length can be reduced down to a few microns, without any loss of accuracy. The results for the self-diffusion of polystyrene in benzene solution

$$D_{\text{self}} \sim c^{-1.7 \pm 0.1} N^{-2 \pm 0.1}$$

are fully consistent with the scaling plus reptation prediction.

SEMILOCAL MOTIONS

A number of important dynamical effects take place on spatial scales that are smaller than the coil size (but still larger than the monomer size).

Neutron Studies on Stretched Chains

In a typical experiment (18), a polystyrene sample ($M = 650.000$) is pulled rapidly in the fluid state, reaching elongation ratios λ of order 3. It is then allowed to relax during a time θ (10 sec $< \theta <$ 30 min), at a fixed temperature slightly above the glass point. Finally, the sample is quenched at room temperature and observed by neutron diffraction. A certain fraction of the polymer chains has been deuterated; in this case, by suitable difference procedures, one can extract the diffraction pattern $S_\theta(\mathbf{q})$ of one chain in a partially elongated state. Here $\mathbf{q}$ is the scattering wave vector; it can lie either parallel to the stretch direction ($q_\parallel$) or perpendicular to it ($q_\perp$). We shall restrict our attention here to the semilocal range $qR_o > 1$ (where R_o is the coil radius).

Before stretching, the scattering is isotropic and described by Debye's now classical law (19)

$$S_i(\mathbf{q}) \sim \frac{1}{q^2}.$$

Immediately after stretching, the chain is deformed by an affine transformation (of eigen values $\lambda, \lambda^{-1/2}, \lambda^{-1/2}$)

$$S(\mathbf{q}) \to \frac{1}{\lambda^2 q_\parallel^2 + \lambda^{-1} q_\perp^2} = S_a(q)(qd < 1).$$

What happens after relaxation during a time θ? For strongly entangled chains, assuming $d \ll q^{-1} \ll R_o$ (where d is the tube diameter), one can provide a very simple prediction. After a time θ, a fraction $M(\theta)$ of the chain is still trapped in the original tube, and gives an anisotropic scattering $S_a(q)$. A fraction $1 - M(\theta)$ has moved out of the original tube, and has returned to isotropy. The intensities from the two parts simply add up: there are no significant interference terms between them for $qR_o > 1$. Thus, we expect

$$S_\theta(q) = M(\theta)S_a(\mathbf{q}) + [1 - M(\theta)]S_i(\mathbf{q}).$$

This type of law is roughly compatible with the data, but many complications occur:

1. The angular range of the available neutron spectrometers limits the data acquisition to $q_\perp$ rather than $q_\parallel$.
2. The chains are not very strongly entangled (d is of order $R_o/3$).
3. Polydispersity can play a role.

However, on the whole, these experiments appear very promising.

Spinodal Decomposition of Blends

Most polymers do not mix; however, certain exceptional polymer pairs A+B are compatible in the melt (20), at least in a certain temperature range. By a rapid change of temperature, one can favor segregation and sometimes induce a "spinodal decomposition," where a fluctuation of the relative concentration A/B, with a well-defined wavelength λ^*, becomes amplified (21). For binary systems made with small molecules, λ^* is usually comparable to the thickness of the interface between the A-rich and the B-rich phases. Does this still hold for polymer alloys? In an early paper, one of us concluded that it does (13). However, after some more reflection, it appeared that the scaling assumptions of (13) were oversimplified. The necessary corrections have been worked out by Pincus (22), and the conclusion is $\lambda^* \gtrsim R_o$. The optimal wavelength is thus much larger than the thickness of an interface: spinodal decompositions are not a very good probe of semilocal scales. Experiments at Bell Laboratories (23) have indeed shown that in the early stages of growth, the basic units have a size comparable to R_o.

Polymer/Polymer Welding

Recent experiments in Lausanne (24) measured the fracture energy for two identical blocks of polymethylmethacrylate put into close contact, during a certain healing time t, at a temperature slightly above the glass point. The fracture energy G_{IC} increases like $t^{1/2}$ at small times. Microscopically, the chains from the two blocks begin to interdigitate, and the healing process is essentially terminated when t becomes larger than the reptation time τ. For $t < \tau$, we are indeed dealing with an interdigitated region of thickness $l(t) < R_o$. Thus, the healing is semilocal.

One essential parameter is the number of bridges per cm² $p(t)$ established between the two halves during the time t. This was analyzed first by a simple argument (25), and later more quantitatively by Prager & Tirrell (26). The result of both papers is a square root growth

$$p(t) \cong p_{\max}(t/\tau)^{1/2}.$$

Thus, it appears that $p(t)$ may control the fracture energy $G_{IC}(t)$. The proportionality between these two quantities is clearly nontrivial! But a simple model, based on plastic deformation at the onset of fracture, does lead to $G_{IC} \sim p(t)$ as observed (27a). Related considerations have been presented by Wool & O'Connor for crack healing (27b).

A special case deserves mention: a dissymmetric interface A/B where the species A and B, making each block, are different but compatible. Here the kinetics of healing could be accelerated by the negative enthalpy of mixing; but the experiments still give $G_{IC} \sim t^{1/2}$, and this result is not fully understood (27a).

Dynamics by Inelastic Neutron Scattering

It is now possible to study chain motions to rather low frequencies ($\omega \sim 10^8$ sec^{-1}) by a very elegant "spin echo" technique using a neutron beam (28a, b). This has led to very good results in the study of nonentangled chains (29a, b). Unfortunately, there is not presently much hope for obtaining results on entanglements (30). The main limitation here is spatial: the scattering wave vectors **q** available on the spin echo machine (and leading to measurable inelastic line widths) are of order 3.10^{-2} Å^{-1} or larger. This means that the largest distances r that can be probed are $q^{-1} \sim 30$ Å. From the systematic analysis of Graessley we know that the chemical distance between consecutive entanglements along one chain is $N_e \sim 300$ monomers. This implies a tube diameter d of order $N_e^{1/2}b$ (where b is a monomer size). Thus, $d \sim 50$ Å. The spin echo experiments do not probe space distances much larger than d; entanglements are not visible! We must wait for the next generation of spectrometers to get significant results on reptation by inelastic neutron scattering.

Computer Experiments

It is possible to study the dynamics of a small number of densely spaced chains by computer methods. This, in principle, can tell us a lot about polymer melts (31a, b). Unfortunately, these computer experiments are not very adequate for reptation studies: the limitation is very similar to that for the spin echo technique. To reach significant results on reptation in melts, we would need chains that are much longer than N_e; this is far beyond present computer capacities.

DIFFUSION-CONTROLLED REACTIONS

Let us think of a dense polymer solution, where a certain reactant X is attached to one chain, while another reactant Y is attached to another chain. We are interested in the reaction $X + Y \underset{k}{\rightarrow} XY$, when this reaction is diffusion-limited. These processes are important in many practical cases: radical polymerizations, condensation polymerizations, cross-linking by irradiation, etc (32). Added to this list are also the purely physical processes in which an optically excited species X* is deactivated or quenched by a functional group Y.

When X and Y are not linked to long chains, the second-order rate constant k is given by a classical formula of von Smoluchowski's (33)

$$k = 4\pi h(D_x + D_y)$$

where D_x, D_y are the diffusion coefficients, and h is the "capture radius," i.e. the distance at which the XY reaction takes place without any delay. How is this modified when we bind X and Y to long chains in a dense solution? One tentative answer proposed recently (34) is to keep the same formula for k, but to use the reptation diffusion coefficients for the two chains. For chains of an equal degree of polymerization (N), this would then lead to $k \sim N^{-2}$. But, as we shall see, this idea is not quite correct: the reaction takes place at distances $r \sim h$ that are much smaller than the coil size R_o, and we are really dealing with a semilocal process.

The complete discussion of reactant motions is somewhat complex (35). In this review, we use a different (and very primitive) approach, based on the notion of a diffusion coefficient $D(r)$ dependent on the distance (r) between X and Y. If $g(r)$ is the pair correlation (probability of finding X and Y at the distance r), the flux of particles converging toward reaction is

$$J = 4\pi r^2 \left[-D(r) \frac{\partial g}{\partial r} \right].$$

In stationary states, this J is constant, and the above equation fixes the profile $g(r)$. The boundary conditions are

$$g(h) = 0 \quad g(\infty) = n_X n_Y.$$

From this one arrives at an explicit form for the rate constant $k \equiv J/n_X n_Y$

$$\frac{4\pi}{k} = \int_h^\infty \frac{dr}{r^2 D(r)}.$$

$D(r)$ is the sum of two coefficients relative to each of the two carrier chains. We assume that the two chains are identical and incorporate this effect by changing $D(r) \to 2D(r)$.

Let us now discuss the r dependence of $D(r)$. Of course for $r > R_o$, we must recover macroscopic diffusion $D(r) \to D \sim N^{-2}$. But for $r < R_o$, we must study motions on time scales $t < \tau$. We shall still call s the curvilinear displacement of one chain in its tube ($s^2 = 2D_{tube}t$). The displacement "as the crow flies," r, associated with s is given by

$$r^2 = sd$$

where d is the tube diameter. This equation expresses the fact that the tube itself has the geometry of a random walk. We see that $r^4 = d^2 D_{tube} t$, and we can now define a diffusion coefficient $D(r)$ through

$$D(r) \cong \frac{r^2}{t} = D_{tube}\left(\frac{d}{r}\right)^2 \quad (r < R_o).$$

Inserting this into the above integral for the rate constant gives a result independent of the capture radius h:

$$k \cong 8\pi D R_o \sim N^{-3/2}.$$

A vast body of experimental data on overall reaction rates is available for radical polymerizations, but their analysis is complex:

1. The reactions are often much more complex than the XY scheme displayed here.
2. The systems are intrinsically polydisperse; the laws for a short chain reacting with a long chain are very different.
3. Even with our simple reaction model, many different regimes are expected, depending on the time scales.

Nevertheless, we may see in the next few years a coherent interpretation scheme for semilocal transport processes and their chemical consequences.

OTHER LINES OF RESEARCH

The reptation ideas can also provide some guidelines for more complex polymer problems. Some of these are listed below:

1. What happens when the test chain (N monomers) is surrounded by a matrix of chemically identical, but shorter, chains (P monomer per chain)? This is being studied experimentally (L. Monnerie, private communication) and leads to rather delicate theoretical discussions (36).

2. The effects involved in polymer crystallization from the melt are still disputed [see section on crystallization in Ref. (37)]. It may be, however, that in certain regimes of fast crystallization, the chains are extracted from the melt by a "suction" process closely related to reptation (38, 39a, b).
3. For branched chains in melts, the reptation process is probably strongly hindered (40), although analog calculations made with computers do not show this hindrance (41) (as explained above, computer methods are not adequate for studies on reptation in melts).
4. When the test chain is rigid, but is embedded in a melt of flexible chains, or in a gel, one expects a very special type of Brownian motion (42). The drift properties of DNA fragments in gel electrophoresis are closely linked to these equations.

On the whole, the reptation idea has been helpful from the early days, when the experiments were mainly concerned with the snake motion of a single flexible chain in a fixed gel (43), to the present, many-chain problems. However, as always in science, we should always be aware of the limitations of our current scheme and search for more profound approaches to polymer dynamics. Perhaps the difference between $a = 3$ and $a = 3.3$ in the exponent for the reptation time $\tau = \tau_o N^a$ conceals a very deep theoretical problem.

Literature Cited

1. De Gennes, P. G. 1971. *J. Chem. Phys.* 55:572–79
2. Ferry, J. D. 1970. *Viscoelastic Properties of Polymers.* New York: Wiley
3. Graessley, W. W. 1974. *Adv. Polym. Sci.* 16:1–179
4. Edwards, S. F. 1967. *Proc. Phys. Soc.* 92:9–16
5. Graessley, W. W. 1980. *J. Polym. Sci. Polym. Phys.* 18:27–34
6. Doi, M. 1981. *J. Polym. Sci. Polym. Lett. Ed.* 19:265–73
7. Doi, M., Edwards, S. F. 1978. *J. Chem. Soc. Faraday Trans.* 2 74:1789–1801; 1802–17; 1818–32
8. Bueche, F. 1968. *J. Chem. Phys.* 48:1410–11
9. Kumagai, Y., Watanabe, H., Miyasaka, K., Hata, T. 1979. *J. Chem. Eng. Jpn.* 12:1–4
10. Klein, J., Briscoe, B. J. 1979. *Proc. R. Soc. London Ser. A* 365:53–73
11. Tanner, J. E. 1971. *Macromolecules* 4:748–50
12. Gilmore, P. T., Falabella, R., Laurence, R. L. 1980. *Macromolecules* 13:880–83
13. De Gennes, P. G. 1980. *J. Chem. Phys.* 72:4756–63
14a. Daoud, M., Cotton, J. P., Farnoux, B., Jannink, G., Sarma, G., Benoit, H., Duplessix, R., Picot, C., De Gennes, P. G. 1975. *Macromolecules* 8:804–18
14b. De Gennes, P. G., 1979. *Scaling Concepts in Polymer Physics.* Ithaca, NY: Cornell Univ. Press.
15a. Callaghan, P. T., Pinder, D. N. 1980. *Macromolecules* 13:1085–92
15b. Callaghan, P. T., Pinder, D. N. 1981. *Macromolecules* 14:1334–40
16. Hervet, H., Urbach, W., Rondelez, F. 1978. *J. Chem. Phys.* 68:2725–29
17a. Léger, L., Hervet, H., Rondelez, F. 1982. *Macromolecules.* 14:1732–38
17b. Hervet, H., Léger, L., Rondelez, F. 1979. *Phys. Rev. Lett.* 42:1681–84
18. Boué, F. 1982. PhD Thesis. Université d'Orsay. In Press

19. Debye, P. 1947. *J. Phys. Colloid Chem* 51:18–28
20. Krause, S. 1972. *J. Macromol. Sci. Rev. Macromol. Chem.* C7:251–300
21a. Cahn, J., Hilliard, J. 1958. *J. Chem. Phys.* 28:258–68
21b. Cahn, J., Hilliard, J. 1959. *J. Chem. Phys.* 31:668–78
22. Pincus, P. 1981. *J. Chem. Phys.* 75:1996–2000
23. Nishi, T., Wang, T., Kwei, T. 1975. *Macromolecules* 8:227–34
24. Jud, K., Dausch, H., Williams, J. G. 1981. *J. Mat. Sci.* 16:204–10
25. De Gennes, P. G. 1980. *C. R. Acad. Sci. Paris B* 291:219–21
26. Prager, S., Tirrell, M. 1981. *J. Chem. Phys.* 75:5194–98
27a. De Gennes, P. G. 1981. *Proc. Paris Conf. Adhesion and Lubrication*, ed. C. Troyanovski. Amsterdam: Elsevier. In press
27b. Wool, R. P., O'Connor, K. M. 1981. *J. Appl. Phys.* 52:5953–63
28a. Mezei, F. 1972. *Z. Phys.* 255:146–60
28b. Mezei, F. 1980. *Neutron Spin Echo.* Berlin/Heidelberg/New York: Springer Verlag
29a. Higgins, J. S., Nicholson, L. K., Hayter, J. B. 1981. *Polymer* 22:163–67
29b. Richter, D., Ewen, B., Hayter, J. B. 1980. *Phys. Rev. Lett.* 45:2121–24
30. Richter, D., Baumgartner, A., Binder, K., Ewen, B., Hayter, J. B. 1981. *Phys. Rev. Lett.* 47:109–13
31a. Baumgartner, A., Binder, K. 1981. *J. Phys. Chem.* 75:2994–3005
31b. Bishop, M., Ceperley, D., Frisch, H., Kalos, M. 1981. *J. Chem. Phys.* 75:5538–42
32. Driscoll, O. 1982. *Pure Appl. Chem.* In press
33. von Smoluchowski, M. 1917. *Z. Phys. Chem.* 92:129–68
34. Tulig, T. J., Tirrell, M. 1981. *Macromolecules* 14:1501–11
35. De Gennes, P. G. 1982. *J. Chem. Phys.* In press
36. Daoud, M., De Gennes, P. G. 1979. *J. Polym. Sci. Polym. Phys. Ed.* 17:1971–81
37. See the ensemble of 1979, *Faraday Discuss. Chem. Soc.* 68
38. Klein, J., Ball, R. C. 1979. *Faraday Discuss. Chem. Soc.* 68:198–209
39a. DiMarzio, E. A., Guttman, C. M., Hoffman, J. D. 1979. *Faraday Discuss. Chem. Soc.* 68:210–17
39b. Hoffman, J. D., Guttman, C. M., DiMarzio, E. A. 1979. *Faraday Discuss. Chem. Soc.* 68:177–97
40. De Gennes, P. G. 1975. *J. Phys. Paris* 36:1199–1203
41. Evans, K. E., Edwards, S. F. 1981. *J. Chem. Soc. Faraday Trans.* 2 10:1891–1938
42. De Gennes, P. G. 1981. *J. Phys. Paris* 42:473–77; 1982. *C. R. Acad. Sci. Paris.* In press
43. Kramer, O., Greco, R., Neira, R. A., Ferry, J. D. 1974. *J. Polym. Sci. Polym. Phys. Ed.* 12:2361–74

DYNAMICS OF MOLECULES IN CONDENSED PHASES: PICOSECOND HOLOGRAPHIC GRATING EXPERIMENTS

M. D. Fayer

Department of Chemistry, Stanford University, Stanford, California 94305

INTRODUCTION

The development of laser equipment that can operate routinely in the subnanosecond (and more recently in the subpicosecond) time regimes has made possible the investigation of a wide variety of fast chemical, physical, and biophysical processes. Most of the successful picosecond time scale experiments, although very sophisticated in technique, have utilized the basic approaches that have been applied on slower time scales. One method involves monitoring time-resolved fluorescence following picosecond excitation (1–3). Since the time scale of interest is very short, techniques such as single photon counting, streak cameras, and fluorescence mixing have been employed to provide the necessary time resolution of the fluorescence. In many other experiments, the picosecond probe pulse technique has been used to examine changes in a systems absorption following picosecond optical excitation (4, 5).

In this article I discuss a different approach to the application of subnanosecond laser pulses to the investigation of molecular and excited state dynamics. This involves the optical generation of a transient holographic diffraction grating in a sample, and the observation of various time and frequency dependent phenomena via subsequent Bragg diffraction from the induced grating. The basic experiment works in the manner illustrated in Figure 1. Two time coincident picosecond laser pulses of the same wavelength are crossed inside of the sample to set up an optical interference pattern. The fringe spacing, d, of the interference pattern is determined by the angle between the beams, θ, and the wavelength, λ, of the excitation pulses, i.e.

$$d = \lambda/2\sin(\theta/2). \qquad 1.$$

The interaction of the radiation field with the sample can produce a number of different changes in the sample, depending on the nature of the sample and the wavelength, λ. As discussed in the examples below, in which the samples are solids or liquids, electronic excited states can be produced (6), internal molecular vibrations can be excited (7), or acoustic waves, i.e. phonons, the collective vibrations of the medium, can be generated (7–11). In some experimental situations, more than one of the above types of excitations are simultaneously produced (11).

In all cases, the excitations generated in the sample have a spatial periodicity that mimics the periodicity of the optical interference pattern used to excite the sample. Excitation results in a spatially periodic change in the physical properties of the system. This in turn produces a periodic variation in the sample's complex index of refraction, $\mathbf{n}$ (11),

$$\mathbf{n} = n + iK. \qquad 2.$$

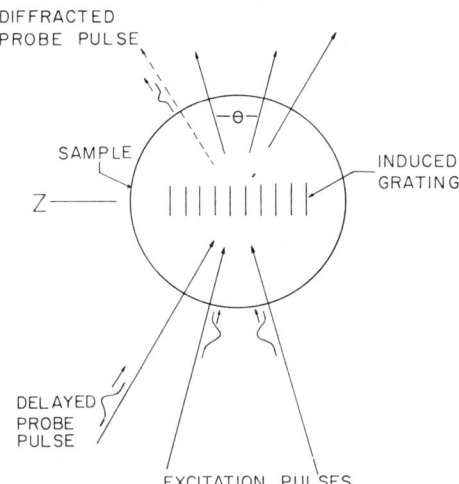

Figure 1 Schematic illustration of the transient grating experiment. Interference between the incoming excitation pulses results in an oscillatory density of excited states, which Bragg-diffracts the subsequent probe pulse. The diffracted probe is the signal, which reflects the time evolution of the excited state population.

The periodic variation in **n** acts as a Bragg diffraction grating for a picosecond probe pulse (11) (see Figure 1). The probe pulse is brought into the sample to meet the Bragg diffraction condition for the holographic transient grating produced by the excitation beams. A part of the probe pulse is diffracted and leaves the sample in a unique direction as a collimated beam. The intensity of the diffracted beam is the observable in the transient grating experiment. The probe pulse can be delayed various amounts in time, and the intensity of the diffracted beam as a function of probe pulse delay can be related to the system's dynamics (6). In addition, the probe pulse can be brought in at a fixed delay time, and the wavelength of either the probe (11) or excitation beams (7) can be varied. In this manner various types of spectroscopic measurements can be made.

Examples of the various types of experiments are presented below. These experiments include measurements of excited state dispersion relations (11), electronic excited state energy transport and trapping (6), the optical generation and detection of ultrasonic waves (8–11), vibrational overtone spectra (7), crystalline elastic constants (8, 9), and photoelastic constants (7).

EXPERIMENTAL SETUP

All of the experiments discussed below were performed with some variation of the transient grating experimental setup illustrated in Figure 2. The laser is a continuously pumped Nd:YAG system that is acousto-optically mode-locked and Q-switched to produce high repetition rate (400 Hz), high power infrared (1.06 μm) picosecond pulses. The laser output is a train of about 40 mode-locked pulses, 5.7 nsec apart, with ~1.4 mJ total energy. A large pulse from the train is selected by a Pockels cell with avalanche transistor driver. The single pulse is frequency doubled using CD*A to give a 20 μJ, 80 psec, transform-limited, TEM$_{00}$ pulse at 532 nm. This passes through a 50% beam-splitter to create the two excitation pulses, which travel equal distances and are focused into the sample.

The unused IR pulse train comes off a reflecting polarizer into another CD*A doubler crystal, and the 532 nm light is used to synchronously pump a dye laser that is spectrally narrowed and tuned by two intracavity etalons. The dye laser is cavity dumped using another Pockels cell with avalanche transistor driver to give a 10 μJ, 30 psec pulse with a spectral width of ~1 cm^{-1}. Synchronization of the two Pockels cells is obtained by a single avalanche transistor, which itself is triggered by the IR pulse train. The dye laser output travels a variable distance controlled

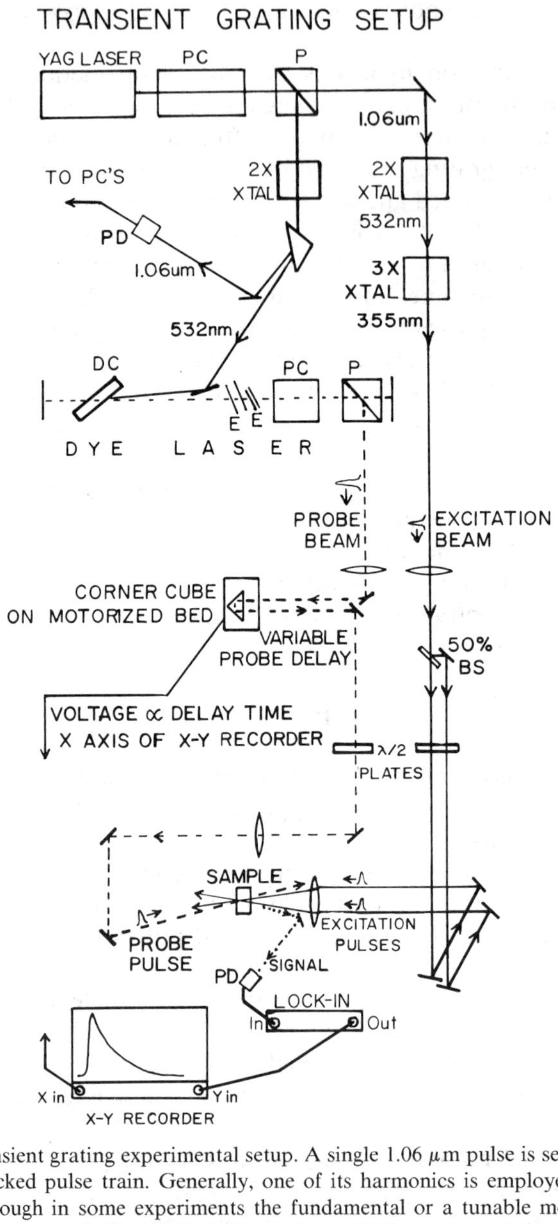

Figure 2 Transient grating experimental setup. A single 1.06 μm pulse is selected from the YAG mode-locked pulse train. Generally, one of its harmonics is employed (shown here with 3×), although in some experiments the fundamental or a tunable mode-locked dye laser pulse may be used. The single pulse is then split into two excitation pulses. These excitation pulses are recombined at the sample, creating the transient grating. The remainder of the pulse train is frequency doubled to synchronously pump a tunable dye laser whose output probes the grating after a variable delay. In some experiments a YAG harmonic is used as a probe. The Bragg-diffracted part of the probe pulse is the transient grating signal. PC ≡ Pockels cell; P ≡ polarizer; PD ≡ photodiode; DC ≡ dye cell; E ≡ etalon; BS ≡ beamsplitter.

by a motorized delay line consisting of a corner cube drawn along a precision optical rail. It probes the grating at an angle satisfying the Bragg diffraction condition. In some of the experiments described below, the dye laser was used for tunable excitation and the probe wavelength was 532 nm. In other experiments, a single color, either dye laser or YAG laser (or a harmonic), was used for both excitation and probing.

A large area photodiode and a lock-in amplifier are used to detect the diffracted signal beam. For time independent experiments, the probe pulse is delayed an appropriate fixed amount and either the probe or excitation wavelength is tuned. The diffracted intensity is recorded as a function of wavelength. For time dependent measurements the lock-in amplifier output drives the y-axis of an x-y recorder. The x-axis is driven by a variable voltage derived from a ten-turn potentiometer connected to the delay line motor, providing the time scale. When the delay line is run, the time dependent diffracted signal is recorded directly on the x-y recorder. If significant data manipulation is required, a minicomputer using analog to digital converters is used to record the time dependent signal for subsequent data analysis.

ELECTRONIC EXCITED STATE PHASE AND AMPLITUDE GRATINGS: PROBE WAVELENGTH DEPENDENCE NEAR A STRONG TRANSITION

As discussed above, the diffraction of the probe pulse from the optically induced grating is due to a spatially periodic variation in the complex index of refraction, **n** (Eq. 2). The diffraction efficiency is proportional to the difference in **n** at the grating peaks and nulls (11). In this section I consider the diffraction efficiency from a grating that arises from the production of electronic excited states by the excitation pulses (11). Generation of electronic excited states causes changes in both the real and the imaginary parts of the index of refraction and, in general, both the spatial variation in n and K give rise to diffraction efficiency (12).

The change, $\Delta K_{ex}(\omega)$, in the imaginary part of **n** due to the excited states is related to the peak-null difference in absorption at the probe wave length, ω. $\Delta K_{ex}(\omega)$ gives rise to an amplitude grating. For ω on or near an absorption peak, $\Delta K_{ex}(\omega)$ can be substantial. Thus the probe pulse experiences alternating regions of high and low optical transmission. This periodic variation in absorption produces, in effect, a multislit diffraction, i.e. an amplitude grating.

The change, $\Delta n_{ex}(\omega)$, in the real part of **n** due to excited states is related to the peak-null difference in optical dispersion at the probe wavelength, ω. $\Delta n_{ex}(\omega)$ gives rise to a phase grating. In the vicinity of an absorption peak, the probe pulse experiences a periodic variation in the velocity of light-producing alterations in the probe wavefront's phase. This phase variation gives rise to diffraction.

For a sinusoidal volume grating, the diffraction efficiency, $\eta(\omega)$, is (11)

$$\eta(\omega) = e^{-\frac{2.3 D_{av}(\omega)}{\cos\theta}} \left[\sinh^2 \frac{\Pi T \Delta K(\omega)}{\lambda \cos\theta} + \sin^2 \frac{\Pi T \Delta n(\omega)}{\lambda \cos\theta} \right]. \qquad 3.$$

The first term in brackets determines the amplitude grating diffraction, $\eta_a(\omega)$, and the second determines the phase grating diffraction, $\eta_p(\omega)$. T is the sample thickness. $D_{av}(\omega)$ is an average optical density adjusted such that the exponential term describes the absorptive loss. In most experimental situations, the total diffraction efficiency is not large, ($\eta < 0.01$) and Eq. 3 can be approximated by (11)

$$\eta(\omega) = e^{-\frac{2.3 D_{av}(\omega)}{\cos\theta}} \left(\frac{\Pi T}{\lambda \cos\theta} \right)^2 \{[\Delta K(\omega)]^2 + [\Delta n(\omega)]^2\}. \qquad 4.$$

The grating diffraction efficiency given by Eq. 4, which depends on $\Delta K(\omega)$ and $\Delta n(\omega)$, must now be related to the excited state concentrations produced by transient grating excitation. This is accomplished by describing the system with a damped harmonic oscillator model for the Kramers-Kronig relations (13). For the probe wavelength, ω, near a single strong optical transition (11), e.g. the probe is tuned near the S_0 to S_1 absorption of a solute molecule in a transparent solvent (as discussed below),

$$\Delta n_{ex}(\omega) = -\frac{N_1}{N_0} \frac{2(\omega_0 - \omega)}{\gamma_0} K_0(\omega) \qquad 5a.$$

$$\Delta K_{ex}(\omega) = -\frac{N_1}{N_0} K_0(\omega) \qquad 5b.$$

with

$$K_0(\omega) = K_0(\omega_0) \frac{\gamma_0^2}{4(\omega_0 - \omega)^2 + \gamma_0^2} \qquad 6a.$$

and

$$K_0(\omega_0) = \frac{1}{2n_0} N_0 \frac{e^2}{3\varepsilon_0 m} \frac{f_0}{\gamma_0 \omega_0}. \qquad 6b.$$

N_0 is the number density of absorbing molecules and N_1 is the number density of excited states at the grating peaks. ω_0 is the frequency of the absorption maximum of the transition being probed and γ_0 is its line width (FWHH). n_0 is the bulk index of refraction (excluding the contribution from the transition being probed), and f_0 is the oscillator strength of the transition being probed. e, m, and ε_0 are the electron charge, electron mass, and the permitivity of free space, respectively.

From Eq. 4 with Eqs. 5 and 6 it can be seen that the signal in a transient grating experiment depends on changes in both the optical density of the sample ($\Delta K_{ex}(\omega)$) and the dispersion of the sample ($\Delta n_{ex}(\omega)$) due to the production of excited states. A conventional picosecond probe pulse experiment only depends on the changes in the optical density upon excitation. If this were the case for a transient grating experiment, the diffraction efficiency (amplitude grating only) would have a probe wavelength dependence proportional to the absorption spectrum squared, i.e. $[\Delta K_{ex}(\omega)]^2$. The amplitude-grating contribution is peaked at the absorption maximum and falls off rapidly at longer and shorter wavelengths. The phase grating contribution, $[\Delta n_{ex}(\omega)]^2$, is zero at the absorption maximum and is peaked at the half heights of the absorption line on both sides of the absorption maximum.

The above considerations are demonstrated experimentally in Figure 3 (11). The sample is a mixed molecular crystal (solid solution) of pentacene in the host *p*-terphenyl. The *inset* in the figure shows the absorption spectrum. The transient grating excitation employed doubled Nd:YAG pulses at 532 nm. The diffraction efficiency $\eta_{ex}(\omega)$ due to the resulting excited state grating was measured at fixed time delay (500 psec) as a function of probe wavelength using tunable dye laser pulses as the probe. The *solid line* through the experimental points (Figure 3a) represents the experimentally measured diffraction efficiency in the vicinity of the pentacene S_0 to S_1 absorption. The *dashed line* shows the theoretically predicted diffraction efficiency from Eq. 4 with Eqs. 5 and 6. The agreement between the two curves is excellent, especially on the red side of the origin, which is spectrally isolated. The blue side of the origin is influenced by the next absorption band (see inset spectrum), which was not included in the calculation.

The diffraction efficiency is not adequately described by amplitude grating effects alone (the *dash-dot curve* in Figure 3a). At the absorption peak, the phase-grating contribution vanishes and thus the amplitude grating, η_a, accounts for all of the observed signal. η_a scales as the square of the absorption strength, and the *dash-dot curve* in Figure 3a was plotted from the inset absorption spectrum. The difference between this curve and the observed signal gives the diffraction resulting from phase-grating effects. This is plotted in Figure 3b (the *solid curve* through the experimental points). The phase-grating contribution rises from zero at the absorption peak to maxima at the halfwidths, then gradually decreases, as theoretically predicted (the *dashed curve*). The asymmetry in

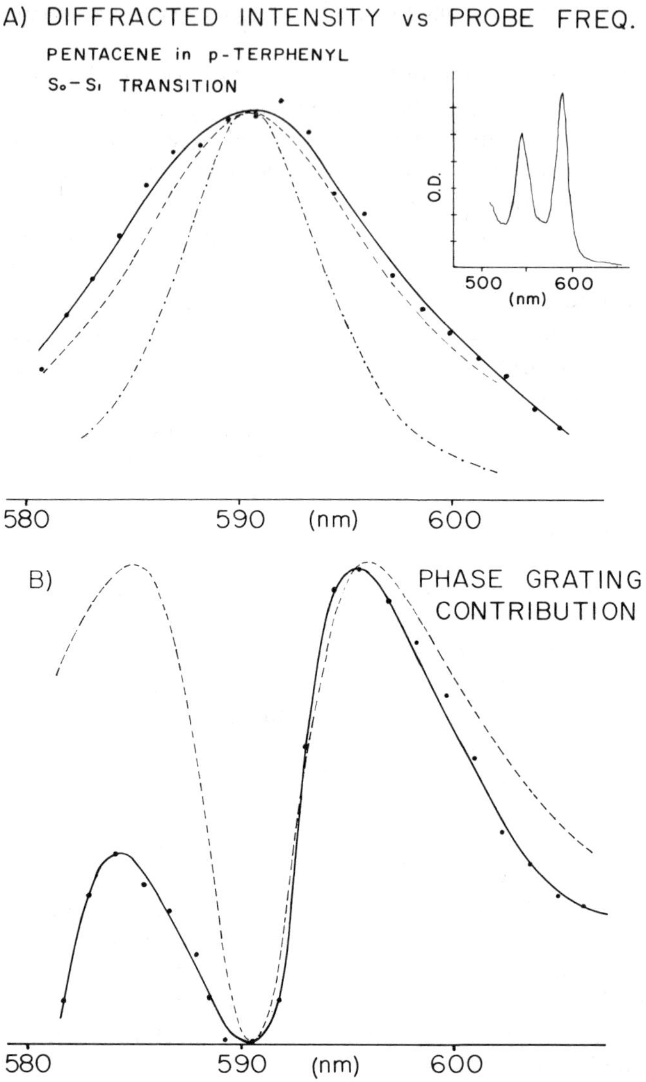

A) DIFFRACTED INTENSITY vs PROBE FREQ.
PENTACENE in p-TERPHENYL
$S_0 - S_1$ TRANSITION

580   590 (nm)   600

B)   PHASE GRATING
                                CONTRIBUTION

580   590 (nm)   600

Figure 3 (*a*) The circles with the solid line are the experimentally measured excited state grating diffraction intensity as a function of probe wavelength near the S_0 to S_1 transition of pentacene in a crystal of *p*-terphenyl. The *dashed curve* is theoretically calculated from the absorption spectrum (*inset*). The *dash-dot curve* is the calculated amplitude grating contribution to the diffraction efficiency. This demonstrates the significant contribution of phase grating effects. (*b*) The points with the *solid line* are the phase-grating contribution to the diffraction intensity obtained by subtracting the curves in (*a*). The predicted *m* shape curve associated with excited state phase grating diffraction is clearly observed. The *dashed curve* is theoretically calculated from the absorption spectrum. On the red side, where the transition is isolated, the agreement is good. On the blue side, interference from the next spectral peak (see inset), which was not included in the calculation, influences the dispersion effect.

$\eta_p(\omega)$ results from the influence of the absorption band (see inset spectrum) on the blue side of the origin. This band was not included in the theoretically calculated curve.

These results confirm the theoretical predictions of the contributions of excited state amplitude and phase gratings to the transient grating diffraction efficiency. Although excited state phase gratings have been discussed by a number of authors (14), experimental observations have been sketchy. The results presented here provide the clearest characterization of the wavelength dependence of excited state phase-grating diffraction. In many experimental situations, failure to account properly for both phase- and amplitude-grating effects can lead to erroneous interpretations of data. This can be true in more general four-wave mixing experiments as well as in transient grating experiments.

TIME DEPENDENT EXCITED STATE GRATING EXPERIMENTS: EXCITED STATE TRANSPORT AND TRAPPING IN CONCENTRATED DYE SOLUTIONS

As with picosecond probe pulse experiments, picosecond transient grating experiments can be used to examine time dependent dynamics of excited state systems. As can be seen from Eqs. 4, 5, and 6, the signal, i.e. the diffraction efficiency, depends on $(N_1)^2$, the square of the number of excited states at the grating peaks. In the absence of any other process, the time dependent signal, $S(t)$, at a fixed probe wavelength will be determined by the excited state lifetime,

$$S(t) = S(0)e^{-2Kt}. \qquad 7.$$

K is the rate constant for the decay of the excited states. The factor of two in the exponential arises because the signal depends on $[N_1(t)]^2$. Thus for an exponential decay of the excited state population, the transient grating signal will decay exponentially with twice the rate constant.

Virtually any experiment that can be performed with a picosecond probe pulse approach can also be performed with a picosecond transient-grating experiment. In many instances there are significant advantages to using the transient-grating approach. For some types of experiments, e.g. measurement of long-range transport of electronic excitations, the transient-grating method can be applied where probe pulse or fluorescence experiments are not applicable. My research group is currently employing transient-grating experiments involving excited states to examine rotational reorientation of molecules in solution, electronic excited state energy transport in dilute solutions, excited state transport in biological systems related to photosynthesis, excited state transport in polymer systems, excited state transport in mixed and pure molecular crystals, and transport and trapping of excited states in polymer systems and concentrated dye solutions. To illustrate the basic ideas, excited transport and trapping in concentrated dye solutions, which are responsible for fluorescence quenching, are discussed here (6).

The experimental evidence suggests that three radiationless processes govern the disposition of electronic excited state energy in concentrated dye solutions. These three processes are energy transfer between dye molecules (1, 15), trapping by dimers (16, 17), which have states of lower energy, and radiationless relaxation (18) of the dimer excited state. A simple model provides a microscopic dynamical picture of fluorescence quenching (19) in concentrated dye solutions. The results given here directly relate to concentration quenching in dye lasers, an important limiting effect (19). In addition, the phenomena under consideration are the important initial steps in photosynthesis, i.e. electronic excitation transfer between chlorophyll chromophores and trapping on reaction centers (dimers) (20–22). Qualitatively, the concentration dependent processes that combine and result in fluorescence quenching work in the following manner. At very low concentration, a dye solution absorbs light and fluoresces. At moderate concentrations, electronic excited state energy transport occurs due to dipole-dipole interactions between the dye molecules (23). The energy transport causes fluorescence depolarization effects (3) but does not affect the fluorescence quantum yield. As the concentration is increased further, ground state dimer formation begins (24, 25) and the rate of energy transport continues to increase. By dimers we mean aggregates of two dye molecules that have distinct spectral and other characteristics. Rapid transport among the monomers allows an excitation to find a dimer and become trapped on it. The experiments indicate that back transfer from the excited dimer to monomers is negligible. Once the excitation is trapped on a dimer, rapid radiationless relaxation to the ground state occurs, and the fluorescence is quenched.

The concentration dependence of the fluorescence quenching is determined by the concentration dependence of the trapping. The trapping rate depends on both the dimer concentration and the concentration dependent rate of energy transport. The model predicts that the trapping rate varies approximately as the cube of the dye concentration. Therefore the onset of fluorescence quenching with increasing concentration is very rapid.

Experimentally, the onset of trapping by dimers manifests itself as an apparent reduction in the excited state lifetime. In the limit that energy transport becomes extremely rapid, the trapping occurs on a time scale that is short relative to the dimer lifetime, and the excited state population decays with the dimer lifetime. In the two systems studied, Rhodamine 6G (R6G) in glycerol and R6G in ethanol, the dimer lifetimes are 830 psec and <50 psec, respectively. Presumably, the dimers have faster radiationless relaxation rates than the monomers because of the loose nature of the dimer complexes. The dimers undergo rapid configurational changes that enhance the radiationless relaxation rates. This is consistent with the longer dimer lifetime in the glycerol solvent. Since glycerol is much more viscous than ethanol, it will "hold" the dimer complex more rigidly and therefore slow radiationless relaxation.

A formally correct, accurate treatment of this problem involves the solution of the Master Equation for excited state transport and trapping in an ensemble of randomly distributed dye molecules and dimer traps in solution. A solution to the Master Equation has recently been obtained using a diagrammatic self-consistent approach to obtain the Green function describing the transport and trapping problem (16). Description of this formalism is beyond the scope of this discussion; rather a heuristic, qualitatively correct, analysis (6) in terms of a simple set of rate equations is presented. These rate equations employ a trapping rate constant. The full diagrammatic Green function treatment confirms the results presented here; however, trapping is not governed by a rate constant, but involves a time dependent trapping rate function.

In the simple model the rate equations governing the excited state populations are:

$$dM^*/dt = -K_M M^* - K_T M^* \qquad 8a.$$

$$dD^*/dt = -K_D D^* - K_T M^*. \qquad 8b.$$

M^* is the concentration of excited monomers, and D^* is the concentration of excited dimers. K_M is the rate constant for decay of excited monomers to the ground state by radiative and nonradiative processes, and K_D is the analogous rate constant for decay of dimers to the dimer ground state. K_T is the trapping rate constant.

As discussed above, the transient grating signal depends on the square of the excited state concentrations, $[N_1(t)]^2$, i.e.

$$S(t) = A[N_1(t)]^2 \qquad 9.$$

where A contains all of the time independent parameters such as beam geometries and $N_1(t) = M^* + D^*$, the sum of the excited monomer and dimer concentrations. Solution of the rate equations yields

$$N_1(t) = M_0^* [\exp[-(K_M + K_T)t] + [K_T/(K_M + K_T - K_D)] \\ \times \{\exp(-K_D t) - \exp[-(K_M + K_T)t]\}]. \qquad 10.$$

$N_1(t)$ is the time-dependent function determined experimentally from $S(t)$. It is informative to note some special cases of Eq. 10. If K_T is very small, trapping is negligible and $N_1(t)$ decays exponentially with the monomer rate constant, K_M. If $K_D \gg K_M, K_T$, then $N_1(t)$ decays exponentially with a rate constant $(K_M + K_T)$. And if $K_T \gg K_M, K_D$, excitations are immediately trapped by dimers and $N_1(t)$ decays exponentially with the dimer rate constant, K_D.

In general, trapping is characterized by a time-dependent trapping rate function (16, 26). Trapping occurs when an excitation has visited enough distinct sites so that on the average it has sampled one trap species. For a random walk on an isotropic three dimensional lattice, the number of distinct sites visited increases linearly with time (27). Therefore, trapping can be characterized by a trapping rate constant which depends on the site-to-site hopping time. I assume that trapping can also be characterized by a trapping rate constant in the solution systems discussed here. At high concentration, this is reasonable since transport is basically diffusive and isotropic in three dimensions. Although we do not have a periodic lattice, the randomness in spatial distribution of the sites (dye molecules) associated with a solution is taken into consideration in the calculation of the hopping time.

The above considerations lead to the following expression for the trapping rate constant (6), K_T:

$$K_T = \left(\frac{Pq}{h_1 M_1^2}\right) M^3. \qquad 11.$$

K_T depends on the cube of the dye concentration, M. The other parameters are constants. q is the monomer-dimer equilibrium constant. h_1 is the site-to-site hopping time and M_1 is the dye concentration, both for a solution with the Forster unitless concentration, $C = 1$ (6). P is the probability that on any step a distinct site is visited. It corrects for the return to previously visited sites. For an isotropic three dimensional random walk on a lattice, $P \approx 0.7$, and I use this value here.[1] Equation 11

---

[1] Our P corresponds to Montroll's $1/u_0$, which is 0.65946 for a simple cubic lattice. This value was misprinted originally, but the correct value $u_0 = 1.5164$ was used in later works. See, for example, Ref. (28).

shows that this model predicts that the trapping rate constant depends on the concentration cubed. Thus, trapping increases very rapidly with concentration.

Examination of $N_1(t)$, Eq. 10, which gives the time-dependent signal, shows that in general the decays are nonexponential. In the data analysis the following procedure is employed. The experimental decays are plotted on log paper and a decay constant is determined for each concentration. These are then compared to a theoretical effective decay constant, K_{eff}, obtained from Eq. 10 by finding the time required for $N_1(t)$ to fall to $1/e$. Thus

$$K_{\text{eff}} = 1/t^\dagger \qquad 12.$$

with $t^\dagger$ obtained from Eq. 10 by

$$N_1(t^\dagger) = 1/e[N_1(0)]. \qquad 13.$$

Transient grating experiments were performed on a series of solutions of Rhodamine 6G in glycerol ranging in concentration from 8.7×10^{-4} m/l to 0.05 m/l. A typical result and log plot are shown in Figure 4. In all cases the data appeared to decay exponentially for several lifetimes. Thus, the decay could be characterized by an effective rate constant, K_{eff}, as discussed above. A plot of K_{eff} vs R6G concentration is shown in Figure 5. First consider the qualitative features of the concentration dependence. At low concentration, K_{eff} is concentration independent and

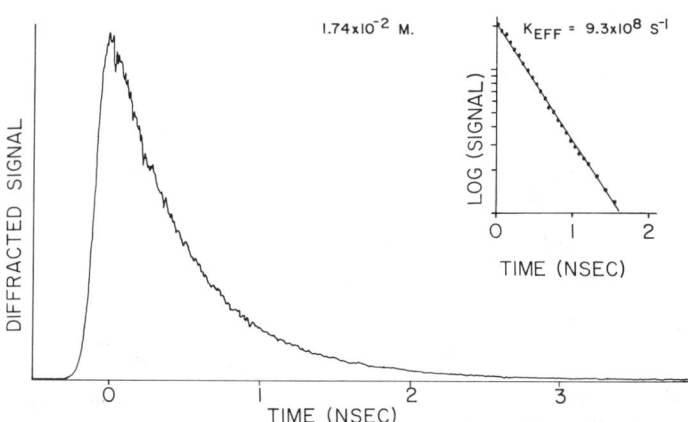

Figure 4 Transient grating results for Rhodamine 6G in glycerol. Probe wavelength = 560 nm. Inset shows the log of the data vs time. The effective decay constant for this data set is $K_{\text{eff}} = 9.3 \times 10^8 \text{ s}^{-1}$.

given by the monomer decay rate: $K_{eff} = K_M = 3.3 \times 10^8 s^{-1}$. This represents the limit $K_T = 0$, i.e. no trapping, since there are few dimers and transport is relatively slow. At high concentration, K_{eff} is essentially concentration independent and given by the dimer decay rate: $K_{eff} \approx K_D = 1.2 \times 10^9 s^{-1}$ and the dimer lifetime is 830 psec. This corresponds to the limiting case $K_T \gg K_D, K_M$ (instantaneous trapping), which occurs at high concentration since the dimer population is substantial and energy transfer is fast.

In addition to affecting excited state dynamical processes, dimer formation should give rise to changes in the ground state absorption spectra of the solutions (24, 25). Spectra of solutions of many concentrations were examined, and it was found that the onset of spectral changes coincides with the onset of changes in the excited state decay rate. This clearly demonstrates that the concentration dependence of the decay rate is due to changes in the ground state molecules and not to processes such as excimer formation that only affect the excited states.

Detailed comparison of the model and the experimental data is given in Figure 5. The monomer and dimer decay rates were determined from the transient grating (TG) data at low and high concentration, respec-

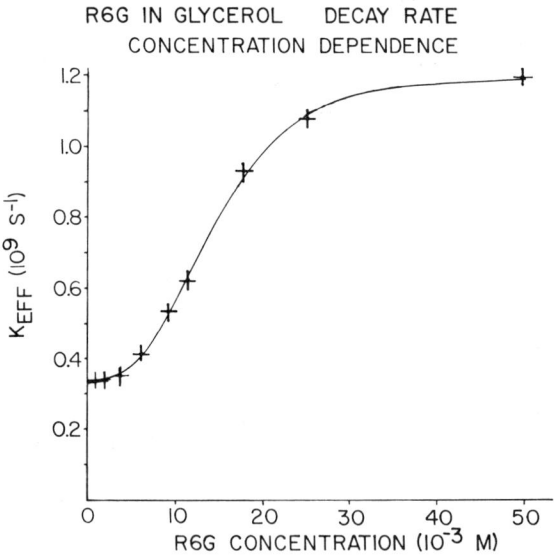

Figure 5 Effective decay constant, K_{eff} vs concentration of Rhodamine 6G in glycerol. + indicates experimental data. As the R6G concentration increases, excited state transport and trapping by R6G dimers becomes increasingly rapid. Fast radiationless relaxation by the dimers decreases the excited state lifetime and quenches fluorescence. The *solid curve* is calculated.

tively. Since the equilibrium constant, q, is not known, TG data at a single intermediate concentration was used to determine K_T. Decay constants K_{eff} were then calculated at other concentrations by scaling K_T as the concentration cubed and using Eq. 10. The calculated values of K_{eff} as a function of concentration yielded the curve (*solid line*) shown in Figure 5. The curve fits the experimentally measured decay constants over the range of concentrations, indicating that the microscopic model is basically correct.

Knowing K_T allows calculation of the equilibrium constant for dimer formation from Eq. 10 since the other parameters are known. An equilibrium constant $q = 9.7$ 1/m was obtained. Equilibrium constants for various solutions of R6G in glycerol-water mixtures have been determined from concentration dependent absorption spectra to range from 28 1/m in the solution with the most water to 11 1/m in the solution with the least water (25). The equilibrium constant that resulted from the time-dependent measurements is consistent with these values. This provides additional support for the basic model.

Similar experiments were performed for the system R6G in ethanol (6). At low concentration the rate constant is determined by the monomer decay rate: $K_{eff} = K_M = 2.7 \times 10^8$ s^{-1}. As the concentration rises, the decay rate rapidly increases. The measurement at the highest concentration is instrumentally limited by the laser pulse duration. The excited state decay constant at high concentration is at least 2×10^{10} s^{-1}, i.e. the lifetime is less than 50 psec. This is in marked contrast to the glycerol solutions, in which the dimer lifetime is 830 psec.

Clearly the radiationless relaxation rates of the loosely bound dimer are influenced by the solvent viscosity. The low viscosity of ethanol permits rapid configurational fluctuations that lead to very fast radiationless relaxation. The fluctuations occur more slowly in glycerol, and thus the dimer lifetime is longer.

These results directly apply to concentration-dependent fluorescence quenching in dye solutions. Trapping on dimers, which increases as the cube of the dye concentration, leads to fast radiationless relaxation and thus quenches fluorescence. The solvent-dependent dimer lifetime also influences fluorescence quenching. In high concentration R6G in ethanol solutions, fluorescence is completely quenched since the dimer radiationless relaxation rate is extremely fast. In high concentration glycerol solutions, fluorescence is only partially quenched since the dimer decay rate is only four times faster than the monomer decay rate. This allows some radiative relaxation to occur.

Transient grating experiments were used in these measurements for two reasons. First, a TG experiment is inherently more sensitive than a probe pulse experiment, although in principle both could provide the same information about the processes under consideration here. In a probe pulse experiment, small changes in an intense beam provide the information. In a grating experiment there is a dark background. The entire diffracted beam is the signal. In a probe pulse experiment in which F is the fraction change in the probe, roughly F^2 of the probe would be diffracted in a grating experiment. In a probe pulse experiment it is extremely difficult to see a change of $F = 0.001$. However, in a grating experiment it is straightforward to detect 10^{-6} diffraction of the probe since it is against a dark background. We found in probe pulse experiments that the highly concentrated samples required very large excitation power densities to achieve sufficient bleaching of the ground state population to give reasonable signal. These very high power densities resulted in anomalous power-dependent decays. In the grating experiments it was possible to use low power densities and still retain good signal-to-noise ratios, making the experiments possible. The second advantage of the grating method is a reduction in the problem of reabsorption of fluorescence in concentrated samples. In a grating experiment, the relevant reabsorption path length is a few fringe spacings, i.e. a few microns. In a probe pulse experiment, reabsorption must not occur in a length corresponding to the probe spot size, ~ 100 μm. In addition, there are a number of types of experiments, such as the measurement of long-range transport processes (29) or the experiments described in the next section, which are only possible using the transient grating approach.

OPTICAL GENERATION AND DETECTION OF ULTRASONIC WAVES AND APPLICATIONS TO PHYSICAL MEASUREMENTS

Optical generation of ultrasonic waves has been of interest to workers in nonlinear optics, acoustics, and condensed matter spectroscopy for some time (30–36). The interaction between light and material acoustic fields is of interest in its own right; furthermore, efficient light to ultrasound conversion holds promise for a variety of scientific and practical applications. Here, a convenient method for optical excitation of coherent acoustic waves in transparent or light-absorbing liquids and solids is described. The acoustic frequency can be continuously and easily varied from about 3 MHz to 30 GHz with our experimental apparatus, and a considerably wider range should be possible. In anisotropic media (crystals, liquid crystals, stretched films, etc) any propagation direction can be selected.

The technique, called Laser Induced Phonons (LIPS), is based on the transient grating experiment (Figure 1) and works as follows. Two time-coincident laser pulses of approximately 100 psec duration intersect inside the sample, setting up an optical interference pattern, i.e. alternating intensity peaks and nulls. Energy deposited into the system via optical absorption or stimulated Brillouin scattering results in the launching of counterpropagating ultrasonic waves (phonons) whose wavelength and orientation match the interference pattern geometry. The acoustic wavelength is given by Eq. 1. With the Nd:YAG laser system in our laboratory we can vary d between 1 mm and 0.1 μm, i.e. over a range of four orders of magnitude. In most materials this corresponds to tunable acoustic frequencies from about 3 MHz to 30 GHz.

The acoustic wave propagation, which continues long after the excitation pulses leave the sample, causes time-dependent, spatially periodic variations in the material density, and since the sample's optical properties (real and imaginary parts of the index of refraction) are density-dependent, the irradiated region of the sample acts as a Bragg diffraction grating. This propagation of the optically excited ultrasonic waves can be optically monitored by time-dependent Bragg diffraction of a variably delayed probe laser pulse.

Diffraction from an ultrasonic wave grating can in principle arise from two distinct mechanisms: (a) changes in the number density of molecules, and (b) changes in the positions of resonances (density-dependent spectral shifts). Both of these could result in amplitude and phase grating diffraction. Under most conditions, phase grating diffraction due to changes in number density predominates (11), and in fact accounts for all acoustic diffraction devices and experiments realized to date. However, it appears that spectral shift effects could be observed in a LIPS experiment under the proper circumstances (11).

The effects of changes in number density are readily calculated. If the peak-null variation in strain is S, then the corresponding peak-null variation of the real and imaginary parts of **n** are, respectively (11),

$$\Delta n_d = -S\frac{(n_0^2 - 1)}{2n_0} \qquad \text{14a.}$$

$$\Delta K_d = -SK_0(\omega). \qquad \text{14b.}$$

These expressions are used with Eq. 4 to calculate the diffracted signal from an ultrasonic grating. It is clear that there will be a phase grating contribution due to density changes at any probe wavelength. However, there will only be amplitude grating diffraction when the probe frequency, ω, is on or near an absorption. Furthermore, it is found that the amplitude contribution will always be small unless the transition is very strong. In the experiments discussed below, all of the ultrasonic gratings are phase gratings.

The mechanism by which LIPS ultrasonic waves are generated depends upon whether the sample is optically absorbing or transparent at the excitation wavelength. If the excitation pulses are absorbed into high-lying vibronic levels, rapid radiationless relaxation and local heating at the interference maxima (the transient grating peaks) occurs. Thermal expansion then drives material in phase away from the grating peaks and toward the grating nulls, setting up counterpropagating waves. The acoustic response to absorbed excitation pulses, in terms of relative material displacement (strain), has been shown to be (8–11)

$$S_{zz} = A\{\cos kz - \tfrac{1}{2}[\cos(\omega t + kz) + \cos(\omega t - kz)]\}$$
$$= A \cos kz (1 - \cos \omega t), \qquad 15.$$

where S_{zz} is the compressional strain along the z-direction, A is the amplitude of the acoustic disturbance, k is the grating (phonon) wave vector, and ω is the acoustic frequency. ($\omega/k = v_z$, the longitudinal speed of sound along the z-direction in the medium.) The excitation geometry is as shown in Figure 1. The acoustic disturbance in Eq. 15 can be viewed as a steady-state expansion (the dc term) plus a transient response (the counterpropagating waves). For simplicity we assume propagation of a single, longitudinal wave, although we (8–11) have demonstrated, and illustrate below, that in anisotropic media quasilongitudinal and quasitransverse waves can also be generated. Local changes in material density, ρ, are given by

$$\delta\rho = -\rho_0 S_{zz}, \qquad 16.$$

where ρ_0 is the normal density.

The excitation pulses are taken to be instantaneous and cross inside the sample at $t = 0$. At this time $\delta\rho = 0$ everywhere, i.e. the density is uniform. Material then moves away from the peaks and toward the nulls, causing the density at the peaks to decrease and the density at the nulls to increase. The density excursion is largest at $t = \pi/\omega$ (and $t = 3\pi/\omega$, $5\pi/\omega$, etc), and returns to normal at $t = 2\pi/\omega$, $4\pi/\omega$, etc (after each acoustic cycle). Thus the density at the grating peaks oscillates between normal and a reduced value, while the density at the grating nulls oscillates between normal and an increased value. The variably delayed probe pulse undergoes no diffraction at $t = 0$, $2\pi/\omega$, etc (uniform density), and is most strongly diffracted at $t = \pi/\omega$, $3\pi/\omega$, etc (maximum density excursion).

In samples that are transparent at the excitation wavelength, optical energy is coupled directly into the sample's acoustic field via stimulated Brillouin scattering (9, 10). This process takes advantage of the inherent spectral line width in the 100 psec excitation pulses. Higher frequency photons from each pulse are annihilated to create lower frequency photons in the opposite pulse and phonons of the difference frequency and wave vector in the medium. Counter-propagating waves (a standing wave) are thus produced. The acoustic response has been predicted earlier to be (9, 10)

$$S_{zz} = -\frac{B}{2}[\sin(\omega t + kz) + \sin(\omega t - kz)]$$
$$= -B\cos kz \sin \omega t, \qquad 17.$$

where B is the acoustic wave amplitude. Experimental observations of this acoustic response are presented below (7, 10). Notice that since there is no radiationless relaxation or heating, no static thermal expansion occurs. (Compare with Eq. 15.) The density at any point in the sample oscillates both above and below normal, and the density is normal everywhere ($\delta\rho \equiv 0$) at $t = 0$, π/ω, $2\pi/\omega$, etc (twice each acoustic cycle). Thus the variably delayed probe pulse undergoes no diffraction twice each acoustic cycle. The different time dependences of the diffracted signal arising from the two acoustic wave-generating mechanisms allow simple determination of the mechanism of phonon excitation in a LIPS experiment.

Figure 6 shows LIPS data from pure ethanol and from solutions of malachite green in ethanol. The 532 nm excitation light is absorbed by malachite green but not by ethanol. The grating fringe spacing is 2.47 μm and the speed of sound in ethanol is 1.16×10^5 cm/sec (37). The acoustic frequency is therefore $\omega = 2.95 \times 10^9$ s^{-1}, and the acoustic period $\tau_{ac} = 2.13$ nsec. (The acoustic period is the time required for the ultrasonic waves to travel one fringe spacing.) Figure 6a shows LIPS data from pure ethanol. The important feature is the frequency of oscillations in the diffracted intensity. Signal vanishes every 1.07 nsec, or exactly twice each acoustic cycle. This indicates that the acoustic disturbance is as given in Eq. 17, and that simulated Brillouin scattering is responsible for acoustic wave production. This is to be expected since the green excitation light is not absorbed by pure ethanol.

Figure 6c shows data from a solution of 5×10^{-5} M malachite green in ethanol. This concentration is such that the acoustic disturbance due to optical absorption completely obscures that due to electrostriction. The diffracted intensity vanishes once each acoustic cycle, indicating that the

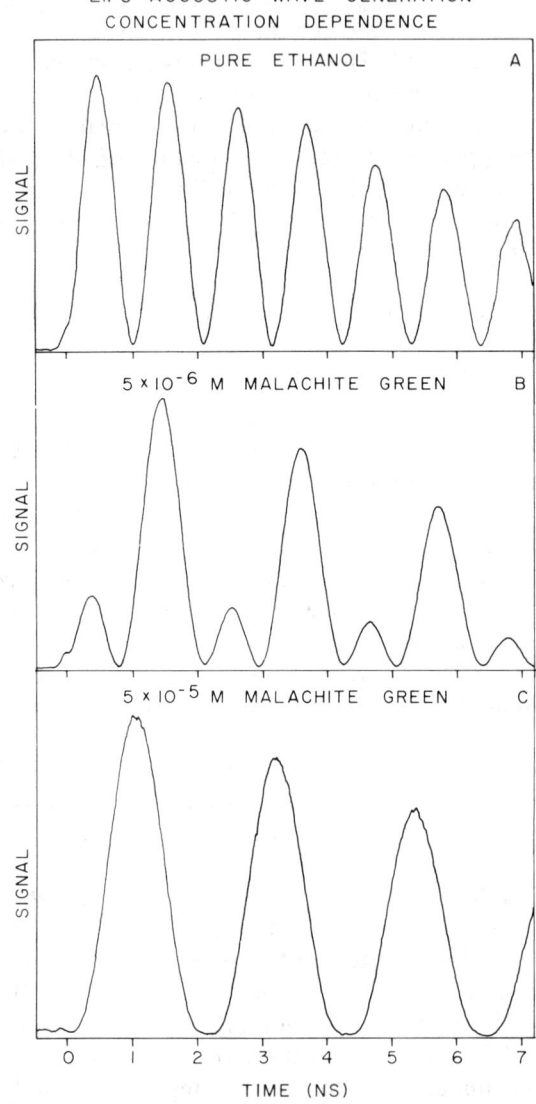

Figure 6 Laser Induced Phonons (LIPS) transient grating data from pure ethanol and solutions of malachite green in ethanol. Excitation $\lambda = 532$ nm; probe $\lambda = 566$ nm. Fringe spacing $d = 2.47$ μm; acoustic cycle $\tau_{ac} = 2.13$ nsec. Experimental conditions other than sample were identical throughout. (*a*) Pure ethanol. Electrostrictively generated standing wave causes diffraction intensity to oscillate twice each acoustic cycle. (*b*) 5×10^{-6} M malachite green in ethanol. Electrostriction and optical absorption produce comparable responses. (*c*) 5×10^{-5} M malachite green in ethanol. Optical absorption effects dominate. Diffracted signal oscillates once each acoustic cycle.

acoustic response is as given in Eq. 15. Figure 6b shows data from a solution of intermediate malachite green concentration. Optical absorption and electrostriction excite acoustic responses of comparable magnitudes, and the effects of both mechanisms are observed. The total acoustic disturbance is simply the sum of Eqs. 15 and 17 with appropriately weighted amplitudes A and B. Diffracted signal at any time goes as the square of the density excursion inside the material, calculated from Eq. 16. Equations 15 through 17 predict the positions and relative amplitudes of the signal maxima for any concentration of absorbing species, and a gradual transition from data of the form in Figure 6a to that in Figure 6c is predicted and was observed for a large number of intermediate concentration solutions. The amplitude of the acoustic response to optically absorbed excitation pulses (A in Eq. 15) has been quantitatively related to the absorption strength and to thermal expansion and elastic stiffness coefficients of the sample (7, 8). The amplitude of the electrostrictively generated waves (B in Eq. 17) has been similarly related to the sample's electrostrictive (photoelastic) constants (7, 9). Comparison of the acoustic response due to a known amount of absorption to the electrostrictively generated response, using data such as that in Figure 6b, permits the accurate measurement of photoelastic constants (7).

It is of interest to note that only weak absorption is needed to produce a detectable "heating" acoustic response. For example, with YAG fundamental (1.06 μm) excitation pulses, absorption into vibrational overtones of water produces a response that obscures the electrostrictively generated response. With deuterated water (or with H_2O and 532 nm excitation), electrostriction effects dominate. Similar results have been obtained with benzene and deuterated benzene. By tuning the excitation wavelength, an overtone absorption spectrum can be taken (7). This is illustrated in Figure 7. In this experiment (7), the dye laser was used for excitation, and a green ($2 \times$ Nd:YAG) probe pulse was employed. The dye laser was tuned to various points across the $0 \rightarrow 6$ transition of a benzene C-H stretching vibrational mode centered at 607 nm. At each wavelength the diffracted intensity of the first and second maxima in the LIPS signal from pure benzene was recorded. The data sets have the appearance of Figure 6b. From this data the absorption strength is determined. The sample path length was only 1 mm, and although the spectrum is quite noisy, it is clear that the technique has the potential for great sensitivity.

Finally, I wish to illustrate the results of LIPS ultrasonic wave generation in anisotropic media. Figure 8 shows LIPS data from the monoclinic molecular crystal α-perylene. This has been discussed in detail previously

(8, 38). Excitation and probe wavelengths were 532 nm and were weakly absorbed by the sample, leading to one oscillation per acoustic cycle. In Figures 7a and 7b, the grating was aligned along the $\bar{b}$ and $\bar{a}$ crystallographic axes, respectively, and single longitudinal waves were generated. With the grating aligned between the $\bar{a}$ and $\bar{b}$ axes in the $\overline{ab}$ plane, quasilongitudinal and quasitransverse waves of different frequencies were generated. The theory of LIPS acoustic wave generation in anisotropic media has been detailed in Ref. (8), and predicts the observed results. In anisotropic media, pure longitudinal waves or quasilongitudinal and quasitransverse waves will be generated depending upon grating orientation. Anisotropic acoustic parameters (velocities and attenuations) can thus be measured with this technique.

The LIPS technique is an extremely versatile tool for controlled optical generation of ultrasonic waves in condensed media. LIPS experiments have been performed on transparent and absorbing solutions, organic and inorganic crystals, glasses, and plastics. The effects have been observed from liquid helium temperatures to room temperature. Acoustic waves have been generated by stimulated Brillouin scattering and by optical absorption. With a mode-locked Nd:YAG system, the acoustic frequency can be varied between about 3 MHz and 30 GHz, and above 60 GHz in some materials. Any propagation direction can be selected,

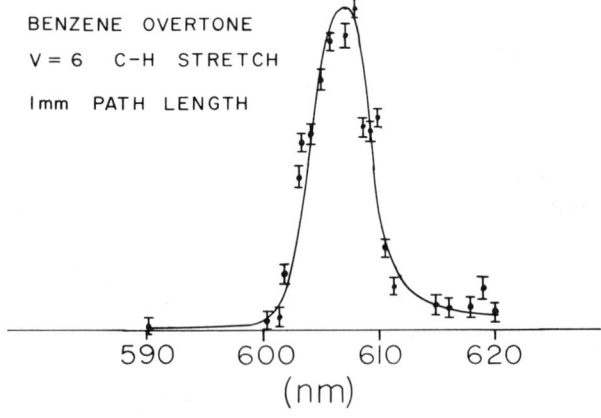

Figure 7 A spectrum of the $v = 6$ transition of a C-H stretch of benzene in a 1 mm path of pure benzene. The spectrum is obtained from the LIPS transient grating experiment as a function of excitation wavelength for fixed probe pulse wavelength. The data has an appearance like that of Figure 6b. At each wavelength the relative height of the first and second peak is determined. This relative height is proportional to the extent of absorption. Thus, the technique is a type of coherent acoustic spectroscopy.

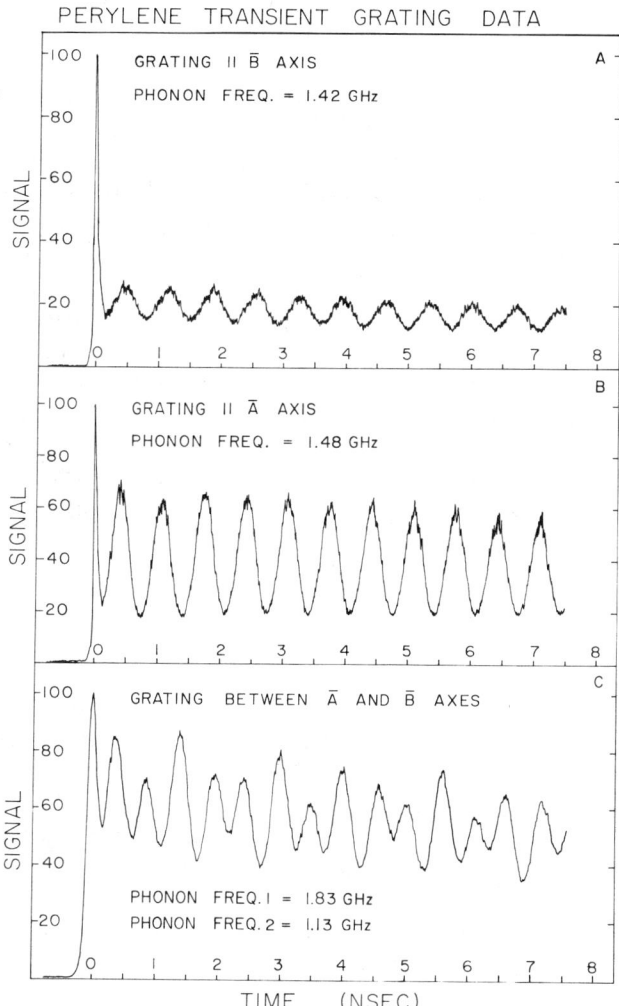

Figure 8 LIPS results for pure α-perylene. Acoustic wavelength is 1.73 μm in all cases. (*a*) Acoustic waves propagate along the $\bar{B}$ (symmetry) axis. Single-frequency modulation is observed. (*b*) Acoustic waves propagate along the $\bar{A}$ axis; single-frequency modulation is observed. (*c*) Waves propagate between $\bar{A}$ and $\bar{B}$ axes, in $\bar{A}\bar{B}$ plane. Beating is due to generation of quasilongitudinal and quasitransverse waves of identical wavelength but having different frequencies.

and in anisotropic materials quasilongitudinal and quasitransverse waves can be generated. The optically generated acoustic waves can be optically amplified, cancelled, or phase shifted (10).

There are a wide variety of applications for the LIPS effect (8–11). It has been used to measure anisotropic elastic constants. Acoustic attenuation measurements have been made. After the traveling waves leave the excitation region, only the "static" acoustic response arising from the absorbing mechanism remains. This decays slowly because of thermal diffusion, and measurement of its decay in crystals as a function of orientation can yield thermal diffusion tensors. LIPS could find possible applications in the field of optical communications. The diffracted probe beam could be both amplitude and frequency (phase) modulated to carry information. The applicability of LIPS to measure photoelastic constants and absorption spectra of weakly absorbing materials has been demonstrated. Finally, LIPS may be of great use in the nondestructive acoustic testing of a wide variety of materials for which conventional techniques are not possible. We are currently involved in making structural measurements on phospholipid bilayers and on single crystal optical fibers using these optically generated and detected acoustic effects.

CONCLUDING REMARKS

In our studies of condensed phase molecular dynamics we have found picosecond transient grating experiments to be versatile tools with a wide range of applications. Here three examples have been discussed in some detail: observation of the dispersive as well as the absorptive characters of excited states, examination of electronic excited state transport and trapping in concentrated dye solutions, and optically generated and detected acoustic wave experiments. A wide variety of other experiments were mentioned. Picosecond transient grating experiments provide new observables that will continue to contribute to an enhanced understanding of condensed phase systems.

Acknowledgment

The material presented in this article is based on extensive work performed in collaboration with a number of investigators. I would like to take this opportunity to thank Professor Keith A. Nelson, Dr. R. Casalegno, R. J. Dwayne Miller, Dr. C. R. Gochanour, and Dr. D. R. Lutz. Also I would like to thank the National Science Foundation, Division of Material Research for supporting the bulk of the research described in this article and the Department of Energy for support of the work on excited state dynamics in concentrated dye solutions.

Literature Cited

1. Hetherington, W. M., Michaels, R. H., Eisenthal, K. B. 1979. *Chem. Phys. Lett.* 66:230–33
2. Tredwell, C. J., Porter, G. 1978. *Chem. Phys. Lett.* 56:278–82
3. Gochanour, C. R., Fayer, M. D. 1981. *J. Phys. Chem.* 85:1989–94
4. Busch, G. E., Jones, R. P., Rentzepis, P. M. 1973. *Chem. Phys. Lett.* 18:178–85
5. Chuang, T. J., Eisenthal, K. B. 1971. *Chem. Phys. Lett.* 11:368–70
6. Lutz, D. R., Nelson, K. A., Gochanour, C. R., Fayer, M. D. 1981. *Chem. Phys.* 58:325–34
7. Miller, R. J. D., Casalegno, R., Nelson, K. A., Fayer, M. D. 1982. *Chem. Phys.* In press
8. Nelson, K. A., Fayer, M. D. 1980. *J. Chem. Phys.* 72:5202–18
9. Nelson, K. A., Lutz, D. R., Fayer, M. D. 1981. *Phys. Rev. B* 24:3261–75
10. Nelson, K. A., Miller, R. J. D., Lutz, D. R., Fayer, M. D. 1982. *J. Appl. Phys.* 53:1144–49
11. Nelson, K. A., Casalegno, R., Miller, R. J. D., Fayer, M. D. 1982. *J. Chem. Phys.* In press
12. Kogelnik, H. 1969. *Bell Syst. Tech. J.* 48:2909–13
13. Born, M., Wolf, E. 1965. *Principles of Optics.* Oxford: Pergamon. 3rd ed.
14. Scrivner, G. P., Tubbs, M. R. 1974. *Opt. Commun.* 10:32
15. Gochanour, C. R., Andersen, H. C., Fayer, M. D. 1979. *J. Chem. Phys.* 4254–71
16. Loring, R. F., Andersen, H. C., Fayer, M. D. 1982. *J. Chem. Phys.* 76:2015–27
17. Imbusch, G. F. 1967. *Phys. Rev.* 153:326–35
18. Birks, J. B. 1970. *Photophysics of Aromatic Molecules,* pp. 142–92. New York: Wiley-Interscience
19. Schäfer, F. P. 1977. In *Topics in Applied Physics,* ed. F. P. Schäfer, 1:21–24, 158–60. New York: Springer-Verlag. 2nd ed.
20. Berlman, I. 1973. *Energy Transfer Parameters of Aromatic Compounds,* p. 62. New York: Academic
21. McElroy, J. D., Feher, G., Mauzerall, D. C. 1972. *Biochim. Biophys. Acta* 267:363–67
22. Norris, J. R., Druyan, M. E., Katz, J. J. 1973. *J. Am. Chem. Soc.* 95:1680–85
23. Berlman, I. 1973. See Ref. 20, pp. 27–47, and references therein
24. Selwyn, J. E., Steinfeld, J. I. 1972. *J. Phys. Chem.* 76:762–71
25. Bojarski, C., Kuśba, J., Obermueller, G. 1975. *Acta Phys. Polon. A* 48:85–92
26. Wieting, R. D., Fayer, M. D., Dlott, D. D. 1978. *J. Chem. Phys.* 69:1996–2010
27. Montroll, E. W. 1964. *Proc. Symp. Appl. Math.* 16:193. Providence, R.I: Am. Math. Soc.
28. Montroll, E. W. 1969. *J. Math. Phys.* 10:753–59
29. Fayer, M. D. 1982. In *Modern Problems in Solid State Physics: Molecular Solids,* ed. A. A. Maradudin. Amsterdam: North Holland. In press
30. Quate, C. F., Wilkinson, D. W., Winslow, D. K. 1965. *Proc. IEEE* 53:1604–13
31. Kroll, N. M. 1965. *J. Appl. Phys.* 36:34–46
32. Cachier, G. 1971. *J. Acoust. Soc. Am.* 49:974–85
33. Yang, K. H., Richards, P. L., Shen, Y. R. 1973. *J. Appl. Phys.* 44:1417–23
34. Grill, W., Weis, O. 1975. *Phys. Rev. Lett.* 35:588–91
35. Meltzer, R. S., Rives, J. E. 1977. *Phys. Rev. Lett.* 38:421–25
36. Eichler, H., Stahl, H. 1973. *J. Appl. Phys.* 44:3429–42, and references therein
37. Altenburg, K. 1972. *Z. Phys. Chem. Leipzig* 250:399–411
38. Nelson, K. A., Dlott, D. D., Fayer, M. D. 1979. *Chem. Phys. Lett.* 64:88–93

MOLECULAR FEATURES OF METAL CLUSTER REACTIONS

E. L. Muetterties, R. R. Burch, and A. M. Stolzenberg

Department of Chemistry, University of California, Berkeley, California 94720

INTRODUCTION

Molecular metal clusters now represent one of the largest families of metal complexes.[1] Until quite recently, the major thrust in metal cluster research had been synthesis and structural characterization. Presently, the emphasis is shifting slowly to a characterization of cluster chemistry. Essential to an understanding of clusters and their chemistry and to a logical exploitation of the chemistry is the delineation of the molecular features of important and typical cluster reactions.

Inorganic and organometallic complexes with a single metal atom, an extensively studied area, display in their chemistry various classes of reactions of general and systematic importance (9). For example, a classic reaction is ligand substitution that has as the net reaction the replacement of one ligand by another and that mechanistically may be associative, dissociative, or interchange in character (10).[2] Electron transfer reactions, of inner or outer coordination sphere mechanistic form, comprise another important reaction class. The complementary reaction processes of oxidative-addition and reductive-elimination, as exemplified in hydrogen (H_2) addition to and elimination from metal complexes, is

---

[1] There has been no general review of molecular metal cluster chemistry. Most reviews have been directed to metal carbonyl clusters. In any case, see Refs. 1–8 for extant reviews.

[2] Respectively, reactions that proceed through an intermediate in which the metal atom has a higher coordination number than in the metal reactant, an intermediate of lower coordination number, and no discernable intermediate. With these definitions, the classic nucleophilic substitution reaction at tetrahedral carbon is an interchange reaction, although it has the intimate character of associative reactions.

another class reaction of great significance in organometallic and catalytic chemistry. Formally, these reaction classes should have their analogs in metal cluster chemistry, and there presently are ample data to support this statement. In each of these class reactions, the metal clusters may prove to be far more diverse with respect to reaction mechanism than the mononuclear metal complexes. Metal-metal bond energies in clusters are typically small, 15–50 kcal/mol, and there may be fragmentation in the course of a cluster class reaction, although the nuclearity may remain the same for cluster reactant and isolated cluster product (8, 11). Our present review focuses on reaction mechanism in cluster class reactions; the review is critical, not comprehensive. Reaction mechanisms of cluster reactions have not been extensively studied; to date, the majority of such studies have addressed one class of reactions, the ligand substitution reaction, and this comprises the major part of the discussion. Hydrogen addition and elimination reactions represent another set. Catalytic reactions of clusters are relatively few in number and still fewer have been studied mechanistically. Accordingly, we do not address the subject of cluster initiated catalytic reactions as a separate topic.

The family of known metal clusters comprises many classes, which range from those with metal atoms of relatively high formal oxidation states, such as the cluster halides, sulfides, and oxides, to those with zero-valent metal atoms, as in the largest class of clusters, the metal carbonyl clusters (1–8). It is the latter class that primarily has been studied in the context of reaction mechanism and is the cluster class discussed in this review with respect to ligand substitution reactions. For oxidative-addition/reductive-elimination reactions, the most definitive data are derived from a set of coordinately unsaturated metal clusters in which the ligands are hydrides and trivalent phosphorous compounds, although some data are available for metal carbonyl clusters.

We attempt in our review to identify key experiments for a more incisive identification of reaction mechanisms in this relatively complex area of molecular metal cluster chemistry.

LIGAND SUBSTITUTION REACTIONS

General Considerations

Ligand substitution reactions are the set of reactions in which a ligand, L, bound to one or more metal centers, is replaced by a new ligand, L', as generally represented in Eq. 1:

$$M_xL_y + L' \rightarrow M_xL_{(y-1)}L' + L. \qquad 1.$$

That this set of reactions is central to the rational synthesis of specific metal complexes should be clear from the definition. Less apparent is the importance of this class of reactions to the development of an understanding of both electron transfer reactions and the variation of structure and reactivity with electronic structure of mononuclear metal complexes. There is every reason to believe that ligand substitution reactions of metal clusters will have a similar importance in the expanding knowledge of these multinuclear systems.

In this discussion of reaction mechanism of cluster reactions, specifically for ligand substitution reactions, we adapt the precise definitions of Langford & Gray (10) for substitution reactions of mononuclear metal complexes to the area of cluster reactions. Reaction mechanism is divided into *stoichiometric* and *intimate* mechanisms. The stoichiometric mechanism is the set of elementary steps. In elaborating the set of elementary steps for a reaction, the crucial issue to be resolved is whether there is an intermediate(s). The study of the elementary steps is the study of the intimate mechanism. For cluster substitution reactions, there are three classes of reaction pathways or stoichiometric mechanism, namely:

1. Associative, A, paths whereby the cluster substitution reaction proceeds through a demonstrable intermediate in which there are more ligands bound to cluster metal atoms than in the reactant cluster (the net process in the formation of the intermediate is bond-making).
2. Dissociative, D, paths whereby the cluster substitution reaction proceeds through a demonstrable intermediate in which either there are less ligands bound to the cluster metal atoms than in the reactant cluster or a metal-metal bond has been cleaved (the net process in the formation of the intermediate is bond-breaking).
3. Interchange, I, paths wherein there is no *demonstrable* intermediate.

Establishment of the stoichiometric reaction form is the most difficult experimental task, and most ligand substitution reactions assume accurately or by default the interchange connotation. For this intermediate, or indeterminate, case, the intimate reaction mechanism can be established by conventional kinetic studies in which the leaving and entering group kinetic dependencies are elucidated. Virtually all the distinctions in mechanistic character of cluster reactions that we present below relate primarily to the intimate rather than the stoichiometric character of the cluster ligand substitution reactions; no intermediates have been demonstrated for these cluster ligand substitution reactions. Hence, all the cluster reactions studied to date are interchange, I, in character[3] and intimate distinctions are between I_a, interchange with substantial bond-

---

[3] In fact, they may be associative or dissociative, but unless intermediates are demonstrated, the reactions by definition (10) are interchange.

ing to both leaving and entering groups in the transition state, and I_d, interchange with only weak bonding to leaving and entering groups in the transition state.

The concept of coordination saturation has been quite successful in rationalizing the reactivity of mononuclear metal complexes. It is a corollary of the 18-electron rule and reflects the large energy separation, in most complexes, between closely spaced *ns*, *np*, and $(n-1)$d orbitals and orbitals of higher principle quantum number (12). Coordinately saturated, i.e. 18 electron complexes, tend to undergo ligand substitution via a dissociative pathway (Eqs. 2 and 3). However, unsaturated

$$ML_n \underset{k_{-1}}{\overset{k_1}{\rightleftharpoons}} ML_{(n-1)} + L \qquad\qquad 2.$$

$$ML_{(n-1)} + L' \overset{k_2}{\to} ML_{(n-1)}L' \qquad\qquad 3.$$

complexes have energetically accessible empty orbitals. These low-lying orbitals permit, but do not necessitate, associative or interchange reaction mechanisms (Eq. 4):

$$ML_n + L' \to ML_n L' \to ML_{(n-1)}L' + L. \qquad\qquad 4.$$

One might expect that the concept of coordination saturation and the related expectations pertaining to reaction mechanism could be extended, in a formal sense, to metal clusters. Unfortunately, there is no simple way to determine accurately the energy levels of a given cluster, and thus its coordination saturation, short of performing a gas phase photoelectron spectroscopic study and a molecular orbital calculation.[4] In particular, the energy gap between the highest occupied molecular orbital and the lowest unoccupied orbital must be known to determine the effective coordination saturation of the cluster.[4] If the gap is very large, the cluster effectively may be considered saturated, and ligand substitution will tend to be dissociative in character. When a small gap exists, the cluster may be considered unsaturated. In this event, the nature of the low-lying orbitals will determine the precise character of the reaction mechanism. Where the lowest unoccupied orbital has metal-metal antibonding character, scission of a metal-metal bond may be the dominant first reaction step. Nevertheless, the reaction sequence cannot be as easily or clearly anticipated as the above discussion suggests. For example, the $M_2(CO)_{10}$ clusters of group VIIB metals, which must be considered saturated on

---

[4]J. Lauher (13, 14) has developed a general characterization of metal carbonyl clusters based on a qualitative extended Hückel molecular orbital analysis, but this effective structural prognosticator does not directly yield the key information about energy level separations.

any qualitative or quantitative basis, could react through metal-metal bond scission (see below) resulting from population of the lowest unoccupied molecular orbital, which has metal-metal antibonding character, either through thermal population of this molecular orbital or by overlap of this orbital with an occupied orbital of an entering ligand. Such metal-metal bond scission preequilibria thus could have the character of a dissociative or an associative reaction.

Experimental definition of an associative or dissociative reaction path for a mononuclear metal complex requires the demonstration, by a kinetics experiment or by spectroscopy, of an intermediate. The same requirements apply to cluster reactions. Kinetic demonstration of an intermediate could comprise the relatively incisive set of experiments showing the accumulation of an intermediate. Presence of an intermediate derived from a dissociative step could be *suggested* by rate inhibition with some added species. If such reaction path diversion can be demonstrated, then the complex derived from the putative intermediate and the added species should be detectable by spectroscopic means.

The intimate character of the reaction steps can be established by kinetic studies. Rates of associative reactions should have a first-order dependence on the concentration of both the metal complex and the entering ligand. Excluding the concentration regime wherein $k_{-1}[ML_{(n-1)}][L]$ (Eqs. 2 and 3) is not insignificant compared with $k_2[ML_{(n-1)}][L']$,[5] the rates of dissociative reactions are independent of the concentration of entering ligand. Typically, the rate of substitution is determined as a function of metal complex and of entering ligand concentration and the data are fitted to a rate expression. The ubiquitous rate expression for all ligand substitution reactions, both for mononuclear and cluster metal complexes, is of the form: rate $= (k_1 + k_2[L'])$ [complex]. Such a derived rate expression does not necessarily imply that there are two distinct reaction pathways—namely a dissociative pathway represented by k_1 and an associative pathway represented by k_2. In fact, the rate expression provides little information about the stoichiometric reaction (is there an intermediate or not?). The rate expression can, however, provide information about the intimate character of the reaction if solvent is varied, if the steric and electronic nature of the entering ligand is varied, and if the nature of the leaving ligand (where feasible) is similarly varied. For example, if a reaction is fundamentally associative in character, the k_1 term in the rate expression may reflect a competing

---

[5] In this regime, the rate will exhibit a dependence upon entering ligand concentration. It will, however, be less than a first-order dependence.

solvent participation in the ligand substitution reaction. If a reaction rate is essentially insensitive to the nature of the entering ligand, then the reaction has a dissociative (intimate mechanism) character. Dependence of rate upon the steric character of the leaving ligand also can be a relatively incisive indicator of a dissociative (intimate) reaction mechanism.

Inevitably, the mechanistic interpretation of rate data for cluster reactions will be less unequivocal than for reactions of mononuclear metal complexes. This statement is based on the recognition of the far greater number of reaction pathways available to metal clusters. Consider simply some of the reaction pathways that in principle are plausible for ligand substitution in a trinuclear metal carbonyl wherein the three metal atoms are in a triangular array and bonded together by single metal-metal bonds. These plausible pathways are enumerated below:

1. Simple ligand dissociation, D:

$$M_3(CO)_{12} \rightleftharpoons M_3(CO)_{11} \xrightarrow{L} M_3(CO)_{11}L$$

2. Ligand interchange, I_a or I_d:

$$M_3(CO)_{12} + L \rightleftharpoons M_3(CO)_{12}\ldots L \rightarrow M_3(CO)_{11}L$$

3. Associative substitution, A:

$$M_3(CO)_{12} + L \rightleftharpoons M_3(CO)_{12}L \rightarrow M_3(CO)_{11}L$$

4. Cluster fragmentation, D:

$$M_3(CO)_{12} + L \rightleftharpoons LM(CO)_3 \xrightarrow{M_3(CO)_{12}} M_3(CO)_{11}L$$

5. Intramolecular metal-metal bond breaking, D^6:

$$M_3(CO)_{12} \rightleftharpoons (OC)_4M - \overset{\overset{\displaystyle O\ \ O}{\overset{\displaystyle C\ \ C}{\diagup\diagdown\diagup\diagdown}}}{\underset{\underset{\displaystyle C\ \ C}{\underset{\displaystyle O\ \ O}{\diagdown\diagup\diagdown\diagup}}}{M}} - M(CO)_3 \xrightarrow{L} M_3(CO)_{11}L$$

6. Ligand assisted intramolecular metal-metal bond breaking, A or I_a[6]:

$$M_3(CO)_{12} + L \rightleftharpoons (OC)_5M - \underset{\underset{\underset{O}{C}}{\diagup}}{\overset{L}{\underset{|}{M}}} {\overset{}{\underset{\underset{O}{C}}{\diagdown}}} - M(CO)_5 \rightleftharpoons M_3(CO)_{11}L$$

The above is not intended to be a comprehensive list of possibilities.

Mechanistic studies largely advance through elimination of mechanistic possibilities. Since many of the above pathways will have identical rate expressions in some concentration regimes, it is clear that a simple kinetic study of a single substitution reaction for $M_3(CO)_{12}$ with a specific ligand can make *no* mechanistic distinctions. Additionally, it is uncertain whether concentration regimes that would permit differentiation between rate expressions for some of the different pathways will be experimentally accessible. What is certain is that the ingenuity, skill, and exactitude of the experimentalist will be rigorously taxed in the area of cluster reaction mechanisms. The occurrence of cluster fragmentation to yield mononuclear intermediates could be addressed by a challenging double labeling study using pure metal isotopes, i.e. reaction of a mixture of $^xM_3(CO)_{12}$ and $^yM_3(CO)_{12}$. Distinction between an intramolecular metal-metal bond breaking mechanism and a dissociative substitution mechanism (or ligand interchange, I_d) could be even more difficult to achieve. Differentiation should be possible based upon the values of the activation parameters $\Delta S^\ddagger$ and $\Delta V^\ddagger$, but these quantities are difficult to determine or interpret accurately.

Cis and trans effects are well-established ligand-metal-ligand interactions that can determine the stereochemical outcome of ligand substitution reactions of mononuclear metal complexes. Similar effects should exist in the analogous reactions of metal clusters. These effects may pertain to metal-ligand bond as well as to metal-metal bond activation. In addition, there is the possibility of the transmission of ligand effects to metal sites other than those bound directly to the ligand. This transmission could be mediated through electronic or steric effects to the remote sites or could be the consequence of stereochemical nonrigidity of intermediates with open coordination sites.

[6] The stereochemistry of the intermediate or transition state is only suggestive for the general case and is not intended to imply a favored form.

Some, but not all, of the issues raised above can be addressed with the mechanistic data presently available. These are discussed in the following sections.

Binuclear Metal Carbonyl Complexes

Binuclear metal complexes represent the simplest class of cluster complexes. This class may be subdivided into two groups: complexes containing a simple metal-metal bond and those containing a metal-metal bond supported by bridging ligands. Examples of the former are the $M_2(CO)_{10}$ complexes of group VIIB metals (Mn, Tc, Re). Such complexes as $Co_2(CO)_8$ typify the latter group.

REACTIONS OF $M_2(CO)_{10}$ COMPLEXES Structural characterization of the $M_2(CO)_{10}$ complexes (15–17) show that they can be regarded as two octahedra sharing a common apex. The four-fold axes of these octahedra are coincident, but the basal CO groups of the two halves are rotated 45° with respect to each other. This results in a staggered configuration of approximately D_{4d} symmetry. Interestingly, the metal-metal bonds in these clusters are significantly longer (about 0.5 Å) than the sum of the "normal" covalent radii of the metal. Furthermore, the basal M–CO bonds are bent toward the other half of the molecule.

The above observations might suggest that these complexes are likely candidates for ligand substitution reactions that proceed through metal-metal bond scission. This type of reactivity is well established for the photochemically activated reactions of these complexes. An absorption band in the near UV spectra of these complexes has been assigned as a $\sigma-\sigma^*$ transition characteristic of the metal-metal bond (18). Irradiation at or near this absorption results in homolysis of this bond to form radical intermediates, as expected. When the photolysis is conducted in CCl_4, mononuclear metal carbonyl chlorides result (Eq. 5) (19):

$$M_2(CO)_{10} \xrightarrow[CCl_4]{h\nu} 2M(CO)_5Cl. \qquad 5.$$

On the other hand, photolysis in alkane solvents in the presence of phosphines results in the formation of a variety of phosphine-substituted products (20, 21). Of these, the disubstituted derivative, $M_2(CO)_8P_2$, is the initial product. These results can be rationalized by the following scheme (Eqs. 6–9).

$$M_2(CO)_{10} \rightleftharpoons 2M(CO)_5\cdot \qquad 6.$$

$$M(CO)_5\cdot + L \rightleftharpoons M(CO)_4L\cdot + CO \qquad 7.$$

$$M(CO)_4L\cdot + M(CO)_5\cdot \rightleftharpoons M_2(CO)_9L \qquad 8.$$

$$2M(CO)_4L\cdot \rightleftharpoons M_2(CO)_8L_2 \qquad 9.$$

As predicted by this scheme, flash photolysis of a mixture of $Mn_2(CO)_{10}$ and $Re_2(CO)_{10}$ yields $MnRe(CO)_{10}$, photolysis of $MnRe(CO)_{10}$ yields both $Mn_2(CO)_{10}$ and $Re_2(CO)_{10}$, and photolysis of $Mn_2(CO)_9[P(C_6H_5)_3]$ yields both $Mn_2(CO)_{10}$ and $Mn_2(CO)_8[P(C_6H_5)_3]_2$.

Substantial disagreement remains about the reaction mechanism, which could vary with the metal,[7] of the thermal ligand substitution reactions of $M_2(CO)_{10}$ complexes. Two opposing schools advocate different limiting mechanisms. Atwood and others propose a mechanism involving CO dissociation (Eqs. 10–13).

$$M_2(CO)_{10} \underset{k_{-1}}{\overset{k_1}{\rightleftharpoons}} M_2(CO)_9 + CO \qquad 10.$$

$$M_2(CO)_9 + L \overset{k_2}{\rightarrow} M_2(CO)_9L \qquad 11.$$

$$M_2(CO)_9L \rightleftharpoons M_2(CO)_8L + CO \qquad 12.$$

$$M_2(CO)_8L + L \rightarrow M_2(CO)_8L_2 \qquad 13.$$

Poë, however, favors a radical mechanism similar to the photochemically initiated one (Eqs. 6–9). An unequivocal resolution of this reaction mechanism issue(s) cannot be achieved with the experimental data presently available.

Determination of the rate expression for a reaction often permits one to distinguish between possible mechanisms based upon the agreement of predicted and observed rate behavior. Unfortunately, both of the above mechanisms predict rate expressions for ligand substitution that are first order in $[M_2(CO)_{10}]$ and independent of $[L]$, when limiting concentrations of L are employed. For less than limiting rates of reaction, however, a radical mechanism requires the dependence on $[M_2(CO)_{10}]$ to decrease from first to half order. No evidence as yet has been adduced that any ligand substitution reaction of the $M_2(CO)_{10}$ complexes exhibits less than a first-order dependence upon the concentration of complex. No conclusion may be drawn from this failure to observe a less than first-order dependency in that it could merely reflect achievement of limiting rates for these reactions at exceedingly low concentrations of entering ligand. The observed independence of the reaction rates upon the concentration of entering ligand does rule out a third possibility, an associative (or I_a) mechanism (Eq. 14).

$$M_2(CO)_{10} \rightleftharpoons M_2(CO)_{10}L \rightleftharpoons M_2(CO)_9L + CO \qquad 14.$$

[7] The metal-metal and metal-ligand bond energies in dimanganese decacarbonyl and dirhenium decacarbonyl differ by a factor of about three. Estimated bond enthalpy contributions to the enthalpy of disruption (11) are Mn–Mn, 16 kcal/mol, Mn–CO, 24 kcal/mol, Re–Re, 45 kcal/mol, and Re–CO, 31 kcal/mol.

Although no investigator has found less than first order kinetic behavior during ligand substitution, Poë has demonstrated such behavior for decomposition reactions of $M_2(CO)_{10}$ complexes. In particular, half-order dependences have been found for the thermal decompositions of $Mn_2(CO)_{10}$ under an inert atmosphere (22) and of $[M(CO)_4L]_2$ in the presence of free ligand, L, with $M = Mn$, $L = P(OC_6H_5)_3$, and $M = Re$, $L = P(C_6H_5)_3$ (23, 24).[8] The decomposition of $Mn_2(CO)_{10}$ and of $MnRe(CO)_{10}$ under varied concentrations of O_2 has been shown, under certain conditions, to undergo a transition from first to half-order dependence upon the concentration of complex (22). One of the most unusual experimental findings is the observation of identical limiting rates for oxidation and for substitution of $MnRe(CO)_{10}$ over a wide temperature range (25).

The less than first-order kinetic behavior observed for the oxidation and decomposition reactions is not inconsistent with a CO dissociative mechanism.[9] However, the failure to observe inhibition of the decomposition of $Mn_2(CO)_{10}$ by CO is inconsistent and is strongly suggestive of a radical process.

The foregoing observations are relevant to ligand substitution reactions only if both classes of reactions share common rate-limiting steps. If, in fact, the decomposition reaction and the ligand substitution reaction are not mechanistically related, the precise identity of reaction rates over a wide temperature range is a most unusual coincidence. Still, Atwood maintains that the decomposition reactions are ill-defined and irrelevant to the substitution reactions (26). In contrast, Poë holds that the decomposition reactions establish a precedent for the existence of radicals in the thermal reactions of $M_2(CO)_{10}$ complexes and that the identical limiting rates of these two types of reactions suggest the involvement of radicals in the substitution reactions as well (27). Furthermore, Poë states that the evidence presented for these reactions must either "be refuted or reinterpreted...when any case is...made for a totally different mechanism" (25).

[8] The relevance of the mechanistic features of the latter reaction to the mechanism of the reactions of the parent $M_2(CO)_{10}$ must be considered dubious at best.

[9] Half-order kinetics will be observed for CO dissociative mechanisms when the CO deficient complex undergoes a first- or pseudo-first-order reaction (Reaction 11) to form products at such a low rate that the equilibrium with CO and undissociated complex (Reaction 10) remains undisturbed. In the absence of added CO, the concentration of CO and the CO deficient complex will be equal, at least initially. Both concentrations will have a square root dependence on the concentration of undissociated complex. In the presence of excess free CO, the concentration of the CO deficient complex will be decreased and will vary linearly with the concentration of undissociated complex.

Another possible kinetic differentiation between the two mechanisms hinges on the rate of appearance of bis-phosphine substituted product. This product can form directly from $M_2(CO)_{10}$, given the radical mechanism, but forms only after production of $M_2(CO)_9L$, given the CO dissociative mechanism. Once again, the available observations are equivocal if not contradictory. Evidence has been presented that suggests $Mn_2(CO)_8[P(C_6H_5)_3]_2$ is an initial product during substitution of $Mn_2(CO)_{10}$ (28). Similarly,

$$\{[(C_6H_5)_3P](CO)_4MnRe(CO)_4[P(C_6H_5)_3]\}$$

is reported to appear in the initial stages of the substitution of $MnRe(CO)_{10}$, given a "sufficiently high," but unspecified, phosphine concentration (29). Others claim that this disubstituted complex appears only later (30). Lacking the knowledge of what phosphine concentration is "sufficiently high," it is not feasible to determine whether these two observations are contradictory. Another suggestive observation pertains to CO inhibition: Total suppression of the initial formation of disubstituted product from $Re_2(CO)_{10}$ and $P(C_6H_5)_3$ is effected by a carbon monoxide pressure of one atmosphere (31). This observation, however, can either be interpreted as being inconsistent with the radical pathway or as being a reflection of the requirement of a very large triphenylphosphine concentration for direct formation of the bis-phosphine product in a carbon monoxide atmosphere.

As anticipated in the introductory discussion, even for this prototypical class of metal clusters, mechanistic probes other than kinetic analysis must be considered in order to establish the mechanism of ligand substitution. One such probe, the examination of the product distribution from the substitution of $MnRe(CO)_{10}$, has raised questions about the involvement of the radical mechanism (30). At no point during these substitution reactions (29, 30) or during thermal decomposition of $MnRe(CO)_{10}$ under O_2 (22) are any homodimetallic species, $Mn_2(CO)_{10-x}L_x$ or $Re_2(CO)_{10-y}L_y$, observed. These observations are not totally inconsistent with the involvement of radicals, but their rationalization requires one to postulate either a much higher rate for recombination of radicals containing unlike metals than those of like metals or a very rapidly established equilibrium that greatly favors the mixed metal species. The validity of the first postulate might be questionable given the measurement of near-diffusion-controlled rates for the recombination of $M(CO)_5$ radicals (32, 33). Assuming the presence of radicals, failure to observe any thermal formation of $MnRe(CO)_{10}$ from $Mn_2(CO)_{10}$ and

$Re_2(CO)_{10}$ under a variety of conditions,[10] both in our laboratory and in others (26, 31), would seem to invalidate the second postulate. Further examination of the $MnRe(CO)_{10}$ system has shown that at 130°C under N_2, small amounts of $Mn_2(CO)_{10}$ and $Re_2(CO)_{10}$ can be formed very slowly. This reaction can be completely quenched, however, by the presence of CO (31). Taken as a whole, the results of these crossover experiments imply the absence of significant concentrations of metal radical intermediates in the reactions of $Re_2(CO)_{10}$ and $MnRe(CO)_{10}$, if not $Mn_2(CO)_{10}$ as well.

The most direct approach to resolve this apparent dilemma of reaction mechanism would be to isotopically label the metal centers. The greater metal-metal bond strength of $Re_2(CO)_{10}$ compared with other complexes in the manganese group (11, 34–37) suggests that it, of any, is the one most likely to follow a CO dissociative pathway. We are currently preparing isotopically pure samples of $^{185}Re_2(CO)_{10}$ and $^{187}Re_2(CO)_{10}$ and will look for label crossover during ligand substitution reactions of physical mixtures of these complexes.

Questions of mechanism aside, substitution of phosphines or phosphites for CO on $M_2(CO)_{10}$ have a noticeable effect on further reactivity. Reaction rates increase upon substitution by phosphine, but not by phosphite, as evidenced by the decrease in the temperature required for further substitution (24, 29, 38, 39). In addition, for as yet undetermined reasons, substitution on one metal center directs further substitution to the other metal center.

REACTIONS OF $M_2(CO)_8$ COMPLEXES The structural features of the d^9 cobalt dinuclear metal carbonyl complex, $Co_2(CO)_8$, differ somewhat from those of the complementary d^7 manganese group metal carbonyl complexes. In addition to an isomeric form of D_{3d} symmetry (40, 41) that is analogous in structure to the D_{4d} symmetry manganese group dinuclear carbonyls, $Co_2(CO)_8$ has two other isomeric forms. Of the three isomers, the lowest energy form can be approximated as two $Co(CO)_5$ square pyramids sharing a basal edge. This C_{2v} symmetry form with two bridging carbonyls is the one found in the solid state (42). The third isomeric form is detected only in solution and has an as yet undetermined structure (43, 44).

---

[10] Substitution reactions of dimanganese decacarbonyl and dirhenium decacarbonyl take place at much different rates. One must take care, therefore, to select conditions such that suitable concentrations of both types of radicals should be present. Of the two, the manganese complex is the more reactive. Based on estimates of metal-metal and metal-CO bond energies,[7] this higher reactivity is consistent with either a CO ligand dissociation or a metal-metal bond cleavage reaction path.

The reactivity of $Co_2(CO)_8$ in ligand substitution reactions, like its structural features, is diverse and divergent from that of the manganese group carbonyls (45). Reaction with less basic ligands, such as $As(C_6H_5)_3$, or with more sterically demanding ligands such as $P(t-C_4H_9)_3$, leads sequentially to $Co_2(CO)_7L$ and, at a lower rate, to $Co_2(CO)_6L_2$. The reaction rate is first order in carbonyl complex and independent of ligand concentration. The exchange of $Co_2(CO)_8$ with ^{13}CO proceeds at a rate approximately equal to the rate of reaction with $As(C_6H_5)_3$, calculated by extrapolation to the same temperature. Thus, the mechanism presumably involves rate-limiting CO dissociation.[11]

With basic and less sterically demanding ligands like $P(n-C_4H_9)_3$, a rapid reaction takes place that leads to the formation of

$$\{Co(CO)_3[P(n-C_4H_9)_3]_2^+\}[Co(CO)_4^-].$$

The dependence of the rate of this reaction upon $Co_2(CO)_8$ concentration is complex, the order decreasing from about 1.5 to 1.0 as temperature increases. Traces of oxygen greatly inhibit the reaction. A complicated chain reaction mechanism involving electron transfer has been devised to reproduce reasonably the observed behavior when the rate equations predicted by the mechanism are numerically integrated. This proposed scheme entails the initial formation of

$$Co_2(CO)_8[P(n-C_4H_9)_3],$$

which then fragments to generate chain-carrying species.

The reaction of $Co_2(CO)_7[P(n-C_4H_9)_3]$ and $P(n-C_4H_9)_3$ proceeds at a rate much lower than that observed for the reaction of $Co_2(CO)_8$ with $P(n-C_4H_9)_3$ and yields only

$$Co_2(CO)_6[P(n-C_4H_9)_3]_2.$$

Neither of these substituted species are observed during the reaction of the parent, $Co_2(CO)_8$, with tributylphosphine. Thus, both would seem to be mechanistically unimportant in the reaction of the parent complex, $Co_2(CO)_8$, with $P(n-C_4H_9)_3$.

Trinuclear Metal Carbonyl Clusters

The simplest set of trinuclear metal carbonyls is the trimetal dodecacarbonyls, $M_3(CO)_{12}$, which are derived from the iron group elements. Both $Ru_3(CO)_{12}$ and $Os_3(CO)_{12}$ have structures of D_{3h} symmetry with all

---

[11] These data are also consistent with a rate-limiting homolysis of the Co–Co bond to form a bridged biradical. The large positive entropy of activation measured for this reaction makes such a step unlikely, however.

terminal carbonyl groups (46, 47). The coordination sphere about each metal atom is octahedral with two metal-metal bonds at *cis* positions and perpendicular to the major three-fold axis. Triiron dodecacarbonyl has only C_{3v} symmetry because one metal-metal vector has two bridging carbonyl ligands, above and below the Fe_3 plane (48). Within this set of trinuclear metal carbonyls, only the ruthenium complex has been sufficiently studied that the reaction mechanism can be discussed objectively.

Reaction of $Ru_3(CO)_{12}$ with a wide variety of ligands leads directly to the formation of tri-substituted $Ru_3(CO)_9L_3$ complexes (49–51). The failure to observe less substituted cluster products, with the exception of the reaction with $P(OC_6H_5)_3$ (51), contrasts with observations for substitution reactions of $Os_3(CO)_{12}$ (52) and suggests that replacement of carbonyl ligands by phosphorus ligands accelerates the rate of further reactions. This rate enhancement has been confirmed in a number of kinetic investigations. The rate expression for substitution reactions of the parent molecule, $Ru_3(CO)_{12}$, has terms both independent of and linearly dependent upon entering ligand concentration (49–51). It has been proposed that these two terms correspond to carbonyl dissociative and ligand associative pathways, but no intermediates have been demonstrated. The first-order pathway could just as well be based on an intramolecular metal-metal bond-breaking process. The second-order term becomes more significant with increasing ligand nucleophilicity. In contrast, reactions of both

$$Ru_3(CO)_{11}[P(C_6H_5)_3] \quad \text{and} \quad Ru_3(CO)_{10}[P(C_6H_5)_3]_2$$
$$\text{with} \quad P(C_6H_5)_3$$

exhibit rate expressions independent of $P(C_6H_5)_3$ concentration (53).[12] After allowance for statistical factors, introduction of one $P(C_6H_5)_3$ substituent into $Ru_3(CO)_{12}$ was found to increase the rate of the assumed CO dissociation roughly 60-fold while substitution of a second $P(C_6H_5)_3$ further increased the rate by only a factor of 1.3, Table 1. No explanation of this ligand effect affording rate enhancement was offered save the suggestion that a more complex mechanism might be involved.

The reaction of $Ru_3(CO)_{12}$ with $P(n-C_4H_9)_3$ differs from those of other ligands in that both trinuclear and mononuclear products were observed (54). The relative yield of

$$Ru_3(CO)_9[P(n-C_4H_9)_3]_3 \quad \text{to both} \quad Ru(CO)_4[P(n-C_4H_9)_3]$$
$$\text{and} \quad \textit{trans}\text{-}Ru(CO)_3[P(n-C_4H_9)_3]_2$$

---

[12] These complexes can undergo ligand substitution reactions by the second-order pathway at rates as fast as the second-order process that occurs with $Ru_3(CO)_{12}$. Such reactions would be undetectable due to the much greater rate of the first-order pathway.

depends upon the original ratio of $Ru_3(CO)_{12}$ to phosphine, a high ratio favoring trisubstituted trinuclear product. The ratio of monophosphine to bisphosphine mononuclear complexes appears to be an invariant 2:1. These results have been interpreted by Poë, for the sake of simplicity, in terms of a cleavage of $Ru_3(CO)_{11}[P(n-C_4H_9)_3]$ into reactive mononuclear fragments (54). Addition of phosphine to these fragments affords the observed mononuclear products, whereas trimerization of these fragments results in formation of the $Ru_3(CO)_9[P(n-C_4H_9)_3]_3$ product. This proposed mechanistic scheme, as set out in Eqs. 15–20, at least in principle illustrates that starting and ending with intact clusters does not necessarily imply the intermediacy of intact clusters.

$$Ru_3(CO)_{12} + P(n-C_4H_9)_3 \rightarrow Ru_3(CO)_{11}[P(n-C_4H_9)_3] + CO \quad 15.$$

$$Ru_3(CO)_{11}[P(n-C_4H_9)_3] \rightarrow 2Ru(CO)_4 + Ru(CO)_3[P(n-C_4H_9)_3] \quad 16.$$

$$3Ru(CO)_4 \rightarrow Ru_3(CO)_{12} \quad 17.$$

$$3Ru(CO)_3[P(n-C_4H_9)_3] \rightarrow Ru_3(CO)_9[P(n-C_4H_9)_3]_3 \quad 18.$$

$$Ru(CO)_4 + P(n-C_4H_9)_3 \rightarrow Ru(CO)_4[P(n-C_4H_9)_3] \quad 19.$$

$$Ru(CO)_3[P(n-C_4H_9)_3] + [P(n-C_4H_9)_3] \rightarrow Ru(CO)_3[P(n-C_4H_9)_3]_2 \quad 20.$$

Implicit in this proposed reaction scheme is the fully selective reformation of

$$Ru_3(CO)_{12} \quad \text{and} \quad Ru_3(CO)_9[P(n-C_4H_9)_3]_3$$

from $Ru(CO)_4$ and $Ru(CO)_3[P(n-C_4H_9)_3]$,

Table 1 Comparison of relative first order rate constants upon successive phosphorus ligand substitution of metal clusters

Cluster n	$Ru_3(CO)_{12-n}[P(C_6H_5)_3]_n$[a]		$Co_4(CO)_{12-n}[P(OCH_3)_3]_n$[b]		$Ir(CO)_{12-n}[P(C_6H_5)_3]_n$[c]	
	Raw	Corrected[d]	Raw	Corrected[d]	Raw	Corrected[d]
0[e]	1	1	1	1	1	1
1	55	60	0.9	1.7	220	1320
2	40	78	1.3	3.9	3500	42000

[a] Refs. (53, 78).
[b] Refs. (65, 73, 74, 78).
[c] Refs. (69, 78).
[d] Corrected for the assumed number of dissociable[a] CO groups present assuming that the first order pathway comprises CO ligand dissociation.
[e] Initial rates of the three clusters are not equal.

respectively. Such selective reformation of cluster molecules from mononuclear fragments is not necessarily an expected property of the mononuclear fragments. Furthermore, the complete fragmentation of the trinuclear clusters to these mononuclear fragments is another somewhat implausible postulate on thermochemical grounds. Poë, himself, recognizes some of these difficulties attendant to the scheme and raises the possibilities of phosphine-induced fragmentations and of binuclear intermediates in the fragmentation and reassembly processes. Still, these discussions continue to imply that trisubstituted products result from recombination of mononuclear fragments. It is possible, however, to propose alternative schemes, fully consistent with the experimental data, that allow formation of the trisubstituted product from intact or open-chain Ru_3 clusters. These alternatives require phosphine-dependent fragmentation reactions. We plan an unambiguous test of this question by means of a mass spectral analysis of the trinuclear cluster products of the reaction of tri-n-butylphosphine with a physical mixture of $^{101}Ru_3(CO)_{12}$ and $^{104}Ru_3(CO)_{12}$.

Further insight into the tendency of these cluster complexes to fragment has been afforded by studies of several reactions of $Ru_3(CO)_9L_3$ complexes (55–57). These reactions include such varied processes as the replacement of $P(C_6H_5)_3$ by $P(n-C_4H_9)_3$, replacement of phosphine ligands by CO, and thermal decomposition. Three distinct reaction pathways have been identified for these complexes. The lowest energy path is replacement of the phosphorus ligand, presumably by means of a reversible phosphorous ligand dissociation. Suppression of this pathway with a high concentration of phosphorus ligands permits detection of a pathway presumed to involve reversible carbonyl ligand dissocation. In the presence of sufficient phosphorus ligand and CO concentration to suppress both reaction pathways, yet a third pathway is observed: fission into mono- and binuclear fragments. These observations suggest that fragmentation is not as facile as other processes for these clusters. No similar statement can be made about reversible scission of one metal-metal bond in such complexes, though. A scheme, consistent with the experimental evidence, has been proposed in which just such a scission is an initial step common to all three of the above pathways.

Experimental verification of reversible scission of one metal-metal bond will be difficult to obtain. As discussed above, such quantities as entropies and volumes of activation could be informative. Another possible approach would be a ^{13}C NMR spin saturation transfer study of the P_{CO} dependence of the rate of exchange between $M_3(CO)_{12}$ and free CO. Both approaches could fail when metal-metal bond scission occurs as a preequilibrium to a dissociative step.

Since $Os_3(CO)_{12}$ and its derivatives should be much more resistant to metal-metal bond-breaking processes than either $Ru_3(CO)_{12}$ or $Fe_3(CO)_{12}$, a study of substitution reactions of the osmium cluster system may provide basic benchmarks for the reaction pathways of intact $M_3(CO)_{12}$ and $M_3(CO)_{12-x}L_x$ clusters. Establishment of the rate expressions from a widely varying set of experimental conditions for an electronically and sterically diverse set of ligands should serve as one aspect of such a study; complementary experiments would be spectroscopic probes, e.g. EPR and chemically induced dynamic polarization (CIDNP) experiments, use of radical traps to seek evidence for radical intermediates, kinetic demonstration of the accumulation of an intermediate, and stereochemical (retention or loss of optical activity) studies.

Tetranuclear Metal Carbonyl Clusters

In the cobalt group, all three metals—cobalt, rhodium, and iridium—form tetranuclear dodecacarbonyl clusters, $M_4(CO)_{12}$. Only $Ir_4(CO)_{12}$ has full T_d symmetry both in the solid (58) and solution states (59) (Figure 1). Each iridium atom in this cluster has octahedral coordination with three terminal carbonyl ligands and three Ir–Ir bonds, each of which is *trans* to an Ir–C carbonyl bond (the angles are less than the 180° for regular octahedral geometry). Analogous is $Ir_4(CO)_{11}[CNC(CH_3)_3]$, which has one unique terminal isocyanide ligand (60). In contrast, $Co_4(CO)_{12}$ and $Rh_4(CO)_{12}$ have a C_{3v} form in both the solid (61, 62) and solution states (63, 64) (Figure 2). In this form, there is a unique apical metal atom that has three terminal carbonyl ligands, each of which is *trans* to a metal-metal bond. The three metal atoms in the basal plane have three edge-bridging carbonyl ligands roughly in this basal plane, and each metal atom has two terminal carbonyl ligands, one equatorially and one axially placed. All phosphine derivatives of $Co_4(CO)_{12}$, $Rh_4(CO)_{12}$, and $Ir_4(CO)_{12}$ have the bridged (idealized C_{3v}

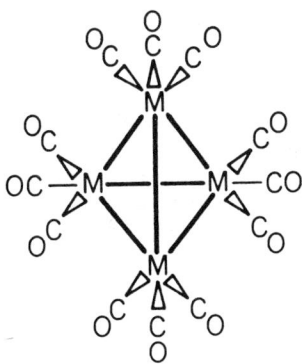

Figure 1 The $M_4(CO)_{12}$ structure of idealized T_d symmetry found for $Ir_4(CO)_{12}$. In this structure, all of the carbonyl groups are equivalent and all are terminally bound to single iridium atoms. The structure of isocyanide derivatives of $Ir_4(CO)_{12}$ are fully analogous.

symmetry) structure. Substitution of $Co_4(CO)_{12}$ appears to occur with sequential replacement of the axial CO groups in the basal plane (65–68). The phosphine-substituted iridium clusters differ in stereochemistry from the cobalt analogs in that after the replacement of one axial CO ligand, subsequent substitutions occur at the equatorial sites of the other basal plane iridium atoms (69–71). Only on substitution with a fourth phosphine does the unique apical iridium atom bear a phosphine substituent. With unidentate phosphines, no metal atom in the cluster is doubly substituted by phosphine ligands in $Ir_4(CO)_{12-x}(PR_3)_x$ ($x = 1-4$) complexes. In all these clusters, there is fairly facile, intramolecular carbonyl site exchange. All of this site exchange in the parent molecules and some of the site exchange in the phosphine derivatives can be explained by a relatively fast interchange between the idealized T_d form (all terminal ligands) and the idealized C_{3v} form (bridging CO ligands). A full discussion of these dynamic processes is presented in a review (72).

Clearly, the structure and stereochemistry of the $M_4(CO)_{12}$ clusters and their derivatives is a relatively sensitive function of metal atom and of the character of the ligands. The energy ordering of, and the energy gap between, the idealized C_{3v} and T_d structural forms as well as the stereochemistry of phosphine ligand placement are all affected by these changes in metal atom and in the nature of the ligands. The cluster structural form of excited state(s) that could undergo ligand substitution reactions is unknown–the structural form could be quite different from the aforementioned, idealized T_d and C_{3v} forms. Plausible alternative forms are a butterfly structure (scission of one metal-metal bond) and an idealized D_{2d} form in which a tetrahedral array of metal atoms has four of the six edges bridged by carbonyl ligands.

In the initial reaction of $Co_4(CO)_{12}$ with phosphites and phosphines, e.g. $P(OCH_3)_3$ and $P(C_6H_5)_3$, the rate of the first substitution is too high for study of the reaction by conventional techniques (73). However,

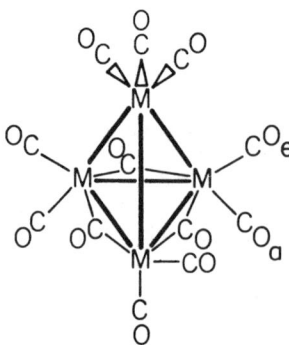

Figure 2 The $M_4(CO)_{12}$ structure of idealized C_{3v} symmetry found for $Co_4(CO)_{12}$ and $Rh_4(CO)_{12}$. Three of the tetrahedral edges are bridged by carbonyl groups. Phosphorus ligand derivatives of all cobalt group tetrametal dodecacarbonyls are based upon this type of carbonyl bridged framework.

carbonyl exchange with ^{13}CO is much slower. These observations suggest that the fast reaction of $Co_4(CO)_{12}$ with phosphorous ligands does not proceed through a $Co_4(CO)_{11}$ intermediate.

In the much slower second step of ligand substitution, specifically the reaction of $Co_4(CO)_{11}[P(OCH_3)_3]$ with $P(OCH_3)_3$, the kinetics are largely first order in complex and first order in phosphite, yet this reaction is inhibited by carbon monoxide. An additional observation that seemingly compounds the mechanistic complexity here is that the reaction of $Co_4(CO)_{11}[P(OCH_3)_3]$ with triphenylphosphine to give

$$Co_4(CO)_{10}[P(OCH_3)_3][P(C_6H_5)_3]$$

proceeds at rates independent of the phosphine concentration. Semiquantitative data for ^{13}CO incorporation into $Co_4(CO)_{12}$ indicate that this process occurs with a rate constant comparable to that observed for CO dissociation in $Co_4(CO)_{11}[P(OCH_3)_3]$ (73, 73a).

Finally, the third substitution, namely the reaction of

$$Co_4(CO)_{10}[P(OCH_3)_3]_2$$

with $P(OCH_3)_3$ to give $Co_4(CO)_9[P(OCH_3)_3]_3$, proceeds at a rate independent of phosphite concentration (74).

In this series of phosphite ligand introduction into $Co_4(CO)_{12}$, the comparison of the values for k_1, the first-order rate constant, shows little substantial change with increasing degree of phosphite substitution in the cluster (Table 1). One interpretation of these data is that little enhancement of the rate of CO dissociation occurs upon substitution of $Co_4(CO)_{12}$, in sharp contrast to the findings reported below for $Ir_4(CO)_{12}$. However, the experimental k_1 parameters do not necessarily relate to a ligand dissociative, D, process—an alternative would be an intramolecular metal-metal bond scission preequilibrium. With regard to the values of k_2, the second-order rate constant,[13] these constants appear to decrease with increasing phosphite substitution in the cluster.

In summary, a set of important rate data (73, 74) are available for the $Co_4(CO)_{12}/P(OCH_3)_3$ reaction system but they do not allow any definitive statements concerning general mechanisms of ligand substitution.

Ligand substitution reactions in $Ir_4(CO)_{12}$ appear to be formally analogous to those of $Co_4(CO)_{12}$ in that the initial substitutions, exclud-

---

[13]An interpretation offered for the second term entails the formation of an intermediate that results from ligand addition at a basal cobalt, cleavage of the metal-metal bond between that cobalt and the apical cobalt atom, and shift of one bridging CO from the substituted cobalt to the apical cobalt (Figure 3). Such an intermediate has been suggested in the reaction of $Co_4(CO)_{12}$ and CO to form $Co_2(CO)_8$ (75), but its presence has not been demonstrated in either case.

ing the $As(C_6H_5)_3$ reaction, show largely a second-order rate behavior, but the subsequent substitutions occur at rates independent of the entering ligand (69, 76–79). There are exceptions to these observations: reactions (76) of $P(n-C_4H_9)_3$ with $Ir_4(CO)_{11}L$ show second-order kinetic behavior as do all reactions of t-butyl isocyanide (77) with tetrairidium clusters with any degree of substitution.

Despite the formal similarities in the substitution reactions between the cobalt and iridium clusters, the basis for the change in the form of the rate expression is quite different. Successive substitutions of $Ir_4(CO)_{12}$ become increasingly rapid due to a large increase in the first-order rate constants (69–76). (See Table 1 for comparison of the cobalt and iridium first-order rate constants.) This first-order step has been interpreted in terms of a CO ligand dissociation step, D, but an equally acceptable interpretation is a preequilibrium step comprising an intramolecular metal-metal bond scission. The rate of the second-order pathway is so small relative to the first-order path that an accurate determination of k_2 for substitutions beyond the first ligand is very difficult. It is believed, however, that the rate of this pathway does not change with degree of substitution (78). Supporting this contention is the small variation in the rates of substitution with t-butyl isocyanide (77).

The substantial increase in the first-order pathway was initially ascribed to the structural change between $Ir_4(CO)_{12}$ and $Ir_4(CO)_{11}L$ (69). It was noted, however, that the additional increase upon further substitution implies that ligand effects can be transmitted by means other than the structural change. Further studies have supported this by demonstrating that the variation of the identity of the substituted ligand can itself modulate the first-order rate by 1–2 orders of magnitude (78, 79). Rate enhancement could occur either through destabilization of the ground state or stabilization of the transition state for either CO dissociation or for an intramolecular metal-metal bond scission or both. Neither crystal

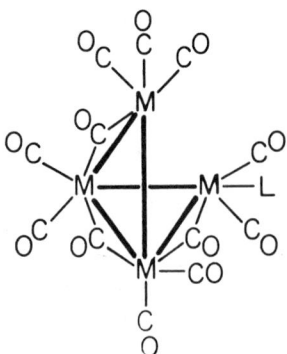

Figure 3 A representation of an intermediate proposed to account for the second-order term in the rate expression for substitutions of members of the $Co_4(CO)_{12}$ series. Note the cleavage of one metal-metal bond and the shift of a bridging CO ligand concomitant to ligand addition.

structure determinations nor infrared spectra provide support for destabilization of the ground state. The trends observed for the relative labilizing ability of the various ligands parallel that of cis-labilization of ligands of mononuclear complexes (78, 79). Unsaturation created by CO dissociation or intramolecular metal-metal bond scission at a substituted metal site could easily be transferred to an unsubstituted site by fluxional processes. The increase in lability due to the second substituent is greater than that expected merely from doubling the number of cis-labilized sites. Thus, the equatorial geometry of this second substituent and possibly some further cooperative effect must also effect the rate of CO dissociation or metal-metal bond scission. The role of steric effects in labilization of the mono- and disubstituted iridium clusters has been suggested to be unusual (78, 79), assuming that, in fact, the rate-determining step is CO dissociation. In contrast, CO dissociation (presumed rate determining process) in $Ir_4(CO)_8L_4$ has been found to be 3300 times greater for $L = P(C_2H_5)_3$ than $P(CH_3)_3$, an electronically similar but sterically less demanding ligand (80).

Metal-metal bond strengths in clusters generally increase within a periodic group as the atomic number of the metal increases (8, 11). Thus, we may expect that reaction rates for ligand substitution by a metal-metal bond-breaking process should decrease in going, for example, from $Co_4(CO)_{12}$ to $Ir_4(CO)_{12}$. Yet it is the first-order pathway that is of greatest significance for iridium. Perhaps, then, the proposed (69, 76–80) ligand (CO) dissociative pathway is the more reasonable one to explain the first-order pathways (intimate mechanism), although the M–CO bond strengths also increase in going from cobalt to iridium.

OXIDATIVE-ADDITION AND REDUCTIVE-ELIMINATION REACTIONS

There presently are many examples of reversible oxidative-addition and reductive-elimination reactions of complexes in which there is a single transition metal atom. Most commonly, such processes are observed in the chemistry of square planar d^8 complexes of the transition metals (81) (Eqs. 21 and 22), but examples can be found involving metals in nearly all parts of the transition metal series. For example, $(\eta^5-C_5H_5)_2MH_3$ complexes (M = Ta, Nb) can exist in equilibrium in solution with H_2 and $(\eta^5-C_5H_5)_2MH$ (82).

21.

$$\text{L}_2\text{ML}_2 + RX \rightleftharpoons \text{L}_4\text{M}(R)(X) \text{ or } \text{L}_4\text{M}(R)(X) \qquad 22.$$

Among cluster compounds, most oxidative-additions occur with loss of other ancillary ligands. A prototype is $Os_3(CO)_{12}$, which oxidatively adds hydrogen under vigorous conditions to form $H_2Os_3(CO)_{10}$ with loss of two carbon monoxide ligands (83). Other transition metal compounds interconvert monometal complexes and cluster compounds in undergoing processes of oxidative-addition and reductive-elimination. For example, $Co_2(CO)_8$ reacts with hydrogen to form $HCo(CO)_4$ (84, 85). Conversely, $Rh(CNR)_4^+$ complexes react with halogens to form $X_2[Rh(CNR)_4]_y^{y+}$ compounds (86, 87). These attendant processes, ligand lability and cluster fragmentation or condensation, are features that effectively preclude facile, reversible oxidative-addition reactions to cluster compounds. Generally, these competing processes are a consequence of the coordination saturation of the majority of known cluster compounds.

A new class of coordinately unsaturated cluster compounds that undergoes reversible oxidative-additions has recently emerged (88, 89) and these are well-defined structurally (90, 91). The class comprises $[HRh(PY_3)_2]_n$ molecules where PY_3 = trialkyl phosphite, *tris*(dialkylamino)phosphine, or trifluorophosphine. Their coordination unsaturation is best demonstrated by their ability to undergo rapid addition of ligands like carbon monoxide (92) and alkynes (93) to expand their coordination spheres but with retention of cluster form. The best-defined members of this class include dimers,

$$\{(\mu\text{-H})Rh[P(O-i-C_3H_7)_3]_2\}_2 \text{ and } \{(\mu\text{-H})Rh[P(N(CH_3)_2)_3]_2\}_2,$$

and trimers,

$$\{(\mu\text{-H})Rh[P(OCH_3)_3]_2\}_3 \text{ and } \{(\mu\text{-H})Rh[P(OC_2H_5)_3]_2\}_3.$$

The molecular structure of both the dimer and the trimer derived from isopropyl and methyl phosphites, respectively, have been precisely determined by low temperature neutron and x-ray diffraction (90, 91). Actually, the hydride ligands in these compounds are the only ones in rhodium cluster chemistry that have been located with relatively high precision. The structures are represented in Formula 1 and Figure 4.

In both structures, the local coordination geometry, H_2RhP_2, can be described as square planar for each of the formally d^8 rhodium(I) centers provided the rhodium-rhodium interactions are ignored. In fact, the rhodium-rhodium separations are similar to those in rhodium metal (2.69 Å at 20°C) with a shorter separation of 2.65 Å prevailing in the dimer

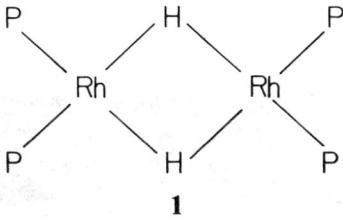

1

and a longer one of 2.80 Å in the trimer (90, 91). In the dimer, the nearly square H_2RhP_2 planes share the H···H edge and all eight of the atoms in the immediate coordination sphere, as represented in Formula **1**, are coplanar. The analogous dimer derived from the aminophosphine is not crystallographically defined, but NMR studies established it to be isostructural with the dimer derived from triisopropyl phosphite (93). In the trimer based on trimethyl phosphite, there again are nearly square coordination planes of H_2RhP_2, and the hydrogen atom vertices are shared (see Figures 4 and 5). The H_2RhP_2 planes are not mutually

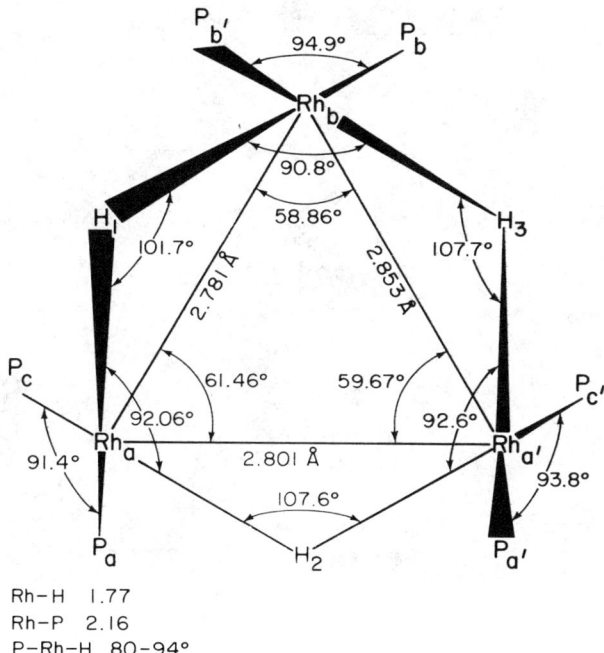

Figure 4 A stereochemical representation of the molecular structure of $\{(\mu-H)Rh[P(OCH_3)_3]_2\}_3$ adapted from an ORTEP drawing. Important bond distances and angles are included.

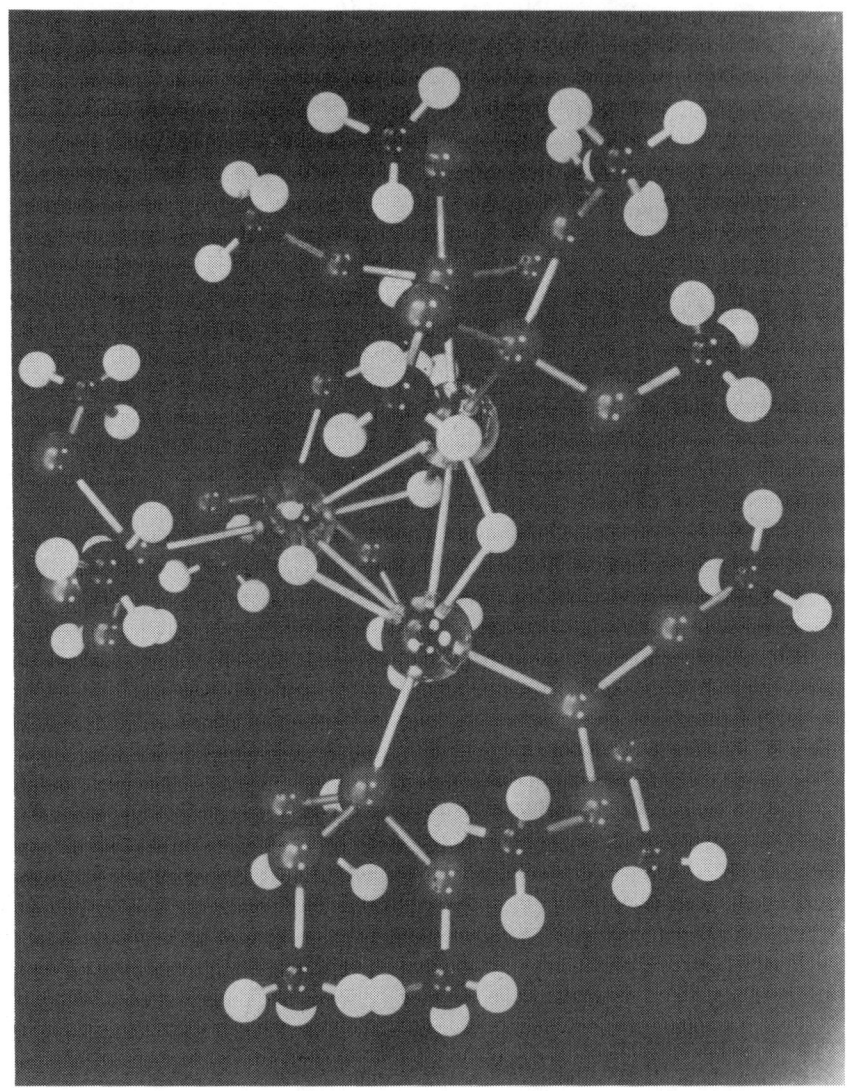

Figure 5 A ball-and-stick representation of {(μ–H)Rh[P(OCH$_3$)$_3$]$_2$}$_3$ elaborating its three-dimensional structure.

coplanar; each is aligned with respect to the plane of the three rhodium atoms so as to generate a two-fold axis of symmetry passing through one rhodium atom and the midpoint of the bond between the two other rhodium atoms (91). The dimers and the trimers rapidly and reversibly add one equivalent of H_2 according to Eqs. 23 and 24 (89).

$$\{(\mu-H)Rh[PR_3]_2\}_2 + H_2 \rightleftharpoons H_4Rh_2[PR_3]_4 \qquad 23.$$

$$\{(\mu-H)Rh[PR_3]_2\}_3 + H_2 \rightleftharpoons H_5Rh_3[PR_3]_6 \qquad 24.$$

Spectrally, there was no evidence of further reaction with hydrogen to generate mononuclear metal complexes (89). X-ray crystallography established a remarkable structure for $H_4Rh_2\{P[N(CH_3)_2]_3\}_4$ in which there is an octahedral Rh(III) center joined to a square planar Rh(I) center as shown below in Formula **2** (93). Again there are two hydride hydrogen atoms between the two rhodium atoms.

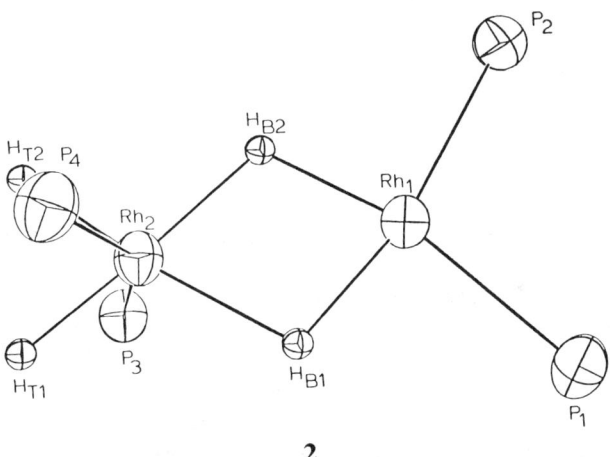

2

The Rh(III) center bears two terminal hydride ligands, and all four of the hydride ligands together with the two phosphorus atoms form an approximately octahedral local coordination geometry about this rhodium atom. All four of the hydrides, the two rhodium atoms, and the two phosphorus atoms on the Rh(I) are very nearly coplanar. In solution, the same structural form prevails and the structure is stereochemically rigid on the NMR time scale (93).

NMR spectroscopy has established that the ground state structure of $H_4Rh_2[P(O-i-C_3H_7)_3]_4$ differs from the triaminophosphine analog in that it has three bridging hydride ligands and a unique terminal hydride ligand (89). The molecule further exists in an NMR time scale equilibrium with an excited state that has a structural form similar to the aminophosphine complex shown above (93). The structures of the ground and excited state forms of the molecule, as originally proposed, are depicted in Formulas 3 and 4 (89).

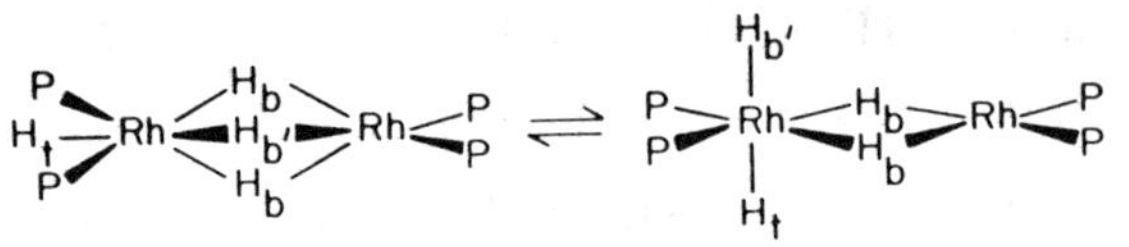

3   4

In this original representation of the excited state form, the basic structure was analogous to 3 but all four phosphorus atoms were placed in a common plane. An alternative representation of the hydrogen adduct and the excited state form in the phosphite system is as shown in 5 and 6 (93).

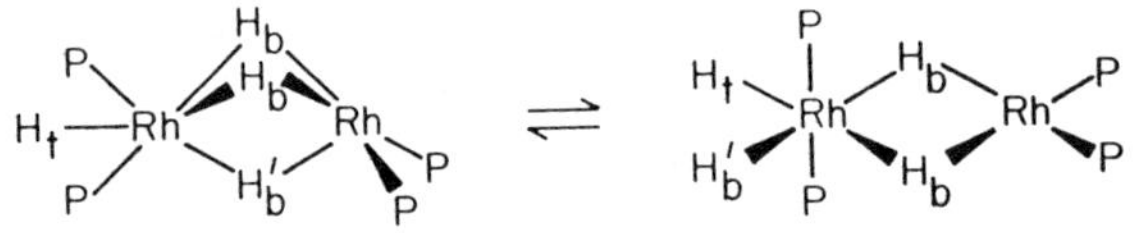

5   6

A crystallographic determination of the molecular structure for

$$[(C_3H_7-i-O)_3P]_2(H)Rh(\mu-H)_3Rh[P(O-i-C_3H_7)_3]_2$$

is being attempted.

Although ambiguities remain in the stereochemical interpretation of the NMR spectra of $H_5Rh_3[PR_3]_6$, the structure appears to consist of a trimeric cluster of rhodium atoms again in different oxidation states and with a high degree of hydride hydrogen atom mobility within the cluster (89).

All these hydrogen adducts based on d^8 Rh(I) square planar aggregates share the common and important feature that they react rapidly with small unsaturated organic molecules, like olefins or acetylenes, to bring about hydrogenation and regeneration of the $\{HRh(PY_3)_2\}_x$ parent molecule (89). In the presence of excess hydrogen gas and excess substrate molecules, catalytic cycles prevail. In these catalytic cycles, the polynuclear species remain intact; no mononuclear species appear to be formed in any step of the catalytic cycles (see Figure 6) (89). However, the only unequivocal demonstration that there is no fragmentation of a cluster in a catalytic cycle requires a double or multiple labeling experiment. This is not feasible for the rhodium clusters because there is only one stable rhodium isotope.

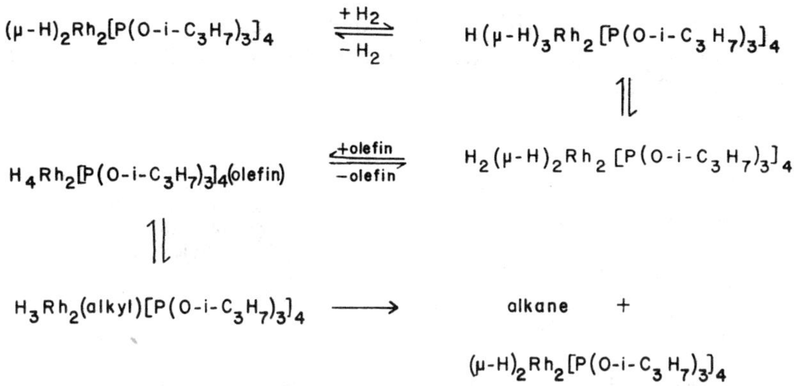

$(\mu-H)_2Rh_2[P(O-i-C_3H_7)_3]_4$ $\underset{-H_2}{\overset{+H_2}{\rightleftharpoons}}$ $H(\mu-H)_3Rh_2[P(O-i-C_3H_7)_3]_4$

$\updownarrow$

$H_4Rh_2[P(O-i-C_3H_7)_3]_4(\text{olefin})$ $\underset{-\text{olefin}}{\overset{+\text{olefin}}{\rightleftharpoons}}$ $H_2(\mu-H)_2Rh_2[P(O-i-C_3H_7)_3]_4$

$\updownarrow$

$H_3Rh_2(\text{alkyl})[P(O-i-C_3H_7)_3]_4$ $\longrightarrow$ alkane +

$(\mu-H)_2Rh_2[P(O-i-C_3H_7)_3]_4$

Figure 6 Pathway for the catalytic hydrogenation of olefins by $\{(\mu-H)Rh[P(O-i-C_3H_7)_3]_2\}_2$. $H_3Rh_2(\text{alkyl})[P(O-i-C_3H_7)_3]_4$ is a postulated species in this catalytic cycle.

There now have been a number of other detailed studies of cluster-based homogeneous catalytic reactions (94–112). Some studies conclude that the cluster remains intact through the catalytic cycle, but the nuclearity of the active catalyst in many cases is not known. In other cases, the coordination saturation of the cluster results in generation of a catalytically active monomer. This is a consequence of the fact that the metal-ligand bond strengths are often greater than those of the metal-metal bonds, resulting in cluster fragmentation under vigorous conditions. Cluster catalysts have the potential for a degree of stereo- and chemoselectivity not available to their monomeric counterparts. But it is clear that such reactive cluster compounds will not become commonplace until the synthesis and chemistry of coordinately unsaturated cluster compounds is more fully developed.

The scientific beauty of the $\{(\mu-H)Rh(PY_3)_2\}_x$ clusters is their well-defined equilibria with hydrogen wherein both the oxidative-addition and reductive-elimination reactions can be studied. A near term objective in our studies is the establishment of the forward (oxidative-addition) and backward (reductive-elimination) rate constants and the enthalpic and entropic thermodynamic parameters for the hydrogen equilibria. Also, the clusters and the hydrogen adducts can be structurally defined by x-ray and neutron diffraction and many of these are, in fact, so defined as noted above. Finally, the hydrogen adducts are catalytically important as they generally represent the first intermediate (Figure 6) in hydrogenations of organic molecules as catalyzed by these clusters.

ACKNOWLEDGMENTS

This analysis was supported by the National Science Foundation. We also wish to thank Professors J. D. Atwood, D. J. Darensbourg, and A. J.

Poë for providing us with preprints of their recent research papers and the National Science Foundation for a Graduate Fellowship for R. R. Burch (1979–1982).

Literature Cited

1. King, R. B. 1972. *Progr. Inorg. Chem.* 15:287–473
2. Chini, P. 1980. *J. Organomet. Chem.* 200:37–61
3. Johnson, B. F. G. 1980. *Transition Metal Clusters.* New York: Wiley
4. Chini, P., Longoni, G., Albano, V. G. 1976. *Adv. Organomet. Chem.* 14:285–344
5. Corbett, J. D. 1976. *Prog. Inorg. Chem.* 21:129–58
6. Schäfer, H., Schnering, H. G. 1964. *Angew. Chem.* 76:833–49
7. Tachikawa, M., Muetterties, E. L. 1981. *Prog. Inorg. Chem.* 28:203–38
8. Muetterties, E. L., Rhodin, T. N., Band, E., Brucker, C. F., Pretzer, W. R. 1979. *Chem. Rev.* 79:91–137
9. Basolo, F., Pearson, R. G. 1967. *Mechanism of Inorganic Reactions.* New York: Wiley. 2nd ed.
10. Langford, C. H., Gray, H. B. 1965. *Ligand Substitution Processes*, Chap. 1. New York: Benjamin
11. Connor, J. A. 1977. *Topics Curr. Chem.* 71:71–110
12. Green, M. L. H. 1972. *Organomettallic Compounds*, Vol. 2: *The Transition Elements*, pp. 2–6. London: Chapman & Hall
13. Lauher, J. W. 1978. *J. Am. Chem. Soc.* 100:5305–15
14. Lauher, J. W. 1979. *J. Am. Chem. Soc.* 101:2604–7
15. Dahl, L. F., Ishishi, E., Rundle, R. E. 1957. *J. Chem. Phys.* 26:1750–51
16. Dahl, L. F., Rundle, R. E. 1963. *Acta Crystallogr.* 16:419–26
17. Bailey, M. F., Dahl, L. F. 1965. *Inorg. Chem.* 4:1140–45
18. Levenson, R. A., Grey, H. B., Ceasar, G. P. 1970. *J. Am. Chem. Soc.* 92:3653–58
19. Wrighton, M. S., Bredesen, D. 1973. *J. Organomet. Chem.* 50:C35–C38
20. Wrighton, M. S., Ginley, D. S. 1975. *J. Am. Chem. Soc.* 97:2065–72
21. Kidd, D. R., Brown, T. L. 1978. *J. Am. Chem. Soc.* 100:4095–103
22. Fawcett, J. P., Poë, A., Sharma, K. R. 1976. *J. Am. Chem. Soc.* 98:1401–7
23. Chowdhury, D. M., Poë, A., Sharma, K. R. 1977. *J. Chem. Soc. Dalton Trans.*, pp. 2352–55
24. DeWit, D. G., Fawcett, J. P., Poë, A. J. 1976. *J. Chem. Soc. Dalton Trans.*, pp. 528–33
25. Poë, A. 1981. *Inorg. Chem.* 20:4029–31
26. Atwood, J. D. 1981. *Inorg. Chem.* 20:4031–32
27. Poë, A. 1981. *Inorg. Chem.* 20:4032–33
28. Haines, L. I. B., Hopgood, D., Poë, A. J. 1968. *J. Chem. Soc. A*, pp. 421–28
29. Fawcett, J. P., Poë, A. 1976. *J. Chem. Soc. Dalton, Trans.*, pp. 2039–44
30. Sonnenberger, D., Atwood, J. D. 1980. *J. Am. Chem. Soc.* 102:3484–89
31. Schmidt, S., Trogler, W. C., Basolo, F. 1982. *Inorg. Chem.* 21:1698–99
32. Wegman, R. W., Olsen, R. J., Gard, D. K., Faulkner, L. R., Brown, T. L., 1981. *J. Am. Chem. Soc.* 103:6089–92
33. Hughey, J. L. 4th, Anderson, C. P., Meyer, T. L. 1977. *J. Organomet. Chem.* 125:C49–C52
34. Quicksall, C. O., Spiro, T. G. 1969. *Inorg. Chem.* 8:2363–67
35. Spiro, T. G., 1970. *Progr. Inorg. Chem.* 11:1–51
36. Good, W. D., Fairbrother, D. M., Waddington, G. 1958. *J. Phys. Chem.* 62:853–56
37. Connor, J. A., Skinner, H. A., Virmani, Y. 1972. *J. Chem. Soc. Faraday Trans. 1* 68:1754–63
38. Wawersick, H., Basolo, F. 1969. *Inorg. Chim. Acta* 3:113–20
39. Haines, L. I. B., Poë, A. J. 1969. *J. Chem. Soc. A*, pp. 2826–33
40. Noack, K. 1964. *Helv. Chim. Acta* 47:1064–67
41. Noack, K. 1964. *Helv. Chim. Acta* 47:1555–63
42. Sumner, G. G., Klug, H. P., Alexandra, L. E. 1964. *Acta Crystallogr.* 17:732–42
43. Bor, G., Noack, K. 1974. *J. Organomet. Chem.* 64:367–72
44. Sweany, R. L., Brown, T. L. 1977. *Inorg. Chem.* 16:415–21
45. Forbes, N. P., Brown, T. L. 1981. *Inorg. Chem.* 20:4343–47
46. Churchill, M. R., Hollander, F. J., Hutchinson, J. P. 1977. *Inorg. Chem.* 16:2655–59

47. Churchill, M. R., DeBoer, B. G. 1977. *Inorg. Chem.* 16:878–84
48. Cotton, F. A., Troup, J. M. 1974. *J. Am. Chem. Soc.* 96:4155–59
49. Candlin, J. P., Shortland, A. C. 1969. *J. Organomet. Chem.* 16:289–99
50. Poë, A. J., Twigg, M. V. 1973. *J. Organomet. Chem.* 50:C39–C42
51. Poë, A. J., Twigg, M. V. 1974. *J. Chem. Soc. Dalton Trans.*, pp. 1860–66
52. Deeming, A. J., Johnson, B. F. G., Lewis, J. 1970. *J. Chem. Soc. A*, pp. 897–901
53. Malik, S. K., Poë, A. 1978. *Inorg. Chem.* 17:1484–88
54. Poë, A., Twigg, M. V 1974. *Inorg. Chem.* 13:2982–85
55. Keeton, D. P., Malik, S. K., Poë, A. 1977. *J. Chem. Soc. Dalton Trans.*, pp. 233–39
56. Keeton, D. P., Malik, S. K., Poë, A. 1977. *J. Chem. Soc. Dalton Trans.* pp. 1392–97
57. Malik, S. K., Poë, A. 1979. *Inorg. Chem.* 18:1241–45
58. Churchill, M. R., Hutchinson, J. P. 1978. *Inorg. Chem.* 17:3528–35
59. Quicksall, C. O., Spiro, T. G. 1969. *Inorg. Chem.* 8:2011–13
60. Churchill, M. R., Hutchinson, J. P. 1979. *Inorg. Chem.* 18:2451–54
61. Carre, F. H., Cotton, F. A., Frenz, B. A. 1976. *Inorg. Chem.* 15:380–87
62. Wei, C. H. 1969. *Inorg. Chem.* 8:2384–97
63. Aime, S., Osella, D., Milone, L., Hawkes, G. E., Randall, E. W. 1981. *J. Am. Chem. Soc.* 103:5290–92
64. Evans, J., Johnson, B. F. G., Lewis, J., Norton, J. R., Cotton, F. A. 1973. *J. Chem. Soc. Chem. Commun.* pp. 807–8
65. Darensbourg, D. J., Incorvia, M. J. 1981. *Inorg. Chem.* 20:1911–18
66. Cohen, M., Kidd, D. R., Brown, T. L. 1975. *J. Am. Chem. Soc.* 97:4408–9
67. Aime, S., Milone, L., Osella, D., Poli, A. 1978. *Inorg. Chim. Acta* 30:45–49
68. Huie, B. T., Knobler, C. B., Kaesz, H. D. 1975. *J. Chem. Soc. Chem. Commun.* pp. 684–85
69. Karel, K. J., Norton, J. R. 1974. *J. Am. Chem. Soc.* 96:6812–13
70. Malatesta, L., Caglio, G. 1967. *J. Chem. Soc. Chem. Commun.* pp. 420–421
71. Albano, V., Bellon, P., Scatturin, V. 1967. *J. Chem. Soc. Chem. Commun.* pp. 730–31
72. Band, E., Muetterties, E. L. 1978. *Chem. Rev.* 78:639–58
73. Darensbourg, D. J., Incorvia, M. J. 1980. *Inorg. Chem.* 19:2585–90
73a. Darensbourg, D. J., Peterson, B. S., Schmidt, R. E. Jr. 1982. *Organometallics* 1:306–11
74. Darensbourg, D. J., Incorvia, M. J. 1979. *J. Organomet. Chem.* 171:89–96
75. Bor, G., Dietler, U. K., Pino, P., Poë, A. 1978. *J. Organomet. Chem.* 154:301–15
76. Sonnenberger, D., Atwood, J. D. 1981. *Inorg. Chem.* 20:3243–46
77. Stuntz, G. 1978. PhD thesis. Univ. Ill., Urbana
78. Sonnenberger, D. C., Atwood, J. D. 1982. *J. Am. Chem. Soc.* In press
79. Sonnenberger, D. C., Atwood, J. D. 1982. *Organometallics.* In press
80. Darensbourg, D. J., Baldwin-Zuschke, B. J. 1981. *Inorg. Chem.* 20:3846–50
81. Collman, J. P., Roper, W. R. 1968. *Adv. Organomet. Chem.* 7:53–94
82. Tebbe, F. N., Parshall, G. W. 1973. *J. Am. Chem. Soc.* 93:3793–95
83. Johnson, B. F. G., Lewis, J., Kilty, P. A. 1968. *J. Chem. Soc. A* pp. 2859–64
84. Wegman, R. W., Brown, T. L. 1980. *J. Am. Chem. Soc.* 102:2494–95
85. Alemdaroglu, N. H., Penninger, J. M. L., Oltay, E. 1976. *Monatsh. Chem.* 107:1043–53
86. Olmstead, M. M., Balch, A. L. 1978. *J. Organomet. Chem.* 148:C15–C18
87. Balch, A. L., Olmstead, M. M. 1979. *J. Am. Chem. Soc.* 101:3128–29
88. Day, V. W., Fredrich, M. F., Reddy, G. S., Sivak, A. J., Pretzer, W. R., Muetterties, E. L. 1977. *J. Am. Chem. Soc.* 99:8091–93
89. Sivak, A. J., Muetterties, E. L. 1979. *J. Am. Chem. Soc.* 101:4878–87
90. Brown, R. K., Williams, J. M., Sivak, A. J., Muetterties, E. L. 1980. *Inorg. Chem.* 19:370–74
91. Teller, R. G., Williams, J. M., Koetzle, T. F., Burch, R. R., Gavin, R. M., Muetterties, E. L. 1981. *Inorg. Chem.* 20:1806–11
92. Burch, R. R., Muetterties, E. L., Schultz, A. J., Gebert, E. G., Williams, J. M. 1981. *J. Am. Chem. Soc.* 103:5517–22
93. Meier, E. B., Burch, R. R., Muetterties, E. L., Day, V. W. 1982. *J. Am. Chem. Soc.* 104:2661–63
94. Smith, A. K., Bassett, J. M. 1977. *J. Mol. Catal.* 2:229–41

95. Demitras, G. C., Muetterties, E. L. 1977. *J. Am. Chem. Soc.* 99:2796–97
96. Band, E., Pretzer, W. R., Thomas, M. G., Muetterties, E. L. 1977. *J. Am Chem. Soc.* 99:7380–1
97. Keister, J. B., Shapley, J. R. 1976. *J. Am. Chem. Soc.* 98:1056–57
98. Agapion, A., Jordan, R. F., Zyzyck, L. A., Norton, J. R. 1977. *J. Organomet. Chem.* 141:C35–C39
99. Fredianai, P., Matteoli, U., Bianchi, M., Piacenti, F., Menchi, G. 1978. *J. Organomet. Chem.* 150:273–78
100. Graff, J. L., Sanner, R. D., Wrighton, M. S. 1979. *J. Am. Chem. Soc.* 101:273–75
101. Adams, R. D., Golembeski, N. M. 1979. *J. Am. Chem. Soc.* 101:2579–87
102. Vidal, J. L., Walker, W. E. 1980. *Inorg. Chem.* 19:896–903
103. Pittman, C. U., Ryan, R. C., McGee, J., O'Connor, J. P. 1979. *J. Organomet. Chem.* 178:C43–49
104. Pittman, C. U., Wilemon, G. M., Wilson, W. D., Ryan, R. C. 1980. *Angew. Chem. Int. Ed. Engl.* 19:478–79
105. Thomas, M. G., Pretzer, W. R., Beier, B. F., Hirsekorn, F. J., Muetterties, E. L. 1977. *J. Am. Chem. Soc.* 99:743
106. Muetterties, E. L., Pretzer, W. R., Thomas, M. G., Beier, B. F., Thorn, D. L., Day, V. W., Anderson, A. B. 1978. *J. Am. Chem. Soc.* 100:2090–96
107. Muetterties, E. L., Band, E., Kokorin, A., Pretzer, W. R., Thomas, M. G. 1980. *Inorg. Chem.* 19:1552–60
108. Slater, S., Muetterties, E. L. 1980. *Inorg. Chem.* 19:3337–42
109. Slater, S., Muetterties, E. L. 1980. *Inorg. Chem.* 20:1604–6
110. Wang, H.-K., Choi, H. W., Muetterties, E. L. 1981. *Inorg. Chem.* 20:2661–63
111. Choi, H. W., Muetterties, E. L. 1981. *Inorg. Chem.* 20:2664–67
112. Schunn, R. A., Demitras, G. C., Choi, H. W., Muetterties, E. L. 1981. *Inorg. Chem.* 20:4023–25

PHOTOFRAGMENT ALIGNMENT AND ORIENTATION

Chris H. Greene

Department of Physics and Astronomy, Louisiana State University, Baton Rouge, Louisiana 70803

Richard N. Zare

Department of Chemistry, Stanford University, Stanford, California 94305

INTRODUCTION

Early measurements demonstrated long ago the wealth of chemical information that can be extracted from photofragmentation experiments in which a target system is dissociated or ionized following the absorption of a sufficiently energetic photon (1-8). For many years experimentalists measured primarily *total* photofragmentation cross sections or rate constants. More recently, a relatively few studies have determined some measure of the *anisotropy* characterizing the photofragmentation process (9-20). In contrast to isotropic rate constant measurements that ignore all directional information, measurements of fragment anisotropy give a much more detailed picture of the dynamics of the photoejection process. In the past it has been common to consider separately photodissociation of isolated molecules and photoionization of free atoms or molecules. In what follows we shall call photodissociation/photoionization by the single name, *photofragmentation*, for, as we show, both processes can be treated together in a unified manner.

The anisotropy receiving most attention to date has been the photofragment angular distribution. For photofragmentation by a beam of linearly polarized light this distribution takes the form

$$\frac{d\sigma}{d\Omega} = \frac{\sigma}{4\pi}[1+\beta P_2(\cos\theta)] \qquad 1.$$

for an electric dipole transition, where θ is the angle between the final recoil direction of the fragments and the electric vector of the light beam. Note that Eq. 1 applies to a single-photon process whereby the target breaks up into only two fragments. Measurements of the anisotropy parameter β in photoionization experiments (6–8, 21–33) have been instrumental in testing various theories of electron correlation (20, 34–37) and have helped unravel the complicated couplings of electronic, vibrational, and rotational motions in molecules (38–45). Measurements of β in photodissociation experiments have yielded information on the symmetry nature of the dissociative molecular state and the time required for the fragments to escape from the excited complex (10–14, 46).

In this article we consider primarily another anisotropy, namely the alignment or orientation of individual fragments. Angular momentum is transferred from the light beam to the ensemble of fragments. Not only does each fragment quantum state have a given energy E but also a definite angular momentum **J** defined in magnitude and quantized in space. To make the concepts of alignment and orientation concrete, let us consider the case of an ensemble of fragments with total angular momentum $J=2$. If we need to account only for the occupation probabilities of the five magnetic sublevels, $M=2,1,0,-1,-2$, then the five-dimensional vector

$$\mathbf{N} = \begin{pmatrix} N_2 \\ N_1 \\ N_0 \\ N_{-1} \\ N_{-2} \end{pmatrix} \qquad 2.$$

suffices, where N_M is the number of fragments in the state $|JM\rangle$.

Note: Equation 2 assumes that the system has cylindrical symmetry. More generally, the excited state ensemble must be represented by a $(2J+1)\times(2J+1)$ density matrix ρ whose diagonal elements $\rho_{MM} = \langle JM|\rho|JM\rangle$ represent occupation numbers (populations) in the levels M and whose off-diagonal elements $\rho_{M'M} = \langle JM'|\rho|JM\rangle$ represent coherence terms (containing phase information) between the M' and M

levels. It is useful to introduce the spherical tensor operators (47–50)

$$T(J)_{LM_L} = \sum_{M,M'} (-1)^{J-M} (JM', J-M | LM_L) |JM'\rangle\langle JM|$$

which satisfy the orthonormality conditions

$$Tr\left[T(J)^\dagger_{LM_L} T(J)_{L'M'_L}\right] = \mathbf{T}(J)^*_{LM_L} \cdot \mathbf{T}(J)_{L'M'_L} = \delta_{LL'}\delta_{M_L M'_L}$$

and behave under rotation like the spherical harmonics Y_{LM_L}. The restriction on the Clebsch-Gordan coefficient $(JM', J-M|LM_L)$ shows that L ranges from 0 to $2J$ and M_L ranges in unit steps from $-L$ to L. Hence the $\mathbf{T}(J)_{LM_L}$ span the $(2J+1)\times(2J+1)$ space of the density matrix. Specifically, we may decompose ρ as

$$\rho = \sum_{L,M_L} \rho_{LM_L} \mathbf{T}_{LM_L}$$

where the expansion coefficients ρ_{LM_L}, given by

$$\rho_{LM_L} = Tr\left[\mathbf{T}^\dagger_{LM_L}\rho\right] = \mathbf{T}^*_{LM_L} \cdot \rho = \langle T^\dagger_{LM_L}\rangle Tr[\rho],$$

are the multipole moments of the system. When $M_L = 0$ we recover the $\hat{\mathbf{T}}_L \equiv \langle T^\dagger_{L0}(J)\rangle$ whose $2J+1$ elements of the form $(-1)^{J-M}(JM, J-M|L0)$ are the same as displayed in Eq. 4 for the case of a $J=2$ system.

This vector can of course be expanded as a linear combination of the "Cartesian" basis vectors:

$$\begin{pmatrix}1\\0\\0\\0\\0\end{pmatrix}, \begin{pmatrix}0\\1\\0\\0\\0\end{pmatrix}, \begin{pmatrix}0\\0\\1\\0\\0\end{pmatrix}, \begin{pmatrix}0\\0\\0\\1\\0\end{pmatrix}, \begin{pmatrix}0\\0\\0\\0\\1\end{pmatrix}. \qquad 3.$$

However, to bring out the symmetries of the system it is much more convenient to introduce the "spherical" basis vectors:

$$\hat{\mathbf{T}}_0 = \frac{1}{\sqrt{5}}\begin{pmatrix}1\\1\\1\\1\\1\end{pmatrix}, \quad \hat{\mathbf{T}}_1 = \frac{1}{\sqrt{10}}\begin{pmatrix}2\\1\\0\\-1\\-2\end{pmatrix}, \quad \hat{\mathbf{T}}_2 = \frac{1}{\sqrt{14}}\begin{pmatrix}2\\-1\\-2\\-1\\2\end{pmatrix},$$

$$\hat{\mathbf{T}}_3 = \frac{1}{\sqrt{10}}\begin{pmatrix}1\\-2\\0\\2\\-1\end{pmatrix}, \quad \hat{\mathbf{T}}_4 = \frac{1}{\sqrt{70}}\begin{pmatrix}1\\-4\\6\\-4\\1\end{pmatrix}, \qquad 4.$$

which satisfy the orthonormality condition

$$\hat{\mathbf{T}}_i \cdot \hat{\mathbf{T}}_j = \delta_{ij}. \qquad 5.$$

Then **N** may be expanded in terms of the $\hat{\mathbf{T}}_L$ basis set as

$$\mathbf{N} = \sum_{L=0}^{4} n_L \hat{\mathbf{T}}_L \qquad 6.$$

where the coefficients $n_L = \hat{\mathbf{T}}_L \cdot \mathbf{N}$ represent the 2^L multipole moments of **N**.

For $L = 0$

$$n_0 = \frac{1}{\sqrt{5}}(N_2 + N_1 + N_0 + N_{-1} + N_{-2}) \qquad 7.$$

is proportional to the total number of $J = 2$ fragments. It is called the monopole component of **N**. For $L = 1$

$$n_1 = \frac{1}{\sqrt{10}}(2N_2 + N_1 - N_{-1} - 2N_{-2})$$

$$= \frac{1}{\sqrt{2}} n_0 \langle J_z \rangle \qquad 8.$$

is proportional to the magnetic dipole moment of the ensemble and is known as the *orientation*. For $L = 2$

$$n_2 = \frac{1}{\sqrt{14}}(2N_2 - N_1 - 2N_0 - N_{-1} + 2N_{-2})$$

$$= \frac{1}{3}\sqrt{\frac{5}{14}} n_0 \langle 3J_z^2 - \mathbf{J}^2 \rangle \qquad 9.$$

is proportional to the quadrupole moment of **N** and is known as the *alignment*. Similarly

$$n_3 = \frac{1}{\sqrt{10}}(N_2 - 2N_1 + 2N_{-1} - N_{-2})$$

$$= \frac{1}{6\sqrt{2}} n_0 \langle J_z(5J_z^2 + 1 - 3\mathbf{J}^2) \rangle \qquad 10.$$

and

$$n_4 = \frac{1}{\sqrt{70}}(N_2 - 4N_1 + 6N_0 - 4N_{-1} + N_{-2})$$

$$= \frac{1}{12\sqrt{14}} n_0 \langle 3\mathbf{J}^2[\mathbf{J}^2 - 2(5J_z^2 + 1)] + 5J_z^2(7J_z^2 + 5) \rangle \qquad 11.$$

are the octopole and hexadecapole moments of **N**, respectively. The M state population is completely described by the n_L coefficients. For arbitrary (integral or half-integral) J there are in general $2J+1$ different multipole moments. As J becomes large we approach the classical correspondence limit and the n_L become the average value of the Legendre polynomials $\langle P_L(\hat{\mathbf{J}}\cdot\hat{\mathbf{z}})\rangle$, i.e. the classical multipole moments of the angular momentum distribution. Indeed the orientation and the alignment are strictly proportional to

$$\langle P_1(\hat{\mathbf{J}}\cdot\hat{\mathbf{z}})\rangle = \langle J_z\rangle/|\mathbf{J}| \quad \text{and} \quad \langle P_2(\hat{\mathbf{J}}\cdot\hat{\mathbf{z}})\rangle = \langle \tfrac{1}{2}(3J_z^2 - \mathbf{J}^2)/\mathbf{J}^2\rangle,$$

respectively, for all J. However, this identification of n_L with $\langle P_L(\hat{\mathbf{J}}\cdot\hat{\mathbf{z}})\rangle$ only holds for $L \geqslant 3$ in the high J classical limit.

Figure 1 illustrates the form of these distributions for a $J=2$ system. If the only moment present is that of the monopole, then all the magnetic

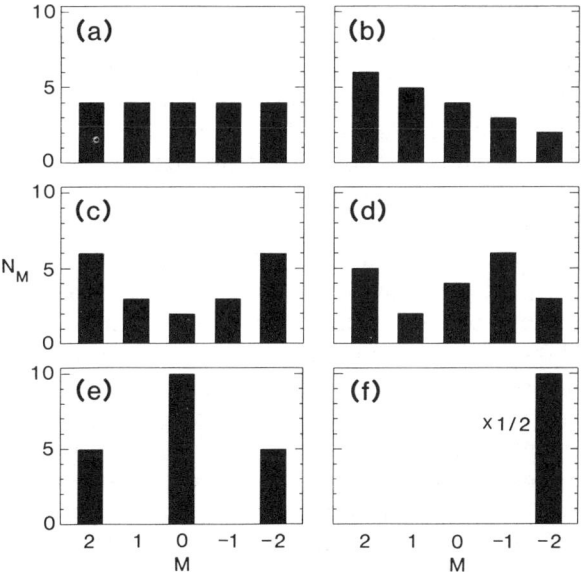

Figure 1 Population versus M state for a $J=2$ system showing the distribution for (a) a pure monopole moment, $\mathbf{N} = 4\sqrt{5}\,\hat{\mathbf{T}}_0$, ($b$) a monopole plus dipole moment, $\mathbf{N} = 4\sqrt{5}\,\hat{\mathbf{T}}_0 + \sqrt{10}\,\hat{\mathbf{T}}_1$, ($c$) a monopole plus quadrupole moment, $\mathbf{N} = 4\sqrt{5}\,\hat{\mathbf{T}}_0 + \sqrt{14}\,\hat{\mathbf{T}}_2$, ($d$) a monopole plus octopole moment, $\mathbf{N} = 4\sqrt{5}\,\hat{\mathbf{T}}_0 + \sqrt{10}\,\hat{\mathbf{T}}_3$, ($e$) a monopole plus hexadecapole moment, $\mathbf{N} = 4\sqrt{5}\,\hat{\mathbf{T}}_0 + \sqrt{70}\,\hat{\mathbf{T}}_4$, and ($f$) a special mixed state in which all the population is in the $M = -2$ sublevel. The latter may be decomposed into a linear superposition of multipole moments, specifically, $\mathbf{N} = 4\sqrt{5}\,\hat{\mathbf{T}}_0 - 4\sqrt{10}\,\hat{\mathbf{T}}_1 + 20\sqrt{2/7}\,\hat{\mathbf{T}}_2 - 2\sqrt{10}\,\hat{\mathbf{T}}_3 + 2\sqrt{10/7}\,\hat{\mathbf{T}}_4$. It is natural to define the polarization of the system as the difference between the actual multipole moment distribution and the distribution of an unpolarized ensemble, corresponding to equal occupation of all the magnetic sublevels [shown in (a)]. Thus (b) describes pure orientation while (c) describes pure alignment.

sublevels are equally populated. We also see that for orientation (and other odd multipoles) the population of $+M$ and $-M$ sublevels differ while for alignment (and other even multipoles) the population may be unequal only for different $|M|$ values.

Experimentally, the orientation or alignment of an excited state fragment is most easily determined by measuring the polarization of its fluorescence (14, 51, 52). The orientation or alignment of a ground state fragment (as well as the octopole and hexadecapole moments) can alternatively be extracted using laser-induced fluorescence (53), but we do not discuss this topic here. To date, only a handful of such emission polarization measurements have been carried out, with the result that fragment polarization is observed in almost every case (54–72). The mere observation of nonzero polarization is not surprising of course, since the incident photon is itself highly anisotropic and the subsequent fragmentation is such a violent event. Now that this polarization has been demonstrated, we can begin to use such measurements as a comprehensive tool for learning about the dynamics of the photofragmentation process and the resulting fragment charge distributions. The remainder of this article addresses the problem of extracting useful dynamical information from polarization measurements.

This is far from the first study of this problem. Starting in the 1950s, the physico-mathematical methods required were developed within the context of perturbed angular correlations of nuclei (48, 73–76). Using density matrix techniques and the powerful methods of Racah algebra, the polarization of light emitted from a fragment was related to the multipole moments of the fragment excited state. This approach derived essentially all the relevant formulae necessary to invert a measurement of polarized emission (or resonance fluorescence) and determine the desired fragment multipole moments. But owing to the heavy use of angular momentum coupling machinery, these results have not been immediately accessible to most experimentalists interested in probing molecular dynamics. Significant progress toward dispelling this obscurity was achieved in the 1973 review article of Fano & Macek (FM) (77). That article obtains the general expression for the intensity I of polarized light emitted in any direction following an arbitrary excitation process, expressing I in terms of simple *geometrical* factors and a few *dynamical* parameters describing the excited fragment multipole moments. Fano & Macek thus provided a unified framework for disentangling dynamical from geometrical information based on emission polarization measurements performed after some type of "collisional" excitation. Here the collision refers to excitation by bombardment with photons, electrons, or even heavy particles. A first goal for the present article is to summarize the main results of FM, both because we hope to broaden its application in the chemical domain, and because the FM framework provides a convenient basis for expressing the results of what follows.

The second goal of this article is to present an interpretation of the measured orientation or alignment, indicating what the fragment multipole moments imply about the photofragmentation process. At the heart of this treatment is the angular momentum transfer formulation of Fano & Dill (78–80). This formulation demonstrates how only three dynamical amplitudes, or more precisely, their two mutual ratios, determine the orientation or alignment of a photofragment, in contrast to the (often) enormous number of continuum amplitudes that enter the more usual formulations of photofragmentation scattering dynamics. This treatment is carried out for arbitrary angular momentum but special attention is given to the high J limit characteristic of many molecular processes.

THE FANO-MACEK FORMULATION OF ORIENTATION AND ALIGNMENT

We begin by considering in Figure 2 the simplest and most common experimental geometry in which a photon beam $h\nu$ impinges on a gas of randomly oriented target molecules or atoms. After absorbing a photon, the target species dissociates or ionizes, leaving, for example, a fragment A (atom or molecule) in an excited state with angular momentum J_i. At 90° with respect to the photon symmetry axis $\hat{z}$, a fluorescence photon (emitted in the transition $J_i \to J_f$) is detected using a linear polarizer set at an angle χ relative to the z axis. Two experimental quantities are usually of interest:

1. The total intensity I_0 of light with energy $h\nu'$ emitted in all directions, which is proportional to the population (monopole moment n_0) of the excited state J_i.
2. The linear polarization of this light, defined as $P = (I_\parallel - I_\perp)/(I_\parallel + I_\perp)$, with $I_\parallel$ and $I_\perp$ the intensities observed when the linear polarizer is set at $\chi = 0$ and $\chi = 90°$, respectively, which is a measure of the alignment (quadrupole moment n_2) of the excited state J_i.

Here the total intensity I_0 is most usually of interest when the partial cross section for creating fragments $A(J_i)$ is desired. One obvious, but impractical, way of measuring I_0 is to detect all photons emitted into all directions with arbitrary polarizations. This measurement is independent of the collision-produced anisotropy. On the other hand, if the more usual experimental geometry of Figure 2 is adopted, care must be

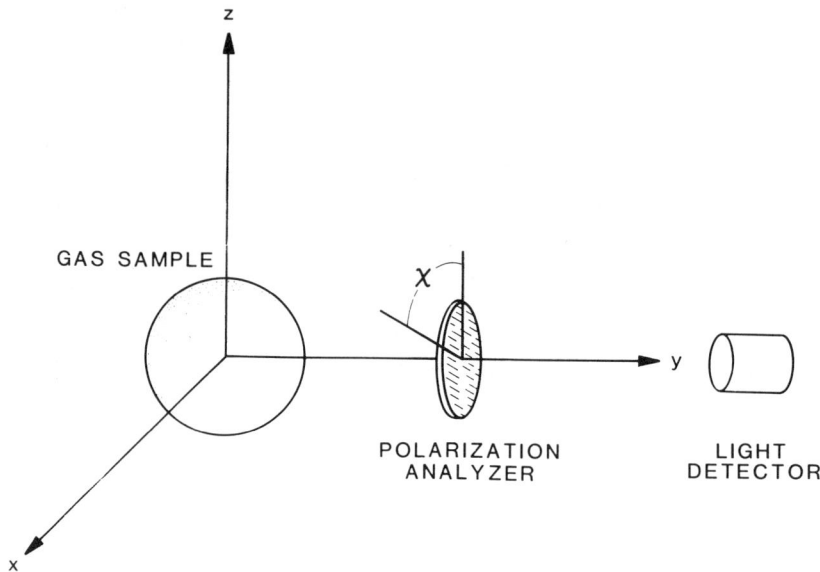

Figure 2 Traditional right-angle geometry for measuring emission polarization. The z axis is the photon symmetry axis directed along the electric vector for a plane polarized light beam or along the propagation direction for an unpolarized light beam. The polarization-analyzer light-detector combination lies in the xy plane and its direction is taken to define the y axis. The transmission axis of the linear polarizer makes the angle χ with respect to the z axis such that the setting $\chi = 0$ corresponds to the transmission of the $I_\parallel$ light signal and $\chi = 90°$ to $I_\perp$.

exercised since the intensity of polarized or even of unpolarized light depends on the alignment and is thus not proportional to I_0 alone. The second experimental quantity of interest, P, depends directly on the alignment and vanishes if there is no alignment. Yet the measurement of P with the experimental setup of Figure 2 may prove difficult, for example, if the fluorescent photons lie in the vacuum ultraviolet spectrum where suitable polarizing materials are not readily available. One major motivation for presenting the general formulation of light emission here is to show the possibility of alternative experimental methods of obtaining the same information about the atomic or molecular system.

Excitation-Detection Geometry

The right-angle setup of Figure 2 is a specialized case of the more general arrangement shown in Figure 3, where there are two coordinate frames of interest. The first, or "collision frame" of coordinates (x, y, z), is adapted to the symmetry of the exciting collision, with the z axis chosen to be its dominant symmetry axis. When no other axis is singled out by the collision, the collision frame is cylindrically symmetric and the choice of the x and y coordinates is arbitrary. For example, in a molecular photodissociation experiment using unpolarized incident photons,

$$h\nu + AB \rightarrow A^*(J_i) + B$$
$$\hookrightarrow A(J_f) + h\nu'$$

the incident photon beam axis serves as the z axis. But if B is detected along a new axis $\hat{w}$ in coincidence with the fluorescence photon, the cylindrical symmetry is destroyed. For this more complicated coincidence experiment, a plane of symmetry still exists, namely the wz plane. By

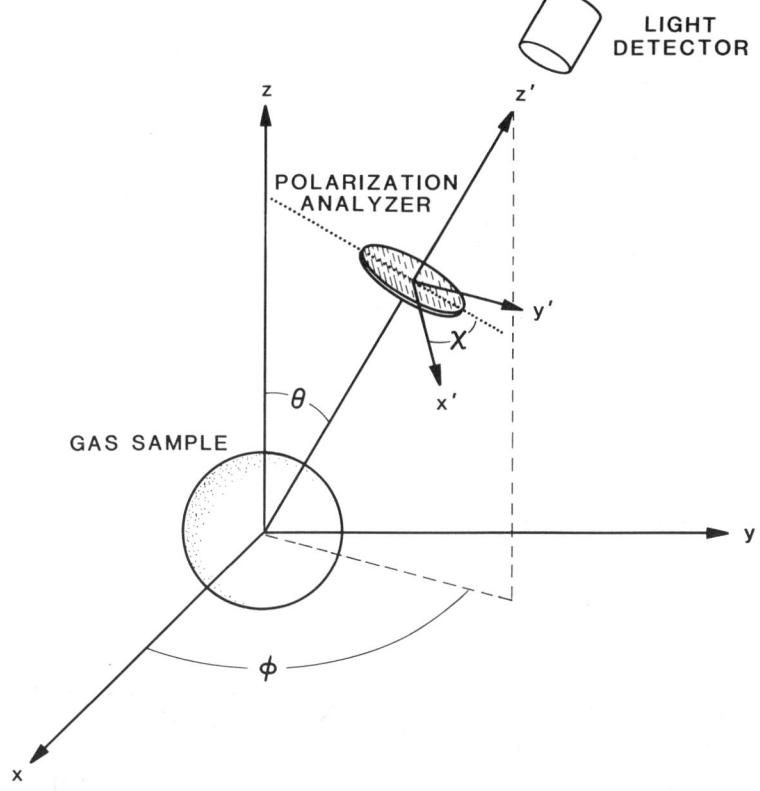

Figure 3 The collision frame x, y, z and the detector frame x', y', z' defining the Euler angles ϕ, θ, χ. The angles θ and ϕ are the polar angles of the axis of the light detector (the z' axis) referred to the collision frame. The angle χ is measured between the transmission axis of the polarization analyzer (the x' axis) and the zz' plane, i.e. between the x' axis and a line drawn normal to the detector axis which intersects the z axis.

convention the x and y coordinates of the collision frame in this problem are chosen so that $\hat{y}$ is parallel to $\hat{z} \times \hat{w}$.

Although the physics of the collision is described most naturally in the collision frame, the detection of the fluorescence photon is better described in a "detector frame" of coordinates (x', y', z'). The direction of observation of the photon is taken as the z' axis, so its polarization vector $\hat{\mathcal{E}}$ is confined to the x'y' plane. In the detector frame the detected polarization vector $\hat{\mathcal{E}}$ can be written

$$\hat{\mathcal{E}} = (\cos\beta, i\sin\beta, 0) \qquad 12.$$

Thus $\beta = 0$ represents the observation of light which is linearly polarized along x', while $\beta = \pi/4$ represents the detection of left-circularly polarized light (81). The multipole moments in the two coordinate frames are related using rotation matrices, whose arguments (ϕ, θ, χ) are the Euler angles required to rotate (x, y, z) into (x', y', z'), as shown in Figure 3.

Source Anisotropy and the Emitted Light

We present and discuss the results of Fano & Macek that relate the excited state fragment multipole moments to the distribution of its emitted light. For details regarding derivations, their original article should be consulted. The first key result is the intensity of fluorescence from the transition $J_i \to J_f$ which is observed by a light detector sensitive to the polarization characterized by β as in Eq. 12:

$$I = \tfrac{1}{3} I_0 \{ 1 - \tfrac{1}{2} h^{(2)}(J_i, J_f) \mathcal{Q}_0^{\text{det}} + \tfrac{3}{2} h^{(2)}(J_i, J_f) \mathcal{Q}_{2+}^{\text{det}} \cos 2\beta$$
$$+ \tfrac{3}{2} h^{(1)}(J_i, J_f) \mathcal{O}_0^{\text{det}} \sin 2\beta \}. \qquad 13.$$

Here the light intensity detected at a particular location is expressed in terms of two new kinds of quantities. First, the $h^{(k)}(J_i, J_f)$ ($k = 1$ or 2) are functions that depend only on the angular momentum quantum numbers of the initial (J_i) and final (J_f) fragment states. Thus the $h^{(k)}(J_i, J_f)$ are *geometrical* quantities which are independent of the fragmentation dynamics. Second, the quantities $\mathcal{Q}_0^{\text{det}}, \mathcal{Q}_{2+}^{\text{det}}, \mathcal{O}_0^{\text{det}}$ are the expectation values of multipole moment operators of the fragment excited state, which thus contain all the *dynamical* information about the excited state that can be learned from a fluorescence measurement. More explicitly, these multipole moments are the expectation values of certain combinations of angular momentum operators in the "primed" detector frame of Figure 3:

$$\mathcal{Q}_0^{\text{det}} = \langle (J_i | 3J_{z'}^2 - \mathbf{J}^2 | J_i) \rangle / J_i(J_i + 1),$$
$$\mathcal{Q}_{2+}^{\text{det}} = \langle (J_i | J_{x'}^2 - J_{y'}^2 | J_i) \rangle / J_i(J_i + 1), \qquad 14.$$
$$\mathcal{O}_0^{\text{det}} = \langle (J_i | J_{z'} | J_i) \rangle / \sqrt{J_i(J_i + 1)}.$$

The expectation values of operators in Eq. 14 require some explanation. The matrix element $\langle(J_i|J_{z'}|J_i)\rangle$, for example, is an average over several excited state matrix elements of the usual type found in quantum mechanics $(J_i M_i|J_{z'}|J_i M_i')$, but weighted according to the distribution of M_i quantum numbers populated in the excited fragment state. In explicit density matrix language (50, 82), we say that a collision creates an excited state density matrix $\rho_{M_i'M_i}$. Then the brackets $\langle\ \rangle$ in Eq. 14 imply the usual trace operation:

$$\langle(J_i|J_{z'}|J_i)\rangle = \sum_{M_i,M_i'} \rho_{M_i'M_i}(J_i M_i|J_{z'}|J_i M_i'). \qquad 15.$$

From an experimental or operational point of view it is not necessary to introduce density matrices, but in the language of density matrices the measurement of orientation or alignment corresponds to the measurement of the first and second rank multipole moments of the density matrix.

Comparison of Eq. 12 and Eq. 13 shows how the term $1-\frac{1}{2}h^{(2)}(J_i,J_f)\mathcal{Q}_0^{\text{det}}$ is observed by an unpolarized detector $(I_\parallel + I_\perp)$, proportional to the sum of $\beta=0$ and $\pi/2$, the term $\frac{3}{2}h^{(2)}(J_i,J_f)\mathcal{Q}_{2+}^{\text{det}}\cos 2\beta$ is observed when linear polarization $(I_\parallel - I_\perp)$ is detected, proportional to the difference of $\beta=0$ and $\pi/2$, and finally the term $\frac{3}{2}h^{(1)}(J_i,J_f)\mathcal{O}_0^{\text{det}}\sin 2\beta$ is observed only through a circular polarizer, $\beta=\pm\pi/4$. Thus $\mathcal{Q}_0^{\text{det}}$ and $\mathcal{Q}_{2+}^{\text{det}}$ are two irreducible elements of the quadrupole-moment matrix (of rank $k=2$) of the excited state, while $\mathcal{O}_0^{\text{det}}$ is the z'-component of an excited state magnetic dipole moment (rank $k=1$).

The functions $h^{(k)}(J_i,J_f)$ are important in that they isolate the sole effect of the final state $|J_f\rangle$ on the distribution of emitted polarized light. They reflect the amount of information lost when light is emitted and no further observations are made on the final state. Derived by Fano & Macek (77) as a ratio of Wigner $6j$ coefficients, they are given explicitly in Table 1. The definitions of the alignment tensor $\mathcal{Q}^{\text{det}}$ and depolarization coefficient $h^{(2)}(J_i,J_f)$ given here coincide with those of Fano & Macek (77). We have found it convenient to modify their definitions of $\mathcal{O}^{\text{det}}$ and $h^{(1)}(J_i,J_f)$, however. Our orientation vector $\mathcal{O}^{\text{det}}$ is $\sqrt{J_i(J_i+1)}$ times that of FM, while the coefficient $h^{(1)}(J_i,J_f)$ is $1/\sqrt{J_i(J_i+1)}$ times that of FM, which ensures that both $\mathcal{O}^{\text{det}}$ and $h^{(1)}(J_i,J_f)$ remain finite in the limit $J_i \to \infty$.

In writing the anisotropy tensor elements $(\mathcal{Q}_0^{\text{det}}, \mathcal{Q}_{2+}^{\text{det}}, \mathcal{O}_0^{\text{det}})$ in a "detector"-based coordinate system (x',y',z'), we have not utilized the known symmetry of the exciting collision, which is clearly better described in the "collision" frame (x,y,z) of Figure 3. Although only three multipole elements in the detector frame are relevant to light emission, there may be more moments in the collision frame or possibly less, depending on

Table 1 The geometrical factors $h^{(k)}(J_i, J_f)$ for the $J_i \to J_f$ transition

J_f	Emission branch	$h^{(1)}(J_i, J_f)$	$h^{(2)}(J_i, J_f)$
$J_i + 1$	$P \downarrow$	$-\dfrac{J_i}{\sqrt{J_i(J_i+1)}}$	$-\dfrac{J_i}{2J_i+3}$
J_i	$Q \downarrow$	$\dfrac{1}{\sqrt{J_i(J_i+1)}}$	1
$J_i - 1$	$R \downarrow$	$\dfrac{J_i+1}{\sqrt{J_i(J_i+1)}}$	$-\dfrac{J_i+1}{2J_i-1}$

the symmetry of the collision process. Rather than introducing the tensorial considerations required to formulate this problem in complete generality, we proceed to treat the most common and simplest symmetry case first.

Cylindrically Symmetric Collision Frame

Collisions like the one depicted in Figure 2, in which only one axis $\hat{z}$ is singled out, have cylindrical symmetry about that axis. Consequently only axially symmetric tensor elements $T_q^{(k)}$ with $q = 0$ can be nonzero in this frame. For the present problem the collision frame alignment tensor $\mathcal{A}$ has only the element $\mathcal{A}_0$ (denoted $\mathcal{A}_0^{col}$ in FM) and the orientation vector $\mathcal{O}$ has only a nonzero z-component $\mathcal{O}_0$ (denoted $\mathcal{O}_0^{col}$ in FM). Moreover $\mathcal{O}$ is the mean value of a *pseudovector* **J**, and must accordingly vanish unless any of the collision partners have an initial net orientation or helicity themselves. For the experiment of Figure 2, where the target molecules AB are randomly oriented, we then have three possibilities:

1. The incident photons are unpolarized and incident along $\hat{z}$. Then $\mathcal{O}$ vanishes identically and $\mathcal{A}_0$ is the only nonzero anisotropy in the collision frame.
2. The incident photons are linearly polarized along $\hat{z}$. Here also $\mathcal{O}$ vanishes and $\mathcal{A}_0$ is the only nonzero anisotropy. The measured value of $\mathcal{A}_0$ for linearly polarized photons is equal to -2 multiplied by the value of $\mathcal{A}_0$ obtained with unpolarized photons.
3. The incident photons are circularly polarized along the incidence axis $\hat{z}$. Now both $\mathcal{A}_0$ and $\mathcal{O}_0$ are nonzero. The alignment $\mathcal{A}_0$ obtained with circularly polarized light is the same as that obtained with unpolarized light. Note that the values of $\mathcal{O}_0$ obtained with right- and

with left-circularly polarized light have the same magnitude but opposite signs.

The results 1–3 do not rely on the fact that we are discussing photodissociation. They apply equally to any photofragmentation experiment in which the target molecules or atoms are randomly aligned and oriented.

To utilize the collision frame symmetries of $\mathcal{A}$ and $\mathcal{O}$ we must relate them to the detector frame quantities $\mathcal{A}^{det}$ and $\mathcal{O}^{det}$ which determine the light distribution through Eq. 13. This relationship is obtained by FM using rotation matrices.[1] In terms of the Euler angles (ϕ, θ, χ) defined in Figure 3, FM obtain for a cylindrically symmetric collision frame the results:

$$\mathcal{A}_0^{det} = \mathcal{A}_0 P_2(\cos\theta), \quad \mathcal{A}_{2+}^{det} = \mathcal{A}_0 \left(\tfrac{1}{2}\sin^2\theta \cos 2\chi\right),$$

$$\mathcal{O}_0^{det} = \mathcal{O}_0 \cos\theta, \qquad\qquad 16.$$

where $P_2(\cos\theta) = \tfrac{3}{2}\cos^2\theta - \tfrac{1}{2}$ is the second rank Legendre polynomial, and where $\mathcal{A}_0$ and $\mathcal{O}_0$ are given by Eq. 14 with the primes removed. Inserting these into Eq. 13 gives the distribution of light emitted as a function of the polar coordinates (θ, ϕ) of the light detector and the orientation χ of its polarization analyzer for a collision frame having cylindrical symmetry. The result is

$$I(\phi, \theta, \chi) = \tfrac{1}{3}I_0\{1 - \tfrac{1}{2}h^{(2)}(J_i, J_f)\mathcal{A}_0 P_2(\cos\theta)$$
$$+ \tfrac{3}{4}h^{(2)}(J_i, J_f)\mathcal{A}_0 \sin^2\theta \cos 2\chi \cos 2\beta$$
$$+ \tfrac{3}{2}h^{(1)}(J_i, J_f)\mathcal{O}_0 \cos\theta \sin 2\beta\}. \qquad 17.$$

In the most common experiments, using either linearly polarized or unpolarized incident light, $\mathcal{O}_0$ vanishes and this result simplifies for linearly polarized detection ($\beta = 0$) to

$$I(\phi, \theta, \chi) = \tfrac{1}{3}I_0\{1 - \tfrac{1}{2}h^{(2)}(J_i, J_f)\mathcal{A}_0[P_2(\cos\theta) - \tfrac{3}{2}\sin^2\theta \cos 2\chi]\}. \quad 18.$$

Equation 18 can be utilized to design an experiment that measures the alignment $\mathcal{A}_0$ of an excited atomic or molecular fragment.

---

[1] We caution that the real second rank tensor $\mathcal{A}$ was defined by FM with a nonstandard normalization (83). For details of the transformation between collision and detector frames using real rotation matrices see Hertel & Stoll (84). We choose to retain the nonstandard normalization so that $\mathcal{A}_0$ ranges between 2 and -1 for any J_i (as does the asymmetry parameter β), and $\mathcal{A}_{1+}$ as well as $\mathcal{A}_{2+}$ range between -1 and 1. If a standard normalization is adopted, then the range of $\mathcal{A}_{1+}$ and $\mathcal{A}_{2+}$ must be multiplied by $\sqrt{3}$.

DETECTION OF LINEARLY POLARIZED LIGHT   The most common method of determining the alignment is to detect light emitted at $\theta = 90°$ with respect to the collision axis z. The azimuthal angle ϕ is irrelevant in problems having a cylindrically symmetric collision frame, as may be seen in Eq. 18. The light intensities at two settings χ of the linear polarizer are required; these are most usually $I_\parallel = I(\phi, \frac{1}{2}\pi, \chi = 0)$ and $I_\perp = I(\phi, \frac{1}{2}\pi, \chi = \frac{1}{2}\pi)$. The "linear polarization" is then defined by

$$P = (I_\parallel - I_\perp)/(I_\parallel + I_\perp), \qquad 19.$$

and is related to the alignment by

$$P = 3h^{(2)}(J_i, J_f)\mathcal{A}_0 / [4 + h^{(2)}(J_i, J_f)\mathcal{A}_0]. \qquad 20.$$

Thus the two quantities of interest, namely I_0 and $\mathcal{A}_0$, are determined by

$$I_0 = I_\parallel + 2I_\perp; \quad \mathcal{A}_0 = \frac{1}{h^{(2)}(J_i, J_f)} \frac{4P}{3 - P}. \qquad 21.$$

Extraction of $\mathcal{A}_0$ is now possible, provided the light detected corresponds to a specific transition (branch) $J_i \to J_f$, using the values of $h^{(2)}(J_i, J_f)$ given in Table 1.

Often the experimentalist is not interested in the alignment, and wishes to measure I_0 alone. When this is the case, I_0 can be extracted from an intensity measurement at a *single* setting χ_M of the linear polarization analyzer; this is clearly an improvement on the method suggested by Eq. 21, requiring two measurements (of $I_\parallel$ and $I_\perp$). The so-called "magic angle" χ_M can be found by requiring the quantity multiplying $\mathcal{A}_0$ in Eq. 18 to vanish (85). Setting $\theta = \frac{1}{2}\pi$ for right angle photon observations as in Figure 2, the magic angle is given by the condition $\cos 2\chi_M = -\frac{1}{3}$, i.e. by

$$\chi_M = \cos^{-1}\left(\frac{1}{\sqrt{3}}\right) = 54.7°. \qquad 22.$$

At this setting the observed intensity is $I(\phi, \frac{1}{2}\pi, \chi_M) = \frac{1}{3}I_0$. If the right-angle detection geometry ($\theta = \frac{1}{2}\pi$) is inconvenient for some reason, a different magic angle $\chi_M(\theta)$ can be found for other observation directions θ. The general expression for the polarizer setting at which I is independent of $\mathcal{A}_0$ is then given by

$$\cos 2\chi_M(\theta) = \frac{(\cos^2\theta - \frac{1}{3})}{\sin^2\theta}, \qquad 23.$$

which has a solution only for $162.4° \geq \theta \geq 17.6°$.

DETECTION OF UNPOLARIZED LIGHT   In some experiments, notably those in which the fluorescence photons are relatively energetic, it may not prove practical to detect linearly polarized light. But the same information (I_0 and $\mathcal{Q}_0$) can be found by detecting only unpolarized light, provided two detector positions θ are utilized. The intensity observed by a detector responding equally to arbitrary light polarization, i.e. having no polarization bias, is simply the sum of the $\chi = 0$ and $\chi = \frac{1}{2}\pi$ intensities in Eq. 18,

$$I_{unp}(\theta) = \tfrac{2}{3} I_0 \left[1 - \tfrac{1}{2} h^{(2)}(J_i, J_f) \mathcal{Q}_0 \left(\tfrac{3}{2} \cos^2\theta - \tfrac{1}{2} \right) \right]. \qquad 24.$$

The intensity is independent of the alignment when θ is taken to be the magic angle 54.7°. While an intensity measurement at this value of θ suffices to determine I_0, one additional measurement is obviously needed to extract $\mathcal{Q}_0$.

QUALITATIVE INTERPRETATION OF $\mathcal{Q}_0$ To conclude this discussion of the cylindrically symmetric collision frame, we consider briefly the physical meaning of the alignment parameter $\mathcal{Q}_0$. For definiteness, we denote the collision cross sections for producing excited states $|J_i M_i\rangle$ by $\sigma(J_i M_i)$. Then $\mathcal{Q}_0$ can be expressed as

$$\mathcal{Q}_0 = \frac{\sum_{M_i} [3M_i^2 - J_i(J_i+1)] \sigma(J_i M_i)}{J_i(J_i+1) \sum_{M_i} \sigma(J_i M_i)}. \qquad 25.$$

Clearly $\mathcal{Q}_0$ measures the relative populations of small $|M_i|$ states (with negative contributions to $\mathcal{Q}_0$) versus large $|M_i|$ states (with positive contributions). If $\sigma(J_i M_i)$ is independent of M_i, then Eq. 25 shows that $\mathcal{Q}_0 = 0$.

Thus $\mathcal{Q}_0$ provides information on the nature of the spatial distribution of angular momentum vectors $\mathbf{J}_i$. This information is readily related to the shape of the excited state charge distribution. To illustrate this we treat in detail the example $J_i = 1$. Noting that $\sigma(J_i M_i) = \sigma(J_i, -M_i)$, and for convenience taking the $\sigma(J_i M_i)$ to be normalized, we have

$$\sigma(10) + 2\sigma(11) = 1. \qquad 26.$$

The alignment can now be evaluated using Eq. 25:

$$\mathcal{Q}_0 = -1 + 3\sigma(11). \qquad 27.$$

By inverting Eq. 26 and Eq. 27, we can express the $\sigma(1 M_i)$ in terms of $\mathcal{Q}_0$ (this is only possible for $J_i = 1$, since the quadrupole moment is the highest multipole for a state with $J_i = 1$),

$$\sigma(1,0) = \tfrac{1}{3} - \tfrac{2}{3} \mathcal{Q}_0; \quad \sigma(1, \pm 1) = \tfrac{1}{3} + \tfrac{1}{3} \mathcal{Q}_0. \qquad 28.$$

Now, the $\sigma(1 M_i)$ also determine the average excited state charge density of, for example, a one-electron atom through

$$\rho \propto \sum_{M_i} |Y_{J_i M_i}(\theta, \phi)|^2 \sigma(J_i M_i). \qquad 29.$$

For our example with $J_i = 1$ then, the angular variation of the charge distribution is given by

$$\rho \propto 1 - 2\mathcal{A}_0 P_2(\cos\theta). \qquad 30.$$

Thus if $\mathcal{A}_0 = 0$, the charge distribution is isotropic. If $\mathcal{A}_0 < 0$, ρ becomes elongated along the collision axis z. Finally if $\mathcal{A}_0 > 0$, ρ becomes flattened along the collision axis. These three alternatives are depicted qualitatively in Figure 4.

Breakdown of Cylindrical Symmetry

Next we consider experiments of greater complexity, possessing a *plane* of symmetry only (53, 83, 84, 86–89). This plane is taken by convention to be the xz-plane in the collision frame. All tensor elements (multipole moments) not symmetric under the transformation $y \to -y$ vanish identically. Restricting ourselves again to the usual class of experiments in which the collision partners have no net helicity (preferred spin direction), this leaves three elements of the alignment tensor and one element of the

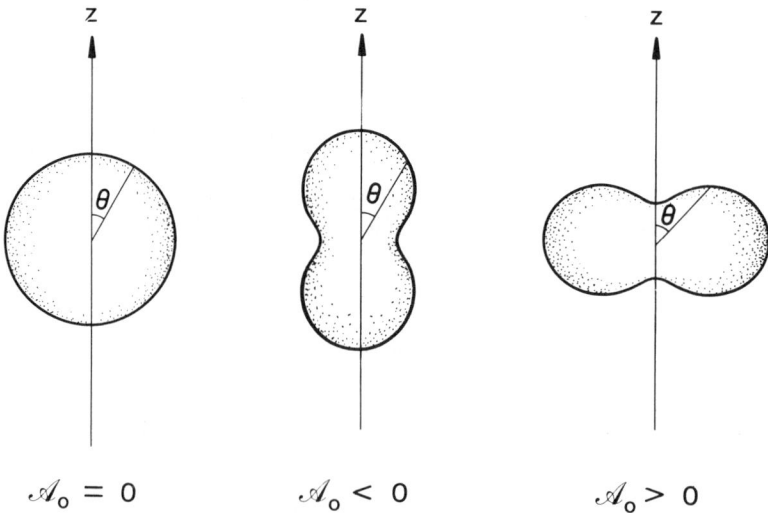

Figure 4 Form of the charge distribution $\rho(\theta)$ as a function of the alignment parameter $\mathcal{A}_0$.

orientation vector:

$$\mathcal{Q}_0 = \langle (J_i | 3J_z^2 - \mathbf{J}^2 | J_i) \rangle / J_i(J_i+1),$$
$$\mathcal{Q}_{1+} = \langle (J_i | J_x J_z + J_z J_x | J_i) \rangle / J_i(J_i+1),$$
$$\mathcal{Q}_{2+} = \langle (J_i | J_x^2 - J_y^2 | J_i) \rangle / J_i(J_i+1),$$
$$\mathcal{O}_{1-} = \mathcal{O}_y = \langle (J_i | J_y | J_i) \rangle / \sqrt{J_i(J_i+1)}. \qquad 31.$$

Here $\mathcal{O}_{1-}$ is simply the standard tensor notation for the y-component of the pseudovector $\mathcal{O}$. Note that $\mathcal{Q}_0$, $\mathcal{Q}_{1+}$, and $\mathcal{Q}_{2+}$ are manifestly symmetric under the transformation $y \to -y$. By this reasoning it might appear that $\mathcal{O}_y$ is odd, but we must be careful to properly account for the pseudovector nature of $\mathcal{O}$. One simple way to do this is to rewrite the components of $\mathcal{O}$ as quadratic operators in $\mathbf{J}$, utilizing the commutation rules $i\hbar \mathbf{J} = \mathbf{J} \times \mathbf{J}$. Then we see explicitly that

$$J_y = (J_z J_x - J_x J_z)/i\hbar, \qquad J_x = (J_y J_z - J_z J_y)/i\hbar,$$
$$J_z = (J_x J_y - J_y J_x)/i\hbar, \qquad 32.$$

which makes it clear that only $\mathcal{O}_y$ is even under the transformation $y \to -y$.

RELATIONSHIP BETWEEN DETECTOR AND COLLISION FRAME MULTIPOLES
As discussed in FM, these four anisotropy elements describe the distribution of polarized fluorescence. While the intensity expression (Eq. 13) in the detector frame is unchanged, the relationship between the detector- and collision-frame alignment and orientation is more complicated than for a cylindrically symmetric collision frame. This relationship, corrected for some misprints in the original article of Fano & Macek, is given by:

$$\mathcal{Q}_0^{det} = \mathcal{Q}_0 P_2(\cos\theta) + \mathcal{Q}_{1+}(\tfrac{3}{2}\sin 2\theta \cos\phi) + \mathcal{Q}_{2+}(\tfrac{3}{2}\sin^2\theta\cos 2\phi);$$
$$\mathcal{Q}_{2+}^{det} = \mathcal{Q}_0(\tfrac{1}{2}\sin^2\theta\cos 2\chi)$$
$$\qquad + \mathcal{Q}_{1+}(\sin\theta\sin\phi\sin 2\chi - \sin\theta\cos\theta\cos\phi\cos 2\chi)$$
$$\qquad + \mathcal{Q}_{2+}[\tfrac{1}{2}(1+\cos^2\theta)\cos 2\phi\cos 2\chi - \cos\theta\sin 2\phi\sin 2\chi];$$
$$\mathcal{O}_0^{det} = \mathcal{O}_{1-}\sin\theta\sin\phi. \qquad 33.$$

By inserting Eq. 33 into Eq. 13 the general expression for spatial variations of the polarized fluorescence is obtained (we do not bother to write it down here).

EXPERIMENTAL DETERMINATION OF $\mathcal{C}$ AND $\mathcal{O}$ It is instructive to compare the preceding analysis with the more common method of analyzing polarized light. Any beam of light can be completely characterized, as is well known (90, 91), by four Stokes parameters I, M, C, and S. The emitted light distribution predicted by Eq. 13 and Eq. 33 depends on the total intensity I_0 and on four anisotropy elements ($\mathcal{C}_0, \mathcal{C}_{1+}, \mathcal{C}_{2+}, \mathcal{O}_{1-}$), making a total of five independent parameters. It should not be surprising that this treatment requires more dynamical parameters than the Stokes treatment, since Eq. 13 and Eq. 33 essentially predict the Stokes parameters associated with a light beam emitted into *any* direction (θ, ϕ) of space. Consequently we reach an important conclusion:

> all anisotropy parameters cannot be determined by polarization measurements made with a detector fixed in only one direction of space (θ, ϕ).

This implies that determination of $\mathcal{C}_0$, $\mathcal{C}_{1+}$, $\mathcal{C}_{2+}$, and $\mathcal{O}_{1-}$ is far more difficult experimentally than is the determination of $\mathcal{C}_0$ in cylindrically symmetric problems, which requires only two intensity measurements with a linear polarizer and a fixed detector. Even so, this remains far simpler than a collision frame without even a plane of symmetry, which can have nine nonvanishing quantities, five elements of $\mathcal{C}$, three elements of $\mathcal{O}$ and, of course, I_0.

It should be pointed out that the expressions 13 and 33 given above for the emitted light distribution apply to any type of excitation process having a plane of symmetry (89). However, when the excitation *is* produced by photofragmentation of nonoriented (92–95) targets using either linearly polarized or unpolarized incident light, $\mathcal{O}_{1-}$ is identically zero owing to symmetry under $\hat{\mathbf{z}} \to -\hat{\mathbf{z}}$. A consideration of chiral targets can be found elsewhere (96–98).

Depolarization Caused by Unresolved Structure

The preceding formulation has treated light emission by an atom or molecule from a state of well-defined angular momentum $\mathbf{J}_i$. When light is emitted from two or more closely lying levels that are coherently excited, this formulation must be modified. Figure 5 illustrates qualitatively the reason for this added complexity. That is, the two or more paths leading from the initial photofragmentation to the final photon detection are indistinguishable, at least when the detector resolution cannot distinguish $h\nu_{ac}$ from $h\nu_{bc}$. Accordingly, the intensity is the absolute square of a coherent summation over the amplitudes for all indistinguishable pathways a, b, ..., and the intensity may exhibit "quantum beats": cross terms which oscillate in the time t of photon emission after the fragmentation as $\cos[(E_a - E_b)t/\hbar]$ (99).

The quantitative treatment of such nonstationary states requires, in general, the solution to the full time-dependent Schroedinger equation as a function of t. While this may prove in general to be a difficult problem, Fano & Macek have found a very simple formulation that is applicable when the energy splittings of interest are caused by hyperfine structure (77, 100, 101). They show that the hyperfine structure induces a time dependence in all of the multipole moments according to

$$\mathcal{Q}(t) = \mathcal{Q}(0)g^{(2)}(t),$$
$$\mathcal{O}(t) = \mathcal{O}(0)g^{(1)}(t), \qquad 34.$$

where

$$g^{(k)}(t) = \sum_{F,F'} \frac{(2F+1)(2F'+1)}{(2I+1)} \begin{Bmatrix} F & F' & k \\ J_i & J_i & I \end{Bmatrix}^2 \cos \omega_{F'F} t. \qquad 35.$$

Here $\omega_{F'F}$ is the frequency splitting between two hyperfine levels whose total angular momenta are F' and F, J_i is the excited state angular momentum quantum number, and I is the nuclear spin.

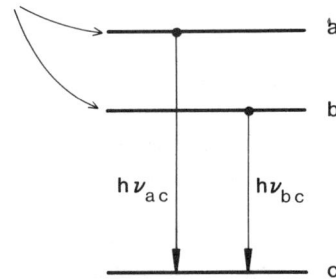

Figure 5 Diagram of two indistinguishable paths whose amplitudes add coherently and cause quantum beats. In the photofragmentation process pictured, two levels a and b are populated and locked together in phase (coherently excited). They both radiate to the same final level c. Then the intensity is proportional to the square of the sum of the probability amplitudes associated with each individual path, i.e. to $|\langle c|\hat{\mathcal{E}}\cdot\mathbf{r}|a\rangle + \langle c|\hat{\mathcal{E}}\cdot\mathbf{r}|b\rangle|^2$. The cross term may contribute constructively or destructively to the total transition rate, causing an interference effect, called quantum beats, to appear in the intensity observed at some fixed location in space by a detector that responds to both $h\nu_{ac}$ and $h\nu_{bc}$. If $|a(t)\rangle = |a(0)\rangle\exp(-iE_a t/\hbar)$ and $|b(t)\rangle = |b(0)\rangle\exp(-iE_b t/\hbar)$, then the cross term shows a time dependence proportional to $\cos(E_a - E_b)t/\hbar$ with a beat frequency equal to the energy difference between the levels divided by Planck's constant.

Despite the somewhat complicated appearance of Eq. 35, it has a simple physical interpretation. At $t=0$, the fragmentation process is usually assumed to align and orient the electronic charge cloud only. As the time t increases, the angular momentum operator $\mathbf{J}$, whose components determine $\mathcal{C}$ and $\mathcal{O}$, are no longer constants of the motion. Instead $\mathbf{J}$ and $\mathbf{I}$ precess about $\mathbf{F}$, causing the electronic and nuclear multipole moments to oscillate back and forth. This qualitative view is supported by the fact that $g^{(k)}(t)$ attains its maximum value, unity, at $t=0$. Accordingly, we interpret $\mathcal{C}(0)$ and $\mathcal{O}(0)$ as the *electronic* contributions to the alignment and orientation.

Most experiments do not achieve the time resolution required to observe the oscillations implied by Eq. 35. To use this result in the context of time-unresolved measurements, we integrate over all time including an exponential decay factor $\exp(-t/\tau)$ for the total intensity. This gives

$$\langle \mathcal{C}(t) \rangle = \mathcal{C}(obs) = \mathcal{C}(0) \bar{g}^{(2)},$$
$$\langle \mathcal{O}(t) \rangle = \mathcal{O}(obs) = \mathcal{O}(0) \bar{g}^{(1)}, \qquad 36.$$

where the time-averaged value of $g^{(k)}(t)$ is given by

$$\bar{g}^{(k)} = \sum_{F, F'} \frac{(2F+1)(2F'+1)}{(2I+1)} \left\{ \begin{matrix} F & F' & k \\ J_i & J_i & I \end{matrix} \right\}^2 \frac{1}{1+\omega_{F'F}^2 \tau^2} \qquad 37.$$

in which each factor $\cos \omega_{F'F} t$ is replaced by $(1+\omega_{F'F}^2 \tau^2)^{-1}$. The "depolarization factor" $\bar{g}^{(k)}$ lies between two extremes:

1. $|\omega_{F'F}| \ll \tau^{-1}$. In this limit the precession of $\mathbf{J}$ about $\mathbf{F}$ is so slow that essentially no precession occurs before the light is emitted. Then $\bar{g}^{(k)} \approx 1$ and the unresolved oscillations have no depolarizing effect.

2. $|\omega_{F'F}| \gg \tau^{-1}$. In this limit the precession is rapid and $\mathbf{J}$ precesses about $\mathbf{F}$ many times before the light is emitted. Then all terms with $F' \neq F$ vanish and the observed alignment $\mathcal{C}(obs)$ or orientation $\mathcal{O}(obs)$ is reduced from the actual electronic alignment or orientation by the factor

$$\bar{g}^{(k)} \approx \sum_F \frac{(2F+1)^2}{(2I+1)} \left\{ \begin{matrix} F & F & k \\ J_i & J_i & I \end{matrix} \right\}^2. \qquad 38.$$

But if $J_i \gg I$, as is typical of molecular problems, this depolarization factor is still close to unity because the precession has little effect on the direction of $\mathbf{J}_i$. In this limit the squared 6j-coefficient is approximately given in terms of a Legendre polynomial and

$$\bar{g}^{(k)} \xrightarrow[J_i, F \gg I]{} \sum_F \frac{2F+1}{2I+1} \frac{[P_k(\hat{\mathbf{J}}_i \cdot \hat{\mathbf{F}})]^2}{(2J_i+1)}, \qquad 39.$$

which approaches 1 as $\hat{\mathbf{J}}_i \cdot \hat{\mathbf{F}}$ approaches unity. When $I \approx J_i$ a sizeable depolarization can occur, as demonstrated for atoms by Hafner, Kleinpoppen & Krüger (102).

Often, in light emission by molecules, the hyperfine precession is negligibly slow on the time scale of the radiative lifetime τ. Then case 1 applies and the hyperfine depolarization effect can be neglected. It remains possible, however, that the light emission from members of a fine structure multiplet is unresolved. In such a case the preceding formulation still applies, provided we make the substitutions $I \to S$, F, $F' \to J_i$, J'_i, and $J_i \to N$, where N is the orbital angular momentum of the molecule and S is its spin. Then $\mathcal{Q}(0)$ and $\mathcal{O}(0)$ are purely *orbital* multipoles.

ANGULAR MOMENTUM TRANSFER TREATMENT OF PHOTOFRAGMENTATION

We turn to a detailed analysis of the dynamical information contained in the alignment and orientation of photofragments. In the section above we emphasize the importance of separating the fragmentation dynamics, contained solely in $\mathcal{Q}_0$ and $\mathcal{O}_0$ for cylindrically symmetric problems, from geometrical aspects of dipole radiation emission. In this section we develop this point of view one step further, by showing how $\mathcal{Q}_0$ and $\mathcal{O}_0$ can themselves be analyzed into the following:

1. a geometrical contribution reflecting only the dipole nature of the photofragmentation process;
2. a few amplitudes containing all dynamical information that it is possible to learn from fluorescence polarization measurements.

Angular Momentum Transfer Formalism

The most obvious way to specify the probability for any given photofragmentation process is in terms of continuum amplitudes specified in *jj*-coupling. In the following we consider only processes of the type

$$h\nu(j_{ph}=1) + AB(J_0) \to [A(j_i) + B(s)](\ell). \qquad 40.$$

Here we have indicated a photon carrying one unit of angular momentum incident on a molecule AB (not necessarily a diatomic) with angular momentum J_0. The fragments A and B so produced have internal angular momenta $\mathbf{j}_i$ and $\mathbf{s}$, respectively, and the orbital angular momentum of the A–B composite system is ℓ. For photodissociation experiments, in which B can be an atom or molecule, $\mathbf{s}$ is its total angular momentum, whereas in photoionization experiments B is a photoelectron with spin quantum number $s = \frac{1}{2}$. The partial cross section for producing fragments A in some state with angular momentum $\mathbf{j}_i$ is determined by a reduced dipole matrix element between the initial (bound) state and the final (continuum) state with total angular momentum $\mathbf{J}$,

$$\sigma(j_i) = C_0 \sum_{s,l,j_s,J} \left|\left(j_i(s\ell)j_s J \| r^{(1)} \| J_0\right)\right|^2. \qquad 41.$$

In Eq. 41, C_0 is a constant proportional to the photon energy, which is irrelevant for our analysis, and $\mathbf{j}_s = \boldsymbol{\ell} + \mathbf{s}$ is the sum of all fragment angular momenta in the final state that are not observed. The final state wave function is assumed to be energy-normalized, and to satisfy the incoming-wave boundary condition at large fragment separations $r_{AB} \to \infty$.

Recall that the polarization of light emitted by fragment A in a dipole transition $j_i \to j_f$ depends on the partial cross section $\sigma(j_i m_i)$ for leaving A in a state with magnetic quantum number m_i. Unlike Eq. 41 for $\sigma(j_i)$, however, $\sigma(j_i m_i)$ involves a coherent summation over J, (i.e. cross terms with $J \neq J'$) even though the summations over s, ℓ, and j_s remain incoherent. Physically, this expresses the fact that the operators $\mathbf{J}^2$ and j_{iz} do not commute.

The main idea of the angular momentum transfer formulation, as used long ago by Fano (103), and as extensively developed more recently by Dill & Fano (78–80), is that any fragment anisotropy should be analyzed in terms of a new set of continuum amplitudes. The new amplitudes are characterized not by the total angular momentum $\mathbf{J}$, but by $\mathbf{j}_t$, the unobserved angular momentum transferred from the target to the photofragments. When so defined, $\mathbf{j}_t^2$ commutes with the angular momenta associated with the observed anisotropy, and hence the contributions to the anisotropy from each $\mathbf{j}_t$ add incoherently.

Applications of this approach have been restricted until very recently to angular distributions of photofragments, in which case the observed angular momentum is $\boldsymbol{\ell}$ and the angular momentum transfer is given by $\mathbf{j}_i + \mathbf{s} - \mathbf{J}_0 = \mathbf{j}_{ph} - \boldsymbol{\ell}$. Dill et al (72) first applied this method to our present problem of photofragment alignment, defining a different angular momentum transfer according to

$$\mathbf{j}_t = \boldsymbol{\ell} + \mathbf{s} - \mathbf{J}_0 = \mathbf{j}_{ph} - \mathbf{j}_i. \qquad 42.$$

Klar (56, 57) has also applied an angular momentum transfer formalism to treat this problem but with a different definition of $\mathbf{j}_t$. As discussed by D. Dill in unpublished correspondence with R. N. Zare in 1977, and as amplified by Greene & Zare (59), the new continuum amplitudes S are related to the old ones by

$$S(J_0, s, \ell, j_s, j_i; j_t) = \sum_J \left(\frac{2j_t+1}{3}\right)^{1/2} \begin{Bmatrix} j_i & j_s & J \\ J_0 & \ell & j_t \end{Bmatrix}$$
$$\times (j_i(s\ell)j_s J \| r^{(1)} \| J_0). \qquad 43.$$

To within normalization, this results in the incoherent summation

$$\sigma(j_i m_i) = \sum_{s,\ell,j_s,j_t} |S(J_0, s, \ell, j_s, j_i; j_t)|^2 (j_i m_i, j_t q - m_i | 1q)^2. \qquad 44.$$

Here q is the component of incident photon angular momentum along the quantization axis $\hat{z}$, i.e. the photon helicity. Linearly polarized incident light thus has $q = 0$, unpolarized light is the average of $q = \pm 1$, and left-circularly polarized light has $q = +1$.

Equation 44 shows how the m_i dependence of $\sigma(j_i m_i)$ for each j_t arises only through Wigner coefficients (Clebsch-Gordan coefficients). Consequently, the alignment $\mathcal{C}_0$ and orientation $\mathcal{O}_0$ contributed by each j_t can be evaluated in closed form. We have found that

$$\mathcal{C}_0(j_i) = \sum_{j_t} |S(j_i; j_t)|^2 \mathcal{C}_0(j_i; j_t) \Big/ \sum_{j_t} |S(j_i; j_t)|^2,$$

$$\mathcal{O}_0(j_i) = \sum_{j_t} |S(j_i; j_t)|^2 \mathcal{O}_0(j_i; j_t) \Big/ \sum_{j_t} |S(j_i, j_t)|^2, \qquad 45.$$

where

$$|S(j_i; j_t)|^2 = \sum_{s,\ell,j_s} |S(J_0, s, \ell, j_s, j_i; j_t)|^2, \qquad 46.$$

and the summation over j_t is restricted to the three values $j_i - 1$, j_i, and $j_i + 1$. In Eq. 45 we have introduced two universal functions describing the alignment $\mathcal{C}_0(j_i; j_t)$ and orientation $\mathcal{O}_0(j_i; j_t)$ arising from each transfer j_t. These are given in Table 2, where the dependence on q is shown explicitly. Note that $\mathcal{C}_0$ for circularly polarized ($q = \pm 1$) or for unpolarized incident light is simply minus one-half the value of $\mathcal{C}_0$

---

[2] More generally, elliptically polarized incident radiation can be shown to excite two nonzero alignment parameters, $\mathcal{C}_0$ and $\mathcal{C}_{2+}$. These are referred to a coordinate system in which the photon beam is incident along $\hat{z}$ and the major and minor axes of the ellipse coincide with $\hat{x}$ and $\hat{y}$ in Figure 3. If the alignment obtained using linearly polarized incident radiation is denoted $\mathcal{C}_0^{\text{lin}}$, then the alignment parameters obtained using elliptically polarized radiation are given by $\mathcal{C}_0 = -\frac{1}{2}\mathcal{C}_0^{\text{lin}}$, $\mathcal{C}_{1+} = 0$, $\mathcal{C}_{2+} = -\frac{1}{2}p\mathcal{C}_0^{\text{lin}}$. Here $p = (I_y - I_x)/(I_y + I_x)$ is the linear polarization of the incident radiation.

Table 2 The universal orientation and alignment factors $\mathcal{O}_0(j_i; j_t)$ and $\mathcal{A}_0(j_i; j_t)$, where the dependence on the incident photon helicity, q, has been made explicit

j_t	Absorption branch	$\mathcal{O}_0(j_i; j_t, q)$	$\mathcal{A}_0(j_i; j_t, q)$
$j_i + 1$	$P\uparrow$	$-\dfrac{\frac{1}{2}qj_i}{\sqrt{j_i(j_i+1)}}$	$-(-1)^q(1-\frac{1}{2}q^2)\dfrac{2j_i-1}{5(j_i+1)}$
j_i	$Q\uparrow$	$\dfrac{\frac{1}{2}q}{\sqrt{j_i(j_i+1)}}$	$(-1)^q(1-\frac{1}{2}q^2)\dfrac{(2j_i-1)(2j_i+3)}{5j_i(j_i+1)}$
$j_i - 1$	$R\uparrow$	$\dfrac{\frac{1}{2}q(j_i+1)}{\sqrt{j_i(j_i+1)}}$	$-(-1)^q(1-\frac{1}{2}q^2)\dfrac{2j_i+3}{5j_i}$

obtained with linearly polarized light ($q = 0$).² These results for the alignment were presented before (59), but have not been given previously for the orientation.

The three alternative values of $h^{(2)}(j_i, j_f)$ in Table 1, coupled with the three values of $\mathcal{A}_0(j_i; j_t)$ in Table 2, determine nine limiting cases for the linear polarization in the high j_i limit through Eq. 20. These yield exactly the same results as found by Macpherson, Simons & Zare (14).

Dependence of the Universal Alignment and Orientation Functions on Parity-Favoredness

The values of the universal alignment function $\mathcal{A}_0(j_i; j_t)$ are plotted in Figure 6, while those of the universal orientation function $\mathcal{O}_0(j_i; j_t)$ are shown in Figure 7. The alignment contributions associated with angular momentum transfers $j_t \neq j_i$ are negative, while those with $j_t = j_i$ are positive. In addition, the two alignment curves with $j_t \neq j_i$ converge as $j_i \to \infty$ to the same value $-2/5$, while the curve $\mathcal{A}_0(j_i; j_i)$ approaches $4/5$ in the same limit. These results suggest strongly that contributions to the fragmentation having transfers $j_t \neq j_i$ differ qualitatively from contributions from transfers $j_t = j_i$. This qualitative difference can be traced to the *parity-favoredness quantum number* π_f, which is either $+1$ or -1 for each angular momentum transfer (104). In general π_f is defined by

$$\pi_f = (-1)^{j_t - j_{obs} + j_{ph}} = (-1)^{j_t - j_{obs} + 1}, \qquad 47.$$

which reduces in the present problem with $j_{obs} = j_i$ to

$$\pi_f = (-1)^{j_t - j_i + 1} = \begin{cases} +1, & j_t \neq j_i \quad \text{(parity-favored)} \\ -1, & j_t = j_i \quad \text{(parity-unfavored)}. \end{cases} \qquad 48.$$

The sign of the alignment is thus solely determined by the parity-favoredness.

An analogous result was first derived for the photofragment angular distribution asymmetry parameter β by Dill & Fano (79). [The definition of β is given in Eq. 1, and is not the same β of Eq. 12]. In particular, for photofragmentation experiments using linearly polarized incident photons, the parity-unfavored contributions to β are identically -1. Parity-favored contributions to β can instead range anywhere from -1 to 2, depending on the specific dynamics of the problem, with experience showing that they tend to be positive. We should point out that whereas the alignment contribution of each transfer can be evaluated in closed form (Table 2) independently of dynamical considerations, the same is not true for β. This stems from the fact that j_i, the "observed" angular momentum quantum in alignment measurements, has only one value, while the observed angular momentum in angular distribution measurements is ℓ, which can range over many values.

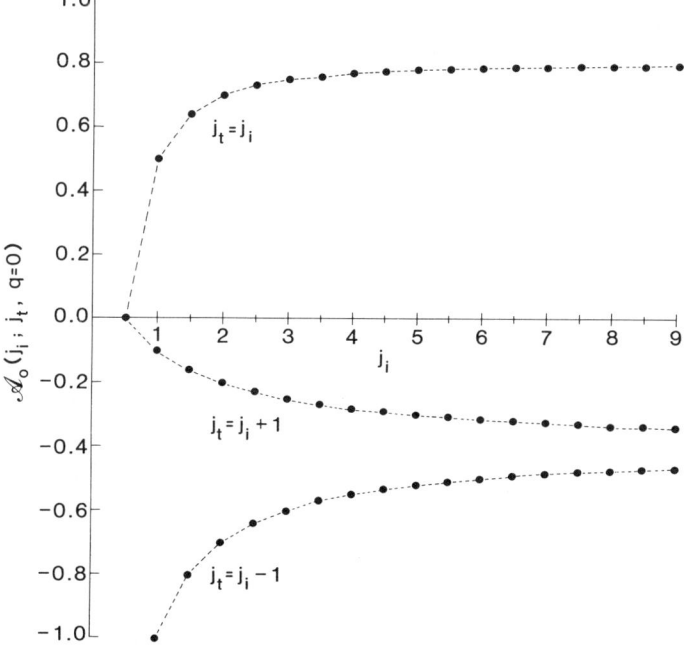

Figure 6 Plot of the universal alignment function $\mathcal{A}_0(j_i; j_t)$ for linearly polarized incident light ($q = 0$) as a function of the photofragment angular momentum quantum number j_i and the angular momentum transfer quantum number j_t. Contrast the behavior of the parity-favored ($j_t \neq j_i$) and parity-unfavored ($j_t = j_i$) contributions. For circularly polarized ($q = \pm 1$) or unpolarized incident light, the values are to be multiplied by $-\frac{1}{2}$.

In the case of photoionization in which the ejected particle (electron) is the same as the collection of particles the electric vector of the light beam acts upon, the sign of either second rank anisotropy (i.e. β or $\mathcal{Q}_0$) provides an indication of whether the fragmentation is a simple, direct break-up process. To see this connection, we consider the two possible limits for the coefficient of $P_2(\cos\theta)$, which is β in Eq. 1 and $-2\mathcal{Q}_0$ in Eq. 30. The upper limit for β is 2, for which the photofragmentation axis $\hat{\mathbf{k}}$ peaks along the incident polarization axis $\hat{\mathcal{E}}$:

$$d\sigma/d\Omega \propto |\hat{\mathcal{E}} \cdot \hat{\mathbf{k}}|^2. \qquad 49.$$

Similarly, Eq. 30 and Figure 4 show that when $-2\mathcal{Q}_0 = 2$, the charge density is elongated along the polarization vector also. This is the pattern expected for *direct* processes, which tend to pull the fragments apart along the direction of the force that is exerted on the target by the photon electric field. These are the parity-favored processes. In the opposite extreme, the coefficient of $P_2(\cos\theta)$ is -1, and the photofrag-

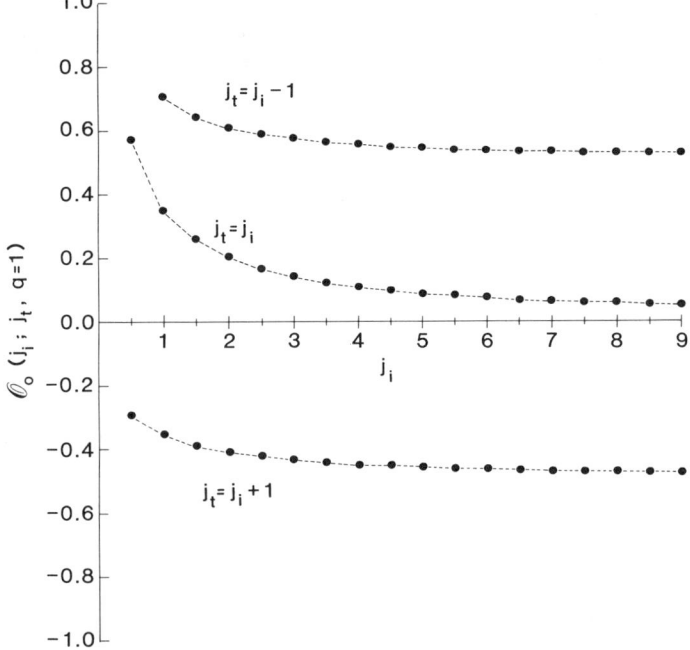

Figure 7 Plot of the universal orientation function $\mathcal{O}_0(j_i; j_t)$ for left-circularly polarized incident light ($q = 1$) as a function of the photofragment angular momentum quantum number j_i and the angular momentum transfer quantum number j_t. For right-circularly polarized light the values are multiplied by -1, while for linearly polarized light or unpolarized light they vanish.

mentation peaks at right angles to $\hat{\mathscr{E}}$:

$$d\sigma/d\Omega \propto |\hat{\mathscr{E}} \times \hat{\mathbf{k}}|^2. \qquad 50.$$

Similarly, the charge density is elongated orthogonally to $\hat{\mathscr{E}}$ when $-2\mathcal{Q}_0 = -1$. The observation of such an angular distribution or alignment is far more interesting and less common than the pattern in Eq. 49, and is the signature of screw-type interactions between the fragments (79, 105–107). A simple classical analogy of each type of fragmentation process is provided by an impulsive force **F** delivered radially to an object resting on a merry-go-round (11, 108). If the merry-go-round is stationary or rotating slowly, the object is ejected axially along **F** (axial recoil), but if the merry-go-round initially is rotating rapidly, then the object is ejected tangentially, perpendicular to **F** (transverse recoil). The latter type of fragmentation process is generally associated with parity-unfavored angular momentum transfers.[3]

In the case of photodissociation, the initial recoil direction of the nuclei has in general some fixed orientation with respect to the electron cloud, for example, along the electronic transition moment for a parallel-type transition or at right angles to the electronic transition moment for a perpendicular-type transition. With suitable modification, the connection between parity favoredness or unfavoredness and axial or transverse recoil exactly carries over.

The manner in which these ideas apply to fragment orientation is perhaps less obvious, as the parity-favored and unfavored terms of the universal function $\mathcal{O}_0(j_i; j_t)$ plotted in Figure 7 do not have opposite signs. It is apparent, however, that parity-favored and parity-unfavored contributions to $\mathcal{O}_0(j_i)$ are qualitatively different. That is, $\mathcal{O}_0(j_i; j_t) \to 0$ as $j_i \to \infty$ for the parity-unfavored transfer $j_t = j_i$, while $\mathcal{O}_0(j_i; j_t) \to \pm 1/2$ for the parity-favored transfers $j_t = j_i \mp 1$.

Extraction of Dynamical Information from Measurements of $\mathcal{Q}_0$ and $\mathcal{O}_0$

Equations 45 and 46, along with Table 2, represent a focal point of this article. They show how the fragment alignment and orientation are generally the incoherent average of three alternative values of the transfer j_t. Moreover, Eq. 45 fully disentangles the fragmentation dynamics, contained in $S(j_i; j_t)$, from the geometrical considerations, contained in

---

[3] While this is strictly true of the *alignment* pattern, there are examples of parity-favored angular momentum transfers which nonetheless give rise to *angular distributions* of the orthogonal type shown in Eq. 50. These exceptions are sometimes termed "dynamically unfavored." They originate from the coherent contribution to the anisotropy of more than one value of j_{obs}. See Greene (106) and Fano & Greene (107).

the universal functions for the alignment $\mathcal{A}_0(j_i; j_t)$ and the orientation $\mathcal{O}_0(j_i; j_t)$.

Let us introduce the three normalized quantities

$$S_+ = \frac{|S(j_i; j_i+1)|^2}{\sum_{j_t}|S(j_i; j_t)|^2},$$

$$S_0 = \frac{|S(j_i; j_i)|^2}{\sum_{j_t}|S(j_i; j_t)|^2},$$

and

$$S_- = \frac{|S(j_i; j_i-1)|^2}{\sum_{j_t}|S(j_i; j_t)|^2}, \qquad 51.$$

which may be interpreted as the fraction of each angular momentum transfer contributing to the photofragmentation process. We call the S the angular momentum transfer channel probabilities. Because $S_+ + S_0 + S_- = 1$, only two of these dynamical quantities are independent. Then Eqs. 45 and 46 may be rewritten as

$$\mathcal{A}_0(j_i) = S_+\mathcal{A}_0(j_i; j_i+1) + S_0\mathcal{A}_0(j_i; j_i) + S_-\mathcal{A}_0(j_i; j_i-1);$$
$$\mathcal{O}_0(j_i) = S_+\mathcal{O}_0(j_i; j_i+1) + S_0\mathcal{O}_0(j_i; j_i) + S_-\mathcal{O}_0(j_i; j_i-1). \qquad 52.$$

Thus the three channel probabilities S represent *all* the dynamical information one can learn about a given photofragmentation process by measuring the polarization of light emitted by one fragment.

In the most common photofragment emission studies only the alignment $\mathcal{A}_0(j_i)$ is measured. Although it is not possible to deduce the individual channel probabilities, this does provide some measure of the ratio of parity-unfavored to parity-favored contributions. Indeed in limit of large j_i, Figure 6 shows how $\mathcal{A}_0(j_i; j_i+1)$ and $\mathcal{A}_0(j_i; j_i-1)$ both approach the same value ($-2/5$). Thus in the high j_i limit, a measurement of $\mathcal{A}_0$ alone gives directly the ratio

$$\gamma = \frac{S_0}{S_+ + S_-} = \frac{|S(j_i; j_i)|^2}{|S(j_i; j_i+1)|^2 + |S(j_i; j_i-1)|^2}. \qquad 53.$$

This parametric dependence of $\mathcal{A}_0$ on γ is illustrated in Figure 8, which may facilitate the interpretation of experimental data in the limit of large j_i.

Knowledge of $\mathcal{A}_0$ and $\mathcal{O}_0$ provides sufficient information to extract all three channel probabilities. For example, a measurement of $\mathcal{A}_0$ and $\mathcal{O}_0$

using left circularly polarized incident light can be inverted, giving

$$\mathcal{S}_+ = \frac{5}{3}\frac{j_i}{2j_i+1}\mathcal{A}_0(j_i) - \frac{(2j_i+3)\sqrt{j_i(j_i+1)}}{(j_i+1)(2j_i+1)}\mathcal{O}_0(j_i) + \frac{2j_i+3}{3(2j_i+1)};$$

$$\mathcal{S}_- = \frac{5}{3}\frac{j_i+1}{2j_i+1}\mathcal{A}_0(j_i) + \frac{(2j_i-1)\sqrt{j_i(j_i+1)}}{j_i(2j_i+1)}\mathcal{O}_0(j_i) + \frac{2j_i-1}{3(2j_i+1)};$$

$$\mathcal{S}_0 = -\frac{5}{3}\mathcal{A}_0(j_i) - \frac{\mathcal{O}_0(j_i)}{\sqrt{j_i(j_i+1)}} + \frac{1}{3}. \qquad 54.$$

Note the simple form the channel probabilities attain in the high j_i limit, but that for smaller values of j_i Eq. 54 must be used in full. We regard the

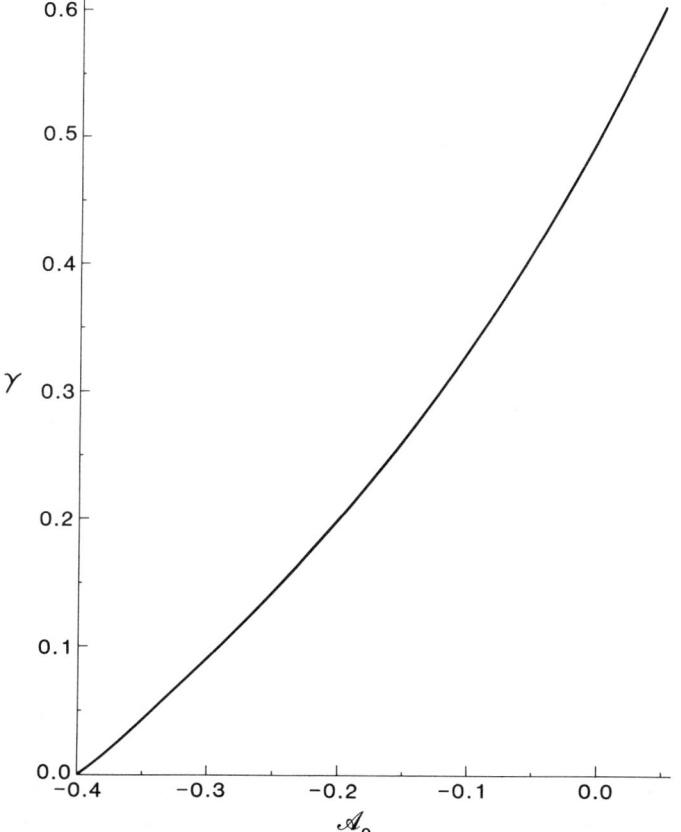

Figure 8 The dependence of γ, the ratio of parity-unfavored to the sum of parity-favored channel probabilities, on $\mathcal{A}_0$, the value of the alignment parameter in the high j_i limit. The actual function plotted is $\gamma = (5\mathcal{A}_0 + 2)/(4 - 5\mathcal{A}_0)$.

extraction of these angular momentum transfer channel probabilities as the primary goal for this class of experiments.

ACKNOWLEDGMENTS

We thank R. Bersohn, P. J. Brucat, C. D. Caldwell, and G. W. Loge for critically reading earlier drafts of this manuscript. This work was supported by grants from the National Science Foundation.

Literature Cited

1. Simons, J. P. 1977. *Gas Kinetics and Energy Transfer*, *Chem. Soc. Spec. Period. Rep.* 2:58–95
2. Gelbart, W. M. 1977. *Ann. Rev. Phys. Chem.* 28:323–48
3. Okabe, H. 1978. *Photochemistry of Small Molecules*. New York: Wiley. 431 pp.
4. Leone, S. R. 1982. *Adv. Chem. Phys.* 50:255–324
5. Berkowitz, J. 1979. *Photoabsorption, Photoionization, and Photoelectron Spectroscopy*. New York: Academic. 469 pp.
6. Fano, U., Cooper, J. W. 1968. *Rev. Mod. Phys.* 40:441–507
7. Pratt, R. H., Ron, A., Tseng, H. K. 1973. *Rev. Mod. Phys.* 45:273–325, 663
8. Samson, J. A. R. 1982. *Handb. Phy.* 31: In press
9. Bersohn, R., Lin, S. H. 1969. *Adv. Chem. Phys.* 16:67–100
10. Jonah, C. 1971. *J. Chem. Phys.* 55:1915
11. Zare, R. N. 1972. *Mol. Photochem.* 4:1–37
12. Yang, S. C., Bersohn, R. 1974. *J. Chem. Phys.* 61:4400–7
13. Dzvonik, M. J., Yang, S., Bersohn, R. 1974. *J. Chem. Phys.* 61:4408–21
14. Macpherson, M. T., Simons, J. P., Zare, R. N. 1979. *Mol. Phys.* 38:2049–55
15. Band, Y. B., Freed, K. F., Kouri, D. J. 1981. *Chem. Phys. Lett.* 79:233–37
16. Band, Y. B., Freed, K. F. 1981. *Chem. Phys. Lett.* 79:238–43
17. Band, Y. B., Freed, K. F., Kouri, D. J. 1981. *J. Chem. Phys.* 74:4380–94
18. Balint-Kurti, G. G., Shapiro, M. 1981. *Chem. Phys.* 61:137–55
19. Shapiro, M., Bersohn, R. 1982. *Ann. Rev. Phys. Chem.* 33: 409–42
20. Starace, A. F. 1982. *Handb. Phys.* 31: In press
21. Berkowitz, J., Erhardt, H. 1966. *Phys. Lett.* 21:531–32
22. Niehaus, A., Ruf, M. W. 1972. *Z. Phys.* 252:84–94
23. Morgenstern, R., Niehaus, A., Ruf, M. W. 1970. *Chem. Phys. Lett.* 4:635–38
24. Dehmer, J. L., Chupka, W. A., Berkowitz, J., Jivery, W. T. 1975. *Phys. Rev. A* 12:1966–73
25. Carlson, T. A., Jonas, A. E. 1971. *J. Chem. Phys.* 55:4913–24
26. Karlsson, L., Mattson, L., Jadrny, R., Siegbahn, K., Thimm, K. 1976. *Phys. Lett. A* 58:381–84
27. Harrison, H. 1970. *J. Chem. Phys.* 52:901–5
28. Schönhense, G. 1981. *J. Phys. B* 14:L187–92
29. Flügge, S., Mehlhorn, W., Schmidt, V. 1972. *Phys. Rev. Lett.* 29:7–9
30. Oh, S. D., Pratt, R. H. 1974. *Phys. Rev. A* 10:1198–1203
31. Hall, J. L., Siegel, M. W. 1968. *J. Chem. Phys.* 48:943–45
32. Cooper, J., Zare, R. N. 1968. *J. Chem. Phys.* 48:942–43
33. Hotop, H., Lineberger, W. C. 1975. *J. Phys. Chem. Ref. Data* 4:539–76
34. Cooper, J., Zare, R. N. 1969. *Lectures in Theoretical Physics*, ed. S. Geltman, K. T. Mahanthappa, W. E. Brittin, 11-C:317–37. New York: Gordon & Breach
35. Tully, J. C., Berry, R. S., Dalton, B. J. 1968. *Phys. Rev.* 176:95–105
36. Buckingham, A. D., Orr, B. J., Sichel, J. M. 1970. *Philos. Trans. R. Soc. London Ser. A* 268:147–57
37. Peshkin, M. 1970. *Adv. Chem.Phys.* 18:1–14
38. Dehmer, J. L., Dill, D. 1979. *Electron Molecule and Photon Molecule Collisions*, ed. V. McKoy, T. Rescigno, B. Schneider, pp. 225–65. New York: Plenum

39. Dill, D. 1972. *Phys. Rev. A* 6:160–72
40. Raoult, M., Jungen, Ch., Dill, D. 1980. *J. Chim. Phys.* 77:599–604
41. Dehmer, J. L., Dill, D., Wallace, S. 1979. *Phys. Rev. Lett.* 43:1005–8
42. Stockbauer, R., Cole, B. E., Ederer, D. L., West, J. B., Parr, A. C., Dehmer, J. L. 1979. *Phys. Rev. Lett.* 43:757–61
43. West, J. B., Parr, A. C., Cole, B. E., Ederer, D. L., Stockbauer, R., Dehmer, J. L. 1980. *J. Phys. B* 13:L105–8
44. Cole, B. E., Ederer, D. L., Stockbauer, R., Codling, K., Parr, A. C., West, J. B., Poliakoff, E. D., Dehmer, J. L. 1980. *J. Chem. Phys.* 72:6308–10
45. Swanson, J. R., Dill, D., Dehmer, J. L. 1981. *J. Phys. B* 14:L207–11
46. Loge, G. W., Zare, R. N. 1981. *Mol. Phys.* 43:1419–28
47. Fano, U. 1957. *Rev. Mod. Phys.* 29:74–93
48. Fano, U., Racah, G. 1959. *Irreducible Tensorial Sets*. New York: Academic. 171 pp.
49. Brink, D. M., Satchler, G. R. 1968. *Angular Momentum*. Oxford: Clarendon. 160 pp. 2nd ed.
50. Blum, K. 1981. *Density Matrix Theory and Applications*. New York: Plenum. 217 pp.
51. Percival, I. C., Seaton, M. J. 1958. *Philos. Trans. R. Soc. London Ser. A* 251:113–38
52. Van Brunt, R. J., Zare, R. N. 1968. *J. Chem. Phys.* 48:4304–8
53. Case, D. A., McClelland, G. M., Herschbach, D. R. 1978. *Mol. Phys.* 35:541–73
54. Caldwell, C. D., Zare, R. N. 1977. *Phys. Rev. A* 16:255–62
55. Mauser, W., Mehlhorn, W. 1980. *Extended Abstracts of the 6th Int. Conf. Vacuum Ultraviolet Rad. Phys.*, Charlottesville, VA, Vol. 2, Pap. 7, pp. 1–3
56. Klar, H. 1979. *J. Phys. B* 12:L409–12
57. Klar, H. 1980. *J. Phys. B* 13:2037–49
58. Theodosiou, C. E., Starace, A. F., Tambe, B. R., Manson, S. T. 1981. *Phys. Rev. A* 24:301–7
59. Greene, C. H., Zare, R. N. 1982. *Phys. Rev. A* 25:2031–37
60. Grader, R. J., Oliver, A. J., Ebert, P. J. 1977. *Phys. Rev. A* 16:2388–91
61. Scofield, J. H. 1976. *Phys. Rev. A* 14:1418–20
62. Vigué, J., Grangier, P., Roger G., Aspect, A. 1981. *J. Phys. Paris* 42:L531–35
63. Rothe, E. W., Krause, U., Düren, R. 1980. *Chem. Phys. Lett.* 72:100–3
64. Chamberlain, G. A., Simons, J. P. 1975. *Chem. Phys. Lett.* 32:355–58
65. Chamberlain, G. A., Simons, J. P. 1975. *J. Chem. Soc. Faraday Trans. 2* 71:2043–50
66. Macpherson, M. T., Simons, J. P. 1977. *Chem. Phys. Lett.* 51:261–64
67. Macpherson, M. T., Simons, J. P. 1978. *J. Chem. Soc. Faraday Trans. 2* 74:1965–77
68. Macpherson, M. T., Simons, J. P. 1979. *J. Chem. Soc. Faraday Trans. 2* 75:1572–92
69. Poliakoff, E. D., Southworth, S. H., Shirley, D. A., Jackson, K. H., Zare, R. N. 1979. *Chem. Phys. Lett.* 65:407–9
70. Husain, J., Wiesenfeld, J. R., Zare, R. N. 1980. *J. Chem. Phys.* 72:2479–83
71. Loge, G. W., Wiesenfeld, J. R. 1981. *Chem. Phys. Lett.* 78:32–35
72. Poliakoff, E. D., Dehmer, J. L., Dill, D., Parr, A. C., Jackson, K. H., Zare, R. N. 1981. *Phys. Rev. Lett.* 46:907–10
73. Alder, K. 1952. *Helv. Phys. Acta* 25:235–58
74. Frauenfelder, H., Steffen, R. M. 1968. *Alpha-, Beta-, and Gamma-Ray Spectroscopy*, ed. K. Siegbahn, Vol. 2, Chap. 19A, pp. 997–1198. Amsterdam: North-Holland. 2nd ed.
75. Matthias, E., Olsen, B., Shirley, D. A., Templeton, J. E., Steffen, R. M. 1971. *Phys. Rev. A* 4:1626–58
76. Steffen, R. M., Alder, K. 1975. *The Electromagnetic Interaction in Nuclear Spectroscopy*, ed. Hamilton, pp. 505–643. Amsterdam: North Holland
77. Fano, U., Macek, J. H. 1973. *Rev. Mod. Phys.* 45:553–73
78. Fano, U., Dill, D. 1972. *Phys. Rev. A* 6:185–92
79. Dill, D., Fano, U. 1972. *Phys. Rev. Lett.* 29:1203–5
80. Dill, D. 1973. *Phys. Rev. A* 7:1976–87
81. Jackson, J. D. 1975. *Classical Electrodynamics*, p. 274. New York: Wiley. 848 pp. 2nd ed.
82. Blum, K. 1978. *Progress in Atomic Spectroscopy* ed. W. Hanle, H. Kleinpoppen, pp. 71–110. New York/London: Plenum
83. Macek, J., Hertel, I. V. 1974. *J. Phys. B* 7:2173–88
84. Hertel, I. V., Stoll, W. 1977. *Adv. At. Mol. Phys.* 13:113–228
85. Samson, J. A. R. 1969. *J. Opt. Soc. Am.* 59:356–57

86. King, G. C. M., Adams, A., Read, F. H. 1967. *J. Phys. B* 5:L254–57
87. Macek, J., Jaecks, D. H. 1971. *Phys. Rev. A* 4:2288–300
88. Blum, K., Kleinpoppen, H. 1979. *Phys. Rep.* 52:203–61
89. Hermann, H. W., Hertel, I. V. 1982. *Comments. At. Mol. Phys.* In press
90. Shurcliff, W. A. 1962. *Polarized Light: Production and Use*. Cambridge: Harvard Univ. Press. 207 pp.
91. Fano, U. 1949. *J. Opt. Soc. Am.* 39:859–63
92. Dill, D. 1976. *J. Chem. Phys.* 65:1130–33
93. Dill, D., Siegel, J., Dehmer, J. L. 1976. *J. Chem. Phys.* 65:3158–60
94. Davenport, J. W. 1976. *Phys. Rev. Lett.* 36:945–49
95. Smith, R. J., Anderson, J., Lapeyre, G. J. 1976. *Phys. Rev. Lett.* 37:1081–84
96. Ritchie, B. 1976. *Phys. Rev. A* 13:1411–15
97. Ritchie, B. 1976. *Phys. Rev. A* 14:359–62
98. Ritchie, B. 1976. *Phys. Rev. A* 14:1396–1401
99. Zare, R. N. 1971. *Acc. Chem. Res.* 4:361–67
100. Macek, J., Burns, D. 1976. *Topics in Current Physics*. Vol. 1: *Beam Foil Spectroscopy*, ed. S. Bashkin, pp. 237–64. New York: Springer-Verlag
101. Andrä, H. J. 1974. *Phys. Scr.* 9:257–80
102. Hafner, H., Kleinpoppen, H., Krüger, H. 1965. *Phys. Lett.* 18:270–71
103. Fano, U. 1957. *Nuovo Cimento* 5:1358–60
104. Fano, U. 1964. *Phys. Rev.* 135:B863–64
105. Alder, K., Winther, A. 1962. *Nucl. Phys.* 37:194–200
106. Greene, C. H. 1980. *Phys. Rev. Lett.* 44:869–71
107. Fano, U., Greene, C. H. 1980. *Phys. Rev. A* 22:1760–63
108. Zare, R. N., Herschbach, D. R. 1963. *Proc. IEEE* 51:178–82

ELECTRONIC STRUCTURE CALCULATIONS USING THE Xα METHOD

David A. Case

Department of Chemistry, University of California, Davis, California 95616

INTRODUCTION

During the past decade, the Xα method has become increasingly popular as an approximate approach to determining molecular electronic structures, especially in the areas of inorganic chemistry and solid state and surface physics. It is a descendent of the Thomas-Fermi-Dirac theory, which postulates an exchange energy proportional to the cube root of the electron density (1). In 1951, Slater (2) proposed using such a term as an effective potential, thus reducing the many-electron problem to a series of one-electron differential equations (in atomic units):

$$\left[-1/2\nabla^2 + V_c + V_{X\alpha\uparrow}\right]\phi_{i\uparrow} = \varepsilon_{i\uparrow}\phi_{i\uparrow}. \qquad 1.$$

Here $\phi_{i\uparrow}$ is an orbital for an electron with $M_S = 1/2(\uparrow)$, and $\varepsilon_{i\uparrow}$ is its one-electron energy. V_c is the classical Coulomb potential (including electron self-interaction terms), and $V_{X\alpha\uparrow}$ is an approximate potential representing the effects of electron exchange. In Slater's model, this is related to $\rho_\uparrow$, the local density of electrons of the same spin:

$$V_{X\alpha\uparrow} = -3\alpha(3\rho_\uparrow/4\pi)^{1/3}. \qquad 2.$$

Here α is an adjustable parameter, which can be determined in a variety of ways (see below).

Eqs. 1 and 2 define what is usually termed the Xα method. It has much in common with density functional methods, which invoke an exchange term like that in Eq. 2, but which also express the kinetic energy as a functional of the electron density (3–5). Here I shall consider only calculations in which the kinetic energy is represented as a differential operator. Density functional theory has also inspired the development of more elaborate potentials that reflect the effects of electron correlation as well as exchange (6–9). In the cases in which these potentials are of a local form, the same computational methods may be used as for the $X\alpha$ potential. For this reason, molecular calculations incorporating these potentials will be reviewed here, even though, strictly speaking, they are not $X\alpha$ calculations.

When Keith Johnson reviewed the $X\alpha$ method for this series seven years ago (10), it was feasible to attempt a comprehensive review of all molecular calculations. This is no longer the case, and I restrict my coverage in a number of ways, emphasizing "practical" aspects of the method as it applies to inorganic chemistry. I make no attempt to review recent advances in density functional theory, and I discuss atomic results only to the extent that they shed light on current approaches to molecular problems. In surveying applications, I attempt to place in perspective the use of effective exchange potentials in relation to other approaches to the many-electron problem. Because of space limitations and my background, I do not discuss applications to problems in solid state and surface physics; these have been reviewed elsewhere (8, 11–13).

Earlier reviews of the $X\alpha$ method have appeared, mostly emphasizing applications using the multiple scattering approximations. Those with the most chemical emphasis include reviews by Johnson (10, 14), Connolly (15), Weinberger & Schwarz (16), and Rösch (17). Density functional theory has also received a number of recent reviews (3–5, 18). The historical development of the $X\alpha$ method has been outlined by Slater (19) and by Connolly (15).

LOCAL EXCHANGE AND CORRELATION POTENTIALS

Slater's statistical approximation to the exchange potential, Eq. 2, has been the one most widely adopted for practical calculations. The classic survey of atoms by Herman & Skillman (20) used this form with $\alpha = 1$, as originally suggested by Slater. Most recent calculations use a smaller value of α, between 0.67 and 0.80. The reasons for this change have been discussed by Slater (18, 19). "Optimal" values for the ground states of atoms may be determined by requiring the $X\alpha$ statistical energy to be the same as the Hartree-Fock total energy (21, 22). These are most commonly used in multiple scattering calculations. Since there is an ambiguity in any heteronuclear molecule about which α factor should be used, many calculations adopt an average value, usually 0.7, for all atoms. This has been a common procedure in applications that use linear combination of atomic orbitals (LCAO) approximations. In related work, Gopinathan et al (23) have used a parametrization of the Fermi hole that yields values dependent only upon the total number of spin-up and spin-down electrons. Some studies have adopted empirical adjustments of the αs to fit certain molecular properties (24), while others have advocated using calculated charges to interpolate between the optimal atomic and ionic values, which may be quite different for light atoms (25).

Considerable effort has been extended to determine potentials that have a more firm theoretical foundation, and which incorporate the effects of electron correlation as well as exchange. The most widely used approach has been the local density approximation, in which the contribution to the exchange and correlation energy at each point is assumed to be the same as that in a homogeneous electron gas of the same density. For the non-spin-polarized ("paramagnetic") case, the parametrization of Hedin & Lundqvist (6) has been widely adopted. Extensions to the "ferromagnetic" case in which spin polarization is included have been made by several groups (7–9). The "local spin density" (LSD) parametrization of Gunnarsson & Lundqvist (7) has been the most popular for molecular calculations.

For many applications these refinements may be unimportant, since the approximations involved in solving the one-electron equations are often greater than the errors in the potentials themselves. The Hedin-Lundqvist (6) function has the form of the $X\alpha$ potential, but with a density-dependent α that resembles that of the "optimal" α factors, and it has been found that ground state potential energy curves are only slightly dependent upon the potential used (26, 27). Spin-dependent properties may show larger differences. The LSD potential, for example, has a smaller splitting between spin-up and spin-down potentials than does the $X\alpha$ potential, especially for large amounts of spin polarization (7). The practical consequences of this difference for calculations of spin-dependent molecular properties have not yet been made clear.

Local exchange and correlation potentials have also been proposed for relativistic calculations (28–31). These exhibit important differences from their nonrelativistic counterparts in regions of high electron density, but such effects appear to be relatively unimportant for the valence orbitals in atoms (30). All of the molecular calculations reviewed here have used nonrelativistic local potentials.

Atomic calculations are very useful in evaluating exchange and correlation potentials since the one-electron equations may be solved to high precision by numerical integration procedures. One test of the quality of the resulting orbitals is to determine the energy of a single configuration constructed from them. Orbitals obtained from the $X\alpha$ potential turn out to be as good as or better than Hartree-Fock orbitals using a double-zeta basis set (32). For example, results for iron (33) are superior to all but the largest of the Gaussian orbital expansions proposed by Hay (34). This indicates the advantages of the numerical integration technique for transition metal atoms, where radial flexibility is very important.

Other atomic calculations have focused on the ability of local density functionals to determine excited states and multiplet splittings. Ziegler, Rauk & Baerends (35) have argued that these statistical approximations should only be applied to single determinants. In many cases the diagonal sum rule (36) may be used to calculate multiplet splittings from these (35, 37–39). Von Barth (37) and Wood (38) show that excitation energies using the LSD potential closely approximate experimental results, whereas results with the $X\alpha$ potential (using $\alpha = 2/3$) agree well with Hartree-Fock. Hence it appears that electron correlation effects are present to some extent in the LSD potential.

Larger systematic deviations from experiment are found for transition metals, in general favoring configurations with larger numbers of $3d$ or $4d$ electrons. As an example, the stability of the $3d^{n-1}4s^1$ configuration in the iron series atoms, compared with the $3d^{n-2}4s^2$ configuration, is overestimated by about 1 eV in both the LSD and $X\alpha$ potentials (33, 39–41). Although the overall results are considerably better than Hartree-Fock (39), the remaining errors are large enough to make calculations of charge transfer energies on molecules containing transition metal ions suspect.

MULTIPLE SCATTERING THEORY

One popular method of solving Eq. 1 approximately is to use multiple scattering, or scattered wave (SW), theory. In this approach, the potential is assumed to be spherically symmetric inside a cell surrounding each atom, and outside an "outer sphere" that surrounds the entire molecule. In the remaining, "intersphere," region a constant potential is assumed. With these assumptions, the wavefunction may be determined essentially exactly, subject only to errors in the radial numerical integrations inside the atomic spheres and to truncation of angular momentum expansions. In practice, it is usually feasible to include enough mesh points and spherical harmonic expansion functions to obtain convergence, even for sizeable molecules. Detailed expositions of this theory have been presented in several places (14, 16, 17, 42–44).

Overlapping Spheres

A useful survey of proposed modifications to the simple multiple scattering theory is given by Herman (45). Most appear to require a substantial increase in computation time. The only extension that has been widely adopted is the use of atomic cells that are allowed to overlap. These were first introduced in an empirical manner (46, 47), showing that considerable improvements in orbital energies are obtained when larger sphere radii are used. A derivation can be formulated in terms of projection operators (48), but this does little to explain why the procedure seems to work. The justification usually cited is that allowing the spheres to overlap removes charge from the intersphere region, where the assumption of a constant potential is particularly unrealistic.

Two principal conceptual problems arise when overlap is allowed. First, in the overlap regions the wavefunction is not uniquely defined. This does not appear to be a major problem for modest amounts of overlap, since tests show that the wavefunctions are sensibly the same no matter which atomic cell they are calculated from (45, 48, 49). Second, the charge in these overlap regions is effectively "double-counted," i.e. is assigned to more than one atomic center in the usual normalization process. It is even possible for the calculated charge inside the atomic spheres to integrate to more than unity, so that the intersphere charge appears to be negative. Some investigators make no corrections for such "problems," while others renormalize negative intersphere charges to zero. A simple double-counting correction proposed by Herman(45) has also been widely adopted. More elaborate truncated sphere models have been proposed (49, 50), but have not yet been tested numerically.

Whether or not the atomic spheres are allowed to overlap, the difficult task of choosing sphere radii must be faced. The most widely adopted and successful procedure has been that suggested by Norman (51), which begins from a superposition of atomic charge densities in the molecular geometry. Relative sphere sizes are determined by finding the "atomic number radii" that enclose an electron charge equal to the nuclear charge. Absolute radii are then chosen by scaling these values uniformly to satisfy the virial theorem. This procedure makes sphere sizes dependent upon the molecular environment in a chemically intuitive way, and has been shown to yield good results for ionization energies and molecular properties in several studies in which (limited) variations in sphere radii were considered (33, 51–54). Because of its dependence upon the virial ratio, Norman's method is best suited to molecules near their equilibrium geometries.

Other schemes have been proposed as well. Many early calculations chose sphere radii ratios based on empirical atomic or ionic radii. More recently, Bloor & Sherrod (55) have used radii that enclose a fixed number of electrons (slightly less than the nuclear charge) in the isolated atoms. For studies of potential energy curves, a method based on matching potentials at the sphere radii and minimizing the exchange term appears to give good results (25, 56).

In many cases, no great differences are encountered for fairly wide variations in sphere radii (45, 48, 54–55, 57). Changes in sphere radii often move orbital energies uniformly up or down (58), which may be of little consequence unless absolute ionization energies are desired. It is prudent to treat the results of multiple scattering calculations with appropriate caution. Nevertheless, the wide range of chemically useful results that have been obtained (outlined below) testifies to the general validity of the procedures that have been used to choose sphere radii.

Rydberg and Continuum States

One of the most attractive features of multiple scattering theory is its numerical flexibility in the outer sphere region, which allows diffuse orbitals and continuum states to be described in a particularly convenient fashion. Through the matching procedure at the outer sphere boundary, a single-center, partial wave expansion at large distances is merged smoothly to a multicenter description at short distances. This allows a simple description of many processes that can be handled only with difficulty using basis set expansions.

Although Rydberg states in molecules are defined by a variety of criteria, this category generally includes diffuse states with density maxima at distances greater than that typical of valence orbitals. Roberge & Salahub (59) studied 36 singlet and triplet excited states of H_2S, yielding excellent agreement both with experiment and with large-scale configuration interaction (CI) calculations; the average absolute energy difference between the $X\alpha$ and CI results was only 0.24 eV. The majority of these states are clearly Rydberg in character, having more than 90% of their charge in the outer sphere region. Other multiple scattering studies of Rydberg states include water (60) and the radical anions of some simple organic molecules (61).

The modifications required to incorporate scattering boundary conditions into the multiple scattering formalism have been given by Dill & Dehmer (62, 63). These unbound orbitals can then be used as final states in calculations of photoionization intensities. Most calculations have concentrated on diatomic and triatomic molecules (63–69), including calculations of partial photoionization cross sections from particular orbitals, fixed-molecule angular distributions, and asymmetry parameters for freely rotating molecules. In general the results are significantly better than previous approaches using plane-wave approximations for the final states. Extensions of this approach to the study of K-shell x-ray absorption spectra of transition metal complexes have also been reported (70, 71).

These continuum orbitals can also be used to study problems in electron-molecule scattering (63, 72–77). For reasons of space, I shall not review this material here; two other recent reviews exist for the interested reader (63, 72).

Molecular Properties

One measure of the quality of wavefunctions determined by the multiple scattering method is the determination of molecular properties from them. The straightforward evaluation of expectation values is complicated by the irregular shape of the intersphere region, which makes numerical integration difficult. To date such calculations have been limited to diatomic and triatomic molecules (78–83). These have achieved good results for the total energy and for some one-electron properties, while other properties are of more limited accuracy. A vexing problem arises when overlapping spheres are allowed, since the wavefunction and normalization procedures are not uniquely defined, as is mentioned above.

Case & Karplus (84) have proposed an approximate procedure to circumvent these difficulties and to allow calculations to be made on large molecules. This consists of partitioning the intersphere charge among the atoms and representing it by extending the atomic radial functions beyond the sphere radii. Comparisons to numerical integration results for LiH show that the errors introduced by this procedure are no greater than those of the multiple scattering approximations themselves (53). The quality of results for small molecules (53, 54, 84) is intermediate between that of minimum basis and double-zeta Hartree-Fock calculations. Certain systematic deficiencies have been identified: in general the charge distributions are too compact (i.e. expectation values of $\langle r^2 \rangle$ are too small), and field gradients at hydrogen atoms are often not well determined. Nevertheless, the results, even for large molecules, can be of useful accuracy for the interpretation of Mossbauer and magnetic resonance data (33, 85–87). Other applications include calculations of molecular hyperpolarizabilities (88) and (with a slightly different method) carbon-13 NMR chemical shifts (89).

Some general conclusions about the effects of sphere radii may be drawn from these results. Increasing the radius of one atomic cell relative to others in the molecule will draw charge into that cell, as if it had become more electronegative. A uniform scaling of all radii to increase overlap will make the charge distribution more uniform, as if differences in electronegativities among the atoms had been reduced. Hydrogen appears to be the most difficult atom to represent well. There is some evidence that the hydrogen radii predicted by the Norman procedure may be too large; nevertheless, this algorithm still appears to be the best *general* guide to the choice of sphere radii.

Very recently Cook & Karplus (90) have extended these procedures to two-electron operators. As an example, Coulomb and exchange integrals for ozone were calculated in a molecular orbital basis. These were found to be insensitive to the choice of sphere radii and to be in good agreement with Hartree-Fock values. This result opens the way to include configuration interaction starting from the multiple-scattering orbitals, in analogy to the early atomic calculations of Zare (91). Total energies may also be determined in this way, although at present the calculations are too time-consuming to be practical.

An alternative approach to calculating molecular properties from multiple scattering wavefunctions is to project the numerical orbitals onto a basis set (92), at which point conventional programs may be used. This procedure can also provide a population analysis for the scattered wave orbitals.

Another measure of wavefunction quality may be obtained by comparing electron density maps to those determined by x-ray, neutron and electron diffraction. Multiple scattering deformation densities have been calculated for cyclic octasulfur (93, 94) and for eight first and second row homonuclear diatomics (95; B. N. McMaster, V. H. Smith, Jr., and D. R. Salahub, manuscript submitted). These demonstrate that provided enough partial waves are used the multiple scattering results compare well with large basis set calculations and with experiment. Overlapping spheres give uniformly better results than touching spheres, primarily by increasing the electron density near the midpoint of bonding σ orbitals, with a consequent reduction of density behind the atoms. This is achieved through a change in hybridization, with *p*-character increasing relative to *s*-character when sphere overlap is included. These deformation densities provide further evidence that in spite of the muffin-tin approximations to the potential the orbitals obtained in a scattered wave calculation are surprisingly accurate.

BASIS SET EXPANSION METHODS

The Discrete Variation Method

Multiple scattering theory has its origins in solid state physics and in a sense is quite foreign to quantum chemistry, which typically uses basis set expansion methods to obtain approximate solutions to wave equations. In 1973, Baerends, Ellis & Ros (96, 97) proposed the use of a numerical integration procedure for the Hamiltonian matrix elements, coupled with a fitting of the Coulomb potential to a series of subsidiary functions. This enables calculations to be done rapidly even for fairly large basis sets, since the two-electron integrals required for ab initio theories are not needed. This procedure is called the discrete variational method (DVM) or the Hartree-Fock-Slater, HFS-LCAO method.

Two general methods of choosing basis sets have evolved from this original work. One approach uses numerical basis functions calculated for an atom embedded in a potential well (98), which may be chosen by a "self-consistent charge" procedure to reflect the calculated molecular charges (99). This procedure utilizes the numerical freedom of the discrete variational method, but yields a basis whose size is that of a minimum basis set, or slightly larger.

A second approach has been to adopt a large Slater-type orbital basis, of double- or triple-zeta quality, including polarization functions where necessary. A careful study of basis set effects for CO and its transition metal complexes has been reported (27, 100), showing that the Xα results are as close to experiment as Hartree-Fock results using the same basis set. An important advantage of the HFS-LCAO method over the usual semiempirical approaches [such as extended Huckel or Intermediate Neglect of Differential Overlap (INDO)], lies in its ability to study basis set effects even for fairly large molecules. A pseudopotential scheme has been developed (101), but in some cases there can be substantial differences between these and the more usual "frozen core" results (102).

One difficulty with the HFS-LCAO method has been that the number of sample points required to evaluate the total energy is much greater than that needed to determine the orbitals and one-electron energies. The incorporation of analytical integration techniques wherever possible has improved the situation substantially, allowing the calculation of potential energy curves for small molecules (27). An alternative approach to calculating bond dissociation energies is the transition state method of Ziegler & Rauk (103). They expand the total energy in a Taylor series, about a density half-way between the true molecular density and a reference density created from a superposition of fragments. As with the original Slater transition state, the necessary derivatives for the Taylor expansion are easily calculated. The final expression relates the binding energy of two fragments to changes in the bond order matrix, and to integrals over the reference, transition state, and final molecular charge densities. This scheme may also be used to decompose the interaction energy into "steric" and "electronic" components (104, 105).

Other Basis Set Approaches

Sambe & Felton (106) were the first to use Gaussian orbitals and analytical integrations within the $X\alpha$ method. Dunlap et al (107, 108) have extended their fitting procedures to allow accurate evaluations of the total energies. Similar calculations in which the Coulomb potential is evaluated exactly have also been reported (109). Sambe & Felton themselves have recently proposed several modifications to this scheme in an attempt to develop an inexpensive method for transition metal clusters (110, 111).

Following the work of Anderson & Wooley (112), Gunnarsson, Harris & Jones have developed the "linear muffin-tin orbital" (LMTO) method (26, 41, 113, 114). This is something of a cross between multiple scattering and basis set methods. The orbitals are chosen from a muffin-tin procedure, and the Hamiltonian is diagonalized in this basis. This effectively linearizes the energy dependence of multiple scattering theory, so that all of the orbital energies may be determined from a single matrix diagonalization. Since this basis set is not simply related to those commonly used by quantum chemists, it is difficult to judge its quality. Results for small molecules suggest that it is approximately equivalent to a double-zeta Slater basis (27).

RELATIVISTIC CALCULATIONS

In view of the promise the $X\alpha$ method holds for practical calculations on molecules containing heavy atoms, it is not surprising that several approaches to include relativistic effects have been tried. I discuss here recent progress in this area. Earlier work is described in a review by Pyykkö (115).

Multiple Scattering Theory

Within multiple scattering theory, two approaches have been used. The first is a "quasirelativistic" theory in which the radial functions inside each atomic sphere satisfy an average Dirac equation that includes the Darwin and mass-velocity corrections, but not the spin-orbit term (116–118). Such calculations are no more difficult or expensive than nonrelativistic multiple scattering calculations, yet they include important direct and indirect level shifts (118–123). A second step may follow in which the spin-orbit operator is diagonalized in the space of the valence orbitals (119, 120). In this way complex wavefunctions that transform according to the molecular double point group are obtained.

An alternative procedure is to use the full four-component Dirac formalism throughout (124–126). This reduces exactly (both in theory and in practice) to the usual nonrelativistic multiple scattering theory in the limit $c \to \infty$. The self-consistent version of this procedure has been coded by Case & Yang (127, 128), and results for UF_6 have been compared in detail to those of other computational methods. For valence levels this method appears to be practically equivalent to the quasirelativistic scheme. The additional complexity of the Dirac theory makes the calculations several times more expensive than quasirelativistic ones. In exchange for this, one obtains an explicit representation of the small components of the wavefunction, which may be of importance in understanding magnetic properties and in making connections to relativistic crystal field theories (129, 130). Both of these methods may be expected to have an accuracy comparable to that found in nonrelativistic scattered wave calculations.

Basis Set Expansions

Computations employing the Dirac equation must take care to avoid contamination from states of negative energy, which satisfy the differential equation but are not of physical interest for chemical problems. As with numerical atomic calculations, the multiple scattering theory achieves this goal through the imposition of the proper boundary conditions. This is not so easy to do with basis set expansions, and other special procedures are generally necessary (131). Two ways to avoid this problem have been used in molecular $X\alpha$ calculations. Rosén & Ellis (132) proposed using a linear combination of numerical atomic spinors, each of which satisfies the Dirac equation for an atomic reference problem. In conjunction with the discrete variational method of calculating matrix elements, this procedure provides a straightforward way of obtaining simple molecular orbital descriptions incorporating relativistic effects (115, 132–136).

An alternative approach is to use Foldy-Wouthuysen transformations to derive an effective equation for the large components of the positive energy solution. This yields corrections to the nonrelativistic Hamiltonian that may be used as perturbations. Snijders, Baerends & Ros (137, 138) have applied this scheme, using the $X\alpha$ potential and numerical integration techniques. Since relativistic effects are large at high densities, a frozen core approximation is used in which relativistic atomic cores are assumed to be unchanged upon molecule formation. The relativistic corrections to the charge density also change the effective Coulomb and exchange potentials. These contributions are sometimes referred to as indirect relativistic corrections, and are determined by a self-consistent perturbation scheme. This total approach, when implemented with a flexible basis set, appears to give a very good interpretation of molecular ionization potentials (138, 139). When combined with the transition state method of calculating binding energies (140, 141), it may be used to analyze relativistic effects on potential energy curves in a way that allows separate evaluation of the various relativistic corrections involved.

To date, both the multiple scattering and basis set methods have concentrated on molecules that can be reasonably represented by a single configuration. Many heavy molecules fall into an "intermediate coupling" regime in which this is no longer true (142), and multiconfiguration methods will be required. Whether these can be fit into the framework of $X\alpha$ theory remains to be seen.

SELECTED APPLICATIONS

Ozone

Ozone has been an interesting case for study, since the Hartree-Fock method predicts a triplet ground state and places the true singlet ground state more than 2 eV higher in energy. Messmer & Salahub (143) showed that $X\alpha$ scattered wave calculations indeed predict the correct ground state. They went on to calculate vertical excitation energies to 15 singlet and triplet excited states, in general achieving results that compared favorably to ab initio CI calculations. Cook & Karplus (90, 144) have shown that some of the remaining error arises from the muffin-tin approximations, which leave out angular interactions between orbitals that are important in calculations of singlet-triplet splittings; these can be retained by performing numerical integrations over the scattered wave orbitals (90, 121), or by use of the basis set expansion methods (35). The latter procedure has recently been applied to ozone (145), showing that a principal difference between $X\alpha$ and Hartree-Fock calculations is that the former gives a larger exchange contribution for the $1a_2(\pi)$ orbital. This increase tends to make the exchange contributions from the valence orbitals more nearly equal and is in the direction required to achieve the correct ordering of states and ionization potentials.

Similar improvements over Hartree-Fock theory may be expected in other molecules with significant biradical character; it has been recognized for many years that the $X\alpha$ method generally allows molecules to dissociate correctly into radical atoms or fragments. One approach to understanding this feature is to recognize that the $X\alpha$ energy expression depends only upon the one-particle density, which molecular orbital methods may treat fairly well, even in cases in which the Hartree-Fock two-particle density is significantly in error. For example, the Hartree-Fock energy expression is quite sensitive to the presence of ionic terms in molecular wavefunctions, while density functional methods respond to ionic character only if it is present in the total charge density. The correlation present in Xα calculations on excited states of ozone has been analyzed in these terms by Cook (144). An important caveat follows, however: Since the orbitals from the Xα method are very similar to those of Hartree-Fock theory (32, 33, 90, 145), the single determinant wavefunctions built from them may have substantial errors, even when their statistical energies are qualitatively correct. Furthermore, prediction of geometry is still a source of difficulty: HFS-LCAO calculations on ozone, for example, show a D_{3h} conformation to be lower in energy than the open structure (146).

One approach to obtaining improved wavefunctions within local density methods is to relax the spin and space restrictions on the orbitals, yielding "broken symmetry" solutions to Eq. 1. This technique has been used for some time in ab initio calculations for core ionizations or excitations from equivalent lone pairs. Similar applications involving the Xα method have also appeared (147–149). The same general techniques may be applied to transition metal systems in cases in which there is an instability in the restricted solution (150–153). Because the resulting orbitals are similar to those of the generalized valence bond method, Noodleman & Norman (150) have called this the "Xα valence bond method," although no explicit superposition of configurations is involved. Rather, the energies of space and spin eigenstates are recovered through approximate projection techniques (150, 151). This promises to be an important approach to calculations on multinuclear transition metal complexes.

Transition Metal Carbonyls

This important class of molecules has received extensive study by Xα methods, from both the multiple scattering and basis set approaches. The neutral $Cr(CO)_6$ and $Ni(CO)_4$ species have received close attention and have been the subject of some controversy.

For $Cr(CO)_6$ we may compare results from multiple scattering calculations (154) with those from large basis set expansions (27, 100, 155). Two principal differences appear in the one-electron energies.

1. Two of the levels which derive from the 5σ orbital of CO are about 2 eV lower in the SW calculation than in the LCAO one, relative to the levels derived from the CO 1π orbitals.
2. The basis set results predict a 1.5 eV splitting of the orbitals derived from the 4σ orbital of CO, while the SW results show very little splitting.

These differences may arise in part from the different α values used in the SW and LCAO calculations, as well as from differences in the errors introduced by the muffin-tin and finite basis set approximations. In the absence of firm assignments of the excitation and photoionization spectra, it is difficult to decide which set of results is the more accurate. For most purposes the two are equivalent; the ionization potentials of the SW calculations, for example, are much closer to the extended basis set results than to the single-zeta results.

On the basis of their scattered wave calculations, Johnson & Klemperer (154) argued that the metal-carbon bonding interactions in $Cr(CO)_6$ are predominantly σ in character with only a small π contribution. They found both bonding and antibonding π interactions, allowing back donation to occur with little net bonding interaction. Bursten et al (156) projected the scattered wave results onto a basis set in order to analyze the bonding characteristics in a more conventional language. The numbers obtained in this way are close to those from the HFS-LCAO calculations (155) and give greater importance to the π component of the bond. Nevertheless, the differences in interpretation among the various groups appear to rest more on the importance given to distinguishing back-bonding from back-donation than on intrinsic differences in the computational approaches.

A similar situation exists for $Ni(CO)_4$, for which an early scattered wave calculation was cited as evidence for lack of back-bonding to CO (157). The opposite conclusion may be reached from the same calculation if charge densities are analyzed in a way that emphasizes the orbital behavior inside the atomic spheres (158–160). Calculations using basis set expansions (104, 155) or overlapping spheres (156, 161) also support the existence of important backbonding in this complex. It is thus important to realize that the charge distributions obtained from multiple scattering calculations will not necessarily be in agreement with Mulliken populations or with traditional pictures. Aids in the interpretation of these distributions may come by projection onto basis sets (92), by partitioning the intersphere charge among the atoms (84), or through analysis of the atomic wavefunctions (158, 159) or potentials (85, 87). No single scheme will be appropriate in all cases (and none should be expected to eliminate arguments among inorganic chemists!).

Molecules with Metal-Metal Bonds

The study of dinuclear transition metal species with metal-metal bonds formed from the d-orbitals has accelerated since the first characterizations nearly 20 years ago (162). A variety of Xα-SW calculations have been reported, beginning with the study of $Mo_2Cl_8^{2-}$ by Norman &

Kolari (163). For molecules such as this, with a formal quadruple metal bond, the multiple scattering calculations (163–169) are generally in agreement with the level orderings predicted from qualitative molecular orbital theory, but may be quite different from the results of single configuration Hartree-Fock theory (170, 171), expecially for the Cr dimers. Inclusion of configuration interaction yields results more in line with the $X\alpha$ calculations (170–172). The predictions of the multiple scattering theory have been substantiated by subsequent spectroscopic studies, especially using polarization information from single crystals (173–176). This usefulness as an aid in assignments is impressive, but it does not mean that all excitation energies are in quantitative agreement with experiment. A particular problem arises in the prediction of $\delta \to \delta^*$ transitions in the quadruply bonded species (173). Noodleman & Norman (150) have shown that substantial improvement is achieved with a broken symmetry calculation that allows left-right correlation to be described within the $X\alpha$ model. The importance of such effects has also been noted in ab initio calculations (177–179).

The assignment of the photoionization spectrum of these quadruply bonded species has been a subject of some controversy, with the $X\alpha$ results suggesting a higher ionization potential for the σ component of the metal-metal bond than is found in the ab initio calculations (163, 164, 168–171). [A similar difference is seen in comparisons of $X\alpha$ (180, 181) and Hartree-Fock (182) results for Rh(II) dimers.] While the interpretation of the photoelectron spectra of these complexes is not unambiguous (183), studies of homologous series definitely favor the assignments suggested by the $X\alpha$ calculations (168).

For Tc, Ru, and Rh, stable odd-electron dimers can be prepared that formally have a mixed valence character. For both the Ru and Rh species, the spin properties predicted by $X\alpha$-SW calculations have been shown to be in good agreement with conclusions from ESR spectra (181, 184). The predicted electronic transitions have also been confirmed by polarized single crystal spectroscopy (185–188). These results, in combination with those for the quadruply bonded dimers, make a convincing case for the usefulness (if not the absolute accuracy) of $X\alpha$-SW calculations for complex inorganic species.

Transition Metal Halides and Oxohalides

Transition metal halides and oxohalides offer several examples of comparisons of $X\alpha$ calculations to more traditional methods of quantum chemistry. Lee et al (189) have compared ab initio CI and $X\alpha$-SW calculations for the dihalides from Cr to Ni. Neither Koopman's theorem nor ΔSCF ab initio methods are reliable for metal ionizations; the scattered wave calculations (189, 190), although not without errors, lead to assignments closest to those of the CI calculations. Incorporation of spin polarization effects has also been shown to improve the $X\alpha$ results (191). Scattered wave calculations (without sphere overlap) have been compared to ab initio SCF and intermediate neglect of differential overlap (INDO) results for copper halides (192, 193). The ab initio and $X\alpha$-SW methods appear to have about the same level of error, while the (much faster) INDO method is impressive in its ability to rationalize optical spectra.

Similar comparisons have been made for transition metal oxohalides (194), in which $X\alpha$-SW calculations are shown to be in good agreement with experiment and with ab initio CI calculations; again Koopman's theorem and ΔSCF calculations give an incorrect ordering of ionization energies. The scattered wave model also has been impressive in correlating electron spin resonance data for this class of compounds (195, 196), in line with a general ability of such calculations to predict spin distributions in transition metal complexes [see for example (33, 85, 86, 181, 197)]. The understanding of the optical spectra of these complexes is still uncertain, however, with $X\alpha$-SW results leading to assignments different from those obtained from ab initio calculations (198).

As increasingly detailed spectroscopic examinations are made of transition metal complexes, it will become more important to be able to estimate term splittings arising from interelectronic repulsions. Two recent calculations on the tetrahedral chlorides of V and Cr illustrate results based on fits to ligand field theories (199) or on direct calculation of two electron integrals (144).

CONCLUDING REMARKS

The applications of the $X\alpha$ method cited above represent only a small fraction of those published in the last five years. While it is clear that this approach is generally sufficiently accurate to be useful to inorganic chemists, it may still be too early to make confident generalizations about its absolute accuracy. This is due in part to the diversity of structures of interest to the inorganic chemist, and in part to the lack of reliable ab initio calculations with which to make secure comparisons. For a fairly wide variety of problems, the $X\alpha$ results appear to be distinctly superior to those obtained from Hartree-Fock theory, but no clear understanding of the origins and limitations of this generalization has yet emerged. Both the exchange averaging implicit in a local density functional approach (143, 144) and the explicit averaging of the muffin-tin approximations

(144, 200) have been suggested as sources of this behavior. Fundamentally, however, the Xα method is a single configuration approach, one that must inevitably break down in cases in which qualitatively different configurations are strongly mixed together, or in which two states of the same symmetry are in close proximity. As more careful comparisons to good quality ab initio wavefunctions are made, the limits of applicability of the Xα method should become clearer.

One such limit may arise in applications to transition metal complexes. It was noted above that Xα calculations on these atoms display a marked bias in favor of configurations containing extra d-electrons. A similar trend has been seen in a number of molecular calculations, with ligands such as halides (201–203), cyanide (33), porphyrins (85–87), and cyclopentadienyl (204), leading to low predictions for the energies of ligand-to-metal charge transfer transitions. It is not yet clear how widespread this type of error might be, but it should be considered as a possible source of error in other calculations of charge transfer transitions.

Most of the calculations reviewed here have been concerned with molecules at their equilibrium geometries. Increasing attention will be paid to structural predictions and correlations. The initial calculations along these lines have been quite promising [see e.g. (26–27, 103–109, 140, 141)] and may represent the first steps toward a practical and quantitative molecular orbital theory for inorganic chemistry.

ACKNOWLEDGMENTS

I owe a great debt to Joe Norman, who helped me assemble the references and provided much useful advice. Mike Cook, Keith Johnson, Martin Karplus, Louis Noodleman, and Cary Yang were also most helpful. While this review was being written, the author's research was supported by a grant from the National Institutes of Health.

Literature Cited

1. Dirac, P. A. M. 1930. *Proc. Cambridge Philos. Soc.* 26:376–85
2. Slater, J. C. 1951. *Phys. Rev.* 81:385–90
3. Parr, R. G., Levy, M. L. 1983. *Ann. Rev. Phys. Chem.* 34: In preparation
4. Rajagopal, A. K. 1980. *Adv. Chem. Phys.* 41:59–193
5. Bamzai, A. S., Deb, B. M. 1981. *Rev. Mod. Phys.* 53:95–126
6. Hedin, L., Lundqvist, B. I. 1971. *J. Phys. C* 4:2064–83
7. Gunnarsson, O., Lundqvist, B. I. 1976. *Phys. Rev. B* 13:4274–98
8. Moruzzi, V. L., Janak, J. F., Williams, A. R. 1978. *Calculated Electronic Properties of Metals.* New York: Pergamon. 188 pp.
9. von Barth, U., Hedin, L. 1972. *J. Phys. C* 5:1629–42
10. Johnson, K. H. 1975. *Ann. Rev. Phys. Chem.* 26:39–57
11. Simonetta, M., Gavezzotti, A. 1980. *Adv. Quantum Chem.* 12:103–58
12. Messmer, R. P. 1981. *Surf. Sci.* 106:225–38
13. Johnson, K. H. 1978. *CRC Crit. Rev. Solid State Mat. Sci.* 7:101–27

14. Johnson, K. H. 1973. *Adv. Quantum Chem.* 7:143–85
15. Connolly, J. W. D. 1977. *Semiempirical Methods of Electronic Structure Calculation, Part A: Techniques,* ed. G. A. Segal, pp. 105–32. New York: Plenum. 274 pp.
16. Weinberger, P., Schwarz, K. 1975. *International Review of Science, Physical Chemistry, Ser. Two, Vol. 1, Theoretical Chemistry,* ed. A. D. Buckingham, C. A. Coulson, pp. 257–84. London: Butterworth. 396 pp.
17. Rösch, N. 1977. *Electrons in Finite and Infinite Structures,* ed. P. Phariseau, L. Scheire, pp. 1–143. New York: Plenum. 443 pp.
18. Slater, J. C. 1972. *Adv. Quantum Chem.* 6:1
19. Slater, J. C. 1974. *Quantum Theory of Molecules and Solids,* Vol. 4. New York: McGraw-Hill. 573 pp.
20. Herman, F., Skillman, S. 1963. *Atomic Structure Calculations.* Englewood NJ: Prentice-Hall
21. Schwarz, K. 1972. *Phys. Rev. B* 5:2466–68
22. Schwarz, K. 1974. *Theor. Chim. Acta* 34:225–31
23. Gopinathan, M. S., Whitehead, M. A., Bogdanovic, R. 1976. *Phys. Rev. A* 14:1–10
24. McAdon, M. H., Konowalow, D. D. 1979. *J. Chem. Phys.* 71:3089–98
25. Michels, H. H., Hobbs, R. H., Wright, L. A., Connolly, J. W. D. 1978. *Int. J. Quantum Chem.* 13:169–87
26. Gunnarsson, O., Jones, R. O. 1977. *J. Chem. Phys.* 67:3970–79
27. Baerends, E. J., Ros, P. 1978. *Int. J. Quantum Chem. Symp.* 12:169–90
28. Rajagopal, A. K. 1978. *J. Phys. C* 11:L943–48
29. McDonald, A. H., Vosko, S. H. 1979. *J. Phys. C* 12:2977–90
30. Das, M. P., Ramana, M. V., Rajagopal, A. K. 1980. *Phys. Rev. A* 22:9–13
31. Ramana, M. V., Rajagopal, A. K. 1981. *Phys. Rev. A* 24:1689–95
32. Schwarz, K., Connolly, J. W. D. 1971. *J. Chem. Phys.* 55:4710–14
33. Aizman, A., Case, D. A. 1981. *Inorg. Chem.* 20:528–33
34. Hay, P. J. 1977. *J. Chem. Phys.* 66:4377–84
35. Ziegler, T., Rauk, A., Baerends, E. J. 1977. *Theor. Chim. Acta* 43:261–71
36. Slater, J. C. 1960. *Quantum Theory of Atomic Structure.* New York: McGraw-Hill. 502 pp.
37. von Barth, U. 1979. *Phys. Rev. A* 20:1693–1703
38. Wood, J. H. 1980. *J. Phys. B* 13:1–14
39. Gunnarsson, O., Jones, R. O. 1980. *J. Chem. Phys.* 72:5357–62
40. Harris, J., Jones, R. O. 1978. *J. Chem. Phys.* 68:3316–17
41. Harris, J., Jones, R. O. 1979. *J. Chem. Phys.* 70:830–41
42. Williams, A. R. 1974. *Int. J. Quantum Chem. Symp.* 8:89–108
43. Faulkner, J. S. 1979. *Phys. Rev. B* 19:6186–6206
44. Williams, A. R., Morgan, J. van W. 1974. *J. Phys. C* 7:37–60
45. Herman, F. 1977. *Electrons in Finite and Infinite Structures,* ed. P. Phariseau, L. Scheire, pp. 382–410. New York: Plenum. 443 pp.
46. Herman, F., Batra, I. P. 1974. *Phys. Rev. Lett.* 33:94–97
47. Rösch, N., Klemperer, W. G., Johnson, K. H. 1973. *Chem. Phys. Lett.* 23:149–54
48. Herman, F., Williams, A. R., Johnson, K. H. 1974 *J. Chem. Phys.* 61:3508–22
49. Yang, C. Y., Johnson, K. H. 1976. *Int. J. Quantum Chem. Symp.* 10:159–65
50. Chia-Chung, S., Chan, S. 1978. *Sci. Sin.* 21:327–46
51. Norman, J. G. Jr. 1976. *Mol. Phys.* 31:1191–98
52. Weber, J. 1977. *Chem. Phys. Lett.* 45:261–64
53. Cook, M., Karplus, M. 1980. *J. Chem. Phys.* 72:7–19
54. Case, D. A., Cook, M., Karplus, M. 1980. *J. Chem. Phys.* 73:3294–3313
55. Bloor, J. E., Sherrod, R. E. 1980. *J. Am. Chem. Soc.* 102:4333–40
56. Michels, H. H., Hobbs, R. H., Wright, L. A. 1978. *J. Chem. Phys.* 69:5151–61
57. Salahub, D. R., Messmer, R. P., Johnson, K. H. 1976. *Mol. Phys.* 31:529–34
58. Kai, A. T., Larsson, S. 1978. *Int. J. Quantum Chem.* 13:367–74
59. Roberge, R., Salahub, D. R. 1979. *J. Chem. Phys.* 70:1177–86
60. Boring, A. M., Wood, J. H., Moskowitz, J. W., Connolly, J. W. D. 1973. *J. Chem. Phys.* 58:5163–66
61. Bloor, J. E., Paysen, R. A., Sherrod, R. E. 1979. *Chem. Phys. Lett.* 60:476–82

62. Dill, D., Dehmer, J. L. 1974. *J. Chem. Phys.* 61:692–99
63. Dehmer, J. L., Dill, D. 1979. *Electron-Molecule and Photon-Molecule Collisions*, ed. T. Rescigno, V. McKoy, B. Schneider, pp. 225–63. New York: Plenum. 355 pp.
64. Dehmer, J. L., Dill, D. 1976. *J. Chem. Phys.* 65:5327–34
65. Wallace, S., Dill, D., Dehmer, J. L. 1979. *J. Phys. B* 12:L417–20
66. Roche, M., Salahub, D. R., Messmer, R. P. 1980. *J. Electron Spectrosc. Relat. Phenom.* 19:273–84
67. Grimm, F. A., Carlson, T. A., Dress, W. B., Agron, P., Thomson, J. O., Davenport, J. W. 1980. *J. Chem. Phys.* 72:3041–48
68. Carlson, T. A., Krause, M. O., Grimm, F. A., Allen, J. D. Jr., McLaffy, D., Keller, P. R., Taylor, J. W. 1981. *J. Chem. Phys.* 75:3288–92
69. Grimm, F. A. 1980. *Chem. Phys.* 53:71–75
70. Kutzler, F. W., Natoli, C. R., Misemer, D. K., Doniach, S., Hodgson, K. O. 1980. *J. Chem. Phys.* 73:3274–88
71. Kutzler, F. W., Scott, R. A., Berg, J. M., Hodgson, K. O., Doniach, S., Cramer, S. P., Chang, C. H. 1981. *J. Am. Chem. Soc.* 103:6083–88
72. Lane, N. F. 1980. *Rev. Mod. Phys.* 52:29–119
73. Siegel, J., Dehmer, J. L., Dill, D. 1980. *Phys. Rev. A* 21:85–94
74. Dehmer, J. L., Siegel, J., Dill, D. 1978. *J. Chem. Phys.* 69:5205–6
75. Gyemant, I., Varga, Z. S., Benedict, M. G. 1980. *Int. J. Quantum. Chem.* 17:255–63
76. Bloor, J. E., Sherrod, R. E., Grimm, F. A. 1981. *Chem. Phys. Lett.* 78:351–56
77. Loomba, D., Wallace, S., Dill, D., Dehmer, J. L. 1981. *J. Chem. Phys.* 75:4546–52
78. Danese, J. B., Connolly, J. W. D. 1974. *J. Chem. Phys.* 61:3063–70
79. Danese, J. B. 1974. *J. Chem. Phys.* 61:3071–80
80. Danese, J. B. 1977. *Chem. Phys. Lett.* 45:150–54
81. Li, C. H. 1976. *Int. J. Quantum Chem. Symp.* 10:193–202
82. Woodruff, S. B., Wolfsberg, M. 1976. *J. Chem. Phys.* 65:3687–97
83. Woodruff, S. B., Wolfsberg, M. 1978. *Chem. Phys. Lett.* 56:125–29
84. Case, D. A., Karplus, M. 1976. *Chem. Phys. Lett.* 39:33–38
85. Case, D. A., Karplus, M. 1977. *J. Am. Chem. Soc.* 99:6182–94
86. Sontum, S. F., Case, D. A. 1982. *J. Phys. Chem.* 86:1596–1606
87. Case, D. A., Huynh, B. H., Karplus, M. 1979. *J. Am. Chem. Soc.* 101:4433–53
88. Bergman, J. G., Ginsberg, A. P., Maurin, M. 1980. *J. Am. Chem. Soc.* 102:118–22
89. Freier, D. G., Fenske, R. F., Xiao-Zeng, Y. 1982. *J. Chem. Phys.* In press
90. Cook, M., Karplus, M. 1981. *Chem. Phys. Lett.* 84:565–70
91. Zare, R. N. 1966. *J. Chem. Phys.* 45:1966–78
92. Bursten, B. E., Fenske, R. F. 1977. *J. Chem. Phys.* 67:3138–45
93. Salahub, D. R., Foti, A. E., Smith, V. H. Jr. 1977. *J. Am. Chem. Soc.* 99:8067–68
94. Salahub, D. R., Foti, A. E., Smith, V. H. Jr. 1978. *J. Am. Chem. Soc.* 100:7847–59
95. Mrozek, J., Smith, V. H. Jr., Salahub, D. R., Ros, P., Rosendaal, A. 1980. *Mol. Phys.* 41:509–19
96. Baerends, E. J., Ellis, D. E., Ros, P. 1973. *Chem. Phys.* 2:41–51
97. Baerends, E. J., Ros, P. 1973. *Chem. Phys.* 2:52–59
98. Averill, F. W., Ellis, D. E. 1973. *J. Chem. Phys.* 59:6412–18
99. Rosen, A., Ellis, D. E., Adachi, H., Averill, F. W. 1976. *J. Chem. Phys.* 65:3629–34
100. Heijser, W., Baerends, E. J., Ros, P. 1980. *J. Mol. Struct.* 63:109–20
101. Snijders, J. G., Baerends, E. J. 1977. *Mol. Phys.* 33:1651–62
102. Geurts, P. J. M., Gosselink, J. W., Van der Avoird, A., Baerends, E. J., Snijders, J. G. 1980. *Chem. Phys.* 46:133–48
103. Ziegler, T., Rauk, A. 1977. *Theor. Chim. Acta* 46:1–10
104. Ziegler, T., Rauk, A. 1979. *Inorg. Chem.* 18:1755–59
105. Ziegler, T., Rauk, A. 1979. *Inorg. Chem.* 18:1558–65
106. Sambe, H., Felton, R. H. 1975. *J. Chem. Phys.* 62:112–16
107. Dunlap, B. I., Connolly, J. W. D., Sabin, J. R. 1979. *J. Chem. Phys.* 71:3396–3402
108. Dunlap, B. I., Connolly, J. W. D., Sabin, J. R. 1979. *J. Chem. Phys.* 71:4993–99

109. Kitaura, K., Satoko, C., Morokuma, K. 1979. *Chem. Phys. Lett.* 65:206–11
110. Sambe, H. 1981. *Chem. Phys.* 59:315–27
111. Sambe, H., Felton, R. H. 1981. *Chem. Phys.* 59:329–39
112. Anderson, O. K., Wooley, R. G. 1973. *Mol. Phys.* 26:905–27
113. Gunnarsson, O., Harris, J., Jones, R. O. 1977. *Phys. Rev. B* 15:3027–38
114. Jones, R. O. 1979. *J. Chem. Phys.* 71:1300–8
115. Pyykkö, P. 1978. *Adv. Quantum Chem.* 11:353–409
116. Cowan, R. D., Griffin, D. C. 1976. *J. Opt. Soc. Am.* 66:1010
117. Koelling, D. D., Harmon, B. N. 1977. *J. Phys. C* 10:3107–14
118. Wood, J. H., Boring, A. M. 1978. *Phys. Rev. B* 18:2701–11
119. Boring, M., Wood, J. H. 1979. *J. Chem. Phys.* 71:32–41
120. Boring, M., Wood, J. H. 1979. *J. Chem. Phys.* 71:392–99
121. Wood, J. H., Boring, M., Woodruff, S. B. 1981. *J. Chem. Phys.* 74:5225–33
122. Thornton, G., Edelstein, N., Rösch, N., Egdell, R. G., Woodwark, D. R. 1979. *J. Chem. Phys.* 70:5218–21
123. Cotton, F. A. 1980. *J. Mol. Struct.* 59:97–108
124. Yang, C. Y., Rabii, S. 1975. *Phys. Rev. A* 12:362–69
125. Yang, C. Y. 1978. *J. Chem. Phys.* 68:2626–29
126. Cartling, B. G., Whitmore, D. M. 1976. *Int. J. Quantum Chem.* 10:393–412
127. Case, D. A., Yang, C. Y. 1980. *J Chem. Phys.* 72:3443–48
128. Case, D. A., Yang, C. Y. 1980. *Int. J. Quantum Chem.* 18:1091–99
129. Chatterjee, R., Newman, D. J., Taylor, C. D. 1973. *J. Phys. C* 6:706–14
130. Lewis, W. B., Mann, J. B., Liberman, D. A., Cromer, D. T. 1970. *J. Chem. Phys.* 52:809–20
131. Ishikawa, Y., Malli, G. L. 1981. *Chem. Phys. Lett.* 80:111–13
132. Rosén, A., Ellis, D. E. 1975. *J. Chem. Phys.* 62:3039–49
133. Gubanov, V. A., Rosén, A., Ellis, D. E. 1979. *J. Inorg. Nucl. Chem.* 41:975–86
134. Gubanov, V. A., Rosén, A., Ellis, D. E. 1979. *J. Phys. Chem. Solids* 40:17–28
135. Ellis, D. E., Rosén, A. 1977. *Z. Phys. A* 283:3–10
136. Adachi, H., Rosén, A., Ellis, D. E. 1977. *Mol. Phys.* 33:199–205
137. Snijders, J. G., Baerends, E. J. 1978. *Mol. Phys.* 36:1789–1804
138. Snijders, J. G., Baerends, E. J., Ros, P. 1979. *Mol. Phys.* 38:1909–29
139. Jonkers, J. G., DeLange, C. A., Snijders, J. G. 1980. *Chem. Phys.* 50:11–20
140. Ziegler, T., Snijders, J. G., Baerends, E. J. 1981. *J. Chem. Phys.* 74:1271–84
141. Pyykkö, P., Snijders, J. G., Baerends, E. J. 1981. *Chem. Phys. Lett.* 83:432–37
142. Lee, Y. S., Ermler, W. C., Pitzer, K. S. 1980. *J. Chem. Phys.* 73:360–66
143. Messmer, R. P., Salahub, D. R. 1976. *J. Chem. Phys.* 65:779–84
144. Cook, M. R. 1981. PhD thesis. Harvard Univ., Cambridge, Mass.
145. Salahub, D. R., Lamson, S. H., Messmer, R. P. 1982. *Chem. Phys. Lett.* In press
146. Laidlaw, W. G., Trisc, M. 1979. *Chem. Phys.* 36:323–25
147. Connolly, J. W. D., Siegbahn, H., Gelius, U., Nordling, C. 1973. *J. Chem. Phys.* 58:4265–77
148. Banna, M. S., Frost, D. C., McDowell, C. A., Noodleman, L., Wallbank, B. 1977. *Chem. Phys. Lett.* 49:213–17
149. Noodleman, L., Post, D., Baerends, E. J. 1982. *Chem. Phys.* In press
150. Noodleman, L., Norman, J. G. Jr. 1979. *J. Chem. Phys.* 70:4903–6
151. Noodleman, L. 1981. *J. Chem. Phys.* 74:5737–43
152. Norman, J. G. Jr., Ryan, P. B., Noodleman, L. 1980. *J. Am. Chem. Soc.* 102:4279–82
153. Aizman, A., Case, D. A. 1982. *J. Am. Chem. Soc.* In press
154. Johnson, J. B., Klemperer, W. G. 1977. *J. Am. Chem. Soc.* 99:7132–37
155. Baerends, E. J., Ros, P. 1975. *Mol. Phys.* 30:1735–47
156. Bursten, B. E., Freier, D. G., Fenske, R. F. 1980. *Inorg. Chem.* 19:1810–11
157. Johnson, K. H., Wahlgren, U. 1972. *Int. J. Quantum Chem. Symp.* 6:243–55
158. Larsson, S. 1978. *Theor. Chim. Acta* 49:45–53
159. Larsson, S., Braga, M. 1979. *Int. J. Quantum Chem.* 15:1–5
160. Braga, M., Larsson, S., Leite, J. R. 1979. *J. Am. Chem. Soc.* 101:3867–73
161. McIntosh, D. F., Ozin, G. A., Messmer, R. P. 1981. *Inorg. Chem.* 20:3640–50

162. Cotton, F. A. 1978. *Acc. Chem. Res.* 11:225–32
163. Norman, J. G. Jr., Kolari, H. J. 1975. *J. Am. Chem. Soc.* 97:33–37
164. Norman, J. G. Jr., Kolari, H. J., Gray, H. B., Trogler, W. C. 1977. *Inorg. Chem.* 16:987–93
165. Cotton, F. A., Kalbacher, B. J. 1977. *Inorg. Chem.* 16:2386–96
166. Cotton, F. A., Stanley, G. G. 1977. *Inorg. Chem.* 16:2668–71
167. Mortola, A. P., Moskowitz, J. W., Rösch, N., Cowman, C. D., Gray, H. B. 1975. *Chem. Phys. Lett.* 32:283–86
168. Bursten, B. E., Cotton, F. A., Cowley, A. H., Hanson, B. E., Lattman, M., Stanley, G. G. 1979. *J. Am. Chem. Soc.* 101:6244–49
169. Bursten, B. E., Cotton, F. A. 1980. *Faraday Discuss. Chem. Soc.* 14:180–93
170. Guest, M. F., Garner, C. D., Hillier, I. H., Walton, I. B. 1978. *J. Chem. Soc. Faraday Trans. 2* 74:2092–98
171. Guest, M. F., Garner, C. D., Hillier, I. H., MacDowell, A. A., Walton, I. B. 1979. *J. Chem. Soc. Faraday Trans. 2* 75:485–93
172. Benard, M., Veillard, A. 1977. *Nouv. J. Chim. Paris* 1:97–99
173. Trogler, W. C., Gray, H. B. 1978. *Acc. Chem. Res.* 11:232–39
174. Fanwick, P. E., Martin, D. S. Jr., Cotton, F. A., Webb, T. R. 1977. *Inorg. Chem.* 16:2103–6
175. Cotton, F. A., Fanwick, P. E. 1979. *J. Am. Chem. Soc.* 101:5252–55
176. Martin, D. S. Jr., Newman, R. A., Fanwick, P. E. 1979. *Inorg. Chem.* 18:2511–20
177. Hay. P. J. 1978. *J. Am. Chem. Soc.* 100:2897–98
178. Benard, M. 1978. *J. Am. Chem. Soc.* 100:2354–62
179. Benard, M. 1979. *J. Chem. Phys.* 71:2546–56
180. Norman, J. G. Jr., Kolari, H. J. 1978. *J. Am. Chem. Soc.* 100:791–99
181. Norman, J. G. Jr., Renzoni, G. E., Case, D. A. 1979. *J. Am. Chem. Soc.* 101:5256–67
182. Nakatsuji, H., Ushio, J., Kanda, K., Onishi, Y., Kawamura, T., Yonezawa, T. 1981. *Chem. Phys. Lett.* 79:299–304
183. Coleman, A. W., Green, J. C., Hayes, A. J., Seddon, E. A., Lloyd, D. R., Niwa, Y. 1979. *J. Chem. Soc. Dalton Trans.* 1057–64
184. Bursten, B. E., Cotton, F. A. 1981. *Inorg. Chem.* 20:3042–48
185. Cotton, F. A., Fanwick, P. E., Gage, L. D., Kalbacher, B. J., Martin, D. S. Jr. 1977. *J. Am. Chem. Soc.* 99:5642–45
186. Martin, D. S. Jr., Webb, T. R., Robbins, G. A., Fanwick, P. E. 1979. *Inorg. Chem.* 18:475–78
187. Martin, D. S. Jr., Newman, R. A., Vlasnik, L. M. 1980. *Inorg. Chem.* 19:3404–7
188. Clark, R. J. H., Ferris, L. T. H. 1981. *Inorg. Chem.* 20:2759–66
189. Lee, E. P. F., Potts, A. W., Doran, M., Hillier, I. H., Delaney, J. J., Guest, M. F. 1980. *J. Chem. Soc. Faraday Trans 2* 76:506–19
190. Berkowitz, J., Streets, D. G., Garritz, A. 1979. *J. Chem. Phys.* 70:1305–11
191. MacNaughton, R. M., Bloor, J. E., Sherrod, R. E., Schweitzer, G. K. 1981. *J. Electron Spectrosc. Relat. Phenom.* 22:1–25
192. DeMello, P. C., Hehenberger, M., Larsson, S., Zerner, M. 1980. *J. Am. Chem. Soc.* 102:1278–88
193. Larsson, S., Hehenberger, M., DeMello, P. C. 1980. *Int. J. Quantum Chem.* 18:1271–8
194. Doran, M., Hawksworth, R. W., Hillier, I. H. 1980. *J. Chem. Soc. Faraday Trans. 2* 76:164–71
195. Sunil, K. K., Rogers, M. T. 1981. *Inorg. Chem.* 20:3283–87
196. Sunil, K. K., Harrison, J. F., Rogers, M. T. 1982. *J. Chem. Phys.* 76:3078–97
197. Larsson, S. 1975. *Theor. Chim. Acta* 39:173–83
198. Weber, J., Garner, C. D. 1980. *Inorg. Chem.* 19:2206–9
199. Weber, J., Daul, C. 1980. *Mol. Phys.* 39:1001–11
200. Bursten, B. E., Jensen, R. J., Gordon, D. J., Treichel, P. M., Fenske, R. F. 1981. *J. Am. Chem. Soc.* 103:5226–31
201. Larsson, S., Connolly, J. W. D. 1974. *J. Chem. Phys.* 60:1514–21
202. Kambali, U., Güdel, H. U., Weber, J. 1982. *Inorg. Chem.* In press
203. Bursten, B. E., Cotton, F. A., Green, J. C., Seddon, E. A., Stanley, G. G. 1980. *J. Am. Chem. Soc.* 102:955–68
204. Weber, J., Goursot, A., Pénigault, E., Ammeter, J. H., Bachmann, J. 1982. *J. Am. Chem. Soc.* 104:1491–1506

COLLISIONS OF RYDBERG ATOMS WITH MOLECULES

F. B. Dunning and R. F. Stebbings

Rice University, Department of Space Physics and Astronomy, and The Rice Quantum Institute, Houston, Texas 77001

Introduction

In recent years there has been an explosion of interest in the study of collision processes involving atoms in highly excited Rydberg states, and a wide range of reaction processes have been observed. In this review we concentrate on two of these, namely near-resonant collisional energy transfer and Rydberg electron transfer to targets that attach free thermal electrons.

Collisional transfer of internal energy is expected to be particularly efficient if the collision is resonant, i.e. if the internal energy lost by one collision partner equals that gained by the other. It is difficult to achieve energy resonance in studies using low-lying states since these are relatively few in number and, moreover, they are typically widely separated in energy. However, with Rydberg species, effects due to energy resonance are readily observable because Rydberg states are closely spaced in energy, thus enabling precise choice of the excitation energy. In addition, the high density of Rydberg states allows study of resonance effects in collisions in which only small amounts of energy are transferred. In the present article we focus on interactions in which there is a resonant interchange between molecular rotational or vibrational energy and energy of Rydberg electronic excitation.

Theoretical discussions of Rydberg atoms frequently make use of a model in which it is assumed that the separation of the Rydberg electron from its associated ion core is so large that the electron can be viewed as an independent, "essentially free" particle. Collisions are then analyzed in terms of the separate interactions between the Rydberg electron, the ionic core, and the target particle. Comparisons between theoretical and experimental results have shown that many Rydberg atom-molecule collision processes can be successfully described by considering only the binary electron-target interaction. This suggests that, since the time-averaged kinetic energy of a Rydberg electron is typically only a few millielectron volts, the study of Rydberg collisions can provide data on electron-molecule scattering in an energy regime virtually inaccessible using alternate techniques.

Experimental Considerations

The study of Rydberg atom collision processes must be approached with caution because, as experimental techniques have become more sophisticated, it has become apparent that many effects not normally of concern in studies involving atoms in ground or low-lying excited states become important. For example, because Rydberg states are closely spaced in energy and are only weakly bound, thermal energy collisions can lead to rapid state mixing and ionization. Thus, even though laser-induced excitation may result in the production of atoms in a single, well-defined Rydberg state, collisional mixing can lead to the rapid evolution of a complex, time dependent population distribution. Rydberg atoms are readily perturbed by external fields (1, 2). Weak electric fields can, for example, have marked effects on Rydberg excitation and collision processes (M. P. Slusher, private communication.). Effects due to interactions with background 300 K blackbody radiation must also be considered (3, 4). In addition, since Rydberg atoms have relatively short lifetimes, collision experiments must typically be completed within a few microseconds following excitation.

Rydberg atoms may be produced using a variety of techniques, including electron impact excitation, electron capture, and photoexcitation. Only the latter technique, however, affords the resolution needed to confine the excitation to a single, well-defined Rydberg state. Indeed, it is the availability of tunable dye lasers that has been responsible for many of the recent experimental advances in the study of Rydberg species.

Rydberg atoms are frequently detected either by field ionization or by observing the radiation they emit in spontaneous decay. With appropriate optical filtering, fluorescence measurements enable state-selective detection of Rydberg atoms with values of $n \lesssim 20$. Atoms with larger values of n are difficult to study using fluorescence techniques because of their long natural lifetimes. State-selective detection at high n is, however, possible using field ionization which, moreover, also provides absolute detection.

Rydberg state population distributions are frequently analyzed by use of selective field ionization (SFI) in which the Rydberg atoms are ionized in a time-dependent electric field (4–13). Since Rydberg atoms in different quantum states ionize at different field strengths, measurement of the field dependence of the ionization signal enables, in principle, the different states initially present to be identified. However, caution must be exercised in interpreting such data since a given SFI feature can only be correlated with a particular initial (zero-field) Rydberg state if the nature of the path to ionization is known. Previous detailed studies of field ionization (14–21) have shown that in an increasing electric field Rydberg atoms typically follow either predominantly adiabatic or predominantly diabatic paths to ionization. The probability of diabatic passage increases with $|m_\ell|$ and n. For $n \sim 30$ it appears that in an ionizing field having a slew rate of $\sim 10^9$ V cm^{-1} sec^{-1}, xenon atoms with $|m_\ell| \leq 3$ ionize predominantly adiabatically, while those with $|m_\ell| \geq 4$ ionize predominantly diabatically. Models of field ionization have been developed that permit estimation of the range of values of electric fields over which ionization is expected, following either adiabatic or diabatic passage to ionization (7, 20a). These models also enable synthesis of the SFI profiles to be expected following ionization of a mixture containing Rydberg atoms having a distribution of different initial states (12, 20b).

Resonant Rotational Energy Transfer

In a collision with a Rydberg atom, rotational deexcitation of a polar molecule can provide the energy necessary to further excite or ionize the Rydberg atom. Collisions of Xe(nf) atoms with a variety of simple polar targets, including HF, HCl, and NH$_3$, have been studied by the Rice University group (6, 7, 9, 10), using the apparatus shown in Figure 1. As indicated in the inset, the Xe(nf) atoms are produced in a two-step process: electron impact excitation to the metastable $^3P_0$ level followed by laser-induced optical excitation to Rydberg states. The metastable atoms are contained in a low density thermal energy beam and are photoexcited by a pulsed dye laser, typically in the presence of target gas, in a region located between two parallel grids. After allowing collisions to occur for a selected time interval, typically a few microseconds, the states of the excited atoms present are determined by use of SFI, for which purpose a ramped potential is applied to the lower grid. The electrons liberated at ionization are detected by a particle multiplier whose output is fed to a time-domain multichannel analyzer (MCA). The MCA is started at the beginning of the ionizing voltage ramp and is stopped by the first electron pulse subsequently registered by the multiplier. For sufficiently low count rates ($\lesssim 0.1$ per laser pulse) the MCA stores a signal proportional to the probability of a field-ionization event per unit time as the ionizing field is increased. Measurement of the time depen-

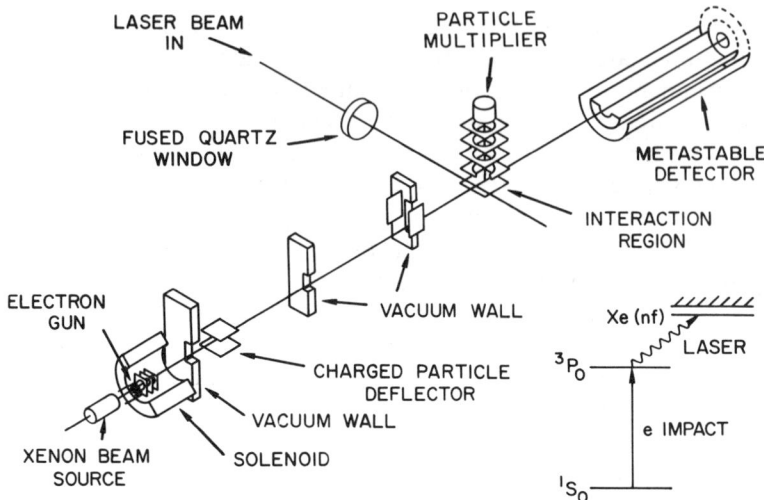

Figure 1 Schematic diagram of the apparatus used by the Rice University group to study collisions involving Xe(nf) atoms.

dence of the voltage ramp then permits determination of the field strengths at which ionization events occur and hence the field ionization signal per unit field increment.

SFI data obtained both for parent $27f$ atoms and following Xe($27f$)–HF collisions are shown in Figure 2a (9). Collisions result in the appearance of several ionization features that can be identified with the resonant transfer of HF rotational energy. The rotational energy levels of HF are given approximately by

$$E_J = BJ(J+1) \qquad \qquad 1.$$

where J is the rotational quantum number and B is the rotational constant. In dipole allowed $J \to J-1$ rotational deexcitations, the energies released are

$$\Delta E_J = E_J - E_{J-1} = 2BJ. \qquad \qquad 2.$$

The results of transferring these amounts of energy to a Xe($27f$) atom are indicated in the inset in Figure 2a, which includes a partial term diagram for xenon, together with a series of arrows whose lengths ($2BJ$) correspond to the energies released in the indicated rotational transitions. The widths of the arrows are proportional to the room-temperature populations of the upper rotational levels involved.

Rotational deexcitation from levels with $J \geqslant 4$ provides sufficient energy to cause collisional ionization via the reaction

$$\text{Xe}(27f) + \text{HF}(J) \to \text{Xe}^+ + \text{HF}(J-1) + e \quad J \geqslant 4. \qquad 3.$$

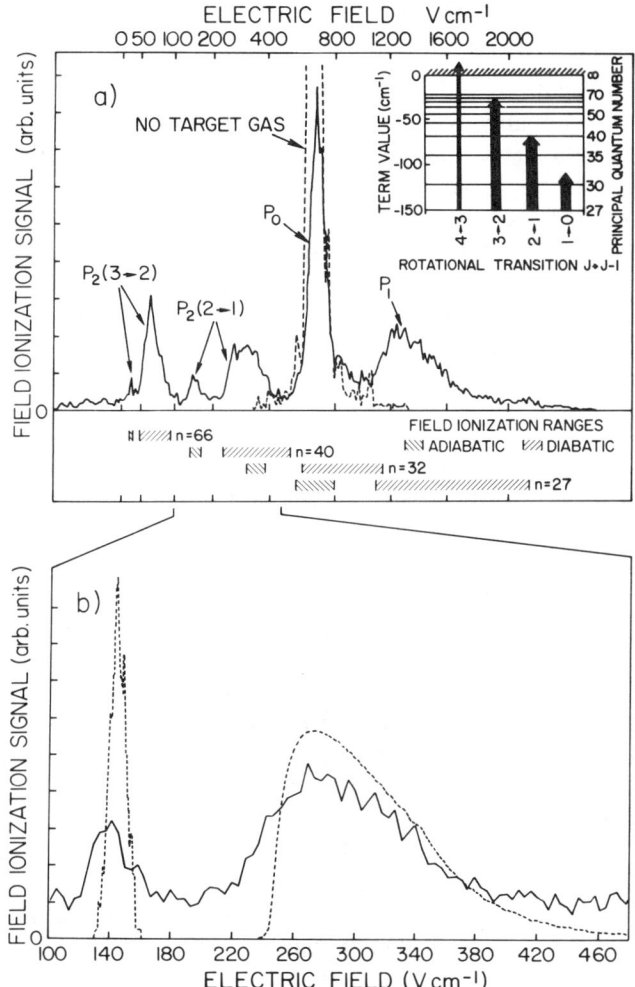

Figure 2 Interpretation of SFI features observed following Xe(27f)–HF collisions (9). (*a*) SFI spectrum obtained after collisions with HF. The features can be identified, as discussed in the text, by reference to the partial xenon term diagram shown in the *inset* and the *horizontal bars* beneath the data which indicate the field ionization ranges for states populated in n- and ℓ-changing collisions. The *dashed line* shows the SFI profile obtained in the absence of target gas. (*b*) Comparison of the SFI signal that results primarily from ionization of the products of Reaction 5 with the profile calculated for ionization of a mixture containing equal numbers of atoms in each $n = 40$ Stark state.

Rotational deexcitation from levels with $J \leq 3$ will lead to excitation to selected groups of higher-lying states in n-changing collisions of the type

$$\text{Xe}(27f) + \text{HF}(J) \rightarrow \text{Xe}(n', \ell') + \text{HF}(J-1) \quad J \leq 3. \qquad 4.$$

The range of field strengths over which the products of such collisions should ionize adiabatically and diabatically are indicated by the horizontal bars below the data. It is evident that the peaks labeled P_2 result from the ionization of the products of n-changing collisions. That these peaks are discrete, points to the near-resonant nature of the transfer of rotational to electronic energy. Since collisions populate only a few groups of excited states, it is possible, in certain cases, to identify separately the adiabatically and diabatically ionizing collision products. For example the second two peaks in Figure 2a, labeled $P_2(2 \rightarrow 1)$, result primarily from ionization of states produced via the reaction

$$\text{Xe}(27f) + \text{HF}(J=2) \rightarrow \text{Xe}(n', \ell') + \text{HF}(J=1) \qquad 5.$$

where $n' \sim 40$. This portion of the spectrum is shown on an expanded scale in Figure 2b, together with the SFI profile calculated for ionization of a mixture containing equal numbers of atoms in each $n = 40$ Stark state, assuming states with $|m_\ell| \leq 3$ ionize adiabatically, $|m_\ell| \geq 4$ diabatically. The generally good agreement between the experimental and theoretical profiles indicates that collisions result in n-mixed products having a wide range of values of ℓ and $|m_\ell|$. The experimentally observed adiabatic ionization feature has a somewhat greater width than is calculated. This may be a consequence both of instrumental effects and of the fact that not all low-$|m_\ell|$ atoms ionize completely adiabatically. However, the uncertainties are such that collisional population of $n = 41$ states cannot be entirely ruled out, although the magnitude of the energy defect associated with excitation to $n = 41$ levels (~ 2.1 cm^{-1}) is larger than that (~ 1.2 cm^{-1}) for excitation to $n = 40$ levels. No significant production of $n = 39$ or $n = 42$ states is evident; this indicates that n-changing requires energy resonance to better than the energy defects (~ 5 cm^{-1}) associated with transitions to these states.

SFI data obtained following $\text{Xe}(23f)$–HF interactions provide evidence of the reaction

$$\text{Xe}(23f) + \text{HF}(J=0) \rightarrow \text{Xe}(21, \ell') + \text{HF}(J=1) \qquad 6.$$

in which electronic energy is transferred from the Rydberg atom to excite rotationally the target molecule. Failure to observe the analogous reaction in collisions involving $26f$ and $27f$ atoms may be indicative of the near-resonant nature of such energy transfer processes, because for $\text{Xe}(23f)$ the energy defect for this process is ~ 1.5 cm^{-1}, whereas for $\text{Xe}(26f, 27f)$ it is ~ 2.6 cm^{-1}.

The peak labeled P_0 in Figure 2a results from adiabatic ionization of the remaining 27f atoms together with the low-$|m_\ell|$ products of ℓ-changing collisions of the type

$$\text{Xe}(27f) + \text{HF}(J) \rightarrow \text{Xe}(27, \ell') + \text{HF}(J) \qquad 7.$$

in which J and the principal quantum number of the Rydberg atom are unchanged. P_1 results from the diabatically ionizing products of such collisions. The small amount of energy required for ℓ-changing transitions is provided by the energy of relative motion.

SFI data obtained following Xe(31f)–HCl collisions (10) are shown in Figure 3. These data can be interpreted by reference to the partial term diagram shown in the inset and to the horizontal bars shown beneath the

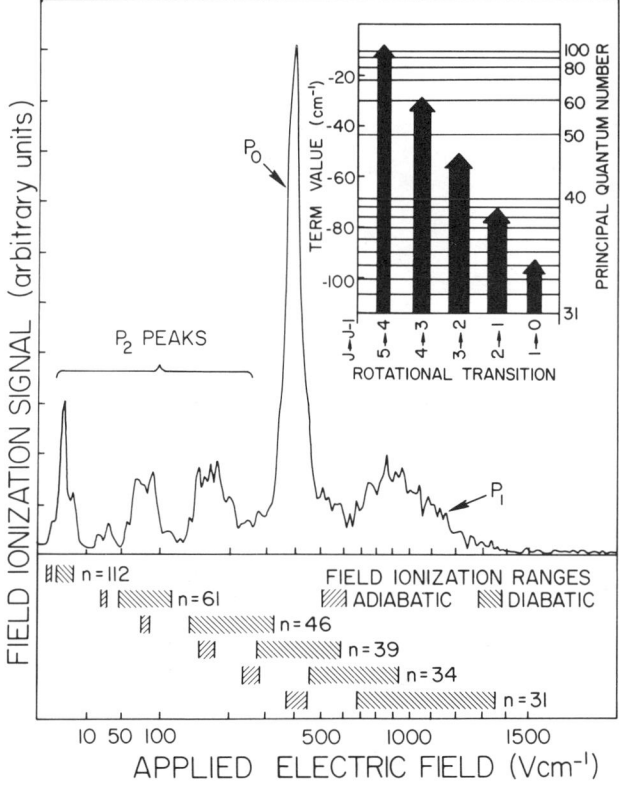

Figure 3 SFI spectrum obtained following Xe(31f)–HCl collisions (10). The inset shows a partial term diagram for Xe, with arrows indicating the energies available from rotational deexcitation of HCl. The widths of the arrows are proportional to the room temperature populations in the upper rotational levels involved. The *horizontal bars* beneath the data give the expected field ionization ranges for states populated in n- and ℓ-changing collisions.

data. SFI features resulting from ionization of the products of near-resonant n-changing collisions are once again clearly evident. Similar results have also been obtained for NH_3 (7).

The rate constants for n- and ℓ-changing can be determined from measurement of the time dependence of the corresponding SFI features. For example, data pertaining to $Xe(31f)$–HCl collisions, obtained at a low slew rate to better resolve the P_2 peaks, is presented in Figure 4 (10).

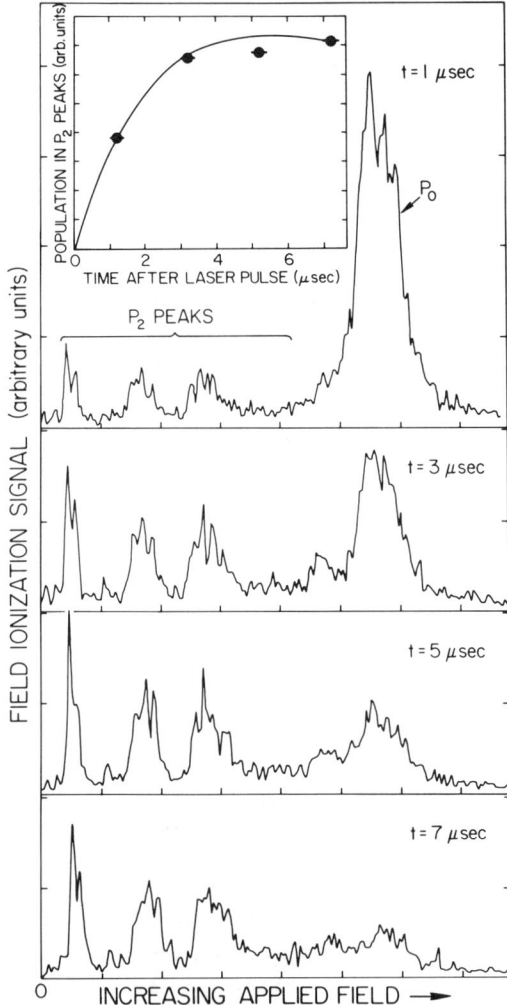

Figure 4 Dependence of the SFI features observed following $Xe(31f)$–HCl collisions upon collision time t (10). The time dependence of the population comprising the P_2 peaks is shown in the *inset* together with the result of a model fit to the data.

The build-up of the P_2 peaks coupled with the decay of the parent $31f$ population, is clearly evident. The time development of the P_2 peaks is shown in the inset together with the results of a model fit to the data. Representative measured rate constants are presented in Table 1. The rate constants for n-changing, i.e. rotational to electronic energy transfer, are very large.

Also included in Table 1 are rate constants for collisional ionization determined from measurement of the number of Xe^+ ions produced in known time intervals following laser excitation. Because the n- and ℓ-changing rate constants are large, part of the observed Xe^+ signal may result from ionization of the products of such collisions. This can be taken into account, however, by postulating that state-changing collisions populate a "reservoir-state" and then including ionization from this reservoir state in analysis of the data. Rate constants for collisional ionization in $Xe(nf)$–NH_3 collisions are shown in Figure 5, together with the results of several theoretical calculations (22a, b, 23). The theoretical results were obtained by considering only the Rydberg electron-target interaction utilizing electron-NH_3 scattering parameters obtained on the basis of a simple dipole interaction. The step-like structure in the calculated values results from the discrete nature of rotational energy transfer. Thus, for the range of n included in Figure 5, only a fraction of the NH_3 atoms, which are at room temperature, are in rotational states with a sufficiently large value of J that a $J \rightarrow J-1$ transition can provide enough energy ($2BJ$) to cause ionization. However, as n is increased, the

Table 1 Rate constants pertaining to $Xe(nf)$ collisions with polar targets[a]

		Rate constant in units of 10^{-7} cm^3 sec^{-1}		
Target species	n	ℓ-changing k_ℓ	n-changing k_n	Collisional ionization k_i
NH_3[b]	31	~13	~4	2.3 (1.0)
HCl[c]	31	4.8 (2.4)[e]	5.5 (2.5)[f] $k_n^{2\rightarrow1}$ 3.0 (1.5)[g]	0.9 (0.4)
HF[d]	27	4.3 (2.2)[e]	$k_n^{3\rightarrow2}$ 2.2 (1.1)[g]	1.5 (0.8)

[a] The numbers in parentheses are the experimental uncertainties.
[b] Ref. (7).
[c] Ref. (10).
[d] Ref. (9).
[e] Included in these rate constants are the small effects of those n-changing collisions that lead to ionization in the vicinity of P_0 and P_1.
[f] For all n-changing collisions that lead to peaks P_2 in Figure 4.
[g] These data pertain to n-mixing resulting from the rotational transition specified by the superscript.

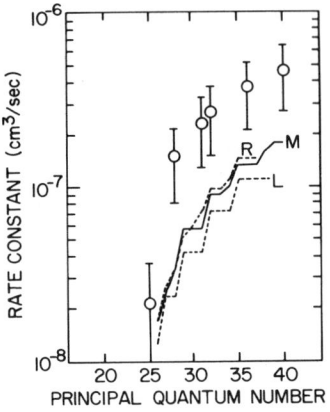

Figure 5 Rate constants for collisional ionization in Xe(nf)–NH$_3$ interactions. ϕ — experimental data (7). Calculated values, R—Rundel (7); M—Matsuzawa (22a,b); L—Latimer (23).

energy required to cause ionization decreases and more of the NH$_3$ population is able to cause ionization. Since the number of rotational transitions that can lead to ionization increases discontinuously as n is increased, this leads to step-like increases in the rate constant. Although the experimental data in Figure 5 follow the trend of the theoretical calculations, i.e. the collisional ionization cross section increases with n, the quantitative agreement is poor. In addition, the experimental uncertainties are such that no step-like structure can be discerned. However, step-like structure has been reported in studies of collisional ionization of krypton Rydberg atoms by HCl and HF (24).

Resonant Vibrational Energy Transfer

Sharply resonant collisional energy transfer from electronic excitation of Na(ns) states to vibrational excitation in CH$_4$ or CD$_4$ has been observed by Gallagher et al (25). The sodium atoms and target gas were contained in a heated pyrex cell. The ns states were excited by two-step laser-induced photoexcitation via the intermediate $3p$ state, as illustrated in the inset of Figure 6. The ns population was monitored by observing the $ns \to 3p$ fluorescence. Velocity-averaged cross sections $\sigma_d(ns)$ for collisional depopulation were determined from measurements of the decay of the ns population as a function of CH$_4$ or CD$_4$ number density; these are shown in Figure 6.

The depopulation cross sections expected in the absence of effects due to resonant energy transfer are shown by the dotted line in Figure 6. For certain values of n the measured depopulation cross sections exceed these values. As indicated in Table 2, these large cross sections can be explained as arising from close energy resonances between electronic

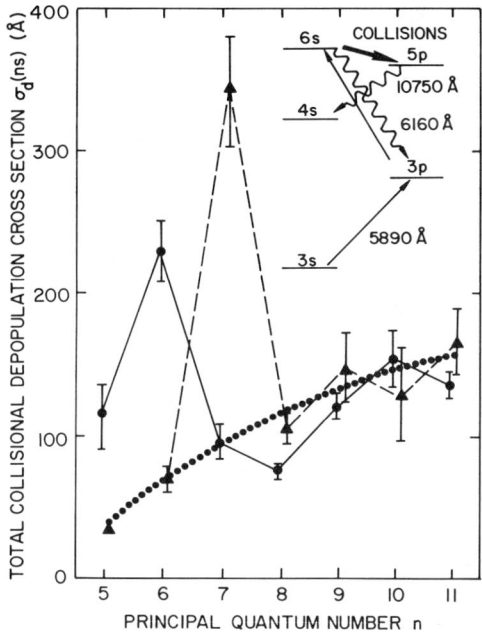

Figure 6 Cross sections for collisional depopulation in Na(ns)–CH$_4$ (●, ———) and Na(ns)–CD$_4$ (▲, – – –) collisions (25). The *smooth dotted curve* shows the depopulation cross section expected in the absence of resonant $e-v$ transfer. The inset shows the energy levels pertinent to the study of $6s$ and $5p$ states. The *straight arrows* indicate laser excitation steps; the *heavy arrow*, collisional transfer; and the *wavy arrows*, the observed fluorescence.

transitions from these states and vibrational transitions in CH$_4$ or CD$_4$. Resonances are observed at different values of n for CH$_4$ and CD$_4$ because these molecules have different vibrational energy level spacings. To confirm that the transitions listed in Table 2 were responsible for the enhanced values of $\sigma_d(ns)$, the time dependence of the lower state populated in each transition was determined. In the case of Na($6s$)–CH$_4$

Table 2 Parameters pertaining to transitions resulting in enhanced depopulation cross sections in Na(ns)–CH$_4$, CD$_4$ collisions

Na transition	Energy (cm^{-1})	CH$_4$, CD$_4$ transitions			State specific cross sections (Å^2)[a]	
		Species	Mode, branch	Branch center frequency (cm^{-1})	CH$_4$	CD$_4$
$5s \rightarrow 4p$	2930	CH$_4$	ν_3,P	2940	103 (22)	10 (14)
$6s \rightarrow 5p$	1331	CH$_4$	ν_4,R	1340	135 (22)	9 (4)
$7s \rightarrow 5d$	975	CD$_4$	ν_4,P	965	12 (4)	215 (33)

[a] Numbers in parentheses are experimental uncertainties.

collisions, for example, measurements of the time dependence of the $5p \rightarrow 4s$ fluorescence indicated that collisionally induced $6s \rightarrow 5p$ transitions were indeed responsible for the large value of $\sigma_d(6s)$ and yielded the state-specific cross section $\sigma(6s \rightarrow 5p)$ for this process. This, and other, state-specific cross sections are included in Table 2. As might be expected, the magnitudes of these cross sections are approximately equal to the amount by which the total depopulation cross section exceeds that expected in the absence of resonant electronic to vibrational energy transfer.

Resonant energy transfer is observed in cases in which the orbital angular momentum quantum number of the Rydberg electron changes by both $\Delta \ell = 1$ and $\Delta \ell = 2$. Since the corresponding cross sections are comparable, this suggests that the interaction does not occur at long range, relative to radius of the sodium atom, as this would favor $\Delta \ell = 1$ transitions. This, coupled with the size of the cross sections, implies that the interaction must take place at about the mean radius of the electron orbit and can thus be described qualitatively in terms of Rydberg electron-molecule scattering.

Not all energy resonances or near-resonances lead to enhanced depopulation cross sections. For example, no resonance enhancement of $\sigma_d(8s)$ was observed even though the $8s \rightarrow 7s$ transition is resonant with a ν_4 transition in CD_4. This can be explained qualitatively by considering the scattering of the Rydberg electron by the target molecule. Since, classically, the motion of an electron in an s state is purely radial, only scattering angles of 0° or 180° can result in transitions to a final s state. However, low-energy electron-molecule scattering leading to vibrational excitation is generally isotropic. Thus, $s \rightarrow s$ transitions are not expected to lead to enhanced depopulation cross sections. Other resonances that produce no enhancement are associated with large changes in n. The lack of enhancement can then be attributed to the poor overlap between the radial wavefunctions of the initial and final states that leads to weak coupling of the atomic states during the collision.

Attaching Targets

The majority of experimental studies concerning collisions with molecules that attach thermal energy electrons have focused on those reactions that result in negative ion formation. Such processes have been analyzed theoretically (26a–c) using the "essentially free" electron model in which negative ion formation is viewed as resulting from attachment of the Rydberg electron to the target molecule, with the Rydberg core playing the role of spectator. On the basis of this model, the rate constants for attachment of Rydberg electrons and of free electrons having the same velocity distribution should be equal. Studies of both Rydberg and free electron collisions with a variety of attaching targets, including SF_6, CCl_4, C_7F_{14}, CH_3I, and C_6F_6, have been reported (27–35). In the present discussion we concentrate on SF_6, as this has been most widely investigated.

Absolute measurements of the rate constant $k_i(nf)$ for collisional ionization via the electron transfer reaction

$$Xe(nf) + SF_6 \rightarrow Xe^+ + SF_6^- \qquad 8.$$

have been undertaken by the Rice University group (30, 32), using the apparatus shown in Figure 1. The measured rate constants are presented in Table 3. Also included are rate constants for Rydberg electron attachment to SF_6 derived using the relationship

$$k = \int_0^\infty v \sigma_a(v) f(v) \, dv \qquad 9.$$

where v is the electron velocity; $\sigma_a(v)$ is the calculated cross section for free electron attachment (36), derived assuming that the interaction is dominated by the polarization potential and that the process is purely s-wave capture; and $f(v)$ is the classical velocity distribution appropriate to electrons in hydrogenic orbits of $\ell = 3$. The agreement between the calculated rate constants and the Rydberg data lends credence to the "essentially free" electron model.

Velocity-averaged cross sections σ_{Ra} for Rydberg electron attachment derived from the data in Table 3 are presented in Figure 7. These are obtained using the expression

$$\sigma_{Ra} = \frac{k_i(nf)}{\bar{v}_n} \qquad 10.$$

Table 3 Rate constants $k_i(nf)$ for collisional ionization via electron transfer in $Xe(nf)$–SF_6 collisions

	Rate constant in units of 10^{-7} cm^3 sec^{-1}	
n	$k_i(nf)$[a]	k[b]
25	3.7 (0.7)	3.4
26	4.3 (0.9)	3.5
28	4.2 (0.8)	3.5
31	3.8 (0.8)	3.6
33	3.5 (0.7)	3.6
35	4.3 (0.9)	3.7
38	4.2 (0.8)	3.7
40	4.0 (0.8)	3.7

[a] Ref. (32).
[b] Calculated using Eq. 9 and data in Ref. (36).

where $\bar{v}_n$ is the velocity corresponding to the time-averaged kinetic energy of a Rydberg electron in the nf state. This kinetic energy is used to locate the Rydberg data on the electron energy axis. Also included in Figure 7 are free electron attachment cross sections, obtained using the threshold photoelectron spectrum by electron attachment (TPSA) technique (37) and from analysis of swarm data (38, 39). The good agreement between the Rydberg and free electron data again indicates that the Rydberg electron interacts with the target molecule much as would a free electron of comparable kinetic energy.

The primary negatively charged species produced in Rydberg-SF_6 collisions at room temperature is SF_6^-. Free electrons, which could result either from Rydberg collisions or from SF_6^- autodetachment, are not observed. Thus the SF_6^- ions formed in Rydberg collisions have a much longer lifetime against autodetachment than do the excited SF_6^- ions that result from free electron attachment. Klots (31) suggested that such observations can be reconciled if the SF_6^- ions formed in the initial Rydberg electron-SF_6 encounter are stabilized by an interaction with the

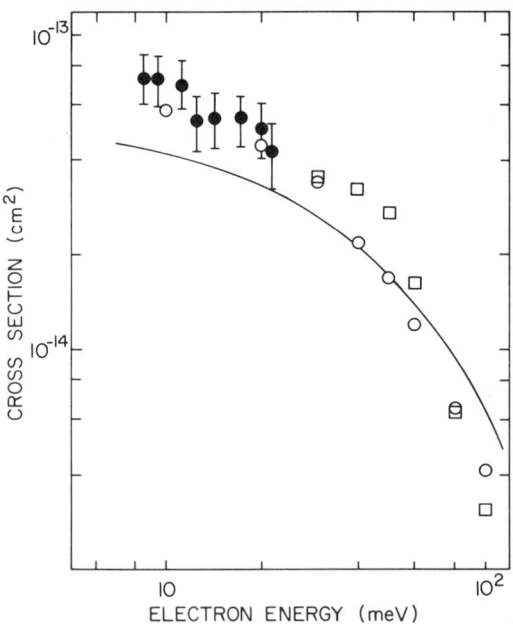

Figure 7 Cross sections for electron attachment to SF_6. ●, values derived from Rydberg data using Eq. 10; ———— data obtained with the TPSA technique (37); □, ○, swarm-unfolded measurements (38, 39) obtained using two different electron energy distributions [the points at low energies were taken from ref. (37)].

core in which the negative ion loses part of its internal energy to kinetic energy.

Ion production as a function of temperature in Ar**–SF_6 collisions has been investigated in detail by Astruc et al (34). The product ions were analyzed mass spectroscopically. The Ar^+ signal, and hence the electron transfer rate, was found to be independent of temperature in the range 25–270°C. At room temperature, only SF_6^- ions were observed. Increasing the SF_6 temperature resulted in a marked decrease in SF_6^- production and in the growth of a small SF_5^- signal. A similar increase in SF_5^- formation with increasing temperature has been noted in free electron studies. The Rydberg data are interpreted using a model that assumes that Rydberg electron attachment initially results in the formation of SF_6^- in an excited state. This excited SF_6^- may be subsequently stabilized by an interaction with the core in which part of its internal energy is converted to kinetic energy, or it may undergo autodetachment, or if it has sufficient energy it may dissociate, yielding SF_5^-. The data indicate that stabilization becomes less effective, and autodetachment increasingly important, as the SF_6 temperature is raised because, although the electron transfer rate is independent of temperature, both the SF_6^- and the total $SF_6^- + SF_5^-$ signals decrease markedly. The observation of stable SF_6^- ions, however, shows that post-attachment interactions with the Rydberg ion core may be important.

Studies with other attaching targets (32, 33, 35) have further demonstrated that free electron attachment and Rydberg collisions result in formation of the same negative ion species and that the rate constants for their formation are comparable, as expected on the basis of the "essentially free" electron model. For example, in the case of CCl_4 (32, 40, 41), collisions with both Rydberg atoms and free electrons result in dissociation, i.e.

$$e + CCl_4 \rightarrow CCl_3 + Cl^-$$
$$Xe(nf) + CCl_4 \rightarrow Xe^+ + CCl_3 + Cl^-. \qquad 11.$$

In contrast to the case of $Xe(nf)$–SF_6 collisions, the rate constant for electron transfer in $Xe(nf)$–CCl_4 collisions increases markedly with increasing n, i.e. decreasing Rydberg electron energy. This behavior is, however, consistent with that expected from theoretical free electron attachment cross sections (32, 36). The opposite behavior is noted in $Xe(nf)$–C_7F_{14} collisions (33)—the rate constant for electron transfer decreasing with increasing n, although this is again consistent with free electron data.

Thus, investigations of collisions involving attaching targets point to further application of Rydberg atoms to study low-energy ($\lesssim 10$ meV) electron-molecule scattering. Although such investigations can be undertaken using low-ℓ states, high-ℓ states are to be preferred as they have narrower Rydberg electron momentum distributions and hence afford better energy resolution. A number of low energy electron scattering resonances should be accessible to study in this manner (42).

Conclusion

Rydberg atoms display a variety of novel and unusual collisional properties. In particular they provide an excellent means to study effects of energy resonance in collisions and make possible the study of electron scattering processes at very low electron energies. In addition, the study of Rydberg atom-molecule collisions holds the promise of providing information on thermal-energy ion-molecule collisions (43), since products resulting from the core-target interaction should be separately identifiable, especially at high n. Indeed, continued study of Rydberg atom collisions will yield much new information concerning a variety of collision processes. If past experiments provide any guide, many new and interesting phenomena remain to be discovered.

Acknowledgments

The research by the authors and their colleagues described in this article was supported by the National Science Foundation under contract PHY81-08452 and by The Robert A. Welch Foundation.

Literature Cited

1. Zimmerman, M. L., Littman, M. G., Kash, M. M., Kleppner, D. 1979. *Phys. Rev. A* 20:2251–75
2. Zimmerman, M. L., Castro, J. C., Kleppner, D. 1978. *Phys. Rev. Lett.* 40:1083–86
3. Farley, J. W., Wing, W. H. 1981. *Phys. Rev. A* 23:2397–2424
4. Hildebrandt, G. F., Beiting, E. J., Higgs, C., Hatton, G. J., Smith, K. A., Dunning, F. B., Stebbings, R. F. 1981. *Phys. Rev. A* 23:2978–82
5. Gallagher, T. F., Cooke, W. E., Edelstein, S. A. 1978. *Phys. Rev. A* 17:904–8
6. Smith, K. A., Kellert, F. G., Rundel, R. D., Dunning, F. B., Stebbings, R. F. 1978. *Phys. Rev. Lett.* 40:1362–65
7. Kellert, F. G., Smith, K. A., Rundel, R. D., Dunning, F. B., Stebbings, R. F. 1980. *J. Chem. Phys.* 72:3179–90
8. Kellert, F. G., Higgs, C., Smith, K. A., Hildebrandt, G. F., Dunning, F. B., Stebbings, R. F. 1980. *J. Chem. Phys.* 72:6312–13
9. Higgs, C., Smith, K. A., McMillian, G. B., Dunning, F. B., Stebbings, R. F. 1981. *J. Phys. B* 14:L285–89
10. Stebbings, R. F., Dunning, F. B., Higgs, C. 1981. *J. Electron Spectrosc. Relat. Phenom.* 23:333–38
11. Higgs, C., Smith, K. A., Dunning, F. B., Stebbings, R. F. 1981. *J. Chem. Phys.* 75:745–48
12. Kellert, F. G., Jeys, T. H., McMillian, G. B., Smith, K. A., Dunning, F. B., Stebbings, R. F. 1981. *Phys. Rev. A* 23:1127–33
13. MacAdam, K. B., Rolfes, R., Crosby, D. A. 1981. *Phys. Rev. A* 24:1286–98
14. Gallagher, T. F., Humphrey, L. M., Hill, R. M., Edelstein, S. A. 1976. *Phys. Rev. Lett.* 37:1465–68
15. Gallagher, T. F., Humphrey, L. M., Cooke, W. E., Hill, R. M., Edelstein, S. A. 1977. *Phys. Rev. A* 16:1098–1108

16. Littman, M. G., Kash, M. M., Kleppner, D. 1978. *Phys. Rev. Lett.* 41:103–7
17. Gallagher, T. F., Cooke, W. E. 1979. *Phys. Rev. A* 19:694–99
18. Vialle, J. L., Duong, H. T. 1979. *J. Phys. B* 12:1407–23
19. Komarov, I. V., Grozdanov, T. P., Janev, R. K. 1980. *J. Phys. B* 13:L573–76
20a. Jeys, T. H., Foltz, G. W., Smith, K. A., Beiting, E. J., Kellert, F. G., Dunning, F. B., Stebbings, R. F. 1980. *Phys. Rev. Lett.* 44:390–93
20b. Jeys, T. H., McMillian, G. B., Smith, K. A., Dunning, F. B., Stebbings, R. F. 1982. *Phys. Rev. A.* In press
20c. McMillian, G. B., Jeys, T. H., Smith, K. A., Dunning, F. B., Stebbings, R. F. 1982. *J. Phys. B.* In press
21a. Rubbmark, J. R., Kash, M. M., Littman, M. G., Kleppner, D. 1981. *Phys. Rev. A* 23:3107–17
21b. Neijzen, J. H. M., Dönszelmann, A. 1982. *J. Phys. B* 15:L87–91
22a. Matsuzawa, M. 1971. *J. Chem. Phys.* 55:2685–89
22b. Matsuzawa, M. 1974. *J. Electron Spectrosc. Relat. Phenom.* 4:1–12
23. Latimer, C. J. 1977. *J. Phys. B* 10:1889–95
24. Matsuzawa, M., Chupka, W. A. 1977. *Chem. Phys. Lett.* 50:373–76
25. Gallagher, T. F., Ruff, G. A., Safinya, K. A. 1980. *Phys. Rev. A* 22:843–48
26a. Matsuzawa, M. 1972. *J. Phys. Soc. Jpn.* 32:1088–92; 33:1108–19
26b. Matsuzawa, M. 1975. *J. Phys. B* 8:L382–86; 8:2114–22
26c. Matsuzawa, M. 1977. *J. Phys. B* 10:1543–55
27. Hotop, H., Niehaus, A. 1967. *J. Chem. Phys.* 47:2506–7
28. Sugiura, T., Arakawa, K. 1971. *Recent Developments in Mass Spectroscopy, Proc. Int. Conf. Mass-Spectrosc. Kyoto 1969*, ed. K. Ogata, T. Kayakawa, pp. 848–52
29. Stockdale, J. A., Davis, F. J., Compton, R. N., Klots, C. E. 1974. *J. Chem. Phys.* 60:4279–85
30. West, W. P., Foltz, G. W., Dunning, F. B., Latimer, C. J., Stebbings, R. F. 1976. *Phys. Rev. Lett.* 36:854–58
31. Klots, C. E. 1977. *J. Chem. Phys.* 66:5240–41
32. Foltz, G. W., Latimer, C. J., Hildebrandt, G. F., Kellert, F. G., Smith, K. A., West, W. P., Dunning, F. B., Stebbings, R. F. 1977. *J. Chem. Phys.* 67:1352–59
33. Hildebrandt, G. F., Kellert, F. G., Smith, K. A., Dunning, F. B., Stebbings, R. F. 1978. *J. Chem. Phys.* 68:1349–54
34. Astruc, J. P., Barbé, R., Schermann, J. P. 1979. *J. Phys. B* 12:L377–81
35. Dimicoli, I., Botter, R. 1981. *J. Chem. Phys.* 74:2346–54; 74:2355–60
36. Klots, C. E. 1976. *Chem. Phys. Lett.* 38:61–64
37. Chutjian, A. 1981. *Phys. Rev. Lett.* 46:1511–14
38. Christodoulides, A. A., Christophorou, L. G., Pai, R. Y., Tung, C. M. 1979. *J. Chem. Phys.* 70:1156–68
39. McCorkle, D. L., Christodoulides, A. A., Christophorou, L. G., Szamrej, I. 1980. *J. Chem. Phys.* 72:4049–57
40. Christophorou, L. G., Stockdale, J. A. D. 1968. *J. Chem. Phys.* 48:1956–60
41. Schultes, E., Christodoulides, A. A., Schindler, R. N. 1975. *Chem. Phys.* 8:354–65
42. Sasano, K., Matsuzawa, M. 1981. *J. Phys. B* 14:L91–96
43. Kocher, C. A., Smith, A. J. 1977. *Phys. Rev. Lett.* 39:1516–19

POLYELECTROLYTE THEORIES AND THEIR APPLICATIONS TO DNA

Charles F. Anderson and M. Thomas Record, Jr.

Department of Chemistry, University of Wisconsin, Madison, Wisconsin 53706

INTRODUCTION

Perspective and Summary

Fundamental advances in the description of cylindrical polyelectrolytes and their interactions with ions in solution, and the application of these theories to calculate the physical properties of locally cylindrical chain polyelectrolytes such as DNA, have occurred in the last 15 years. This resurgence of theoretical and experimental interest in solutions of cylindrical polyelectrolytes has been catalyzed by four interrelated developments:

1. the development [principally by Manning (1–6)] of counterion condensation (CC) theory as a simple, analytic alternative to the Poisson-Boltzmann (PB) theory of cylindrical polyions (7, 8);
2. the rigorous statistical mechanical examination of the PB equation, demonstrating its applicability to dilute solutions of highly charged cylindrical polyelectrolytes at (excess) salt concentrations approaching the usual experimental range (9);
3. the derivation, from either the CC (1) or the PB (10, 11) theories, of analytical or closed-form expressions for many of the thermodynamic and molecular properties of solutions of cylindrical polyelectrolytes;
4. the resulting ability to incorporate polyelectrolyte effects into the theoretical description of the thermodynamic, kinetic, hydrodynamic, and molecular properties of DNA and other sufficiently rod-like polyions.

Previous reviews of various aspects of polyelectrolyte solutions have usually adopted a single unified theoretical approach, based on either the cylindrical PB equation (7, 12–14) or the concept of counterion condensation (3, 5, 6, 15, 16). In contrast, the focus of this review is the comparison of these two theoretical models as well as the comparison of the thermodynamic and molecular descriptions of DNA and its interactions provided by these models. Other reviews covering experimental investigations (17–22) and theoretical interpretations (5, 22, 23) of the physical properties and conformational and binding equilibria of DNA have recently appeared. The general field of polyelectrolytes is too broad to survey in an article of this length. Among the many important areas not covered here is the application of polyelectrolyte theory to transport properties; the complexities of these nonequilibrium phenomena continue to command the interest of theoreticians and experimentalists (24–34). Certain other physical properties of polyelectrolyte solutions have also been surveyed recently (35).

We begin with an overview of the PB and CC theories and then compare the basic models, theoretical developments, thermodynamic results, and molecular pictures of the two theories. A brief summary of remaining theoretical uncertainties and possible approaches to their resolution is also given. The next section explains how the analytic or closed-form expressions that can be obtained from each theory have been used to describe the thermodynamic and physical binding of counterions to cylindrical polyions. The final section reviews the application of each theory to analyze and interpret the salt-dependence of equilibria involving polyions such as double helical DNA.

General Remarks on the Poisson-Boltzmann and Counterion Condensation Theories

Although flexible coil chain polyions, such as high molecular weight DNA, occupy a roughly spherical volume in solution, they often are locally rod-like, especially if the density of structural charges is high. For such polyions, in a solution containing enough added salt, the average length of a locally rod-like segment exceeds the Debye length, a measure of the distance over which electrostatic interactions are not effectively screened. If this condition is met, the cylindrical model is an appropriate basis for theoretical calculations of those properties that depend on electrostatic interactions between the polyion and small ions (36). Accordingly, in both the PB and CC theories a chain polyion is modeled as an isolated, infinitely long cylinder whose charge density is expressed by the nondimensional parameter $\xi = e^2/\varepsilon k T b$, where b is the average axial distance between polyion structural charges, ε is the dielectric constant of pure solvent, and the other symbols are conventional. In the PB theory the magnitude of ξ has a pronounced effect on the radial and salt-dependence of the potential derived from the cylindrical PB equation, and hence on the radial distribution(s) of counterions near the polyion. For highly charged polyions ($\xi > 1$) this local counterion gradient can be very steep, especially at low salt concentrations, and it is relatively insensitive to changes in the bulk salt concentration (37, 38). In the CC theory the magnitude of ξ determines whether counterion condensation (association with the polyion) occurs, and, if so, to what extent. For polyions having $\xi > 1$ in a solution containing only univalent counterions, the number of condensed counterions per polyion charge, $1 - \xi^{-1}$, is nearly invariant to changes in the bulk salt concentration, unless it approaches 1 M (4).

COMPARISON OF POISSON-BOLTZMANN AND COUNTERION CONDENSATION POLYELECTROLYTE THEORIES

Overview of the Development of the Two Theories

Manning has published two detailed theories of counterion condensation. The first (1, 2) was formulated primarily for the purpose of describing the effects of condensation on the limiting laws for thermodynamic properties reflecting the interactions of a cylindrical polyion with small ions in solution. Limiting laws for certain transport properties were also derived (39, 40). By 1972 it had been established that expressions derived from condensation theory provide a simple and useful means of analyzing the results of a diverse group of experimental investigations (15). Since then numerous additional studies of the equilibrium (41–50) and transport (51–61) properties of solutions of chainlike polyions have been analyzed in terms of Manning's limiting laws (or extensions thereof), and the consequences of counterion condensation have been incorporated into thermodynamic analyses of the conformational transitions (62–64) and binding equilibria (23, 65) of nucleic acids. In general, the utility of theoretical descriptions based on the concept of counterion condensation has continued to be confirmed (16), often at salt concentrations well above its expected range of applicability. Moreover, the extent of binding predicted by the condensation model has been detected by various relatively direct spectroscopic methods (6). Some of the basic assumptions underlying Manning's model have been subjected to further theoretical scrutiny (66–70); various ways of extending the theoretical validity of the limiting laws have been proposed (71–76). Counterion condensation has also been incorporated into theoretical treatments of the effect of electrostatic interactions on polyion configurations in solution (77, 78).

Manning's second theory of condensation was formulated to account for the observed persistence, at moderate salt concentrations, of the extent of counterion binding predicted for the limit of infinite dilution. From a simple two-phase model, counterion condensation is derived as the result of a free energy minimization (4). In some respects this new theoretical approach is related to Manning's earlier development of the condensation model in the same way that Bjerrum's ion-pairing theory is related to the Debye-Hückel theory of simple electrolyte solutions. However, the analogy is imperfect in the sense that Manning's later formulation was not intended as a means of extending the limiting laws, but rather as an accurate molecular description of condensation as a binding phenomenon. A similar treatment of counterion condensation in terms of a free energy minimization was presented independently (73); some aspects of the relationship between this theory and Manning's limiting laws were examined (75). Manning's molecular theory of condensation has been used in analyses of competitive binding equilibria (5, 79–83), potentiometric titration (84), transport properties (85, 86), conformational transitions (5, 87), and other salt-dependent phenomena peculiar to nucleic acids, such as collapse (88), unwinding (89), and bending (90–92).

For many years theoretical descriptions of the thermodynamic and hydrodynamic properties of polyelectrolyte solutions have been based on the Poisson-Boltzmann (PB) equation, which, unlike Manning's condensation theories, is not confined to polyions of cylindrical symmetry. In this article only the cylindrical PB equation and its thermodynamic consequences are considered. The exact analytic solution of this equation is known only for solutions containing no added salt, in which all the counterions have the same valence (93, 94). [For salt-free solutions containing two types of univalent counterions of different size, the PB equation can still be solved analytically (95); numerical solutions have been computed for salt-free solutions containing various proportions of uni- and bivalent counterions (96).] The linearized (or "Debye-Hückel") form of the cylindrical PB equation can be solved analytically (97), but the radial dependence of the resulting "Debye-Hückel" potential is inaccurate near the polyion unless its charge density is relatively low or the concentration of added salt is high, or both. These conditions are rarely met in systems that have been investigated experimentally. In solutions having a high enough ratio of added salt to polyion charges, the

Debye-Hückel potential always is valid at large enough distances from the polyion, whereas in the region of solution around the polyion where the concentration of coions can be neglected, the functional form of the analytic solution for the salt-free case is approximately valid. Thus, it is plausible to attempt matching appropriately parameterized salt-free and Debye-Hückel potentials to construct a "hydrid" potential that is approximately valid both near and far from the polyion. Various criteria for matching these potentials have been proposed (98–102) and other ways of approximating the solution of the cylindrical PB equation have been devised for the case of excess uniunivalent salt (103–106). Accurate numerical solutions of this equation have been computed for a wide range of polyion charge densities and salt concentrations (8, 107). The cylindrical PB equation appropriate for polyions in solutions with excess salt consisting of a mixture of uni- and bivalent counterions has also been solved numerically (106, 108, 109).

For a salt-free solution of cylindrical polyions, expressions for the thermodynamic properties (osmotic coefficient, counterion activity coefficient) can be derived analytically because the solution of the PB equation is known (110). The lack of an analytic solution of the cylindrical PB equation for the case of added salt complicates its use in analyzing thermodynamic measurements. Analyses of this kind have been extensively reviewed (7, 12, 35, 111). More recently the chemical potentials of the electrolyte and polyelectrolyte components of a solution containing an isolated cylinder in the presence of excess uniunivalent salt were computed, and their thermodynamic self-consistency was demonstrated (8). The computations required numerical integrations of functions of the PB potential, which also was computed numerically. A simpler alternative way to evaluate the Donnan coefficient for this system has been presented (10). This approach, valid only for cylindrical geometry, is based on a simple analytic relationship between the known salt concentrations on each side of the semipermeable membrane and the PB potential at the polyion surface (obtained numerically). By introducing this relationship into some thermodynamic results of the PB cell model (112), relatively simple closed-form expressions have been derived for the mean ionic activity coefficient and osmotic coefficient characteristic of a solution containing cylindrical polyions and excess uniunivalent salt (11). The evaluation of these expressions is as straightforward as the calculation of a potentiometric titration curve, because only the polyion surface potential is required.

Despite the widespread utilization of the cylindrical PB equation, there have been few attempts to examine any features of the model upon which the solution of the equation is based. For cylindrical polyions having a periodically varying continuous surface charge density with average value in the range $10^{-4}-10^{-2}$ charges per $Å^2$, approximate solutions of the PB equation were computed using series expansions (103). For salt concentrations in the range 0.01–0.2 M, the *average* potential at the polyion surface shows little departure from the value for the corresponding uniformly charged cylinder. Marked deviations about this average are manifested at charge densities lower than $10^{-3}/Å^2$. (The surface charge density of double-helical DNA is $\sim 10^{-2}/Å^2$.) The effect of spatial variations in dielectric saturation on the polyion surface potential has been computed by incorporating a semiempirical function for the dielectric tensor into the cylindrical PB equation (113). For highly charged polyions the magnitude of this correction appears significant, but its salt-dependence at a fixed charge density is small.

Before 1979 there had been no rigorous investigation of the statistical mechanical validity of the cylindrical PB equation. For solutions of low molecular weight electrolytes, the spherical PB equation is known to be theoretically unsatisfactory except for highly dilute solutions, where it can be linearized (114). However, in polyelectrolyte solutions the natural asymmetry resulting from the large number of like charges on a single molecule affords some reason to expect that the PB equation, in the reference frame of the polyion, may remain valid well above the limit of infinite dilution. If the polyion charge density is high enough and the concentration of added salt not too high, interactions of the field due to polyion charges with the small ions in solution may make more important contributions to the excess electrostatic free energy than do interactions among the small ions (25). The first analytic approach to testing the effects of small ion interactions on the accuracy of the PB equation was presented by Fixman (9). In view of the significance of his conclusions, and the complexity of the demonstration, a somewhat detailed summary is appropriate.

A rigorous generalized form of the PB equation is derived for a polyelectrolyte of any geometry in a solution containing added salt. This equation allows for the possibility of a variable dielectric tensor, external electric fields, and non-Coulombic interparticle forces. The effect of interactions (or correlations) among the small ions is incorporated in terms of an integral over the difference between their Coulombic interaction potential and the suitably normalized direct correlation function. The magnitude of this integral is estimated in the region near the surface of a typical highly charged polyion, where the local counterion concentration is ≥ 1 M even if the bulk salt concentration is ≤ 0.1 M. (The estimate is obtained by adapting the mean spherical approximation for the restricted primitive model of an electroneutral solution (0.10 M) of small ions with hard sphere diameters of 4 Å.) Comparing the estimated correction with the PB potential itself indicates that small ion interactions do indeed introduce significant error into the "PB approximation," whereby the potential of mean force in the Boltzmann factor is replaced by the mean potential (115). The failure of this approximation seems to invalidate the usual (uncorrected) form of the PB equation. However, Fixman goes on to show that, at least for dilute solutions of highly charged polyions, the solution of the usual PB equation, and the corresponding ion distributions, may remain accurate at surprisingly high salt concentrations.

To estimate the range of accuracy of the usual PB equation for a uniformly charged cylindrical polyion, Fixman obtains a formal solution of this equation in terms of an integral over the excess charge density. The integrand is parameterized in a way that conforms with a simple physical picture of counterion condensation. The parameters are chosen so that close agreement is maintained, over a wide range of salt concentration and polyion charge densities, between the approximate solution (in terms of a transcendental equation) and the exact numerical solution of the PB equation at the polyion surface. One of the parameters in the approximate solution reflects the local counterion charge density at the surface. This parameter is correlated with the correction that would result from properly accounting for interactions among the small ions. From the insensitivity of the approximate solution to large changes in the charge density parameter, it is inferred that the solution of the generalized PB equation cannot be very sensitive to the significant failure of the PB approximation caused by interactions among the small ions. (For example, if the charge density parameter is doubled in the "approximate solution," the corresponding value of the local counterion concentration at the polyion surface is only about 25% greater. In this calculation the polyion is assumed to have the radius and charge density of DNA, and the solution is 0.1 M in uniunivalent salt. Smaller discrepancies occur for lower salt concentrations.) Fixman concluded that if the mean potential at the polyion surface is high enough, it is effectively "buffered" against contributions from small ion interactions by the functional form of the generalized PB equation.

To establish more precise bounds on the applicability of the usual cylindrical PB equation, numerical solution of Fixman's generalized PB equation should be useful, especially as a means of examining more realistic models of the polyelectrolyte solution. At present the theoretical basis for applying the uncorrected PB equation to moderately concentrated salt solutions containing cylindrical polyions is as well developed as is the CC theory for such systems. Therefore, it is appropriate to consider the various ways in which the two theories can be compared. At the experimental level their relative merits have been amply discussed (1, 15). It is clear that results derived from the CC theory have the advantages of being simpler, devoid of empirical or unmeasurable parameters, and generally capable of providing agreement with experiment. However, there are applications for which the PB theory has proved to be at least equally well-suited (107, 113, 116–118). At the theoretical level it is pertinent to consider the simple model for a polyelectrolyte solution upon which each theory is based and then to examine how each theoretical development proceeds from this model to derive expressions that can be used to analyze experimental measurements. In the following sections we discuss the model assumptions and the two theoretical approaches to the derivation of limiting law expressions for the thermodynamic properties. Perhaps the most fundamental basis of comparison between the two theories is at the molecular level of the counterion radial distribution function.

In the PB theory the radial distribution functions are obtained directly from the appropriate Boltzmann factors of the mean potential. Although the counterion radial distribution function does not enter explicitly into either of Manning's formulations of the CC theory, the later one (4) does quantify the solution volume within which the condensed counterions are located. (We discuss this theory in greater detail in a following section.) Whether the counterion distributions predicted by the two theories agree even qualitatively in the limit of infinite dilution is currently a matter of controversy. A mathematical analysis of the characteristics of an upper bound for the exact solution of the cylindrical PB equation led to the conclusion that the Debye length, κ^{-1}, diverges more strongly than the radius enclosing the predicted number of condensed counterions (119). However, the limiting behavior of the latter radius was not clarified. For an infinite *line* of continuous charge density having $\xi > 1$, the radius enclosing the condensed counterions has been shown to collapse to an "encapsulated delta function" in the limit of infinite dilution in solutions containing either excess (120) or no (121) added salt. For a highly charged ($\xi > 1$) infinite cylinder in salt-free solution, an analytic expression for the counterion radial distribution function in the limit of infinite dilution has been derived (102). This expression does not agree quantitatively with the volume of the condensed layer predicted by Manning's theory (4), but indicates the persistence of a high counterion charge density in the vicinity of the polyion as the concentration goes to zero. However, counterion radial distribution functions computed from the cylindrical PB equation for the case of excess salt give no indication that a finite charge density near the surface of a highly charged polyion remains in the limit of infinite dilution (38). A mathematical approach to bringing the counterion radial distributions predicted by the two theories into closer correspondence has been examined (122). At moderate salt concentrations (.01–.3 M) the two theories are in qualitative agreement in predicting a large and relatively salt-independent counterion accumulation close to a highly charged polyion ($\xi > 1$). The CC theory generally predicts a higher density of counterions within a given volume surrounding the polyion.

Spectroscopic methods of various kinds provide the most direct experimental means of gaining information about the counterion radial distribution near a polyion (6, 35). Changes in the ^{23}Na NMR quadrupolar relaxation rate as a function of the extent of protonation (charge density) of poly(acrylic acid) have been analyzed in terms of the PB (123) and CC (124) theories. The latter permits a simpler and more self-consistent molecular interpretation. More recently ion distributions of both ^{23}Na (or ^{7}Li) and Mn(II) were computed for solutions of poly(acrylic acid) (109). These computations were found to give an accurate simulation of the NMR results (125). The applicability of Manning's "two-variable" theory (5) for the competitive condensation of uni- and bivalent counterions was not tested. Another direct approach to the radial distribution function is afforded by Monte Carlo simulations. Such computations are appearing regularly now for charged sheets. They have been used to investigate the accuracy of the planar (Gouy-Chapman) PB equation (131–134), and of a more rigorous version of this equation (135). For cylindrical polyions in the absence of added salt, the counterion radial distribution predicted by the analytic solution of the PB equation (94) has been compared with Monte Carlo simulations for hard sphere counterions of different radii (102). This PB equation was found to give accurate ion distributions for counterions of diameter <6 Å; however, interpolyion interactions were not included in the simulation. It has recently been demonstrated that Monte Carlo simulations of ion and solvent distributions are feasible for highly detailed and realistic models of polyions such as helical DNA (136). At the inescapable expense of complexity, Monte Carlo methods do have the distinct advantage of allowing the consequences of many model assumptions to be checked, singly, and in groups. Although interpolyion interactions may be difficult to simulate, Monte Carlo computations probably are the most promising means of accounting accurately for interactions among small ions. A Monte Carlo approach to treating the effects of ion distributions on polyion configurations in solution has been proposed (137).

Comparison of the Models and Their Thermodynamic Consequences

Both the Poisson-Boltzmann (PB) and counterion condensation (CC) theories are based on essentially the same model for the components of a polyelectrolyte solution. The solvent is assumed to be a continuous medium having a spatially uniform dielectric constant, ε, that does not vary with solution composition (local or bulk). The small mobile ions (counterions and, if added salt is present, coions) are assumed to form a continuous charge density, or ionic atmosphere, which is highly nonuniform in the reference frame of the polyion. Interactions between the polyion charges and its ionic atmosphere are purely Coulombic. Differences in the size, polarizability, and solvation of the small ions are (usually) ignored. The screening effect of the ionic atmosphere is characterized by the Debye length, κ^{-1}. In Manning's CC theory (1), κ is proportional to the square root of the ionic strength determined by the concentration of all the small ions (including those originating from the polyelectrolyte component). In the Poisson-Boltzmann cell model κ has a different definition (112), though it often is nearly identical to Manning's in magnitude (for example, in solutions containing excess uniunivalent salt). In both theories the Debye length is a measure of the distance, in the reference frame of the polyion, over which electrostatic interactions remain significant. If the Debye length is less than the local radius of curvature of the polyion chain, then knowledge of the configurations of the chain in solution may be unnecessary for the purpose of computing properties that depend on polyion interactions with the small ions. On the basis of this assumption the polyion is modeled as an infinitely long straight line (1, 15) or impenetrable cylinder (36). The true structural charge distribution on the polyion is idealized as a linear array of regularly spaced discrete charges (4, 66, 68, 69), or as a line of uniform, continuous charge density (1, 119), or as a cylindrical surface of uniform, continuous charge (93, 94). For each of these models the polyion charge density can be characterized by the nondimensional parameter $\xi = e^2/\varepsilon kTb$.

The charge density parameter ξ enters into the solution of the PB equation by an application of the Gauss' Law boundary condition at the surface of a uniformly charged impenetrable cylinder. The PB equation for this model can be solved for *any* value of ξ, at *any* salt concentration.

The PB potential at the surface, $y_a = e|\psi(a)|/kT$, does diverge in the limit of infinite dilution, but the local concentration(s) of small ions at the polyion surface remains finite (10). The analytic solution of the cylindrical PB equation for the salt-free case takes on different functional forms, in the limit of infinite dilution, depending on the magnitude of ξ relative to unity (93). There is a concomitant dependence of the analytic form of the limiting-law expression for the osmotic coefficient on the magnitude of ξ; the possible correspondence between this effect and the concept of counterion condensation has been discussed (7). From numerical solutions of the PB equation for the case of excess added salt, analytic limiting law expressions for the Donnan salt exclusion coefficient were deduced (126). Again, a striking dependence on the magnitude of ξ relative to unity is encountered.

The significance of ξ in determining the limiting law expressions for thermodynamic properties derived from the PB equation was noted in Manning's first paper on counterion condensation (1). But his theoretical development was not founded on any explicit consideration of the characteristics of the solution (exact or approximate) of the cylindrical PB equation. Some confusion has arisen concerning the relevance of the Debye-Hückel solution of the linearized PB equation to one of Manning's basic postulates: "that the uncondensed ions can be treated in the Debye-Hückel approximation." In connection with this postulate an earlier investigation of the cylindrical PB equation was cited (100). The principal conclusion of this mathematical analysis was that the Debye-Hückel solution to the PB equation remains an "asymptotically valid" approximation to the exact (numerical) solution of the PB equation, in the limit of infinite dilution. The implication that in this limit the Debye-Hückel potential provides a numerically accurate representation of the counterion radial distribution function has subsequently been disputed, on the basis of both analytic (101, 102) and numerical (9, 107) investigations of the PB equation. However, these findings do not necessarily undermine Manning's use of the "Debye-Hückel approximation," which refers to a way of deriving an expression for the contribution to the excess electrostatic free energy of an infinitely long, continuously charged line, in the limit of infinite dilution. The derivation proceeds by expressing the potential acting on any point on the line as a superposition of screened Coulombic potentials arising from all of its charged segments. This potential is then subjected to a (Guntelberg) charging process to derive the "Debye-Hückel approximation" to the excess electrostatic free energy due to the polyion. This expression is rigorous in the limit of infinite dilution because it is identical to the lead (ring) term in the appropriate MacMillan-Mayer cluster series expansion (2, 67, 127).

For moderately charged cylindrical polyions ($\xi<1$), the dominance of the Debye-Hückel term over higher order terms in the cluster expansion has been demonstrated numerically at low salt (67). However, in the experimentally accessible range of salt concentrations, the Debye-Hückel approximation has been found inadequate for polyions with $\xi<1$, in the sense that higher order terms in the free energy cluster expansion make nonnegligible contributions to the thermodynamic properties of the solution (72, 74). For highly charged polyions ($\xi>1$) Manning did not explicitly derive limiting law expressions from an expression for the excess electrostatic free energy. Instead, the value of ξ in each of the limiting law expressions derived from the Debye-Hückel form of the excess electrostatic free energy was assigned the value unity, and an appropriate adjustment in the number of unbound counterions was made. A justification for this procedure (2) was based on the salt-dependence of some higher order cluster integrals which, in the limit of infinite dilution, diverge if ξ exceeds unity. [Some aspects of the cluster series expansion appropriate for a finite polyion have recently been clarified (70).] Above the limit of infinite dilution, there is as yet no demonstration, in terms of a cluster series representation of the radial distribution function, that counterion condensation is a physical binding phenomenon. The experimental evidence indicating the existence of a highly concentrated zone of mobile counterions close to a cylindrical polyion has been accounted for by Manning's later theory of counterion condensation (6).

The object of Manning's first condensation theory was the derivation of limiting law expressions for the polyelectrolyte contribution to the "colligative properties"—the Donnan salt exclusion coefficient, the osmotic coefficient, the mean and single ion activity coefficients. These expressions are derived by taking appropriate derivatives of the excess electrostatic free energy, formulated in the "Debye-Hückel" approximation discussed above. Interactions among the small ions and interpolyion interactions are assumed to make negligible contributions to the excess electrostatic free energy. The former approximation can be justified in the limit of infinite dilution, but, even in this limit, the latter approximation appears to require sufficient excess salt to ensure the proper behavior of the cluster expansion (127). The physical relevance of the limit of infinite dilution may appear dubious, not only because it is experimentally inaccessible, but also because the cylindrical model for the polyion must fail when its length is exceeded by the Debye length. Consideration of this limit is justified by assuming that there exists a range of salt concentrations over which the rod-like model is still valid, and at which the low-salt asymptotic functional form of the excess electrostatic free energy has effectively been attained. The limiting laws derived by Manning are simple analytic expressions depending only on ξ, whose magnitude relative to unity determines their functional forms. In applying them to experimental data, two further significant approximations are commonly employed. Even though polyion concentrations are typically high, and the ratio of added salt to polyion monomer not much different from unity, interpolyion interactions are neglected. Interactions among the small ions in the usual experimental range of salt concentrations cannot be neglected. Instead, their contribution to the excess electrostatic free energy is treated as an independent additive term, whose magnitude is estimated semiempirically by measurements on comparable salt concentrations containing no polyions. A theoretical justification for this procedure has been given, in terms of an appropriate mode-mode expansion (71).

Recently a set of expressions for the Donnan coefficient (10), osmotic coefficient, and mean ionic activity coefficient (11) characteristic of a solution of cylindrical polyions containing excess uniunivalent salt was derived on the basis of the PB cell model (112), specialized to cylindrical geometry. This model, as developed by Marcus, bypasses the explicit details of charging the PB potential at the polyion surface and, without evaluating the excess electrostatic free energy, arrives directly at simple expressions for the polyion contribution to the thermodynamic properties in terms of the local concentrations of the small ions at the boundary of an electroneutral cell surrounding the polyion. At this finite boundary it is assumed that the electric field in the reference frame of the central polyion vanishes. This outer boundary condition in some sense accounts for interpolyion interactions, but the exact nature of the approximation involved has not been elucidated. (Interactions between highly charged cylindrical polyions have been treated by using the Debye-Hückel PB potential to compute the appropriate virial coefficients (128).) In solutions containing a moderate excess of added salt over polyion monomer concentration, the neglect of interpolyion interactions appears justified (12) and this condition ensures that the concentrations of counterions and coions at the outer boundary are very nearly equal. The difference between $C_3(R)$, the small ion concentration at the outer boundary, and C_3, the bulk salt concentration, is *not* negligible in the context of the PB cell model, unless a huge excess of salt is present. For polyions of cylindrical symmetry, $C_3(R)$, C_3, and S_a where:

$$S_a \equiv (1/4)(x_a)^2 [\exp[y_a] + \exp[-y_a]]$$

$$x_a^2 \equiv 8\pi e^2 a^2 C_3(R)/\varepsilon kT$$

are related by a simple expression (10). Introducing this relationship into the general expressions for the thermodynamic properties derived by Marcus produces simple closed form expressions that are explicit functions of y_a. At very low salt concentrations the asymptotic forms of these expressions are dominated by S_a, whose limiting values are the following (in the absence of physical binding of counterions—see section below):

If $\xi \geqslant 1$, $\lim_{x_a \to 0} S_a = (\xi - 1)^2$ 1a.

If $\xi \leqslant 1$, $\lim_{x_a \to 0} S_a = 0$ 1b.

Because S_a is directly proportional to the local mobile ion concentration(s) at the polyion surface, Eq. 1a indicates that *even* in the limit of infinite dilution, this local concentration remains finite, if $\xi > 1$. For more weakly charged polyions ($\xi \leqslant 1$), the local ion concentration vanishes in the limit of infinite dilution. The persistence of a finite local counterion concentration at the surface of a highly charged polyion is strikingly reminiscent of the condensation hypothesis, which, interpreted in terms of binding, requires that if $\xi > 1$ the fraction $1 - \xi^{-1}$ counterion per polyion charge remains bound in the limit of infinite dilution. However, in the thermodynamics of the PB cell model a local ion concentration is equivalent to a local chemical activity (112). Thus, S_a is, in general, only a crude measure of the number of ions enclosed within a *finite* solution volume. Although the low salt limiting forms of S_a are not *direct* measures of the extent of physical binding, they ensure that the limiting law expressions derived from the PB cell model for the Donnan coefficient (10), osmotic coefficient and mean ionic activity coefficient (11) are identical to the corresponding limiting laws obtained from the CC theory (1), at least for the case of excess added salt. This striking correspondence between the thermodynamic consequences of the PB and CC theories must ultimately be due to an exact agreement between the asymoptotic salt dependences of the excess electrostatic free energies predicted by the two theories. [Further clarification of this point is being developed (B. K. Klein, M. T. Record, Jr., and C. F. Anderson, unpublished)].

The characteristic salt dependences of the thermodynamic properties discussed thus far are all determined primarily by the PB potential at the polyion surface. Thus, it is appropriate to examine the salt-dependence of y_a. In view of the controversy concerning Manning's use of the "Debye-Hückel" approximation (1), it is of particular interest that the solution of the linearized PB equation can be used (or modified) to provide an accurate measure of the salt dependence of y_a, both in very concentrated salt solutions and in the limit of infinite dilution. In the first column of

Table 1 are numerical computations of y_a for an isolated cylinder in solution with excess uniunivalent salt (130). The next three columns tabulate approximate values of the surface potential obtained in alternative ways: (a) by assuming that the analytic solution of the linearized (Debye-Hückel) PB equation, y_a (DH), is valid at the polyion surface for any value of the potential; (b) by assuming that the low salt asymptotic functional form of the exact PB equation, y_a (A), is valid at any salt concentration; (c) by inverting the low salt limiting expression for S_a, valid for $\xi > 1$ (10). The tabulated potentials cover values of ξ that are representative of most biopolyelectrolytes (for DNA, $\xi = 4.2$). The range of the variable $x_a = 0.33a[C_3]^{1/2}$ (a in Å, C_3 in moles/liter), corresponds to 6×10^{-6} M $\leq C^3 \leq 6 \times 10^{-2}$ M for polyions with $a = 13$ Å, the value often assumed for helical DNA. At salt concentrations lower than 10^{-4} M, there are few reliable means of studying a polyelectrolyte solution, whereas at salt concentrations higher than 0.1 M, the theoretical accuracy of the PB equation is unsubstantiated (9).

The analytic solution of the linearized PB equation is

$$y_a (\text{DH}) = 2\xi K_0(x_a)/x_a K_1(x_a)$$

where K_0 and K_1 are modified Bessel functions (97). At salt concentra-

Table 1 The dependence of the surface potential on polyion charge density and salt concentration

	y_a	y_a (DH)	y_a (A)	y_a (S_a)	S_a	ΔS_a
$\xi = 0.5$						
$x_a = 0.01$	4.54	4.72	4.60		2.34×10^{-3}	2.34×10^{-3}
$= 0.10$	2.36	2.46	2.30		2.67×10^{-2}	2.67×10^{-2}
$= 1.0$	.691	.700	0		.624	.624
$\xi = 1.0$						
$x_a = 0.01$	7.88	9.44	9.21		6.61×10^{-2}	6.61×10^{-2}
$= 0.10$	4.21	4.93	4.60		.168	.168
$= 1.0$	1.34	1.40	0		1.02	1.02
$\xi = 2.0$						
$x_a = 0.01$	10.78	18.9	9.21	10.6	1.20	.20
$= 0.10$	6.36	9.85	4.60	5.99	1.44	.44
$= 1.0$	2.41	2.80	0	1.39	2.81	1.81
$\xi = 4.2$						
$x_a = 0.01$	12.95	39.7	9.21	12.92	10.52	.28
$= 0.1$	8.37	20.7	4.60	8.32	10.79	.55
$= 1.0$	3.93	5.88	0	3.71	12.73	2.49

tions high enough to ensure that $y_a < 1$, the PB equation evidently can be linearized and y_a (DH) should be a good approximation to y_a. Inspection of the tabulated values confirms that for all ξ, agreement between y_a (DH) and y_a improves with increasing x_a. Even at low values of x_a for $\xi = 0.5$, y_a (DH) remains a good approximation to y_a, although the latter is significantly greater than unity. A still better approximation to y_a (PB) for $\xi = 0.5$ at low values of x_a is provided by the asymptotic PB potential, y_a (A) $= -2\xi \ln x_a$ (valid for any $\xi < 1$ (100). This asymptotic functional form of the exact solution to the PB equation *coincides* with the asymptotic functional form of y_a (DH), because as $x_a \to 0$, $-K_0(x_a) \sim \ln x_a$ and $x_a K_1(x_a) \sim 1$. This coincidence ensures that y_a (DH) and y_a provide the *same* limiting laws for the polyion contribution to thermodynamic properties, even though the PB equation evidently cannot be linearized as $x_a \to 0$ (because $y_a \to \infty$).

For highly charged polyions ($\xi > 1$), a comparison of the tabulated values of y_a and y_a (DH) indicates their poor agreement at low salt concentrations. For any value of ξ, the low salt asymptotic form of y_a (DH) is $-2\xi \ln x_a$, but for $\xi \geq 1$, the low-salt asymptotic form of the exact (numerical) potential at the polyion surface is y_a (A) $= -2 \ln x_a$ (119). This logarithmic salt-dependence can be obtained simply by assigning ξ the *effective* value of unity in the low-salt asymptotic expression for y_a (DH). This modification of the Debye-Hückel potential at the surface of a highly charged polyion does provide the same limiting laws of the polyion contributions to solution nonideality as does the exact PB surface potential. Moreover, these limiting laws are in agreement with those derived by Manning from the hypothesis of counterion condensation (1, 11). In the context of the PB equation, it is not possible to attach a direct molecular interpretation to the reduction of ξ in y_a (DH) that ensures the correct asymptotic salt dependence as $x_a \to 0$. To arrive at an interpretation in terms of counterion association, it is necessary to examine the *radial distribution* predicted by the PB equation in the limit of infinite dilution (119, 120).

For $\xi > 1$, the tabulated values of y_a are not very close to those of y_a (A), even at $x_a = 0.01$. However,

$$y_a(S_a) = \ln\left[4(\xi - 1)^2\right] + y_a(A)$$

is a good approximation to y_a, especially for $\xi = 4.2$ [The simplicity of this analytic expression is striking. More accurate analytic approximations are available, but they require the solution of transcendental equations (9, 99, 102) or introduce arbitrary parameters (106)]. The physical significance of the close agreement between $y_a(S_a)$ and y_a for $\xi = 4.2$ is that the latter maintains an approximately linear dependence on $-2\ln x_a$ at salt concentrations well into the typical experimental range (130). Some molecular and thermodynamic implications of this logarithmic salt dependence are discussed in the section below. In view of the importance of divergences in Manning's formulation of condensation theory it is interesting to note that if, in the limit of infinite dilution, y_a for $\xi > 1$ did have the "Debye-Hückel" asymptotic form, $-2\xi \ln x_a$, S_a would diverge! The correct asymptotic form of y_a prevents a nonphysical divergence of the local counterion concentration at the polyion.

The salt dependence of y_a is directly relevant in PB analyses of competitive binding (see final section) and potentiometric titrations; the salt dependences of other thermodynamic properties of a solution of cylindrical polyions in the presence of excess uniunivalent salt are determined by S_a (10, 11). In the PB expressions for these properties the contribution of S_a is scaled by the molar volume of the polyion. Therefore, the relative magnitudes of deviations of these properties from their limiting law values cannot be directly inferred by comparing the tabulated values of

$$\Delta S_a = \left(S_a - \lim_{x_a \to 0} S_a \right)$$

to those of S_a. However, the Table does provide a useful indication of the sensitivity of ΔS_a to trends in ξ and x_a. At fixed x_a, $\Delta S_a / S_a$ decreases steadily if $\xi \geq 1$; but ΔS_a is a monotonically increasing function of ξ. Because of the latter dependence, the PB theory predicts that at moderate salt concentrations deviations of the thermodynamic coefficients from their low salt limiting values exhibit greater salt-dependences for more highly charged polyions.

COUNTERION BINDING IN POLYELECTROLYTE SOLUTIONS

The interactions between a counterion and a cylindrical polyion in solution have been described in terms of two types of dissociation: physical and thermodynamic. The fraction of counterions in a polyelectrolyte component, $M_N^+ P^{N-}$, that are physically dissociated from the polyion in solution is designated α. Incomplete physical dissociation of counterions reduces the structural charge density ξ to the effective value $\lambda \equiv \alpha \xi$. The nature of the interaction between the $N(1-\alpha)$ bound counterions and N polyion charges has been interpreted differently in the two theories. In the PB theory the $N(1-\alpha)$ ions usually are considered site-bound (112, 126) (but see below). In the CC theory (4), condensed ions are considered to be (predominantly) mobile near the polyion surface. The thermodynamic degree of dissociation, designated i, refers to the fraction of counterions from $M_N^+P^{N-}$ which can be considered dissociated in order to avoid explicit consideration of the nonideality arising from small ion-polyion interactions in analyzing thermodynamic measurements. The rigorous definition of i in terms of the appropriate thermodynamic derivative is given below (138). (The symbol i has sometimes been used nonrigorously in conjunction with additivity rules (12).) In general i < α because i incorporates not only physical dissociation but also the effects of long-range electrostatic interactions. Some of the misunderstandings that have arisen concerning the description of counterion-polyion interactions are due to the failure to use a thermodynamically rigorous definition of i and the failure to distinguish i from α. In this section the relationship between these quantities is discussed in terms of expressions derived from both the PB and CC theories.

In experimental and theoretical investigations of cylindrical polyions in solution with a single type of electrolyte, the principal independent variables, in addition to ξ and α, are (a) the polyelectrolyte concentration on the molar (C_2) or molal (m_2) scales (or equivalent concentrations $C_u = NC_2$ or $m_u = Nm_2$, where N is the number of structural charges on the polyion), and (b) the electrolyte concentration on the molar (C_3) or molal (m_3) scale. The partial molar volume of the polyelectrolyte may make a very important contribution to the theoretical thermodynamic properties expressed on the molar scale. Therefore, electrostatic contributions to these properties are conveniently isolated by use of the molal scale (B. K. Klein, M. T. Record, Jr., and C. F. Anderson, unpublished work). However, in discussing local concentrations in the context of the PB theory, the molar scale is appropriate. For example, the quantity S_a defined above can be expressed in terms of the local molar concentrations of counterions and coions at the polyion surface:

$$S_a = 2\pi\xi a^2 b[C_+(a) + C_-(a)].$$

The Physical Degree of Dissociation, α

PB ANALYSIS PB theory predicts that the local molar concentration of counterions, $C_+(a) = C_3(R)\exp[y_a]$, at the surface ($r = a$) of a cylindrical polyanion greatly exceeds the bulk value (C_3), as a consequence of the Boltzmann distribution of ions in the electric field surrounding the polyion. For highly charged polyions ($\lambda > 1$) in excess salt ($C_3 \gg C_u$), $C_+(a)$ is relatively insensitive to changes in C_3, if C_3 is not too high (37, 38). Even in the limit $C_3 \to 0$, $C_+(a)$ retains a large (limiting-law) value (10):

$$\lim_{C_3 \to 0} C_+(a) = (\lambda - 1)^2 / 2\xi \overline{V}_u \qquad 2.$$

where $\overline{V}_u$ is the molar volume of the polyion monomer. Typical values of $C_+(a)$ under limiting-law conditions are of order 1 M.

The existence of such a high local concentration of counterions will promote a significant extent of physical counterion binding, $(1-\alpha)$, if any region of the polyion (not necessarily the fixed charges) possesses even a weak chemical affinity (i.e. an equilibrium constant K of order 1 M^{-1}) for the counterion. [An example may be the interaction between tetramethyl or tetraethyl ammonium cations and the AT base pairs of DNA (139).] Minimization of the total free energy, using the PB expression for the electrostatic free energy, yields the local mass action law (in excess salt) (112):

$$K = (1-\alpha)/\alpha C_+(a) = (1-\alpha)/\alpha C_3(R)\exp(y_a) \qquad 3.$$

where $C_3(R)$, the molar electrolyte concentration at the outer boundary of the cylindrical cell, in this context may be approximated by C_3. From Eqs. 2 and 3 an analytical expression is obtained for α as a function of K and ξ, under limiting law conditions. At finite C_3, values of y_a and $C_3(R)$ can be obtained by numerical solution of the PB equation in order to calculate α from Eq. 3. Since $C_+(a)$ is found to be buffered against changes in C_3, α also decreases only slowly with increasing C_3.

In the context of the PB approach, α may also be interpreted as a parameter that corrects for failures in the model near the polyion surface (e.g. the assumptions of a uniform charge density, a uniform dielectric constant, and a regular cylindrical geometry). Alternatively, $1-\alpha$ may include those ions contained within the cylindrical volume of the polyion (e.g. ions in the grooves of helical DNA).

CC ANALYSIS   The central tenet of CC theory is that the effective charge density $\lambda = \alpha\xi$ of a cylindrical polyion ($\xi > 1$) is reduced to the critical value $\lambda = 1$ by the phenomenon of counterion condensation. In Manning's original formulation of the theory, condensation was introduced as a mathematical device necessary to prevent the divergence of the phase integral of an infinite line charge under limiting law conditions; no physical interpretation of condensation in terms of a mode of counterion binding was provided (1). Nevertheless, the logical physical inference that the $1-\xi^{-1}$ condensed ions per polyion monomer are closely associated with the polyion (and may be considered to be bound) was utilized in applications of the theory (23, 62). Numerous experiments have strongly suggested the approximate validity of this picture, not just under limiting-law conditions but over a wide range of salt concentrations (6). Recently Manning reformulated the condensation model to provide (a) a molecular picture of counterion condensation; (b) an explanation of the apparent validity of the condensation criterion ($\alpha = \xi^{-1}$) at finite salt concentrations; and (c) a theoretical framework for the analysis of the competitive binding of counterions of different valence. The condensed counterions are assumed to occupy a coion-free cylindrical shell of volume V_p surrounding the polyion. Though trapped radially at a uniform local concentration C_{loc} by the polyion field, the counterions are assumed to be free to translate on the polyion surface, and consequently are referred to as territorially bound. Three key assumptions are employed in developing the consequences of this model:

1. The real charge distribution on the polyion is replaced by a line of evenly spaced discrete charges.
2. The territorially bound ions are assumed to reduce the magnitude of each discrete charge from unity to the effective value α.
3. The volume V_p of the coion-free region occupied by the condensed ions is assumed to be a structural feature of the model, and hence independent of salt concentration.

Assumptions 1 and 2 motivate the use of a modified Debye-Hückel free energy, which may be considered to result from a superposition of the screened Coulomb potentials of a linear array of charges, separated by the distance b and assigned the effective value α. (Use of the equally plausible alternative assumption of a uniformly charged cylinder of effective charge density λ would provide the same result for the electrostatic free energy under limiting-law conditions (130) but would not retain the logarithmic dependence on C_3 over as wide a range of salt concentration.) Minimization of the total free energy (electrostatic and mixing contributions) under limiting law conditions demonstrates that $\alpha = \xi^{-1}$ and provides an analytic expression for V_p. If V_p is assumed to be independent of C_3, then the result $\alpha = \xi^{-1}$ is shown to be independent of C_3.

COMPARISON OF THE PB AND CC PREDICTIONS FOR THE LOCAL COUNTERION CONCENTRATION AND FOR α The PB and CC theories provide qualitatively similar pictures of the ion distribution near a highly charged polyion. This local counterion concentration is very large and is relatively insensitive to changes in the bulk electrolyte concentration as a result of the buffering action of the polyion field. Coions are virtually absent from the solution volume near the polyion. In the CC theory, the counterions in the local region are considered to be territorially bound to the polyion (which necessitates their inclusion in α as we have defined it). In the PB theory, these ions are treated as being unbound, though their high local concentration may drive binding.

The CC theory, applied to helical DNA ($\xi = 4.2$), predicts that the charge fraction after condensation is $\alpha = \xi^{-1} = 0.24$ in a uniunivalent salt solution. The territorially bound counterions (the fraction $1 - \xi^{-1} = 0.76$ per phosphate) are predicted to occupy a region extending radially from the surface of the DNA molecule (radius 10 Å) for a distance of 7 Å. In this region the local counterion concentration $C_{loc} = 1.2$ M, independent of C_3 as long as $C_{loc} \gg C_3 \gg C_u$. The predictions of the PB theory depend on the choice of α. For $\alpha = 1, C_+(a) = 2.4$ M under limiting-law conditions (assuming $a = 13$ Å); $C_+(a)$ increases slowly with increasing C_3. If a weak intrinsic binding constant $K = 1$ M^{-1} is assumed in Eq. 3, then α decreases from 0.63 at $C_3 \to 0$ (where $C_+(a) \simeq 0.6$ M), to 0.50 at $C_3 = 0.2$ M (where $C_+(a) \simeq 1$ M).

The Thermodynamic Degree of Dissociation i

DEFINITION OF i IN TERMS OF THE PREFERENTIAL INTERACTION PARAMETER AND THE DONNAN COEFFICIENT The thermodynamic degree of dissociation (i) includes contributions from the physical degree of dissociation (α) and from local ion concentration gradients in the vicinity of the polyion (screening or nonideality contributions). Since i is a thermodynamic parameter characterizing the extent of ion-polyion interactions, its rigorous theoretical definition is obtained from the electrolyte-polyelectrolyte preferential interaction parameter Γ_{3u}°, expressed on the basis of polyion charges (u) (138). The corresponding experimental definition of i is obtained from the molal Donnan (coion) distribution coefficient, Γ_m°:

$$\Gamma_m^{\circ} \equiv \lim_{m_u \to 0} (m_3 - m_3')/m_u \qquad 4.$$

where m_3' is the salt concentration in the polyion-free solution.

Thermodynamic (colligative) measurements on solutions of highly charged polyions ($\xi > 1$, or $b < 7.1$ Å in H$_2$O at 25°C) invariably demonstrate extreme deviations from ideal behavior. This topic has been frequently reviewed (7, 15, 111), and we cite only a single relevant result here. The experimental Donnan coefficient of helical DNA approaches a value $\Gamma_m^{\circ} \simeq -0.1$ at low salt concentrations (126). If the solution were thermodynamically ideal (no small ion-polyion interactions) and yet the polyion were fully dissociated ($\alpha = 1$), then $\Gamma_m^{\circ} = -0.5$; at the other extreme, if the polyelectrolyte were completely undissociated ($\alpha = 0$), then $\Gamma_m^{\circ} = 0$. Clearly, even at high dilution of both polyelectrolyte and salt, the DNA solution is highly nonideal; the polyelectrolyte behaves thermodynamically as if a large fraction of the counterions remained associated with the polyion.

The origin of the unusual degree of nonideality in a polyelectrolyte solution is the local electrostatic potential resulting from the fixed array of like charges on the polyion. The local gradients in the concentrations of small ions surrounding the polyion are the microscopic manifestation of this electrostatic effect. The local nonuniformity of ion concentrations gives rise to large corrections to the ideal entropic mixing terms in the formulation of the free energy of the solution. This microscopic nonuniformity is mirrored at a macroscopic level by the nonuniform electrolyte distribution in the Donnan dialysis equilibrium. The Donnan coefficient is equal to the preferential interaction parameter, $\Gamma^{\circ}_{3u} = N^{-1}\Gamma^{\circ}_{32}$, which provides a complete thermodynamic description of electrolyte-polyelectrolyte interactions (138):

$$\Gamma^{\circ}_{3u} \equiv \lim_{m_u \to 0} \left(\frac{\partial m_3}{\partial m_u}\right)_{\mu_3, T, P} = -N^{-1} \lim_{m_u \to 0} \left(\frac{\partial \ln \gamma_2}{\partial \ln a_3}\right)_{m_u, T, P} \qquad 5.$$

where γ_2 is the activity coefficient of the electroneutral polyelectrolyte component ($M_N^+ P^{N-}$), defined to include all deviations from an uncharged ($\alpha = 0$), unhydrated reference state. The preferential interaction parameter can be shown to be equal to the derivative of the logarithm of the binding polynomial of the polyelectrolyte monomer with respect to the logarithm of the activity of the electrolyte. From this relationship, the contributions to Γ°_{3u} from counterion dissociation, hydration, and screening effects have been resolved. For a uniunivalent electrolyte ($M^+ X^-$), the anion of which does not interact with the polyanion, a good approximation to Γ°_{3u} is

$$\Gamma^{\circ}_{3u} = -\tfrac{1}{2}\left[\alpha + 2m_3\nu_1/m_1 + \lim_{m_u \to 0} N^{-1}[\partial \ln \gamma_{2,\alpha}/\partial \ln a_{\pm}]_{T, p, m_u}\right] \qquad 6.$$

where ν_1 is the amount of bound solvent (moles solvent/mole polyelectrolyte monomer), $a_{\pm}$ is the mean ionic activity of the electrolyte (evaluated in the absence of the polyelectrolyte), and $\gamma_{2,\alpha}$ is the activity coefficient of the electroneutral polyelectrolyte component with $N\alpha$ dissociated counterions and $N\nu_1$ molecules of bound solvent ($\gamma_{2,\alpha} \neq \gamma_2$). The term in brackets in Eq. 6 is a thermodynamic degree of counterion dissociation: it expresses the corrections to the physical degree of dissociation per monomer (α) arising from hydration ($2m_3\nu_1/m_1$) and screening $[N^{-1}(\partial \ln \gamma_{2,\alpha}/\partial \ln a_{\pm})]$ effects. This quantity is designated i, so that Eq. 6 may be more simply written: $\Gamma^{\circ}_{3u} = -\mathrm{i}/2$. The relationship between i and the Donnan coefficient has been considered previously in the context of the PB equation (126).

The fundamental significance of the Donnan distribution in polyelectrolyte solution thermodynamics generally, and in the thermodynamics of the cell model in particular, cannot be overemphasized. The equivalence of Γ_m^o and Γ_{3u}^o provides a direct route to the measurement and interpretation of the extent of interaction of the electrolyte with the polyelectrolyte, in the context of the thermodynamic binding parameter. In addition the Donnan ion distributions are a macroscopic representation of the local ion distributions. In excess salt, $m_3' = m_3(R)$; the salt concentration in the polyion-free solution is equal to that at the outer boundary of the cylindrical cell model (99). Moreover, the concentration difference $m_3 - m_3'$ is a measure of the local concentration of ions at the surface of the polyion (10).

EVALUATION OF i FROM THE PB AND CC THEORIES In the range of salt concentrations where the PB and CC theories are presumed applicable ($m_3 < 0.1m$), the hydration term in Eq. 6 is usually negligible. Both the PB and CC theories provide closed-form expressions for the activity coefficient derivative, from which i can be evaluated. For a polyion with $\lambda > 1$, the PB result (obtained from the statistical thermodynamics of the PB cylindrical cell model) is (130)

$$i_{PB} = (2\xi)^{-1}(1+\delta) \qquad 7.$$

where δ is a measure of the extent to which the local gradients in ion concentrations (the difference between the sum of the surface concentrations and the sum of the outer boundary concentrations of counterions and coions) differ from their limiting-law values. The correction term δ, determined by a, ξ, α, and m_3, can be evaluated numerically from the PB equation. The explicit dependence of i on α has disappeared from Eq. 7 because this term is exactly cancelled by one in the PB expression for the activity coefficient derivative (see Eq. 6). At low salt concentrations, $i_{PB} = (2\xi)^{-1}$ is independent of α. Even if the physical degree of dissociation is unity ($\alpha = 1$), the polyion will behave thermodynamically as a weak electrolyte. For helical DNA ($\xi = 4.2$), the low salt limit of i_{PB} is 0.12. To rephrase the meaning of this result, if DNA is modeled as an incompletely dissociated polyanion with 88% of its counterions bound (under limiting-law conditions), then all polyelectrolyte-electrolyte contributions to nonideality can be neglected. The actual nonideality resulting from the screening effect and physical binding of counterions is thermodynamically equivalent to the binding of 0.88 counterion per DNA nucleotide under limiting law conditions. Above limiting law conditions, the dependence of i_{PB} on m_3 is less than that of α, for any value of the counterion binding constant K. For no choice of K, however, does the PB equation predict i_{PB} to be independent of m_3; even if α is held constant, i_{PB} varies with m_3 (130).

To evaluate i from the CC theory under limiting-law conditions, one assumes that $\alpha = \xi^{-1}$, and obtains the activity coefficient derivative for this value of α using an expression for the electrostatic free energy which is consistent with PB theory under limiting-law conditions (130). The activity coefficient derivative is $-(2\xi)^{-1}$, so that $i = (2\xi)^{-1}$, in agreement with the PB result. Outside of the limiting law region, Manning's molecular thermodynamic formulation of CC theory predicts that α remains fixed at ξ^{-1}, and consequently it must follow that $i_{CC} = (2\xi)^{-1}$ over the range of salt concentrations where the CC model is applicable (up to at least 0.1 M) (4).

Since i is directly measurable from the Donnan equilibrium distribution, it is unfortunate that only limited Donnan data are available on DNA or other locally rodlike polyions of known ξ (126, 140). Furthermore, interpretation of these data is complicated by the experimental difficulty of imposing the condition of excess salt at low salt concentrations, and by the dominant contribution of the polyion partial molar volume to measurements of the molar Donnan coefficient made at high salt concentrations (130). At low univalent salt concentrations, the DNA data are in semiquantitative agreement with predictions of the CC and PB limiting-law equations ($\Gamma_m^\circ = -i/2 = -(4\xi)^{-1} = -0.06$). Although the data obtained at high salt concentrations should in principle allow one to distinguish between the predictions of the PB and CC theories, in practice this has not yet been possible.

Physical Interpretations of Thermodynamically Bound Counterions

Under limiting-law conditions, CC theory predicts that $1 - \xi^{-1}$ counterions are territorially bound per polyion structural charge, and both CC and PB theories predict that the thermodynamic ion binding parameter $1 - i$ equals $1 - (2\xi)^{-1}$ in this limit. The thermodynamic binding parameter $1 - i$ is the hypothetical extent of counterion association required to account for all deviations from ideality due to polyion-small ion interactions. Since both counterion and coion gradients contribute to this parameter, one might not expect it to have a simple molecular interpretation. However, it has been observed that application of the Bjerrum criterion for ion association to the PB radial distribution of counterions about a cylindrical polyion defines a *net* fractional extent of neutralization of the polyion by small ions equal to $1 - (2\xi)^{-1}$ at any concentration of excess salt (141). (The fractional excess of counterions over coions, per polyion charge, within a cylindrical shell defined by the radius at which the counterion radial distribution function passes through a minimum is $1 - (2\xi)^{-1}$.) There may be no rigorous connection between this ion-association parameter and the thermodynamic binding parameter $1-i$ (the two are only equal, in the PB analysis, at infinite dilution), but each of these parameters reflects interactions involving both counterions and coions. Other definitions of counterion association based on the PB equation have been proposed (12).

THEORIES OF POLYELECTROLYTE EFFECTS ON EQUILIBRIA INVOLVING DNA

This section reviews some applications of the PB and CC theories to the interpretation of polyelectrolyte effects on equilibria involving DNA (or other highly charged, locally rodlike polyelectrolytes). In general, processes in which the structural charge density (ξ) of DNA is modified will be affected by changes in the electrolyte activity a_3. The theoretical framework for analysis of these effects is based on the preferential interaction parameter Γ_{3u}°, which in both PB and CC theories has the limiting-law value $\Gamma_{3u}^\circ = -(4\xi)^{-1}$ for a highly charged polyelectrolyte in solution with excess uniunivalent electrolyte. For any equilibrium (conformational, binding, solubility, etc) involving DNA, the thermodynamic equilibrium constant K_T (a function of temperature and pressure only) may be written $K_T = K_{obs} K_\gamma$, where K_{obs} is the equilibrium macromolecular concentration ratio and K_γ is the corresponding ratio of macromolecular activity coefficients. At low macromolecule concentrations, the salt-dependence of K_{obs} is given by

$$\left(\frac{\partial \ln K_{obs}}{\partial \ln a_\pm}\right)_{T,P,m_u \to 0} = -\left(\frac{\partial \ln K_\gamma}{\partial \ln a_\pm}\right)_{T,P,m_u \to 0} = 2\Delta\Gamma_{32}^\circ \qquad 8.$$

where the second equality follows from Eq. 5. In Eq. 8, $\Delta\Gamma_{32}^\circ$ is the stoichiometrically-weighted difference in preferential interaction parameters of the product and reactant macromolecular components. The quantity $2\Delta\Gamma_{32}^\circ$ is the difference in the thermodynamic degree of dissociation of reactant and product polyelectrolytes (since for a uniformly charged polyion $2\Delta\Gamma_{32}^\circ = -Ni$). If $2\Delta\Gamma_{32}^\circ$ is negative, then the thermodynamic degree of dissociation of the products is greater than that of the reactants, and the process of converting reactants to products is accompanied by a release of thermodynamically bound ions. This process of ion release, which may involve both physically bound ions and screening (ion gradient) effects, provides an entropic driving force for conversion of reactants to products as the bulk electrolyte concentration is reduced. In particular, this entropic effect causes DNA to denature and cationic ligands to bind to DNA as the salt concentration is reduced (5, 23).

Conformational Equilibria

A cooperative order-disorder transition between conformational states of DNA or polynucleotides can be described thermodynamically as a two-state equilibrium between cooperatively-denaturing units, characterized by an equilibrium constant K_{obs} and a midpoint transition temperature T_m. Provided that the cooperative unit is sufficiently large so that end effects are negligible, and that the size of the cooperative unit is not a function of salt concentration, then the effect of uniunivalent salt activity on T_m is expressed as

$$-\frac{d(1/T_m)}{d\ln a_\pm} = \frac{1}{T_m^2}\frac{dT_m}{d\ln a_\pm} = -\frac{R(2\Delta\Gamma_{3u}^\circ)}{\Delta H_{obs}^\circ} = \frac{R\Delta i}{\Delta H_{obs}^\circ} \qquad 9.$$

where ΔH_{obs}° is the observed transition enthalpy (per mole of nucleotides denatured) at T_m, and Δi (or $-2\Delta\Gamma_{3u}^\circ$) is the stoichiometrically weighted difference between the thermodynamic degrees of dissociation of the product and reactant states (130). Calorimetric determinations of ΔH_{obs}° have shown it to increase approximately in proportion to T_m^2, when T_m is varied by changing $a_\pm$; consequently T_m (rather than $1/T_m$) is plotted as a function of $\ln a_\pm$ (or $\ln C_3$, if the effects of electrolyte-electrolyte interactions are neglected). Plots of T_m vs $\ln C_3$ are usually linear (within experimental error) over the range 10^{-3} M $< C_3 < 10^{-1}$ M; inclusion of activity coefficients ($\gamma_\pm$) extends the range of linearity to higher salt concentrations (142). For an AT-rich DNA at pH 7 in NaCl, $dT_m/d\ln a_\pm = 8.9$; T_m increases by approximately 19° per decade increase in the univalent salt concentration (64). For $C_3 > 0.5$ M, salt-specific effects (presumably differential hydration effects) are observed; T_m passes through a maximum and then decreases with increasing C_3 in the molar range (23, 143).

Data on the effects of base composition, strandedness, pH and oligocations (e.g. Mg^{2+}) on $dT_m/d\ln a_\pm$ have been recently reviewed (5, 23). Consistent theoretical interpretations of these effects have been achieved using Eq. 9 (or analogs thereof) with the PB/CC limiting value of $i((2\xi)^{-1})$, where ξ for the helical forms is known from structural studies, and where ξ for the denatured single strands is used as a fitting parameter. [Attempts to equal or improve the quality of the theoretical fit using the more general expression for i derived from the PB equation (Eq. 7) have as yet proved unsuccessful (130).]

Recently, much attention has been given to the transition from right to left-handed forms of helical double stranded polynucleotides containing alternating GC base pairs [poly(dG-dC)·poly(dG-dC)], which occurs at 2.5 M NaCl or 0.7 M $MgCl_2$, independent of temperature (144). Methylation of either C (145) or G (146) dramatically reduces the concentration of electrolyte required to effect the transition. For poly (dG-m⁵dC)·poly(dG-m⁵dC), the transition midpoint is reduced to 0.7 M NaCl or 0.05 M NaCl, 6×10^{-4} M $MgCl_2$ (145). Analysis of the high and low salt transitions of these polymers will require consideration of both electrostatic and hydration effects on $\Delta\Gamma^\circ_{32}$ (cf Eq. 6), but should provide valuable insights into the contributions of these factors to the relative stability of the B and Z helices.

Structural Equilibria

Aspects of the double helical structure of DNA in solution are sensitive to the type and concentration of electrolyte. Work in this area has recently been reviewed (5, 22). Both the persistence length a and the winding angle θ are linear functions of the logarithm of the univalent salt concentration over an extended range; a (and, therefore, the bending force constant to which it is related) decreases with increasing [NaCl]. One expression characterizing this dependence is $a(\text{Å}) = 317 - 502 \ln[\text{NaCl}]$, valid at $[\text{NaCl}] \leq 1.0$ M (92). In NaCl (and also KCl, LiCl, and NH_4Cl), $d\theta/d\ln[\text{MCl}] = 0.37$; in RbCl and CsCl, $d\theta/d\ln[\text{MCl}] = 0.55$ (147). Absolute values of θ were not determined.

If the dependences of a and θ on univalent salt concentration are reflections of a polyelectrolyte effect, they must indicate that the processes of bending and of winding DNA are accompanied by slight increases in ξ (axial compression) (5, 22, 89–92). An increase in ξ will decrease the thermodynamic degree of dissociation i (recall $i = -2\Gamma^\circ_{3u} = (2\xi)^{-1}$ under limiting-law conditions). Therefore an increase in ξ upon bending or winding DNA would result in an uptake of counterions in the thermodynamic sense. This entropic contribution to the free energy of the bending or winding process would necessarily vary with the logarithm of the bulk electrolyte concentration, and would be less unfavorable at high bulk salt concentrations. (Molecular models for these effects have been proposed (89–92).)

Binding of Oligocations

Binding constants K_{obs} for the interactions of oligocations with DNA are extremely sensitive functions of the univalent salt concentration. Generally $\ln K_{obs}$ decreases linearly with $\ln C_3$, with a slope $d\ln K_{obs}/d\ln C_3$ which is proportional to the valence of the oligocation (5, 23, 65, 148). For oligocations such as Mg^{2+}, polyamines, and oligolysines, which might be expected to interact with DNA by a predominantly electrostatic (nonspecific) mechanism, the extrapolated K_{obs} at $C_3 = 1$ M is generally of order of magnitude 1 M^{-1} (5, 23, 65, 148). Although in principle electrolyte effects on these equilibria may be analyzed using Eq. 8, this approach suffers from the lack of information concerning the preferential interaction parameter of the complex and the activity coefficient of the oligocationic ligand.

Record and co-workers (23, 65, 148) found that values of $-(d\ln K_{obs}/d\ln C_3)$ for the binding of oligolysines of valence Z in the range $3 \leqslant Z \leqslant 8$ and of polyamines with $2 \leqslant Z \leqslant 4$ were all in the range $0.82Z$–$0.95Z$, centered on the value $0.88Z$. This result was derived theoretically (65) from an analysis of ligand binding to a rodlike polyion in which counterion binding and small ion screening interactions were described on the basis of Manning's original condensation model (1). In terms of the present development, this analysis involved the evaluation of α and the activity coefficient derivative in Eq. 6 for the free and complexed polyion (23). With the further assumption that the interactions of an oligocation with electrolyte ions in solution are equivalent to the nonideality of Z univalent cations, $a_\pm$ in Eq. 8 can be replaced by C_3 and the equation can be written:

$$-\frac{d\ln K_{obs}}{d\ln C_3} = Z(1+2\Gamma^o_{3u}) = Z(1-i) = 0.88Z. \qquad 10.$$

On the basis of Manning's later condensation model (5), the last equality is expected to hold at all salt concentrations lower than ~ 1 M. As in the application of Eq. 9 to cooperative conformational equilibria, the limiting law value of i provides good agreement with ligand-binding experiments over a wide range of ionic conditions (23). The evaluation of $-d\ln K_{obs}/d\ln C_3$ from the PB cell model is currently being investigated (B. K. Klein, C. F. Anderson, and M. T. Record, unpublished).

Other molecular thermodynamic theories of the interaction of oligocations with a polyelectrolyte such as DNA have been developed, based either on CC theory or PB theory. In the application of the later CC model developed by Manning (5), the oligocations of valence Z are assumed to displace Z univalent counterions from the volume V_p in which the univalent ions are territorially bound. The extent of territorial binding of the oligocation is obtained by a free energy minimization procedure similar to that used to obtain the extent of territorial binding ($\alpha = \xi^{-1}$) of a univalent counterion. The theory predicts that $-(d\ln K_{obs}/d\ln C_3) = Z$ and that the standard free energy of binding approaches zero (or $K_{obs} \to 1$) as the bulk salt concentration approaches the counterion concentration in the local phase ($C_{loc} = 1.2$ M for DNA). These predictions are in reasonable agreement with data on the interactions of a variety of oligocations with DNA and other chain polyelectrolytes (5). The theory has been extended to treat high binding densities of the territorially bound oligocation by introducing individual values of V_p

for each counterion valence, evaluated from the free energy minimization for that counterion alone (122). Although the physical interpretation of these different volumes is unclear, the extended theory has been useful in analyzing data on the oligocation-induced transition of helical DNA to a compact form (149–151). An analogous molecular thermodynamic theory of the binding of a Z-valent oligocation to a polyanion in the presence of excess salt is readily formulated in the context of the PB model by postulating the existence of a local equilibrium between oligocations at the polyion surface (at a local concentration specified by the Boltzmann distribution law) and bound ions. In its simplest form, the theory predicts that $-d\ln K_{obs}/d\ln C_3 \simeq Z$. Such an analysis has been applied to discuss the interactions of oligocations with DNA (152, 153) and with ionic bilayers (154, 155).

The binding of proteins to DNA often exhibits a dramatic salt-dependence (65, 22, 23). For the nonspecific interaction of *E. coli* lac repressor with DNA, $-d\ln K_{obs}/d\ln C_3 = 11 \pm 1$ (156, 157); for example, K_{obs} decreases by a factor of 10 when the bulk electrolyte concentration is increased from 0.15 to 0.20 M. Not only do these large electrolyte effects indicate the importance of electrostatic interactions and counterion release in stabilizing the repressor-DNA complex at low salt, but they also suggest that small variations in the in vivo ionic environment could have a significant effect on the distribution and stability of protein-DNA complexes involved in the control of gene expression.

ACKNOWLEDGMENTS

We acknowledge with thanks the many individuals who supplied us with preprints or reprints of their work. We also thank Sandra Shaner and Sharlyn Mazur for their assistance in the preparation of this article, and Gretchen Van Zile for typing the manuscript. Work from the authors' laboratory was supported by grants from the NSF (PCM 79-04607) and NIH (GM23467).

Literature Cited

1. Manning, G. S. 1969. *J. Chem. Phys.* 51:924–33
2. Manning, G. S. 1969. *J. Chem. Phys.* 51:3249–52
3. Manning, G. S. 1974. In *Polyelectrolytes*, ed. E. Sélégny, pp. 9–37 Dordecht: Reidel. 533 pp.
4. Manning, G. S. 1977. *Biophys. Chem.* 7:95–102
5. Manning, G. S. 1978. *Q. Rev. Biophys.* 11:179–246
6. Manning, G. S. 1979. *Acc. Chem. Res.* 12:443–49
7. Katchalsky, A. 1971. *Pure Appl. Chem.* 26:327–73
8. Stigter, D. 1975. *J. Colloid Interface Sci.* 53:296–306
9. Fixman, M. 1979. *J. Chem. Phys.* 70:4995–5005
10. Anderson, C. F., Record, M. T. Jr. 1980. *Biophys. Chem.* 11:353–60
11. Klein, B. K., Anderson, C. F., Record, M. T. Jr. 1981. *Biopolymers* 20:2263–80

12. Katchalsky, A., Alexandrowicz, Z., Kedem, O. 1966. In *Chemical Physics of Ionic Solutions*, ed., B. E. Conway, R. G. Barradas, pp. 295–346. New York: Wiley. 622 pp.
13. Nagasawa, M. 1971. *Pure Appl. Chem.* 26:519–36
14. Nagasawa, M. 1975. *J. Polym. Sci. Polym. Symp.* 49:1–29
15. Manning, G. S. 1972. *Ann. Rev. Phys. Chem.* 23:117–40
16. Ise, N., Okubo, T. 1978. *Macromolecules* 11:439–47
17. Krey, A. K. 1980. In *Progress in Molecular and Subcellular Biology*, ed. F. E. Hahn, pp. 42–87. Berlin: Springer-Verlag
18. Wells, R. D., Goodman, T. C., Hillen, W., Horn, G. T., Klein, R. D., Larson, J. E., Müller, U. R., Neuendorf, S. K., Panayotatos, N., Stirdivant, S. M. 1980. *Progr. Nucl. Acids Res. Mol. Biol.* 24:167–267
19. Wilson, W. D., Jones, R. L. 1981. In *Intercalation Chemistry*, ed. M. S. Whittingham, A. J. Jacobson. New York: Academic
20. Helene, C., Maurizot, J. C. 1981. *CRC Crit. Rev. Biochem.* 10:213–58
21. Sarma, R. H., ed. 1981. *Biomolecular Stereodynamics*. New York: Adenine. 472 pp.
22. Record, M. T. Jr., Mazur, S. J., Melançon, P., Roe, J.-H., Shaner, S. L., Unger, L. 1981. *Ann. Rev. Biochem.* 50:997–1024
23. Record, M. T. Jr., Anderson, C. F., Lohman, T. M. 1978. *Q. Rev. Biophys.* 11:103–78
24. Schmitt, A., Meullenet, J. P., Varoqui, R. 1978. *Biopolymers* 17:413–23, 1249–55
25. Schellman, J. A., Stigter, D. 1977. *Biopolymers* 16:1415–34
26. Yoshida, N. 1978. *J. Chem. Phys.* 69:4867–71
27. Stigter, D. 1978. *J. Phys. Chem.* 82:1417–23, 1424–29
28. Stigter, D. 1979. *J. Phys. Chem.* 83:1663–70, 1670–75
29. Fixman, M. 1980. *Macromolecules* 13:711–16
30. Imai, N., Sasaki, S. 1980. *Biophys. Chem.* 11:361–67
31. Minakata, A., Shimizu, T., Nakamura, H., Wada, A. 1980. *Biophys. Chem.* 11:403–10
32. Okubo, T. 1980. *Biophys. Chem.* 11:425–31
33. Joshi, Y. M., Kwak, J. C. T. 1980. *Biophys. Chem.* 12:323–28
34. Fixman, M., Jagannathan, S. 1981. *J. Chem. Phys.* 75:4048–59
35. Anderson, C. F., Morawetz, H. 1982. In *Encyclopedia of Chemical Technology*, Vol. 18, ed. R. Kirk, D. Othmer, pp. 495–530. New York: Wiley
36. Strauss, U. P., Ander, P. 1958. *J. Am. Chem. Soc.* 80:6494–99
37. Stigter, D. 1978. *Prog. Colloid Polym. Sci.* 65:45–52
38. Guêron, M., Weisbuch, G. 1980. *Biopolymers* 19:353–82
39. Manning, G. S. 1969. *J. Chem. Phys.* 51:934–38
40. Manning, G. S. 1970. *Biopolymers* 9:1543–46
41. Rinaudo, M., Milas, M. 1976. *Chem. Phys. Lett.* 41:456–59
42. Kwak, J. C. T., Morrison, N. J., Spiro, E. J., Iwasa, K. 1976. *J. Phys. Chem.* 80:2753–61
43. Pass, G., Phillips, G. O., Wedlock, D. J. 1977. *Macromolecules* 10:197–201
44. Diakun, G. P., Edwards, H. E., Wedlock, D. J., Allen, J. C., Phillips, G. O. 1978. *Macromolecules* 11:1110–14
45. Kowblansky, M., Tomasula, M., Ander, P. 1978. *J. Phys. Chem.* 82:1491–96
46. Joshi, Y. M., Kwak, J. C. T. 1978. *Biophys. Chem.* 8:191–201
47. Joshi, Y. M., Kwak, J. C. T. 1979. *J. Phys. Chem.* 83:1978–83
48. Rinaudo, M., Karimian, A., Milas, M. 1979. *Biopolymers* 18:1673–83
49. Rochas, C., Rinaudo, M. 1980. *Biopolymers* 19:1675–87
50. Kowblansky, M., Zema, P. 1981. *Macromolecules* 14:166–70
51. Tuffile, F. M., Ander, P. 1975. *Macromolecules* 8:789–92
52. Szymczak, J., Holyk, P., Ander, P. 1975. *J. Phys. Chem.* 79:269–72
53. Holyk, P., Szymczak, J., Ander, P. 1976. *J. Phys. Chem.* 80:1626–28
54. Kowblansky, M., Ander, P. 1977. *J. Phys. Chem.* 81:2024–32
55. Kowblansky, A., Sasso, R., Spagnuolo, V., Ander, P. 1977. *Macromolecules* 10:78–83
56. Ander, P., Gangi, G., Kowblansky, A. 1978. *Macromolecules* 11:904–8
57. Wingrove, D. E., Ander, P. 1979. *Macromolecules* 12:135–40
58. Trifiletti, R., Ander, P. 1979. *Macromolecules* 12:1197–1203

59. Ander, P., Leung-Louie, L., Silvestri, P. 1979. *Macromolecules* 12:1204–7
60. Lubas, W., Ander, P. 1980. *Macromolecules* 13:318–21
61. Ander, P., Lubas, W. 1981. *Macromolecules* 14:1058–61
62. Manning, G. S. 1972. *Biopolymers* 11:937–49, 951–55
63. Manning, G. S. 1976. *Biopolymers* 15:2385–90
64. Record, M. T. Jr. 1975. *Biopolymers* 14:2137–58
65. Record, M. T. Jr., Lohman, T. M., deHaseth, P. L. 1976. *J. Mol. Biol.* 107:145–58
66. Bailey, J. M. 1973. *Biopolymers* 12:559–74
67. Bailey, J. M. 1973. *Biopolymers* 12:1705–8
68. Skolnick, J., Fixman, M. 1978. *Macromolecules* 11:867–71
69. Soumpasis, D. 1978. *J. Chem. Phys.* 69:3190–96
70. Woodbury, C. P. Jr., Ramanathan, G. V. 1982. *Macromolecules* 15:82–88
71. Iwasa, K., Kwak, J. C. T. 1976. *J. Phys. Chem.* 80:215–16
72. Iwasa, K., Kwak, J. C. T. 1977. *J. Phys. Chem.* 81:408–12
73. Iwasa, K. 1977. *J. Phys. Chem.* 81:1829–33
74. Iwasa, K., McQuarrie, D. A., Kwak, J. C. T. 1978. *J. Phys. Chem.* 82:1979–85
75. Iwasa, K. 1979. *Biophys. Chem.* 9:397–404
76. Skolnick, J. 1979. *Macromolecules* 12:515–21
77. Bailey, J. M. 1977. *Macromolecules* 10:725–30
78. Soumpasis, D. M., Bennemann, K. H. 1981. *Macromolecules* 14:50–54
79. Kwak, J. C. T., Joshi, Y. M. 1981. *Biophys. Chem.* 13:55–75
80. Mattai, J., Kwak, J. C. T. 1981. *Biochim. Biophys. Acta* 667:303–12
81. Mattai, J., Kwak, J. C. T. 1982. *Biophys. Chem.* 14:55–64
82. Mattai, J., Kwak, J. C. T. 1982. *J. Phys. Chem.* 86:1026–30
83. Shimizu, T., Minakata, A., Imai, N. 1982. *Biophys. Chem.* 14:333–39
84. Manning, G. S. 1981. *J. Phys. Chem.* 85:870–77
85. Manning, G. S. 1980. *J. Phys. Chem.* 84:3331–32
86. Manning, G. S. 1981. *J. Phys. Chem.* 85:1515–26
87. Manning, G. S. 1977. *Biophys. Chem.* 9:189–92
88. Wilson, R. W., Bloomfield, V. A. 1979. *Biochemistry* 18:2192–96
89. Manning, G. S. 1981. *Biopolymers* 20:2337–50
90. Manning, G. S. 1979. *Biopolymers* 18:2929–42
91. Manning, G. S. 1980. *Biopolymers* 19:37–52
92. Manning, G. S. 1981. *Biopolymers* 20:1261–70, 1751–55
93. Fuoss, R. M., Katchalsky, A., Lifson, S. 1951. *Proc. Natl. Acad. Sci. USA* 37:579–89
94. Alfrey, T. Jr., Berg, P. W., Morawetz, H. 1951. *J. Polymer Sci.* 7:543–47
95. Gregor, H. P., Gregor, J. M. 1977. *J. Chem. Phys.* 66:1934–39
96. Dolar, D., Peterlin, A. 1969. *J. Chem. Phys.* 50:3011–15
97. Hill, T. L. 1955. *Arch. Biochem. Biophys.* 57:229–39
98. Alexandrowicz, Z. 1962. *J. Polym. Sci.* 56:97–114, 115–32
99. Alexandrowicz, Z., Katchalsky, A. 1963. *J. Polym. Sci. Part A* 1:3231–60
100. MacGillivray, A. D., Winkleman, J. J. Jr. 1966. *J. Chem. Phys.* 45:2184–88
101. Philip, J. R., Wooding, R. A. 1970. *J. Chem. Phys.* 52:953–59
102. LeBret, M., Zimm, B. H. 1982. *Biopolymers* 21:In press
103. Sugai, S., Nitta, K. 1973. *Biopolymers* 12:1363–76
104. Rinaudo, M., Loiseleur, B. 1973. *J. Chim. Phys.* 70:1305–8, 1697–701 (In French)
105. Delville, A. 1980. *Chem. Phys. Lett.* 69:386–88
106. Weisbuch, G., Guéron, M. 1981. *J. Phys. Chem.* 85:517–25
107. Stigter, D. 1978. *J. Phys. Chem.* 82:1603–6
108. Ishikawa, M. 1979. *Macromolecules* 12:502–5
109. Westra, S. W. T., Leyte, J. C. 1979. *Ber. Bunsenges. Phys. Chem.* 83:672–77
110. Lifson, S., Katchalsky, A. 1954. *J. Polym. Sci.* 13:43–53
111. Armstrong, R. W., Strauss, U. P. 1969. In *Encyclopedia of Polymer Science and Technology, Vol. 10*, ed. N. M. Bikales, pp. 781–853. New York: Interscience
112. Marcus, R. A. 1955. *J. Chem. Phys.* 23:1057–68

113. Ishikawa, M. 1979. *Macromolecules* 12:498–502
114. Friedman, H. L. 1977. *J. Electrochem. Soc.* 124:421
115. Rice, S. A., Nagasawa, M. 1961. *Polyelectrolyte Solutions.* New York: Academic
116. Rinaudo, M. 1974. See Ref. 3, pp. 157–93
117. Manning, G. S. 1978. *J. Phys. Chem.* 82:2349–51
118. Guéron, M., Weisbuch, G. 1979. *J. Phys. Chem.* 83:1991–98
119. MacGillivray, A. D. 1972. *J. Chem. Phys.* 56:80–85; 57:4071–78
120. Lampert, M. A., Crandall, R. S. 1980. *Chem. Phys. Lett.* 72:481–86
121. Lampert, M. A. 1982. *Biopolymers* 21:159–67
122. Wilson, R. W., Rau, D. C., Bloomfield, V. A. 1980. *Biophys. J.* 30:317–24
123. Van der Klink, J. J., Zuiderweg, L. H., Leyte, J. C. 1974. *J. Chem. Phys.* 60:2391–99
124. Manning, G. S. 1975. *J. Chem. Phys.* 62:748–49
125. Westra, S. W. T., Leyte, J. C. 1979. *Ber. Bunsenges. Phys. Chem.* 83:678–82
126. Gross, L. M., Strauss, U. P. 1966. See Ref. 12, pp. 361–89
127. Manning, G. S., Zimm, B. H. 1965. *J. Chem. Phys.* 43:4250–59
128. Stigter, D. 1977. *Biopolymers* 16:1435–48
129. Deleted in Proof
130. Klein, B. K. 1980. PhD thesis. Univ. Wis., Madison
131. Jönsson, B., Wennerström, H., Halle, B. 1980. *J. Phys. Chem.* 84:2179–85
132. Torrie, G. M., Valleau, J. P. 1980. *J. Chem. Phys.* 73:5807–16
133. Van Megan, W., Snook, I. 1980. *J. Chem. Phys.* 73:4656–62
134. Snook, I., Van Megan, W. 1981. *J. Chem. Phys.* 75:4104–6
135. Outhwaite, C. W., Bhuiyan, L. B., Levine, S. 1981. *Chem. Phys. Lett.* 78:413–15
136. Clementi, E., Corongiu, G. 1981. See Ref. 21, pp. 209–59
137. Brender, C., Lax, M., Windmer, S. 1981. *J. Chem. Phys.* 74:2576–81
138. Eisenberg, H. 1976. *Biological Macromolecules and Polyelectrolytes in Solution*, pp. 48–59. Oxford: Clarendon. 279 pp.
139. Shapiro, J. T., Stannard, B. S., Felsenfeld, G. 1969. *Biochemistry* 8:3232–41
140. Strauss, U. P., Helfgott, C., Pink, H. 1967. *J. Phys. Chem.* 71:2550–56
141. Kotin, L., Nagasawa, M. 1962. *J. Chem. Phys.* 36:873–79
142. Gruenwedel, S. W., Hsu, C. -H. 1969. *Biopolymers* 7:557–70
143. Hamaguchi, K., Geiduschek, E. P. 1962. *J. Am. Chem. Soc.* 84:1329–38
144. Pohl, F. M., Jovin, T. M. 1972. *J. Mol. Biol.* 67:375–96
145. Behe, M., Felsenfeld, G. 1981. *Proc. Natl. Acad. Sci. USA* 78:1619–23
146. Moller, A., Nordheim, A., Nichols, S. R., Rich, A. 1981. *Proc. Natl. Acad. Sci. USA* 78:4777–81
147. Anderson, P., Bauer, W. 1978. *Biochemistry* 17:594–601
148. Braunlin, W. H., Strick, T. J., Record, M. T. Jr. 1982. *Biopolymers* 21: In press
149. Wilson, R. W., Bloomfield, V. A. 1979. *Biochemistry* 18:2192–96
150. Bloomfield, V. A., Wilson, R. W., Rau, D. C. 1980. *Biophys. Chem.* 11:1339–43
151. Widom, J., Baldwin, R. L. 1980. *J. Mol. Biol.* 144:431–53
152. Clement, R. M., Sturm, J., Daune, M. P. 1973. *Biopolymers* 12:405–21
153. Daune, M. P. 1972. *Eur. J. Biochem.* 26:207–11
154. Woolley, P., Teubner, M. 1979. *Biophys. Chem.* 10:335–50
155. Hauser, H., Darke, A., Phillips, M. C. 1976. *Eur. J. Biochem.* 62:335–44
156. deHaseth, P. L., Lohman, T. M., Record, M. T. Jr. 1977. *Biochemistry* 16:4783–90
157. Revzin, A., von Hippel, P. H. 1977. *Biochemistry* 16:4769–76

COMPLEX COORDINATES IN THE THEORY OF ATOMIC AND MOLECULAR STRUCTURE AND DYNAMICS

William P. Reinhardt

Department of Chemistry, University of Colorado and Joint Institute for Laboratory Astrophysics, University of Colorado and National Bureau of Standards, Boulder, Colorado 80309

INTRODUCTION

During the past ten years there has been rapid development and application of a theory variously known as *complex scaling*, *complex coordinates*, *coordinate-rotation*, and *dilatation analyticity*, to problems of resonances in atomic and molecular physics and in chemistry. Resonances are ubiquitous. A few examples:

1. an excited species radiates;
2. a multiply excited atom autoionizes;
3. an atomic or molecular system in an external field is subject to field or multiphoton ionization;
4. an appropriately excited molecule dissociates unimolecularly;
5. electrons attach to molecules and the resulting quasibound molecular ions dissociate into stable ionic and neutral subsystems.

Each of these processes is typified by the formation and decay of an intermediate state, or *resonance*, which is a nonstationary (or quasibound) state with a lifetime long enough to be well characterized, and long enough to make its explicit recognition of experimental and theoretical importance. The simplest, and most naive, mathematical description of such states is that they resemble bound stationary states in that they are "localized" in space (at $t = 0$), and their time evolution is given by (1–3)

$$\psi_R(t) = \exp(-iE_R t/\hbar)\psi_R(0) \qquad 1.$$

which is the usual stationary state time dependence, except that now the energy, E_R, of the resonant state is complex:

$$E_R = E_{res} - i\Gamma/2 \qquad 2.$$

where E_{res} and Γ are real, and $\Gamma \geq 0$. The presence of the "$-i\Gamma/2$" forces exponential decay, if we accept Eq. 1, thus allowing description of a decaying state. If we assume for the moment that such a simple description of the time evolution is adequate (it very often is), a priori calculation of the real and imaginary parts of E_R allows prediction of the formation energies and lifetimes of intermediate species, and thus is an important task. However, an immediate question arises: If the Hamiltonian for the system at hand is Hermitian, how can a "complex eigenvalue" such as E_R ever occur? Hermitian operators have real eigenvalues! This observation has led to almost continuous discussion and dispute since 1930. Refs. (1–7) contain sensible discussions. Many others do not. However, even without detailing any of this discussion, it is straightforward to state that one of the major purposes of the introduction of complex coordinates in nonrelativistic quantum theory is to produce exactly those (non-Hermitian) operators which have the complex energies of Eq. 2 among their actual eigenvalues. That the procedure for constructing these new operators is, at first glance, totally trivial is certainly a reason for the popularity and utility of the method. A pleasant bonus is that the eigenfunctions associated with these complex resonance eigenvalues are square integrable (henceforth, L^2) and thus satisfy our feeling that resonances are localized, at least at complex values of the coordinates.

Thus, one result of the use of complex coordinates is calculation of the energies and lifetimes of decaying systems, be they atoms or molecules, and for all types of decay mechanisms. As the formalism is developed, it will also emerge that use of complex coordinates by no means implies a particular a priori assumption as to the actual time evolution: it might be exponential for all times, t, of physical interest, $\tau_{short} \lesssim t \lesssim \tau_{long}$ [for example, (5) and (6) contain discussions of the origin of the mandatory existence of long-time departures from pure exponential decay] or it might be strongly nonexponential, and not at all well described by a single complex eigenvalue for any reasonable length of time. Another use of the technique is the production of rigorous mathematical results governing what we can expect Schrödinger theory to predict for bound state and scattering dynamics of atoms and molecules. The third major use of the method is in the area of computational scattering theory. In the flurry of successful activities relating to calculation of resonance properties, many workers have overlooked an original motivation for interest in the method: namely, direct calculation of scattering and photoabsorption cross sections *without detailed enforcement of boundary conditions*. The ability to develop computational methods that flout the usual channel-by-channel accounting of multichannel scattering theory is essential as the number of channels increases (or becomes infinite as in impact ionization or collisional dissociation) and was a dominant interest of many of the early workers attempting to apply complex coordinates in atomic and molecular physics (8–18). These three areas—resonances, results of rigorous theory, and potential for development of techniques that ignore usual scattering boundary conditions—form the body of the review, and are interwoven with discussion of applications.

Before proceeding a few additional comments are in order. The status of theory and computation as of early 1978 are well represented in the proceedings (19) of a March 1978 Sanibel Workshop. An extensive discursive review of computational aspects relating mainly to atomic electronic resonance structure, and, in particular, use of complex basis functions has been prepared by Junker (20). The present review is intended for physical chemists, which is taken to imply that there is usually no need for the great mathematical precision that can be found in abundance in much of the cited literature. Thus, for example, the *spectrum* of an operator is taken to consist of the set of its eigenvalues; we assume that square integrable bound state eigenfunctions correspond to discrete eigenvalues, and that δ-function normalized scattering, or continuum, eigenfunctions correspond to eigenvalues in the continuous spectrum: this is the usual language of physicists.

CONCEPTUAL AND MATHEMATICAL BACKGROUND

Before stating the results of the theory of many-particle analytically continued Hamiltonians developed by Aguilar & Combes (21), Balslev & Combes (22), Reed & Simon (7), Simon (23, 24), and van Winter (25), it is useful to give a little qualitative background and motivation. (Experts may skip the next two subsections.)

How Can Coordinates Be Complex?

We take the view of Heisenberg: Only observables are subject to physical interpretation, and observables are matrix elements and eigenvalues.

Consider the hydrogen atom Hamiltonian (in a.u.) for s-states ($l=0$)

$$\mathcal{H} \equiv \mathcal{H}(r) = -\frac{1}{2}\frac{1}{r^2}\frac{d}{dr}r^2\frac{d}{dr} - \frac{1}{r} \qquad 3.$$

r being the radial variables in spherical polar coordinates. The expected value of the energy is thus

$$\langle E \rangle = \int_0^\infty R(r)\left[-\frac{1}{2}\frac{1}{r^2}\frac{d}{dr}r^2\frac{d}{dr} - \frac{1}{r}\right]R(r)r^2\,dr \bigg/ \int_0^\infty R^2(r)r^2\,dr \qquad 4.$$

assuming that the state function is $\Psi(r,\theta,\phi) = R(r)/\sqrt{4\pi}$, and that $R(r)$ is real, as it may always be taken (s-states are nondegenerate). It is usually assumed that the integration in Eq. 4 is over the real range of r from 0 to ∞, and that the radial coordinate of Eq. 3 is real. However, if $R(r)$ is analytic in r (e.g. a combination of actual hydrogenic functions, or Slater functions, or Gaussian radial functions) we can rewrite, using Cauchy's theorem, Eq. 4 in terms of contour integrals

$$\langle E \rangle = \int_C R(\rho)\left(-\frac{1}{2}\frac{1}{\rho^2}\frac{d}{d\rho}\rho^2\frac{d}{d\rho} - \frac{1}{\rho}\right)R(\rho)\rho^2\,d\rho \bigg/ \int_C R^2(\rho)\rho^2\,d\rho \qquad 5.$$

without changing the value of $\langle E \rangle$. Three possible contours, C, are shown in Figure 1. If we take $R(r) = 2e^{-r}$ (the exact radial $1s$ function), the contour need not even return to the real axis at ∞, as long as ρ has a positive real part at ∞ insuring convergence of the integral. Thus, for

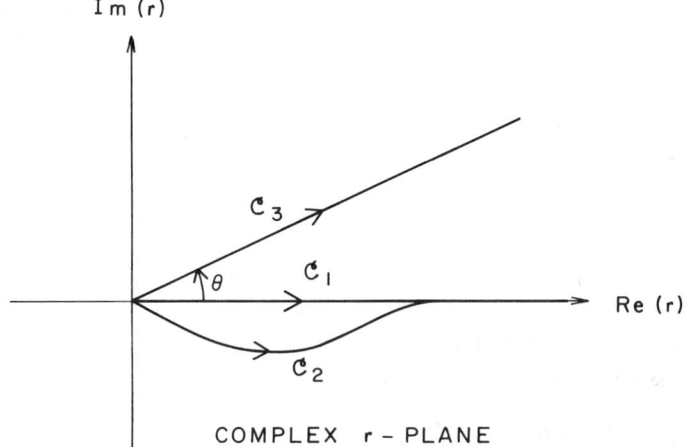

Figure 1 Complex r integration contours. The integral in Eq. 6 has the same value on each of these. C_1 is the usual real axis contour, C_2 is asymptotically real, and C_3 heads off into the complex plane even as $\rho \to \infty$.

example, taking[1] $\rho = re^{i\theta}$ (θ real, for the moment), which "scales" r by the complex factor $e^{i\theta}$, we have

$$\langle E \rangle = \frac{\int_0^\infty e^{-re^{i\theta}}\left(-\frac{e^{-2i\theta}}{2}\frac{1}{r^2}\frac{d}{dr}r^2\frac{d}{dr} - \frac{e^{-i\theta}}{r}\right)e^{-re^{i\theta}}(e^{i\theta}r)^2 e^{i\theta}\,dr}{\int_0^\infty (e^{-re^{i\theta}})^2 (e^{i\theta}r)^2 e^{i\theta}\,dr}, \qquad 6.$$

which may be directly evaluated, giving the usual value, $-1/2$. Thus the expected energy is not altered if we use the complex scaled wave function

$$R(re^{i\theta}) = 2e^{-re^{i\theta}} \qquad 7.$$

along with the scaled operator

$$\mathcal{H}(\theta) \equiv \mathcal{H}(re^{i\theta}) = -e^{-2i\theta}\frac{1}{2}\frac{1}{r^2}\frac{d}{dr}r^2\frac{d}{dr} - \frac{e^{-i\theta}}{r} \qquad 8.$$

integrated with the scaled measure $e^{3i\theta}r^2\,dr$. Note also that just as $R(r) = 2e^{-r}$ is an eigenfunction of the $\mathcal{H}(r)$ of Eq. 3, with eigenvalue $-1/2$, $2e^{-e^{i\theta}r}$ is an eigenfunction of $\mathcal{H}(re^{i\theta})$, with the same eigenvalue, as is easily checked by substitution, or more laboriously, by usual power series techniques (26, 27). As Eqs. 6–8 give the same matrix element and eigenvalue as $2e^{-r}$ with Eqs. 3 and 4, no predictions are changed. Whether we prefer to think of the coordinate r in $R(r)$ as real, or as a scaled complex coordinate $\rho = e^{i\theta}r$, makes no difference in this example as long as the following hold:

1. r is not itself an observable. (This is usually true—think about it, spectroscopy and scattering experiments do not give "r" directly).
2. The operators are also appropriately scaled.
3. We note that as Eqs. 5–8 arose from Eq. 4 by changing the radial integration contour via Cauchy's theorem, there is no complex conjugation of the $e^{i\theta}$ arising from the scale transformation, whether it occurs on the left or right of the operator. On the other hand, if $R(r)$ for r real (i.e. $\theta = 0$) is intrinsically complex (e. g. $R(r) = re^{-r} + ie^{-2r}$) then the complex conjugation of the usual definition of expectation value does apply to (but only to) the intrinsically complex part of the function. Thus, for example, the spherical harmonics, $Y_{l,m}$, are conjugated as usual.

Thus, without changing any observable properties of the system, we can work with a complex radial variable. The only care that need be taken is

---

[1] The "angle" θ defining the complex scaling factor $e^{i\theta}$ is not to be confused with the θ of usual (r, θ, ϕ) coordinates. This will always be clear in context.

with regard to Condition 3 above, as not all "i"s are treated alike in forming expectation values—an odd rule unless it turns out to be useful. Finally, we note that the $e^{3i\theta}$, arising as $r^2\,dr \to e^{3i\theta}r^2\,dr$ when $r \to re^{i\theta}$, is often included with the scaled functions, leaving the "measure" $r^2\,dr$ unchanged thus

$$\langle E \rangle = \frac{\int_0^\infty (e^{i3\theta/2}e^{-re^{i\theta}})\left(-\frac{e^{-2i\theta}}{2}\frac{1}{r^2}\frac{d}{dr}r^2\frac{d}{dr} - \frac{e^{-i\theta}}{r}\right)(e^{i3\theta/2}e^{-re^{i\theta}})r^2\,dr}{\int_0^\infty (e^{i3\theta/2}e^{-re^{i\theta}})^2 r^2\,dr}.$$

9.

In this case the $e^{i3\theta/2}$, as well as the $re^{i\theta}$, are not complex conjugated. But, again, any intrinsically complex functions are treated as usual. The reason for this last rearrangement becomes clear below, when a unitary transformation is introduced to effect the scale transformation: Everyone knows that operators don't "act" on the integration measure, so the $e^{i3\theta/2}$'s have to be stuck in! The ideas of this subsection are taken far more seriously in Refs. (8, 10, 28–32), where they are referred to as contour distortion techniques.

So far, much complexity, a peculiar set of operations for performing expectation values (not all is are equivalent!) and no new results, but:

What About Boundary Conditions?

Boundary conditions determine whether an operator has eigenvalues, and whether the corresponding eigenfunctions are L^2 (bound states) or non-L^2, δ-function-normalized, scattering states. What happens to boundary conditions as $r \to re^{i\theta}$? For our example of $R(r) = 2e^{-r}$, which is certainly L^2, there is no problem. $R(re^{i\theta}) = 2\exp(-r\exp(i\theta))$ is also L^2, unless $|\theta| \geq \pi/2$, a restriction that must be observed. Thus the boundary condition of square integrability is preserved. Conversely, no new square integrable eigenfunctions with real eigenvalues suddenly appear as $r \to re^{i\theta}$ and $\mathcal{H}(r) \to \mathcal{H}(re^{i\theta})$. That is $\mathcal{H}(r)$ and $\mathcal{H}(re^{i\theta})$ have the same real bound state eigenvalues.

What about scattering states? This is where things begin to happen. Scattering wave functions are non-normalizable, but they remain finite as $r \to \infty$. Thus, unless the potentials are too long ranged, radial scattering solutions look like linear combinations $(e^{+ikr})/r$ and $(e^{-ikr})/r$ as $r \to \infty$. k is the momentum (in units of $\hbar$) and $E = k^2/2$. To preserve this (bounded) asymptotic form as $r \to re^{i\theta}$ we must simultaneously take $k \to ke^{-i\theta}$. If we don't, one of the exponentials will grow exponentially at ∞, violating the boundary condition of everywhere finite wave functions.

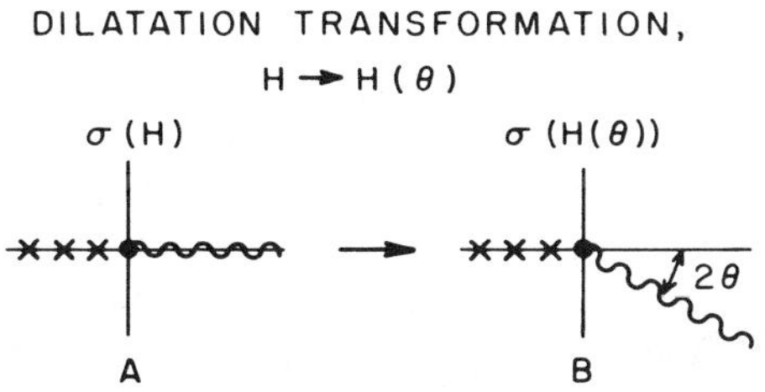

DILATATION TRANSFORMATION,
$$H \rightarrow H(\theta)$$

Figure 2 Effect of the transformation $r \rightarrow r\exp(i\theta)$ on the spectrum "σ" of a one-body problem. Bound states are invariant as is the threshold where the continuum begins. The continuous spectrum "rotates" about the threshold by -2θ, exposing a higher Riemann sheet of the resolvent.

If $k \rightarrow ke^{-i\theta}$, then $E = k^2/2 \rightarrow e^{-2i\theta}k^2/2$, for the allowed scattering eigenenergies. This suggests immediately: the complex scaled Hamiltonian $\mathcal{H}(\theta)$ of Eq. 8 has the same bound state eigenvalues as does the original of $\mathcal{H}$ (Eq. 3), but the scattering states have energies $Ee^{-2i\theta}(0 \leq E < \infty)$. The continuum is then *rotated* into the lower half complex energy plane, as shown in Figure 2. This is the correct result even though for the Coulomb problem the asymptotic forms aren't quite $(e^{\pm ikr})/r$. It is the fact that the continuous spectrum of $\mathcal{H}(\theta)$ is different from that of $\mathcal{H}$ which is the key to the utility of the $r \rightarrow re^{i\theta}$ transformation. With this very qualitative motivation, we now simply state the results of the full theory.

Spectral Theory of Dilatation Analytic Hamiltonians

This section contains the mathematical results of the theory of complex scaling. The basic result is that if, for an N-body Coulomb system, we take (θ real and postive, to simplify the discussion)

$$\mathcal{H}(\theta) = -e^{-2i\theta}\frac{\nabla_N^2}{2} + e^{-i\theta}V_N^{coul}, \qquad 10.$$

where $e^{-2i\theta}, e^{-i\theta}$ are complex numbers which scale the ordinary N-body kinetic energy $-\nabla_N^2/2$, and N-body Coulomb potential energy V_N^{coul}, $\mathcal{H}(\theta)$ has complex eigenvalues with L^2 eigenfunctions (for certain ranges of θ) which we *associate* [a carefully chosen word, see (33) and below] with resonances. Those wishing to see applications before investing the energy to follow a fuller statement of the theory may look at Figures 2 and 3, and then proceed to the applications sections, having noted that Eq. 10 is just Eq. 8 scaled up to N-particles.

The rigorous theory (7, 21–25) that follows gives results for eigenvalues of complex scaled N-body nonrelativistic Coulomb Hamiltonians, and their behavior under the complex scale transformation $r \rightarrow re^{i\theta}$. The parallel results for the analytic structure of Hilbert space matrix elements of the *resolvent* $(z - \mathcal{H})^{-1}$ are also stated. Rather than by direct use of Cauchy's theorem and the contour distortions of the previous sections, the rigorous theory is compactly formulated in terms of unitary transformations called dilatation transformations (dilate ≡ dilatate ≡ stretch), conveying the idea that stretching scales the system but does not change angles. We thus define a unitary operator $U(\theta)$ whose action on *wave functions* is defined in configuration space for three dimensions

$$U(\theta)\psi(\vec{r}) \equiv e^{i3\theta/2}\psi(\vec{r}e^{i\theta}) \qquad 11.$$

for a two-body system, $\vec{r}$ being the relative center-of-mass coordinate vector (see Eq. 9 and following equations). For N-body systems, if $\psi^N(\vec{r})$ is the corresponding function $\vec{r} \equiv (\vec{r}_1, \vec{r}_2, \vec{r}_3 \ldots \vec{r}_N)$, where only $N-1$ of the $\vec{r}_i$ are independent in center-of-mass coordinates, we have

$$U(\theta)\psi_N(\vec{r}) \equiv e^{i3(N-1)\theta/2}\psi(\vec{r}e^{i\theta}). \qquad 12.$$

The corresponding transformation on the N-body *operators*, exemplified by the Hamiltonian, is

$$\mathcal{H}_N(\theta) \equiv \mathcal{H}_N(\vec{r}e^{i\theta}) \equiv U(\theta)\mathcal{H}_N(\vec{r})U(\theta)^{-1} \qquad 13.$$

For the case of the N-body Coulomb problem

$$\mathcal{H}_N \equiv \mathcal{H}_N(\vec{r}) = -\frac{\nabla_N^2}{2} + V_N^{coul}(\vec{r}) \qquad 14.$$

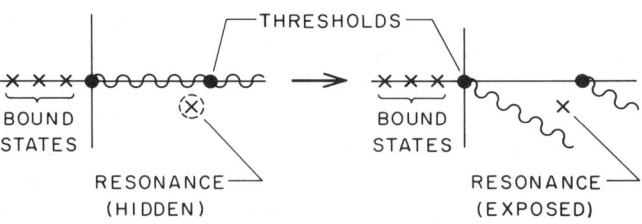

Figure 3 Effect of dilatation transformation on a many-body Hamiltonian. Again bound states and thresholds are invariant. However, as the continua rotate, complex resonance eigenvalues may be exposed. Such eigenvalues correspond to poles of the resolvent $R_\phi(z)$, but are "hidden" on a higher sheet if $\theta = 0$, and will be exposed if the cuts are appropriately moved.

and

$$\mathcal{H}_N(\theta) \equiv U(\theta)\mathcal{H}_N(\vec{r})U(\theta)^{-1} = \mathcal{H}(\vec{r}e^{i\theta})$$

$$= -e^{-2i\theta}\frac{\nabla_N^2}{2} + V_N^{coul}(\theta)$$

$$= -e^{-2i\theta}\frac{\nabla_N^2}{2} + e^{-i\theta}V_N^{coul}. \qquad 15.$$

That is, for N-body systems with only pair-wise Coulomb interactions, the scaled Hamiltonian $\mathcal{H}(\theta)$ is obtained by simply multiplying the total relative kinetic energy by $e^{-2i\theta}$, and the potential energy by $e^{-i\theta}$. Note that this is precisely the analog of Eq. 8, as obtained by contour distortion. Not only does the Coulomb potential scale in a simple manner; the Coulomb interaction belongs to a special class of potentials called *dilatation analytic*, a more restrictive condition than simple analyticity in the interparticle coordinates, which the Coulomb potential has (except at $r_{ij} = 0$, which causes no difficulty). The precise definition, and determination of which potentials are or are not dilatation analytic, is a nontrivial mathematical technicality (see Refs. 7, 21–25, 34–36), into which we shall not delve. That the N-body Coulomb potential is dilatation analytic allows us to state at once (suppressing the "N" in $\mathcal{H}_N$):

1. Bound state eigenvalues of $\mathcal{H}(\theta)$ are independent of θ, and identical to those of $\mathcal{H}$ for $|\theta| \leq \pi/2$.

2. Scattering thresholds corresponding to the possibility of fragmentation of different subsystems in differing states of excitation are also independent of θ, $|\theta| \leq \pi/2$.

3. The segments of continua beginning at each scattering threshold rotate by an angle 2θ into the lower half plane ($\theta \geq 0$), each about its individual threshold. (This follows from the fact that the continua are related to the kinetic energy, not the potential.)

Points 1, 2, and 3 mimic the simplistic analysis of the preceding subsection (see Figure 2); now, additionally:

4. New, complex, discrete eigenvalues of $H(\theta)$ may appear in the lower half complex energy plane, in the sector $0 > \arg(z - E_0^{\text{thresh}}) \geq -2\theta$ where z is the complex energy, and E_0^{thresh} is the lowest energy-scattering threshold. These are the complex eigenvalues that we associate with resonances. The corresponding eigenfunctions $\chi(\theta)$ are L^2, as befits their association with discrete eigenvalues. Note that the $\chi(0)$ obtained by taking $\lim_{\theta \to 0} \chi(\theta)$ will not be L^2, unless its eigenvalues were real to begin with, and it represents a bound state. The discrete complex eigenvalues of $H(\theta)$ are independent of θ as long as they remain isolated from parts of the continuum. As the continua rotate as a function of θ,

discrete complex eigenvalues may appear or disappear as the continua sweep by, and:

5. All discrete eigenvalues of $\mathcal{H}(\theta)$ are of finite multiplicity, and can accumulate only at thresholds (i.e. only near a threshold is it possible to have an infinite number of discrete, bound state, or resonance eigenvalues, in an arbitrarily small region of the complex plane).

These changes in the spectrum (the set of eigenvalues) "σ" of $\mathcal{H}(\theta)$ as a function of θ, are illustrated in Figures 2 and 3.

An important reinterpretation of Figures 2 and 3 is in terms of the resolvent. Consider a matrix element of the resolvent $(z - \mathcal{H})^{-1}$:

$$R_\phi(z) \equiv \langle \phi, (z - \mathcal{H})^{-1} \phi \rangle \qquad 16.$$

as a function of the complex variable z. For all but exceptional L^2 vectors ϕ, in our N-body Hilbert space, $R_\phi(z)$ has well-known properties: it is analytic in z everywhere off the real axis [$\mathcal{H} = \mathcal{H}(\vec{r})$ is Hermitian]; it has (simple) poles corresponding to bound state eigenvalues; and, has branch points corresponding to thresholds, with the segments of continua running to $+\infty$ from the various thresholds acting as branch cuts. The branch cuts force $R_\phi(z)$ to be carefully interpreted as z approaches the real axis if $Re(z) \geq E_0^{\text{thresh}}$, the lowest scattering threshold energy. Thus we associate points of nonanalyticity of $R_\phi(z)$ with the spectrum (eigenvalues) of the operator. Thus $R_\phi(z)$ is analytic for $Im(z) \neq 0$, as the spectrum of $\mathcal{H}$ is real.

What about the resolvent $(z - \mathcal{H}(\theta))^{-1}$? How is it related to $(z - \mathcal{H})^{-1}$? This is where the utility of defining the complex scale transformations via the unitary transformations of Eqs. 12 and 13 becomes apparent. Consider

$$R_\phi^\theta(z) \equiv \langle \phi, U(\theta)^{-1} U(\theta)(z - \mathcal{H})^{-1} U(\theta)^{-1} U(\theta) \phi \rangle$$
$$= \langle \phi(\theta), (z - \mathcal{H}(\theta))^{-1} \phi(\theta) \rangle. \qquad 17.$$

Using our association of the singularity structure of a resolvent with the spectrum of the operator, we see that the analytic structure of $R_\phi^\theta(z)$ is the same as that of $R_\phi(z)$ except that the *branch cuts* of $R_\phi^\theta(z)$ are rotated into the lower half plane, by angle 2θ. As $R_\phi^\theta(z) = R_\phi(z)$ in the upper half z-plane, these two functions are simply different integral representations of the same more general multisheeted analytic function. Each representation has its own domain of validity, as the cut locations give the natural boundaries of the representation. Thus in Figures 2 and 3, $R_\phi(z)$ is valid for $0 \leq \arg(z) \leq 2\pi$; $R_\phi^\theta(z)$ is valid for $-2\theta \leq \arg(z) \leq (2\pi - 2\theta)$. The two representations are not equal on the segment $0 \leq \arg(z) \leq -2\theta$, where $R_\phi^\theta(z)$ provides the (unique) analytic continuation of

$R(z)$ into the lower half z-plane. This geometric interpretation involving the Riemann sheet structure makes the statements 1, 2, and 4, above, transparent. All follow from the fact that poles (bound states or resonances) and branch points (thresholds) are intrinsic singularity properties of an analytic function. As long as a representation restricted to a cut plane "exposes" these singularities, they are independent of cut location; but, as soon as the cuts move in such a way as to hide singularities, they are (suddenly) no longer seen in a specific representation. The cuts themselves, of course, are not singularities of the analytic function, but only boundaries of a representation, and we may attempt to place them as we wish. Those who prefer to see these properties exemplified in terms of explicitly soluble model problems, rather than as following from analytic function theory, should look at such solved problems in (20, 37, 38a,b), where complex coordinates have been explicitly included, and (39), where it would be a useful exercise to do so.

Finally, many computational results, such as those involving atoms in fields and molecular structure and dynamics (discussed below), indicate that similar properties of spectra and resolvents exist for systems that interact via nondilatation analytic potentials [see also Yaris et al (40)]. The restrictive requirements of dilatation analyticity are thus too strong. What is the appropriate generalization of the concept of dilatation analyticity? Where will results differ from those for dilatation analytic potentials? If the past continues to be a guide to the future, positive results of computation will be an important guide to mathematicians as to the potential existence of theorems waiting to be discovered.

Theoretical Applications of Spectral Theory

Many quite strong additional theoretical results that apply to the dynamics of atomic and molecular electronic structure have been demonstrated using the dilatation analyticity of the N-body Coulomb Hamiltonian. Briefly, dilatation analyticity has been used in the following cases:

1. In combination with its cousin Boost Analyticity to establish dispersion relations (41, 42) in e^+-H scattering, and to suggest their nonutility in e^--H scattering, contrary to earlier conjecture (43 and references therein).
2. To prove (24, 44, 45) nonexistence of discrete eigenvalues (real or complex) of $\mathcal{H}(\theta)$ with $Re(E) \geq 0$. In this theorem $E = 0$ is the threshold for total breakup. A very simple proof was later found (33, 46) and warrants inspection.

3. In derivation and discussion of the usual Born-Oppenheimer approximation (47). Are diatomic potentials analytic in the internuclear distance, R?
4. In establishing fundamental results about convergence and summability of Rayleigh-Schrödinger perturbation theory for resonances (48 and references therein) and time-dependent perturbation theory (24, 49).
5. In establishing the possible behavior of $E(1/Z)$, the ground state energy of an atomic system in $1/Z$ perturbation theory (50), forcing the conclusion that conjectured (50, 51 and references therein) bound states in the continuum are likely to occur only on sets of measure zero in $1/Z$.
6. In setting rigorous bounds on the rate of fall off in coordinate space of bound state wave functions (52, 53) and on the rate of fall off in time of resonant time evolution (54).

Other more technical mathematical results are discussed by Reed & Simon (7) and Simon (33). However, it is important to note at this point that for the N-body Coulomb problem the relationship of poles of $\langle \phi(\theta),(z-\mathcal{H}(\theta))^{-1}\phi(\theta)\rangle$ to poles of scattering amplitudes, and thus to observables, is not fully established. This general point is discussed by Simon (33). More recently Nuttall & Singh (55) have given a proof of the connection for e^- atom scattering for energies below the lowest three-body (the collisional dissociation or impact ionization) threshold. Thus there are open theoretical questions, in contrast to the two-body case (e.g. 56 and references therein) and the N-body case with shorter ranged potentials (e.g. 57 and references therein). The importance of such a gap in the mathematical foundations is well illustrated by an experiment of Peart & Dolder (58) observing a resonance (albeit very broad) in H^{2-} above $E = 0$. If observed resonances are in one-to-one correspondence with complex eigenvalues of $\mathcal{H}(\theta)$, this observation contradicts 2, above, and we could conclude the experiment to be an artifact. This puzzling situation is discussed by Doolen (59) and Nuttall (60). It is especially unclear how to interpret the stabilization calculation of an H^{2-} resonance with E well above zero (61) as this relates directly to the original Hamiltonian, as do eigenvalues of $\mathcal{H}(\theta)$, not even to the (remote) possibility that $\mathcal{H}(\theta)$ has no eigenvalue corresponding to an S-matrix pole.

COMPUTATIONAL APPLICATIONS

Applications to atomic and molecular autoionization, atomic structure in ac and dc fields, molecular predissociation, and calculation of cross sections are discussed under this heading. The discussion of atomic electronic structure is sufficiently detailed to reflect the changes in computational strategy as they have developed since 1972.

Atoms: The Direct Approach

From a purely computational point of view the simplest way to implement the theoretical developments for N-body systems is to use standard configuration interaction or other variational techniques which produce a real symmetric matrix representation of the usual electronic Hamiltonian,

$$(\overline{\mathcal{H}})_{ij} = \langle \chi_i(\vec{r}) | \mathcal{H}(\vec{r}) | \chi_j(\vec{r}) \rangle.$$

Then

$$\overline{\mathcal{H}} = \overline{KE} + \overline{PE} \qquad 18.$$

where $\overline{KE}$ and $\overline{PE}$ are the real symmetric representations of the kinetic and potential energies in the variational basis. Rayleigh-Ritz theory now directs us to find the eigenvalues of $\overline{\mathcal{H}}$ as equivalent to a linear variation in our space of trial functions. The extension to $\mathcal{H}(\theta)$, and thus to direct determination of resonance eigenvalues as first suggested by J. Nuttall in 1972 (private communication), is to take (as follows at once from Eq. 18)

$$\overline{\mathcal{H}}(\theta) = e^{-2i\theta} \overline{KE} + e^{-i\theta} \overline{PE} \qquad 19.$$

where $\overline{KE}$ and $\overline{PE}$ are the same real symmetric matrices as in Eq. 18, although we might expect to choose the trial function space differently. Solution of the complex symmetric matrix eigen-problem

$$\overline{\mathcal{H}}(\theta) \bar{c}_i = E_i \bar{c} \qquad 20.$$

then yields the complex eigenvalues E_i, and the complex vectors $\bar{c}_i$. All of the burden of trying to represent a complex valued wave function is placed on these vectors of expansion coefficients. We note at once that as $\overline{\mathcal{H}}(\theta)$ is not Hermitian, its spectral resolution involves both its left and right eigenfunctions (62). However, elementary manipulation gives the results that the left eigenfunction $\bar{d}_i$ corresponding to eigenvalue E_i is just $(\bar{c}_i)^T$ where T is the transpose. (Note that the transpose is not the usual Hermitian conjugate, which is the transpose complex conjugate.) Thus

$$\overline{\mathcal{H}}(\theta) = \sum_i E_i (\bar{c}_i)(\bar{c}_i)^T.$$

This is a bi-orthogonal expansion (62, 63) and $(\bar{c}_i)^T (\bar{c}_j) = \delta_{ij}$ (rather than the usual $(\bar{c}_i)^\dagger (\bar{c}_j) = \delta_{ij}$ for Hermitian problems).

The ansatz of Eq. 19 is very attractive as it implies that standard, existing variational codes may be used to generate $\overline{\mathcal{H}}(\theta)$, leaving only the problem of solution of Eq. 20, which is almost always straightforward. This is, in spite of the limitations discussed in the following subsection, still an attractive feature for exploratory calculations. One can determine whether the method will be of any utility without much work on code development. We refer to the use of Eqs. 19 and 20 with real symmetric representations of $\overline{KE}$ and $\overline{PE}$ as the *Direct Approach*.[2]

The most important question about the Direct Approach is: Does it work? This is answered affirmatively by Doolen et al (64, 65). Next, how can we optimize the choice of θ, given that in a finite variational expansion the resonance eigenvalues are not independent of θ (although in a complete basis they would be)? This is answered in an empirical manner by Doolen (66). Doolen observed, for a fixed L^2 expansion basis, that approximate complex resonance eigenvalues, when plotted as a function of θ (note that in the Direct Approach no new matrix elements need be calculated as θ is changed!), followed trajectories (henceforth θ-trajectories) that *paused* for certain values of θ, suggesting some sort of stationary property. This is illustrated in Figure 4 in a calculation of E_{res} and $\Gamma/2$ for the lowest ^1S resonance in helium $(2s^2)$. The approximate stationarity, easily observed in the figure, is even more dramatic for larger basis sets (66). The analytic theory of the morphology of these θ-trajectories has been given by Moiseyev et al (67), and the existence and interpretation of such stationarity analyzed in terms of an extension of the usual virial theorem to complex scalings by Brändas et al (68, 69), Canuto & Goscinski (70), Winkler & Yaris (71, 72), Yamabe et al (63, 73), Certain (74), and Moiseyev et al (75–78). Application of the virial results require that θ in the $e^{i\theta}$ scaling be itself a complex number, thus allowing optimization of $Re\theta$ and $Im\theta$, to ensure simultaneous stationarity of both the real and imaginary parts of the resonance eigenvalue. This is completely consistent with the general theory of dilatation transformations; the restriction to real θ in our earlier discussions was simply for convenience.

A very large number of workers have used the θ-trajectory techniques as the basis of computational algorithms to locate resonances. Winkler & Yaris (79, 80) and Weinhold (81a) have shown that low-order perturbation theory may sometimes be used to avoid repetition of eigenvalue computations as a function of θ, although Moiseyev & Certain (81b) have demonstrated that such a theory may not be expected to have a large radius of convergence.

Applications of the Direct Approach to resonance structure of few body atomic systems have been notably successful. A highly selective

---

[2] Ironically, the first published results (112) ostensibly using complex coordinates to locate a resonance (following Nuttall's suggestion) were by a different method, to be discussed below.

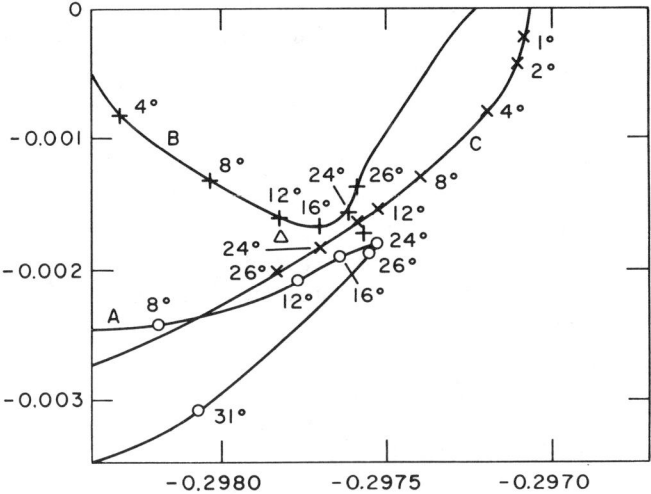

Figure 4 θ-trajectories for location of the complex eigenvalue corresponding to the lowest ^1S Feshbach resonance in He. Trajectories show behavior of complex eigenvalue as a function of the complex scaling angle θ, for three choices of nonlinear parameters in a Hylleraas-type basis. The trajectories kink, or pause near values of θ approximately consistent with the Virial Theorem. Figure reproduced from (66), with permission.

sampling of recent papers (containing references to earlier work) follows: Ho, in an extensive series of papers, has looked at autoionizing resonance structure in He and H^- associated with higher ($n > 2$) excitation thresholds in He^+ and H (82–85). Ho (86) and Moiseyev & Weinhold (87), have examined a large number of isoelectronic series of resonances from $Z = 1$ to 10; the latter authors, using data of Ho (82), have examined correlation in the resonances as a function of $1/Z$. Doolen et al have established the existence of a very narrow Feshbach resonance in e^+-H scattering (88). Similarly, resonances in the (e^-, e^+, e^-) system (89) and in the scattering of positronium (e^+, e^-) from H-atoms have been found (90, 91).

Atoms in Fields: The Direct Approach

The Direct Approach has been applied to problems involving the dc Stark (electric field) and Zeeman (magnetic field) problems, and to multiphoton ionization. The dc Stark Hamiltonian (for hydrogen)

$$\mathcal{H}^{\text{Stark}} = -\frac{\nabla^2}{2} - \frac{1}{r} + \vec{F} \cdot \vec{r} \qquad 21.$$

has no bound states, but for small fields ($\vec{F}$) has long-lived states which may be treated as resonances. Using the Direct Approach ansatz,

$$\mathcal{H}(\theta) = e^{-2i\theta}\overline{KE} + e^{-i\theta}\overline{PE} + e^{+i\theta}\overline{\vec{F} \cdot \vec{r}}, \qquad 22.$$

Reinhardt (92) obtained excellent positions and widths of Stark broadened hydrogenic states for $n=1$, with subsequent extensions to $n=2$ states (93). Many others have subsequently used the Stark problem as a test case, for example (68, 94–96). The same techniques were then applied to the Stark broadening of autoionizing states by Wendoloski & Reinhardt (97); to the combined Stark-Zeeman effect by Chu (98); and to the multiphoton (ac Stark) ionization by Chu & Reinhardt (99), Reinhardt (100), Chu (101, 102, and references therein), and Preobrazhensky & Rapoport (103) for the case of an oscillatory classical field, and for quantized fields by Grossman & Tip (104). Many of these results were obtained before there was any real theoretical justification for the use of the dilatation transformation. The field coupling $\vec{F} \cdot \vec{r}$ is not well behaved at ∞, and is not encompassed by the standard theory of Refs. (7, 21–25). Qualitative mathematical results were obtained by Cerjan et al (105) and the whole theory finally put on a firm mathematical footing by Herbst (106, 107) and Herbst & Simon (108a, b, 109). The clear lesson is that there are many things to be discovered via computations, and that in particular the numerical utility of complex scale transformations is often ahead of rigorous mathematical foundation.

Critique and Extensions of Computational Technique

The Direct Approach has the following advantages:

1. Boundary conditions are ignored—often a nontrivial fact: for example, for resonances near the $n=4$ threshold (85) in, say, He, where a sufficiently large number of open and almost open channels are present as to make use of other methods problematic; or for the problem of Stark broadening of the $^1P2s2p$ shape resonance (97) in H^-, which lies above the $n=2$ threshold, and in the presence of the field a minimum of 12 important open channels are strongly coupled, ignoring the problem of enforcing Stark boundary conditions. In these cases, both of which were motivated by concurrent experimental work, no other computational techniques have even yet been quantitatively applied.
2. It is possible to take existing codes and put them into the service of the direct method without reprogramming integrals or integral transformations. This means that exploratory calculations are easily undertaken.

But there are problems. The Direct Approach very often requires solution of large eigenproblems: in (85), Ho used Hylleraas bases of up to 165 terms, in (88) Doolen et al used bases ranging from 286 to 680 Hylleraas terms. This latter is an extreme case, and quantitative results were obtained for a very narrow resonance ($\Gamma/2 \sim 7\times 10^{-5}$ a.u.), very slightly below the $n=2$ threshold in e^+-H scattering, and where other computational methods were not able to come to decisive conclusions. In general, that widths are usually quite small implies the need for fairly high precision. However, use of a large basis is aesthetically offensive to some, and one can ask: Can some asymptotic information be included to speed convergence? More fundamentally, the Direct Approach simply runs out of steam for e^--atom resonances involving several shells of atomic electrons. In the Direct Approach the *inner shells* cause grave difficulties. The origin of this difficulty, illustrated by our elementary discussion of the H atom, is that for a bound state, if

$$\mathcal{H}(r)R(r) = ER(r), \quad \text{then} \quad \mathcal{H}(re^{i\theta})R(re^{i\theta}) = ER(re^{i\theta}).$$

Thus, if an inner shell is orbital in, say, neon is well represented by $2e^{-Zr}$, with an effective Z of about 9, the corresponding 1s orbital for the scaled Hamiltonian $\mathcal{H}(\theta)$, will be $2e^{-9re^{i\theta}}$. This is highly oscillatory and will be quite difficult to represent in the basis of real radial functions of the Direct Approach, where, again, the expansion coefficients themselves bear the whole burden of the fact that $\mathcal{H}(\theta)$ is not $\mathcal{H}$. This is all the more frustrating as in e^--Ne scattering resonance, the 1s orbital probably plays no real role in the resonance formation. Apparently somewhat different, but in fact closely related, methods for solving this inner-shell problem have been introduced. The simplest to describe is:

METHOD OF COMPLEX BASIS FUNCTIONS   A matrix element of the scaled Hamiltonian

$$I = \langle \chi_i(\vec{r}) | \mathcal{H}(\theta) | \chi_j(\vec{r}) \rangle,$$

which might occur in the Direct Approach, might well employ basis functions of the form

$$\chi_i(\vec{r}) = A(\phi_{i(1)}(\vec{r}_1)\phi_{i(2)}(\vec{r}_2)\ldots\phi_{i(N+1)}(\vec{r}_{N+1})) \qquad 23.$$

in scattering of an electron from an N electron atom, A being an appropriately normalized antisymmetrizer. In the Direct Approach the $\phi_{i(1)}(\vec{r}_1)$ are of the form $R_i(r_1)Y_{lm}(\Omega_1)$ with $R_i(r_1)$ a real radial function. A moment's reflection indicates that the variable change $\vec{r} \to \vec{r}e^{-i\theta}$ in the matrix element leaves it invariant:

$$I = \langle \chi_i(\vec{r}e^{-i\theta}) | \mathcal{H} | \chi_j(\vec{r}e^{-i\theta}) \rangle$$

and thus, it is possible to reinterpret the Direct Approach as keeping the unscaled (normal) Hamiltonian, $\mathcal{H}$, but working with the complex radial functions $R_j(re^{-i\theta})$ implicit in

$$\chi_i(\vec{r}e^{-i\theta}) = A\left(\phi_{i(1)}(\vec{r}_1 e^{-i\theta})\phi_{i(2)}(\vec{r}_2 e^{-i\theta})\dots\phi_{i(N+1)}(\vec{r}_{N+1} e^{-i\theta})\right). \qquad 24.$$

However, as realized in a slightly different context by Rescigno & Reinhardt (15, 16, 18), the use of complex basis functions with the unscaled Hamiltonian is more flexible than the Direct Approach if we scale the different $\phi_j(\vec{r}_j)$ differently. The method is in general no longer equivalent to the spectral theory of dilatation analyticity, as it may well not correspond to an easily derived variable scaling of the operator. Thus, in introducing the method of complex basis functions for resonance determination, Rescigno & McCurdy (110, 111) have chosen to take $\mathcal{H}$ unscaled and to take

$$\chi_i = A\left(\phi_{i(1)}(\vec{r}_1)\phi_{i(2)}(\vec{r}_2)\dots\phi_{i(k)}(\vec{r}_k)\phi_{i(k+1)}\right.$$
$$\left.\times(\vec{r}_{k+1}e^{-i\theta})\dots\phi_{i(N+1)}(\vec{r}_{N+1}e^{-i\theta})\right) \qquad 25.$$

where orbitals $i(1)$ through $i(k)$ correspond to a subspace of orbitals important for the static interaction, polarization, and correlation, and $i(k+i)$ through $i(N+1)$ correspond to the orbitals involved in resonance formation and scattering. Thus, inner shells are simply not scaled, thus avoiding the aforementioned inner-shell problem. This idea of not scaling "tight" orbitals and scaling "loose, or scattering" type orbitals is intuitively related to the concept of exterior scaling discussed below. The method has been applied to e^--Be scattering in the static exchange approximation (111) and works well. More recently, K. T. Chung and B. Davis (private communication, 1981) have used an explicit multichannel version of this technique: They use $\mathcal{H}(\theta)$ and a trial function of the form

$$\Psi = A\left(\psi_1(\vec{r}_1 e^{i\theta},\dots,\vec{r}_N e^{i\theta})\right.$$
$$\left. + \sum_i^{\text{open channels}} \phi_i^{\text{Target}}(\vec{r}_1 e^{i\theta},\dots,\vec{r}_{N-1}e^{i\theta})\phi_N(\vec{r}_N)\right) \qquad 26.$$

where $\psi_1(\vec{r}e^{i\theta},\dots,\vec{r}_N e^{i\theta})$ is just an L^2 eigenfunction of $\mathcal{H}(\vec{r})$, which approximates the resonance, with the indicated variable changes, and $\phi_i(\vec{r}_N)$, which represents the scattering electron, is expanded in Slaters. While this appears to differ from the Rescigno-McCurdy ansatz, note that the change of variables $\vec{r} \to \vec{r}e^{-i\theta}$, in both $\mathcal{H}(\theta)$ and Ψ, yields the former type ansatz. K. T. Chung and B. Davis have obtained results of exceptional stability for doubly excited states of He using this method.

METHOD OF COMPLEX SIEGERT FUNCTIONS Bardsley et al (112, 113), Junker & Huang (27, 114, 115), Junker (20, 116, 117), and Nicolaides et al (118–121) have implemented a method closely related to the Siegert method (3). The Bardsley & Junker paper (112) was the first computational application of any type of complex scaling to a resonance calculation. The use of complex coordinates was suggested to them by Nuttall, who was rather surprised by the way they used his suggestion: he had expected them to attempt a Direct Approach calculation.

Siegert proposed that one solve $\mathcal{H}\psi(r) = E\psi(r)$ with the boundary condition $\psi(r) \to e^{+ikr}/r$. With this unusual boundary condition, resonance solutions with complex E and k are found, but, k has a negative imaginary part implying that $\psi(r)$ diverges as $r \to \infty$, making the method problematic for computations. However, simply looking for formal solutions of the Schrödinger equation, if

$$\mathcal{H}\psi(r) = E\psi(r), \quad \text{then} \quad \mathcal{H}(\theta)\psi(re^{i\theta}) = E\psi(re^{i\theta})$$

by simply regarding r as a dummy variable. Thus applying the Siegert condition amounts to solving

$$\mathcal{H}(\theta)\psi(re^{i\theta}) = E\psi(re^{i\theta})$$

with the rotated Siegert condition

$$\psi(re^{i\theta}) \to e^{ikre^{i\theta}}/(re^{i\theta})$$

which, if θ is large enough, is a decaying function even for k complex in the lower half k plane: a far more palatable boundary condition for actual applications. Thus if $k = pe^{-i\alpha}$ (p, α, real and positive), as long as $\theta > \alpha$, an L^2 Siegert function may be found. Thus, an L^2 eigenfunction in a sense suddenly "appears" as θ increases from 0 and finally becomes larger than α; this is highly reminiscent of the dilatation analyticity spectral theory, where a discrete eigenvalue "suddenly" appears as a cut rotates far enough into the lower half plane.

In fact, both methods have found the same L^2 eigenfunction, a fact recognized (and often exploited) by numerous workers, in addition to the above: for example, Nuttall (60); Atabek et al (122–124), the latter of these (124) containing an interesting discussion of optimization of θ; Isaacson et al (125, 126); Simons (127, 128). In particular, McCurdy & Rescigno (129a), Yaris & Taylor (129b), and Bačić & Simons (130) have been able to determine partial resonance widths (i.e. fractional decay into different open channels) via relationships involving Siegert boundary conditions, an alternative method to that of Noro & Taylor (131). Thus, excellent results for many atomic resonances are obtained using a trial function of the approximate form (written here for a single open channel)

$$\Psi = A\left(\sum_i a_i \chi_i(\vec{r}_1 e^{i\theta}, \vec{r}_2 e^{i\theta}, \ldots, \vec{r}_{N+1} e^{i\theta}) + \phi^{\text{target}}(\vec{r} e^{i\theta}) f(r_{N+1} e^{i\theta})\right) \quad 27.$$

where $f(re^{i\theta}) \to e^{ikre^{i\theta}}/re^{i\theta}$ asymptotically, is used in diagonalization of $\mathcal{H}(\theta)$ (not $\mathcal{H}$). In some cases k is determined self-consistently, in others (27) the method is generalized by using k as a complex variational parameter.

The two methods (complex coordinates and complex Siegert functions) are not as dissimilar as they appear. The trial function, Ψ, of Eq. 27 is employed with $\mathcal{H}(\theta)$. If both Ψ and $\mathcal{H}(\theta)$ are subject to the transformation $r_j \to r_j e^{-i\theta}$ (for all j) and $k \to ke^{-i\theta}$, a trial function of the form of Eq. 25 is obtained, which is to be used with $\mathcal{H}$, just as in the Rescigno-McCurdy ansatz. The only difference is the expansion basis used for the term $f(\vec{r}_{N+1})$, which in both methods will be an L^2 function for θ large enough. Thus both methods solve the inner-shell problem in the same way. Moiseyev et al (132) have carried out a study of the two methods using a newly defined criterion for stabilized complex eigenvalues, with the conclusion that neither method works well. This is in strong contrast to the findings of other workers, and the origins of the differences should be resolved as soon as possible.

COMPLEX VARIATIONAL THEORY   The observation that (Eqs. 22, 24)

$$\langle \chi_i(\vec{r}) | \mathcal{H}(\theta) | \chi_j(\vec{r}) \rangle = \langle \chi_i(\vec{r}e^{-i\theta}) | \mathcal{H} | \chi_j(re^{-i\theta}) \rangle$$

for all (i, j) implies that scaling all basis coordinates $\vec{r} \to \vec{r}e^{-i\theta}$ and leaving $\mathcal{H}$ unscaled is identically equivalent to a dilatation transform and has an identical spectral theory, provided only that the basis functions analytically continue. Thus, if a Direct Approach calculation gives a stabilized complex eigenvalue $E_R(\theta)$ at $\theta = \tilde{\theta}$, and

$$\Psi_R(\tilde{\theta}) = \sum_i c_i(\tilde{\theta}) \chi_i(\vec{r}) \quad 28.$$

$\bar{c}(\tilde{\theta})$ being an eigenfunction of the matrix representation of $\mathcal{H}(\theta)$ in the $\chi_i(\vec{r})$ basis, then the corresponding function in the rotated basis is

$$\tilde{\Psi}_R(\tilde{\theta}) \equiv \sum_i c_i(\tilde{\theta}) \chi_i(\vec{r}e^{-i\tilde{\theta}}) \quad 29.$$

[with the identical $c_i(\tilde{\theta})$] as $\bar{c}(\tilde{\theta})$ is automatically an eigenfunction [with the same $E_R(\tilde{\theta})$] of the matrix representation unscaled Hamiltonian $\mathcal{H}$, in the $\chi_i(\vec{r}e^{-i\tilde{\theta}})$ basis. As the spectral theory tells us to expect $\Psi_R(\tilde{\theta})$ to be L^2, $\tilde{\Psi}_R(\tilde{\theta})$ is also L^2, establishing the complete equivalence of the pairs $\{\mathcal{H}(\theta)$ with basis $\chi_i(\vec{r})\}$ and $\{\mathcal{H}$, with basis $\chi_i(\vec{r}e^{-i\theta})\}$ as far as computations are concerned. This is the old rotation group (physical rotations in this single occurrence!) duality: Do we rotate objects, or do we rotate the coordinates?

If coordinates corresponding to different electrons are scaled differently, however, [as in Rescigno et al (111)] a new flexibility is obtained. Further, if even coordinates in the radial expansion of a *single electron* are scaled differently

$$\left(\text{e.g.} \quad R(r) = re^{-\alpha r} + r^2 e^{-\beta x}, \quad \text{with} \quad \alpha \to \alpha e^{-i\theta_1}, \beta \to \beta e^{-i\theta_2}\right)$$

use of complex basis functions gives still more flexibility. Note that even in this most general case, as only radial (rather than angular) coordinates are scaled, a stretching of coordinates (albeit nonuniform) is occurring without any angular distortion. We thus call this most general type of transformation a generalized dilatation transformation (GDT). It is defined by its action on the wave function. How are we to use this flexibility? Stillinger et al (133a,b) and Herrick et al (133c, 134, 135), beginning in the mid 1970s, have implicitly used a variational principle for resonances based on locating complex energy stationary points of the GDT functional (note that $\mathcal{H}$ is unscaled)

$$E^{GDT}(\Psi) \equiv \frac{\int (\Psi)^{GDT(*)} \mathcal{H} \Psi \, d\tau}{\int (\Psi)^{GDT(*)} \Psi \, d\tau}. \qquad 30.$$

The new quantity $\Psi^{GDT(*)}$ we define as the generalized dilatation transformation complex conjugate: GDT(*) implies that only those quantities (such as spherical harmonics) not involved in the GDT are conjugated. That is, only those quantities that would be conjugated in the absence of a GDT. The new GDT notation is introduced to indicate the generality of the variational theory implied. A simple example will make this clear. Given a one-dimensional radial Hamiltonian $\mathcal{H}(r)$, which might support both bound states and resonances, we write

$$R(r, \{\alpha_i\}) = \sum_i c_i r e^{-\alpha_i r}$$

and take

$$E^{GDT}\{c_i \alpha_i\} = \frac{\int_0^\infty R(r, \{\alpha_i\}) \mathcal{H}(r) R(r, \{\alpha_i\}) r^2 \, dr}{\int_0^\infty R(r, \{\alpha_i\})^2 r^2 \, dr} \qquad 31.$$

with no conjugates at all in this one-dimensional problem. We now vary all of the c_i and α_i independently, including independent complex variations. As the (possible) complex values of the c_i and α_i are associated with the GDT they are not involved in complex conjugation. This is clearly far more general than simply assuming, as would be the case in application of the method of complex coordinates [as in (111)] to a one-dimensional problem, that the trial function be of the form

$$R(re^{i\theta}, \{\alpha_j\}) = \sum_j c_j(re^{-i\theta}) e^{-\alpha_j re^{-i\theta}},$$

with real α_j, which would then be made stationary without conjugation of the $e^{-i\theta}$'s. It is also more general than the variational theories discussed in (68, 70, 73), for example. Such a variation can only give stationarity, rather than the usual variational upper bound (which would be regained were ordinary *'s reinserted) even for the (real) ground state. Herrick attempted to publish an explicit discussion of these points (D. R. Herrick, 1978 preprint, 1981 private communication), but the complex-coordinate community was not yet prepared to understand the real import of his remarks: his 1978 paper was rejected. Quite independently, Junker (20, 96, 136, 137 and references therein) intuitively rediscovered the utility of Eq. 30 and applied it to several systems. Junker's broad grasp of the many forms of complex basis functions led him to the correct conclusions, even though one is confused by his implication (20, 96) that $\tilde{\Psi}_R(\tilde{\theta})$ of Eq. 29 is the continuation of $\Psi_R(\tilde{\theta})$ to $\theta \to 0$. It simply isn't, unless all the c_i's are independent of θ: they are not. This misinterpretation does not affect his conclusions. Junker calls the method of Eq. 30 *complex stabilization*, from its similarity to ordinary stabilization (138), and he cogently points out that the method can wreak havoc with boundary conditions. Additionally, Junker has, in a very successful series of calculations (20, 96), applied the method to a model problem, a e^--Be resonance, and to the Stark effect, and has pointed out that an earlier e^--Be calculation by Donnelly & Simons (140) also exemplifies the method. [Taylor & Yaris (139a) and Simons et al (139b,c) have discussed aspects of the relationship of various complex coordinate methods to the usual stabilization techniques.] Perhaps the most spectacular example is that of McCurdy et al (141), in which, in an SCF calculation of a Ca^- resonance employing real basis functions and a real Hamiltonian, use of the scalar product implicit to Eq. 30 with respect to the radial expansion coefficients (which are part of the GDT) gave a complex eigenvalue from the otherwise ordinary SCF equations. This is clearly a very powerful method, although its limitations are not yet defined. What is the distribution of eigenvalues for a large basis? The GDT(*) conjugation has clear origins in contour distortion and dilatation transformation theory, but its use as outlined in this section is of far greater generality.

Molecular Electronic Structure: Real Axis Clamped Nuclei

The spectral theory of Refs. (7, 21–25) is immediately valid for the full molecular problem, treating all nuclei and electrons as mobile particles. It is also immediately consistent with the Born-Oppenheimer approximation, but with a twist. Consider the H_2^+ Hamiltonian, with nuclei at $\vec{R}_\alpha$ and $\vec{R}_\beta$ in the Born-Oppenheimer approximation (no nuclear kinetic energies) and $\vec{r}_1$ denoting the electron coordinate

$$\mathcal{H} = -\frac{\nabla_1^2}{2} - \frac{1}{|\vec{r}_1 - \vec{R}_\alpha|} - \frac{1}{|\vec{r}_1 - \vec{R}_\beta|}. \qquad 32.$$

Can we simply take

$$\mathcal{H}(\theta) = -e^{-2i\theta}\frac{\nabla_1^2}{2} - e^{-i\theta}\left(\frac{1}{|\vec{r}_1 - \vec{R}_\alpha|} + \frac{1}{|\vec{r}_1 - \vec{R}_\beta|}\right) \qquad 33.$$

and, if so, what is its interpretation? This is transparently answered by inspection of the Hamiltonian in confocal elliptic coordinates (e.g. 142). If $r_\alpha = |\vec{r}_1 - \vec{R}_\alpha|$ and $r_\beta = |\vec{r}_1 - \vec{R}_\beta|$, and R is the scalar distance between nuclei, then

$$\mathcal{H} = -\frac{2}{R^2(\xi^2 - \eta^2)}\left[(\xi^2-1)\frac{\partial^2}{\partial\xi^2} + 2\xi\frac{\partial}{\partial\xi} + (1-\eta^2)\frac{\partial^2}{\partial\eta^2} - 2\eta\frac{\partial}{\partial\eta}\right.$$
$$\left.+ \left(\frac{1}{\xi^2-1} + \frac{1}{1-\eta^2}\right)\frac{\partial}{\partial\phi^2}\right] - \frac{2}{R(\xi+\eta)} - \frac{2}{R(\xi-\eta)} \qquad 34.$$

where ϕ is the azimuthal angle and $\xi = (r_\alpha + r_\beta)/r$ and $\eta = (r_\alpha - r_\beta)/R$ are dimensionless. In these variables the expected scaling $e^{-i2\theta}KE, e^{-i\theta}PE$ can only arise if ξ, η are unscaled, and $R \to Re^{i\theta}$. That is, under the simultaneous scalings $r_\alpha \to r_\alpha e^{i\theta}, r_\beta \to r_\beta e^{i\theta}$ and $R \to Re^{i\theta}$. The conclusion is that, for an N-electron molecule with arbitrarily many nuclei, the naive transformation

$$\mathcal{H} \to \mathcal{H}(\theta) = e^{-2i\theta}KE + e^{-i\theta}PE$$

is consistent with the Born-Oppenheimer Approximation, but at complex internuclear distances. We call this the CCBOA (Complex Coordinate Born-Oppenheimer Approximation). There is absolutely nothing wrong with this, as observables are always matrix elements of transition operators over appropriate vibrational functions, and we can distort the contours! Thus, for a diatom, a transition moment between vibrational states $\psi_i(R), \psi_j(R)$ might be written in the form

$$\left| \int_0^\infty \psi_i(Re^{i\theta}) M(Re^{i\theta}) \psi_j(Re^{i\theta}) d(Re^{i\theta}) \right|^2. \qquad 35.$$

The electronic transition moment $M(Re^{i\theta})$ in Eq. 35 is exactly what would follow (automatically) from the use of $e^{-2i\theta}KE + e^{-i\theta}PE$. These points have been noted by Bardsley (143), Junker (136), and informally by the author and others (private communications from T. N. Rescigno, C. W. McCurdy, V. McKoy, J. Simons). However, no calculations of the CCBOA type have been carried out. There is an incredibly strong (and largely irrational, in most cases) resistance to thinking of complex R. What is done instead is to clamp the nuclei on the real axis, and only scale the electronic coordinates. We refer to this as the CCRACNA (Complex Coordinate Real Axis Clamped Nuclei Approximation); it initially caused no end of difficulty. Note carefully that the CCRACNA is not simply the Born-Oppenheimer approximation, which is not at all inconsistent with the simple scaling

$$\mathcal{H}(\theta) = e^{-2i\theta}KE + e^{-i\theta}PE.$$

The problems in the CCRACNA arise from the fact that if nuclear coordinates are left real, and an electronic coordinate $\vec{r}_i$ is scaled $\vec{r}_i \to \vec{r}_i e^{i\theta}$, the nuclear-electron interaction

$$\sum_\alpha |\vec{r}_i e^{i\theta} - \vec{R}_\alpha|^{-1}$$

is nonanalytic, (33), as the argument of the absolute value can vanish for a continuous range of values such that

$$|\vec{r}_i| = |\vec{R}_\alpha|, \hat{r}_i \cdot \hat{R}_\alpha = \cos\theta,$$

giving rise to a continuous line of square root branch points. McCurdy & Rescigno (144), and independently Moiseyev & Corcoran (145), nevertheless established that the CCRACNA could be made to work. In (144) in application to H_2^+ and a model problem this was done by the method of complex basis functions discussed above. Diffuse basis functions (only) were scaled $r \to re^{-i\theta}$, in a sense performing the scaling for $|\vec{r}_i| > |\vec{R}_\alpha|$ and avoiding the singularities. In (145) in applications to autoionizing states of H_2 and H_2^-, matrix elements of the RACNA Hamiltonian were analytically continued [see especially footnote 12 of (145) for an important remark on stability of this method for the Gaussian bases used in both (144) and (145)]. The success of these calculations prompted a quick response by Simon (146), who suggested use of an exterior scaling, as illustrated in Figure 5, where the coordinates are kept on the real r-axis long enough to get past any interior nonanalyticities. This is an old idea [Figure 5 is reproduced from (32)] but one that fills the bill, in that it allows formulation of a solid mathematical foundation for the CCRACNA. McCurdy (147) and Morgan & Simon (148) have analyzed the relationship between actual computations and the complex exterior scaling concept. Qualitatively, phrased in the language of GDT's, if we scale the diffuse basis functions, but not tight ones, a relationship (not an identity) can be established between the idea of Figure 5 and actual computations. This idea is implicit in the Siegert work of Isaacson & Miller (126). More recently, Deguchi & Nishikawa (149, 150), using the generator coordinate formulation of Lathouwers (151), have suggested a related approach, which has the very attractive feature that through generator coordinates nonadiabatic effects might well be included, although this has not yet been explored. Actual computational applications to molecules is a developing field. At present the only real electronic structure results are the following: the He + H Penning ionization calculations of (126); the H_2 and H_2^- work of (145); a calculation by Rescigno et al (152) of an N_2^- shape resonance [using a complex SCF formalism developed by McCurdy et al (153, 154) and concurrently by Froelich (see 155)]; and the work on competition between dissociation and autoionization of Moiseyev (156, 157).

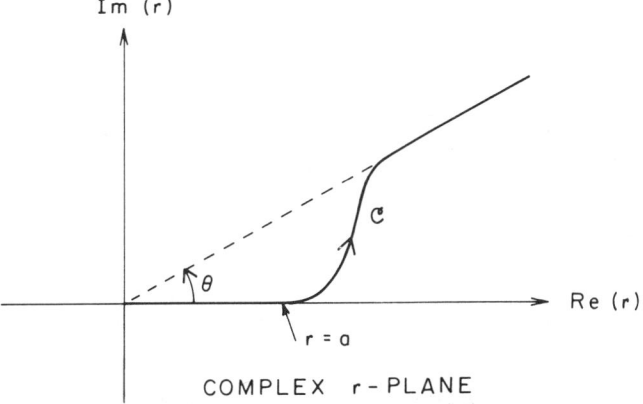

Figure 5 Exterior scaling contour for r distortion appropriate to a situation where for $r < a$ the potential is nonanalytic. As $r \to re^{i\theta}$ as $r \to \infty$, the spectral theory is unchanged, as the spectrum of $\mathcal{H}(\theta)$ is determined by asymptotic boundary conditions. Reproduced from (32), with permission.

Molecular Predissociation

All applications discussed so far (except the CCRACNA) have consisted of systems with one- and two-body interactions that scale as $PE \to e^{-i\theta}PE$ (or $PE \to e^{+i\theta}PE$ for the Stark problem); however, the general techniques have much greater applicability. For nuclear motion on a Born-Oppenheimer potential surface $V(\vec{r}_1,\ldots,\vec{r}_N)$

$$\mathcal{H}(\theta) = e^{-2i\theta}KE(\vec{r}) + V(\vec{r}_1 e^{i\theta}, \vec{r}_2 e^{i\theta} \ldots \vec{r}_N e^{i\theta}). \qquad 36.$$

But where does one find $V(\vec{r}e^{i\theta})$? In a model this is simple; however, for an ab initio surface it perhaps seems a formidable problem. It isn't: We note with amusement that precisely this potential function of complex internuclear distance $(\vec{r}_1 e^{i\theta},\ldots,\vec{r}_N e^{i\theta})$ is that which would be immediately obtained from the simple ansatz of the CCBOA, rather than from the CCRACNA. Given such an $\mathcal{H}(\theta)$, one can, for example, look at rotational predissociation of diatoms (e.g. 158, 159) or at various predissociating channels in a tri-atom, say Ar(H$_2$). Chu (160) and Chu & Datta (161) have looked at rotational predissociation of van der Waals com-

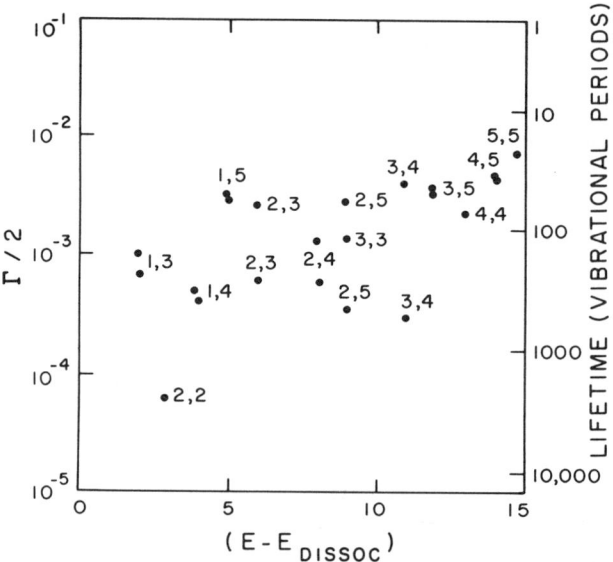

Figure 6 Imaginary parts of complex eigenvalues as a function of energy above dissociation for two coupled Morse oscillators. Parameters have been chosen to mimic interacting symmetric and antisymmetric stretches in H$_2$O. The states where lifetimes are shown are resonances corresponding to doubly excited states of the unperturbed oscillators. The lifetimes of these states are certainly nonstatistical. The complex eigenvalues were obtained by R. Hedges and W. P. Reinhardt using a spline basis.

plexes via $\mathcal{H}(\theta)$ of Eq. 36 as a Direct Application, i.e. diagonalization in a basis with no boundary conditions except square integrability. Chu has subsequently extended the method to consider molecular dynamics in strong laser fields (162).

Atabek & Lefebvre (163–165) and Bačić & Simons (130, 166) have developed multichannel Siegert-like formalisms which when used with complex coordinates allow use of square integrable boundary conditions using an L^2 expansion (130, 166) or numerical solutions (163–165) of the multichannel problem (see also 167). Model problems, usually involving linear triatoms or van der Waals predissociation, have been solved by these authors. Waite & Miller (168) and R. Hedges and W. P. Reinhardt (unpublished), using the Direct Approach, have examined resonance eigenvalue distributions for vibrational predissociation of very strongly coupled triatom systems and investigated the effect of potential surface on RRKM or non-RRKM (i.e. mode specific) decay rates. Typical complex eigenvalues obtained by R. Hedges and W. P. Reinhardt for a coupled Morse oscillator problem, modeling the symmetric and antisymmetric stretch in H_2O, are shown in Figure 6, where strongly nonstatistical behavior is seen for a sequence of multiply excited states above the dissociation limit. These results contrast with the results of (168) for the Henon-Heiles model problem, but are in rough consonance with a second model problem considered by Waite & Miller, where some mode specificity was observed.

CONCLUSIONS: THE FUTURE

Use of complex coordinate techniques in various guises is now widespread for determination of complex resonance eigenvalues in all types of physical situations: electronic autoionization, field ionization, molecular predissociation. Progress has been made in sorting out the amplitudes for decay into each of several decay channels when competing mechanisms exist. The computational theory for carrying out such calculations is becoming clearly defined. New computational variants, such as use of complex coordinates with many-body optical potentials (169–172), are continually appearing. What lies beyond?

Time Dependences

In many applications exponential decay is not a reasonable approximation. General time dependences are given by

$$|t\rangle = e^{-i\mathcal{H}(\theta)t/\hbar}|0\rangle \qquad 37.$$

where $\mathcal{H}(\theta)$ may be replaced by its biorthogonal spectral resolution, and thus

$$e^{-i\overline{\mathcal{H}}(\theta)t/\hbar} \cong \sum_j e^{-iE_j t/\hbar} \overline{c}_j \overline{c}_j^T \qquad 38.$$

where the E_i are complex. Equation 38 is a quadrature of continuum time dependence and exponential decay, as appropriate. This would seem to require that all vectors/eigenvalues of $\mathcal{H}(\theta)$ be found, a prohibitive task for large systems. However, this is not necessarily the case. Figure 7, prepared by the author for this discussion, shows time evolution of the probability of an H atom leaving the 1s state, on sudden application of a strong dc field. There is much nonexponential structure due to the sudden turn on. However, the most important feature of the results is that they have been obtained by a sequence of simple power expansion of Eq. 37 and Padé resummations: no eigenvalues or eigenvectors of $\overline{\mathcal{H}}(\theta)$ were even estimated. Converged results for short times were even found for $\theta = 0$! These results illustrate the capability of complex coordinate methods to handle complex time evolution, and indicate that systems of

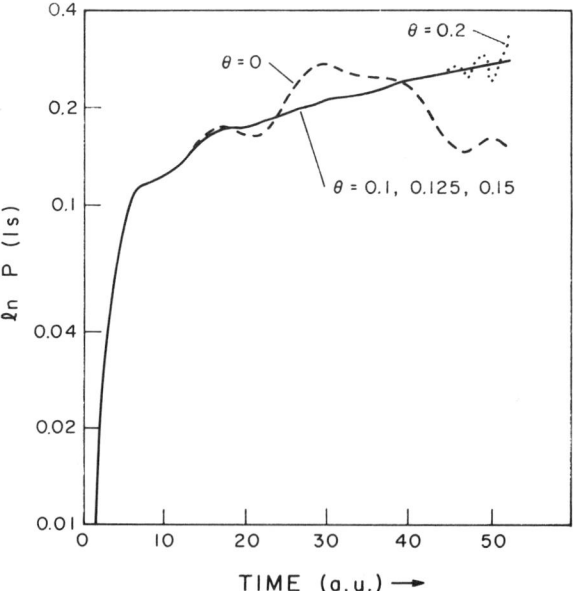

Figure 7 Time evolution of the probability of leaving the 1s state of atomic hydrogen under the influence of a suddenly applied dc electric field as computed by expansion of the exponential $\exp(-i\mathcal{H}(\theta)t/\hbar)$, without explicit determination of eigenparameters. This technique allows the use of very large matrix representations, and thus should allow extension to calculation of time evolution (and thus spectra) in molecular systems. The method has no great difficulty in coping with a multi-timescale decay process.

many degrees of freedom may well be approached, as time evolution can be calculated directly, without solution of large complex eigenproblems.

Scattering

As mentioned in the introduction, early motivation for moving branch cuts off the real axis was to make discretized solution of many-particle scattering problems tractible. Except for application to photoabsorption (173–176), a problem which can be reformulated as a complex eigenvalue problem in any case (99), this aspect has not been actively pursued. Why? Simply because scattering amplitudes of the form

$$\langle k|V(r)(z-H)^{-1}V(r)|k\rangle,$$

the $|k\rangle$'s being plane wave states, do not survive insertion of our unitary transform $U(\theta)U(\theta)^{-1}$ unless $V(r)$ is exponentially damped, apparently excluding the long-range interactions of atoms, molecules, and especially charged systems such as ions and electrons. However, if mathematical difficulties of this type had been taken too seriously many of the computations discussed in this review would not have been even attempted. One can only agree with the remarks of Taylor & Yaris (139a) that there are many ways to by-pass this apparent obstacle. Thus, a parting conjecture: Ways will soon be found to harness the power of complex coordinate techniques for solution of atomic and molecular scattering problems, as well as for determination of resonance parameters, and of time evolution of initially bound systems.

ACKNOWLEDGMENTS

The author acknowledges fruitful conversations, over a period of many years, with J. Nuttall (who introduced him to these techniques), B. Simon, C. W. McCurdy, and with his several collaborators. The technical bibliographic assistance of the Atomic Collisions Information Center at JILA (P. Krog and N. Lewis) and assistance in manuscript preparation from JILA Scientific Reports Office (L. Volsky and G. Romey) have been invaluable, and are most gratefully acknowledged. This work has been supported, in part, by Grants PHY79-04928, CHE80-11442 to the University of Colorado by the National Science Foundation.

Literature Cited

1. Weisskopf, V. F., Wigner, E. P. 1930. *Z. Phys.* 63:54–73
2. Gamow, G. 1931. *Constitution of Atomic Nuclei and Radioactivity*. Oxford: Oxford Univ. Press. 114 pp.
3. Siegert, A. F. S. 1939. *Phys. Rev.* 56:750–52
4. Zumino, B. 1956. *NY Univ. Inst. Math. Sci., Res. Rep.* CX-23. 31 pp.
5. Schwinger, J. 1960. *Ann. Phys. NY* 9:169–93
6. Goldberger, M. L., Watson, K. 1964. *Collision Theory*, pp. 424–509. New York: Wiley. 919 pp.

7. Reed, M., Simon, B. 1978. *Methods of Modern Mathematical Physics*, 4:51–60. New York: Academic. 396 pp.
8. Nuttall, J. 1967. *Phys. Rev.* 160:1459–67
9. Nuttall, J., Cohen, H. L. 1969. *Phys. Rev.* 188:1542–44
10. Nuttall, J. 1972. In *The Physics of Electronic and Atomic Collisions, VII ICPEAC, Invited Papers and Progress Reports*, ed. T. R. Govers, F. J. deHeer, pp. 265–76. Amsterdam: North-Holland. 496 pp.
11. McDonald, F. A., Nuttall, J. 1972. *Phys. Rev. C* 6:121–25
12. Nuttall, J. 1973. *Comput. Phys. Commun.* 6:331–35
13. Hendry, J. A. 1972. *Nucl. Phys. A* 198:391–96
14. Raju, S. B., Doolen, G. 1974. *Phys. Rev. A* 9:1965–68
15. Rescigno, T. N., Reinhardt, W. P. 1973. *Phys. Rev. A* 8:2828–34
16. Rescigno, T. N., Reinhardt, W. P. 1974. *Phys. Rev. A* 10:158–67
17. Baumel, R. T., Crocker, M. C., Nuttall, J. 1975. *Phys. Rev. A* 12:486–92
18. Reinhardt, W. P. 1973. *Comput. Phys. Commun.* 6:303–15
19. Proc. 1978 Sanibel Workshop Complex Scaling. 1978. *Int. J. Quantum Chem.* 14:343–542
20. Junker, B. R. 1982. *Adv. Atom. Molec. Phys.* 18:In press
21. Aguilar, J., Combes, J. M. 1971. *Commun. Math. Phys.* 22:269–79
22. Balslev, E., Combes, J. M. 1971. *Commun. Math. Phys.* 22:280–94
23. Simon, B. 1972. *Commun. Math. Phys.* 27:1–9
24. Simon, B. 1973. *Ann. Math.* 97:247–74
25. van Winter, C. J. 1974. *J. Math. Anal. Appl.* 47:633–70
26. Junker, B. R. 1978. *Int. J. Quantum Chem.* 14:371–82
27. Junker, B. R., Huang, C. L. 1978. *Phys. Rev. A* 18:313–23
28. Wick, G. C. 1954. *Phys. Rev.* 96:1124–34
29. Regge, T. 1959. *Nuovo Cimento* 14:951–76
30. Eden, R. J. 1962. In *Lectures in Theoretical Physics*, 1:1–101. New York: Brandeis Summer Inst. 385 pp.
31. Newton, R. G. 1966. *Scattering Theory of Waves and Particles*, pp. 337–38. New York: McGraw-Hill. 681 pp.
32. Taylor, J. R. 1972. *Scattering Theory*, pp. 220–27. New York: Wiley. 477 pp.
33. Simon, B. 1978. *Int. J. Quantum Chem.* 14:529–42
34. Babbitt, D., Balslev, E. 1974. *Commun. Math. Phys.* 35:173–79
35. Babbitt, D., Balslev, E. 1975. *J. Funct. Anal.* 18:1–14
36. Balslev, E. 1974. *Rep. Math. Phys.* 5:393–413
37. Doolen, G. D. 1978. *Int. J. Quantum Chem.* 14:523–28
38a. Simons, J. 1980. *Int. J. Quantum Chem. Symp.* 14:113–21
38b. Lefebvre, R. 1980. *Chem. Phys. Lett.* 70:430–33
39. Stillinger, F. H. 1979. *J. Math. Phys.* 20:1891–95
40. Yaris, R., Bendler, J., Lovett, R. A., Bender, C. M., Fedders, P. A. 1978. *Phys. Rev. A* 18:1816–25
41. Tip, A. 1978. *Comments At. Mol. Phys.* 7:137–51
42. Tip, A. 1977. *J. Phys. B* 10:L11–16
43. Gerjuoy, E., Lee, C. M. 1978. *J. Phys. B* 11:1137–55
44. Simon, B. 1974. *Math. Ann.* 207:133–38
45. Balslev, E. 1975. *Arch. Ration. Mech. Anal.* 59:343–57
46. Hunziker, W. 1977. In *The Schrödinger Equation*, ed. W. Thirring, P. Urban, pp. 43–72. New York: Springer. 224 pp.
47. Aventini, P., Seiler, R. 1975. *Commun. Math. Phys.* 41:119–34
48. Harrell, E., Simon, B. 1980. *Duke Math. J.* 47:845–902
49. Simon, B. 1971. *Phys. Lett. A* 36:23–24
50. Reinhardt, W. P. 1977. *Phys. Rev. A* 15:802–5
51. Stillinger, F. H. 1977. *Phys. Rev. A* 15:806
52. Deift, P., Hunziker, W., Simon, B., Vock, E. 1978. *Commun. Math. Phys.* 64:1–34
53. Combes, J. M., Thomas, L. 1973. *Commun. Math. Phys.* 34:251–70
54. Rauch, J. 1978. *Commun. Math. Phys.* 61:149–68
55. Nuttall, J., Singh, S. R. 1980. *Commun. Math. Phys.* 72:1–13
56. Perry, P. A. 1981. *Commun. Math. Phys.* 81:243–59
57. Balslev, E. 1980. *Ann. Inst. Henri Poincare Sect. A* 32:125–60
58. Peart, B., Dolder, K. 1973. *J. Phys. B* 6:1497–1502

59. Doolen, G. 1975. *Phys. Rev. A* 12:1121–22
60. Nuttall, J. 1979. *Comments At. Mol. Phys.* 9:15–21
61. Taylor, H. S., Thomas, L. D. 1972. *Phys. Rev. Lett.* 28:1091–92
62. Morse, P. M., Feshbach, H. 1953. *Methods of Theoretical Physics*, 1:884–86. New York: McGraw-Hill. 997 pp.
63. Yamabe, T., Tachibana, A., Fukui, K. 1978. *Adv. Quantum Chem.* 11:195–221
64. Doolen, G., Hidalgo, M., Nuttall, J., Stagat, R. 1973. In *Atomic Physics*, ed. S. J. Smith, G. K. Walters, 3:257–60. New York: Plenum, 676 pp.
65. Doolen, G. D., Nuttall, J., Stagat, R. W. 1974. *Phys. Rev. A* 10:1612–15
66. Doolen, G. D. 1975. *J. Phys. B* 8:525–28
67. Moiseyev, N., Friedland, S., Certain, P. R. 1981. *J. Chem. Phys.* 74:4739–40
68. Brändas, E., Froelich, P. 1977. *Phys. Rev. A* 16:2207–10
69. Brändas, E., Froelich, P., Hehenberger, M. 1978. *Int. J. Quantum Chem.* 14:419–41
70. Canuto, S., Goscinski, O. 1978. *Int. J. Quantum Chem.* 14:383–91
71. Winkler, P. 1977. *Z. Phys. A* 283:149–60
72. Yaris, R., Winkler, P. 1978. *J. Phys. B* 11:1475–80
73. Tachibana, A., Yamabe, T., Fukui, K. 1979. *Mol. Phys.* 37:1045–57
74. Certain, P. R. 1979. *Chem. Phys. Lett.* 65:71–72
75. Moiseyev, N., Certain, P. R., Weinhold, F. 1978. *Int. J. Quantum Chem.* 14:727–36
76. Moiseyev, N., Certain, P. R., Weinhold, F. 1978. *Mol. Phys.* 36:1613–30
77. Moiseyev, N., Friedland, S., Certain, P. R. 1981. *J. Chem. Phys.* 74:4739–40
78. Moiseyev, N. 1981. *Phys. Rev. A* 24:2824–25
79. Winkler, P., Yaris, R. 1978. *J. Phys. B* 11:1481–95
80. Winkler, P., Yaris, R. 1978. *J. Phys. B* 11:4257–70
81a. Weinhold, F. 1979. *J. Phys. Chem.* 83:1517–20
81b. Moiseyev, N., Certain, P. R. 1979. *Mol. Phys.* 37:1621–32
82. Ho, Y. K. 1979. *J. Phys. B* 12:387–99
83. Ho, Y. K. 1979. *J. Phys. B* 12:L543–46
84. Ho, Y. K. 1980. *Phys. Lett. A* 77:147–49
85. Ho, Y. K. 1980. *Phys. Lett. A* 79:44–46
86. Ho, Y. K. 1981. *Phys. Rev. A* 23:2137–49
87. Moiseyev, N., Weinhold, F. 1979. *Phys. Rev. A* 20:27–31
88. Doolen, G. D., Nuttall, J., Wherry, C. J. 1978. *Phys. Rev. Lett.* 40:313–15
89. Ho, Y. K. 1979. *Phys. Rev. A* 19:2347–52
90. Drachman, R. J., Houston, S. K. 1975. *Phys. Rev. A* 12:885–90
91. Ho, Y. K. 1978. *Phys. Rev. A* 17:1675–78
92. Reinhardt, W. P. 1976. *Int. J. Quantum Chem. Symp.* 10:359–67
93. Cerjan, C., Hedges, R., Holt, C., Reinhardt, W. P., Scheibner, K., Wendoloski, J. J. 1978. *Int. J. Quantum Chem.* 14:393–418
94. Atabek, O., Lefebvre, R. 1981. *Int. J. Quantum Chem.* 19:901–6
95. Froelich, P., Hehenberger, M., Brandas, E. 1977. *Int. J. Quantum Chem. Symp.* 11:295–99
96. Junker, B. R. 1982. In *Electronic and Atomic Collisions, XII ICPEAC, Invited Talks*, ed. S. Datz, pp. 491–501. Amsterdam: North-Holland. 872 pp.
97. Wendoloski, J. J., Reinhardt, W. P. 1978. *Phys. Rev. A* 17:195–220
98. Chu, S.-I. 1978. *Chem. Phys. Lett.* 58:462–66
99. Chu, S.-I., Reinhardt, W. P. 1977. *Phys. Rev. Lett.* 39:1195–98
100. Reinhardt, W. P. 1980. *Electronic and Atomic Collisions, XI ICPEAC, Invited Lectures and Progress Reports*, ed. N. Oda, K. Takayanagi, pp. 729–39. Amsterdam: North Holland. 844 pp.
101. Chu, S.-I. 1978. *Chem. Phys. Lett.* 54:367–72
102. Chu, S.-I. 1979. *Chem. Phys. Lett.* 64:178–82
103. Preobrazhensky, M. A., Rapoport, L. P. 1980. *Zh. Eksp. Teor. Fiz.* 51:929–35 (*Sov. Phys. -JETP* 51:468–71)
104. Grossmann, A., Tip, A. 1980. *J. Phys. A* 13:3381–97
105. Cerjan, C., Reinhardt, W. P., Avron, J. E. 1978. *J. Phys. B* 11:L201–5
106. Herbst, I. W. 1979. *Commun. Math. Phys.* 64:279–98

107. Herbst, I. W. 1980. *Commun. Math. Phys.* 75:197–205
108a. Herbst, I. W., Simon, B. 1978. *Phys. Rev. Lett.* 41:67–69
108b. Herbst, I. W., Simon, B. 1978. *Phys. Rev. Lett.* 41:1759 (Erratum)
109. Herbst, I. W., Simon, B. 1981. *Commun. Math. Phys.* 80:181–216
110. Rescigno, T. N., McCurdy, C. W. 1978. *Bull. Am. Phys. Soc.* 23:1104
111. Rescigno, T. N., McCurdy, C. W., Orel, A. E. 1978. *Phys. Rev. A* 17:1931–38
112. Bardsley, J. N., Junker, B. R. 1972. *J. Phys. B* 5:L178–80
113. Bain, R. A., Bardsley, J. N., Junker, B. R., Sukumar, C. V. 1974. *J. Phys. B* 7:2189–2202
114. Junker, B. R., Huang, C. L. 1976. *Bull. Am. Phys. Soc.* 21:1252
115. Junker, B. R., Huang, C. L. 1977. In *X ICPEAC, Abstracts*, ed. M. Barat, J. Reinhardt, pp. 614–15. Paris: Commis. Energ. At. 1320 pp.
116. Junker, B. R. 1978. *Int. J. Quantum Chem.* 14:371–82
117. Junker, B. R. 1978. *Phys. Rev. A* 18:2437–42
118. Nicolaides, C. A., Beck, D. R. 1978. *Int. J. Quantum Chem.* 14:457–513
119. Nicolaides, C. A., Komninos, Y., Mercouris, T. 1981. *Int. J. Quantum Chem. Symp.* 15:355–67
120. Nicolaides, C. A., Komninos, Y., Mercouris, T. 1981. *Phys. Lett. A* 84:421–23
121. Komninos, Y., Nicolaides, C. A. 1981. *Chem. Phys. Lett.* 78:347–50
122. Atabek, O., Lefebvre, R. 1981. *Chem. Phys.* 56:195–201
123. Atabek, O., Lefebvre, R. 1980. *Phys. Rev. A* 22:1817–19
124. Atabek, O., Lefebvre, R., Requena, A. 1981. *Chem. Phys. Lett.* 78:13–15
125. Isaacson, A. D., McCurdy, C. W., Miller, W. H. 1978. *Chem. Phys.* 34:311–17
126. Isaacson, A. D., Miller, W. H. 1979. *Chem. Phys. Lett.* 62:374–77
127. Simons, J. 1980. *Int. J. Quantum Chem. Symp.* 14:113–21
128. Simons, J. 1980. *J. Chem. Phys.* 73:992–93
129a. McCurdy, C. W., Rescigno, T. N. 1979. *Phys. Rev. A* 20:2346–51
129b. Yaris, R., Taylor, H. S. 1979. *Chem. Phys. Lett.* 66:505–10
130. Bačić, Z., Simons, J. 1981. *Int. J. Quantum Chem.* 21:727–34
131. Noro, T., Taylor, H. S. 1980. *J. Phys. B* 13:L377–81
132. Moiseyev, N., Certain, P. R., Weinhold, F. 1981. *Phys. Rev. A* 24:1254–59
133a. Stillinger, F. H., Stillinger, D. K. 1974. *Phys. Rev. A* 10:1109–21
133b. Stillinger, F. H., Weber, J. A. 1974. *Phys. Rev. A* 10:1122–30
133c. Herrick, D. R., Stillinger, F. H. 1975. *J. Chem. Phys.* 62:4360–65
134. Herrick, D. R. 1976. *J. Chem. Phys.* 65:3529–35
135. Sherman, P. R., Herrick, D. R. 1981. *Phys. Rev. A* 23:2790–93
136. Junker, B. R. 1980. *Phys. Rev. Lett.* 44:1487–90
137. Junker, B. R. 1980. *Int. J. Quantum Chem. Symp.* 14:53–66
138. Hazi, A., Taylor, H. S. 1970. *Phys. Rev. A* 1:1109–20
139a. Taylor, H. S., Yaris, R. 1980. *Comments At. Mol. Phys.* 9:73–85
139b. Simons, J. 1981. *J. Chem. Phys.* 75:2465–67
139c. Bačić, Z., Simons, J. 1982. *J. Phys. Chem.* 86:1192–99
140. Donnelly, R. A., Simons, J. 1980. *J. Chem. Phys.* 73:2858–66
141. McCurdy, C. W., Lauderdale, J. G., Mowrey, R. C. 1981. *J. Chem. Phys.* 75:1835–42
142. Levine, I. N. 1974. *Quantum Chemistry*, pp. 294–95. Boston: Allyn & Bacon. 406 pp. 2nd ed.
143. Bardsley, J. N. 1978. *Int. J. Quantum Chem.* 14:343–52
144. McCurdy, C. W., Rescigno, T. N. 1978. *Phys. Rev. Lett.* 41:1364–68
145. Moiseyev, N., Corcoran, C. 1979. *Phys. Rev. A* 20:814–17
146. Simon, B. 1979. *Phys. Lett. A* 71:211–14
147. McCurdy, C. W. 1980. *Phys. Rev. A* 21:464–70
148. Morgan, J. D., Simon, B. 1981. *J. Phys. B* 14:L167–71
149. Deguchi, K., Nishikawa, K. 1980. *J. Phys. B* 13:L511–14
150. Deguchi, K., Nishikawa, K. 1980. *J. Phys. B* 13:L515–18
151. Lathouwers, L. 1978. *Phys. Rev. A* 18:2150–58
152. Rescigno, T. N., Orel, A. E., McCurdy, C. W. 1980. *J. Chem. Phys.* 73:6347–48
153. McCurdy, C. W. 1981. In *Quantum Mechanics in Mathematics, Physics and Chemistry*, ed. K. E. Gustafson, W. P. Reinhardt, pp. 383–406. New York: Plenum. 506 pp.

154. McCurdy, C. W., Rescigno, T. N., Davidson, E. R., Lauderdale, J. G. 1980. *J. Chem. Phys.* 73:3268–73
155. Mishra, M., Ohrn, Y., Froelich, P. 1981. *Phys. Lett. A* 84:4–8
156. Moiseyev, N. 1981. *Int. J. Quantum Chem.* 20:835–42
157. Moiseyev, N. 1981. *Mol. Phys.* 42:129–39
158. Brändas, E., Elander, N., Froelich, P. 1978. *Int. J. Quantum Chem.* 14:443–56
159. Atabek, O., Lefebvre, R. 1980. *Chem. Phys.* 52:199–210
160. Chu, S.-I. 1980. *J. Chem. Phys.* 72:4772–76
161. Chu, S.-I., Datta, K. K. 1982. *J. Chem. Phys.* 76:5307–20
162. Chu, S.-I. 1981. *J. Chem. Phys.* 75:2215–21
163. Atabek, O., Lefebvre, R. 1981. *Chem. Phys.* 55:395–406
164. Atabek, O., Lefebvre, R. 1981. *Chem. Phys.* 56:195–201
165. Atabek, O., Lefebvre, R. 1980. *Phys. Rev. A* 22:1817–19
166. Bačić, Z., Simons, J. 1980. *Int. J. Quantum Chem. Symp.* 14:467–75
167. Schneider, B. I. 1981. *Phys. Rev. A* 24:1–3
168. Waite, B. A., Miller, W. H. 1981. *J. Chem. Phys.* 74:3910–15
169. Winkler, P. 1979. *Z. Phys. A* 291:199–202
170. Winkler, P., Yaris, R., Lovett, R. 1981. *Phys. Rev. A* 23:1787–94
171. Donnelly, R. A., Simons, J. 1980. *J. Chem. Phys.* 73:2858–66
172. Mishra, M., Froelich, P., Ohrn, Y. 1981. *Chem. Phys. Lett.* 81:339–46
173a. Rescigno, T. N., McKoy, V. 1975. *Phys. Rev. A* 12:522–25
173b. Rescigno, T. N., McCurdy, C. W., McKoy, V. 1976. *J. Chem. Phys.* 64:477–80
174. Sukumar, C. V., Kulander, K. C. 1978. *J. Phys. B* 11:4155–65
175. McCurdy, C. W., Rescigno, T. N. 1980. *Phys. Rev. A* 21:1499–1505
176. Semenov, V. E. 1980. *Opt. Spectrosc. (USSR)* 48:398–401. (1980. *Opt. Spektrosk.* 48:723–27)

LIGHT AND RADICAL IONS

Terry A. Miller

Bell Laboratories, Murray Hill, New Jersey 07974

Introduction

If an electron is removed from most neutral molecules, a radical cation is formed. Until recently, the gas phase, or isolated molecule, study of radical cations has largely been the domain of the mass spectroscopist. However, in the last three to five years, there has been great interest in the interaction of radiation and molecular ions. In the last issue (1981) of the *Annual Review of Physical Chemistry*, two articles dealt directly with the subject, "High Resolution Spectroscopy of Molecular Ions" (1) and "Fast Ion Beam Photofragment Spectroscopy," (2) with another two indirectly addressing it (3, 4). It is certainly a measure of the vitality of the area that there exists anything left to review one year later.

Limiting the scope of this review, however, rather than finding new material, turned out to be the larger problem. This review is confined to work involving the interaction of light (visible and near UV and IR radiation) with molecular ions. Because the spectroscopy of diatomic ions was treated in detail last year (1), only polyatomic ion work is covered, and its spectroscopic aspects emphasized. Furthermore, this review is restricted to positive ions and does not consider the large body of literature (5) on photodetachment and photoelectron spectroscopy of negative ions.

When an isolated molecule, ion or not, absorbs a photon and undergoes a transition to an excited electronic state, it may decay radiatively or nonradiatively. If the radiative decay pathway is negligible, then absorption spectroscopy is the usual means for probing the electronic transition. However, absorption spectroscopy of ions has been mostly ineffectual. Gas phase ion densities exceeding $10^8/cm^3$ ($10^{10}/cm^3$ in plasmas) can only be obtained with great difficulty in the laboratory, and absorption spectroscopy is just not sensitive enough to be useful at these low concentrations. In some cases, the density problem can be overcome by isolating ions in an inert gas matrix. Correspondingly, in the gas phase, if the electronic transition changes the ion's mass via dissociation, nonradiating electronic transitions can be studied indirectly by monitoring the ions mass spectrometrically. Experiments on nonradiating ions are reviewed below; however, the coverage is rather succinct because such experiments have been the subject of recent reviews (2, 6) and because the spectroscopic content of many of the studies is limited.

Ions with excited electronic states that radiatively decay generally yield more information about themselves. The section, Experimental Techniques—Radiative, discusses, with particular reference to their strengths and weaknesses, recent advances in experimental techniques that have made possible the acquisition of extensive information about this class of molecular ions. The section, Light-emitting Ions, surveys the types of polyatomic ions for which radiative transitions are known. Next is discussed the information learned about vibrational structure and Jahn-Teller effects in organic ions. The final section summarizes the present status of polyatomic ion work and looks to the future. Several reviews (7–12), not mentioned above, and a couple of books (13, 14) touch on some aspects of the interaction of radiation and molecular ions.

Experimental Techniques — Nonradiative

While molecular ions that decay radiatively often reveal the most information about themselves, considerable information can be obtained about the interaction of light and polyatomic molecular ions, even when the excited ions decay nonradiatively.

The simplest expedient for gaining information about nonradiating states is usually to record their absorption spectrum; however, as mentioned above, because of the low densities obtainable, there are essentially no reports of absorption spectra of polyatomic ions in the gas phase. One method of circumventing this problem has been to examine the absorption spectra of ions in inert gas matrices where concentrations $\gtrsim 10^{14}$ ions/cm^3 can be obtained. In the next section the criteria are discussed in more detail for obtaining unperturbed ion spectra in matrices, but suffice it to say here that if one uses a Ne, or in some cases an Ar matrix, unperturbed spectra are obtainable.

Whether or not the state decays radiatively, many excited ionic states are sufficiently long lived so that lifetime broadening does not prohibit spectral information from being obtained from their matrix absorption spectra. Examples of nonradiating polyatomic molecular ions whose absorption spectra have been obtained in inert gas matrices include the benzyl (15), tropylium (15), toluene (16, 17), cycloheptatriene (16), napthalene (18), benzyl chloride (19), fluorotoluene (19), monofluorobenzene (20, 21), monochlorobenzene (21), monobromobenzene (21), and 1,4-difluorobenzene (20) cations.

Another means of studying the spectroscopy of nonradiating ions is to detect the appearance of photofragment ions, or the disappearance of parent ions, upon the absorption of light. In one version of this technique, an ion cyclotron resonance (ICR) spectrometer is used as the ion detector (6, 22–24). The ions in the ICR cell are irradiated by a tunable frequency light source and the concentration of ions of a given mass continuously monitored by the ICR spectrometer. Light sources utilized include wavelength selected arc lamp, Ar^+ laser, pulsed or CW dye laser, and CO_2 laser. ICR photodissociation spectroscopy can be used as an indirect means of detecting the optical absorption spectrum of an ion. It can also be used to differentiate between structural isomers (which are equivalent to ICR alone) and to characterize rearrangement reactions. The method even provides some information on photofragmentation dynamics by roughly characterizing kinetic energy release and angular distributions of fragments, given the known trapping potential.

Some of the large number of polyatomic cations studied by ICR photodissociation spectroscopy include methyl (25, 26), alkyl- (27), and halogen-substituted (28) benzenes, chain and cycloalkanes (29, 30), and halogen-substituted (31) versions thereof, as well as dienes, trienes (32–34), and unsaturated aldehydes and ketones (35).

Besides the ICR experiments, other apparatuses for doing ion photofragment spectroscopy include fast ion beams, drift tubes, and tandem quadrupole spectrometers. In all cases loss of parent ions, or fragment ion production, is detected as a function of irradiating wavelength. Species studied in this way include CH_3I^+ (36), $C_2H_5I^+$ (37), $C_3H_7I^+$ (37), N_2O^+ (38, 39), $(O_2)_2^+$ (40), $(NO)_2^+$ (40), $(CO_2)_2^+$ (40), CO_4^+ (41), and $O_2^+ \cdot H_2O$ (41).

In general, all the photofragment studies for polyatomic ions have been of relatively low resolution and do not reveal detailed structural information. Nonetheless, they are quite valuable in that they constitute our only spectroscopic information on these nonradiating ions.

Experimental Techniques — Radiative

The simplest experimental technique for studying radiating electronic transitions is classical emission spectroscopy. The largest optical spectrographs yield spectra of ions that are nearly Doppler-limited. Smaller spectrographs, of course, have lower resolution, but can often have significantly greater optical throughput, an important advantage when dealing with emitting species at low concentrations.

Overall, the basic design and performance of optical spectrographs have not changed greatly over recent years. What has been improved considerably are the optical emission sources. The traditional source for emission studies has been a discharge tube. Recent work (42–45) by the group at Orsay, headed by Leach, shows that nice results can still be obtained by that method, even for relatively large organic ions. However, the technique has a couple of important limitations. The processes occurring in the discharge are largely uncontrolled, but always rather severe, leading simultaneously to dissociation, ionization, and multiple step reactions. This makes it difficult to generate ions selectively, particularly those of large fragile organic molecules. Even when such ions are successfully generated, they may well be vibrationally and rotationally hot, leading to quite congested spectra.

The recent effort in the development of ion sources has been toward producing more selective ionization and less internal energy in the ion. The use of controlled electron impact ionization has been very effective in reaching the former, though not perhaps the latter, goal. This technique has been pioneered by Maier and co-workers (10, 11, 46–48).

Experiments of this sort utilize standard optical equipment; however, they replace the traditional discharge with a source using a beam of nominally monochromatic electrons. The idea is to match roughly the energy of the electrons to the peak of the ionization cross section. This maximizes ionization while minimizing fragmentation. In a typical experiment, electron beams of 20–40 eV, with 0.1–1.0 mA current, are used to excite the sample compound at a pressure of 10^{-5}–10^{-2} torr. The ion spectra are recorded with a scanning monochromator with moderate resolution (~ 0.1–1 nM). Data from photoelectron spectra of the corresponding neutral greatly aid such experiments. They delineate the spectral region that needs to be searched and provide an initial criterion by which ion spectra can be recognized.

These electron beam excited emission experiments are relatively rapid and provide a most convenient way of surveying for ionic emissions. In the last few years, emissions from roughly 100 organic ions have been detected (11) by such experiments. The limitation of such experiments is that the resolution, and hence the information content of such spectra, is often less than one desires. Occasionally, the resolution problem is instrumental; however, usually the problem lies in the inherent congestion present in the spectrum of polyatomic ions, caused by the population of many vibrational and rotational levels at room or elevated temperature.

A number of recent experiments have been performed to try to circumvent this problem by obtaining the spectra of "cold" or selectively excited ions. One experimental technique has been to continue to use controlled electron bombardment excitation and to record emission spectra, but to replace the room temperature sample of neutrals with a very low temperature sample. Since polyatomic molecules generally have quite low vapor pressures at low temperatures, this work has utilized the free jet expansion technique (49) to reach very low temperatures. The sample gas is entrained in an inert gas carrier and expanded through a small nozzle (typically 0.01–1 mm diameter). Adiabatic cooling leaves the seed molecule at a translational, and usually also internal, temperature of a few degrees Kelvin or less.

Controlled electron impact excitation then produces excited, emissive ionic states that are likewise internally quite cold. Experiments (50) on N_2^+, in which the ion's excited state rotational temperature could be accurately measured, indicated rotational temperatures $\lesssim 30$ K for excitation by electrons with energies of 100 eV or more. Obviously, conservation of angular momentum dictates little rotational warming upon ionization. Final state vibrational distributions will largely be governed by Franck-Condon overlap considerations; however, particularly in large organic molecules, the removal of one electron may not drastically change the electronic potential; thus, $\Delta v = 0$ transitions will be favored. These techniques have yielded very cold, and greatly simplified, emission spectra of several fluorobenzene cations (51, 52a, b).

The most important innovation in obtaining simplified, and so interpretable, spectra of polyatomic ions probably has been to employ lasers, combined with temperature control. Since radical cations are open-shell species, their electronic spectra usually fall into accessible spectral regions for lasers, even when the corresponding closed shell neutrals do not. For example, neutral N_2 has no electronic absorptions at wavelengths longer than 2000 Å, yet N_2^+ has both a red and violet electronic absorption. Similarly, in neutral benzenoid compounds the first electronic transition is typically ~ 2600 Å, while in the benzenoid cations it falls into the 4000–6000 Å region. Additionally, laser-induced fluorescence techniques have a sensitivity approaching 10^5 ions/cm^3. This compares favorably with the $\sim 10^8$ ions/cm^3, which, as noted above, represents the highest densities typically obtainable in the laboratory.

Two general kinds of laser-induced fluorescence spectra can be recorded. The first, an *excitation* spectrum (53–60), closely resembles, in information content, the classical optical absorption spectrum. The laser frequency is varied, and essentially the total fluorescence is detected as a function of laser frequency. This type of spectrum typically does a very good job of characterizing the excited electronic state in the transition; however, the only ground state levels interrogated are those thermally populated.

Laser excited, wavelength-resolved emission spectra characterize (61) mainly the ground state. In this experiment, the laser frequency is fixed so that it coincides with a feature in the excitation spectrum, and the resulting emission is wavelength-resolved with a monochromator.

The advantage of this experiment over a conventional emission experiment is that only a single excited state level fluoresces at a time (neglecting relaxation processes that can usually be controlled). This means the spectrum is usually simple enough to be easily interpretable. By changing excitation frequencies, different emission spectra can be recorded so that the ground state can be completely characterized.

As in nonlaser experiments, temperature control of the ionic species is also important. In the initial laser-induced fluorescence experiments, the ions were produced by Penning ionization. Although Penning ionization is a relatively gentle means of producing ions, it nonetheless often produces vibrationally and rotationally "hot" ions. The initial LIF experiments, however, with few exceptions, detected ions equilibrated to an ambient room temperature distribution. The reason was that the ions generally suffered many collisions with the buffer gas before interrogation by the laser. The net result was that the room temperature LIF spectra were typically less congested than those of the same ion obtained under discharge or controlled electron bombardment conditions.

In later Penning ionization experiments (7, 62–64) the entire discharge and flow tube apparatus for delivering the metastable atoms to the reaction chamber was immersed in liquid N_2. In this way the ions were equilibrated to a much lower temperature. Experiments on N_2^+ indicated (8) an internal temperature of ~ 100 K. By careful mixing, LIF spectra of ions whose parents have relatively low vapor pressures, e.g. trichlorobenzene, were obtained under these same low temperature conditions. These spectra were greatly simplified and more resolved than any previous gas phase LIF spectra of large ions.

The quest for even simpler and more interpretable large ion spectra led to the low temperature limit, matrix isolation spectra of ions. It was first thought that the ion-host interaction might be too strong to obtain unperturbed ion spectra in matrices. The criterion for the matrix to be sufficiently unperturbing so that the matrix spectra can be considered the low temperature limit of the gas phase spectra seems to be that the electron affinity of the excited ionic state (the ionization plus excitation energy) be small compared to the energy of the host's conduction bond (7, 8, 65). The electron affinity of the lower excited states of organic ions are usually $\lesssim 13$ eV, thus Ne with a conduction band energy of only 1–2 eV below its ionization potential of 21.6 eV clearly satisfies this criterion. Other inert gas matrices satisfy this criterion less well, if at all (for example the ionization potential of Ar is only 15.7 eV).

Numerous comparisons are now available between Ne matrix and gas phase ion spectra. Almost without exception, the vibrational frequencies recorded in the matrix and gas phase for both ground and excited states of ions are identical within experimental error (65–67). This correspondence holds even when dealing with subtle interactions like Jahn-Teller effects (68–70). In a similar way even relative band intensities are reproduced. The one clearly measurable difference (67, 71, 72) between the gas phase and a Ne matrix is a shift of the entire spectrum, i.e. T_e, to a lower frequency by an amount of $\lesssim 1\%$.

Both LIF excitation and wavelength-resolved emission spectra for a large number of ions have been obtained in Ne matrices. The experimental techniques (7, 8) are, overall, very like those described earlier for gas-phase LIF spectra. The sample preparation is of course rather different. Typically, a mixture of 1 part in 10^3 or 10^4 of the parent neutral is deposited on a substrate held at liquid He temperature. The sample is exposed to VUV radiation, e.g. Lyman-α at 1216 Å, until the ion concentration ceases to increase, typically 0.1–5 minutes. Unlike in the gas phase, laser excitation spectra are typically taken with the light dispersed through a monochromator. This technique greatly increases spectral resolution, usually to the monochromator bandwidth, by eliminating different matrix sites and inhomogeneous broadening.

Recently, the advantages of extremely low temperature in matrices, and isolation in the gas phase, have been combined in a single laser-induced fluorescence experiment (76). As in the electron excited emission spectra of ions from a free jet expansion, very cold, organic parent neutral molecules are produced by a supersonic jet. However, now higher transient ion concentrations are produced by two-photon ionization with a pulsed excimer laser. These ions are probed downstream by a tunable dye laser. In this way, gas phase LIF excitation spectra have been obtained for large organic ions essentially "frozen out" in their lowest internal states. However, perhaps most exciting about this experiment is that not only have cold bare ion spectra been obtained but spectra for ions, clustered by several inert gas atoms, have also been seen (76).

While we have made much of the importance of being able to control temperature (i.e. control internal state population), there are some LIF experiments on ions where such control is not possible, or even perhaps desirable. Recently there have been several reports (73–75) of LIF studies of ions, particularly small diatomic and triatomic ones, created by electron impact and held in a three-dimensional rf quadrupole ion trap. These ions are produced and held under essentially collision-free conditions so that the distribution of internal energy characterizing the ion's formation remains essentially unchanged.

To conclude this section on radiative experimental techniques we discuss briefly two techniques for obtaining lifetime and quantum yield information for ions with detectable emissions. If one is interested in lifetimes only, one can measure directly time-resolved emission profiles. The main requirement here is a pulsed excitation source. Pulsed electron beam excitation has enjoyed considerable popularity (9, 11, 77). Its principal advantage is its near universality. Its principal drawback is lack of selectivity, making it difficult to detect lifetime changes within a given electronic state. One must even take care to insure that the results are not contaminated by contributions from electronic states other than the one being measured.

Laser excitation generally avoids the problem of selectivity, since single vibronic levels can be populated. Several measurements of ion lifetimes have recently been reported (78–81) using laser excitation. The principal limitation here has so far been in detecting laser-excited, fluorescence decay curves from ions in a perturbation-free environment. In gas phase experiments, lifetimes usually have to be extrapolated to zero pressure; in matrices, small corrections need to be made for the medium's index of refraction, and lifetime measurements are limited to the vibrationless level due to rapid vibrational relaxation.

The third technique that has enjoyed considerable success in providing ion lifetimes is coincidence measurements (9–11, 82–87). Suppose an ion is produced by photoionization (inelastic electron excitation has also been used) into an excited state ($\tilde{A}$), then

$$M + h\nu_{UV} \rightarrow M^+(\tilde{A}) + e_{KE}$$
$$\downarrow h\nu_R$$
$$M^+(\tilde{X}). \qquad\qquad 1.$$

On the right hand side of Eq. 1 are three quantities: the photo-ion M^+; an electron, e_{KE}, with well-determined kinetic energy which, given the energy of the photoionizing photon $h\nu_{UV}$ defines rather precisely the excited state energy $M^+(\tilde{A})$; and finally the photon, $h\nu_R$, emitted in the radiative decay of $M^+(\tilde{A})$ to the ground (or lower) state $M^+(\tilde{X})$. These three quantities can be detected in pair-wise coincidence. If the fast photoelectrons are collected in coincidence with the photo-ions, a time of flight mass spectrum is obtained, which will reveal the extent, if any, of photofragment ion production in the initial step. If the photoelectrons, e_{KE}, of specific kinetic energy and the photons $h\nu_R$ are collected in coincidence, this yields a lifetime for the specified excitation energy of $M^+(\tilde{A})$. If the photo-ions and photons $h\nu_R$ are collected, once again a fluorescence decay curve can be obtained, but without specific energy specification within $M^+(\tilde{A})$.

If C is the number of coincidence photons and photo-electrons or ions (depending on the technique) and T is the total number of photo-electrons or ions, then

$$C/T = f_{h\nu_R}\phi \qquad \qquad 2.$$

where $f_{h\nu_R}$ is the overall collection efficiency for photons, $h\nu_R$, including both molecular branching ratio and instrumental effects; ϕ is the quantum yield for emission, defined as the ratio of the radiative decay rate to the sum of the radiative and nonradiative decay rates. The coincidence methods appear to be the only means yet used to measure quantum yields for polyatomic ions. As might be expected, the principal limitation on overall accuracy is the absolute determination of $f_{h\nu_R}$. Some, though not complete, specification of $f_{h\nu_R}$ can be made by calibrating the apparatus with ions like $N_2^+(\tilde{B})$ for which ϕ is unity.

Light-emitting Ions

As can be gleaned from the previous section, electronically excited states that decay radiatively generally provide considerable information about themselves. This gives rise to the important question of which polyatomic molecular ions may be expected to have excited states that decay radiatively. Work in the past few years has identified several such classes of polyatomic ions.

The simplest class of (photon) emitting ions is triatomic cations. (Interestingly, there is no known case of a polyatomic anion emitting in the gas phase. Generally speaking, the electron affinities of anions are insufficient to support bound excited states.) For triatomic species, once placed in a bound, excited electronic state, emission is virtually mandatory, because the density of states is too low to permit internal conversion in the isolated molecule. (One can, of course, excite nonbound states, or predissociating states, which lead to fragmentation rather than radiation, or at best a competition between the two.) There are now a number of triatomic ions that are known to emit, and Table 1 lists these ions along with the band systems observed. It also includes the lifetimes of the excited states, as well as the experimental techniques that have been used to probe these ions.

If one moves to polyatomic ions with four or more atoms, one has, as well as radiation and fragmentation, the possibility of internal conversion as a decay route. The density of states may now be high enough so that electronic excitation energy can be converted into vibrational excitation of the ground (or other lower) electronic state. These highly vibrationally excited ions can subsequently fragment or isomerize.

Nonetheless, there have been three categories of larger, organic ions for which emissions have been observed. These categories are olefinic, acetylenic, and benzenoid cations. Although these groupings are quite broad and the species structurally quite different, there is some commonality with regards to the observed electronic transitions. In all cases

Table 1 Triatomic molecular ions for which spectra have been observed[a]

Cation	States	Lifetime (nsec)	Experimental techniques
ClCN	$\tilde{B}^2\Pi \leftrightarrow \tilde{A}^2\Pi, \tilde{A}^2\Sigma^+ \leftrightarrow \tilde{X}^2\Pi(C_{\infty v})$	150–205 ($\tilde{B}$), 4400 ($\tilde{A}$)	EEEB (47), TP (47, 82)
BrCN	$\tilde{B}^2\Pi \leftrightarrow \tilde{A}^2\Pi, \tilde{A}^2\Sigma^+ \leftrightarrow \tilde{X}^2\Pi(C_{\infty v})$	270–300 ($\tilde{B}$), 3000 ($\tilde{A}$)	EEEB (47), LIFG (74), TP (47, 82)
ICN	$\tilde{B}^2\Pi \leftrightarrow \tilde{A}^2\Pi, \tilde{A}^2\Sigma^+ \leftrightarrow \tilde{X}^2\Pi(C_{\infty v})$	300 ($\tilde{B}$), 1200 ($\tilde{A}$)	EEEB (47), TP (47)
H_2O	$\tilde{A}^2A_1 \leftrightarrow \tilde{X}^2B_1(C_{2v})$	3000, 10500	EEEB (88–90), IBOP (91), TP (92, 93)
H_2S	$\tilde{A}^2A_1 \leftrightarrow \tilde{X}^2B_1(C_{2v})$	4300	EED (94), TP (95)
N_2O	$\tilde{A}^2\Sigma^+ \leftrightarrow \tilde{X}^2\Pi(C_{\infty v})$	220–260	EED (96), EEEB (97), SJES (52), TP (82, 86, 98–100, 120)
CO_2	$\tilde{B}^2\Sigma_u^+, \tilde{A}^2\Pi_u \leftrightarrow \tilde{X}^2\Pi_g(D_{\infty h})$	117–140 ($\tilde{B}$), 102–124 ($\tilde{A}$)	EED (101–105), LIFG (54), SJES (52), TP (82, 85, 106–108, 120)
CS_2	$\tilde{B}^2\Sigma_u^+, \tilde{A}^2\Pi_u \leftrightarrow \tilde{X}^2\Pi_g(D_{\infty h})$	294–360 ($\tilde{B}$), 2400–4000 ($\tilde{A}$)	EED (109–113), CHLE (114), PIE (115), LIFM (80, 116), LIFG (117), TP (82, 99, 116, 120)
OCS	$\tilde{A}^2\Pi \leftrightarrow \tilde{X}^2\Pi(C_{\infty v})$	<30–175	EED (118), TP (82, 99, 120)
HCP	$\tilde{A}^2\Sigma^+ \leftrightarrow \tilde{X}^2\Pi(C_{\infty v})$	1190	EEEB (119), TP (119)

[a]Under *States* are indicated the states involved in the observed electronic transition(s). The symbol in parenthesis gives the assumed molecular symmetry. The legend for the experimental techniques is as follows: EEEB = Emission Excited by Electron Bombardment; EED = Emission Excited by a Discharge; LIFG = Laser Induced Fluorescence in the Gas phase; LIFM = Laser Induced Fluorescence in an inert gas Matrix; IBOP = Ion Beam, Optical spectroscopy; PIE = Photo-Ionization induced Emission; CHLE = Chemiluminescence Emission; SJES = Supersonic Jet Emission Spectrum; TP = lifetime (τ) or quantum yield (ϕ) determination. For the listed lifetimes, where several different experiments gave similar but slightly different lifetimes, a range is given. Where discrepant values have been reported, if possible selection of the most reliable value has been made. Where such selection has not been possible, all values are listed. All lifetime values are in nanoseconds and for the vibrationless level of the indicated state.

they are $\pi - \pi$ transitions of a "hole-promotion" type (32, 163) not found in neutral molecules. For all these ions, in the parent neutrals at least the last two occupied molecular orbitals are π orbitals.

For example, in benzene, these filled π orbitals are the doubly degenerate e_{1g} and the nondegenerate a_{2u}. In diacetylenes, the two highest filled orbitals (122) are $2p\pi_u$ and $2p\pi_g$. For the olefinic compounds, the n highest filled orbitals are again π orbitals where n is the number of double bonds in the neutral ($n > 2$ is required for observed emission from the cation). Roughly speaking, the ground state of all the radical cations is obtained by removing an electron from the highest filled π orbital. The excited, emitting electronic state is formed by promoting this "hole" to the next deeper π orbital.

While this general scheme seems to hold for all the observed ionic transitions, it requires a little elaboration. In asymmetrically substituted benzenes the doubly degenerate $^2E_{1g}$ ground state splits into nondegenerate $\tilde{A}$ and $\tilde{X}$ states. Emission from the "hole promotion" $\tilde{B}$ state into both $\tilde{A}$ and $\tilde{X}$ has been observed, but not emission between $\tilde{A}$ and $\tilde{X}$. In some benzenoid cations, including benzene itself, a σ state slips below the "hole promotion" π state. In these cations, emission is generally not detectable. In the olefinic compounds, detailed MO calculations (163) have been made on 1,3,5-hexatriene, and it appears (32, 165, 166) that the emitting state is a mixture of the "hole promotion" state and one formed by excitation of the unpaired electron to the highest unoccupied π orbital of the neutral.

Table 2 lists, in a fashion analogous to Table 1, the acetylenic cations for which emission has been observed. The cation of acetylene itself (with the four π electrons of the triple bond in a doubly degenerate π orbital) does not emit. However, as Table 2 shows, if one or both H's are replaced by a halogen or the pseudo-halogen, CN, emission is detected.

As Table 2 further shows, the combination of two triple bonds (and hence two filled π orbitals) into a diacetylene unit gives rise to many emitting cations. While quantum yield measurements are not widely available (see Table 2) ethyl- and dimethyldiacetylene radical cations, $C_6H_6^+$, give readily observable spectra whereas the isomeric benzene radical cation appears to have a vanishingly small fluorescence quantum yield (48). On the other hand, although diacetylene cation itself appears to have near unity quantum yield for emission from its vibrationless level, in a number of the substituted diacetylene cations in Table 2, fragmentation is competitive with radiation.

In a similar way, for the olefinic cations listed in Table 3, internal conversion and fragmentation are competitive with emission. The case of the 1,3,5-hexatriene cations is probably the best studied (136–138). Things are slightly complicated by the existence of *cis* and *trans* isomers, but there seems now good agreement that the quantum yield for emission from the vibrationless level (137) of the $\tilde{A}$ state for the *trans* cation is $\sim 7\%$. The competing nonradiative pathway is believed to be rapid internal conversion to high vibrational levels of the ground state. This species then isomerizes to a cyclic $C_6H_8^+$, which finally fragments to $C_6H_7^+ + H$.

Table 4 lists the known emitters among the benzenoid cations. As can be seen from Table 4, generally speaking if one has three or more F substituents on the ring, the "hole promoting" π state apparently is the

Table 2 Acetylenic cations known to have observable fluorescence emission[a]

Cation	Electronic transition	Lifetime	Experimental technique
$Cl-C\equiv C-Cl$	$\tilde{A}^2\Pi_g \leftrightarrow \tilde{X}^2\Pi_u (D_{\infty h})$	21	EEEB (46), TP (46)
$Br-C\equiv C-Br$	$\tilde{A}^2\Pi_g \leftrightarrow \tilde{X}^2\Pi_u (D_{\infty h})$	≤ 6	EEEB (46), TP (46)
$I-C\equiv C-I$	$\tilde{A}^2\Pi_g \leftrightarrow \tilde{X}^2\Pi_u (D_{\infty h})$	≤ 6	EEEB (46), TP (46)
$H-(C\equiv C)-Cl$	$\tilde{A}^2\Pi \leftrightarrow \tilde{X}^2\Pi (C_{\infty v})$	17	EEEB (77), TP (77)
$H-(C\equiv C)-Br$	$\tilde{A}^2\Pi \leftrightarrow \tilde{X}^2\Pi (C_{\infty v})$	12	EEEB (77), TP (77)
$H-C\equiv C-I$	$\tilde{A}^2\Pi \leftrightarrow \tilde{X}^2\Pi (C_{\infty v})$	15	EEEB (77), TP (77)
$H-(C\equiv C)_2-H$	$\tilde{A}^2\Pi_g \leftrightarrow \tilde{X}^2\Pi_u (D_{\infty h})$	72(0.72)	EED (122), EEEB (123), LIFM (66), LIFG (8), TP (123, 126, 127)
$F-(C\equiv C)_2-F$	$\tilde{A}^2\Pi_g \leftrightarrow \tilde{X}^2\Pi_u (D_{\infty h})$	21	EEEB (121), TP (121)
$Cl-(C\equiv C)_2-Cl$	$\tilde{A}^2\Pi_g \leftrightarrow \tilde{X}^2\Pi_u (D_{\infty h})$	21(0.47)	EEEB (128), LIFG (124), TP (124, 128)
$Br-(C\equiv C)_2-Br$	$\tilde{A}^2\Pi_g \leftrightarrow \tilde{X}^2\Pi_u (D_{\infty h})$	12	EEEB (128), TP (128)
$I-(C\equiv C)_2-I$	$\tilde{A}^2\Pi_g \leftrightarrow \tilde{X}^2\Pi_u (D_{\infty h})$	≤ 6	EEEB (128), TP (128)
$CH_3-(C\equiv C)_2-CH_3$	$\tilde{A}^2E_u \leftrightarrow \tilde{X}^2E_g (D_{3d})$	24	EEEB (126, 130), LIFG (7, 60, 125), LIFM (7, 65), TP (126, 130, 133)
$CF_3-(C\equiv C)_2-CF_3$	$\tilde{A}^2E_u \leftrightarrow \tilde{X}^2E_g (D_{3d})$	46	EEEB (121), TP (121)
$H-(C\equiv C)_3-H$	$\tilde{A}^2\Pi_g \leftrightarrow \tilde{X}^2\Pi_u (D_{\infty h})$	17	EEEB (123), TP (123)
$H-(C\equiv C)_4-H$	$\tilde{A}^2\Pi_g \leftrightarrow \tilde{X}^2\Pi_u (D_{\infty h})$	≤ 6	EEEB (123), TP (123)
$H-(C\equiv C)_2-CH_3$	$\tilde{A}^2E \leftrightarrow \tilde{X}^2E (C_{3v})$	50(0.94)	EEEB (126), LIFG (125), TP (133)
$H-(C\equiv C)_2-C_2H_5$	$\tilde{A}^2A'' \leftrightarrow \tilde{X}^2A'' (C_S)$	$\leq 6(0.004)$	EEEB (130), TP (130, 133)
$C_2H_5-(C\equiv C)_2-C_2H_5$	$\tilde{A}^2A'' \leftrightarrow \tilde{X}^2A'' (C_S)$	8(0.19)	EEEB (126), LIFG (132), TP (126, 132)
$Cl-(C\equiv C)-CH_3$	$\tilde{A}^2E \leftrightarrow \tilde{X}^2E (C_{3v})$	19	EEEB (134), TP (134)
$Br-(C\equiv C)-CH_3$	$\tilde{A}^2E \leftrightarrow \tilde{X}^2E (C_{3v})$	13	EEEB (134), TP (134)
$Cl-(C\equiv C)_2-H$	$\tilde{A}^2\Pi \leftrightarrow \tilde{X}^2\Pi (C_{\infty v})$	41(0.79)	EEEB (134), LIFG (124), TP (124, 134)

Table 2 *Continued*

Cation	Electronic transition	Lifetime	Experimental technique
Br–(C≡C)$_2$–H	$\tilde{A}^2\Pi \leftrightarrow \tilde{X}^2\Pi(C_{\infty v})$	27	EEEB (134), TP (134)
Cl–(C≡C)$_2$–CH$_3$	$\tilde{A}^2E \leftrightarrow \tilde{X}^2E(C_{3v})$	22	EEEB (134), TP (134)
Br–(C≡C)$_2$–CH$_3$	$\tilde{A}^2E \leftrightarrow \tilde{X}^2E(C_{3v})$	10	EEEB (134), TP (134)
F–(C≡C)$_2$–CF$_3$	$\tilde{A}^2E \leftrightarrow \tilde{X}^2E(C_{3v})$	30	EEEB (121), TP (121)
N≡C–C≡C–C≡N	$\tilde{A}^2\Sigma_g^+ \leftrightarrow \tilde{X}^2\Pi_u(D_{\infty h})$	13(0.12)	EEEB (129), LIFG (129), TP (129)
N≡C–(C≡C)$_2$–C≡N	$\tilde{A}^2\Pi_u \leftrightarrow \tilde{X}^2\Pi_g(D_{\infty h})$	≤6	EEEB (131), TP (131)
H–(C≡C)$_2$–C≡N	$\tilde{A}^2\Pi \leftrightarrow \tilde{X}^2\Pi(C_{\infty v})$	15	EEEB (135), TP (135)
CH$_3$–(C≡C)$_2$–C≡N	$\tilde{A}^2E \leftrightarrow \tilde{X}^2E(C_{3v})$	8	EEEB (135), TP (135)
C$_2$H$_5$–(C≡C)$_2$–C≡N	$\tilde{A}^2A'' \leftrightarrow \tilde{X}^2A''(C_S)$	≤6	EEEB (135), TP (135)

[a] The assumed molecular symmetry is given in parenthesis following the symbols for the electronic transition. The legend for the experimental techniques is as follows: EEEB = Emission Excited by Electron Bombardment; EED = Emission Excited by a Discharge; LIFG = Laser Induced Fluorescence in the Gas phase; LIFM = Laser Induced Fluorescence in an inert gas Matrix; TP = lifetime (τ) or quantum yield (ϕ) determination. The lifetimes for the $\tilde{A}$ state vibrationless level are given in nsecs. Where measured the quantum yield ϕ is given in parenthesis after the lifetime.

lowest excited electronic state in the cation and fluorescence is detected with near unity quantum yield, at least from the vibrationless level of the $\tilde{B}$ state. With two or less fluorines the σ state appears to drop below the π excited state and low or vanishing quantum yields are noted. Since these states are below the fragmentation thresholds in these cations, this energy must remain stored in the isolated ion or its isomers. As Table 4 shows,

Table 3 Olefinic cations known to have observable fluorescence emission[a]

Cation	Electronic transition	Lifetime	Experimental technique
trans-1,3,5-Hexatriene	$\tilde{A}^2B_g \leftrightarrow \tilde{X}^2A_u(C_{2h})$	17(0.07)	EEEB (136, 137), LIFM (138), TP (136, 137)
cis-1,3,5-Hexatriene	$\tilde{A}^2A_2 \leftrightarrow \tilde{X}^2B_1(C_{2v})$	≤6(0.003)	EEEB (136, 137), TP (136, 137)
trans-1,3,5-Heptatriene	$\tilde{A}^2A'' \leftrightarrow \tilde{X}^2A''(C_S)$	9	EEEB (137), TP (137)
trans-1,3,5,7-Octatetraene	$\tilde{A}^2A_u \leftrightarrow \tilde{X}^2B_g(C_{2h})$	≤6	EEEB (139), TP (139)
cis-1,2-Difluoroethylene	$\tilde{A}^2A_1 \leftrightarrow \tilde{X}^2B_1(C_{2v})$	320(0.067)	EEEB (140), TP (140, 141)

[a] The legend for experimental techniques is as follows: EEEB = Emission Excited by Electron Bombardment; LIFM = Laser Induced Fluorescence in an inert gas Matrix; TP = lifetime (τ) or quantum yield (ϕ) determination. The assumed symmetry is given in parenthesis following the symbols for the electronic transitions. The quantum yield, ϕ, for the vibrationless level is given in parenthesis after the lifetime, if measured. The lifetimes for the $\tilde{A}$ states' vibrationless level are in nsecs.

Table 4 Benzenoid cations known to have observable fluorescence emissions[a]

Cation[b]	Electronic transition	Lifetime	Experimental Technique
Hexafluorobenzene	$\tilde{B}^2A_{2u} \leftrightarrow \tilde{X}^2E_{1g}(D_{6h})$	48–56(1.00)[b]	EEEB (48, 142, 143), EED (42), LIFG (58, 64, 70), LIFM (67, 69, 70, 144, 146), ABM (69, 145), SJES (51, 52), LIFJ (76), TP (48, 81, 84, 87)
Pentafluorobenzene	$\tilde{B}^2B_1 \leftrightarrow \tilde{X}^2A_2, \tilde{A}^2B_1(C_{2v})$	45–52(0.98)	EEEB (48, 143), EED (42), LIFG (58), LIFM (67, 71), LIFJ (76), TP (48, 81, 84, 87), SJES (52)
1,2,3,5-Tetrafluorobenzene	$\tilde{B}^2B_1 \leftrightarrow \tilde{X}^2A_2, \tilde{A}^2B_1(C_{2v})$	50–57(0.99)[c]	EEEB (48, 143), EED (42), SJES (52), LIFG (58), LIFM (67, 71), TP (48, 81, 84, 87)
1,2,4,5-Tetrafluorobenzene	$\tilde{B}^2B_{3u} \leftrightarrow \tilde{X}^2B_{2g}, \tilde{A}^2B_{1g}(D_{2h})$	30–36(0.61)[c]	EEEB (48, 143), EED (42), SJES (52), LIFG (58), LIFM (67, 71), TP (48, 81, 84, 87)
1,2,3,4-Tetrafluorobenzene	$\tilde{B}^2B_1 \leftrightarrow \tilde{X}^2A_2, \tilde{A}^2B(C_{2v})$	50–57(0.90)[c]	EEEB (48, 143), EED (42), SJES (52), LIFG (58), LIFM (67, 71), TP (48, 81, 84, 87)
1,3,5-Trifluorobenzene	$\tilde{B}^2A_2'' \leftrightarrow \tilde{X}^2E''(D_{3h})$	57–64(1.00)[c]	EEEB (48, 143), EED (42–44), LIFG (57, 58, 61, 63, 147, 148), LIFM (67, 63, 146, 149), SJES (52), LIFJ (76), TP (48, 81, 84, 87)
1,2,4-Trifluorobenzene	$\tilde{B}^2A'' \leftrightarrow \tilde{X}^2A'', \tilde{A}^2A''(C_S)$	10–16(0.14)[c]	EEEB (48, 143), LIFM (72), TP (48, 84, 87), SJES (52)
1,2,3-Trifluorobenzene	$\tilde{B}^2A'' \leftrightarrow \tilde{X}^2A'', \tilde{A}^2A''(C_S)$	48–58(0.74)	LIFG (150, 151), LIFM (150, 151), TP (87, 150)
1,3-Difluorobenzene	$\tilde{B}^2B_1 \leftrightarrow \tilde{X}^2A_2, \tilde{A}^2B_1(C_{2v})$	≤6(0.006)	EEEB (48, 143), ABM (145), LIFM (153), TP (48, 84, 87)
1,3,5-Trichlorobenzene	$\tilde{B}^2A_2' \leftrightarrow \tilde{X}^2E''(D_{3h})$	22(0.41)	EEEB (161), EED (45), LIFG (59, 147, 148), LIFM (146–149), LIFJ (76), TP (81, 161, 162)

Compound	Transition	Value	References
1,4-Dichlorobenzene	$\tilde{B}^2B_{3u} \leftrightarrow \tilde{X}^2B_{2g}, \tilde{A}^2B_{1g}(D_{2h})$	$\leqslant 6(0.005)$	EEEB (161), TP (161, 162)
1,3-Dichlorobenzene	$\tilde{B}^2B_1 \leftrightarrow \tilde{X}^2A_2, \tilde{A}^2B_1(C_{2v})$	$\leqslant 6(0.004)$	EEEB (161), TP (161, 162)
1,3,5-Trichloro-2,4,6-trifluorobenzene	$\tilde{B}^2A_2'' \leftrightarrow \tilde{X}^2E''(D_{3h})$	33–34(0.66)	EEEB (154), LIFG (68, 148), LIFM (148, 155), TP (162)
1,2,4,5-Tetrachloro-3-fluorobenzene	$\tilde{B}^2B_1 \leftrightarrow \tilde{X}^2A_2(C_{2v})$	—	EEEB (154), TP (154)
1,3,5-Trichloro-2-fluorobenzene	$\tilde{B}^2B_1 \leftrightarrow \tilde{X}^2B_1, \tilde{A}^2A_2(C_{2v})$	22	EEEB (154), TP (154)
1,4-Dichloro-2,5-difluorobenzene	$\tilde{B}^2A_u \leftrightarrow \tilde{X}^2B_g(C_{2h})$	$\leqslant 6$	EEEB (154), TP (154)
1,4-Dichloro-2-fluorobenzene	$\tilde{B}^2A'' \leftrightarrow \tilde{X}^2A''(C_S)$	$\leqslant 6$	EEEB (154), TP (154)
1,3-Dichloro-2,4,6-trifluorobenzene	$\tilde{B}^2B_1 \leftrightarrow \tilde{X}^2B_1(C_{2v})$	38	EEEB (154), TP (154)
1,3-Dichloro-2,5-difluorobenzene	$\tilde{B}^2B_1 \leftrightarrow \tilde{X}^2B_1(C_{2v})$	38	EEEB (154), TP (154)
1,3-Dichloro-2,4-difluorobenzene	$\tilde{B}^2A'' \leftrightarrow \tilde{X}^2A'', \tilde{A}^2A''(C_S)$	29	EEEB (154), TP (154)
1,3-Dichloro-5-fluorobenzene	$\tilde{B}^2B_1 \leftrightarrow \tilde{X}^2B_1, \tilde{A}^2A_2(C_{2v})$	8	EEEB (154), LIFM (72), ABM (72), TP (154)
1,3-Dichloro-4-fluorobenzene	$\tilde{B}^2A'' \leftrightarrow \tilde{X}^2A''(C_S)$	8	EEEB (154), TP (154)
1,3-Dichloro-2-fluorobenzene	$\tilde{B}^2B_1 \leftrightarrow \tilde{X}^2B_1(C_{2v})$	14	EEEB (154), TP (154)
1-Chloro-pentafluorobenzene	$\tilde{B}^2B_1 \leftrightarrow \tilde{X}^2A_2, \tilde{A}^2B_1(C_{2v})$	42–43	EEEB (154), LIFM (156), ABM (156), TP (154, 156)
1-Chloro-2,3,5,6-tetrafluorobenzene	$\tilde{B}^2B_1 \leftrightarrow \tilde{X}^2A_2, \tilde{A}^2B_1(C_{2v})$	52	EEEB (154), TP (154)
1-Chloro-2,3,4,5-tetrafluorobenzene	$\tilde{B}^2A'' \leftrightarrow \tilde{X}^2A''(C_S)$	27	EEEB (154), TP (154)
1-Chloro-2,4,5-trifluorobenzene	$\tilde{B}^2A'' \leftrightarrow \tilde{X}^2A''(C_S)$	$\leqslant 6$	EEEB (154), TP (154)
1-Chloro-2,3,6-trifluorobenzene	$\tilde{B}^2A'' \leftrightarrow \tilde{X}^2A''(C_S)$	21	EEEB (154), TP (154)
1-Chloro-3,5-difluorobenzene	$\tilde{B}^2B_1 \leftrightarrow \tilde{X}^2A_2, \tilde{A}^2B_1(C_{2v})$	$\leqslant 6$	EEEB (154), ABM (72), TP (154)
1,3,5-Tribromo-2,4,6-trifluorobenzene	$\tilde{B}^2A_2' \leftrightarrow \tilde{X}^2E''(D_{3h})$	$\leqslant 6$	EEEB (157), LIFM (172), TP (157, 162)
1,3-Dibromotetrafluorobenzene	$\tilde{B}^2B_1 \leftrightarrow \tilde{X}^2B_1, \tilde{A}^2A_2(C_{2v})$	$\leqslant 6$	EEEB (157), TP (157)
1,4-Dibromotetrafluorobenzene	$\tilde{B}^2B_{3u} \leftrightarrow \tilde{X}^2A_2, \tilde{A}^2B_1(D_{2h})$	$\leqslant 6$	EEEB (157), TP (157)
Pentafluorophenol	$\tilde{B}^2A'' \leftrightarrow \tilde{X}^2A''(C_S)$	31–35	EEEB (158), LIFM (159), LIFG (163) TP (81, 158)
2,3,5,6-Tetrafluorophenol	$\tilde{B}^2A'' \leftrightarrow \tilde{X}^2A'', \tilde{A}^2A''(C_S)$	41	EEEB (158), TP (158)

Table 4 *Continued*

Cation[b]	Electronic transition	Lifetime	Experimental Technique
2,4,5-Trifluorophenol	$\tilde{B}^2A'' \leftrightarrow \tilde{X}^2A', \tilde{A}^2A''(C_s)$	⩽6	EEEB (158), LIFG (163), TP (158)
2,3,4-Trifluorophenol	$\tilde{B}^2A'' \leftrightarrow \tilde{X}^2A', \tilde{A}^2A''(C_s)$	26	EEEB (158), LIFG (163), TP (158)
3,5-Difluorophenol	$\tilde{B}^2A'' \leftrightarrow \tilde{X}^2A', \tilde{A}^2A''(C_s)$	36	EEEB (158), LIFG (163), TP (158)
2,5-Difluorophenol	$\tilde{B}^2A'' \leftrightarrow \tilde{X}^2A', \tilde{A}^2A''(C_s)$	13	EEEB (158), LIFG (163), TP (158)
Methyl-pentafluorobenzene	$\tilde{B}^2B_1 \leftrightarrow \tilde{X}^2A_2, \tilde{A}^2B_1(C_{2v})$	43–52	EEEB (164), LIFM (81, 156), ABM (156), TP (81, 156)
Perfluoro-1,4-xylene	$\tilde{B}^2B_1 \leftrightarrow \tilde{X}^2A_2(D_{2h})$	11	EEEB (164), TP (164)
Perfluorotoluene	$\tilde{B}^2B_1 \leftrightarrow \tilde{X}^2A_2(C_{2v})$	⩽5, 45	EEEB (164), LIFM (156), ABM (156), TP (156)
2,3,5,6-Tetrafluorotoluene	$\tilde{B}^2B_1 \leftrightarrow \tilde{X}^2A_2(C_{2v})$	43	EEEB (164), TP (164)
1,3,5-Trimethyl-2,4,6-trifluorobenzene	$\tilde{B}^2A'_2 \leftrightarrow \tilde{X}^2E''(D_{3h})$	37	EEEB (164), LIFM (81), ABM (81), TP (81)
1-Methyl-2,4,6-trifluorobenzene	$\tilde{B}^2B_1 \leftrightarrow \tilde{X}^2A_2(C_{2v})$	47	LIFM (81), ABM (81), TP (81)
2,4-Difluoro-1,3,5-trimethylbenzene	$\tilde{B}^2A'' \leftrightarrow \tilde{X}^2A', \tilde{A}^2A''(C_s)$	31	LIFM (81), ABM (81), TP (81)
2,5-Dichlorotoluene	$\tilde{B}^2A_1 \leftrightarrow \tilde{X}^2A''(C_s)$	⩽6	EEEB (164), TP (164)
3,5-Dichlorotoluene	$\tilde{B}^2B_1 \leftrightarrow \tilde{X}^2A_2(C_{2v})$	10	EEEB (164), TP (164)
Perfluorobenzonitrile	$\tilde{B}^2B_1 \leftrightarrow \tilde{X}^2A_2(C_{2v})$	⩽6	EEEB (164), TP (164)
Trifluoroborazine	$\tilde{A}^2A'_2 \leftrightarrow \tilde{X}^2E''(D_{3h})$	⩽6	EEEB (160), TP (160)

[a] The assumed molecular symmetry is given in parentheses following the symbols for the electronic transition. The legend for the experimental techniques is as follows: EEEB = Emission Excited by Electron Bombardment; EED = Emission Excited by a Discharge; LIFG = Laser Induced Fluorescence in the Gas phase; LIFM = Laser Induced Fluorescence in an inert gas Matrix; ABM = (optical) Absorption spectra in an inert gas Matrix; SJES = Supersonic Jet Emission Spectrum; LIFJ = Laser Induced Fluorescence in a supersonic Jet; TP = lifetime (τ) or quantum yield (ϕ) determination. Where several similar determinations of the lifetimes (in nsecs for the $\tilde{B}$ state vibrationless level) exist, a range is given. Where determined, quantum yields (ϕ) for the vibrationless level are given in parentheses following the lifetime.

[b] Very weak fluorescence, with essentially no spectral information, has been reported for a few other benzenoid cations. Ref. (164) reports very weak gas phase emissions from the cations of pentafluoronitrobenzene, pentafluoroaniline, pentafluorobenzoic acid, pentafluoromethoxybenzene, pentafluorobenzaldehyde, and 2,5-dichloro-1,4-xylene. Ref. (81) reports weak LIFM from 1-fluoro-2,4,6-trimethylbenzene. The situation with respect to fluorescence from the benzene cation is unclear, Ref. (152b) reports LIF in an Ar matrix from the lowest vibrational levels, but in the gas phase the quantum yield for fluorescence is believed to be $< 10^{-5}$ [Refs. (9, 48)].

[c] The quoted values of ϕ are from Ref. (87), the values from Ref. (84) are generally $\sim 1/2$ as large. Lifetimes (τ) from Ne matrix studies (e.g. 81) are converted to gas phase values by multiplying by 1.135 to correct for the medium's index of refraction.

various combinations of methyl or heavier halogen substituents also give rise to emitting species. Likewise a number of fluorophenols are known to emit.

Vibronic Structure and Jahn-Teller Effects

One of the most important results of the recent rapid advances in the study of molecular ions is the accumulation of detailed information about the vibronic structure of polyatomic cations in both their ground and excited electronic states. For practically every ion listed in Tables 1–4, one can consult the original references and find relatively precise information about a number of the ground and excited state vibrational frequencies.

It is obviously impossible to present all this detailed vibrational information here, but several general conclusions can be drawn. For the ions studied, there have been only small changes in vibrational frequencies going from the parent neutral to either the ground or excited state of the ion. Indeed a recent tabulation (148) of 72 independent ground and excited state frequencies for six asymmetrically substituted halobenzene cations showed that with respect to the corresponding vibrations in the neutral there was an average change of only 3%. In no case was there a well-founded vibrational assignment in the ion that differed from the neutral frequency by more than 10%.

Perhaps the most interesting vibronic structure yet studied is found in the ground electronic states of the substituted benzene cations of nominal D_{3h} or D_{6h} symmetry and so doubly degenerate electronic states. Cations in this category for which spectra have been observed include $C_6F_6^+$ (70) and sym- $C_6F_3H_3^+$ (146), $C_6F_3Cl_3^+$ (148), $C_6F_3Br_3^+$ (172), and $C_6Cl_3H_3^+$ (146).

All of these ions have been studied in more or less detail; for illustrative purposes, the results obtained for the most symmetrical, $C_6F_6^+$ are reviewed. The electronic transition, $\tilde{B}^2A_{2u} \leftrightarrow \tilde{X}^2E_{1g}$, has been studied by a variety of spectroscopic techniques. Its emission spectrum was first obtained by controlled electron impact excitation (48, 143) and later from a discharge source (42). More recently the emission spectrum has been greatly simplified by cooling the parent in a free jet expansion (51, 52). Its laser excitation spectrum has been obtained at room temperature (58), liquid N_2 temperature (64), and at near 0 K in a free jet expansion (76), and in 5 K solid Ar (67, 144) and Ne matrices (69, 70, 146). Its laser excited, wavelength-resolved emission spectrum has been observed at room and liquid N_2 temperature in the gas phase (64) and at 5 K in a Ne matrix (69, 70, 146). These extensive observations have well and redundantly characterized the vibronic structure of both the $\tilde{B}$ and $\tilde{X}$ states with no significant discrepancies reported among the various experiments.

If $C_6F_6^+$ conformed in both electronic states to the D_{6h} symmetry of its parent neutral, the vibrational structure of the $\tilde{B} \leftrightarrow \tilde{X}$ transition should be very simple. Selection rules prohibit vibrational progressions in any but the totally symmetric vibrational modes for this allowed electronic transition. In C_6F_6 there are only two such a_{1g} modes, ν_1 and ν_2, with approximate frequencies of 1500 and 550 cm^{-1}, respectively (58). The very first laser excitation spectra (58) showed the vibronic structure of the $\tilde{B} - \tilde{X}$ transition to be much more complicated than could be accounted for on the basis of only two active modes, ν_1 and ν_2.

The excitation spectrum could be explained if one assumed that some of the e_{2g} modes, ν_{15}, ν_{16}, ν_{17}, and ν_{18}, were also active in the spectrum. Theory predicted that these would be precisely the modes that would become active if a Jahn-Teller distortion affected the ground $^2E_{1g}$ state (167–170). Subsequent experiments (64, 69, 70, 146) established the vibronic structure of the $\tilde{X}\ ^2E_{1g}$ state. The e_{2g} modes are indeed active. What's more, unlike all the observed modes in nondegenerate states of benzenoid cations, the spacing of the progressions are not harmonic, but very irregular. This provides confirmation that the $\tilde{X}$ state is indeed distorted by a Jahn-Teller effect. Figure 1 shows the positions that were finally, and usually redundantly, established for the Jahn-Teller active vibronic levels.

The remaining problem of how to interpret these levels to obtain useful information about the Jahn-Teller effect was solved in the following way (70, 147). The Hamiltonian for a single, doubly degenerate Jahn-Teller active mode i with orthogonal coordinates $Q_\pm$, can be written (70, 147, 169),

$$\hat{\mathcal{H}}_i^e = \hat{\mathcal{H}}_T + (2\pi^2\omega_i^2)Q_+Q_- + 2\omega(D_i h\omega)^{1/2}Q_-$$
$$+ K_i(2\pi^2\omega^2)Q_+^2 + \text{H.C.} \qquad 3.$$

Figure 1 Observed and calculated Jahn-Teller active vibronic energy level structure of the $\tilde{X}^2E_{1g}$ state of $C_6F_6^+$. The observed level structure is derived from a variety of experiments; M = matrix, laser excited, emission; H = hot bands in the gas phase laser excitation spectrum, R = wavelength resolved emission in the gas phase, and U = unrelaxed, laser excited, matrix emission. The calculated energy levels were obtained as described in the text. Levels shown by *dashed lines* are calculated to exist at the positions shown, but correspondingly calculated transition intensities indicate that these levels would not have been observed in the experiments performed, and have not been. (The three numbers below many of the calculated levels indicate the relative probabilities for transitions to the indicated level from $v' = 0$, 1, and 2 respectively of the excited electronic state.) The columns are labeled by the symmetry quantum number j. Energy level structure in the $j = 3/2, 5/2$ states is calculated below 1000 cm^{-1} above the vibrationless level, that in the $j = 1/2$ state below 1750 cm^{-1}.

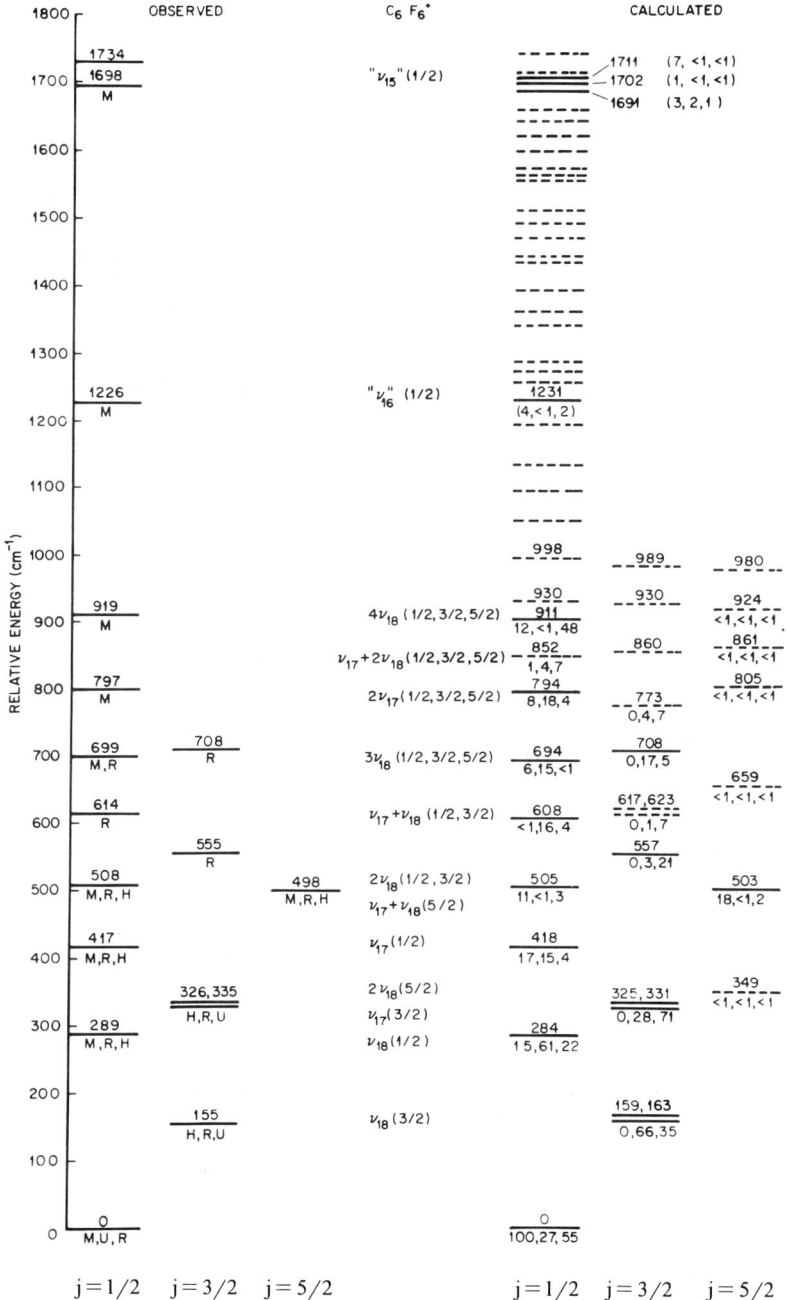

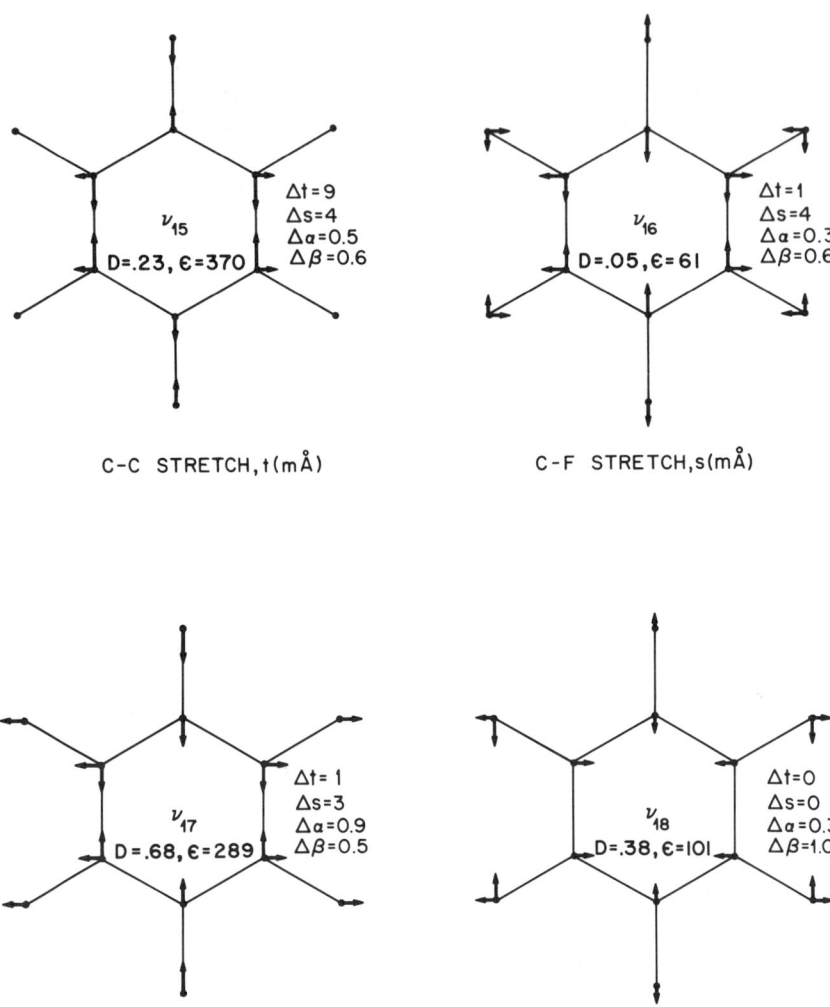

Figure 2 Schematic representation of the experimentally determined Jahn-Teller distortion of $C_6F_6^+$. Each of the four rings represent one component of the four e_{2g} doubly degenerate normal modes of $C_6F_6^+$. Although the motions given, and the information summarized, is appropriate for the normal coordinates, which are linear combinations of the symmetry coordinates, we label the representations by the (unnormalized) symmetry coordinates that make the dominate contribution to each normal mode, i.e. ν_{15}: C–F stretch: $(2s_1 - s_2 - s_3 + 2s_4 - s_5 - s_6)$; ν_{16}: C–C stretch: $(t_1 - 2t_2 + t_3 + t_4 - 2t_5 + t_6)$; ν_{17}: C–C–C bend: $(2\alpha_1 - \alpha_2 - \alpha_3 + 2\alpha_4 - \alpha_5 - \alpha_6)$; ν_{18}: C–F bend: $(\beta_2 - \beta_3 + \beta_5 - \beta_6)$. To obtain a representation of the shift $\delta\rho$ of the minimum of the Jahn-Teller distorted potential from the symmetrical position for a given mode, one inserts the bond length and angle changes given to the right of each ring into the above expressions for the symmetry coordinates. For

In the above H.C. stands for Hermitian conjugate and $\hat{\mathcal{H}}_T$ = kinetic energy operator for the nuclear vibration; ω_i = unperturbed frequency of the vibration; D_i = linear Jahn-Teller distortion parameter; K_i = quadratic Jahn-Teller distortion parameter. The parameters ω_i, D_i, and K_i can be related (147) algebraically to the Jahn-Teller stabilization energy, ε_i, and the change in the normal coordinate, $\delta\rho_i$, needed to take one from the symmetrical (D_{6h} or D_{3h}) position to the new minimum of the Jahn-Teller distorted potential.

It was found that the four e_{2g} modes could not be treated separately and so the final Hamiltonian became (70, 147, 171)

$$\hat{\mathcal{H}} = \sum_{i=1}^{4} \hat{\mathcal{H}}_i^e. \qquad 4.$$

The matrix elements of this Hamiltonian can be computed in a basis set consisting of products of the two degenerate electronic wavefunctions and a two-dimensional harmonic oscillator wavefunction for each of the e_{2g} modes. Taking maximum advantage of symmetry blocking (169), the resulting Hamiltonian matrices that required diagonalization were of order, ~7000.

---

example, for mode 17,

$$\delta\rho_{17} = \sum_i (\Delta s_i + \Delta t_i + \Delta\alpha_i + \Delta\beta_i)$$

$$= (6\hat{s}_1 - 3\hat{s}_2 - 3\hat{s}_3 + 6\hat{s}_4 - 2\hat{s}_5 - 3\hat{s}_6) \text{ milli-}\text{Å}$$

$$+ (1\hat{t}_1 - 2\hat{t}_2 + 1\hat{t}_3 + 1\hat{t}_4 - 2\hat{t}_5 + 1\hat{t}_6) \text{ milli-}\text{Å}$$

$$+ (1.8\hat{\alpha}_1 - 0.9\hat{\alpha}_2 - 0.9\hat{\alpha}_3 + 1.8\hat{\alpha}_4 - 0.9\hat{\alpha}_5 - 0.9\hat{\alpha}_6) \text{ degrees}$$

$$+ (0.5\hat{\beta}_1 - 0.5\hat{\beta}_3 + 0.5\hat{\beta}_4 - 0.5\hat{\beta}_6) \text{ degrees}.$$

$\beta = (\phi_i - \phi_{i+1})/2$, ϕ_i = C-C-F bond angle and the hatted characters indicate unit vectors in the given directions. Note that if the symmetry coordinates were true normal coordinates for ν_{17}, $\Delta t = \Delta s = \Delta\beta = 0$, but in the actual normal coordinate representation of ν_{17} there are nonzero contributions from Δt, Δs, and $\Delta\beta$. Nonetheless, the change in the C-C-C bond still plays a major role in determining the shift to the new distorted minimum for mode 17. Noting that in D_{6h} symmetry the C-C-C bond angles are all 120°, we see that this representation of the distorted potential minimum corresponds to a benzene ring with these C-C-C angles, $\alpha_1 = 121.8°$, $\alpha_2 = 119.1°$, $\alpha_3 = 119.1°$, $\alpha_4 = 121.8°$, $\alpha_5 = 119.1°$, and $\alpha_6 = 119.1°$. Using the remaining information on $\Delta\alpha$, $\Delta\beta$, Δs, and Δt in the figure, and standard C-C, and C-F bond lengths, representations of the distortion in all the bond lengths and angles for any of the four modes may be simply worked out. In the center of each ring, the stabilization energy, ε_i, (in cm^{-1}) and linear Jahn-Teller distortion parameter, D_i, is given. For the e_{2g} modes in $C_6F_6^+$, the quadratic distortion parameter K_i is negligibly small, except for mode 17 where it is found to be 0.006.

The results of such diagonalizations are shown (70) on the right-hand side of Figure 1. As one can see, there is excellent agreement between calculated and all observed vibronic levels. More reassuring, the eigenvectors from the same calculation predict extremely well the observed intensities (70).

From the parameters obtained from the calculation and a normal modes analysis, the distorted geometry of the C_6F_6 cation at its Jahn-Teller distorted equilibrium position is obtained and shown in Figure 2. As can be seen from Figure 2, major distortions include an unequal C–C–C internal ring bond angle ranging from 119 to 122° and unequal C–C bond lengths differing by ~ 0.03 Å. Similar distortions were found for the other symmetrical cations investigated (148, 172).

The Jahn-Teller stabilization energies, ε_i, for each mode are also given in Figure 2. They add up to a total Jahn-Teller stabilization energy of 821 cm^{-1} for $C_6F_6^+$. For all the benzenoid cations studied (148, 169) the stabilization energies were in the range of 500–1000 cm^{-1}. This similarity of geometric distortion and stabilization energy leads to the conclusion that there is a characteristic Jahn-Teller effect for any benzenoid compound when one electron is removed from the highest doubly degenerate π orbital.

Conclusions

The past few years have seen the spectroscopy of polyatomic radical cations advance from a state of virtual nihility to one of vigorous activity. Several techniques have been developed to obtain some spectral and structural information about ions whose states decay nonradiatively. At the same time even more impressive studies have been made for those ions with excited states that decay radiatively. There are now roughly 100 polyatomic ions in the latter category whose spectra have been characterized in more or less detail. Detailed vibrational, and in some cases rotational, structure has been observed for many of these ions. Subtle effects, like Jahn-Teller distortion, have been observed and analyzed in great detail for several symmetric ions. New techniques for obtaining cold ions and selective excitation have already demonstrated the ability to interpret the spectra of relatively large organic ions, and they hold more promise for the future. In a very real sense the ions of traditional condensed-phase chemistry are more and more becoming accessible to the detailed analysis and understanding of the isolated molecule spectroscopist.

Literature Cited

1. Saykally, R. J., Woods, R. C. 1981. *Ann. Rev. Phys. Chem.* 32:403–31
2. Moseley, J., Durup, J. 1981. *Ann. Rev. Phys. Chem.* 32:53–76
3. Johnson, P. M., Otis, C. E. 1981. *Ann. Rev. Phys. Chem.* 32:139–57
4. Green, S. 1981. *Ann. Rev. Phys. Chem.* 32:103–38
5. Corderman, R. R., Lineberger, W. C. 1979. *Ann. Rev. Phys. Chem.* 30:347–78
6. Dunbar, R. C. 1982. *Specialist Period. Rep.* 6: In press
7. Miller, T. A., Bondybey, V. E. 1980. *J. Chim. Phys.* 77:695–704
8. Miller, T. A., Bondybey, V. E. 1982. *Appl. Spectrosc. Rev.* 18:105–69
9. Maier, J. P. 1979. In *Kinetics of Ion-Molecule Reactions*, ed. P. Ausboos, pp. 437–62. New York: Plenum
10. Maier, J. P. 1980. *Chimia* 34:219–31
11. Maier, J. P. 1981. *Angew. Chem.* 20:638–46
12. Andrews, L. 1979. *Ann. Rev. Phys. Chem.* 30:79–101
13. Berkowitz, J., ed. 1980. *Molecular Ions Geometric and Electronic Structure.* Dordrecht, Neth.: Reidel
14. Miller, T. A., Bondybey, V. E. 1982. *Physical Chemistry Advances in the Study of Molecular Ions.* Amsterdam: North-Holland
15. Andrews, L., Keelan, B. W. 1981. *J. Am. Chem. Soc.* 103:99–103
16. Andrews, L., Keelan, B. W. 1980. *J. Am. Chem. Soc.* 102:5732–36
17. Andrews, L., Miller, J. H., Keelan, B. W. 1980. *Chem. Phys. Lett.* 71:207–10
18. Andrews, L., Blankenship, T. A. 1981. *J. Am. Chem. Soc.* 103:5977–79
19. Andrews, L., Keelan, B. W. 1981. *J. Am. Chem. Soc.* 103:822–29
20. Bondybey, V. E., Miller, T. A., English, J. H. 1980. *J. Chem. Phys.* 72:2193–94
21. Keelan, B. W., Andrews, L. 1981. *J. Am. Chem. Soc.* 103:829–32
22. Dunbar, R. C. 1981. In *Physical Methods of Modern Chemical Analysis*, Vol. 2, ed. T. Kuwana. New York: Academic
23. Dunbar, R. C. 1979. In *Kinetics of Ion-Molecule Reactions*, ed. P. Ausboos, pp. 463–85. New York: Plenum
24. Dunbar, R. C. 1979. In *Gas Phase Ion Chemistry.* ed. M. T. Bowen, 2:181–220. New York: Academic
25. Teng, H. H., Dunbar, R. C. 1978. *J. Chem. Phys.* 68:3133–38
26. Dunbar, R. C. 1978. *J. Chem. Phys.* 68:3125–32
27. Dunbar, R. C. 1979. *J. Phys. Chem.* 83:2376–78
28. Dunbar, R. C., Teng, H. H., Fu, E. W. 1979. *J. Am. Chem. Soc.* 101:6506–10
29. van Dishhoek, E. F., van Velzen, P. N., Van der Hart, W. J. 1979. *Chem. Phys. Lett.* 62:135–38
30. Benz, R. C., Dunbar, R. C. 1979. *J. Am. Chem. Soc.* 101:6363–66
31. Morgenthaler, L. N., Eyler, J. R. 1979. *J. Chem. Phys.* 71:1486–91
32. Dunbar, R. C., Teng, H. H. 1978. *J. Am. Chem. Soc.* 100:2279–83
33. Dunbar, R. C. 1976. *Anal. Chem.* 48:723–26
34. Benz, R. C., Dunbar, R. C., Claspy, P. C. 1981. *J. Am. Chem. Soc.* 103:1799–1802
35. Honovich, J. P., Dunbar, R. C. 1981. *J. Phys. Chem.* 85:1558–67
36. McGilvery, D. C., Morrison, J. D. 1977. *J. Chem. Phys.* 67:368–69
37. Goss, S. P., McGilvery, D. C., Morrison, J. D., Smith, D. L. 1981. *J. Chem. Phys.* 75:1820–28
38. Thomas, T. F., Dale, F., Paulson, J. F. 1977. *J. Chem. Phys.* 67:793–800
39. Larzilliere, M., Carre, M., Gaillard, M. L., Rostas, J., Horani, M., Velghe, M. 1980. *J. Chim. Phys. Phys. Chim. Biol.* 77:689–93
40. Smith, G. P., Lee, L. C. 1978. *J. Chem. Phys.* 69:5393–99
41. Smith, G. P., Lee, L. C., Moseley, J. T. 1977. *J. Chem. Phys.* 67:3818–28
42. Cossart-Magos, C., Cossart, D., Leach, S. 1979. *Mol. Phys.* 37:793–830
43. Cossart-Magos, C., Cossart, D., Leach, S. 1978. *J. Chem. Phys.* 69:4313–14
44. Cossart-Magos, C., Cossart, D., Leach, S. 1979. *Chem. Phys.* 41:345–62
45. Cossart-Magos, C., Cossart, D., Leach, S. 1979. *Chem. Phys.* 41:363–72
46. Allan, M., Kloster-Jensen, E., Maier, J. P. 1977. *J. Chem. Soc. Faraday Trans. 2* 73:1417–24
47. Allan, M., Maier, J. P. 1976. *Chem. Phys. Lett.* 41:231–35

48. Allan, M., Maier, J. P., Marthaler, O. 1977. *Chem. Phys. Lett.* 26:131–40
49. Levy, D. H. 1980. *Ann. Rev. Phys. Chem.* 31:197–225
50. De Koven, B. M., Levy, D. H., Harris, H. H., Zegarski, B. R., Miller, T. A. 1981. *J. Chem. Phys.* 74:5659–68
51. Miller, T. A., Zegarski, B. R., Sears, T. J., Bondybey, V. E. 1980. *J. Phys. Chem.* 84:3154–56
52a. Carrington, A., Tuckett, R. P. 1980. *Chem. Phys. Lett.* 74:19–23
52b. Tuckett, R. 1981. *Chem. Phys.* 58:151–62
53. Engelking, P. C., Smith, A. L. 1975. *Chem. Phys. Lett.* 36:21–22
54. Bondybey, V. E., Miller, T. A. 1977. *J. Chem. Phys.* 67:1790–92
55. Miller, T. A., Bondybey, V. E. 1977. *Chem. Phys. Lett.* 50:275–77
56. Cook, J. M., Miller, T. A., Bondybey, V. E. 1978. *J. Chem. Phys.* 69:2562–68
57. Miller, T. A., Bondybey, V. E. 1978. *Chem. Phys. Lett.* 58:454–56
58. Bondybey, V. E., Miller, T. A. 1979. *J. Chem. Phys.* 70:138–46
59. Miller, T. A., Bondybey, V. E., English, J. H. 1979. *J. Chem. Phys.* 70:2919–25
60. Miller, T. A., Bondybey, V. E., Zegarski, B. R. 1979. *J. Chem. Phys.* 70:4982–85
61. Sears, T. J., Miller, T. A., Bondybey, V. E. 1980. *J. Chem. Phys.* 72:6749–54
62. Sears, T. J., Miller, T. A., Bondybey, V. E. 1980. *J. Am. Chem. Soc.* 102:4864–66
63. Bondybey, V. E., Sears, T. J., English, J. H., Miller, T. A. 1980. *J. Chem Phys.* 73:2063–68
64. Sears, T. J., Miller, T. A., Bondybey, V. E. 1981. *J. Am. Chem. Soc.* 103:326–29
65. Bondybey, V. E., English, J. H., Miller, T. A. 1979. *J. Chem. Phys.* 70:1765–68
66. Bondybey, V. E., English, J. H. 1979. *J. Chem. Phys.* 71:777–82
67. Bondybey, V. E., Miller, T. A., English, J. H. 1979. *J. Am. Chem. Soc.* 101:1248–53
68. Sears, T. J., Miller, T. A., Bondybey, V. E. 1980. *J. Am. Chem. Soc.* 102:4864–66
69. Bondybey, V. E., Miller, T. A. 1980. *J. Chem. Phys.* 73:3053–59
70. Sears, T. J., Miller, T. A., Bondybey. V. E. 1981. *J. Chem. Phys.* 74:3240–48
71. Bondybey, V. E., English, J. H., Miller, T. A. 1980. *J. Mol. Spectrosc.* 81:455–72
72. Bondybey, V. E., Vaughn, C. R., Miller, T. A., English, J. H., Shiley, R. H. 1981. *J. Chem. Phys.* 74:6584–91
73. Mahan, B. H., O'Keefe, A. 1981. *J. Chem. Phys.* 74:5606–12
74. Grieman, F. J., Mahan, B. H., O'Keefe, A. 1981. *J. Chem. Phys.* 74:857–61
75. Mahan, B. H., O'Keefe, A. 1981. *J. Chem. Phys.* 74:5606–12
76. Heaven, M., Miller, T. A., Bondybey, V. E. 1982. *J. Chem. Phys.* 76:3831–32
77. Allan, M., Kloster-Jensen, E., Maier, J. P. 1977. *J. Chem. Soc. Faraday Trans. 2* 73:1406–16
78. Bondybey, V. E., Miller, T. A. 1978. *J. Chem. Phys.* 69:3597–602
79. Katayama, D., Welsh, J. A. 1981. *J. Chem. Phys.* 75:4224–30
80. Bondybey, V. E., English, J. H., Miller, T. A. 1979. *J. Chem. Phys.* 70:1621–25
81. Bondybey, V. E., Vaughn, C., Miller, T. A., English, J. H., Shiley, R. H. 1981. *J. Am. Chem. Soc.* 103:6303–7
82. Eland, J. H. D., Devoret, M., Leach, S. 1976. *Chem. Phys. Lett.* 43:97–101
83. Leach, S., Devoret, M., Eland, J. H. D. 1978. *Chem. Phys.* 33:113–21
84. Dujardin, G., Leach, S., Taieb, G. 1980. *Chem. Phys.* 46:407–21
85. Schlag, E. W., Frey, R., Gotchev, B., Peatman, W. B., Pollak, H. 1977. *Chem. Phys. Lett.* 51:406–8
86. Frey, R., Gotchev, B., Peatman, W. B., Pollak, H., Schlag, E. 1978. *Chem. Phys. Lett.* 54:411–14
87. Maier, J. P., Thommen, F. 1981. *Chem. Phys.* 57:319–32
88. Lew, H., Heiber, I. 1973. *J. Chem. Phys.* 58:1246–47
89. Wehinger, P. A., Wyckoff, S., Herbig, G. H., Herzberg, G., Lew, H. 1974. *Astrophys. J.* 190:L43–46
90. Lew, H. 1976. *Can. J. Phys.* 54:2028–49
91. Carrington, A., Milverton, D. R. J., Roberts, P. G., Sarre, P. J. 1978. *J. Chem. Phys.* 68:5659–61
92. Curtis, L. S., Erman, P. 1977. *J. Opt. Soc. Am.* 67:1218–30

93. Möhlmann, G. R., Bhutani, K. K., de Heer, F. J., Tsurubuchi, S. 1978. *Chem. Phys.* 31:273–80
94. Duxbury, G., Horani, M., Rostas, J. 1972. *Proc. Roy. Soc. Ser. A* 331:109–37 and references therein
95. Möhlmann, G. R., de Heer, F. J. 1975. *Chem. Phys. Lett.* 36:353–56
96. Callomon, J. H., Creutzberg, F. 1974. *Philos. Trans. R. Soc. London* 277:157–89
97. Van Sprang, H. A., Möhlmann, G. R., de Heer, F. J. 1978. *Chem. Phys.* 33:65–72
98. Bloch, M., Turner, D. W. 1975. *Chem. Phys. Lett.* 30:344–61
99. Smith, W. H. 1969. *J. Chem. Phys.* 51:3410–12
100. Fink, E. H., Welge, K. H. 1968. *Z. Naturforsch. Teil A* 23:358–76
101a. Mrozowski, S. 1941. *Phys. Rev.* 60:730–38
101b. Mrozowski, S. 1947. *Phys. Rev.* 72:682–98
102. Gauyacq, D., Horani, M., Leach, S., Rostas, J. 1975. *Can. J. Phys.* 53:2040–59
103. Bueso-Sanllehi, F. 1941. *Phys. Rev.* 60:556–70
104. Johns, J. W. C. 1964. *Can. J. Phys.* 42:1004–7
105. Gauyacq, D., Larcher, C., Rostas, J. 1979. *Can. J. Phys.* 57:1634–49
106. Smith, A. J., Read, F. H., Imhof, R. E. 1975. *J. Phys. B* 8:2869–79
107. Hesser, J. E. 1968. *J. Chem. Phys.* 48:2518–35
108. Erman, P., Brzozowski, J., Sigfridson, B. 1973. *Nucl. Instrum Methods* 110:471–76
109. Price, W. C., Simpson, D. W. 1932. *Proc. R. Soc. London Ser. A* 165:272–90
110. Laird, R. K., Barrow, R. F. 1950. *Proc. Phys. Soc. London Sect. A* 63:412
111. Callomon, J. H. 1958. *Proc. R. Soc. London Ser. A* 244:220–44
112. Leach, S. 1970. *J. Chim. Phys.* 67:74
113. Balfour, W. J. 1976. *Can. J. Phys.* 54:1969–78
114. Coxon, J. A., Marcoux, P. J., Setser, D. W. 1976. *Chem. Phys.* 17:403–15
115. Lee, L. C., Judge, D. L., Ogawa, M. 1975. *Can. J. Phys.* 53:1861–68
116. Bondybey, V. E., English, J. H. 1980. *J. Chem. Phys.* 73:3098–3102
117. Miller, T. A., Bondybey, V. E. 1982. To be published
118. Horani, M., Leach, S., Rostas, J., Berthier, G. 1966. *J. Chim. Phys.* 63:1015–25
119. King, M. A., Kroto, H. W., Nixon, J. F., Klapstein, D., Maier, J. P., Marthaler, O. 1981. *Chem. Phys. Lett.* 82:543–45
120. Maier, J. P., Thommen, F. 1980. *Chem. Phys.* 51:319–27
121. Allan, M. Maier, J. P., Marthaler, O., Stadelmann, J. P 1979. *J. Chem. Phys.* 70:5271–75
122. Callomon, J. H. 1956. *Can. J. Phys.* 34:1046–69
123. Allan, M., Kloster-Jensen, E., Maier, J. P. 1976. *Chem. Phys.* 7:11–18
124. Maier, J. P., Marthaler, O., Misev, L., Thommen, F. 1981. *Discuss. Faraday Soc.* 71:181–89
125. Maier, J. P., Misev, L. 1980. *Chem. Phys.* 51:311–18
126. Maier, J. P., Marthaler, O., Kloster-Jensen, E. 1980. *J. Chem. Phys.* 72:701–8
127. Maier, J. P., Thommen, F. 1980. *J. Chem. Phys.* 73:5616–19
128. Allan, M., Kloster-Jensen, E., Maier, J. P., Marthaler, O. 1978. *J. Electron Spectrosc. Relat Phenom.* 14:359–70
129a. Maier, J. P., Marthaler, O., Thommen, F. 1979. *Chem. Phys.* 60:193–96
129b. Maier, J. P., Misev, L., Thommen, F. 1982. *J. Phys. Chem.* 86:514–18
130. Allan, M., Maier, J. P., Marthaler, O., Kloster-Jensen, E. 1978. *Chem. Phys.* 29:331–37
131. Kloster-Jensen, E., Maier, J. P., Marthaler, O., Mohraz, M. 1979. *J. Chem. Phys.* 71:3125–28
132. Maier, J. P., Misev, L., Thommen, F. 1981. *Helv. Chim. Acta* 64:1985–90
133. Foster, P., Maier, J. P., Thommen, F. 1981. *Chem. Phys.* 59:85–90
134. Maier, J. P., Marthaler, O., Kloster-Jensen, E. 1980. *J. Electron Spectrosc. Relat. Phenom.* 18:251–65
135. Bieri, G., Kloster-Jensen, E., Kvisle, S., Maier, J. P., Marthaler, O. 1980. *J. Chem. Soc. Faraday Trans. 2* 76:676–84
136. Allan, M., Maier, J. P. 1976. *Chem. Phys. Lett.* 43:94–96
137. Allan, M., Dannacher, J., Maier, J. P. 1980. *J. Chem. Phys.* 73:3114–22
138. Bondybey, V. E., English, J. H., Miller, T. A. 1980. *J. Mol. Spectrosc.* 80:200–8
139. Jones, T. B., Maier, J. P. 1979. *Int. J. Mass Spectrosc. Ion Phys.* 31:287–91

143. Allan, M., Maier, J. P. 1975. *Chem. Phys. Lett.* 34:442–45
144. Bondybey, V. E., English, J. H., Miller, T. A. 1978. *J. Am. Chem. Soc.* 100:5251–52
145. Bondybey, V. E., Miller, T. A., English, J. H. 1980. *J. Chem. Phys.* 72:2193–94
146. Bondybey, V. E., Miller, T. A., English, J. H. 1980. *Phys. Rev. Lett.* 44:1344–47
147. Sears, T. J., Miller, T. A., Bondybey, V. E. 1980. *J. Chem. Phys.* 72:6070–80
148. Sears, T. J., Miller, T. A., Bondybey, V. E. 1981. *Discuss. Faraday Soc.* 71:175–80; 341–46
149. Bondybey, V. E., Miller, T. A., English, J. H. 1979. *J. Chem. Phys.* 71:1088–1100
150. Bondybey, V. E., English, J. H., Miller, T. A., Shiley, R. H. 1980. *J. Mol. Spectrosc.* 84:124–31
151. Bondybey, V. E., English, J. H., Miller, T. A. 1981. *J. Mol. Spectrosc.* 90:592–95
152a. Miller, J. H., Andrews, L., Lund, P. A., Schwartz, P. N. 1980. *J. Chem. Phys.* 73:4932–39
152b. Miller, J. H., Andrews, L. 1980. *Chem. Phys. Lett.* 72:90–93
153. Bondybey, V. E., English, J. H., Miller, T. A. 1979. *Chem. Phys. Lett.* 66:165–68
154. Maier, J. P., Marthaler, O., Mohraz, M., Shiley, R. H. 1980. *Chem. Phys.* 47:295–305
155. Bondybey, V. E. 1979. *J. Chem. Phys.* 71:3586–91
156. Bondybey, V. E., Miller, T. A., English, J. H. 1980. *J. Chim. Phys.* 77:667–72
157. Maier, J. P., Marthaler, O., Mohraz, M. 1980. *Chem. Phys.* 47:307–12
158. Maier, J. P., Marthaler, O., Mohraz, M., Shiley, R. H. 1980. *J. Electron Spectrosc. Relat. Phenom.* 19:11–20
159. Bondybey, V. E., English, J. H., Miller, T. A., Vaughn, C. B. 1981. *J. Phys. Chem.* 85:1667–70
160. Jones, T. B., Maier, J. P., Marthaler, O. 1979. *Inorg. Chem.* 18:2140–42
161. Maier, J. P., Marthaler, O. 1978. *Chem. Phys.* 32:419–27
162. Maier, J. P., Thommen, F. 1982. *J. Chem. Phys.* In press
163. Maier, J. P., Misev, L., Shiley, R. H. 1980. *Helv. Chim. Acta* 63:1920–25
164. Maier, J. P., Marthaler, O., Mohraz, M. 1980. *J. Chim. Phys.* 77:661–65
165. Shida, T., Iwata, S. 1973. *J. Am. Chem. Soc.* 95:3473–83
166. Zahradnik, R., Carsky, P. 1970. *J. Phys. Chem.* 74:1240
167. Jahn, H. A., Teller, E. 1937. *Proc. Roy. Soc. London Ser. A* 161:220–35
168. Moffett, W., Liehr, A. D. 1956. *Phys. Rev.* 106:1195–1200
169. Longuet-Higgins, H. C. 1961. *Adv. Spectrosc.* 2:429–72
170. Herzberg, G. 1966. *Electronic Spectra of Polyatomic Molecules*. Princeton, N.J: Van Nostrand
171. Sloane, C. S., Silbey, R. 1972. *J. Chem. Phys.* 56:6031–43
172. Bondybey, V. E., Sears, T. J., Miller, T. A., Vaughn, C., English, J. H., Shiley, R. H. 1981. *Chem. Phys.* 61:9–16

DYNAMICS OF PROTEINS

P. G. Debrunner and H. Frauenfelder

Department of Physics, University of Illinois at Urbana-Champaign, Urbana, IL 61801

INTRODUCTION

It has long been recognized that proteins are dynamic systems. Although the three-dimensional structures, which have been determined for an increasing number of proteins, provide some basic insight into the architecture of these molecules, taken alone they cannot explain the observed properties and functions in any mechanistic sense. To quote an example, Perutz & Mathews first noted that the equilibrium structure of hemoglobin leaves no room for a ligand to reach the buried heme binding site (1); the same observation applies to the O_2 storage protein myoglobin (Mb), as revealed by x-ray diffraction at 2 Å resolution (2). There is ample evidence that the substrate gains access to the interior of Mb as a result of the thermal motion of the polypeptide chain (3). Chain mobility is obviously essential in protein folding (4, 5), but the question that concerns us here is the dynamics of the folded, native form. Conformational transitions (6–8), allostery (9), active transport (10), enzymatic activity (6, 11), and motility (12, 13) are processes that obviously depend on internal mobility. To delineate the driving forces and pathways of protein reactions is still a distant goal (14). Much progress has been made, however, in the description of the internal dynamics of proteins.

While theory makes important predictions about the fluctuations in the thermodynamic parameters of macromolecules (15), a deeper understanding of protein dynamics has come from experiments and model calculations (16, 17) on several globular, water soluble proteins. There is hope that the results are representative for this class of proteins, but the methods used do not yet provide a comprehensive picture of protein dynamics. The problem is that the characteristic frequencies of the interatomic forces have an enormously wide spectrum, from 10^{-13} sec to

10^{-3} sec (14), and no single approach to modeling or measuring is likely to cover the whole range of interest. After a brief review of the evidence for protein dynamics, we focus on a specific approach, the study of ligand binding to heme proteins. We show how a reaction that is seemingly simple under physiological conditions can be resolved into processes occurring on vastly different time scales by observations at low temperatures and in different media.

Several aspects of protein dynamics have been covered in recent reviews (18, 19) summarizing in particular the results of nuclear magnetic resonance studies of proton exchange (20) and molecular dynamics calculations (16, 17). Here we only consider globular proteins in the native state and ignore the intriguing questions of protein folding and motility.

STATIC VERSUS DYNAMIC STRUCTURE

Because x-ray crystallography has become a major source of information about proteins (21, 22), it is important to know to what extent the structure and dynamics in a crystal are representative of these properties in solution. While a specific answer can be given after a case by case study only, there are several lines of evidence that suggest close structural similarity between a protein in a crystal and in solution. An important, yet qualitative argument is the success of the concept of a structure-function relationship in rationalizing the active site structures deduced by protein crystallographers (8, 22–24). If these rationalizations are valid, the active site conformation in solution must be close to the one observed in a crystal. In proteins that crystallize in different space groups, the atomic coordinates typically show only minor deviations, in spite of different crystal-packing effects (25, 26). In a broader sense, proteins related by function (27) or evolution (28) show striking similarity or complementarity (23) of their active sites.

Among the methods that allow the study of protein structure in solution, diffuse x-ray (29) or neutron scattering (30) is sensitive to the overall shape of the molecule, while NMR is sensitive to the local arrangement of the resonant nuclei (31, 32). Both methods confirm the overall similarity of the structure in the crystal and in solution (29, 32, 33), but small differences that can be rationalized have been noted (32, 34).

Temperature Factor in X-ray Diffraction

One of the problems facing the protein crystallographer is that protein crystals typically show a high degree of disorder (22). Even if good diffraction patterns are observed and an electron density map can be constructed, there may be segments in the amino acid chain that assume more than one conformation (35) or that cannot be located at all (36–38). A number of refinement schemes have been elaborated that adapt methods used for small molecules (39) and may locally restrain the fits to the established stereochemistry of the individual amino acids (40, 41). These refinement routines typically assign an adjustable temperature factor $\exp[-B_i \sin^2\theta/\lambda^2]$ to each nonhydrogen atom to account for the deviation of atoms from their ideal position in a given scattering plane (22, 42). The temperature or Debye-Waller factor has a well-defined meaning for x-ray scattering on regular, harmonic solids, where $B = 8\pi\langle x^2 \rangle$ is proportional to the mean square displacement $\langle x^2 \rangle$ of an atom perpendicular to the scattering plane. According to the Debye model, $\langle x^2 \rangle$ is constant at low temperatures T and increases linearly with T in the classical, high temperature limit (42). For a protein, the validity and significance of an (isotropic) temperature factor is questionable, but the simplicity of the expression and its success in reducing the residual index

$$R = \sum_{hkl} |F_0 - F_c| \Big/ \sum_{hkl} F_0$$

(22) in the refinement process justifies its use.

Since x-ray scattering is a fast process ($\Delta t \sim 10^{-18}$ sec) compared to vibrational periods ($1/\nu \sim 10^{-13}$ sec), it does not differentiate between dynamic disorder resulting from thermal motion and static disorder resulting from crystal imperfections, radiation damage, etc. The parameter B thus contains contributions from lattice disorder as well as from dynamics (43). The latter part must include a mean square displacement due to high-frequency, interatomic vibrations (42), which is expected to be of the same order of magnitude as that found in small organic molecules, $\langle x^2 \rangle \sim 0.05$ Å^2, and a potentially larger contribution due to low-frequency segmental motion (44, 45). Shifting the emphasis significantly, one may focus attention on the temperature factor as an important source of information about protein dynamics (25, 46–48). The present state of the art in protein crystallography indeed allows one to determine individual B-values—in well-behaved cases even anisotropic B-values (45)—for the resolved atoms. A successful refinement requires a large data set and a good structural model. Accordingly, the method has been applied to such well-known, small to medium-size proteins as myoglobin (46, 49, 50), lysozyme (25, 47, 48), rubredoxin (45), cytochrome c (51), and a serine protease (52). In the brief discussion of the results that follows, we call $B/(8\pi^2) = \langle x^2 \rangle$ a mean square displacement (msd), keeping in mind, however, that this interpretation deserves further scrutiny.

Several methods have been used to differentiate between the msd due to lattice disorder, $\langle x^2 \rangle_{ld}$, and the dynamic, temperature dependent msd $\langle x^2 \rangle_{cv}$, which is assumed to consist of a vibrational part, $\langle x^2 \rangle_v$, and a conformational part $\langle x^2 \rangle_c$. If the various terms are statistically independent we have (46)

$$\langle x^2 \rangle = \langle x^2 \rangle_{ld} + \langle x^2 \rangle_v + \langle x^2 \rangle_c.$$

Lacking evidence to the contrary, it is reasonable to assume that $\langle x^2 \rangle_{ld}$ is a constant throughout the molecule. In the case of myoglobin it is then possible to estimate $\langle x^2 \rangle_{ld} \simeq 0.045$ Å^2 for the heme iron, since the dynamic msd of the latter has been measured independently by Mössbauer spectroscopy (53–55). With these simplifications it is found

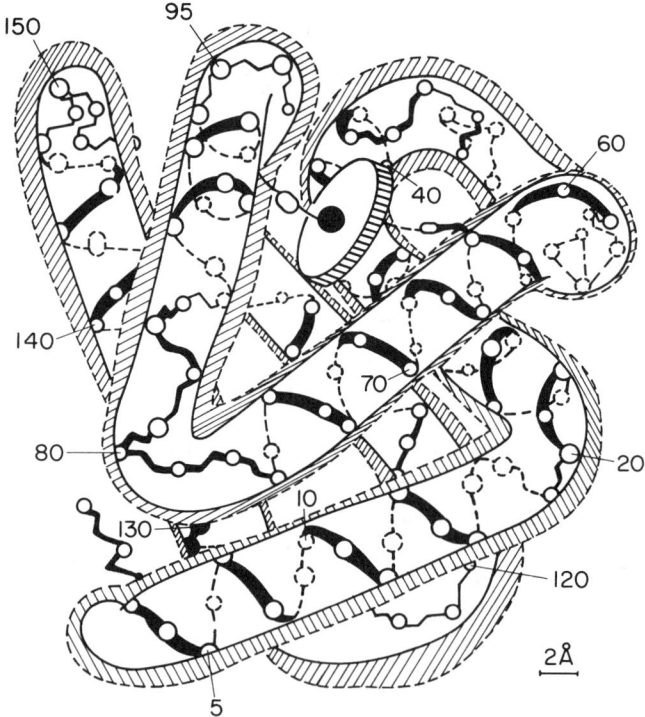

Figure 1 Main chain structure of Mb. *Solid lines* indicate the static structure, the shaded area the region reached by conformational fluctuations with a 99% probability. *Circles* denote the α carbons. [Reprinted with permission from *Nature* 280(5723):558–63. Copyright (c) 1979, Macmillan Journals Limited.] (Ref. 46.)

that the msd of several atoms in the protein backbone is of the same order of magnitude as that found in small organic molecules, whereas it approaches 0.3 Å^2 near residue 122 (Figure 1) and exceeds this value at the N-terminus. Atoms in the side groups quite generally tend to have larger msd, and the msd increases with increasing distance from Cα. Another general trend is an increase in $\langle x^2 \rangle$ from the center of the protein toward the surface. Of 44 residues with dynamic msd larger than 0.15 Å^2, almost all are on the surface and only seven are hydrophobic. Similar conclusions can be drawn from Table 1: Coulomb interaction with the solvent causes large msd for the charged and polar groups on the surface. Hydrophobic groups, however, are less strongly coupled to the fluctuating Coulomb field of the solvent "bath."

The mobility near the active center is of particular interest. As noted above, some surface atoms have to move away from their equilibrium positions before the substrate, O_2, can penetrate to the heme binding site (Figure 1) (1–3, 35). A likely path for O_2 diffusion is provided by residues 43–45 with average msd $\langle x^2 \rangle_{av} \sim 0.15$ Å^2. Significant differences in mobility are observed for the groups in contact with the heme. Those on the distal, O_2-binding side are exceptionally rigid, $\langle x^2 \rangle_{av} \sim 0.04$ Å^2, while those on the proximal side are unusually mobile, $\langle x^2 \rangle_{av} \sim 0.12$ Å^2. The proximal side undergoes conformational changes on O_2-binding (49, 50); in the tetrameric hemoglobin these changes have been associated with the stereochemical "trigger" of the cooperativity observed among the subunits (9). The pattern of the msd deduced from the crystallographic refinement of myoglobin thus correlates well with the stereochemistry and function of the protein.

The temperature dependence of the msd should provide information about the conformational modes. In myoglobin very few atoms show the classical behavior $\langle x^2 \rangle \propto T$; in fact, if the msd is roughly parametrized by a power law, $\langle x^2 \rangle = AT^\nu$, ν ranges from -1.5 to 10. Residues with small $\langle x^2 \rangle$ tend to have large ν values and thus large fractional change, $d\ln\langle x^2 \rangle/dT$, with temperature, whereas residues with large $\langle x^2 \rangle$ have small ν and small fractional change. This behavior can be modeled by

Table 1 Average side-chain displacements in metmyoglobin at 250 K

	$\langle x^2 \rangle_{cv}$(Å^2)	
Residue type	Outside	Inside
Charged	0.20 ± 0.02	0.10 ± 0.02
Neutral polar	0.16 ± 0.04	0.08 ± 0.04
Nonpolar	0.07 ± 0.02	0.05 ± 0.01

suitable anharmonic potentials (56), but a more realistic approach is called for to study the correlation of the atomic motions in space and time. Other, crystallographic aspects of low temperature x-ray scattering are discussed in Ref. (57).

As mentioned above, the msd of the iron in Mb can be determined from Mössbauer measurements (53–55). The temperature dependence of $\langle x^2 \rangle$ reveals several distinctly different dynamic processes (55). Below 160 K, the msd increases linearly with temperature, as expected classically for the vibrational part $\langle x^2 \rangle_v$ of a harmonic system. Above 160 K, conformational modes become important, and the $\langle x^2 \rangle_c$ values deduced from the Mössbauer intensity are compatible with the x-ray diffraction data. Above 220 K the msd is larger in frozen aqueous solutions than in polycrystalline samples; this excess msd correlates with line broadening and must be of diffusional character.

The significance of the B-parameters was confirmed beyond doubt by two independent refinements of hen egg white and human lysozyme, respectively (25, 47, 48). Despite differences in crystal-packing effects, the B-parameters of corresponding atoms turned out to be very similar. Two regions of high mobility line the surfaces of the substrate binding site; these regions assume a different, more tightly packed conformation in the inhibitor complex. It appears that high mobility is a typical characteristic of binding sites for large substrates, as it is also found in serine proteases (52). Mobility presumably allows for efficient sampling and matching of corresponding molecular surfaces in preparation of the more rigid enzyme substrate complex.

An interesting comparison of empirical B-values (51) with molecular dynamics simulations was made in the case of ferrocytochrome c (58). If allowance is made for a uniform lattice disorder term in the experimental msd, $\langle x^2 \rangle_{ld} \simeq 0.047$ Å^2, obtained by matching the dynamic contribution for the protein core from both methods, a satisfactory correlation is found between the results of the simulation and the refinement.

Fluorescence Quenching and Depolarization

While x-ray diffraction provides no information about the time scale of conformational fluctuations, an important benchmark on the rate of specific modes comes from fluorescence experiments. Lakowicz & Weber observed dynamic quenching by O_2 in various globular proteins; they concluded that O_2 must get access to tryptophan residues on a nanosecond time scale, apparently as a result of conformational fluctuations (59). Making the admittedly unrealistic assumptions of (a) isotropic diffusion of O_2 within the protein and (b) a partition coefficient of unity, the diffusion coefficient is found to be two to five times smaller than in water, with activation enthalpies ranging from 12 to 17 kJ mol^{-1}.

The tryptophan fluorescence of globular proteins is also efficiently quenched by acrylamide; activation enthalpies are from 8 to 45 kJ mol^{-1} (60). Fluorescence also allows measurement of the reorientation of the fluorophore if the depolarization is observed with high time resolution. Subnanosecond motions of tryptophans have been observed by this method (61).

NMR

NMR is probably the most versatile technique for studying internal motions in proteins (18, 31, 62–64). Typically, specific assignments of individual resonances can be made. For ^{13}C the quantities of interest are the longitudinal and transverse relaxation times, which depend on the motion of the nearby nuclei. Effective correlation times ranging from 10^{-5} sec to 10^{-12} sec may be important (18, 63). For protons, the change in resonance frequency due to exchange between different environments is typically observed; the relevant time range is 1 sec to 10^{-5} sec. A well-documented example is the motion of the aromatic side groups in the trypsin inhibitor (65). The motion of some of these rings is limited to small librations, while others undergo 180° flips at the same temperatures. Activation enthalpies from 70 to 155 kJ mol^{-1} have been observed. Model calculations yield rotational barriers of comparable height if the displaced sidegroups are allowed to relax to their minimum energy position (66–68). A very powerful method for the study of the local geometry and flexibility of a protein is two-dimensional NMR, since it allows comprehensive measurements of shifts and coupling constants (64).

An entirely different application of NMR is the study of amide proton exchange in D_2O (20). The best known example again is the trypsin inhibitor, for which the kinetics of exchange of individually assigned protons has been studied as a function of temperature, pH etc (69–72). The main conclusion is that water can penetrate into the densely packed protein in much the same way as O_2 and acrylamide are found to permeate through proteins and collide with tryptophans. Qualitatively, Lumry's model of mobile defects can account for these observations (73, 74). Even in the extensively studied case of the trypsin inhibitor, however, the mechanistic details of the processes involved are still under discussion (75).

LIGAND BINDING TO HEME PROTEINS

The biological function of a protein must be determined largely by dynamic features. To explore the functional aspects of protein dynamics, the kinetics of a suitable reaction is studied under various external conditions. Considering that physical chemists still argue about the proper description of bimolecular processes (76), it is reasonable to focus on a simple, yet biologically significant reaction such as oxygen binding to myoglobin (Mb), $Mb + O_2 \leftrightarrow MbO_2$. Generalizing slightly to include binding processes of other small ligands to the heme iron, and considering heme complexes in solution as well as in proteins, the experimenter has a model system with well-characterized reactants and numerous controllable variables at his disposal. Moreover, the reaction is readily initiated by photodissociation of the ligand-heme complex, and rebinding can be monitored optically (77). Extensive work has allowed the resolution of the ligand binding reaction into a number of elementary steps (78–88). The changes in the electronic state of the chromophore (89), interesting as they are, do not concern us here. Rather, we concentrate on the results that best illustrate the dynamic nature of proteins.

Kinetics at Low Temperatures

Under physiological conditions, O_2 and CO binding to Mb (77) follows ordinary second order kinetics (pseudo first order at high ligand concentration) in spite of the ligand having no direct access from the solvent to the buried binding site at the heme iron (1–3). Entirely different, more complex kinetics of rebinding are observed, however, after photodissociation at low temperatures (79), as illustrated in Figure 2. Here the fractions $N(t)$ of Mb molecules that have not rebound CO after a saturating flash applied at $t = 0$ are plotted against time. Processes occurring over an enormous time range from 10^{-6} sec to 10^3 sec, are significant, hence the log-log scale employed in Figure 2. Data sets of the type shown have been obtained for temperatures down to 4.2 K[1] for different viscosities (87) and various combinations of ligands and heme proteins (80, 82, 85). The kinetics observed for given variables are reproducible; ie for identical initial conditions, each flash produces the same kinetics, but they vary systematically with temperature, viscosity,

---

[1] No adverse effect of low temperatures on Mb and similar proteins is detectable. Specifically, thermal cycling leaves the room temperature kinetics unchanged, and there is no evidence of a phase transition associated with freezing or crystallization. This crucial and perhaps surprising feature must be a result of the bound water (90) and the glassy nature of the medium. It is well known that native proteins have a sizable amount of bound water associated with them, roughly equivalent to a monolayer, which shows no sign of a phase transition upon cooling, but which gradually loses its mobility as judged by NMR (91) and dielectric relaxation measurements (92, 93). The glycerol or other glass-forming agent, which was used in all samples to keep them transparent, should enhance the amorphous nature of the protein environment.

ligand, and protein. In all cases the data can be parametrized by a model of sequential barriers that the ligand encounters as it moves along some reaction coordinate (79). From the covalently bound state A, the ligand first moves to a weakly bound state B, then back to A or on to state C, etc, and it may or may not reach the solvent, represented as a state S, before rebinding occurs. The transition rate k_{ij} from state (i) to state (j) is assumed to follow an Arrhenius relation

$$k_{ij} = A_{ij}\exp(-H_{ij}/RT) \qquad 1.$$

where the preexponential A_{ij} is independent of T, and H_{ij} is an enthalpy of activation. A sequential model then leads to a system of coupled differential equations, the solution of which produces the solid lines in Figure 2. A stochastic model of ligand migration was discussed by Hänggi (83, 94). For CO binding to Mb, five states, A, B, C, D, S, and four barriers can be resolved, for MbO$_2$ only four states, and for CO binding to free heme only three states and two barriers are seen. The results agree, at room temperature, with the second order kinetics noted above; they are also compatible, however, with the picture emerging from trajectory calculations for Mb (3), which show that a small ligand released from the distal side of the heme encounters obstacles in all directions and can escape only with concomitant displacement of the residues surrounding the heme. In low temperature flash experiments, the rebinding rates can be slowed sufficiently to reveal individual barriers,

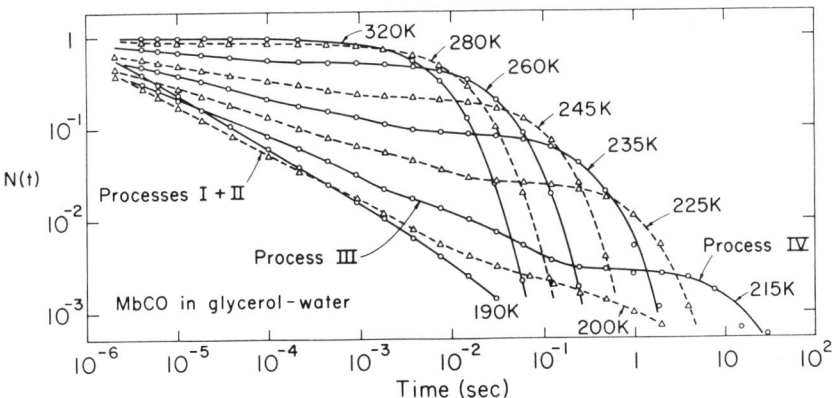

Figure 2 Rebinding of CO after photodissociation of MbCO in 3:1 v/v glycerol-water. Empirically, four processes labeled I–IV can be distinguished. I–III are independent of, IV is proportional to (CO). In IV, rebinding occured after CO reached the solvent S, in III after CO reached the third well D but not S, etc. (Reprinted with permission from *Biochemistry* 14:5355. Copyright 1975, American Chemical Society.) (Ref. 79.)

and there is hope, in principle, that the observed potential can be correlated with structural features of the protein.

At temperatures below ~200 K the photodissociated ligand can no longer escape from the protein and, being trapped next to the heme iron, it will eventually rebind (78). At very low temperatures, $T < 40$ K, the Arrhenius rate for over-the-barrier transitions is negligible, and molecular tunneling becomes dominant (81, 86, 88). Tunneling has been clearly identified in MbCO on the basis of the temperature and mass dependence of the binding rate (88).

The kinetics observed in the temperature range 40 K $< T <$ 180 K is of particular interest (78). Although the analysis of Figure 2 and of similar data suggests that only process I, the transition across the innermost barrier B → A is involved, the fraction $N(t)$ follows a power law in time rather than an exponential. This observation can be rationalized by assuming a distribution $g(H)$ of barrier heights with the normalization $\int g(H)dH = 1$. The fraction $N(t)$ is then given by the expression

$$N(t) = \int g(H)\exp(-kt)dH; \quad k = A\exp(-H/RT). \qquad 2.$$

It is indeed possible to find a unique spectral function $g(H)$ that reproduces the kinetics observed over the temperature range 40 K $< T <$

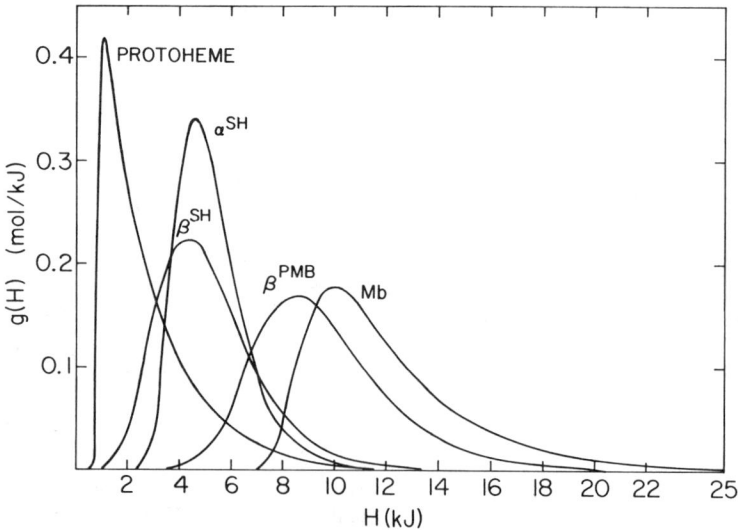

Figure 3 Activation enthalpy spectrum of the innermost barrier for CO binding to protoheme, separated α and β chains of hemoglobin, the p-mercuribenzoate of the latter, and of myoglobin. (Reprinted with permission from *Biochemistry* 17:43. Copyright 1978, American Chemical Society.) (Ref. 82.)

180 K. All heme protein adducts studied so far exhibit nonexponential low temperature kinetics; the spectral functions $g(H)$ differ from case to case and are thus a characteristic of each system (78–80, 82). The most plausible explanation for a distribution in barrier heights is the assumption that the active site geometry differs slightly from molecule to molecule. Although the detailed mechanism of bond formation $B \rightarrow A$ is not known (95, 96), it is clear that sizable atomic displacements and a spin change from $S=2$ to $S=0$ are involved (97). The energy of the transition state must depend on the coordinates of the complex and thus be sensitive to the local geometry. Each protein can assume a range of conformations; and at the temperatures considered, the fluctuation rate among substates is very slow, so that different conformations are frozen in. A spread in the low temperature heme geometries is evident also from the g-strain observed in EPR (98) and Mössbauer spectra (99) of ferric heme proteins.

Influence of Solvent Viscosity

There is no reason to believe that the other barriers BC, CD,... are intrinsically sharp. They appear sharp, however, if the thermal fluctuation rates between conformational substates become fast compared to the rates k_{ij} (Eq. 1). The fast fluctuation limit is approached at temperatures above 200 K. A notable exception is MbCO embedded in solid polyvinyl alcohol (PVA) (79), in which the ligand is always confined to the protein, and the three barriers observed are all distributed. The reason for this exceptional behavior must be the rigidity of the matrix, which entails limited conformational fluctuations, large local distortions, or both. The question then arises as to what extent the internal motions of the protein are dynamically coupled to the solvent matrix. In order to answer this fundamental question, we compare the kinetics measured at fixed temperatures but different solvent viscosities η. Specifically, the MbCO rebinding rates in aqueous solutions with added glycerol, sucrose, or ethylene glycol are analyzed in terms of the four barrier model described above, and the results shown in Figure 4 are obtained (87). The rate across the innermost barrier $B \rightarrow A$ is assumed to be independent of solvent and is not shown, but all other rates k_{ij}, $i, j = B, C, D, S$, are seen to vary systematically with solvent viscosity. Judging from the smooth variation of the rates with η and T, no specific interactions with glycerol etc are involved. In fact, for fixed η the log k values are to a large extent linear in T^{-1}, i.e. the rates follow an Arrhenius type relation,

$$k_{ij}^{(\eta)}(T) \propto \exp\left(-H_{ij}^{(\eta)}/RT\right). \qquad 3.$$

In addition, for fixed T, the $\log k_{ij}^{(T)}(\eta)$ values decrease approximately linearly with $\log \eta$, suggesting a power law

$$k_{ij}^{(T)}(\eta) \propto \eta^{-\kappa_{ij}},\qquad 4.$$

where κ_{ij} is a number that can be deduced from the data. Again, MbCO in solid PVA is an exception in that it shows finite rates despite the high viscosity η of the matrix, which should make the rate approach zero according to Eq. 4.

These empirical observations have far-reaching consequences. They demonstrate that the dynamic coupling of the medium with the protein has a controlling influence on all rates except for the rate k_{BA} of covalent bond formation. The viscosity of the solvent appears to be the crucial physical parameter. Standard transition state theory (100) predicts rate constants of the form of Eq. 1, with

$$A = \nu \exp(S^\ddagger/R),\qquad 5.$$

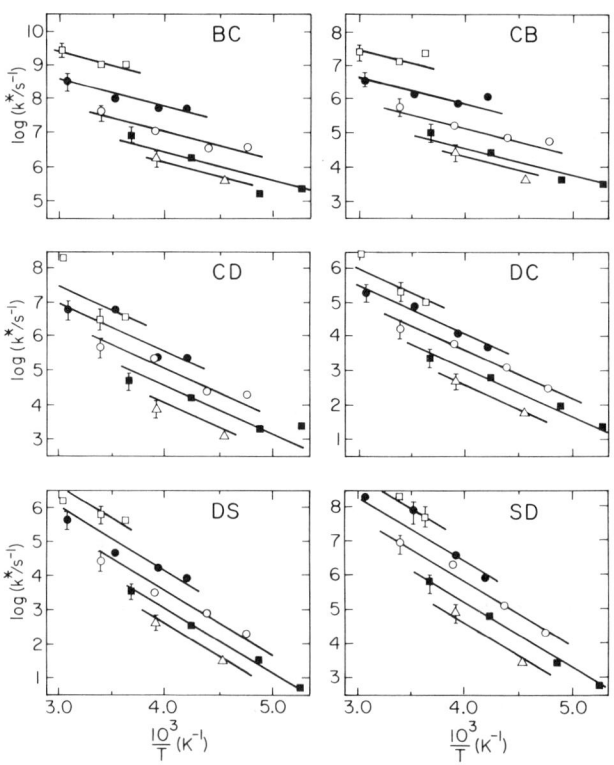

Figure 4 Rate coefficients $k^{(\eta)}$ vs $10^3/T$ for MbCO at the following fixed viscosities, □ 10cP, ● 1P, ○ 10P, ■ 100P, △ 1000P. The *solid lines* are explained in the text. The units for $k_{SD}^{(\eta)}$ are M^{-1} sec^{-1}. (Reprinted with permission from *Biochemistry* 19:5147. Copyright 1980, American Chemical Society.) (Ref. 87.)

where the frequency $\nu \sim 10^{13}$ sec^{-1} is supposed to be constant or at most weakly temperature dependent, while $S^{\ddagger}$ and $H = H^{\ddagger}$ are the temperature independent entropy and enthalpy of activation.

Although transition state theory is able to reproduce all the kinetic data observed for a given system with a relatively small number of thermodynamic parameters $H_{ij}^{\ddagger}$ and $S_{ij}^{\ddagger}$, this description has a great flaw: $H_{ij}^{\ddagger}$ and $S_{ij}^{\ddagger}$ depend on the solvent viscosity η as well as on the protein and the ligand, yet the η-dependence is not explicitly shown. Because of the dynamic coupling between solvent and protein, it is not possible to separate the measured enthalpies and entropies into contributions from the solvent and from the protein-ligand system alone. Based on the empirical relation Eq. 3, one may write the rates $k_{ij}^{(\eta)}(T)$ evaluated at constant viscosity η in the form

$$k_{ij}^{(\eta)}(T) = \nu \exp\left(-H_{ij}^{(\eta)}/RT + S_{ij}^{(\eta)}/R\right). \qquad 6.$$

While this parametrization achieves the desired separation of solvent properties from the properties of the protein-ligand system, it does so at the price of allowing $H^{(\eta)}$ and $S^{(\eta)}$ to be implicitly temperature dependent through $\eta(T)$. Not surprisingly, the values $H^{(\eta)}$ and $S^{(\eta)}$ are greatly reduced compared to the corresponding values $H^{\ddagger}$ and $S^{\ddagger}$, as illustrated in Table 2. In a plot of $H^{(\eta)} - TS^{(\eta)}$ versus reaction coordinate, the potential minima corresponding to the states B, C, D, S remain relatively constant, while the outer barriers increase with viscosity. That these barriers are dynamic in origin is suggestive of solvent-coupled fluctuations.

The empirical relation Eq. 4 is reminiscent of Kramers' equation (101)

$$k = (\nu_0 \nu_0' \rho / 3\eta) \exp(-H/RT) \qquad 7.$$

Table 2 Parameters deduced from the rates k_{BC} and k_{CB} of MbCO according to different models

	Eqs. 1,5 [a]		Eq. 6 [b]			Eqs. 8,9 [c]		
Transition	$H_{ij}^{\ddagger}$ [d]	$S_{ij}^{\ddagger}/R$	$H_{ij}^{(\eta)}$ [d]	$S_{ij}^{(\eta)}/R$	$\log A_{ij}$ [e]	$\log A_{ij}^{0}$ [f]	κ	H [d]
B → C	54	11.4	16	−4.3	12.5	9.1	0.8	15
C → B	45	3.8	15	−9.2	10.6	7.4		

[a] 79% glycerol.
[b] $\eta = 2$ P, $\nu = 10^{13}$ sec^{-1}.
[c] Best fit for 10^{-1} P $\leq \eta \leq 10^3$ P, 190 K $\leq T \leq$ 320 K.
[d] In kJ mol^{-1}.
[e] A_{ij} in units of cP$^{\kappa}$ sec^{-1}.
[f] A_{ij}^{0} in units of sec^{-1}.

for the transition rate of a particle across a barrier of height H in one dimension as a result of Brownian motion in a highly viscous medium. The particle is initially in a harmonic potential with undamped frequency ν_0; ν'_0 is the equivalent frequency of the parabolic barrier; and ρ is the mass divided by a linear dimension. At low viscosity, the Kramers Eq. 7 accounts well for the outer barrier of CO binding to free heme, where the transition D → S can be described with the reasonable parameters $\nu_0\nu'_0 = 2.4 \cdot 10^{27}$ sec^{-2}, $\rho = 10^{-14}$ g cm^{-1}, and $H = 20$ kJ mol^{-1}. At high viscosities an additional term $A^0 \exp(-H/RT)$, $A^0 = 5 \cdot 10^{10}$ sec^{-1}, can be added to account for the fact that k_{DS} becomes independent of η. For the data of Figure 4 a similarly modified Kramers equation with an additional parameter $\kappa_{ij}, 0 < \kappa_{ij} < 1$,

$$k_{ij}(T) = \left(A^0_{ij} + A_{ij}\eta^{-\kappa_{ij}}\right)\exp\left(-H_{ij}/RT\right) \qquad 8.$$

is found to work well. In fact the solid lines in Figure 4 use the further, simplifying condition

$$H_{ij} = H_{ji}, \qquad \kappa_{ij} = \kappa_{ji} \qquad 9.$$

with the parameters listed in Table 2. Condition 9 can be interpreted by picturing the barrier as a gate opened and closed by conformational fluctuations. The enthalpy $H_{ij} = H_{ji}$ then is the energy of activation of the "open" or transmission state, and $\kappa_{ij} = \kappa_{ji}$ measures the extent to which the particular fluctuations involved are coupled to the solvent. Similar viscosity dependences of reaction rates have been observed in catalysis (102) and in the photocycle of bacteriorhodopsin (103), and it is clear that dynamic coupling to the solvent is a general phenomenon for proteins.

SUMMARY AND OUTLOOK

The experiments and the molecular dynamics computations discussed here imply that proteins are fluctuating systems and that the fluctuations are, at least in the binding of ligands to heme proteins, essential for biological function. Despite the rapid progress made during the past few years, the field is only at a beginning, and more exciting results can be expected. Progress is likely in at least four directions. Experiments, particularly x-ray diffraction, NMR, fluorescence, Raman scattering, and Mössbauer effect, will yield more detailed information on the spatial and temporal properties of protein motions. Molecular dynamics will give the corresponding theoretical understanding of these processes. The theoretical basis underlying calculations and data evaluation, such as the application of a modified Kramers equation to protein reactions, will probably be strengthened and generalized. And finally, experiment and theory will move on to biologically more interesting processes such as enzyme catalysis and transport of information, charge, and matter.

Literature Cited

1. Perutz, M. F., Mathews, F. S. 1966. *J. Mol. Biol.* 21:199–202
2. Takano, T. 1977. *J. Mol. Biol.* 110:569–84
3. Case, D. A., Karplus, M. 1979. *J. Mol. Biol.* 132:343–68
4. Nemethy, G., Scheraga, H. A. 1977. *Q. Rev. Biophys.* 10:239–352
5. Levitt, M. 1976. *J. Mol. Biol.* 104:59–107
6. Citri, N. 1973. *Adv. Enzymol. Relat. Areas. Mol. Biol.* 37:397–648
7. Banks, R. D., Blake, C. C. F., Evans, P. R., Haser, R., Rice, D. W., Hardy, G. W., Merrett, M., Phillips, A. W. 1979. *Nature* 279:773–77
8. Anderson, C. M., Zucker, F. H., Steitz, T. A. 1979. *Science* 204:375–80
9. Perutz, M. F. 1979. *Ann. Rev. Biochem.* 48:327–86
10. Hobbs, A. S., Albers, R. W. 1980. *Ann. Rev. Biophys. Bioeng.* 9:259–91
11. Careri, G., Fasella, P., Gratton, E. 1979. *Ann. Rev. Biophys. Bioeng.* 8:69–97
12. Squire, J. M. 1975. *Ann. Rev. Biophys. Bioeng.* 4:137–63
13. Highsmith, S., Kretzschmar, K. M., O'Konski, C. T., Morales, M. F. 1977. *Proc. Natl. Acad. Sci. USA* 74:4986–90
14. Careri, G., Fasella, P., Gratton, E. 1975. *CRC Crit. Rev. Biochem.* 3:141–64
15. Cooper, A. 1976. *Proc. Natl. Acad. Sci. USA* 73:2740–41
16. McCammon, J. A., Karplus, M. 1980. *Ann. Rev. Phys. Chem.* 31:29–45
17. Karplus, M., McCammon, J. A. 1981. *CRC Crit. Rev. Biochem.* 9:293–349
18. Gurd, F. R. N., Rothgeb, T. M. 1979. *Adv. Protein Chem.* 33:74–165
19. Cooper, A. 1980. *Sci. Prog.* 66:473–97
20. Woodward, C. K., Hilton, B. D. 1979. *Ann. Rev. Biophys. Bioeng.* 8:99–127
21. Matthews, B. W. 1976. *Ann. Rev. Phys. Chem.* 27:493–523
22. Blundel, T. L., Johnson, L. N. 1976. *Protein Crystallography*. London: Academic
23. Poulos, T. L., Kraut, J. 1980. *J. Biol. Chem.* 255:10322–30
24. Weber, I. T., Johnson, L. N., Wilson, K. S., Yeates, D. G. R., Wild, D. L., Jenkins, J. A. 1978. *Nature* 274:433–37
25. Artymiuk, P. J., Blake, C. C. F., Grace, D. E. P., Oatley, S. J., Phillips, D. C., Sternberg, M. J. E. 1979. *Nature* 280:563–68
26. Robertus, J. D., Alden, R. A., Birktoft, J. J., Kraut, J., Powers, J. C., Wilcox, P. E. 1972. *Biochemistry* 11:2439–49
27. Delbaere, L. T. J., Hutcheon, W. L. B., James, M. N. G., Thiessen, W. E. 1975. *Nature* 257:758–63
28. Dickerson, R. E., Timkovich, R., Almassy, R. J. 1976. *J. Mol. Biol.* 100:473–91
29. Stuhrmann, H. B. 1973. *J. Mol. Biol.* 77:363–69
30. Engelman, D. M., Moore, P. B. 1975. *Ann. Rev. Biophys. Bioeng.* 4:219–41
31. Wüthrich, K. 1976. *Nuclear Magnetic Resonance in Biological Research — Peptides and Proteins*. Amsterdam: North Holland
32. Campbell, I. D., Dobson, C. M., Williams, R. J. P. 1978. *Adv. Chem. Phys.* 39:55–95
33. Timchenko, A. A., Ptitsyn, O. B., Dolgikh, D. A., Fedorov, B. A. 1978. *FEBS Lett.* 88:105–8
34. Timchenko, A. A., Ptitsyn, O. B., Troitsky, A. V., Denesyuk, A. I. 1978. *FEBS Lett.* 88:109–13
35. Steigemann, W., Weber, E. 1979. *J. Mol. Biol.* 127:309–38
36. Huber, R., Deisenhofer, J., Colman, P. M., Matsushima, M., Palm, W. 1976. *Nature* 264:415–20
37. Harrison, S. C., Olson, A. J., Schutt, C. E., Winkler, F. K. 1978. *Nature* 276:368–73
38. Wiegand, G., Kukla, D., Scholze, H., Jones, A., Huber, R. 1979. *Eur. J. Biochem.* 93:41–50
39. Watenpaugh, K. D., Sieker, L. C., Herriott, J. R., Jensen, L. H. 1973. *Acta Crystallogr. B* 29:943–56

40. Jack, A., Levitt, M. 1978. *Acta Crystallogr. A* 34:931–35
41. Konnert, J. H., Hendrickson, W. A. 1980. *Acta Crystallogr. A* 36:344–50
42. Willis, B. T. M., Pryor, A. W. 1975. *Thermal Vibrations in Crystallography.* London: Cambridge Univ. Press
43. Parak, F., Formanek, H. 1971. *Acta Crystallogr. A* 27:573–78
44. Peticolas, W. L. 1979. *Methods Enzymol.* 61:425–58
45. Watenpaugh, K. D., Sieker, L. C., Jensen, L. H. 1980. *J. Mol. Biol.* 138:615–33
46. Frauenfelder, H., Petsko, G. A., Tsernoglou, D. 1979. *Nature* 280:558–63
47. Artymiuk, P. J., Blake, C. C. F., Oatley, S. J. 1979. *J. Chim. Phys.* 76:813–15
48. Sternberg, M. J. E., Grace, D. E. P., Phillips, D. C. 1979. *J. Mol. Biol.* 130:231–53
49. Frauenfelder, H., Petsko, G. A. 1980. *Biophys. J.* 32:465–78
50. Phillips, S. E. V. 1980. *J. Mol. Biol.* 142:531–54
51. Takano, T., Dickerson, R. E. 1980. *Proc. Natl. Acad. Sci. USA* 77:6371–75
52. Sielecki, A. R., Hendrickson, W. A., Broughton, C. G., Delbaere, L. T. J., Brayer, G. D., James, M. N. G. 1979. *J. Mol. Biol.* 134:781–804
53. Dwivedi, A., Pederson, T., Debrunner, P. G. 1979. *J. Phys.* 40 (C2):531–32
54. Parak, F., Frolov, E. N., Mössbauer, R. L., Goldanskii, V. I. 1981. *J. Mol. Biol.* 145:825–33
55. Keller, H., Debrunner, P. G. 1980. *Phys. Rev. Lett.* 45:68–71
56. Gavish, B. 1981. *Proc. Natl. Acad. Sci. USA* 78:6868–72
57. Singh, T. P., Bode, W., Huber, R. 1980. *Acta Crystallogr. B* 36:621–27
58. Northrup, S. H., Pear, M. R., McCammon, J. A., Karplus, M., Takano, T. 1980. *Nature* 287:659–60
59. Lakowicz, J. R., Weber, G. 1973. *Biochemistry* 12:4171–79
60. Eftink, M. R., Ghiron, C. A. 1981. *Anal. Biochem.* 114:199–227
61. Munro, I., Pecht, I., Stryer, L. 1979. *Proc. Natl. Acad. Sci. USA* 76:56–60
62. Andrew, E. R., Bryant, D. J., Cashell, E. M. 1980. *Chem. Phys. Lett.* 69:551–54
63. Jardetzky, O. 1981. *Acc. Chem. Res.* 14:291–98
64. Nagayama, K. 1981. *Adv. Biophys.* 14:139–44
65. Wagner, G., Wüthrich, K., Tschesche, H. 1978. *Eur. J. Biochem.* 86:67–76
66. Gelin, B. R., Karplus, M. 1975. *Proc. Natl. Acad. Sci. USA* 72:2002–6
67. Hetzel, R., Wüthrich, K., Deisenhofer, J., Huber, R. 1976. *Biophys. Struct. Mechanism* 2:159–80
68. McCammon, J. A., Wolynes, P. G., Karplus, M. 1979. *Biochemistry* 18:927–42
69. Hilton, B. D., Woodward, C. K. 1978. *Biochemistry* 17:3325–32
70. Wüthrich, K., Wagner, G. 1979. *J. Mol. Biol.* 130:1–18
71. Hilton, B. D., Woodward, C. K. 1979. *Biochemistry* 18:5834–41
72. Wüthrich, K., Eugster, A., Wagner, G. 1980. *J. Mol. Biol.* 144:601–4
73. Lumry, R., Rosenberg, A. 1975. *Colloq. Int. CNRS* 246:53–61
74. Richards, F. M. 1979. *Carlsberg Res. Commun.* 44:47–63
75. Hilton, B. D., Trudeau, K., Woodward, C. K. 1981. *Biochemistry* 20:4697–4703
76. Walker, R. B., Light, J. C. 1980. *Ann. Rev. Phys. Chem.* 31:401–33
77. Antonini, E., Brunori, M. 1971. *Hemoglobin and Myoglobin in Their Reactions with Ligands.* Amsterdam: North-Holland
78. Austin, R. H., Beeson, K., Eisenstein, L., Frauenfelder, H., Gunsalus, I. C., Marshall, V. P. 1974. *Phys. Rev. Lett.* 32:403–5
79. Austin, R. H., Beeson, K. W., Eisenstein, L., Frauenfelder, H., Gunsalus, I. C. 1975. *Biochemistry* 14:5355–73
80. Alberding, N., Austin, R. H., Chan, S. S., Eisenstein, L., Frauenfelder, H., Gunsalus, I. C., Nordlund, T. M. 1976. *J. Chem. Phys.* 65:4701–11
81. Alberding, N., Austin, R. H., Beeson, K. W., Chan, S. S., Eisenstein, L., Frauenfelder, H., Nordlund, T. M. 1976. *Science* 192:1002–4
82. Alberding, N., Chan, S. S., Eisenstein, L., Frauenfelder, H., Good, D., Gunsalus, I. C., Nordlund, T. M., Perutz, M. F., Reynolds, A. H., Sorensen, L. 1978. *Biochemistry* 17:43–51
83. Alberding, N., Frauenfelder, H., Hänggi, P. 1978. *Proc. Natl. Acad. Sci. USA* 75:26–29

84. Alberding, N., Austin, R. H., Chan, S. S., Eisenstein, L., Frauenfelder, H., Good, D., Kaufmann, K., Marden, M., Nordlund, T. M., Reinisch, L., Reynolds, A. H., Sorensen, L. B., Wagner, G. C., Yue, K. T. 1978. *Biophys. J.* 24:319–34
85. Beece, D., Eisenstein, L., Frauenfelder, H., Good, D., Marden, M., Reinisch, L., Reynolds, A. H., Sorensen, L. B., Yue, K. T. 1979. *Biochemistry* 15:3421–23
86. Frauenfelder, H. 1979. *Tunneling in Biological Systems*, ed. B. Chance, D. C. DeVault, H. Frauenfelder, R. A. Marcus, J. R. Schrieffer, N. Sutin, pp. 627–49. New York: Academic
87. Beece, D., Eisenstein, L., Frauenfelder, H., Good, D., Marden, M. C., Reinisch, L., Reynolds, A. H., Sorensen, L. B., Yue, K. T. 1980. *Biochemistry* 19:5147–57
88. Alben, J. O., Beece, D., Bowne, S. F., Eisenstein, L., Frauenfelder, H., Good, D., Marden, M. C., Moh, P. P., Reinisch, L., Reynolds, A. H., Yue, K. T. 1980. *Phys. Rev. Lett.* 44:1157–60
89. Gouterman, M. 1978. *The Porphyrins*, ed. D. Dolphin, 3(Pt. A):1–165. New York: Academic
90. Cooke, R., Kuntz, I. D. 1974. *Ann. Rev. Biophys. Bioeng.* 3:95–126
91. Kuntz, I. D., Brassfield, T. S., Law, G. D., Purcell, G. V. 1969. *Science* 163:1329–31
92. Takashima, S. 1970. *Physical Principles and Techniques of Protein Chemistry*, Pt. A, ed. S. J. Leach, pp. 291–333. New York: Academic
93. Singh, G. P., Parak, F., Hunklinger, S., Dransfeld, K. 1981. *Phys. Rev. Lett.* 47:685–88
94. Hänggi, P. 1978. *J. Theor. Biol.* 74:337–59
95. Jortner, J., Ulstrup, J. 1979. *J. Am. Chem. Soc.* 101:3744–54
96. Hopfield, J. J. 1979. See Ref. 86, p. 646
97. Gerstman, B., Austin, R. H., Hopfield, J. J. 1981. *Phys. Rev. Lett.* 47:1636–39
98. Brill, A. S. 1979. See Ref. 86, pp. 561–68
99. Dwivedi, A., Toscano, W. A. Jr., Debrunner, P. G. 1979. *Biochim. Biophys. Acta* 576:502–8
100. Glasstone, S., Laidler, K. J., Eyring, H. 1941. *The Theory of Rate Processes*. New York: McGraw-Hill
101. Kramers, H. A. 1940. *Physica* 7:284–304
102. Gavish, B., Werber, M. M. 1979. *Biochemistry* 18:1269–79
103. Beece, D., Bowne, S. F., Czégé, J., Eisenstein, L., Frauenfelder, H., Good, D., Marden, M. C., Marque, J., Ormos, P., Reinisch, L., Yue, K. T. 1981. *Photochem. Photobiol.* 33:517–22

RECENT DEVELOPMENTS IN ELECTRON PARAMAGNETIC RESONANCE: TRANSIENT METHODS

S. I. Weissman

Department of Chemistry, Washington University, St. Louis, Missouri 63130

INTRODUCTION

The most recent article in *Annual Review of Physical Chemistry* under the title "Electron Spin Resonance" is Freed's (1), which appeared in 1972. In the intervening years several thousand papers have been abstracted in *Chemical Abstracts* under the title "Electron Spin Resonance." In most of these papers "conventional electron paramagnetic resonance" (EPR) spectroscopy, i.e. recording of the absorption spectrum under slow passage steady state conditions, has been used. The principles and practice of such EPR spectroscopy are so well known that the results that it yields are not, I believe, appropriate for a review dealing with recent developments in EPR. Many of the findings are novel and of great interest to chemists, but would be more appropriately included in reviews of the particular chemical phenomena with which they deal. (The reader may note that Electron Spin Resonance has been metamorphosed to Electron Paramagnetic Resonance. I make the change to avoid irritating rigorous colleagues who point out that no pure electron spin resonance, completely devoid of orbital effects, exists.)

This review deals for the most part with the various methods in which time evolution of transient magnetization is observed. The book edited by Kevan & Schwartz on *Time Domain Electron Spin Resonance Spectroscopy* (2) contains much of the material of this review. Many important and elegant developments are not included herein.

Despite restriction of this review to transient phenomena, some comments on the extensive use of EPR may be appropriate. Most commonly used spectroscopic techniques, such as optical and infrared absorption spectroscopy, or nuclear magnetic resonance (NMR), are applicable to the majority of chemical systems. EPR, on the other hand, is applicable to only a very small fraction of the millions of chemical compounds that have been characterized. Thus, all compounds containing hydrogen, with the exception of para hydrogen and possibly a few others that might be isolated in singlet nuclear spin states, have proton NMR spectra. Why the widespread use of EPR, despite the fact that most stable compounds are analogous with respect to their electronic properties to the nuclear spin properties of para hydrogen and thus have no EPR responses? Aside from the addiction of some investigators (including me) to viewing well-resolved hyperfine patterns, it is the role of paramagnetic substances as intermediates in chemical processes and the rich yield of structural and kinetic information obtainable from EPR that is responsible, I believe, for its voluminous literature. Electron spin pairing and the absence of orbital degeneracy in the majority of chemical species are responsible for their not having EPR responses. But in chemical processes, intermediates are often not so restricted. In redox reactions, in the passage from one spin-paired species to another by one-electron transfer steps, odd electron intermediates must occur. In other processes, particularly photochemical ones, intermediates with unpaired spins often are involved. Such intermediates, some sufficiently long lived for study by conventional EPR, have been receiving increasing attention.

TRANSIENT METHODS

Let us consider first transient magnetization in systems in which the primary time dependence is produced chemically: A chemical process is initiated by an appropriate pulse, typically a pulse of light or of high energy electrons, and the subsequent evolution is monitored through response to continuous irradiation at one of the EPR transitions of a product. This method is analogous to the optical ones in which time evolution of products is monitored by evolution of optical absorption from a continuous beam. As we shall see, in the magnetic case the evolution of the absorption is determined not only by the chemical contribution but by interaction with the probing radiation field; that simple extrapolation to the behavior under low intensity monitoring field does not yield directly the chemical evolution.

Among the earliest transient EPR experiments of the type described in the preceding paragraph are those of Fessenden et al (3–5) on the radiolysis of alkanes. They found that under continuous electron irradiation of liquid alkanes in the cavity of a conventional EPR spectrometer, the spectra of odd electron species appeared. Irradiation of liquid methane produced a normal absorption spectrum of methyl radicals and an abnormal spectrum of hydrogen atoms, the low field member of the hydrogen doublet being emissive, the high field one enhanced absorptive. Observation of the transient behavior led to an understanding of the steady state spectra. The transient experiments were carried out by irradiating with short electron pulses and recording the time evolution of the absorption of a continuous microwave field tuned to one of the transitions. From the nature of the transient absorption the authors were able to conclude that the hydrogen atoms are born with almost equal populations of their four eigenstates, but that the ensuing chemical processes deplete two of the eigenstates, the ones in which electron and nuclear spins are antiparallel, more rapidly than the two with parallel nuclear and electron spins. Fessenden & Verma have applied the method to a variety of other systems (6, 7).

I discuss next the nature of the transients appearing in such experiments and how they are interpreted. Consider the following idealized situation: Paramagnetic species with a single sharp resonance are created instantaneously while under constant irradiation by an exactly resonant microwave field whose rotating component in the resonant direction has amplitude B_1. They are created with one of their eigenstates substantially more populated than the other. Their chemical lifetimes are long, their relaxation times T_1 and T_2 are equal, and the ambient temperature is high. A close approach to these idealizations is achieved in the photolysis of rubidium and cesium anions to produce solvated electrons (8, 9) and in the formation of semiquinones by photolysis of quinones in alcohols (10, 11).

How does the rate of absorption of microwave energy evolve in time? It is given by M_x, the component of magnetization out of phase with the driving field: $A \exp - t/T_2 \sin \gamma B_1 t$, where t is the time measured from the instant of creation of the spins and γ is their magnetogyric ratio. Note that at the instant of creation, when the population difference is greatest, the rate of absorption is zero. It reaches its maximum value at $t = \pi/2\gamma B_1$. The oscillatory response has been well known since the work of Torrey (12, 13) on transient nutation in the early days of NMR. Torrey observed the oscillatory responses in systems in which the number of spins is independent of time but in which the probing field is turned on suddenly; in our case the probing field is always present and the spins are created suddenly. The reason for the oscillatory response has long been known (14). γB_1 is the frequency with which the probing field carries the system through the succession of stages: initial population → coherent superposition → inverted population → coherent superposition → initial population. The rate of transfer of energy between probing field and spin systems is greatest when there is maximum coherence between the spin eigenstates.

If the irradiating field at frequency ω is not equal to the resonant frequency ω_0 the signal is given by

$$M_x(t) = \gamma B_1 \{(\omega - \omega_0)^2 + (\gamma B_1)^2\}^{-1/2}$$
$$\times \exp{-t/T_2} \sin\{(\omega - \omega_0)^2 + (\gamma B_1)^2\}^{1/2} t$$

when $\gamma B_1 \ll 1/T_2$; i.e. when the amplitude of the driving field is smaller than the homogeneous line breadth, the oscillations are lost and the on-resonance response becomes approximately $A\gamma B_1 t \exp{-t/T}$. When $\gamma B_1 \gg 1/T_2$ the amplitude of the first maximum is independent of B_1. The transient response is not impaired by a power level, which would saturate the resonance in a CW experiment.

Our idealized system demonstrates a feature of all time-resolved spectroscopic observation that has often been ignored: the dynamics of the interaction of the probing radiation field with the chemical system. In the studies of transient optical absorption in condensed systems, the effects are usually justifiably ignored because the characteristic times for interaction with the radiation field are almost always shorter than the resolving time of detection. In gaseous systems where narrow absorptions are monitored, or even in condensed systems to which subpicosecond methods are applied, the effects may be significant.

Let us examine now some of the effects that occur in most real systems. For a single homogeneously broadened line with $T_1 \neq T_2$, the responses are generally triexponential:

$$M_x(t) = \sum_{m=1}^{m=3} Cml^{\lambda mt},$$

the λs being solutions of the cubic

$$\lambda^3 + A\lambda^2 + B\lambda + C = 0$$

$$A = \frac{2}{T_2} + \frac{1}{T_1}$$

$$B = \left(\frac{1}{T_2}\right)^2 + \frac{2}{T_1 T_2} + (\omega - \omega_0)^2 + 4\gamma^2 B_1^2$$

$$C = \frac{1}{T_1 T_2^2} + \frac{(\omega - \omega_0)^2}{T_1} + \frac{4\gamma^2 B_1^2}{T_2}.$$

At resonance ($\omega = \omega_0$), the three solutions are

$$\lambda_1 = -1/T_2,$$

$$\lambda_{2,3} = -\frac{\left(\frac{1}{T_1}+\frac{1}{T_2}\right)}{2} \pm \left\{\left(\frac{\frac{1}{T_1}-\frac{1}{T_2}}{2}\right)^2 - 4\gamma^2 B_1^2\right\}^{1/2}.$$

If $\left(\frac{\frac{1}{T_1}-\frac{1}{T_2}}{2}\right)^2 > 4(\gamma B_1)^2$

there are no oscillatory solutions. Numerical solutions for a wide variety of cases are given by Hore & McLauchlan (15). Their solutions are for situations in which a system is created with large population differences between eigenstates and relaxes toward equality of population, i.e. $h\omega_0 \ll kT$. Hore & McLauchlan have applied the results to determination of relaxation rates (16, 17). R. A. Smith (18a–c) has worked out (the optical regime) solutions for cases with $h\omega_0 \gg kT$. Smith's language is that of atomic spectroscopy but is easily translated to that of magnetic resonance. His rate of transition-inducing collisions is $1/T_1$ and of dephasing collisions $1/T_2$.

We now drop the restriction of a single homogeneously broadened line. Consider an inhomogeneously broadened line for which γB_1 is considerably smaller than the inhomogeneous line breadth and in which the phase memory time T is equal to T_1. The time dependence of the signal is given by

$$M_x(t) \sim \int_{-\infty}^{\infty} d\omega_0 g(\omega_0) \gamma B_1 \{(\gamma B_1)^2 + (\omega - \omega_0)^2\}^{-1/2}$$

$$\times \exp -t/T \sin\{(\gamma B_1)^2 + (\omega - \omega_0)^2\}^{1/2} t$$

where $g(\omega_0)$ is the shape function of the resonance. As is well known, the rise time of the signal now becomes of the order of the reciprocal line breadth. An example of this effect is given in the work of Furrer et al (19). The reason for the rapid rise is that the off-resonance isochromats nutate with increasing frequency and decreasing amplitude as the frequency deviation $\omega - \omega_0$ increases. Their sum in the limit of an infinitely broad line has its maximum amplitude at $t = 0$. We see the justification for the remark above that in optical experiments in condensed media, in which inhomogeneous line breadths of hundreds of wave numbers are often encountered with corresponding rise times shorter than a fraction of a picosecond, that the assumption that the instantaneous rate of light absorption is proportional to concentration of absorbing species is valid.

To return to magnetic resonance, the predicted rapid rise followed by a damped oscillation has been observed in some but not all the transitions of newly created photoexcited triplets in single crystal matrices (20, 21). Thus, in the triplet state of pentacene the low field emissive transition exhibits no oscillations in its transient response, the high field absorptive exhibits one well-defined oscillation. The inhomogeneous line breadths are about 10^8 Hz, γB_1 about 10^6 Hz. The differences in the two cases are probably caused by differences in radiation-induced spectral diffusion.

The aim of most of the work on time-resolved EPR has been to determine the course of the chemical processes that are monitored. The complications induced by the dynamics of interaction with the probing radiation are probably considered by some to be at best an interesting nuisance. How to sift out the chemical time evolution? Fessenden's and McLauchlan's solutions (6, 16) have been to record the time evolution at varying probing amplitudes, solve the kinetic equations—Bloch equations for the magnetic time evolution coupled to assumed chemical kinetic equations—and seek a fit to the observed transients. The following features are considered: initial populations, relaxation rates, and state-dependent rates of reactions. The latter feature, important in chemically induced electron and nuclear polarization (CIDEP and CIDNP), is detectable in time-resolved EPR experiments but not in conventional kinetic experiments.

Other uses of the method are exemplified by the studies of Wan (22) and co-workers on a variety of photochemical processes, of Doetschman (23) and Chisholm (24) on photoproduction of methylenes and biradicals, and of Kothe (25) on detection of a short-lived photoexcited quartet state. The method is applicable to randomly oriented molecules in rigid glasses and has succeeded in determining the path of intersystem crossing for hexahelicene, a molecule for which no suitable crystalline host has been found (26).

The use of randomly oriented molecules in rigid glasses, although it affords lower sensitivity than is obtainable from well-oriented systems, has obvious advantages. For triplet and quartet molecules the response at each field comes only from molecules at a limited set of orientations. Through study of dependence of transients on static magnetic field it is possible to deduce the selectivity in a population of eigenstates, and in favorable cases not only the magnitude but the sign of spin-spin coupling parameter D. To my knowledge, no oscillations in the transient nutation responses in glassy media have yet been observed.

PULSING THE RADIATION FIELD: DELAYED TRANSIENT NUTATION, FREE INDUCTION DECAYS, AND ELECTRON SPIN ECHOES

Attempts to obtain the "pure" chemical kinetics, uncontaminated by the dynamics of interaction with the probing radiation field, have been made through the use of pulsed microwave radiation. Simplest of these is the delayed transient nutation method of Furrer et al (19). Following initiation by a photolytic or radiolytic pulse, the system evolves in the absence of microwave fields (but in the presence of the static field); the microwave radiation is then switched on and evokes a transient response. The amplitude of the earliest maximum in the transient is interpreted as being proportional to the population difference between the pair of states in resonance. The method is, I believe, reliable and useful in many cases. It may be difficult to apply when the species are very short-lived owing to the instrumental transient introduced upon switching on the microwaves, but with appropriate signal-averaging and subtraction of the off-resonance response from the resonance response, this method should be widely applicable.

The free induction method of Trifunac & Lawler (27) is related to the delayed transient nutation method: a $\pi/2$ microwave pulse is applied subsequent to the initiating radiolytic pulse. The maximum amplitude of the free induction decay is interpreted as yielding the population difference of resonant levels. The method is applicable only when lines are sufficiently narrow to give free induction decays longer than the dead time of the instrument. For a dead time of 100 nsec, the lines must be narrower than ~ 0.4 Oe. I see little advantage in the method over delayed transient nutation; the latter is useable for broad lines and will yield an oscillatory transient nutation for narrow lines—the first maximum of which represents the intensity of the free induction decay immediately after a $\pi/2$ pulse.

We come finally to the electron spin echo (ESE) method. Because of its increasing use I discuss its applications, not only to systems that evolve in time in the absence of probing radiation, but to stable ones in thermal equilibrium as well.

After a few early experiments (28–31) and little development for some years, the use of electron spin echoes has recently increased. I do not discuss here the experimental arrangements that are used in obtaining electron spin echoes: they are presented along with an illuminating exposition of the physical principle in the article by Mims (32). A recent article describes the most powerful echo instrument (5 kW coherent pulses produced by a magnetron) yet advertised (33).

The physical principles involved in ESE are no different from those which have long been understood in NMR (34, 35). But owing to the magnetogyric ratio of the electron—some three orders of magnitude greater than the largest nuclear ones—the experimental methods in ESE are different and at present seem to be more demanding than those used in NMR. Mims' paper (32) presents an exposition of the methods, their limitations and range of application. I summarize some of the points here.

Consider first the possibility of obtaining an EPR spectrum via the free induction decay—the method now used almost universally in NMR. Many of the spectra of interest to chemists have hyperfine patterns spanning some 50 Oe. If the free induction decay is to be a faithful Fourier transform of the entire spectrum, the tilt angle should not vary greatly across the spectrum; the amplitude of the rotating field in the pulse should then be at least 100 Oe. The power required to produce the 100 Oe amplitude at X band is of the order of 20 kW, the exact value depending on the nature of the structure enclosing the sample. A $\pi/2$ pulse of 100 Oe amplitude has a duration of about 1 nsec. The power and switching requirements and the dead time associated with the high power would appear to make such EPR Fourier transform spectroscopy presently out of reach. However, if the requirement that the entire spectrum be acquired in each free induction decay is relaxed, or if the paramagnetic species have narrow spectral distribution, the free induction decay is useful, and with presently available rapid digitizing devices and data processing, sensitivities greater than those afforded by CW methods are possible. Thus CW methods for paramagnetic species with relaxation times of the order of 1 msec are troublesome because of saturation; free induction decays (FID) at pulse repetition rates of 1 kHz are practical and would serve very well in such cases. One million FIDs in 20 min is a figure that should appeal to practitioners of NMR.

We consider next the spin echo methods that have been used in EPR. To my knowledge the only pulse sequences that have received widespread use in EPR are ones very familiar from NMR.

$$\frac{\pi}{2} - \tau - \frac{\pi}{2}, \quad \frac{\pi}{2} - \tau - \pi, \quad \text{and} \quad \frac{\pi}{2} - \tau_1 - \frac{\pi}{2} - \tau_2 - \frac{\pi}{2}.$$

A few others such as Carr-Purcell trains and rotary echoes have had limited use; I do not discuss them here. (The elaborate sequences that have been used in NMR, such as a recent one of 58 pulses, have not yet been attempted in EPR. I do not speculate as to how many of them will become appliable or useful.)

Discussions of pulse and echo phenomena often treat the observed time evolution as resulting from tilting, dephasing, or refocusing of magnetization that had been present prior to application of pulses. Such representation is applicable to ESE of systems that had been in thermal equilibrium in a strong static field before imposition of the pulses. But many of the systems for which ESE has been done have zero magnetization before the pulses are applied. A triplet system at zero field whose density matrix in the eigenbase of the Hamiltonian is diagonal, or a triplet system in strong external static field born with equal populations in the $M_s = \pm 1$ states, i.e. with density matrix

$$\begin{array}{c} \begin{array}{ccc} +1> & |0> & |-1> \end{array} \\ \begin{array}{c} +1> \\ |0> \\ -1> \end{array} \left(\begin{array}{ccc} f & 0 & 0 \\ 0 & 1-2f & 0 \\ 0 & 0 & f \end{array} \right) \end{array}$$

has zero expectation value of all components of magnetization. A pulse at a frequency resonant with a pair of levels establishes coherence between them and an observable time dependent magnetization. The ideas used in discussion of "fictitious two level systems," i.e. multilevel systems in which only one pair of levels is close to resonance with an oscillating field, are applicable. One might think of a fictitious magnetization corresponding to the population difference between the levels that are brought into a coherent superposition by the oscillating field as executing the motions described in much of the literature on pulses, but it is not a way of thinking that I have found to be particularly useful. What then do we mean by a $\pi/2$ pulse or a π pulse for a system which prior to the pulse had no magnetization to be tilted? By a $\pi/2$ pulse we mean one that produces a coherent superposition of a pair of states with equal magnitudes of their amplitudes; or in other words one with maximum off-diagonal element of the density matrix. A π pulse is one that interchanges a pair of diagonal elements. We should also note that the amplitude and duration of oscillating field required for a particular kind of pulse depends on the nature of the states that are mixed by it. For an allowed transition between a pair of triplet Zeeman levels, the required pulse at fixed amplitude is shorter by $\sqrt{2}$ than for a doublet system, owing to the stronger coupling of the two spin system to the radiation field.

We now look into the kind of information obtainable from the simplest pulse sequence, $\pi/2, \tau, \pi$, in EPR. An echo appears at time τ, following the π pulse, via the refocusing, which has been described in many ways: runners on a racetrack reversing directions, ants circling, a flipped pancake. The property of interest is the dependence of the amplitude of the echo at its peak on the time τ. The experiment is carried out by imposing repeated pairs of $\pi/2$ and π pulses, the time τ being slowly advanced, and the aperture in a box car integrator simultaneously advanced to occur at time τ following the second pulse. The dependence of the peak amplitude of the echo on τ is the echo envelope. The echo envelope always decays to zero, not necessarily monotonically, at sufficiently long τ. Clearly no echo will be observed with the simple $\pi/2, \tau, \pi$ scheme if the decay time is shorter than or of the order of the dead time of the instrument. Unfortunately, many paramagnetic materials, particularly ones of biological interest, such as iron and copper compounds in liquid environments, have decay times at temperatures at which they are biologically active too short to permit their observation by echo methods. But useful information concerning biologically active molecules may be obtained under conditions in which they are not biologically active, as the extensive and valuable x-ray diffraction studies of crystalline biomolecules demonstrates.

Let us consider first studies of paramagnetic materials in solution in nonviscous fluids. The first observation of echoes from such a system is that of Blume, who studied metals in liquid ammonia (28). He used an NMR spectrometer operating at a field of about 10 Oe. Many such systems have spectra of resolved homgeneously broadened lines. To obtain echoes from them, inhomogeneous broadening is introduced either by residual inhomogeneity of the magnetic field or by intent (appropriately placed paper clips or hairpins are useful) in order to shorten the free induction decay. In liquids the echo envelope is a monotomically decaying function, often a simple exponential. The relaxation parameters obtainable from the echo envelope bear on the dynamics of the liquid system. An exposition of interpretation of echoes in liquids is given by Stillman & Schwartz (36). In principle the echo method could be used as it has been used in NMR for measurement of diffusion rates. To my knowledge, however, no successful use of ESE measurement of diffusion rates of paramagnetic species has been reported. We have failed in our laboratory, even in a favorable case—electrons in liquid ammonia—owing to an inability to achieve large enough field gradients.

In single crystals, glasses, and polycrystalline materials, the echo behavior is richer than in liquids. Both the $\pi/2, \tau, \pi$ sequence and the $\pi/2, \tau_1\pi/2, \tau_2\pi/2$ sequences usually yield echo envelopes that are not monotonic in time. They may have complicated oscillatory behavior whose Fourier transform contains frequencies of coherent processes that are induced by the pulses. These frequencies are hyperfine splittings and sums and the differences of them. Mims has given a full analysis of the oscillatory behavior of the echo envelope (32). I present here a pallid restatement of his arguments. Consider a four-level system: an electron spin and a nucleus of spin $1/2$ in interaction with an external magnetic field. The four eigenstates and their energies are

I: $\quad |\alpha\rangle|a_U\alpha' + b_U\beta'\rangle \left[\dfrac{\omega_0 + \delta_U}{2}\right]$

II: $\quad |\alpha\rangle|b_U\alpha' - a_U\beta'\rangle \left[\dfrac{\omega_0 - \delta_U}{2}\right]$

III: $\quad |\beta\rangle|a_L\alpha' + b_L\beta'\rangle \left[-\dfrac{\omega_0 + \delta_L}{2}\right]$

IV: $\quad |\beta\rangle|b_L\alpha' - a_L\beta'\rangle \left[-\dfrac{\omega_0 - \delta_L}{2}\right]$.

$|\alpha\rangle$ and $|\beta\rangle$ are the electron spin functions, $|\alpha'\rangle$ and $|\beta'\rangle$ are the nuclear spin functions, and a_U, b_U, a_L, b_L are amplitudes of the nuclear spin functions in the upper and lower states. ω_0 is the electron spin Larmor frequency; δ_U and δ_L are the hyperfine splittings in the two electronic states. The anisotropies in the system are such that at $a_L \neq a_U$; $b_L \neq b_U$; the nuclear spins are quantized in different directions in the two electronic states.

Suppose the system is originally in the lower pair of states; its density matrix is

$$\rho(0) = \begin{matrix} & \text{I:} & \text{II:} & \text{III:} & \text{IV:} \\ & \begin{pmatrix} 0 & 0 & 0 & 0 \\ 0 & 0 & 0 & 0 \\ 0 & 0 & \tfrac{1}{2} & 0 \\ 0 & 0 & 0 & \tfrac{1}{2} \end{pmatrix} \end{matrix}.$$

What is the state of the system immediately after a $\pi/2$ pulse? The $\pi/2$ pulse is characterized by $\gamma B_1 \tau = \pi/2$ where B_1 is its amplitude and τ its duration. If a weak pulse is used, i.e. one in which the Larmor frequency in the rotating frame is less than the hyperfine frequencies — $\gamma B_1 \ll \delta_U, \delta_L$, or the same statement $\tau \gg 1/\delta_U, 1/\delta_L$, a $\pi/2$ pulse can be produced only when the frequency of the irradiating field is on resonance with one of the four ESR transitions, say the I, III one. Subsequent to the pulse the density matrix evolves as (neglecting relaxation processes)

$$\begin{matrix} & & \text{I:} & \text{II:} & \text{III:} & \text{IV:} \\ \text{I:} & & \tfrac{1}{4} & 0 & \tfrac{1}{4}\exp^{i(\omega_0+\delta_L)t} & 0 \\ \text{II:} & \Bigg(& 0 & 0 & 0 & 0 \\ \text{III:} & & \exp^{-i(\omega_0+\delta_L)t} & 0 & \tfrac{1}{4} & 0 \\ \text{IV:} & & 0 & 0 & 0 & \tfrac{1}{2} \end{matrix} \Bigg).$$

Although the nuclear spin eigenfunctions in the two states mixed by the pulse are not orthogonal, because of the slowness of the process accompanying the pulse ($\tau \gg 1/\delta_U, 1/\delta_L$), the nuclear magnetization evolves adiabatically to its new state. Consider next the behavior under a strong pulse, $\tau \ll 1/\delta_U, 1/\delta_L$. If the nuclear eigenstates in the two electronic states are the same, a condition that occurs when the hyperfine field is parallel to the external field, only two electronic transitions are permitted and the density matrix subsequent to the pulse is

$$\begin{array}{c} \\ \text{I:} \\ \text{II:} \\ \text{III:} \\ \text{IV:} \end{array} \begin{pmatrix} \text{I:} & \text{II:} & \text{III:} & \text{IV:} \\ \tfrac{1}{4} & 0 & \tfrac{1}{4}\exp^{i(\omega_0+\delta_L)t} & 0 \\ 0 & \tfrac{1}{4} & 0 & \tfrac{1}{4}\exp^{i(\omega_0+\delta_U)t} \\ \tfrac{1}{4}\exp^{-i(\omega_0+\delta_L)t} & 0 & \tfrac{1}{4} & 0 \\ 0 & \tfrac{1}{4}\exp^{-i(\omega_0+\delta_U)t} & 0 & -\tfrac{1}{4} \end{pmatrix}.$$

A refocusing π pulse produces an echo whose intensity is independent of the hyperfine splittings δ_L and δ_U; the two frequencies $\omega_0 + \delta_L$ and $\omega_0 + \delta_U$ behave just as they would if they had been produced by a field inhomogeneity. If, however, all four transitions are allowed (i.e. if there is "partitioning," in the language of the trade), off-diagonal elements between all states will be produced; the zeros in the density matrix are replaced by $\exp \pm \delta_L \tau$ and $\exp \pm i\delta_U \tau$. These terms represent coherent superpositions of the two hyperfine states in each of the electron spin states. They oscillate at the frequencies of the hyperfine splittings. The coherent superpositions are produced only when both of the following conditions are fulfilled: the pulse duration is short, $\tau < 1/\delta$, and the directions of the effective fields at the positions of the nuclei in the two electronic states are not parallel. Prior to the pulse, the nuclei are in their eigenstates, quantized along the direction of their local fields; the rapid microwave pulse, through its reorientation of electron spin, produces a nonadiabatic change in the direction of the local field, following which the nuclei that are no longer in stationary states begin a free induction decay. The electron spin now evolves under an oscillating local field arising from the precessing nuclei. The amplitude of the echo depends on the phase of the local field at the instant of application of the refocusing pulse, hence an oscillatory echo envelope. The $\pi/2, \tau, \pi$ sequence yields an echo envelope that finally decays with characteristic phase memory time, T_m. The $\pi/2, \tau_1, \pi/2, \tau_2, \pi/2$ sequence decays with characteristic time of order of T_1, the spin lattice relaxation time, often much longer than T_m. T_m is related to the familiar T_2 of NMR; it is the characteristic time for true irreversible decay of coherence in a single isochromat.

Interpretation of an echo envelope is usually done by making a plausible model of the system under study, calculating its echo envelope, and adjusting its parameters to make an acceptable fit to the data. (There is no universal agreement on acceptability of fit.) The calculations may be fairly complicated. For instance, for a three-pulse sequence on an electronic doublet system in interaction with only one spin 1/2 nucleus, one transforms the four-by-four density matrix representing the initial state of the system by the matrix representing the $\pi/2$ pulse, evaluates the matrix at time τ, after the first pulse as it evolves under the Hamiltonian of the system, transforms again for the second pulse, evaluates at time τ_2, transforms by the matrix representing the third pulse, and finally evaluates the intensity of the echo at time τ_1 after the third pulse by taking the trace of product of the density matrix with the matrix representing the measured component of magnetization. For systems with many nuclei, in particular ones with quadrupole moments, the calculations become lengthy.

The bulk of the work on ESE of stable paramagnetic systems has been done by two groups—those of Mims and Kevan. Mims and co-workers, in addition to having made many of the important experimental advances, have published a series of papers displaying the power of the ESE method for detecting structural features of complicated molecular aggregates. They have successfully exploited Mims' method for measuring shifts in resonance frequencies linear in applied static electric fields (37, 38). The effect is especially valuable in structural determinations, since its magnitude depends on departures from centrosymmetric symmetry about the paramagnetic center. The method has been applied to rare earths (39), cytochromes (40, 41), hemoproteins (42, 43), to the copper-containing complexes laccase and ceruplasim (44), iron-sulfur proteins (45), copper complexes with imidazole (46), to the copper in a cytochrome heart oxidase (47), to copper, cobalt, and iron complexes of bleomycin (48), iron imidazole complexes (49), to the copper in superoxide dismutase (50), to the non-blue copper protein galactose oxidase, entirely different from the blue copper protein (51), and to many others. In addition to the electric field shifts, they have studied, via the echo modulations, the hyperfine interactions with ^{14}N and ^{15}N, ^{1}H and ^{2}H, as well as those of the paramagnetic ion. A review of the work is given in Ref. (52).

Mims' work has not been confined to molecules with biological-sounding names. Copper in binary oxide glasses (53) and rare earths in frozen water, alcohol, or glycerol (54, 55) have also been studied.

Kevan's group has used ESE echo modulation for probing the environments of various paramagnetic species. They have studied, via modulation of the echo envelope, electrons formed by radiolysis of frozen media. They have studied electrons in frozen deuterated sodium hydroxide in deuterated water (56), in methyl tetrahydrofuran (57), the aqueous matrices (58), and in ethanol (59). They have investigated the environments of silver atoms formed by radiolysis of silver salts in various matrices (60, 61). One of their studies demonstrates how the environment of silver atoms formed by radiolysis at 4.2 K changes on annealing at 77 K and recooling to 4.2 K (62). The environment immediately after radiolysis is interpreted as representing that of the silver ion prior to radiolysis. After annealing, the environment adjusts itself to the equilibrium environment for neutral silver atoms. One of the studies by the Kevan group reveals the sensitivity of the echo modulation pattern to small structural differences. The conventional EPR spectra of the radical CH_2OH on various zeolites are indistinguishable, yet their modulation patterns are dramatically different (62). One of the most detailed studies of the structure of the medium around a trapped electron has been that of Kevan and co-workers on electrons in glasses of the deuterated ethylalcohols: CH_3CH_2OD, CH_3CD_2OH, and CD_3CH_2OH (63).

In the analysis of the echo envelope one tries to disentangle the oscillatory features from the decay. A method for analysis of the modulation independent of the decay function has been devised by Ichikawa, Kevan, Bowman, Dikanov, and Tsevtkov (64). Effects of quadrupole couplings have also been discussed (65).

ESE spectroscopy has shown some promise for obtaining information concerning molecular motions. It has been applied to nitroxide spin labels (66) in an attempt to obtain rotational correlation times from phase memory times. It has also been used in a version of saturation transfer spectroscopy to investigate molecular motions with characteristic times in the range 10^{-5} sec (67).

We now consider the application of ESE spectroscopy to the study of transient species. Although diphenylmethylene may not qualify as "transient" because of its long lifetime in rigid media at low temperatures, some of the studies of it merit consideration. As mentioned above, the nature of its photolytic birth has been investigated by ESE and transient nutation methods. When it is formed in crystalline benzophenone at helium temperatures, it displays modulations of its echo envelope (68). Doetschman et al have argued that part of the modulation arises from interaction with nitrogen nuclei in the dinitrogen that had been expelled from the parent diphenyldiazomethane (69). T. S. Lin (private communication) has confirmed Doetschman's conjecture; annealing the crystals and bringing them back to helium temperature produces changes in the modulation consistent with the nitrogen having diffused away from the methylene during the annealing. In benzophenone glass, however, there is no observable coupling to nitrogen (71).

Among the studies of ESE of transient systems are those carried out at the Argonne Laboratory, mainly on photosynthetic systems. The usefulness of the ESE method was demonstrated in a study of CIDEP in the paramagnetic species produced by the photolytic reaction between quinones and alcohols. No modulation of the echo envelope was found; the dependence of the echo intensity on the time between photolytic flash and a pair of microwave pulses was recorded (72). The Argonne apparatus has been described (73) and reviews of the work published (74, 75). One of the most striking findings has been that the phase of the echo from one of the members of the radical pair that is produced in photosynthesis is shifted from the normal one. The echo experiments are carried out with coherent systems—the pulses are chopped out of a continuously running local oscillator. The peculiar phase shift was explained as arising from modulation of the evolution of one member of the radical pair, the longer-lived one, by scalar coupling to its shorter-lived partner (76).

The final applications of ESE to transient species that I cite are to studies of photoexcited triplets. A study of a short-lived (~ 100 μsec) triplet—that of pentacene in a single crystal host of p-terphenyl—at room temperature has given a rich harvest of data: structural information from echo envelope modulation and kinetic data from phase memory times and from dependence of echoes on time between the photolyzing pulse and application of the microwave pulses (77). A vivid example of the effect of partitioning on the echo envelope is seen in the differences between the echo envelopes of the two triplet transitions at fixed orientation of the molecules. In the $M_s = 0$ state of the triplet, the effective field at the nuclei is the external field. In the $M_s = \pm 1$ state, it is the vector sum of the external field and the hyperfine fields. The hyperfine fields have opposite orientation in $M_s = +1$ and $M_s = -1$. In one they combine to yield an effective field turned far from the direction of the external field, in the other the fields are closer to parallel, which in turn leads to a high amplitude nuclear spin procession in one and not in the other.

Finally, I discuss two works that do not fit into the arbitrary classification I have made:

The first is the observation by the workers at Leiden that a light flash alone is capable of evolving a precessing magnetization and free induction decay (78, 79). The system is tetramethyl pyrazine in a crystalline durene host. At zero external magnetic field, two of the triplet eigenstates are populated following photoexcitation. Are they formed via a single channel in a coherent superposition, or by separate incoherent channels? The single channel alternative is described by a density matrix with nonvanishing off-diagonal elements; the two separate channels alternative is described by a diagonal density matrix of the system immediately after the flash. The experimental results decide in favor of the first possibility. Another study of anthracene that also has two zero field levels populated after the flash showed no coherence (80).

The second work, from Novosibirsk, I cite because of its demonstration of the sensitivity that can be achieved in some cases. Anion-cation radical pairs formed by radiolysis recombine to leave one member of the radical pair in an excited singlet state that radiates its characteristic fluorescence. The intensity of the fluorescence is monitored as the system is swept through resonance in an EPR apparatus. At resonance, the rate of production of singlets from the original singlet-triplet mixture of the radical pair is changed by the resonant microwave field with a corresponding change in fluorescence intensity. The steady number of radical pairs in the cavity was about 20 and they were detected with a signal to noise ratio of about 10 (81).

Literature Cited

1. Freed, J. 1972. *Ann. Rev. Phys. Chem.* 23:265–311
2. Kevan, L., Schwartz, R. N. 1979. *Time Domain Electron Spin Resonance*. New York: Wiley
3. Fessenden, R. W., Schuler, R. H. 1973. *J. Chem. Phys.* 39:2147
4. Verma, N. C., Fessenden, R. W. 1973. *J. Chem. Phys.* 58:2501
5. Fessenden, R. W. 1973. *J. Chem. Phys.* 58:2489
6. Verma, N. C., Fessenden, R. W. 1976. *J. Chem. Phys.* 65:2139
7. Fessenden, R. W., Verma, N. C. 1976. *J. Am. Chem. Soc.* 98:243
8. Glarum, S. H., Marshall, H. 1970. *J. Chem. Phys.* 52:5555
9. Kim, S. S., Weissman, S. I. 1978. *Chem. Phys. Lett.* 58:326
10. Wong, S. K., Wan, J. K. S. 1972. *J. Am. Chem. Soc.* 94:7197
11. Atkins, P. W., Dobbs, A. J., McLauchlan, K. A. 1974. *Chem. Phys. Lett.* 25:105
12. Torrey, H. C. 1949. *Phys. Rev.* 76:1059
13. Abragam, A. 1961. *The Principles of Nuclear Magnetism*, p. 68, London: Oxford Univ. Press
14. Rabi, I. I., Zacharias, J. R., Millman, S., Kusch, P. 1938. *Phys. Rev.* 53:318
15. Hore, P. J., McLauchlan, K. A. 1979. *J. Magn. Res.* 36:129
16. Hore, P. J., McLauchlan, K. A. 1981. *Mol. Phys.* 42:533
17. Hore, P. J., McLauchlan, K. A. 1981. *Mol. Phys.* 42:1009
18a. Smith, R. A. 1978. *Proc. Roy. Soc. London Ser. A* 362:13
18b. Smith, R. A. 1979. *Proc. R. Soc. London Ser. A* 368:163
18c. Smith, R. A. 1980. *Proc. R. Soc. London Ser. A* 371:319
19. Furrer, R., Fujara, F., Lange, C., Stehlik, D., Vieth, H. M., Vollman, W. 1980. *Chem. Phys. Lett.* 75:323
20. Kim, S. S., Weissman, S. I. 1979. *Rev. Chem. Intermed.* 3:107
21. Furrer, R., Fujara, F., Lange, C., Stehlik, D., Vieth, H. M., Vollman, W. 1980. *Bull. Magn. Res.* 2:114
22. Wan, J. K. S., Wong, S. K., Hutchinson, D. A. 1974. *Acc. Chem. Res.* 7:58
23. Doetschman, D. 1976. *J. Phys. Chem.* 80:2167

24. Chisholm, W. P., Weissman, S. I., Burnett, M. N., Pagni, R. M. 1980. *J. Am. Chem. Soc.* 102:7104
25. Kothe, G., Kim, S. S., Weissman, S. I. 1980. *Chem. Phys. Lett.* 71:445
26. Kim, S. S., Weissman, S. I. 1979. *J. Am. Chem. Soc.* 101:5863
27. Trifunac, A. D., Lawler, R. G. 1981. *Chem. Phys. Lett.* 84:515
28. Blume, R. V. 1958. *Phys. Rev.* 109:1867
29. Mims, W. B., Nassau, K., McGee, J. D. 1961. *Phys. Rev.* 123:2059
30. Brown, I. M. 1979. See Ref. 2, Ch. 6
31. Brändle, R., Kruger, C. J., Müller-Warmuth, W. Z. 1970. *Naturforschung A* 24:1
32. Mims, W. B. 1972. In *Electron Paramagnetic Resonance*, ed. S. Geschwind, pp. 263–351. New York: Plenum
33. Semenov, A. G., Schirov, M. D., Zhidkov, V. D., Khmelinskii, V. E., Dvornikov, Z. V. 1980. *Inst. Chem. Kinet. Combust.*, Novosibersk, Preprint 3
34. Hahn, E. L., Maxwell, D. E. 1952. *Phys. Rev.* 88:1070
35. Slichter, C. P. 1978. *Principles of Magnetic Resonance*. New York: Springer Verlag
36. Stillman, A. E., Schwartz, R. N. 1981. *J. Phys. Chem.* 85:3031
37. Mims, W. B. 1974. *Rev. Sci. Instrum.* 45:1583
38. Mims, W. B. 1976. *The Linear Electric Field Effect in Paramagnetic Resonance*. Oxford: Clarendon
39. Mims, W. B., Mashur, G. J. 1972. *Phys. Rev. B* 5:3605
40. Peisach, J., Mims, W. B. 1973. *Proc. Natl. Acad. Sci. USA* 70:2972
41. Mims, W. B., Peisach, J. 1974. *Biochemistry* 13:3346
42. Mims, W. B., Peisach, W. J. 1975. *Proc. 18th Congr. Ampere*. Amsterdam: North Holland
43. Mims, W. B., Peisach, J. 1976. *J. Chem. Phys.* 64:1074
44. Mondovi, B., Graziani, M. T., Mims, W. B., Oltzik, R., Peisach, J. 1977. *Biochemistry* 16:4198
45. Peisach, J., Orme-Johnson, N. R., Mims, W. B., Orme-Johnson, W. H. 1977. *J. Biol. Chem.* 252:5643
46. Mims, W. B., Peisach, J. 1978. *J. Chem. Phys.* 69:4921
47. Mims, W. B., Peisach, J., Shaw, R. W., Beinert, H. 1980. *J. Biol. Chem.* 255:6843
48. Burger, R. M., Adler, A. D., Horwitz, S. B., Mims, W. B., Peisach, J. 1981. *Biochemistry* 20:1701
49. Peisach, J., Mims, W. B. 1977. *Biochemistry* 16:2795
50. Fee, J. A., Peisach, J., Mims, W. B. 1981. *J. Biol. Chem.* 256:1910
51. Kosman, D. J., Peisach, J., Mims, W. B. 1980. *Biochemistry* 19:1304
52. Mims, W. B., Peisach, J. 1979. In *Biological Applied Magnetic Resonance*, ed. R. Shulman. New York: Academic
53. Mims, W. B., Petersen, G. E., Kurkjian, C. R. 1978. *Phys. Chem. Glasses* 19:14
54. Mims, W. B., Davis, J. L. 1976. *J. Chem. Phys.* 65:3266
55. Mims, W. B., Davis, J. L. 1976. *J. Chem. Phys.* 64:4836
56. Bowman, M. K., Kevan, L., Brown, I. M. 1973. *Chem. Phys. Lett.* 22:16
57. Kevan, L., Bowman, M. K., Narayana, P. A., Boeckman, R. K., Yudanov, V. F., Tsevtkov, Y. D. 1975. *J. Chem. Phys.* 63:409
58. Narayana, P. A., Bowman, M. K., Kevan, L., Yudanov, V. F., Tvetskov, Y. D. 1975. *J. Chem. Phys.* 63:3365
59. Narayana, M., Kevan, L. 1980. *J. Chem. Phys.* 72:2891
60. Narayana, M., Li, A. S. W., Kevan, L. 1981. *J. Phys. Chem.* 85:132
61. Ichikawa, T., Kevan, L., Narayana, P. A. 1979. *J. Chem. Phys.* 71:3792
62. Ichikawa, T., Kevan, L. 1980. *J. Am. Chem. Soc.* 102:2650
63. Narayana, M., Kevan, L. 1981. *J. Am. Chem. Soc.* 103:1618
64. Ichikawa, T., Kevan, L., Bowman, M. K., Dikanov, S. A., Tsvetkov, Y. D. 1979. *J. Chem. Phys.* 71:1167
65. Narayana, P. A., Kevan, L. 1977. *J. Magn. Res.* 26:437
66. Madden, K., Kevan, L. 1980. *J. Phys. Chem.* 84:2691
67. Dzuba, S. A., Sailikhov, K. M., Tsevtkov, Y. D. 1981. *Chem. Phys. Lett.* 79:568
68. Cheng, C. P., Lin, T. S., Sloop, D. J., 1976. *Chem. Phys. Lett.* 44:576
69. Doetschman, D., Fierstein, D. E., Michaelis, J., Desantolo, A. M., Utterback, S. C. 1980. *Chem. Phys. Lett.* 79:539
70. Deleted in proof
71. Lin, T. S., Bowman, M. K., Norris, J. R., Closs, G. L. 1981. *Chem. Phys. Lett.* 78:283
72. Trifunac, A. D., Norris, J. R. 1978. *Chem. Phys. Lett.* 59:140

73. Norris, J. R., Thurnauer, M. C., Bowman, M. K., Trifunac, A. D. 1979. *Energy Res. Abstr.* 4: Abstr. No. 8888
74. Norris, J. R., Shipman, L., Thurnauer, M., Bowman, M. K. 1981. *Bull. Magn. Reson.* 2:52
75. Norris, J. R., Thurnauer, M. C., Bowman, M. K. 1980. *Act. Biol. Mol. Phys.* 17:365
76. Thurnauer, M. C., Norris, J. R. 1980. *Chem. Phys. Lett.* 76:557
77. Sloop, D. J., Yu, H. L., Lin, T. S., Weissman, S. I. 1981. *J. Chem. Phys.* 75:3746
78. Nonhof, C. J., Plantenga, F. L., Schmidt, J., Varma, C. A. G. O., Van der Waals, J. H. 1979. *Chem. Phys. Lett.* 60:353
79. Van der Waals, J. H. 1980. *J. Mol. Struct.* 59:259
80. Felix, C. C., Weissman, S. I. 1975. *Proc. Natl. Acad. Sci. USA* 72:4203
81. Anisimov, O. A., Grigoryants, V. M., Molchanov, V. K., Molin, Y. N. 1979. *Chem. Phys. Lett.* 66:265

SOME PHYSICAL STUDIES OF THE CONTRACTILE MECHANISM IN MUSCLE

Manuel F. Morales, Julian Borejdo, Jean Botts, Roger Cooke, Robert A. Mendelson, and Reiji Takashi

Cardiovascular Research Institute, University of California, San Francisco, California 94143

Introduction

Research of the 1940s and 1950s (1–4) established that, on the μm scale, muscles shorten because an exergonic reaction—hydrolysis of adenosine triphosphate (ATP)—causes filament assemblies of myosin and of actin to translate by one another. In the myosin filaments the stems ["light meromyosin" (LMM)] of Y-shaped myosin molecules are aggregated to form the filament cores, but the arms of the Ys issue radially from the filaments, ending near the adjacent actin filaments that are in parallel array. Each myosin arm—a "cross bridge"—consists of a long slender "S-2" segment proximal to the filament core and a distal globular "S-1" segment. After it was discovered that the S-1 segment bears a binding site for actin and an ATPase site (5), and that relative to the filament (or muscle fiber) axes the cross bridges assume different attitudes in different physiological states (6), the idea arose that a cross bridge, bearing on the adjacent actin and supplied with ATP, is the unitary engine of muscle.

In muscle shortening, myosin and actin filaments translate past one another, but within their respective filaments, myosin and actin molecules hold their places; so excluding long-range elasticity, it is reasonable to guess that rotation will be a component of whatever motion the unitary engine executes. Furthermore, while the filaments translate distances of order 1 μm the "throws" of the cross bridges are of order 0.01 μm. For these reasons it is plausible that the impulsive cycle of a cross bridge is a repetitive rotation. Evidence for repetitiveness has indeed appeared (7). H. E. Huxley (8) discussed models of the unitary engine, including one in which myosin S-1 segments "rolled" on the adjacent actins. A. F. Huxley and R. M. Simmons selected this particular model, expressed it in more molecular terms, and supported it with mechanical data (9). In 1972, we (10, 11) suggested that ATPase chemistry could be incorporated into the model by assuming a correspondence between enzymatic states (in the degradation of ATP by S-1) and attitudinal states of S-1 relative to the fiber axis. Of course more fundamental than these attitudes are the angles relating S-1 and actin (12). Also, although it is easier to speculate about the motions of rigid bodies, protein chemistry suggests that myosin S-1 and actin may also undergo local deformations during their interactions. With these extensions and provisos understood, the Huxley-Simmons model of the cycling cross bridge, or variants thereof, continues to stimulate mechanistic investigations of contractility. In the following pages we review some essential features of the unitary engine, emphasizing physical approaches that our laboratory has employed to investigate it.

Protein Morphology

Because the force-*generating* (as distinct from force-transmitting) interactions occur between actin and myosin S-1[1], we ignore the morphology of the S-2 and LMM segments, both of which appear to be "coiled coils" of largely α-helical strands.

Although lens aberrations reduce its resolution (from 0.004 to 1 or 2 nm) and staining or shadowing artifacts arise in specimen preparation, much of the present morphology of actin and S-1 comes from electron microscopy. A recent development (14, 15) is the inference of 3-D structure from appropriate micrographs, capitalizing on the great depth of focus of the EM. By using the symmetry of repeating units within the specimen, one estimates the number of views required for deducing the structure of the repeating unit. The output of a density scan of the specimen is digitized, then Fourier analyzed into a set of phases and amplitudes. Density not predicted by the symmetry is removed, and several micrographs may be averaged (16). Fourier inversion is then performed using the symmetry of the object much as in x-ray diffraction analysis. Unlike the latter, phases are in the original digitized image.

Recent 3-D reconstructions (17–19) suggest that in the "thin" (actin) filament, monomers are arranged in two intertwined right-handed helical strands with a 70–76 nm pitch and a near repeat of 35–38 nm. The monomers are also on a "genetic" helix of 5.9 nm pitch and a unit axial

---

[1]For a divergent view see Ueno & Harrington (13).

translation of 2.75 nm. The latter helix has ca. 13 monomers in 6 repeats so that successive monomers rotate by 166°. From reconstruction work the actin monomer is now thought to be a pear-like object (20).

Independent information comes from 2-D actin sheets with crystalline order (21, 22). In these the monomer in projection is 5.6×3.3 (nm)^2 with a pronounced (projected) mass asymmetry along the long axis and a major and minor mass peak at each end. Also, crystallographic analysis of actin-DNAase crystals (23) at 0.6 nm resolution shows asymmetries along the long axis, but yields somewhat different dimensions. These newer results do not give the absolute orientation of the monomer in the filament helix, so functionally interesting regions (the binding sites) are not yet located on the monomer.

Crystallographic observations of S-1 have not been made, so its shape is known only at low (>3 nm) resolution from EM, hydrodynamic, and solution x-ray scattering observations. Individual myosin molecules were visualized earlier (24), but the definition of S-1 shape has been improved by 3-D reconstruction of "decorated" thin filaments (many S-1s attached to a filament) (17). S-1 appeared like a cupped hand wrapped around the filament axis at an inclination of about 45°. The inclination and the wrap-around feature combined to generate successive "arrowheads" along the filament when viewed in projection. With the adoption of an independently established shape for actin, the attitude of bound S-1 and the boundary of S-1 in the acto-(S-1) complex could be located (17).

Measurements of the rotational diffusion coefficient by fluorescence methods (see below) also suggested that the S-1 particle was asymmetric and had an axial ratio greater than 3.5 if considered a prolate ellipsoid of revolution (25). From combining several hydrodynamic data it was argued (26) that S-1 was more oblate than prolate. Less simple shapes began to emerge from rotary shadowed (27) and negatively stained (28) preparations. In these, S-1 appeared pear-shaped, 19 nm in longest dimension, with mass concentrated in the distal region.

The best present estimates of S-1 shape come from combining x-ray scattering observations from solutions of native, active S-1 and from the EM reconstructions of others. From the small angle scattering it was found (29) that the S-1 radius of gyration is 3.24 nm. To have such a radius of gyration and a mass of the order of 10^5 daltons the S-1 particle has to be quite asymmetric. Subsequently, larger angle data were obtained (30) and Fourier-transformed to get the distribution of chords. The maximum chord is 12 nm. Various shapes were tested in regard to how well they predicted the larger angle data. The best fit was given by an S-1 shape deduced from an EM image reconstruction (J. Seymour and E. J. O'Brien, unpublished).

On recent reexamination (32, and Figure 1) it was discovered that previous work (17) had misplaced the boundary of S-1 in the acto-(S-1) complex. When this defect was corrected, the S-1 shape and position were significantly changed. The S-1 shape deduced by Taylor & Amos (32) also fits the solution scattering data rather well (33). There are other reconstructed acto-(S-1) images (34). In these images S-1 appears to have regions of variable electron density, interpreted by the authors as distinct domains (see below). A second novel feature of these images is that S-1 is not tilted relative to the filament axis, even though "arrowheads" are seen in conventional projection. It is argued that "arrowheads" arise solely from S-1 bending or slewing around the axis, and that the fundamental motion of S-1 during impulsion is not tilting, but a largely azimuthal movement. These contentions are subjects of lively debate. The reconstructions of Taylor & Amos (32) certainly retain the tilt feature. Also, by using the new low-temperature fixation technique, decorated filament images have been produced devoid of arrowheads; this leads to suggestions that arrowheads are an artifact of negative staining (35).

In summary, S-1 probably has the shape deduced in image reconstructions (Figure 1)—an elongate (ca. 12 nm), somewhat oblate object with more of its mass nearer the actin binding site. Some important consequences follow. If such an object does tilt during impulsion, its "throw" would be of the order of 10 nm, as mechanical experiments (36) seem to require. An object of this dimension could nearly span the interfilament distance, and be within thermal vibration distance of a binding site on actin. A special, Ca^{2+} induced, "reach out" process (37) need not be invoked (38). Finally, since nucleotide binding has little effect on the rotational diffusion coefficient of S-1 (39), and since the shape of native S-1 free in solution (x-ray scattering data) can be very plausibly reconciled with the shape of S-1 bound to actin (EM images), morphological investigations at the present resolution give no support to theories of impulsion (see below) in which S-1 suffers significant changes in shape. States in which both actin and nucleotides are bound to S-1 have not been studied extensively, and present resolution cannot exclude changes in local conformation.

Segmental Flexibility of Myosin

Because various proteolytic enzymes first cut myosin at the same two places—at the region between S-1 and S-2, and between S-2 and LMM —it has long been surmised that these are noncompact, and therefore potentially flexible places. Flexibilities at these places have played important roles in theories of myosin function (40), and thus their experimental demonstration has been correspondingly significant.

Examination for (S-1)-(S-2) flexibility has been conducted (25) using the technique of time-resolved fluorescence anisotropy decay, or TRFAD (see 41 for a fuller account).

In TRFAD experiments, the sample is a population of rotating fluorophores at the origin of coordinates, O. The fluorophores (intrinsic like tryptophane of a protein, or extrinsic like a dye attached to a protein) are

Figure 1 A 3-D model of the complex of F-actin ("A") decorated with S-1, deduced by reconstructing images from electron micrographs of uranyl acetate-stained acto-(S-1), according to Taylor & Amos (32).

imagined to be rigidly attached to macromolecules. In classical parlance, each fluorophore, say the nth, has imbedded absorption and emission axes. Along one of these directions ($\hat{\alpha}_n$), dipolar oscillations are excited by the electric vector of the incident exciting light; along the other (direction $\hat{\varepsilon}_n$) occur the dipolar oscillations that generate fluorescence emission. In general, $\hat{\alpha}_n \neq \hat{\varepsilon}_n$, but for simplicity we equate them here. Suppose that at $t=0$ such a population of "labeled macromolecules" receives a pulse of exciting light traveling from $+X$ to 0. The pulse is polarized, with its electric vector along a unique direction in the normal plane, say $\hat{k}$. The probability that the nth attached fluorophore will be excited is $[\hat{k}\cdot\hat{\alpha}_n(0)]^2$. Obviously, labeled macromolecules whose $\hat{\alpha}$s were at or near the Z-direction will be preferentially excited. Now consider the observation of the anticipated fluorescence at $+Y$, supposing that, at will, photons polarized along the $\hat{k}$- or $\hat{i}$-directions can be detected. If the nth macromolecule was excited at $t=0$, it can emit a photon at any later time, t_n (at the "excited state lifetime", τ, for the "average" macromolecule). The probability that it will contribute to observed intensity in the Z-direction ($I_\parallel$) will be $[\hat{k}\cdot\hat{\varepsilon}_n(t_n)]^2$, and the probability that it will contribute to observed intensity in the X-direction ($I_\perp$) will be $[\hat{i}\cdot\hat{\varepsilon}_n(t_n)]^2$. If we invoke the simple case, $\hat{\alpha}_n = \hat{\varepsilon}_n$, then the contribution of the nth macromolecule to $I_\parallel$ will be $[\hat{k}\cdot\hat{\alpha}_n(t_n)]^4$, and that to $I_\perp$ will be $[\hat{i}\cdot\hat{\alpha}_n(t_n)]^4$. Between $t=0$ and $t=t_n$ the nth macromolecule will rotate in Brownian motion, and $\hat{\alpha}_n$, initially quasiparallel to k, will change in a succession of random displacements. If t_n is short, however, $\hat{\alpha}_n(t_n)$ will not be too different from $\hat{\alpha}_n(0)$, and $[\hat{k}\cdot\hat{\alpha}_n(t_n)]^2$ will be large compared to $[\hat{i}\cdot\hat{\alpha}_n(t_n)]^2$. But if t_n is very long the two contributions will be very nearly the same. The same conclusion holds on summing over all the labeled macromolecules: If the excitation of the sample is initially in the Z-direction, soon after the flash $I_\parallel$ will be large compared to $I_\perp$, but as time passes they will tend to equalize. Realizing that, by symmetry, oscillations in the (unobserved) Y-direction must equal those in the X-direction, one surmises that the total intensity, I_{tot}, of emission must be $I_\parallel + 2I_\perp$. Furthermore, a very useful quantity for describing the observations in a TRFAD experiment is the "anisotropy,"

$$r = (I_\parallel - I_\perp)/(I_\parallel + 2I_\perp).$$

As a function of time after the flash, one expects $r(t)$ to decay monotonically to zero. But deduction of the kinetics of the decay is a problem in the rotational diffusion of the macromolecule bearing the fluorophore, dependent on the geometrical relations between the $\hat{\alpha}$ and $\hat{\varepsilon}$ axes and the principal rotational axes of the macromolecule. For a sphere,

$$r(t) = r(0)\exp(-t/\Phi); \quad \Phi = V\eta/kT = (6D)^{-1}, \qquad 1.$$

where Φ is the "rotational correlation time," V, the volume, and η, the ambient viscosity. In a general ellipsoid the three principal axes can be used as a moving coordinate system, and the $\hat{\alpha}$ and $\hat{\varepsilon}$ directions can be positioned in the system by direction cosines, α_1, α_2, α_3, and ε_1, ε_2, ε_3, respectively. For this case, it has been shown (42, 43) that $r(t)$ is the sum of five exponential terms. For an ellipsoid of revolution, $r(t)$ reduces to

$$r(t) = a(\alpha_1, \varepsilon_1)e^{-At} + b(\alpha_1, \alpha_3, \varepsilon_1, \varepsilon_3)e^{-Bt}$$
$$+ c(\alpha_1, \alpha_3, \varepsilon_1, \varepsilon_3)e^{-Ct}, \qquad 2.$$

where $A = D_\perp$, $B = 5D_\perp + D_\|$, and $C = 2D_\perp + 4D_\|$, the Ds being rotational diffusion coefficients about the minor ($D_\perp$) and major ($D_\|$) axes. The a, b, and c are constants dependent only on the direction cosines. The TRFAD apparatus records the $I_\|(t)$ and $I_\perp(t)$ required to calculate $r(t)$, but, as explained above, one can from the same experiment construct

$$I_{tot}(t) = I_{tot}(0)\exp(-t/\tau) \qquad 3.$$

where τ is the aforementioned excited state lifetime.

The two S-1 segments of myosin can each be uniquely spin-labeled (44) at their reactive thiols, "SH_1" (45) by labels of iodoacetamide reactivity. Analogously, one can uniquely fluorescence-label the segments using N-iodoacetylamino-1-naphthalene-5-sulfonate ("1,5 IAEDANS") (46, 47). The LMM segment of myosin—and therefore any structure containing it, such as intact myosin or joined S-2 and LMM (so-called "rod")—is soluble only in special salt concentrations (eg 0.6 M KCl), but self-aggregates in dilute salt or in water. The S-1 segment in the absence of nucleotides or polyphosphates forms very stable complexes with high polymers of actin. Studies of S-1 motion (25) began by recording $r(t)$ from various systems in which the S-1 segment had been labeled: myosin, HMM, S-1; also HMM + excess F-actin, and myosin aggregated in low salt concentration. When Eq. 2 was applied to S-1 data, only the first exponential term could be isolated, and therefore $D_\perp$ could be evaluated. By using Perrin's formulas and the molecular weight and specific volume of S-1, the lengths of the principal axes of the prolate ellipsoid of revolution hydrodynamically equivalent to S-1 could be estimated. The length of the major axis could not greatly exceed the known interfilament distance (in order to comply with EM pictures). This constraint could only be satisfied by assuming large direction cosines, α_1 and ε_1, i.e. by assuming that the axes of the 1,5 IAEDANS were quasiparallel to the major axis of the S-1 ellipsoid. The answer to the question of whether there was flexibility at the (S-1)-(S-2) junction emerged from comparing the rotational correlation time, Φ, when S-1 was rotating freely in solution and when it was part of HMM. It turned out that $\Phi_{HMM} > \Phi_{S-1}$, but not much greater, as would be expected from assuming an HMM molecule with various rigid connections between its segments. If the two S-1s of HMM are envisioned as attached to the stem of HMM by joints, then the motion of the two S-1 segments is akin to the motion of the two ligated pieces of a "once-broken stick" (48). In fact Φ_{HMM} and Φ_{S-1} fit the Yu-Stockmayer ratio fairly well, suggesting that the attachments of the S-1s to the stem are like molecular swivels. Recently it has been pointed out (48a) that the neglect of translational motion in the Yu-Stockmayer treatment may alter the expected $D_\perp$ somewhat; in the present case calculations show this neglect is unimportant (W. A. Wegener, private communication).

These early conclusions about (S-1)–(S-2) flexibility based on TRFAD measurements (25) have since been confirmed in various ways, for example by electric birefringence studies (49) and by visualization of a "hinge" in EM pictures (27). Also it has been found (38) that Φ for "single-headed" HMM is the same as Φ for the normal HMM with both heads attached, indicating that as the S-1 segments rotate they do not "bump" to any extent. However, the most important confirmation has come from the application of the new saturation-transfer electron paramagnetic resonance (STEPR) method by Thomas and his associates (50). Like TRFAD, STEPR indicates that the S-1 segments have independent rotational motion; but because it can sense slower motions than can TRFAD, it strengthens the latter's conclusions about the immobilization of S-1 on binding to F-actin or upon aggregation in low salt concentrations.

As remarked at the outset, early proteolytic studies (as well as guesses concerning the possible function of cross-bridges) suggested that there is a point, or short region of flexibility, somewhere between S-2 and LMM, i.e. in the isolatable fragment (S-1s removed) known as "myosin rod." This possibility was examined using the method of electric birefringence (51), which we anticipated would be best suited to sense motions with correlation times of hundreds of microseconds as might be executed by very long rods (52). To solubilize rod and LMM and yet maintain the low ionic strength enabling the electric field to operate, the fragments were dissolved in dilute solutions of ATP or pyrophosphate (53). In making the requisite preparations of rod, S-2, and LMM, it was found (54, 52) that S-2 is considerably longer than had been previously thought—the earlier estimation overlooking the fact that proteolysis not only severs S-2 from LMM but rapidly digests a substantial portion of S-2. Operationally, this leads to preparations of "long" and "short" S-2. Since both S-2 and LMM have "coiled coil" structures, direct estimates of their lengths can be made from electron microscopy and from the mass-per-unit-length measurements (55). The experiments consisted in orienting the appropriate material in a field of ca. 50 kv cm^{-1}, then suddenly annihilating the field and recording the relaxation curve, $\Delta n(t)$, of the difference of refractive indices photometrically. The data fitted best the simple relation,

$$\Delta n = \Delta n(0)\exp(-t/\tau). \qquad 4.$$

In the case of a rigid long rod, the relaxation time, τ, is related to the rotational coefficient (about its minor axis), $D_\perp$, by $\tau = (6D_\perp)^{-1}$ (56); in turn, Broersma's equation (57) relates $D_\perp$ to the dimensions of the rod. From the electric birefringence data, an LMM length of 79.1 nm could be calculated, in excellent agreement with an EM length of 78.5 nm. On the other hand, the calculated lengths for the rod (121.9 nm) and for long S-2 (58.7 nm) were statistically significantly less than the corresponding EM lengths (136.0 nm and 65.0 nm, respectively). This was taken to mean that there is in the S-2 portion of the rod a region in which bending motion is possible (52). Presumably this region lies in the "difference piece" between long and short S-2. (A recent correction (57a) of the Broersma equation gives a rod length slightly longer than the one calculated here.) It has been found (58) that the amino acid sequences near the N-termini of long and short S-2 are very similar, showing that the difference piece fits between short S-2 and LMM (58). The question of whether there are bend points or regions in myosin rod has also been investigated by electron microscopy (27). From examining a large number of micrographs it has been found that the most frequently observed bend point in a rod occurs at 42.4 nm from the (S-1)-(S-2) junction. Using the length-per-unit-mass value for a coiled coil one can calculate its length to be 41 nm; therefore, as expected, the modal bend point (27) lies in the difference piece.

The foregoing studies directly support the thought that some sort of flexibility also exists at the (S-2)-LMM junctional region[2]; however, the stiffness of this joint, or the amplitude through which LMM can turn when S-1 and S-2 are fixed, remains unknown. As is seen in subsequent sections, what has been learned about flexibility in the (S-1)-(S-2) and

[2] Very recently the systems studied by electric birefringence have also been studied by dynamic light scattering (59). The qualitative conclusion that LMM is stiff but that some region of S-2 can bend has been confirmed. But the new study includes modeling of myosin rod as a "once-broken stick," and concludes that the two segments sweep in cones whose minimum angle is 128°. NMR investigation, however, provided no evidence for extensive regions of random coil character. The problem has also been investigated by labeling myosin rod with dansyl sulfonyl chloride and studying its TRFAD. By this method flexibility has been found at low pH, but not (or nearly not) at neutral pH (60).

(S-2)-LMM regions is consistent with theories of cross-bridge function in which the bending of S-2 away from the thick filament allows (S-1)-actin contact despite variable interfilament spacing, and in which S-1 rotates in impulsion.

Measuring the Binding of Myosin Fragments to Actin, Using TRFAD

The thermodynamics and kinetics of actin-myosin binding processes are highly important subjects in contractility research, but theoretically sound methods, applicable to μM concentrations of the participants, have been hard to develop. Binding is detectable by viscosity and light scattering, or even simple turbidimetric techniques, but it is hard to extract stoichiometry from such observations because the theories of these methods for very large, often polydisperse, macromolecular systems are very complex. Sedimentation analysis, using sensitive UV-optics (61) or radioactively labeled constituents (62), has proved to be a reliable method, though rather laborious and—in company with all transport methods—open to the question of whether binding and sedimentation "couple." The aformentioned experiences with TRFAD, in particular the observation that S-1 rotation ceases ($\Phi \to \infty$) upon addition of excess F-actin, led to experimentation with TRFAD as a means of studying binding equilibria in dilute solutions of muscle proteins. Of course, the use of "steady-state" fluorescence depolarization effects has often been made in the past, but a time-resolving apparatus renders the approach much more convenient.

In the preceding section it is rationalized that for a pure component, namely S-1,

$$r_f(t) = r_f(0)\exp(-t/\Phi_f), \qquad\qquad 5.$$

where the subscript f means "free." In a binary mixture some of the S-1 will be bound to unlabeled (and thus unseen) F-actin. Bound component (subscript b) will generate quite a different decay curve, say $r_b(t)$. The observed anisotropy of the mixed system, $r_{sys}(t)$, will not in general be separable into r_f and r_b. However, as first shown by Wahl (63), separation is possible if throughout the decay the total molar fluorescence intensity emitted by the two components is the same. In other connections, changes in excited state lifetime, τ, of IAEDANS-labeled S-1 upon ligating actin or nucleotide have been measured, and found to be about 1% (39). If these negligibly small differences hold throughout the decay, it is valid to write,

$$r_{sys}(t) = x_f r_f(t) + x_b r_b(t), \qquad\qquad 6.$$

where x_f and x_b are mole fractions, so that $x_f + x_b = 1$. Unlike the case of sedimentation, it is certain that decay rate, Φ^{-1}, is very much faster than the association or dissociation rates, so in Eq. 6 it is safe to insert for r_f and r_b the functions observed in the absence of interactions. Φ_f is of the order of 10^2 nsec, but because of the length and mass of F-actin, $\Phi_b > 10^4$ nsec, so $r_b \cong r_b(0)$. In principle, $r_b(0) = r_f(0)$. With these acceptable approximations, Eq. 6 was used to find x_f and thus the association constant of S-1 to F-actin in very dilute solution (64). Alternatively, from Eq. 6 the expression for $\ln[r_{sys}(t)]$ can be differentiated to give

$$x_b^{-1} = 1 + \left\{ \left(\Phi_{sys}^{-1} - \Phi_b^{-1} \right) / \left(\Phi_f^{-1} - \Phi_{sys}^{-1} \right) \right\} \exp\left\{ \left(\Phi_f^{-1} - \Phi_b^{-1} \right) t_m \right\}. \qquad 7.$$

where $-\Phi_{sys}^{-1}$ is the measurable slope over an essentially linear portion of the logarithmic decay curve at a representative time, t_m, in that linear range (65).

We stress in this review that an omnipresent danger in methods employing labeled proteins is that the labeling itself may affect the intended measurement. It has been suggested that in affinity measurements this danger can be examined by investigating whether varying the percent labeling affected estimates of x_f (64); it does not. This problem was further pursued (66), and it was concluded that the test could detect per se effects of labeling when the affinity constants of labeled and unlabeled participants differ by a factor greater than two. A further reassurance of this and nonlabeling sedimentation methods is that they yield results in quantitative agreement with each other (61, 67).

The foregoing method of estimating myosin-actin affinity, K, using TRFAD has yielded several results pertinent to the study of contractile mechanisms. For example, under quasiphysiological conditions, the free energies of A-binding and N-binding ("A" for actin and "N" for nucleotide) are roughly comparable (64); K decreases with increasing ionic strength (64, 68); K increases with temperature (69). Some special problems have also been investigated: Since myosin is duplex and F-actin has many binding sites, the question has arisen whether the two S-1 moieties of myosin (as HMM) bind to the same or to different actin filaments (both in solution and in the organized muscle structure), and what is the functional significance of the arrangement. TRFAD is one of the methods that has been used to investigate the issue in solution, and it has been found (69) that in dilute solution, where "two-filament binding" is improbable, the second moiety binds to actin less tightly than the first.[3] Certain modifications of the S-1 particle have been suspected of

---

[3] Quantitatively, the affinities depend on the statistical treatment (70, 71), but also different affinities have been deduced for HMM using different experimental methods (cf 72).

affecting (S-1)-actin affinity. It was found using TRFAD that it did not matter for affinity whether light-chain 1 or light-chain 3 (polypeptides attached to the heavy chain of S-1) were attached to S-1 (68); however, it was found that blocking the more reactive thiols of S-1 decreases K (65). Finally, a particularly important result of using TRFAD was the demonstration that ternary, A-(S-1)-N, complexes exist (73). In Highsmith's work, N = ADP; subsequently complexes with other nucleotides have also been shown (74); it is possible that ternary complexes are significant in energy transduction (see below).

Orientation of Myosin Segments in Organized Muscle

As suggested in the Introduction, it is a popular conjecture that the *articulated* myosin cross bridge delivers to the adjacent actin filament repeated mechanical impulses timed with ATPase events. But what is the nature of the cross-bridge transition that imparts the momentum to actin? Especially following the EM work (6) and the mechanical investigations (9), attention has focused on the possibility that segmental rotation of the actin-bound S-1 moiety is the impulsive transition. In an isolated myosin molecule the S-1 moiety can rotate around the (S-1)-(S-2) junction; the question addressed in this section is whether this reorientation occurs in the organized lattice of muscle, and whether it invariably accompanies activity.

X-ray diffraction (XRD) analysis of isolated live muscles has advanced the general idea that cross bridges are involved in impulsion. Since both the myosin and actin filaments are helical arrays of the corresponding monomers, the diffraction pattern of muscle has thick filament-related and thin filament-related layer lines. In "relaxed" muscle the former are enhanced, while in "rigor" muscle the latter are enhanced at the expense of the former. That during activity cross bridges produce little diffracted intensity has been interpreted to mean that the cross bridges are in motion and therefore in axial disarray. X-ray diffraction, however, has not shed light on the EM-based proposal that rotation is the most likely motion, or on the expectation from mechanics that a rotating cross-bridge would have a "throw" of the order of 10 nm. The obstacles encountered by the classical techniques are partly technical (the EM is not applicable to dynamic systems and XRD has to be interpreted through quite elaborate models) and partly inherent in features of the specimen. One of the architectural difficulties is a mismatch between the spacings of the actin filament assembly and those of the myosin filament assembly. An inherent complexity in the dynamic system is that molecular transitions are stochastic; while this characteristic is sometimes more tractable in forced transients and oscillations or (see below) in fluctuations from a steady state, it overlays the detection problem with an analytic problem. The alternative of investigating cross-bridge orientation by using small, direction-reporting groups imbedded in the segments suspected of motion has offered means of overcoming some of these obstacles. These "probes" have either fluorescence or electron paramagnetic (EPR) properties. Of course, investigations with probes have specific difficulties of their own, to be mentioned in context. One general problem, already introduced, is the effect of the probe on the activity of the system; another is that only intrinsic probes can be used with live muscle. Work with extrinsic probes is mostly on so-called "glycerinated," "skinned," or "chemically skinned" muscle fibers. Basically, the membrane structure of the fibers must be destroyed to provide access to the contractile system. Such fibers develop tension cm^{-2} roughly comparable to live muscle when they are provided with MgATP, but legitimate questions can be raised about preservation of order in their protein assemblies. Some probe work has also been carried out with the myofibrils obtained from "live" fibers by mincing, but these specimens are very small and fragile, and do not lend themselves to simultaneous mechanical analysis.

The monomers in the filaments of a sarcomere are in helical arrays, with opposite polarities to either side of the central normal plane. If the probes attached to each monomer—be they intrinsic or extrinsic—are attached uniquely, they too will be in helical arrays of circular symmetry. To understand how the field of polarized fluorescence around the fiber reports the geometric orientation of the attached probes, the geometrical discussion above must be continued. Considering only the static (ignoring t-dependence) case and considering the "sample" to be a muscle fiber oriented along the Z-axis, one recognizes that the plane of excitation may include the $\hat{k}$-direction or the $\hat{j}$-direction. If, as in the former example, it includes the $\hat{k}$-direction, then the probability that the nth fluorophore will be excited, and also be detected along the observer's $\hat{k}$-direction, will be $[\hat{k} \cdot \hat{\alpha}_n]^2 [\hat{k} \cdot \hat{\varepsilon}_n]^2$; the contribution of the nth fluorophore to $I_{\hat{k}\hat{k}}$ (or, more conventionally, $I_\parallel$) will therefore be proportional to this expression. The unit vectors, $\hat{\alpha}_n$ and $\hat{\varepsilon}_n$ can also be written in terms of their components expressed in spherical coordinates, θ_n, the inclination to the Z-axis, and φ_n, the azimuth measured from $+X$. Because of the circular symmetry of the helix, θ_n will be the same for all the probes in one half-sarcomere. In one perfect helix, φ_n may assume only values corresponding to, say, hexagonal packing, but in a muscle fiber with $\sim 10^{13}$ S-1s the density of azimuthal directions is essentially even, so, on integrating over all the probes in a half-sarcomere, one applies $(1/2\pi)\int_0^{2\pi} d\varphi$ to the foregoing expression; in this averaging φ disappears. This integration must respect the facts that $\hat{\alpha} \cdot \hat{\varepsilon}$ is a physical constant of the fluorophore, and, in combining sarcomere halves, one half is the mirror image of the other. By these simple methods are obtained the geometric factors to which $I_{\|\perp}$, $I_{\|\|}$, $I_{\perp\|}$ and $I_{\perp\perp}$ are proportional. Assuming that the physical factors (lamp intensity, optics, etc) are the same for all directions, ratios such as $I_{\|\|}/I_{\|\perp}$ are simple trigonometric expressions from which θ can be found. A summary of this analysis appears in the Appendix to Ref. (75). It is also important to predict theoretically the effects of static and of dynamic disorder. It is important to note that except in unusual cases [such a case might be S-1 labeled with IAEDANS in which (see above) there is reason to think that $\hat{a}$ and $\hat{e}$ of the bound fluorophore are roughly parallel to the major axis of S-1] one is deriving the orientation, not of the labeled macromolecule, but of the probe (76).

The idea arose in a study of the polarized tryptophane fluorescence of muscle proteins and fibers (77) that fluorescence polarization effects provide a means of detecting differences in cross-bridge attitudes in different "physiological states" (i.e. in rigor, wherein because of the absence of ATP, most or all myosin-actin contacts are formed and the filaments thereby "cross-linked," in *relaxation*, wherein myosin ATPase is ongoing but no myosin-actin contacts exist, and in *activity*, wherein actomyosin ATPase is ongoing and the cross bridges may be in motion). In that study it was found that $P_\perp$, the polarization of fluorescence

$$[(I_{\perp\|} - I_{\perp\perp})/(I_{\perp\|} + I_{\perp\perp})]$$

correlated with physiological state, and it was surmised that this resulted from changing cross-bridge attitudes. The phenomenon was later studied in greater depth (78, 10). $P_\perp$ (activity) was found to be intermediate between distinct values $P_\perp$ (relaxation) and $P_\perp$ (rigor). Also, $P_\perp$ was studied as a function of various nucleotides, and it was shown that the phenomenon also occurred in live fibers. Aware that there is no tryptophane in S-2, these authors concluded that changes in $P_\perp$ resulted from reorientations of the S-1 segments. They and others (11) formulated the notion of "one nucleotide ligand binding—one θ." Although the tryptophane signal is strong and its use does not require chemical perturbation of the proteins, it is hard to interpret in detailed terms, basically because S-1 contains several tryptophanes in unknown arrangement and subject to local deformation by nucleotide binding. Extrinsic labeling of fibers with IAEDANS, in analogy with the myosin labeling discussed above, offered a promising alternative (79), and was soon used in a full-scale fiber study (75). Labeling whole fibers uniquely is, of course, much less certain than labeling isolated myosin, so the finding (75) that the rigor tension achieved by labeled fibers is close to that of nonlabeled fibers is technologically important. The technique of specifically and uniquely labeling the reactive thiol of S-1 in fibers was improved by the knowledge (80) that specificity is much better in relaxation than in rigor because actin-myosin contacts activate actin Cys-373. By using the analysis sketched above it was estimated (75) that in rigor, $\theta_\alpha \cong 40°$. If for IAEDANS α and ε are roughly parallel to the principal axis of S-1, 40° is approximately the inclination of S-1 itself. The observations in relaxation could not be rationalized in terms of a single θ value, however, and it was concluded that relaxation involved disordering. The problem has recently been reinvestigated by a different fluorescence technique (81). Now θ is not inferred by manipulating the 4 orthogonal Is (see above), but from the record of total emission intensity (collected by a wide aperture lens), $I_{\omega\|} + 2I_{\omega\perp}$, as a function of ω—the angle that the excitation plane makes with the Z-axis. Thus, in effect one studies the linear dichroism of the fiber at the excitation wavelength. Although corrections must be made for the collecting lens, total intensity is more accurately measured than the intensity ratios. Also, these observations have employed the new fluorophore, iodoacetyl tetramethyl rhodamine (IATR). As before, the rhodamine dipole axis is clearly placed (ca. 80°) in rigor, but in relaxation, total disordering is indicated. The activity state also seems to correspond to total disorder. A new state is seen, however, when the fiber equilibrates with MgADP. Now a distinct dichroic pattern indicates that the rhodamine dipoles are at 40° to the fiber axis. This "ADP state" is seen less well with iodoacetyl fluoresceine (IAF), and it is not seen with IAEDANS or with spin labels (see below). Finally, little tension change is recorded in the transition, rigor → ADP. Preliminary exploration by TRFAD of IATR-labeled S-1 in solution shows that ADP addition causes no change in τ. Because τ is short, measuring Φ is difficult, but no major decrease of Φ ("loosening" of the label) is evident. Observations have also been made (82) using as a fluorophore either ε-ADP or ε-ATP bound to the N-site. When ε-ADP is used, the probe dipole is at ca. 50° to the fiber axis. When ε-ATP is used (with Ca^{2+} to produce activity; with chelator to produce relaxation) disorder increases, and the center of the angular distribution shifts to ca. 60°. These results are consistent with those of dichroism in the sense that ε-ADP is a defined state, and relaxation as well as activity are disordered states; results with spin labels (below) are similarly consistent. Speculative interpretation of the foregoing depends on whether S-1 is assumed rigid or segmented. If S-1 were a rigid cylinder (or ellipsoid of revolution) whose attitude is described by φ, θ, and a "torsional" angle, ψ (describing rotation about the cylinder axis), the optic dipole of IATR may be roughly parallel to a cylinder radius while the dipoles of all the other probes are roughly parallel to the cylinder axis. From binding work (above) it is known that the A-M affinity is different from the A-M-ADP affinity; this could correspond to a difference in attitude (specifically, ψ) of attachment that IATR could sense but the other probes couldn't. Also explained would be the minimal changes in τ and Φ when ADP is added to IATR-labeled S-1 in solution. The disorder observed in relaxation (or, for a time, during the active cycle) could be partially in φ and θ, but mainly in ψ, so that layer lines (somewhat lower in intensity) of the diffraction pattern (which is insensitive to torsion) could remain. On the other hand, were S-1 segmented so that one piece maintains attitude as long as S-1 is bound to actin while the other moves under the influence of N-binding, most probe ring structures could lie on the static piece and that of IATR or IAF on the moving piece. Then, regardless of the orientation of the initial probe dipoles only IATR or IAF would sense ADP-binding to a fiber. In this model the disorder of relaxation (or activity) could be freely assigned to all three angular coordinates. As discussed above, S-1 is now known to be a "cylinder" or an "ellipsoid" only in a general sense, so the present interpretations have to be tentative, but they will be useful in considering active state data (below).

Nitroxyl-type spin labels can also sense attitude with respect to the imposed magnetic field, H (W. L. Hubbel, unpublished), for in the experimental situation H can be made to coincide with the muscle fiber axis. The interactions generating the EPR spectral lines arising in this placement of the specimen are described in the "spin-Hamiltonian" operator (83),

$$\mathcal{H} = \beta \mathbf{S} \cdot \bar{\bar{g}} \cdot \mathbf{H} + g_e \beta \mathbf{S} \cdot \bar{\bar{T}} \cdot \mathbf{I}. \qquad 8.$$

The interaction between the electron spin, $\mathbf{S}$, and $\mathbf{H}$ is described by the tensor, $\bar{\bar{g}}$, and that between the nuclear spin, $\mathbf{I}$, and $\mathbf{S}$, by the tensor, $\bar{\bar{T}}$. β is the Bohr magneton, and g_e is the g-factor for the electron. In these spin labels the electron interacts mainly with the nuclear spin of ^{14}N; for this nucleus $\mathbf{I}$ is known to be unitary $\hat{I}$. When $\mathcal{H}$ operates on the spin eigenfunctions of the electron and nucleus, the energy of the spin interactions is obtained. These interactions give rise to three energy levels, corresponding to $I_z = -1, 0, +1$, hence to three transitions in the EPR spectrum, one for each value of I_z. Both the splitting between these three spectral lines (determined by $\mathbf{I}$) and the position of their range along $\mathbf{H}$ (determined by g) are functions of the relative orientations of the spin label and $\mathbf{H}$ (or the fiber axis). The density of the unpaired electron is (approximately) cylindrically symmetric about an axis perpendicular to the plane of the label ring. We thus assume that the interactions can be described by cylindrically symmetric tensors, $T_\parallel$ and $T_\perp$ and $g_\parallel$ and $g_\perp$, where the directions are relative to the axis of symmetry of the spin label. The values of the magnetic field at which resonance occurs, H_{res}, are functions of I_z and θ, the angle between **H** and the principal axis of the spin label:

$$H_{res}(\theta, I_z) = H_{res}(0,0)\left[1 - \left[(g_\| - g_\perp)/g\right]\sin^2\theta\right]$$
$$- I_z\left[T_\|^2\cos^2\theta + T_\perp \sin^2\theta\right]^{1/2}. \qquad 9.$$

Here the gs and Ts are scalar quantities obtained from the diagonal terms of the corresponding tensors. $H_{res}(\theta, I_z)$ is three-valued, so the corresponding three peaks in the EPR spectrum are determined by θ (84).

It is important to note that if the probes are of two distinct θ-classes, the spectrum consists of two overlapping patterns, each of three lines; this is easily recognized in the spectrum; this argument can be extended to distributions. In principle, if one knows experimentally the $H_{res}(\theta, I_z)$, Eq. 9 can be inverted to find the θs or the distribution of the θs. Because the assumption of cylindrical symmetry is not precise, it is in practice easier to fit experimental spectra (more precisely, peak positions along H) by comparing with positions generated by substituting proposed θ-distributions into Eq. 9. The inference of θ-distributions, using thiol-directed spin labels (iodoacetamide reactivity, "IASL," or maleimide reactivity, "MSL") has been an important contribution of EPR work (86).

In rigor, θ(IASL) = 68° and θ(MSL) = 82°, and the distribution around these values is narrow (ca. 15°) full width at half maximum (86). In relaxation the EPR spectra indicate orientational disorder with either probe. As with some fluorescence probes, no "ADP state" is evident. Thus, qualitatively, the EPR analysis agrees closely with the foregoing fluorescence work. It adds, however, information about the disorder. The disorder in relaxation is not only extensive (the θ-distributions are roughly rectangular), but also dynamic, i.e. the saturation transfer technique (STEPR) shows that in relaxation the probes are moving with frequencies of the order of 10^5 Hz (87). A further result of EPR analysis is that contrary to some other techniques it establishes that aPP(NH)P-, or pyrophosphate-binding, produces mixed states (mixture of rigor + relaxation), not new states (86).

The Active State

As remarked in the introduction, it is currently speculated that muscle contraction results from repetitive reorientation of cross bridges, and that such motion—in contrast to the much faster Brownian motion—is ATPase-driven and slow [in a glycerinated isometrically held single fiber, of the order of 1 Hz (88, 89); in a live fiber perhaps 10 times faster (90), but probably no component process is faster than 1 kHz]. Consequently it has been a major immediate goal to obtain unequivocal experimental evidence proving that cross bridges move, and a more distant goal to show that their movement is coupled to ATPase.

Studies of transient situations have figured prominently in these attempts. Podolsky (91) first observed that if the tension of a contracting muscle is "step-changed," its shortening speed "hunts" for its new steady state value; he interpreted this in terms of moving cross bridges. A. F. Huxley & Simmons (9) found that if the length of a muscle developing isometric tension is step-changed, the tension hunts for its new value in a sequence of "phases." They suggested that in the first and fastest phase the tension drops rapidly due to a passive elasticity, possible in S-2, but that in the second phase it recovers more slowly due to a reorientation of S-1 along the θ-coordinate, under the influence of a torque generated at the A-M interface. Their data fitted these interpretations very well, even though they were not directly observing S-1. The idea that a torque on S-1 was generated at the A-M interface and not, for example, at the (S-1)-(S-2) junction received direct support when it was shown (92) that as a single fiber is stretched away from full (actin-myosin filament) overlap, its ability to transit from the relaxation value of its (tryptophane) $P_\perp$ to the activity or to the rigor value of $P_\perp$ diminished linearly with overlap. That is, actin is required for generating the torque, so it must be generated at the A-M interface.

The time resolution in taking x-ray diffraction patterns (intensity vs angle plots) has steadily improved (93–98). Studies with this technique have shown, generally speaking, that the transient onset of activity is associated with the formation of actin-myosin unions and a decay of activity with their severance; however, the diffraction pattern, while information-rich, cannot yet be interpreted so as to discern rotations. A technologically remarkable study with time resolution (ca. 1 kHz) comparable to that of the mechanical measurements was recently reported (98) in which features of the pattern and tension were recorded following step-changes in length. Somewhat mystifyingly, step-lengthening and step-shortening the muscles produced pattern changes in the same direction. Although it was not clear that the reflections followed were sensitive to the θ-variable, the authors suggested that perhaps cross bridges were being displaced symmetrically from the same distribution center to a new distribution. At this writing the knowledge emerging from the study of transients by mechanical or by x-ray methods shows that cross bridges may move, but the nature of the motion is hard to infer. In anticipation of remarks below, we note that the analytic aspect, e.g. Fourier analysis, of transient experiments has not yet been exploited.

In recent years it has become possible to observe the behavior of spin labels in the active state because of an advance (99) in which the fiber bundle is first labeled in a rather nonspecific way with the rigidly fastened MSL probe, then upon spin-annihilation with a reductant such as $Fe_3(CN)_6^{3-}$, nonspecifically placed probes are destroyed, leaving a specifically labeled fiber-bundle. When isometrically held fiber bundles of this kind are activated, the resulting spectra are well explained by assuming that during steady state activity there exist two cross bridge populations: 20–30% with orientations corresponding to rigor, and 80–70% in a situation of dynamic disorder (frequencies of 10^6 Hz) (86). These results agree with those of dichroism (81) and of fluorescence polarization (82) in the resemblance between the active and relaxed states. None of the three methods, however, is well suited to studying motions in the $0-10^3$ Hz range, such as ATPase-driven motion is likely to be.

This handicap is overcome in "correlation spectroscopy" (for reviews see 100–105). In this approach, as usual, the measured variable—e.g. tension or fluorescence intensity—is a summation of effects from molecular elements; also, as usual, the value of the variable is recorded at a succession of discrete times during a transient. But in this case, in contrast to other methodologies, the sampling rate is fast enough that it "catches" contributions from the same molecular elements more than once during the transient. If there is a priori a notion of the frequencies characterizing the phenomenon, and the sampling rate is set substantially higher than these frequencies, the approach will certainly detect at least the slower frequencies of the phenomenon. The transient situation can be experimentally arranged, for instance by "step-changing" state variables of the system, as in the aforementioned dynamic x-ray diffraction work, but it is in some ways better to make use of the "micro-transients" (random thermal fluctuations) spontaneously and continuously produced while a fiber is in "steady state" activity. To be specific, if $X_0, X_1,...,X_n$ are successive enzymatic states of the N-site of S-1, in the steady state of activity there will occur steady state or time-averaged values of the concentrations, $[X_0],[X_1],...,[X_n]$. If in addition to chemical composition these states differ from each other in another way, say by some property Ω we may expect that an external measurement of the system will record

$$([X_0]\Omega_0 + \cdots + [X_n]\Omega_n)/(\Sigma[X_i]),$$

a "steady state Ω". Clearly, number or concentration fluctuations in the $[X_i]$ due to the stochastic nature of the chemical reactions that connect the X_i into a chemical network will be reflected in fluctuations in the measured Ω. The displacements causing the fluctuations in the $[X_i]$ or the Ω_i are, of course, very small, but the "relaxation" back to the steady state that follows each such displacement carries the same empirical information (since in simple cases it has the same mathematical form) as the relaxation curve in a macrotransient. This empirical information is the "power spectrum," $S(\omega)$, a display giving for each component Fourier frequency $\nu = \omega/2\pi$ the square of the corresponding coefficient or weighting. If the experimental record of the fluctuations is $\delta(t) = \Omega(t) - \langle\Omega\rangle$, where $\langle\Omega\rangle$ is the time-average of $\Omega(t)$, then over an arbitrary time interval a computer can calculate the "auto-correlation function," $G(\tau) = \langle\delta(t)\delta(t+\tau)\rangle$. The Wiener-Khintchine theorem asserts that the Fourier transform of $G(t)$ is $S(\omega)$. In the first applications of correlation spectroscopy to the active state, Ω was the intensity of visible light scattered by an activated whole muscle (106); in our first application it was the isometric tension of an active glycerinated single fiber (107). In both cases it was hard to ascribe the frequencies recorded to motions of particular molecular entities. In the tension case, however, it was shown to be plausible to assign them to cross bridges by simulating on a computer the stochastic behavior of a model (108) (hypothesizing that tension results from cross-bridge motion) whose constants had been obtained from quite different experiments; the empirical and generated $S(\omega)$s were found to be similar. Superior identification of molecular origin became possible by fluorescence-labeling the entity of interest and using as Ω some fluorescence property. Since fractional number fluctuations fall as the inverse square of the number of elements observed, this approach requires a compromise (with detectability) number of fluorophores in the field of observation ($<10^6$) and intense (laser) illumination (with attendant problems of photoinstability). Comparison between experiment and hypothesis can proceed in much the same way, viz simulation of the supposed underlying chemical system employing separately measured rate constants in either Monte-Carlo or Poissonian assumptions (104). The first application to the contractile system was a measurement of the translational diffusion coefficient of S-1 in the presence of actin and ATP, and an effort to obtain an A-M affinity in this active situation (109). More recently, the question of whether cross bridges execute repetitive rotational motion during activity has been investigated (110). For this purpose the reactive thiols of the S-1 moieties of single glycerinated muscle fibers were labeled with (the more photostable) iodoacetylrhodamine to the degree that gave the requisite number of fluorophores in the field. The ratio of two orthogonally polarized fluorescence intensities, Ω, was very sensitive to rotational movement but minimally sensitive to translational movement. The power spectra inferred from the fluctuation traces obtained in either rigor or relaxation were essentially those of instrumental noise. In isometric activity, however, the fluctuations were larger and generated a distinctly non-flat power spectrum dominated by frequencies in the 0–5 Hz range. These frequencies were thus of the order of the ATPase frequencies observed in the same system (89). This is an experimental indication that during activity, cross bridges, wholly or in part, rotate repetitively. Taken together with spin label studies of the active state (above), one must suppose that these rotations occur during the rigor-like attached state (of the two states sensed by EPR) but are too slow to be detected by EPR. From the static state dichroism studies with the same label, the rotation, if S-1 is rigid, would be mainly along the ψ-coordinate, or it would be the rotational motion of a portion of S-1. While the indication of cross-bridge rotation is experimental, its coupling with ATPase is merely a surmise. This surmise could be experimentally examined by asking whether fluctuations in the concentrations of ATPase intermediates are cross-correlated with orientational fluctuations. Along this line, some theoretical (111) and experimental (T. Ando, J. A. Duke, Y. Tonomura, M. F. Morales, unpublished) steps toward detecting stochastic behavior in steady state ATPase have been taken.

The Transduction Mechanism

In the foregoing sections the concern has been with how cross bridges may impel filaments and thus produce muscle contraction. Tacitly, it has been supposed that these impelling motions are "coupled" to ATPase, but nothing has been said about the mechanism of coupling except that it resides on S-1 and that both actin and nucleotides bind to S-1. The object now is to outline a hypothesis of energy transduction in S-1, to describe how certain observations support it, and to speculate about future research.

The demonstrated existence of ternary complexes, "AMN" (73, 113, 114) confirms the early assertion (115) that S-1 has two spatially distinct sites. Furthermore, very early observations (2, 116) showed that binding of N to the N-site reduces the affinity of A to the A-site, and vice versa, i.e. that there is intersite communication. The researches of Tonomura (117), Trentham et al (118), and Taylor (119) uncovered several sequential intermediates in ATPase. These intermediates can be regarded as Ns [distinct from substrate ($N_1 =$ MgATP) or final product ($N_n =$ MgADP)] with high affinity for the N-site. Of course it is the large, negative ΔG^0 of ATP hydrolysis (120) that "drives" the temporally sequential transmutations, $N_1 \rightarrow N_2 \rightarrow \cdots N_n$. Because of intersite communication it is reasonable to think that each N_i imposes its own influence at the A-site, thereby causing a temporal sequence of A-M affinities, but it is further postulated that specifiable relations besides affinity are also changing sequentially, among them angular relations in the AMN complex (12); it is this latter postulate that creates a "machine." Above, it was left unsettled which of the three attitudinal angles (φ, θ, ψ) relating M to A actually changes in cross-bridge impulsion. But the fluctuation experiments (110) did show that some "slow" (order of the ATPase rate) rotation occurs. So, to be suitably noncommital, it will be said that some angular function, Ω, pertaining to AM attitude in AMN changes sequentially, along with affinity. So, this is the hypothesis: There is a temporal sequence of occupants ("enzymatic intermediates") of the N-site, driven by the large, $-\Delta G(\text{ATP} \rightarrow \text{ADP} + \text{P})$. Because of intersite communication, this imposes a corresponding sequence at the A-site, in regard to affinity and the Ω-property.

What might be the mechanism of intersite communication? Long-range field effects are probably insufficiently specific, so communication by means of propagated structural distortion seems more likely. But this latter option can have two versions. The two sites could be directly linked by polypeptide chains with suitable properties; in this case "paths of influence" should be experimentally traceable. Alternatively, the situation could resemble a popular view of hemoglobin (121), i.e. two or more relational states of M could preexist, and the equilibrium between them could be shifted by ligand binding; in this case there would be no path of influence connecting the sites, but there might be a "fault line" along which the interstate transition occurs, and along this line there might be groups responsive to ligands. The simpler first version is initially pursued here, and the second is presented with the NMR work that inspired it.

Since there is no crystallographic information about S-1, the reference frame for this particle continues to be its primary amino acid sequence (M. Elzinga, personal communication), and the methods for acquiring mechanistic information are varied and indirect:

1. Residues have been found on the proteins that react more or less specifically with defined small molecules ("probes"), having useful properties once attached.
2. By using cross-linking reagents under special conditions and examining the results on polyacrylamide gel electrophoresis (PAGE), it has been shown that certain primary sequences are probably associated in the native state.
3. By applying proteolytic enzymes under special conditions, then analyzing the results on PAGE, selective cuts in the primary sequence have been achieved.

Each technique has spawned various strategies. Again, probes have been of special utility. Single, environment-sensing fluorescence and EPR probes have reported useful information [the attitude and motion probes (see above) are special cases of this category]; but probes and reagents have also been used in tandem. For example, the reactivity of a probe toward its target site may depend on whether another probe is in place at another site, or, after a probe is in place, application of annihilating reagents (acrylamide, ascorbate, etc) may test for access to the bound probe.

The most informative probe pairs are those between which energy transfer can be detected. While such probes can be paramagnetic, most applications to actin and S-1 have been with fluorescence probes, and resonance energy transfer has been investigated. When transfer occurs, the distance between the two probes can be estimated (123, 124) from the measured E, "the fraction of energy transferred." The basic strategy follows. One selects two attachable probes, a "donor" ($\mathcal{D}$) with absorbance at a "short" wavelength and whose emission at a longer wavelength can be represented (relative to an imbedded framework) by $\hat{\varepsilon}_\mathcal{D}$, and an "acceptor" ($\mathcal{C}$) whose absorbance, represented by $\hat{\alpha}_\mathcal{C}$, must to some extent overlap in wavelength the emission of $\mathcal{D}$. The rates at which $\mathcal{D}$ can lose absorbed energy are k_f (fluorescence), k_t (transfer to $\mathcal{C}$, if $\mathcal{C}$ is present), and k_q (all other mechanisms, e.g. collisional, dielectric relaxation, etc). The experiment consists in measuring the quantum yield of $\mathcal{D}$ attached alone [$F_\mathcal{D} = k_f/(k_f + k_q)$], or of $\mathcal{D}$ when $\mathcal{C}$ is also attached

$$[F_{\mathcal{D}\mathcal{C}} = k_f/(k_f + k_q + k_t)];$$

alternatively it consists of measuring the corresponding excited state lifetimes

$$[\tau_\mathcal{D} = (k_f + k_q)^{-1} \quad \text{and} \quad \tau_{\mathcal{D}\mathcal{C}} = (k_f + k_q + k_t)^{-1}].$$

Now $E = k_t/(k_f + k_q + k_t)$, so, experimentally, E can be obtained as $1 - (F_{\mathcal{D}\mathcal{C}}/F_\mathcal{D})$ or as $1 - (\tau_{\mathcal{D}\mathcal{C}}/\tau_\mathcal{D})$. von Förster (125) showed that if R is the $\mathcal{D} - \mathcal{C}$ distance, $k_t = (\text{const})R^{-6}$. Substituting into the expression for E and rearranging,

$$E = R_0^6/(R^6 + R_0^6), \quad \text{where} \quad R_0^6 = (\text{const})/(k_f + k_q) = (\text{const})/\tau_\mathcal{D}.$$

In classical terms one might expect k_t to be proportional to the electrical influence that $\hat{\varepsilon}_\mathcal{D}$ has on $\hat{\alpha}_\mathcal{C}$ and to that which $\hat{\alpha}_\mathcal{C}$ has on $\hat{\varepsilon}_\mathcal{D}$ (66). That of $\hat{\varepsilon}_\mathcal{D}$ on $\hat{\alpha}_\mathcal{C}$ is

$$\hat{\alpha}_\mathcal{C} \cdot \mathcal{E}_\varepsilon = [3(\hat{\varepsilon}_\mathcal{D} \cdot \hat{R})(\hat{\alpha}_\mathcal{C} \cdot \hat{R}) - \hat{\varepsilon}_\mathcal{D} \cdot \hat{\alpha}_\mathcal{C}]R^{-3},$$

where $\mathcal{E}_\varepsilon$ is the electrical field of $\hat{\varepsilon}_\mathcal{D}$ at $\mathcal{C}$, and **R** is the vector distance between $\mathcal{D}$ and $\mathcal{C}$. This expression is symmetrical, and is also the influence of $\hat{\alpha}_{\mathcal{C}}$ on $\hat{\varepsilon}_{\mathcal{D}}$, so k_t should be proportional to

$$\left[\hat{\varepsilon}_{\mathcal{D}} \cdot \hat{\alpha}_{\mathcal{C}} - 3(\hat{\varepsilon}_{\mathcal{D}} \cdot \hat{R})(\hat{\alpha}_{\mathcal{C}} \cdot R)\right]^2 R^{-6}.$$

This is the origin of the factor, R^{-6}, in k_t and E. The expression in brackets, usually called κ, a function of the cosines of the indicated angles, specifies the angular relations between $\hat{\varepsilon}_{\mathcal{D}}$ and $\hat{\alpha}_{\mathcal{C}}$; it is a factor of "const," and therefore of R_0. Satisfactory values of the other experimental factors of "const," viz overlap integral and refractive index, can be found, but typically κ^2 is indeterminate, and precludes estimation of R from experimental knowledge of E. The usual circumvention of this difficulty is to appeal to angular randomization of $\hat{\varepsilon}_{\mathcal{D}}$ and/or $\hat{\alpha}_{\mathcal{C}}$ by high frequency (picosecond) movement about a symmetry axis that remains essentially rigid on the nanosecond time scale, thus permitting use of the same or similar probes in TRFAD (above) (66). The effect of such randomization on the anisotropy manifests itself experimentally as a reduced "zero-time anisotropy," $r(0)$, i.e. one less than expected from immobile, randomized (congruent) dipoles, viz $2/5$; invariably, the $r(0)$ obtained by extrapolation from the usable TRFAD range is less than $2/5$, but technical uncertainties keep this observation from being conclusive. More convincing is steady state data showing that the anisotropy of immobilized (by high concentrations of sugar) IAEDANS-labeled S-1 (66, 126) is well under that of frozen IAEDANS (46). Further indications that randomization occurs are that interchanging the locations of $\mathcal{D}$ and $\mathcal{C}$, and immersing them in chaeotropic solutions to loosen surface structure, leave E unchanged, as would be expected if the probes were randomized before the maneuvers are instituted. A systematic method of examining for high-frequency randomization has been proposed (128). Full application has not yet been attempted, but anisotropy measurements have indicated that probes attached to Cys 373 are indeed randomized (129). So, while the randomization assumption is not as well validated as it will be, it is universally made, in which case (125) $\kappa^2 \to 2/3$; thus E becomes a function of R only, and the distance estimate can be made.

As discussed above, clever use of cross-linking reagents has elicited information about the "micro-mechanics" of cross bridges; it has also been applied to the transducer itself. The assignment of the A-binding site to a particular stretch of the S-1 heavy chain was first made on the basis of cross-linking results (130, 131). A more detailed study of A-M binding has now been made (132). A different application concerns the SH_1-SH_2 distance. It had been initially shown that two particular thiols of S-1 ("SH_1" and "SH_2"; see below) could be cross-linked by p-phenylene dimaleimide, thus establishing that in the final product the two sulfur atoms were 1.4 nm apart. It has now been shown (133), however, that analogous reagents ranging in separation distance from .1 nm to 1.4 nm could also satisfactorily link the two atoms; thus the conclusion is that the S-1 region enveloping the two thiols is very flexible.

Proteolytic techniques as a means of relating structure to function were introduced into biochemistry by Gergely (134) and were immediately used to elucidate the segmental nature of myosin (5, 135). Recently, a new advance was made with this method (136). Conditions have been found wherein trypsin will cleave S-1 into three major fragments (27K, 50K, and 20K) which remain associated under mild conditions but are separable by PAGE. The fragmented S-1 retains the ability to hydrolyze ATP and bind actin. These three fragments have become a valuable framework in which to assign specific groups and functionalities (130). It is unclear whether these fragments are *domains*, and, if so, whether they relate to morphological domains (137). The proteolysis locations, however, are significant landmarks for they identify heavy chain segments at which trypsin effects considerable eversions (138, 139). Moreover, changes in the feasibility of these eversions are quantitatively expressed as changes in proteolytic rate. From the attack of either trypsin or chymotrypsin on actin, a fragment of 34K, retaining the nucleotide binding site, can be isolated (140); the fragment no longer binds Ca^{2+}, polymerizes, or activates myosin ATPase. Thus in actin, too, proteolysis separates functionalities.

Various workers using these methods have located several groups, residues, and proteolytic cuts, as diagrammed in Figure 2, and in some instances the distances between points. The conformation of polypeptide chains in Figure 2, however, is purely schematic, as is the manner in which the chains have been fitted into the gross shapes discussed above. Encircled groups react with solute reagents and must have surface locations. Not all groups that have been detected can be placed in such a diagram. For example, a Trp residue perturbable by N-binding (so of value in kinetics) has been well studied (152), and is thought to be within charge-transfer distance from the ring of bound N, but the chain fragment on which it resides is unknown. Similarly, there exist definite binding sites for ANS and bis-ANS, and these bound fluorophores can accept energy from a vicinal Trp (153, 154); but, again, they have not been located on the primary sequence.

Now we consider what present chemical knowledge of S-1 suggests about internal dynamics, and how our hypothesis accommodates these suggestions. The separation of A-site and N-site is further confirmed by emerging structural work. The 50K fragment has been implicated in actin-binding (130, 131). It has also been shown (130) that the bound

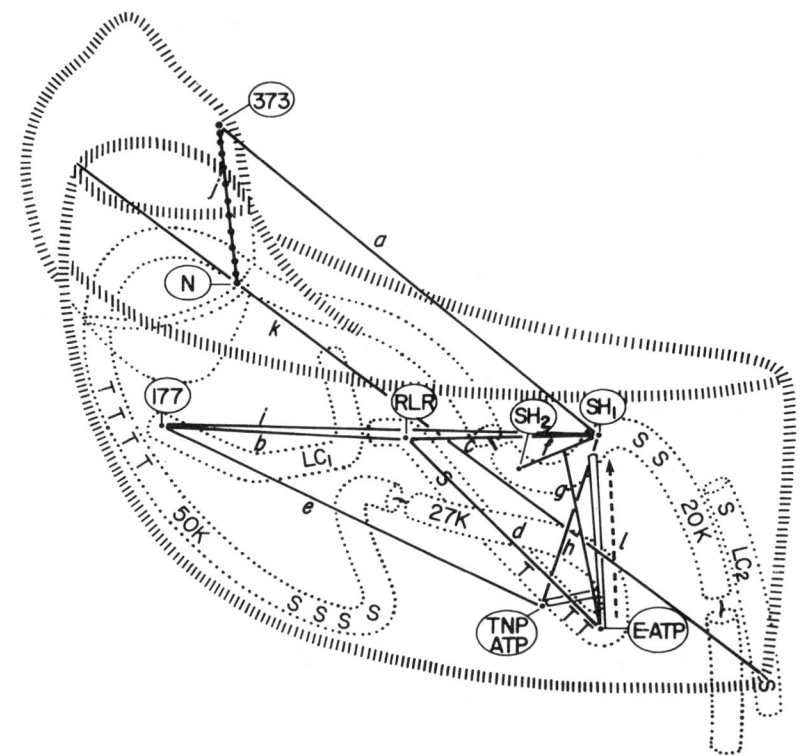

Figure 2 Composite diagram indicating landmark points, interpoint distances, and certain residues located on acto-(S-1). Distances in nm are: $a = 5.8$ (126); $b = 3.2$ (R. Takashi, unpublished); $c = 2.6$ (142); $d = 4.2$ (R. Takashi, unpublished); $e = 5.5$ (144); $f = 1.4$ (145); $g = 2.5$ (146); $h = 3.8$ (147); $i = 4.0$ (148); $j = 3.0$ (149); $k = 12.0$ (30); $l = 2.0$ (stretched length of ATP from space-filling models). Light chains are indicated as "LC_1" and "LC_2," and heavy chain fragments as "20K" (which attaches to S-2), "50K," and "27K." Thiols are either identified by their conventional designations ("SH_1", "SH_2," actin-Cys 373, LC_1-Cys 177) or by "S" (T. Hozumi, unpublished). Tryptophanes are indicated by "T" (151).

actin protects against the 50K/20K tryptic cut[4]. The N-site, on the other hand, seems to be associated with other fragments. There are currently two photoaffinity analogs of ATP: in one (155) the photoreactive group is linked to the ribose moiety, and in another (R. G. Yount, personal communication) it is linked to a ring. Both analogs attach to 27K (157; R. G. Yount, personal communication). MgATP or Mg-pyrophosphate

---

[4]Actually, Mornet et al concluded from their solution work that (a pair of) actin is bound to both 50K and 20K regions. That one S-1 can bind to two actins, however, is controversial in the light of some reconstructed images of acto-(S-1) (32).

when bound to the N-site protects SH_1 and SH_2 from reaction with sulfhydryl reagents (158); this suggests that the polyphosphate moiety of ATP may be near the SH_1 region of 20K. Therefore, as first suggested by Kassab, the N-site may lie across 27K and 20K, but have no direct connection with 50K. The length of extended ATP (calculated from models) and the distance from the fluorophoric ring of ε-ATP to crosslinked $(SH_1)-(SH_2)$ (146) are in harmony with this assignment.

If the simpler idea that there may be "paths of influence" connecting the A-site and N-site is followed, the overall influence of bound actin on the N-site is shown by its activation of S-1 MgATPase (2) and its diminution of (S-1)-ADP affinity (116). Initially, the discovery (130) that the 20K/50K cut abolished actin activation of myosin Mg^{2+}ATPase, without affecting myosin ATPases, seemed an indication that the path of influence had been proteolytically severed, but the observation had been made at a single actin concentration. It has since been shown (159) that if the actin concentration is increased enough the full ATPase is nearly restored: hence, the 20K/50K cut appears to diminish the actin affinity to certain enzymatic intermediates of S-1. The observation (159) is presumably related to the inhibition of the 20K/50K cut by A-binding. Nevertheless, the effect of bound actin is also felt at other points. For example, actin-binding depresses the reactivity of SH_1 (80) and at the 27K/50K cut it reverses certain suppressive effects of N-binding, without causing N-dissociation (160). However, actin-binding is not felt globally; for example, it has no effect on the reactive lysyl residue ("RLR") (161). Thus, the influence of A-binding on N-binding seems to pass through both cuts and SH_1, but does not reach RLR.

The influence of bound N on the actin-affinity of the A-site is shown by the actin dissociative effects of most nucleotides and polyphosphates. An attempt to discover a path of influence can begin by considering the interactions of the N-site with its immediate environment. One set of these interactions is an extensively studied phenomenon sometimes called "modification."

The original facts about modification are that at neutral pH and room temperature, (a) reacting SH_1 with common sulfhydryl reagents (45), or (b) adding EDTA (162), or (c) substituting ITP for ATP (163) results in a much faster myosin ATPase rate when Ca^{2+} is the coactivator, and a much slower rate if K^+ or NH_4^+ is the coactivator. After much subsequent research, it is clear that these observations are a manifestation of a general pattern. When the Ca^{2+}-activated room temperature activity is measured as a function of pH, the resulting curve has crests near 6 and 9, and a trough near neutrality (164). Any one of the maneuvers (a–c) converts this pH-dependence to a sigmoidal, titration-like curve with inflection near neutrality (158). A similar sigmoidal curve can also be produced, without any maneuver, by lowering the temperature below 15°C. Another generality emerges from studying the influence of structure-disrupting ions on Ca^{+2} ATPase activity (165). As expected, activity falls as concentrations of these ions increase, but it falls much less if maneuvers (*a*) or (*c*) are first made [(*b*) has not been tested]. Because of the implications of these effects, it was suggested (166) that the N-site can exist in either a (native) "α-state", or a (modified) "β-state". What actually happens in modification? Several authors (166–168) found that modification of Ca^{2+} ATPase can also be produced by moderate concentrations of organic solvents. It was found (169, 170) that modification could be sensed by an accompanying increase in fluorescence intensity of ANS bound to myosin, or by a decrease in energy transferred to ANS by a vicinal Trp. It was shown (171, 172) that modification can be produced by attaching a trinitrophenyl group to RLR. Finally, recent work (65) concludes, as regards maneuver (a), that what counts is not the SH-blocking but the bulk of the group introduced. Collectively, modification studies now suggest that a region surrounding the N-site can be placed in one or another conformation, and that this can be done by actions taken at different points in the region.

Reciprocal to modification are the effects of ligating the N-site on surrounding points; invariably in these effects it matters whether the N-site is ligated with ATP or with ligands incapable of forming enzymatic intermediates [e.g. ADP, pyrophosphate, aPP(NH)P]. These reciprocal effects were observed early (152, 170). Unfortunately, both observations involved tryptophanyl residues, known to be near the N-site (173) but not yet located on the primary sequence. Subsequently, other such effects appeared, viz a reduction in the formation of a 29K precursor in the 27K/50K cut (174, 175), a reduction in the reactivity of RLR to trinitrobenzene sulfonate (160), and an increase in the reactivity of SH_2 to thiol reagents (176). From our present viewpoint the most interesting reciprocal effects are at SH_1 because, in turn, effects at SH_1 may plausibly cause effects at the A-site. Seidel & Gergely (177) discovered the effect of N-binding on the mobilization of spin labels attached to SH_1 (and deduced thereby the existence of the intermediate complex now known as "$M^{**}ADP, P_i$"). Subsequently, it was found (80) by following an in vivo lead (178) that the reactivity of SH_1 to iodoacetamide and its analogs is decreased by ATP and increased by ADP, pyrophosphate, and aPP(NH)P. That influence can travel beyond SH_1 to actin is suggested by the fact that reaction of SH_1 and other reactive thiols of the region reduces actin-affinity (65). Once at the interface, the influence spreads in actin, at least to Cys-373, for it is known that the reactivity of this thiol to iodoacetamide analogs is enhanced in myosin-ligated F-actin over what it is in free F-actin (80, 126). Once again, available evidence suggests that, as in the $A \to N$ direction, the $N \to A$ influence passes through SH_1 and nearby points such as the 20K/50K cut.

In summary, the simple idea that there is direct communication between the A-site and N-site of S-1 is encouraged by some available data. But the data are insufficiently precise and extensive to permit a definitive conclusion, so we must also examine the alternative, "hemoglobinoidal" view. Shriver & Sykes (179) were the first to obtain ^{31}P-NMR spectra from M·ADP. They noted that the signal from the β-phosphate was clearly a doublet, and found furthermore that the ratio of the populations corresponding to the peaks could be shifted in a regular way by varying the temperature between 0° and 20°C. In a rather bold but stimulating speculation, they have proposed that not only M·ADP, but all M·N, exist in two states—indeed, that the two states globally preexist in M, and correspond to the two extremal values of Ω at which actin can bind. The Shriver-Sykes view calls for a more elaborate chemical scheme, and would make it unnecessary to search for a direct path of influence between binding sites, but it does not change the principle of transducer operation proposed early in the preceding section. That the two Shriver-Sykes states are related, or even identical, to the α and β states postulated from biochemical observations is an intriguing possibility. Another recent NMR investigation that may relate to the conformational instability discussed here (180) has concluded that there is an S-1 region of much internal motion and that this motion is quenched by actin binding.

In the absence of any experimental evidence we can only speculate about two remaining features of our transducer hypothesis: Does the temporal sequence of states at the A-site include a change in relations, Ω, among the transitions between bound states? If changes in the configuration of polypeptide chains stretching between the N-site and A-site are the means of communication, what is the nature of the forces that send the "messages"? The first question may be approachable by measuring changes in the distances between several noncollinear points on actin and on S-1. Since all the participants (proteins, ATP, Mg^{2+}) are very polar, it is natural to wonder whether electrostatic interactions at the ends of the intervening chains ultimately generate the "contractile force."

ACKNOWLEDGMENTS

This review emphasizes research conducted by a group successively encamped at the Naval Medical Research Institute, Dartmouth, and the

University of California at San Francisco. On this account we acknowledge with special thanks the cited and uncited contributions of many Fellows and Visiting Scientists who are no longer in the group authorship. J. Borejdo and M.F.M. are, respectively, Established Investigator and Career Investigator of the American Heart Association. The UCSF research cited was supported by grants HL-16683, PCM-7922174, and CI-8.

Literature Cited

1. Engelhardt, V. A., Ljubimova, M. N. 1939. *Nature* 144:669-71
2. Szent-Gyorgyi, A. 1947. *Chemistry of Muscular Contraction*. New York: Academic. 150 pp.
3. Huxley, A. F., Niedergerke, R. 1954. *Nature* 173:971-73
4. Huxley, H. E., Hanson, J. 1954. *Nature* 173:973-76
5. Mueller, H., Perry, S. V. 1962. *Biochem. J.* 85:431-39
6. Reedy, M. K., Holmes, K., Tregear, R. T. 1965. *Nature* 207:1276-80
7. Gordon, A. M., Huxley, A. F., Julian, F. J. 1966. *J. Physiol.* 184:170-92
8. Huxley, H. E. 1969. *Science* 164:1356-66
9. Huxley, A. F., Simmons, R. M. 1971. *Nature* 233:533-38
10. Dos Remedios, C. G., Yount, R. G., Morales, M. F. 1972. *Proc. Natl. Acad. Sci. USA* 69:2542-46
11. Botts, J., Cooke, R., Dos Remedios, C. G., Duke, J. Mendelson, R. A., Morales, M. F., Tokiwa, T., Viniegra, G. 1972. *Cold Spring Harbor Symp. Quant. Biol.* 37:195-200
12. Morales, M. F., Botts, J. 1979. *Proc. Natl. Acad. Sci. USA* 76:3857-59
13. Ueno, H., Harrington, W. F. 1981. *Proc. Natl. Acad. Sci. USA* 78:6101-05
14. De Rosier, D. J., Klug, A. 1968. *Nature* 217:130-34
15. Klug, A., Amos, L. A. 1975. *J. Mol. Biol.* 99:51-64
16. Amos, L. A. 1975. *J. Mol. Biol.* 99:65-73
17. Moore, P. H., Huxley, H. E., De Rosier, D. J. 1970. *J. Mol. Biol.* 50:279-95
18. Wakabayashi, T., Huxley, H. E., Amos, L. A., Klug, A. 1975. *J. Mol. Biol.* 93:477-97
19. O'Brien, E. J., Bennett, P. M., Hanson, J. 1975. *J. Mol. Biol.* 99:461-75
20. O'Brien, E. J., Morris, E. P., Seymour, J., Couch, J. 1980. *Muscle Contraction: Its Regulatory Mechanisms*, pp. 147-64. Tokyo: Springer Verlag
21. Dos Remedios, C. G., Dickens, M. 1978. *Nature* 276:731-33
22. Aebi, U., Smith, P. R., Isenberg, G., Pollard, T. 1980. *Nature* 276:731-33
23. Suck, D., Kabsch, W., Mannherz, H. G. 1981. *Proc. Natl. Acad. Sci. USA* 78:4319-23
24. Slayter, H. S., Lowey, S. 1967. *Proc. Natl. Acad. Sci. USA* 58:1611-18
25. Mendelson, R. A., Morales, M. F., Botts, J. 1973. *Biochemistry* 12:2250-55
26. Yang, J. T., Wu, C. C. 1977. *Biochemistry* 16:5785-89
27. Elliott, A., Offer, G. 1978. *J. Mol. Biol.* 123:505-19
28. Takahashi, T. 1978. *J. Biochem.* 83:905-8
29. Kretzschmar, K. M., Mendelson, R. A., Morales, M. F. 1978. *Biochemistry* 17:2314-18
30. Mendelson, R. A., Kretzschmar, K. M. 1980. *Biochemistry* 19:4103-8
31. Deleted in proof
32. Taylor, K. A., Amos, L. A. 1981. *J. Mol. Biol.* 147:297-324
33. Mendelson, R. A., Giniger, E. 1982. *Biophys. Soc. Abstr.* 37:54a
34. Toyoshima, C., Wakabayashi, T. 1979. *J. Biochem.* 86:1887-90
35. Heuser, J. E., Cooke, R. 1982. *J. Mol. Biol.* In press
36. Ford, L. E., Huxley, A. F., Simmons, R. M. 1977. *J. Physiol.* 269:441-515
37. Morimoto, K., Harrington, W. F. 1974. *J. Mol. Biol.* 88:693-709
38. Mendelson, R. A., Cheung, P. C. 1976. *Science* 194:190-92
39. Mendelson, R. A., Putnam, S., Morales, M. F. 1975. *J. Supramol. Struct.* 3:162-68
40. Hanson, J. 1968. *Q. Rev. Biophys.* 1:177-216

41. Yguerabide, J. 1972. *Methods Enzymol.* 26C:498–578
42. Belford, G. G., Belford, R. L., Weber, G. 1972. *Proc. Natl. Acad. Sci. USA* 69:1392–93
43. Ehrenberg, M., Rigler, R. 1972. *Chem. Phys. Lett.* 14:539–44
44. Quinlivan, J., McConnell, H. M., Stowring, L., Cooke, R., Morales, M. F. 1969. *Biochemistry* 8:3644–47
45. Kielley, W. W., Bradley, L. B. 1956. *J. Biol. Chem.* 218:653–59
46. Hudson, E. N., Weber, G. 1973. *Biochemistry* 12:4154–61
47. Takashi, R., Duke, J., Ue, K., Morales, M. F. 1976. *Arch. Biochem. Biophys.* 175:279–83
48. Yu, H., Stockmayer, W. H. 1967. *J. Chem. Phys.* 47:1369–71
48a. Wegener, W. A. 1981. *Biopolymers* 20:303–26
49. Kobayashi, S., Tatsuka, T. 1975. *Biochim. Biophys. Acta* 47:1369–71
50. Thomas, D. D., Seidel, J. C., Hyde, J. S., Gergely, J. 1975. *Proc. Natl. Acad. Sci. USA* 72:1729–33
51. O'Konski, C. T., Krause, S. 1976. *Molecular Electro-Optics*, pp. 63–120. New York: Dekker
52. Highsmith, S., Kretzschmar, K. M., O'Konski, C. T., Morales, M. F. 1977. *Proc. Natl. Acad.Sci.USA* 74:4986–90
53. Brahms, J., Brezner, J. 1961. *Arch. Biochem. Biophys.* 95:219–28
54. Weeds, A. G., Pope, B. 1977. *J. Mol. Biol.* 111:129–57
55. Lowey, S., Slayter, H. S., Weeds, A. G., Baker, H. 1969. *J. Mol. Biol.* 42:1–29
56. O'Konski, C. T., Zimm, B. H. 1950. *Science* 111:113–16
57. Broersma, S. 1960. *J. Chem. Phys.* 32:1626–31
57a. Tirado, M., de la Torre, G. 1980. *J. Chem. Phys.* 73:1986–93
58. Lu, R. C. 1980. *Proc. Natl. Acad. Sci. USA* 77:2010–13
59. Highsmith, S., Wang, C., Zero, K., Pecora, R., Jardetzky, O. 1982. *Biochemistry* 21:1192–97
60. Harvey, S. C., Cheung, H. C. 1977. *Biochemistry* 16:5181–87
61. Margossian, S. S., Lowey, S. 1975. *Fed. Proc.* 34:671–93
62. Marston, S., Weber, A. 1975. *Biochemistry* 14:3868–73
63. Wahl, P. 1969. *Biochim. Biophys. Acta* 175:55–64
64. Highsmith, S., Mendelson, R. A., Morales, M. F. 1976. *Proc. Natl. Acad. Sci. USA* 73:133–37
65. Botts, J., Ue, K., Hozumi, T., Samet, J. 1979. *Biochemistry* 18:5157–63
66. Morales, M. F., Botts, J. 1980. *Muscle Contraction—Its Regulatory Mechanism*, pp. 133–43. Tokyo: Springer Verlag
67. Margossian, S. S., Lowey, S. 1978. *Biochemistry* 17:5431–39
68. Wadzinski, L., Botts, J., Wang, A., Woodward, J., Highsmith, S. 1979. *Arch. Biochem. Biophys.* 198:397–402
69. Highsmith, S. 1977. *Arch. Biochem. Biophys.* 180:404–8
70. Peller, L. 1975. *J. Supramol. Struct.* 3:169–74
71. Hill, T. L. 1978. *Nature* 274:825–26
72. Greene, L. E., Eisenberg, E. 1980. *J. Biol. Chem.* 255:543–48
73. Highsmith, S. 1976. *J. Biol. Chem.* 251:6170–72
74. Greene, L. E., Eisenberg, E. 1978. *Proc. Natl. Acad. Sci. USA* 75:54–58
75. Borejdo, J., Putnam, S. V. 1977. *Biochim. Biophys. Acta* 459:578–95
76. Mendelson, R. A., Wilson, M. 1982. *Biophys. J.* In press
77. Aronson, J. F., Morales, M. F. 1969. *Biochemistry* 8:4517–22
78. Dos Remedios, C. G., Millikan, R. G., Morales, M. F. 1972. *J. Gen. Physiol.* 59:103–20
79. Nihei, T., Mendelson, R. A., Botts, J. 1974. *Biophys. J.* 14:236–42
80. Duke, J., Takashi, R., Ue, K., Morales, M. F. 1976. *Proc. Natl. Acad. Sci. USA* 73:302–6
81. Borejdo, J., Assulin, O., Ando, T., Putnam, S. V. 1982. *J. Mol. Biol.* In press
82. Yanagida, T. 1981. *J. Mol. Biol.* 146:539–60
83. Carrington, A., McLachlan, A. 1967. *Introduction to Magnetic Resonance.* New York: Harper. 266 pp.
84. McCalley, R. C., Shimshick, E. J., McConnell, H. M. 1972. *Chem. Phys. Lett.* 23:115–19
85. Deleted in proof
86. Thomas, D. D., Cooke, R. 1980. *Biophys. J.* 32:891–906
87. Thomas, D. D., Ishiwata, S., Seidel, J. C., Gergely, J. 1980. *Biophys. J.* 32:873–90
88. Loxdale, H. D. 1976. *J. Physiol.* 260:4–5
89. Takashi, R., Putnam, S. V. 1979. *Anal. Biochem.* 92:375–82

90. Julian, F. 1969. *Biophys. J.* 9:547–70
91. Podolsky, R. J. 1960. *Nature* 188:666–68
92. Nihei, T., Mendelson, R. A., Botts, J. 1974. *Proc. Natl. Acad. Sci. USA* 71:274–77
93. Matsubara, I., Yagi, N., Hashizume, H. 1975. *Nature* 255:728–29
94. Podolsky, R. J., St. Onge, R., Yu, L. C., Lymn, R. W. 1976. *Proc. Natl. Acad. Sci. USA* 73:813–17
95. Huxley, H. E. 1979. *The Molecular Basis of Force Development in Muscle*, pp. 1–13. Palo Alto, Calif: Palo Alto Med. Res. Found.
96. Yu, L. C., Hartt, J. E., Podolsky, R. J. 1979. *J. Mol. Biol.* 132:53–68
97. Huxley, H. E., Faruqi, A. R., Bordas, J., Koch, M. H. J., Milch, J. R. 1980. *Nature* 284:140–43
98. Huxley, H. E., Simmons, R. M., Faruqi, A. R., Kress, M., Bordas, J., Koch, M. H. J. 1981. *Proc. Natl. Acad. Sci. USA* 78:2297–2301
99. Graceffa, P., Seidel, J. C. 1980. *Biochemistry* 19:33–39
100. Pecora, R. 1972. *Ann. Rev. Biophys. Bioeng.* 1:257–76
101. Webb, W. B. 1976. *Q. Rev. Biophys.* 9:49–68
102. Madge, D. 1977. *Chemical Relaxation in Molecular Biology*, pp. 43–83. Berlin: Springer Verlag
103. Feher, G. 1978. *Trends Biochem. Sci.* 3:111–13
104. Borejdo, J. 1980. *Curr. Top. Bioenerg.* 10:1–40
105. Chen, Y. 1978. *Adv. Chem. Phys.* 37:67–97
106. Carlson, F. D., Bonner, R., Frazer, A. 1972. *Cold Spring Harbor Symp. Quant. Biol.* 37:389–96
107. Borejdo, J., Morales, M. F. 1977. *Biophys. J.* 20:315–34
108. Huxley, A. F. 1957. *Prog. Biophys. Biophys. Chem.* 7:255–318
109. Borejdo, J. 1979. *Biopolymers* 18:2807–20
110. Borejdo, J., Putnam, S. V., Morales, M. F. 1979. *Proc. Natl. Acad. Sci. USA* 76:6346–50
111. Morales, M. F. 1982. *Proc. Natl. Acad. Sci. USA* 79:1126–28
112. Deleted in Proof
113. Tonomura, Y., Morita, F. 1960. *J. Am. Chem. Soc.* 82:5172–77
114. Beinfeld, M. C., Martonosi, A. N. 1975. *J. Biol. Chem.* 250:7871–78
115. Barany, M., Barany, K. 1959. *Biochim. Biophys. Acta* 35:293–309
116. Kiely, B., Martonosi, A. N. 1968. *J. Biol. Chem.* 243:2273–78
117. Tonomura, Y. 1972. *Muscle Proteins, Muscle Contraction, and Cation Transport.* Tokyo: Univ. Tokyo Press
118. Trentham, D. R., Eccleston, J. F., Bagshaw, C. R. 1976. *Q. Rev. Biophys.* 9:217–81
119. Taylor, E. W. 1979. *CRC Crit. Rev. Biochem.* 6:103–14
120. Podolsky, R. J., Morales, M. F. 1956. *J. Biol. Chem.* 218:945–59
121. Perutz, M. F. 1978. *Sci. Am.* 239:92–125
122. Deleted in proof
123. Schiller, P. W. 1975. *Biochemical Fluorescence*, pp. 285–303. New York: Dekker
124. Stryer, L. 1978. *Ann. Rev. Biochem.* 47:819–46
125. von Förster, T. 1948. *Ann. Phys.* 2:55–75
126. Takashi, R. 1980. *Biochemistry* 18:5164–69
127. Deleted in proof
128. Dale, R., Eisinger, J. 1979. *Biophys. J.* 26:161–93
129. Wahl, P., Mihashi, K., Auchet, J. C. 1975. *FEBS Lett.* 60:164–67
130. Mornet, D., Bertrand, R., Pantel, P., Audemard, E., Kassab, R. 1979. *Biochem. Biophys. Res. Commun.* 89:925–32
131. Yamamoto, K., Sekine, T. 1979. *J. Biochem.* 86:1855–62
132. Mornet, D., Bertrand, R., Pantel, P., Audemard, E., Kassab, R. 1981. *Nature* 293:301–6
133. Wells, J. A., Knoeber, C., Sheldon, M. C., Werber, M. M., Yount, R. G. 1980. *J. Biol. Chem.* 255:11135–40
134. Gergely, J. 1950. *Fed. Proc.* 9:176
135. Mihalyi, E., Szent-Gyorgyi, A. G. 1953. *J. Biol. Chem.* 201:184–96
136. Balint, M., Wolf, I., Tarcsafalvi, A., Gergely, J., Sreter, F. A. 1978. *Arch. Biochem. Biophys.* 190:793–99
137. Katayama, E., Wakabayashi, T. 1981. *J. Biochem.* 90:703–14
138. Stroud, R. M. 1974. *Sci. Am.* 231:74–89
139. Stroud, R. M., Krieger, M., Koeppe, R. E., Kossiakoff, A. A., Chambers, J. L. 1975. *Proteases and Biological Control*, pp. 13–32. Cold Spring Harbor, NY: Cold Spring Harbor Lab.
140. Jacobson, G. R., Rosenbusch, J. P. 1976. *Proc. Natl. Acad. Sci. USA* 73:2742–46

141. Deleted in proof
142. Takashi, R., Muhlrad, A., Botts, J. 1982. *Biochemistry*. In press
143. Deleted in proof
144. Moss, D. J., Trentham, D. R. 1980. *Fed. Proc.* 39:1935
145. Reisler, E., Burke, M., Himmelfarb, S., Harrington, W. F. 1974. *Biochemistry* 13:3837-40
146. Perkins, J., Wells, J. A., Yount, R. G. 1980. *Fed. Proc.* 39:1937
147. Tao, T., Lankin, M. 1981. *Biochemistry* 20:5051-55
148. Marsh, D. J., Lowey, S. 1980. *Biochemistry* 19:774-84
149. Miki, M., Mihashi, K. 1978. *Biochim. Biophys. Acta* 533:163-72
150. Deleted in proof
151. Hozumi, T. 1981. *J. Biochem.* 90:785-88
152. Morita, F. 1967. *J. Biol. Chem.* 242:4501-6
153. Cheung, H. C., Morales, M. F. 1969. *Biochemistry* 8:2177-82
154. Takashi, R., Tonomura, Y., Morales, M. F. 1977. *Proc. Natl. Acad. Sci. USA* 74:2334-38
155. Jeng, S. F. Guillory, R. J. 1975. *J. Supramol. Struct.* 3:448-68
156. Deleted in proof
157. Szilagyi, L., Balint, M., Sreter, F. A., Gergely, J. 1979. *Biochem. Biophys. Res. Commun.* 87:936-45
158. Morales, M. F., Hotta, K. 1960. *J. Biol. Chem.* 235:1979-86
159. Botts, J., Muhlrad, A., Takashi, R., Morales, M. F. 1982. *Biochemistry*. In press
160. Muhlrad, A., Hozumi, T. 1982. *Proc. Natl. Acad. Sci. USA*. 79:958-62
161. Muhlrad, A., Fabian, F. 1970. *Biochim. Biophys. Acta* 216:422-27
162. Friess, E. T. 1954. *Arch. Biochem. Biophys.* 51:17-23
163. Blum, J. J. 1955. *Arch. Biochem. Biophys.* 55:486-511
164. Gilmour, D. 1960. *Nature* 186:295-98
165. Warren, J. C., Stowring, L., Morales, M. F. 1966. *J. Biol. Chem.* 241:309-16
166. Rainford, P., Hotta, K., Morales, M. F. 1964. *Biochemistry* 3:1213-20
167. Tonomura, Y., Sekiya, K., Imamura, K. 1963. *Biochim. Biophys. Acta* 69:296-305
168. Yasui, T., Watanabe, S. 1965. *Molecular Biology of Muscular Contraction*, pp. 97-108. Tokyo: Igaku Shoin
169. Duke, J. A., McKay, R., Botts, J. 1966. *Biochim. Biophys. Acta* 126:600-3
170. Cheung, H. C. 1969. *Biochim. Biophys. Acta* 194:478-85
171. Kubo, S., Tokura, S., Tonomura, Y. 1960. *J. Biol. Chem.* 235:2835-39
172. Fabian, F., Muhlrad, A. 1968. *Biochim. Biophys. Acta* 162:596-603
173. Onishi, H., Tonomura, Y. 1973. *J. Biochem.* 74:435-50
174. Hozumi, T., Muhlrad, A. 1981. *Biochemistry* 20:2945-50
175. Miyanishi, T., Tonomura, Y. 1981. *J. Biochem.* 89:831-39
176. Sekine, T., Yamaguchi, M. 1963. *J. Biochem.* 54:196-98
177. Seidel, J. C., Gergely, J. 1971. *Biochem. Biophys. Res. Commun.* 44:826-30
178. Barany, M., Barany, K., Gaetjen, E. 1971. *J. Biol. Chem.* 246:3241-49
179. Shriver, J. W., Sykes, B. D. 1981. *Biochemistry* 20:6357-62
180. Highsmith, S., Akasaka, K., Konrad, M., Goody, R., Holmes, K., Wade-Jardetzky, N., Jardetzky, O. 1979. *Biochemistry* 18:4238-44

RESONANCE RAMAN SCATTERING: The Multimode Problem and Transform Methods

P. M. Champion

Department of Chemistry, Worcester Polytechnic Institute, Worcester, Massachusetts 01609

A. C. Albrecht

Department of Chemistry, Cornell University, Ithaca, New York 14853

INTRODUCTION

During the last several years an enormous amount of experimental and theoretical work has been directed toward a better understanding of the resonance Raman effect. As a result, a variety of comprehensive review articles has recently appeared (1–4) that encompass a large body of the current work. In light of the substantial nature of these previous reviews, it seems appropriate to permit a new review to concentrate upon only one particular aspect of the topic. The choice of focus here is the highly significant (but often suppressed) role played by the many dimensional vibrational subspace of a medium-to-large polyatomic molecule (and even its "surroundings") when it is coupled to an electronic excitation that is active in resonance Raman scattering (RRS). This is known as the multimode problem. It happens that this choice of topic is particularly timely. New transform techniques have recently emerged that explicitly link the conventional absorption band, which naturally contains a great deal of multimode information, to the Raman excitation profile expected for resonance scattering throughout the transition. Thus, the immense complexity of the multimode problem, which originally led it to be either ignored or suppressed, can now be approached directly with quite powerful new techniques.

Although multimode effects are formally present in most of the general theories of resonance Raman scattering, in the process of transcription to an applicable theory, the true multimode description is quite often dropped in favor of a single mode (or 2–3 mode) theory that is more amenable in practice. One serious consequence of this limitation is that Raman excitation profiles (and corresponding absorption bands), which are normally broad because of multimode effects, are analyzed, instead, in terms of artificially large damping factors. Thus, exceptionally short relaxation times are erroneously attributed to the resonant excited molecular states. In this review we try to outline some of the recent work that has been directed toward an understanding of the effects of the full multimode vibrational subspace in large polyatomic molecules. Some rather interesting phenomena arise when the full subspace is included in the actual calculations; such phenomena are not predicted by theories that rely on the superposition of single mode intensity calculations.

One aspect of this problem concerns the relationship between the absorption intensity (band shape) and the resonance Raman intensity (Raman excitation profile, REP). That there is a deep theoretical relationship between these quite different experimental observations has been known for many years, but has emerged only recently in the form of useful transform methods for determining REPs from absorption bands. The explicit participation of the full multimode subspace in this relationship we also discuss in this review.

Terminology can be a confusing aspect of an explicit approach. Here, in order to give a concise description of the various efforts that have been directed toward an understanding of multimode effects, we choose in the Basic Theory section to partition the problem along the lines of traditional vibronic spectroscopy with its own somewhat standard terminology. This approach, although not completely general, exposes the vibrational subspace explicitly through the zeroth-order Born-Oppenheimer, or adiabatic approximation. The various nuclear coordinate dependent effects that give rise to inelastic scattering phenomena can then be considered via a Taylor's series expansion of the molecular Hamiltonian in terms of the nuclear coordinates. This explicit nuclear coordinate dependence can, in some cases, be treated exactly (e.g. the Franck-Condon effect). In other cases a perturbation framework may be useful, although not always successful (Herzberg-Teller vibronic coupling). In the Specific Approaches section we review papers that deal directly with the multimode question in RRS and, on occasion, even touch upon breakdown of the adiabatic approximation. We then discuss transform techniques, which are able to formally bypass multimode complexities.

BASIC THEORY

The explicit dependence of the molecular Hamiltonian on the nuclear coordinates, Q_i is traditionally exposed via a Taylor's expansion as given by Eq. 1:

$$\hat{H}_m(q,Q) = \hat{H}_e(q,Q) + \hat{T}_N \qquad \text{1a.}$$

$$\hat{H}_e(q,Q) = \hat{H}_e^0(q,Q^0) + \sum_i \left(\frac{\partial V(q,Q)}{\partial Q_i}\right)_{Q^0} Q_i \qquad \text{1b.}$$

$$+ \frac{1}{2} \sum_{ij} \left(\frac{\partial^2 V(q,Q)}{\partial Q_i \partial Q_j}\right)_{Q^0} Q_i Q_j + \cdots$$

where T_N is the nuclear kinetic energy operator, q represents electronic coordinates, and Q^0 is a convenient choice of origin for the nuclear displacement coordinates (usually taken to be the ground state equilibrium configuration). The crude adiabatic wavefunctions (ψ_m^0) generated by H_e^0 are then taken as an electronic basis set. The remaining Q-dependent terms in Eq. 1 can be used to generate the vibrational basis functions as well as to describe the perturbing influence of nuclear motion on the crude B-O basis states.

The situation is depicted schematically in Figure 1 where $\{v_m\}$ represents a multimode vibrational basis set that is associated with the crude adiabatic Born-Oppenheimer (CABO) electronic state, ψ_m^0. The matrix elements coupling the CABO electronic states are noted explicitly up to the quadratic terms in Q. In the figure the subscripted Vs represent the various derivatives in Eq. 1. The electronic matrix elements of the derivatives shown on the diagonal blocks are used, along with $\hat{T}_N$, to find the respective multimode vibrational basis sets, $\{v_m\}$. The V_{Q_i} give rise to shifts in the nuclear equilibrium positions with respect to the ground state ($\Delta_i^{(m)}$). The $\langle \psi_m^0 | V_{Q_i Q_i} | \psi_m^0 \rangle$ are the force constants ($k_i^{(m)}$), and the $\langle \psi_m^0 | V_{Q_i Q_j} | \psi_m^0 \rangle$ result in Duschinsky mixing of the coordinate space (*wrt* the ground state). The off-diagonal blocks of Figure 1 contain matrix elements that are involved in the mixing of the electronic states due to nuclear motion (i.e. vibronic coupling). Within a Herzberg-Teller scheme, these matrix elements serve to mix the CABO states as well as to introduce explicit nuclear coordinate dependence into the electronic wavefunctions. Under certain circumstances the nuclear kinetic energy operator can mix the resulting adiabatic B-O states giving rise to non-adiabatic effects. This situation is much more complex and such mixing of the electronic basis functions is not indicated in Figure 1. The γ in Figure 1 represents radiative coupling between the ground and excited states. Vibronic effects between these states are suppressed, since their energy separation usually considerably exceeds the energy separation between the resonant state and other excited states.

The fact that there are, in general, many normal modes, Q_i, which participate in the nuclear motion means that a variety of situations may be represented by Figure 1. For example, if Q_i is a non-totally symmetric mode, the linear derivative, V_{Q_i}, will vanish from the diagonal elements. On the other hand, if Q_i is a totally symmetric mode, the linear terms will remain on the diagonal; this results in a shift of the excited state potential energy surface with respect to the ground state and gives rise to the well known Franck-Condon effect. Under certain circumstances (e.g. when ψ_1^0 and ψ_2^0 have the same symmetry) a totally symmetric mode can have nonzero linear elements in both the diagonal and off-diagonal positions. As mentioned above, the terms $V_{Q_i Q_j}$ give rise to Duschinsky mixing of the vibrational basis set and are notoriously difficult to treat in a general way. This difficulty is magnified in large multimode systems since, in principle, a given mode may be coupled to many other modes.

	$\psi_0^0 \{v_0\}$	$\psi_1^0 \{v_1\}$	$\psi_2^0 \{v_2\}$	
$\psi_0^0 \{v_0\}$	$V_{Q_i} \rightarrow \Delta_i^{(0)} \equiv 0$ $V_{Q_i Q_i} \rightarrow k_i^{(0)}$ $V_{Q_i Q_j} \rightarrow 0$	γ	γ	....
$\psi_1^0 \{v_1\}$	γ	$V_{Q_i} \rightarrow \Delta_i^{(1)}$ $V_{Q_i Q_i} \rightarrow k_i^{(1)}$ $V_{Q_i Q_j}$	$V_{Q_i} \rightarrow h_i^{(1,2)}$ $V_{Q_i Q_i}$ $V_{Q_i Q_j}$	....
$\psi_2^0 \{v_2\}$	γ	$V_{Q_i} \rightarrow h_i^{(2,1)}$ $V_{Q_i Q_i}$ $V_{Q_i Q_j}$	$V_{Q_i} \rightarrow \Delta_i^{(2)}$ $V_{Q_i Q_i} \rightarrow k_i^{(2)}$ $V_{Q_i Q_j}$	....
	⋮	⋮	⋮	

Figure 1 A schematic representation of the crude adiabatic Born-Oppenheimer approach to molecular eigenstates. The purely electronic part of the wavefunctions, ψ_m^0, is separated from the multimode vibrational state, $\{v_m\}$. (When dealing with simple quadratic potential energy surfaces, the $\{v_m\}$ are usually taken as a product of independent harmonic oscillators.) The mth diagonal block contains the electronic matrix elements that are needed to construct the nuclear potential energy surface for state m and, along with the nuclear kinetic energy operator, can be used to find the $\{v_m\}$. The off-diagonal blocks contain terms that cause mixing of the CABO states and lead to explicit nuclear coordinate dependence in the transition moment (i.e. breakdown of the Condon approximation). The γ represent the radiative coupling between the ground and excited states.

Our purpose for the presentation of Figure 1 is to help describe some of the various methods by which one can construct the excited molecular states, Φ_m, associated with a given resonant transition. These states are needed if one is to calculate explicitly either a direct single-molecule absorption spectrum, or a resonance Raman scattering excitation profile. The $T=0K$ absorption cross-section at frequency ν, can be written as

$$I_A(\nu) \propto \left(\frac{\nu}{c}\right) \sum_m \frac{|\langle \psi_0\{0\}|\mu|\Phi_m\rangle|^2 \Gamma_m}{(E_m - h\nu)^2 + \Gamma_m^2} \qquad 2.$$

where μ represents the electric dipole operator and E_m is the energy of state Φ_m (Γ_m is the associated damping factor).

Similarly, we can write the cross-section for the ($T=0K$) resonance Raman scattering as (4, 5):

$$\frac{d\sigma}{d\Omega} \propto \left(\frac{\nu}{c}\right)^4 |\sum_{\rho,\sigma} \alpha_{\rho\sigma}|^2 \quad \text{where} \qquad 3.$$

$$\alpha_{\rho\sigma} = \sum_m \frac{\langle \psi_0\{0\}|\mu_\rho|\Phi_m\rangle\langle\Phi_m|\mu_\sigma|\psi_0\{f\}\rangle}{E_m - h\nu - i\Gamma_m} + (\text{nonresonant term}) \qquad 4.$$

is the ρ,σ th component of the polarizability tensor and $\{f\}$ represents the final vibrational state of the system.

It is topical to note how the interaction of light with matter can take on different forms depending on the gauge within which the radiation field is expressed. Though a correct treatment of spectroscopies must be gauge invariant, most applications deal with a limited basis set of wavefunctions, which, furthermore, are usually approximate. Under these conditions the question of which gauge is the "best" becomes a significant one. Recently Robinson et al (6, 7) found that the momentum form (Coulomb gauge) for the field in the dipole approximation gives significantly different results for Raman scattering near resonance than does the length form (Lamb gauge), when a limited basis set is considered. Raman experiments are too few and not sufficiently refined to say definitely which is "correct" on experimental grounds alone. Long ago, however, W. E. Lamb, Jr. (8) pointed out how for ordinary absorption the momentum operator, as conventionally applied, does not give the observed Lorentzian lineshape for individual transitions as in Eq. 2. K-H. Yang (9a, see also 9b) has discussed this question from a general point of view and shows how for any spectroscopy (up to the dipole approximation in the field) any choice of gauge requires the length operator, not the momentum operator, for the mixing of exact dark eigenstates. The length form has been the common choice in most spectroscopies, certainly in Raman scattering, and must remain so (as here in Eq. 4), with any exception reserved for some unusual choice of approximate wavefunctions.

One of the main difficulties with Eqs. 2–4 in the multimode limit is that the Σ_m becomes prohibitive (in practice) as the number of vibrational modes increases. Usually it is possible to limit the sum to one electronic state in the B-O approximation (i.e. the resonant state), but the remaining vibronic manifold still presents formidable problems, especially if the energy gap between the vibronic states is smaller than (or on the order of) the damping factor Γ_m. The number of vibronic states, of course, greatly exceeds the number of normal modes in the problem since each individual mode has a complete set of associated eigenstates and energies that are needed to span the coordinate space. The density of vibronic states in energy space, $\rho(\varepsilon)$, is thus a sensitive (and explosive) function of the number of normal modes associated with the problem.

Theories of RRS can be distinguished by the level of approximation used in defining Φ_m. The simplest approach to Φ_m takes it to be a product of a crude adiabatic electronic function such as ψ_1^0, and the multimode vibrational function $\{v_1\}$. In the language of Figure 1, only diagonal vibronic coupling terms are considered. This is known as the "Condon" or "A-term" approximation in RRS (no variation of electronic transition moment with nuclear coordinates). Usually $\{v_1\}$ is written as a product of harmonic oscillator wavefunctions in the potential energy surface of the excited electronic state. Thus, diagonal vibronic coupling terms up to second order only are included, and mixed derivatives at the second order are excluded (no Duschinsky rotation). Often changes in force constants upon electronic excitation are taken to be absent. The only vibronic effect, then, is the shift of the equilibrium position of the excited state surface from that of the ground state via the diagonal linear vibronic coupling terms ($\Delta_i^{(1)}$). When, in addition, only one excited electronic state is present in the resonance region, this set of approximations has come to be called the "standard assumptions." Deviations from these include the variation of the electronic transition moment with nuclear coordinates (non-Condon or B- or C-term effects) as prescribed by the *off-diagonal* vibronic coupling terms (linear and higher) in Figure 1. Then there are the so-called "quadratic vibronic coupling" effects, which are concerned with force constant changes and mode mixing (Duschinsky rotations) upon electronic excitations. These are diagonal at the quadratic level in Figure 1, but exclude that part which directly recovers the ground state force field. Anharmonic corrections appear at higher order. In addition to the above, more than one electronic transition might act within the region of resonance. Finally, at one of the most refined levels, nonadiabatic effects, outside of Figure 1, must be considered.

In the following section we discuss some of the techniques that have been used to approximate the Φ_m and Σ_m in Eqs. 2–4 when a number of normal modes are assumed to be active (i.e. the matrix elements depicted in Figure 1 are nonzero for more than one of the Q_i). The recently developed transform techniques, which relate Eqs. 2 and 4 in a general way, and are particularly relevant in the multimode limit, we examine separately in the Transform Methods section.

SPECIFIC APPROACHES

A number of authors have recognized the difficulties inherent in Eqs. 1–4 and have utilized a variety of techniques in the calculation of absorption bands and Raman excitation profiles. Generally speaking, the calculations involve two related, yet separable, problems: (a) obtaining a good approximation to the excited molecular wavefunctions (Φ_m) and energies (E_m), and (b) performing the sum over excited states. In practice it appears that one must find a reasonable balance between the number of modes to be treated and the accuracy of the wavefunctions. As the calculation of Φ_m becomes more refined (e.g. inclusion of Duschinsky and nonadiabatic effects) it becomes more difficult to handle large numbers of modes. Conversely, the calculations that comfortably handle many degrees of freedom are most likely to suffer from simple approximations to Φ_m.

Calculation of Φ_m

An extensive amount of work on the problem of calculating Φ_m (with emphasis on nonadiabatic effects) has been previously reviewed by Siebrand & Zgierski (3). Here we discuss briefly some of the key contributions; particularly those that involve multimode calculations.

The role of nonadiabatic coupling in the calculation of excited molecular states and its application to RRS has been explored explicitly by Small & Yeung (10), Friedman & Hochstrasser (11), and Gregory et al (12). These theories are much more complex than the simple CABO picture outlined above. These approaches are generally more appropriate for the calculation of the Φ_m when the electronic energy spacings are small (i.e. on the order of the vibrational quanta). The excited states in these more elaborate theories are usually taken as linear combinations of the CABO basis set. Unfortunately, the complications involved in the nonadiabatic calculations become magnified as the number of participating modes increases.

Usually, practical calculations are performed with only one or two modes considered active. For example, early work by Friedman &

Hochstrasser (13) has treated the case of two-mode interference phenomena in Raman scattering. The single mode nonadiabatic theory of Gregory et al (12) has been developed in terms of a unitary transformation to a diabatic basis set. Detailed discussion and example calculations based on this model can be found in papers by Siebrand (14), Henneker et al (15, 16), and Siebrand & Zgierski (3).

Recently, the model has been expanded by Henneker et al (17) to include two Raman active modes; one totally symmetric and one nontotally symmetric. A variety of computer calculations are presented in this work that pass from the limits of weak coupling

$$\langle \psi_m^0 | V_{Q_i} | \psi_n^0 \rangle \ll (E_n - E_m)$$

to a strong coupling (i.e. Jahn-Teller) regime. Explicit discussion of intermediate cases is also included. Special attention is given to the totally symmetric coordinate, which is allowed to assume different equilibrium positions in the two vibronically coupled states (i.e. $\Delta_{Q_s}^{(1)} \neq \Delta_{Q_s}^{(2)}$ in Figure 1). Breakdown of the Condon approximation is investigated in a variety of theoretical situations.

In a similar vein, Köppel et al (18a–d) have explored multimode effects of a totally symmetric vibrational subspace on the vibronic coupling induced by a nontotally symmetric mode. Although their work is focused primarily on the calculation of photoelectron spectra of small molecules, the generality of their approach makes it applicable to a variety of problems. When the electronic energy separations are small (or if the number of symmetric modes is very large), they have shown that the modulation of the electronic energy levels by the totally symmetric modes can strongly affect the vibronic coupling. This leads to nonadiabatic behavior between the two states over a wider range of normal coordinate space than might otherwise be expected from a single mode calculation. The spectra presented in Ref. (18) are calculated using numerical methods and two to four active modes are treated explicitly.

Using a different approach, Shelnutt et al (19–22) have considered a variety of problems involved in the calculation of Raman excitation profiles. This approach also tends to focus on various techniques for obtaining the excited molecular wavefunctions. The method relies on a perturbation theory that utilizes crude Born-Oppenheimer wavefunctions as a basis set. However, the ground state harmonic oscillator functions are consistently used as the vibrational basis to calculate the excited state wavefunctions. Unfortunately, the convergence of this theory is limited as a result of this choice of basis functions. Simple Franck-Condon effects are difficult to treat since the diagonal matrix elements (the $\Delta_i^{(m)}$ in Figure 1) are treated as perturbations on an equal footing with the off-diagonal elements ($h_i^{(n,m)}$). In principle, several normal modes can be considered simultaneously with Shelnutt's computer approach. Nonadiabatic, interference, and Duschinsky effects as well as "helping mode" phenomena can be treated with this theory as long as the number of modes and their Franck-Condon activity are (implicitly) assumed to be small enough to allow for the convergence of the perturbation expansion. Some consequences of this approach, and its application to the Raman excitation profile data for cytochrome c, have been previously discussed by Champion & Albrecht (23).

Another difficulty arising in the calculation of the Φ_m concerns mode-mixing or the Duschinsky effect. The problem of Duschinsky rotation of the normal coordinate space is, by its very nature, a multimode phenomenon. Several authors have recently addressed this question in the context of Raman scattering. Perhaps the most comprehensive (and comprehensible) treatment is that given by Hassing & Mortensen (24). These authors carry out the Herzberg-Teller expansion in the crude adiabatic basis set and collect all terms in the excited state vibrational potential that have a $Q_i Q_j$ dependence. Two such terms arise. The most obvious is the $V_{Q_i Q_j}$ "intrastate" term that is found in the diagonal positions of Figure 1. A more subtle "interstate" contribution arises from the Herzberg-Teller expansion in conjunction with the linear nuclear potential energy terms. Sharf & Honig (25) and Small (26) have treated this second contribution in great detail and have shown how an "induced" Duschinsky rotation must occur whenever two or more normal modes are active in vibronic coupling. Hassing & Mortensen (24) have investigated the interplay of both of the Duschinsky terms and have presented explicit multimode calculations of the polarization dispersion and the Raman excitation profiles. Moreover, they have shown that the Duschinsky terms and the nonadiabatic terms are of the same order of magnitude and that these effects may cancel (or reinforce) each other in the Raman excitation profile. They suggest that polarization dispersion measurements may be of great help in the separation of the Duschinsky and nonadiabatic contributions to the Raman scattering tensor.

Siebrand & Zgierski (27) have also treated the Duschinsky effect using the diabatic transformation theory discussed earlier (14-17). The case of two non-totally symmetric modes involved in vibronic coupling is treated explicitly at the non-adiabatic level and a variety of excitation profiles and polarization dispersion curves are displayed. In a further development, Zgierski & Pawlikowski (28) have recently applied Duschinsky mixing effects (two totally symmetric modes) along with earlier vibronic interference models (29) in order to account for the Raman excitation profiles of copper tetraphenylporphyrin.

Sum-Over-States

The second main problem in the evaluation of Eqs. 2–4 involves the sum-over-states. As the number of normal modes increases, the number of possible states, Φ_m, increases dramatically since each additional mode brings with it a complete set of basis states (these are often taken to be the harmonic oscillator basis functions). It becomes quite difficult to focus on the sum-over-states problem without using a simple form for the Φ_m. The very elaborate and general calculations of Φ_m discussed in the previous section become difficult to apply when many modes are considered simultaneously. As a result, most authors who face the sum-over-states problem in a true multimode sense (i.e. more than two or three modes) must utilize the simple crude adiabatic Born-Oppenheimer wavefunctions discussed in the Basic Theory section. As we have indicated, often a simple product of harmonic oscillator basis functions is taken to represent the vibrational component of both the ground and excited molecular states. Further calculations (often aided by computer summation techniques) are then facilitated by the well-defined mathematical properties of the Hermite polynomials.

Some of the earliest multimode calculations involving Raman spectra concentrate on the changes in molecular geometry that accompany electronic excitation. These calculations lend themselves naturally to a normal mode analysis of the molecular system. Suzuki et al (30) studied the relative Raman intensities of several active modes of tetrachlorobutadiene and used this information to predict both the excited state molecular structure and the ultraviolet absorption spectrum. Warshel & Dauber (31) have derived explicit expressions for the multimode contributions to the Raman scattering. Their approach also relies on simplifying assumptions to calculate Φ_m (e.g. adiabatic Born-Oppenheimer approximation and separable harmonic oscillator vibrational basis states). These authors predict relative Raman intensities through explicit calculation of electronic transition moments and their derivatives with respect to the normal coordinates. Both Condon and non-Condon effects are thereby considered. The actual calculations presented, however, are simplified (one- or two-mode) versions of the more general multimode expressions.

More recently, Peticolas et al (32) have explored multimode calculations of relative Raman intensities in order to aid in the determination of molecular force fields. The usual crude adiabatic and harmonic oscillator approximations are employed; however, only totally symmetric Franck-Condon modes are considered in the calculations. The relative intensities of the Raman lines are related to changes in the excited state equilibrium geometry and compared to molecular orbital bond order calculations. In molecules of low symmetry, where the number of interaction constants greatly exceeds the number of observed frequencies, Peticolas et al suggest that this approach may be of great help in finding the true set of molecular force fields.

In addition to the above work, which focuses on the relative intensities of Raman scattered lines in a Raman spectrum excited at one frequency, there has been a great deal of effort directed at the calculation of the dispersion of Raman intensities throughout the entire resonance region (i.e. calculation of the REP) in the multimode limit. Some of the specific problems inherent in such calculations have been outlined by Siebrand & Zgierski (33). These authors have focused on Franck-Condon effects in the multimode limit and have examined multiple electronic state contributions and mode mixing (Duschinsky effect) in the analysis. Crude adiabatic Born-Oppenheimer product functions with harmonic oscillator basis states are used in their computer sum-over-states approach. Their work also tries to incorporate another important factor in RRS, that of inhomogeneous sources of broadening. For example the distribution of molecules over different sites in a sample (ensemble effect) can cause broadening of an absorption band that is fundamentally different from multimode phenomena (single molecule effect). Inhomogeneous broadening effects cannot be folded into RRS theory at the amplitude level, where multimode effects directly appear. Thus, while both sources of broadening appear at the modulus square level in the theory of absorption, they must be explicitly separated in order to arrive at a unified theory RRS and absorption. The actual calculations (33) turn out to be limited to two or three modes and are applied to chromate ion and β-carotene. In this latter example, rather large values for the homogeneous damping factors and the inhomogeneous broadening are obtained in a multiparameter fit to the REP data.

Hoskins (34) has also explored a multimode model in an attempt to understand the REP's of lycopene, a molecule closely related to β-carotene. Again the crude adiabatic Born-Oppenheimer approximation along with the harmonic oscillator basis and a computer sum-over-states is employed. This development also considers mode mixing and uses two to three modes in the actual calculations. The homogeneous damping factors found for lycopene ($\Gamma_{FWHM} \approx 900$ cm^{-1}) in this analysis appear to be excessively large.

More recently, Kodama & Bandrauk (35) have explored a "true" multimode theory that takes into account more than two to three modes. In particular, they have considered low frequency phonon modes in the analysis of the Raman scattering of β-carotene in order to reduce the large size of the homogeneous damping factors (lifetime broadening) found in previous analyses of the data (33, 34). A simple Einstein oscillator model is used in order to simplify the computer sum-over-states.

This approach is quite similar in spirit to the work of Champion & Albrecht (36), discussed below. Here, the multimode effects due to the low frequency phonon subspace are condensed into a single Einstein oscillator that can have its excited state displacement varied in order to fit the experimental data. This results in a broadening of the calculated spectral features without having to resort to large values for the damping factor (i.e. ultrashort lifetimes).

Hassing & Mortensen (37) have also discussed an explicit multimode model involving two modes. They have shown how interference between a totally symmetric and non-totally symmetric mode can mimic the same type of dispersion (REP) that has been attributed to nonadiabatic effects (38). Such results expose some of the basic ambiguities inherent in the present state of Raman theory. (More experiments are clearly needed to resolve some of these questions.)

As an example of *single*-mode, *multi*-state summation theories we should also mention the elaborate Green's function approach of Hong (39, 40). This is an excellent example of how the sum-over-states for even a single mode can be quite complicated. Strong vibronic coupling of two electronic states as well as Franck-Condon effects are treated with a continued fraction solution for the exact Green's function. A variety of absorption spectra, REPs, and quantum yields are calculated and displayed.

A more formally powerful multimode Green's function treatment has been developed in a series of papers by Fujimura & Lin (41–43). These authors present the general multimode expression for the Raman scattering cross-section in the displaced harmonic oscillator model. The actual model calculations presented are confined to only two modes, however. Both strong and weak coupling are considered and the effect of nonadiabatic corrections is assessed and found to appear only in the high order approximations. Extension of this theory to nonzero temperatures (42) is also discussed in the multimode limit and Stokes/anti-Stokes ratios are calculated. Example single mode calculations of the temperature dependence of the Raman excitation profile are also presented. Finally, this very general theory is expanded (43) to produce analytical expressions for the resonance Raman scattering and resonance fluorescence cross-sections (derived in the displaced harmonic oscillator model).

Working in another direction, Korenowski et al (44) and Champion & Albrecht (36) have explored analytical models that encompass a large multimode Franck-Condon subspace. Closure of the sum-over-states by integration techniques has proven quite successful from a phenomenological point of view. Korenowski et al have used a semiempirical approach to the problem. They have duplicated the observed absorption spectrum of benzene by convolving a Gaussian distribution function with the Franck-Condon progression of the Raman active mode under investigation. [A similar use of a Gaussian distribution function to account for multimode effects has also been discussed by Lukashin & Frank-Kamenetskii (45).] The use of this distribution function in the calculation of the REP then allows for much smaller (and more reasonable) excited state damping factors.

The simultaneous modeling of the absorption band shape and REP of cytochrome c (36) is accomplished by using closed form integral solutions for Eqs. 2–4. (See Figure 2.) In this model the homogeneous damping factor is found to be much smaller than the absorption bandwidth. Moreover, the calculated REP is much more sensitive to the size of the damping factor than is the absorption bandshape. The greatly reduced damping factor seems to be a general property of calculations that embrace the full vibrational subspace (35, 44). [The small damping factor is also found to be consistent with absolute quantum yield measurements that give estimates of the excited state lifetime in cytochrome c (46, 47).] In fact, the Franck-Condon/Density-of-States distribution function (36, 48) serves to "fill out" the absorption spectrum and REP. The parameters that describe this effect condense our lack of detailed knowledge about the complicated FC/DOS function into phenomenological variables that can be used to "fit" the experimental data. The analytical model presented by Champion & Albrecht (36) has been subsequently shown (49) to be an exact representation of the transform theory discussed in the next section.

Further work on the role of librational broadening in molecular transitions has been presented by Korenowski & Albrecht (50) where particular attention has been given to the C-H local mode overtones of benzene. Two- and three-mode calculations have also been performed using a computer sum-over-states technique to assess the multimode interference effects in smaller molecules (51, 52).

The relationship between multimode effects and vibronic borrowing in the calculation of absorption spectra has been recently explored by Champion (53). This work focuses on the Herzberg-Teller (H-T) approach to weakly allowed transitions in the multimode limit and shows how vibronic intensity can be induced at the 0-0 transition frequency, contrary to the traditional assumptions. Explicit multimode expressions are derived for the 0-0 and 0-1 transition amplitudes that can be used in a computer sum-over-states calculations of the vibronically induced band shape. Breakdown of the H-T approach for multimode systems is also discussed.

Champion & Albrecht (48) have also used a computer modeling approach in order to elucidate the effects of a large multidimensional Franck-Condon subspace on the absorption band shape of a strongly allowed electronic transition. These calculations use an exact counting density-of-states approach that captures the essential character of the observed spectral band shapes. Nevertheless, these calculations are flawed in that the computer is unable to consider all the possible final states of the true multimode system within a reasonable computation time (20–40 modes implies ca. 10^{10} states in which one or more modes carries one quantum of excitation in the upper electronic state). As a result, calcu-

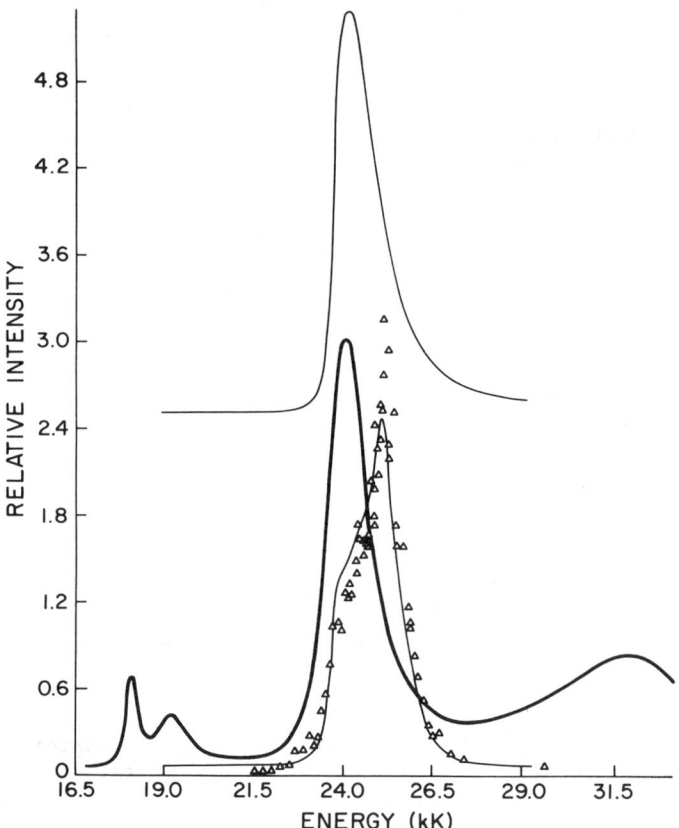

Figure 2 The resonance Raman excitation profile of 1362 cm^{-1} mode of ferrocytochrome c in the Soret band (S$_2$) region. The *solid dark line* is the observed absorption spectrum. The small *triangles* represent the REP data points and the *thin solid lines*, I_A (*top*) and REP (*bottom*), are the results of a calculation described in Ref. (36). The calculation results in "transform pairs" for the absorption and REP that depend upon three main parameters. Two of these are needed to simulate the Franck-Condon/Density-of-States function. The third is the damping factor, Γ_m, which can be varied between 10–100 cm^{-1} and still generate reasonable fits to the data. In this figure the damping factor is taken to be 50 cm^{-1}.

lated spectral band shapes underestimate the multimode effects. Purely theoretical arguments help us to approximate these errors by predicting the exponential form of the ratio of the 0-0 transition intensity to the total band intensity (this ratio is analogous to the "recoiless fraction" in Mossbauer spectroscopy). Thus, we can make accurate estimates of the errors inherent in the computer sum-over-states technique (48).

When one adds to this intramolecular multimode picture the additional complications of very low frequency modes arising from intermolecular potentials in the condensed phase, the problem appears to be overwhelming. Yet, an encouraging fact is that the vast majority of both intra- and intermolecular multimode effects must already be contained in the observed absorption band shape. Blazej & Peticolas recognized this point in their analysis of the observed REPs from the pyrimidine mononucleotides (54). They reasoned that were one to (formally) remove the FC subspectrum of the Raman scattered vibration from the absorption spectrum, one could apply the conventional Kramers-Kronig transform to the remainder of the spectrum in order to obtain the real part of the molecular polarizability (the imaginary part of the polarizability is related to the absorption spectrum via the optical theorem). The Raman active vibration is reintroduced explicitly under the usual set of "standard assumptions" to obtain REPs in which the multimode problem in $3N-7$ space has been subsumed by employing the transform. It was also noticed how the absolute measure of the REP determines the FC coupling parameter [S_k in Ref. (48), $\Delta^2 j/2$ in Ref. (54)] of the scattered vibration. Thus, through this transform method, excited state geometries of the pyrimidine nucleotides are deduced from the REP and the related absorption data. As it turns out, this approximate approach to the multimode problem is in the vein of an exact general transform method for determining REPs that was discovered more than a decade ago by a group of Estonian physicists. The general realm of powerful transform techniques for determining REPs constitutes the subject of the next section.

TRANSFORM METHODS

A major development in the theory of RRS has roots in work from the Institute of Physics and Astronomy of the Estonian Academy of Sciences. Nearly fifteen years ago the theory of "secondary radiation" was presented in the language of the time correlation formalism—the time integral form of the Kramers-Heisenberg expression for the linear polarizability. Here the concept of "secondary radiation" embraces all possible one-photon forms of radiation that emerge from a sample that has been, or is being, illuminated. In this work, principally by Hizhnyakov & Tehver, the distinction between scattering and luminescence was clearly drawn (55–59). [See also Toyozawa (60) for a clear discussion of this question.] The former consists of both Rayleigh and Raman scattering, in which the secondary photon is emitted prior to the dephasing of the intermediate state. All light appearing after the dephasing of the intermediate state is termed "luminescence" and consists of "hot luminescence" when it originates from the (dephased) intermediate state (resonance fluorescence), or states not yet in thermal equilibrium. "Ordinary" luminescence occurs when light originates from thermally equilibrated states less energetic than the intermediate level. The latter luminescence would include both conventional fluorescence and phosphorescence.

Of special consequence to multimode RRS, however, is a transform law for the off-diagonal (i.e. $\{i\} \neq \{f\}$) molecular polarizability (for RRS) very much akin to the transform law, already established, for the diagonal Rayleigh polarizability ($\{i\} = \{f\}$) (55, 56, 61, 62). How the diagonal linear molecular polarizability can be obtained from the absorption band is well known. The Kramers-Kronig transform of the band gives the real part of the polarizability, while the optical theorem shows how the absorption function itself represents the imaginary part. Thus, the complex molecular polarizability at frequency ν is directly related to a function given by:

$$\phi(\nu) = P \int d\nu' I_A(\nu') [\nu'(\nu' - \nu)]^{-1} - \pi i I_A(\nu)/\nu \qquad 5.$$

(where P denotes principal value). The excitation profile of Rayleigh scattering is just proportional to $|\phi(\nu)|^2$ and is thus fully determined by the absorption band (via Eq. 5). The important discovery by Hizhnyakov & Tehver is that the REP of a Raman line can be similarly related to the absorption band. They have shown how the REP is simply proportional to the square modulus of the difference between the molecular polarizability at the incident frequency, ν, and the molecular polarizability at the scattered frequency, ν_s. That is,

$$I_R(\nu) \propto S |\phi(\nu) - \phi(\nu_s)|^2 \qquad 6.$$

where S is the linear coupling strength of the scattered mode. Thus, the off-diagonal polarizability component important to RRS has as its real part the difference of the Kramers-Kronig transform between the observed absorption band and the identical band displaced by the frequency of the Raman scattered vibration (note that the linear ν dependence of Eq. 2 is factored out); the imaginary part is this difference itself. It is important to note that the transform is expressed in terms only of the incident frequency, the scattered frequency, and the absorption band; the absolute magnitude scales with the coupling strength. The complete transform prescription developed by Hizhnyakov & Tehver appears as an infinite series in different orders of phonon (or vibrational quanta) interaction, of which Eq. 6 is the leading odd numbered (non-Rayleigh) term. As such, it corresponds to the Stokes scattering of fundamentals at 0 K. The complete series, however, in which each term has its own transform prescription, is very general and holds for Stokes and anti-Stokes scattering of fundamentals as well as overtones and at any temperature!

The transform method for determining REPs is the epitome of a multimode approach. It contains within it all of the multimode information that is carried by the absorption band. This technique formally bypasses the explicit single and multimode sum-over-states approach to the modeling of an observed REP. Simply said, the transform of the absorption band is the theoretically expected REP. When disagreement with experiment arises, one can search for its origins in having more than one electronic transition (with different vibronic coupling parameters) in the same resonance region, in inhomogeneous effects related to site broadening, in neglected contributions from the nonresonant terms of Eq. 4, or, ultimately, in the failure of the "standard assumptions."

Though the transform method introduced by Hizhnyakov & Tehver has long been in print, it did not come to the attention of most practicing spectroscopists until a paper by Tonks & Page in 1979 (63). Here the transform law is clearly presented; the set of "standard assumptions" under which it holds is set forth. By way of testing these assumptions the method is used to predict REPs in β-carotene and cyanocobalamin with encouraging agreement with experiment (see Figure 3a). Hassing & Mortensen (64) derived the same transform law under the set of "standard assumptions" but used the conventional optical theorem and the usual Kramers-Heisenberg formulation (Eqs. 3, 4) as the starting point. Through model calculations they emphasized the importance of temperature effects upon calculated REP, which are not contained in the leading term of the transform law. Their approach is straightforward and illustrates some of the details of the transform approach, especially when polarization of the scattered light and asymmetric polarizability tensors are considered. More recently, Tonks & Page (65) have presented a helpful, comprehensive reexamination of the time correlator based derivation of the transform law. They show that it indeed formally accounts for temperature effects, though only after higher order terms in the Hizhnyakov & Tehver expansion are included (as we mention above).

Thus the formal transform prescription becomes complicated, though still explicit, at temperatures when FC active modes can be thermally activated.

Deviations from the "standard assumptions" within the time correlator approach have been examined. Non-Condon effects, or B and C-term corrections (variation of the electronic transition moment with nuclear coordinates) were formally included long ago by Tehver (61); however, quadratic vibronic effects (frequency shifts upon electronic excitation, mode mixing, or Duschinsky rotation) have been examined only more recently by Hizhnyakov & Tehver (62) and by Tonks & Page (66). The latter, in a model calculation valid at 0 K, show how the REP of a weakly

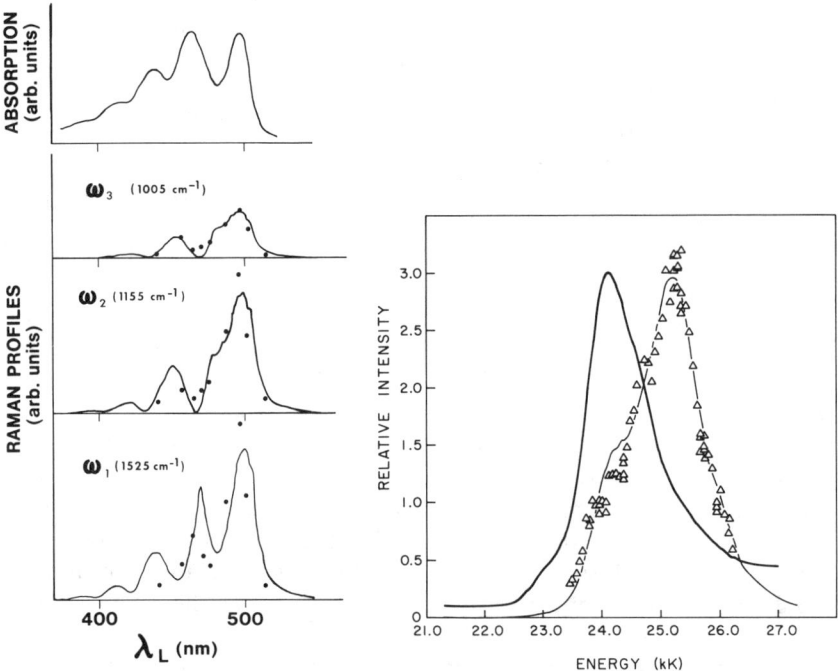

Figure 3 Applications of the transform method: A. The absorption and REP data from β-carotene (Inagaki, D., Tasumi, M., Miyazawa, T. 1974. *J. Mol. Spectrosc.* 50:286–93). The transform (Eq. 6) of the absorption spectrum using the three scattered frequencies leads to reasonably good fits to the three observed REPs (solid circles in Figure). Both the absorption and REP were accumulated at low temperature. [Figure from Ref. (63).] B. Applications of the transform method: The absorption (*heavy line*) and REP data (Δ) (1362 cm^{-1} mode) from ferrocytochrome c along with the corresponding REP (*light line*) determined by the transform of the absorption. The observations are made at 77 K. In this case we have included some weak non-Condon effects due to vibronic borrowing, leading to a slightly altered form of Eq. 6 for the transform law. The details of these calculations can be found in Ref. (70).

coupled vibration can be strongly modified by mixing with a mode that is strongly FC coupled to the electronic transition, a point already made earlier (33).

In 1979 Lee & Heller (67) also explored the general theory of RRS in the framework of time evolution formalism. Their work differs from that of Hizhnyakov & Tehver, not in its theoretical basis, but at the interpretive and calculational level. RRS is analyzed in terms of wave-packet propagation in the excited potential energy surface. RRS is dealt with at different levels of approximations extending from the "standard assumptions" to non-Condon effects, and the quadratic vibronic coupling parameters corresponding to frequency shifts and mode-mixing Duschinsky rotations. Their work necessarily formally contains the same generality found in the transform laws developed by Hizhnyakov & Tehver since both proceed from the same starting point, though the latter explicitly includes temperature effects. Differences will appear at the calculational level, particularly as numerical techniques are required when working outside of the "standard assumptions." Tannor & Heller (68) have presented model REP and absorption band calculations and compared their results to cases in which sum-over-states methods have been used (e.g. see 48) within the context of the "standard assumptions." The results for REP calculations are comparable to what can be obtained when using the transform law of Eq. 6. Their approach to multimode absorption band calculations, however, is clearly superior to a brute force sum-over-states.

A somewhat different approach to the transform theory has been outlined by Champion & Albrecht (49). This work focuses on the modeling of experimental data using a "transform compatible" theory. It is argued that the band shapes of electronic transitions and the REP should be modeled simultaneously in order to extract the maximum information from the experimental data. When absolute intensity measurements are available, however, the transform theory by itself can give an estimate of the coupling strength. This was pointed out previously (45, 54), is implicit in the work of Warshel & Dauber (31), and has been applied by Stallard et al (69) for cytochrome c (see Figure 3b). The transform theory as developed by Champion & Albrecht (49) relies on the "standard assumptions" but allows for the phenomenological modeling of both the REP and absorption band shapes simultaneously. Unlike the previous transform theories, this approach explicitly exposes the excited state damping factor and (implicitly) depends upon the Franck-Condon/Density-of-States function (48) associated with the electronic transition. Calculations of absorption-REP transform pairs are discussed using an analytic form (36) of the theory. Some examples of these transform pairs are presented in Figure 4.

One might be left with the impression that a correct transform method for REPs must have as its basis the time evolution approach to scattering. In fact, since the Kramers-Heisenberg sum-over-states expression, itself, is just the consequence of the time integration, the transform theory must follow directly from it, as well. That this is the case is already indirectly evident from the optical-theorem approach (64). With the K-H expression as the starting point, we have found (70) that not only does the simple transform theory follow from the sum-over-states expression, but

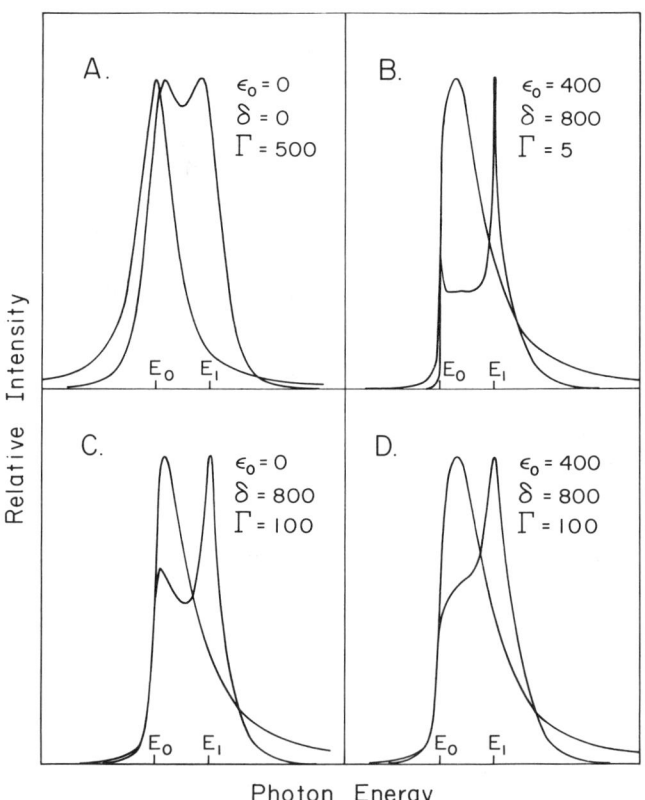

Figure 4 Several I_A-REP transform pairs generated using the analytic forms of Ref. (36). The Franck-Condon/Density-of-States function is described by a truncated Lorentzian (width is δ, displacement is ε_0). For cases B–D, the damping factor, Γ, is much smaller than the absorption bandwidth. The absorption curves, I_A, have one maximum while the REPs have two maxima, one located close to the 0-0 transition energy, E_0, and the other, one Raman quantum away, at E_1. The Raman quantum is 1362 cm^{-1}. In the calculation A, the FC/DOS function approaches a delta function limit representing a single mode model. Comparison of B to D shows how subtle variations in the absorption band shape can lead to drastic differences in the REP. Details regarding this figure can be found in Ref. (49).

refinement to include non-Condon effects in the scattered mode is straightforward. In fact, the transform can be formulated under conditions in which the adiabatic approximation is introduced only in the coordinate of the scattered vibration. The remaining nuclear-electronic space can be treated outside of the adiabatic approximation in the language of genuine molecular eigenstates. Though certain approximations are required, they are apparently less restrictive than those normally required in the full-adiabatic approximation.

Perhaps the most serious failure of the transform method will not lie in its theoretical structure but in difficulties arising when the individual molecular states attached to the resonant absorption band carry significantly different vibronic coupling parameters in the scattered mode. A resonance absorption consisting of two nearly degenerate, very different electronic transitions would be one example; however, if the concept of the inverse transform (49) [REP → $I_A(\nu)$] proves to be a viable one, a new tool becomes available for resolving complex absorption bands into their electronic subspectra.

In any case, when applying the transform method to an experimental absorption band, even when it can be characterized by a single set of vibronic coupling parameters, it can be essential to separate multimode broadening phenomena from inhomogeneous broadening. The transform approach is essentially a single-molecule theory in which multimode broadening is automatically (and correctly) taken into account. Ensemble effects such as site broadening must be deconvoluted from an observed spectrum so that the single molecule absorption spectrum, $I_A(\nu)$ can be found. If the full transform theory is not used (i.e. higher orders), temperature effects are to be treated together with site distribution phenomena in a similar fashion. Specific ground state and specific site absorption bands must be individually transformed to REPs that are then ensemble-averaged by the Boltzmann distribution function or a site distribution function to give an observed REP. In this respect, the multimode broadening can be considered as "homogeneous" in that it is associated with a single molecule [see Kodama & Bandrauk (35) for example]. On the other hand, since multimode broadening is clearly not a lifetime effect, we have encountered just the opposite terminology, i.e. that multimode broadening is an inhomogeneous effect! This view can be reinforced by a mathematically appealing argument that stems from consideration of the absorption cross-section as a sum of individual transitions, each having a (Lorentzian) width determined by lifetime and pure dephasing effects. These individual transitions can be taken as the manifold of multimode transitions belonging to a single molecule in addition to the transitions arising at different energies due to ensemble effects. The fallacy in this approach is not apparent when the absorption process is considered independently from the Raman scattering, and for this reason it is tempting to label multimode broadening as simply another form of "inhomogeneous" broadening. That this is misleading can be seen by a simple consideration of the expression for the Raman scattering intensity. Equation 4 shows clearly that the sum over the intermediate states of an individual molecule must occur at the amplitude level. The general connection between absorption and Raman scattering via the transform (Eq. 6) then shows that multimode broadening must be considered as separate and distinct from inhomogeneous (ensemble) effects.

SUMMARY

We have briefly reviewed some of the recent developments in resonance Raman scattering theory that attempt to treat the multimode nature of scattering phenomena in complex molecular systems. Approaches that focus on the detailed calculation of the excited state wavefunctions appear to be excessively difficult to apply in practice when the number of vibrational modes exceeds two or three. These difficulties arise most consistently when perturbational approaches are used and nonadiabatic effects, Duschinsky rotations, etc are considered.

The sum-over-states problem has met with more success in the multimode limit when simple wavefunctions are used. There is promise that brute force summation techniques can be adequately replaced by analytic approximations. Moreover, if applied with care, multimode analysis techniques may allow us to gain reliable information about the excited state geometries and the normal modes of complex molecules.

In a more general sense, however, the transform approach comes to the heart of resonance scattering theory. No matter whether the transform prescription is cast explicitly in the time domain or in the frequency domain, it applies as well to the one-dimensional problem of diatomics as to the many dimensional, multimode space of large polyatomic molecules. The complexities of the multimode problem are naturally "built-in" to the theory of both absorption and scattering. Appropriate modeling techniques will allow us to treat multiple excited states and quadratic as well as linear coupling. Absolute intensity measurements along with careful theoretical analysis should yield values for the coupling constants themselves. Absorption band shape and REP calculations can give us reasonable estimates of the excited state damping factors (lifetimes) as well as approximations to the Franck-Condon/Density-of-States function, which is a manifestation of the large (linearly coupled) multimode subspace associated with complex molecules.

There are signs that under appropriate conditions the transform techniques continue to be valid even outside of the adiabatic approximation, at least for those nuclear coordinates not directly involved in the scattering. It remains to be seen how much substance there is to this promise of increased generality. For the immediate future, in any case, there is an urgent need for much quantitative REP work of the best quality in order to put the transform methods to experimental testing. In complex molecules the questions of inhomogeneous broadening and multiple electronic transitions within a resonance region are likely to cloud the issue. As we develop more confidence in the transform methods, however, these very techniques may offer an entirely new way of examining such questions.

ACKNOWLEDGMENTS

A. C. A. wishes to thank Professor John Page for several very helpful discussions regarding the transform method. We acknowledge support through grants from the National Institutes of Health (AM-30714) and the National Science Foundation (CHE-8016526) and the Materials Science Center of Cornell University.

Literature Cited

1. Johnson, B., Peticolas, W. 1976. *Ann. Rev. Phys. Chem.* 27:465–91
2. Warshel, A. 1977. *Ann. Rev. Biophys. Bioeng.* 6:273–300
3. Siebrand, W., Zgierski, M. 1979. In *Excited States*, ed. E. C. Lim, 4:1–132. New York: Academic
4. Sonnich Mortensen, O., Hassing, S. 1980. In *Advances in Infrared and Raman Spectroscopy*, ed. R. Clark, R. Hester, 6:1060. London: Heyden
5. Tang, J., Albrecht, A. C. 1970. In *Raman Spectroscopy*, ed. H. Szymanski, 2:33–68. New York: Plenum
6. Berg, J. O., Robinson, G. W. 1977. *Isr. J. Chem.* 16:235–40
7. Robinson, G. W., Auerbach, R. A. 1980. *J. Chem. Phys.* 74:2083–90
8. Lamb, W. E. Jr. 1952. *Phys. Rev.* 85:259–76
9a. Yang, K-H. 1976. *Ann. Phys. NY* 101:61–96
9b. Yang, K-H. 1982. *J. Phys. A* 15:437–50
10. Small, G., Yeung, E. 1975. *Chem. Phys.* 9:379–83
11. Friedman, H., Hoschstrasser, R. 1973. *Chem. Phys.* 1:457–67
12. Gregory, A., Henneker, W., Siebrand, W., Zgierski, M. 1976. *J. Chem. Phys.* 65:2071–87
13. Friedman, J., Hochstrasser, R. 1975. *Chem. Phys. Lett.* 32:414–19
14. Siebrand, W. 1977. *Chem. Phys. Lett.* 51:5–7
15. Henneker, W., Penner, A., Siebrand, W. Zgierski, M. 1978. *J. Chem. Phys.* 69:1884–96
16. Henneker, W., Penner, A., Siebrand, W., Zgierski, M. 1978. 69:1704–21
17. Henneker, W., Siebrand, W., Zgierski, M. 1981. *J. Chem. Phys.* 74:6560–79
18a. Köppel, H., Domcke, W., Cederbaum, L. S., von Niessen, W. 1978. *J. Chem. Phys.* 69:4252–63
18b. Köppel, H., Cederbaum, L. S., Domcke, W., von Niessen, W. 1979. *Chem. Phys.* 37:303–17
18c. Domcke, W., Köppel, H., Cederbaum, L. S. 1981. *Mol. Phys.* 43:851–75
18d. Köppel, H., Cederbaum, L. S., Domcke, W. 1982. *J. Chem. Phys.* In press
19. Shelnutt, J. Cheung, L., Chang, R., Yu, N-T., Felton, R. 1977. *J. Chem. Phys.* 66:3387–98
20. Shelnutt, J., O'Shea, D. 1978. *J. Chem. Phys.* 69:5361–74
21. Shelnutt, J. 1980. *J. Chem. Phys.* 72:3948–58
22. Shelnutt, J. 1981. *J. Chem. Phys.* 74:6644–57

23. Champion, P. M., Albrecht, A. C. 1981. *J. Chem. Phys.* 75:3211–14
24. Hassing, S., Mortensen, O. S. 1981. *J. Mol. Spectrosc.* 87:1–17
25. Sharf, B., Honig, B. 1970. *Chem. Phys. Lett.* 7:132–36
26. Small, G. 1971. *J. Chem. Phys.* 54:3300–6
27. Siebrand, W., Zgierski, M. 1979. *Chem. Phys. Lett.* 62:3–8
28. Zgierski, M., Pawlikowski, M. 1981. *Chem. Phys. Lett.* 78:451–55
29. Zgierski, M., Shelnutt, J. Pawlikowski, M. 1979. *Chem. Phys. Lett.* 68:262–66
30. Suzuki, E., Hamoguchi, H., Haroda, I., Matsuura, H., Shimanouchi, R. 1976. *J. Raman Spectrosc.* 5:119–34
31. Warshel, A., Dauber, P. 1977. *J. Chem. Phys.* 66:5477–88
32. Peticolas, W., Strommen, D., Lakshminarayanan, V. 1980. *J. Chem. Phys.* 73:4185–91
33. Siebrand, W., Zgierski, M. 1979. *J. Chem. Phys.* 71:3561–69
34. Hoskins, L. C. 1981. *J. Chem. Phys.* 74:882–85
35. Kodama, K., Bandrauk, A. 1981. *Chem. Phys. Lett.* 80:248–52
36. Champion, P. M., Albrecht, A. C. 1979. *J. Chem. Phys.* 71:1110–21
37. Hassing, S., Mortensen, O. S. 1977. *Chem. Phys. Lett.* 47:115–18
38. Zgierski, M. 1975. *Chem. Phys. Lett.* 36:390–98
39. Hong, H. 1977. *J. Chem. Phys.* 67:801–12
40. Hong, H. 1977. *J. Chem. Phys.* 67:813–23
41. Fujimura, Y., Lin, S. H. 1979. *J. Chem. Phys.* 70:247–62
42. Fujimura, Y., Lin, S. H. 1979. *J. Chem. Phys.* 71:3733–43
43. Fujimura, Y., Kono, H., Nakajima, T., Lin, S. H. 1981. *J. Chem. Phys.* 75:99–106
44. Korenowski, G., Ziegler, L., Albrecht, A. C. 1978. *J. Chem. Phys.* 68:1248–52
45. Lukashin, A. V., Frank-Kamenetskii, M. D. 1978. *Chem. Phys.* 35:469–76
46. Champion, P. M., Lange, R. 1980. *J. Chem. Phys.* 73:5947–57
47. Champion, P. M., Perreault, G. J. 1981. *J. Chem. Phys.* 75:490–91
48. Champion, P. M., Albrecht, A. C. 1980. *J. Chem. Phys.* 72:6498–6506
49. Champion, P. M., Albrecht, A. C. 1981. *Chem. Phys. Lett.* 82:410–13
50. Korenowski, G. M., Albrecht, A. C. 1979. *Chem. Phys.* 38:239–44
51. Champion, P. M., Korenowski, G. M., Albrecht, A. C. 1979. *Solid State Commun.* 32:7–12
52. Korenowski, G. M. 1979. PhD thesis, Cornell Univ., Ithaca, NY
53. Champion, P. M. 1982. *Chem. Phys. Lett.* 86:231–34
54. Blazej, E., Peticolas, W. 1980. *J. Chem. Phys.* 72:3134–42
55. Rebane, K., Hizhnyakov, V., Tehver, I. 1967. *ENSV TA Toimet. Fuus. Matem.* 16:202–31 (in English). ("*Proc. Acad. Sci. Estonian SSR Phys. Math.*")
56. Hizhnyakov, V., Tehver, I. 1967. *Phys. Status Solidi* 21:755–68
57. Hizhnyakov, V., Tehver, I. 1970. *Phys. Status Solidi* 39:67–78
58. Hizhnyakov, V., Tehver, I. 1977. *Phys. Status Solidi B* 82:k89–k93
59. Hizhnyakov, V., Tehver, I. 1979. *J. Lumin.* 18/19:673–77
60. Toyozawa, Y. 1976. *J. Phys. Soc. Jpn.* 41:400–11
61. Tehver, I. 1968. *ENSV TA Toimet. Fuus. Matem.* 17:235–38 (in English). ("*Proc. Acad. Sci. Estonian SSR Phys. Math.*")
62. Hizhnyakov, V. V., Tehver, I. J. 1980. *Opt. Commun.* 32:419–21
63. Tonks, D. L., Page, J. B. 1979. *Chem. Phys. Lett.* 66:449–53
64. Hassing, S., Mortensen, O. S. 1980. *J. Chem. Phys.* 73:1078–83
65. Tonks, D. L., Page, J. B. 1981. *J. Chem. Phys.* 75:5694–5708
66. Tonks, D. L., Page, J. B. 1981. *Chem. Phys. Lett.* 79:247–52
67. Lee, S.-Y., Heller, E. J. 1979. *J. Chem. Phys.* 71:4777–88
68. Tannor, D. J., Heller, E. J. 1982. *J. Chem. Phys.* In press
69. Stallard, B. R., Champion, P. M., Callis, P. R., Albrecht, A. C. 1982. *J. Chem. Phys.* In press
70. Stallard, B. R., Callis, P. R., Champion, P. M., Albrecht, A. C. 1982. *J. Chem. Phys.* In press

IONIZATION IN SOLUTION BY PHOTOACTIVATED ELECTRON TRANSFER

D. Mauzerall and S. G. Ballard

The Rockefeller University, New York, New York 10021

INTRODUCTION

Electron transfer reactions have held a special place in chemistry because of their apparent simplicity. Both theoretical and experimental work have uncovered a world of detail beneath this plain exterior. The progress in understanding these reactions in photosynthetic systems and the relevance of these processes to efficient utilization of solar energy have contributed to the momentum of advancement. In this article we restrict ourselves to a subtopic of this field: that of photochemical ion formation in liquids. Reviews in the general area of electron transfer reactions are available: for inorganic complexes, see (1); for photochemistry, see (2); for photosynthesis, see (3), and for radiation chemistry, see (4). We also set aside the topics of direct electron photoejection (5–7) and of multiphoton events (8).

The basic problem of photogeneration of ions in solution is how do the initially formed, highly reactive radical ions ever escape? It is argued that, aside from electron spin selection rules, a finite escape requires a finite jump to a distance greater than the contact or collapse radius of the initial ion pair. It is also argued that the escape and geminate recombination of the ions cannot be consistently described by the usual steady state kinetics but require an appropriate set of time dependent diffusion equations. We begin the review with a somewhat selective literature survey of photogenerated ions from charge transfer states and from bimolecular encounters. We choose work showing direct evidence for free ion formation from kinetic analysis of absorbance transients or from conductivity measurements. The theory of ion escape is developed and applied to examples from the literature. Finally, we suggest what could be done to improve our understanding of these processes, so fundamental and so deceptively simple in appearance.

CHARGE-TRANSFER COMPLEXES

Relevant Properties of Charge-Transfer States

Mulliken's theory of charge-transfer (C-T) complexes has been greatly elaborated and refined since its first appearance (9–12). The expected sandwich-type geometry is generally confirmed in the few cases for which crystal structures are known (13–15). As indicated by their dipole moments and enthalpies of formation (12), the majority of C-T complexes show only a small degree of transfer of charge in their ground states. However, the donor and acceptor redox potentials may be such that free ions are formed in the ground state in polar solvents (16). There is great interest in crystalline charge-transfer complexes that behave as "organic metals" (17).

Although the considerable polarity of the Franck-Condon first electronic excited state is shown by electrochromic shifts of the C-T absorption spectra (18–20), the data indicate incomplete electron transfer (21). Most C-T complexes fluoresce only weakly even in nonpolar solvents. Exceptions are the tetracyanobenzene (TCNB) aromatic hydrocarbon complexes (22). Temperature dependence of the fluorescence Stokes shifts of these molecules indicate that charge transfer in the relaxed excited state is almost complete (23).

Detection of Charge Transfer Ionization

Early examples of ion radical formation from irradiated C-T complexes are solutions of chloranil (24), tetracyanoethylene (TCNE), and pyromellitic dianhydride (PMDA) (25, 26) in donor solvents, eg tetrahydrofuran.

Interest in the ionization of C-T excited states quickened with the advent of pulsed lasers as actinic sources. In 1969, kinetically resolved (millisecond) growing-in of absorption due to $PMDA^-$ was observed following C-T excitation of the PMDA/mesitylene complex in fluid ether/isopentane solutions at 117 K (27). Phosphorescence of the C-T triplet state decayed with the same first-order rate constant (800 sec^{-1}) as the growth of $PMDA^-$ absorption. The ionic character of the exciplex in the TCNB-benzene and TCNB-toluene systems was shown by the similarity of the decay times of $TCNB^-$ absorption and C-T fluorescence (28, 29).

Unambiguous detection of free ions came with use of polar solvents and application of fast kinetic conductance measurements, first reported by Mataga and co-workers for the TCNB-toluene complex dissolved in a 1:2 toluene/acetonitrile mixture at room temperature (30). The important parameters in these experiments are the kinetics of the ionization process and the yield of ions. The latter, and particularly its variation with solvent over an extended range of dielectric constant, has been rather neglected. As we indicate (Theory section), crucial information can be obtained by this basically simple measurement. Quantum yields of free ions were obtained for the TCNB/benzene (0.1), TCNB/toluene (0.098), TCNB/mesitylene, (0.046) and TCNB/hexamethylbenzene (0.005) complexes, excited in their C-T bands in acetonitrile solution. The 20-fold decrease of yield across this series parallels a monotonic ten-fold increase in ground-state association constant (31). A roughly linear relationship was noted between the donor ionization potential and relative ion yield in this series (30). However, this is far from being generally true (32). Absolute yields of free ion formation also were determined for the TCNB/benzene complex as a function of solvent dielectric constant, using a series of mixtures of benzene (donor solvent) and various polar solvents in a 1:2 volume ratio (30). The results were: CH_3CN (0.1); acetone (0.035); 1,2-dichloroethane (0.001); diethyl ether (0); and benzene (0). Similarly, yields were compiled for the TCNB/toluene complex, in 1:2 mixtures of toluene with acetonitrile (0.089), isopropanol (0.029), dichloroethane (0.001), and diethyl ether (0).

The kinetics of these processes were studied in laser photoconductivity experiments (33). The photocurrent rise-time on excitation of the TCNB-benzene complex in 1,2-dichloroethane was shorter than the convolution of the C-T fluorescence lifetime (20 nsec) with the flash envelope. Thus, it was concluded that ion formation occurred principally from the unrelaxed first excited singlet state of the complex. However, since the "escape time" in dichloroethane is ~ 20 nsec even if the electron transfer distance is 8 Å, much greater than the C-T contact distance, the fast ionization kinetics require additional explanation. The high acceptor concentrations (up to 0.1 M) used may provide the means of nanosecond escape by charge exchange:

$$(D^+A^-)\cdots A \to (D^+A)\cdots A^-.$$

The rise-times of photocurrent (free ions) and optical absorbance changes ($TCNB^-$ or $PMDA^-$, free ions and D^+A^- pairs indistinguishable) for the complexes of methyl-tetrahydrofuran (MTHF) and the acceptors TCNB and PMDA were less than 10 nsec in the polar solvents butyronitrile, acetonitrile, and ethanol at room temperature (34). With pure MTHF as solvent, the photocurrent rose more slowly (~300 nsec). Similarly, slow "growing in" of photoconduction occurred in the TCNB-MTHF complexes in 1,2-dichloroethane, where photoconduction rise-times of up to several microseconds were reported. The rise-time of optical absorbance in all cases was <10 nsec. These results show the expected slow separation of the initial geminate ion pair in solvents of low dielectric constant. The quantum efficiencies of free ion production were not obtained for any of the above complexes. For the case of the TCNB/naphthalene complex, ion yields relative to a "standard" [the TCNB-toluene-acetonitrile system; $\phi = 0.098$ (35)] in several solvents were quoted. Absolute yields, which we obtain by scaling the relative yields to the standard are: acetonitrile (0.057); butyronitrile (0.024); 1,2-dichloroethane (0.0095), and diethyl ether (0.00019).

Some of this data was recompiled (36), but curious discrepancies appear. The reported yields, particularly in nonpolar solvents, vary among different papers by factors of up to 100. Only the high yield values in polar solvents were consistent. However, this was pioneering work, and improvements in technique have led to better data. A recent review of the work of Mataga's group is available (37).

Ion Formation from Charge-Transfer Triplet States

In contrast to the fast C-T ionizations discussed above, radical anion formation from the TCNE-tetrahydrofuran complex in dichloroethane irradiated in the C-T band requires 300 μsec (38) and is correlated with decay of the C-T triplet state. Although the decay kinetics of the TCNE$^-$ signal were not reported, they appear from the data to be slow compared to the time scale of formation; this points to the likelihood of free ions. The growing-in of PMDA$^-$ absorbance in the PMDA-dioxane system at room temperature (39) required 70 μsec. Thus, it was concluded that the C-T triplet state is involved in this case also. Partial quenching of ionization of the TCNB/α-methylstyrene complex in amyl alcohol by molecular oxygen was interpreted as indicating ion formation from both singlet and triplet excited C-T states (40). Quenching was complete at low temperatures. The insensitivity of the anion radical signal to molecular oxygen on excitation of the C-T complex of TCNE and acetonitrile was interpreted as showing that ion formation is from the singlet state (41). A similar lack of oxygen quenching had been noted earlier with the complexes of PMDA with triphenylene, anthracene, and pyrene (42). However, these conclusions must be tempered by the complicated effects possible when O_2 is used as a quencher. Not only is it a good electron acceptor itself, but, because it is in a ground triplet state, collision with an excited singlet state will greatly enhance intersystem crossing. In fact, it may be possible to obtain both singlet oxygen and the aromatic triplet from the encounter of excited aromatic singlet and oxygen (43). A second point is that in several cases (39, 40, 42), the decay of the ions do not follow second order kinetics, and are often anomalously slow. Thus, relaxation cannot be attributed to simple bimolecular recombination of the primary (photo) ions. This demonstrates the need for more careful choice of experimental systems. This same view was expressed by Ottolenghi in his useful 1973 review of the photochemistry of C-T complexes (44).

Variety of Initial Ion Pairs

The ionic photodissociation of the perylene-PMDA and perylene-TCNE complexes in acetonitrile was studied in 1976 by Hentzchel & Watkins (45), using conventional flash photolysis. They observed that free ions resulted from both photodissociation of the C-T complex and from free acceptor quenching of the excited donor (perylene, $^1P^*$) by electron transfer in bimolecular encounters. The efficiencies of the two ionogenic processes in the two systems studied were Perylene-PMDA: bimolecular quenching (0.07); C-T complex (0.015); Perylene-TCNE: bimolecular quenching (0.11); C-T complex (0.024); i.e. in each case the C-T yield is lower than the yield of free ions from the bimolecular quenching reaction by a factor of about four. This is reasonably interpreted as indicating that the radius for electron transfer between $^1P^*$ and the quencher in bimolecular encounter is greater than the "contact" radius of the corresponding C-T complex in its ground state.

There is now considerable evidence for a structural variety of initial ion pairs, i.e. for the view that electron transfer reactions can occur without prior contact or exciplex formation. This evidence is particularly clear when donor and acceptor are linked by three methylene groups. Szwarc and co-workers (46) have shown that conformational effects on electron exchange are maximal for this chain length. The time dependence of the fluorescence of 1-anilino-3-(4-anthryl)-propane showed that noncoplanar ion pairs could form, i.e. long distance electron transfer occurs, even in nonpolar solvents (47). Poorly resolved changes in absorption spectra following picosecond excitation of 1-dimethylanilino-3-(1-pyrenyl)-propane have been interpreted as charge transfer in a loose structure followed by a collapse to a sandwich-type exciplex (48). The formation of exciplexes by 1-dimethylanilino-3-(4-anthryl)-propane occurs in about 10 psec in dimethyl-aniline but about in 1 nsec in hexane (49). In acetonitrile, two photo processes can be seen in (50). The exciplex is formed from the closed conformer in <2 psec and lives 580 psec. The anthracene fluorescence is quenched in 7 psec, representing the electon transfer time in the open conformer. Thus, the electron not only can jump a finite distance (>5 Å) but can do so very fast. The picosecond kinetics of the reaction of singlet anthracene in neat diethylaniline and when diluted by hexane or acetonitrile could be fit by a diffusion equation, including transient terms, with a reaction constant of 10^{11} M^{-1} sec^{-1} and a reaction radius of 8 Å (51). In our view, the rate of electron transfer by tunneling will be largely independant of the solvent static dielectric constant but will depend on the optical dielectric constant (refractive index). The observed insensitivity of charge transfer to solvent is in agreement with this view. The yield of ions, however, will be strongly dependent on the static dielectric constant parameter because of the collapse to the ground state from the singlet ion pair at close distance. A review article (52), implies that the ion yield in hexane is low. The pulse energies used in usual psec experiments, 1–10 mJ cm^{-2}, are dangerously close to that resulting in two-photon effects. It is hoped that future experiments will pay as much attention to yields and excitation energies as to third-order terms in the diffusion equation.

It is unfortunate that so much of the exciplex work has been done with alkyl anilines and their variants as donors. Many studies using a variety of approaches [solvent effects (53, 54) and sterically twisted molecules (55)] have shown beyond rational doubt that the angular position of the unshared pair of electrons (n orbital) on the nitrogen has a dramatic effect on the photophysics of these molecules, e.g. multiple fluorescence states exist. This internal degree of freedom severely complicates any detailed interpretation of complexes containing these molecules.

BIMOLECULAR ENCOUNTER

Fluorescence Quenching by Electron Transfer

The literature on quenching of excited states by electron transfer extends back at least to the early 1930s (56, 57), and is now voluminous. Linschitz was the first to adduce strong evidence that the common mechanism of quenching of excited states was via electron transfer (57a). Here we review only work that bears directly on the formation of free ions in such quenching processes.

Flash photolysis of solutions of perylene and amines in polar solvents showed formation of the perylene anion radical (58). In nonpolar solvents only the perylene triplet state was identified. The fluorescence quenching constant k_q for each amine donor was found to depend dramatically on solvent polarity (59). For polar solvents, $k_q \sim 2 \times 10^{10}$ M^{-1} sec^{-1} (encounter limited rate for reaction radius of 7 Å, i.e. greater than the sum of the molecular radii of perylene and quencher), whereas in nonpolar solvents $k_q \sim 10^8 - 10^9$ M^{-1} sec^{-1} and quenching of the perylene fluorescence was observed to give rise to a new, structureless, red-shifted "exciplex" emission. Its intensity is about 1% of the original ^{1}P* fluorescence (60). In polar solvents the triplet state is also formed, but in lower yield than in nonpolar solvents.

From the ten-fold stronger variation of exciplex fluorescence intensity than of lifetime with solvent dielectric constant, Knibbe et al (61) concluded that not only does the C-T exciplex decay more quickly in polar solvents, but its initial formation probability is less also. Their model invoked two parallel quenching reactions of ^{1}P*. The first gives the fluorescent C-T exciplex, the lifetime of which declines with increasing dielectric constant because of increased dissociation into free ions. The competing second reaction was supposed to produce a solvent-shared nonfluorescent ion pair that dissociates with high probability. Rapid increase in its formation rate constant with increasing solvent polarity was reasonably invoked to explain the observed decline in exciplex fluorescence intensity. This model was elaborated in a subsequent paper (62), where a more complete kinetic scheme for fluorescence quenching, free ion formation and triplet state production was presented. The quantum yields of free ions in CH$_3$CN were measured for excited anthracene, perylene, tetracene, and coronene quenched by DEA. Values of 0.04, 0.06, 0.02, and 0.02, respectively, were quoted.

The first of a series of now classic papers by Weller's group on the energetics of fluorescence quenching by electron transfer was published in 1969 (63). The second-order rate constants (k_q) for quenching of the singlet-excited state of aromatic molecules by alkylamino- and methoxybenzenes with polarographic oxidation potentials spanning the range 0.16 V–1.49 V were measured in acetonitrile. Strong correlation was observed between k_q and the energy change ΔG involved in the electron transfer step, taken as

$$\Delta G = E^0_{D/D^+} - E^0_{A^-/A} - \Delta E_{A*} - e^2/\varepsilon r_0, \qquad 1.$$

where E^0_{D/D^+} and $E^0_{A^-/A}$ are, respectively, the standard equilibrium donor oxidation and acceptor reduction potentials, ΔE_{A*} is the molar enthalpy of the excited state acceptor obtained from the fluorescence 0,0-transition, and $e^2/\varepsilon r_0$ is an electrostatic term intended to take into account the interionic potential energy at distance r_0. For $\Delta G < -0.5$ eV, k_q was observed to be at or close to the encounter limit in all cases ($\sim 2 \times 10^{10}$ M^{-1} sec^{-1}), contrary to the parabolic relationship predicted by the Marcus theory (64). For $\Delta G > -0.5$ eV, k_q decreases monotonically and finally becomes proportional to $\exp(-\Delta G/RT)$. Thus, in this limit, and possibly only in this limit, one can consider electron transfer to be an activated state process.

An even more extensive set of data conforming to this picture was published later by Rehm & Weller (65), and numerous other examples have been forthcoming from other laboratories, eg (66, 67). With regard to the use of equations such as Eq. 1, however, we note that E^0s are Gibbs free energies, while E_A^* and the questionable electrostatic terms are enthalpies. The important entropy changes that must accompany formation of charged species in solution are usually completely neglected.

The dipolar nature of the exciplex was demonstrated by the increase in wavelength of the weak emission band with increasing solvent dielectric constant (68). The anthracene/DEA exciplex was assigned a dipole moment of 10 D in nonpolar solvents, identical with the value for the pyrene-diethyl, aniline (DEA) exciplex (69). The ionic character of C-T exciplexes increases with increasing solvent polarity (70).

Detection of Free Ions

Free ion formation was inferred by Grellman & Watkins (71) from absorbance transients seen in the bimolecular quenching of the first excited singlet state of perylene by the ground state in acetonitrile. Interestingly, they claim in this paper that no ionization resulting from triplet-triplet encounter is detectable, despite its apparently more favorable energetics as compared to the first excited singlet-ground state reaction ($\Delta E_{1_{p^*}} = 2.85$ eV; $2\Delta E_{3_{p^*}} = 3.1$ eV). In the case of pyrene (72), in nonpolar solvents the singlet-ground state reaction becomes slow and ion formation then proceeds via the more strongly exothermic triplet-triplet reaction ($\Delta E_{1_{p^*}} = 3.3$ eV; $2\Delta E_{3_{p^*}} = 4.2$ eV. In acetonitrile, $E^0_{p/p^+} - E^0_{p^-/p} = 3.16$ eV).

The most compelling and useful direct evidence for formation of free ions in photoredox reactions came with the application of fast electrical conductance measurements. Kawada & Jarnagin (73) observed photocurrents following bimolecular encounter of triplet states of phenanthrene and anthracene in tetrahydrofuran solution: $^3A^* + {}^3A^* \to A^+ + A^-$. This work was later extended to naphthalene and pyrene (74). Variation of free ion yields with solvent dielectric constant was systematically studied for the first time. A review of this work, with compilation of the T-T yield data, is available (6). Jarnagin's data is replotted in Figure 2; for analysis, see Theory section below.

Photo-conductance measurement was first applied to the problem of dynamic fluorescence quenching by Mataga's group (35). Formation and decay of free ions were followed conductometrically in the bimolecular quenching reaction of the fluorescent state of pyrene $^1P^*$ by dimethyl aniline (DMA). In hydrocarbon solvents, exciplex fluorescence paralleled formation and decay of absorbance changes attributable to $\dot{P}^-$, but no free ions were detectable (instrument sensitivity not stated). Thus, quenching of $^1P^*$ produces an exciplex that decays without dissociation into free ions. For solvents with $\varepsilon > 10$, photocurrents were observed, and relative yields of free ions obtained. Decay of the conductance was second order. It was later shown (75) that the primary component of ion formation in the pyrene/DMA system in polar solvents develops more rapidly (<10 nsec) than the convolution of the flash profile with the exciplex fluorescence decay time (33 nsec for pyrene/DMA, 25 nsec for anthracene/DMA). Thus, ionic dissociation was inferred to occur from an "incompletely relaxed" state of the ion pair intermediate. It is certain, however, that thermal relaxation of excited states occurs in picosecond times. Either charge exchange or the complexities of the photophysics of DMA described above (C-T section) may account for the rapid ion formation. Free ion quantum yields in acetonitrile were 0.5 for the pyrene/DMA reaction and 0.3 for anthracene/DMA, i.e. substantially greater than Weller's figures for anthracene/DEA in acetonitrile (62). Variation of the pyrene/DMA reaction yield ($\phi_\pm$) with solvent was also tabulated:

$\phi_\pm(\text{acetonitrile}) = 0.5$; $\phi_\pm(\text{acetone}) = 0.17$;

$\phi_\pm(1,2 \text{ dichloroethane}) = 0.01$.

Dilute (10^{-3} M) solutions of vinyl carbazole and ethyl carbazole quenched by the electron acceptors TCNB and TCNE (10^{-2} M; no C-T complex detectable) in a variety of organic solvents were also shown to produce free ions upon flash irradiation (76). In acetonitrile and tetrahydrofuran, the ion current rise-time was not resolvable, consistent with rapid ion uncorrelation following bimolecular-encounter quenching at the diffusion-controlled rate. However, in dichloromethane, a rise-time of ~ 200 nsec was seen. This is probably the first observation of the slow electrostatic uncoupling of photo-produced ions in nonpolar solvents (see Theory section). Photocurrent decay kinetics were shown to be clean second-order in all three solvents. Quenching of the triplet state of benzophenone by DEA in acetonitrile was found by Arimitsu & Masuhara using conventional flash photolysis to yield free ion radicals with near unity yield (77).

Nanosecond flash photolysis of the pyrene/DEA system in methanol at 233 K showed biphasic decay kinetics of the flash-induced $\dot{P}^-$ absorbance (78). An initial fast phase ($t_{1/2} \sim 40$ nsec, independent of flash intensity) was attributed to geminate recombination within the initial ion pair $\dot{P}^- \cdots \text{DE}\dot{A}^+$. It was followed by a much slower ($t_{1/2} \sim 25$ μsec, intensity dependent) second-order phase, corresponding to bimolecular recombination of the escaped ions. Slowing of the geminate phase to tens of nanoseconds was attributed to high viscosity of the solvent at low temperature. However, we estimate the time scale of geminate recombination at ~ 2 nsec in this system, hence well below the resolving time of the instrument used. Possibly, the 40 nsec absorbance changes are due to protonation of $\dot{P}^-$.

Yields of free ions formed in electron transfer quenching of the $^1P^*$ state of pyrene by both amine electron donors and by nitrile, anhydride and phthalate ester acceptors spanning a large redox range were measured, using both conventional nanosecond flash photolysis and transient photocurrent measurement (79). No systematic correlation was found between ΔG and yield of free ions. The systems studied may not be simple, however. For example, the anhydrides may react with solvent water leading to high background conductances, as observed, and to possible proton transfer reactions. The quantum yields of free ions in the quenching of pyrene by DMA span the range 0.03 (in dichloromethane) to 0.5 (in acetonitrile) (80). This data is replotted (Figure 2) and analyzed in the Theory section.

The conductance rise kinetics in the reaction of excited pyrene with p-dicyanobenzene become progressively slower as the solvent dielectric constant is decreased, whereas the absorbance transient rise-time remains less than the 20 nsec flash time (81). In acetonitrile and acetone, growth of conductance is not kinetically resolvable, but in pyridine, 1,2-dichloroethane, and dichloromethane, approximately exponential rise kinetics are seen, with first-order constants 9, 0.9 and 1.3×10^7 sec^{-1} respectively. These were interpreted as electrostatic uncorrelation times, and correspond approximately to the Eigen equation values (see Theory). Nanosecond spectroscopic separation of the geminate (~ 3 nsec) and bimolecular (~ 1 μsec) phases of photo-ion recombination has been described by Schulten et al (82) for excited pyrene quenched by 3,5-dimethoxydimethylaniline in methanol. They used an adequate description of the process involving the diffusion of charged species.

The second-order rate constants for electron-transfer quenching of the triplet-state of zinc uroporphyrin by acceptors in aqueous solution with simultaneous production of the porphyrin cation radical were determined, using low energy flash photolysis (83). Rates were shown to be controlled by reactant charge and by extent of electrostatic screening from added electrolyte ions, not by quencher redox potential. Isolation of the Coulomb interaction term gave the critical radius for electron transfer $r_0 = 22$ Å ± 4 Å, i.e. twice the sum of the radii of porphyrin and quencher, 11 Å.

Second-generation transient conductance methodology employing pulsed biphasic rather than the usual static cell polarization (84) was applied to investigation of photochemical ionization reactions of lumiflavin in acetonitrile solution (85). Following the actinic flash a positive, kinetically first-order conductance change on the microsecond time scale was attributed to unimolecular release of flavin ion radicals following electron transfer in a triplet-state flavin aggregate species. Subsequently, fast proton transfer reactions of the photo-produced ions revealed themselves in complex, multiphasic conductance transients, exquisitely sensitive to solvent pH. The kinetic constants of these reactions were determined. Photoionization reactions of the protonated flavin at low pH were also elucidated, as was the reaction between triplet flavin and indole. This work demonstrated the power of the conductance method to reveal reactions involving charged species that are practically invisible to optical detection (e.g. $F^- + H^+ \rightleftarrows FH$). Much simpler systems for studying ion formation following photo-activated electron transfer in solution are the triplet-triplet (T-T) and triplet-ground state (T-P) reactions of porphyrins and metalloporphyrins:

$$T + T \stackrel{k_{TT}}{\to} {}^1(P^+ \cdots P^-)_{r_0} \leftrightarrow {}^3(P^+ \cdots P^-)_{r_0} \stackrel{\phi'^{\pm}}{\to} \dot{P}^+ + \dot{P}^-$$

$$T + P \stackrel{k_{TP}}{\to} {}^3(P^+ \cdots P^-)_{r_0} \leftrightarrow {}^1(P^+ \cdots P^-)_{r_0} \stackrel{\phi^{\pm}}{\to} \dot{P}^+ + \dot{P}^-.$$

We have employed the pulsed conductivity method to study these reactions, which have provided an unusually complete and easily analyzed set of experimental data on photoactivated electron transfer and on the electrostatic uncorrelation and recombination of ion radicals in solution (86–89). Ionization kinetics and yields were measured over a wide range of solvent dielectric constant ($\varepsilon = 2.3$ to $\varepsilon = 172$). Coherent averaging of the conductance transients, mandatory in low-ε solvents, was made possible by the quantitatively cyclical nature of the porphyrin excited state redox reactions in oxygen-free ($[O_2] < 10^{-9}$ M) solution, and by the electrochemically gentle low-voltage pulsed measurement system. Ion yields as low as 10^{-5} (i.e. [free ion] $< 10^{-12}$ M) were measurable with time resolution of some tens of microseconds. The encounter-limited rate constant for the T-T ionogenic reaction, and analysis of the variation of the absolute yield of free ions with solvent dielectric constant (Theory section, Figure 3) independently fixed the radius of forward electron transfer (r_0) at 20 ± 2 Å. The same r_0 could also be used to model the T+P yield (Figure 3), but satisfactory fit to the data requires an additional parameter, the electron spin dephasing time in the geminate pair, either directly or in a model based on equilibration of the porphyrin triplet state and ions (see Theory section). Bimolecular, encounter limited ion recombination following both the T-T and T-P reactions was found to follow closely the Debye-Smoluchowski equation (90), from which the hydrodynamic radius of the porphyrin $r_h = 7.1 \pm 0.4$ Å was obtained. This value is close to the 7.0 Å radius of the equivalent hydrodynamic sphere estimated from crystallographic dimensions within the uncertainties of hydrodynamic theory applied to molecular motion in liquids (91, 92). This figure has been directly confirmed by measurement of the electrical mobility of a porphyrin ion in solution (A. C. Albrecht, personal communication). The mobility must be known with confidence since it scales the observed photocurrents to absolute ion yields, via the cell conductance equation. Direct proof of the $T+T$ and $T+P$ mechanisms of ion formation in porphyrin solutions has recently been obtained by simultaneously monitoring the growth of electrical conductance and the decay of delayed fluorescence from the triplet state following an actinic flash (S. G. Ballard, 1981, submitted to *Chem. Phys. Lett.*). Figure 1 shows data obtained for the $T+P$ reaction of magnesium octaethyl porphyrin in acetonitrile. Curve *b* shows the decay to baseline of the delayed fluorescence signal obtained using a gated photomultiplier system, and curve *a*

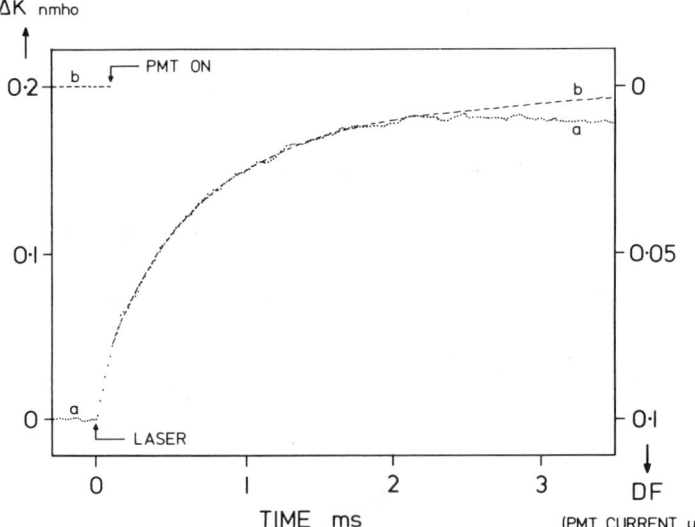

Figure 1 Kinetics of formation of MgOEP ion radicals $\dot{P}^+ + \dot{P}^-$ (*curve a*, left-hand scale) and simultaneous decay of the triplet state T (*curve b*, right-hand scale) following weak flash irradiation (500 nm, 5 nsec pulse; $P^*/P = 10^{-4}$) of 5 μM solution of MgOEP in CH_3CN. Ions (produced by T-P reaction) are detected by transient conductimetry, and T is followed via delayed fluorescence at 610 nm. Data was obtained at room temperature (20° C), and is taken from S. G. Ballard (submitted to *Chem. Phys. Lett.*).

shows the corresponding growth of photocurrent. The two quantities, ΔK and DF, track precisely until truncation of the conductance signal by bimolecular ion recombination becomes significant. Similar correspondence is seen for the T+T reaction.

Abrupt (<300 nsec instrument resolution) ionization following light absorption in the porphyrin 440 nm interband trough was also reported, and attributed to electron ejection following sequential 2-photon excitation (89). This ionization is now known to be due to electron transfer involving microcrystals, which in the case of the poorly soluble symmetrical porphyrins deposit from polar solvents onto the electrodes of the conductance cell (S. G. Ballard, P. Kirshman, unpublished results). The energetics of the T+T and T+P reactions have been briefly discussed (93). In contrast to the fluorescence quenching data of Rehm & Weller (65) and others (66, 67), the rate constants for the porphyrin T-T and T-P reactions were found to be independent of solvent dielectric constant. This poses no problem for the encounter-limited T-T reaction, which is highly exothermic in all solvents; however, the T-P reaction appears from the standard redox potentials of the ion products to be marginal or slightly endothermic, even in polar solvents $\{E^0_{P/P^+} - E^0_{P^-/P} = 2.2$ eV $[E^0_{P/P^+}$ measured potentiometrically in MeOH (94), $E^0_{P^-/P}$ measured polarographically in DMSO (95)]; $E_T = 1.8$ eV$\}$. Lowering the solvent polarity raises the redox potentials of the ions even further. This reveals itself as a decline in yield of free ions. We argue that the irreversible positive entropy flow accompanying escape to high dilution is what allows any escape at all. These effects are contained in the diffusion equation describing these processes (Theory section). The substantial independence of k_{TP} from solvent dielectric constant indicates that the rate is controlled by extra-thermodynamic factors such as spin dephasing (see Theory section). Ion formation from the triplet-triplet annihilation reaction of chlorophyll-*a* in acetonitrile has also been reported (96) in various solvents ranging from diethyl ether to acetonitrile. Oxidation of the triplet state of chlorophyll-*a* by *p*-benzoquinone in a series of aliphatic alcohol solvents has been shown to produce free ions (97). It was concluded that the reaction radius r_0 was approximately the sum of the porphyrin and quinone radii. Our analysis of this work requires a finite spin dephasing time of the triplet ion pair and an r_0 of up to 20 Å. A comprehensive overview of recent work on the energetics of electron transfer in chlorophylls and related pigments is available (98). We point out, however, that the much-used electrostatic charging models are very crude approximations and should not be taken too seriously. Further, experimental redox potentials are themselves subject to substantial errors, particularly when derived polarographically. Thus, much caution is needed in dealing with ion thermodynamics, particularly when extrapolating to weakly polar solvents. The intrinsically irreversible photoreactions are a case in which no equilibrium model, however derived, is admissible.

THEORY OF ION ESCAPE

Diffusion of Ions

The equilibrium thermodynamics of ion pairs have been reviewed (99). The problem of the escape of thermodynamically unstable species from a reaction complex is of wider occurrence. The yield of escaped species has often been calculated by assuming that escape and collapse can be represented by first-order constants (63, 100). Thus the yield is given by:

$$\phi = {}^{k}esc/({}^{k}col + {}^{k}esc). \qquad 2.$$

The situation is aggravated by the tendency to fit the data with an arbitrary collection of rate constants (63). We believe this approach is incorrect, and is clearly wrong when the collapse is encounter limited. In these cases a purely stochastic approach is needed. Basically, not all configuration space is explored by any given pair, as is required by steady-state chemical kinetics: the escaped and collapsed groups have different histories. The latter have encountered the critical reaction radius, which we represent by r_m, and the former have not. The motion of the isolated ion pair can be described, to the approximation of a continuum solvent and spherical symmetry, by a Smoluchowski equation with a Coulomb term:

$$\frac{\partial u}{\partial t} = D\left[\frac{\partial^2 u}{\partial r^2} + \left(\frac{2}{r} + \frac{r_c}{r^2}\right)\frac{\partial u}{\partial r}\right] \qquad 3.$$

where u is the probability density of ion pairs separated by distance r, D is the sum of the diffusion constants of the ions, and r_c is the Coulomb radius: $e^2/\varepsilon kT$, where ε is the dielectric constant of the solvent. The time-dependent probability distribution is complex, but has been analyzed (101–103) and can be computed. An analytic solution has been obtained for the special case of an initial Boltzmann distribution of ion pairs (103a). An analytic expression can be obtained for the distribution as $t \to \infty, r \to \infty$ with the boundary conditions:

$$u = \delta(r - r_0) \quad \text{at } t = 0 \qquad 1.$$

and

$$u = 0 \text{ at } r = r_m \quad \text{for all } t \qquad 2.$$

Condition **1** states that all ions are initially at r_0 and Condition **2** states that a perfect sink occurs at r_m. The yield of escaped ions, defined as the total flux out of the sphere $4\pi r^2$ as $r \to \infty, t \to \infty$ (101) is:

$$\phi = \frac{\exp(-r_c/r_0) - \exp(-r_c/r_m)}{1 - \exp(-r_c/r_m)} = \frac{e^{-x} - e^{-y}}{1 - e^{-y}}. \qquad 4.$$

Note that the diffusion constant D (and thus solvent viscosity) does not appear in the solution. The yield is independent of D, which only establishes the time scale of the escape. If there is a distribution of initial separation distances, one simply averages Eq. 3 over this distribution. Because of the sink condition, when $r_0 = r_m$ the yield is zero. As $r_c \to 0$,

$$\phi_0 \to 1 - r_m/r_0. \qquad 5.$$

It is often thought that the Onsager equation (104) $\phi = \exp(-r_0/r_c)$ represents the yield of escaped ions. This is mistaken for two reasons. First, the Onsager equation is based on the assumption that r_m is zero and thus does not apply to the problem with collapse at a finite radius. As $r_c \to 0$, the Onsager yield $\to 1$, whereas a finite r_m gives Eq. 5. An encounter radius of measure zero can only lead to *no* collapse whatsoever. Second, and more basically, the Onsager equation is a steady state solution of Eq. 3. It assumes continual replenishment at r_0 and thus gives the equilibrium distribution of unreacted ions. The equivalent transient problem is to solve Eq. 3, but with a reflecting boundary at r_m (which can shrink to 0) instead of a sink. The escaped yield is then unity for any r_c. This is because infinite dilution has been assumed (i.e. independent collapse and escape events), and it is only a matter of time (10^{-9} to 10^{-4} sec for $\varepsilon = 100$ to 2, respectively) that the entropy flow (first two terms on the right of Eq. 3) driven by thermal fluctuations (D) will overcome the Coulomb binding energy (third term on right of Eq. 3). For the case of present interest, infinite sink at r_m, the steady-state solution of Eq. 3 with finite concentration of ion pairs following the transient collapse and escape, leads to the Debye equation (90), i.e. the encounter-limited rate of reaction of the now randomized charged species.

In the misguided spirit of the steady state approach, the Eigen equation (105) for k_{esc}, and the Debye equation (90) for k_{col} have often been assumed

$$k_{esc} = \frac{3r_c}{r_0^3 - r_m^3} \cdot \frac{D}{\exp(r_c/r_0) - 1} \qquad 6.$$

$$k_{coll} = \frac{4\pi N r_c D}{1 - \exp(-r_c/r_m)} \qquad 7.$$

where we distinguish between the formation radius, r_0, and the reaction radius, r_m. Thus

$$\phi = [1 + C(e^x - 1)/(1 - e^{-y})]^{-1} \qquad 8.$$

where we have replaced r_c/r_0 by x and r_c/r_m by y, and C is a concentration ratio term $(4\pi N(r_0^3 - r_m^3)/3$, N = molecules per unit volume) which normalizes the first-order (Eigen) and second-order (Debye) rate constants. If it is further assumed that $r_0 = r_m$, but that the volume term is $4\pi N r_e^3/3$; Eq. 8 becomes:

$$\phi = [1 + Ce^x]^{-1}. \qquad 9.$$

It is amusing that the value of $r_e = 7$ Å almost universally used in the description of photoformed ions in solution (106) is that value of r_0 (7.35 Å) which leads to $C = 1$ for the chemists' 1 mole per liter of concentration units. This is equivalent to assuming the "standard state" of the ion pairs is 1 M, thus defining away the influence of entropy, which is the ultimate origin of these concentration terms. This factor C simply does not arise in the correct treatment of the transient problem. Thus, even with favorable assumptions, the true equation for the yield (Eq. 4) cannot be regained from Eq. 2, but only approximated. It is also clear that no Arrhenius or absolute reaction rate theory expressions for k_{esc} and k_{col} could ever generate the actual yield equation. The transient solution is simply discarded in the steady-state arguments, and with it goes the heart of the problem. Putting it another way, the escape and collapse fluxes are time dependent, and one must integrate these fluxes over time to obtain yields. The short cut of time-independent fluxes is a crude approximation.

Boundary Conditions

Before discussing the choice of the boundary conditions, we must comment on the definition of the yield itself, which involves both $r \to \infty$ and $t \to \infty$. Clearly, to define escape, r must be much larger than r_c. Its practical limit is that the escaping ions from another pair must not cross into the first pairs' escaping volume. Thus, the total concentration of ions formed must be much less than $3 \times 10^{27}/4\pi \times 6 \times 10^{23} \times r_c^3$, i.e. $\ll 0.1$ M in acetonitrile and $\ll 2 \times 10^{-5}$ M in benzene. Even the latter condition is met by conductometric measurements. The factor 10^{27} converts liter to $Å^3$. The time required for this escape (τ) will be ~ 3 nsec (acetonitrile) to $\sim 10^{-4}$ sec (benzene). To avoid encounter with neutral molecules over this time span requires their concentrations to be $\ll 3 \times 10^4/4\pi 8(6D\tau)^{3/2}$, i.e. $\ll 2 \times 10^{-3}$ M and $\ll 3 \times 10^{-8}$ M, respectively, in acetonitrile and benzene. To meet the latter requirement requires some care even with the conductometric (89) or delayed fluorescence methods (107). The concentrations used in the study of weak charge-transfer complexes (~ 0.1 M, see C-T complexes) are totally inappropriate. How realistic are the boundary Conditions 1 and 2?:

1. Formation of ions at $r_0 > r_m$. This would be a direct consequence of charge transfer by electron tunneling. The basic idea was discussed by Gurney (108), and there is now considerable evidence for electron tunneling in molecular systems (109–112, 113). Our view of electron transfer via tunneling is based on the weak overlap of the relevant molecular orbitals. For large molecular π systems the abundance of possible energy states (essentially a continuum) and the small reorganization energies will allow the tunneling probability to become rate limiting, and sufficiently large to allow encounter-limited reactions at greater than molecular radii. The transfer probability to and from excited states will be larger than that between ground state molecules because the tunneling parameter is larger, i.e. the tail of the orbital extends further into the medium (114, 115). The tunneling parameter is thought to be 1–2 Å^{-1} but is not yet known with any certainty. Experiments to obtain this parameter from systems where donor and acceptor are rigidly held in three-dimensional space are in progress. A rigid co-facial porphyrin-quinone complex with 10 Å spacing has been recently synthesized in high yield by entropically favored macropolycyclization (J. Lindsey and D. Mauzerall, submitted to J. Am. Chem. Soc.) Both singlet and triplet states of the zinc derivative are observed to be quenched indicating rapid (<2 ns) electron transfer.

A second possible mechanism of forming reactive species at greater than contact distance is by thermalization of the initial hot species formed in an exothermic reaction. This mechanism would probably yield a broader or Gaussian distribution of r_0's than would the tunneling mechanism.

Evidence for electron transfer at greater than molecular contact distances was obtained by analyzing the rate constants of the second-order reaction of negatively charged porphyrin triplets with variously charged acceptors (83). These experiments also showed the expected high yield of ions from "repulsive encounters" (83, 115), i.e. of similarly charged donor and acceptor species. These concepts have been adopted by investigators of reactions in charged micelles to show photochemical reactions with high yield (116). Borsenberger et al (117) studied the photoionization of triphenylamine in bisphenol-A-polycarbonate film as a function of electric field strength. Using the Onsager theory, they concluded that r_0 was 22–27 Å and independent of temperature. This is very good evidence for electron tunneling in a photochemical system, as is the previously mentioned result of Crawford et al (50) on a tethered donor-acceptor pair.

2. A perfect sink exists at r_m. For those reactions where the products can be shown to react at the encounter-limited rate, this condition is simply a statement of fact. If r_m is not a perfect sink, the boundary conditions first stated by Collins & Kimball (118) must be used. This leads to a smooth transition from the diffusion-limited rate constant to the regular second order rate constant. Experimentally, an inefficient sink can be recognized from the less than linear slope of recombination rate constant versus reciprocal viscosity. This slope tends to zero as the sink becomes inefficient and "normal" second-order kinetics are observed. For the case of interest in the presence of Coulomb potential, the solution of Eq. 3 with finite flux at r_m, following Hong & Noolandi (102) or Berlin et al (103), gives for the escape yield:

$$\phi = \frac{e^{-x} + \left(\frac{D \cdot y}{Kr_m} - 1\right)e^{-y}}{1 + \left(\frac{D \cdot y}{Kr_m} - 1\right)e^{-y}} \qquad 10.$$

and, for $r_c \to 0$:

$$\phi_0 = \frac{1 - \frac{r_m}{r_0} + \frac{D}{Kr_m}}{1 + \frac{D}{Kr_m}} \qquad 11.$$

where K is the surface rate constant at r_m, with units of cm sec^{-1}. Berlin et al swallow r_m in their definition of K. At $r_c = 0$ and $r_m = r_0$, we recover the yield equivalent to the Collins-Kimball case. At $K \to \infty$ we recover the infinite sink yield, Eq. 4. A crucial point is that for the case of finite K, the yield does depend on the diffusion constant and thus on viscosity. The dependence is not simple, and is inextricably mixed with r_c and r_m. For the trivial case of $K \to 0$, the yield is unity, and *again* independent of viscosity. With appropriate care, this viscosity dependence can be used to determine the "perfection" of the sink. A nice example of this unexpectedly complex dependence of observable on viscosity has been worked out by Sterna et al (119) for the case of ^{13}C enrichment by photolysis of a ketone. The mechanism involves bond scission and reformation controlled by spin states—a morass that we soon must enter.

Equation 10 is an excellent solution to the problem of proton escape from molecules that are strong acids in their excited states. Thus, the data on the dissociation yield of hydroxypyrenetrisulfonate (0.4 to 0.7 yield for lifetimes of 1.5 to 5 nsec) is simply fit with $r_0 = 5$ Å since D, r_c, and $K(r_m/\text{lifetime of excited state})$ are known. r_0 is 1 to 2 Å beyond the edge

(r_m) of the molecule. The analogous, more complex attempt by Haar et al (120) to explain this data leads to a similar conclusion.

Rate of Escape

Whereas the collapse rate is well defined, having a transient and steady-state (Debye, Eq. 7) regions, the escape rate is poorly defined. The escape to infinity used to define the yields obviously gives an infinitely slow rate. The flux from r_m is complex when molecules originate at $r_0 > r_m$. A simple definition is the outward net flux of ions through the surface $r = r_c$. The rate then has a lag period for $r_c > r_0$ (the time to diffuse from r_0 to r_c), a transient rate and a steady state rate. Even the seemingly rational definition of escape as the flux beyond r_c fails when $r_c < r_0$ or r_m. Thus the problem is ill poised. A reasonable estimate of the escape rate is the flux beyond $r_c + r_0$. Any such defined flux can be readily computed (Eq. 3) and the steady-state value is approximated by the Eigen equation (Eq. 6). The measured escape rate necessarily includes the rate of reaction at r_m.

Spins

The preceding discussion has treated the reactive species in a purely classical way. Quantum mechanics adds a powerful selection rule. Radicals can only recombine to form a chemical bond, or electron transfer occur to singlet ground (or singlet excited) states, if the electron spins are paired. If the electron spins are unpaired, only excited molecular triplet states can be formed, i.e. the pair collapse is prevented. In the homogeneous recombination of radicals, the relative phases of the two electron spins are random, and thus 3/4 of the time they will be triplet-like and 1/4 of the time singlet-like. Since many such reactions leading to bound or ground states are measured to be encounter limited, the spin flip must occur on a time scale comparable to diffusion over the reaction radius, i.e. $r_m^2/6D \sim 10^{-10}$ sec. On the other hand, the initial pair in a radical dissociation may have a well-defined spin state. It is a singlet for usual thermal chemical bond breaking and for electron transfer to and from singlet states. These pairs will have a high probability of recombining. Many molecular singlet states rapidly inter-system cross to triplets, which have lower energy. Electron transfer (or bond scission) from these states produces pairs that cannot recombine to ground states until spin flips have occurred. Thus the probability of these species escaping is larger than for the singlet pairs. Note that this argument is completely extra-thermodynamic and is totally unrelated to the energy dissipated by the pair collapse. We assume the electron overlap of the pair is sufficiently small that the interaction energy is negligible compared to kT ($\sim 10^{-2}$

eV) yet sufficient for electron tunneling ($\sim 10^{-4}$ eV). The pair escape probability will depend on the spin dephasing time. It is tempting to use Eq. 10 as a solution to this problem, i.e. to reduce the time dependence to a time-variable boundary condition. However, the interconversion of singlets and triplets occurs at all distances, intermingled with the Coulomb-prejudiced random walk. A realistic formulation of the problem is a separate diffusion equation (Eq. 3) for singlet and triplet species, and a kinetic equation, derived from quantum mechanics, for singlet-triplet interconversion. Solutions for the coupled system of partial differential equations are not available, but can be computed. This radical pair interaction is the chief mechanism responsible for Chemically Induced Dynamic Nuclear Polarization (121). The case of uncharged species has been considered by Freed & Pedersen (122) using finite difference equations. Schulten & Schulten have computed the ion pair case using a density matrix approach (123) and Schulten & Epstein a Monte Carlo-path integral method (124). We have computed solutions using finite difference equations based on those used in simple diffusion problems (125). The transparency of this method far outweighs its relative inefficiency. Addition of more realistic conditions such as the distance dependence of spin flip rate by exchange and of electron tunneling rate, and of diffusion constant at close encounters, etc which cause enormous complications to the already near unsolvable analytic equations are relatively easily incorporated into the finite difference program. There are three critical parts to these computations:

1. the number of diffusive steps or channels,
2. the boundary condition at r_m, and
3. the singlet-triplet interconversion rate.

1. Our finiteness requires that the diffusion to infinity be truncated rather shortly. Schulten & Schulten (123) have used the stratagem of placing a perfect sink or reflector at r_f (the final channel). While this is satisfactory for initial kinetics on the nanosecond time scale, the yield of ions (i.e., ions with $r \gg r_c$) is strongly affected by finite r_f. We have varied r_f and extrapolated the yield to $r \to \infty$. This procedure is adequate but tedious. The use of a "Virtual Boundary" eliminates this difficulty. The last channel is simply clamped at the known long time, large r solution to Eq. 3, thus assuring an asymptotically correct answer. In practice an empirical equation that reproduced the yields given by Eq. 4 to within 1% was found useful. A dramatic condensation of the number of channels (Δr) is possible. Since the number of computational cycles increases as $\sim (\Delta r)^3$ in the finite difference method, this reduction of channel number is most helpful.

2. The rate of reaction at the encounter distance r_m determines the relative importance of diffusion on the yield (cf. Eqs. 4 and 10). We adopt the simplest picture: r_m is a perfect sink for singlets and a perfect reflector for triplets. The latter is the straightforward requirement of quantum mechanics. The former is required if the products do follow encounter-limited kinetics in the homogeneous phase of the reaction. This is an experimentally determinable fact. It is inconsistent to allow reflections at r_m if the products in the random spin state react at the encounter limit. Schulten & Schulten (123) treat the rate at r_m as an adjustable parameter. They also assume that the rates of singlet and triplet collapse are the same. This assumption leads them to the erroneous conclusion that the effect of magnetic fields on the ratio of singlet to triplet escape is independent of solvent viscosity and dielectric constant. Our calculations show otherwise (unpublished work).

3. The most widely quoted mechanism for electron spin dephasing is that of coupling to magnetic nuclei (126). The interaction is coherent, but most molecules contain a large number of hydrogen (spin 1/2) and even some nitrogen nuclei (spin 1). Thus the decay of the initial electron spin state is best described as dispersion into a rather large sea of spin levels. An element of irreversibility thus enters even on the quantum scale. The relaxation is by dispersion, not dissipation. This process is just what makes up the electron spin resonance spectrum of the free radicals. For the spin one half case of Gaussian line shape and equal widths for cation and anion the inter-conversion probabilities (J. Geronimo, unpublished work) are:

low field $\quad \rho_{T-T} = \frac{7}{9} + \frac{1}{9}(1 - t^2\omega)$

$$\times [\exp(-t^2\omega/2) + (1 - t^2\omega)\exp(-t^2\omega)] \quad \quad 12.$$

high field $\quad \rho_{T-T} = \frac{5}{6} + \frac{1}{6}\exp(-t^2\omega) \quad \quad 13.$

and the Brocklehurst relation:

$$\rho_{S-S} = 3\rho_{T-T} - 2 \quad \quad 14.$$

where $\omega = $ (linewidth)^2 in terms of frequency (2.8 Mc/G). The case of unequal cation-anion linewidths is somewhat more complex. The initial spin state decays smoothly on the nanosecond time scale (few gauss linewidths) to a steady state, the low field case showing a shallow undershoot to the statistical probability at $t^2\omega = 3$. The steady state values differ slightly from the random ratio of three triplets to one singlet. This further relaxation occurs on the microsecond-millisecond time scale for organic radicals in solution (127). These time scales are those of the T_2 (spin-spin) and T_1 (spin-lattice) relaxations of magnetic resonance terminology. Because of the $\exp(-t^2\omega)$ dependence of the rate of spin diffusion, a "lag" is seen in the plot of triplet-singlet interconversion with time, which translates into a lag in the effect of external magnetic fields on this interconversion. This effect can be seen in the data of Schulten et al (82) and in our data (unpublished). Magnetic field effects arise in general from splitting of the magnetic sublevels of the triplet state, so that ultimately only T_0 can communicate with S.

Schulten et al (82) observed a 15% decrease in the yield of pyrene triplet formed during geminate recombination of pyrene anion with 3,5-dimethoxyaniline cation in methanol in the presence of a 100 G magnetic field. Michel-Beyerle et al (128) observed both an 8% increase in ion yield and a smaller decrease in yield of pyrene triplet from the reaction of excited pyrene and diethylaniline in methanol in the presence of a 40 G magnetic field. No magnetic field effects, even at 200 G were observed in isopropanol. However, Frankevich & Fedotova (129) did observe a 0.5% decrease of ion yield with magnetic field in this isopropanol system, but the effect was highly solvent-dependent. The ion yield actually increased by 5% on adding 5% acetonitrile. These complex effects may be caused by the interaction of spin dephasing time, viscosity and dielectric constant, exemplified by Eq. 10, and aggravated by the idiosyncracies of anilines.

Analysis of Yield Data

The work of Jarnagin and his co-workers on the photoionization of aromatic hydrocarbons in solutions (6, 73, 74) provided clear evidence for the formation of aromatic anion and cation pairs at the encounter-limited rate of two triplet states. The equation used to fit the ion yields was derived by the steady-state assumptions, criticized above, and moreover is valid only at large r_c. Their data for phenanthrene is replotted in Figure 2. If only singlet ion pairs are formed, an r_0 of 12 to 13 Å fits the data. This is somewhat larger than twice the molecular radius. The data cannot be fit by assuming only triplet ion pairs are formed at $r_0 = r_m$, regardless of spin dephasing time. Thus the conclusions reached are similar to those obtained in the porphyrin triplet-triplet reaction (see below).

Selecting the most extensive set of solvent variation data from the work of Mataga and co-workers, we show in Figure 2 the ion yield from the toluene-tetracyanobenzene complex (36) as a function of Coulomb radius of the solvent. The data can be fit by assuming singlet ion pairs (as is assumed by these workers) and by use of Eq. 4, with $r_0 \sim 25$ Å and $r_m \sim 11$ Å. However, it is unlikely that C-T absorption would be observable at such a distance (r_0) and a more reasonable fit is made by assuming ion escape from the triplet state. If the initial ion yield is unity, the escaped ion yield data can be fit with a rather rapid spin dephasing time, equivalent to a linewidth of ~1000 G. This is similar to the porphyrin-triplet case (see below). Masuhara & Mataga (37) fit their data with an empirical equation. We believe their use of steady-state kinetics and free energy relations is not justified.

Our results with zinc octaethyl porphyrin are shown in Figure 3. The yield of ions from the T-T reaction, assuming only singlet-state ions, is fit by Eq. 4 with $r_0 = 19 \pm 1$ Å and $r_m = 14 \pm 3$ Å. The yield of ions from the ground-state triplet-state reaction is fit with $r_0 = 20$ Å, $r_m = 15$ Å and a spin dephasing time of about 0.6 nsec. The latter would correspond to a hyperfine linewidth of ~500 G, whereas the measured value is ~5 G.

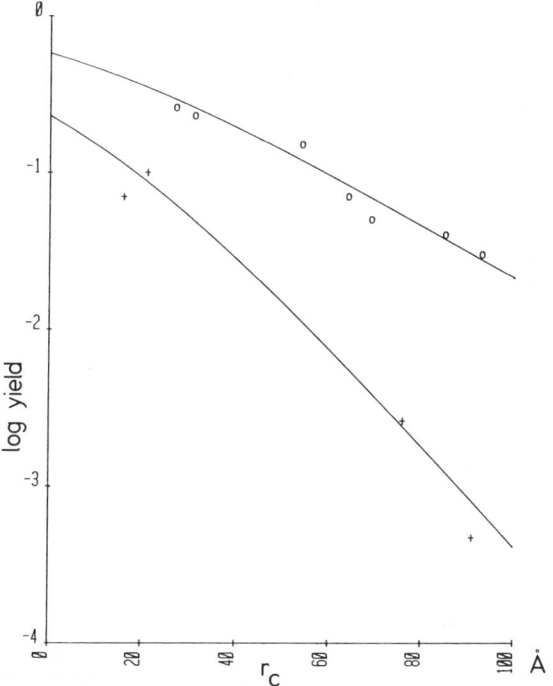

Figure 2 The log of the yield of escaped ions is plotted versus the Coulomb radius, r_c, of the solvent. The crosses are the data of Gary et al (74) from their Table 4, p. 1583. They concern the yield of ions from the reaction of two phenanthrene triplets. The data are fit (*solid line*) by Eq. 4 with $r_0 = 13$ Å, $r_m = 10$ Å. The circles are the data of Masuhara et al (36) from their Table 1, p. 996. They concern the yield of ions from the photoexcited tetracyanobenzene-toluene complex. The data are fit (*solid line*) by Eq. 4 with $r_0 = 26$ Å, $r_m = 11$ Å, or by the computed solution to the triplet pair escape with a spin dephasing time of about 0.5 nsec.

The decrease in yield at low r_c could be explained if the spin-dephasing time were shortened in solvents of high dielectric constant. Our recent experiments suggest that porphyrins may also aggregate more readily in these highly polar solvents and thus the effect could be an artifact.

The inefficiency of the singlet-triplet reaction of the porphyrin ($k \sim 10^8$ M^{-1} sec^{-1}) suggests a simple model. Since over 100 collisions are required to cause quenching of the triplet state, the radial distribution of triplets about the ground state can be taken as the equilibrium distribution $4\pi r^2 g(r)$, where $g(r)$ is $\exp(-r_c/r)$ for ions and 1 for neutral radicals. If reversible electron transfer occurs at r_0, ions will be present at

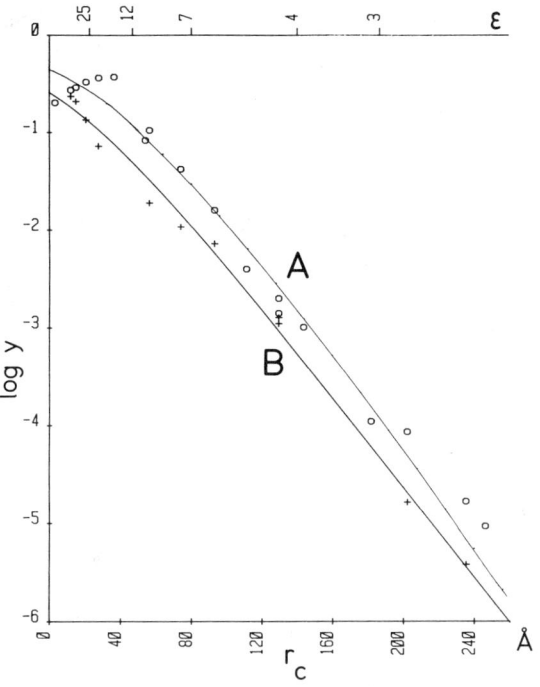

Figure 3 Ion yield data for the T-P (o) and T-T (+) reactions in solutions of zinc octaethyl porphyrin, as a function of solvent Coulomb radius. *Line B* is the least squares fit of the T-T data, using Eq. 4. The parameters used are $r_0 = 19 \pm 1$ Å, $r_m = 14 \pm 3$ Å. *Line A* is a fit of the T-P data by computed solution to the triplet pair escape. The parameters are $r_0 = 20$ Å, $r_m = 15$ Å and the spin dephasing time is 0.6 nsec. Equivalently, the T-P data is fit by Eq. 15 with $r_0 = 23$ Å and $r_m = 12$ Å. The data is taken from Ref. (89). Note that the slopes of these lines are determined mainly by r_0 and thus are well defined at large r_c. The y-intercepts and curvatures of the lines are largely determined by r_m and the spin dephasing time, the derived values of which are therefore more dependant on absolute calibration of yield data.

$r < r_0$ but neutral triplet + ground state at $r > r_0$. Only when the spins of the ion pair are in the singlet state will there be a possibility of escape or collapse. Thus, a distribution of singlet-state ions will form between r_0 and r_m if the spin flip is independent of distance. The yield of escaped ions is:

$$\phi = \frac{\int_{r_m}^{r_0} Y(r) 4\pi r^2 \exp(-r_c/r) \, dr}{\int_{r_m}^{r_0} \pi r^2 \exp(-r_c/r) \, dr} \qquad 15.$$

where $Y(r)$ is the yield of escaped ions originating at r (Eq. 4). The dependence of spin dephasing on distance is easily incorporated into the average. A simple approximation of the exchange interaction is that spin flips are prohibited close to r_m. One then simply changes the lower limit of the above integrals, leaving r_m in Y untouched. The effect is small. Freezing the spin state for 2 Å beyond r_m increases the yield only 4%, and even freezing spins to within 2 Å of r_0 increases the yield by only 30% for r_m and r_0, about 10 Å and 20 Å respectively. The results of such calculations show that the singlet-triplet yield data can be fit with $r_0 = 23 \pm 1$ Å and $r_m = 12 \pm 2$ Å. As expected, r_0 must be somewhat larger than before because the increased yield caused by spin dephasing delay has been replaced by a distribution of radii. However, the troublesome rapid spin dephasing time has been diluted by a factor of $\sim 10^2$. That is, the total time spent on spin dephasing will be the number of encounters multiplied by the time spent in diffusing between r_0 and r_m. This is

$$100 \cdot (\Delta r)^2 / 6D = 10^2 \times 121 \text{ Å}^2 / 6.3 \times 10^{11} \text{ Å}^2 \text{ sec}^{-1} \equiv 70 \text{ nsec}.$$

This corresponds to hyperfine linewidth of ~ 5 G, as is observed. Thus the extraordinarily weak dependence of the second-order rate constant ($\sim 10^8$ M^{-1} sec^{-1}) on solvent properties is explained. It is determined by the spin properties of the radical. This mechanism also predicts only weak effects of magnetic field and viscosity on the yield of ions, as is observed. We have further shown that the yield of ions from triplet plus ground state porphyrin is not sensitive to the heavy atom effect (replacing Zn by Mg) or to change of molecular symmetry from D4h to D2h (Zn porphyrin to Zn chlorin) (S. G. Ballard, unpublished work). The small increase of ion yield seen in the presence of magnetic fields may be caused by an increased yield of triplet state.

The equilibration of triplet states and triplet ion pairs and the resulting inefficient quenching of the triplet state will occur when the free energy change of electron transfer process is near zero. This possible mechanism may explain the data of Vogelmann et al (130) on the quenching of excited acridine orange by various aromatic amines and ethers. They fit their data to a Weller-type scheme, but must postulate that the triplet state is less reactive than the singlet by 0.12 eV more than the known 0.34 eV singlet-triplet energy gap.

Our view predicts that the bimolecular reaction of excited donor or acceptor with its partner will always produce a higher yield of ions than from excited CT complex. The data of Hentzchel & Watkins (45) on perylene acceptors are in agreement with this prediction, as are the data of Carapelluci & Mauzerall (83). It is the nonzero lifetime of the CT state that allows separation to the ion pair, which can then continue to escape, return to the CT state, or electron transfer (tunnel!) to the ground state. It has often been assumed that all photoelectron transfers proceed through the C-T state, but we believe this hypothesis is without support. In our view, the direct electron transfer to ground state, when encounter-limited, is simply that occuring at r_m.

What Does r_0 Mean?

The measured value of r_0 includes several effects. The parsimonious assumption of a unique r_0 to fit the data in all solvents could be relaxed if necessary. A distribution of r_0 values is to be expected, e.g. from the exponential dependence of electron tunneling rate on distance, or from the distribution model used to explain the porphyrin ground-triplet state data. The average yield over any distribution is readily calculated from the δ-function response (eg. Eq. 15). The average r_0 value so determined will be underestimated if there are steric or orientational hindrances to the reaction. It will also be underestimated if the probability of initial formation of ions decreases with r_c. If this is so, as is usually assumed, then for a given yield of ions a *larger* r_0 will be necessary. Thus our use of a unique r_0 is a lower bound. Only the hydrodynamic slowing of two closely approaching spheres would tend to cause an overestimation of r_0. The possible slowing of molecular diffusion at near-encounter distances is related to the cage effect in liquids. This effect has been much discussed from the view of the structure of the liquid state. Recent numerical calculations (131, 132) on nonpolar solutes in water have given evidence of a double minimum in the plot of interaction energy versus solute-solute distance: that in contact and that at a separation of about one solvent molecule. The increased time spent at this solvent-separated position would translate into a larger r_0 for ion escape if this effect were present in polar organic solvents, and if the lifetime were sufficiently long, 10 to 100 psec.

OUTLOOK

Although a few straws have been plucked from the wind, more solid conclusions require more substantial effort. The homogeneous monophotonic reactions are of interest because the energy of the ions formed on reaction of a ground and excited state of the same molecular species is about equal to that of the excited state. Thus, conversion of photon energy to chemical energy is very efficient in this resonant case. Clearly it is the long triplet-state lifetime and the spin selection rule that make the triplet plus ground state reaction so efficient in forming free ions. The long life permits all triplets to be trapped even in dilute solution and the spin rule enforces the delay that enhances ion pair escape. Note that both of these effects are entropic and are a measure of the entropy flow which characterizes the basic irreversibility of these processes. There remains to pin down the mechanism(s) of the spin-dephasing process. The required experiments are clear: the effect of magnetic fields combined with isotopic substitution, heavy atom, and symmetry changes. Their execution and interpretation are more problematical.

In contrast, the case of the reaction between excited singlet and ground state molecules is in disarray. Claims of free ion yields ranging from zero to substantial exist, although often on somewhat different systems. From our viewpoint this variety in yield reflects a variation in reaction radius (r_0) from contact (r_m) to substantially greater than r_m. Less ambiguous experiments than fluorescence quenching measurement are sorely needed. The same comment can be made for the reaction of two excited singlet states that must be experimentally distinguished (e.g. by double pulse techniques) from two photon processes.

The formation of ions from the reaction of two triplet states is well documented, but the spin state(s) of the resulting ion pair is not unambiguously known. The effects of magnetic fields and other relevant parameters should allow resolution of this uncertainty. A central problem is to distinguish between contact or tight ion pairs and solvent-separated or loose ion pairs. We propose an operational definition: If the absorption spectra of the initial product is the sum of the spectra of the anion and cation, then the product is the solvent-separated pair. If not, then one must invoke appreciable electronic interaction, and the product is the contact ion pair.

It is unlikely that the question of r_0 will be completely settled by studies in solution with molecules allowing free diffusion. Molecules capable of hindered diffusion (the tethered or limp-noodle models) are even less likely to provide clear answers. What is needed are molecules rigidly held in three-dimensional space so both distance and angular coordinates are known. Such species are under study in several laboratories, including our own (J. Lindsey, unpublished work).

The electrostatic effects of ion movement described by the diffusion equation(s) predict negative temperature dependence of ion yield, and even of escape kinetics, under some conditions. This is because the dielectric constant of many solvents increases faster than the reciprocal of the absolute temperature. This effect will be particularly evident in the ion yields, since they are often independent of viscosity. The ion escape kinetics will, in addition, be inversely proportional to viscosity and thus these effects may be empirically separated by a careful study of both yields and kinetics. The use of the coulomb radius, r_c, to characterize the electrostatic effects of a solvent has much to recommend it. The linear component of the variation of the dielectric constant with temperature neatly cancels, so r_c is more closely a constant of the system. It is analogous to the kineticist's use of viscosity divided by temperature to characterize viscous effects.

There are two valuable experimental techniques to be used in the study of the photoionization problems, ideally in combination. The most direct approach is to follow the very rapid geminate pair collapse and escape through picosecond spectroscopy. A complication is the short singlet state lifetime and the ensuing high concentration of acceptor or donor needed to observe appreciable reaction. Charge transfer complexes offer the possibility of specific excitation, but again their weak binding constants require high concentrations. Study over a careful region of concentration would be needed to allow a suitable time window before secondary encounters confuse the interpretation of data. Careful choice of the system and solvents is necessary for any hope of success. We recommend the porphyrins. They have been chosen by the fine comb of evolution precisely for the electron transfer function in cells. It simply is not true that complex molecules yield complex results. As we have argued, just the opposite is often the case. The attractiveness of the conductivity technique is that it specifically measures the yield, and the kinetics, of the free ions. The optical absorption methods can rarely distinguish contact ion pairs, let alone those solvent separated pairs, from free ions. The ion escape kinetics themselves may be of considerable interest. These are best measured by a fast conductivity apparatus.

In summary, we believe the emerging view of ion formation in liquids rescues the subject from the static, thermodynamic picture and allows the fascinating richness of dynamic detail to appear.

Acknowledgment

We acknowledge the support of the National Science Foundation (Grants PCM-80-11485 and CHE 77-25152) and the Ferry Fund.

Literature Cited

1. Connors, R. D. 1980. *Electron Transfer Reactions.* London: Butterworths. 351 pp.
2. Sutin, N. 1979. *J. Photochem.* 10:19–40
3. Blankenship, R. E., Parson, W. W. 1978. *Ann. Rev. Biochem.* 47:635–53
4. Hummel, A. 1974. *Adv. Radiat. Chem.* 4:1–102
5. Albrecht, A. C. 1970. *Acc. Chem. Res.* 3:238–48
6. Jarnagin, R. C. 1971. *Acc. Chem. Res.* 4:420–27
7. Gerischer, H., Kolb, D. M., Sass, J. K. 1978. *Adv. Phys.* 27:432–98
8. Friedrich, D. M., McClain, W. M. 1980. *Ann. Rev. Phys. Chem.* 31:559–77
9. Mulliken, R. S. 1950. *J. Am. Chem. Soc.* 72:600–8
10. Mulliken, R. S., Person, W. B. 1962. *Ann. Rev. Phys. Chem.* 13:107–26
11. Beens, H., Weller, A. 1975. See Ref. 21, Ch. 4, Sect. 2
12. Foster, R. 1969. *Organic Charge-Transfer Complexes*, pp. 13–16. London: Academic. 470 pp.
13. Powell, H. M., Huse, G., Cooke, P. W. 1943. *J. Chem. Soc.* 153–57
14. Kuroda, H., Ikemoto, I., Akamatu, H. 1966. *Bull. Chem. Soc. Jpn.* 39:1842–49
15. Prout, C. K., Wright, J. D. 1968. *Angew. Chem.* 80:688–97
16. Davis, K. M. C., Symons, M. C. R. 1965. *J. Chem. Soc.* 2079–83
17. Torrance, J. B. 1979. *Acc. Chem. Res.* 12:79–86
18. Bergmann, K., Eigen, M., de Mayer, L. 1963. *Ber. Bunsenges. Phys. Chem.* 67:819–26
19. Chan, R. K., Liao, S. C. 1970. *Can. J. Chem.* 48:299–305
20. Varma, C. A. G. O., Oosterhof, L. J. 1971. *Chem. Phys. Lett.* 9:406–11
21. Birks, J. B., ed. 1975. *Organic Molecular Photophysics*, 2:519, Tbl. 9.31. London: Wiley. 653 pp.
22. Mataga, N., Murata, Y. 1969. *J. Am. Chem. Soc.* 91:3144–52
23. Kobayashi, T., Yoshihara, K., Nagakura, S. 1971. *Bull. Chem. Soc. Jpn.* 44:2603–10
24. Lagercrantz, C., Yhland, M. 1962. *Acta Chem. Scand.* 16:1043–45
25. Ward, R. L. 1963. *J. Chem. Phys.* 39:852–53
26. Ilten, D. F., Calvin, M. 1965. *J. Chem. Phys.* 42:3760–66
27. Potashnik, R., Goldschmidt, C. R., Ottolenghi, M. 1969. *J. Phys. Chem.* 73:3170–71
28. Potashnik, R., Ottolenghi, M. 1970. *Chem. Phys. Lett.* 6:525–28
29. Masuhara, H., Mataga, N. 1970. *Chem. Phys. Lett.* 6:608–10
30. Masuhara, H., Shimada, M., Tsujino, N., Mataga, N. 1971. *Bull. Chem. Soc. Jpn.* 44:3310–16
31. Iwata, S., Tanaka, J., Nagakura, S. 1966. *J. Am. Chem. Soc.* 88:894–902
32. Hinatu, J., Yoshida, F., Masuhara, H., Mataga, N. 1978. *Chem. Phys. Lett.* 59:80–83
33. Shimada, M., Masuhara, H., Mataga, N. 1972. *Chem. Phys. Lett.* 15:364–65
34. Shimada, M., Masuhara, H., Mataga, N. 1973. *Bull. Chem. Soc. Jpn.* 46:1903–9
35. Taniguchi, Y., Nishina, Y., Mataga, N. 1971. *Bull. Chem. Soc. Jpn.* 45:764–69
36. Masuhara, H., Hino, T., Mataga, N. 1975. *J. Phys. Chem.* 79:994–1000
37. Masuhara, H., Mataga, N. 1981. *Acc. Chem. Res.* 14:312–18
38. Achiba, Y., Katsumata, S., Kimura, K. 1972. *Chem. Phys. Lett.* 13:213–16
39. Achiba, Y., Kimura, K. 1973. *Chem. Phys. Lett.* 19:45–48
40. Irie, M., Tomimoto, S., Koichiro, H. 1972. *J. Phys. Chem.* 76:1419–24
41. Kimura, K., Achiba, Y., Katsumata, S. 1973. *J. Phys. Chem.* 77:2520–23
42. Pilette, Y. P., Weiss, K. 1971. *J. Phys. Chem.* 75:3805–14

43. Orbach, N., Novros, J., Ottolenghi, M. 1973. *J. Phys. Chem.* 77:2831–36
44. Ottolenghi, M. 1973. *Acc. Chem. Res.* 6:153–60
45. Hentzchel, P., Watkins, A. R. 1976. *J. Phys. Chem.* 80:494–500
46. Shimada, K., Szwarc, M. 1975. *J. Am. Chem. Soc.* 97:3321–23
47. Pragst, F., Hamann, H. J., Teuchner, K., Daehne, S. 1978. *J. Lumin.* 17:425–37
48. Migita, M., Kawai, M., Mataga, N., Sakata, Y., Misumi, S. 1978. *Chem. Phys. Lett.* 53:67–70
49. Chuang, T. J., Cox, R. J., Eisenthal, K. B. 1974. *J. Am. Chem. Soc.* 96:6828–31
50. Crawford, M. K., Wang, Y., Eisenthal, K. B. 1981. *Chem. Phys. Lett.* 79:529–33
51. Chuang, T. J., Eisenthal, K. B. 1975. *J. Chem. Phys.* 62:2213–22
52. Eisenthal, K. B. 1977. *Ann. Rev. Phys. Chem.* 28:207–32
53. Kosower, E. M., Dodink, H. 1978. *J. Am. Chem. Soc.* 100:4173–79
54. Kosower, E. M., Dodink, H., Kanety, H. 1978. *J. Am. Chem. Soc.* 100:4179–88
55. Grabowski, Z. R., Rotkiewicz, K., Siemiarczuk, A., Cowley, D. J., Baumann, W. 1979. *Nouv. J. Chim.* 3:443–54
56. Baur, E. 1932. *Z. Phys. Chem. Abt. B* 16:465
57. Weiss, J., Fishgold, H. 1936. *Z. Phys. Chem. Abt. B* 32:135
57a. Linschitz, H., Pekkarinen, L. 1960. *J. Am. Chem. Soc.* 82:2411–16
58. Leonhardt, H., Weller, A. 1961. *Z. Phys. Chem. N.F.* 29:277–80
59. Leonhardt, H., Weller, A. 1963. *Ber. Bunsenges. Phys. Chem.* 67:791–95
60. Mataga, N., Okada, T., Ezumi, K. 1965. *Mol. Phys.* 10:203–4
61. Knibbe, H., Röllig, K., Schäfer, F. P., Weller, A. 1967. *J. Chem. Phys.* 47:1184–85
62. Knibbe, H., Rehm, D., Weller, A. 1968. *Ber. Bunsenges. Phys. Chem.* 72:257–63
63. Rehm, D., Weller, A. 1969. *Ber. Bunsenges. Phys. Chem.* 73:834–39
64. Marcus, R. A. 1964. *Ann. Rev. Phys. Chem.* 15:155–96
65. Rehm, D., Weller, A. 1970. *Isr. J. Chem.* 8:259–71
66. Traber, R., Vogelmann, E., Schreiner, S., Werner, T., Kramer, H. E. A. 1981. *Photochem. Photobiol.* 33:41–48
67. Vogelmann, E., Rauscher, W., Kramer, H. E. A. 1979. *Photochem. Photobiol.* 29:771–76
68. Beens, H., Knibbe, H., Weller, A. 1967. *J. Chem. Phys.* 47:1183–84
69. Mataga, N., Okada, T., Yamamoto, N. 1966. *Bull. Chem. Soc. Jpn.* 39:2562
70. Mataga, N., Okada, T., Yamamoto, N. 1967. *Chem. Phys. Lett.* 1:119–21
71. Grellman, K. H., Watkins, A. R. 1971. *Chem. Phys. Lett.* 9:439–43
72. Watkins, A. R. 1976. *J. Phys. Chem.* 80:713–17
73. Kawada, A., Jarnagin, R. C. 1966. *J. Chem. Phys.* 44:1919–28
74. Gary, L. P., de Groot, K., Jarnagin, R. C. 1968. *J. Chem. Phys.* 49:1577–87
75. Taniguchi, Y., Mataga, N. 1972. *Chem. Phys. Lett.* 13:596–99
76. Taniguchi, Y., Nishina, Y., Mataga, N. 1973. *Bull. Chem. Soc. Jpn.* 46:1646–49
77. Arimitsu, S., Masuhara, H. 1973. *Chem. Phys. Lett.* 22:543–46
78. Goodall, D. M., Orbach, N., Ottolenghi, M. 1974. *Chem. Phys. Lett.* 26:365–68
79. Hino, T., Akazawa, H., Masuhara, H., Mataga, N. 1976. *J. Phys. Chem.* 80:33–37
80. Masuhara, H., Hino, T., Mataga, N. 1975. *J. Phys. Chem.* 79:994–1000
81. Hino, T., Masuhara, H., Mataga, N. 1976. *Bull. Chem. Soc. Jpn.* 49:394–96
82. Schulten, K., Staerk, H., Weller, A., Werner, H. J., Nickel, B. 1976. *Z. Phys. Chem. N.F.* 101:371–90
83. Carapellucci, P. A., Mauzerall, D. 1975. *Ann. NY Acad. Sci.* 244:214–38
84. Ballard, S. G. 1976. *Rev. Sci. Instrum.* 47:1157–62
85. Ballard, S. G., Mauzerall, D. C., Tollin, G. 1976. *J. Phys. Chem.* 80:341–51
86. Ballard, S. G., 1977. PhD thesis. The Rockefeller Univ., New York
87. Ballard, S. G., Mauzerall, D. 1978. *Biophys. J.* 24:335–45
88. Ballard, S. G., Mauzerall, D. 1979. In *Tunneling in Biological Systems*, ed. B. Chance, D. deVault, pp. 581–89. Proc. Symp. Nov. 2–5, 1977, Johnson Found., Philadelphia. New York: Academic. 758 pp.

89. Ballard, S. G., Mauzerall, D. C. 1980. *J. Chem. Phys.* 72:933–47
90. Debye, P. 1942. *Trans. Electrochem. Soc.* 82:265–72
91. Alwattar, A. H., Lumb, M. D., Birks, J. B. 1973. In *Organic Molecular Photophysics*, ed. J. B. Birks, Vol. 1, Ch. 8, pp. 403–56. London: Wiley. 600 pp.
92. Stiles, P. J. 1981. *Chem. Phys. Lett.* 80:73–75
93. Ballard, S. G., Mauzerall, D. C. 1980. *J. Electrochem. Soc.* 80:495–96
94. Fuhrhop, J.-H., Mauzerall, D. C. 1969. *J. Am. Chem. Soc.* 91:4174–81
95. Fuhrhop, J.-H., Kadish, K. M., Davis, D. G. 1973. *J. Am. Chem. Soc.* 95:5140–47
96. Imura, T., Furutsuka, T., Kawabe, K. 1975. *Photochem. Photobiol.* 22:129–34
97. Gudkov, N. D., Stolovitskii, Yu. M., Yevstigneyev, V. B. 1975. *Biophysics* 20:217–21
98. Seely, G. R. 1978. *Photochem. Photobiol.* 27:639–54
99. Huyskens, P. L. 1980. *Bull. Soc. Chim. Belg.* 89:937–49
100. Gouterman, M., Holten, D. 1977. *Photochem. Photobiol.* 25:85–92
101. Monchick, L. 1956. *J. Chem. Phys.* 24:381–85
102. Hong, K. M., Noolandi, J. 1978. *J. Chem. Phys.* 68:5163–71
103. Berlin, Yu. A., Cordier, P., Delaire, J. A. 1980. *J. Chem. Phys.* 73:4619–27
103a. Flannery, M. R. 1981. *Phys. Rev. Lett.* 47:163–66
104. Onsager, L. 1934. *J. Chem. Phys.* 2:599–615
105. Eigen, M. 1954. *Z. Phys. Chem. N.F.* 1:176–200
106. Weller, A. 1961. *Prog. React. Kin.* 1:187–214
107. Feitelson, J., Mauzerall, D. 1982. *J. Phys. Chem.* 86:1623–28
108. Gurney, R. 1931. *Proc. R. Soc. London Ser. A* 134:137–54
109. Brocklehurst, B. 1973. *Chem. Phys.* 2:6–18
110. Miller, J. R. 1975. *Science* 189:221–23
111. Libby, W. F. 1977. *Ann. Rev. Phys. Chem.* 28:105–10
112. Chance, B., DeVault, D., eds. 1979. *Tunneling in Biological Systems*. New York: Academic. 758 pp.
113. DeVault, D. 1980. *Q. Rev. Biophys.* 13:387–564
114. Mauzerall, D. 1976. *Brookhaven Symp. Biol.* 28:64–73
115. Mauzerall, D. 1978. In *The Porphyrins*, ed. D. Dolphin, 5:29–52. New York: Academic. 548 pp.
116. Brugger, P. A., Infelta, P. P., Braun, A. M., Gratzel, M. 1981. *J. Am. Chem. Soc.* 103:320–26
117. Borsenberger, P. M., Contois, L. E., Hoesterey, D. C. 1978. *J. Chem. Phys.* 68:637–41
118. Collins, F. C., Kimball, G. E. 1949. *J. Colloid Sci.* 4:425–37
119. Sterna, L., Ronis, D., Wolfe, S., Pines, H. 1980. *J. Chem. Phys.* 73:5493–99
120. Haar, H. P., Klein, U. K. A., Hauser, M. 1978. *Chem. Phys. Lett.* 58:525–30
121. Lepley, A. R., Closs, G. L., ed. 1973. *Chemically Induced Magnetic Polarization*. New York: Wiley. 416 pp.
122. Freed, J. H., Pedersen, J. B. 1976. *Adv. Magn. Res.* 8:1–84
123. Schulten, Z., Schulten, K. 1977. *J. Chem. Phys.* 66:4616–34
124. Schulten, K., Epstein, I. R. 1979. *J. Chem. Phys.* 71:309–16
125. Carslaw, H. S., Jaeger, J. C. 1959. *Conductivity of Heat in Solids*, pp. 466–78. Oxford: Clarendon. 510 pp. 2nd ed.
126. Brocklehurst, B. 1976. *J. Chem. Soc. Faraday Trans. 2* 72:1869–83
127. Hore, P. J., Joslin, C. G., McLauchlan, K. A. 1979. *Chem. Soc. Rev.* 8:29–62
128. Michel-Beyerle, M. E., Haberkorn, R., Bube, W., Steffens, E., Schröder, H., Neusser, H. J., Schlag, E. W., Seidlitz, H. 1976. *Chem. Phys.* 17:139–45
129. Frankevich, E. L., Fedotova, E. Ya. 1980. *High Energy Chemistry*, pp. 355–59. New York: Plenum
130. Vogelmann, E., Rauscher, W., Traber, R., Kramer, H. E. A. 1981. *Z. Phys. Chem. N.F.* 124:13–22
131. Swaminathan, S., Beveridge, D. L. 1979. *J. Am. Chem. Soc.* 101:5832–33
132. Pangali, C., Rao, M., Berne, B. J. 1979. *J. Chem. Phys.* 71:2982–90

THEORIES OF THE DYNAMICS OF PHOTODISSOCIATION

M. Shapiro[1]

Department of Chemical Physics, Weizmann Institute of Science, 76100 Rehovot, Israel

R. Bersohn[1,2]

Department of Chemistry, Columbia University, New York, New York 10027

INTRODUCTION

This review is an account of some theoretical methods in photodissociation dynamics with which the authors are familiar, and a brief comparison of experiments with theory. For reasons of space, numerous topics have been omitted, such as a survey of experimental results (1–3), infrared multiphoton dissociation (e.g. see 4), van der Waals complex dissociation (e.g. see 4), and complementary aspects of photodissociation of polyatomics, reviewed by Gelbart (5).

The fascinating aspect of photodissociation dynamics, particularly of a triatomic $ABC \to A + BC$, is the possibility of solution of the quantum equations of motion, on the ground and excited state potential surfaces, and eventual comparison with the experimental cross sections. This is much more difficult for a three body exchange reaction $A + BC \to AB + C$, mainly because of the different boundary conditions and because a wide range of impact parameters must be considered. The observables in photodissociation and reaction dynamics are very similar—the disposal

[1] Work supported by the U S–Israel Binational Science Foundation Grant No. 2430.
[2] Work supported by the U S National Science Foundation.

of the available energy into translation, vibration, and rotation, and the angular distribution of the products. The available energy is "tuned" in reactive scattering by changing the reactant velocities or sometimes their internal states. In photodissociation the dissociating wavelength is changed.

OBSERVABLES

Consider a triatomic molecule ABC undergoing an electronic excitation due to the action of radiation and breaking apart into two fragments,

$$ABC \xrightarrow{\hbar\omega} A + BC. \qquad 1.$$

If not forbidden by energy requirements, there are, in principle, three "fragment channels" of dissociation denoted by α (A, B, or C) in which

$$\begin{array}{ll} ABC \rightarrow A+BC & \alpha = A \\ B+AC & \alpha = B \\ C+AB & \alpha = C. \end{array} \qquad 2.$$

The radiation can be represented as

$$\varepsilon(t) = \int d\omega \varepsilon(\omega) \cos(\omega t + \phi_\omega) \qquad 3.$$

where ϕ_ω is a phase whose randomness would reflect the incoherence of the radiation, and

$$\varepsilon(\omega) \equiv \hat{\varepsilon} \cdot \varepsilon(\omega) \qquad 4.$$

is the radiation electric field at a given frequency ω. In the following the radiation-matter interaction is treated semiclassically, i.e. the molecular energy levels are quantized, but not those of the photons. Assuming a dipolar transition, we can write the time dependent Hamiltonian as

$$\hat{H} = H - \mu_\varepsilon \int d\omega \varepsilon(\omega) \cos(\omega t + \phi_\omega) \qquad 5.$$

where $\mu_\varepsilon = \boldsymbol{\mu} \cdot \hat{\varepsilon}$ is the component of the transition-dipole operator in the direction of $\hat{\varepsilon}$, and H is the molecular (matter) Hamiltonian.

The eventual breaking up of the molecule implies that the molecule is promoted, by the radiation, to an unbound (continuum) part of the excited state Hamiltonian. We therefore expand the solution of the time dependent Schrödinger equation

$$\partial \psi / \partial t = -\frac{i\hat{H}\psi}{\hbar} \quad \text{as,} \qquad 6.$$

$$\psi(t) = a(t)|E_i\rangle\exp(-iE_i t/\hbar)$$
$$+ \sum_{\mathbf{n}} \int dE\, b(E,\mathbf{n}|t)|E,\mathbf{n}^-\rangle\exp(-iEt/\hbar), \qquad 7.$$

where $a(t)$ and $b(E,\mathbf{n}|t)$ are coefficients to be determined. In Eq. 7, $|E_i\rangle$ is the initial (bound) wavefunction in the ground electronic state and $|E,\mathbf{n}^-\rangle$ are the "incoming" scattering wavefunctions in the excited electronic state. $|E_i\rangle$ satisfies the time independent Schrödinger equation for the ground state,

$$(E_i - H_{gr})|E_i\rangle = 0 \qquad 8a.$$

and $|E,\mathbf{n}^-\rangle$ satisfies the "incoming" Lippmann-Schwinger equation in the excited state,

$$|E,\mathbf{n}^-\rangle = \left(I + (E - i\varepsilon - H_{ex})^{-1} V_{ex}\right)|E,\mathbf{n}^0\rangle \qquad 8b.$$

where H_{gr} and H_{ex} are the molecular Hamiltonians in the ground and excited states, respectively, and V_{ex} is the interaction potential in the excited state.

As explained below we employ the "incoming" solution $|E,\mathbf{n}^-\rangle$ rather than the "outgoing" solution $|E,\mathbf{n}^+\rangle$, more customary in collision problems, because $|E,\mathbf{n}^-\rangle$ is the scattering state that evolves in the *distant future* to a well-defined state of the fragments $|E,\mathbf{n}^0\rangle$ [6, 7]. In contrast, $|E,\mathbf{n}^+\rangle$ corresponds to a well-defined fragment state $|E,\mathbf{n}^0\rangle$ in the *remote past*. Such a state would be appropriate for a radiative recombination of the fragments, but not for photodissociation, where for $t<0$ the wave function is $|E_i\rangle$, a molecular ground state. In the above, the collective fragment index $\mathbf{n}$ incorporates the e, α, v, j, m_j, and $\hat{k}$ quantum numbers, where e is the fragment electronic state, $\alpha(=A,B,C)$ is the fragment channel index, (Eq. 2), v, j, m_j are the diatomic vibrational, rotational, and magnetic quantum numbers, $\hat{k}$ is the direction of recoil of the ejected atom in the center of mass frame; $\mathbf{n}$ spans all accessible (open) channels at a given energy E.

By energy conservation we have that, given the internal energy $\varepsilon_\mathbf{n}$, the magnitude of the momentum carried by the ejected atom in the c.m. system is given as,

$$\hbar k_\mathbf{n} = \{2\mu_\alpha(E - \varepsilon_\mathbf{n})\}^{1/2}, \qquad 9.$$

where μ_α is the reduced mass in the α fragment channel. For $\alpha = A$,

$$\mu_A = m_A(m_B + m_C)/(m_A + m_B + m_C), \qquad 10.$$

with similar definitions holding for μ_B and μ_C. A spectroscopic probe will measure the internal energy distribution directly, from which, by Eq.

9, we can determine the k_n distribution. A time-of-flight experiment measures k_n from which it is sometimes possible to infer (by Eq. 9) the various fragment states.

In order to evaluate $a(t)$ and $b(E,\mathbf{n}|t)$ of Eq. 7, we substitute that expansion in the Schrödinger equation, (Eq. 6), making use of the orthonormality relations,

$$\langle E_i | E_j \rangle = \delta_{ij}, \qquad \text{11a.}$$

$$\langle E_i | E, \mathbf{n}^- \rangle = 0, \qquad \text{11b.}$$

$$\langle E', \mathbf{m}^- | E, \mathbf{n}^- \rangle = \delta(E - E')\delta_{\mathbf{n},\mathbf{m}}. \qquad \text{11c.}$$

We obtain

$$\dot{a}(t) = \frac{i}{\hbar} \int d\omega\, \varepsilon(\omega)\cos(\omega t + \phi_\omega)\Big\{ a(t)\langle E_i|\mu_\varepsilon|E_i\rangle$$

$$+ \sum_{\mathbf{n}} \int dE\, b(E,\mathbf{n}|t)\langle E_i|\mu_\varepsilon|E,\mathbf{n}^-\rangle \exp(-i\omega_{E_i}t)\Big\} \quad \text{and} \quad \text{12a.}$$

$$\dot{b}(E',\mathbf{m}|t) = \frac{i}{\hbar}\int d\omega\,\varepsilon(\omega)\cos(\omega t + \phi_\omega)\Big\{ a(t)\langle E'\mathbf{m}^-|\mu_\varepsilon|E_i\rangle\exp(i\omega_{E'_i}t)$$

$$+ \sum_{\mathbf{n}} \int dE\, b(E,\mathbf{n}|t)\langle E',\mathbf{m}|\mu_\varepsilon|E,\mathbf{n}^-\rangle\cdot \exp(i\omega_{E'E}t)\Big\},$$

where 12b.

$$\omega_{Ei} = (E - E_i)/\hbar \qquad \text{13a.}$$

$$\omega_{E'E} = (E' - E)/\hbar. \qquad \text{13b.}$$

In the spirit of the Rotating Wave Approximation (RWA), we neglect all the highly oscillatory terms on the r.h.s. of Eqs. 12 and we retain only those terms, containing $\exp[i(\omega - \omega_{Ei})t]$ or $\exp[-i(\omega - \omega_{Ei})t]$, where cancellation of oscillation is at all possible. The resulting equation for $\dot{b}(E',\mathbf{m}|t)$ is

$$\dot{b}(E',\mathbf{m}|t) = \frac{i}{2\hbar}\int d\omega\,\tilde{\varepsilon}(\omega)a(t)\langle E,\mathbf{m}^-|\mu_\varepsilon|E_i\rangle\exp\big[i(\omega_{E'_i} - \omega)t\big], \quad 14.$$

where

$$\tilde{\varepsilon}(\omega) = \varepsilon(\omega)\exp(-i\phi_\omega). \qquad 15.$$

If we assume that initially $(t = 0)$ the molecule is in state $|E_i\rangle$ i.e. $a(0) = 1$, $b(E,\mathbf{n}|0) = 0$, and that $a(t)$ is changing slowly i.e. $a(t) \simeq 1$ we can write the first-order result for $b(E', \mathbf{m}|t)$ as

$$b(E',\mathbf{m}|t) = \frac{i}{2\hbar}\int d\omega\, \tilde{\varepsilon}(\omega)\langle E\mathbf{m}^-|\mu_\varepsilon|E_i\rangle$$
$$\times \int_0^t dt' \exp\left[i(\omega_{E_i'}-\omega)t'\right]. \qquad 16.$$

For sufficiently long times such that $(\omega_{E_i}-\omega)t \gg 2\pi$ we find

$$\int_0^t dt' \exp\left[i(\omega_{E_i'}-\omega)t'\right] \simeq 2\pi\delta(\omega_{E_i'}-\omega) = 2\pi\hbar\delta(E-E_i-\hbar\omega),$$
$$\qquad 17.$$

and it follows that,

$$\lim_{t\to\infty} b(E,\mathbf{m}|t) = \pi i \int d\omega\, \hat{\varepsilon}(\omega)\langle E,\mathbf{m}^-|\mu_\varepsilon|E_i\rangle \delta(E-E_i-\hbar\omega)$$
$$= \frac{\pi i}{\hbar}\varepsilon(\omega_{E_i})\langle E,\mathbf{m}^-|\mu_\varepsilon|E_i\rangle \qquad 18.$$

Using Eqs. 7 and 18 we obtain the following form for the wavefunction $\psi(t)$ at $t \gg 2\pi/(\omega_{E_i}-\omega)$

$$\psi(t) = |E_i\rangle\exp(-iE_i t/\hbar)$$
$$+ \pi i \sum_\mathbf{n} \int d\omega\, \tilde{\varepsilon}(\omega)|E,\mathbf{n}^-\rangle\langle E,\mathbf{n}^-|\mu_\varepsilon|E_i\rangle\exp(-iEt/\hbar) \qquad 19.$$

where E, by Eq. 18, is

$$E = E_i + \hbar\omega. \qquad 20.$$

We now wish to calculate the probability of observing asymptotically (as $t \to \infty$) the fragments in a given recoil direction $\hat{k}$ and given set of internal quantum numbers $\mathbf{n}$. The relevant state in the asymptotic region is $|E,\mathbf{n}^\circ\rangle$, which satisfies the asymptotic Schrödinger equation

$$(E-H_\alpha^\circ)|E,\mathbf{n}^\circ\rangle = 0 \qquad 21.$$

where H_α°, the free Hamiltonian, is defined as

$$H_\alpha^\circ \equiv \lim_{R_\alpha\to\infty} H_{ex} = H_{ex} - V_{ex}^\alpha, \quad \alpha = A, B, C \qquad 22.$$

R_α being the separation of the α atom from the c.m. of the complementary diatomic. The free Hamiltonian is a sum of internal Hamiltonians for the fragments and the kinetic energy operator in R_α. For example, for $\alpha = A$,

$$H_A^\circ = K(\mathbf{R}_A) + h_A + h_{BC}(\mathbf{r}_{BC}). \qquad 23.$$

We know from Eqs. 8, 21, and 22 that

$$\langle E, \mathbf{n}^\circ | E_i \rangle = 0. \qquad 24.$$

Hence the amplitude to observe state $|E', \mathbf{m}^\circ\rangle$ at time t, which is simply $\langle E', \mathbf{m}^\circ | \psi(t) \rangle$ is, by Eqs. 19 and 24,

$$\langle E', \mathbf{m}^\circ | \psi(t) \rangle = i\pi \sum_\mathbf{n} \int d\omega\, \tilde{\varepsilon}(\omega) \langle E', \mathbf{m}^\circ | E, \mathbf{n}^- \rangle$$

$$\cdot \langle E, \mathbf{n}^- | \mu_\varepsilon | E_i \rangle \exp(-iEt/\hbar). \qquad 25.$$

At this point, we make full use of the asymptotic properties of the "*incoming*" solutions, because we know (6, 7) that

$$\lim_{t \to \infty} \langle E', \mathbf{m}^\circ | E, \mathbf{n}^- \rangle \exp(-iEt/\hbar) = \delta(E - E')\delta_{\mathbf{n},\mathbf{m}}, \qquad 26.$$

i.e. as mentioned above $|E, \mathbf{n}^-\rangle \exp(-iEt/\hbar)$ approaches in the distant future a *single* well-defined asymptotic state $|E, \mathbf{n}^\circ\rangle$. From Eqs. 20, 25, and 26 we obtain that

$$\lim_{t \to \infty} \langle E', \mathbf{m}^\circ | \psi(t) \rangle = \frac{\pi i}{\hbar} \tilde{\varepsilon}(\omega_{E'i}) \langle E' \mathbf{m}^- | \mu_\varepsilon | E_i \rangle \qquad 27.$$

and therefore

$$\psi(t) \underset{t \to \infty}{=} |E_i\rangle \exp(-iE_it/\hbar) + \sum_\mathbf{n} \int dE\, \frac{\pi i}{\hbar} \tilde{\varepsilon}(\omega_{E_i}) | E, \mathbf{n}^\circ \rangle$$

$$\cdot \langle E, \mathbf{n}^- | \mu_\varepsilon | E_i \rangle \exp(-iEt/\hbar). \qquad 28.$$

The probability, per unit energy, to make a transition to the $|E, \mathbf{m}^\circ\rangle$ state of the products is therefore,

$$\lim_{t \to \infty} |b(E, \mathbf{m} | t)|^2 = \frac{\pi^2}{\hbar^2} \varepsilon^2(\omega_{E_i}) |\langle E, \mathbf{m}^- | \mu_\varepsilon | E_i \rangle|^2. \qquad 29.$$

Using Eq. 29 we now define the photodissociation cross section for making a specific transition as

$$\sigma_{\mathbf{m}i}(\hbar\omega_{E_i}) = \frac{\text{radiation energy absorbed due to a transition}}{\text{incident radiation intensity}}$$

The radiation energy absorbed due to a transition to a final state at energy $E \to E + dE$ is $\hbar\omega \times$ probability of absorption per energy $\times dE$ which is given by

$$\hbar\omega_{E_i} \frac{\pi^2}{\hbar^2} \varepsilon^2(\omega_{E_i}) |\langle E, \mathbf{m}^- | \mu_\varepsilon | E_i \rangle|^2\, dE.$$

The incident intensity (radiation energy impinging on a unit area at unit time) in the frequency range ($\omega_{E_i} \to \omega_{E_i} + dE/\hbar$) is

$\varepsilon^2(\omega_{E_i})c/8\pi \cdot \dfrac{dE}{\hbar}$, hence

$$\sigma_{mi}(\hbar\omega_{E_i}) = \dfrac{\hbar\omega_{E_i}(\pi^2/\hbar^2)\varepsilon^2(\omega_{E_i})|\langle E,\mathbf{m}^-|\mu_\varepsilon|E_i\rangle\, dE}{\varepsilon^2(\omega_{E_i})c/8\pi \cdot (dE/\hbar)}$$

$$= \dfrac{8\pi^3}{c}\omega_{E_i}|\langle E,\mathbf{m}^-|\mu_\varepsilon|E_i\rangle|^2. \qquad 30.$$

Due to the normalization of Eqs. 11a–c this cross section has units of area. The total absorption cross section, i.e. the absorption spectrum, is then given as

$$I_i(\hbar\omega_{E_i}) = \sum_{\mathbf{m}} \sigma_{mi}(\hbar\omega_{E_i}). \qquad 31.$$

Eqs. 30 and 31 constitute the starting point for almost all the dynamical theories of photodissociation reviewed below. A notable exception is the wavepacket propagation method (8–10) in which the starting point is Eq. 19. Equation 19 is also a suitable starting point for all classical and semiclassical methods using time dependent trajectories. Due to lack of space the classical and semiclassical techniques are not reviewed here.

THE FRANCK-CONDON, FORCED OSCILLATOR AND IMPULSIVE COLLISION MODELS

The act of photodissociation in classical terms takes place in two distinct steps—molecular excitation followed by fragment separation. In quantum mechanical terms the two processes are not incoherent and therefore amplitudes for the different intermediate states can interfere. At a certain level of approximation one can prove (11–14) that the cross section for making a transition from the initial molecular state i to the final state $\mathbf{n}$ of the fragments is given by

$$\sigma_{ni} = \left|\sum_{\bar{\mathbf{n}}} S^H_{\mathbf{n}\bar{\mathbf{n}}} A_{\bar{\mathbf{n}}i}\right|^2 \qquad 32.$$

where $A_{\bar{\mathbf{n}}i}$ is the relative probability amplitude for generation by photon excitation of a state which correlates with the final state $\bar{\mathbf{n}}$ of the fragments. $S^H_{\mathbf{n}\bar{\mathbf{n}}}$ is the amplitude for making a transition from a fragment state $\bar{\mathbf{n}}$ to the fragment state $\mathbf{n}$ during the motion on the excited potential surface (15). The set of such amplitudes $S^H_{\mathbf{n}\bar{\mathbf{n}}}$ is collectively the $\mathbf{S}^H$ matrix for the half (H) collision which begins with the fragments in contact and ends with them far apart. The effect of the S^H matrix is variously described as $V-T$ relaxation on the excited surface, final state interaction or vibrationally nonadiabatic effects of the half collision.

Before considering the problem in all its generality it is helpful to consider two extreme models. In one, the so-called Franck-Condon model (13, 14a, b, 16–20a, b) the final state interactions are neglected and the S^H matrix is therefore taken to be the unit matrix. Let us begin with the rigorous expression for the cross sections (Eq. 30) in a more explicit form

$$\sigma_{ni} = \frac{8\pi^3\omega}{c} |\langle \chi^-(E, \mathbf{n}, Q_{ex}) | \mu_e(Q_{ex}) | \chi(E_i, Q_{gr}) \rangle|^2 \quad \text{where} \qquad 33.$$

$$\chi^-(E, \mathbf{n}, Q_{ex}) = \langle Q_{ex} | E, \mathbf{n}^- \rangle \quad \text{and} \quad \chi(E_i, Q_{gr}) = \langle Q_{gr} | E_i \rangle \qquad 34.$$

and Q_{gr} and Q_{ex} are the sets of nuclear coordinates of the ground and excited states, respectively, and the energies are defined by Eq. 20.

For a molecule with N atoms there are just $3N-6$ internal nuclear coordinates Q_{gr} ($3N-5$ if it is linear). Thus the $3N-6$ excited state coordinates Q_{ex} must be a function of Q_{gr}. The reason for distinguishing them is that they correspond to different types of motion. The set of coordinates Q_{gr} are most conveniently chosen to be the normal vibrational coordinates of the ground state. The set Q_{ex} can be written $(\mathbf{R}, \mathbf{r})$ where the coordinate $\mathbf{R}$ is singled out because it is the displacement between the centers of masses of the two fragments. The requirement that there be no relaxation on the upper state is satisfied by imposing the mathematical condition

$$V(Q_{ex}) = V_1(R) + V_2(\mathbf{r}). \qquad 35.$$

The potential V_1 repels the two fragments, but its exact form is not usually known. The potential V_2 may be taken to be the potential for the vibrations of the fragments and is therefore often known from spectroscopy. The physical meaning of Eq. 35 is that the internal motions of the fragments are completely decoupled from their relative transitional motion. For example, in the dissociation of CH_3I the normal modes of the CH_3 group would be assumed to be independent of the distance from the dissociating I atom. Thus, the vibrational distribution of the fragments is generated solely by the excitation. It then follows from Eq. 35 that

$$\chi^-(E, \mathbf{n}, Q_{ex}) = \chi_\mathbf{k}(\mathbf{R})\phi_\mathbf{n}(\mathbf{r}) \qquad 36.$$

where $\mathbf{k}$ is the momentum of the recoiling atom in the center of mass frame.

Because in most cases we deal with molecules with low vibrational energy it is usually safe to ignore any dependence of the dipole operator

μ_ε on the nuclear coordinates (indeed we seldom have the necessary information). The cross sections can then be written as

$$\sigma_{ni} = \frac{8\pi^3\omega}{c}|f_{ni}|^2 = \frac{8\pi^3\omega}{c}|\mu_\varepsilon|^2|\langle\chi^-(E,\mathbf{n},Q_{ex})|\chi(E_i,Q_{gr})\rangle|^2. \qquad 37.$$

Thus, the internal state distribution of the fragments is generated by the projection of the initial internal state onto the final internal states.

The calculation of the Franck-Condon factors of Eq. 37 requires the knowledge of the wave function $\chi_k(R)$, which in turn depends on the potential $V_1(R)$. The form of this potential is seldom known and must be guessed. A guess leads to a certain final state distribution which in principle can be compared with experiment. The guess can then be refined until agreement is obtained between theory and experiment. Thus if the theoretical approximations are valid, the experiment can be thought of as determining $V_1(R)$. In practice $V_1(R)$ is often represented by a straight line with a certain slope passing through a point on the upper surface obtained by vertical excitation by an amount $\hbar\omega$ from the ground state. The only parameter then is the slope of the line.

Few repulsive potentials yield an analytic solution to the Schrödinger equation. The linear repulsion yields Airy functions whose overlap integral with harmonic eigenfunctions are known analytically (21). Band & Freed have calculated the overlap integral of Eq. 37 for the special case of a linear triatomic dissociation in which rotation and the degenerate bending vibration were neglected and a linear repulsion assumed. The ground state and excited state wavefunctions are, respectively, $\phi_{n_1}(Q_1)\phi_{n_2}(Q_2)$ and $\chi_{E_k}(R)\phi_n(r)$ where Q_1 and Q_2 are the normal mode coordinates in the ground state, n_1, n_2 and n are vibrational quantum numbers, and E_k is the relative kinetic energy obtained from the conservation of energy condition, Eq. 2.

$$E_k = \hbar\omega - \varepsilon(n) - D_0 + \varepsilon(n_1) + \varepsilon(n_2) = \frac{\hbar^2 k^2}{2\mu}. \qquad 38.$$

In Eq. 38 the εs are vibrational energies, D_0 is the minimum energy required to dissociate the molecule from its lowest quantum state, and μ is the reduced mass. The flavor of the results is given by the probability of making a transition from the ground state ($n_1 = n_2 = 0$) of the parent triatomic to the state $\chi_{E_k}(R)\phi_n(r)$,

$$P(n) = \frac{N}{2^n n!} \sum_{r=0}^{n} \binom{n}{r} C_r Ai^{(n-r)}(f). \qquad 39.$$

In this expression N is a normalization constant; the coefficients C_r and the argument f of the Airy function derivatives, $Ai^{(n-r)}$ are functions of the force constants, the atomic masses, the equilibrium bond distances, and the photon energy.

The Franck-Condon model of Freed & Band in its simplest form neglects all final state interactions. However, there are other types of Franck-Condon models. In an early, perceptive paper (19) Mitchell & Simons argued that product vibrational excitation depends on the difference between the free BC oscillator and the BC oscillator imbedded in the excited ABC molecule located vertically above the ground state (rather than in the ground state of ABC). Figure 1 gives examples of schematic excited potential surfaces of NOX and XCN (X is a halogen) which do and do not give rise to vibrational excitation. OCS in its excited $^1\Sigma^+$ state (to be discussed in the section below, Comparison with Experiment) is similar to NOX. A superior Franck-Condon method would presumably involve the instantaneous "equilibrium" BC distance on the upper state surface vertically above the ground state, but this is usually unknown.

The forced oscillator model (22–26) assumes that all of the fragment vibrational excitation arises from the final state interactions, i.e. that the repulsive force that separates the fragments also does work on their internal coordinates. For example, in a molecule like ICN the CN group is assumed to have the same bond distance and force constant as the free

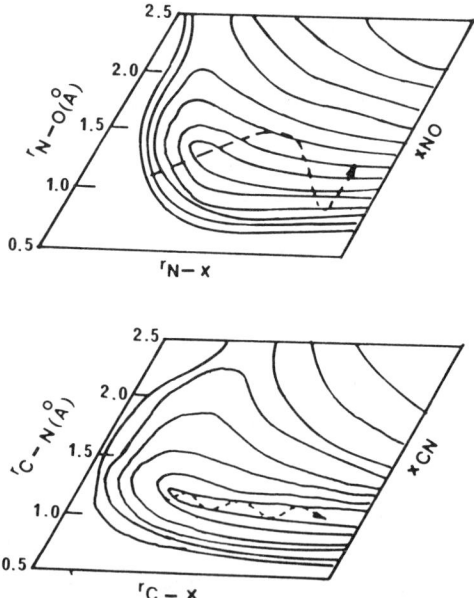

Figure 1 Excited potential energy surfaces of XNO and XCN [from Ref. (19)].

CN radical. Product vibrational excitation, if any will, arise from the external force of the departing I atom.

At this point when discussing the photodissociation of a linear triatomic, ABC contact can be made with the great body of calculations (15, 27–29) on the linear collisions of an atom A with the B end of the diatomic molecule BC. An exponential repulsion between the A and B atoms is invariably assumed, i.e. $V = V_o e^{-\alpha R_{AB}}$. The transformation

$$R_{AB} = R - \frac{M_C R_{BC}^o}{M_{BC}} - \frac{M_C}{M_{BC}}(R_{BC} - R_{BC}^o) \qquad 40.$$

where R is the distance between the A atom and the center of mass of the BC molecule and R_{BC}^o is the equilibrium BC bond distance suggests an expansion in powers of the displacement $R_{BC} - R_{BC}^o$:

$$V = V_o e^{\alpha M_C R_{BC}^o / M_{BC}} e^{-\alpha R} \left(1 + \alpha \frac{M_C}{M_{ABC}}(R_{BC} - R_{BC}^o) + \cdots \right)$$

$$= V_{\text{elastic}} + V_{\text{inelastic}}. \qquad 41.$$

The first term can only lead to elastic scattering; the second and higher terms can produce either elastic or inelastic scattering.

For small vibrational amplitudes we can assume that the second term is much smaller than the first. The classical equation of motion for a particle moving in an exponentially repulsive potential can then be solved analytically, i.e. an analytic function $R(t)$ can be found. The small perturbing potential $V(R(t), R_{BC} - R_{BC}^o)$ can be thought of as an external potential driving the oscillator. Suppose an oscillator with amplitude $R_{BC} - R_{BC}^o$ is subjected to its restoring force $-k(R_{BC} - R_{BC}^o)$ and an additional force

$$F(t) = -\left[\frac{dV(t)}{dR_{BC}}\right]_{R_{BC} = R_{BC}^o} = \alpha V_o e^{-\alpha R(t)} \qquad 42.$$

which lasts for a time T. The average gain in energy of the oscillator is

$$\frac{1}{2\mu}\left| \int F(t)e^{i\omega_o t} dt \right|^2 = \frac{|f(\omega_o)|^2}{2\mu} = \langle \Delta E \rangle \qquad 43.$$

where $f(\omega_o)$ is the Fourier transform of the force evaluated at the vibrational frequency $\omega_o = (k/\mu)^{1/2}$ and $\mu = M_B M_C / M_{BC}$.

Exactly the same expression as Eq. 43 can be derived from quantum mechanical perturbation theory (22). The distribution over quantum states v has been obtained by solving exactly the time dependent forced harmonic oscillator (23). Specifically, if the oscillator was initially in its ground state, the probability that it will be found in the state v after the external force has ceased to act is given by the Poisson distribution

$$P(v) = \frac{1}{v!}\left(\frac{\Delta E}{\hbar\omega_o}\right)^v \exp\left(-\frac{\langle\Delta E\rangle}{\hbar\omega_o}\right). \qquad 44.$$

Extension of this calculation to a polyatomic molecule with more than three atoms is immediate. The repulsive force can be written

$$F(t) = \sum_{i=1}^{N} \left(\frac{\partial V}{\partial Q_i}(R(t), Q_2, \ldots, Q_N)\right)_o \qquad 45.$$

because, just as with the linear triatomic, the repulsive force depends on interatomic distances which can in turn be expressed in terms of the distance R between the centers of masses of the fragments and the normal mode coordinates.

In first order, certain modes Q_i would not contribute to Eq. 45. For example, in the dissociation of SO_3 the departing O atom would be expected to remain more or less in the plane bisecting the OSO angle; if so, the symmetric stretch and the bending mode might be excited but not the asymmetric stretch. Similarly in the dissociation of CH_3X, if the X atom departs along the three-fold axis, in first order a_1 but not e type vibrations could be excited.

The generalized force $F(t)$ of Eq. 45 in a polyatomic molecule generates a Poisson distribution in each vibrational mode for whose coordinate there is a nonzero term in Eq. 45. The joint probability $P_{n_1,\ldots,n_N}$ for having varying numbers n_i of quanta in each of the N modes can be written

$$P_{n_1,\ldots,n_N} = \prod_{i=1}^{N} \frac{\left(\frac{\Delta E_i}{\hbar\omega_i}\right)^{n_i}}{n_i!} \exp - \sum_{j=1}^{N}\left(\frac{\Delta E_j}{\hbar\omega_j}\right) \qquad 46.$$

and using the multinomial theorem we have

$$P_{n_1,\ldots,n_N} = \frac{\left(\sum_i \frac{\Delta E_i}{\hbar\omega_i}\right)^{\sum n_k}}{\left(\sum_i n_i\right)!} \exp\left(-\sum_j \Delta E_j/\hbar\omega_j\right) \qquad 47.$$

$$= \frac{1}{v!}\left(\frac{\Delta E}{\hbar\omega}\right)^v \exp(-\Delta E/\hbar\omega). \qquad 48.$$

Thus, a collection of oscillators, each of which has a Poisson distribution in its quantum states, has an overall distribution of vibrational energy which is also given by a Poisson distribution with a parameter $\Delta E/\hbar\omega$. If all the frequencies ω_i were the same, then ΔE would be the average energy transferred to the vibrating molecule.

Unfortunately the verification that a fragment distribution is close to Poissonian often proves little. In the first place the Poisson distribution is very elastic. By a suitable choice of the parameter ΔE or $\Delta E/\hbar\omega$ in Eqs. 44 and 48 one can fit a distribution which peaks at $v=0$ as well as a strongly inverted distribution. Second, one can show (30, 31) that a linear forced harmonic oscillator is exactly equivalent to a displaced harmonic oscillator of the same frequency. Thus, in simple cases both the Franck-Condon and the forced harmonic oscillator models can lead to a Poisson distribution.

Wilson & Levine (32) introduced a compact impulsive collision model. For an ABC molecule dissociating into A+BC, the atom C was assumed to have a momentum q times the momentum of the B atom just after dissociation. Setting $q=0$ yields the spectator model and for $q=1$ the BC molecule is rigid. The parameter q gives the net effect of the sudden approximation when the B-C relative motion is slow compared to the A-B motion.

The Franck-Condon and Final State Interaction models can be unified by the use of Eq. 32 (11). This approach has been developed by Atabek, Beswick, Lefebvre, Mukamel & Jortner [12, 33-35], and by Halavee & Shapiro (36). These methods vary mainly in the way the half collision S^H matrix is evaluated. In principle it is possible to evaluate it exactly (15); however, usually the final state interaction is small enough to allow the use of a unitarized diabatic distorted-waves approximation (12, 33-35) or a first order adiabatic distorted-waves approximation (36). These approaches were mainly applied to model collinear photodissociation of XCN (X = H, Cl, Br, I) (33-36).

GENERAL TREATMENT OF PHOTODISSOCIATION DYNAMICS

In this section we review some of the general methods now available for calculating the photodissociation *amplitude*

$$f_{\mathbf{m},i} = \langle E, \mathbf{m}^- | \mu_\varepsilon | E_i \rangle \qquad 49.$$

from which, using Eq. 30, one can obtain the photodissociation cross sections. Using different functional forms for the operator in this equation the same methods are applicable to all kinds of dissociation processes including predissociation and unimolecular decay. The photodissociation amplitude is different in structure from the scattering amplitude,

$$t_{m,n} = \langle E, \mathbf{m}^- | V | E, \mathbf{n}^\circ \rangle \qquad 50.$$

because the initial state $|E_i\rangle$ of Eq. 49 is an interacting bound state whereas $|E, \mathbf{n}^\circ\rangle$ of Eq. 50 is a non-interacting (free) continuum state.

The need to evaluate the scattering component $|E, \mathbf{m}^-\rangle$ makes the "brute force" evaluation of $f_{m,i}$ very difficult because it is nearly impossible to generate $\langle \mathbf{R}, \mathbf{r} | E, \mathbf{m}^- \rangle$, which is a highly oscillatory function of $\mathbf{R}$ over a sufficient number of points necessary to calculate the $f_{m,i}$ integral, except possibly for a diatomic case. For this reason we would like to make the connection with time-independent scattering theory where efficient techniques exist to generate directly the $t_{m,n}$ integrals (27–29).

The key is to relate the dissociation problem to a scattering problem with a source term. This means that instead of solving a set of *homogeneous* differential equations, known in scattering theory as "coupled channels" or "close coupling" equations (6, 7, 29), one solves a set of *inhomogeneous* differential equations with $|E_i\rangle$ acting as the source. The method of solution can be explicit, i.e. $|E_i\rangle$ is generated separately and then used as a source term, or implicit. In the implicit method both bound and continuum subspaces are included in a unified set of equations. The implicit method has the advantage that one can easily handle a general ground state potential or high initial vibration of the triatomic molecule. It has a slight disadvantage in that the resulting channel potential matrix is nonsymmetric, which reduces somewhat the efficiency of the numerical calculations. The first method developed, called the "artificial channel method" (11, 37–39), utilized the *implicit* approach. Later a number of time independent *explicit* (40, 41) and implicit (42–46) methods were introduced. In addition the problem has been treated by time-dependent propagation schemes (8–10).

In the artificial channel method one reexpresses the inhomogeneous set of differential equations as a set of homogeneous differential equations. This is done by incorporating together the differential equations for the source and the scattering process. In doing so one encounters a very serious problem: Because the source is a bound wavefunction its direct integration from $R = 0$ to some asymptotic value of R is difficult, except at energies exactly equal to one of the eigenenergies. This is so, because at any other energy, the solutions of the bound manifold diverge asymptotically. The problem is solved by adding one extra "artificial" continuum channel, which serves as a source for the bound manifold and as shown below, stabilizes the solutions.

In order to demonstrate the procedure we consider, for simplicity, a three-channel problem one representing the continuum (excited) state

$|E, \mathbf{m}^-\rangle$, one the bound (ground) state $|E_i\rangle$, and the third—the so-called artificial channel. The differential equations to be solved are,

$$\left[E - \varepsilon_1 + \frac{\hbar^2}{2\mu} \frac{d^2}{dR^2} - V_{ex}(R) \right] \psi_1(R) = \mu(R)\psi_2(R) \qquad 51a.$$

$$\left[E - \varepsilon_2 + \frac{\hbar^2}{2\mu} \frac{d^2}{dR^2} - V_{gr}(R) \right] \psi_2(R) = W(R)\psi_3(R) \qquad 51b.$$

$$\left[E - \varepsilon_3 + \frac{\hbar^2}{2\mu} \frac{d^2}{dR^2} - V_a(R) \right] \psi_3(R) = 0. \qquad 51c.$$

where $\varepsilon_3, W(R)$ and $V_a(R)$ are chosen arbitrarily. Because ψ_2 corresponds to a bound manifold and ψ_1 and ψ_3 to continua, channels 1 and 3 are "open" and channel 2 is "closed." This means that

$$E - \varepsilon_1 > 0 \qquad 52a.$$
$$E - \varepsilon_2 < 0 \qquad 52b.$$
$$E - \varepsilon_3 > 0. \qquad 52c.$$

We now impose the usual scattering boundary conditions. Because there are three components, there are three independent ways in which this can be done. Choosing channel 3 as our incoming channel, we find

$$\psi_3^{+(3)}(R) \xrightarrow[R \to \infty]{} \{\sin k_3 R + t_{33} \exp(ik_3 R)\} \left(\frac{\mu}{\pi k_3} \right)^{1/2} \qquad 53a.$$

$$\psi_2^{+(3)}(R) \xrightarrow[R \to \infty]{} 0 \qquad 53b.$$

$$\psi_1^{+(3)}(R) \xrightarrow[R \to \infty]{} 0 + t_{13} \exp(ik_1 R) \left(\frac{\mu}{\pi k_1} \right)^{1/2} \qquad 53c.$$

where

$$k_1 = \{2\mu(E - \varepsilon_1)\}^{1/2}/\hbar$$
$$k_3 = \{2\mu(E - \varepsilon_3)\}^{1/2}/\hbar. \qquad 54.$$

The coefficients t_{13} and t_{33} are elements of the scattering transition-amplitudes as defined in Eq. 50. They are determined numerically (27) by propagating the solutions of Eqs. 51a–c from $R = 0$ where they must vanish to some sufficiently large value of R. It is much easier to determine the **t** matrix than the wavefunctions themselves because in doing so it is not necessary to follow all the oscillations of the continuum wavefunctions. As pointed out above this is the main reason for making the connection with the inhomogeneous scattering problem.

Supposing that we have computed t_{13}, it is now possible to show (37) that it is given by

$$t_{13} = \langle E, 1^- | \mu | \psi_2^{+(3)} \rangle \qquad 55.$$

where $|E, 1^-\rangle$ is the "incoming" scattering state of Eq. 8b, for a single channel case. In the limit $\varepsilon \to 0 | E, 1^- \rangle$ is therefore a solution of the homogeneous part of Eq. 51a, i.e.

$$\left[E - \varepsilon_1 + \frac{\hbar^2}{2\mu} \frac{d^2}{dR^2} - V_{ex}(R) \right] \langle R | E, 1^- \rangle = 0 \qquad 56.$$

$|\psi_2^{+(3)}\rangle$ of Eq. 55 is a solution of the full inhomogeneous Eq. 51b, subject to the boundary conditions embodied in Eq. 53b. It follows from Eqs. 51b and 53b that it may be written as,

$$|\psi_2^{+(3)}\rangle = \sum_i (E - E_i)^{-1} | E_i \rangle \langle E_i | W | \psi_3^{+(3)} \rangle$$

$$+ \int_{\varepsilon_2}^{\infty} dE' (E - E')^{-1} | E' \rangle \langle E' | W | \psi_3^{+(3)} \rangle \qquad 57.$$

where E_i and $|E_i\rangle$ are the discrete eigenenergies and eigenstates of H_{gr} defined in Eq. 8a and $|E'\rangle$ are the continuum eigenstates of the same ground state Hamiltonian. In our simple three-channel example,

$$H_{gr} = \varepsilon_2 - \hbar^2 / 2\mu \, d^2/dR^2 + V_{gr}(R). \qquad 58.$$

Substitution of Eq. 57 in Eq. 55 yields for t_{13},

$$t_{13}(E) = \sum_i \langle E, 1^- | \mu | E_i \rangle \langle E_i | W | \psi_3^{+(3)} \rangle / (E - E_i) + A(E) \qquad 59.$$

where

$$A(E) = \int_{\varepsilon_2}^{\infty} \frac{dE' \langle E, 1^- | \mu | E' \rangle \langle E' | W | \psi_3^{+(3)} \rangle}{E - E'}. \qquad 60.$$

Because $E < \varepsilon_2$, Eq. 52b, and $E' > \varepsilon_2$, the integral defining $A(E)$, Eq. 60, is an analytic function of E. However, the first term in Eq. 59 has a simple pole whenever $E \to E_i$. The desired photodissociation amplitude for the three-channel problem, $f_{1,i}$, (defined in Eq. 49) is obtained from the residues of $t_{13}(E)$

$$f_{1i}(E_i) \equiv \langle E_i, 1^- | \mu | E_i \rangle = \lim_{E \to E_i} \left[\frac{t_{13}(E)(E - E_i)}{\langle E_i | W | \psi_3^{+(3)} \rangle} \right] \qquad 61.$$

at the pole positions, $E = E_i$.

The normalization integral $\langle E_i | W | \psi_3 \rangle$ is obtained by performing a small calculation in which we set $\mu = W$, $V_{gr} = V_{ex} = V_a$ and $\varepsilon_1 = \varepsilon_3$.

Under these conditions $|E,1^-\rangle = |\psi_3^{-(3)}\rangle$ and we have from Eq. 61 that

$$|\langle\psi_3^{+(3)}|W|E_i\rangle|^2 = \lim_{E \to E_i} |t_{13}(E)(E - E_i)|. \qquad 62.$$

Once E_i and $\langle E_i|W|\psi_3^{+(3)}\rangle$ have been determined, we can calculate $\langle E,1^-|\mu|E_i\rangle$ at energies $E \neq E_i$ by adding $(E - E_i)$ to ε_2 of Eq. 51b, calculating $t_{13}(E)$, and using Eq. 61. In particular, for photodissociation, we calculate $\langle E,1^-|\mu|E_i\rangle$ at different photon energies by adding $\hbar\omega_{E,i} = E - E_i$ to ε_2.

The multichannel analogue of the above method has been applied to both collinear [N_2O (47), HCN, DCN (48), and CH_3I (49)] and three-dimensional [XeD_2 (37), H_2O (50), ArN_2 (51)] photodissociation problems. In addition, it has been used as an accurate alternative to the large-basis variational methods for determining vib-rotational energy levels of triatomic molecules, e.g. H_2O (52). The method is based on finding the pole positions of the multichannel analogue of $t_{13}(E)$ (Eq. 59), which yield directly the desired eigenvalues.

As some comparisons with experimental results are reviewed in the section entitled Comparison with Experiment, we confine ourselves here to a brief discussion of the first realistic case to be studied by this method, namely the VUV dissociation of N_2O. We discuss this calculation because it helped clarify the respective roles of Franck-Condon vs final states interactions (see section above). Essentially, the photodissociation process was shown to be a combined result of both: While the overall features resemble expectations based on the Franck-Condon model, the detailed final state distribution deviates considerably from that model. This is so, even if the probabilities for energy transfer in a full collision on the excited surface are small.

In order to see why this is so, we note from Eq. 32 that even if $|S_{n\bar{n}}^H|^2 \approx 0.01$, $|S_{n\bar{n}}^H| \approx 0.10$ and $|f_{ni}|^2$ may be changed by roughly that fraction, if $S_{n\bar{n}}^H$ and $A_{\bar{n}i}$ interfere constructively.

Explicit time independent methods have been late in development, mainly due to the numerical difficulties associated with generating the continuum wavefunction $\langle\mathbf{R},\mathbf{r}|E,\mathbf{n}^-\rangle$. These difficulties were recently overcome by Kulander & Light (40) by using an R-matrix propagation technique (53). In their method, which has so far been applied to collinear photodissociation only, one divides the $f_{n,i}$ integral into sectors, such that $V_{ex}(R)$ (Eq. 51a) can be represented as a piecewise collection of flat potential steps. It is then possible to express $\langle Rr|E,\mathbf{n}^-\rangle$ as a piecewise analytic solution. Thus in the ith sector, centered around $R = R_i$, one writes,

$$\langle R,r|E,\mathbf{n}^-\rangle = \sum_m \phi_m(r|R_i)(a_m^n(R_i)\sin k_m R + b_m^n(R_i)\cos k_m R) \qquad 63.$$

where a_m^n, b_m^n are, to be determined, coefficients and $\phi_m(r|R_i)$ are sector dependent harmonic oscillator eigenfunctions. In the case of a two fragment channel problem (see Eq. 2) it is more convenient to use other coordinates in the interaction region. These are the reaction coordinate u and a coordinate orthogonal to it, v. The expansion (Eq. 63) is then performed in these coordinates.

Assuming that $|E_i\rangle$, the initial bound state (Eq. 8a), is accurately represented, say as a solution of a harmonic potential, one can obtain, using Eq. 63, a piecewise analytic expression to the $f_{\mathbf{n},i}$ integral in each sector. The overall $f_{\mathbf{n},i}$ integral is then obtained by propagating $|E,\mathbf{n}^-\rangle$ from one sector to another using the R-matrix propagation technique.

Kulander & Light have applied their method to the photodissociation of a collinear CO_2 molecule. In this case either one of the oxygen atoms can be removed. As a result one needs to solve a reactive problem on the excited state surface. In principle one could use the artificial channel idea for such processes as well, but this, so far, has not been done. One drawback of the Kulander & Light method is the need to know $|E_i\rangle$ prior to the R matrix propagation step. For this reason, the method has so far been applied to the case where V_{gr} could be assumed harmonic. In principle, one could, however, extend the method to treat anharmonic ground state potentials. For the one-fragment-channel dissociation this method can, in principle, be extended to three dimensions. For two or three fragment channels this is much more difficult because the techniques for performing accurate three-dimensional quantum mechanical calculations for reactive scattering, for systems other than H_3, are still in the development stage.

Other promising techniques for solving the triatomic photodissociation problem are time dependent propagation schemes developed mainly by Kulander & Bottcher (8) and Heller (9, 10) the R-matrix method of Numrich & Kay (42), and the multichannel complex-scaling methods as developed by Atabek & Lefebvre (43a,b) and Chu (44).

The time dependent technique is based on considering an ultrashort coherent broad band excitation process in which one can approximate $\tilde{\varepsilon}(\omega)$, as defined by Eqs. 3 and 15 as a constant,

$$\tilde{\varepsilon}(\omega) \cong \bar{\varepsilon} \qquad 64.$$

Under this condition Eq. 25 can be rewritten as

$$\langle E',\mathbf{m}^o|\psi(t)\rangle = \frac{i\pi}{\hbar}\bar{\varepsilon}\langle E',\mathbf{m}^o|\left\{\sum_\mathbf{n}\int dE|E,\mathbf{n}^-\rangle\langle E,\mathbf{n}^-|\exp(-iEt/\hbar)\right\}$$

$$\cdot \mu_\varepsilon|E_i\rangle. \qquad 65.$$

Recognizing that the term in curly brackets is simply $\exp(-iH_{ex}t/\hbar)$, one obtains that,

$$\langle E', \mathbf{m}^0 | \psi(t) \rangle = \frac{i\pi}{\hbar} \bar{\varepsilon} \langle E', \mathbf{m}^0 | \exp(-iH_{ex}t/\hbar) \mu_\varepsilon | E_i \rangle \qquad 66.$$

and the probability of observing state $|E', \mathbf{m}^\circ\rangle$ as $t \to \infty$ is simply given as

$$\lim_{t \to \infty} |b(E, \mathbf{m}|t)|^2 = \frac{\pi^2}{\hbar^2} \bar{\varepsilon}^2 |\langle E, \mathbf{m}^\circ | \lim_{t \to \infty} \exp(-iH_{ex}t/\hbar) | \phi_i \rangle|^2 \quad \text{where}$$
$$\qquad 67.$$

$$|\phi_i\rangle \equiv \mu_\varepsilon |E_i\rangle. \qquad 68.$$

The process is envisioned as follows. A wavepacket given by $|\phi_i\rangle$ is formed by the photon at $t=0$ in the excited state. For $t>0$ $|\phi_i\rangle$ propagates by the excited state evolution operator $\exp(-iH_{ex}t/\hbar)$, and our detection system probes this wavepacket at $t = \infty$.

As pointed out above, this approach represents a physically realizable situation provided that we have a completely coherently ultrashort white pulse. For direct photodissociation, e.g. CH_3I, where the absorption lineshape is typically ~ 4000 cm^{-1}, Eq. 64 implies that ε must be constant over this range. If we impose condition 64 over a frequency range of ~ 4000 cm^{-1} we get typically a pulse whose duration is $10^{-14} - 10^{-15}$ sec, which is very short indeed. However, as a computational technique this method can provide the exact photodissociation amplitudes for all types of excitations because, $\exp(iEt/\hbar)b(E, \mathbf{m}|t)/\bar{\varepsilon}$ as obtained from Eq. 66 is formally identical to the photodissociation amplitude (see Eqs. 49 and 27). Heller has developed both a semiclassical and an exact numerical procedures of performing the temporal propagation (9, 10).

The complex rotation method [43a, b, 44] is not discussed here because it is reviewed in this volume by W. P. Reinhardt.

NONLINEAR TRIATOMICS

Morse, Freed & Band (54–56) as well as Beswick & Gelbart (57) have made Franck-Condon calculations of the vibrational and rotational distributions resulting from the photodissociation of a nonlinear triatomic. Balint-Kurti & Shapiro (58) have given a formal exact scattering theory solution to the problem. It is instructive to compare these treatments.

First we consider the scattering theory solution. The bound state wave function of the molecule in the state $J_i M_i$ is expanded in a set of basis functions whose coordinates are appropriate to the fragments:

$$\Psi^{J_iM_i}(\mathbf{R},\mathbf{r},E_i) = \left(\frac{2J_i+1}{4\pi}\right)^{1/2} \times \sum_{vJ\lambda} t_\lambda \chi_{vj}(r) \frac{\Phi_{vj\lambda}^{J_ip_i}(R)}{R}$$

$$\{D^{J_i}_{\lambda M_i}(\phi_R,\theta_R,0)Y_{j\lambda}(\gamma,\psi) + p_i D^{J_i}_{-\lambda M_i}(\phi_R,\theta_R,0)Y_{j-\lambda}(\gamma,\psi)\}, p_i = \pm 1$$

69.

where $t_\lambda = \frac{1}{2}$ for $\lambda = 0$, $\frac{1}{\sqrt{2}}$ for $\lambda > 0$

and p_i determines the parity of the state given as $p_i(-1)^{J_i}$. The angles θ_R, ϕ_R describe the orientation with respect to a fixed frame of the vector **R** from the center of mass of the diatomic to the leaving atom and γ, ψ are the polar coordinates of the diatomic internuclear vector in a molecule fixed system whose z axis is in the direction of **R**. The $\chi_{vj}(r)$ basis functions describe the vibrational motion of a diatomic fragment in the bound triatomic system. This fragment will eventually become a free fragment once the photodissociation is complete.

The excited state wave function in the space fixed system normalized on a wave number scale becomes asymptotically

$$\psi^{-(\hat{k}vjm_j)}(\mathbf{R},\mathbf{r},E) \underset{R\to\infty}{\sim} \left(\frac{\mu k_{vj}}{\hbar^2(2\pi)^3}\right)^{1/2}$$

$$\cdot \left\{ e^{i\mathbf{k}_{vj}\cdot\mathbf{R}} \chi_{vj}(r) Y_{jm_j}(\hat{r}) + \sum_{v',j',m_{j'}} f_{v'j'm_{j'},vjm_j}(\hat{k};\hat{R}) \right.$$

$$\left. \cdot \frac{e^{-ik_{v'j'}R}}{R} \chi_{v'j'}(r) Y_{j'm_{j'}}(\hat{r}) \right\}. \qquad 70.$$

When these wave functions are substituted into the Schrödinger equation for the ground and the excited state, respectively, with appropriate coupling terms representing the radiation matter interaction, we find for the photofragmentation amplitude

$$f(\hat{k}Evjm_j|E_iJ_iM_ip_i) = \left(\frac{\mu k_{vj}}{2\pi^2\hbar^2}\right)^{1/2} (-1)^{M_i+j-m_j}$$

$$\times \sum_{J,\lambda} \begin{pmatrix} J & 1 & J_i \\ -M_i & 0 & M_i \end{pmatrix} (2J+1)^{1/2}$$

$$\cdot D^J_{\lambda M_i}(\phi_k,\theta_k,0) D^j_{-\lambda-m_j}(\phi_k,\theta_k,0)$$

$$\times t(EJvj\lambda p|E_iJ_ip_i). \qquad 71.$$

The angles θ_k, ϕ_k describe the orientations of the observation direction $\hat{k}$. The differential photofragmentation cross section is given by

$$\sigma(\hat{k}Evjm_j|E_iJ_iM_ip_i) = \frac{8\pi^3\omega}{c}|f(\hat{k}Evjm_j|E_iJ_iM_ip_i)|^2. \qquad 72.$$

The double sum over J and λ in the scattering amplitude is limited to $J = J_i, J_i \pm 1$ and $0 < \lambda < \min(J, j)$. It leads to a quadruple sum over J, J', λ, and λ' in the cross section. However, most of the complexity of the problem has arisen because we are photodissociating a rotating molecule with $J_i > 0$. The use of such states in an experiment introduces difficulties that are mainly kinematic. In short, little physics is lost by dissociating a supersonically cooled beam of largely nonrotating molecules with $J_i = 0$. Moreover, it follows from Eq. 71 that the angular distribution of the fragment is far more intricate than the usual expression $(1 + \beta P_2(\cos\theta_k))$. However, if we do not resolve the individual m_j states and therefore average over them and let $J_i = 0$, the measured angular distribution for a given final vibrational state is

$$\sigma(\theta_k vE) \equiv \sum_j \sigma(\theta_k Evj|E_iJ_i = 0, p_i = 1) = \frac{2\omega\mu}{3\hbar^2c}$$

$$\cdot \left[\left\{ \sum_j k_{vj} |t(EJ = 1, vj\lambda = 0, p = 1|E_iJ_i = 0, p_i = 1)|^2 \right\} \right.$$

$$\cdot (1 + 2P_2(\cos\theta_k)) \qquad 73.$$

$$+ \left\{ \sum_j k_{vj} |t(EJ = 1, vj\lambda = 1, p = 1|E_iJ_i = 0, p_i = 1)|^2 \right\}$$

$$\left. \cdot (1 - P_2(\cos\theta_k)) \right].$$

For any given final state v, j of the diatomic there are two t-matrix amplitudes to be calculated ($\lambda = 0, 1$). Depending on their relative magnitudes, the angular distribution of the fragments varies continuously from a purely parallel distribution ($\beta = 2$) to a purely perpendicular distribution ($\beta = -1$). The $\lambda = 0$ and 1 components correspond respectively to the transition dipole moment being along and perpendicular to the fragment separation vector **R**. Equation 73 shows that for $J_i = 0$ the fragment angular distribution has the classical form $(1 + \beta P_2(\cos\theta_k))$. The same is true for any J_i (58). We see from Eq. 73 that if we sum over all final m_j states for a given value of λ, all the partial cross sections have the same dependence on θ_k. An m_j-averaged angular distribution cannot be decomposed to its various j dependent partial cross sections $|t(Evj J \lambda p)|^2$. However, if we perform an m_j selection, by some polarization experiment, we can in principle extract the j-partial cross sections. In order to see this we take the simple case of $J_i = 0$. For a parallel transition $\lambda = 0$ and we have from Eq. 71 that (58),

$$\sigma(\theta_k Evm_j) \equiv \sum_j \sigma(\theta_k Evjm_j) = g(\theta_k) \sum_j \sigma(Evj)|Y_{jm_j}(\theta_k, 0)|^2 \qquad 74.$$

where the partial photodissociation cross sections are

$$\sigma(Evj) = |t(EJ = 1, vj\lambda = 0, p)|^2 \quad \text{and} \qquad 75.$$

$$g(\theta_k) = C_0 + C_2 P_2(\cos \theta_k) \qquad 76.$$

with C_0 and C_2 given by simple 3-j and 6-j coefficients. Because each j-dependent partial cross section has a different angular dependence, by fitting the

$$\sigma(\theta_k Evm_j)/g(\theta_k) \quad \text{to a sum of} \quad |Y_{jm_j}(\theta_k, 0)|^2$$

we can obtain the partial photodissociation cross sections $\sigma(Evj)$. The extraction of the partial photodissociation cross sections is possible because contrary to crossed molecular beams experiments we add probabilities and not amplitudes.

Equation 71 is an exact but formal solution to the problem of the nonlinear triatomic. The numerical magnitudes of the t-matrix elements,

$$t(EJv j \lambda p | E_i J_i p_i)$$

constitute the real solution to any specific problem. Nevertheless without any numerical calculations it is possible to derive the form of the angular distribution for both oriented and unoriented fragments and to see why it is so desirable to do experiments on nonrotating molecules.

Beswick & Gelbart (57) developed a Franck-Condon approach to the rotational distributions equivalent in spirit to that of Morse, Freed & Band (54–56). Their approach is simpler to describe. In this model the bending vibration wave-function of the electronic ground state of a molecule with total angular momentum J is expanded in terms of rotational wave functions. The coefficients in this expansion yield the desired probabilities.

The Hamiltonian is expressed in terms of dissociation coordinates, R, r and the angle γ between the vectors $\mathbf{R}$ and $\mathbf{r}$. The bending vibration is assumed to be a pure bend depending only on γ and not on the bond distances. R and r are therefore replaced by average values $\bar{R}, \bar{r}$ and the

Hamiltonian for rotation and bending becomes just

$$H = \frac{\hbar^2}{2\mu_{A,BC} \bar{R}^2} l^2 + \frac{\hbar^2}{2\mu_{B,C} \bar{r}^2} \mathbf{j}^2 + V(\gamma) \qquad 77.$$

where $\mu_{i,j} = m_i m_j/(m_i + m_j)$, l is the orbital angular momentum of the atom A around the diatomic BC and **j** is the rotational angular momentum of the diatomic. The symmetric top wave functions are

$$\langle \gamma, \psi, \theta_R, \phi_R | jJM\lambda \rangle = \left(\frac{2J+1}{4\pi}\right)^{1/2} D^J_{\lambda M}(\phi_R, \theta_R, 0) Y_{j\lambda}(\gamma, \psi) \qquad 78.$$

where the quantum numbers are identical to those defined in Eq. 69. If the bending potential $V(\theta)$ is expanded in terms of Legendre polynomials, then the wave function for bending and total rotation can be readily expanded

$$|\chi^{\text{bound}}_{nJM}\rangle \equiv \sum_{j,\lambda} C^{(nJM)}_{j\lambda} | jJM\lambda \rangle \qquad 79.$$

where n is the quantum number of the bending vibration. The squares of the coefficients summed over all helicities, λ, are the rotational populations:

$$P^J_j = \sum_\lambda |\langle \chi^{\text{bound}}_{nJM} | jJM\lambda \rangle|^2 = \sum_{\lambda = -\min(j,J)}^{+\min(j,J)} |C^{nJM}_{j\lambda}|^2. \qquad 80.$$

The essential physical feature of this model is that it is independent of the upper potential surface; all final state interactions are neglected. It is a sudden approximation in which the excited states are those of a free rotor. The calculations of Eq. 79 are greatly simplified when $J = 0$.

A sample result is given in Figure 2. For ICN the rotational distribution over j of the CN radical is seen to be virtually independent of J. That is, it arises almost entirely from the bending vibration. The reason is that the light CN rotates around the very heavy I atom; the greater the initial J, the greater will be the orbital angular momentum. In this Franck-Condon treatment, the stiffer the triatomic the more energy is put into rotational motion. In classical terms this is paradoxical but in quantum mechanics it follows that the stiffer the molecule the more energy will be in the zero-point vibration.

Useful physical insights have already emerged from the two theoretical approaches. Perhaps the most important is that the j distribution is sensitive to the initial J_i state and that comparison of theory with experiment is far easier when $J_i = 0$. The classical angular distribution $(1 + \beta P_2(\cos \theta))$ is confirmed by a quantum mechanical calculation (58).

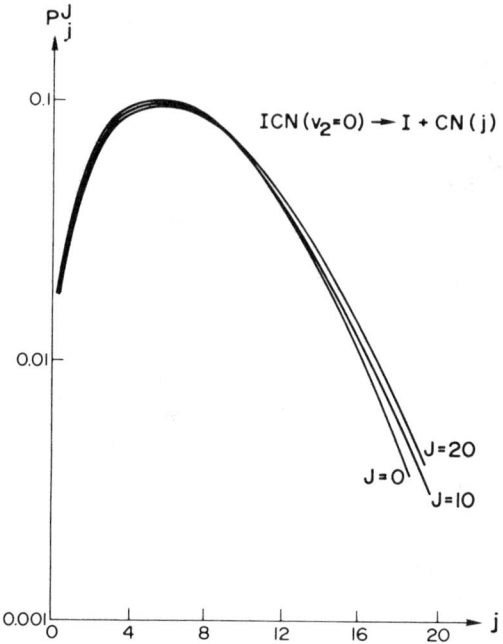

Figure 2 Distribution of CN rotational states, j, for a given initial state, J, of ICN [from Ref. (57)].

COMPARISON WITH EXPERIMENT

ICN

ICN is one of the few stable linear triatomics which near room temperature can be dissociated with UV (as contrasted to VUV) light. For this reason its photodissociation has been extensively studied experimentally and therefore theoretically. This is particularly true of its first (A) absorption band which peaks at 250 nm. This structureless absorption bond was originally believed to be a single transition and many theoretical studies were based on this assumption. The complexity of the real molecule is shown by Figure 3 (59), which proves that this simple looking band consists of (at least) three absorptions, two leading to I atoms in the ground $^2P_{3/2}$ state and one leading to I atoms in the excited $^2P_{1/2}$ state. At 266 nm virtually all of the CN radicals are produced in the $v=0$ level of the ground $X^2\Sigma$ electronic state. The rotational distribution at 266 nm has been interpreted as a sum of three Boltzmann distributions with "temperatures" of 37 K, 489 K, and 6134 K.

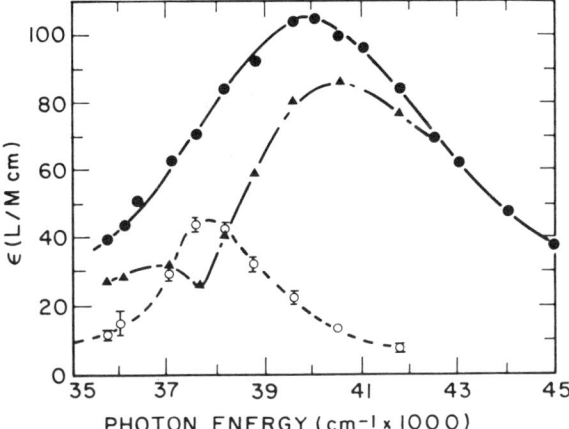

Figure 3 Experimental results for the production of $I(^2P_{1/2})$ from ICN [from Ref. (59)]. Solid dots, absorption spectrum of ICN $-\varepsilon[ICN]$; open circles, partial absorption due to production of $I(^2P_{1/2}) - \phi[I(^2P_{1/2})] \times \varepsilon[ICN]$; triangles, $\varepsilon[ICN] \times \{1 - \Phi[I(^2P_{1/2})]\}$.

The theoretical paper which has so far come closest to experimental reality is that by Morse, Freed & Band (55), who do assume two excited state potential surfaces. They show that some rotational excitation of the CN fragment is excited just because of the bending mode zero point motion. However, a rotational temperature of 6134 K goes far beyond their model. A powerful bending force is acting on one of the three excited state potential surfaces, in other words the final state interactions are dominant. ICN, BrCN, and HCN have been studied extensively in the vacuum ultraviolet (60–63), because excited electronic states of CN are thereby generated.

CO_2

Walsh's rules state that a triatomic molecule with 16 valence electrons (e.g. ICN, CO_2, N_2O) will have a linear ground state, but if there are 17 (NO_2, CO_2^-) or 18 (SO_2, O_3) valence electrons, the ground state will be nonlinear. The basis of this prediction is that the ninth valence molecular orbital has a lower energy in a bent configuration. A corollary is that the lower excited states of the 16 valence electron molecules that correspond to promotions from the eighth (antibonding π) to the ninth (antibonding σ) molecular orbitals will be bent. Using spectroscopic terminology, a $^1\Pi \leftarrow {}^1\Sigma$ transition necessarily, by the Renner-Teller theorem, produces a bent upper state. Absorption of still higher energy photons brings about transitions to excited $^1\Sigma$ states. Some of these may also bend but they do not have to.

There are absorptions in CO_2, N_2O, and OCS that peak at 111.9, 129.1, and 155 nm, respectively, which are thought to involve a $^1\Sigma^+$ upper state, e.g.

$$N_2O(\chi^1\Sigma^+) + h\nu \to N_2O(^1\Sigma^+) \to N_2(^1\Sigma) + O(^1S).$$

At least the necessary conditions for a completely collinear dissociation are satisfied. In contrast, a dissociation

$$CO_2(^1\Sigma) + h\nu \to CO(a'^3\Sigma^+) + O(^3P)$$

might involve an excited Σ or a Π state.

The $\Sigma \leftarrow \Sigma$ dissociation of N_2O has been given an "exact" treatment by Shapiro (47), i.e. accurate cross sections were obtained for a repulsive potential derived from the absorption spectrum. As yet no data exist to test this calculation.

When photons in the 901–923 Å range are absorbed by CO_2, the $a'^3\Sigma^+ \to a^3\Pi$ fluorescence of CO is observed and presumably an $O(^3P)$ atom is formed simultaneously. The CO produced is observed (64) to have a monotonic decrease of vibrational population between $v = 5$ and 12; Freed & Band (14) were able to fit the data with the Franck-Condon model. On the experimental side, measurements of fluorescence polarization of the CO fragments (see the article by Greene & Zare in this volume) and their rotational distribution would be useful in testing collinearity. The observed diffuse band structure in the VUV spectrum of CO_2 has been given an elegant qualitative explanation by Pack (65). Kulander & Light (40) have tried to verify Pack's mechanism by an accurate calculation. The band structure obtained by Kulander & Light is, however, twice as large as that observed experimentally.

OCS

When OCS absorbs a 157 nm photon, it has unit quantum yield for producing sulfur atoms in the ^1S state. From the translational energy distribution measured at 157 nm, if one assumes negligible rotational energy one finds a strongly inverted vibrational population peaking at the maximum, $v = 7$ (66). The distribution can be fitted by the expression

$$P_v = P_v^o N' e^{-\lambda v} = N(E_{AVL} - \varepsilon(v))^{1/2} e^{-\lambda v}$$

where P_v^o is the prior distribution weighting equally all states with no rotational energy, N, N' are normalization constants, $\varepsilon(v)$ is the vibrational energy of CO with respect to the zero point energy, and $\lambda = -6.5$ is the surprisal parameter (67).

The CO distances in free CO and OCS are 1.128 and 1.160 Å, respectively. Using these data, the ground state force constants, and reasonable values for the repulsive force between the CO and S, a Franck-Condon calculation using the Freed & Band theory (66) yields a vibrational distribution peaking at $v=0$.

A reasonable explanation for the inverted distribution has already been given by Pack (65). The absorption band of OCS between 160 and 145 nm has considerable vibrational structure, which is ascribed to a symmetric type vibration in the excited state. In other words instead of executing an asymmetric vibration and falling apart at once, the molecule executes a symmetric type vibration with a lengthy trajectory on the upper surface before it dissociates, forming a "hot" diatomic. This is perhaps the simplest example of a vibrational predissociation. It is an extreme case in which final state interactions dominate.

CH_3I

CH_3I is an example of a quasilinear triatomic. If we consider only $^1A_1 \leftarrow {}^1A_1$ spectroscopic transitions, in first-order only the two a_1 vibrations of the CH_3 radical would be excited by the act of dissociation. One of these, the symmetric C–H stretch, is not excited because the C–H bond has the same length in the ground and in the first excited state of CH_3I. Thus, we are left with excitation of the symmetric bend and a Hamiltonian depending on only two coordinates that is isomorphic to the Hamiltonian of a linear X–C–I molecule. (Similarly, if C_{2v} symmetry is preserved during dissociation and C–H stretching modes are ignored, ethylene oxide is a quasilinear tetraatomic and iodobenzene is a quasilinear pentaatomic).

An essentially exact calculation has been carried out of the partial (to a specific vibrational state of the symmetry bending mode of CH_3) and the total absorption cross sections as a function of photon energy through the first absorption band (49, 68). An excited state potential

$$V_{ex}(R, r_{C-H_3}) = 9.618 \exp(-1.40\,R)$$
$$+ 2.604 \exp(-1.20\,R + 0.24\,r_{C-H_3}) + \tfrac{1}{2}(0.0362) r_{C-H_3}^2$$

was assumed where R is the distance between centers of masses of the fragments and r_{C-H_3} is the distance of the carbon nucleus from the plane of the three protons. Figures 4 and 5 are contour plots of the vibrational distribution $P(v, E)$ (which is interpolated between physical integral values of v) obtained by exciting the ground vibrational state and the state with one quantum in the C–I stretch. With this potential, the **S** matrix for the half collision has moderately large off-diagonal elements.

For a 266 nm photon, both the predicted and the experimental distribution (69) peak at $v = 2$, although the latter is narrower. A stiffer test of the theory will come from data obtained at a different wave length.

H_2O

The VUV photodissociation of H_2O at ~ 130 nm has been a source of fascination for dynamicists and spectroscopists for more than a decade (2). The reason for this is two-fold: First, the OH($A^2\Sigma$) produced, with a quantum yield at 8–10% (70), displays an "abnormal" (i.e. partly inverted) rotational state distribution (71). Second, in the absorption

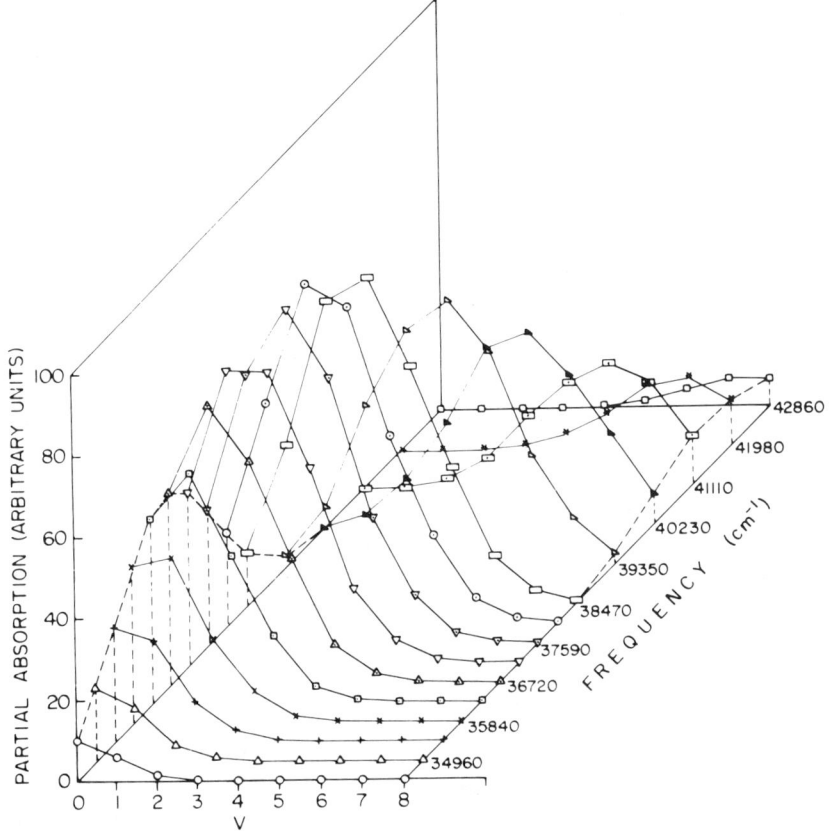

Figure 4 A three-dimensional plot of the detailed cross sections $\sigma_{v,0}(\hbar\omega_{E0})$ for the photodissociation of CH_3I in its ground (**0**) vibrational state as a function of photon frequency ($\hbar\omega_{E0}$) and final vibrational quantum number (v) of the CH_3 fragment [from Ref. (49)].

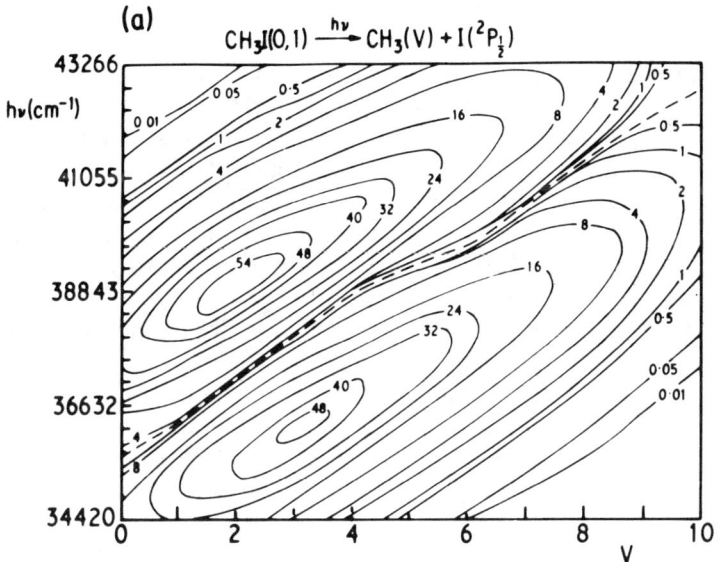

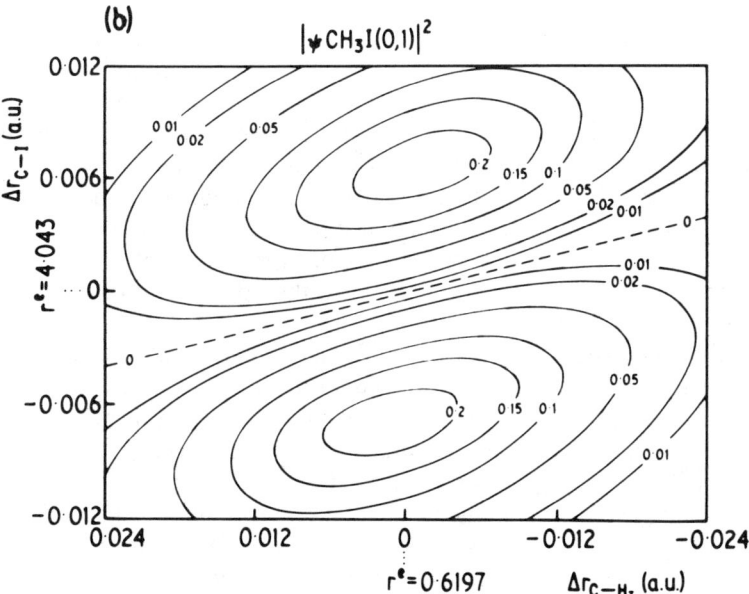

Figure 5 (a) Contour plot of the detailed cross sections $\sigma_{v,1}(\hbar\omega_{E1})$ for the photodissociation of CH_3I in its first excited (**1**) vibrational state as a function of $\hbar\omega_{E1}$ and v. (b) Contour plot of the probability-density $|\psi_1(r_{C-I}, r_{C-H_3})|^2$ of CH_3I in its first excited (**1**) vibrational state [from Ref. (68)].

spectrum (72) there appears to be a progression of diffuse bands, which has been difficult to assign (73). These two observations were correlated by the studies of Macpherson & Simons (74). They showed that the excitation spectrum for the emission of OH($^2\Sigma$) photodissociated from H$_2$O also has a band structure, very similar to the absorption spectrum. The appearance of the diffuse bands is therefore probably due to the $\tilde{B}^1A'$ ($^1A_1$ in C_{2v}) mixed Rydberg and valence state that correlates at bent configurations with H+OH($^2\Sigma$). The $\tilde{B}^1A'$ is, however, mainly repulsive (75), and it was not clear whether a rotational or vibrational predissociation on a single surface could account for the structure.

Segev & Shapiro (50, 76) performed a full three-dimensional calculation on the 130 nm photodissociation of water. In this calculation the ground state (X^1A') potential was that given by Sorbie & Murrell (77) and the $\tilde{B}^1A'$ excited state potential was that given by Flouquet & Horsley (75).

In Figure 6 (76) the calculated absorption spectrum and the OH($^2\Sigma$) rotational state distribution are shown at different wavelengths. The agreement with experiment is very good, especially regarding the positions of the calculated resonances ("diffuse bands"). The calculation showed clearly that most of the diffuse bands arise from a rotational predissociation on a single ($\tilde{B}^1A'$) potential surface, although inhomogeneous predissociation (due to other electronic states) cannot be ruled out. The molecule does not dissociate directly because of a small barrier for dissociation which was found by Flouquet & Horsley to exist only in the bent configuration. In order to dissociate, the molecule has to straighten somewhat; this induces a torque on the OH fragment and excites the bending motion at selected ("resonance") energies. At these energies, the molecule can exist for a number of vibrations, thus greatly enhancing the photoabsorption cross section; hence the appearance of the "diffuse bands".

The torque exerted on the molecule in the course of dissociation has long been recognized (71) as a possible source for the OH($^2\Sigma$) "abnormal" rotation. The calculation of Segev & Shapiro (50, 76), Figure 6 indeed shows that the OH($^2\Sigma$) recoils with a considerable rotational excitation, which, at certain photon energies, is even inverted. Figure 6 shows that at photon energies "on-resonance," the OH($^2\Sigma$) rotational "temperature" is much cooler than at off-resonance photon energies. We see here a fine interplay between a Franck-Condon type distribution at off-resonance, where the molecule is very short-lived and hence cannot redistribute its energy, and final state interactions that dominate the scene on-resonance.

The OH($^2\Sigma$) rotational distribution is in reasonable agreement with the experiment of Lee et al (78) at 133.6 nm. The vibrational distribution is

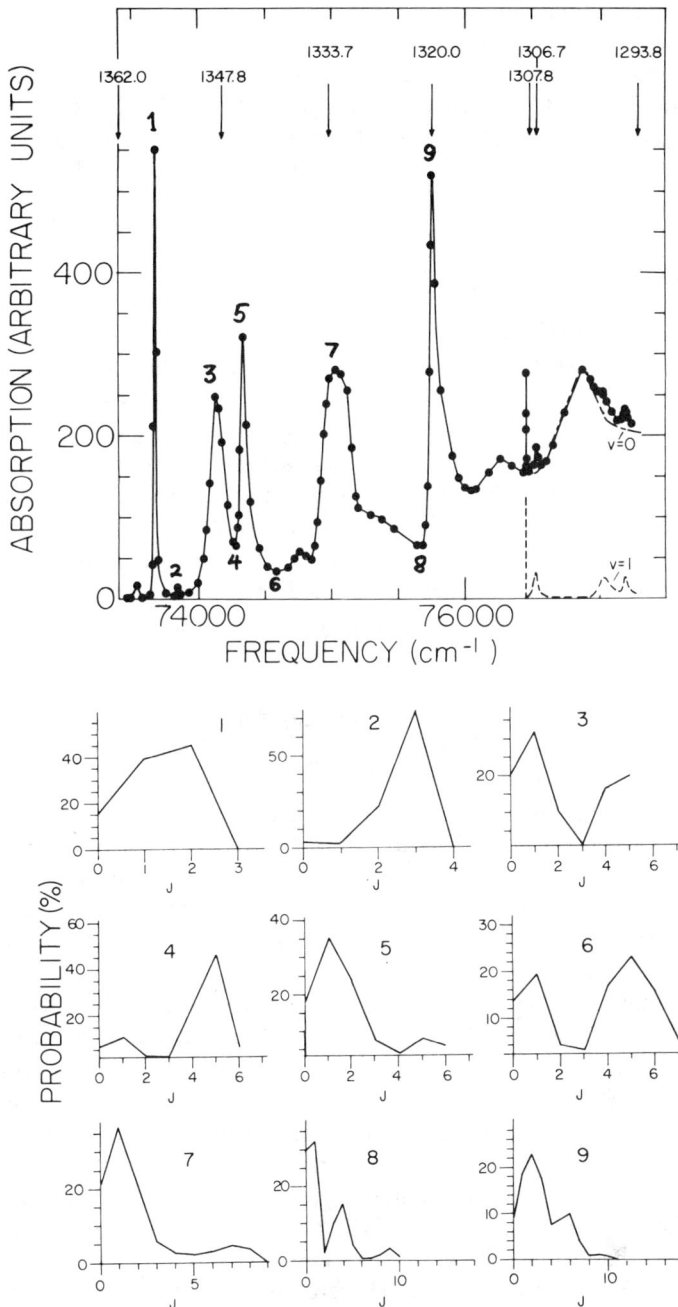

Figure 6 H$_2$O absorption spectrum and rotational distribution of the OH($^2\Sigma$) fragment at selected frequencies [from Ref. (76)]. The arrows indicate the positions of the observed diffused bands [from Ref. (78)].

in good agreement with experiment (71); the vibrational Franck-Condon approximation seems to apply well to this case (79), probably because the equilibrium OH distance remains roughly the same both in ground (X^1A') and excited ($\tilde{B}^1A'$)H_2O states, as well as in the OH($^2\Sigma$) and OH($^2\Pi$) fragment states.

SUMMARY

In this article, we have discussed two distinct approaches to the photodissociation problem: the Franck-Condon factor and the exact coupled channels solution. The Franck-Condon school, notably Mitchell & Simons, Freed, Band & Morse, compute $|\langle f(Q_{ex})|i(Q_{gr})\rangle|^2$, the overlap between the initial ground state and final fragment states where Q_{ex}, Q_{gr} are the natural coordinates for the excited and ground states, respectively. The exact solution school, notably Shapiro & co-workers, Atabek, Beswick & Lefebvre, and Kulander & Light, use the techniques of scattering theory to derive a set of coupled channels equations for the ground state and the excited state together. In these calculations the same set of coordinates Q_{ex} is used to describe both the excited and the ground state.

The Franck-Condon approach has the major advantage in that having omitted the final state interactions, it is much simpler and leads to analytical expressions with a ready physical interpretation. However, the model is incomplete and in at least one case, CH_3I, the effects of the half-collision were proven to be substantial.

The exact solution approach suffers from possibly requiring a large number of "channels," i.e. basis set wave functions. Therefore, solutions of the equations for systems more complicated than triatomics may be prohibitively difficult.

The dynamics of photodissociation is still a young field. Ideally what we would wish from an experiment on a triatomic molecule is an accurate final state distribution of the product diatomic as a function of photon energy, $P(v, j; E)$ when the triatomic is initially in the $J = 0$ rotational state and in a known (usually the ground) vibrational state. Data of such quality and completeness are available for very few systems, but the next decade should be witness to many more. While the Franck-Condon method is easier to apply, really good data deserve an exact calculation to test, among other things, how well we can rely on the simpler model. Indeed, if the Franck-Condon model were always correct, the only part of the upper potential surface that we could learn anything about would be the part directly above the limited region of vibration of the electronic ground state.

Literature Cited

1. Leone, S. R. 1981. *Adv. Chem. Phys.*
2. Okabe, H. 1978. *Photochemistry of Small Molecules*. New York: Wiley
3. Simons, J. P. 1977. In *Gas Kinetics and Energy Transfer*, ed. P. G. Ashmore, R. J. Donovan, 2:58. London: Burlington
4. Jortner, J., Levine, R. D., Rice, S. A., eds. 1981. *Adv. Chem. Phys.* 47
5. Gelbart, W. M. 1977. *Ann. Rev. Phys. Chem.* 28:323
6. Levine, R. D. 1969. *Quantum Mechanics of Molecular Rate Processes*. Oxford: Clarendon
7. Taylor, J. R. 1972. *Scattering Theory*. New York: Wiley
8. Kulander, K. C., Bottcher, C. 1978. *Chem. Phys.* 29:141
9. Heller, E. J. 1978. *J. Chem. Phys.* 68:2066
10. Kulander, K. C., Heller, E. J. 1978. *J. Chem. Phys.* 69:2439
11. Shapiro, M. 1973. *Israel J. Chem.* 11:691
12. Atabek, O., Beswick, J. A., Lefebvre, R., Mukamel, S., Jortner, J. 1976. *Mol. Phys.* 31:1
13. Band, Y. B., Freed, K. F. 1974. *Chem. Phys. Lett.* 28:328
14a. Freed, K. F., Band, Y. B. 1977. In *Excited States*, ed. E. C. Lim, 3:109 and references therein
14b. Caplan, C. E., Child, M. S. 1972. *Mol. Phys.* 23:249
15. Shapiro, M., Levine, R. D. 1970. *Chem. Phys. Lett.* 5:499
16a. Berry, M. J. 1974. *Chem. Phys. Lett.* 27:73
16b. Berry, M. J. 1974. *Chem. Phys. Lett.* 29:329
16c. Berry, M. J. 1974. *J. Chem. Phys.* 61:3114
17. West, G. A., Berry, M. J. 1974. *J. Chem. Phys.* 61:4700
18. Abgrall, M., Fiquet-Fayard, F. 1974. *J. Chem. Phys.* 60:4497
19. Mitchell, R. C., Simons, J. P. 1967. *Faraday Discuss. Chem. Soc.* 44:208
20a. Simons, J. P., Tasker, P. W. 1973. *Mol. Phys.* 26:1267
20b. Simons, J. P., Tasker, P. W. 1974. *Mol. Phys.* 27:1691
21. Stueckelberg, E. C. G. 1932. *Phys. Rev.* 42:518
22. Rapp, D., Kassal, T. 1969. *Chem. Rev.* 69:61
23. Pechukas, P., Light, J. C. 1966. *J. Chem. Phys.* 44:3897
24. Holdy, K. E., Klotz, L. C., Wilson, K. R. 1970. *J. Chem. Phys.* 52:4588
25. Heidrich, F. E., Wilson, K. R., Rapp, D. 1971. *J. Chem. Phys.* 54:3885
26. Busch, G. E., Wilson, K. R. 1972. *J. Chem. Phys.* 56:3626, 3638, 3655
27. Secrest, D., Johnson, B. R. 1966. *J. Chem. Phys.* 45:4556
28. Gordon, R. G. 1969. *J. Chem. Phys.* 51:14
29. Bernstein, R. B., ed. 1979. *Atom-Molecule Collision Theory*. New York: Plenum
30. Levine, R. D. 1971. *Chem. Phys. Lett.* 10:510
31. Gislason, E. A., Kleyn, A. W., Los, J. 1981. *Chem. Phys.* 59:91
32. Wilson, A. D., Levine, R. D. 1974. *Mol. Phys.* 27:1197
33. Atabek, O., Beswick, J. A., Lefebvre, R., Mukamel, S., Jortner, J. 1976. *J. Chem. Phys.* 65:4035
34a. Mukamel, S., Jortner, J. 1974. *J. Chem. Phys.* 60:4760
34b. Mukamel, S., Jortner, J. 1976. *J. Chem. Phys.* 65:3735
35. Beswick, J. A., Jortner, J. 1977. *Chem. Phys.* 24:1
36. Halavee, U., Shapiro, M. 1977. *Chem. Phys.* 21:105
37. Shapiro, M. 1972. *J. Chem. Phys.* 56:2582
38. Atabek, O., Lefebvre, R. 1977. *Chem. Phys.* 23:51
39. Atabek, O., Lefebvre, R., Jacon, M. 1978. *Chem. Phys. Lett.* 58:196
40. Kulander, K. C., Light, J. C. 1980. *J. Chem. Phys.* 73:4337
41. Band, Y. B., Freed, K. F., Kouri, D. J. 1981. *J. Chem. Phys.* 74:4380
42. Numrich, R. W., Kay, K. G. 1979. *J. Chem. Phys.* 70:4343
43a. Atabek, O., Lefebvre, R. 1981. *Chem. Phys.* 55:395
43b. Atabek, O., Lefebvre, R. 1981. *Chem. Phys.* 56:195
44. Chu, S. 1980. *J. Chem. Phys.* 72:4772
45. Grabensteller, J. E., LeRoy, R. J. 1979. *Chem. Phys.* 42:41
46. Kodama, K., Bandrauk, A. D. 1981. *Chem. Phys.* 57:461
47. Shapiro, M. 1977. *Chem. Phys. Lett.* 46:442
48. Beswick, J. A., Shapiro, M., Sharon, R. 1977. *J. Chem. Phys.* 67:4045

49. Shapiro, M., Bersohn, R. 1980. *J. Chem. Phys.* 73:3810
50. Segev, E., Shapiro, M. 1980. *J. Chem. Phys.* 73:2001
51. Beswick, J. A., Shapiro, M. 1982. *Chem. Phys.* 64:333
52. Shapiro, M., Balint-Kurti, G. G., 1979. *J. Chem. Phys.* 71:1461
53. Light, J. C., Walker, R. B. 1976. *J. Chem. Phys.* 65:4272
54. Morse, M. D., Freed, K. F., Band, Y. B. 1979. *J. Chem. Phys.* 70:3604
55. Morse, M. D., Freed, K. F., Band, Y. B. 1979. *J. Chem. Phys.* 70:3620
56. Morse, M. D., Freed, K. F., 1981. *J. Chem. Phys.* 74:4395
57. Beswick, J. A., Gelbart, W. M., 1980. *J. Phys. Chem.* 84:3148
58. Balint-Kurti, G. G., Shapiro, M. 1981. *Chem. Phys.* 61:137
59. Tipps, W. M., Baronavski, A. T., 1980. *Chem. Phys. Lett.* 71:395
60. Simons, J. P., Tasker, P. W., 1974. *Mol. Phys.* 27:1691
61. Ashfold, M. N. R., Simons, J. P. 1977. *Chem. Phys. Lett.* 47:65
62a. Ashfold, M. N. R., Simons, J. P. 1978. *J. Chem. Soc. Faraday Trans. 2* 73:858
62b. Ashfold, M. N. R., Simons, J. P. 1978. *J. Chem. Soc. Faraday Trans. 2* 74:280
63. Ashfold, M. N. R., Macpherson, M. T., Simons, J. P. 1979. *Top. Current Chem.* 86:1
64. Lee, L. C., Judge, D. L. 1973. *Can. J. Phys.* 51:378
65. Pack, R. T. 1976. *J. Chem. Phys.* 65:4765
66. Kanfer, S., 1981. PhD thesis. Columbia Univ., New York
67. Levine, R. D., Kinsey, J. L. 1979. In *Atom Molecule Collision Theory*, ed. Bernstein, p. 693. New York: Plenum
68. Shapiro, M. 1981. *Chem. Phys. Lett.* 81:521
69. Sparks, R. K., Shobatake, K., Carlson, L. R., Lee, Y. T. 1981. *J. Chem. Phys.* 75:3838
70. Lee, L. C., 1980. *J. Chem. Phys.* 72:4334
71. Carrington, T., 1964. *J. Chem. Phys.* 41:2012
72. Watanabe, K., Zelikoff, M. 1953. *J. Opt. Soc. Am.* 43:753
73. Wang, H., Felps, W. S., McGlynn, S. P. 1977. *J. Chem. Phys.* 67:2614
74. Macpherson, M. T., Simons, J. P. 1977. *Chem. Phys. Lett.* 51:261
75. Flouquet, F., Horsley, J. A., 1974. *J. Chem. Phys.* 60:3767
76. Segev, E., Shapiro, M. 1982. *J. Chem. Phys.* In press
77. Sorbie, K. S., Murrell, J. N. 1976. *Mol. Phys.* 31:905
78. Lee, L. C., Oren, L., Phillips, E., Judge, D. L. 1978. *J. Phys. B* 11:47
79. Akamatsu, R., O-Ohata, K. 1977. *J. Phys. Soc. Jpn.* 43:264

POLYACETYLENE, $(CH)_x$:
The Prototype Conducting Polymer

S. Etemad and A. J. Heeger

Department of Physics, University of Pennsylvania,
Philadelphia, Pennsylvania 19104

A. G. MacDiarmid

Department of Chemistry, University of Pennsylvania,
Philadelphia, Pennsylvania 19104

INTRODUCTION

The emergence of conducting polymers as a new class of electronic materials has attracted considerable attention: the study of these systems has generated entirely new scientific concepts as well as potential for new technology. As polymers, these materials have a highly anisotropic quasi-one-dimensional structure, which makes such systems fundamentally different from conventional inorganic semiconductors. First, their chain-like structure leads to strong coupling of the electronic states to conformational excitations (solitons) peculiar to a one-dimensional (1-d) system. Second, the relatively weak interchain binding allows diffusion of dopant molecules into the structure (between chains), while the strong intrachain carbon-carbon bonds maintain the integrity of the polymer. The prototype example of these conducting polymers is polyacetylene, $(CH)_x$. In this review we introduce the novel concepts associated with this new class of electronic materials by describing some important properties of this well studied example.

Polyacetylene is the simplest conjugated polymer. It consists of weakly coupled chains of CH units forming a pseudo-one-dimensional lattice.

Three of the four carbon valence electrons are in sp^2 hybridized orbitals; two of the σ-type bonds construct the 1-d lattice while the third forms a bond with the hydrogen side group. The 120° bond angle between these three electrons can be satisfied by two possible arrangements of the carbons, *trans*-(CH)$_x$ and *cis*-(CH)$_x$, with two and four CH monomers per unit cell, respectively (see Figure 1). In either isomer the remaining valence electron has the symmetry of a $2p_z$ orbital with its charge density lobes perpendicular to the plane defined by the other three. In terms of an energy-band description, the σ-bonds form low-lying completely filled bands, while the π-bond leads to the partially filled energy band structure responsible for the important electronic properties.

If all the bond lengths were equal, pure *trans*-(CH)$_x$ would be a quasi-1-d *metal* with a half-filled band. Such a system is unstable with respect to a dimerization distortion, the Peierls instability (1, 2), in which adjacent CH groups move toward each other, forming alternately short (or double) bonds and long (or single) bonds, thereby lowering the energy of the system (see Figure 2). Clearly, by symmetry arguments, one could interchange the double and single bonds without changing the energy. Thus, there are two lowest energy states, *L* and *R*, having two distinct bonding structures. This two-fold degeneracy leads to the existence of nonlinear topological excitations, bond-alternation domain walls or solitons, which appear to be responsible for many of the remarkable properties of (CH)$_x$ (3–8).

Both the *cis*- and *trans*-forms (see Figure 1) can be prepared as silvery, flexible films, which can be made either free-standing or on a variety of

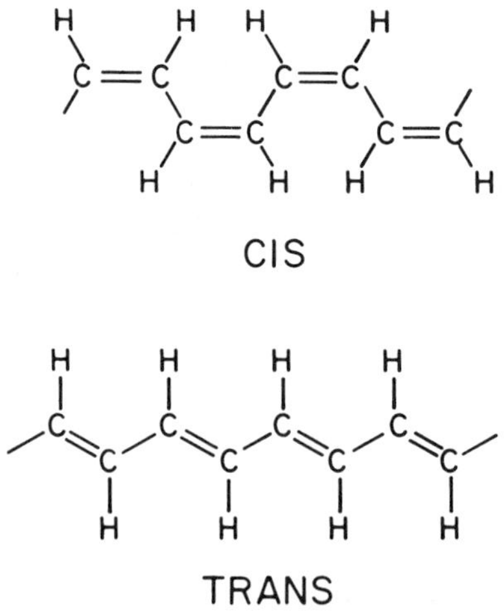

Figure 1 Polymer chain structures of *cis*- and *trans*-(CH)$_x$.

substrates, such as glass or metal, with thicknesses varying from 10^{-5} cm to 0.5 cm[9]. The *trans*-isomer is the thermodynamically stable form. Any *cis-trans* ratio can be maintained at low temperatures; but complete isomerization from *cis*- to *trans*-$(CH)_x$ can be accomplished after synthesis by heating the film to temperatures above 150° C for a few minutes (10, 11).

Electron microscopy studies (9, 12, 13) show that the as-grown $(CH)_x$ films consist of randomly oriented fibrils (typical fibril diameter ~ 200 Å, see Figure 3). The fibril diameter can be varied significantly with different polymerization conditions. The films can be stretch-oriented in excess of three times their original length with concomitant partial alignment of the fibrils (12, 14, 15). The bulk density is about 0.4 gm/cm^3 compared with 1.2 gm/cm^3 as obtained by flotation techniques (16). Therefore, the polymer fibrils fill only about one-third of the total volume and the effective surface area is quite high (~ 60 m^2/gm). X-ray studies (17, 18) show that the $(CH)_x$ films are highly crystalline. A description of the crystal structure and chain packing is given below.

Recent studies have demonstrated that after synthesis, $(CH)_x$ films can be doped chemically (19–21) or electrochemically (22) at room temperature with a variety of donors or acceptors to form n- or p-type semiconductors; i.e. $[(D^+)_y CH]_x$ or $[CH(A^-)_y]_x$ where D and A represent donor and acceptor species, respectively, and y is the dopant concentration. Doping to high levels (above $\sim 1\%$) results in a semiconductor-metal transition (20, 23, 24), giving a whole new class of metals with a wide range of electronegativity. Moreover, the existing experimental data already show that these materials have potential for use in a number of areas of future technology. The electrical conductivity of polyacetylene can be varied in a controlled manner over twelve orders of magnitude through chemical or electrochemical doping. Values greater than 3×10^3 (Ω-cm^{-1}) have already been achieved with only partially crystalline

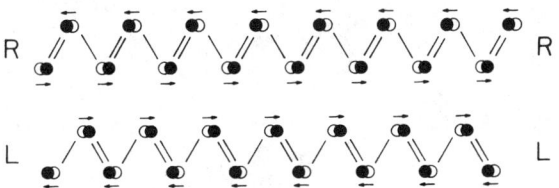

TRANS-$(CH)_x$

Figure 2 The degenerate ground state of *trans*-$(CH)_x$, an example of a broken symmetry. The double and single bonds can exist in two different, but equivalent, configurations.

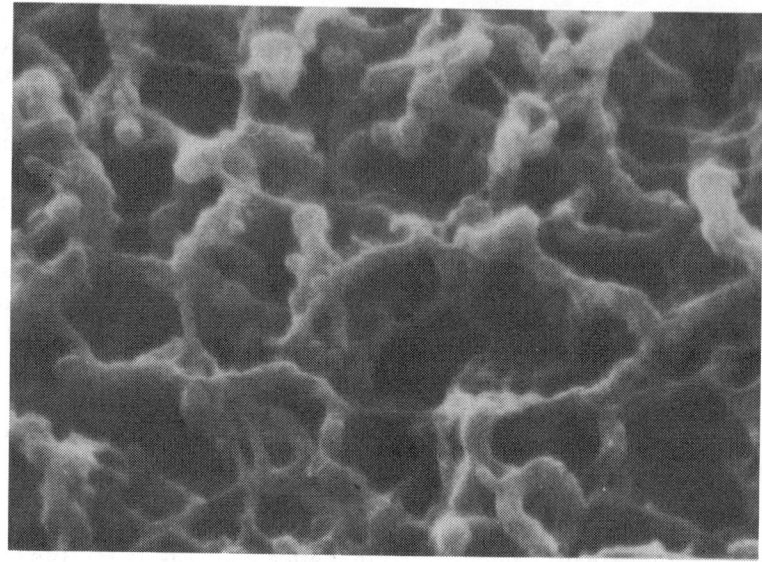

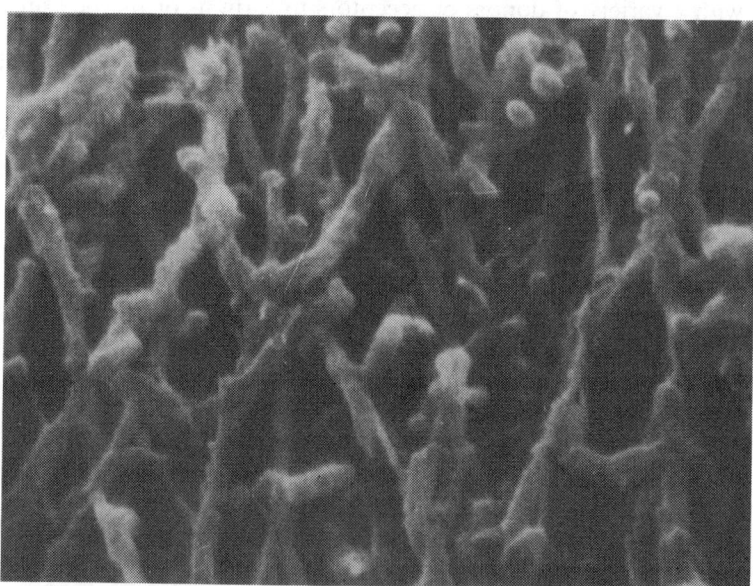

Figure 3 Electron micrographs of polyacetylene (dull side). The fibril diameter is approximately 200 Å. The upper panel is from an as-grown film; the lower panel was obtained after stretch-orientation ($l/l_0 \simeq 3$).

samples (23), and analysis of the transport (23) and optical (25) data implies that a further increase of at least one order of magnitude should be possible. Controlled electrochemical doping and "undoping" have been demonstrated (22), and prototype rechargeable batteries have been constructed using $(CH)_x$ as both the active cathode and anode (26, 27). Photovoltaic phenomena have been observed in heterojunctions (28), Schottky barrier junctions (29, 30), and photoelectrochemical (31) junctions. Thus, although these and other potential applications will require considerable future work before ultimate technological value can be determined, the properties appear promising.

A sound understanding of the electronic properties of polyacetylene is essential to any future development of this polymer or related polymers as useful electronic materials. In the following sections we describe the general electronic properties of polyacetylene with emphasis on potential applications. Then we outline the present understanding of the electronic excitations of this simplest of all conjugated polymers.

ELECTRONIC PROPERTIES: THE POTENTIAL FOR NEW TECHNOLOGY

General Features

Simple estimates lead to a picture of $(CH)_x$ as a broad band, quasi-one-dimensional semiconductor. The overall bandwidth associated with electronic motion along the chain, W, can be estimated from tight-binding theory; $W = 2zt_0$ where z is the number of nearest neighbors and t_0 is the inter-carbon transfer integral for π-electrons. From theoretical and spectroscopic studies of aromatic ring systems and short chain polyenes, t_0 can be estimated as 2–2.5 eV, so that $W \simeq 8$–10 eV. Because of the bond-alternation (see below) $trans$-$(CH)_x$ is a semiconductor with an energy gap of about 1.5 eV (25, 32). As a result of the large bandwidth and unsaturated π-system, $(CH)_x$ is fundamentally different from either the traditional organic semiconductors made up of weakly interacting molecules (e.g. anthracene, etc) or the saturated polymers with monomeric units of the form $\left(\begin{smallmatrix} R & R \\ & C \end{smallmatrix}\right)$ where there are no π-electrons (e.g. polyethylene). Polyacetylene is therefore electronically more nearly analogous to the traditional inorganic semiconductors; however, the transverse bandwidth due to interchain coupling is much less. The large nearest neighbor interchain spacing (~ 4 Å) implies a transverse bandwidth which is much smaller, and is comparable to the longitudinal bandwidth in molecular crystals (2) such as TTF-TCNQ; i.e. of order 0.1 eV. Weak interchain coupling is therefore implied, and the system may be regarded as quasi-one-dimensional.

The discovery that polyacetylene can be doped after synthesis (19–21), at room temperature, and with a variety of dopants makes it fundamentally different from conventional covalent semiconductors. The ability to dope $(CH)_x$ *after synthesis* using chemical or electrochemical techniques is due to a combination of its open morphology (9, 12, 13) with associated high surface area and the weak interchain binding that allows diffusion of the dopant ions between the polymer chains. The 200 Å diameter fibrillar morphology (Figure 3) ensures that a dopant molecule has to diffuse less than ~ 100 Å to reach any chain. On the other hand, the strong intrachain bonding maintains the integrity of the polymer during the diffusion process and thereby ensures the reversibility of the doping.

The carriers generated by the doping of $(CH)_x$ result from charge transfer [e.g. see (33)]. Charge transfer occurs from polymer to acceptor (A) with the polymer chain acting as a poly(cation) in the presence of an A^- species. For a donor (D), the polymer chain acts as a poly(anion) in the presence of D^+ species. The A^- or D^+ ions reside between polymer chains. The principal electrical properties (23, 34–36) associated with the doping process in *trans*-$(CH)_x$ are sketched in Figure 4. Figure 4a is a plot of the logarithm of the conductivity ($\log \sigma$) together with the thermoelectric power (S) for n- and p-type doped *trans*-$(CH)_x$. For dopant concentrations below the semiconductor-metal transition, the

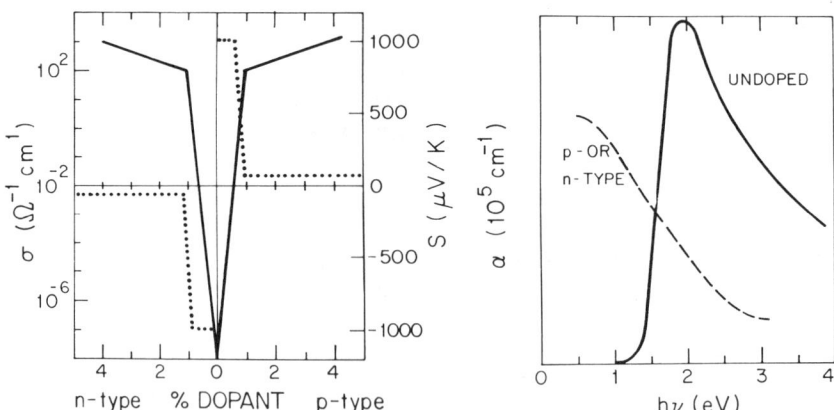

Figure 4 Schematic diagrams of transport and optical properties as a function of doping. (*a*) Conductivity (σ) and thermopower (S) as a function of dopant concentration for both n- and p-type doping. (*b*) Optical absorption of *trans*-$(CH)_x$ before doping (showing the interband transition) and after doping to metallic levels (showing the free carrier absorption).

conductivity increases exponentially with y. In the same region ($y < 10^{-3}$), S is large and characteristic of a semiconductor. The sign of S changes with little change in its magnitude on crossing the boundary between n- and p-type doping. The fact that the magnitude of S behaves similarly for n- and p-type doping as a function of y, but opposite in sign (36), reflects the electron-hole symmetry of the band structure, and is an indication that the carriers responsible for the high conductivity of doped $(CH)_x$ have been transferred onto the polymer from the dopant. At higher concentration, the conductivity begins to saturate. Similarly, at dopant levels above a few tenths of a percent, S becomes small and linear in T, characteristic of delocalized carriers (23).

The reversibility of the doping and undoping process, as monitored through the absorption coefficient (α) measurements (37–41), is shown in Figure 4b. For undoped trans-$(CH)_x$, α resembles that of a 1-d semiconductor with band gap $E_g \simeq 1.5$ eV. The 1-d divergence in the joint density of states at the band edge is smeared by a combination of 3-d and disorder effects. The results from n- and p-type doped polymers are similar and resemble free carrier absorption at high doping levels ($y \simeq 0.07$). Furthermore, chemical compensation (39, 40, 41) or electrochemical undoping (37) of either n- or p-type doped samples gives back the spectrum of the undoped trans-$(CH)_x$. These results demonstrate that doping reversibly transforms the electronic structure of the polymer from a semiconductor to a metal.

Schottky Diodes, Heterojunctions and Solar Cells

The possibility of using $(CH)_x$ in active electronic devices is appealing. Experiments have been reported where metallically doped $(CH)_x$ was used as a Schottky-barrier contact to Si and GaAs (28); also p-$(CH)_x/n$-ZnS and p-$(CH)_x/n$-CdS heterojunctions have been fabricated (28) and shown to exhibit photoresponse to visible light.

The near perfect match of the trans-$(CH)_x$ band gap to the solar energy spectrum has also been exploited in photovoltaic devices using the Schottky-barrier configuration (29, 30) formed with inexpensive metals such as aluminum. Although a polyacetylene photovoltaic device has the correct ingredients (low cost, high α, matched E_g, relatively easy large area production) for large scale energy conversion applications, the highest internal conversion efficiency achieved so far is 0.3% for an Al:p-$(CH)_x$ under low illumination (30). Figure 5 shows an I-V characteristic for such a device (42) with a back-to-forward resistance ratio of about 1100 and a back-biased impedance of about 90 MΩ. Detailed studies of the I-V characteristics indicate complex competing processes (recombination, trapping, diffusion etc), also implied by the high value

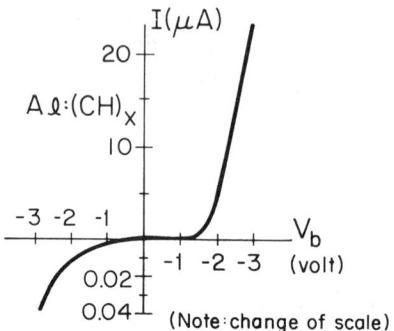

Figure 5 I-V characteristic for an A : p-$(CH)_x$ Schottky-barrier diode; back-to-forward resistance of about 1100 and a back-biased resistance of 90 MΩ.

($\geqslant 3$) of the diode quality factor. Significant improvement will be required for efficient device operation. Nevertheless, the results are encouraging at this early stage of development.

The mechanism of Schottky-barrier formation on polyacetylene has been investigated by in situ monitoring the x-ray photoemission during junction formation (43). A 0.6 eV shift in energy of the carbon core level electrons has been interpreted as due to band bending at the Mg/$(CH)_x$ interface, resulting from Schottky barrier formation in a p-type semiconductor via a space-charge (carrier depletion) layer and an associated potential barrier region in $(CH)_x$. No band bending was observed at the Au/$(CH)_x$ interface upon contact formation. The associated I-V curves were in agreement with expectations: the Mg/$(CH)_x$ contact was rectifying, whereas the Au/$(CH)_x$ contact was ohmic (30, 43). Analysis of the results showed that $(CH)_x$ energy bands are initially flat (or minimally bent) at a free surface and that Schottky-barrier formation is most likely due to the electrostatics of work function differences at the metal-$(CH)_x$ interface region.

Electrochemical Doping and Battery Applications

Polyacetylene can act either as an electron source or an electron sink according to whether it is, respectively, oxidized (doped p-type) or reduced (doped n-type). Thus, a whole class of "metallic plastics" can be made from this conjugated polymer, ranging in electrochemical potential from highly electropositive to highly electronegative. Because the doping can be carried out after synthesis using electrochemical techniques, and because the doping is reversible, electrodes fabricated from polyacetylene may find use in a variety of electrochemical applications (22, 26, 27, 37).

Lightweight, high power-density, high energy-density rechargeable organic storage batteries, some of which involve no metal ion or free metal in either their charge or discharge processes, have been developed using polymer electrodes. The concepts involved represent an entirely new approach to battery technology.

Batteries using $(CH)_x$ for both electrodes have the $(CH)_x$ in different oxidation states; e.g. for the charged state $(CH^{+a})_x$ and $(CH^{-b})_x$. The charged state may be simply attained by passing a d.c. current between two pieces of $(CH)_x$ film immersed in a suitable electrolyte. During discharge, electrons flow from the less oxidized state (the anode of the battery) to the more oxidized state (the cathode of the battery) until both electrodes have the same oxidation state, i.e. the same chemical composition. In order to maintain electrical neutrality in the system, counter anions and cations must also be present. These, however, do not undergo any electrochemical reaction; only the conductive polymer is oxidized or reduced during the battery charging or discharging processes.

For example, if two strips of $(CH)_x$ film are placed in a solution of $Li^+(ClO_4)^-$ in propylene carbonate and are attached to the positive and negative terminals, respectively, of a battery or a d.c. power source (~ 4 volts), oxidation of the $(CH)_x$ occurs at the positive electrode (anode) and reduction occurs at the negative electrode (cathode) during this "charging" operation. The charging reactions are:

at anode (+): $(CH)_x + xy(ClO_4)^- \rightarrow [(CH)^{+y}(ClO_4^-)_y]_x + xye^-$

at cathode (−): $(CH)_x + xyLi^+ + xye^- \rightarrow [Li_y^+(CH)^{-y}]_x$

net reaction: $2(CH)_x + xyLi^+(ClO_4)^- \rightarrow [(CH)^{+y}(ClO_4^-)_y]_x$
$$+ [Li_y^+(CH)^{-y}]_x.$$

The discharge reactions are the reverse of the above. Other counter anions, e.g. PF_6^- can be used in place of ClO_4^-.

The battery system so far investigated to the greatest extent involves a strip of $(CH)_x$ film and a strip of Li, or Al, foil immersed in a solution of $LiClO_4$ in propylene carbonate (26, 27). The net charging reaction, accomplished by attaching the $(CH)_x$ to the positive terminal and the Li or Al to the negative terminal, respectively, of a d.c. power source is,

$$(CH)_x + xyLi^+(ClO_4)^- \rightarrow [(CH)^{+y}(ClO_4^-)_y]_x + xyLi.$$

The discharge reaction is the reverse of the above.

The initial open circuit voltages, (V_{oc}) and short circuit currents (I_{sc}) obtained from the cells are interestingly large. For example, the $[(CH)^{+y}(ClO_4)_y]_x/Li$ battery gives $V_{oc} = 3.7$ volts and I_{sc} in excess of 0.1 amp/cm² of $(CH)_x$ film (~ 4 mg). Definitive energy and power densities have not been obtained for the batteries because these parameters are affected greatly by the weight of the packaging material, etc employed; however, the available data from experimental cells indicate that the energy and power densities may well exceed those of, for example, conventional lead/acid batteries by a considerable margin. To the accuracy of the experimental measurements, the $(CH)_x$ electrode in this system appears to be electrochemically reversible with no observable degradation after a number of complete charge-discharge cycles.

The large currents obtained from the small weights of films is related to the morphology of $(CH)_x$. As described above (see Figure 3), $(CH)_x$ film consists of an interwoven network of approximately 200 Å $(CH)_x$ fibrils that fill only about 1/3 of the volume of the film. For example, 1 cm² piece of film, 0.01 cm in thickness, has an effective surface area in contact with the electrolyte of approximately 2.5×10^3 cm²!

These initial results indicate that electrochemical studies of $(CH)_x$ and other organic metals represent an extensive area for further research that is not only of fundamental scientific interest but also of potential technological value.

SOLITONS IN *trans*-$(CH)_x$

General Concepts

The recent progress in developing an understanding of the physical properties of polyacetylene results principally from two factors. First, the simplicity of the $(CH)_x$ chain structure is the key to its success as the prototype conjugated polymer. This is clearly reflected in the variety of parallel theoretical and experimental studies that have been carried out successfully within the past few years. Second, the thermodynamically stable *trans*-$(CH)_x$ has the unusual broken symmetry degenerate ground state, described in Figure 2. As a result, the polymer can sustain free stable solitons as natural nonlinear excitations. The possibility of experimental studies of such solitons in polyacetylene, therefore, not only promises a deeper understanding of this important conducting polymer, but also represents a unique opportunity to explore nonlinear phenomena in condensed matter.

The mathematical theory of solitons in $(CH)_x$ has been investigated in detail in several recent papers which show that the coupling of these conformational excitations to the π-electrons leads to unusual electrical and magnetic properties (3-6, 44, 45). In the following paragraphs we describe the concepts and theoretical results in simple, schematic terms. Where necessary we refer to specific theoretical results; however, the fundamental origins of these novel concepts can be at least qualitatively understood in elementary terms.

Because of the Peierls instability (1, 2), adjacent CH groups move toward each other, forming alternately short (or double) bonds and long (or single) bonds, thereby, lowering the electronic energy by opening an energy gap at the Fermi level. One can describe the resulting ground state in terms of the inequality of the transfer (resonance) integrals for the single and double bonds (t_1 and t_2) that cause the opening of a $\pi - \pi^*$ gap, $2\Delta = 2(t_1 - t_2)$, separating the highest occupied MO states from the lowest unoccupied MO state.

The existence of bond alternation in *trans*-$(CH)_x$ is fundamental to the soliton concept. Only recently, however, has direct experimental evidence of the dimerization distortion been obtained from X-ray scattering studies by Fincher et al (18). They find a crystal structure similar to that proposed earlier (46) (space group $P2_1/n$ with $a = 4.25$ Å, $b = 7.33$ Å, $c = 2.46$ Å, $\beta = 91.5$ Å, and two chains per unit cell). The structure is dimerized as shown schematically in Figure 6 with $u_0 \simeq 0.03$ Å.

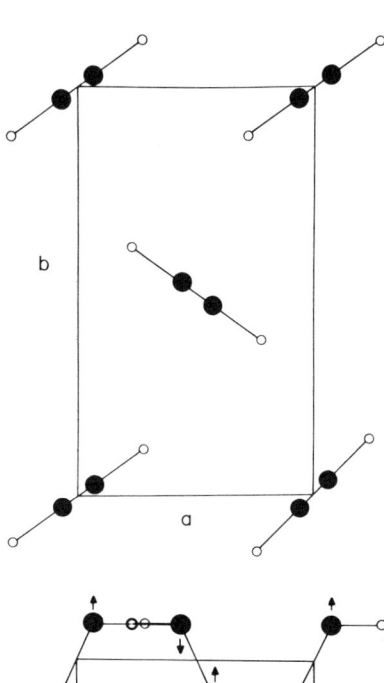

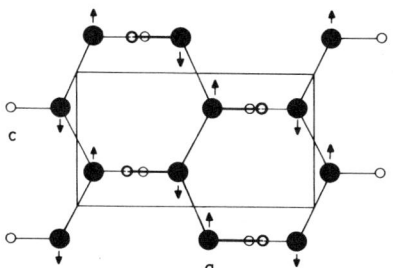

Figure 6 Schematic diagram of crystal structure of *trans*-$(CH)_x$ (see 16). The black dots represent carbon atoms; the open circles represent hydrogen. The small arrows (a–c plane projection) imply a dimerization distortion with $u_0 \sim 0.03$ Å.

The dimerization distortion leads to an energy gap in the electronic excitation spectrum. Using the Hamiltonian adopted from the Hückel theory by Su, Schrieffer & Heeger (SSH) (3) to describe $trans$-$(CH)_x$, the value of the symmetry breaking one-dimensional energy gap, appropriate for k parallel to the chain axis in reciprocal space, can be written as

$$2\Delta = 8\beta u_0 \qquad 1.$$

where β, the electron-phonon coupling constant, describes the modulation of π-electron transfer integral due to the atomic motion. Analysis of the optical absorption data (25, 32, 47) leads to a value for $2\Delta \simeq 1.6-1.8$ eV. This value, when substituted into Eq. 1, leads to a value for the electron-photon coupling constant of about $\beta = 7.1$ eV/Å, in agreement with estimates obtained from studies of lattice dynamics in polyacetylene (48, 49) (see below). Thus, the existence of bond alternation in long chain polyenes has been confirmed. Moreover, the good agreement between the measured energy gap and that predicted by the SSH (3) electron-lattice Hamiltonian (using the measured distortion) implies that the effective electron-electron interation is relatively weak and does not dominate the physics.

As described above, one could interchange the double and single bonds in $trans$-$(CH)_x$ without changing the energy. Thus, there are two degenerate lowest-energy states, L and R, having two distinct bonding structures as shown in Figure 2. This two-fold ground state degeneracy leads to the existence of nonlinear topological excitations, bond-alternation domain walls or solitons.

When a single chain of cis-$(CH)_x$ begins to isomerize, the isomerization process can, in principle, commence at different parts of the chain, one having configuration (L), the other configuration (R). When these two

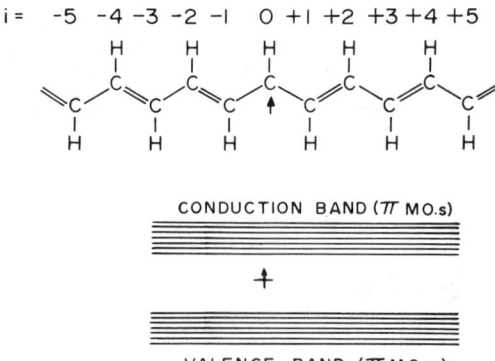

Figure 7 Diagrammatic representation of a neutral soliton (free radical located on a nonbonding molecular orbital) in $trans$-$(CH)_x$. The index (i) labels the C-atoms relative to the soliton center (see text).

different configurations meet, a free radical is produced, as shown schematically in Figure 7. This has been confirmed experimentally; when pure $cis\text{-}(CH)_x$, possessing no free spins, undergoes isomerization, approximately one in 3000 of the (CH) units in the resulting $trans\text{-}(CH)_x$ is in the form shown in Figure 7 with an unpaired spin and a Curie law magnetic susceptibility (11, 50). This species is actually a valid excited state of $trans\text{-}(CH)_x$, which could be formed (in pairs) if $trans\text{-}(CH)_x$ could be heated to a sufficiently high temperature (4000 K!) without thermal decomposition.

Associated with the kink is an electronic bound state with energy at mid-gap by virtue of electron-hole symmetry. Alternatively, the molecular orbital associated with a kink is a nonbounding state (51); i.e. halfway between the bonding valence band and the antibonding conduction band (Figure 7). If the localized state contains one electron, the soliton is neutral, with spin 1/2, and therefore is paramagnetic. When the electron in the localized state is removed, for example by acceptor doping, the soliton is positively charged, with spin zero and nonmagnetic. This has indeed been observed. Thus, if $(CH)_x$ is carefully and slowly oxidized by, for example, AsF_5, positive solitons are formed; the number of Curie spins originally present decreases by more than a factor of a thousand, to levels below one part per million (52).

The positive soliton (Figure 7) is equivalent to a stabilized (by delocalization) carbonium ion on the $(CH)_x$ chain. Similarly, double occupancy induced by donor doping would lead to a spin-zero negatively charged state. The negative soliton is thus equivalent to a stabilized carbanion as shown in Figure 8.

For simplicity, the solitons are shown in Figures 2, 7, and 8 as if localized on one lattice site, whereas detailed calculations (3–6) have shown that minimization of the energy spreads the kink over a region of

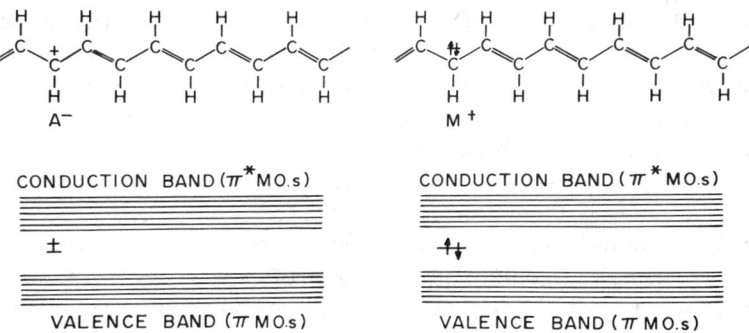

Figure 8 Diagrammatic representations of positive and negative solitons; i.e. carbonium ion and carbanion, located in a nonbonding molecular orbital in $trans\text{-}(CH)_x$.

about 15 (CH) units. Thus, in the case of a positive soliton, although the maximum charge density is adjacent to the counter anion, A^-, $\sim 85\%$ of the charge is spread out symmetrically on either side over about 15 (CH) units. The spatial spread of the soliton can be roughly estimated from an inspection of the molecular orbital wave functions of the localized nonbonding state in Figure 7. If c_is are the MO coefficients of the nonbonding state on the ith carbon atom, by symmetry one can see $c_i = c_{-i}$. Hence $c_i = 0$ for all odd sites, and for the even sites the c_i satisfy the relation (51)

$$c_{2i}t_2 + c_{2(i+1)}t_1 = 0 \qquad \qquad 2.$$

where t_1 and t_2 are the transfer integrals between two carbon atoms on the ends of a single and double bond, respectively. Eq. 2 shows that the MO wavefunction of the mid-gap state peaks at the center and decreases as $c_{2i}/c_0 = (t_1/t_2)^i$. Defining $2t_0 = (t_1 + t_2)$ and recalling that the $\pi - \pi^*$ gap is $2 = 2(t_2 - t_1)$, one can estimate the spread of the nonbonding states to be about $2t_0/\Delta \sim 7$ sites. This underestimates the actual size of the spread by about a factor of two since the simple MO formalism does not allow relaxation of the bond lengths near the soliton center (3).

The location of the center of the soliton on the chain is arbitrary since all sites are equivalent. To move the soliton involves minor shifts ($\sim .01$ Å) of the C−H units in the vicinity consistent with moving the boundary between the two domains of opposite bond alternation. The soliton in trans-$(CH)_x$, therefore, is like a "particle" consisting of a localized structural distortion intimately coupled to a self-trapped nonbonding localized state at the center of the gap. Su, Schrieffer & Heeger showed that the kinetic mass of the soliton could be written as (3)

$$m_s \sim M_{CH} \frac{u_0^2}{al}$$

where M_{CH} is the mass of the C − H fragment, l is the half-width of the soliton wavefunction ($l \simeq 7a$) in units of the C − C distance along the chain (a), and u_0 is the symmetry breaking dimerization distortion. Using $u_0 \simeq 0.03$ Å, one obtains $m_s \sim m_e$ where m_e is the electron mass. As a result, one expects neutral solitons in trans-$(CH)_x$ to be highly mobile in agreement with experiment (50, 54).

The intimate coupling of the structural distortion and nonbonding electronic state not only determines the creation energy and spatial spread of a solition, but also dramatically alters the dynamics of a soliton pair. For example, recombination of a pair of highly mobile solitons on a given chain, as well as bonding of two solitons on neighboring chains (to form a cross-link), is fundamentally different from the expected behavior of two moving free radical nonbonding M.O. states. The spatial spread of the soliton wavefunction ($l \simeq 7a$) plus the well-known robust nature of the nonlinear conformational excitations (55) inhibits soliton recombination, leading to relatively long recombination times. Mathematical models, for example, show that solitons on model chains can bounce off one another many times before finally recombining to return the chain to the ground state (56).

The presence of neutral solitons induced as defects during isomerization is a fortunate accident, for the number of thermally induced neutral soliton excitations in undoped $(CH)_x$ would be far too small to be observable. In this context, however, we stress that it is not necessary to first have neutral solitons in order to obtain positive (or negative) solitons after doping. Since the nonbonded localized state appears at the center of the energy gap, i.e. the chemical potential, the relevance of the solitons to the doping of $(CH)_x$ depends on the energy for creation of a soliton, E_s, as compared with Δ, the energy required for adding an electron or a hole. If $\Delta < E_s$, charge-transfer doping would occur by creating free band excitations; if $\Delta > E_s$, soliton formation would be favored. Both numerical calculations (3) using a discrete lattice model and analytical results for a continuum model (5) indicate that, within the one-electron approximation, $\Delta > E_s$. The continuum model (5) yields

$$E_s = 2\Delta/\pi. \qquad 3.$$

Thus, for example, if a perfect *trans*-$(CH)_x$ sequence were p-doped, removing electrons from the system, the most stable configuration for the resulting positive charge would be the nonmagnetic charged soliton (Figure 8). This has been confirmed experimentally. On careful p-doping with AsF_5 one finds that (52)[1] (*a*) the original free Curie spins in $(CH)_x$ decrease as indicated above and (*b*) even though the amount of dopant species is far in excess of that necessary to remove the original free spins, no additional Curie spins appear.

The fact that $2E_s < 2\Delta$ ensures that solitons are the primary electronic excitations of the *trans*-$(CH)_x$ lattice. The topological nature of the solitons, however, makes them fundamentally different from the conventional electronic excitations of three-dimensional semiconductors. In the following sections we first present evidence for soliton formation in *trans*-$(CH)_x$. The antisymmetric nature of a soliton wave function and its associated mid-gap state have been observed directly in studies of the

[1] Note that although there is some disagreement in the literature concerning the magnitude of the temperature independent Pauli susceptibility at intermediate doping levels, the dramatic reduction in Curie spins ($\chi \sim 1/T$) is a general feature of all reported data. See for example Figure 3 of (53).

lattice dynamics and absorption spectra upon dilute doping. Then we discuss the dynamics of soliton formation as indicated in the experiments that involve an initial photoinjection of an electron-hole pair.

Effects on Lattice Dynamics

Specific evidence for the formation of charged solitons upon doping has emerged from infrared studies of the donor and acceptor states in lightly doped polyacetylene (49, 57, 58, 59). In a series of experiments with various dopants in both $(CH)_x$ and $(CD)_x$ it was found that upon dilute doping new absorption modes appeared in the IR region, with remarkable intensity. The doping-induced modes are primarily polarized parallel to the chain direction and their intensities grow in proportion to the dopant level, becoming comparable to any of the intrinsic IR lines of the undoped polymer at about 0.1%. Thus, the doping-induced IR modes have oscillator strengths enhanced by approximately 10^3; such a large enhancement must arise from coupling of the new vibrational modes (induced by doping) to the electronic oscillator strength of the polyene chain. These results are quite general; the same modes can be observed for iodine and AsF_5 (p-type), and for Na doping (n-type).

Mele & Rice (58) have been able to interpret successfully these results in terms of a theory of the lattice dynamics of solitons in $trans$-$(CH)_x$ or $trans$-$(CD)_x$. In the frequency range of the vibrational modes they found several internal modes peculiar to the soliton antisymmetric structure. In

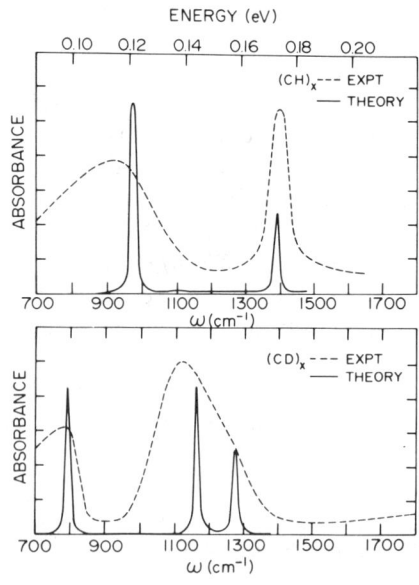

Figure 9 Comparison of the experimentally derived and the theoretically calculated additional absorption due to the infrared active vibrational modes (IAVM) of solitons in $trans$-$(CH)_x$ (49).

particular, three of these modes were found to be strongly infrared active, deriving their oscillator strength from interactions with the π-electrons. They have shown that the dominant motions associated with the infrared active vibrational modes (IAVM) of a soliton involve an antisymmetric contraction of the single (or double) bonds on one side of the soliton center and an expansion on the other, thus driving charge back and forth across the soliton center. Because of the spatial spread of the soliton over about 15 sites, they find that antisymmetric oscillation of the electronic charge generates a large oscillating dipole moment consistent with the experimentally observed enhanced oscillator strength of the dopant-induced IAVM.

The force field model constructed by Mele & Rice (48, 49, 58) relied on a *single* adjustable parameter, the electron-lattice coupling constant β, which leads to nonlocal coupling of the atomic displacements through their interaction with the extended π-electronic states. The value of β was determined by a fit of the calculated zone center frequencies of gerade modes to the observed Raman lines in unperturbed *trans*-$(CH)_x$ and $(CD)_x$ (49). A comparison of the IAVM of solitons in *trans*-$(CH)_x$ and *trans*-$(CD)_x$ is shown in Figure 9. The *dashed curves* are the experimental results for a concentration of approximately 0.1% and the *solid curves* are the result of the calculations with no adjustable parameter (the value of $\beta = 6.9$ eV/Å was already fixed by the fit to the Raman data).

As shown in Figure 9, the calculated infrared absorption associated with the IAVM of solitons is in remarkably good agreement with the experimental results. First, the frequencies are in agreement to an accuracy of a few percent for all the IAVM for both $(CH)_x$ and $(CD)_x$. Second, the increased width and asymmetry of the higher frequency mode in $(CD)_x$ appears to be accounted for by the blue shifting of one of the degenerate modes in the low frequency line of *trans*-$(CH)_x$. Finally, the calculated absolute integrated oscillator strengths correspond to approximately 20% of the total associated with the added charge, and they agree with the experimental values to within a factor of 2–3. The relative oscillator strengths of the different modes are even more accurate. In $(CH)_x$, the lower frequency mode is more intense, whereas in $(CD)_x$ we find that the higher frequency absorption is more intense. In both cases the experimental and theoretical intensity ratios are in close agreement.

The Mid-gap State

In contrast to the case of conventional semiconductors, we find that donor (acceptor) doping does not result in generation of localized states near the conduction (valence) band. Instead, independent of the nature of the dopant, addition or removal of electrons to or from the polymer chain is accommodated by formation of soliton-induced localized states near mid-gap (37, 38, 60–63). Calculations of the absorption coefficient (α) show that a soliton kink on a chain suppresses the interband transition, whereas transitions involving the soliton level are found to have a significantly enhanced absorption cross-section (60–63). The results are in agreement with the experimental absorption spectra obtained from trans-$(CH)_x$ lightly doped with a variety of dopants (37, 38, 60).

The absorption spectra of pristine and lightly doped (a few tenths of a percent AsF_5) trans-$(CH)_x$ are compared in Figure 10 (60). The doping was carried out in situ using extreme care so that the results could be directly and quantitatively compared. The appearance of a mid-gap absorption band concurrent with the uniform suppression of the entire interband ($\pi - \pi^*$) transition has been quantitatively reproduced in a variety of n- and p-type doping experiments. The insensitivity of the energy of the mid-gap absorption to the sign of the transferred charge or the dopant species demonstrates that the doping-induced mid-gap transi-

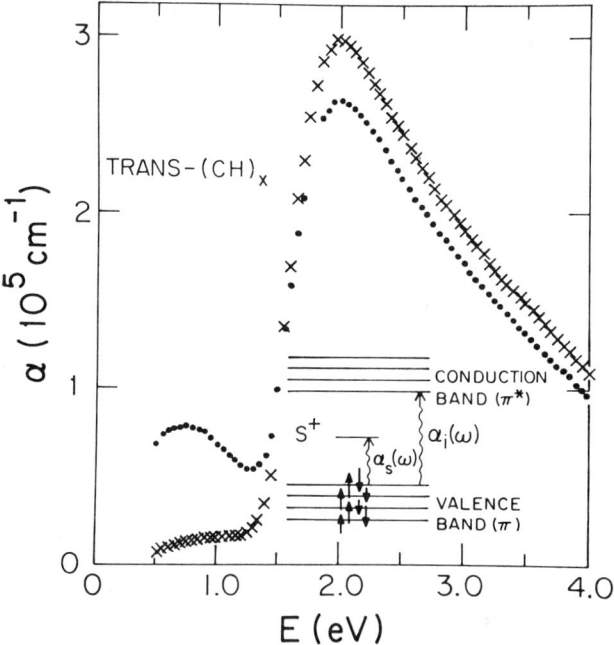

Figure 10 Absorption coefficient of trans-$(CH)_x$: xxx before doping; ... after doping to about 1%. The inset shows the interband transition (i) and the mid-gap transition (α_s) in the context of a band diagram (60).

tion is a feature of the doped *trans*-$(CH)_x$ chain and that the Coulomb interaction causes at most minor shifts in transition energy. Furthermore, in a series of opto-electrochemical experiments (37), which are ideal for in situ monitoring the doping and undoping process, the two curves similar to Figure 10 have been cycled back and forth assuring the reversibility of the doping process.

The strong absorption band with threshold at 1.4 eV and peak at 1.95 eV has been attributed to the direct interband transition in a one-dimensional (1-d) band structure, and can be viewed as arising from a transition from the 1-d peak in the density of states in the valence band to that in the conduction band. The rounding appears to shift the position of the peaks in the VB and CB densities of states by about 0.2 eV (25, 37, 60).

The characteristic features of the "mid-gap" and main absorption bands can be explained in detail if we assume that the doping proceeds through formation of positive charged solitons and that the low energy absorption band is associated with the transition from the valence band into the mid-gap level to form a neutral soliton. As the number of charged solitons increases with doping, the strength of the low energy transition grows proportionally. Further, detailed calculations (60–63) predict that the interband transition will be uniformly suppressed with the introduction of solitons, again in precise agreement with the experimental results. From comparison of the strength of the mid-gap transition and the associated reduction in the interband transition to the theoretical expressions for $\alpha(\omega)$, a value for the soliton half-width of about 7–9 (CH) units is obtained, in excellent agreement with theory.

Photoexcitations of Polyacetylene

Interest in photoexcitation studies (64–67) of $(CH)_x$ has been stimulated by the recent calculations of Su & Schrieffer (67), who considered direct injection of an e-h pair and studied the time evolution of the system. Their principal result was the conclusion that in *trans*-$(CH)_x$ a photoinjected e-h pair evolves to a soliton-antisoliton pair in a time of order of the reciprocal of an optical phonon frequency. This is the central experimental question addressed in this section: Are solitons photogenerated in polyacetylene? The proposed photogeneration of charged solitons has clear implications that can be checked through studies of photoluminescence and photoconductivity (64–66). We therefore discuss in this section the experimental results obtained with these two complementary techniques in *cis*- and *trans*-$(CH)_x$.

In the scattered light spectrum from *cis*-$(CH)_x$, there is a relatively broad luminescence structure peaking at 1.9 eV, near the interband absorption edge, together with a series of multiple order Raman lines (66, 68). Through measurements of the excitation spectrum (66), it is found that the luminescence turns on sharply for excitation energies greater than 2.05 eV, implying a Stokes' shift of about 0.15 eV. Isomerization of the same sample to *trans*-$(CH)_x$ quenches the luminescence; there is no indication of luminescence near the interband absorption edge of *trans*-$(CH)_x$ even at temperatures as low as 7 K.

The photoconductive response of the two isomers, on the other hand, behaves oppositely upon isomerization (64–66). Whereas no detectable steady state photoconductivity is observed for *cis*-$(CH)_x$, a relatively large signal appears upon in situ *cis-trans* isomerization. The upper limit on the steady state photocurrent for $h\nu \simeq 2.0$ eV photons in *cis*-$(CH)_x$ is more than three orders of magnitude smaller than that in *trans*-$(CH)_x$. Considering that upon *cis-trans* isomerization, the carrier mobility is expected to decrease as a result of increased disorder, the enhancement of the photocurrent implies a large increase in the lifetime of the photogenerated carriers. Furthermore, in *trans*-$(CH)_x$ the photoconductivity has a threshold at 1.0 eV, well below the interband absorption edge at 1.5 eV, implying the presence of states deep inside the gap (64–66). The free carrier generation efficiency rises exponentially above the threshold and changes to a slow increase above the onset of the interband transition.

In traditional semiconductors, photoconductivity and recombination luminescence are intimately related, and both may be observed after photoexcitation. Photoconductivity indicates the presence of free carriers generated by the absorbed photons. Although the subsequent recombination of these photogenerated carriers can take place either radiatively or nonradiatively, recombination luminescence is commonly observed, at least at low temperatures. The fundamental differences between such traditional data and those obtained from polyacetylene can be seen by comparison of the results obtained from $(CH)_x$ and cadmium sulfide (CdS). The luminescence and multiple order Raman scattering data (69) from CdS are similar to the results obtained from *cis*-$(CH)_x$. However, in CdS, a strong photoconductive response is observed for photon energies just above the band edge (70), whereas in *cis*-$(CH)_x$ significant photogeneration of free carriers is not observed even for photon energies 1 eV above the band edge. Isomerization to *trans*-$(CH)_x$ quenches the luminescence at all temperatures, but turns on the photoconductivity. In neither isomer is the traditional combination of effects observed.

The unusual combination of photoconductivity and luminescence results in *cis*- and *trans*-$(CH)_x$ exists in spite of the close similarity of their microscopic structure; within the next-nearest-neighbour approximation, the two isomers are *identical*. In particular, the absorption spectra of the

Table 1 A comparison of some physical properties of the two isomers of (CH_x)

	Cis-$(CH)_x$	trans-$(CH)_x$
Degenerate ground state	No	Yes
Free stable solitons	No	Yes
Highly mobile neutral spin	No	Yes
Photoconductivity	No	Yes
Band edge luminescence	Yes	No

two isomers are quite similar (25). We, therefore, do not expect any major change in the electronic interaction upon isomerization. Table 1 compares the relevant physical properties of the two isomers.

The unusual changes in the photoresponse and scattered light spectrum upon isomerization of cis-$(CH)_x$ to trans-$(CH)_x$ are the result of the change in symmetry of the polymer and not to some extrinsic effects. In trans-$(CH)_x$, the degenerate ground state leads to free soliton excitations, absence of the band edge luminescence, and photoconductivity. In cis-$(CH)_x$ the nondegenerate ground state leads to confinement of the photogenerated carriers, absence of photoconductivity, and to the observed recombination luminescence.

A schematic diagram of the photogeneration of a charged soliton-antisoliton pair in trans-$(CH)_x$ is shown in Figure 11. The incident photon (for $\hbar\omega > 2\Delta$) generates an e-h pair within the rigid lattice (Figure 11a). the system rapidly evolves to a soliton pair (Figure 11b) as shown by Su & Schrieffer (67). After a time of order 10^{-13} sec, their

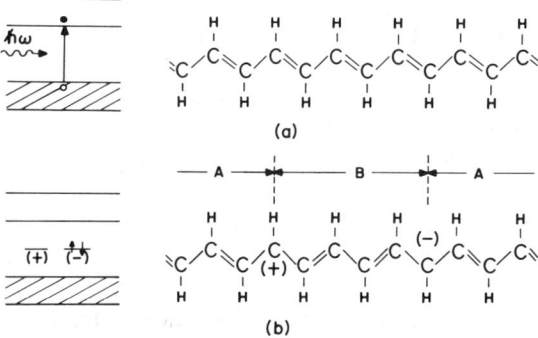

Figure 11 Band diagram and chemical structure diagram for trans-(CH). (*a*) The band diagram shows schematically the absorption of a photon and the creation of an e-h pair. (*b*) A trans-$(CH)_x$ chain containing a charged solition-antisoliton pair. Since A and B are degenerate, the solitons are free and can move apart with no cost in energy. The corresponding band diagram shows the mid-gap states associated with the two solitons; one empty (+) and the other double occupied (−).

results imply the presence of two kinks separating degenerate regions. Because of the precise degeneracy of the A and B phases, the two charged solitons are free to move in an applied electric field and contribute to the photoconductivity (64–65). The simultaneous absence of band edge luminescence in trans-$(CH)_x$ even at the lowest temperatures is consistent with the proposed photogeneration of charged soliton-antisoliton pairs. In this case, band edge luminescence cannot occur since there are no electrons and holes. The charged carrier pair consists of two kinks with associated mid-gap states that cannot give rise to band edge luminescence (66).

The ground state degeneracy in trans-$(CH)_x$ is not present in cis-$(CH)_x$, so that soliton photogeneration would not lead to photoconductivity in the cis-isomer. Since the cis-transoid configuration has a lower energy than the trans-cisoid configuration, domain walls would separate nondegenerate regions. The energy required to make a pair of kinks would be

$$E_{tot} = 2E_s + n\Delta E_0 \qquad \qquad 4.$$

where E_s is the energy for creation of a single soliton, analogous to the soliton creation energy in trans-$(CH)_x$, n is the number of CH monomers separating the two kinks, and E_0 is the energy difference between cis-transoid and trans-cisoid configurations—expected to be a small fraction of the full gap $2\Delta \sim 2.0$ eV in cis-$(CH)_x$. Thus, as they begin to form, the two solitons would be "confined," or bound into a polaron-like entity; the farther apart, the greater the energy. As a result, one expects the photogenerated pair to quickly recombine, thereby quenching the photoconductivity in cis-$(CH)_x$ (44, 64, 65).

SUMMARY AND CONCLUSIONS

The essence of the soliton theory is that a rigid band picture for trans-$(CH)_x$ is simply and fundamentally not valid. Thus, unlike traditional semiconductors, the trans-$(CH)_x$ lattice is inherently unstable to the presence of a pair of electrons (holes) and/or an electron-hole pair. The results summarized in this review indicate that solitons are the principal electronic excitations of trans-$(CH)_x$. There are further indications that solitons are responsible for the conductivity of undoped polyacetylene (36, 71, 72). For the concentration range 0.2% to 6% in the highly conducting regime, comparison of transport (23, 24), optical (37), and spin susceptibility data (52, 73) implies that the current carrying delocalized carriers are spinless. Thus, a generalization of the soliton theory to high densities may be required for an understanding of the high conductivity of the doped polymer.

That solitons are the primary electronic excitations in *trans*-(CH)$_x$ may have important consequences for the potential use of this polymer in device applications. Unlike electrons or holes, solitons by their topological constraints cannot easily cross an interface. Thus, for example, the observation of reduced solar energy conversion efficiency in Al:(CH)$_x$ Schottky devices at high light intensities (31) could possibly be caused by the photogeneration of solitons inside the depletion depth. Since subsequent injection of charge across the interface is inhibited, the characteristics of the device may change once their concentration exceeds a nominal threshold. Detailed studies of charge injection into polyacetylene at interfaces should be carried out to clarify such questions.

The broken symmetry, degenerate ground state of *trans*-(CH)$_x$ is quite special. However, there exist many analogous conjugated polymers with a nearly degenerate ground state. Examples are *cis*-polyacetylene (Figure 1) and poly(paraphenylene) (74). The latter is particularly easy to visualize: poly(paraphenylene) can be imagined in two nearly equivalent structures, benzenoid (Figure 12a) and quinoid (Figure 12b). Although neither of these limiting forms is literally correct, we may think of the ground state as essentially equivalent to Figure 12a, with the structure of Figure 12b being higher in energy. The energy difference is estimated to be $\Delta E_0 \sim 0.35$ eV per phenyl monomer. Consider then the excitation of a neutral soliton-antisoliton pair, as shown schematically in Figure 12c. The energy required is similar to Eq. 4, where E_s is now the energy for creation of a single soliton (analogous to the soliton creation in (CH)$_x$) and n is the number of phenyl monomers separating the two kinks. The two solitons would be "confined," or bound into a polaron-like entity; the farther

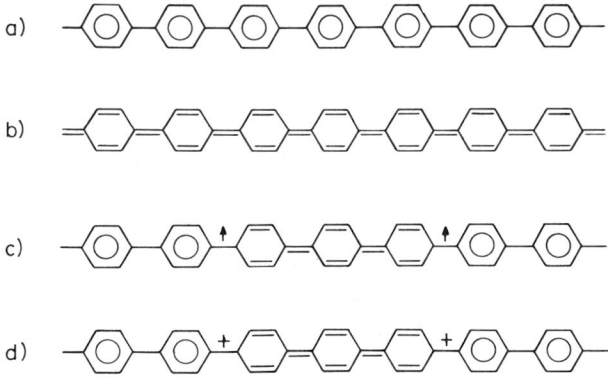

Figure 12 Poly(paraphenylene): (*a*) the lowest-energy benzenoid structure; (*b*) the higher-energy quinoid structure; (*c*) chain containing a confined pair of neutral soliton; (*d*) chain containing a confined pair of charged solitons with compensating A$^-$ ions nearby.

apart, the greater the energy. Detailed theoretical analyses of systems with near-degenerate ground states have not been carried out. However, arguing by analogy with the degenerate trans-$(CH)_x$ case, for each soliton a nonbonded localized electronic state will appear in the gap. Such a state can be readily ionized by charge transfer to give charged bound solition-antisoliton pairs of charge $2|e|$. We note that such bound pairs would act as spinless charge carriers. Because of the distinct energy difference, $\Delta E_0 > 0$, E_{tot} will be greater than the corresponding energy with $\Delta E_0 = 0$. Nevertheless, in the presence of sufficiently strong acceptors, oxidation will occur. It appears that charge transfer, with the concomitant formation of soliton-antisoliton conformational changes in the polymer, may occur with the smallest energy cost. In this context, Peo et al (75) have reported magnetic susceptibility results for doped poly(paraphenylene); they find that any Pauli contribution is small, $\chi_p < 10^{-7}$ emu/mole, consistent with these ideas.

The field of conducting polymers has been firmly established. The rapid progress in the last few years has demonstrated that the potential exists for the discovery of new concepts and phenomena as well as the development of new technology. The future of the field relies on creative synthesis of new systems. Keeping in mind the results and concepts described in this review, one may attempt to clarify the essential features required for conducting polymers. Clearly, conjugated systems with a π-electron band structure make up the most promising class of polymers. Such polymers can undergo charge transfer reactions either electrochemically or with suitable donors or acceptors to provide potential carriers. The mobility of these carriers will be determined by a variety of effects, including chain perfection, crystallinity, etc (76, 77); however, a delocalized π-system with relatively broad energy bands is certainly an advantage for effective electrical transport. Moreover, in such conjugated polymers, the π-electron charge transfer leaves the σ-bonds intact, thus ensuring the structural integrity of the doped polymer chains.

Future synthesis might attempt to move in two important directions: (a) new systems with a broken symmetry degenerate ground state and (b) poly(heterocycles) (74, 79). The emphasis on broken symmetry systems is clear in the context of the important concepts developed in the study of trans-$(CH)_x$. Other systems with multiply degenerate ground states can be envisioned. Theoretical considerations of such systems have already suggested that important new phenomena might be observed, e.g. fractionally charged solitons (78). The suggestion of synthesis of poly(heterocycles) is an attempt to go in another direction. Although the broad bandwidth and low density of states of $(CH)_x$ offer many advantages, traditionally, collective phenomena (e.g. magnetism, superconductivity etc) in condensed matter systems are enhanced in *narrow* band materials with high state densities at the Fermi energy. Indeed, the longstanding interest in transition metals and the more recent exciting advances in charge transfer molecular crystals have resulted at least in part from the relatively narrow band width and high density of states. Poly-(heterocycles) might, therefore, offer an opportunity for achieving such phenomena in polymers.

ACKNOWLEDGMENT

We thank all of our colleagues whose work contributed to the body of knowledge that we have reviewed briefly in this paper. Because of length constraints, our focus was on particular aspects of the conducting polymer field. We, therefore, did not include or refer to many important contributions; for these omissions we apologize. The work at the University of Pennsylvania that contributed to this review was supported by the Office of Naval Research, the Defense Advanced Research Project Agency, the National Science Foundation, the National Science Foundation-Material Research Laboratory program, the Army Research Office, and the Department of Energy (Advanced Energy Projects). Without this support, clearly, the significant progress described in this review could not have been achieved.

Literature Cited

1. Peierls, R. E. 1955. *Quantum Theory of Solids*, p. 108. London: Oxford Univ. Press
2. Devreese, J. T., Evrard, R. P., van Doren, V. E., eds. 1978. *Highly Conducting One-Dimensional Solids*. NY: Plenum. (Contains several reviews on this topic)
3a. Su, W. P., Schrieffer, J. R., Heeger, A. J. 1979. *Phys. Rev. Lett.* 42:1698
3b. Su, W. P., Schrieffer, J. R., Heeger, A. J. 1981. *Phys. Rev. B* 22:2099
4. Rice, M. J. 1979. *Phys. Lett. A* 71:152
5. Takayama, H., Lin-Liu, Y. R., Maki, K. 1980. *Phys. Rev. B* 21:2388
6a. Brazovskii, S. 1978. *JETP Lett.* 28:656
6b. Brazovskii, S. 1980. *JETP* 78:677
7a. Heeger, A. J., MacDiarmid, A. G. 1981. *Proc. Int. Conf. Low Dimensional Synthetic Metals, Helsingor, Denmark. Chem. Scripta* 17:115; 1981. *Proc. Int. Conf. Low Dimensional Conductors, Boulder, Colo., Mol. Cryst. Liquid Cryst.* 77:1
7b. Alcacer, L., ed. 1980. *The Physics and Chemistry of Low Dimensional Solids*, pp. 353, 393. Dordrecht: Reidel
8. Etemad, S., Heeger, A. J., Lauchlan, L., Chung, T.-C., MacDiarmid, A. G. 1981. *Proc. Int. Conf. Low Dimensional Conductors, Boulder, Colo., Mol. Cryst. Liquid Cryst.* 77:43
9. Ito, T., Shirakawa, H., Ikeda, S. 1974. *J. Polym. Sci. Polym. Chem. Ed.* 12:11
10. Ito, T., Shirakawa, H., Ikeda, S. 1975. *J. Polym. Sci.* 13:1943
11. Shirakawa, H., Ito, T., Ikeda, S. 1978. *Makromol. Chem.* 179:1565
12. Shirakawa, H., Ikeda, S. 1979/1980. *Synth. Metals* 1:175
13. Karasz, F. E., Chien, J. C. W., Galkiewicz, R., Wnek, G. E., Heeger, A. J., MacDiarmid, A. G. 1979. *Nature* 282:286
14. Park, Y. W., Druy, M. A., Chiang, C. K., MacDiarmid, A. G., Heeger, A. J., Shirakawa, H., Ikeda, S. 1979. *J. Polym. Sci. Polym. Lett. Ed.* 17:628
15. Druy, M. A., Tsang, C.-H., Brown, N., Heeger, A. J., MacDiarmid, A. G. 1980. *J. Polym. Sci. Polym. Phys. Ed.* 18:429
16. Ito, T., Shirakawa, H., Ikeda, S. 1976. *Kobunshi Ronbunshu* 33:339; 1976. *Chem. Abstr.* 85:78449c
17. Akaishi, T., Miyasaka, K., Ishikawa, K., Shirakawa, H., Ikeda, S. 1980. *J. Polym. Sci. Polym. Phys. Ed.* 18:745

18. Fincher, C., Chen, C. E., Heeger, A. J., MacDiarmid, A. G., Hastings, J. B. 1982. *Phys. Rev. Lett.* 48:100
19a. Shirakawa, H., Louis, E. J., Macdiarmid, A. G., Chiang, C. K., Heeger, A. J. 1977. *Chem Commun.* 578
19b. Chiang, C. K., Druy, M. A., Gau, S. C., Heeger, A. J., Shirakawa, H., Louis, E. J., MacDiarmid, A. G., Park, Y. W. 1978. *J. Am. Chem. Soc.* 100:1013
20. Chiang, C. K., Fincher, C. R., Park, Y. W., Heeger, A. J., Shirakawa, H., Louis, E. J., Gau, S. C., MacDiarmid, A. G. 1977. *Phys. Rev. Lett.* 39:1098
21. MacDiarmid, A. G., Heeger, A. J. 1979. *Synth. Metals* 1:101
22. Nigrey, P. J., MacDiarmid, A. G., Heeger, A. J. 1979. *J. Chem. Soc. Chem. Commun.* p. 594
23a. Park, Y. W., Heeger, A. J., Druy, M. A., MacDiarmid, A. G. 1980. *J. Chem. Phys.* 73:946
23b. Park, Y. W., Denenstein, A., Chiang, C. K., Heeger, A. J., MacDiarmid, A. G. 1979. *Solid State Commun.* 29:747
24. Moses, D., Denenstein, A., Chen, J., McAndrew, P., Woerner, T., Heeger, A. J., MacDiarmid, A. G., Park, Y. W. 1982. *Phys. Rev. B.* In press
25. Fincher, C. R., Ozaki, M., Tanaka, M., Peebles, D., Lauchlan, L., Heeger, A. J., MacDiarmid, A. G. 1979. *Phys. Rev. B* 20:1589
26a. Nigrey, P. J., McInnes, D., Nairns, D. P., MacDiarmid, A. G., Heeger, A. J. 1981. *J. Electrochem. Soc.* 128:1651
26b. McInnes, D., Druy, M. A., Nigrey, P. J., Nairns, D. P., MacDiarmid, A. G., Heeger, A. J. 1981. *J. Chem. Soc. Chem. Commun.* p. 317
27. Nigrey, P. J., MacDiarmid, A. G., Heeger, A. J. 1981. *Proc. Int. Conf. Low Dimensional Conductors, Boulder, Colo., Mol. Cryst. Liquid Cryst.* 77
28a. Ozaki, M., Peebles, D., Weinberger, B. R., Heeger, A. J., MacDiarmid, A. G. 1980. *J. Appl. Phys.* 51:4252
28b. Ozaki, M., Peebles, D., Weinberger, B. R., Chiang, C. K., Gau, S. C., Heeger, A. J., MacDiarmid, A. G. 1979. *Appl. Phys. Lett.* 35:83
29a. Tani, T., Grant, P. M., Gill, W. D., Street, G. B., Clarke, T. C. 1980. *Solid State Commun.* 33:499
29b. Tani, T., Grant, P. M., Gill, W. D., Street, G. B., Clarke, T. C. 1980. *Synth. Metals* 1:301
29c. Grant, P. M., Tani, T., Gill, W. D., Krounbi, M., Clarke, T. C. 1981. *J. Appl. Phys.* 52:869
30a. Weinberger, B. R., Gau, S. C., Kiss, Z. 1981. *Appl. Phys. Lett.* 38:555
30b. Weinberger, B. R., Akhtar, M., Gau, S. C. 1982. *Synth. Metals* 4:187
31. Chen, S. N., Heeger, A. J., Kiss, Z., MacDiarmid, A. G., Gau, S. C., Peebles, D. L. 1980. *Appl. Phys. Lett.* 36:96
32. Shirakawa, H., Ito, T., Ikeda, S. 1973. *Polym. J.* 4:460
33a. Chiang, C. K., Park, Y. W., Heeger, A. J., Shirakawa, H., Louis, E. J., MacDiarmid, A. G. 1978. *J. Chem. Phys.* 69:5098
33b. Hsu, S. L., Signorelli, A. J., Pez, G. P., Baughman, R. H. 1978. *J. Chem. Phys.* 69:106
33c. Street, G. B., Clarke, T. C. 1980. *Adv. Chem. Ser.* 186:177
34. Chiang, C. K., Gau, S. C., Fincher, C. R., Park, Y. W., MacDiarmid, A. G., Heeger, A. J. 1978. *Appl. Phys. Lett.* 33:188
35a. Kwak, J. F., Clarke, T. C., Greene, R. L., Street, G. B. 1979. *Solid State Commun.* 31:355
35b. Seeger, K., Gill, W. D., Clarke, T. C., Street, G. B. 1978. *Solid State Commun.* 28:873
36. Moses, D., Chen, J., Denenstein, A., Kaveh, M., Chung, T.-C., Heeger, A. J., MacDiarmid, A. G. 1981. *Solid State Commun.* 40:1007
37. Feldblum, A., Kaufman, J., Etemad, S., Heeger, A. J., Chung, T.-C., MacDiarmid, A. G. 1982. *Phys. Rev. B.* In press
38. Tanaka, M., Watanabe, A., Tanaka, J. 1980. *Bull. Chem. Soc. Jpn.* 53:645; Tanaka, M., Watanabe, A., Tanaka, J. 1980. *Bull. Chem. Soc. Jpn.* 53:3430
39. Clarke, T. C., Street, G. B. 1980. *Synth. Metals* 1:119
40. Chung, T. C., Feldblum, A., Heeger, A. J., MacDiarmid, A. G. 1981. *J. Chem. Phys.* 74:5504
41. Francois, B., Bernard, M., Andre, J. J. 1981. *J. Chem. Phys.* 75:4142
42. Schlesinger, Y., and Kaufer, J. (unpublished results)
43. Waldrop, J. R., Cohen, M. J., Heeger, A. J., MacDiarmid, A. G. 1981. *Appl. Phys. Lett.* 38:53
44. Brazovskii, S. A., Kirova, N. N. 1981. *JETP Lett.* 33:6
45. Mele, E. J., Rice, M. J. 1981. *Phys. Rev. B* 23:5397
46. Baughman, R. H., Hsu, S. L., Anderson, L. R., Pez, G. P., Signorelli, A. J. 1979. *Molecular Metals*, p. 189. New York: Plenum

47a. Moses, D., Feldblum, A., Denenstein, A., Chung, T.-C., Heeger, A. J., MacDiarmid, A. G. 1981. *Proc. Int. Conf. Low Dimensional Conductors, Boulder, Colo., Mol. Cryst. Liquid Cryst.* 77;
47b. Moses, D., Feldblum, A., Denenstein, A., Ehrenfreund, E., Chung, T.-C., Heeger, A. J., MacDiarmid, A. G. 1982. *Phys. Rev. B.* In press
48. Mele, E. J., Rice, M. J. 1980. *Solid State Commun.* 34:339
49. Etemad, S., Pron, A., Heeger, A. J., MacDiarmid, A. G., Mele, E. J., Rice, M. J. 1981. *Phys. Rev. B* 23:5137
50a. Goldberg, I. B., Crowe, H. R., Newman, P. R., Heeger, A. J., MacDiarmid, A. G. 1979. *J. Chem. Phys.* 70:1132
50b. Weinberger, B. R., Ehrenfreund, E., Pron, A., Heeger, A. J., MacDiarmid, A. G. 1980. *J. Chem. Phys.* 72:4749
51. Pople, J. A., Walmsley, J. H. 1962. *Mol. Phys.* 5:15
52. Ikehata, S., Kaufer, J., Woerner, T., Pron, A., Druy, M. A., Sivak, A., Heeger, A. J., MacDiarmid, A. G. 1980. *Phys. Rev. Lett.* 45:1123
53. Tomkiewicz, Y., Schultz, T. D., Brom, H. B., Taranko, A. R., Clarke, T. C., Street, G. B. 1981. *Phys. Rev. B* 24:4348
54. Nechtschein, M., Devreux, F., Greene, R. L., Clarke, T. C., Street, G. B. 1980. *Phys. Rev. Lett.* 44:356
55. Bishop, A. R., Schenider, T., eds. 1978. *Solitons and Condensed Matter Physics, Springer Ser. Solid State Sci.* Berlin: Springer-Verlag
56. Su, W. P. 1981. *Proc. Int. Conf. Low Dimensional Conductors, Boulder, Colo., Mol. Cryst. Liquid Cryst.* 77:265
57. Fincher, C. R., Ozaki, M., Heeger, A. J., MacDiarmid, A. G. 1979. *Phys. Rev. B* 19:4140
58. Mele, E. J., Rice, M. J. 1980. *Phys. Rev. Lett.* 45:926
59. Rabolt, J. F., Clarke, T. C., Street, G. B. 1979. *J. Chem. Phys.* 71:4614
60. Suzuki, N., Ozaki, M., Etemad, S., Heeger, A. J., MacDiarmid, A. G. 1980. *Phys. Rev. Lett.* 45:1209; erratum, *Phys. Rev. Lett.* 45:1483
61. Gammel, J., Tinka, Krumhansl, J. A. 1980. Preprint
62. Kivelson, S., Lee, T.-K., Lin-Liu, Y. R., Peschel, I., Yu, L. 1982. *Phys. Rev. B.* In press
63. Horovitz, B. 1981. *Solid State Commun.* 41:593
64. Etemad, S., Ozaki, M., Heeger, A. J., MacDiarmid, A. G. 1981. *Proc. Int. Conf. Low Dimensional Synthetic Metals, Helsingor, Denmark*, 1980. *Chem. Scripta* 17:159
65. Etemad, S., Mitani, M., Ozaki, M., Chung, T.-C., Heeger, A. J., MacDiarmid, A. G. 1981. *Solid State Commun.* 40:75
66. Lauchlan, L., Etemad, S., Chung, T.-C., Heeger, A. J., MacDiarmid, A. G. 1981. *Phys. Rev. B* 24:1
67. Su, W. P., Schrieffer, J. R. 1980. *Proc. Natl. Acad. Sci. USA* 77:5626
68. Lichtman, L. S., Sarhangi, A., Fitchen, D. C. 1980. *Solid State Commun.* 36:869
69a. Leite, R. C. C., Scott, J. C., Damen, T. C. 1969. *Phys. Rev. Lett.* 22:780
69b. Klein, M. V., Porto, S. P. S. 1969. *Phys. Rev. Lett.* 22:782
70. Bube, R. H. 1960. *Photoconductivity of Solids*, pp. 230, 391. New York: Wiley
71. Kivelson, S. 1981. *Phys. Rev. Lett.* 46:1344
72. Epstein, A. J., Rommelmann, H., Abkowitz, M., Gibson, H. W. 1981. *Phys. Rev. Lett.* 47:1549
73. Epstein, A. J., Rommelmann, H., Druy, M. A., Heeger, A. J., MacDiarmid, A. G. 1981. *Solid State Commun.* 38:683
74. Baughman, R. H., Bredas, J. L., Chance, R. R., Eckhardt, H., Elsenbaumer, R. L., Ivory, D. M., Miller, G. G. Preziosi, P. F., Shacklette, L. W. 1981. *Conductive Polymers*, ed. R. B. Seymour, p. 137. New York: Plenum
75. Peo, M., Roth, S., Dransfeld, K., Tieke, B., Hocker, J., Gross, H., Grupp, A., Sixl, H. 1980. *Solid State Commun.* 35:119
76. Haberkorn, H., Naarmann, H., Penzien, K., Schlag, J. 1982. *Synth. Metals.* In press
77a. Diets, W., Cukor, P., Rubnes, M., F., Jonson, H. 1981. *J. Electron. Mat.* 10:683
77b. Diets, W., Cukor, P., Rubnes, M. F. 1981. See Ref. 74
78. Su, W. P., Schrieffer, J. R. 1981. *Phys. Rev. Lett.* 46:738
79. Kanazawa, K. K., Diaz, A. F., Gill, W. D., Grant, P. M., Street, G. B., Gardini, G. P., Kwak, J. R. 1980. *Synth. Metals* 1:329

TIME-RESOLVED RESONANCE RAMAN STUDIES OF HEMOGLOBIN

J. M. Friedman, D. L. Rousseau, and M. R. Ondrias

Bell Laboratories, Murray Hill, New Jersey 07974

INTRODUCTION

The binding of small molecular ligands to the hemes in hemoglobin (Hb) is a highly localized perturbation. Nonetheless, this localized binding initiates a sequence of propagating structural events that culminates in a change in quaternary structure, proceeding from the low affinity deoxy T state to the high affinity liganded R state (1). Although these two equilibrium species have been well characterized, both the mechanism by which ligation triggers structural destabilization and the dynamic pathways relating ligation to change in quaternary structure remain as yet undetermined. Moreover, these questions are interdependent, i.e. the dynamics are influenced by the ligation-sensitive structures surrounding the heme. Understandably, it is of fundamental interest to probe the ligation-dependent interactions that affect the stability of the heme environment (hemepocket) in order to establish the relationship between the structural variations of this environment and kinetic phenomena.

In the equilibrium structures of Hb the functionally relevant energies associated with the coupling of ligation and quaternary structure may be delocalized. However, in metastable species, generated immediately after ligation or deligation, these energies must be manifested, at least transiently, at the interface between the binding site, i.e. the iron porphyrin, and the surrounding protein. Hence, appropriate time resolved studies of these metastable species would enhance the likelihood of detecting these hitherto unidentified, physiologically important heme-protein interactions. Photolysis of liganded Hb's has been used extensively to prepare nonequilibrium populations of deoxy Hb. In turn, transient absorption spectroscopy has yielded extensive data on the kinetics of these nonequilibrium species. Such studies (2–5) have shown that photodissociation and the subsequent recovery to the ground electronic state of the deoxyheme occur on a picosecond time scale. However, transient absorption studies (5, 6, 7) also reveal differences between the spectrum of the stable T state deoxy Hb and that of the photolyzed species. In all likelihood, differences in spectra observed on these time scales (psec) originate from environmental differences about the porphyrin chromophore. Thus, these spectral properties offer a measure of the above mentioned heme-protein interactions. Absorption spectroscopy, unfortunately, cannot be used to determine the structural origins of these interactions because of the numerous contributions to a porphyrin absorption spectrum. In contrast, resonance Raman spectroscopy provides a more sensitive probe of localized structure.

The frequency of the resonance Raman bands of deoxy and liganded hemes have been correlated with numerous structural parameters [for a recent review see (8)]. Thus, by utilizing transient Raman spectroscopy, it is possible to follow the dynamics associated with specific structural features (9), as well as to compare the influence of transient and stable protein environments upon the electronic and nuclear structure of the porphyrin. The extreme sensitivity of several Raman spectral frequencies to the state of ligation also permits Raman spectroscopy to be used to monitor ligand recombination subsequent to photolysis. At room temperature the photodissociated ligands can undergo an ultrafast ($\leqslant 100$ nsec) germinate recombination (5, 9–12), the yield of which is directly related to the properties of the hemepocket about the porphyrin. In the following account we review the recent time resolved resonance Raman (TRRR) studies of Hb.

PICOSECOND STUDIES

Most of the large scale rearrangements of protein structure that occur subsequent to photolysis are expected to and are observed (6, 7, 9) to occur on time scales that are considerably longer than picoseconds. There are, however, tertiary structural changes that are expected to occur on the picosecond time scale and thus kinetic and structural questions may be addressed by Raman investigations in this time regime. To date the studies that have been reported focus upon the electronic and nuclear rearrangements of the heme that occur subsequent to photolysis of HbCO and HbO$_2$ on the 30–50 psec time scale. Photolysis transforms the heme from a ligated low spin Fe^{+2} species to a five coordinate high spin Fe^{+2} species. The TRRR studies (13–16) confirm the findings based upon transient absorption results (2–5) that within picoseconds subsequent to photolysis of ligated Hb, the heme spectroscopically resembles a five coordinate high spin heme. Although the picosecond spectra closely resemble that of equilibrium deoxy Hb, small differences are evident in several of the Raman peaks. Some of these differences persist into the nanosecond regime. In contrast to HbCO, the photolyzed HbO_2 species at 30 psec manifests (17) a much greater deviation from the deoxy spectrum. Within 50 psec, however, the photolyzed HbO_2 spectrum looks very much like that associated with photolyzed HbCO (15).

In the picosecond TRRR studies, the only Raman lines that have been monitored are those that are sensitive to the size of the central core of the porphyrin macrocycle. Coppey et al (16) studies the photolysis of HbCO and HbO_2 with 30 psec pulses. Within their 4 cm^{-1} precision, they found that in 30 psec the core-size sensitive depolarized band at 1554 cm^{-1} was at the same frequency as in deoxy Hb. They obtained the same results for HbCO and HbO_2. In a study of 30 psec photolysis of HbCO with better sensitivity, Terner et al (13, 14) reported a transient spectrum with depolarized modes at 1603 and 1542 cm^{-1} and an anomolously polarized mode at 1552 cm^{-1}. Similar results were obtained by Nagumo et al (15) for 50 psec photolysis of HbO_2, namely frequencies of 1602, 1543, and 1548–1554 cm^{-1}, respectively. These compare to the corresponding deoxy values of 1607 (ν_{10}), 1549 (ν_{11}), and 1558 (ν_{19}) cm^{-1}.

The observation that those Raman bands sensitive to the porphyrinato core size are shifted by a few cm^{-1} to lower frequency in the transient species relative to deoxy Hb has been cited (13–15) as evidence that the porphyrinato core in the transient is expanded relative to its size in deoxy Hb. From these data it was inferred that the iron atom is nearly in plane in these photolyzed transients (13, 14). For reasons discussed in subsequent sections, this inference has now been questioned, and it has been proposed that for these transient species the iron atom has relaxed to an out of plane position and that the changes in the Raman spectrum originate from other perturbations.

Recently, Terner et al (17) reported the transient Raman spectrum of HbO_2 photolyzed with high intensity 30 psec pulses. As may be seen in Figure 1, they found frequencies substantially lower than those reported in the earlier studies in which longer and lower intensity laser pulses were generally used. They resolved their spectra into lines at 1590(ν_{10}), 1538(ν_{11}), and 1550(ν_{19}) cm^{-1}. These larger shifts from the deoxy values are too great to be ascribed to core expansion and, since ν_{10} and ν_{11} are sensitive to the porphyrin π-electron density, they were interpreted as arising from an electronically excited high-spin heme. The highly perturbed spectrum associated with the photolyzed HbO_2 species at 30 psec

(17) may correspond to the spectral intermediate observed in a picosecond absorption study (5). In both cases the features of the spectra were ascribed to an excited electronic state associated with a bottleneck in the relaxation of the photodissociated state in HbO_2.

NANOSECOND STUDIES

The picosecond studies reveal that for both HbO_2 ($t > 50$ psec) and HbCO, the photolyzed species have Raman spectra that are slightly shifted in frequency relative to the equilibrium deoxy species. Nanosecond studies (13, 14, 18, 19) reveal that these same differences in the core size marker bands persist for at least tens of nanoseconds. The picosecond studies have been limited to resonant excitation sources that selectively enhance the core size marker bands; however, the nanosecond

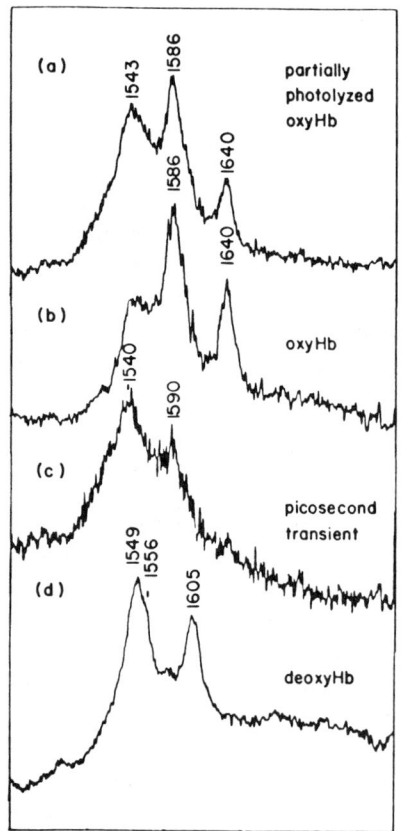

Figure 1 ~ 30 psec frequency-doubled (5306 Å) YAG laser excited-Raman spectra of (*a*) partially photolyzed HbO_2, obtained with tight focusing of the laser (~100 μm spot); (*b*) unphotolyzed HbO_2, obtained with diffuse focusing; (*c*) HbO_2 photoproduct, obtained by subtracting spectrum (*b*) from (*a*), with a scale factor adjusted to blank out the 1640 cm^{-1} band; (*d*) deoxyHb. Reprinted with permission from Ref. (17).

sources cover a much greater excitation regime. Thus nanosecond experiments utilizing Soret band resonances have allowed for equilibrium versus transient comparisons that involve the electron density marker line (ν_4) and the iron-proximal histidine (Fe–His) stretching mode.

Electron Density Marker Line

The electron density marker line (ν_4), which occurs at ~1355 cm^{-1} for deoxyhemes and in the 1370s for liganded ferrous hemes, has been shown to display a sensitivity in its frequency (for deoxyheme) to variations in the π^* electron distribution originating from perturbations occurring either through the iron (20) or through the periphery of the porphyrin (21). Hence the frequency of this band can be used as an antenna that responds to perturbations altering the π^* distribution in the ring. Recently (22) TRRR has been used to probe the influence of ligation and quaternary structure on the binding as reflected in changes in the frequency of ν_4. Because changes in the state of ligation or of oxidation produce large shifts in the Raman spectrum of the heme, it is desirable to compare hemes in identical states of oxidation and ligation in order to isolate the effects due to the protein environment. These comparisons are made by generating and comparing the resonance Raman spectrum of stable deoxy Hb's and metastable deliganded Hb's produced within 10 nsec after photodissociating a particular liganded species. Since the electronic state relaxation subsequent to photodissociation occurs on a picosecond (2–5) time scale and the protein response to deligation does not begin until well after 10 nsec (6, 7, 9), the protein structure about the two electronically relaxed deliganded (deoxy) hemes are qualitatively dissimilar; in one case the structure of the surrounding protein is at equilibrium, whereas in the other, the structure of the surrounding protein is still that characteristic of a liganded heme and is thus metastable. Differences in the two spectra can reflect both transient and static differences in the environments of the deliganded hemes.

The static differences in environment that appear under steady state conditions have been determined from Raman difference spectroscopy studies. It has been demonstrated (23–25) that in equilibrium forms of deoxy Hb, the frequency of ν_4 is sensitive to quaternary structure. Quaternary structure dependent differences occur in chemically modified (des-Arg) human hemoglobins, mutant hemoglobins (Kempsey) and in a hemoglobin of a different species (carp). The magnitudes of the differences are sensitive to allosteric effectors such as IHP and in the des-Arg Hb system the frequency differences display a dependence upon the presence of inorganic phosphates as well. Differences have also been observed between human deoxy hemoglobin and its isolated chains or sperm whale myoglobin. Data obtained using Raman difference spectroscopy show that a general trend exists in comparisons between the spectra of high affinity and low affinity globins in the 1200–1700 cm^{-1} region. Among the various human hemoglobins, all of these π^* sensitive lines respond in a concerted manner, shifting to lower frequency upon conversion to the high affinity forms. The ν_4 mode shifts to lower frequency by as much as 1.5 cm^{-1} in going from a deoxy T to a deoxy R structure.

Before considering the transient forms of Hb, we first consider the comparison between Hb and myoglobin (Mb). The steady state resonance Raman spectra of the liganded and deoxy forms of Hb and Mb reveal that the response of the porphyrin π system to ligation is very nearly identical for these two proteins. Consequently, for these two proteins, a comparison of the frequency shifts of the ν_4 vibrational mode between the deoxy form and the metastable species (≤ 10 nsec after photolysis) should provide a direct measure of the responsiveness of the respective hemepockets to ligation. Figure 2 depicts the results of such a spectral comparison. The difference between deoxy Hb and photolyzed HbCO is displayed in the upper portion of the figure. Also shown is the computer subtracted difference between the two bands. Both a nonlinear least squares analysis and a computation based on the shape of the difference curve (26) reveal that ν_4 in the metastable species is 2.5 cm^{-1} lower in frequency than that of the stable deoxy form. The lower portion of the figure reveals that there is a much smaller shift between the corresponding spectra in Mb. The nonlinear least squares analysis indicates that there is a barely detectable ($\approx .25$ cm^{-1}) shift for Mb that is in the same direction as for Hb. To determine whether the reduced response in Mb is related to its being monomeric and/or noncooperative, a similar set of spectra were generated from the isolated subunits of Hb. Both the isolated β^{SH} chains (tetrameric and noncooperative) and α^{SH} chains (monomeric and noncooperative) manifest a 1.6 cm^{-1} shift to lower frequency in a comparison of the deoxy form and the metastable species. Based on the responsiveness of the isolated chains and of the intact Hb, it is concluded that compared to Mb the protein structure about the heme is inherently more responsive to ligation-induced changes at the porphyrin. The question remains as to what role quaternary structure plays in influencing this response of the protein to ligation.

The difference in frequency observed between deoxy Hb (T state) and the photolyzed R state Hb at 10 nsec exceeds the steady state differences described above, indicating that ligation induces additional changes. A systematic study (22) of the effect of ligation for each quaternary structure was carried out using mutant (Hb$_{KANSAS}$) forms of Hb as well as nitrosyl Hb. Both COHb$_{KANSAS}$ and NOHb in the presence of IHP at low pH can be switched to the T structure. It was observed that for photolyzed $COHb_{KANSAS}(T)$ and $NOHbA(T)$, ν_4 is a few tenths (>0.6) of a cm^{-1} higher in frequency than for photolyzed HbCO(R). Other R state carboxy Hb's derived from $Hb_{KEMPSEY}$ (pH 6.5 and 9) and Hb_{KANSAS} (pH 9) yielded the same frequency as photolyzed HbCO (pH 6.5 and 9). The addition of IHP to HbCO shifted ν_4 a few tenths of a cm^{-1} to higher frequency.

A pattern associated with the shifts of ν_4 with respect to ligation and quaternary structure can be ascertained from the summary of the behavior of this mode shown in Figure 3. The vertical axis of Figure 3 represents the frequency shift of ν_4 with respect to deoxy Hb(T), which is taken as the zero point. Descending the vertical axis corresponds to a

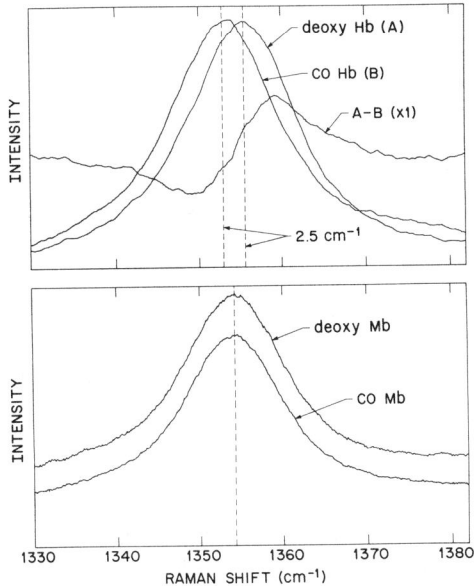

Figure 2 A comparison between Hb (human adult) and Mb (whale) with respect to the shifting of ν_4 in going from the stable deoxy form to the metastable species occurring within 10 nsec of the photolysis of the corresponding carboxy form. The 10 nsec long excitation pulses, from the 4200 Å (bis-MSB) output of a molectron UV 1000 nitrogen laser pumped dye laser operating at 10 Hz, were focused (500 mm f.l. lens) into a cuvette in which the 2° C solution was recirculated. No differences were observed between these spectra and those derived from noncirculating samples maintained at the same temperature. For these spectra and the ones shown in the subsequent figures, the 90° scattered light was dispersed by a 1-meter F/8 J. Y. Ramanor HG-23 spectrometer fitted with an RCA C31034-02HQ photomultiplier tube. The output of the photomultiplier tube was gated (1 nsec) and averaged on a PAR #163 box car integrator. The resulting signal was stored and processed on a Nicolet 1174 signal averager that allowed for repetitive scans of a given spectrum.

lowering of the frequency of ν_4. The left side of the figure contains the T state Hb's whereas the right side contains the R state Hb's along with some monomeric species. A comparison of the right and left sides of Figure 3 reveals the differing response of Hb to ligation as a function of quaternary structure. In every instance examined, [including $Hb_{KEMPSEY}$ at pH 6.5 and 9 (22) and $Hb_{KANSAS}(R)$ which are not shown in the figure], the metastable deoxy heme associated with the photolyzed R quaternary structure, is at lower frequency than those of the corresponding T structure.

The above results demonstrate that the R-T differences in the photolyzed Hb mimick those in the steady state species insofar as the high

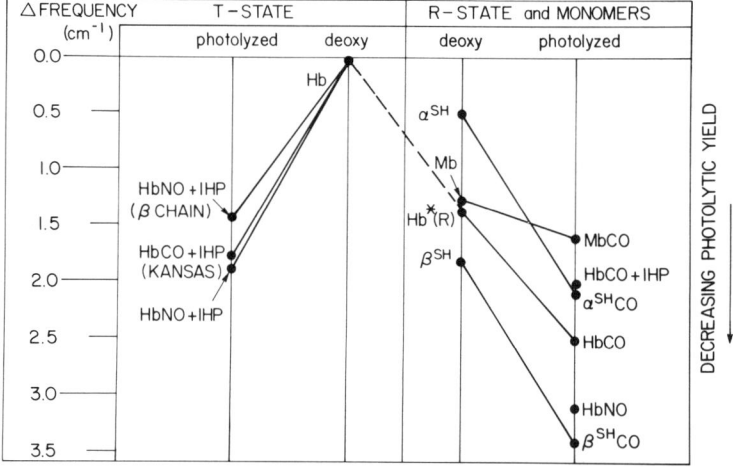

Figure 3 Frequency changes in hemoglobins on photolysis. The points along the vertical lines represent the relative frequencies of ν_4 for deligated (deoxy) ferrous hemes in a variety of stable and metastable protein configurations. The frequencies are referenced to that of deoxy Hb(T) (1357 cm^{-1}) which is taken as the zero point. Points below the zero indicate lower frequencies (e.g. $2.0 = 1357 - 2.0$ cm^{-1}). The points designated liganded proteins, such as HbCO and HbNO, refer to the frequency associated with the deoxy heme in the metastable protein occurring within 10 nsec of photolysis of the corresponding ligated Hb. Protein configurations that decrease the frequency of ν_4 of the deoxy heme are associated with porphyrin macrocycles that have a higher π^* electron density. The Mb is from the skeletal muscle of whale. Except for the Hb_{KANSAS} ($\gg 100~\mu M$), the heme concentrations for the above solutions were $\approx 50~\mu M$. The frequency used for deoxy Hb in the R state (Hb(R)) is derived from two independent experiments. One measurement, derived from a Raman difference study, (23), is associated with deoxy (NES-des-ArgHb(R), a chemically modified Hb which remains stabilized in the R structure when deoxygenated. This same frequency is observed (9) for the transient deoxy Hb species occurring a few μsec after photodissociating HbCO with a 10 nsec pulse. At this temporal point the hemepocket has undergone a deligation related relaxation ($T \approx 1~\mu$sec), leaving the system in a deoxy R state configuration which subsequently decays ($T \gg 1~\mu$sec) to the stable deoxy T structure.

affinity forms have lower frequencies for ν_4. It is also clear from these studies that for both the R and the T structures ligation induces additional changes in the protein structure that cause ν_4 in the photolyzed protein to be at lower frequency. These results indicate the following ordering with respect to the decreasing frequency of ν_4: deoxy T > deoxy R ≈ photolyzed T > photolyzed R.

None of the above experiments distinguish between the α and β subunits. Recent continuous wave (cw) studies (27) on Fe–Co hybrids have been used to determine the effect of quaternary structure upon ν_4, for the individual subunits. Owing to the optical absorption spectral differences between the iron heme and the cobalt substituted heme, the iron heme can be selectively examined by a suitable choice of laser frequency. Taking advantage of this it was found that for the deoxy T state, ν_4 for the α subunit is about 0.9 cm^{-1} higher in frequency than the composite peak, while the β subunit is 1.4 cm^{-1} lower. Upon switching to the R state, ν_4 of the α subunit shifts 2.3 cm^{-1} to lower frequency, whereas the β heme shifts 0.5 cm^{-1} to lower frequency. Transient Raman studies on Fe–Co hybrids (J. M. Friedman, M. Ikeda-Saito, and T. Yonetani, to be published) reveal that for the photolyzed R state at high pH in the absence of phosphates there is no detectable difference in the frequency of ν_4 between α and β subunits. Both subunits give rise to values for ν_4 that are 2.5 cm^{-1} lower than for deoxy HbA. Addition of IHP to the ligated hybrids at low pH produces dramatic chain specific shifts. These are shown in Figure 4. The addition of IHP shifts ν_4 of the photolyzed α(Fe)β(Co) hybrid to values that are characteristic of the T structure, whereas for the α(Co)β(Fe) hybrid the shift to higher frequency is not as great.

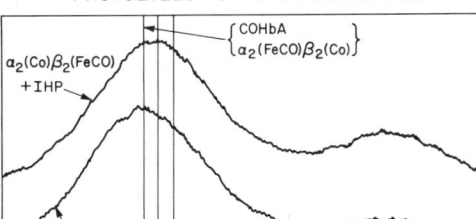

Figure 4 Band I (ν_4) for the carboxy Fe–Co hybrids in the presence and absence of IHP.

Iron-Histidine Stretching Mode

Although ν_4 is very strongly enhanced for Soret band excitation, the size of the shifts and the uncertain origin of the shifts make detailed analysis difficult. In contrast, the low frequency region of the Soret enhanced Raman spectra has features that are more easily studied.

Systematic changes are evident (25, 28, 29) in the low frequency (50–500 cm^{-1}) region of the cw spectrum for a wide series of hemoglobins. Although changes have been detected in several modes, by far the largest quaternary structure dependent differences occur in the mode at 216–225 cm^{-1}. This mode is of particular interest because it has been assigned as the iron-histidine stretching frequency from isotopic substitution studies of the iron in model compounds, hemoglobin, and myoglobin and deuteration studies of the coordinated imidazole (30). The changes detected in this mode in the chemically modified, mutant, and carp hemoglobins indicate a quaternary structure induced change in the iron-histidine bond in which the frequency of this mode is lower in the T structure than in the R structure. For deoxy T structures, the Fe–His stretching mode is centered at ~ 216 cm^{-1} and is very asymmetric, whereas for R state species it is shifted to ≈ 222 cm^{-1} and is quite symmetric.

The behavior of the α and β subunits of the Hb tetramer can be selectively examined by the investigation of valency or metal hybrid hemoglobins. Utilizing excitation frequencies that enhance only the ferrous subunits the sensitivity of α and β subunits to quaternary structure can be determined. Changes in the low frequency region of valency hybrid hemoglobins were reported by Nagai & Kitagawa (29). They concluded that the asymmetric shape observed from HbA (T structure) is a consequence of subunit distinguishability, the α-subunit having a peak near 203 cm^{-1} and the β-subunit having a peak near 218 cm^{-1}. A recent study of the cobalt-iron hybrids (27) has confirmed these results. It was found that in the α-subunits the Fe–His stretching mode occurs as a doublet with maxima at 201 and 212 cm^{-1} whereas in the β-subunits it appears at 218 cm^{-1}.

Upon conversion to the R-state, the differences between subunits are markedly diminished. Des-Arg (R-state) derivatives of both valency and metal hybrid preparations yield Fe–His stretching modes at 220–224 cm^{-1} for both the α and β subunits. This strongly suggests that the α-subunits are more responsive to quaternary structure than the β-subunits.

Studies complementary to the cw work on the Fe–His stretching mode have been conducted recently (31, 32). The shift in the frequency of the Fe–His stretching mode in going from deoxy Hb to photolyzed COHb is shown in Figure 5. No such change is seen for Mb, as seen in Figure 6. The effect of IHP on the spectra of photolyzed ligated species is seen in Figure 7. For NOHb, where the addition of IHP is known to switch the system to the T state, the frequency shifts from 231 to ~ 222 cm^{-1}. For the carboxy derivative that does not switch, the effect of IHP is smaller but in the same direction. The carboxy derivative of Hb$_{KANSAS}$ that does switch to the T state in the presence of IHP manifests the same frequency shift as the NO derivative (32). The corresponding subunit specific responses are shown in Figure 8 (J. M. Friedman, M. Ikeda-Saito, and T. Yonetani, to be published). As with ν_4, there are no detectable subunit dependent differences in the Fe–His stretching mode for the R state ligated species in the absence of IHP. It can be seen that the addition of IHP clearly has a different affect on the two subunits, with the Fe–His stretching mode of the α(Fe)β(Co) hybrid becoming the more T-like, i.e. lower frequency. A summary of the shifts in the Fe–His stretching mode is shown in Figure 9. As with ν_4, the effect of ligation is to shift the Fe–His mode of the T system toward R-like values and to shift the R systems still further.

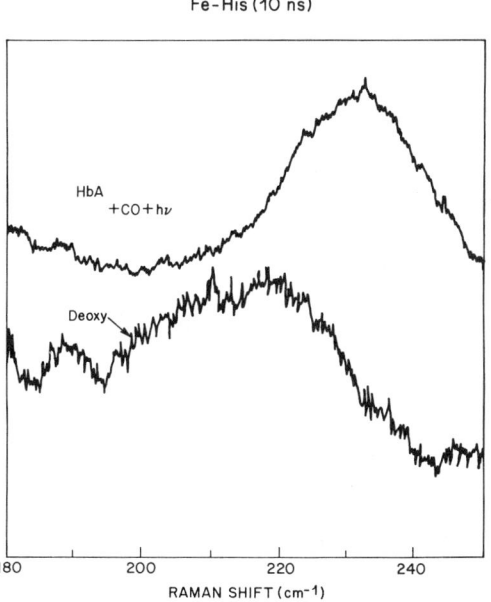

Figure 5 The iron-histidine stretching mode for photolyzed (10 nsec) carboxy HbA at pH 8.7 (*upper spectrum*) and deoxy HbA (*lower*). The spectra were generated using the output of a Lambda Physics excimer pumped dye laser at 4340 Å (Silbene 3, .7 mJ, 10 nsec duration).

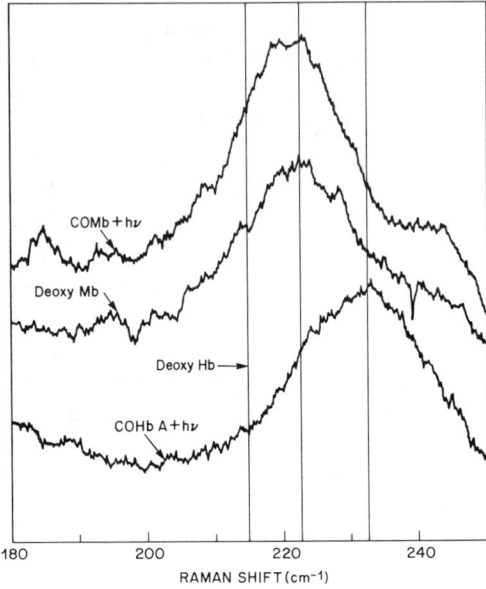

Figure 6 A comparison of the Fe–His stretching mode between Hb and Mb. The *upper curve* is the spectrum of photolyzed carboxy sperm whale Mb; the *middle spectrum* is of the corresponding dexoy Mb; and the *bottom spectrum* is of photolyzed COHbA at high pH. The excitation frequency is 4350 Å.

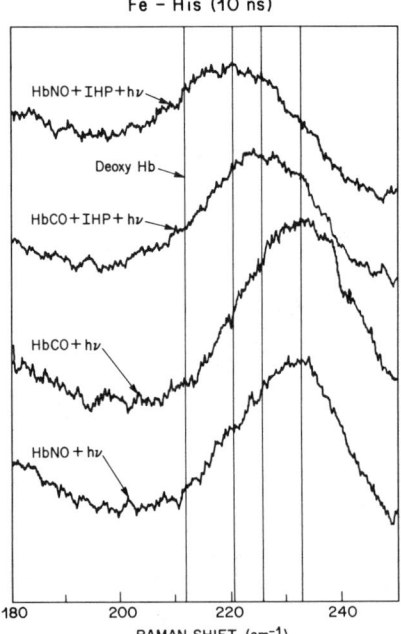

Figure 7 The influence of IHP upon the Fe–His stretching mode in NOHbA and COHbA. For NOHbA, IHP at low pH switches the quaternary structure to the T state, whereas COHbA in the presence of IHP remains in a modified R state.

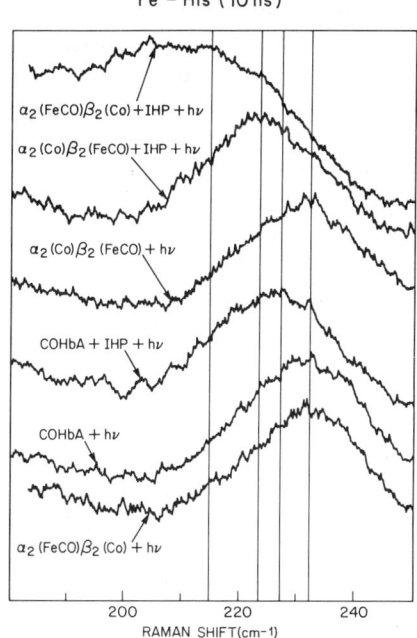

Figure 8 The influence of IHP upon the carboxy Fe–Co hybrids as reflected in the behavior of the Fe–His stretching mode.

Based on the behavior of the Fe–His stretching modes, it follows that the transient species possess a stronger Fe–His bond than the deoxy counterparts. Specifically, within each quaternary structure the Fe–His stretching mode is at higher frequency for the transient. These frequencies demonstrate that for both quaternary structures, ligation induces substantial structural changes about the heme. Baldwin and Chothia (33) recently concluded that the positioning of the histidine is primarily a consequence of the quaternary structure, irrespective of the presence of bound ligand. However, the data discussed here in which the frequency difference between the Fe–His stretching mode in an R–T comparison and a photolyzed-deoxy comparison are roughly equal indicate that the positioning of the histidine is controlled both by quaternary structure and by ligation.

Although the various ligands are known to generate slightly different structures (34), for a given quaternary structure in the transient species at 10 nsec, the Fe–His frequencies are indistinguishable (32) (R-state Hbs in the absence of IHP have an Fe–His stretching mode at ~ 231 cm^{-1} and T-state Hbs have a frequency of ~ 222 cm^{-1}). Especially dramatic are the results from NOHb+IHP in which there is strong evidence that the Fe–His bond is either severed or severely weakened (35, 36). However, at 10 nsec the Fe–His frequency is the same as that of photolyzed

COHb$_{KANSAS}$ in which there is no evidence for an anomalous Fe–His bond. Furthermore, despite substantial differences in the iron-to-center distance in COHb between the α and β subunits, there are no chain specific differences in the transients. These results indicate that although the protein retains memory of the ligated state, the ligand specific features have relaxed within 10 nsec of photolysis.

The question remains as to whether, within 10 nsec, the Fe–His bond length has achieved its equilibrium "deoxy" value, which would then allow for a direct structural comparison of the Fe–His linkage between the equilibrium and the transient species. Central to this question is when the in-plane-iron of the ligated heme reassumes an out of plane position characteristic of the deoxy heme. Theoretical (37–40) and experimental (41) evidence demonstrates that in a five-coordinated heme, nonbonded interactions between the proximal histidine and the pyrrole nitrogens of the prophyrin macrocycle strongly favor an out-of-plane iron. Calculations (40) reveal that the nonbonded repulsive energy drops from 15 kcal

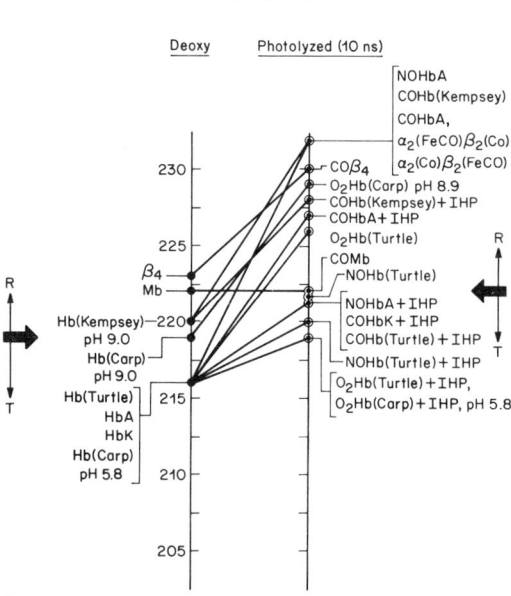

Figure 9 The change in frequency of the Fe–His stretching mode in going from equilibrium deoxy species (*left side*) to transient deoxy species (*right side*) occurring within 10 nsec of photolysis. Both T and R quaternary structures are represented on either side of the figure. Hb(T) refers to turtle (Chelonia mydas) Hb, which has an unstable ligated R structure that is easily switched to the T structure by the addition of phosphates at low pH (J. M. Friedman, to be published). Hb K refers to hemoglobin Kansas, which when ligated can also be switched to the T structure.

to 6 kcal when the iron atom moves from in-plane to 0.3 Å out-of-plane. Thus, in the absence of some protein-induced counter force, the potential surface of the Fe–His stretching motion can be expected to be highly perturbed by this strong repulsive force. Simple potential function considerations (32) readily demonstrate that the greater the influence of the repulsive term, the lower the vibrational frequency becomes. Thus, for an in-plane iron with stronger nonbonded interactions, the Fe–His mode would be at lower frequency than it would be for an out-of-plane iron. The observation that the frequency of the Fe–His stretching mode in the photolyzed protein is substantially higher than that of the deoxy protein necessitates that substantial motion of the iron has occurred and the nonbonded interactions have been largely relieved.

If, in the transient, the iron remained in plane, as has been suggested (13, 14), then in order to account for the increase in frequency, it would be necessary to postulate that the protein generates a force counter to the sizable repulsive force discussed above. Without this counter force the iron would be expected to assume an out-of-plane configuration within a few vibrational periods, i.e. < psec. The observed increase in frequency would have to be explained in terms of this postulated restraining force inducing a shortening of the Fe–His bond. Such rigidity as would be necessary for these putative forces is not evidenced in either crystallographic (33) or theoretical studies (37, 38). Furthermore, if the iron were to maintain the same position in the transient species as in the six-coordinate hemoglobin, then the Raman frequency would be expected to correlate with the known displacements in the ligated species. Recent analysis (32) indicates that there is no such correlation. This is especially notable in the comparison of COMb and the β subunits of COHb, which have identical iron displacements (0.2 Å) but very different transient Raman frequencies (222 and 231 cm^{-1}, respectively).

Even though it has been inferred from transient data that in 10 nsec the iron has relaxed to an out-of-plane position, the data still have significant implications regarding the forces in the ligated protein. The R–T differences in the transient spectra indicate that there are quaternary structure-dependent protein forces acting on the iron-histidine bond in the ligated hemoglobins. This follows because the protein quaternary structure is known (6, 7, 9) to relax on a much longer time scale. In order to translate these differences into energies in the six-coordinate case, it is necessary to know the R–T dependent forces responsible for the variation in Raman frequency. As a first step Friedman et al (32) have correlated known structural features with the observed frequencies.

A relationship was found (32) between the frequency of the Fe–His stretching mode and the tilt of the histidine (in its own plane) with respect to the porphyrin plane. This tilt results in a difference in the distances between the two pairs of imidazole carbons and pyrrole nitrogens. For myoglobin, in going from ligated to deoxy, the histidine retains its symmetric orientation. In contrast, for hemoglobin, comparing ligated to deoxy, the histidine goes from a symmetric to a tilted orientation. Through the increase in nonbonded interactions (38), the tilting is expected to weaken the Fe–His bond and thereby lower the Raman frequencies. It follows that if the change in Raman frequency is interpreted as being due to a change in nonbonded interactions, no difference between the deoxy and photolyzed myoglobin is expected, whereas in hemoglobin the frequency decreases in going from the photolyzed conformation with a symmetric histidine to the deoxy conformation with a tilted histidine. Given this relationship it appears that the Raman frequency of the Fe–His stretching mode is modulated by changes in the nonbonded interactions due to tilting of the histidine.

Correlation Between ν_4 and Fe–His Stretching Mode

It has been demonstrated (25) that for the equilibrium species there is a linear inverse correlation between the frequencies of ν_4 and the Fe–His stretching mode. Recent TRRR studies (J. M. Friedman, to be published) indicate that this correlation holds as well for the transients at 10 nsec. A composite of cw and transient frequencies is shown in Figure 10. From these data it appears that there is a direct link between the behavior of the Fe–His stretching mode and the distribution of electrons within the π system of porphyrin.

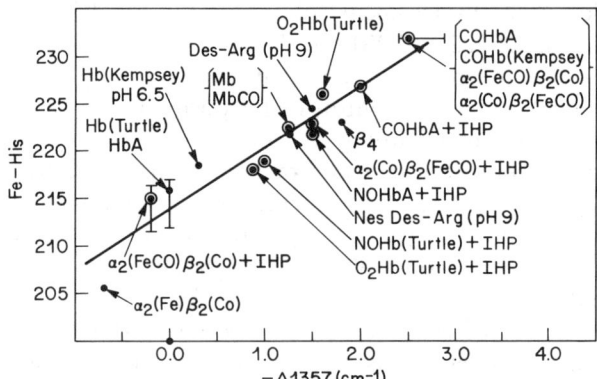

Figure 10 The inverse correlation between the frequency of the Fe–His stretching mode and ν_4 for both steady state and photolyzed ligated species of Hb. Species without a ligand designation are deoxy and are indicated by points that are not encircled.

These data bear significantly on the mechanism by which the protein interacts with the porphyrin. From equilibrium Raman difference studies on a variety of hemoglobins it has been found that only the Fe–His stretching mode and ν_4 reliably change frequency with quaternary structure (25). Since the frequency changes in these modes have now been found to correlate under a variety of conditions, it appears that there is only one significant protein-heme interaction, rather than multiple interactions. However, the specific interaction that leads to the correlated behavior of these two important modes cannot at present be identified. It is tempting to speculate that the changes in the porphyrin electron density result from changes in back-donation from the iron atom due to orientational differences in the proximal histidine. On the other hand, other routes are certainly possible, such as a direct $\pi-\pi$ interaction between the histidine and the porphyrin macrocycle. Whatever the mechanism of coupling between these modes, for complete evaluation of the cooperative energy the electronic energy changes in the porphyrin must be considered in addition to the changes in energy of the iron-histidine bond.

TIME EVOLUTION OF RAMAN MODES

Most transient Raman experiments reported to date have been carried out with single pulses, i.e. the same pulse both photolyzes the ligand and gives rise to the Raman scattering. To probe the dynamics of the time evolution of the transient species, two pulses are necessary—one for photolysis and one for the Raman probe. It would be extremely valuable to plot the continuous variation of the complete Raman spectrum from the picosecond regime in which only the heme is reorganizing to the millisecond regime in which the entire protein structure has fully relaxed. From such studies one should be able to separate readily influences resulting from quaternary structure and those resulting from ligation.

The only reported two-pulse experiment is that of Lyons & Friedman (9). They followed the frequency changes of ν_4 from 10 nsec to 1 msec in order to resolve ligation and quaternary structure effects by virtue of their different time evolution. The experiment is summarized in Figure 11. Because of geminated recombination (10–12), $\approx 50\%$ of the hemes remain ligated after ≈ 100 nsec subsequent photolysis. Thus, if the system is probed with a very low intensity source, the resulting ν_4 band will arise from those hemes that are evolving as unligated species. If a high intensity source is used as the second pulse, the resulting spectrum reflects the properties of the hemes that have been evolving as both ligated and deligated sites. If all the heme sites evolve in the same fashion, independent of the degree of ligation, the high and low intensity spectra should be identical. The results are shown in Figure 12. This experiment demonstrates that the photolyzed sites that remain deligated undergo a microsecond relaxation to a value comparable to those of R state deoxy Hb's observed in the cw experiments (23–25). The R to T transition, which occurs much more slowly, causes both types of sites to relax to higher frequencies. The lower curve (dashed) in Figure 12 is a result of the deconvolution that yields the time evolution of the recombined sites. The observed increase in frequency after many tens of microseconds is consistent with a shift in equilibrium toward the T-state. The results of this experiment demonstrate that there is an initial local relaxation about the deligated hemes that precedes the more global changes associated with the change in quaternary structure. In a recent experiment, J. M. Friedman (unpublished results) found that the time variation of the Fe–His stretching mode was the same as that of ν_4; this upholds the generality of the correlation between these modes.

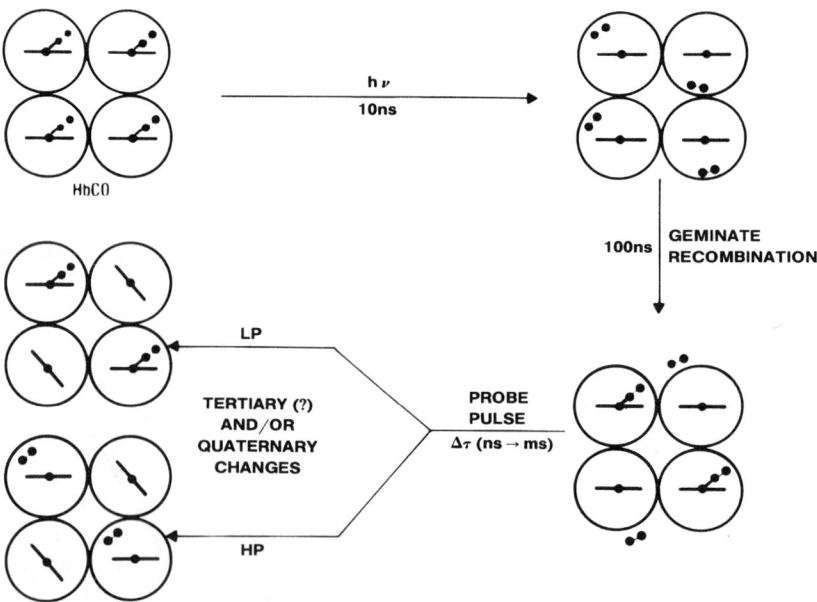

Figure 11 Schematic of the two pulse experiment reported by Lyons & Friedman (9) depicting the two populations of hemes that occur subsequent to geminate recombination. If a low intensity probe pulse is used (for $t > 100$ nsec), ν_4 has contributions only from those sites that have not undergone geminate recombination. When the probe pulse has sufficient intensity to rephotolyze the recombined sites ($t > 100$ nsec), ν_4 has contributions from both heme sites.

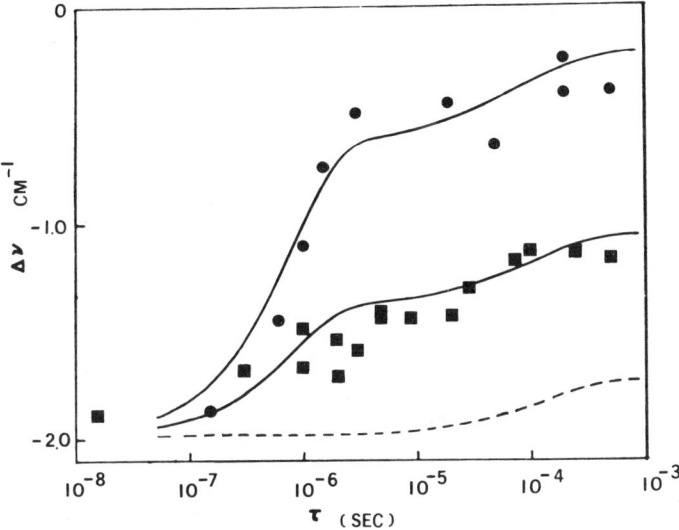

Figure 12 The behavior of the deoxy peak positions of ν_4 as a function of delay after photolysis. The positions for the peak obtained at low power (*solid dots*) were extracted from spectra excited at 4060 Å with a pulse energy of 50 μJ. Results obtained with a high power (300 μJ) probe pulse at 4200 Å are shown by the *squares*. On the *vertical scale*, the Raman shifts are referred to an origin that is the Raman frequency of a steady-state deoxy species, obtained in the same apparatus. (That is, $\Delta\nu = 0$ means $\nu = 1357$ cm^{-1}.) The CO pressure is 50 mm Hg, the temperature is 12° C, and the concentration is 44 μM (heme) for the low power experiment, and 68 μM for the high power. Similar results are obtained for concentrations in the range 30–100 μM. The *dashed line* shows the position of the deoxy peak associated with the rephotolyzed sites evolving ligated after the geminate recombination.

CONCLUSIONS

Resonance Raman scattering from transient species of Hb's are clearly very rich in information. Modes sensitive to the porphyrin core size, the heme electron density, and the Fe–His stretching frequency have already been studied in considerable detail in specific time domains. Further studies in the picosecond range will be most useful to determine exactly the state of the heme before it has had time to display significant structural and electronic relaxation. By following the time evolution of these changes, the molecular dynamics of the propagation of changes from the local ligand binding site at the iron to the protein subunit interfaces can be elucidated. Further comparisons of the magnitude of the Raman frequency shifts and their temporal behavior for proteins stabilized in the high (R) and low (T) affinity quaternary structures will also yield important results concerning the energetics of cooperativity and the kinetics of ligand association. The importance of identifying the structural determinant of these frequency shifts is indicated by the observation that for both stable deoxy and transient deoxy species the frequency of either the Fe–His stretching mode of ν_4 is an indicator of the stability of the quaternary structure.

Literature Cited

1. Perutz, M. F. 1970. *Nature* 228:726–34
2. Shank, C. V., Ippen, E. P., Bersohn, R. 1976. *Science* 193:50–51
3a. Noe, L. J., Eisert, W. G., Rentzepis, P. M. 1978. *Proc. Natl. Acad. Sci. USA* 75:573–77
3b. Reynolds, A. H., Rand, S. D., Rentzepis, P. M. 1981. *Proc. Natl. Acad. Sci. USA* 78:2292–96
4. Greene, B. I., Hochstrasser, R. M., Weissman, R. B., Eaton, W. A. 1978. *Proc. Natl. Acad. Sci. USA* 75:5255–59
5. Chernoff, D. A., Hochstrasser, R. M., Steele, A. W. 1980. *Proc. Natl. Acad. Sci. USA* 77:5606–10
6. Sawicki, C. A., Gibson, Q. H. J. 1974. *Biol. Chem.* 251:1533–42
7. Lindqvist, L., El Mohsni, S., Tfibel, F., Alpert, B. 1980. *Nature* 288:729–30
8. Asher, S. A. 1981. *Methods Enzymol.* 76:371–13
9. Lyons, K. B., Friedman, J. M. 1982. In *Interactions Between Iron and Proteins in Oxygen and Electron Transport*, ed. C. Ho. Amsterdam: Elsevier-North Holland. In press
10. Duddell, D. A., Morris, J. R., Muttucuraru, N. J., Richards, D. A. 1980. *Photochem. Photobiol.* 31:479–84
11. Alpert, B., El Mohsni, S., Lindqvist, L., Tfibel, F. 1979. *Chem. Phys. Lett.* 64:11–16
12. Friedman, J. M., Lyons, K. B. 1980. *Nature* 5756:570–79
13. Terner, J., Spiro, T. G., Nagumo, M., Nicol, M. F., El-Sayed, M. A. 1980. *J. Am. Chem. Soc.* 102:3238–39
14. Terner, J., Spiro, T. G., Nagumo, M., Nicol, M. F., El-Sayed, A. J. 1981. *Proc. Natl. Acad. Sci. USA* 78:1313–17
15. Nagumo, M., Nicol, M., El-Sayed, M. A. 1981. *J. Phys. Chem.* 85:2435–38
16. Coppey, M., Tomberg, H., Valat, P., Alpert, B. 1980. 284:568–70
17. Terner, J., Voss, D., Paddock, C., Miles, R. B., Spiro, T. G. 1982. *J. Phys. Chem.* 86:859–61
18. Lyons, K. B., Friedman, J. M., Fleury P. A. 1978. *Nature* 275:565–66
19. Srivastawa, R. B., Schuyler, M. W., Dosser, L. R., Purcell, F. J., Atkinson, G. H. 1978. *Chem. Phys. Lett.* 56:595–98
20. Spiro, T. G., Burke, J. M. 1976. *J. Am. Chem. Soc.* 98:5482–89
21. Tsubaki, M., Nagai, K., Kitagawa, T. 1980. *Biochemistry* 19:379–85
22. Friedman, J. M., Stepnoski, R. A., Stavola, M., Ondrias, M. R., Cone, R. L. 1982. *Biochemistry* 21:2022–27
23. Shelnutt, J. A., Rousseau, D. L., Friedman, J. M., Simon, S. R. 1979. *Proc. Natl. Acad. Sci. USA* 76:4409–13
24. Rousseau, D. L., Shelnutt, J. A., Ondrias, M. R., Friedman, J. M., Henry, E. R., Simon, S. R. 1982. In *Interactions Between Iron and Proteins in Oxygen and Electron Transport*, ed. C. Ho. Amsterdam: Elsevier-North Holland. In press
25. Ondrias, M. R., Rousseau, D. L., Shelnutt, J. A., Simon, S. R. 1982. *Biochemistry*. In press
26. Rousseau, D. L. 1981. *J. Raman Spectrosc.* 10:94–99
27. Ondrias, M. R., Rousseau, D. L., Kitagawa, T., Ikeda-Saito, M., Inubushi, T., Yonetani, T. 1982. *J. Biol. Chem.* In press
28. Nagai, K., Kitagawa, T., Morimoto, H. J. 1980. *J. Mol. Biol.* 136:271–89
29. Nagai, K., Kitagawa, T. 1980. *Proc. Natl. Acad. Sci. USA* 77:2033–37
30. Kitagawa, T., Nagai, M., Tsubaki, M. 1979. *FEBS. Lett.* 104:376–79
31. Irwin, M. J., Atkinson, G. H. 1981. *Nature* 293:317–18
32. Friedman, J. M., Rousseau, D. L., Ondrias, M. R., Stepnoski, R. A. 1982. *Science*. In press
33. Baldwin, J. M., Chothea, C. 1979. *J. Mol. Biol.* 129:175–201

34. Moffat, J., Detherage, F., Seybert, D. W., 1979. *Science* 206:1035-42
35. Maxwell, J., Caughey, W. S. 1976. *Biochemistry* 15:388-96
36. Nagai, K., Welborn, C., Dolphin, D., Kitagawa, T. 1980. *Biochemistry* 19:4755-761
37. Warshel, A. 1977. *Proc. Natl. Acad. Sci. USA* 74:1789-93
38. Gelin, B. R., Karplus, M. 1977. *Proc. Natl. Acad. Sci. USA* 74:801-5
39. Olafson, B. D., Goddard, W. A. III. 1977. *Proc. Natl. Acad. Sci. USA* 74:1315-19
40. Goddard, W. A. III, Olafson, B. D. 1972. In *Biochemical and Clinical Aspects of Oxygen*. pp. 87-123. New York: Academic
41. Scheidt, W. R., Frisse, M. E. 1975. *J. Am. Chem. Soc.* 97:17-21

HYDROCARBON BOND DISSOCIATION ENERGIES

Donald F. McMillen and David M. Golden

Department of Chemical Kinetics, SRI International, Menlo Park, California 94025

INTRODUCTION

This review considers the best available values for homolytic bond dissociation energies (BDEs) of various classes of neutral compounds. (BDEs in ionic species is a legitimate subject that we touch on briefly and could easily be included in a longer review. The same can be said for heterolytic BDEs, which are not reviewed as such, although some of the ionic thermochemical data discussed yield values for these processes.)

Our major emphasis is on hydrocarbons and their nitrogen, oxygen, sulfur, halogen, and silicon-containing derivations, but we include limited data for inorganic molecules. We focus particularly on prototypical radicals whose heats of formation, formerly thought to be well in hand, have recently been called into serious question. We intend to include all the major types of sigma bonds, if not all specific cases where known or estimatable heats of formation allow bond dissociation energies to be generated. BDEs are defined in the usual way, from Reaction 1, as:

$$A - B \rightarrow A + B$$
$$DH(A-B) \equiv \Delta H^\circ_{298} = \Delta H^\circ_{f,298}(A,g) + \Delta H^\circ_{f,298}(B,g) - \Delta H^\circ_{f,298}(AB,g).$$

1.

In the short space available we refrain from reminding the reader of all the reasons that BDE information is fundamental to chemistry. There exist so many reviews and compilations of BDEs that we refer the still curious to those in Table 1 and to older reviews such as those of Benson (240) and Szwarc (241) for intellectual succor if required!

We attempt in this review to acknowledge all the standard techniques for measuring BDEs in polyatomic molecules and to offer our critical analysis of selected portions of the literature. This leaves us with values that we recommend as the most likely to be correct at the time of this writing. We reserve both the right to be wrong and to remain good friends with those people with whose conclusions we may disagree!

We draw heavily from the reviews listed in Table 1, without the benefit of which this task would have been more time consuming. Of particular help has been the compilation by Kerr, including the updated listing he was kind enough to send us in advance of its publication in 1983. In cases in which no new critical assessment has been made, the values accepted by the authors listed in Table 1 have been accepted here also; where the values listed differ, the differences are the result either of reevaluation of original data, changes in thermochemical values used in converting that data into ΔH°_{298} values, or changes in the assumption about activation energy for radical combination reactions, as discussed below.

Table 1 Important bond dissociation energy reviews

Title	Author, year	Comments	Ref.
Bond strengths in polyatomic molecules	Kerr 1981	Handy abstraction, most recent comprehensive	(1)
Free-radical and molecule thermochemistry from studies of gas-phase iodine atom reactions	Golden & Benson 1969	Detailed review of iodination method	(2)
Bond dissociation energy values in silicon-containing compounds and some of their implications	Walsh 1981	Recent, critical	(3)
Thermochemistry of nitro compounds, amines, and nitroso compounds	Batt & Robinson 1981	Recent, critical	(4)
Bond dissociation energies by kinetic methods	Kerr 1966	Discusses individual results in many halogenation studies	(5)
Thermochemistry of fluorocarbon radicals	Rodgers 1978	Critical, based on reassessment of complete thermodynamic functions for certain fluorocarbons	(6)

Table 1 *Continued*

Title	Author, year	Comments	Ref.
Thermochemistry and kinetics of sulfur-containing molecules and radicals	Benson 1978	Recent, critical	(7)
Kinetic data on gas-phase unimolecular reactions	O'Neal & Benson 1970	Detailed, conveniently presented; critical summary of published rate studies	(8)
The thermochemistry of CN and NC groups	Batt 1981		(9)
Pyrolysis studies of main group metal-alkyl bond dissociation energies: VLPP of GeMe$_4$, SbE$_3$, and PhEt$_4$	Smith & Patrick 1982	Recent, includes reanalyses of related literature data	(10)
Homopolar and heteropolar bond dissociation energies and heat of formation of radicals and ions in the gas phase. I. Data on organic molecules	Egger & Cocks 1973	Most extensive compilation of experimental and group additivity-generated energies bond dissociation	(11)
Thermochemistry of peroxides	Baldwin 1982	Includes derived group values for peroxides and polyoxides	(12)

BDE DETERMINATION SCHEMES

BDEs may, in principle, be determined directly if an experiment is designed to measure the enthalpy change in Reaction 1 [strictly speaking, at 298 K, but heat capacities can be estimated with sufficient accuracy to allow for measurements at other temperatures (13).] BDEs are also obtained if the three heats of formation in the definition are known from independent experiments. Since values of $\Delta H^\circ_{f,298}(AB, g)$ are often (but not always) known through calorimetric and/or empirical correlation techniques (13–16), the problem is often one of determining $\Delta H^\circ_{f,298}(A, g)$ and $\Delta H^\circ_{f,298}(B, g)$. Furthermore, if A or B are atoms, we can accept the spectroscopically determined values (14) as given, thus reducing the BDE to a value of the previously unknown radical heat of formation.

Since most radicals cannot be handled in such a way as to make feasible the direct (calorimetric) determination of their absolute heats of formation, techniques are used to determine relative values, which can be anchored with accessible absolute values such as heats of formation of atoms.

Halogenation Kinetics

The reactions $RH + X_2 \rightleftarrows RX + HX$ and $RH + X_2 \rightleftarrows olefin + 2HX$ have been shown (2) to proceed via the following mechanism:

$$X_2 \rightleftarrows 2X$$
$$X + RH \underset{-1}{\overset{1}{\rightleftarrows}} R + HX$$
$$R + X_2 \underset{-2}{\overset{2}{\rightleftarrows}} RX + X$$
$$(RX \underset{-3}{\overset{3}{\rightleftarrows}} olefin + HX) \quad .$$

Measurements of the kinetics of these reactions, or their reverse for $X = I$, generally lead to Arrhenius values for the rate constants k_1 and/or k_{-2}. (It is usual to obtain the value of k_{-1}/k_2 as well.) If k_{-1} and/or k_2 Arrhenius parameters could be determined absolutely, the temperature dependence thus provided for the equilibrium constants $K_1 = k_1/k_{-1}$ and/or $K_2 = k_2/k_{-2}$ would yield directly ΔH and ΔS values for the respective processes. In any case, the only unknown heat of formation is likely to be that of R, which is thus determined. The assumption, at first weakly based, that $E_{-1} = 1 \pm 1$ kcal mol^{-1} and $E_2 = 0 \pm 1$ kcal mol^{-1} (2) has been "confirmed" in direct determinations for CH_3 (17), t-C_4H_9 (18), $PhCH_2$, and C_3H_5 (19, 22). Actually, higher values ($E_{-1} \approx 4$ kcal mol^{-1}) were measured (19) for C_3H_5 and $PhCH_2$ at ca. 1000 K. Reasonable transition state models predict sufficient curvature (21) to accommodate much (at least 2 kcal mol^{-1}) lower values at ca. 500 K where most iodinations are measured. In addition, the heat of formation of C_3H_5 determined (22) in an equilibrium study confirms the assumed E_{-1} value for iodination. t-Butyl + DI has been shown (18) to have an activation energy of ~ 2 kcal mol^{-1} in the range 644 to 722 K and its heat of formation has been confirmed in equilibrium studies (S. W. Benson, unpublished data) as has that of CH_3 (17).

The actual reported BDEs are evaluated by assuming that $E_{-1} = 1 \pm 1$ at reaction temperature, usually about 500 K. Then, $\Delta H_T^\circ = E_1 - E_{-1}$ is corrected to 298 K by $\Delta C_p^\circ(T-298)$ using estimates for C_p° of the radical (13). This technique, where X is the I-atom, has been reviewed by Golden & Benson (2), and extended to many Si compounds by Walsh (3) and to

F-containing compounds by Rodgers (6). Most recent values are compiled by Kerr (1).

In the case where X = Br, data is available on both the direct bromination of specific RH compounds, and on relative brominations at low yields of RH and a reference compound R'H. These have been tabulated through 1971 by Benson & O'Neal (24), and more recent values are taken into account in Kerr's compilation (1) and here. (Recent 298 K measurements (25, 242) of the absolute values of the rate constants for the reaction of Br_2 with CF_3, C_2F_5, and n-C_3F_7 have been compatible with $E_2 \leq 2$ kcal mol^{-1}.)

For X = Cl, both types of data, competitive and direct, are also available, but the chlorination reactions are so exothermic that the individual rate constants are difficult to extract from these chain reactions. When they have been extracted [an excellent older review by Fettis & Knox (26) lists many] the activation energies for Reaction 1 turn out to be 0–1 kcal mol^{-1}, and thus all the information content is in the more difficult to obtain k_{-1}, which is endothermic for most Rs.

Halogenation Equilibria

Recently, Benson and co-workers (17, 27) have used the very low-pressure pyrolysis (VLPP) technique (28) (referred to for these experiments as VLP Reactor) to measure directly the equilibrium constants for the reactions:

$$Cl + CH_4 \rightleftharpoons HCl + CH_3$$
$$Cl + c\text{-}C_3H_6 \rightleftharpoons HCl + c\text{-}C_3H_5$$
$$Br + i\text{-}C_4H_{10} \rightleftharpoons HBr + t\text{-}C_4H_9.$$

In the first (17a, b) and third (S. W. Benson, unpublished data) cases, Benson and co-workers have verified the iodination results and in the second (27) have determined the elusive value of DH(c-C_3H_5-H). It is our opinion that the best values come from the iodination kinetics and, of course, from the equilibrium VLPR studies. Some good values also can be obtained from bromination studies, especially for the stronger bonds (24).

Bond Scission Reactions — Treatment of Data

A large number of reported values for BDEs come from direct measurement of the high-pressure value of the activation energy for bond scission. This method has caused some difficulties since the relationship between BDE and the activation energy for bond scission ($E_{d,T}$) involves the value of $E_{r,T}$, which is not often accurately known. In fact, there are good reasons to suspect from both theory and experiment that the value of $E_{r,T}$ can be a little ($\sim$1-2 kcal mol^{-1}) negative. There is general agreement that there does not exist a potential energy barrier to combination of radicals [$E^o_{0,r} = 0$ (see below)], so we suggest the following convention (23):

$$AB \underset{r}{\overset{d}{\rightleftharpoons}} A + B,$$

$\Delta E^o_0 = E^o_{0,d} - E^o_{0,r} \equiv E^o_{0,d} \equiv \Delta E^{\ddagger}_{0,d}$ = the critical energy.

The subscript "0" refers to K and the superscript "o" to the standard state. The desired bond dissociation energy (ΔH^o_{298}) can be related to the measured activation energy $E_{T,d}$ (assuming high-pressure limit), in the following ways:

$$E_{T,d} = \Delta E^{\ddagger}_{T,d} + RT = \Delta E^{\ddagger}_{0,d} + \langle \Delta C^{\ddagger}_{v,d} \rangle^T_0 T + RT$$

(from the transition state theory, where $k = kT/h \exp(\Delta S^{\ddagger}_T/R - \Delta E^{\ddagger}_{T,d}/RT)$)

$$E_{T,d} = \Delta E^o_0 + (R + \langle \Delta C^{\ddagger}_{v,d} \rangle^T_0)T$$
$$= \Delta E^o_T - \langle \Delta C_v \rangle^T_0 T + (R + \langle \Delta C^{\ddagger}_{v,d} \rangle^T_0)T$$
$$= \Delta H^o_T + \langle \Delta C^{\ddagger}_{v,r} \rangle^T_0 T$$

or

$$E_{T,d} = \Delta H^o_{298} + \langle \Delta C^o_p \rangle^T_{298}(T - 298) + \langle \Delta C^{\ddagger}_{v,r} \rangle^T_0 T$$

$$BDE = \Delta H^o_{298} = E_{T,d} - \langle \Delta C^o_p \rangle^T_{298}(T - 298) - \langle \Delta C^{\ddagger}_{v,r} \rangle^T_0 T.$$

Thus, BDE values require both the overall heat capacity change and enough information about the transition state to write values for $\Delta C^{\ddagger}_{v,r}$ as a function of temperature. Where we have reevaluated data to arrive at the values recommended herein, the above convention has been used. Knowledge of the transition state may appear to be a requirement that is not readily met. However, if the Gorin rotational transition state model is used for bond dissociation processes, $\Delta C^{\ddagger}_{v,r}$ will be $-\frac{1}{2}R$ for all cases (13) at all temperatures above $\sim$ 20 K, and

$$BDE = \Delta H^o_{298} = E_{T,d} - \langle \Delta C^o_p \rangle^T_{298}(T - 298) + \frac{1}{2}RT.$$

This results in BDE values that are lower by $\frac{1}{2}RT$ than values based on the assumption that the activation energy for recombination (from rate constants measured in concentration units) is 0 at the reaction temperature. For studies up to 1000 K, this change amounts to 1 kcal mol^{-1} or less. Given other uncertainties in bond energy measurements, this change does not greatly change the implications of any single BDE measurement. However, in attempts to improve consistency by closing thermochemical cycles consisting of several BDE values, the difference can become important.

Measurement of Bond Scission Reaction Rates

It is well recognized that pyrolysis of even small organic molecules can be enormously complicated. Therefore, the extraction of accurate rate parameters or even absolute rates from pyrolysis results (192, 197) is often supportive rather than definitive, unless specific measures are taken to simplify the reaction sequence. These measures include determination of reaction rates at very small extents of reaction (235), and/or use of specific techniques, some of the more widely used of which are discussed briefly below.

CARRIER TECHNIQUES Pyrolysis in the presence of a large excess of a radical scavenger should, in principle, eliminate secondary reactions. However, many examples in which systematic errors have substantially perturbed the Arrhenius parameters (though generally not the rate constants) obtained from toluene carrier data can be seen by perusing Ref. (8).

SHOCK TUBE The single-pulse shock tube (SPST) has provided one of the simplest, most convenient and surface-interaction-free techniques for bringing a reaction mixture quickly to very high temperatures, followed by rapid cooling (29, 30). It thus is a convenient tool for studying the initial decomposition steps (i.e. unimolecular, high activation energy reactions) in complex decomposition processes in which the bulk of the secondary reactions exhibit substantially lower activation energies. The reaction time is generally short enough to eliminate all secondary reactions of the substrate except those involving H-atom chains. These can be satisfactorily suppressed by use of a scavenger such as toluene (29).

One of the principal difficulties has been satisfactory temperature definition. This difficulty appears to be essentially eliminated by use of the SPST in the "comparative rate" mode as developed by Tsang, where comparison of the fractional decomposition of the substrate with that of a "temperature standard"—a substrate whose Arrhenius parameters are already well known—allows determination of the substrate Arrhenius parameters with high precision (30). Typically, activation energies in the range 50–75 kcal mol^{-1} are determined with a standard deviation of 0.3 kcal mol^{-1} or less. Treatment of the errors involved in the comparative rate technique indicates the following:

1. Uncertainties in reaction time contribute no error to the determined activation energy and essentially none to the A-factor (29).

2. For substrates and standards with similar activation energies (e.g. within ~10%), maintenance of the fractional decomposition below ~20% ensures that variations in temperature with time and with space within the reaction volume contribute no significant error to the measured parameters (31).

In all cases in which a test substrate was available, the comparative rate single-pulse shock tube has proved (29, 30) to be at least as precise and accurate as other methods by which the test substrate parameters were originally determined. This has included a number of bond scission rate measurements, for example, bond homolysis rates of alkynes (98, 231). However, in certain cases of interest in this review (e.g. hexamethylethane and isobutyl- and neopentyl-benzene) (32, 69, 237), unexplainable differences exist between shock tube and flow-tube data and other bond energy sources. In a subsequent section of this paper, we discuss discrepancies in the specific case of homolyses that generate t-butyl radical.

VERY LOW-PRESSURE PYROLYSIS The Very Low-Pressure Pyrolysis (VLPP) technique, which consists essentially of pyrolysis in a Knudsen cell reactor (22, 28) (typical pressure 0.1 to 10 mTorr), is comparably effective to shock-tube techniques in minimizing the impact of secondary bimolecular reactions in otherwise complex reaction systems. Mass spectrometric detection of a modulated molecular beam formed from the VLPP reactor effluent allows detection of reactive products without any gas-gas or gas-wall collisions after leaving the reactor. Measurement of the reactor wall temperature (and therefore the reaction temperature) is unambiguous.

Because VLPP is a flow technique, because the inlet system can readily be made small enough for convenient heating, and because molecular beam detection enables measurement that involves no surface collisions downstream of the reactor exit aperture, VLPP is more readily applicable to high molecular weight, low vapor pressure substrates (33) than is the inherently batch-reactor shock tube technique. The principal limitation of the VLPP technique stems from the fact that, at mTorr pressures and at the temperatures required to observe substantial cleavage of 70–90 kcal mol^{-1} bonds, unimolecular reactions of small molecules are generally not at their high pressure limits (28). When the degree of falloff is substantial ($k/k_\infty \ll 1.0$), the temperature dependence is increasingly attenuated for increasing values of the A-factor. This means that a precise determination of the activation energy generally requires that the A-factor be specified, or vice versa. Thus, under these conditions, VLPP does not readily provide an independent determination of both Arrhenius parameters. This could actually be considered an advantage in the sense that it saves the researcher from the illusion that a precisely determined Arrhenius plot is necessarily correspondingly accurate. In any case, the reader must always be aware that Arrhenius parameter measurements attributed to VLPP measurements are almost always one parameter determinations.

The frequently raised spectre of complicating surface reactions in a technique where the gas-surface collision frequency is much higher than the gas-gas collision frequency is in some cases a real limitation. However, in many cases, particularly with substrates of low and moderate polarity, surface contributions have, in fact, presented no problem. Two of the principal reasons that can be cited for the unimportance of surface reaction in these cases are low residence time on the reaction surface at high temperature (28) and the fact that bond scission reactions, as such, are generally not accelerated by the influence of polar media.

EVANS-POLANYI RELATIONSHIPS Correlations between the rates of atom abstraction reactions $R_1\cdot + R_2-X \rightarrow R_1-X + R_2\cdot$ and the BDEs of the bonds R_2-X offer a convenient, but not unambiguous, method for determining bond strengths. The advantages are that the correlations can generally be made in terms of absolute rates rather than activation energies, and that the absolute rates can be determined in a competitive manner. Thus, difficulties with random and systematic experimental errors are minimized. However, the disadvantage is that the required correlations can be expected to hold only within a closely related series, i.e. Evans-Polanyi relationships are most reliable for interpolation within a series of closely related compounds for which the answer is already pretty well known anyway.

An example of the difficulties that can be encountered in the application of Evans-Polanyi relationships is provided by use of the correlation shown by Kerr & Parsonage (34) to hold with good precision for $CH_3 + RH \rightarrow CH_4 + R\cdot$, where RH are alkanes. The correlation leads to a predicted activation energy for $CH_3 + PhCH_2-H \rightarrow CH_4 + PhCH_2\cdot$ of ~ 6 kcal/mol, but the measured value (34) is ~ 10 kcal/mol. The above cautions notwithstanding, Evans-Polanyi relationships can be helpful when judiciously used. A number of them are shown in a graphical form in Ref. (35), which illustrates in a very usable way the extent of nonlinearity of such "free energy" relationships. Furthermore, a plethora of Evans-Polanyi relationships appear to support equilibrium-based BDEs in alkane (30) and halocarbon (24) systems, as is discussed below.

Chemical Activation

Another method, which can in principle be used for BDE determination, involves that of chemical activation, wherein a species excited with known energy may dissociate via two pathways. The branching ratio, unfolded through unimolecular reaction rate theory, can give relative heats of formation of product radicals (36). This tends to be confirmatory rather than definitive.

Optical Detection of Product Energy

In several experiments it is possible to study reactions such as $X + RY \rightarrow R + XY^*$, where the asterisk represents nonthermal energy distribution. If XY^* is spectroscopically monitored, a lower limit for $DH(R-Y)$ is determined [if $DH(X-Y)$ is known]. Bogan & Setser (37) have used this method where $X = F$ and $Y = H$ to produce limits for several organic species. Zare and co-workers (38) have detected XY^* in many systems.

We can also include, under this general heading, measurement of the threshold energy for photodissociation or optical detection and/or determination of the extent of excess energy in the fragments (e.g. see 58, 80). This can be very precise, but as the complexity of the fragments increases, it becomes increasingly difficult to make the excess energy determination (234). This technique is therefore limited to diatomics, triatomics, and polyatomics with high symmetry.

Thermodynamic Cycles Including Ions

Just as relative values of free radical heats of formation may be measured from equilibrium constants involving neutral species, so may relative heats of formation of ions, both positive and negative, be determined. If some absolute values exist, these relative numbers can be put on an absolute basis.

For negative ions, electron affinities of halogen atoms are known with great accuracy and may be used as a fiduciary. Equilibria, such as $F^- + RH \rightleftharpoons HF + R^-$, which measure relative acidities, can be determined (39, 72) and the acidity of RH compared to that of HF. If values for the electron affinity of R exist (55), the BDE may be determined viz:

$$
\begin{array}{lll}
R-H \rightarrow H^+ + R^- & \Delta H_{\text{acid}} & (a) \\
H^+ + e^- \rightarrow H & -IP(H) & (b) \\
\underline{R^- \rightarrow R + e^-} & \underline{EA(R)} & (c) \\
R-H \rightarrow R + H & BDE(R-H) = \Delta H_{\text{acid}} - IP(H) + EA(R). &
\end{array}
$$

Strictly speaking, adiabatic values for IP and EA are values of ΔH_0° for the reactions written. These should be adjusted by heat capacity values to 298 K. If we take (b) and (c) together, $H^+ + R^- \rightarrow H + R$, $\Delta H_0^\circ =$

$EA(R) - IP(H)$, and

$$\Delta H°_{298} = \Delta H°_0 + \langle \Delta C_p \rangle_0^{298}(298),$$

$$\langle \Delta C_p° \rangle_0^{298} = \langle (C_p°(T)[R] - C_p°(T)[R^-]) + (C_p°(T)[H] - C_p°(T)[H^+]) \rangle_0^{298} \approx 0.$$

This term is small, since the ion and neutral structures and frequencies are fairly similar. Thus, neglect of this correction will normally result in BDEs that are in error by ≤ 1 kcal/mol. In many cases precision is limited by the precision of the EA measurement, since single equilibrium constants should give precise values of $\Delta H°$, and many are cited (39) as having <1 kcal mol^{-1} errors. However, where the equilibrium constants are the product of a number of individually measured values, error accumulation may make the equilibrium constant the principal source of error.

To obtain BDEs from positive ion cycles entails more difficulty than from negative ions. In principle, the problem is the same. If the heats of formation of positive ions can be linked by the study of equilibria, such as $R_1^+ + R_2X \rightleftharpoons R_2^+ + R_1X$, then if the value of the $\Delta H_f(R^+)$ were known for any species, an absolute scale of values could be established. If, in addition, ionization potentials for the radicals are known, the values of $\Delta H_f(R)$ could be ascertained.

The problems here are several-fold. It is particularly difficult to measure the equilibria in many cases, since establishing these in the absence of competing reactions is often difficult (40). Nevertheless, this difficulty is sometimes surmountable and values of hydride transfers to carbonium ions are reported to chemical accuracy (40).

It is also feasible to obtain relative carbonium ion stabilities by measuring equilibria involving proton transfers to olefins:

$$R_2^+ + \text{Olefin-1} \rightleftharpoons \text{Olefin-2} + R_1^+$$

as, for example:

$$t\text{-}C_4H_9^+ + C_3H_6 \rightleftharpoons i\text{-}C_4H_8 + s\text{-}C_3H_7^+$$

It is, however, difficult to find a good (i.e. $< \pm 1$ kcal mol^{-1}) absolute value for any carbonium ion upon which to anchor the scale of proton affinities of olefins and other such bases (i.e. NH_3) (61). No convenient marker analogous to the halogen negative ions exists here, unless one accepts the heat of formation of a radical measured with other techniques and the ionization potential of that radical is known.

IPs of many radicals have been measured using both photoelectron spectroscopy (41) and electron impact appearance potentials (42, 104) but these are difficult experiments that attempt to determine threshold energies in the range of 150–200 kcal mol^{-1} to precisions of <1% in order to provide values of chemical accuracy. (Photoelectron spectroscopy affords the IPs of many species and thus the values of heats of formation of radical cations, since the parent hydrocarbon is known or can be estimated accurately. These radical cations unfortunately cannot be tied to carbonium ions independently of a radical heat of formation.)

Radical and/or ion heats of formation can, in principle, be determined from appearance potential measurements. These are difficult to measure and may often contain errors due to excited state formation. The problems are reviewed in an extensive literature search summarized in Ref. (42).

HEATS OF FORMATION AND BDEs IN PROTOTYPICAL SYSTEMS

Among the earliest organic radicals to be studied were the prototypical primary, secondary, and tertiary systems: ethyl, iso-propyl, and tertiary-butyl, as well as the unique system represented by the allyl radicals.

As might have been expected, values for these heats of formation varied a bit between laboratories, but by the time of the 1969 review of Golden & Benson (2) it seemed safe to say that the values therein were universally agreed upon.

$$[DH(\text{Et-H}) = 98 \pm 1,: DH(\text{iPr-H}) = 95 \pm 1, DH(\text{t-Bu-H}) = 92 \pm 1,$$
$$\text{and } DH(\text{CH}_3\text{-H}) = 104 \pm 1 \text{ kcal mol}^{-1}].$$

In 1978 Tsang showed (44) the analysis of data for reactions of the type, $R-R \underset{k_r}{\overset{k_d}{\rightleftharpoons}} 2R$, where both k_d and k_r were available, could only be quantitatively compatible with BDE values by raising the accepted values of the BDEs by 2–4 kcal mol^{-1}. This relatively small revision, if appropriate, has serious impact on some current understanding of mechanistic organic chemistry [see a particularly colorful discussion by Doering (45)] to say nothing of impugning the accuracy of some recent kinetic studies (18, 19) of R+HX processes and raising problems associated with assigning to such H-atom metathesis reactions (2) the negative activation energies that would be required by substantially increased heats of formation. The above, together with a recent flood of reports (44–49) claiming that ΔH_f°, 298(t-butyl) ≈ 10–11 kcal mol^{-1}, requires special scrutiny.

Thermochemistry of the t-Butyl Radical

We examine here the conflicting evidence concerning the appropriate values of both the entropy and heat of formation of t-butyl radicals.

Much of the same sort of discussion is apropos for ethyl and isopropyl radicals as well.

IODINATION There have been two reported measurements of the rate constant k_1:

$$I + i\text{-}C_4H_{10} \underset{-1}{\overset{1}{\rightleftharpoons}} t\text{-Bu}\cdot + HI$$

These are listed in Refs. (50, 51):

$$\log k_1/M^{-1}s^{-1} = 10.9 \pm 0.2 - (21.3 \pm 0.5)/\Theta; [525 \leqslant T/K \leqslant 583]$$
$$\times (\Theta = 2.303\, RT)$$
$$\log k_1/M^{-1}s^{-1} = 11.3 \pm 0.4 - (22.6 \pm 1.1)/\Theta.$$

The latter authors (Knox & Musgrave) actually prefer a somewhat higher activation energy, but these values are in essential agreement over the temperature range of the experiments. We accept the parameters of Teranishi & Benson (50) as discussed by Golden & Benson (2); however, in recognition of the possibility that unknown systematic errors have substantially perturbed these sets of parameters, we have used not only the activation energies, but also the absolute rates as bases for deriving the t-butyl radical heat of formation.

Application of the assumption that $E_{-1} \approx 1 \pm 1$ kcal mol^{-1} (at reaction temperature), together with estimates of the t-butyl radical heat capacity led Golden & Benson (2) to a value of $\Delta H^\circ_{f,298}(t\text{-butyl}) = 7.6 \pm 1.2$ kcal mol^{-1}, corresponding to $DH(t\text{-butyl-H}) = 91.8 \pm 1.2$ kcal mol^{-1}, which they rounded off to 92 ± 1 kcal mol^{-1}. This fit well with the values they recommended for other BDEs: $DH(CH_3\text{-H}) = 104$, $DH(\text{ethyl-H}) = 98$, $DH(\text{isopropyl-H}) = 95$ kcal mol^{-1}, all to ± 1 kcal mol^{-1} uncertainty.

Rossi & Golden (18) reported a measurement of $\log k_{-1}/M^{-1} = 9.68 - 2.12/\Theta$; $[644 \leqslant T/K \leqslant 722]$ using DI instead of HI. (They suggest that $k_H/k_D \approx 1.5$.) The Arrhenius parameters are fairly uncertain since in this temperature range they represent a very small variation in k. Thus, Rossi & Golden chose to extrapolate the Teranishi & Benson parameters for k_1 up to 700 K in order to establish $K_1 \equiv k_1/k_{-1}$ at 700 K. [They also extrapolated the reported value (52, 53) of k_{-1} down to 550 K with no serious difference in final results.] Then, using the values of entropy and heat capacity for t-butyl radical suggested by Choo et al (53), whose essential assumption is that the three methyl rotors have barriers of 2 kcal/mol^{-1}, and other easily available thermochemical data, they concluded (18) that $\Delta H^\circ_f, 298(t\text{-Bu}) = 8.4$ kcal mol^{-1} and $DH(t\text{-Bu-H}) \approx 92.7$ kcal mol^{-1}.

Recently, Pacansky and co-workers have studied the infrared spectrum and performed theoretical calculations on t-butyl radical (54a,b). An entropy and heat capacities may be computed from these results. The most important outcome is the assignment of barriers of 0.5 kcal mol^{-1} to the three methyl rotors. Benson has suggested (private communication, 1981) that the same evidence may only require that one rotational barrier (in the coupled set of three) is low; the others can be higher. In what follows, we discuss t-butyl thermochemistry for the Pacansky et al suggestion, $S^{\circ}_{300}(t\text{-butyl}) = 76.0$ cal mol^{-1} K^{-1}, and for the Benson suggestion $S^{\circ}_{300}(\text{t-Bu}) = 74.0$, assigning two barriers to have heights of 2.4 kcal mol^{-1}.

Following the Rossi et al prescription (22) with these new entropies and heat capacities, we get $\Delta H^{\circ}_{f,298}(\text{t-Bu}) = 9.7$ or 8.7 kcal mol^{-1} using the higher and lower entropy models, respectively [i.e. $DH(\text{t-Bu-H}) = 94.0$ or 93.0 kcal mol^{-1}. These values imply $E_{-1} = -0.4$ and $E_{-1} = 0.2$ kcal mol^{-1}, respectively (in pressure units). These differ from the Benson & Golden assumption (2) that $E_{-1} = 1 \pm 1$ kcal mol^{-1}, but if $\Delta H^{\circ}_{f,298}(\text{t-Bu})$ is substantially higher, a disconcerting negative value of E_{-1} is required.

BOND SCISSION   There are several reported values of the rate constant for the decomposition of hexamethyl ethane (HME); Tsang (32) reports, using shock tube techniques, $\log k_2/s^{-1} = 16.4 - 68.4/\Theta$ [$1000 \leq T/K \leq 1100$], and using flow-tube techniques (56): $\log k_2/s^{-1} = 17.4 - 72.1/\Theta$ [$725 \leq T/K \leq 900$].

$$\text{Me}_3\text{CCMe}_3 \underset{-2}{\overset{2}{\rightleftharpoons}} 2t\text{-Bu} \cdot$$

Atri et al (57) report, from a study of HME pyrolysis in the presence of oxygen:

$$\log k_2/s^{-1} = 16.8 - 69.4/\Theta; \quad [713 \leq T/K \leq 815].$$

These parameters yield indistinguishable rate constants over the stated range. If the value of k_2 is taken as $\log k_2/s^{-1} = -4.90$ at 700 K and the value of k_{-2}, from the expression of Parkes & Quinn (52) [which agrees exactly with the value of Choo et al (53) at 670 K], is taken to be $\log k_{-2}/\text{cm}^3 \text{ mol}^{-1} \text{ s}^{-1} = 11.8$, we may compute $\Delta H^{\circ}_{f,298}(\text{t-Bu})$ subject to the entropy choice. Using the values above, the results are, $\Delta H^{\circ}_{f,298}(\text{t-Bu}) = 11.8$ and 10.6 kcal mol^{-1}.

An exhaustive (and exhausting) study of the papers on which these conflicting values are based leaves us perplexed. The values above imply a problem with the iodination work. Either Teranishi & Benson (50) and Knox & Musgrave (51) are equally wrong (i.e. k_1 too high by a factor of 10), or k_{-1} has substantial (2-3 kcal mol^{-1}) *negative* activation energy if the "high" values for $\Delta H^{\circ}_{f,298}(\text{t-Bu})$ are accepted. This is in conflict with any theory about H-atom metathesis reactions (13), to say nothing of the experimental results of the Rossi & Golden study (18, 19). Furthermore, the same sorts of errors would have to exist for ethyl and i-propyl radicals, which should scale up with t-butyl. [The methyl radical heat of formation derived from iodination rates (2) has been validated by spectroscopic measurements (49, 58) and in equilibrium studies (17), so the reaction $CH_3 + HI \rightarrow CH_4 + I$ is known to have $E \approx 1$ kcal mol^{-1}.]

If the high heat of formation values from the bond-scission studies (32, 56, 57) are incorrect, either the three studies of k_2 have to be a factor of ten or more too low, or k_{-2} is a factor of ten too high. The former seems unlikely given the different techniques used in the bond scission studies, and the latter seems equally unlikely, given the agreement between Parkes & Quinn (52) and Choo et al (53). Also, k_{-2} cannot be too low or it would not agree with the observation that this process is diffusion controlled in solution (59).

There are several other methods for determination of $\Delta H^\circ_{f,298}(t\text{-Bu})$. Canosa & Marshall (47) review these. In their own study, they rely on measurements of $H + \mathord{>\mkern-5mu=} \underset{-3}{\overset{3}{\rightleftarrows}} t\text{-Bu}\cdot$, both k_3 and k_{-3}, and thus K_3. Using an entropy much like the lower value suggested above, they report $\Delta H^\circ_{f,298}(t\text{-Bu}) = 10.6 \pm 0.5$ kcal mol^{-1}. We find that a simple RRK calculation suggests that k_3 might be pressure-dependent under these conditions. If k_3 were higher, then $\Delta H^\circ_{f,298}(t\text{-Bu})$ would be lower! Other methods depend on values for k_2 or rate constants for $CH_3 + t\text{-Bu}$ recombination, which cannot be said to be known accurately.

Benson and co-workers (unpublished data, 1981) have studied the equilibrium: $Br + i\text{-}C_4H_{10} \rightleftarrows t\text{-Bu} + HBr$ and claim to measure second-law values corresponding to S°_{298} and $\Delta H^\circ_{f,298}(t\text{-Bu})$ of 74 ± 2.0 e.u. and 8.3 ± 0.5 kcal mol^{-1}. This direct method should be accurate, but the work is not published at this date and can only be regarded tentatively.

We are tempted to follow the easy course here of averaging all the possible values and recommending $\Delta H^\circ_{f,198}(t\text{-butyl}) = 10.2 \pm 1.5$ kcal mol^{-1}. We resist this temptation and, instead, offer our opinion that the iodination system result is the most likely to be correct. This is based on the fact that several other BDEs determined this way have been shown to be correct (i.e. in agreement with other methods), and on the extreme unlikelihood that $R + HI$ reactions have negative activation energies. It should be borne in mind by those who disagree with our choice that the t-butyl value cannot be changed in isolation; all the iodination-based values would have to be rethought. The inability to make a hard choice for t-butyl doesn't cause sanguinity about such a course of action!

We also recommend the lower entropy for t-butyl radicals, $S^\circ_{298}(t\text{-butyl}) = 74.0$ cal mol^{-1} K^{-1}. We find the argument compelling that if the H–H nonbonded distance in the t-butyl of Yoshimine & Pacansky (54b) is, in fact, ~2.3 Å as we calculate, then the repulsive interaction is ~1.3 kcal mol^{-1} per H-H interaction [see Figure 1 in (60)]. Thus, one low torsional frequency and two high ones (barrier = 2.4 kcal mol^{-1}) seems reasonable.

We recommend $\Delta H^{\circ}_{f,298}(t\text{-Bu}) = 8.7 \pm 1$ kcal mol^{-1} [$DH(t\text{-Bu-H}) = 93 \pm 1$ kcal mol^{-1}], while remaining unsatisfied with the discrepancy discussed above.

We have in this analysis avoided using data not determined on neutrals in the gas phase. An interesting solution study by Griller and co-workers has concluded (46a,b) that alkyl radicals are 1-2 kcal mol^{-1} less stable than the Golden & Benson values (2). In particular, they recommend $\Delta H^{\circ}_{f,298}(t\text{-Bu}) = 9.4 \pm 1$; their value was obtained using the lower entropy and was derived relative to a value for $\Delta H^{\circ}_{f}(CH_3) = 34.7$, which is lower than that recommended here by 0.7 kcal mol^{-1}. Thus, their value could be as high as 11.0 kcal mol^{-1}, providing that the heats of solution of CH_3 and t-butyl are only trivially different!

There are several values (59) of the heat of formation of $t\text{-Bu}^+$-cation that can be combined with the value of Koenig et al (238) and Houle & Beauchamp (62) for the adiabatic IP of the radical (6.70 eV) to produce a heat of formation, but they vary too much to provide a basis for choosing between the values of ca. 8 kcal/mol^{-1} and ca. 11 kcal/mol^{-1}.

BDEs in Hydrocarbons and Their Derivatives

ALKANES With the qualification discussed above, the BDEs in saturated alkanes (open chain) are well characterized by the values 98, 95, and 93 (all ± at least 1) kcal mol^{-1} for primary, secondary, and tertiary C-H bonds (2). The radical heats corresponding to these BDE values, in conjunction with heats of formation of other species of interest, lead to other BDEs. Values derived in this manner are indicated in Table 2 (and the following tables) by the footnote c. Generally, we have limited these derived values to cases of prototypical systems (i.e. radical center primary, secondary, tertiary, connected to an olefin, etc), in cases in which published heats of formation exist for the parent compounds. The selection of BDE values thus derived is usually sufficient to allow the reader to estimate related bond energies roughly by inspection, and more precisely with the aid of additivity procedures (13). The general presentation of values to a tenth of a kilocalorie is not meant to imply a level of accuracy, but is intended to minimize accumulation of roundoff errors, since a number of values are often combined to obtain a heat of formation that is then used to obtain other BDE values, etc. The commonly invoked uncertainty of ±1 kcal mol^{-1} is often retained here, with the realization that it cannot take into account unforseen systematic error, and is therefore generally optimistic.

HYDROCARBON BOND DISSOCIATION 509

Table 2 BOND DISSOCIATION ENERGIES OF ALKANES

$\Delta H_f^o(R)$	R	(52.1) H	CH$_3$	C$_2$H$_5$	(13.9±1) i-C$_3$H$_7$	(78.6±2) t-C$_4$H$_9$	(47.8) C$_6$H$_5$	(39.1) PhCH$_2$	(9.4) CH$_2$CH:CH$_2$OH
(35.1±0.15) [7]	CH$_3$	105.1 ± 0.2 [17]	85.8 ± 1 c	85.7 c	84.1 ± 1 c	101.8 ± 2 [2,65,66]	75.8 ± 1.5 [19,70-72]	79.4 ± 1 [19,22]	92.3 ± 0.3 c
(25.9 ± 1) [2]	C$_2$H$_5$	98.2 ± 1 c [2]	82.2 ± 1 c	81 c	79.1 ± 1 c	97.4 ± 2 c	71.8 ± 1 c [69]	70.3 ±1	91.5 ± 1 c
(21.0 ± 1) [2]	n-C$_3$H$_7$	86.5 ± 1 c	82.1 ± 1 c	80.4 ± 2 c	78.9 ± 1 c	97.7 ± 2 c	72.1 ± 1 c [69]	70.0 ± 1 c,[210,230]	91.6 ± 1 c
(18.2 ± 1) [2]	i-C$_3$H$_7$	85.7 ± 1 c	81 ± 1 c	79.0 ± 2.5 c	75.6 ± 1 c	95.9 ± 2 c	71.3 ± 1 c [69]	69.6 ± 2	92.7 ± 1 c
(13.0 ± 2) [2]	s-C$_4$H$_9$	85.0 ± 1	80.0 ± 1	78.5 ± 1	--	--	--	--	--
(8.7 ± 1) [2,18,44,46-48,51,235]	t-C$_4$H$_9$	93.2 ± 2	79.1 ± 1 c	75.6 ± 2 c	71.2 ± 1 c	--	69.6 ± 2 a [69]	67.2 ± 1 c	92.8 ± 1 c
(8.7 ± 2) [67]	CH$_2$C(CH$_3$)$_3$ [67]	100 ± 2	82.7 ± 1	--	--	--	--	--	--
(66.9 ± 0.25) [27]	cyclopropyl-H	106.3 ± 0.3 [27,63]							
(51.1 ± 1.6) [225]	cyclopropyl-methyl-H	97.4 ± 1.6 [225]							
(51.2 ± 1) [63]	cyclobutyl-H	96.5 ± 1 [63,226]							
(24.3 ± 1) [63]	cyclopentyl-H	94.5 ± 1 [63]							
(13.9 ± 1) [63]	cyclohexyl-H							95.5 ± 1 [63]	
(12.1 ± 1) [63]	cycloheptyl-H							92.5 ± 1 [63]	
(91.0 ± 1) [63]	spiropentyl-H							98.8 ± 1 [63]	
(32.6 ± 2.5) [68]	norbornyl(3°)-H							96.7 ± 2.5 [68]	

[a] Numbers in brackets denote references.

[b] Values in parentheses near radicals and atom are $\Delta H_{f,298}^o$.

[c] Derived from referenced radical heats of formation and stable species heats of formation from references 13, 14, 15, or 16. The assigned 1 kcal mole^{-1} uncertainties should be read "at least" 1 kcal mole^{-1}.

Cyclic compounds may show variations due to differences in strain energies between the parent compound and the cyclic radical (63). Thus, while DH(cyclopentyl-H) and DH(cyclohexyl-H) differ little from normal secondary C–H bonds, DH(cyclobutyl) is 1 to 2 kcal mol^{-1} higher and DH(cycloheptyl) is 2–3 kcal mol^{-1} lower than secondary (63). (This is surprising, since cycloheptene has only marginally less strain energy than cycloheptane.) It seems reasonable that DH(cyclopropyl-H) should be high, but it is puzzling that DH(spiropentyl-H) (63), at 98.8 kcal mol^{-1} (4 kcal mol^{-1} higher than secondary) is considerably lower than DH(cyclopropyl-H) at 106.3 kcal mol^{-1} (27). This is qualitatively reasonable because in the rigid spiropentyl radical is a cyclopropylmethyl radical in which optimum overlap between the unpaired electron in its p-orbital and the Walsh orbitals in the adjacent ring is assured. However, 7 kcal mol^{-1} of resonance stabilization energy seems too high, since the cyclopropylmethyl radical itself is reported (225) to have little or no (1 ± 1 kcal mol^{-1}) resonance stabilization. In fact a spiropentyl-H BDE of 101–103 kcal mol^{-1} seems likely, since the 98.8 kcal mol^{-1} shown in Table 2 is derived from a series of relative rates of H-atom abstractions by CF$_3$ radical that show cyclopropyl-H to be 101 kcal mol^{-1} rather than the presently accepted 106 kcal mol^{-1} value (27). DH(norbornyl-H) is, as is to be expected, higher than tertiary, because the radical center cannot relieve strain by flattening out (68).

All of the above relationships between primary, secondary, and tertiary cyclic and acyclic alkanes are supported by numerous Evans-Polanyi relationships (23, 24, 34, 35, 63, 64, 66, 105), the dismissing of which would require invoking unforseen and very fortuitous kinetic effects.

ALKENES Linear alkenes have two very different characteristic types of C–H bonds: the vinyl C–H bond, in which the H is bonded to a sp^2 carbon atom, and the allyl C–H bond, in which the H is bonded to an sp^3 carbon atom bonded to an sp^2 site.

In the former instance, evidence (2, 73, 236) points toward a very strong (ca. 110 kcal mol^{-1}) bond, but the exact value has remained elusive. In the latter case, the value of the heat of formation of C_3H_5, the allyl radical, is known to good precision from several experiments (22, 77, 78, 239, 245 and references cited therein). There is also considerable data concerning the effects of substitution on the allyl radical.

The allyl-H bond in propylene is weaker than a primary C–H bond (~ 12 kcal mol^{-1}) (19, 22) because the allyl radical is stabilized by the delocalization of three electrons over three carbon centers (i.e. $\overline{CH_2\text{--}CH\text{--}CH_2}$ is a better description than $CH_2=CH\text{--}CH_2\cdot$). This "resonance stabilization energy" (RSE) (sometimes "allyl resonance energy") may be operationally defined as

$$RSE \equiv DH(\text{primary C–H}) - DH(\text{allyl–H}) \approx 12 \text{ kcal mol}^{-1}.$$

Of course, delocalization can extend over several carbon atoms in a polyenic chain, so that the electron could, in principal, be delocalized over larger odd numbers of carbon atoms [$\cdot CH_2(CH=CH)_nCH=CH_2$]. Data exist (2, 74, 75, 232, 239) that indicate that for $n=1$, $RSE = 19 \pm 3$ kcal mol^{-1}. The RSE is 19 kcal mol^{-1} when it is based on removal of the secondary H in 1,4-pentadiene, but as the values in Table 3 indicate, when the RSE is based on the primary hydrogen in 1,3-pentadiene, it appears to be only 15 kcal mol^{-1}, *unless* the 4.3 kcal mol^{-1} of RSE exhibited by trans-1,3-pentadiene itself (15) is taken into account. The RSE then becomes 19 kcal mol^{-1}, agreeing with the above value, as it should, if it is a property of the radical itself. The same consideration of stabilization in the ground state must be taken into account in the derivation of radical resonance energies from the isomerization kinetics of conjugated olefins (74).

Allylic C–H bonds in cyclic alkenes and dienes show C–H bond weakening much like that in acyclic systems, but the exact amounts are more open to question. The value chosen here for the 1,3-cyclohexadienyl radical is a compromise between the reported BDE [70 kcal mol^{-1} (76, 243, 244)] and the judgment reflected in Ref. 13 that the reported BDE represents an unreasonably high 25 kcal mol^{-1} of stabilization energy. The BDE for cyclopentadienyl-H is adjusted downward from the original iodination values (2) to the same extent that recent bond strength measurements (232) have warranted for iodination-based pentadienyl values. The rationalization for this adjustment arises from the probability that the originally assumed $E_{-1} = 1 \pm 1$ value for reaction -1

$$(HI + \cdot CH_2CH=CHCH=CH_2 \rightleftharpoons I\cdot + H-CH_2CH=CHCH=CH_2)$$

is several kcal mol^{-1} too low for these particular cases when reaction -1 is only slightly exothermic (74). The result of the adjustment seems reasonable, since molecular orbital theory suggests that, apart from strain energy in the sigma-bond system, the stabilization energy will be higher in the closed cyclic system than in either the acyclic pentadienyl or the cyclohexadienyl systems.

Resonance stabilization energy	19	21	22	28	18

Table 3 BOND DISSOCIATION ENERGIES FOR ALKENES, ALKYNES AND AROMATICS

$\Delta H_f^o(R)$	R	H	CH_3
(70.4 ± 2) [2,73,77]	·C=C	110 ± 2 [2,73,77]	100.6 ± 2 c
(78.6 ± 2) [2,43,65,66]	·C$_6$H$_5$	110.9 ± 2 [2,43,65,66]	101.8 ± 2 c
(135 ± 1) [80,227]	·C≡C	132 ± 5 [80,227]	125.7 ± 1 c
(-130.9 ± 2) [154]	·C$_6$F$_5$	113.9 [88]	- - -
(39.1 ± 1.5) [22]	·C-C=C	86.3 ± 1.5 [22,77,245]	74.4 ± 1 c
(30.0 ± 1.5) [82,83]	·C-C=C-C	85.6 ± 1.5 [82]	72.9 ± 0.8 [82,83]
(30.4 ± 1.3) [2]	·C(C)-C=C	82.5 ± 1.3 [2]	- - -
(18.5 ± 1.5) [84]	·C(C)(C)-C=C	77.2 ± 1.5 [84]	68.1 ± 1.5 c,[84]
(9.0 ± 1.5) [85]	·C(C)(C)-C(C)=C	76.3 ± 1.1 [85]	- - -
(9.5 ± 1.5) [85]	·C-C(C)=C(C)	78.0 ± 1.1 [85]	- - -
	·C(Cl)-C=C	88.6 ± 1.4 [101]	- - -
(38.4) [79]	cyclopentenyl	82.3 ± 1 [79]	- - -
(49 ± 3) [2,74,75,232]	·C(C=C)(C=C)	76 ± 3 [2,74,75,232]	- - -
(49 ± 3) [2,74,75,232]	·C=C-C=C	83 ± 3 [2,74,75,232]	- - -
(57.9 ± 1.5) [77,78]	cyclopentadienyl	71.1 ± 1.5 [77,78]	- - -
(47 ± 5) [13,76]	cyclohexadienyl	73 ± 5 [13,76]	- - -
(64.8 ± 2) [77,87]	cycloheptatrienyl	73.0 ± 2 [77,87]	- - -
(105.1 ± 4.1) [77]	cyclopropenyl	90.6 ± 4 [77]	- - -
(61.4 ± 1.5) [86]	diphenylmethyl	75.3 ± 1.5 [86]	- - -

Table 3 (concluded)

$\Delta H_f^o(R)$	R	H	CH_3
(47.8 ± 1.5) [19,70,71,72]	·C-C₆H₅	88.0 ± 1 [19,70,71,72,77]	75.8 ± 1 c
(60.4) [29]	1-naphthyl-C·	85.1 ± 1.5 [33]	72.9 ± 1.5 [33]
(80.7) [29]	9-anthryl-C·	81.8 ± 1.5 [33]	67.6 ± 1.5 [33]
(74.4) [29,103]	9-phenanthryl-C·	85.1 ± 1.5 [33,103]	72.9 ± 1.5 [33,103]
(40.4) [102]	·C-C-phenyl	85.4 ± 1.5 [102]	74.6 ± 1.5 [102]
(33.2) [102]	C-Ċ-C phenyl	84.4 ± 1.5 [102]	73.7 ± 1.5 [102]
- - -	(C₆H₅)₂CH·	84 ± 2 [93]	72 ± 2 [93]
- - -	(C₆H₅)₂C(C)·	81 ± 2 [93]	69 ± 2 [93]
- - -	·furyl	86.5 ± 2 [93]	75 ± 2 [93]
- - -	·indenyl	[84 ± 3] [93]	
(81.4 ± 2) [98,228,229]	·C-C≡C	89.4 ± 2 [98]	76.0 ± 2 [228]
(70.2) [98,99]	·C-C≡C-C	87.2 ± 2 [98,99]	73.7 ± 1.5 [99]
(65.2 ± 2.3) [100]	·C̈-C≡C-C (C)	87.3 ± 2.7 [100]	76.7 ± 1.5 [100]
(53.0 ± 2.3) [100]	·C̈-C≡C-C (C, C)	82.3 ± 2.7 [100]	72.5 ± 1.5 [100]
(61.5 ± 2) c [94]	·C̈-C≡C (C)	81.0 ± 2.3 [94]	70.7 ± 1.5 [94]
(70.5 ± 2.2) [90]	·C̈-C≡C (C)	83.1 ± 2.2 [90]	73.0 [90]

$CH_2=CHCH_2--OC_6H_5$			49.8 ± 2 [81]
$CH≡CCH_2--n-C_3H_7$			73.2 ± 1.5 [98,231]
$CH≡CCH_2--s-C_4H_9$			71.7 ± 1.5 [98]
$CH≡CCH_2--CH_2C_6H_5$			61.4 ± 2 [229]
$CH_2=C=CH_2--CH_3$			76.5 ± 2.2 [90]

[a] Numbers in brackets denote references.
[b] Values in parentheses near radicals and atom are $\Delta H_{f,298}^0$.
[c] Derived from referenced radical heats of formation and stable species heats of formation from references 13, 14, 15, and 16. The assigned 1 kcal/mole^{-1} uncertainties should be read "at least" 1 kcal mole^{-1}.

As a prototypical radical of probable importance in the pyrolysis of certain aromatic systems (89), the cyclohexadienyl radical stability should be verified.

The stability of the cyclohexadienyl (86) radical formed by abstraction of an H-atom from 9,10-dihydroanthracene corresponds to a C–H bond strength in the latter compound of 77 kcal mol^{-1}, or 18 kcal mol^{-1} of RSE. Thus, the ease of hydrogen removal from the dihydro compound is roughly equal to that for 1,4-cyclohexadiene. However, since the total resonance energy in anthracene is far less than three times that of benzene (13, 15), formation of cyclohexadienyl radicals by H-atom addition to the respective aromatic compound is far more favored for anthracene than for benzene ($\Delta H°_{298} = -47$ and -25 kcal mol^{-1}, respectively), and consequently is expected to be more important in pyrolyses of the three-membered aromatic (89). The observed stabilization energy for the cyclohexadienyl radical in the anthracene system (86) demonstrates that a method recently developed for the estimation of benzylic radical stabilization energy (92, 93) is valid also for cyclohexadienyl radicals in polycyclic aromatic systems.

ALKYNES The HC≡C–H BDE has not been determined with high precision, but it seems clear that this bond is at least 15 kcal mol^{-1} stronger than the C–H bonds in ethylene. As the values in Table 3 indicate, the propargyl radical exhibits ~4 kcal/mole less stabilization energy than the allyl radical, consistent with the expectation that nondegenerate canonical forms should result in lower resonance energy. As with allylic radicals, alkyl substitution on the terminal positions of the propargyl radical decreases the bond strength to the same or to a lesser extent than it would in an unstabilized alkyl radical (90), and thus does not appreciably increase the propargylic stabilization energy as commonly defined.

PHENYLIC AND BENZYLIC BONDS As with vinyl C–H bonds, the phenyl C–H bond is very strong, and for that reason, its abstraction is difficult to measure directly without interference from competing reactions, particularly addition-elimination processes (8, 34). Nevertheless, a number of studies clearly place it in the 110–111 kcal mol^{-1} range (2, 43, 65, 66, 154, 236). The C_6F_5–X (where X = H, F) bonds appear to be several kcal mol^{-1} stronger than the corresponding C_6H_5–X bonds (88, 154), as is the case in the fluorinated alkanes (see below).

Recent experiments show the benzyl radical (19, 77) to exhibit slightly less stabilization energy than the allyl radical (22). These measurements decrease the uncertainty of the accepted heat of formation of the benzyl radical. However, the need for verification and improved precision, such as might come from equilibrium studies, still exists, since there is probably no other single radical upon which so many bond energy values depend.

Benzylic radicals in polycyclic aromatic systems can exhibit substantially more stabilization energy than benzyl radical itself (33). In three-ringed systems, this "extra" stabilization reaches a maximum at 8 kcal mol^{-1} in the 9-anthrylmethyl radical (33), in good agreement with the predictions of the simple two-parameter estimation method developed by Stein & Golden (92).

The stabilization energy of the diphenylmethyl radical has not been unambiguously determined in the gas phase. However, relative rates in gas-phase (93) pyrolyses and liquid-phase Evans-Polanyi relationships (95, 96) and stilbene and tetraphenylethylene isomerization activation energies (97) indicate that Ph$_2$CH–H is 3–5 kcal mol^{-1} weaker than PhCH$_2$–H, and 0–2 kcal mol^{-1} weaker than PhCH(CH$_3$)–H. In other words, making the benzyl radical secondary by substitution of a methyl group stabilizes it just about as much as substitution of a second phenyl group; the second phenyl group results in little increase in resonance stabilization energy. The presence of a nitrogen in the aromatic ring has little effect on benzylic bond strengths: The bonds in 2-, 3-, and 4-ethyl pyridine are 2, 0, and 1 kcal mol^{-1} stronger than in ethyl benzene (200). Similarly, the presence of *ortho*-methyl substituents has only a minor (steric) effect on benzylic bond strength, weakening the bond in 2-ethyl toluene by about 1 kcal mol^{-1} compared to ethyl benzene itself (224).

CARBON-CENTERED OXYGEN-CONTAINING RADICALS   The values listed in Table 4 show that the BDE for a hydrogen on a carbon bound to an acyclic or cyclic alcohol or ether-oxygen is lowered by 3±2 kcal mol^{-1}. This is consistent with the results of an extensive study by Malatesta & Ingold (105) of low-temperature, liquid-phase *t*-butoxy radical abstractions of hydrogen from cyclic ethers. In this study (105), the abstraction rate of α-hydrogens was shown to be markedly increased when the dihedral angle between the developing p-orbital and the oxygen p-type orbital is very small. Barring steric or electronic effects that could affect differently the radical product and the transition state leading to it, the enhanced rates suggest that the BDE of the "axial" α-hydrogen in tetrahydrofuran is 3.5 to 7 kcal mol^{-1} lower than that in cyclopentane. This stabilization is associated with a three-center, two-electron interaction involving the p-type nonbonding electrons on the oxygen and the p-type single electron on the carbon (–ĊH–Ö– ↔ –CH=Ȯ–). Clearly, in this case, the stabilizing effect of two additional electrons in a bonding orbital is greater than the destabilizing effect of one electron in an

Table 4 BOND DISSOCIATION ENERGIES OF C,H,O COMPOUNDS

		Oxygen-Centered Radicals				
$\Delta H^0_{f,298}(R)$	R_1	(52.1)[a] H	R_1	(9.4) OH	(35.1) CH_3	(25.9) C_2H_5
(9.43) [14][b]	OH	119 ± 1 [14]	51 ± 1 [14]	51 ± 1 [14]	92.3 ± 1 c	91.5 ± 1 c
(4.2) [120]	OCH_3	104.4 ± 1 [120]	37.6 ± 2 [120,122]	- - -	83.3 ± 1 c	81.8 ± 1 c
(-4.1) [120]	OC_2H_5	104.2 ± 1 [120]	37.9 ± 1 [120]	- - -	82.7± 1 c	82.1 ± 1 c
(-9.9) [120]	$O-n-C_3H_7$	103.4 ± 1 c	37.1 ± 1 [120]	- - -	82.0 ± 1 c	- - -
(-15.0) [4]	$O-n-C_4H_9$	102.9 ± 1 c	- - -	- - -	- - -	- - -
(-12.5) [120]	$O-i-C_3H_7$	104.7 ± 1 c	37.7 [120]	- - -	82.8 ± 1 c	- - -
(-16.6) [123]	$O-s-C_4H_9$	105.5 ± 1 c	36.4 ± 1 [12,124]	- - -	- - -	- - -
(-21.7) [120,123]	$O-t-C_4H_9$	105.1 ± 1 [82,119,120]	38.0 ± 1 [120]	- - -	83.1 ± 1.5 c	- - -
- - -	$O-t-C_5H_{11}$	- - -	39.3 ± 1 [130]	- - -	- - -	- - -
- - -	$OCH_2C(CH_3)_3$	102.3 ± 1.5 [119,82]	36.4 ± 1 [126]	46.3 ± 1.9 [120,121]	- - -	- - -
(11.4) [81,133,134]	OC_6H_5	86.5 ± 2 [81]	- - -	- - -	63.8 ± 1 c	63 ± 1.5 [81]
- - -	OCF_3	- - -	46.2 ± 1 [128]	- -	- - -	- - -
- - -	$OC(CF_3)_3$	- - -	35.5 ± 1.1 [129]	- - -	- - -	- - -
(2.5 ± 0.6) [127]	O_2H	87.2 ± 1.0 [127]	- - -	- - -	- - -	- - -
(-49.6 ± 1) [24]	O_2CCH_3	105.8 ± 2 [24]	30.4 ± 2 [24,119]	- - -	- - -	82.6 ± 1.5 c
(-54.6 ± 1) [24]	$O_2CC_2H_5$	106.4 ± 2 [24]	30.4 ± 2 [24,119]	- - -	- - -	- - -
(-59.6 ± 1) [24]	$O_2C-n-C_3H_7$	105.9 ± 2 [24]	30.4 ± 2 [24,119]	- - -	- - -	- - -

$CH_3CO_2-CH_2C_6H_5$	67 ± 2 [8]
$C_6H_5CH_2-OCH_3$ 70.5 ± 2	c
$CF_3OO-OCF_3$ 30.3 ± 2	[214]
HO_2-NO_2	23 ± 2 [215]
$CH_2C(C)O_2-NO_2$ 26 ± 2	[216]

continued..

Table 4 BOND DISSOCIATION ENERGIES OF C,H,O COMPOUNDS (concluded)

		Carbon-Centered Radicals			
$\Delta H^{o}_{f,298}(R)$	R_1	(52.1) H	(35.1) CH_3	(78.6) C_6H_5	R_1
(8.9 ± 1.2) [2,24]	CHO	87 ± 1 [2]	82.5 ± 1 c	96.3 ± 1.5 c	68.4 ± 2 c
(-5.8 ± 0.4) [2,24]	$COCH_3$	86.0 ± 0.8 [107]	81.2 ± 1 c	93.5 ± 1.5 c	67.4 ± 2.3 [8,118]
(17.3) [108]	$COCH:CH_2$	87.1 ± 1 [108]	- - -	- - -	- - -
(-10.2 ± 1) [111]	COC_2H_5	87.4 ± 1 [111]	80.6 ± 2 c	94.4 ± 2 c	- - -
(26.1 ± 2)	COC_6H_5	86.9 ± 1 [112]	81.9 ± 2 c	90.1 ± 2 c	66.4 [8]
	$COCF_3$	91.0 ± 2 [114]	- - -	- - -	- - -
(-5.7) [109]	CH_2COCH_3	98.3 ± 1.8 [109,110]	86.4 ± 1 c	- - -	- - -
(-16.8 ± 1.7) [115]	$CH(CH_3)COCH_3$	92.3 ± 1.4 [115]	- - -	- - -	- - -
(-6.2 ± 1.5) [2,131]	CH_2OH	94 ± 2 [2]	- - -	96.4 ± 2 c	80.2 ± 3
(-15.2 ± 1) [116]	$CH(OH)CH_3$	93 ± 1 [116]	- - -	- - -	84.9 ± 2 c
(-26.6 ± 1.1) [2,24]	$C(OH)(CH_3)_2$	91 ± 1 [2]	- - -	- - -	- - -
(-2.8 ± 1.2) (2,24]	CH_2OCH_3	93 ± 1 [2]	86.4 ± 1.5 c	- - -	- - -
(-4.3 ± 1.5) [24,113]	tetrahydrofuran-2-yl	92 ± 1 [2]	- - -	- - -	- - -
(0.0) [132]	$CH(OH)CH:CH_2$	81.6 ± 1.8 [132]	- - -	- - -	- - -
(-40.4 ± 1) [24,117]	$COOCH_3$	92.7 ± 1 [117]	- - -	- - -	- - -
(-16.7) [24,106]	$CH_2OCOC_6H_5$	100.2 ± 1.3 [106]	- - -	- - -	- - -
(-53.3) [112,8]	$COOH-CH_2C_6H_5$	67 [8]	- - -	- - -	- - -
(59.2 ± 3) [8]	$(C_6H_5)_2CH-COOH$	59.4 ± 3 [8]	- - -	- - -	- - -
	$C_6H_5CH_2CO-CH_2C_6H_5$	65.4 ± 2 [8]	- - -	- - -	- - -
(47.8 ± 1.5) [19]	$C_6H_5CH_2-OH$	81.2 ± 1.5 c	- - -	- - -	- - -
(11.4 ± 2) [81]	$C_6H_5CO-CF_3$	73.8 ± 2 [8]	- - -	- - -	- - -

[a]Values in parentheses near radicals and atom are $\Delta H^{o}_{f,298}$.

[b]Numbers in brackets denote references.

[c]Derived from referenced radical heats of formation and stable species heats of formation from references 13, 14, 15, and 16. The assigned 1 kcal mole^{-1} uncertainties should be read "at least" 1 kcal mole^{-1}.

antibonding orbital. This phenomenon apparently does not apply to the α-hydrogens of alkyl carboxylates, as the methyl hydrogens of methyl benzoate are reported to have a BDE of 100 kcal mol^{-1}(106). Finally, it appears that the bonding of a hydrogen to a carbonyl carbon, which is weakened by 11 kcal mol^{-1} in acetaldehyde, or an alkoxy carbon is less weakened if those groups are part of a carboxylic acid or ester grouping.

OXYGEN-CENTERED RADICALS Alkoxy radicals, barring nonbonded interactions, "follow" the rules of group additivity: the O–H bonds in CR$_3$OH and the O–O bonds in alkyl peroxides are 104±2 and 38±1 kcal mol^{-1}, regardless of the nature of the R groups. These values are derived from ROOR and RONO pyrolyses. The exception to this generalization about RO–OR bond strength appears to be the CF$_3$O–OCF$_3$ bond, which, at 96 kcal mol^{-1} is substantially stronger (128, 213). If gas phase acidity-electron affinity cycles are used instead, the RO–H bond energies appear to be systematically 2 kcal mol^{-1} lower at 102±1 kcal mol^{-1}. The neopentyl alcohol RO–H bond for which no RONO-based values exist is shown in Table 4 as an ion cycle derived value of 102 kcal mol^{-1} (191). The source of the systematic difference is not obvious.

The O–H bonds in carboxylic acids appear slightly stronger at 106±2 kcal mol^{-1}. As previously compiled (1), these bonds varied from 103 to 112. However, such variation seemed implausible in the case of carboxylic acids where the variation in R is one more atom removed from the radical center than in the case of the alkoxy radicals. Use of published acyl radical (24) and parent acid (15) heats of formation caused the RCO$_2$–H values to converge plausibly to those shown in Table 4. This is consistent with the bonds in diacyl peroxides being constant at 30±1 kcal mol^{-1} (12).

SULFUR-CONTAINING COMPOUNDS The RS–H bonds in thiols are shown in Table 5 as constant at 91±1.5 kcal mol^{-1}. This is 1 kcal mol^{-1} lower than shown in Ref. (7), as a result of using a different recombination activation energy assumption and newer methyl and benzyl radical DH_f°, 298 values in the thermochemical cycle. As in the case of the alcohols, the pyrolysis-derived values selected here are, for reasons that are not apparent, ~2 kcal mol^{-1} higher than those derived from ion thermochemical cycles (227, 246). They are also ~2 kcal mol^{-1} higher than values shown in Ref. (243).

The values in Table 5 show thio-ether bonds to be 5 to 10 kcal mol^{-1} weaker than the corresponding ether bonds and to be more sensitive to the nature of the alkyl radical (i.e. R$_2$ in R$_1$S–R$_2$). Thus, the MeS–Et bond appears to be 8 kcal mol^{-1} weaker than the MeS–Me bond, whereas in the analogous ethers, the difference is only 1.5 kcal mol^{-1}.

Table 5 BOND DISSOCIATION ENERGIES OF SUFLFUR-CONTAINING COMPOUNDS

R_1-R_2	D^o_{298}		$\Delta H^o_{f,298}(R_1)$	Reference
HS-H	91.1 ± 1		33.6 ± 1.1	180,14
CH_3S-H	90.7 ± 2		33.2 ± 1.5	7
RS-H	91 ± 1.5		- - -	a, 7
CH_3-SH	74 ± 1.5		- - -	a, 7
C_2H_5-SH	70.5 ± 1.5		- - -	a, 7
t-Bu-SH	68.4 ± 1.5		- - -	7
C_6H_5-SH	86.5 ± 2		- - -	7
CH_3S-CH_3	77.2 ± 2		- - -	7,150
CH_3S-C_2H_5	73.3 ± 2		- - -	a
CH_3S-n-C_3H_7	73.78 ± 2		- - -	a
PhS_H	83.3 ± 2		- - -	150
PhS-CH_3	69.4 ± 2		54.9 ± 2	150
$PhCH_2$-SCH_3	61.4 ± 2		- - -	150
SF_5-F	91.1 ± 3.2	(D^o_o)	-215.7 ± 3.2	182,185,194
SF_4F	53.1 ± 6	(D^o_o)	-180.9 ± 5.0	182,183
SF_3-F	84.1 ± 3	(D^o_o)	-115.2 ± 5.8	182
SF_2-F	63.1 ± 7.1	(D^o_o)	-70.4 ± 4.0	182,193
SF-F	91.7 ± 9.3	(D^o_o)	2.9 ± 1.5	182,193
S-F	81.2 ± 8.6	(D^o_o)	- - -	182
HS-I	49.4 ± 2		33.6 ± 2	183
CS-S	102.9 ± 3	(D^o_o)	65.0 ± 0.4	190,195
OS-O	130.1 ± .5	(D^o_o)	- - -	149,186,187
OS-I	43		- - -	188
CSC-Cl	63.4 ± 0.5	(D^o_o)	43 ± 1	189
CH_3SO_2-CH_3	66.8		- - -	112,15
CH_3SO_2-$CH_2CH:CH_3$	49.6		- - -	113,5
CH_3SO_2-$CH_2C_6H_5$	52.9		- - -	113,5
SF_5O-OSF_5	37.2		- - -	211
SF_5OO-OCF_3	30.3		- - -	212
RS_2-H	70 ± 1.5		- - -	7
RS_2-CH_3	57 ± 1.5		- - -	7
HS-SH	66 ± 2		- - -	7
RS-SR	72 ± 2		- - -	a

[a]Derived from referenced radical heats of formation and stable species heats of formation from references 13, 14, 15, and 16.

The recently completed (180) series of BDE values for sequential loss of F from SF_6 show a very pronounced alternation illustrating the cost of "promoting" an electron from a nonbonding to a bonding orbital.

AMINES AND NITRILES   In compiling the values in Table 6, we drew heavily on the recent reviews of Batt & Robinson (4) and Batt (9). Nevertheless, the chosen values sometimes differ from theirs, for reasons stated above. The listed values indicate that whereas the N–H bond in NH_3 is about 3 kcal mol^{-1} stronger than the C–H bond in CH_4, the N–H bond in dimethylamine is several kcal mol^{-1} weaker than the corresponding bond in propane. With the exception of the NH_2 radical itself, derivation of the values for all of the radicals, $R_1R_2N\cdot$ depend, in some way or another, on the heat of formation of the benzyl radical. The values chosen here were based as much as possible on relative pyrolysis rates of the isoelectronic hydrocarbon and amine analogs.

The amino methyl radical ($\cdot CH_2NH_2$) is stabilized by ~ 5 kcal mol^{-1} compared to its isoelectronic counterpart, the ethyl radical, and recent values derived from appearance potential measurements by Griller & Lossing (143) suggest that the amount of stabilization increases markedly as the nitrogen becomes alkyl substituted. Such an effect is not seen in other resonance-stabilized radicals (94, 98, 100).

An α-cyano group appears to stabilize carbon-centered radicals by about 5 kcal mol^{-1} (146, 233). This is somewhat surprising, since the contributing structure (e.g. $CH_2=C=N:$) used to rationalize resonance stabilization has lost the benefit of the extremely strong $C\equiv N$ multiple bond (147), a situation which, in the analogous $\cdot CH_2COCH_3$ system, is used to rationalize the fact that there is little or no resonance stabilization in the acetonyl radical (109).

NITRO AND NITROSO COMPOUNDS AND NITRATES   The various C–N bond energies of C-nitroso compounds listed in Ref. (9) indicate that the correct values are not yet firmly established. Those in Table 7 were selected from the range available to conform to trends observed in other alkyl-X series. Thus, methyl-, i-propyl-, and t-butyl-NO are shown as exhibiting values of 40, 36.5, and 34 kcal mol^{-1}, respectively. On the other hand, the situation seems straightforward with the nitrites and nitrates. The RO–NO and RO–NO_2 bonds have been measured as 40.8 ± 1 and 40.7 ± 0.5, irrespective of the substitution on the α-carbon atom (4). As in the case of peroxide (12), this is merely an illustration of "adherence" to the rules of group additivity: the R_3CO–NO or R_3CO–NO_2 bond strengths are not sensitive to changes that are two atoms removed. This view leads to the observation that the C–NO_2 bonds at 60.8 and 58.5 kcal mol^{-1} for the Me–NO_2 and t-butyl–NO_2 show less

HYDROCARBON BOND DISSOCIATION   521

Table 6 BOND DISSOCIATION ENERGIES OF NITROGEN-CONTAINING COMPOUNDS

$\Delta H^o_{f,298}(R)$		(52.1) H	(35.1) CH_3	Amines and Nitriles (25.9) C_2H_5	(47.8) $C_6H_5CH_2$	(78.6) C_6H_5	(44.3) NH_2
(44.3 ± 1.1) [4,140,141]	NH_2	107.4 ± 1.1 [4,140,141]	84.9 ± 1 c	81.6 ± 2 c	71.1 ± 1 [136]	102 ± 2 c	65.8 c, [14]
(42.4 ± 2) [4,136]	$NHCH_3$	100.0 ± 2.5 [4,136]	82.2 ± 2.5 c	79.8 ± 1 c	68.7 ± 2 [136]	100.6 ± 2.5 c	64.1 ± 2 c [113]
(34.7 ± 2) [4,136]	$N(CH_3)_2$	91.5 ± 2 [4,136]	75.5 ± 2.5 c	72.3 ± 2 c	62.1 ± 2 [136]	93.2 ± 2.5 c	59.0 ± 2 c, [113]
(56.7 ± 2) [4,137]	NHC_6H_5	88.0 ± 2 [4,137]	71.4 ± 2 [137]	69.1 ± 2 c	---	81.1 ± 2.5 c	52.3 ± 2 c
(55.8 ± 2) [137]	$N(CH_3)C_6H_5$	87.5 ± 2 [137]	70.8 ± 2 [137]	---	---	---	---
(8 ± 1) [9,138]	NF_2	75.7 ± 2.5 [138]	---	---	---	---	---
(112 ± 5) [139]	N_3	92 ± 5 [139]	---	---	---	---	---
(35.7 ± 2) [41,142,143]	CH_2NH_2	93.3 ± 2 [143]	82.2 ± 2 c	79.4 ± 2 [143]	68.0 ± 2 [150]	93.3 ± 2 c	---
(30 ± 2) [143]	$CH_2NH(CH_3)$	87 ± 2 [143]	76.6 ± 2.5 c	---	---	87.8 ± 2.5 c	---
(26 ± 2) [143]	$CH_2N(CH_3)_2$	84 ± 2 [143]	73.7 ± 2.5 c	---	---	84.2 ± 1.5 c	---
(104 ± 2) [14,144]	CN	123.8 ± 2 [14,144]	121.8 ± 2 c	118.2 ± 2 c	---	131 ± 2 c	---
(58.5 ± 2.5) [146]	CH_2CN	93 ± 2.5 [9,146]	81.3 ± 3 c	76.9 ± 1.7 [146]	---	---	---
(50.0 ± 2.3) [148]	$CH(CH_3)CN$	89.9 ± 2.3 [147]	78.8 ± 2 [147]	---	---	---	---
(39.8 ± 2.0) [148]	$C(CH_3)_2CN$	86.5 ± 2 [148]	74.7 ± 1.6 [148]	---	---	---	---
(59.4) [113,8]	$C(CH_3)(CN)C_6H_5$	---	59.9 [113]	---	---	---	---

$\Delta H^o_{f,298}(R)$		Nitro and Nitroso Compounds and Nitrates (21.6) NO	(7.9) NO_2	(17) ONO_2
(52.1) [14]	H	---	78.3 ± 0.5 [13]	101.2 ± 0.5 [13]
(9.4) [14]	OH	49.3 [1]	49.4 [14]	39 ± 2 [215]
(35.1 ± .15) [17]	CH_3	40.0 ± 0.8 [152]	60.8 [4]	---
(25.9 ± 1) [2]	$i-C_3H_7$	---	58.6 [4]	---
(18.2 ± 1) [2]	$i-C_3H_7$	36.5 ± 3 [155]	59.0 [4]	---
(8.7 ± 1) [18,23]	$t-C_4H_9$	39.5 ± 1.5 [151]	58.5 [4]	---
(-111.7 ± 3.6) [6,135]	CF_3	42.8 ± 2 [153]	---	---
(19 ± 1) [161]	CCl_3	32 ± 3 [156]	71.3 ± 1 [4]	---
(79 ± 2) [2,65,66]	C_6H_5	50.8 ± 1 [154]	---	---
(-130.9 ± 2) [154]	C_6F_5	49.8 ± 1 [154]	---	---
---	$C(NO_2)_3R_2$	---	48.8 ± 2.5 [4]	---
---	$C(NO_2)_2R$	---	43.7 ± 2.5 [4]	---
---	$C(NO_2)_3$	---	40.5 ± 1 [4]	---
---	RO	40.8 ± 1 [4]	40.7 ± 0.5 [4]	---
(7.9) [14]	NO_2	9.7 ± 0.5 [13,14]	13.6 [13,14]	---

[a] Numbers in brackets denote references.
[b] Values in parentheses near radicals and atoms are $\Delta H^o_{f,298}$.
[c] Derived from referenced radical heats of formation and stable species heats of formation from references 10, or 11, 12, 13. The assigned 1 kcal^{-1} uncertainties should be read "at least" 1 kcal mole^{-1}.

Table 7 BOND DISSOCIATION ENERGIES OF HALOCARBONS

$\Delta H^o_{f,298}(R)$	R	(52.1)[a] H	(35.1) CH_3	(18.9) F	(28.9) Cl	(26.7) Br	(25.5) I	(-111.1) CF_3
(35.1 ± .2)[b] [17]	CH_3	105.1 ± 0.2 [17]	90.4 ± 0.2 c	109.9 ± 1 c	84.6 ± 0.2 c	70.9 ± 0.3 c	57.2 ± 0.3 c	101.6 ± 3 c
(25.9) [2]	C_2H_5[2]	98.2 ± 1 [2]	- - -	107.7 ± 1 c	79.9 ± 1 c	67.8 ± 1 c	53.4 ± 1 c	- - -
(18.2 ± 1) [2]	$i-C_3H_7$	95.1 ± 1 [2]	85.7 ± 1 c	106.5 ± 1.1 c	80.7 ± 1.4 c	68.4 ± 1.2 c	53.5 ± 1.8 c	- - -
(47.8 ± 1.5) [19,70,71,72]	$CH_2C_6H_5$	88.0 ± 1 [19,70,71,72,77]	75.8 ± 1 c	- - -	72.2 ± 1.5 c	57.6 ± 1.5 c	48.2 ± 1.5 c	- - -
(-7.8 ± 2) [157,158]	CH_2F	100 ± 2 [157,158]	119 ± 2 c	85.3 ± 3 c	85.3 ± 3 c	- - -	- - -	94.6 ± 4 c
(-59.2 ± 2) [157,158]	CHF_2	101 ± 2 [157,158]	95.6 ± 2.5 c	126 ± 2 c	- - -	69 ± 2 [163]	- - -	- - -
(-111.7 ± 3.6) [6,135]	CF_3	106.7 ± 1 [6,135]	101.6 ± 3 c	130.5 ± 3 c	86.2 ± 3 c [158]	70.6 ± 3 [164,165]	55.0 ± 3 c [166]	98.7 ± 2.5 c,[167]
(28.3) [160]	CH_2Cl	100.9 ± 2 [160]	- - -	- - -	80.1 c	- - -	- - -	- - -
(24.1) [160]	$CHCl_2$	99.0 ± 2 [160]	- - -	- - -	77.6 c	- - -	- - -	- - -
(19 ± 1) [161]	CCl_3	95.8 ± 1 [161]	- - -	101.9 ± 2.3 c	73.1 ± 1.8 c [161]	55.3 ± 1 c, [161]	- - -	- - -
(-64.3 ± 2) [176,179]	CF_2Cl	101.6 ± 1.0 e	- - -	123 ± 2.2 c	76 ± 2 [176]	64.5±1.5 e	- - -	- - -
(-23) [177]	$CFCl_2$	- - -	- - -	110 ± 6 [176]	73 ± 2 [177]	- - -	- - -	- - -
(41.5) [160]	CH_2Br	102.0 ± 2 [160]	- - -	- - -	- - -	- - -	- - -	- - -
(54.3) [160]	$CHBr_2$	103.7 ± 2 [160]	- - -	- - -	- - -	- - -	- - -	- - -
- - -	CBr_3	96.0 ± 1.6 [162]	- - -	- - -	- - -	56.2 ± 1.8 [162]	- - -	- - -
(-213.4 ± 1) [8,171,242]	C_2F_5	102.7 ± 0.5 [135,242]	- - -	126.8 ± 1.8 c	82.7 ± 1.7 [168]	68.7 ± 1.5 [165,168,169]	51.2 ± 1.0 [166,171]	- - -
- - -	$n-C_3F_7$	104 ± 2 [5]	- - -	- - -	- - -	66.5 ± 2.5 [168]	49.8 ±1 [166]	- - -
- - -	$i-C_3F_7$	103 ± 0.6 e	- - -	- - -	- - -	65.5 ± 1.1 e	- - -	- - -
(-72.3 ± 2) [6,135]	CF_2CH_3	99.5 ± 2.5 [6,159]	- - -	124.8 ± 2 [6,159]	- - -	- - -	52.1 ± 1 [172]	- - -
(-123.6 ± 2) [6,135]	CH_2CF_3	106.7 ± 1.1 [6,159]	- - -	109.4 [6,159]	- - -	- - -	56.3 ± 1 [173]	- - -
- - -	$CHClCF_3$	101.8 ± 1.5 e	- - -	- - -	- - -	65.7 ± 1.5 e	- - -	- - -
- - -	$CClBrCF_3$	96.6 ± 1.5 e	- - -	- - -	- - -	60.0 ± 1.5 e	- - -	- - -
(104 ± 2) [14,144]	CN	123.8 ± 2 [14,144]	121.8 ± 2 [14]	112.3 ± 1.2 [145]	100.8 ± 1.2 [145]	87.8 ± 1.2 [145]	72.5 ± 1 [145]	134.1 ± 2 c
(9.4) [14]	OH	119 ± 1 [14]	92.6 ± 0.2 c	- - -	60 ± 3 [5]	56 ± 3 [5]	56 ± 3 [5]	- - -
(79 ± 2) [2,65,66]	C_6H_5	111.3 ± 2 c,[2,65,66]	102.2 ± 2 c	125.7 ± 2 c	95.7 ± 2 c	80.5 ± 2 c,[178]	65.4 ± 2 c,[91]	- - -
(-130.9 ± 2) [154]	C_6F_5	116.5 ± 2	- - -	- - -	91.6 ± 2 c	- - -	~ 66.2 [88]	- - -

continued...

Table 7. BOND DISSOCIATION ENERGIES OF HALOCARBONS (concluded)

R_1-R_2	D^o_{298}	$\Delta H^o_{f,298}$ R_1	References
C_2Cl_5--H	95 ± 2	8.4 ± 2	1,121,174
$CHCl_2CCl_2$--H	94 ± 2	5.6 ± 2	1,121,175
CF_2ClCF_2--Cl	78 ± 2	-164 ± 4	176
CH_2-CCHCl--H	88.6 ± 1.4	- - -	101
n-C_4H_9--I	49.0 ± 1	- - -	166
C_6H_5CO--Cl	74 ± 3	26.1 ± 2	5,8,112
NF_2--NF_2	21 ± 1	8 ± 1	9,138,149
CH_2CN--i-C_3H_7	73 ± 2.5	58.5	146
CH_2I--H	103 ± 2	55.0 ± 1.6	2,8
CHI_2--H	103 ± 2	79.8 ± 2.2	2,24

[a] Values in parentheses near radicals and atom are $\Delta H^o_{f,298}$.

[b] Numbers in brackets denote references.

[c] Derived from referenced radical heats of formation and stable species heats of formation from references 13, 14, 15, or 16. The assigned 1 kcal mole^{-1} uncertainties should be read "at least" 1 kcal mole^{-1}.

[d] Many of the values in this table are determined by competitive methods and are thus linked, for example, to the "known" BDE of CH_4. Upward revisions of the latter will require increases in such linked values of from 0.3 to 1.0 kcal mole^{-1}. If precision of this level is desired, the original papers should be consulted.

[e] E. Whittle, unpublished work.

dependence on the C-substitution than might be expected. A rationalization is that increased alkyl group polarizability results in an increased bond strength to the electronegative NO_2 group that nearly offsets the increased alkyl radical stability.

HALOCARBONS   In Table 7, "derived" BDE values are listed only for halocarbons for which published BDE values exist, since group additivity is known not to hold for halocarbons for which polar effects are most pronounced (6, 159). Perhaps the most striking observation is that whereas the CH_3–H bond is evidently made ~ 2 kcal mol^{-1} stronger by the substitution of three fluorine atoms, it is made 4 kcal mol^{-1} weaker by substitution of one or two fluorine atoms (20). This result has been questioned previously (5), but has been supported recently by an OH-radical/halocarbon Evans-Polanyi relationship (158). It is somewhat surprising for the planar CH_3 radical and the pyramidal CF_3 radical (and those in between) to fall in a single relationship, but these results are, for the time being, taken at face value since they also lend support to the unusual trends that previous data appear to reveal in the fluoroethane series. Evidently, substitution on the β-carbon strengthens the α C–H bonds, with the CF_3CH_2–H bond being as strong as the CF_3–H bond! This can be ascribed to polar effects that are maximized in CF_3CH_2–H, the fluoroethane with the largest dipole moment. To a lesser extent, C–Cl bonds appear to respond in these ways to fluorine substitution, but C–Br and C–I bonds, not at all. The C–C bond strengths apparently exhibit what may be called a similar trend: the CF_3–CF_3 bond at 99 kcal mol^{-1} is ~ 9 kcal mol^{-1} stronger than the CH_3–CH_3 bond, but the CH_3–CF_3 bond at 102 kcal mol^{-1} is the strongest of the C–C bonds in fluoroethanes.

ORGANOSILICON COMPOUNDS As illustrated in Table 8 and discussed by Walsh (3), a striking feature of silane and the methyl silanes is the almost constant SiH bond strength at $\sim 89 \pm 1$ kcal mol^{-1}. The absence of a weakening effect of substituent methyl groups on Si–H bonds (in contrast with the well-known weakening for C–H bonds) can be rationalized (3) by the lower electronegativity of silicon (versus carbon), which provides a poorer acceptor of electron density from the substituent alkyl groups.

Si–H bonds are only ~ 2 to 10 kcal mol^{-1} weaker than their C–H analogs, despite the greater reactivity of Si–H bonds. For sequential removal of H or halogen atoms from silicon compounds, the same type of alternation of BDE values is observed as discussed above for sulfur fluorides (182). Si–H bond weakening by phenyl substitution is only 1–2 kcal mol^{-1}, in contrast to the 10 kcal mol^{-1} observed (18) for phenyl substituted C–H bonds. As pointed out by Walsh (3), this is not unexpected in view of the weakness of π-bonding in sila olefins.

ORGANOMETALLIC COMPOUNDS $M(Me)_n$–Me AND $M(Et)_n$–Et Smith & Patrick (10) have examined the existing data for metal alkyls in conjunction with their reporting of new experimental results for $GeMe_4$, $SbEt_3$, and $PbEt_4$. This has included a reanalysis of the toluene carrier data of Price and co-workers, which forms the bulk of gas-phase rate measurements for metal alkyls (160, 170, 174, 181, 186, 217, 218, 220, 221). The reanalysis involved assessment of the extent of falloff using a

Table 8 BOND DISSOCIATION ENERGIES OF ORGANO-SILICON COMPOUNDS

$\Delta H_f^o(R)$	R_1	(52.1)[a] H	(35.1) CH_3	(46.4) SiH_3	R_1	(18.9) F	(28.9) Cl
(46.4) [196][b]	SiH_3	90.3 [196]	88.2 ± 3 c, [3]	74 c, [3]	74 c, [3]	- - -	- - -
(36.5) [197]	SiH_2Me	99.6 [197]	88.3 ± 3 c, [3]	- - -	- - -	- - -	- - -
(14.3) [197]	$SiHMe_2$	89.4 [197]	88.1 ± 3 c, [3]	- - -	- - -	- - -	- - -
(-0.8) [198]	$SiMe_3$	90.3 [198]	89.4 ± 3 c, [3]	- - -	80.5 [199]	- - -	113 c, [3]
(53.3) [197]	Si_2H_5	86.3 [197]	- - -	71 c, [3]	68 c, [3]	- - -	- - -
	$SiH_2C_6H_5$	88.2 [201]	- - -	- - -	- - -	- - -	- - -
	$Si(C_6H_5)_3$	- - -	- - -	- - -	88 ± 7 [110]	- - -	- - -
(-8.3) c	CH_2SiMe_2	99.2 [202]	- - -	- - -	- - -	- - -	- - -
(58) [203]	SiH_2	64 [203]	- - -	- - -	- - -	- - -	- - -
(90) [14]	SiH	84 c, [3]	- - -	- - -	- - -	- - -	- - -
(108) [14]	Si	70 c, [3]	- - -	- - -	- - -	132 c, [3]	91 c, [3]
(-76) [3]	$SiCl_3$	91.3 [204]	- - -	- - -	- - -	- - -	111 c, [3]
(-39.1) [14]	$SiCl_2$	- - -	- - -	- - -	- - -	- - -	66 c, [3]
(46.8) [14,3],c	$SiCl$	- - -	- - -	- - -	- - -	- - -	114 c, [3]
(-245) c, [3]	SiF_3	100.1 [205]	- - -	- - -	- - -	160 c, [3]	- - -
[-140.5] [111]	SiF_2	- - -	- - -	- - -	- - -	123 c, [3]	- - -
(-4.6) [14]	SiF	- - -	- - -	- - -	- - -	155 c, [3]	- - -

$SiMe_3-Br$		96 c, [3]	$SiMe_3-s-C_4H_9$		99 [206]
$SiMe_3-I$		77 c, [3]	$SiMe_3-CH_2CH:CH_2$		70 [206]
$SiMe_3-OH$		128	$Me_3SiO-O-t-Bu$		47 [235]
$SiMe_3-NHMe$		100 c, [3]			

[a] Values in parentheses near radicals and atom are $\Delta H_{f,298}^o$.

[b] Numbers in brackets denote reference.

[c] Derived from referenced radical heats of formation and stable species heats of formation from references 13, 14, 15, and 16. There are few explicitly assigned uncertainties since most of the values in this table are taken directly from the critical review (3) by Walsh who does not assign individual uncertainties but states that they should be considered to be from 2 to 3 kcal mole^{-1}.

Gorin transition state model, and derivation of activation energies using A-factors assigned by analogy with other tri- and tetra-alkyls.

The BDE values resulting (10) from these data, treated with a Gorin transition model and A-factors of 16.5 ± 0.2, are shown in Table 9, together with values from previous tables for Groups V, VI, and VII elements, using the format of Ref. (10). The uncertainties notwithstanding, clearly the Group IV elements generally form the strongest methyl bonds. The weakest are the Group III trimethyl compouds [except for the unusually strong $B(Me)_3$]. In each Group, the methyl bond energies decrease going down the Group.

Table 9 BOND ENERGIES OF ORGANOMETALLIC COMPOUNDS—$X(Me)_n$-Me and $X(Et)_n$-Et

Li	Mg		B	C	N	O	F
- - -	- - -		- - -	84 ± 1	82.2 ± 2	83.3 ± 2	109.0 ± 0.2
(61 ± 5)d				(79 ± 1)			(108.3 ± 1)
c [15]				c [67]	c [4]	[120,122]	c [13,2]
			Al	Si	P	S	Cl
			- - -	89.4 ± 3	- - -	78.2	84.6 ± 0.2
							(81.5 ± 1)
				c [3]		[150]	c [3,2]
	Zn	Ga		Ge	As	Se	Br
	68 ± 4	63 ± 4		83 ± 4	67 ± 4	- - -	70.9 ± 0.3
	(57 ± 4)	(50 ± 4)					(68.4 ± 1)
	[10,209]	[10,217,218]		[219]	[10,220]		c [13,2]
	Cd	In		Sn	Sb	Te	I
	60 ± 4	49 ± 4		71 ± 4	61 ± 4	- - -	57.2 ± 0.3
				(63 ± 4)	(58 ± 4)		(54 ± 1)
	[221]	[10,222]		[10,208,223]	[10,207]		c [13,2]
	Hg	Tl		Pb	Bi		
	61 ± 4	40 ± 4		57 ± 4	52 ± 4	- - -	- - -
	(49 ± 4)			(55 ± 4)			
	[10,174,186]	[10,181]		[10,170]	[10,160]		

Other Metal-Organic Bond Energies:

R_1-R_2	$D°_{298}$	Ref.
$BCl_2-C_6H_5$	~ 122	208
BF_2-CH_3	~ 113	157
$(Et)_3GeO-OtBu$	46	125
$(Et)_3SnO-OtBu$	46	49

[a] Numbers in brackets denote references.
[b] Values in parentheses are for $X(Et)_n$-Et.
[c] Derived from referenced radical heats of formation and stable species heats of formation from references 13, 14, 15, and 16. The assigned 1 kcal mole^{-1} uncertainties should be read "at least" 1 kcal mole^{-1}.
[d] Reference 15.

Acknowledgments

This review was made possible, in part, by funds from the US Department of Energy, Division of Chemical Sciences, Office of Basic Energy Sciences, Processes and Techniques Branch, under contract No. DE-AC03-79ER10485. We are grateful for the help of colleagues in supplying us with copies of manuscripts and for stimulating helpful discussions, and for the help of Elaine Adkins and Walter Ogier in preparing the manuscript and decreasing the number of errors in the tables.

Literature Cited

1. Kerr, J. A. 1981. In *Handbook of Chemistry and Physics*, pp. F-222. Boca Raton, Fl.: CRC Press
2. Golden D. M., Benson, S. W. 1969. *Chem. Rev.* 69:125
3. Walsh, R. 1981. *Acc. Chem. Res.* 14:246
4. Batt, L., Robinson, G. N. 1981. In *Chemistry of the Functional Groups*, ed. S. Patai, Suppl. F, p. 1035. Chichester: Wiley Ltd.
5. Kerr, J. A. 1966. *Chem. Rev.* 66:465
6. Rodgers, A. S. 1978. *ACS Symp. Ser.* 66:296
7. Benson, S. W. 1978. *Chem. Rev.* 78:23
8. O'Neal, H. E., Benson, S. W. 1970. *Kinetic Data on Gas-Phase Unimolecular Reactions.* Washington DC: Natl. Ref. Data Serv. NSRDS-NBS 21, US Dept. Commerce
9. Batt, L., 1981. See Ref. 4, Suppl. C
10. Smith, G. P., Patrick, R. 1982. *Int. J. Chem. Kinetics.* In press
11. Egger, K. W., Cocks, A. T. 1973. *Helv. Chim. Acta* 56:1516
12. Baldwin, A. C. 1982. In *Chemistry of the Functional Groups*, ed. S. Patai. Chichester: Wiley Ltd. In press
13. Benson, S. W. 1976. *Thermochemical Kinetics.* New York: Wiley
14. *J.A.N.A.F. Thermochemical Tables*, 1971. Natl. Stand. Ref. Data Ser. Washington DC: US Natl. Bur. Stand., NSRDS-NBS 37
15a. Cox, J. D., Pilcher, G. 1970. *Thermochemistry of Organic and Organometallic Compounds.* New York: Academic
15b. Pedley, J. B., Rylance, J. 1977. *Computer Analyzed Thermochemical Data: Organic and Organometallic Compounds.* Sussex: Univ. Sussex
16. Stull, D. R., Westrum, E. F., Jr., Sinke, G. C. 1969. *The Chemical Thermodynamics of Organic Compounds*, New York: Wiley
17a. Baghal-Vayjooee, M. H., Colussi, A. J., Benson, S. W. 1978. *J. Am. Chem. Soc.* 101:3214
17b. Baghal-Vayjooee, M. H., Colussi, A. J., Benson, S. W. 1979. *Int. J. Chem. Kinet.* 11:147
18. Rossi, M. J., Golden, D. M. 1979. *Int. J. Chem. Kinet.* 11:969
19. Rossi, M. J., Golden, D. M. 1979. *J. Am. Chem. Soc.* 101:1230
20. Tarr, A. M., Coomber, J. W., Whittle, E. 1965. *Trans. Faraday Soc.* 61:1182
21. Golden, D. M. 1979. *J. Phys. Chem.* 83:108
22. Rossi, M. J., King, K. D., Golden, D. M. 1979. *J. Am. Chem. Soc.* 101:1223
23. Chenier, J. H. B., Tong, S. B., Howard, J. A. 1978. *Can. J. Chem.* 56:3047
24. Benson, S. W., O'Neal, H. E. 1973. In *Free Radicals*, 2:275 New York: Wiley
25. Amphlett, J. C., Whittle, E. 1966. *Trans. Faraday Soc.* 62:1662
26. Fettis, G. C., Knox, J. H. 1964. *Prog. React. Kinet.* 2:2
27. Baghal-Vayjooee, M. H., Benson, S. W. 1979. *J. Am. Chem. Soc.* 101:2838
28. Spokes, G. N., Golden, D. M., Benson, S. W. 1973. *Angew. Chem. Int. Ed. Engl.* 12:534
29. Tsang, W. 1964. *J. Chem. Phys.* 41:2487
30. Tsang, W. 1981. In *Shock Waves in Chemistry*, ed. A. Lifshitz, pp. 59–129. New York: Dekker
31. McMillen, D. F., Lewis, K. E., Smith, G. P., Golden, D. M. 1982. *J. Phys. Chem.* 86:709

32. Tsang, W. 1966. *J. Chem. Phys.* 44:4283
33. McMillen, D. F., Trevor, P. L., Golden, D. M. 1980. *J. Am. Chem. Soc.* 102:7400
34. Kerr, J. A., Parsonage, M. J. 1976. *Evaluated Kinetic Data on Gas-Phase Hydrogen Transfer Reactions of Methyl Radicals.* London: Butterworths
35. Agmon, N. 1981. *Int. J. Chem. Kinet.* 13:333
36. Robinson, P. J., Holbrook, K. A. 1972. *Unimolecular Reactions.* New York: Wiley
37. Bogan, D., Setser, D. N. 1978. In *Fluorine-Containing Free Radicals: Kinetics and Dynamics of Reactions*, ed. J. W. Root, pp. 237–50. ACS Symp. Ser. Washington DC: Am. Chem. Soc.
38. Gupta, A., Perry, D. S., Zare, R. N. 1980. *J. Chem. Phys.* 72:6250
39. Bartness, J. E., McIver, R. T. Jr. 1979. In *Gas Phase Ion Chemistry*, pp. 88–119. New York: Academic
40. Aui, D. H., Bowers, M. T. 1979. In *Gas Phase Ion Chemistry*, ed. M. T. Bowers, 2:1. New York: Academic
41. Carlson, T. A. 1974. *Ann. Rev. Phys. Chem.* 26:211
42. Rosenstock, H., Draxl, K., Steiner, G., Herron, J. 1977. *J. Phys. and Chem. Ref. Data* 6; Supplement No. 1
43. Rodgers, A. S., Golden, D. M., Benson, S. W. 1967. *J. Am. Chem. Soc.* 89:4578
44. Tsang, W. 1978. *Int. J. Chem. Kinetics* 10:821
45. Doering, W. V. E. 1981. *Proc. Natl. Acad. Sci. USA* 78:5279
46a. Castelhano, A. L., Marriott, P. R., Griller, D. 1981. *J. Am. Chem. Soc.* 103:4262
46b. Castelhano, A. L., Griller, D. 1982. *J. Am. Chem. Soc.*, In press
47. Canosa, C. E., Marshall, R. M. 1981. *Int. J. Chem. Kinetics* 13:295–303
48. Baldwin, R. R., Walker, R. W. 1981. *J. Chem. Soc., Faraday Trans.* 76:825
49. Herzberg, G. 1961. *Proc. Roy. Soc.* (London) A262:291
50. Teranishi, H., Benson, S. W. 1963. *J. Am. Chem. Soc.* 85:2887
51. Knox, J. H., Musgrave, R. G. 1967. *Trans. Faraday Soc.* 63:2201
52. Parkes, D. A., Quinn, C. P. 1976. *J. Chem. Soc., Faraday Trans. I* 72:1952
53. Choo, K. Y., Beadle, P. C., Piszkiewicz, L. W., Golden, D. M. 1976. *Int. J. Chem. Kinetics* 8:451
54a. Pacansky, J., Chang, J. S. 1981. *J. Chem. Phys.* 74:5538
54b. Yoshimine, M., Pacansky, J. 1981. *J. Chem. Phys.* 74:5168
55. Janousek, B. K., Brauman, J. I. 1979. In *Gas Phase Ion Chemistry*, ed. M. T. Bowers, 2:53–86. New York: Academic
56. Walker, J. A., Tsang, W. 1979. *Int. J. Chem. Kinet.* 11:867
57. Atri, G. M., Baldwin, R. W., Evans, G. A., Walker, R. R. 1978. *J. Chem. Soc. Faraday Trans. 1* 74:366
58. Chupka, W. A. 1968. *J. Chem. Phys.* 48:2337
59. Schuh, H., Fischer, H. 1976. *Int. J. Chem. Kinet.* 8:341
60. Benson, S. W., Luria, M. 1975. *J. Am. Chem. Soc.* 97:704
61. Yamdagni, R., Kebarle, P. 1976. *J. Am. Chem. Soc.* 98:132
62. Houle, F. A., Beauchamp, J. L. 1979. *J. Am. Chem. Soc.* 101:4067
63. Ferguson, K. C., Whittle, E. 1971. *Trans. Faraday Soc.* 67:2618
64. Baldwin, R. R., Walker, R. W. 1979. *Trans. Faraday Soc.* 75:140
65. Chamberlain, G. A., Whittle, E. 1971. *Trans. Faraday Soc.* 67:2077
66. Rosenstock, H. M., Stockbauer, R., Parr, A. C. 1980. *J. Chem. Phys.* 73:773
67. Larson, C. W., Hardwidge, E. A., Rabinovitch, B. S. 1969. *J. Chem. Phys.* 50:2769
68. O'Neal, H. E., Bagg, J. W., Richardson, W. H. 1970. *Int. J. Chem. Kinet.* 2:493
69. Robaugh, D. A., Barton, B. D., Stein, S. E. 1981. *J. Phys. Chem.* 85:2378
70. Miller, R. E., Stein, S. E. 1981. *J. Phys. Chem.* 85:580
71. McMillen, D. F., Ogier, W. C., Ross, D. S. 1981. *J. Org. Chem.* 46:3322
72. Cumming, J. B., Kebarle, P. 1978. *Can. J. Chem.* 56:1
73. Steinkruger, F. J., Rowland, F. S. 1981. *J. Phys. Chem.* 85:135
74. Egger, K. W., Jola, M. 1970. *Int. J. Chem. Kinet.* 2:265
75. Frey, H. M., Krantz, A. 1969. *J. Chem. Soc. A*, p. 1159
76. James, D. G. L., Suart, R. D. 1968. *Trans. Faraday Soc.* 64:2752
77. DeFrees, D. J., McIver, R. T. Jr., Hehre, W. J. 1980. *J. Am. Chem. Soc.* 102:3334

78. Furuyama, S., Golden, D. M., Benson, S. W. 1971. *Int. J. Chem. Kinet.* 3:237
79. Furuyama, S., Golden, D. M., Benson, S. W. 1970. *Int. J. Chem. Kinet.* 2:93
80. Okabe, H., Dibeler, V. H. 1973. *J. Chem. Phys.* 59:2430
81. Colussi, A. J., Zabel, F., Benson, S. W. 1977. *Int. J. Chem. Kinet.* 9:161
82. Trenwith, A. B., Wrigley, S. P. 1977. *J. Chem. Soc. Faraday Trans. 1* 73:817
83. Tsang, W. 1973. *Int. J. Chem. Kinet.* 5:929
84. Trenwith, A. B. 1970. *Trans. Faraday Soc.* 66:1970
85. Rodgers, A. S., Wu, M. C. R. 1973. *J. Am. Chem. Soc.* 95:6913
86. Ogier, W. C., McMillen, D. F., Golden, D. M. 1982. *Int. J. Chem. Kinet.* In press
87. Vincow, G., Dauben, H. J., Hunter, F. R., Volland, W. V. 1969. *J. Am. Chem. Soc.* 91:2823
88. Krech, M. J., Price, S. J. W, Yared, W. F. 1974. *Int. J. Chem. Kinet.* 6:257
89. Stein, S. E. 1981. *Carbon* 19:421
90. Nguyen, T. T., King, K. D. 1981. *J. Phys. Chem.* 85:3130
91. Kominar, R. J., Krech, M. J., Price, S. J. W. 1976. *Can. J. Chem.* 54:2981
92. Stein, S. E., Golden, D. M. 1977. *J. Org. Chem.* 42:839
93. Stein, S. E. 1981. In *New Approaches in Coal Chemistry. ACS Symp. Ser.* 169:97
94. King, K. D. 1977. *Int. J. Chem. Kinet.* 9:907
95. Pryor, W. A., Gojon, G., Church, D. F. 1978. *J. Org. Chem.* 43:793
96. Hendry, D. G., Mill, T., Piszkiewicz, L., Howard, J. A., Kigenmann, H. Y. 1974. *J. Phys. Chem. Ref. Data* 3:937
97. Leigh, W. J., Arnold, D. R. 1981. *Can. J. Chem.* 59:609
98. Tsang, W. 1978. *Int. J. Chem. Kinet.* 10:687
99. Nguyen, T. T., King, K. D. 1982. *Int. J. Chem. Kinet.* 14:613
100. King, K. D., Nguyen, T. T. 1981. *Int. J. Chem. Kinet.* 13:255
101. Alfassi, Z. B., Golden, D. M., Benson, S. W. 1973. *Int. J. Chem. Kinet.* 5:67; 155
102. Robaugh, D. A., Stein, S. E. 1981. *Int. J. Chem. Kinet.* 13:445
103. Ogier, W. C., McMillen, D. F., Golden, D. M. 1982. *Int. J. Chem. Kinet.* In press
104. SenSharma, D. K., Franklin, J. L. 1973. *J. Am. Chem. Soc.* 95:6562
105. Malastesta, V., Ingold, K. U. 1981. *J. Am. Chem. Soc.* 103:609
106. Solly, R. K., Benson, S. W. 1971. *Int. J. Chem. Kinet.* 3:509
107. Devore, J. A., O'Neal, H. E. 1969. *J. Phys. Chem.* 73:2644
108. Alfassi, Z. B., Golden, D. M. 1973. *J. Am. Chem. Soc.* 95:319
109. King, K. D., Golden, D. M., Benson, S. W. 1970. *J. Am. Chem. Soc.* 92:5541
110. Calle, L. M., Kana'an, A. S. 1974. *J. Chem. Thermodyn.* 6:935
111. Watkins, K. W., Thompson, W. W. 1973. *Int. J. Chem. Kinet.* 5:791
112. Solly, R. K., Benson, S. W. 1971. *J. Am. Chem. Soc.* 93:1592
113. Meot-Ner, M. 1982. *J. Am. Chem. Soc.* 104:5
114. Amphlett, J. C., Whittle, E. 1970. *Trans. Faraday Soc.* 66:2016
115. Solly, R. K., Golden, D. M., Benson, S. W. 1970. *Int. J. Chem. Kinet.* 2:381
116. Alfassi, Z. B., Golden, D. M. 1972. *J. Phys. Chem.* 76:3314
117. Solly, R. K., Benson, S. W. 1969. *Int. J. Chem. Kinet.* 1:427
118. Knoll, H., Scherker, K., Geiseler, G. 1973. *Int. J. Chem. Kinet.* 5:271
119. Reed, K. J., Brauman, J. I. 1975. *J. Am. Chem. Soc.* 97:1625
120. Batt, L., Christie, K., Milne, R. T., Summers, A. J. 1974. *Int. J. Chem. Kinet.* 6:877
121. Lewis, D. K. 1976. *Can. J. Chem.* 54:581
122. Batt, L., McCulloch, R. D. 1976. *Int. J. Chem. Kinet.* 8:491
123. Batt, L., McCulloch, R. D. 1976. *Int. J. Chem. Kinet.* 8:911
124. Walker, R. F., Phillips, L. 1968. *J. Chem. Soc.*, p. 2103
125. Rabinovich, I. B., Kiparisova, E. G., Alexksandrov, Y. Y. 1971. *Dokl. Akad. Nauk. SSR* 20:1116
126. Perona, M. J., Golden, D. M. 1973. *Int. J. Chem. Kinet.* 5:55
127. Howard, C. J. 1980. *J. Am. Chem. Soc.* 102:6937
128. Descamps, B., Forst, W. 1976. *J. Phys. Chem.* 80:933
129. Ireton, R., Gordon, A. S., Tardy, D. C. 1977. *Int. J. Chem. Kinet.* 9:769
130. Rulz, R. P., Bayes, K. D. 1981. *J. Phys. Chem.* 85:1622
131. Doncaster, A. M., Walsh, R. 1979. *J. Chem. Soc. Chem. Commun.* p. 904

132. Alfassi, Z. B., Golden, D. M. 1973. *Int. J. Chem. Kinet.* 5:295
133. Carson, A. S., Fine, D. H., Gray, P., Laye, P. G. 1971. *J. Chem. Soc. B*, p. 1611
134. Paul, S., Back, M. H. 1975. *Can. J. Chem.* 53:3330
135. Bassett, J. E., Whittle, E. 1972. *J. Chem. Soc., Faraday Trans. 1* 68:492
136. Golden, D. M., Solly, R. K., Gac, N. A., Benson, S. W. 1972. *J. Am. Chem. Soc.* 94:363
137. Colussi, A. J., Benson, S. W. 1978. *Int. J. Chem. Kinet.* 10:1139
138. Pankratov, A. V., Zercheninov, A. N., Chesnokov, V. I., Zhdanova, N. N. 1969. *Zh. Fiz. Khim.* 43:394
139. Pellerite, M. J., Jackson, R. L., Brauman, J. I. 1981. *J. Phys. Chem.* 85:1624
140. Bohme, D. K., Hemsworth, R. S., Rundle, H. W. 1973. *J. Chem. Phys.* 59:77
141. DeFrees, D. J., Hehre, W. J., McIver, R. T., McDaniel, D. H. 1979. *J. Phys. Chem.* 83:232
142. Colussi, A. J., Benson, S. W. 1977. *Int. J. Chem. Kinet.* 9:307
143. Griller, D., Lossing, F. P. 1981. *J. Am. Chem. Soc.* 103:1586
144. Betowski, D., Mackay, G., Payzant, J., Bohme, D. 1975. *Can. J. Chem.* 53:2365
145. Day, J. S., Gowenlock, B. G., Johnson, C. A. F., McInally, I. D., Pfab, J. 1978. *J. Chem. Soc. Perkin Trans. 2* p. 1110
146. King, K. D., Goddard, R. D. 1975. *Int. J. Chem. Kinet.* 7:837
147. King, K. D., Goddard, R. D. 1975. *J. Am. Chem. Soc.* 97:4504
148. King, K. D., Goddard, R. D. 1976. *J. Phys. Chem.* 80:546
149. Darwent, B. de B. 1970. *Bond Dissociation Energies in Simple Molecules*, NSRDS-NBS:31. Washington DC: Natl. Bur. Stand.
150. Colussi, A. J., Benson, S. W. 1977. *Int. J. Chem. Kinet.* 9:295
151. Coo, K. Y., Mendenhall, G. D., Golden, D. M., Benson, S. W. 1974. *Int. J. Chem. Kinet.* 6:813
152. Batt, L., Milne, R. T. 1973. *Int. J. Chem. Kinet.* 5:1067
153. Glanzer, K., Maier, M., Troe, J. 1979. *Chem. Phys. Lett.* 61:175
154. Choo, K. Y., Golden, D. M., Benson, S. W. 1975. *Int. J. Chem. Kinet.* 7:713
155. Carmichael, P. J., Gowenlock, B. G., Johnson, C. A. F. 1972. *Int. J. Chem. Kinet.* 4:339
156. Carmichael, P. J., Gowenlock, B. G., Johnson, C. A. F. 1973. *J. Chem. Soc., Perkin Trans. 2* p. 1853
157. Skinner, H. A. 1965. *Adv. Organometal. Chem.* 2:49
158. Martin, J. P., Paraskevopoulos, G. 1982. *Can. J. Chem.* In press
159. Chen, S. S., Rodgers, A. S., Chao, J., Wilhoit, R. C., Zwolinski, B. J. 1975. *J. Phys. Chem. Ref. Data* 4:455
160. Price, S. J. W., Trotman-Dickenson, A. F. 1958. *Trans. Faraday Soc.* 54:1630
161. Mendenhall, G. D., Golden, D. M., Benson, S. W. 1973. *J. Phys. Chem.* 77:2707
162. King, K. D., Golden, D. M., Benson, S. W. 1971. *J. Phys. Chem.* 75:987
163. Okafo, E. N., Whittle, E. 1974. *J. Chem. Soc., Faraday Trans. 1* 70:1366
164. Ferguson, K. C., Whittle, E. 1972. *J. Chem. Soc., Faraday Trans. 1* 68:295
165. Ferguson, K. C., Whittle, E. 1972. *J. Chem. Soc., Faraday Trans. 1* 68:641
166. Okafo, E. N., Whittle, E. 1975. *Int. J. Chem. Kinet.* 7:287
167. Coomber, J. W., Whittle, E. 1967. *Trans. Faraday Soc.* 63:1394
168. Coomber, J. W., Whittle, E. 1967. *Trans. Faraday Soc.* 63:2656
169. Ferguson, K. C., Whittle, E. 1972. *J. Chem. Soc., Faraday Trans. 1* 68:306
170. Gilroy, K. M., Price, S. J. W., Webster, N. J. 1972. *Can. J. Chem.* 50:2639
171. Wu, E-C., Rodgers, A. S. 1976. *J. Am. Chem. Soc.* 98:6112
172. Pickard, J. M., Rodgers, A. S. 1976. *Int. J. Chem. Kinet.* 8:809
173. Wu, E-C., Rodgers, A. S. 1973. *Int. J. Chem. Kinet.* 5:1001
174. Lalonde, A. C., Price, S. J. W. 1971. *Can. J. Chem.* 49:3367
175. Franklin, J. A., Huybrechts, G. H., Cillien, C. 1969. *Trans. Faraday Soc.* 65:2094
176. Foon, R., Tait, K. B. 1972. *J. Chem. Soc., Faraday Trans. 1* 68:1121
177. Foon, R., Tait, K. B. 1972. *J. Chem. Soc., Faraday Trans. 1* 68:104
178. Kominar, R. J., Krech, M. J., Price, S. J. W. 1978. *Can. J. Chem.* 56:1589
179. Leyland, L. M., Majer, J. R., Robb, J. C. 1970. *Trans. Faraday Soc.* 66:898
180. Hwang, R. J., Benson, S. W. 1979. *Int. J. Chem. Kinet.* 11:579
181. Price, S. J. W., Richard, J. P., Rumfeldt, R. C., Jacko, M. G. 1973. *Can. J. Chem.* 51:1397
182. Kiang, T., Zare, R. N. 1980. *J. Am. Chem. Soc.* 102:4024

183. Hwang, R. J., Benson, S. W. 1979. *J. Am. Chem. Soc.* 101:2615
184. White, J. N., Gardiner, W. C. 1978. *Chem. Phys. Lett.* 58:470
185. Kiang, T., Estler, R. C., Zare, R. N. 1979. *J. Chem. Phys.* 70:5925
186. Kominar, R. J., Price, S. J. W. 1969. *Can. J. Chem.* 47:991
187. Okabe, H. 1971. *J. Am. Chem. Soc.* 93:7095
188. Rafaey, K. M. A., Franklin, J. L. 1976. *J. Chem. Phys.* 65:1994
189. Okabe, H. 1977. *J. Chem. Phys.* 66:2058
190. Okabe, H. 1972. *J. Chem. Phys.* 56:4381
191. Janousek, B. K., Zimmerman, A. H., Reed, K. J., Brauman, J. I. 1978. *J. Am. Chem. Soc.* 100:6142
192. Foucaut, J-F., Martin, R. 1978. *J. Chim. Phys.* 75:132
193. Chase, M. W. Jr., Curnutt, J. L., McDonald, R. A., Syverud, A. N. 1978. *J. Phys. Chem. Ref. Data* 7:897
194. Babcock, L. M., Streit, G. E. 1981. *J. Chem. Phys.* 74:5700
195. Miletic, M., Eres, D., Veljkovic, M., Zmbov, K. F. 1980. *Int. J. Mass Spectrom. Ion Phys.* 35:231
196. Doncaster, A. M., Walsh, R. 1981. *Int. J. Chem. Kinet.* 13:503
197. Corbel, S., Marquaire, P. M., Come, G. M. 1981. *Chem. Phys. Lett.* 80:34
198a. Doncaster, A. M., Walsh, R. 1979. *J. Chem. Soc., Faraday Trans. 1* 75:1126
198b. Doncaster, A. M., Walsh, R. 1965. *J. Chem. Soc., Faraday Trans. 1* 72:100
199. Davidson, I. M. T., Howard, A. V. 1975. *J. Chem. Soc., Faraday Trans. 1* 71:69
200. Barton, B. D., Stein, S. E. 1981. *J. Chem. Soc., Faraday Trans. 1* 77:1755
201. Barber, M., Doncaster, A. M., Walsh, R. 1982. *Int. J. Chem. Kinet.* 14:669
202. Doncaster, A. M., Walsh, R. 1976. *J. Chem. Soc. Faraday Trans. 1* 72:2908
203. John, P., Purnell, J. H. 1973. *J. Chem. Soc. Faraday Trans. 1* 69:1455
204. Walsh, R., Wells, J. M. 1976. *J. Chem. Soc., Faraday Trans. 1* 72:1212
205. Doncaster, A. M., Walsh, R. 1978. *Int. J. Chem. Kinet.* 10:101
206. Davidson, I. M. T., Wood, I. T. 1980. *J. Organomet. Chem.* 202:C-65
207. Price, S. J. W., Richard, J. P. 1972 *Can. J. Chem.* 50:966
208. Johnson, R. P., Price, S. J. W. 1972. *Can. J. Chem.* 50:50
209. Dunlop, A. N., Price, S. J. W. 1970. *Can. J. Chem.* 48:3205
210. Tsang, W. 1978. *Int. J. Chem. Kinet.* 10:1119
211. Czarnowski, J., Schumacher, H. J. 1978. *Int. J. Chem. Kinet.* 10:11
212. Czarnowski, J., Schumacher, H. J. 1979. *Int. J. Chem. Kinet.* 11:613
213. Czarnowski, J., Catellano, E., Schumacher, H. J. 1968. *Chem. Commun.* 1255
214. Czarnowski, J., Schumacher, H. J. 1981. *Int. J. Chem. Kinet.* 13:639
215. Baldwin, A. C., Golden, D. M. 1978. *J. Phys. Chem.* 82:644
216. Hendry, D. G., Kenley, R. 1977. *J. Am. Chem. Soc.* 99:3198
217. Jacko, M. G., Price, S. J. W. 1963. *Can. J. Chem.* 41:1560
218. Paputa, M. C., Price, S. J. W. 1979. *Can. J. Chem.* 57:3178
219. Dzarnoski, J., Ring, M. A., O'Neal, H. E. 1982. *Int. J. Chem. Kinet.* In press
220. Price, S. J. W., Richard, J. P. 1970. *Can. J. Chem.* 48:3209
221. Krech, M., Price, S. J. W. 1965. *Can. J. Chem.* 43:1929
222. Jacko, M. G., Price, S. J. W. 1964. *Can. J. Chem.* 42:1198
223. Baldwin, A. C., Lewis, K. E., Golden, D. M. 1979. *Int. J. Chem. Kinet.* 11:529
224. Barton, B. D., Stein, S. E. 1980. *J. Phys. Chem.* 84:2141
225. McMillen, D. F., Golden, D. M., Benson, S. W. 1971. *Int. J. Chem. Kinet.* 3:359
226. McMillen, D. F., Golden, D. M., Benson, S. W. 1972. *Int. Chem. Kinet.* 4:487
227. Janousek, B. K., Brauman, J. I., Simons, J. 1979. *J. Chem. Phys.* 71:2057
228. King, K. D. 1978. *Int. J. Chem. Kinet.* 10:545
229. King, K. D., Nguyen, T. T. 1974. *J. Phys. Chem.* 83:1940
230. King, K. D. 1979. *Int. J. Chem. Kinet.* 11:1071
231. King, K. D. 1981. *Int. J. Chem. Kinet.* 13:273
232. Trenwith, A. B. 1980. *J. Chem. Soc. Faraday Trans 1* 76:266
233. King, K. D., Goddard, R. D. 1978. *J. Phys. Chem.* 82:1675
234. Roellig, M. P., Houston, P. L., Asscher, M., Haas, Y. 1980. *J. Chem. Phys.* 73:5081
235. Marquaire, P. M., Come, G. M. 1978. *React. Kinet. Catal. Lett.* 9:171
236. Delliste, D., Richard, C., Martin, R. 1981. *J. Chem. Phys. Phys. Chim. Biol.* 78:655

237. Tsang, W. 1969. *Int. J. Chem. Kinet.* 1:245
238. Koenig, T., Belle, T., Snell, W. 1975. *J. Am. Chem. Soc.* 97:662
239. Frey, H. M., Krantz, A. 1969. *J. Chem. Soc. A* p. 1159
240. Benson, S. W. 1965. *J. Chem. Educ.* 42:502
241. Szwarc, M. 1950. *Chem. Rev.* 47:75
242. Evans, B. S., Whittle, E. 1981. *Int. J. Chem. Kinet.* 13:59
243. Shaw, R., Cruickstank, F. R., Benson, S. W. 1967. *J. Phys. Chem.* 71:4538
244. Hendry, D. G., Schuetzle, D. 1975. *J. Am. Chem. Soc.* 97:123
245. Korth, H-G., Trill, H., Sustmann, R. 1981. *J. Am. Chem. Soc.* 103:4483
246. Janousek, B. K., Reed, K. J., Brauman, J. I. 1980. *J. Am. Chem. Soc.* 102:3125

SOLID STATE NMR OF BIOLOGICAL SYSTEMS

Stanley J. Opella

Department of Chemistry, University of Pennsylvania, Philadelphia, Pennsylvania 19104

INTRODUCTION

Biological NMR

NMR spectroscopy is widely used in the study of biological problems. A tremendous variety of biological systems have been studied: samples range in complexity from inert gases interacting with biomolecules to intact eukaryotic organisms. The field has been thoroughly described in monographs (1–4) and review volumes and series (5–11). There are also a large number of individual reviews on the subject, including in the *Annual Review of Physical Chemistry* (12–15). Even though essentially all of the components of cells have been studied by NMR, the amount, as well as the effectiveness, of the research have not always reflected the fundamental interest in the problems. This is because the studies are often closely linked with the available technology (16). Since the instrumentation and methods for high resolution NMR of liquids have been well developed for a long period of time, small and medium sized molecules in solution are well studied. In contrast, even though the properties of intact functional biological complexes are generally of greater interest than those of the isolated constituent subunits, much less is known about structures with mass greater than 10^6 daltons compared to those with mass of 10^4 daltons or less. This discrepancy results from the difficulties associated with chemical and physical studies of large and complex systems. The size, as reflected in the limited rates and amplitudes of reorientation of sites, and complexity, resulting from many similar sites, combine to make NMR spectroscopy of very large molecules and their aggregates particularly difficult. Investigations of crystals of biomolecules and biological supramolecular structures in solution, such as membranes and nucleoprotein complexes, are at a very early stage because of the only relatively recent development of methods and instrumentation for solid state NMR with high resolution and sensitivity. The spectral overlap among resonances from a large number of nuclei with similar properties causes severe problems in both solution and solid state NMR. Solid state NMR offers no advantages over solution NMR in dealing with the difficulties imposed by the chemical complexity of the samples. But, solid state NMR does offer great advantages in dealing with the slow or limited reorientation of sites of the large macromolecular systems. Obviously the most appropriate biochemical systems for solid state NMR are those that are very large by virtue of being constructed from many identical small subunits. Of course, this is also virtually the definition of a single crystal of a molecule.

Regardless of the complexity or physical state of the sample, NMR spectroscopy relies on analyzing the properties of nuclear resonances. The spectroscopic data are in the form of intensities, absorption frequencies and multiplicities, lineshapes, and relaxation parameters of the resonances. The interpretation requires the assignment of individual resonances to specific molecular sites and the comparison of observed and calculated spectral properties. It is in the acquisition, resolution, and characterization of individual resonance signals that the techniques of NMR have a critical role in defining the feasibility of experiments.

A combination of genetic, biochemical, and spectroscopic strategies are used in biological NMR studies. In particular, the use of specific isotopic labels has the effect of simultaneously simplifying the resolution and assignment problems and increasing the sensitivity of the experiment. Regardless of the biochemical finesse applied to the system, if the molecular sites of interest do not reorient rapidly and isotropically, their resonances will be broad and weak. In liquids, narrowing of lines occurs because the large amplitude, rapid motions completely average the static spin interactions. In high resolution solid state NMR, radio-frequency irradiations and mechanical sample spinning replace molecular motions as line narrowing mechanisms. In solid state NMR spectra the angular properties of the spin interactions are manifested, therefore solid state NMR can be used to obtain molecular information that is not present in solution.

Solid State NMR

Several reviews on high resolution solid state NMR are available (17–20), including in the *Annual Review of Physical Chemistry* (21). Solid state NMR methods can be applied to crystalline or amorphous solids, gels, and liquid samples. The differentiation between solution NMR and solid state NMR lies not in the physical state of the sample, but rather in the properties of the resonances being observed. If a definition of solid state NMR can be made that excludes solution NMR, it has to be based on the manifestation of the static spin interactions as tensors in the various spectral features. This is a consequence of the lack of effective motional narrowing and is usually determined by the response of the spin system to experimental manipulations, e.g. if irradiating the ^{1}H spins narrows a ^{13}C resonance, then static ^{1}H–^{13}C dipolar couplings are present.

A useful division in solid state NMR approaches is made by considering abundant spin problems, generally ^{1}H NMR, and dilute spin problems, such as ^{2}H, ^{13}C, ^{15}N, or ^{31}P NMR, separately. High resolution ^{1}H NMR of solids relies on complex multipulse line-narrowing sequences to average out ^{1}H–^{1}H dipolar couplings enabling ^{1}H chemical shift properties to be studied (17). Even when multipulse and sample spinning procedures are combined, the small frequency range of ^{1}H chemical shifts limits the possible biological applications (22); however, the dilute spin double resonance procedures with and without sample spinning provide many opportunities for fruitful biochemical studies (23, 24).

The spin $S=1/2$ nuclei of ^{13}C, ^{15}N, and ^{31}P are available for NMR studies in natural abundance or labeled sites of biopolymers. The ^{13}C and ^{15}N sites can be isotopically enriched significantly above the natural abundance levels. These dilute spins strongly interact with the abundant ^{1}H spins, but not among themselves. They usually have large, anisotropic chemical shielding tensors. Therefore, the dominant spin interactions are heteronuclear dipolar couplings and chemical shift anisotropy. Deuterium, ^{2}H, has spin $S=1$ with a substantial quadrupole coupling constant and a relatively small gyromagnetic ratio, therefore ^{2}H NMR is dominated by the quadrupole interaction. ^{14}N is the abundant form of nitrogen and its large quadrupole coupling has a substantial influence on spectroscopy in many situations.

DIPOLAR INTERACTION In most organic molecules the dipolar couplings are the strongest interactions for spin $S=1/2$ nuclei. Both homonuclear couplings among the abundant ^{1}H spins and heteronuclear couplings between the dilute and nearby ^{1}H spins are present. In high resolution NMR experiments of powder samples the dipolar couplings cause severe and unwanted line-broadening superimposed on lineshapes resulting from the chemical shift interaction. The dipolar coupling between two spin $S=1/2$ nuclei is simply described since it is a through-space interaction; it causes a splitting of the resonance in single crystals and the characteristic Pake powder pattern in unoriented samples (25). The dipolar interactions can be removed or decoupled by irradiation of the ^1H resonances, resulting in ^1H transitions that are rapid compared to the breadth (in frequency units) of the individual dipole-dipole couplings (23). It is now possible to reduce ^1H–^{13}C interactions by a factor of 10^3, effectively eliminating them as a source of line-broadening.

The individual heteronuclear dipolar couplings can be characterized with separated local field spectroscopy (26). These local fields are of interest because their magnitude depends only on the distance between nuclei and the angle the internuclear vector makes with respect to the applied magnetic field. Like the quadrupole and chemical shift interactions, the dipolar couplings reflect the motions at the site.

In addition to contributing to the broadening of resonance lines, the heteronuclear dipolar interaction is a mechanism for connecting abundant and dilute spins. Cross-polarization procedures rely on this mechanism for the transfer of magnetization from the protons in order to provide the enormous sensitivity enhancements necessary for solid state NMR of dilute spins (23). Not only is each transient for the dilute spin larger than for direct pulses at the observed spin frequency, but the experiment can be recycled at a rate determined by the generally short ^1H T_1 rather than the long dilute spin T_1s. Because cross-polarization double resonance procedures rely on a "solid state effect," the presence of static dipolar couplings, to develop spin magnetization, they can be used to distinguish "solid-like" from "liquid-like" spins.

CHEMICAL SHIFT INTERACTION The chemical shift interaction reflects the local electronic environment and provides a means for distinguishing among individual nuclear sites. Isotropic chemical shift frequencies and multiplicities and chemical shift anisotropy powder patterns are sensitive to molecular dynamics. Because dipolar couplings are generally larger than the chemical shift anisotropy, it is necessary to remove the broadening due to the dipolar interactions to observe the lineshape due to the chemical shift interaction.

The resonance frequencies observed in single crystals depend on the orientation with respect to the applied magnetic field. Powder samples have characteristic lineshapes. The chemical shielding is described by a second rank tensor that is diagonal when oriented in the principal axis system of the molecule. The values reflect the asymmetry of the electronic shielding while the angles determine the orientation within the molecule. The isotropic chemical shift is the trace of the tensor and corresponds to the single resonance frequency observed in solution where motional averaging occurs.

In many cases the chemical shift anisotropy merely represents another broadening mechanism rather than a source of information. This occurs when there is overlap of numerous sites or the isotropic shifts are of interest. Magic angle sample spinning can replace isotropic molecular motion as a line-narrowing mechanism for chemical shift anisotropy and dipolar interactions (27). In general, proton decoupling and magic angle sample spinning are combined to effect complete line narrowing to give the isotropic chemical shift spectrum (24).

QUADRUPOLE INTERACTION   The quadrupole spin interaction arises from the electrostatic interaction of the nuclear quadrupole moment with the electric field gradient at the site. The quadrupole interaction is of importance in spectroscopy where nuclei with spin $S > 1/2$ are present. It has its strongest effect on the resonance properties of the quadrupolar nucleus itself, but also can have a substantial influence on the properties of nearby spin $S = 1/2$ nuclei. The two most important quadrupolar nuclei in biological studies are the natural ^{14}N isotope and ^2H when it replaces ^1H at specific sites in the molecules.

^{14}N has a quadrupole coupling constant of about 3 MHz in peptide bonds (28). This is large compared to the Zeeman interaction. The resulting ^{14}N NMR spectra are spread over a correspondingly large frequency range; this makes ^{14}N NMR spectroscopy difficult, but does provide a very sensitive probe of structure and dynamics. The presence of ^{14}N is manifested in ^{13}C NMR spectra through the $^{14}N-^{13}C$ dipolar coupling.

^2H NMR is a useful probe of the dynamics of many biological systems (29). Because $C-^2H$ bonds can be arranged to replace $C-^1H$ bonds by organic synthesis, site specific information is available from ^2H NMR despite limited chemical shift resolution. The quadrupole interaction is much larger than chemical shift or dipolar interactions in deuterons, therefore the ^2H NMR spectra are dominated by the ^2H quadrupole coupling.

Biological Solid State NMR

The initial description of the high resolution double resonance solid state NMR experiments suggested the potential importance of these techniques for the study of biological systems (30). However, there have been relatively few applications of solid state NMR to biological problems in ten years, probably because of the technical difficulties inherent in the experiments. Only one review of the area has appeared (31). Biological solid state NMR is promising because previously unapproachable systems may be studied and in some situations both improved sensitivity and resolution are possible. The techniques and theories of solid state NMR offer a number of opportunities for new types of studies of biological systems. Comparisons between solution and solid state properties of molecules are particularly important for medium sized, highly structured macromolecules like globular proteins or transfer RNA that have been studied extensively in solution, yet require results from diffraction studies on crystalline samples for the interpretation of spectroscopic results. A corollary to the comparisons made possible by a dual phase NMR approach is the capability for studying systems that are completely insoluble or in which solubilization denatures or otherwise alters the properties of interest. Solid state NMR also extends the range of the method to large systems in solution in which the overall and internal reorientation rates are not sufficient for the complete motional averaging required for solution NMR methods and theories to be applicable. Solid state NMR offers a way of overcoming the problems imposed by lack of motional averaging and for extracting valuable information from the anisotropic unaveraged spin interactions.

Severe line-broadening and the attendant loss of spectral resolution in large biological structures occurs when the effective rotational correlation time is slower than about 10^{-8} sec because of efficient nuclear spin relaxation and when it is slower than about 10^{-6} sec because of unaveraged static nuclear spin interactions. Solid state NMR techniques are capable of dealing with static nuclear spin interactions but not relaxation effects. Therefore, there is a range of correlation times between about 10^{-8} and 10^{-6} sec where neither solution nor solid state NMR experiments are going to be very effective.

BIOLOGICAL SYSTEMS

Amino Acids and Peptides

TENSORS The spectroscopy of crystalline amino acids and peptides is complex because of the presence of multiple spin species. In particular for ^{13}C NMR, ^{14}N quadrupole coupling, ^{13}C chemical shift anisotropy, and both ^{1}H–^{1}H and ^{1}H–^{13}C dipolar interactions must be taken into account.

Representative tensors for all these interactions have been determined for a few of the amino acids and small peptides, although in general such information must be taken from related model compounds. The interpretation of protein spectra depends in an important way on the understanding of the spectral properties of the constituent amino acids and peptides. The spin interactions present in amino acids have a wide range of strength, as determined by the frequency breadth of the energy splittings induced by the interactions. The strength of these interactions, in decreasing order, is ^{14}N quadrupole coupling of the peptide bond; ^{14}N quadrupole coupling of the free amino group; ^{2}H quadrupole coupling of labeled sites; ^{13}C–^{1}H, ^{14}N–^{1}H, and ^{15}N–^{1}H dipolar couplings; ^{13}C chemical shift anisotropy of aromatic rings and carbonyl groups; ^{15}N chemical shift anisotropy of peptide bonds; ^{13}C chemical shift anisotropy of aliphatic groups; and ^{15}N chemical shift anisotropy of free amino groups. Both the magnitude and directional properties of these interactions are important in determining the spectral properties of samples. In addition, these interactions determine the time scale of the dynamic properties that are reflected in various spectral parameters; these time scales range from $>10^6$ Hz for the quadrupole interactions to ~ 10 Hz for the isotropic chemical shifts and small chemical shift anisotropy. All of these interactions can induce relaxation by their local fields fluctuating near the nuclear Larmor frequencies ($\sim 10^9$ Hz).

The ^{14}N quadrupole interaction of amino acids and peptides have been studied with both single crystal and powder experiments. Standard double resonance schemes have yielded the quadrupole tensors for a variety of amino acids and peptides (28). Recent direct observation of ^{14}N spectral lines from single crystals have yielded well-characterized ^{14}N quadrupole tensors for glycine (32) and N-acetyl valine (33). The nuclear quadrupole coupling constant is about 1.2 MHz for the free-amino groups of amino acids. There is a large change in the ^{14}N quadrupole interaction when going from the free amino group of the amino acids to the peptide bond, with the quadrupole coupling constant increasing to 3.2 MHz.

The ^{14}N–^{13}C dipolar coupling has been directly observed in single crystal ^{13}C NMR spectra of glycine (32, 34) and alanine (35). C–N bond lengths have been determined using this dipolar coupling (32). As described below, the existence of the ^{14}N–^{13}C dipolar coupling along with the substantial ^{14}N quadrupole coupling results in distinctive lineshapes in the magic angle spinning ^{13}C NMR spectra of powder samples. The spectral manifestation of these two interactions provides a means for obtaining structural information.

Dipolar couplings between hydrogens and nitrogens or carbons have been observed and characterized in amino acids and peptides (33, 36, 37). Generally, high resolution solid state NMR spectroscopy is carried out with irradiation at proton resonance frequency during data acquisition to decouple the heteronuclear dipolar interactions; however, in other experiments this interaction is resolved and used for determination of bond lengths and descriptions of molecular motions (37, 38).

Single crystal studies have been used to determine the ^{13}C chemical shielding tensors of glycine (32) and alanine (35). The ^{13}C chemical shielding tensors for most major functional groups found in biomolecules have been determined and are described in standard references on solid state NMR (17, 18). The only unusual carbon sites in amino acids are the α- and carbonyl carbons and the studies of alanine and glycine show their chemical shift tensors to be similar to those observed in related compounds. Tensors for other aliphatic sites and aromatic sites of amino acids side chains are expected to be very similar to those from the organic crystalline compounds determined previously. The ^{15}N chemical shielding tensors of histidine have recently been characterized (39) and the ring nitrogens were found to have an anisotropy of about 230 ppm while the amino group has a small value of about 10 ppm. As expected, the principal components are oriented along the molecular symmetry directions of the ring.

HIGH RESOLUTION SOLID STATE NMR SPECTRA   Figure 1 illustrates a number of features of ^{13}C NMR spectra of powder samples with crystalline alanine. Without the use of ^{1}H irradiation only very broad nondescript shapes result. The application of proton decoupling yields characteristic chemical shift powder patterns. The principal values of the carbonyl tensor can be easily measured in the spectrum labeled "stationary, ^{1}H decoupled." The methylene carbon chemical shift properties are significantly altered by interaction with the adjacent ^{14}N. Magic angle sample spinning in the presence of proton decoupling produces a spectrum that has single resonances at the isotropic chemical shift positions for the carbonyl and methyl carbons of alanine, but as the inset shows, unusual lineshapes centered at the isotropic chemical shift frequency for the α-carbon. Other spectra in Figure 1 show the effects of changing magnetic field strength and substituting ^{15}N for ^{14}N in alanine on the lineshape of the α-carbon resonance. This feature results from the interaction of ^{14}N with ^{13}C. It is a consequence of the ^{14}N quadrupole coupling interfering with the magic angle spinning elimination of the ^{14}N–^{13}C dipolar coupling (40–48).

High resolution NMR has proved to be a powerful approach to the analysis of conformations of peptides in solution. The capability of obtaining high resolution NMR spectra of solid samples extends the opportunity for analysis to crystalline peptides and peptides that are immobilized as part of biological complexes. Therefore, comparisons can be made of peptide conformations in different environments and physical states. Some of the information obtained from the analysis of solid state NMR spectra is unique, giving insight into both structural and dynamical aspects of peptide conformation. In the study of peptides by solid state NMR many of the aspects of solid state NMR of organic molecules in general are apparent (42, 45). In particular, high resolution ^{13}C spectra of powders are often confusing because of multiple lines for each carbon

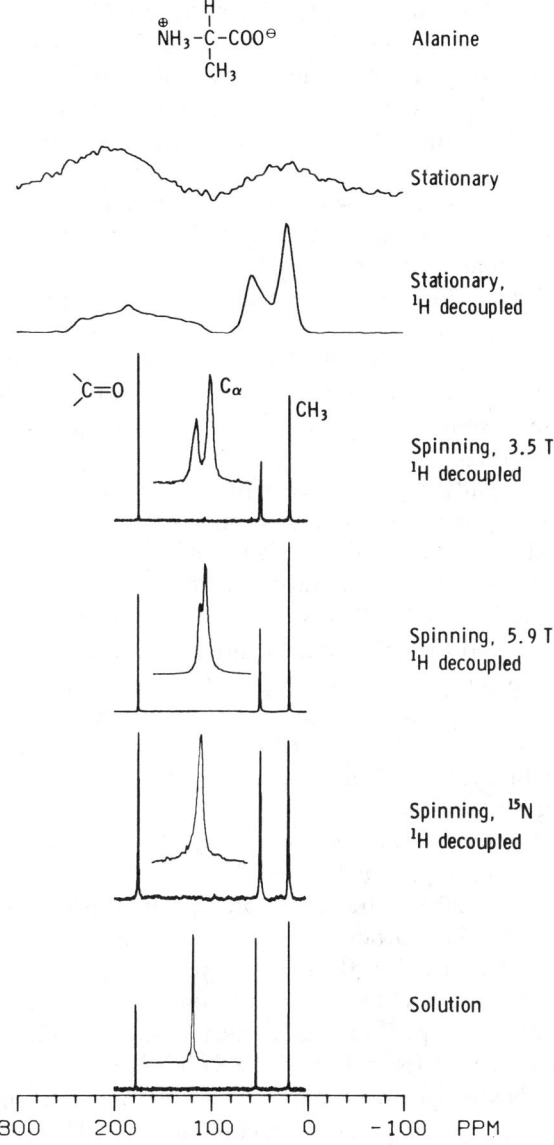

Figure 1 ^{13}C NMR spectra of polycrystalline alanine. All spectra were obtained by cross-polarization and, except the top spectrum labeled "stationary," were obtained with proton decoupling. Field strength and conditions are noted in the Figure. The inset region is of the C_α resonance (M. H. Frey, and S. J. Opella, unpublished results).

site. This can result from a number of factors, in addition to the doublets induced by the $^{13}C-^{14}N$ interactions; other common sources of multiple lines include the presence of different crystal forms and the effect of substituent groups on aromatic rings that result in an asymmetric conformation in the solid state. There are large spectral differences between solutions and powder samples of amino acids and linear small peptides in some cases. The changes in isotropic chemical shift between solid and solution seem to reflect the conformational flexibility of the molecule. Although the actual shifts may result from intermolecular interactions, the accessibility to different environments indicates the capability for changing structure. This is also consistent with the largest changes in chemical shift occurring in a long chain aliphatic amino acids. The initial studies of peptide conformation based on the isotropic chemical shift positions indicate that conformationally flexible peptides, whether linear or cyclic, undergo large changes on going from solution to the solid state, and conformationally rigid cyclic peptides retain the same isotropic chemical shifts in the solid state and solution.

Several model cyclic peptides designed to mimic regions of proteins where hydrogen-bonded reverse turns occur have been studied (49–51). Cyclo-D-Phe-Pro-Gly-D-Ala-Pro is a rigid cyclic pentapeptide that contains in solution and the solid state both a β-turn and a γ-turn. Figure 2 shows that the isotropic ^{13}C chemical shifts for the aliphatic carbons of this crystalline peptide are essentially superimposable on the shifts in solution, thus confirming from NMR data that the peptide has the same conformation in the two states. Only the presence of the asymmetric doublets induced by ^{14}N on the α-carbon carbonyl and Pro-C_δ resonances yield differences between the two spectra. Especially noteworthy is the appearance of an unusually high field proline C_β resonance in both solution and solid state spectra. This upfield-shifted position results from the eclipsing of the Pro-carbonyl and β-methylene groups; this indicates that the local intramolecular effects predominate in determining chemical shifts in both the solid state and in solution where a rigid structure is retained. Other cyclic peptides have given quite different results with varying degrees of shifts between the solid and solution. The high resolution ^{13}C NMR spectra of naturally occurring ferrichrome peptides in solution and as powders have been compared (M. H. Frey, M. Llinas, S. J. Opella, 1982, unpublished results). Isotropic chemical shifts of alumichrome in solution and the solid state show a very high degree of similarity, indicating that the peptide maintains the same conformation in both states. In contrast, the spectra of deferriferrichrome in solution and the solid state show little overlap of resonance frequencies; this is consistent with the high degree of structural flexibility known to be present in the metal free peptide.

DYNAMICS The motions of crystalline amino acids and peptides directly influence the spin interactions. The best characterized motions are the three-fold jumps of methyl groups and the two-fold 180° flips of aromatic rings. The motions can manifest themselves in isotropic chemical shift spectra by averaging of line position. A lack of such averaging yields doublet resonances in aromatic rings that are substituted, since a rapid flip of the ring would give a single line. In addition, dynamics strongly influence powder patterns from chemical shift anisotropy, quadrupole interactions, and dipolar interactions when motions of large amplitude occur at rates comparable to or faster than the relevant spectroscopic time scales. It is possible to calculate the powder patterns in the presence of any type of motion by taking into account the angles, amplitudes, and rates involved, although in general this is difficult. When the rate of

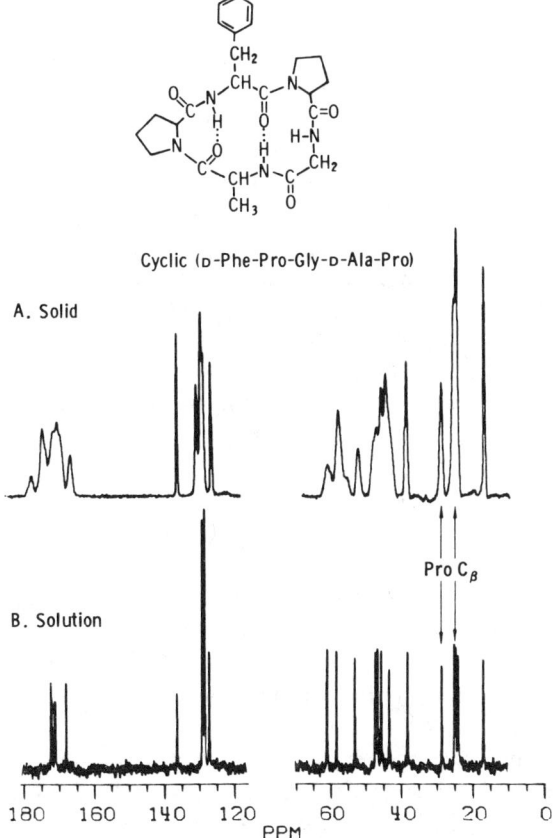

Figure 2 ^{13}C NMR spectra of cyclo-(D-Phe-Pro-Gly-d-Ala-Pro). A. Polycrystalline sample. B. Peptide in CDCl$_3$ solution. The *arrows* point to the Pro C$_\beta$ resonances [from Ref. (49)].

motion is much faster than the spectroscopic time scale, however, the problem is in the fast exchange limit and becomes much simpler. For rapid motions, only the angles involved are needed for calculations of spectra for either rotational or jump motions. Analysis of all the interactions has been carried out for intramolecular dynamics of both aromatic and aliphatic amino acid side chains (52–60).

Proteins

The two major categories of solid state NMR studies can be exploited with proteins. The relatively small globular proteins that are most amenable to high resolution solution NMR are exactly the same ones that crystallize most conveniently for x-ray diffraction analysis. Therefore, solid state NMR studies of crystalline proteins are complementary to diffraction studies of the same samples, and enable comparisons between solution and solid state properties to be made. For the first time, proteins that aggregate or are part of macromolecular complexes can be studied in their native state by NMR. A variety of structural, mechanical, and intrinsic membrane proteins fall into this category; thus, solid-state NMR of proteins can extend the range of the method.

Globular proteins reorient in solution with correlation times around 10^{-8} sec. This results in efficient nuclear spin relaxation and broad lines for many relaxation mechanisms, especially carbons and nitrogens with directly attached protons. Therefore, the ultimate resolution among ^{13}C or ^{15}N sites may be significantly better in the solid state than in solution, because the isotropic chemical shift dispersion is the same and experimental procedures can remove static linebroadening mechanisms, and relaxation-induced linebroadening is absent in immobile samples.

Solid state NMR spectroscopy is well suited for the investigation of protein dynamics. The immobilized protein molecules have only intramolecular motions. This eliminates difficulties inherent in the separation of the intramolecular motions from overall reorientation, especially when they are on similar time scales. In addition, the motion-induced averaging of powder pattern lineshapes is more clearly interpreted than the relaxation measurements used in solution. Therefore, solid state NMR provides an alternative spectroscopic strategy for the study of protein dynamics.

To date, structural proteins that are part of supramolecular structures have been studied more extensively than crystalline proteins. This is partially because of the low resolution available in the early experiments, which provided limited but unique information on the previously unstudied structural proteins. Without atomic resolution, however, compari-

STRUCTURAL PROTEINS OF CONNECTIVE TISSUE   Collagen is the major structural protein of connective tissue. It is found in large amounts in tendon, cartilage, teeth, and other organs where high strength and limited extensibility are important. Collagen exists as a highly ordered triple α-helix that forms very long fibrils about 10^3 Å in diameter. These fibrils are insoluble and clearly cannot undergo rapid isotropic reorientation; as a consequence, the protein cannot be studied in its native state by conventional high resolution solution NMR.

Collagen was one of the first proteins studied by solid state NMR. The initial studies by natural abundance ^{13}C NMR (31, 61–63) showed that the resonances were narrowed by ^{1}H decoupling and magic angle sample spinning, and that cross-polarization of the ^{13}C magnetization was effective. Since collagen has a highly repetitive amino acid sequence with essentially every third residue a glycine, Jelinski & Torchia have biosynthetically ^{13}C labeled the α- and carbonyl carbons of glycine and used the resonance properties to describe backbone dynamics of the protein (64). In solution, collagen behaves like a monomeric protein of 300,000 mol. wt. The interpretation of the α-carbon resonance linewidth in fibrils is complicated by ^{14}N–^{13}C interactions and ^{1}H–^{13}C dipolar relaxation. But the comparison of the 103 ppm carbonyl chemical shift anisotropy in collagen fibrils to polycrystalline model compounds with 144 ppm anisotropy does show some reduction due to limited motional averaging of the protein backbone on the 10^4 Hz time scale. The greatly reduced carbonyl linewidth of 8 ppm for the protein in solution shows nearly complete motional averaging. The longitudinal relaxation parameters, T_1 and NOE, of the glycine backbone sites of collagen were interpreted to show that the protein in solution had rotational diffusion about its long axis as well as some internal reorientation. The discovery of substantial NOE and only somewhat longer T_1 values for collagen in fibrils was taken as evidence that the rapid axial motion is present; the relatively large nonaxially symmetric 103 ppm anisotropy for the carbonyl sites showed that the angular dispersion of motion is limited.

Collagen side chains have been isotopically labeled with ^{2}H and ^{13}C for NMR studies (52–55, 65). The ^{13}C NMR studies of collagen side-chain dynamics used linewidths, lineshapes, NOE, and T_1 values for the alanine, methionine, leucine, and glutamic acid residues. The resonance behavior of all the side-chain sites was somewhat similar to that observed for the glycine sites, in that the chemical shift anisotropy was partially averaged, but the T_1s were short and the NOEs significantly above the minimal values. A definite graduation of motion was seen for the types of side chains, with the long aliphatic side chains having smaller linewidths due to larger amplitude motions than backbone or short chain sites. The ^{13}C NMR studies of collagen have been supplemented by the ^{2}H NMR studies. [3, 3, 3—d_3] alanine and [d_{10}]-leucine were incorporated into separate collagen samples. Powder patterns due to the ^{2}H quadrupole interaction were analyzed. The [d_3]-alanine patterns suggest that the molecule undergoes reorientation about its long axis by jumping between sites ~ 30° apart. The [d_{10}]-leucine patterns indicate that the side-chain jump motions occur rapidly, in addition to the three-fold reorientation of the methyl groups.

Cartilage is a structural matrix in connective tissues. It is made up of about 50% collagen and 50% proteoglycans, which are predominantly polysaccharide with a small amount of protein. The collagen in cartilage is similar, but not identical to that discussed above. The matrix is both structurally and dynamically complex. Conventional solution NMR spectra of cartilage have relatively narrow resonances from the polysaccharides. Only when high power proton decoupling is applied are protein resonances observed (31, 62, 63, 66). These solid state NMR experiments, in which large spectral changes are seen as a function of decoupling power, indicate that the two major structural components have quite different motional properties.

Hard tissues, such as bone or teeth, are a matrix of calcium phosphates, typically hydroxyapatite, and proteins, typically collagen. ^{31}P NMR with magic angle sample spinning has been used to differentiate the organic phosphates in high and low density bone (67) and shows that high density bone is predominantly hydroxyapatite and low density bone is predominantly brushite, based on the analysis of ^{31}P chemical shift tensors as reflected in spinning sideband intensities. Ivory gives high resolution ^{13}C NMR spectra when proton decoupling and magic angle sample spinning are utilized. These spectra correspond to the carbons of the immobilized protein component (24, 62). Enamel contains proteins and highly crystalline hydroxyapatite. In spite of the extreme order of the material, both solution and solid state NMR techniques give ^{13}C resonances (68). Two classes of proteins are indicated, based on their ^{13}C NMR spectra; about 70% of the proteins exhibit rapid, nearly isotropic motion, whereas the other 30% appear to undergo spatially restricted anisotropic motion.

Elastin is the major protein component of elastic fiber connective tissue, such as ligaments. It is a highly cross-linked structural protein similar to collagen in that about one-third of the residues are glycine with a high proline content. It is similar to rubber in its bulk elastic properties.

Elastin is a candidate for solid state NMR studies because in its mature cross-linked form it is insoluble and difficult to study by conventional solution techniques. Solid-state ^{13}C NMR has proven useful in describing this otherwise intractable system (31). The protein is extremely rigid in its unswollen form. In its swollen form a significant fraction of the residues have substantial mobility, since some ^{13}C resonance intensity can be observed by solution NMR methods. Narrower and more intense ^{13}C resonance bands are seen when high-power ^{1}H decoupling is applied; the increase in the aliphatic carbon resonance intensity is as much as 50%. There are substantial restrictions of the motions of even the most mobile elastin sites, since all resonances respond to cross-polarization, which requires at least some unaveraged dipolar coupling to be effective. Analysis of the ^{13}C relaxation parameters indicates that the dynamics of elastin are complex.

NUCLEOPROTEIN COMPLEXES   Viruses are organisms that are so small in size and in genetic content that they can only replicate in an infected cell. The DNA or RNA genome of a virus is packaged within symmetrically arranged protein, and in some cases, lipid components. Although the size of viral particles are small by the standards of biological organisms, they are extremely large by most spectroscopic criteria; in particular they are unable to reorient rapidly enough in solution to narrow resonance lines. Viruses have been studied only very recently by NMR spectroscopy because of the need to use solid state NMR techniques for the detection and resolution of signals. Solution NMR procedures have limited usefulness, but are able to show if any part of the structure has substantial internal mobility.

The coat proteins immobilize the phosphodiester backbone of the RNA in the spherical tomato bushy stunt virus (69) and the DNA in the filamentous bacteriophages fd and Pf1 (70–73) and the head of T4 (74). The DNA in bacteriophage PM2 is packaged by both proteins and lipids and is also immobile (75). The conclusion that the backbone of the nucleic acids are greatly restricted in their motions by encapsidation is based on observing ^{31}P chemical shift anisotropy powder patterns that are asymmetric with the same ~200 ppm breadth observed "model crystalline phosphodiesters." This is quite different from the finding of fully averaged chemical shift anisotropy for nucleic acids in solution in the absence of proteins (76).

^{31}P NMR spectra of the lipid containing PM2 (75) and vesicular stomatitis (77, 78) viruses show that the lipids form membrance bilayers, since the phosphate of the lipid headgroup give axially symmetric powder patterns typical of model and biological membranes (79).

The ^{1}H, ^{13}C, and ^{15}N NMR spectra of a variety of viruses have been obtained by conventional high resolution solution NMR techniques (72, 74, 80–84). Because of the dominant constituent of viruses in the coat protein, these spectra, which contain varying amounts of some parts of the coat proteins, have substantial internal motions. Solid state NMR of viruses enables all coat protein resonances to be observed. The coat proteins of two viruses have been studied by solid state NMR; these are the RNA-containing tobacco mosaic virus and the DNA-containing filamentous bacteriophages. Both types of viruses are long rods of coat protein subunits wrapped helically about the nucleic acid genome packaged inside.

Virus coats illustrate the type of biological problem that is well suited for solid state NMR. The slow reorientation of the subunits results from the overall size of the assembled particle and the local protein folding in protein-protein and protein-nucleic interactions. However, the large size and packing is a consequence of the symmetrical assembly of thousands of identical small subunits and not a single, immense, asymmetric unit. Solid state NMR techniques are capable of narrowing resonances and improving resolution due to the linebroadening effects of the lack of motional narrowing. Resonance bands that are broad due to the overlap of many similar sites are not narrowed by solid state NMR methods. The lack of effective molecular reorientation gives very broad and weak spectra of viruses. This is shown in Figures 3 for both ^{13}C and ^{15}N spectra. Once the linebroadening due to the static spin interactions is removed with spectroscopic techniques, the basic problem remains of studying relatively small protein by NMR.

The filamentous bacteriophage fd infects *E. coli*, therefore there is considerable flexibility in the ability to label isotopically the virus components. Eighty-eight percent of the total mass of the virion is from 2700 copies of the 5000 mol. wt. major coat protein. The total particle is 16×10^6 daltons with nearly all of the nonmajor coat protein mass coming from a single stranded circular DNA packed inside the coat protein shell. The major coat protein of fd has only 50 amino acids, therefore some residues are present only once or a few times in the sequence; this simplifies the spectroscopic problems. For example, there is only one tryptophan residue and it is in position 26 of the sequence. This residue has been labeled in separate samples with ^{13}C at the γ position, ^{15}N at the ε_1 position, and ^{2}H on the five available ring carbon positions. By combining biosynthetic isotopic labeling with solid state NMR for two spin $S = 1/2$ sites, the type of spectra shown in Figure 3B and 3C are obtained (58). The high resolution resonance lines are analogous to those seen for the isolated coat protein in solution where rapid, rotational diffusion is effective in averaging the static spin interactions (72, 85), whereas in the data of Figure 3, spectroscopic irradiations and sample rotations replace molecular rotation as line narrowing mechanisms. Both ^{13}C and ^{15}N labeled Trp 26 fd samples have a single resonance line. This is strong evidence that all 2700 coat protein subunits have the same conformation and environment around residue Trp 26. This information may help distinguish among possible models for the viral architecture based on low resolution diffraction data in model building. Relaxation studies can be useful in characterizing molecular dynamics where there is a single dominant relaxation mechanism such as at the ^{15}N ε_1 site with an attached proton resulting in dipolar relaxation. In general, the high resolution isotropic chemical shift spectra of these proteins obtained with magic angle sample spinning and proton decoupling have molecular information in the frequencies, multiplicities, and relaxation properties of the resonances just as the case for high resolution solution NMR.

Unique to solid state NMR is the information contained in the anisotropic static spin interactions of the individual sites (ie. chemical shift anisotropy, dipolar, and quadrupolar interactions). To extract this information, the NMR experiment must be designed so that one of the

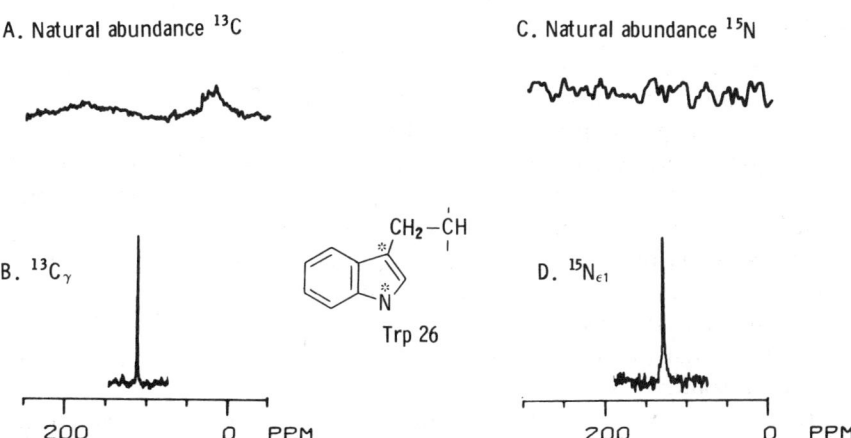

Figure 3 ^{13}C and ^{15}N NMR spectra of fd virus. A. Natural abundance ^{13}C NMR spectrum obtained using solution NMR techniques. B. ^{13}C NMR spectrum of $[^{13}C_\alpha]$-Trp 26 labeled fd obtained with cross-polarization, high power proton decoupling, and rapid magic angle sample spinning. C. Natural abundance ^{15}N NMR spectrum similar to A. D. ^{15}N NMR spectrum of $[^{15}N_\varepsilon]$-Trp-26 labeled fd similar to B [from Ref. (58)].

static spin interactions is not removed by the selective averaging procedures of solid state NMR. The emphasis of experimental design must be on selectivity, because if more than one interaction is left unaveraged to influence the lineshapes, the cumulative broadening effects are so severe that the data are uninterpretable, as seen in Figures 3A and 3C. Such selection procedures have been carried out in a number of cases in the fd coat protein studies, utilizing ^{13}C and ^{15}N chemical shift anisotropy, ^{13}C-^{1}H and ^{15}N-^{1}H dipolar couplings, and ^{2}H quadrupole interactions (37, 57, 58, 84).

The examinations of static spin interactions have been most useful in describing protein dynamics, since the tensors are averaged in specific ways by rotational and jump modes of motion. These studies have shown that the tyrosines and phenylalanine residues of the fd coat protein undergo rapid, 180° flips about their C_β-C_γ bond axis. This is in contrast to the peptide backbone sites and the tryptophan ring, which have no large amplitude motions on the 10^3 Hz time scale, although apparently all sites have small amplitude motions on the 10^9 to 10^{12} Hz time scale.

Tobacco mosaic virus (TMV) infects plants. The viral host has limited isotopic labeling to uniform incorporation of ^{13}C. The virus itself consists of 2200 protein subunits of 17,500 mol. wt. surrounding single-stranded RNA with 6600 nucleotides. The total particle size is 42×10^6 Daltons. High resolution solution ^{1}H and ^{13}C NMR spectra have a substantial amount of relatively narrow resonance intensity from mobile sections of the viral coat protein (80-82).

However, the ^{13}C resonances of TMV narrow substantially and increase in intensity using solid state NMR techniques. With magic angle sample spinning, essentially all of the carbons of TMV coat protein contribute to the resonance intensities (86), whereas with conventional high resolution solution NMR techniques, only the mobile carbons contribute to the spectrum. The large extent of internal mobility seen in the comparison of solution and solid state NMR spectra indicates that the TMV coat protein has extensive internal motions. This is quite different from the coat protein of fd bacteriophage, which has very limited internal motions of sufficient amplitude and rate to give resonances in solution NMR (72).

DNA complexed with histones in soluble chromatin and with protamine in the heads of sperm has been studied by ^{31}P NMR spectroscopy (87). These eukaryotic chromosomal complexes are made up of duplex DNA surrounded by a limited number of relatively small basic proteins. Because the DNA in eukaryotic cell nuclei is condensed much more than the free polymer in solution, significant problems are associated with the extent of folding of the structure. These protein nucleic acid complexes give asymmetric chemical shift anisotropy powder patterns for the ^{31}P resonance of the phosphodiester backbone. However, the total breadth is 132 ppm, which is significantly less than the completely static value of about 200 ppm. Therefore, some reduction occurs by motional averaging of the chemical shift anisotropy powder pattern. This is in contrast to that seen for the viruses, where the full chemical shift anisotropy pattern is observed. Therefore, some motions of the particle or, more likely, local structural fluctuations of limited amplitude must be fast compared to the 10^4 Hz time scale. If this motion were of large amplitude, such as rotation about a single axis, the shape of the spectrum would be an axially symmetric powder pattern, and this is not observed.

GLOBULAR PROTEINS Hemoglobin is the oxygen-carrying protein of red blood cells. It is well studied by NMR and other physical techniques in solution, and an x-ray diffraction structure of the crystalline form has been determined. One of the concepts of biophysics that can be tested by solid state NMR is the comparison of the structure and dynamics of the protein in solution to the crystalline form. One way of showing this, at least implicitly, is by comparing the isotropic chemical shift positions, where samples in solution are obtained by conventional pulse NMR and the same samples as solids are obtained by solid state NMR. This is exactly the sort of comparison described for the cyclic pentapeptides with well-defined stable conformations in both states. Preliminary comparisons of lysozyme and ribonuclease in the solid state and solution have been described (88). Hemoglobin crystals with bound ^{13}CO were found to have the same distinctive chemical shift as seen for bound CO in solution, suggesting that the structure around the binding site of hemoglobin is the same in solution and in the crystalline state (89).

Hemoglobin S is the protein of sickle cell hemoglobin. Hemoglobin S aggregates and forms fibers in red blood cells under conditions of low oxygen tension. This is of course in contrast to the behavior exhibited by normal hemoglobin. The immobilization of the hemoglobin protein molecules in the fibers makes them candidates for solid state NMR. In fact, the requirement of at least some unaveraged static dipolar coupling for cross-polarization means that NMR experiments can be designed that distinguish between aggregates and soluble oligomers of hemoglobin. Such a study has been carried out by natural abundance ^{13}C NMR (90). Since approximately 90% of the carbon in the red blood cell is from hemoglobin, resonance intensity comparisons showed that as much as 80% of the hemoglobin is aggregated in the oxygenated sickled cell.

Myoglobin crystals have been studied by ^{13}C and ^{2}H solid state NMR (91, 92). Myoglobin has two methionine residues, the methyl groups of which can be chemically exchanged easily to perform isotopic labeling.

By taking advantage of the magnetic ordering of paramagnetic myoglobin crystals in the magnetic field used for the NMR experiment, the anisotropic character of the quadrupole or chemical shift anisotropy interactions means that both the spatial orientation of the particular labeled residue within the protein crystal and the rate and types of side-chain motions can be determined. For myoglobin, good agreement has been seen between the structure determined by diffraction methods and the NMR methods, relying on the interpretation of the oriented crystals data as well as the correspondence of isotropic ^{13}C chemical shifts (60).

MEMBRANE PROTEIN COMPLEXES Proteins embedded in membrane bilayers are effectively immobilized. Therefore they can be studied by solid state, but not solution, NMR techniques. Two examples have been examined so far, bacteriorhodopsin and fd coat protein. Both can be isotopically labeled biosynthetically and prepared in large amounts. We can anticipate that many additional examples will be studied.

Bacteriorhodopsin is the major protein constituent of the purple membrane of halobacteria. This protein is being widely studied by physical techniques. Selectively deuterated amino acids have been incorporated into the protein biosynthetically. The shape of the quadrupole powder patterns and their relaxation properties have been used to describe the dynamics of the molecule (55, 56, 59, 60). Rapid 180° flips of phenylalanine and tyrosine have been observed, as well as rapid side-chain motions in aliphatic amino acids.

The coat protein of fd resides in the cell membrane prior to virus assembly. Although the protein is insoluble in water, it can be studied in isolated form by solid state NMR (40). The protein-membrane complex can be reconstituted from pure protein and phospholipids. Therefore, the protein can be studied as an intrinsic membrane protein as well as the structural form found in the virion. Significant differences in the protein dynamics have been seen by solid state NMR between the two forms (G. Leo, C. D. D'Ambrosio, S. J. Opella, unpublished results). This correlates well with studies of the protein in micelles and vesicles (72, 85, 93, 94).

NITROGEN METABOLISM One of the most innovative applications of solid state NMR to biological problems has been developed by Schaefer, Stejskal, and their co-workers in their work on nitrogen metabolism in soybean plants (95–97). In these experiments the incorporation of ^{15}N and ^{13}C labeled nutrients were monitored by the NMR spectra of the materials. In the initial experiments, soybeans were grown on ^{15}N enriched ammonium nitrate fertilizer. Nitrogen incorporation into some or all sites can be determined with simple area measurements on samples that were lyophilized. The double cross-polarization experiment (98) has proved useful in providing detailed information on metabolism using the double labeled compound (4^{13}C-amide-^{15}N asparagine) as the sole nitrogen source. Conventional cross-polarization experiments show that both the ^{13}C and ^{15}N were incorporated nonrandomly into the proteins. The double cross-polarization experiments show that about half the asparagine in the protein went in directly without scrambling of the labels. *Neurospora crassa* nitrogen metabolism has been studied by growing the organism on nitrate (99). The incorporation of label into proteins and amino acids was found to depend upon high levels of nitrate-reducing activity.

Nucleic Acids

It is not possible to overemphasize the importance of solid state NMR to the study of DNA. Both in solution and in the solid state, high molecular weight native duplex DNA has proved to be intractable for conventional NMR studies. Even the first investigations of DNA found the linewidths too great to permit detection of nuclear resonances under high resolution conditions (100). These broad lines are a natural consequence of the stiffness and lack of substantial internal mobility of the polymer. Prior to the availability of solid state NMR technology, it was only possible to study denatured DNA or low molecular weight fragments of DNA in solution.

^{31}P NMR studies have received the most emphasis because of the location of the ^{31}P nuclei in the phosphodiester backbone of the molecule and because of the ease of experimentation, since this nucleus has high sensitivity and a large chemical shift anisotropy. The ^{31}P chemical shift powder patterns for a variety of nucleic acids have been reported and shown to be similar to those seen in other phosphodiesters, including phospholipids (101). Surprisingly, chemical shift tensors of model nucleotides were found to be axially symmetric in the sodium salts but clearly nonaxially symmetric in free acids forms. It is generally assumed that the orientation of the tensor components of phosphodiesters is the same as in model compounds such as barium diethyl phosphate (102).

Highly oriented DNA fibers have been studied by solid state ^{31}P NMR. These samples provide excellent opportunities for structural studies. Naturally occurring DNA (103–105) as well as fibers of poly *d*AT (106) were examined by comparing calculated and observed powder patterns for various angles of the fibers with respect to the applied magnetic field. An excellent match was seen for the orientation of the phosphate groups in the fibers given an A form DNA by x-ray diffraction, as shown Figure 4 (105).

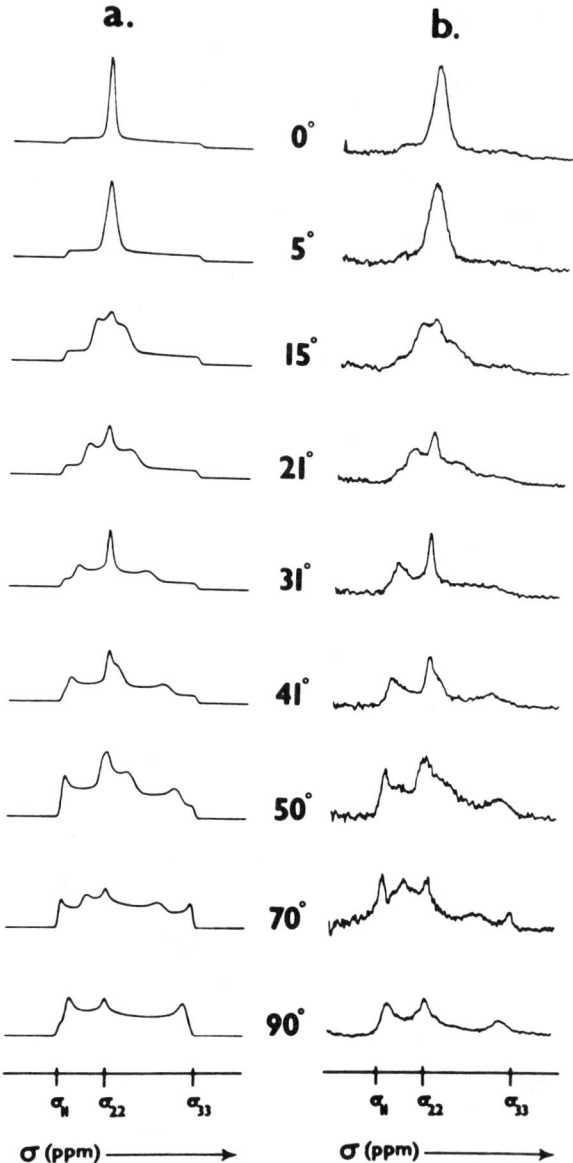

Figure 4 ^{31}P NMR spectra of solid A-form DNA: (a) simulated (b) experimental. These spectra are for various orientations of oriented DNA fibers in the magnetic field [from Ref. (105)].

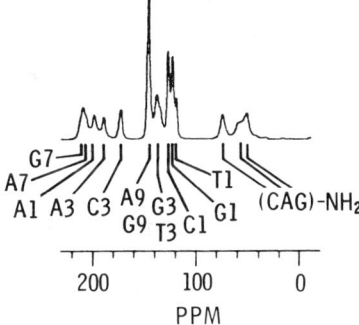

Figure 5 ^{15}N NMR spectrum of solid B-form DNA. The sample is unoriented, but spinning at the magic angle [from Ref. (108)].

The ^{31}P chemical shift anisotropy with DNA backbone provides a means for studying DNA dynamics on the 10^4 Hz time scale. The backbone motions of DNA in the solid state vary significantly as a function of hydration and temperature. A-form DNA has a rigid backbone on this time scale while B-form DNA has substantial motion of the backbone at temperatures above 20°C. In solution, the static chemical shift anisotropy is completely averaged by the bending of the DNA polymer (76). By simultaneously examining the ^{2}H quadrupole pattern for C_8-labeled purine bases of DNA and a ^{31}P chemical shift anisotropy of the backbone in solid B-form DNA, it was possible to show that even under conditions in which substantial backbone motions are present, the bases are rigid (107). ^{15}N NMR isotropic chemical shift spectra of DNA obtained with magic angle sample spinning and proton decoupling have excellent resolution, as shown in Figure 5 (108). The ^{15}N–^{1}H dipolar couplings measured by separated local field spectroscopy are not averaged by motion; thus, rigid bases are indicated (109). Solid state ^{13}C NMR spectra of nucleotides and DNA have been obtained (44, 110, 111); however, the high resolution isotropic chemical shift data are complicated by the presence of the ^{14}N in the bases.

Membranes

The phospholipids of natural and model membrane bilayers have extended long-range order and rapid, local motions. Much of the information on membrane dynamics has come from magnetic resonance studies. Solid state NMR techniques have proved essential because of the lack of complete averaging of the static interactions. Extensive reviews have been written on membrane NMR (14, 29, 78, 112–115), so I describe here only a few illustrative examples of solid state NMR applications.

Phospholipid bilayers are model membranes and have been studied by both dilute spin and abundant spin solid state NMR. Multipulse ^{1}H line-narrowing procedures have been applied to these systems (116, 117) and are essentially the only examples of this type of spectroscopy on biological problems. The initial double resonance ^{13}C NMR studies of model and biological membranes (118, 119) showed that the unsonicated lipid dispersions had some residual linewidth resulting from unaveraged ^{13}C chemical shift anisotropy after the severe broadening effects of the ^{13}C–^{1}H dipolar interactions were decoupled (120–122). Figure 6A is an example of a proton-decoupled ^{13}C NMR spectrum of phospholipid bilayers.

The lineshape caused by chemical shift anisotropy can be analyzed to describe the dynamics of the phospholipids. This has been done in the

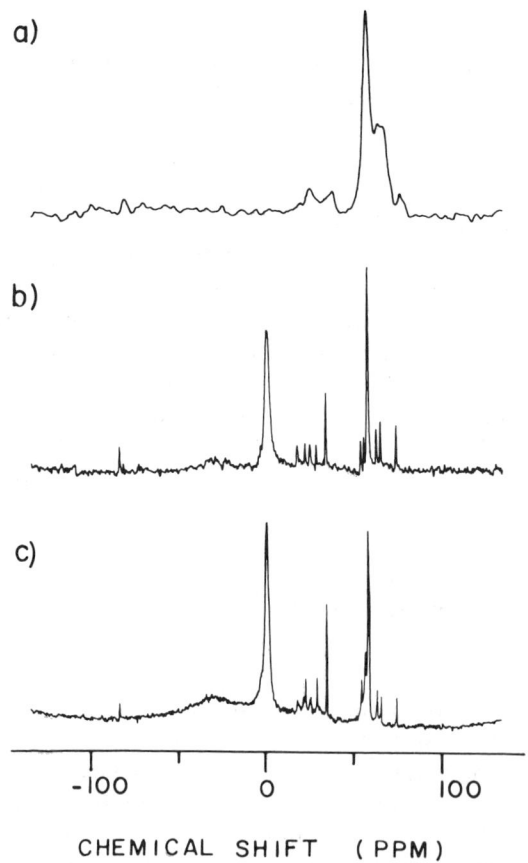

Figure 6 ^{13}C NMR spectra of phospholipid bilayers: (a) DMPC (dimyristoyl phosphatidylcholine) in H$_2$O, stationary; (b) DMPC in H$_2$O with magic angle sample spinning; (c) DPPC (dipalmitoyl phosphatidylcholine) in H$_2$O with magic angle sample spinning [from Ref. (121)].

case of the fatty acid carbonyl resonance (59, 124–126). In particular, the spectra shown in Figure 7 have been used to describe how the conformation and dynamics of the carbonyl group are altered during the phospholipid phase transitions induced by temperature.

Alternative spectroscopic strategies that utilize the tensor properties of the chemical shift anisotropy have been carried out. Magic angle sample spinning gives well-resolved isotropic chemical shift spectra that are equivalent to those seen for the lipids dissolved in organic solvents (121, 122). This is shown in Figure 6. In addition, when multi-bilayers are aligned between glass plates, each site gives a single resonance line, with the frequency a function of the orientation with respect to the magnetic field. Rotational studies of oriented model membranes have been performed for ^{13}C (126), ^{31}P (127, 128), and ^{2}H (129, 130) NMR.

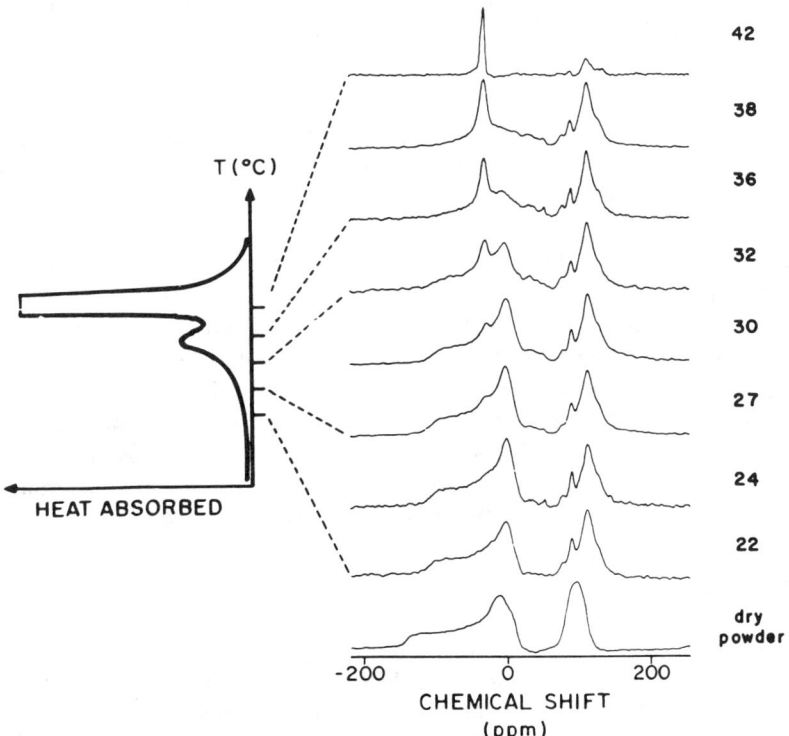

Figure 7 ^{13}C NMR spectra of 2-[1-^{13}C]DPPC in H$_2$O, stationary; except for the bottom spectrum, which is the dry powder. The carbonyl lineshape reflects the lipid dynamics as a function of temperature. A calorimetric trace is on the left. The *dashed lines* show the position of the NMR spectra relative to the temperatures of the phase transitions [from Ref. (123)].

There are extensive studies of the ^{31}P resonance lineshapes of the phosphodiester groups of the membranes (77, 114, 128, 131–135). The powder pattern lineshapes as a function of temperature have been used to describe the phase transitions. The lineshapes have also been used to describe the conformation of the headgroups and their interactions with other molecules.

^{2}H NMR has found its widest use in the study of membranes. There is extensive literature covering all aspects of the results and interpretation of the quadrupole powder patterns from ^{2}H labeled sites of model and biological membranes. This work is well described in reviews (29, 113, 115).

Membranes contain many species in addition to phospholipids. These include proteins, peptides, and steroids. By employing appropriate labels, both the lipids and the other types of molecules can be studied by solid state NMR (115). ^{13}C NMR of labeled cholesterol in bilayers has been used to describe the dynamics and phase transitions of the system (136).

Polysacchrides

Cellulose is the primary constituent of plant cell walls. It is the β-14 polymer of anhydroglucose. The two most common polymorphs are usually identified with the native and regenerated forms of molecules. These two polymorphs can be easily differentiated by solid state NMR spectra (137, 138). Sufficient differences occur among the isotropic chemical shifts to enable one to resolve structures. It is also easy to differentiate amorphous samples from the types 1 and 2 celluloses. Other studies of polysacchrides show that high molecular glucose can adopt a helix form, whereas low molecular weight polysacchrides have considerable random coil form in the solid state (139). In studies of linear and cyclic saccharides, rapid conformational averaging occurs in aqueous solution, with lifetimes much less than the 10^{-3} sec of the conformation dependent ^{13}C chemical shifts obtained in the solid state (140).

SUMMARY

Many biological systems can be studied by solid state NMR. Systems that are intractable for solution NMR methods can be investigated by solid state NMR. Much of the information so gained is unique.

The use of solid state NMR in the study of biological systems is at an early stage. Many of the studies mentioned in this review are the first of their kind, and will undoubtedly be superseded by more definitive work. The widespread adaptation of solid state NMR approaches to chemical and biological problems anticipated from the earliest results in the field has not materialized, probably because of the difficulty in carrying out experiments utilizing high power radiofrequency irradiation and rapid sample spinning. The most active laboratories in the field of biological solid state NMR have emphasized the goal of development of new methods over that of making the experiments routine and reliable.

Areas in which solid state NMR can give results unobtainable in any other way are sure to receive a great deal of interest, perhaps at the expense of more routine studies of a wide range of systems. These areas include structural studies of oriented materials, where single crystal work is impractical, and dynamical studies of the intramolecular motions of molecules in crystals and in immobile macromolecular structures.

ACKNOWLEDGMENTS

The research carried out at the University of Pennsylvania has been supported by the American Chemical Society (PRF 9882-66), the National Science Foundation (PCM 77-05598), the National Institutes of Health (GM-24266 and GM-29754), and the American Cancer Society (NP-255). In addition, this review was written during the tenure (1980–1982) of a Fellowship from the A. P. Sloan Foundation.

Literature Cited

1. Dwek, R. A. 1973. *NMR in Biochemistry: Applications to Enzyme Systems*. London: Oxford Univ. Press. 395 pp.
2. James, T. L., 1975. *Nuclear Magnetic Resonance in Biochemistry*. New York: Academic. 413 pp.
3. Wüthrich, K. 1976. *NMR in Biological Research: Peptides and Proteins*. Amsterdam: Elsevier. 379 pp.
4. Jardetzky, O., Roberts, G. C. K. 1981. *NMR in Molecular Biology*. New York: Academic. 681 pp.
5. Dwek, R. A., Campbell, I. D., Richards, R. E., Williams, R. J. P., eds. 1977. *NMR in Biology*. London: Academic. 381 pp.
6. Agris, P. F., Loeppky, R. N., Sykes, B. D., eds. 1978. *Biomolecular Structure and Function*. New York: Academic. 614 pp.
7. Berliner, L. J., Reuben, J., eds. 1978. *Biol. Magn. Reson.* 1:1–345
8. Opella, S. J., Lu, P., eds. 1979. *NMR and Biochemistry*. New York: Dekker. 434 pp.
9. Shulman, R. G. 1979. *Biological Applications of Magnetic Resonance*. New York: Academic. 595 pp.
10. Cohen, J. S., ed. 1980. *Magn. Reson. Biol.* 1:1–309
11. Bothner-By, A. A., Glickson, J. D., Sykes, B. D., eds. 1982. *Biochemical Structure Determination by NMR*. New York: Dekker. 248 pp.
12. Shulman, R. G. 1962. *Ann. Rev. Phys. Chem.* 13:325–50
13. Dwek, R. A., Richards, R. E. 1967. *Ann. Rev. Phys. Chem.* 18:99–124
14. Bocian, D. F., Chan, S. I., 1978. *Ann. Rev. Phys. Chem.* 29:307–35
15. Patel, D. J. 1978. *Ann. Rev. Phys. Chem.* 29:337–62
16. Opella, S. J., 1977. *Science* 198:158–65
17. Haeberlen, U. 1976. *Adv. Magn. Reson. Suppl.* 1
18. Mehring, M. 1976. *NMR: Basic Principles Prog.* 11:1–246
19. Griffin, R. G., 1977. *Anal. Chem.* 49:A951–62
20. Schaefer, J., Stejskal, E. O. 1979. *Topics in Carbon-13 NMR Spectroscopy*, ed. G. C. Levy, pp. 283–324. New York: Wiley
21. Vaughan, R. W. 1978. *Ann. Rev. Phys. Chem.* 29:397–419
22. Gerstein, B. C. 1981. *Philos. Trans. R. Soc. London Ser. A* 299:521–46

23. Pines, A., Gibby, M. G., Waugh, J. S. 1973. *J. Chem. Phys.* 59:569–90
24. Schaefer, J., Stejskal, E. O. 1976. *J. Am. Chem. Soc.* 98:1031–32
25. Pake, G. E. 1948. *J. Chem. Phys.* 16:327–36
26. Waugh, J. S. 1976. *Proc. Natl. Acad. Sci. USA* 73:1394–97
27. Andrew, E. R. 1981. *Philos. Trans. R. Soc. London Ser. A* 299:505–20
28. Edmonds, D. T. 1977. *Phys. Rep.* 29:233–55
29. Seelig, J. 1977. *Q. Rev. Biophys.* 16:353–418
30. Pines, A., Gibby, M. G., Waugh, J. S. 1972. *J. Chem. Phys.* 56:1176–77
31. Torchia, D. A., VanderHart, D. L., 1979. See Ref. 20, pp. 325–60
32. Haberkorn, R. A., Stark, R. E., van Willigan, H., Griffin, R. G. 1981. *J. Am. Chem. Soc.* 103:2534–39
33. Stark, R. E., Haberkorn, R. A., Griffin, R. G. 1978. *J. Chem. Phys.* 68:1996–97
34. Griffin, R. G., Pines, A., Waugh, J. S. 1975. *J. Chem. Phys.* 63:3676–77
35. Naito, A., Ganopathy, S., Akasaka, K., McDowell, C. A. 1981. *J. Chem. Phys.* 74:3190–97
36. Bodenhausen, G., Stark, R. E., Ruben, D. J., Griffin, R. G. 1979. *Chem. Phys. Lett.* 67:424–27
37. Cross, T. A., Opella, S. J. 1982. *J. Mol. Biol.* In press
38. Frey, M. H., DiVerdi, J. A., Opella, S. J. 1982. Submitted for publication
39. Harbison, C., Herzfeld, J., Griffin, R. G. 1981. *J. Am. Chem. Soc.* 103:4752–54
40. Opella, S. J., Frey, M. H., Cross, T. A. 1979. *J. Am. Chem. Soc.* 101:5856–57
41. Groombridge, C. J., Harris, R. K., Packer, K. J., Say, B. J., Tunner, S. F. 1980. *J. Chem. Soc. Chem. Commun.*, pp. 174–75
42. Frey, M. H., Opella, S. J. 1980. *J. Chem. Soc. Chem. Commun.*, pp. 474–75
43. Hexem, J. G., Frey, M. H., Opella, S. J. 1981. *J. Am. Chem. Soc.* 103:224–26
44. Zumbulyadis, N., Heinrichs, P. M., Young, R. H. 1981. *J. Chem. Phys.*
45. Opella, S. J., Hexem, J. G., Frey, M. H., Cross, T. A. 1981. *Philos. Trans. R. Soc. London Ser. A* 299:665–83
46. Hexem, J. G., Frey, M. H., Opella, S. J. 1982. *J. Chem. Phys.* In press
47. Naito, A., Ganapathy, S., Akasaka, K., McDowell, C. A. 1981. *J. Chem. Phys.* 74:3190–97
48. Naito, A., Ganapathy, S., McDowell, C. A. 1982. *J. Magn. Reson.* 48: In press
49. Pease, L. G., Frey, M. H., Opella, S. J. 1981. *J. Am. Chem. Soc.* 103:467–68
50. Gierasch, L. M., Opella, S. J., Frey, M. H. 1981. *Peptides: Synthesis, Structure, and Function*, ed. D. H. Rich, E. Gross, pp. 267–76. Rockford, Ill.: Pierce Chem. Co.
51. Gierasch, L. M., Frey, M. H., Hexem, J. G., Opella, S. J. 1982. *New Methods and Applications of NMR Spectroscopy*, ed. G. C. Levy. Washington DC: Am. Chem. Soc. In press
52. Jelinski, L. W., Sullivan, C. E., Batchelder, L. S., Torchia, D. A. 1980. *Biophys. J.* 32:515–29
53. Jelinski, L. W., Sullivan, C. E., Torchia, D. A. 1980. *Nature* 284:531–34
54. Batchelder, L. S., Sullivan, C. E., Jelinski, L. W., Torchia, D. A. 1982. *Proc. Natl. Acad. Sci. USA* 79:386–89
55. Kinsey, R. A., Kintanar, A., Tsai, M. D., Smith, R. L., Janes, N., Oldfield, E. 1981. *J. Biol. Chem.* 256:4146–49
56. Kinsey, R. A., Kintanar, A., Oldfield, E. 1981. *J. Biol. Chem.* 256:9028–36
57. Gall, C. M., DiVerdi, J. A., Opella, S. J. 1981. *J. Am. Chem. Soc.* 103:5039–43
58. Gall, C. M., Cross, T. A., DiVerdi, J. A., Opella, S. J. 1982. *Proc. Natl. Acad. Sci. USA* 79:101–5
59. Rice, D. M., Blume, A., Herzfeld, J., Wittebort, R. J., Huang, T. H., DasGupta, S. K., Griffin, R. G. 1981. *Stereodynamics of Molecular Systems*, ed. R. H. Sarma, 2:255–70. New York: Adenine
60. Schramm, S., Kinsey, R. A., Kintanar, A., Rothgeb, T. M., Oldfield, E. 1981. See Ref. 59, pp. 271–86
61. Torchia, D. A., VanderHart, D. L. 1976. *J. Mol. Biol.* 104:315–21
62. Schaefer, J., Stejskal, E. O., Brewer, C. F., Keiser, H. D., Sternlicht, H. 1978. *Arch. Biochem. Biophys.* 190:657–61
63. Danyluk, S. S., Schwartz, H. M. 1979. *Stereodynamics of Molecular Systems*, ed. R. H. Sarma, pp. 439–52. New York: Pergamon
64. Jelinski, L. W., Torchia, D. A. 1979. *J. Mol. Biol.* 133:45–65

65. Jelinski, L. W., Torchia, D. A. 1980. *J. Mol. Biol.* 138:255–72
66. Torchia, D. A., Hasson, M. A., Hascall, V. C. 1977. *J. Biol. Chem.* 252:3617–25
67. Herzfeld, J., Roufasse, A., Haberkorn, R. A., Griffin, R. G., Glimcher, M. J. 1980. *Philos. Trans. R. Soc. London Ser. B* 289:459–69
68. Termine, J. D., Torchia, D. A. 1980. *Biopolymers* 19:741–50
69. Munowitz, M. G., Dobson, C. M., Griffin, R. G., Harrison, S. C. 1980. *J. Mol. Biol.* 141:327–33
70. Cross, T. A., DiVerdi, J. A., Wise, W. B., Opella, S. J. 1979. See Ref. 8, pp. 67–74
71. DiVerdi, J. A., Opella, S. J. 1981. *Biochemistry* 20:280–84
72. Opella, S. J., Cross, T. A., DiVerdi, J. A., Sturm, C. F. 1980. *Biophys. J.* 32:531–48
73. Opella, S. J., DiVerdi, J. A. 1982. *Biochemical Structure Determination by NMR*, ed. A. A. Bothner-By, J. Glickson, B. D. Sykes, pp. 149–68. New York: Dekker
74. deWit, J. L., Dorssers, L. C. J., Schaafsma, T. J. 1979. *Biochem. Biophys. Res. Commun.* 89:435–40
75. Akutsu, H., Satake, H., Franklin, R. M. 1980. *Biochemistry* 19:5264–70
76. Opella, S. J., Wise, W. B., DiVerdi, J. A. 1981. *Biochemistry* 20:284–90
77. Moore, N. F., Patzer, E. J., Wagner, R. R., Yeagle, P. L., Hutton, W. C., Martin, R. B. 1977. *Biochem. Biophys. Acta* 464:234–44
78. Stoffel, W., Bister, K. 1975. *Biochemistry* 14:2841–47
79. Seelig, J. 1978. *Biochim. Biophys. Acta* 515:105–40
80. deWit, J. L., Schaafsma, T. J. 1978. *FEBS Lett.* 92:273–77
81. deWit, J. L., Hemminga, M. A. Schaafsma, T. J. 1978. *J. Magn. Reson.* 31:97–107
82. deWit, J. L., Alma-Zeestraten, N. C. M., Hemminga, M. A., Schaafsma, T. J. 1979. *Biochemistry* 18:3973–76
83. Andree, P. J., Kan, J. H., Mellema, J. E. 1981. *FEBS Lett.* 130:265–68
84. Cross, T. A., Gall, C. M., Opella, S. J. 1981. *Bacteriophage Assembly*, ed. M. S. Dubow, pp. 457–65. New York: Liss
85. Cross, T. A., Opella, S. J. 1981. *Biochemistry* 20:290–97
86. Hemminga, M. A., Veeman, W. S., Hilhorst, H. W. M., Schaafsma, T. J. 1981. *Biophys. J.* 35:463–70
87. DiVerdi, J. A., Opella, S. J., Ma, R.-I., Kallenbach, N. R., Seeman, N. C. 1981. *Biochem. Biophys. Res. Commun.* 102:885–90
88. Jardetzky, O., Wade-Jardetzky, N. G. 1980. *FEBS Lett.* 110:133–35
89. Maciel, G. E., Shatlock, M. P., Houtchens, R. A., Caughey, W. S. 1980. *J. Am. Chem. Soc.* 102:6885–86
90. Sutherland, J. W. H., Egan, W., Schechter, A. N., Torchia, D. A. 1979. *Biochemistry* 18:1797–1803
91. Oldfield, E., Rothgeb, T. M. 1980. *J. Am. Chem. Soc.* 102:3635–37
92. Rothgeb, T. M., Oldfield, E. 1981. *J. Biol. Chem.* 256:1432–46
93. Hagen, D. S., Weiner, J. H., Sykes, B. D. 1978. *Biochemistry* 18:2007–12
94. Dettman, H. D., Weiner, J. H., Sykes, B. D. 1982. *Biophys. J.* 37:243–51
95. Schaefer, J., Stejskal, E. O., Sefcik, M. D., McKay, R. A. 1981. *Philos. Trans. R. Soc. London Ser. A* 299:593–608
96. Schaefer, J., Stejskal, E. O., McKay, R. A. 1979. *Biochem. Biophys. Res. Commun.* 88:274–80
97. Schaefer, J., Skokut, T. A., Stejskal, E. O., McKay, R. A., Varner, J. E., 1981. *Proc. Natl. Acad. Sci. USA* 78:5978–82
98. Schaefer, J., McKay, R. A., Stejskal, E. O. 1979. *J. Magn. Reson.* 34:443–47
99. Jacob, G. S., Schafer, J., Stejskal, E. O., McKay, R. A. 1980. *Biochem. Biophys. Res. Commun.* 97:1176–82
100. McDonald, C. C., Phillips, W. D., Penman, S. 1964. *Science* 144:1234
101. Terao, T., Matsui, S., Akasaka, K. 1977. *J. Am. Chem. Soc.* 99:6136–38
102. Herzfeld, J., Griffin, R. G., Haberkorn, R. A. 1978. *Biochemistry* 17:2711–18
103. Shindo, H., Wooten, J. B., Pheiffer, B. H., Zimmerman, S. B. 1980. *Biochemistry* 19:518–26
104. Shindo, H., Wooten, J. B., Zimmerman, S. B. 1981. *Biochemistry* 20:745–50
105. Nall, B. T., Rothwell, W. P., Waugh, J. S., Rupprecht, A. 1981. *Biochemistry* 20:1881–87
106. Shindo, H., Zimmerman, S. B. 1980. *Nature* 283:690–91
107. DiVerdi, J. A., Opella, S. J. 1981. *J. Mol. Biol.* 149–307–11
108. Cross, T. A., Diverdi, J. A., Opella, S. J. 1982. *J. Am. Chem. Soc.* 104:1759–61

109. DiVerdi, J. A., Opella, S. J. 1982. *J. Am. Chem. Soc.* 104:1761–62
110. Hays, G. R., Huis, R. Coleman, B., Clague, D., Verhoeven, J. W., Rob, F. 1981. *J. Am. Chem. Soc.* 103:5140–46
111. Ganapathy, S., Naito, A., McDowell, C. A. 1981. *J. Am. Chem. Soc.* 103:6011–15
112. Wennerstrom, H., Lindblom, G. 1977. *Q. Rev. Biophys.* 10:67
113. Mantsch, H. H., Saito, H., Smith, I. C. P. 1977. *Prog. Nucl. Magn. Reson. Spectrosc.* 11:211–71
114. Cullis, P. R., Dekruijf, F. B. 1979. *Biochim. Biophys. Acta* 559:399–420
115. Jacobs, R. E., Oldfield, E. 1981. *Prog. NMR Spectrosc.* 14:113–36
116. Seiter, C. H. A., Chan, S. I. 1973. *J. Am. Chem. Soc.* 95:7541–53
117. Silva Crawford, M., Gerstein, B. C., Kwo, A. L., Wade, C. G. 1980. *J. Am. Chem. Soc.* 102:3728–32
118. Urbina, J. S., Waugh, J. S. 1973. *Ann. NY Acad. Sci.* 222:733–39
119. Urbina, J. S., Waugh, J. S. 1974. *Proc. Natl. Acad. Sci. USA* 73:3812–15
120. Cornell, B. A., Keniry, M., Hiller, R. G., Smith, R. 1980. *FEBS Lett.* 115:134–38
121. Haberkorn, R. A., Herzfeld, J., Griffin, R. G. 1978. *J. Am. Chem. Soc.* 100:1296–98
122. Haberkorn, R. A., Herzfeld, J., Griffin, R. G. 1978. *Nuclear Magnetic Resonance Spectroscopy in Molecular Biology*, ed. B. Pullman, pp. 381–91. Dordrecht: Reidel
123. Wittebort, R. J., Schmidt, C. F., Griffin, R. G. 1981. *Biochemistry* 20:4223–28
124. Pope, J. M., Walker, L., Cornell, B. A., Francis, G. W. 1981. *Biophys. J.* 35:509–20
125. Cornell, B. A. 1981. *Chem. Phys. Lipids* 28:69–78
126. Cornell, B. A., Francis, G. W. 1980. *J. Magn. Reson.* 41:175–78
127. McLaughlin, A. C., Cullis, P. R., Hemminga, M. A., Hoult, D. I., Radda, G. K., Ritchie, G. A., Seeley, P. J., Richards, R. E. 1975. *FEBS Lett.* 57:213–18
128. Griffin, R. G., Powers, L., Persham, P. S. 1978. *Biochemistry* 17:2718–22
129. Charvolin, J., Manneville, P., Deloche, B. 1973. *Chem. Phys. Lett.* 23:345–48
130. Seelig, J., Niederberger, W. 1974. *J. Am. Chem. Soc.* 96:2069–72
131. Niederberger, W., Seelig, J. 1976. *J. Am. Chem. Soc.* 98:3704–6
132. Griffin, R. G. 1976. *J. Am. Chem. Soc.* 98:851–53
133. Kohler, S. J., Klein, M. P. 1976. *Biochemistry* 15:967–73
134. Kohler, S. J., Klein, M. P. 1977. *Biochemistry* 16:519–26
135. Kohler, S. J., Ellett, J. D., Klein, M. P. 1976. *J. Chem. Phys.* 64:4451–58
136. Opella, S. J., Yesinowksi, J. P., Waugh, J. S. 1976. *Proc. Natl. Acad. Sci. USA* 73:3812–15
137. Atalla, R. H., Gast, J. C., Sindorf, D. W., Bartuska, V. J., Maciel, G. E. 1980. *J. Am. Chem. Soc.* 102:3249–51
138. Earl, W. L., VanderHart, D. L. 1980. *J. Am. Chem. Soc.* 102:3251–52
139. Saito, H., Tabeta, R. 1981. *Chem. Lett.* 713–16
140. Saito, H., Tabeta, R. Harada, T. 1981. *Chem. Lett.* 571–74

AUTHOR INDEX

(Names appearing in capital letters indicate authors of chapters in this volume.)

A

Abgrall, M., 416
Abkowitz, M., 464
Abragam, A., 303
Achiba, Y., 380, 381
Ackermann, P. G., 22
Adachi, H., 159, 161
Adams, A., 134
Adams, R. D., 115
Adler, A. D., 313
Aebi, U., 321
Agapion, A., 115
Agmon, N., 501, 510
Agris, P. F., 533
Agron, P., 156
Aguilar, J., 225, 229, 231, 238, 245
Aime, S., 105, 106
Aizman, A., 154, 155, 157, 163, 166, 167
Akaishi, T., 445
Akamatsu, R., 440
Akamatu, H., 378
Akasaka, K., 347, 539, 540, 553
Akutsu, H., 547
Albano, V., 106, 108
Albano, V. G., 90
Alben, J. O., 290, 292
Alberding, N., 290-93
Albers, R. W., 283
ALBRECHT, A. C., 353-76; 357, 360, 377
Alcacer, L., 444
Alden, R. A., 284
Alder, K., 124, 145
Alemdaroglu, N. H., 110
Alexandra, L. E., 100
Alexandrowicz, Z., 192, 195, 203, 206, 208, 213, 215
Aleksandrov, Y. Y., 525
Alfassi, Z. B., 512, 516, 523
Alfrey, T. Jr., 194, 199, 200
Allan, M., 260, 264, 266-73
Allen, J. C., 193
Allen, J. D. Jr., 156
Almassy, R. J., 284
Alma-Zeestraten, N. C. M., 548, 550
Alpert, B., 472, 473, 475, 485, 487
Altenburg, K., 81
Alwattar, A. H., 388
Ammeter, J. H., 167
Amos, L. A., 320, 322, 344, 345

Amphlett, J. C., 497, 516
Ander, P., 192, 193, 200
Andersen, H. C., 72-74
Anderson, A. B., 115
ANDERSON, C. F., 191-222; 191-93, 195, 199, 201, 203-9, 213, 215-19
Anderson, C. M., 283, 284
Anderson, J., 136
Anderson, L. R., 453
Anderson, O. K., 160
Anderson, P., 217
Ando, T., 333, 337
Andrä, H. J., 137
Andre, J. J., 449
Andree, P. J., 548
Andrew, E. R., 289, 537
Andrews, L., 258, 259, 272
Anisimov, O. A., 316
Anton, M., 25
Antonini, E., 290
Arakawa, K., 184
Arimitsu, S., 385
Armstrong, R. W., 195, 211
Arnold, D. R., 515
Aronson, J. F., 332
Artymiuk, P. J., 284, 285, 288
Asher, S. A., 472
Ashfold, M. N. R., 433
Aspect, A., 124
Asscher, M., 502
Assulin, O., 333, 337
Astruc, J. P., 184, 187
Atabek, O., 238, 241, 248, 249, 415, 421, 422, 426, 427
Atalla, R. H., 558
Atkins, P. W., 303
Atkinson, G. H., 474, 480
Atri, G. M., 506, 507
Atwood, J. D., 98-100, 103, 108, 109
Auchet, J. C., 342
Audemard, E., 342, 343, 345
Auerbach, R. A., 357
Aui, D. H., 503
Austin, R. H., 290-93
Aventini, P., 234
Averill, F. W., 159
Avron, J. E., 238

B

Babbitt, D., 231
Babcock, L. M., 519
Bachmann, J., 167
Bačić, Z., 241, 249
Back, M. H., 516

Baerends, 153, 154, 159-64, 167
Bagg, J. W., 509, 510
Baghal-Vayjooee, M. H., 496, 497, 506, 509, 510, 521, 522
Bagshaw, C. R., 339
Bailey, J. M., 193, 194, 200-2
Bailey, M. F., 96
Bain, R. A., 241
Baker, H., 327
Balch, A. L., 110
Baldwin, A. C., 495, 516, 520, 521, 525
Baldwin, B. A., 29
Baldwin, J. M., 483, 485
Baldwin, R. L., 219
Baldwin, R. R., 504, 506, 507, 509, 510
Baldwin-Zuschke, B. J., 109
Balfour, W. J., 266
Balint, M., 343, 344
Balint-Kurti, G. G., 119, 425, 427, 429-31
Ball, R. C., 60
BALLARD, S. G., 377-407; 387, 388, 392, 400
Balslev, E., 225, 229, 231, 233, 238, 245
Bamzai, A. S., 152
Band, E., 90, 106, 108, 109, 115
Band, Y. B., 119, 415, 416, 422, 427, 430, 433
Bandrauk, A. D., 422
Banks, R. D., 283
Banna, M. S., 163
Barany, K., 339, 346
Barany, M., 339, 346
Barbé, R., 184, 187
Barber, M., 525
Bardsley, J. N., 236, 241, 246
Baronavski, A. T., 432, 433
Barrow, R. F., 266
Bartness, J. E., 502, 503
Barton, B. D., 500, 509, 515
Bartuska, V. J., 558
Basolo, F., 89, 99, 100
Bassett, J. E., 521, 522
Bassett, J. M., 115
Batchelder, L. S., 544, 545
Batra, I. P., 155
Batt, L., 494, 495, 516, 520, 521, 523, 525
Bauer, W., 217
Baughman, R. H., 453, 465, 466
Baumann, W., 382

563

AUTHOR INDEX

Baumel, R. T., 225
Baumgartner, A., 57
Baur, E., 382
Bayes, K. D., 516
Beadle, P. C., 505-7
Beardslee, R. A., 25, 26, 29
Beauchamp, J. L., 508
Beck, D. R., 241
Beece, D., 290, 292, 293, 296
Beens, H., 359, 378, 384
Beeson, K., 290, 292, 293
Beeson, K. W., 290-93
Behe, M., 216, 217
Beier, B. F., 115
Beinert, H., 313
Beinfeld, M. C., 339
Beiting, E. J., 174, 175
Belford, G. G., 325
Belford, R. L., 325
Belle, T., 508
Bellon, P., 106, 108
Benard, M., 165
Bender, C. M., 233
Bendler, J., 233
Benedict, M. G., 157
Bennemann, K. H., 194
Bennett, P. M., 320
Benoit, H., 53, 54
Benson, S. W., 493-501, 504-14, 516, 519-23, 525
Benz, R. C., 259
Berg, J. M., 157
Berg, J. O., 357
Berg, P. W., 194, 199, 200
Bergman, J. G., 157
Bergmann, K., 360, 378
Berkowitz, J., 119, 120, 132, 166, 258
Berliner, Yu. A., 390, 394
Berliner, L. J., 533
Berlman, I., 72
Bernard, M., 449
Berne, B. J., 402
Bernstein, R. B., 419, 422
Berry, M. J., 416
Berry, R. S., 120
BERSOHN, R., 409-42; 119, 120, 425, 435, 436, 472, 473, 475
Berthier, G., 266
Bertrand, R., 342, 343, 345
Beswick, J. A., 415, 421, 425, 427, 430
Betowski, D., 521, 522
Beveridge, D. L., 402
Bhuiyan, L. B., 199
Bhutani, K. K., 266
Bianchi, M., 115
Bieri, G., 269
Binder, K., 57
Birks, J. B., 72, 360, 378, 388
Birktoft, J. J., 284
Bishop, A. R., 457

Bishop, M., 57
Bister, K., 547, 555
Blake, C. C. F., 283-85, 288
Blankenship, R. E., 377
Blankenship, T. A., 259
Bloch, M., 266
Bloomfield, V. A., 194, 199, 219
Bloor, J. E., 156, 157, 166
Blum, J. J., 345
Blum, K., 121, 129, 134
Blume, A., 544, 552, 557
Blume, R. V., 307, 310
Blundel, T. L., 284, 285
Boatley, S. J., 285, 288
Bocian, D. F., 533, 555
Bode, W., 288
Bodenhausen, G., 539
Boeckman, R. K., 314
Bogan, D., 502
Bogdanovic, R., 153
Bohme, D., 521, 522
Bohme, D. K., 521
Bojarski, C., 72, 76, 77
Bondybey, V. E., 258, 259, 261-64, 266-68, 270-74, 277, 278
Bonner, R., 338
Bor, G., 100, 107, 108
Bordas, J., 336
BOREJDO, J., 319-51; 332, 333, 337, 338, 340
Boring, A. M., 156, 160
Boring, M., 160, 162
Born, M., 21, 68
Borsenberger, P. M., 393
Bothner-By, A. A., 533
Bottcher, C., 415, 422, 426
Botter, R., 184, 187
BOTTS, J., 319-51; 320, 321, 323, 325, 326, 329, 330, 332, 336, 340-42, 344-46
Boué, F., 55
BOUSSEAU, D. L., 471-91
Bowers, M. T., 503
Bowman, M. K., 314, 315
Bowne, S. F., 290, 292, 296
Bradley, L. B., 325, 345
Braga, M., 164
Brahms, J., 326
Brändas, E., 236, 238, 244, 248
Brändle, R., 307
Brassfield, T. S., 290, 291
Brauman, J. I., 502, 512, 516, 521
Braun, A. M., 393
Braunlin, W. H., 217, 218
Brayer, G. D., 285, 288
Brazovskii, S., 444, 452, 455
Brazovskii, S. A., 452, 464
Bredas, J. L., 465, 466
Bredesen, D., 96
Brender, C., 200

Brewer, C. F., 545, 546
Brey, L., 25, 26
Brey, L. A., 31, 32
Brezner, J., 326
Brill, A. S., 293
Brink, D. M., 121
Briscoe, B. J., 52
Brocklehurst, B., 393, 397
Broersma, S., 327
Brom, H. B., 457
Broughton, C. G., 285, 288
Brown, I. M., 307, 314
Brown, N., 445
Brown, R. K., 110, 111
Brown, T. L., 96, 99-101, 106, 110
Brucker, C. F., 90, 109
Brugger, P. A., 393
Brunori, M., 290
Bryant, D. J., 289
Brzozowski, J., 266
Bube, R. H., 462
Bube, W., 398
Buckingham, A. D., 120
Bueche, F., 52
Bueso-Sanllehi, F., 266
BURCH, R. R., 89-118; 110, 111, 113, 114
Burger, R. M., 313
Burke, M., 344
Burnett, M. N., 306
Burns, D., 137
Burr, J. G., 29
Bursten, B. E., 158, 164, 165, 167
Busch, G. E., 63, 418

C

Cachier, G., 78
Caglio, G., 106, 108
Caldwell, C. D., 124
Callaghan, P. T., 54
Calle, L. M., 516, 525
Callomon, J. H., 266, 268, 269
Calvin, M., 378
Campbell, I. D., 284, 533
Candlin, J. P., 102
Canosa, C. E., 504, 507, 509
Canuto, S., 236, 244
Caplan, C. E., 415, 416
Carapellucci, P. A., 386, 393, 402
Careri, G., 283, 284
Carlson, F. D., 338
Carlson, L. R., 436
Carlson, T. A., 120, 156, 503, 521
Carmichael, P. J., 521
Carre, F. H., 105
Carre, M., 259
Carrington, A., 261, 266, 334
Carrington, T., 436, 438, 440

AUTHOR INDEX 565

Carsky, P., 267
Carslaw, H. S., 396
Carson, A. S., 516
Cartling, B. G., 160
Casalegno, R., 64, 65, 67-69, 79-81, 83, 86
CASE, D. A., 151-71; 124, 134, 142, 154-57, 161, 163-67, 283, 287, 290, 291
Cashell, E. M., 289
Castelhano, A. L., 504, 509
Castro, J. C., 174
Catellano, E., 516
Caughey, W. S., 483, 551
Ceasar, G. P., 96
Cederbaum, L. S., 360
Ceperley, D., 57
Cerjan, C., 238
Certain, P. R., 236, 242
Chamberlain, G. A., 124, 509, 512, 514, 521, 522
Chambers, J. L., 343
CHAMPION, P. M., 353-76
Chan, R. K., 360, 378
Chan, S., 155
Chan, S. I., 533, 555, 556
Chan, S. S., 290, 292, 293
Chance, B., 393
Chance, R. R., 465, 466
Chang, C. H., 157
Chang, R., 360
Chao, J., 522, 523
Charvolin, J., 557
Chase, M. W. Jr., 519
Chen, C. E., 445, 453
Chen, J., 445, 448, 449, 464
Chen, S. N., 447, 465
Chen, S. S., 522, 523
Chen, Y., 337
Cheng, C. P., 314
Chenier, J. H. B., 498, 510, 521
Chernoff, D. A., 472-75
Chesnokov, V. I., 521, 523
Cheung, H. C., 327, 328, 343, 346
Cheung, L., 360
Cheung, P. C., 322, 326
Chia-Chung, S., 155
Chiang, C. K., 445, 448
Chien, J. C. W., 445, 448
Child, M. S., 415, 416
Chini, P., 90
Chisholm, W. P., 306
Choi, H. W., 115
Choo, K. Y., 505-7, 512, 514, 521, 522
Chothea, C., 483, 485
Chowdhury, D. M., 98
Christie, K., 516, 525
Christodoulides, A. A., 186, 187
Christophorou, L. G., 186, 187

Chryssomallis, G., 30
Chryssomallis, G. S., 31
Chu, S., 422, 426, 427
Chu, S.-I., 238, 248, 249, 251
Chuang, T. J., 63, 381, 382
Chung, T.-C., 444, 448-50, 460-62, 464
Chupka, W. A., 120, 182, 502, 507
Church, D. F., 515
Churchill, M. R., 102, 105
Chutjian, A., 186
Cillien, C., 523
Citri, N., 283
Clague, D., 555
Clark, R. J. H., 165
Clarke, T. C., 448, 449, 456-58
Claspy, P. C., 259
Clement, R. M., 219
Clementi, E., 199
Closs, G. L., 315, 396
Cocks, A. T., 495, 521
Codling, K., 120
Cohen, H. L., 225
Cohen, J. S., 533
Cohen, M., 106
Cohen, M. J., 450
Cole, B. E., 120
Coleman, A. W., 165
Coleman, B., 555
Collins, F. C., 394
Collman, J. P., 109
Colman, P. M., 285
Colussi, A. J., 496, 497, 506, 509, 516, 519, 521, 522, 525
Combes, J. M., 225, 229, 231, 234, 238, 245
Come, G. M., 499, 509, 525
Compton, R. N., 184
Cone, R. L., 475, 476, 478
Connolly, J. W. D., 152-54, 156, 157, 160, 163, 167
Connor, J. A., 90, 97, 100, 109
Connors, R. D., 377
Contois, L. E., 393
Coo, K. Y., 521
Cook, J. M., 261
Cook, M., 155-58, 162, 163
Cook, M. R., 162, 163, 166, 167
Cooke, P. W., 360, 378
COOKE, R., 319-51; 290, 291, 320, 322, 325, 332, 335, 337
Coomber, J. W., 522, 524
Cooper, A., 283, 284
Cooper, J., 120
Cooper, J. W., 119, 120
Coppey, M., 472, 473
Corbel, S., 499, 525
Corbett, J. D., 90
Corcoran, C., 246, 247

Cordier, P., 390, 394
Cornell, B. A., 556, 557
Corongiu, G., 199
Cossart, D., 260, 270, 273
Cossart-Magos, C., 260, 270, 273
Cotton, F. A., 102, 105, 160, 164, 165, 167
Cotton, J. P., 53, 54
Couch, J., 321
Cowan, R. D., 160
Cowley, A. H., 165
Cowley, D. J., 382
Cowman, C. D., 165
Cox, J. D., 495, 509, 513, 516, 519, 523, 525
Cox, R. J., 381
Coxon, J. A., 266
Cramer, S. P., 157
Crandall, R. S., 198, 206
Crawford, M. K., 381, 393
Creutzberg, F., 266
Crocker, M. C., 225
Cross, T. A., 539, 540, 544, 547-50, 552, 555
Cruickstank, F. R., 511, 516
Cullis, P. R., 555, 557, 558
Cumming, J. B., 502, 509, 513, 522
Curnutt, J. L., 519
Curtis, L. S., 266
Czarnowski, J., 516, 519
Czégé, J., 296

D

Daehne, S., 381
Dahl, L. F., 96
Daintith, J., 270
Dale, F., 259
Dale, R., 342
Dalton, B. J., 120
Danese, J. B., 157
Dannacher, J., 267-69
Danyluk, S. S., 545, 546
Daoud, M., 53, 54, 59
Darensbourg, D. J., 103, 106-9
Darke, A., 219
Darwent, B. de B., 519, 523
Das, M. P., 153
DasGupta, S. K., 544, 552, 557
Datta, K. K., 248
Dauben, H. J., 512
Daul, C., 166
Daune, M. P., 219
Dausch, H., 56
Davenport, J. W., 136, 156
Davidson, E. R., 247
Davidson, I. M. T., 525
Davis, D. G., 388
Davis, F. J., 184
Davis, J. L., 313
Davis, K. M. C., 360, 378

AUTHOR INDEX

Day, J. S., 522
Day, V. W., 110, 111, 113-15
Deb, B. M., 152
DeBoer, B. G., 102
DEBRUNNER, P. G., 283-99; 286, 288, 293
Debye, P., 55, 388, 391
Deeming, A. J., 102
DeFrees, O. J., 510, 512-14, 521, 522
DE GENNES, P. G., 49-61; 49-51, 53, 54, 56-60
de Groot, K., 384, 398, 399
Deguchi, K., 247
deHaseth, P. L., 193, 217-19
de Heer, F. J., 266
Dehmer, J. L., 120, 124, 136, 140, 156, 157, 165
Deift, P., 234
Deisenhofer, J., 285, 289
De Koven, B. M., 261
Dekruijf, F. B., 555, 558
Delabaere, L. T. J., 285, 288
Delaire, J. A., 390, 394
Delaney, J. J., 165, 166
DeLange, C. A., 161
de la Torre, G., 327
Delbaere, L. T. J., 284
Delliste, D., 510, 514
Deloche, B., 557
Delville, A., 195
de Mayer, L., 378
DeMello, P. C., 166
Demitras, G. C., 115
Denenstein, A., 445, 448, 449, 464
Denesyuk, A. I., 284
De Rosier, D. J., 320-22
Desantolo, A. M., 314
Descamps, B., 516
Detherage, F., 483
Dettman, H. D., 552
DeVault, D., 393
Devore, J. A., 516
Devoret, M., 264
Devreese, J. T., 444, 448, 453
Devreux, F., 456
DeWit, D. G., 98, 100
deWit, J. L., 547, 548, 550
Diakun, G. P., 193
Diaz, A. F., 466
Dibeler, V. H., 502, 512
Dickens, M., 321
Dickerson, R. E., 284, 285, 288
Dietler, U. K., 107, 108
Dill, D., 120, 124, 125, 136, 140, 143, 145, 156, 157, 165
DiMarzio, E. A., 60
Dimicoli, I., 184, 187
Dinsdale, R., 270
Dirac, P. A. M., 151, 165
DiVerdi, J. A., 539, 544, 547-50, 552, 555

Dlott, D. D., 74, 84
Dobbs, A. J., 303
Dobson, C. M., 284, 547
Dodink, H., 382
Doering, W. V. E., 504
Doetschman, D., 306, 314
Doi, M., 51
Dolar, D., 194
Dolder, K., 234
Dolgikh, D. A., 284
Dolphin, D., 483
Domcke, W., 360
Doncaster, A. M., 516, 525
Doniach, S., 157
Donnelly, R. A., 244, 249
Dönszelmann, A., 175
Doolen, G., 225, 234, 236
Doolen, G. D., 233, 236-38
Doran, M., 165, 166
Dorssers, L. C. J., 547, 548
Dos Remedios, C. G., 320, 321, 332
Dosser, L. R., 474
Drachman, R. J., 237
Dransfeld, K., 290, 291, 466
Draxl, K, S, 503, 504
Dress, W. B., 156
DRICKAMER, H. G., 25-47; 25-27, 29-34, 37, 38, 41
Driscoll, O., 58
Druy, M. A., 445, 448, 455, 457, 464
Druyan, M. E., 72
Duddell, D. A., 472, 487
Dujardin, G., 264, 270-72
Duke, J., 320, 325, 332, 333, 345-47
Duke, J. A., 346
Dunbar, R. C., 258, 259, 267
Dunlap, B. I., 160, 167
Dunlop, A. N., 525
DUNNING, F. B., 173-89; 174-77, 179-82, 184, 187
Duplessix, R., 53, 54
Düren, R., 124
Durup, J., 257, 258
Duxbury, G., 266
Dvornikov, Z. V., 307
Dwek, R. A., 533
Dwivedi, A., 286, 288, 293
Dzarnoski, J., 525
Dzuba, S. A., 314
Dzvonik, M. J., 119, 120

E

Earl, W. L., 558
Eaton, W. A., 472, 473, 475
Ebert, P. J., 124
Ebisuzaki, Y., 25
Eccleston, J. F., 339
Eckhardt, H., 465, 466

Edelstein, N., 160
Eden, R. J., 228
Ederer, D. L., 120
Edmonds, D. T., 537, 539
Edwards, H. E., 193
Edwards, S. F., 50, 51, 60
Eftink, M. R., 289
Egan, W., 551
Egdell, R. G., 160
Egger, K. W., 495, 511, 512, 521
Ehrenberg, M., 325
Eichler, H., 78
Eigen, M., 360, 378, 391
Eisenberg, E., 329, 330
Eisenberg, H., 208, 211, 212
Eisenstein, L., 290-93, 296
Eisenthal, K. B., 63, 72, 381, 382, 393
Eisert, W. G., 472, 473, 475
Eisinger, J., 342
Eland, J. H. D., 264
Elander, N., 248
Eliason, R. R., 25
Ellett, J. D., 558
Elliott, A., 321, 326, 327
Ellis, D. E., 159, 161
El Mohsni, S., 472, 475, 485, 487
El-Sayed, A. J., 472-74, 485
El-Sayed, M. A., 472-74, 485
Elsenbaumer, R. L., 465, 466
Engelhardt, V. A., 319
Engelking, P. C., 261
Engelman, D. M., 284
English, J. H., 259, 261-64, 266-68, 270-74, 278
Epstein, A. J., 464
Epstein, I. R., 396
Eres, D., 519
Erhardt, H., 120
Erman, P., 266
Ermler, W. C., 162
Estler, R. C., 519
ETEMAD, S., 443-69; 444, 449, 450, 454, 458-62, 464
Eugster, A., 289
Evans, B. S., 497, 522
Evans, D. A., 506, 507
Evans, J., 105
Evans, K. E., 60
Evans, P. R., 283
Evrard, R. P., 444, 448, 453
Ewen, B., 57
Eyler, J. R., 259
Eyring, H., 293
Ezumi, K., 383

F

Fabian, F., 345, 346
Fairbrother, D. M., 100
Falabella, R., 53

Fano, U., 119-21, 124, 125, 129, 136, 137, 140, 142, 143, 145
Fanwick, P. E., 165
Farley, J. W., 174
Farnoux, B., 53, 54
Faruqi, A. R., 336
Fasella, P., 283, 284
Faulkner, J. S., 154
Faulkner, L. R., 99
Fawcett, J. P., 98-100
FAYER, M. D., 63-87; 63-65, 67-69, 72-74, 77-81, 83, 84, 86
Fedders, P. A., 233
Fedorov, B. A., 284
Fedotova, E. Ya., 398
Fee, J. A., 313
Feher, G., 72, 337
Feitelson, J., 393
Feldblum, A., 449, 450, 460, 461, 464
Felix, C. C., 316
Felps, W. S., 438
Felsenfeld, G., 209, 216, 217
Felton, R., 360
Felton, R. H., 160, 167
Fenske, R. F., 157, 158, 164, 167
Ferguson, K. C., 509, 510, 522
Ferris, L. T. H., 165
Ferry, J. D., 49, 50, 60
Feshbach, H., 235
Fessenden, R. W., 302, 303, 306
Fettis, G. C., 497
Fierstein, D. E., 314
Fincher, C., 445, 453
Fincher, C. R., 445, 447, 448, 454, 458, 461, 463
Fine, D. H., 516
Fink, E. H., 266
Fiquet-Fayard, F., 416
Fischer, H., 507, 508
Fishgold, H., 382
Fitchen, D. C., 462
Fixman, M., 191-93, 196, 200, 201, 205, 206
Flannery, M. R., 390
Fleury, P. A., 474
Flouquet, F., 438
Flügge, S., 120
Foltz, G. W., 175, 184, 187
Foon, R., 522, 523
Forbes, N. P., 101
Ford, L. E., 322
Formanek, H., 285
Forst, W., 516
Förster, T. H., 31
Foster, P., 268
Foster, R., 359, 360, 378
Foti, A. E., 158
Foucaut, J-F., 499

Francis, G. W., 557
Francois, B., 449
Frankevich, E. L., 398
Franklin, J. A., 523
Franklin, J. L., 503, 519
Franklin, R. M., 547
FRAUENFELDER, H., 283-99; 124, 285-87, 290-93, 296
Frazer, A., 338
Fredianai, P., 115
Fredrich, M. F., 110
Freed, J. H., 301, 396
Freed, K. F., 119, 415, 416, 422, 427, 430, 433
Freier, D. G., 157, 164
Frenz, B. A., 105
Frey, H. M., 510-12
Frey, M. H., 539, 540, 542, 543, 552
Frey, R., 264, 266
Friedland, S., 236
Friedman, H., 359
Friedman, H. L., 196
Friedman, J., 360
FRIEDMAN, J. M., 471-91; 472, 475, 476, 478, 480, 481, 483, 485, 487, 488
Friedrich, D. M., 377
Frisch, H., 57
Frisse, M. E., 484
Froelich, P., 236, 238, 244, 247-49
Frolov, E. N., 286, 288
Frost, D. C., 163
Fu, E. W., 259
Fuhrhop, J.-H., 388
Fujara, F., 305-7
Fukui, K., 235, 236, 244
Fuoss, R. M., 194, 200, 201
Furrer, R., 305-7
Furutsuka, T., 388
Furuyama, S., 510, 512

G

Gac, N. A., 521
Gaetjen, E., 346
Gage, L. D., 165
Gaillard, M. L., 259
Galkiewicz, R., 445, 448
Gall, C. M., 544, 548-50
Gallagher, T. F., 182, 183
Gammel, J., 460, 461
Ganapathy, S., 540, 555
Gangi, G., 193
Ganopathy, S., 539
Gard, D. K., 99
Gardiner, W. C., 519
Gardini, G. P., 466
Garner, C. D., 165, 166
Garritz, A., 166
Gary, L. P., 384, 398, 399

Gast, J. C., 558
Gau, S. C., 445, 447, 448, 465
Gauyacq, D., 266
Gavezzotti, A., 152
Gavin, R. M., 110, 111, 113
Gavish, B., 288, 296
Gebert, E. G., 110
Geiduschek, E. P., 216
Geiseler, G., 516
Gelbart, W. M., 119, 409, 427, 430
Gelin, B. R., 289, 484-86
Gelius, U., 163
Gergely, J., 326, 335, 343, 344, 346
Gerischer, H., 357, 360, 377
Gerstein, B. C., 535, 556
Gerstman, B., 293
Geurts, P. J. M., 159
Ghiron, C. A., 289
Gibby, M. G., 535-37
Gibson, H. W., 464
Gibson, Q. H. J., 472, 475, 485
Gierasch, L. M., 542
Gill, W. D., 448, 466
Gilmore, P. T., 53
Gilmour, D., 345
Gilroy, K. M., 524, 525
Giniger, E., 322
Ginley, D. S., 96
Ginsberg, A. P., 157
Gislason, E. A., 421
Glanzer, K., 521
Glarum, S. H., 303
Glasstone, S., 293
Glickson, J. D., 533
Glimcher, M. J., 546
Gochanour, C. R., 63-65, 72-74, 77
Goddard, R. D., 520, 521, 523
Goddard, W. A. III, 484
Gojon, G., 515
Goldanskii, V. I., 286, 288
Goldberger, M. L., 224
GOLDEN, D. M., 493-532; 494, 496, 497, 500, 501, 504-16, 520-25
Goldschmidt, C. R., 378
Golembeski, N. M., 115
Good, D., 290, 292, 293, 296
Good, W. D., 100
Goodall, D. M., 385
Goodman, T. C., 192
Goody, R., 347
Gopinathan, M. S., 153
Gordon, A. M., 320
Gordon, A. S., 516
Gordon, D. J., 167
Gordon, R. G., 419, 422
Goscinski, O., 236, 244
Goss, S. P., 259

AUTHOR INDEX

Gosselink, J. W., 159
Gotchev, B., 264, 266
Goursot, A., 167
Gouterman, M. 1, 290, 390
Gowenlock, B. G., 521, 522
Grabensteller, J. E., 422
Grabowski, Z. R., 382
Grace, D. E. P., 284, 285, 288
Graceffa, P., 337
Grader, R. J., 124
Graessley, W. W., 50, 51
Graff, J. L., 115
Grangier, P., 124
Grant, P. M., 466
Gratton, E., 283, 284
Gratzel, M., 393
Gray, H. B., 89, 91, 165
Gray, P., 516
Graziani, M. T., 313
Greco, R., 60
Green, J. C., 165, 167
Green, M. L. H., 92
Green, S., 257
Greene, B. I., 472, 473, 475
GREENE, C. H., 119-50; 124, 140, 145
Greene, L. E., 329, 330
Greene, R. L., 448, 456
Gregg, D. W., 25
Gregor, H. P., 194
Gregor, J. M., 194
Gregory, A., 359, 360
Grellman, K. H., 384
Grey, H. B., 96
Grieman, F. J., 264, 266
Griffin, D. C., 160
Griffin, R. G., 534, 539, 540, 544, 546, 547, 552, 553, 556-58
Grigoryants, V. M., 316
Grill, W., 78
Griller, D., 504, 509, 520, 521
Grimm, F. A., 156, 157
Groombridge, C. J., 540
Gross, H., 466
Gross, L. M., 201, 207, 211, 212, 214
Grossmann, A., 238
Gruenwedel, S. W., 216
Grupp, A., 466
Gubanov, V. A., 161
Güdel, H. U., 167
Gudkov, N. D., 388
Guéron, M., 193, 195, 198, 199, 206, 208
Guest, M. F., 165, 166
Guillory, R. J., 344
Gunnarsson, O., 152-54, 160, 167
Gunsalus, I. C., 290-93
Gupta, A., 502
Gurd, F. R. N., 284, 289
Gurney, R., 393

Guttman, C. M., 60
Gyemant, I., 157

H

Haar, H. P., 395
Haas, Y., 502
Haberkorn, H., 466
Haberkorn, R., 398
Haberkorn, R. A., 539, 546, 553, 556, 557
Haeberlen, U., 534, 535, 540
Hafner, H., 139
Hagen, D. S., 552
Hahn, E. L., 308
Haines, L. I. B., 99, 100
Halavee, U., 421
Hall, J. L., 120
Halle, B., 199
Hamaguchi, K., 216
Hamann, H. J., 381
Hänggi, P., 290, 291
Hanson, B. E., 165
Hanson, J., 319, 320, 322
Hara, K., 27
Hara, Y., 25
Harada, T., 558
Harbison, C., 540
Hardwidge, E. A., 509, 525
Hardy, G. W., 283
Harmon, B. N., 160
Harrington, W. F., 320, 322, 344
Harris, H. H., 261
Harris, J., 154, 160
Harris, R. K., 540
Harrison, H., 120
Harrison, J. F., 166
Harrison, S. C., 285, 547
Hartt, J. E., 336
Harvey, S. C., 327, 328
Hascall, V. C., 546
Haser, R., 283
Hashizume, H., 336
Hassing, S., 353, 357
Hasson, M. A., 546
Hastings, J. B., 445, 453
Hata, T., 52
Hatton, G. J., 174
Hauser, H., 219
Hauser, M., 395
Hawkes, G. E., 105
Hawksworth, R. W., 166
Hay, P. J., 154, 165
Hayes, A. J., 165
Hayes, A. V., 41
Hays, G. R., 555
Hayter, J. B., 57
Hazi, A., 244
Heaven, M., 263, 270, 273
Hedges, R., 238
Hedin, L., 152, 153

HEEGER, A. J., 443-69; 444, 445, 447-55, 457-65
Hehenberger, M., 166, 236, 238
Hehre, W. J., 510, 512-14, 521, 522
Heiber, I., 266
Heidrich, F. E., 418
Heijser, W., 159, 163
Hein, D. E., 25, 29
Heinrichs, P. M., 540, 555
Helene, C., 192
Helfgott, C., 214
Heller, E. J., 415, 422, 426, 427
Helmholz, L., 21
Hemminga, M. A., 548, 550, 557
Hemsworth, R. S., 521
Hendrickson, W. A., 285, 288
Hendry, D. G., 511, 515, 516
Hendry, J. A., 225, 243
Henneker, W., 359, 360
Henry, E. R., 475, 488
Hentzchel, P., 381, 402
Herbig, G. H., 266
Herbst, I. W., 238
Heremans, K., 31
Herman, F., 152, 155, 156
Hermann, H. W., 134, 136
Herrick, D. R., 243
Herriott, J. R., 285
Herron, J., 503, 504
Herschbach, D. R., 124, 134, 142, 145
Hertel, I. V., 131, 134, 136
Hervet, H., 54
Herzberg, G., 266, 274, 504, 507, 525
Herzfeld, J., 540, 544, 546, 552, 553, 556, 557
Hesser, J. E., 266
Hetherington, W. M., 63, 72
Hetzel, R., 289
Heuser, J. E., 322
Hexem, J. G., 540, 542
Hidalgo, M., 236
Higgins, J. S., 57
Higgs, C., 174, 176, 177, 179-81
Highsmith, S., 283, 326, 327, 329, 330, 339, 347
Hildebrandt, G. F., 174, 184, 187
Hilhorst, H. W. M., 550
Hill, T. L., 194, 205, 329, 330
Hillen, W., 192
Hiller, R. G., 556
Hillier, I. H., 165, 166
Hilton, B. D., 284, 289
Himmelfarb, S., 344
Hinatu, J., 379
Hino, T., 380, 386, 398, 399
Hirsekorn, F. J., 115
Ho, Y. K., 237, 238

AUTHOR INDEX 569

Hobbs, A. S., 283
Hobbs, R. H., 153, 156
Hochstrasser, R., 360
Hochstrasser, R. M., 472-75
Hocker, J., 466
Hodgson, K. O., 157
Hoesterey, D. C., 393
Hoffman, J. D., 60
Hoffmann, G., 31
Holbrook, K. A., 502
Holdy, K. E., 418
Hollander, F. J., 102
Holmes, K., 319, 330, 347
Holt, C., 238
Holten, D., 390
Holyk, P., 193
Hong, K. M., 390, 394
Honovich, J. P., 259
Hook, J. W. III, 27, 31
Hopfield, J. J., 293
Hopgood, D., 99
Horani, M., 259, 266
Hore, P. J., 305, 306, 397
Horn, G. T., 192
Horovitz, B., 460, 461
Horsley, J. A., 438
Horwitz, S. B., 313
Hoschstrasser, R., 359
Hotop, H., 120, 184
Hotta, K., 345, 346
Houle, F. A., 508
Hoult, D. I., 557
Houston, P. L., 502
Houston, S. K., 237
Houtchens, R. A., 551
Howard, A. V., 525
Howard, C. J., 516
Howard, J. A., 498, 510, 515, 521
Hozumi, T., 329, 330, 344-46
Hsu, C. -H., 216
Hsu, S. L., 453
Huang, C. L., 227, 241, 242
Huang, T. H., 544, 552, 557
Huber, R., 285, 288, 289
Hudson, E. N., 325, 342
Hughey, J. L., 99
Huie, B. T., 106
Huis, R., 555
Hummel, A., 377
Hunklinger, S., 290, 291
Hunter, F. R., 512
Hunziker, W., 233, 234
Husain, J., 124
Huse, G., 360, 378
Hutcheon, W. L. B., 284
Hutchinson, D. A., 306
Hutchinson, J. P., 102, 105
Hutton, W. C., 547, 558
Huxley, A. F., 319, 320, 322, 330, 336, 338
Huxley, H. E., 319-22, 336
Huybrechts, G. H., 523

Huynh, B. H., 157, 164, 167
Huyskens, P. L., 390
Hwang, R. J., 519, 520
Hyde, J. S., 326

I

Ichikawa, T., 314
Ikeda, S., 445, 447, 448, 453-55
Ikeda-Saito, M., 479, 480
Ikehata, S., 455, 457, 464
Ikemoto, I., 360, 378
Ilten, D. F., 378
Imai, N., 192, 194
Imamura, K., 346
Imbusch, G. F., 72
Imhof, R. E., 266
Imura, T., 388
Incorvia, M. J., 103, 106-8
Infelta, P. P., 393
Ingold, K. U., 510, 515
Inubushi, T., 479, 480
Ippen, E. P., 472, 473, 475
Ireton, R., 516
Irie, M., 380, 381
Irwin, M. J., 480
Isaacson, A. D., 241, 247
Ise, N., 192, 193
Isenberg, G., 321
Ishikawa, K., 445
Ishikawa, M., 195, 196, 198
Ishikawa, Y., 161
Ishishi, E., 96
Ishiwata, S., 335
Ito, T., 445, 447, 448, 453-55
Ivory, D. M., 465, 466
Iwasa, K., 193, 194, 202, 203
Iwata, S., 267, 379

J

Jack, A., 285
Jacko, M. G., 524, 525
Jackson, J. D., 128
Jackson, K. H., 124, 140
Jackson, R. L., 521
Jacob, G. S., 553
Jacobs, R. E., 555, 558
Jacobson, G. R., 343
Jacon, M., 422
Jadrny, R., 120
Jaecks, D. H., 134
Jaeger, J. C., 396
Jagannathan, S., 192
Jahn, H. A., 274
James, D. G. L., 511, 512
James, M. N. G., 284, 285, 288
James, T. L., 533
Janak, J. F., 152, 153
Janes, N., 544, 545, 552
Jannick, G., 53, 54
Janousek, B. K., 502, 512, 516

Jardetzky, O., 289, 327, 347, 533, 551
Jarnagin, R. C., 357, 360, 377, 384, 398, 399
Jelinski, L. W., 544, 545
Jeng, S. F., 344
Jenkins, J. A., 284
Jensen, L. H., 285
Jensen, R. J., 167
Jeys, T. H., 175
Jivery, W. T., 120
John, P., 525
Johns, J. W. C., 266
Johnson, B., 353
Johnson, B. F. G., 90, 102, 105, 110
Johnson, B. R., 419, 422, 423
Johnson, C. A. F., 521, 522
Johnson, J. B., 163, 164
Johnson, K. H., 152, 154-56, 164
Johnson, L. N., 284, 285
Johnson, P. C., 25-27, 30
Johnson, P. M., 257
Johnson, R. P., 525
Jola, M., 511, 512
Jonah, C., 119, 120
Jonas, A. E., 120
Jones, P. F., 30
Jones, R. L., 192
Jones, R. O., 153, 154, 160, 167
Jones, R. P., 63
Jones, T. A., 285
Jones, T. B., 269, 271, 272
Jonkers, J. G., 161
Jönsson, B., 199
Jordan, R. F., 115
Jortner, J., 293, 409, 415, 421
Joshi, Y. M., 192-94
Joslin, C. G., 397
Jovin, T. M., 216
Jud, K., 56
Judge, D. L., 266, 434, 438, 439
Julian, F., 335
Julian, F. J., 320
Jungen, Ch., 120
Junker, B. R., 225, 227, 233, 236, 238, 241, 242, 244, 246

K

Kabsch, W., 321
Kadhim, A. H., 27-30
Kadish, K. M., 388
Kaesz, H. D., 106
Kai, A. T., 156
Kalbacher, B. J., 165
Kallenbach, N. R., 550
Kalos, M., 57
Kambali, U., 167

Kan, J. H., 548
Kana'an, A. S., 516, 525
Kanazawa, K. K., 466
Kanda, K., 165
Kanety, H., 382
Kanfer, S., 434, 435
Karasz, F. E., 445, 448
Karel, K. J., 103, 106, 108, 109
Karimian, A., 193
Karlsson, L., 120
Karplus, M., 155-58, 162-64, 166, 167, 283, 284, 287-91, 484-86
Kash, M. M., 174, 175
Kasha, M., 37
Kassab, R., 342, 343, 345
Kassal, T., 418, 419
Katayama, D., 264
Katayama, E., 343
Katchalsky, A., 191, 192, 194, 195, 200, 201, 203, 206, 208, 211, 213, 215
Katsumata, S., 380
Katz, J. J., 72
Kaufer, J., 449, 455, 457, 464
Kaufman, J., 449, 450, 460, 461, 464
Kaufmann, K., 290
Kaveh, M., 448, 449, 464
Kawabe, K., 388
Kawada, A., 384, 398
Kawai, M., 381
Kawamura, T., 165
Kay, K. G., 422, 426
Kebarle, P., 502, 503, 509, 513, 522
Kedem, O., 192, 195, 203, 208, 215
Keelan, B. W., 259
Keeton, D. P., 104
Keiser, H. D., 545, 546
Keister, J. B., 115
Keller, H., 286, 288
Keller, P. R., 156
Kellert, F. G., 177, 180-82, 184, 187
Keniry, M., 556
Kenley, R., 516
Kerr, J. A., 494, 497, 501, 510, 514, 516, 519, 521-24
Kevan, L., 301, 314
Khmelinskii, V. E., 307
Kiang, T., 519, 524
Kidd, D. R., 96, 106
Kielley, W. W., 325, 345
Kiely, B., 339, 345
Kigenmann, H. Y., 515
Kilty, P. A., 110
Kim, J. J., 25, 26
Kim, S. S., 303, 306
Kimball, G. E., 394
Kimura, K., 380, 381
King, G. C. M., 134

King, K. D., 496, 500, 506, 509, 510, 512-14, 516, 520-23
King, M. A., 266
King, R. B., 90
Kinsey, J. L., 434
Kinsey, R. A., 544, 545, 552
Kintanar, A., 544, 545, 552
Kiparisova, E. G., 525
Kirova, N. N., 452, 464
Kiss, Z., 447, 465
Kitagawa, T., 475, 479, 480, 483
Kitaura, K., 160, 167
Kivelson, S., 460, 461, 464
Klapstein, D., 266
Klar, H., 124, 140
Klein, B. K., 191, 195, 203-7, 210, 213, 214, 216
Klein, J., 52, 60
Klein, M. P., 558
Klein, R. D., 192
Klein, U. K. A., 395
Kleinpoppen, H., 134, 139
Klemperer, W. G., 155, 163, 164
Kleppner, D., 174, 175
Kleyn, A. W., 421
Klick, D. I., 25, 41
Kloster-Jensen, E., 260, 264, 268, 269
Klots, C. E., 184, 186, 187
Klotz, L. C., 418
Klug, A., 320
Klug, H. P., 100
Knibbe, H., 383-85
Knobler, C. B., 106
Knoeber, C., 343
Knoll, H., 516
Knox, J. H., 497, 505, 506, 509
Kobayashi, S., 326
Kobayashi, T., 378
Koch, M. H. J., 336
Kodama, K., 422
Koelling, D. D., 160
Koenig, T., 508
Koeppe, R. E., 343
Koetzle, T. F., 110, 111, 113
Kogelnik, H., 67
Kohler, S. J., 558
Koichiro, H., 380, 381
Kokorin, A., 115
Kolari, H. J., 165
Kolb, D. M., 357, 360, 377
Kominar, R. J., 519, 522, 524, 525
Komninos, Y., 241
Konnert, J. H., 285
Konowalow, D. D., 153
Konrad, M., 347
Köppel, H., 360
Korth, H-G., 510, 512
Kosman, D. J., 313

Kosower, E. M., 382
Kossiakoff, A. A., 343
Kothe, G., 306
Kotin, L., 214
Kouri, D. J., 119, 422
Kowblansky, A., 193
Kowblansky, M., 193
Kramer, H. E. A., 384, 388, 401
Kramer, O., 60
Kramers, H. A., 295
Krantz, A., 510-12
Krause, M. O., 156
Krause, S., 56, 326
Krause, U., 124
Kraut, J., 284
Krech, M., 524, 525
Krech, M. J., 512, 514, 522
Kress, M., 336
Kretzschmar, K. M., 283, 321, 326, 327, 344
Krey, A. K., 192
Krieger, M., 343
Kroll, N. M., 78
Kroto, H. W., 266
Kruger, C. J., 307
Krüger, H., 139
Krumhansl, J. A., 460, 461
Kubo, S., 346
Kukla, D., 285
Kulander, K. C., 251, 415, 422, 425-27, 434
Kumagai, Y., 52
Kuntz, I. D., 290, 291
Kurkjian, C. R., 313
Kuroda, H., 360, 378
Kusba, J., 72, 76, 77
Kusch, P., 303
Kutzler, F. W., 157
Kvisle, S., 269
Kwak, J. C. T., 192-94, 202, 203
Kwak, J. F., 448, 466
Kwei, T., 56
Kwo, A. L., 556

L

Lagercrant, Z. C., 378
Laidlaw, W. G., 163
Laidler, K. J., 293
Laird, R. K., 266
Lakowicz, J. R., 288
Lalonde, A. C., 523-25
Lamb, W. E. Jr., 357
Lampert, M. A., 198, 206
Lamson, S. H., 162, 163
Lane, N. F., 157
Lange, C., 305-7
Langford, C. H., 89, 91
Lankin, M., 344
Lapeyre, G. J., 136
Larcher, C., 266

AUTHOR INDEX 571

Larson, C. W., 509, 525
Larson, J. E., 192
Larsson, S., 156, 164, 166, 167
Larzilliere, M., 259
Lathouwers, L., 247
Latimer, C. J., 181, 182, 184, 187
Lattman, M., 165
Lauchlan, L., 444, 447, 454, 461-64
Lauderdale, J. G., 244, 247
Lauher, J. W., 92
Laurence, R. L., 53
Law, G. D., 290, 291
Lawler, R. G., 307
Lax, M., 200
Laye, P. G., 516
Leach, S., 260, 264, 266, 270-73
LeBret, M., 195, 198, 199, 201, 206
Lee, E. P. F., 165, 166
Lee, L. C., 259, 266, 434, 436, 438, 439
Lee, T.-K., 460, 461
Lee, Y. S., 162
Lee, Y. T., 436
Lefebvre, R., 233, 238, 241, 248, 249, 415, 421, 422, 426, 427
LÉGER, L., 49-61; 54
Leigh, W. J., 515
Leite, J. R., 164
Leone, S. R., 119, 409
Leonhardt, H., 382
Lepley, A. R., 396
LeRoy, R. J., 422
Leung-Louie, L., 193
Levenson, R. A., 96
Levine, I. N., 245
Levine, R. D., 409, 414, 415, 419, 421, 422, 434
Levine, S., 199
Levitt, M., 283, 285
Levy, D. H., 261
Levy, M. L., 152
Lew, H., 266
Lewis, D. K., 516, 523
Lewis, G. N., 11, 14
Lewis, J., 102, 105, 110
Lewis, K. E., 500, 525
Leyland, L. M., 522
Leyte, J. C., 195, 199
Li, A. S. W., 314
Li, C. H., 157
Li, T. M., 31
Liao, S. C., 360, 378
Libby, W. F., 393
Lichtman, L. S., 462
Liehr, A. D., 274
Lifson, S., 194, 195, 200, 201
Light, J. C., 290, 418, 419, 422, 425, 434

Lin, S. H., 119
Lin, T. S., 314, 315
Lindblom, G., 555
Lindqvist, L., 472, 475, 485, 487
Lineberger, W. C., 120
Lin-Liu, Y. R., 444, 452, 455, 457, 460, 461
Linschitz, H., 382
Littman, M. G., 174, 175
Ljubimova, M. N., 319
Lloyd, D. R., 165
Loeppky, R. N., 533
Loge, G. W., 120, 124
Lohman, T. M., 192, 193, 209, 215-19
Loiseleur, B., 195
Longoni, G., 90
Longuet-Higgins, H. C., 274, 277, 278
Loomba, D., 157
Loring, R. F., 72-74
Los, J. R., 421
Lossing, F. P., 520, 521
Louis, E. J., 445, 448
Lovett, R., 249
Lovett, R. A., 233
Lowey, S., 321, 327-29, 344
Loxdale, H. D., 335
Lu, P., 533
Lu, R. C., 327
Lubas, W., 193
Lumb, M. D., 388
Lumry, R., 289
Lundqvist, B. I., 152, 153
Luria, M., 508
Lutz, D. R., 64, 65, 72-74, 77, 80, 81, 83, 86
Lymn, R. W., 336
Lyons, K. B., 472, 474, 475, 485, 487

M

Ma, R.-I., 550
MACDIARMID, A. G., 443-69; 444, 445, 447-51, 453-55, 457-65
MacDowell, A. A., 165
Macek, J., 131, 134, 137
Macek, J. H., 124, 129, 137
MacGillivray, A. D., 195, 198, 200, 206
Maciel, G. E., 551, 558
Mackay, G., 521, 522
MacNaughton, R. M., 166
Macpherson, M. T., 119, 120, 124, 142, 433, 438
Madden, K., 314
Madge, D., 337
Mahan, B. H., 264, 266
Maier, J. P., 258, 260, 264, 266-73

Maier, M., 521
Majer, J. R., 522
Maki, K., 444, 452, 455, 457
Malatesta, L., 106, 108
Malatesta, V., 510, 515
Malik, S. K., 102-4
Malli, G. L., 161
Manneville, P., 557
Mannherz, H. G., 321
Manning, G. S., 191-94, 198-202, 204, 206, 207, 209, 211, 214-18
Manson, S. T., 124
Mantsch, H. H., 555, 558
Marcoux, P. J., 266
Marcus, R. A., 195, 200, 203, 204, 207, 209, 383
Marden, M., 290
Marden, M. C., 290, 292, 293, 296
Margossian, S. S., 328, 329
Marquaire, P. M., 499, 509, 525
Marque, J., 296
Marriott, P. R., 504, 509
Marsh, D. J., 344
Marshall, H., 303
Marshall, R. M., 504, 507, 509
Marshall, V. P., 290, 292, 293
Marston, S., 328
Marthaler, O., 260, 266-73
Martin, D. S. Jr., 165
Martin, J. P., 522, 524
Martin, R., 499, 510, 514
Martin, R. B., 547, 558
Martonosi, A. N., 339, 345
Mashur, G. J., 313
Mastrangelo, C. J., 27
Masuhara, H., 378-80, 385, 386, 398, 399
Mataga, N., 360, 378-81, 383-86, 398, 399
Mathews, F. S., 283, 287, 290
Matsubara, I., 336
Matsui, S., 553
Matsushima, M., 285
Matsuzawa, M., 181, 182, 184, 188
Mattai, J., 194
Matteoli, U., 115
Matthews, B. W., 284
Matthias, E., 124
Mattson, L., 120
Maurin, M., 157
Maurizot, J. C., 192
Mauser, W., 124
MAUZERALL, D., 377-407; 386, 387, 393, 402
Mauzerall, D. C., 72, 387, 388, 392, 400
Maxwell, D. E., 308
Maxwell, J., 483

AUTHOR INDEX

MAYER, J. E., 1-23; 4, 5, 11, 14, 21, 22
Mayer, M. G., 21
Mazur, S. J., 192, 217, 219
McAdon, M. H., 153
McAndrew, P., 445, 464
McCalley, R. C., 335
McCammon, J. A., 283, 284, 288, 289
McClain, W. M., 377
McClelland, G. M., 124, 134, 142
McConnell, H. M., 325, 335
McCorkle, D. L., 186
McCulloch, R. D., 516, 525
McCurdy, C. W., 240, 241, 243, 244, 246, 247, 251
McDaniel, D. H., 521
McDonald, A. H., 153
McDonald, C. C., 553
McDonald, F. A., 225
McDonald, R. A., 519
McDowell, C. A., 163, 539, 540, 555
McElroy, J. D., 72
McGee, J., 115
McGee, J. D., 307
McGilvery, D. C., 259
McGlynn, S. P., 438
McInally, I. D., 522
McIntosh, D. F., 164
McIver, R. T., 521
McIver, R. T. Jr., 502, 503, 510, 512-14, 522
McKay, R., 346
McKay, K. A., 552, 553
McKoy, V., 251
McLachlan, A., 334
McLaffy, D., 156
McLauchlan, K. A., 303, 305, 306, 397
McLaughlin, A. C., 557
MCMILLEN, D. F., 493-532; 500, 509, 510, 512-15, 522
McMillian, G. B., 175-77, 181
McQuarrie, D. A., 194, 202
Mehlhorn, W., 120, 124
Mehring, M., 534, 540
Meier, E. B., 110, 111, 113, 114
Melançon, P., 192, 217, 219
Mele, E. J., 452, 454, 458, 459
Mellema, J. E., 548
Meltzer, R. S., 78
Menchi, G., 115
MENDELSON, R. A., 319-51; 320-23, 325, 326, 328, 329, 332, 336, 344
Mendenhall, G. D., 521, 522
Meot-Ner, M., 516, 519, 521
Mercouris, T., 241
Merrett, M., 283

Messmer, R. P., 152, 156, 162-64, 166
Meullenet, J. P., 192
Mezei, F., 57
Michaelis, J., 314
Michaels, R. H., 63, 72
Michel-Beyerle, M. E., 398
Michels, H. H., 153, 156
Migita, M., 381
Mihalyi, E., 343
Mihashi, K., 342, 344
Miki, M., 344
Milas, M., 193
Milch, J. R, 336
Miles, R. B., 473, 474
Miletic, M., 519
Mill, T., 515
Miller, G. G., 465, 466
Miller, J. H., 259
Miller, J. R., 393
Miller, R. E., 509, 513, 522
Miller, R. J. D., 64, 65, 67-69, 79-81, 83, 86
MILLER, T. A., 257-82; 258, 259, 261-64, 266-68, 270-74, 277, 278
Miller, W. H., 241, 247, 249
Millikan, R. G., 332
Millman, S., 303
Milne, R. T., 516, 521, 525
Milone, L., 105, 106
Milverton, D. R. J., 266
Mims, W. B., 307, 308, 310, 313
Minakata, A., 192, 194
Misemer, D. K., 157
Misev, L., 267-69, 271
Mishra, M., 247, 249
Misumi, S., 381
Mitani, M., 461, 462, 464
Mitchell, D. J., 25-27, 29, 41
Mitchell, R. C., 416, 418
Miyanishi, T., 346
Miyasaka, K., 52, 445
Moffat, J., 483
Moffett, W., 274
Moh, P. P., 290, 292
Möhlmann, G. R., 266
Mohraz, M., 269, 271, 272
Moiseyev, N., 236, 237, 242, 246, 247
Molchanov, V. K., 316
Molin, Y. N., 316
Moller, A., 216
Monchick, L., 390, 391
Mondovi, B., 313
Montroll, E. W., 74
Moore, N. F., 547, 558
Moore, P. B., 284
Moore, P. H., 320-22
MORALES, M. F., 319-51; 283, 320-23, 325-29, 332, 333, 338-43, 345-47

Morawetz, H., 192, 194, 195, 199, 200
Morgan, J. D., 247
Morgan, J. van W., 154
Morgenstern, R., 120
Morgenthaler, L. N., 259
Morimoto, H. J., 480
Morimoto, K., 322
Morita, F., 339, 343, 346
Mornet, D., 342, 343, 345
Morokuma, K., 160, 167
Morris, E. P., 321
Morris, J. R., 472, 487
Morrison, J. D., 259
Morrison, N. J., 193
Morse, M. D., 427, 430, 433
Morse, P. M., 235
Mortola, A. P., 165
Moruzzi, V. L., 152, 153
Moseley, J., 257, 258
Moseley, J. T., 259
Moses, D., 445, 448, 449, 464
Moskowitz, J. W., 156, 165
Moss, D. J., 344
Mössbauer, R. L., 286, 288
Mowrey, R. C., 244
Mrozek, J., 158
Mrozowski, S., 266
Mueller, H., 319, 343
MUETTERTIES, E. L., 89-118; 90, 106, 108-11, 113-15
Muhlrad, A., 344-46
Mukamel, S., 415, 421
Müller, U. R., 192
Müller-Warmuth, W. Z., 307
Mulliken, R. S., 29, 42, 357, 359, 378
Munowitz, M. G., 547
Munro, I., 289
Murata, Y., 360, 378
Murrell, J. N., 438
Musgrave, R. G., 505, 506, 509
Muttucuraru, N. J., 472, 487

N

Naarmann, H., 466
Nagai, K., 475, 480, 483
Nagai, M., 480
Nagakura, S., 378, 379
Nagasawa, M., 192, 197, 214
Nagayama, K., 289
Nagumo, M., 472-74, 485
Naito, A., 539, 540, 555
Nakamura, H., 192
Nakashima, T. T., 25
Nakatsuji, H., 165
Nall, B. T., 553, 554
Narayana, M., 314
Narayana, P. A., 314
Nassau, K., 307
Natoli, C. R., 157

AUTHOR INDEX 573

Nechtschein, M., 456
Neijzen, J. H. M., 175
Neira, R. A., 60
Nelson, K. A., 64, 65, 67-69, 72-74, 77, 79-81, 83, 84, 86
Nemethy, G., 283
Neuendorf, S. K., 192
Neusser, H. J., 398
Newman, R. A., 165
Newton, R. G., 228
Nguyen, T. T., 513, 514, 520
Nichols, S. R., 216
Nicholson, L. K., 57
Nickel, B., 386, 398
Nicol, M., 25, 27, 29, 30, 472, 473
Nicol, M. F., 25, 472-74, 485
Nicolaides, C. A., 241
Niederberger, W., 557, 558
Niedergerke, R., 319
Niehaus, A., 120, 184
Nigrey, P. J., 445, 447, 450, 451
Nihei, T., 332, 336
Nishi, T., 56
Nishikawa, K., 247
Nishina, Y., 380, 384, 385
Nitta, K., 195, 196
Niwa, Y., 165
Nixon, J. F., 266
Noack, K., 100
Noe, L. J., 472, 473, 475
Nonhof, C. J., 315
Noodleman, L., 163, 165
Noolandi, J., 390, 394
Nordheim, A., 216
Nordling, C., 163
Nordlund, T. M., 290, 292, 293
Norman, J. G. Jr., 155, 163, 165, 166
Noro, T., 241
Norris, J. R., 72, 315
Northrup, S. H., 288
Norton, J. R., 103, 105, 106, 108, 109, 115
Novros, J., 381
Numrich, R. W., 422, 426
Nuttall, J., 225, 228, 234, 236-38, 241

O

Oatley, S. J., 284, 285, 288
Obermueller, G., 72, 76, 77
O'Brien, E. J., 320, 321
O'Connor, J. P., 115
O'Connor, K. M., 57
Offen, H. W., 25-30
Offer, G., 321, 326, 327
Ogawa, M., 266
Ogier, W. C., 509, 512-14, 522
Oh, S. D., 120

Ohrn, Y., 247, 249
Okabe, H., 119, 409, 436, 502, 512, 519
Okada, T., 383, 384
Okafo, E. N., 522
Okamoto, B. Y., 25
O'Keefe, A., 264, 266
O'Konski, C. T., 283, 326, 327
Okubo, T., 192, 193
Olafson, B. D., 484
Oldfield, E., 544, 545, 551, 552, 555, 558
Oliver, A. J., 124
Olmstead, M. M., 110
Olsen, B., 124
Olsen, R. J., 99
Olson, A. J., 285
Oltay, E., 110
Oltzik, R., 313
ONDRIAS, M. R., 471-91; 475, 476, 478-81, 483, 485-88
O'Neal, H. E., 495, 497, 499, 501, 509, 510, 514, 516, 521-23, 525
Onishi, H., 346
Onishi, Y., 165
Onsager, L., 391
O-Ohata, K., 440
Oosterhof, L. J., 360, 378
OPELLA, S. J., 533-62; 533, 539, 540, 542-44, 547-50, 552, 555, 558
Orbach, N., 381, 385
Orel, A. E., 240, 243, 244, 247
Oren, L., 438, 439
Orme-Johnson, N. R., 313
Orme-Johnson, W. H., 313
Ormos, P., 296
Orr, B. J., 120
Osella, D., 105, 106
O'Shea, D., 360
Otis, C. E., 257
Ottolenghi, M., 378, 381, 385
Outhwaite, C. W., 199
Ozaki, M., 447, 454, 458, 460-64
Ozin, G. A., 164

P

Pacansky, J., 505, 507
Pack, R. T., 434, 435
Packer, K. J., 540
Paddock, C., 473, 474
Pagni, R. M., 306
Pai, R. Y., 186
Pake, G. E., 535
Palm, W., 285
Panayotatos, N., 192
Pangali, C., 402
Pankratov, A. V., 521, 523
Pantel, P., 342, 343, 345

Paputa, M. C., 524, 525
Parak, F., 285, 286, 288, 290, 291
Paraskevopoulos, G., 522, 524
Park, E. H., 27
Park, Y. W., 445, 448, 464
Parkes, D. A., 505-7
Parr, A. C., 120, 124, 140, 509, 510, 512, 514, 521, 522
Parr, R. G., 152
Parshall, G. W., 109
Parson, W. W., 377
Parsonage, M. J., 501, 510, 514
Pass, G., 193
Patel, D. J., 533
Patrick, R., 495, 521, 524, 525
Patzer, E. J., 547, 558
Paul, S., 516
Paulson, J. F., 259
Paysen, R. A., 156
Payzant, J., 521, 522
Pë, A. J., 99
Pear, M. R., 288
Pearson, R. G., 89
Peart, B., 234
Pease, L. G., 542, 543
Peatman, W. B., 264, 266
Pecht, I., 289
Pechukas, P., 418, 419
Pecora, R., 327, 337
Pedersen, J. B., 396
Pederson, T., 286, 288
Pedley, J. B., 495, 509, 513, 516, 519, 523, 525
Peebles, D. L., 447, 454, 461, 463, 465
Peierls, R. E., 444, 453
Peisach, J., 313
Pekkarinen, L., 382
Peller, L., 329, 330
Pellerite, M. J., 521
Pénigault, E., 167
Penman, S., 553
Penner, A., 360
Penninger, J. M. L., 110
Penzien, K., 466
Peo, M., 466
Percival, I. C., 124
Perkins, J., 344, 345
Perona, M. J., 516
Perry, D. S., 502
Perry, S. V., 319, 343
Persham, P. S., 557, 558
Person, W. B., 359, 378
Perutz, M. F., 283, 287, 290, 293, 340, 471
Peschel, I., 460, 461
Peshkin, M., 120
Peterlin, A., 194
Petersen, G. E., 313
Peterson, B. S., 107, 108
Peticolas, W. L., 285, 353
Petsko, G. A., 285-87

AUTHOR INDEX

Pez, G. P., 453
Pfab, J., 522
Pheiffer, B. H., 553
Philip, J. R., 195, 201
Phillips, A. W., 283
Phillips, D. C., 284, 285, 288
Phillips, D. T., 25, 26
Phillips, E., 438, 439
Phillips, G. O., 193
Phillips, L., 516
Phillips, M. C., 219
Phillips, S. E. V., 285, 287
Phillips, W. D., 553
Piacenti, F., 115
Pickard, J. M., 522
Picot, C., 53, 54
Pilcher, G., 495, 509, 513, 516, 519, 523, 525
Pilette, Y. P., 380, 381
Pincus, P., 56
Pinder, D. N., 54
Pines, A., 535-37, 539
Pines, H., 394
Pink, H., 214
Pino, P., 107, 108
Piszkiewicz, L., 515
Piszkiewicz, L. W., 505-7
Pittman, C. U., 115
Pitzer, K. S., 162
Plantenga, F. L., 315
Podolsky, R. J., 336, 339
Poë, A., 98-100, 102-4, 107, 108
Poë, A. J., 98, 100, 102
Pohl, F. M., 216
Poli, A., 106
Poliakoff, E. D., 120, 124, 140
Politis, T. G., 27, 31, 33, 34
Pollak, H., 264, 266
Pollard, T., 321
Pope, B., 326
Pope, J. M., 557
Pople, J. A., 455, 456
Porter, G., 63
Post, D., 163
Potashnik, R., 378
Potts, A. W., 165, 166
Poulos, T. L., 284
Powell, H. M., 360, 378
Powers, J. C., 284
Powers, L., 557, 558
Prager, S., 57
Pragst, F., 381
Pratt, R. H., 119, 120
Preobrazhensky, M. A., 238
Pretzer, W. R., 90, 109, 110, 115
Preziosi, P. F., 465, 466
Price, S. J. W, 512, 514, 519, 522-25
Price, W. C., 266
Pron, A., 454, 455, 457-59, 464
Prout, C. K., 360, 378

Pryor, A. W., 285
Pryor, W. A., 515
Ptitsyn, O. B., 284
Purcell, F. J., 474
Purcell, G. V., 290, 291
Purnell, J. H., 525
Putnam, S., 322, 328
Putnam, S. V., 332, 333, 335, 337-40
Pyykkö, P., 160-62, 167

Q

Quate, C. F., 78
Quicksall, C. O., 100, 105
Quinlivan, J., 325
Quinn, C. P., 505-7

R

Rabi, I. I., 303
Rabii, S., 160
Rabinovich, I. B., 525
Rabinovitch, B. S., 509, 525
Rabolt, J. F., 458
Racah, G., 121, 124
Radda, G. K., 557
Rafaey, K. M. A., 519
Rainford, P., 346
Rajagopal, A. K., 152, 153
Raju, S. B., 225
Ramana, M. V., 153
Ramanathan, G. V., 193, 202
Rand, S. D., 472, 473, 475
Randall, E. W., 105
Rao, M., 402
Raoult, M., 120
Rapoport, L. P., 238
Rapp, D., 418, 419
Rau, D. C., 199, 219
Rauch, J., 234
Rauk, A., 154, 159, 162, 164, 167
Rauscher, W., 384, 388, 401
Read, F. H., 134, 266
RECORD, M. T. JR., 191-222; 191-93, 195, 201, 203-9, 213, 215-19
Reddy, G. S., 110
Reed, K. J., 516
Reed, M., 224, 225, 229, 231, 234, 238, 245
Reedy, M. K., 319, 330
Regge, T., 228
Rehm, D., 383-85, 388, 390
REINHARDT, W. P., 223-55; 225, 234, 238, 240, 251
Reinisch, L., 290, 292, 293, 296
Reisler, E., 344
Rentzepis, P. M., 63, 472, 473, 475
Renzoni, G. E., 165, 166
Requena, A., 241

Rescigno, T. N., 225, 240, 241, 243, 244, 246, 247, 251
Reuben, J., 533
Revzin, A., 219
Reynolds, A. H., 290, 292, 293, 472, 473, 475
Rhodin, T. N., 90, 109
Rice, D. M., 544, 552, 557
Rice, D. W., 283
Rice, M. J., 444, 452, 454, 455, 458, 459
Rice, S. A., 197, 409
Rich, A., 216
Richard, C., 510, 514
Richard, J. P., 524, 525
Richards, D. A., 472, 487
Richards, F. M., 289
Richards, P. L., 78
Richards, R. E., 533, 557
Richardson, W. H., 509, 510
Richter, D., 57
Rigler, R., 325
Rinaudo, M., 193, 195, 198
Ring, M. A., 525
Ritchie, B., 136
Ritchie, G. A., 557
Rives, J. E., 78
Rob, F., 555
Robaugh, D. A., 500, 509, 513
Robb, J. C., 522
Robbins, G. A., 165
Roberge, R., 156
Roberts, G. C. K., 533
Roberts, P. G., 266
Robertus, J. D., 284
Robinson, G. N., 494, 520, 521
Robinson, G. W., 357
Robinson, P. J., 502
Rochas, C., 193
Roche, M., 156
Rodgers, A. S., 494, 497, 512, 514, 521-23
Rodriguez, S., 29
Roe, J.-H., 192, 217, 219
Roellig, M. P., 502
Roger, G., 124
Rogers, M. T., 166
Röllig, K., 383
Rollinson, A. M., 27, 29
Rommelmann, H., 464
Ron, A., 119, 120
Rondelez, F., 54
Ronis, D., 394
Roper, W. R., 109
Ros, P., 153, 158-61, 163, 164, 167
Rösch, N., 152, 154, 155, 160, 165
Rosen, A., 159, 161
Rosenberg, A., 289
Rosenbusch, J. P., 343
Rosendaal, A., 158
Rosenstock, H., 503, 504

AUTHOR INDEX

Rosenstock, H. M., 509, 510, 512, 514, 521, 522
Ross, D. S., 509, 513, 522
Rossi, M. J., 496, 500, 504-7, 509, 510, 512-14, 521, 522, 524
Rostas, J., 259, 266
Roth, S., 466
Rothe, E. W., 124
Rothgeb, T. M., 284, 289, 544, 551, 552
Rothwell, W. P., 553, 554
Rotkiewicz, K., 382
Roufasse, A., 546
ROUSSEAU, D. L., 471-91; 475, 476, 478-81, 483, 485-88
Rowland, F. S., 510, 512
Rubbmark, J. R., 175
Ruben, D. J., 539
Ruf, M. W., 120
Ruff, G. A., 182, 183
Rulz, R. P., 516
Rumfeldt, R. C., 524, 525
Rundel, R. D., 180-82
Rundle, H. W., 521
Rundle, R. E., 96
Rupprecht, A., 553, 554
Ryan, P. B., 163
Ryan, R. C., 115
Rylance, J., 495, 509, 513, 516, 519, 523, 525

S

Sabin, J. R., 160, 167
Safinya, K. A., 182, 183
Saikhov, K. M., 314
Saito, H., 555, 558
Sakata, Y., 381
Salahub, D. R., 156, 158, 162, 163, 166
Salman, O. A., 37
Sambe, H., 160, 167
Samet, J., 329, 330, 346
Samson, J. A. R., 119, 120, 132
Sanner, R. D., 115
Sarhangi, A., 462
Sarma, G., 53, 54
Sarma, R. H., 192
Sarre, P. J., 266
Sasaki, S., 192
Sasano, K., 188
Sass, J. K., 357, 360, 377
Sasso, R., 193
Satake, H., 547
Satchler, G. R., 121
Satoko, C., 160, 167
Sawicki, C. A., 472, 475, 485
Say, B. J., 540
Scatturin, V., 106, 108
Schaafsma, T. J., 547, 548, 550
Schaefer, J., 534, 535, 537, 545, 546, 552, 553
Schäfer, F. P., 72, 383
Schäfer, H., 90
Schafer, J., 553
Schechter, A. N., 551
Scheibner, K., 238
Scheidt, W. R., 484
Schellman, J. A., 192, 196
Schenider, T., 457
Scheraga, H. A., 283
Scherker, K., 516
Schermann, J. P., 184, 187
Schiller, P. W., 341
Schindler, R. N., 187
Schirov, M. D., 307
Schlag, E., 264, 266
Schlag, E. W., 264, 266, 398
Schlag, J., 466
Schlesinger, Y., 449
Schmidt, C. F., 557
Schmidt, J., 315
Schmidt, R. E. Jr., 107, 108
Schmidt, S., 99, 100
Schmidt, V., 120
Schmitt, A., 192
Schneider, B. I., 249
Schnering, H. G., 90
Scholze, H., 285
Schönhense, G., 120
Schramm, S., 544, 552
Schreiner, S., 384, 388
Schrieffer, J. R., 444, 452, 455, 461, 462, 464, 466
Schröder, H., 398
Schuetzle, D., 511
Schuh, H., 507, 508
Schuler, R. H., 301, 302
Schulten, K., 386, 396-98
Schulten, Z., 396, 397
Schultes, E., 187
Schultz, A. J., 110
Schultz, T. D., 457
Schumacher, H. J., 516, 519
Schunn, R. A, 115
Schuster, G. B., 25-27, 29, 31, 32, 41
Schutt, C. E., 285
Schuyler, M. W., 474
Schwartz, H. M., 545, 546
Schwartz, R. N., 301, 310
Schwarz, K., 152, 154, 163
Schweitzer, G. K., 166
Scofield, J. H., 124
Scott, R. A., 157
Scrivner, G. P., 71
Sears, T. J., 261-63, 270, 271, 273, 274, 277, 278
Seaton, M. J., 124
Secrest, D., 419, 422, 423
Seddon, E. A., 165, 167
Seeger, K., 448
Seeley, P. J., 557
Seelig, J., 537, 547, 555, 557, 558
Seely, G. R., 388
Seeman, N. C., 550
Sefcik, M. D., 552
Segev, E., 425, 438, 439
Seidel, J. C., 326, 335, 337, 346
Seidlitz, H., 398
Seiler, R., 234
Seiter, C. H. A., 556
Sekine, T., 342, 343, 346
Sekiya, K., 346
Selwyn, J. E., 72, 76
Semenov, A. G., 307
Semenov, V. E., 251
Sengupta, P. K., 37
SenSharma, D. K., 503
Setser, D. W., 266, 502
Seybert, D. W., 483
Seymour, J., 321
Shacklette, L. W., 465, 466
Shaner, S. L., 192, 217, 219
Shank, C. V., 472, 473, 475
Shapiro, J. T., 209
SHAPIRO, M., 409-42; 119, 415, 419, 421, 422, 424, 425, 427, 429-31, 434-39
Shapley, J. R., 115
Sharma, K. R., 98, 99
Sharon, R., 425
Shatlock, M. P., 551
Shaw, R., 511, 516
Shaw, R. W., 29, 313
Sheldon, M. C., 343
Shelnutt, J., 360
Shelnutt, J. A., 475, 478, 480, 486-88
Shen, Y. R., 78
Sherman, P. R., 243
Sherrod, R. E., 156, 157, 166
Shida, T., 267
Shiley, R. H., 263, 264, 267, 270-73, 278
Shimada, K., 381
Shimada, M., 379
Shimizu, T., 192, 194
Shimshick, E. J., 335
Shindo, H., 553
Shipman, L., 315
Shirakawa, H., 445, 447, 448, 453-55
Shirley, D. A., 124
Shobatake, K., 436
Shortland, A. C., 102
Shriver, J. W., 347
Shulman, R. G., 533
Shurcliff, W. A., 136
Sichel, J. M., 120
Siebrand, W., 353, 359, 360
Siegbahn, H., 163
Siegbahn, K., 120
Siegel, J., 136, 157
Siegel, M. W., 120

AUTHOR INDEX

Sieker, L. C., 285
Sielecki, A. R., 285, 288
Siemiarczuk, A., 382
Sigfridson, B., 266
Signorelli, A. J., 453
Silbey, R., 277
Silva Crawford, M., 556
Silvestri, P., 193
Simmons, R. M., 320, 322, 330, 336
Simon, B., 224, 225, 229, 231, 233, 234, 238, 245-47
Simon, S. R., 475, 478, 480, 486-88
Simonetta, M., 152
Simons, J., 233, 241, 244, 249, 512, 516
Simons, J. P., 119, 120, 124, 142, 409, 416, 418, 433, 438
Simpson, D. W., 266
Simpson, J. D., 29, 30
Sindorf, D. W., 558
Singh, G. P., 290, 291
Singh, S. R., 234
Singh, T. P., 288
Sinke, G. C., 495, 509, 513, 516, 519, 523, 525
Sivak, A., 455, 457, 464
Sivak, A. J., 110, 111, 113, 114
Siveigart, D. A., 270
Sixl, H., 466
Skillman, S., 152
Skinner, H. A., 100, 522, 525
Skokut, T. A., 552
Skolnick, J., 193, 194, 200
Slater, J. C., 151, 152, 154
Slater, S., 115
Slayter, H. S., 321, 327
Slichter, C. P., 308
Sloane, C. S., 277
Sloop, D. J., 314, 315
Small, G., 359
Smith, A. J., 266
Smith, A. K., 115
Smith, A. L., 261
Smith, D. F., 4, 5
Smith, D. L., 259
Smith, G. P., 259, 495, 500, 521, 524, 525
Smith, I. C. P., 555, 558
Smith, K. A., 174-77, 180-82, 184, 187
Smith, P. R., 321
Smith, R., 556
Smith, R. A., 305
Smith, R. J., 136
Smith, R. L., 544, 545, 552
Smith, V. H. Jr., 158
Smith, W. H., 266
Snell, W., 508
Snijders, J. G., 159, 161, 162, 167

Snook, I., 199
Solly, R. K., 516, 519, 521, 523
Somekh, J., 29
Sonnenberger, D., 99, 108, 109
Sonnenberger, D. C., 103, 108, 109
Sonnich Mortensen, O., 353, 357
Sontum, S. F., 157, 166, 167
Sorbie, K. S., 438
Sorensen, L., 290, 293
Sorensen, L. B., 290, 293
Soumpasis, D., 193, 200
Soumpasis, D. M., 194
Southworth, S. H., 124
Spagbnuolo, V., 193
Sparks, R. K., 436
Spiro, E. J., 193
Spiro, T. G., 100, 105, 472-74, 485
Spokes, G. N., 497, 500, 501
Squire, J. M., 283
Sreter, F. A., 343, 344
Srivastawa, R. B., 474
Stadelmann, J. P., 268, 269
Staerk, H., 386, 398
Stagat, R., 236
Stagat, R. W., 236
Stahl, H., 78
Stanley, G. G., 165, 167
Stannard, B. S., 209
Starace, A. F., 119, 120, 124
Stark, R. E., 539
Stavola, M., 475, 476, 478
STEBBINGS, R. F., 173-89; 174-77, 179-82, 184, 187
Steele, A. W., 472-75
Steffen, R. M., 124
Steffens, E., 398
Stehlik, D., 305-7
Steigemann, W., 285, 287
Stein, S. E., 500, 509, 513-15, 522
Steinfeld, J. I., 72, 76
Steinkruger, F. J., 510, 512
Steitz, T. A., 283, 284
Stejskal, E. O., 534, 535, 537, 545, 546, 552, 553
Stepnoski, R. A., 475, 476, 478, 480, 481, 483, 485
Sterna, L., 394
Sternberg, M. J. E., 284, 285, 288
Sternlicht, H., 545, 546
Stigter, D., 191, 192, 195, 196, 198, 201, 203
Stiles, P. J., 388
Stillinger, D. K., 243
Stillinger, F. H., 233, 234, 243
Stillman, A. E., 310
Stirdivant, S. M., 192
Stockbauer, R., 120, 509, 510, 512, 514, 521, 522

Stockdale, J. A., 184, 187
Stockmayer, W. H., 326
Stoffel, W., 547, 555
Stoll, W., 131, 134
Stolovitskii, Yu. M., 388
STOLZENBERG, A. M., 89-118
St. Onge, R., 336
Stowring, L., 325, 346
Strauss, U. P., 192, 195, 200, 201, 207, 211, 212, 214
Street, G. B., 448, 449, 456-58, 466
Streets, D. G., 166
Streit, G. E., 519
Strick, T. J., 217, 218
Stroud, R. M., 343
Stryer, L., 289, 341
Studebaker, J. F., 28
Stueckelberg, E. C. G., 417
Stuhrmann, H. B., 284
Stull, D. R., 495, 509, 513, 516, 519, 523, 525
Stuntz, G., 108, 109
Sturm, C. F., 547-50, 552
Sturm, J., 219
Su, W. P., 444, 452, 455, 457, 461, 463, 466
Suart, R. D., 511, 512
Suck, D., 321
Sugai, S., 195, 196
Sugiura, T., 184
Sukumar, C. V., 241, 251
Sullivan, C. E., 544, 545
Summers, A. J., 516, 525
Sumner, G. G., 100
Sunil, K. K., 166
Sustmann, R., 510, 512
Sutherland, J. W. H., 551
Sutin, N., 377
Suzuki, N., 460, 461
Swaminathan, S., 402
Swanson, J. R., 120
Sweany, R. L., 100
Sykes, B. D., 347, 533, 552
Symons, M. C. R., 360, 378
Syverud, A. N., 519
Szamrej, I., 186
Szent-Gyorgyi, A., 319, 339, 345
Szent-Gyorgyi, A. G., 343
Szilagyi, L., 344
Szwarc, M., 381, 493
Szymczak, J., 193

T

Tabeta, R., 558
Tachibana, A., 235, 236, 244
Tachikawa, M., 90
Taieb, G., 264, 270-72
Tait, K. B., 522, 523
Takahashi, T., 321

Takano, T., 283, 285, 287, 288, 290
TAKASHI, R., 319-51; 325, 333, 335, 339, 342-47
Takashima, S., 290, 291
Takayama, H., 444, 452, 455, 457
Tambe, B. R., 124
Tanaka, J., 379, 449, 460
Tanaka, M., 447, 449, 454, 460, 461, 463
Tang, J., 357
Taniguchi, Y., 380, 384, 385
Tanner, J. E., 53
Tao, T., 344
Taranko, A. R., 457
Tarcsafalvi, A., 343
Tardy, D. C., 516
Tarr, A. M., 524
Tasker, P. W., 416, 433
Tatsuka, T., 326
Taylor, E. W., 339
Taylor, H. S., 234, 241, 244, 251
Taylor, J. R., 228, 247, 414, 422
Taylor, J. W., 156
Taylor, K. A., 322, 344, 345
Taylor, T. J., 25
Tebbe, F. N., 109
teiner, G., 503, 504
Teller, E., 274
Teller, R. G., 110, 111, 113
Templeton, J. E., 124
Teng, H. H., 259, 267
Teranishi, H., 505, 506
Terao, T., 553
Termine, J. D., 546
Terner, J., 472-74, 485
Teubner, M., 219
Teuchner, K., 381
Tfibel, F., 472, 475, 485, 487
Theodosiou, C. E., 124
Thiessen, W. E., 284
Thimm, K., 120
Thomas, D. D., 326, 335, 337
Thomas, L., 234
Thomas, L. D., 234
Thomas, M. G., 115
Thomas, M. M., 38
Thomas, T. F., 259
Thommen, F., 264, 266, 268-72
Thompson, W. W., 516, 525
Thomson, J. O., 156
Thorn, D. L., 115
Thornton, G., 160
Thurnauer, M., 315
Thurnauer, M. C., 315
Tieke, B., 466
Timchenko, A. A., 284
Timkovich, R., 284
Tip, A., 233, 238
Tipps, W. M., 432, 433

Tirado, M., 327
Tirrell, M., 57, 58
Tokiwa, T., 320, 332
Tokura, S., 346
Tollin, G., 387
Tomasula, M., 193
Tomberg, H., 472, 473
Tomimoto, S., 380, 381
Tomkiewicz, Y., 457
Tong, S. B., 498, 510, 521
Tonomura, Y., 339, 343, 346
Torchia, D. A., 537, 544-47, 551
Torgerson, P. M., 31
Torrance, J. B., 360, 378
Torrey, H. C., 303
Torrie, G. M., 199
Toscano, W. A. Jr., 293
Toyoshima, C., 322
Traber, R., 384, 388, 401
Tredwell, C. J., 63
Tregear, R. T., 319, 330
Treichel, P. M., 167
Treadwell, C. J., 63
Trentham, D. R., 339, 344
Trenwith, A. B., 511, 512, 516
Trevor, P. L., 500, 513, 515
Trifiletti, R., 193
Trifunac, A. D., 307, 315
Trill, H., 510, 512
Trisc, M., 163
Troe, J., 521
Trogler, W. C., 99, 100, 165
Troitsky, A. V., 284
Trotman-Dickenson, A. F., 522, 524, 525
Troup, J. M., 102
Trudeau, K., 289
Tsai, M. D., 544, 545, 552
Tsang, C.-H., 445
Tsang, W., 499-501, 504, 506, 507, 509, 512, 513, 520
Tschesche, H., 289
Tseng, H. K., 119, 120
Tsernoglou, D., 285, 286
Tsubaki, M., 475, 480
Tsujino, N., 379
Tsurubuchi, S., 266
Tsvetkov, Y. D., 314
Tubbs, M. R., 71
Tuckett, R. P., 261
Tuffile, F. M., 193
Tulig, T. J., 58
Tully, J. C., 120
Tung, C. M., 186
Tunner, S. F., 540
Turner, D. W., 266, 270
Twigg, M. V., 102, 103

U

Ue, K., 325, 329, 330, 333, 345-47

Ueno, H., 320
Ulstrup, J., 293
Unger, L., 192, 217, 219
Urbach, W., 54
Urbina, J. S., 556
Ushio, J., 165
Utterback, S. C., 314

V

Valat, P., 472, 473
Valleau, J. P., 199
Van Brunt, R. J., 124
Van der Avoird, A., 159
VanderHart, D. L., 537, 545-47, 558
Van der Hart, W. J., 259
Van der Klink, J. J., 199
Van der Waals, J. H., 315
van Dishhoek, E. F., 259
van Doren, V. E., 444, 448, 453
Van Megan, W., 199
Van Sprang, H. A., 266
van Velzen, P. N., 259
van Willigan, H., 539
van Winter, C. J., 225, 229, 231, 238, 245
Varga, Z. S., 157
Varma, C. A. G. O., 315, 360, 378
Varner, J. E., 552
Varoqui, R., 192
Vaughan, R. W., 534
Vaughn, C., 264, 270-73, 278
Vaughn, C. B., 271
Vaughn, C. R., 263, 270, 271
Veeman, W. S., 550
Veillard, A., 165
Velghe, M., 259
Veljkovic, M., 519
Verhoeven, J. W., 555
Verma, N. C., 302, 303, 306
Vernon, M., 25
Vidal, J. L., 115
Vieth, H. M., 305-7
Vigué, J., 124
Vincow, G., 512
Viniegra, G., 320, 332
Virmani, Y., 100
Visser, A. J. W. G., 31
Vlasnik, L. M., 165
Vock, E., 234
Vogelmann, E., 384, 388, 401
Volland, W. V., 512
Vollman, W., 305-7
von Barth, U., 152-54
von Förster, J., 341, 342
von Hippel, P. H., 219
von Niessen, W., 360
von Smoluchowski, M., 58
Vosko, S. H., 153
Voss, D., 473, 474

AUTHOR INDEX

W

Wada, A., 192
Waddington, G., 100
Wade, C. G., 556
Wade-Jardetzky, N., 347
Wade-Jardetzky, N. G., 551
Wadzinski, L., 329, 330
Wagner, G., 289
Wagner, G. C., 290
Wagner, R. R., 547, 558
Wahl, P., 328, 342
Wahlgren, U., 164
Waite, B. A., 249
Wakabayashi, T., 320, 322, 343
Waldrop, J. R., 449
Walker, J. A., 506, 507
Walker, L., 557
Walker, R. B., 290, 425
Walker, R. F., 516
Walker, R. W., 504, 506, 507, 509, 510
Walker, W. E., 115
Wallace, S., 120, 156, 157
Wallbank, B., 163
Walmsley, J. H., 455, 456
Walsh, R., 494, 496, 516, 524, 525
Walton, I. B., 165
Wan, J. K. S., 303, 306
Wang, A., 329, 330
Wang, C., 327
Wang, H., 438
Wang, H.-K., 115
Wang, T., 56
Wang, Y., 381, 393
Ward, R. L., 378
Warren, J. C., 346
Warshel, A., 353, 484, 485
Watanabe, A., 449, 460
Watanabe, H., 52
Watanabe, K., 438
Watanabe, S., 346
Watenpaugh, K. D., 285
Watkins, A. R., 381, 384, 402
Watkins, K. W., 516, 525
Watson, K., 224
Waugh, J. S., 535-37, 539, 553, 554, 556, 558
Wawersick, H., 100
Webb, T. R., 165
Webb, W. B., 337
Weber, A., 328
Weber, E., 285, 287
Weber, G., 31, 288, 325, 342
Weber, I. T., 284
Weber, J., 155, 166, 167
Weber, J. A., 243
Webster, N. J., 524, 525
Wedlock, D. J., 193
Weeds, A. G., 326, 327
Wegener, W. A., 326
Wegman, R. W., 99, 110

Wehinger, P. A., 266
Wei, C. H., 105
Weinberger, P., 152, 154
Weiner, J. H., 552
Weinhold, F., 236, 237, 242
Weis, O., 78
Weisbuch, G., 193, 195, 198, 199, 206, 208
Weiss, J., 382
Weiss, K., 380, 381
Weisskopf, V. F., 224
Weissman, R. B., 472, 473, 475
WEISSMAN, S. I., 301-18; 303, 306, 315, 316
Welborn, C., 483
Welge, K. H., 266
Weller, A., 359, 378, 382-86, 388, 390, 392, 398
Wells, J. A., 343-45
Wells, J. M., 525
Wells, R. D., 192
Welsh, J. A., 264
Wendoloski, J. J., 238
Wennerström, H., 199, 555
Werber, M. M., 296, 343
Werner, H. J., 386, 398
Werner, T., 384, 388
West, G. A., 416
West, J. B., 120
West, W. P., 184, 187
Westra, S. W. T., 195, 199
Westrum, E. F. Jr., 495, 509, 513, 516, 519, 523, 525
Wherry, C. J., 237, 238
White, J. N., 519
Whitehead, M. A., 153
Whitmore, D. M., 160
Whittle, E., 497, 509, 510, 512, 514, 516, 521, 522, 524
Wick, G. C., 228
Widom, J., 219
Wiegand, G., 285
Wiesenfeld, J. R., 124
Wieting, R. D., 74
Wiget, J. M., 25
Wigner, E. P., 224
Wilcox, P. E., 284
Wild, D. L., 284
Wild, S. M., 27
Wilemon, G. M., 115
Wilhoit, R. C., 522, 523
Wilkinson, D. W., 78
Williams, A. R., 152-56
Williams, J. G., 56
Williams, J. M., 110, 111, 113
Williams, R. J. P., 284, 583
Willis, B. T. M., 285
Wilson, A. D., 421
Wilson, K. R., 418
Wilson, K. S., 284
Wilson, M., 332
Wilson, R. W., 194, 199, 219
Wilson, W. D., 115, 192

Windmer, S., 200
Wing, W. H., 174
Wingrove, D. E., 193
Winkleman, J. J. Jr., 195, 206
Winkler, F. K., 285
Winkler, P., 236, 249
Winslow, D. K., 78
Winther, A., 145
Wise, W. B., 547, 555
Wittebort, R. J., 544, 552, 557
Wnek, G. E., 445, 448
Woerner, T., 445, 455, 457, 464
Wolf, E., 68
Wolf, I., 343
Wolfe, S., 394
Wolfsberg, M., 157
Wolynes, P. G., 289
Wong, S. K., 303, 306
Woo, J. T., 25
Wood, I. T., 525
Wood, J. H., 154, 156, 160, 162
Woodbury, C. P. Jr., 193, 202
Wooding, R. A., 195, 201
Woodruff, S. B., 157, 160, 162
Woods, R. C., 257
Woodward, C. K., 284, 289
Woodward, J., 329, 330
Woodwark, D. R., 160
Wool, R. P., 57
Wooley, R. G., 160
Woolley, P., 219
Wooten, J. B., 553
Wright, J. D., 360, 378
Wright, L. A., 153, 156
Wrighton, M. S., 96, 115
Wrigley, S. P., 512, 516
Wu, C. C., 321
Wu, E-C., 522
Wu, M. C. R., 512
Wüthrich, K., 284, 289, 533
Wyckoff, S., 266

X

Xiao-Zeng, Y., 157

Y

Yagi, N., 336
Yamabe, T., 235, 236, 244
Yamaguchi, M., 346
Yamamoto, K., 342, 343
Yamamoto, N., 384
Yamdagni, R., 503
Yanagida, T., 333, 337
Yancey, J., 27
Yang, C. Y., 155, 160, 161
Yang, J. T., 321
Yang, K. H., 78
Yang, S., 119, 120
Yang, S. C., 119, 120
Yared, W. F., 512, 514, 522

AUTHOR INDEX

Yaris, R., 233, 236, 241, 244, 249, 251
Yasui, T., 346
Yeagle, P. L., 547, 558
Yeates, D. G. R., 284
Yesinowksi, J. P., 558
Yeung, E., 359
Yevstigneyev, V. B., 388
Yhland, M., 378
Yonetani, T., 479, 480
Yonezawa, T., 165
Yoshida, F., 379
Yoshida, N., 192
Yoshihara, K., 378
Yoshimine, M., 505, 507
Young, R. H., 540, 555
Yount, R. G., 320, 332, 343-45
Yu, H., 326
Yu, H. L., 315
Yu, L., 460, 461
Yu, L. C., 336
Yu, N-T., 360
Yudanov, V. F., 314
Yue, K. T., 290, 292, 293, 296

Z

Zabel, F., 516
Zacharias, J. R., 303
Zahradnik, R., 267
ZARE, R. N., 119-50; 119, 120, 124, 137, 140, 142, 145, 158, 502, 519, 524
Zegarski, B. R., 261, 268, 270, 273
Zelikoff, M., 438
Zema, P., 193
Zercheninov, A. N., 521, 523
Zerner, M., 166
Zero, K., 327
Zgierski, M., 353, 359, 360
Zhdanova, N. N., 521, 523
Zhidkov, V. D., 307
Ziegler, T., 154, 159, 162, 164, 167
Zimm, B. H., 195, 198, 199, 201, 202, 206, 327
Zimmerman, A. H., 516
Zimmerman, M. L., 174
Zimmerman, S. B. 1, 553
Zmbov, K. F., 519
Zucker, F. H., 283, 284
Zuiderweg, L. H., 199
Zumbulyadis, N., 540, 555
Zwolinski, B. J., 522, 523
Zyzyck, L. A., 115

SUBJECT INDEX

A

Acetonitrile solutions
  flash irradiation of
    free ion formation and, 385
Acetylacetonate derivatives
  europium chelates with
    energy transfer process in, 41–45
Acetylenic cations
  light-emitting, 266, 268–69
Acridine orange
  quenching of, 401–2
Acrylamide
  tryptophan fluorescence quenching and, 289
Actin
  binding of myosin fragments to, 319, 328–30
Actin-DNase crystals
  long-axis asymmetry of, 321
Actin filaments
  arrangement of, 320–21
  muscle contraction and, 319–20
Actin-myosin cross bridges
  ATPase and, 336–39
Adenosine triphosphate
  hydrolysis of
    muscle contraction and, 319
    polyphosphate moiety of, 345
Alanine
  ^{13}C nuclear magnetic resonance spectra of, 539, 541
Alcohol
  photolytic reaction between quinones and
    electron spin echo spectroscopy and, 315
  radiative rate of indole in effect of dielectric constant on, 34
Aldehydes
  ion cyclotron resonance photodissociation spectroscopy and, 259
Alkanes
  bond dissociation energies and, 508–10
  radiolysis of
    electron paramagnetic resonance and, 302–3
Alkane solvents
  results of photolysis in, 96
Alkenes
  bond dissociation energies and, 510–14

Alkoxy radicals
  bond dissociation energies of, 518
Alkylaminobenzenes
  aromatic molecule singlet-excited state quenching and, 383–84
Alkynes
  bond dissociation energies and, 514
  bond homolysis rates of, 500
Amines
  bond dissociation energies and, 520
  flash photolysis of
    perylene anion radical formation and, 382–83
Amino acids
  nuclear magnetic resonance spectroscopy and, 538–44
Aminophosphine complex
  ground and excited state forms of, 114
Ammonia
  metals in liquid
    electron spin echo method and, 310
Angular momentum transfer
  parity-favoredness quantum number for, 142–43
  of photofragmentation, 139–48
L-Anilino-3-(4-anthryl)-propane
  noncoplanar ion pair formation and, 381
Anisotropic elastic constants
  laser-induced phonons and, 86
Anisotrophy
  of photofragmentation processes, 119
Anthracene
  bimolecular encounter of with phenanthrene, 384
  cyclohexadiene radical formation and, 514
  singlet reactions of
    kinetics of, 382
Anthraldehyde
  fluorescence emission of, 29
Aqueous matrices
  deuterated sodium hydroxide in
    electron spin echo method and, 314
Aromatic carbonyl compounds
  phosphorescence of, 29
Aromatic hydrocarbons

  bond dissociation energies and, 512–13
  excimer formation with pressure of, 30
  fluorescence of, 25–29
  phosphorescence of, 29
  photoionization of
    aromatic anion and cation pair formation and, 398
  tetracyanobenzene complexed with
    emission of, 38–40
Atoms
  complex scale transformations and, 235–37
Atoms in fields
  complex scale transformations and, 237–38
ATPase
  actin-myosin cross bridges and, 336–39
  binding site for in myosin, 319
Auramine-O
  luminescence efficiency of solvent viscosity and, 32
Autoionizing states
  Stark broadening of, 238
Azulene derivative
  S_1-S_0 emission in, 28
Azulene derivatives
  luminescent efficient vs peak energy and, 26–27
  S_2-S_0 emission in, 27

B

Bacteriophage fd
  nuclear magnetic resonance spectroscopy and, 548–51
Bacteriorhodopsin
  nuclear magnetic resonance spectroscopy and, 552
  photocyle of
    viscosity dependences of reaction rates in, 296
Barium diethyl phosphate
  tensor component orientation in, 553
Basis set expansion
  relativistic calculations with, 161–62
  transition metal carbonyls and, 163–64
  wave equation solutions and, 158–60

SUBJECT INDEX 581

Batteries
  polyacetylene and, 450–52
Benzene
  absorption spectrum of,
    364–65
  cyclohexadiene radical
    formation and, 514
  highest filled orbitals of, 267
  ion cyclotron resonance
    photodissociation
    spectroscopy and, 259
  self-diffusion of polystyrene
    chains in, 55
Benzene cations
  vibronic structure of, 273
Benzene radical cation
  fluorescence quantum yield
    of, 267
Benzenoid
  degenerate ground state of,
    465
Benzenoid cations
  first electronic transition in,
    261
  light-emitting, 266, 270
Benzenoid compounds
  first electronic transition in,
    261
Benzoquinone
  chlorophyll-a triplet state
    oxidation and, 389
Benzyl cation
  matrix absorption spectrum
    of, 258–59
Benzyl chloride cation
  matric absorption spectrum
    of, 258–59
Benzylic bonds
  dissociation energies of,
    514–15
Benzylic radicals
  stabilization energy of, 515
Binuclear metal carbonyl
  complexes
  ligand substitution reactions
    of, 96–101
Bisphenol-A-polycarbonate film
  triphenylamine
    photoionization in, 393
Bjerrum's ion-pairing theory
  Debye-Hückel theory and,
    194
Bleomycin
  metal complexes with
    electron spin echo method
      and, 313
Bond dissociation energies
  bond scission reactions and,
    497–99
  measurement of, 499–501
  t-butyl radical and, 504–8
  chemical activation and,
    501–2

determination of, 495–504
halogenation equilibria and,
  497
halogenation kinetics and,
  496–97
heats of formation and,
  504–26
homolytic, 493
in hydrocarbons and their
  derivatives, 508–26
optical detection of, 502
reviews of, 494–95
thermodynamic cycles and,
  502–4
transition state method and,
  159
see also Hydrocarbon bond
  dissociation
Bond scission reactions
  bond dissociation energies
    and, 497–99
  measurement of rates of,
    499–501
Bone
  nuclear magnetic resonance
    spectroscopy and, 546
Boost analyticity
  dilatation analyticity and,
    233
Born-Oppenheimer
  approximation
  complex coordinate, 245–46
  dilatation analyticity and,
    234
Born-Oppenheimer potential
  surface
  nuclear motion on, 248
Bound state wave functions
  dilatation analyticity and,
    234
Bromination
  halogenation kinetics and,
    497
t-Butyl radical
  thermochemistry of, 504–8

C

Calcium phosphates
  nuclear magnetic resonance
    spectroscopy and, 546
Carbon dioxide
  collinear
    photodissociation of, 426
  photodissociation of, 433–34
Carbon monoxide
  basis set effects for, 159
  binding to myoglobin,
    290–92
Carbon tetrachloride
  results of photolysis in, 96
  self-diffusion of polystyrene
    chains in, 54

Carbon-centered
  oxygen-containing radicals
  bond dissociation energies
    and, 515–18
Carbonium ions
  hydride transfers to, 503
Carotene
  Raman excitation profile of,
    369
  resonance Raman scattering
    and, 363
Carr-Purcell trains, 308
Cartilage
  nuclear magnetic resonance
    spectroscopy and, 546
Catalysis
  viscosity dependences of
    reaction rates in, 296
Catalytic reactions
  cluster-based homogeneous,
    115
Cauchy's theorem, 226
Cellulose
  nuclear magnetic resonance
    spectroscopy and, 558
Ceruplasim
  electron spin echo method
    and, 313
Cesium
  photolysis of
    transient absorption and,
      303
Charge-transfer complexes,
  378–82
  charge-transfer ionization
    and, 378–80
  ion formation from triplet
    states in, 380–81
  ion pair variety in, 381–82
Charge-transfer ionization,
  378–80
Chemical activation
  bond dissociation energies
    and, 501–2
Chemically induced electron
  polarization, 306
Chemically induced nuclear
  polarization, 306
Chloranil solutions
  irradiated
    ion radical formation from,
      378
Chlorination
  halogenation kinetics and,
    497
Chlorophyll-a
  triplet state oxidation of, 389
Cholesterol
  nuclear magnetic resonance
    spectroscopy and, 558
Chromate ion
  resonance Raman scattering
    and, 363

SUBJECT INDEX

Chromium
  tetrahedal chlorides of
    ligand field theories and, 166
Clebsch-Gordan coefficient, 121
Cobalt
  tetranuclear dodecacarbonyl cluster formation and, 105
Coincidence measurement
  quantum yields for detectable-emission ions and, 264–65
Collagen
  nuclear magnetic resonance spectroscopy and, 545–56
Competitive binding equilibria
  molecular theory of condensation and, 194
Complex basis functions
  method of, 239–40
Complex coordinate
  Born-Oppenheimer approximation, 245–46
Complex coordinate real axis clamped nuclei approximation, 246–47
Complex coordinates
  see Complex scale transformations
Complex scale transformations, 223–51
  boundary conditions and, 228–29
  computational applications and, 234–49
  mathematical results of, 229–33
  spectral theory of dilatation analytic Hamiltonians and, 229–33
  theoretical applications of, 233–34
Complex scaling
  see Complex scale transformations
Complex Siegert functions
  method of, 241–42
Complex stabilization, 244
Complex variational theory, 242–44
Conformational transitions
  molecular theory of condensation and, 194
Connective tissue
  structural proteins of
    nuclear magnetic resonance spectroscopy and, 545–47
Contractile mechanism in muscle 319–47
    active, 335–39
    binding of myosin fragments to actin and, 328–30
    myosin segment orientation and, 330–35
    protein morphology and, 320–22
    segmental flexibility of myosin and, 322–28
    transduction and, 339–47
Coordinate-rotation
  see Complex scale transformations
Coordination saturation
  mononuclear metal complex reactivity and, 92
Copper complexes
  electron spin echo method and, 313
Copper compounds
  electron spin echo method and, 310
Copper tetraphenylporphyrin
  Raman excitation profile of, 361
Counterion binding
  in polyelectrolyte solutions, 207–15
Counterion condensation
  polyelectrolyte theory
    development of, 193–200
    thermodynamic consequences of, 200–7
    thermodynamic degree of dissociation and, 213–14
Counterions
  thermodynamically bound
    physical interpretations of, 214–15
Crude adiabatic
  Born-Oppenheimer electronic state, 355–56
Crystal violet
  luminescence efficiency of solvent viscosity and, 32
Crystals
  echo behavior in, 310
Cyanocobalamin
  Raman excitation profile of, 369
Cyclic octasulfur
  multiple scattering deformation densities for, 158
Cycloalkanes
  ion cyclotron resonance photodissociation spectroscopy and, 259
Cycloheptatriene cation
  matrix absorption spectrum of, 258–59
Cyclohexadienyl radical
  stability of, 514
Cyclohexane
  pyrene/perylene excimers in, 30
Cyclopropylmethyl
  resonance stabilization in, 510
Cytochrome c
  absorption band shape of, 365
  Raman excitation profile of, 361
  x-ray diffraction and, 285
Cytochromes
  electron spin echo method and, 313

D

dc Stark Hamiltonian
  for hydrogen, 237–38
Debye-Hückel theory
  Bjerrum's ion-pairing theory and, 194
Deferriferrichrome
  ^{13}C nuclear magnetic resonance spectra of, 542
Delayed transient nutation, 307
Density functional theory, 152
Deoxyhemes
  electron density marker line for, 475–79
  resonance Raman band frequencies of, 472
Deuterated ethylalcohols
  electron spin echo method and, 314
Deuterated sodium hydroxide
  electron spin echo method and, 314
Deuterated water
  deuterated sodium hydroxide in
    electron spin echo method and, 314
Diacetylenes
  highest filled orbitals of, 267
Diatomic ions
  laser-induced fluorescence spectra of, 264
Dibenzilmethide
  energy transfer process in, 41–45
Dichloromethane solutions
  flash irradiation of
    free ion formation and, 385
Dicyanobenzene
  reaction of with pyrene
    conductance rise kinetics of, 386
Dienes
  ion cyclotron resonance

SUBJECT INDEX 583

photodissociation
   spectroscopy and, 259
Diffusion rates
   electron spin echo method
   and, 310
1,4-Difluorobenzene cation
   matric absorption spectrum
   of, 258–59
Dilatation analyticity
   of N-body Coulomb
    Hamiltonian, 229–33
   uses of, 233–34
   see also Complex scale
    transformations
Dimethylamine
   bond dissociation energy of,
    520
Dimethyldiacetylene radical
   cation
   fluorescence quantum yield
    of, 267
Dimryistoyl
   phosphatidylcholine
   ^{13}C nuclear magnetic
    resonance spectrum of,
    556
Dipalmitoyl
   phosphatidylcholine
   ^{13}C nuclear magnetic
    resonance spectrum of,
    556
Diphenylmethane dyes
   luminescence efficiency of
    solvent viscosity and,
    31–32
Diphenylmethylene
   electron spin echo
    spectroscopy and,
    314–15
Diphenylmethyl radical
   stabilization energy of, 515
Diphenylpolyenes
   pressure-induced changes in
    peak location for, 26
Dirac theory, 160–61
Discrete variation method,
   158–59
DNA
   A-form
    ^{31}P nuclear magnetic
     resonance spectrum
     of, 554
   B-form
    ^{13}N nuclear magnetic
     resonance spectrum
     of, 555
   conformational states of
    order-disorder transition
     between, 216–17
   Monte Carlo simulations of,
    199
   nuclear magnetic resonance
    spectroscopy and,
    553–54

oligocation interactions with
   binding constants for,
   217–19
polyelectrolyte effects on
   equilibria involving,
   215–19
polyelectrolyte theories of,
   191–219
   counterion binding and,
    207–15
   equilibria and, 215–19
   Poisson-Boltzmann and
    counterion
    condensation, 193–207
viral, 547
Donnan coefficient
   thermodynamic degree of
    dissociation and,
    211–13
Donnan salt exclusion
   coefficient, 201
Duschinsky mixing, 356
Dye solutions
   excited state transport and
    trapping in, 71–78

E

Echo envelopes
   interpretation of, 313
Elastin
   nuclear magnetic resonance
    spectroscopy and,
    546–47
Electric birefringence
   myosin (S-1)-(S-2) flexibility
    and, 326
Electric dipole transition
   photofragment angular
    distribution for, 120
Electrochemical doping
   polyacetylene and, 450–52
Electrolyte-polyelectrolyte
   interactions
   thermodynamic description
    of, 212
Electron capture
   Rydberg atom formation
    and, 174
Electron density
   kinetic energy and, 151–52
Electron donor-acceptor
   complexes
   absorption and fluorescence
    peaks in, 28–29
   tautomer emission in, 37–38
Electronic orbitals
   pressure tuning of, 40–45
Electron impact excitation
   Rydberg atom formation
    and, 174
Electron microscopy
   bend points in myosin rod
    and, 327

Electron paramagnetic
   resonance, 301–16
   myosin segment orientation
    in muscle and, 331
   radiation field pulsing in,
    307–16
   transient methods of,
    302–6
Electron spin dephasing
   mechanism of, 397–98
Electron spin echoes, 307–16
Electron spin resonance
   see Electron paramagnetic
    resonance
Electron transfer
   fluorescence quenching,
    382–84
Emission polarization
   measurement of, 125–28
Entangled polymers, 49–60
   diffusion-controlled reactions
    of, 58–59
   self-diffusion and, 51–55
   semilocal motions of,
    55–58
Ethanol
   deuterated sodium hydroxide
    in
    electron spin echo method
     and, 314
   laser-induced phonon
    transient grating and,
    81–82
   pyrene/perylene excimers in,
    30
   rhodamine 6G in
    decay rate of, 77
    dimer lifetimes of, 73
Ethylalcohols
   deuterated
    electron spin echo method
     and, 314
Ethyl carbazole solutions
   flash irradiation of
    free ion formation and,
     385
Ethyldiacetylene radical cation
   fluorescence quantum yield
    of, 267
Europium chelates
   energy transfer process in,
    41–45
Evans-Polanyi relationships
   equilibrium-based bond
    dissociation energies
    and, 501
   OH-radical/halocarbon, 524
Excimer emission, 30
Exciplex emission, 30
Exciplex formation
   l-dimethylanilino-3-
    (4-anthryl)-propane,
    381–82

584 SUBJECT INDEX

Excited molecular
   wavefunctions
   resonance Raman scattering
      and, 359–61
Excited state transport
   in concentrated dye
      solutions, 71–78

F

F-actin
   3-D model of, 323
Ferrocytochrome c
   Raman excitation profile of,
      366
Feshbach resonance
   e^+-H scattering and, 237
Flavones
   distribution of excitation
      between excited states
      in, 37
Fluorenone
   molecular excitation energy
      of in solution, 27–28
Fluorescence
   of aromatic and related
      hydrocarbons, 25–29
   excited state fragment
      orientation and, 124
   high-pressure studies of,
      25–29
Fluorescence quenching
   electron transfer and, 382–84
   excited state transport and
      trapping in concentrated
      dye solutions and, 72–78
   protein dynamics and,
      288–89
Fluorescence spectra
   laser-induced
      polyatomic ions and,
         261–64
Fluorotoluene cation
   matrix absorption spectrum
      of, 258–59
Foldy-Wouthuysen
   transformations, 161
Forced oscillator
   photodissociation model,
      415–21
Franck-Condon effect, 356
Franck-Condon
   photodissociation model,
      415–21
Free energy minimization
   counterion condensation and,
      194
Free induction decay
   electron paramagnetic
      resonance spectra and,
         308

G

Galactose oxidase
   electron spin echo method
      and, 313
Generalized dilatation
   transformations, 243–44
Glasses
   echo behavior in, 310
Globular proteins
   nuclear magnetic resonance
      spectroscopy and,
         551–52
Glycerol
   effect of pressure on emission
      in, 37
   rhodamine 6G in
      dimer lifetimes of, 73
      effective decay constant vs
         concentration of, 76
      transient grating results
         for, 75
      tautomer emission in, 37
Glycine
   ^{13}C nuclear magnetic
      resonance spectra of,
         539
Ground state energy
   dilatation analyticity and,
      234

H

Halides
   transition metal
      $X\alpha$ calculations and,
         165–66
Halobenzene cations
   vibronic structure of, 273
Halocarbons
   bond dissociation energies
      and, 522–24
Halogenated hydrocarbons
   phosphorescence of, 29–30
Halogenation equilibria
   bond dissociation energies
      and, 497
Halogenation kinetics
   bond dissociation energies
      and, 496–97
Hartree-Fock orbitals, 152–54
Heats of formation
   bond dissociation energies
      and, 504–26
Hedin-Lundqvist function, 153
Heme iron
   dynamic msd of, 286
Heme proteins
   ligands binding to, 289–96
      low-temperature kinetics
         and, 290–93
      solvent viscosity and,
         293–96

Hemes
   liganded
      electron density marker
         line for, 475–79
      resonance Raman band
         frequencies of, 472
Hemoglobin
   iron-histidine stretching
      mode and, 480–86
   nuclear magnetic resonance
      spectroscopy and, 551
   time-resolved resonance
      Raman spectroscopy
         and, 471–90
Hemoglobin S
   nuclear magnetic resonance
      spectroscopy and, 551
Hemoproteins
   electron spin echo method
      and, 313
Heptamethylnonane
   electron donor-acceptor
      complexes in
         emission of, 38–40
Heterocyclic molecules
   high-pressure fluorescence
      studies for, 27
Heterojunctions
   photovoltaic phenomena in,
      447
   polyacetylene and, 449–50
Hexahelicene
   path of intersystem crossing
      in, 306
Hexamethyl ethane
   rate constant for
      decomposition of, 506
Hexane
   radiative rate of indole in
      effect of dielectric constant
         on, 34
High pressure molecular
   luminescence, 25–46
   as a biologic probe, 30–31
   environmental effects of,
      31–40
   excimer and exciplex
      emission and, 30
   fluorescence of aromatic and
      related hydrocarbons
         and, 25–29
   phosphorescence and,
      29–30
   pressure tuning of energy
      levels and, 40–46
Homonuclear diatomics
   multiple scattering
      deformation densities
         for, 158
Huxley-Simmons model, 320
Hydrides
   transfer of to carbonium
      ions, 503

SUBJECT INDEX 585

Hydrocarbon bond dissociation, 493–526
Hydrocarbons
  fluorescence of, 25–29
  see also Aromatic hydrocarbons
Hydrogen
  dc Stark Hamiltonian for, 237–38
  multiple scattering theory and, 158
Hydrogen atoms
  scattering of positronium from, 237
Hydroxyapatite
  nuclear magnetic resonance spectroscopy and, 546
3-Hydroxyflavone
  luminescence of, 37
Hydroxypyrenetrisulfonate
  dissociation yield of, 394–95

I

ICN
  photodissociation of, 432–33
Imidazole
  copper complexes with electron spin echo method and, 313
  iron-histidine stretching mode and, 480
Impulsive collision photodissociation model, 415–21
Indole
  emission behavior as function of pressure in, 33
  radiative rate of in solution effect of dielectric constant on, 34–37
  thermal dissipation of in solution, 34
Inert gas matrices
  absorption spectra of nonradiating polyatomic ions in, 258–59
Inorganic crystals
  luminescence of
  pressure tuning and, 41
Inorganic molecules
  high pressure studies of, 25
Iodination
  t-butyl radical thermochemistry and, 505–8
Iodoacetyl fluoresceine
  myosin segment orientation in muscle and, 333–34
Ion cyclotron resonance spectrometer
  nonradiating ions and, 259
Iridium
  tetranuclear dodecacarbonyl cluster formation and, 105
Iron compounds
  electron spin echo method and, 310
Iron-imidazole complexes
  electron spin echo method and, 313
Iron-sulfur proteins
  electron spin echo method and, 313
Isobutanol
  electron spin echo method and, 313
Isobutanol
  effect of pressure on emission in, 37
  tautomer emission in, 37

J

Jahn-Teller effects, 263
  vibronic structure of polyatomic ions and, 273–78

K

Ketones
  ^{13}C enrichment by photolysis of, 394
  ion cyclotron resonance photodissociation spectroscopy and, 259
Kinetic energy
  electron density and, 151–52
Knudsen cell reactor, 500
Koopman's theorem
  transition metal halides and oxohalides and, 165–66
Kramers-Heisenberg
  sum-over-states expression, 372–73
Kramers-Kronig transform
  resonance Raman scattering and, 368–69

L

Laccase
  electron spin echo method and, 313
Laser-induced fluorescence spectra
  of polyatomic ions, 261–64
Laser-induced phonons, 79–86
Lattice energy
  definition of, 21
Ligaments
  nuclear magnetic resonance spectroscopy and, 546
Ligand-metal-ligand interactions, 95
Ligands
  binding to heme proteins, 289–96
    low-temperature kinetics and, 290–93
    solvent viscosity and, 293–96
  as luminescent probes, 31
Ligand substitution reactions, 90–102
  of binuclear metal carbonyl complexes, 96–101
  of tetranuclear metal carbonyl clusters, 105–9
  of trinuclear metal carbonyl clusters, 101–5
Light absorption
  absorbing species concentration and, 306
Light scattering
  myosin fragment binding to actin and, 328
Linear muffin-tin orbital method, 160
Local spin density potential
  spin-up and spin-down potentials and, 153
Lumiflavin
  photochemical ionization reactions of, 387
Lysozyme
  B-parameters of, 288
  x-ray diffraction and, 285

M

MacMillan-Mayer cluster series expansion, 201
Malachite green in ethanol
  laser-induced phonon transient grating and, 81–83
Membrane protein complexes
  nuclear magnetic resonance spectroscopy and, 552
Membranes
  nuclear magnetic resonance spectroscopy and, 555–58
Metal carbonyl clusters
  ligand substitution reactions of, 96–109
Metal cluster reactions, 89–115
  ligand substitution, 90–109
  oxidative-addition, 109–15
  pathways of, 91
  reductive-elmination, 109–15
Metal complexes
  electron spin echo method and, 313
Metalloporphyrins

SUBJECT INDEX

nonradiative rates for
  fluorescence of, 27
thermal dissipation of in
  solution, 33–34
triplet-triplet and
  triplet-ground state
  reactions of, 387–89
Metal-metal bonds
  $X\alpha$ calculations and, 164–65
Metals
  in liquid ammonia
    electron spin echo method
    and, 310
Methane
  radiolysis of
    electron paramagnetic
    resonance and, 303
Methoxybenzenes
  aromatic molecule
    singlet-excited state
    quenching and, 383–84
5-Methoxyindole
  emission behavior as function
    of pressure in, 33
Methyl benzoate
  methyl hydrogens of
    bond dissociation energy
    of, 518
Methylcyclohexane
  electron donor-acceptor
    complexes in
    emission of, 38–40
  radiative
    effect of dielectric constant
    on, 34
Methylenes
  photoproduction of, 306
Methyl silanes
  bond dissociation energies of,
    524
Methyl tetrahydrofuran
  deuterated sodium hydroxide
    in
    electron spin echo method
    and, 314
Methyl-tetrahydrofuran
  complexes
  photocurrent and optical
    absorbance changes in,
    379–80
Metmyoglobin
  side-chain displacements at
    250 K, 287
Michaelson-Morley experiment,
  14
Microwave energy
  rate of absorption of
    evolution of in time,
    303–4
Molecular electronic structure
  real axis clamped nuclei and,
    245–47
Molecular electronic structures

$X\alpha$ method and, 151–67
  applications of, 162–66
  basis set expansion and,
    158–60
  exchange and correlation
    potentials and, 152–54
  multiple scattering theory
    and, 154–58
  relativistic calculations
    and, 160–62
Molecular ionization potentials
  basis set expansion and,
    161–62
Molecular predissociation,
  248–59
Monobromobenzene
  matrix absorption spectrum
    of, 258–59
Monochlorobenzene cation
  matrix absorption spectrum
    of, 258–59
Monofluorobenzene cation
  matrix absorption spectrum
    of, 258–59
Mononuclear metal complexes
  associative/dissociative
    reaction paths of,
    93–94
  ligand substitution reactions
    of, 95
  reactivity of
    coordination saturation
    and, 92–93
Mossbauer spectroscopy
  dynamic msd of heme iron
    and, 286
Multiphoton (ac Stark)
  ionization, 238
Multiple scattering theory,
  154–58
  metal-metal bonds and,
    165
  molecular properties and,
    157–58
  overlapping spheres and,
    155–56
  relativistic calculations with,
    160–61
  Rydberg and continuum
    states and, 156–57
  transition metal carbonyls
    and, 163–64
Muscle
  myosin segment orientation
    in, 330–35
  see also Contractile
    mechanism in muscle
Myoglobin
  carbon monoxide binding to,
    290–92
  iron-histidine stretching
    mode and, 480–82
  main chain structure of, 286

nuclear magnetic resonance
  spectroscopy and,
  551–52
oxygen binding to, 290
time-resolved resonance
  Raman spectroscopy
  and, 476
x-ray diffraction and, 285–88
Myosin
  binding of to actin, 328–30
  orientation of in organized
    muscle, 330–35
  S-1 segment of
    actin and ATPase binding
    sites of, 319
  segmental flexibility of,
    322–28
Myosin filaments
  muscle contraction and,
    319–20
Myosin S-1
  amino acid sequence of,
    340–47
  asymmetry of, 321
  spin-labeling of, 325
  trypsin cleavage of, 343

N

N-body Coulomb Hamiltonian
  dilatation analyticity of,
    229–33
Naphthalene cation
  matrix absorption spectrum
    of, 258–59
Neutron scattering
  inelastic
    entangled polymers and,
    57
Nitriles
  bond dissociation energies
    and, 520
Nitrogen-containing compounds
  bond dissociation energies
    and, 520–23
Nitrogen metabolism
  nuclear magnetic resonance
    spectroscopy and,
    552–53
Nitroxide spin labels
  electron spin echo
    spectroscopy and, 314
Nonadiabatic coupling
  excited molecular states and,
    359
Nonlinear triatomic molecules
  photodissociation of, 427–31
Nuclear magnetic resonance
  protein dynamics and, 289
  pulsed-field gradient
  self-diffusion coefficient in
    polydimethysiloxane
    melts and, 53

SUBJECT INDEX 587

self-diffusion of polystyrene
chains in carbon,
tetrachloride and, 54
Nuclear magnetic resonance
spectroscopy, 533–58
amino acids and, 538–44
biological, 533–34
biological solid-state, 537–38
membranes and, 555–58
nucleic acids and, 553–55
peptides and, 538–44
polysaccharides and, 558
proteins and, 544–53
solid-state, 534–37
Nucleic acids
conformational transitions
and binding equilibria
of, 193
nuclear magnetic resonance
spectroscopy and,
553–55
Nucleoprotein complexes
nuclear magnetic resonance
spectroscopy and,
547–51

O

OCS
photodissociation of, 434–35
Olefinic cations
internal conversion/
fragmentation vs
emission in, 267–68
light-emitting, 266, 269
Olefinic compounds
highest filled orbitals of, 267
Oligocations
interactions of with DNA
binding constants for,
217–19
Organic molecules
phosphorescence of, 29–30
Organometallic compounds
bond dissociation energies
and, 524–26
Organosilicon compounds
bond dissociation energies
and, 524
Orthonormality, 121–22
Oxidative-addition reactions
of metal clusters, 109–15
Oxohalides
transition metal
$X\alpha$ calculations and, 165–66
Oxygen
binding to myoglobin, 290
Oxygen-containing compounds
bond dissociation energies
and, 518
Ozone
$X\alpha$ scattered wave
calculations and, 162–63

P

Parity-favoredness quantum
number
for angular momentum
transfer, 142–43
Penning ionization, 262
Pentacene
absorption spectrum of in
p-terphenyl, 69–71
triplet state of
inhomogeneous line
breadths of, 306
Peptides
nuclear magnetic resonance
spectroscopy and,
538–44
Perturbation theory
dilatation analyticity and,
234
Perylene
excimer formation and, 30
Perylene complexes
ionic photodissociation of,
381
Perylene solutions
flash photolysis of
perylene anion radical
formation and,
382–83
Phenanthrene
bimolecular encounter of
with anthracene, 384
photoionization of
aromatic anion and cation
pair formation and,
398
Phenylic bonds
dissociation energies of,
514–15
Phospholipid bilayers
nuclear magnetic resonance
spectroscopy and,
555–56
Phospholipids
nuclear magnetic resonance
spectroscopy and,
555–58
Phosphorescence
high-pressure studies of,
29–30
Photodissociation
recoil direction of nuclei in,
145
Photodissociation dynamics,
409–40
amplitude calculation of,
421–27
cross-section for specific
transitions, 414
experimental comparisons of,
432–40

Franck-Condon, forced
oscillator, and impulsive
collision models of,
415–21
nonlinear triatomic, 427–31
observables in, 410–15
Photoelectrochemical junctions
photovoltaic phenomena in,
447
Photoexcitation
polyacetylene, 461–64
Rydberg atom formation
and, 174
Photoexcited triplets
electron spin echo
spectroscopy and,
315–16
Photofragment alignment and
orientation, 119–48
angular momentum transfer
and, 139–48
breakdown of cylindrical
symmetry and, 134–36
cylindrically symmetric
collision frame and,
130–34
dependence of on
parity-favoredness,
142–45
depolarization and, 136–39
emitted light and source
anisotropy and, 128–30
excitation-detection geometry
and, 126–28
Fano-Macek formulation of,
125–39
Photofragment angular
distribution
for electric dipole transitions,
120
Photoionization
anisotrophy parameter in,
120
Photoionization in solution,
377–404
bimolecular, 382–90
charge-transfer complexes
and, 378–82
ion escape and, 390–402
Photolysis
ligand substitution reactions
and, 96–97
Photoredox reactions
free ion formation in, 384
Photosynthesis
electronic excitation transfer
between chlorophyll
chromophores and, 72
Picosecond holographic grating
experiments, 63–86
excited state transport and
trapping in concentrated
dye solutions and, 71–78

optical generation and
ultrasonic wave
detection and, 78–86
probe wavelength dependence
near a strong transition
and, 67–71
schematic illustration of, 64
setup for, 65–67
Planar Poisson-Boltzmann
equation
Monte Carlo simulations of,
199
Poisson-Boltzmann
polyelectrolyte theory
development of, 193–200
thermodynamic consequences
of, 200–7
thermodynamic degree of
dissociation and, 213–14
Polyacetylene, 443–67
crystal structure of, 453
degenerate ground state of,
445
electrochemical doping and
battery applications of,
450–52
electronic properties of,
447–52
electron micrographs of, 446
lattice dynamics of, 458–59
mid-gap state of, 459–61
photoexcitations of, 461–64
polymer chain structures of,
444
Schottky diodes,
heterojunctions, and
solar cells and, 449–50
solitons in, 452–64
Polyatomic ions
light-emitting, 265–73
nonradiative decay of,
258–59
radiative decay of, 259–65
vibronic structure of, 273–78
Polyatomic molecules
bond dissociation energies of,
493–94
Polycrystalline materials
echo behavior in, 310
Polydimethysiloxane melts
self-diffusion coefficient in, 53
Polydispersity
entangled polymers and,
50–51
Polyelectrolyte solutions
counterion binding in,
207–15
physical degree of
dissociation in, 208–11
thermodynamic degree of
dissociation in, 211–14
Polyelectrolyte theories of
DNA, 191–219

counterion binding and,
207–15
equilibria and, 215–19
Poisson-Boltzmann and
counterion condensation,
193–207
Polyion configurations in
solution
electrostatic interactions on,
194
Polymerizations
radical
reaction rates for, 59
Polyisobutylene
fluorenone in
molecular excitation
energy of, 27–28
polyvinyl carbazole excimer
emission in, 30
Polymer melts
self-diffusion in, 52–53
Polymer/polymer welding,
56–57
Polymers
viscoelastic properties of,
49–51
see also Entangled polymers
Polymethylmethacrylate
fluorenone in
molecular excitation energy
of, 27–28
fracture energy for, 56–57
polyvinyl carbazole excimer
emission in, 30
Poly(4-methyl-l-pentene)
fluorenone in
molecular excitation
energy of, 27–28
Polynucleotides
conformational states of
order-disorder transition
between, 216–17
Poly(paraphenylene)
degenerate ground state of,
465
Polysaccharides
nuclear magnetic resonance
spectroscopy and, 558
Polystyrene
fluorenone in
molecular excitation
energy of, 27–28
nonradiative rates for
fluorescence of, 27
polyvinyl carbazole excimer
emission in, 30
Polystyrene chains
neutron studies on, 55–56
self-diffusion of in benzene
solution, 55
self-diffusion of in carbon
tetrachloride, 54
Polyvinyl carbazole

excimer emission in, 30
Porphyrins
triplet-triplet and
triplet-ground state
reactions of, 387–89
Positronium
scattering of from hydrogen
atoms, 237
Potential energy curves
for small molecules, 159
Potentiometric titration
molecular theory of
condensation and, 194
Preferential interaction
parameter
thermodynamic degree of
dissociation and, 211–13
Propylene carbonate
polyacetylene and
battery applications of, 451
Protein conformation
luminescence as biologic
probe of, 30–31
Protein dynamics, 283–97
fluorescence quenching and
depolarization and,
288–89
heme, 289–96
nuclear magnetic resonance
and, 289
x-ray diffraction and, 284–88
Proteins
nuclear magnetic resonance
spectroscopy and,
544–53
Pulsed electron beam excitation
quantum yields for
detectable-emission ions
and, 264
Pyrene
excimer formation and, 30
reaction of with
p-dicyanobenzene
conductance rise kinetics
of, 386
Pyrene/DEA system
flash photolysis of
biphasic decay kinetics of
flash-induced
absorbance and,
385–86
Pyrimidine mononucleotides
Raman excitation profile of,
367
Pyrolysis
very low-pressure
bond scission reaction rate
measurement and,
500–1
halogenation equilibria and,
497
Pyromellitic dianhydride
complexes

oxygen quenching and, 380–81
Pyromellitic dianhydride/mesitylene complex
charge-transfer excitation of, 378
Pyromellitic dianhydride solutions
irradiated
ion radical formation from, 378

Q

Quadratic vibronic coupling effects, 358
Quinoid
degenerate ground state of, 465
Quinones
photolysis of in alcohols
semiquinone formation and, 303
photolytic reaction between alcohols and
electron spin echo spectroscopy and, 315

R

Radical polymerizations
reaction rates for, 59
Raman excitation profiles
calculation of, 360–61
of carotene, 369
of copper tetraphenylporphyrin, 361
of cyanocobalamin, 369
of cytochrome c, 361
of ferrocytochrome c, 366
of pyrimidine mononucleotides, 367
transform method for determining, 369–70
Rare earths
electron spin echo method and, 313
Rayleigh light scattering
forced
self-diffusion of polystyrene in benzene solution and, 55
Rayleigh-Ritz theory
resonance eigenvalues and, 235
Rayleigh-Schrödinger perturbation theory
dilatation analyticity and, 234
Reactive lysyl residue
actin-binding and, 345

reactivity of to trinitrobenzene sulfonate, 346
Redox reactions
electron intermediates in, 302
Red shift
polarizability of the medium and, 26
Reductive-elimination reactions of metal clusters, 109–15
Reptation
entangled polymers and, 49–50
Rescigno-McCurdy ansatz, 240
Resonance Raman scattering, 353–75
excited molecular wavefunction calculation and, 359–61
specific approaches in, 359–67
sum-over-states and, 362–67
theories of, 355–59
transform methods and, 367–74
see also Time-resolved resonance Raman spectroscopy
Resonance states
computational algorithms locating, 236
Resonant rotational energy transfer
Rydberg atom collisions and, 175–82
Resonant states
exponential decay of, 224
time evolution of, 223–24
Resonant vibrational energy transfer
Rydberg atom collisions and, 182–84
Rhodamine 6G in ethanol
decay rate of, 77
dimer lifetimes of, 73
Rhodamine 6G in glycerol
dimer lifetimes of, 73
effective decay constant vs concentration of, 76
transient grating results for, 75
Rhodium
tetranuclear dodecacarbonyl cluster formation and, 105
RNA
viral, 547
Rotating wave approximation, 412
Rubidium
photolysis of
transient absorption and, 303

Rubredoxin
x-ray diffraction and, 285
Rydberg atom collisions, 173–88
attaching targets and, 184–88
experimental considerations and, 174–75
resonant rotational energy transfer and, 175–82
resonant vibrational energy transfer and, 182–84
Rydberg atoms
perturbation of, 174
Rydberg states
multiple scattering theory and, 156–57

S

Saturation-transfer electron paramagnetic resonance
myosin (S-1)-(S-2) flexibility and, 326
Schottky barrier junctions
photovoltaic phenomena in, 447
Schottky diodes
polyacetylene and, 449–50
Sedimentation analysis
myosin fragment binding to actin and, 328
Selective field ionization
Rydberg state population distribution analysis and, 174–75
Self-diffusion
entangled polymers and, 51–55
Semiquinones
photolysis of quinones in alcohols forming, 303
Serine protease
x-ray diffraction and, 285
Shock tube
bond scission reaction rate measurement and, 499–500
Sickle cell hemoglobin
nuclear magnetic resonance spectroscopy and, 551
Sila olefins
bond dissociation energies of, 524
Silanes
bond dissociation energies of, 524
Silver atoms
electron spin echo method and, 314
Single-pulse shock tube
bond scission reaction rate measurement and, 499–500

Solar cells
  polyacetylene and, 449–50
Solitons
  polyacetylene and, 452–64
Solvent viscosiy
  ligands binding to heme
    proteins and, 289–96
Spectral theory of dilatation
    analytic Hamiltonians,
    229–33
  theoretical applications of,
    233–34
Sperm
  nuclear magnetic resonance
    spectroscopy and,
    550–51
Spherical tensor operators, 121
Spinodal decomposition, 56
Stark-Zeeman effect, 238
Sulfur-containing compounds
  bond dissociation energies
    and, 518–20

T

Tautomer emission
  as function of pressure,
    37–38
Teeth
  nuclear magnetic resonance
    spectroscopy and, 546
p-Terphenyl
  pentacene in
    absorption spectrum of,
    69–71
Tetrachlorobutadiene
  excited state molecular
    structure and ultraviolet
    absorption spectrum of,
    362
Tetracyanobenzene
  complexed with aromatic
    hydrocarbons
    emission of, 38–40
Tetracyanobenzene aromatic
  hydrocarbon complexes
    fluorescence of, 378
Tetracyanobenzene complexes
  free ion quantum yields of,
    379
Tetracyanobenzene-toluene
  complex
    free ion detection in, 379
Tetrahydrofuran solutions
  flash irradiation of
    free ion formation and,
      385
Tetramethyl pentadecane
  electron donor-acceptor
    complexes in emission
    of, 38–40
Tetramethyl pyrazine

free induction decay and,
    315–16
Tetranuclear metal carbonyl
  clusters
    ligand substitution reactions
      of, 105–9
Thenoyltrifluoroacetylacetonate
  energy transfer process in,
    41, 45
Thermodynamic degree of
  dissociation
    Donnan coefficient and,
      211–13
    preferential interaction
      paramter and, 211–13
Thomas-Fermi-Dirac theory,
    151
Time-dependent perturbation
  theory
    dilatation analyticity and,
      234
Time-resolved fluorescence
  anisotropy decay
    binding of myosin fragments
      to actin and, 328–30
    myosin (S-1)-(S-2) flexibility
      and, 323–26
Time-resolved resonance
  Raman spectroscopy,
    471–90
  hemoglobin and, 471–90
  iron-histidine stretching
    mode and, 480–87
  nanosecond, 474–87
  picosecond, 472–74
  time evolution of, 487–88
Tobacco mosaic virus
  nuclear magnetic resonance
    spectroscopy and, 550
Toluene
  pyrene/perylene excimers in,
    30
Toluene cation
  matrix absorption spectrum
    of, 258–59
Toluene-tetracyanobenzene
  complex
    ion yield from, 398–99
Tomato bushy stunt virus
  nuclear magnetic resonance
    spectroscopy and, 547
Transition metal carbonyls
  $X\alpha$ scattered wave
    calculations and, 163–64
Transition metal clusters
  linear muffin-tin orbital
    method and, 160
Transition metal complexes
  spin distributions in, 166
Transition metal halides
  $X\alpha$ calculations and, 165–66
Transition metal oxohalides

$X\alpha$ calculations and, 165–66
Transition metals
  K-shell x-ray absorption
    spectra of, 157
  reversible oxidative-addition/
    reductive-elimination
    reactions of, 109
Transport properties
  molecular theory of
    condensation and, 194
Triatomic ions
  laser-induced fluorescence
    spectra of, 264
  light-emitting, 265–66
Triatomic molecules
  nonlinear
    photodissociation of,
    427–31
Trichlorobenzene
  laser-induced fluorescence
    spectra of, 262
Trienes
  ion cyclotron resonance
    photodissociation
    spectroscopy and, 259
Triiron dodecacarbonyl
  symmetry of, 102
Trinitrobenzene sulfonate
  reactive lysyl residue
    reactivity to, 346
Trinuclear metal carbonyl
  clusters
    ligand substitution reactions
      of, 101–5
Triphenylamine
  photoionization of as a
    function of electric field
    strength, 393
Triphenylmethane dyes
  luminescence efficiency of
    solvent viscosity and,
      31–32
Triplet states
  charge-transfer
    ion formation from,
    380–81
tRNA conformation
  pressure effects on, 31
Tropylium cation
  matrix absorption spectrum
    of, 258–59
Trypsin
  myosin S-1 cleavage and, 343
Tryptophan
  emission behavior as function
    of pressure in, 33
  fluorescence quenching of,
    288–89
  as a luminescent probe, 31
Tryptophane fluorescence
  myosin segment orientation
    in muscle and, 332

SUBJECT INDEX 591

Tyrosine
  as a luminescent probe, 31

U

Ultrasonic waves
  optical generation and detection of
    physical measurements and, 78–86
Ultraviolet light
  ICN photodissociation and, 432

V

Vanadium
  tetrahedal chlorides of
    ligand field theories and, 166
van der Waals complexes
  rotational predissociation of, 248–49
Very low-pressure pyrolysis
  bond scission reaction rate measurement and, 500–1
  halogenation equilibria and, 497
Vinyl carbazole solutions
  flash irradiation of
    free ion formation and, 385
Viruses
  nuclear magnetic resonance spectroscopy and, 547–48

W

Water
  dielectric variation of with pressure in, 34
  radiative rate of indole in
    effect of dielectric constant on, 34
  VUV photodissociation of, 436–40

X

Xα method, 151–67
  applications of, 162–66
  basis set expansion and, 158–60
  exchange and correlation potentials and, 152–54
  multiple scattering theory and, 154–58
  relativistic calculations with, 160–62
X-ray diffraction
  A-form DNA phosphate group orientation and, 553–54
  myosin segment orientation in, 330–31
  protein dynamics and, 284–88
Xylenes
  tetracyanobenzene complexed with
    emission of, 38–40

Z

Zinc octaethyl porphyrin
  ion yield from, 399–401
Zinc uroporphyrin
  triplet-state of
    electron-transfer quenching of, 386

CUMULATIVE INDEXES

CONTRIBUTING AUTHORS, VOLUMES 29–33

A

Albery, W. J., 31:227–63
Albrecht, A. C.,
  29:421–40;33:353–76
Alder, B. J., 32:311–29
Anderson, C. F., 33:191–222
Andrews, L., 30:79–101

B

Ballard, S. G., 33:377–407
Barisas, B. G., 29:141–66
Bartlett, R. J., 32:359–401
Bauer, S. H., 30:271–310
Benzinger, J., 29:285–306
Bersohn, R., 33:409–42
Beynon, J. H., 30:51–78
Bocian, D. F., 29:307–35
Borden, W. T., 30:125–53
Borejdo, J., 33:319–51
Botts, J., 33:319–51
Brenton, A. G., 30:51–78
Brochard, F., 32:433–51
Bryant, R. G., 29:167–88
Burch, R. R., 33:89–118

C

Cardillo, M. J., 32:331–57
Case, D. A., 33:151–71
Ceyer, S. T., 29:477–99
Champion, P. M., 33:353–76
Chan, S. I., 29:307–35
Chandler, D., 29:441–71
Chang, J. S., 30:443–69
Cole, R. H., 29:283–300
Cooke, R., 33:319–51
Corderman, R. R., 30:347–78
Cowley, J. M., 29:251–83

D

Davidson, E. R., 30:125–53
Debrunner, P. G., 33:283–99
de Gennes, P. G., 33:49–61
Deslattes, R. D., 31:435–61
Drickamer, H. G., 33:25–47
Duewer, W. H., 30:443–69
Dunning, F. B., 33:173–89

Dunning, T. H. Jr., 30:311–46
Durup, J., 32:53–76
Dykstra, C. E., 32:25–52

E

Ehrlich, G., 31:503–37
Eisenthal, K. B., 29:207–32
Etemad, S., 33:443–69

F

Fanconi, B., 31:265–91
Fayer, M. D., 33:63–87
Frauenfelder, H., 33:283–99
Frenkel, D., 31:491–521
Friedman, H. L., 32:179–204
Friedman, J. M., 33:471–91
Friedrich, D. M., 31:559–77
Frisch, H. L., 32:433–51

G

Gardiner, W. C. Jr., 31:377–99
Gill, S. J., 29:141–66
Goddard, W. A. III, 29:363–96
Golden, D. M., 33:493–532
Green, J. C., 29:161–83
Green, S., 32:103–38
Greene, C. H., 33:119–50
Greer, S. C., 32:233–65
Gutowsky, H. S., 31:1–27

H

Harding, L. B., 29:363–96
Harris, C. B., 29:473–95
Hay, P. J., 30:311–46
Heeger, A. J., 33:443–69
Hildebrand, J. H., 32:1–23
Honig, B., 29:31–57
Hyde, J. S., 31:293–317
Hynes, J. T., 29:301–21

J

Jaynes, E. T., 31:579–601
Johnson, P. M., 32:139–57

Johnson, W. C. Jr., 29:93–114
Jonas, J., 31:1–27

K

Karplus, M., 31:29–45
Kaufman, F., 30:411–42
Kepler, R. G., 39:497–518
Kivelson, D., 31:523–58
Kneba, M., 31:47–79
Koszykowski, M. L.,
  32:267–309
Krajinovich, D. J.,
  30:379–409
Kwok, H. S., 30:379–409

L

Lee, Y. T., 30:379–409
Léger, L., 33:49–61
Levine, R. D., 29:59–92
Levy, D. H., 31:197–225
Light, J. C., 31:401–33
Lineberger, W. C., 30:347–78
Lytle, F. W., 30:215–38

M

MacDiarmid, A. G.,
  33:443–69
Madden, P. A., 31:523–58
Madix, R. J., 29:295–306
Marcus, R. A., 32:267–309
Mauzerall, D., 33:377–407
Mayer, J. E., 33:1–23
McCammon, J. A., 31:29–45
McClain, W. M., 31:559–77
McDonald, J. D., 30:29–50
McMillen, D. F., 33:493–532
McTague, J. P., 31:491–521
Mendelson, R. A., 33:319–51
Miller, T. A., 33:257–82
Miller, W. G., 29:519–35
Moldover, M. R., 32:233–65
Morales, M. F., 33:319–51
Morgan, R. P., 30:51–78
Moseley, J., 32:53–76

CONTRIBUTING AUTHORS 593

Muetterties, E. L., 33:89–118
Mulliken, R. S., 29:1–30

N

Nachtrieb, N. H., 31:131–56
Nagle, J. F. 31:157–95
Noid, D. W., 32:267–309
Nozik, A. J., 29:189–222

O

Olson, D. B., 31:377–99
Ondrias, M. R., 33:471–91
Opella, S. J., 33:533–62
Otis, C. E., 32:139–57
Oxtoby, D. W., 32:77–101

P

Parkhurst, L. J., 30:503–46
Parks, E. K., 30:179–213
Patel, D. J., 29:337–62
Pechukas, P., 32:159–77
Philpott, M. R., 31:97–129
Pocker, Y., 30:579–95
Pollock, E. L., 32:311–29

R

Record, M. T. Jr., 33:191–222
Reinhardt, W. P., 33:223–55
Rousseau, D. L., 33:471–91

S

Sandstrom, D. R., 30:215–38
Saner, K., 30:155–78
Saykally, R. J., 32:403–31
Schulz, P. A., 30:379–409
Schwendeman, R. H., 29:527–58
Scoles, G., 31:81–96
Shapiro, M., 33:409–42
Shen, Y. R., 30:379–409
Shoemaker, R. L., 30:239–70
Stebbings, R. F., 33:173–89
Stolt, K., 31:603–37
Stolzenberg, A. M., 33:89–118
Sudbø, Aa. S., 30:379–409
Swofford, R. L., 29:421–40

T

Takashi, R., 33:319–51
Thomas, D. D., 31:293–317
Troe, J., 29:223–50
Tully, J. C., 31:319–43

V

Vaughan, R. W., 29:397–419
Vaughan, W. E., 30:103–24
Vilches, O. E., 31:463–90

W

Wadt, W. R., 30:311–46
Walker, R. B., 31:401–33
Weinberg, W. H., 29:115–39
Weissman, M. B., 32:205–32
Weissman, S. I., 33:302–18
Wertheim, M. S., 30:471–501
Wexler, S., 30:179–213
Williams, C., 32:433–51
Wilson, E. B., 30:1–27
Wolfrum, J., 31:47–79
Wolynes, P. G., 31:345–76
Woods, R. C., 32:403–31

Y

Yip, S., 30:547–77

Z

Zare, R. N., 33:119–50
Zwemer, D. A., 29:473–95

CHAPTER TITLES, VOLUMES 29–33

BIOPHYSICAL CHEMISTRY

Light Energy Transduction in Visual Pigments and Bacteriorhodopsin	B. Honig	29:31–57
Circular Dichroism Spectroscopy and the Vacuum Ultraviolet Region	W. C. Johnson Jr.	29:93–114
Microcalorimetry of Biological Systems	B. G. Barisas, S. J. Gill	29:141–66
NMR Studies of Membrane Structure and Dynamics	D. F. Bocian, S. I. Chan	29:307–35
High Resolution NMR Studies of the Structure and Dynamics of tRNA in Solution	D. J. Patel	29:337–62
Hemoglobin and Myoglobin Ligand Kinetics	L. J. Parkhurst	30:503–46
Simulation of Protein Dynamics	J. A. McCammon, M. Karplus	31:29–45
Theory of the Main Lipid Bilayer Phase Transition	J. F. Nagle	31:157–95
Saturation-Transfer Spectroscopy	J. S. Hyde, D. D. Thomas	31:293–317
Polyelectrolyte Theories and Their Applications to DNA	C. F. Anderson, M. T. Record, Jr.	33:191–222
Dynamics of Proteins	P. G. Debrunner, H. Frauenfelder	33:283–99
Some Physical Studies of the Contractile Mechanism in Muscle	M. F. Morales, J. Borejdo, J. Botts, R. Cooke, R. A. Mendelson, R. Takashi	33:319–51
Time-Resolved Resonance Raman Studies of Hemoglobin	J. M. Friedman, D. L. Rousseau, M. R. Ondrias	33:471–91
Solid State NMR of Biological Systems	S. J. Opella	33:533–62

CHEMICAL KINETICS—GAS PHASE

Atom and Radical Recombination Reactions	J. Troe	29:223–50
Unimolecular Ion Decomposition	A. G. Brenton, R. P. Morgan, J. H. Beynon	30:51–78
Molecular Beam Studies of Collisional Ionization and Ion-Pair Formation	S. Wexler, E. K. Parks	30:179–213
Four Center Metathesis Reactions	S. H. Bauer	30:271–310
Kinetics of Thermal Gas Reactions With Application to Stratospheric Chemistry	F. Kaufman	30:411–42
Modeling Chemical Processes in the Stratosphere	J. S. Chang, W. H. Duewer	30:443–69
Chemical Kinetics of High Temperature Combustion	W. C. Gardiner Jr., D. B. Olson	31:377–99
Transition State Theory	P. Pechukas	32:159–77
Collisions of Rydberg Atoms with Molecules	F. B. Dunning, R. F. Stebbings	33:173–89

CHEMICAL KINETICS—PHOTOCHEMISTRY AND RADIATION CHEMISTRY

Photosynthesis—The Light Reactions	K. Sauer	30:155–78

CHEMICAL KINETICS—REACTION DYNAMICS

Information Theory Approach to Molecular Reaction Dynamics	R. D. Levine	29:59–92
Creation and Disposal of Vibrational Energy in Polyatomic Molecules	J. D. McDonald	30:29–50
Reactive Molecular Collisions	R. B. Walker, J. C. Light	31:401–33
Fast Ion Beam Photofragment Spectroscopy	J. Moseley, J. Durup	32:53–76

Quasiperiodic and Stochastic Behavior in Molecules	D. W. Noid, M. L. Koszykowski, R. A. Marcus	32:267–309
Photofragment Alignment and Orientation	C. H. Greene, R. N. Zare	33:119–50
Theories of the Dynamics of Photodissociation	M. Shapiro, R. Bersohn	33:409–42

CHEMICAL KINETICS—SOLUTIONS (CONDENSED PHASE)

Homogeneous Catalysis in Solution	Y. Pocker	30:579–95
The Application of the Marcus Relation to Reactions in Solution	W. J. Albery	31:227–63

ELECTROCHEMISTRY

Photoelectrochemistry: Applications to Solar Energy Conversion	A. J. Nozik	29:189–222
Dynamics of Electrolyte Solutions	P. G. Wolynes	31:345–76
Ionization in Solution by Photoactivated Electron Transfer	D. Mauzerall, S. G. Ballard	33:377–407

GEOCHEMISTRY AND COSMOCHEMISTRY

Interstellar Chemistry: Exotic Molecules in Space	S. Green	32:103–38

LASER CHEMISTRY, ENERGY TRANSFER AND RELAXATION

Theoretical Studies of Molecular Electronic Transition Lasers	P. J. Hay, W. R. Wadt, T. H. Dunning Jr.	30:311–46
Multiphoton Dissociation of Polyatomic Molecules	P. A. Schulz, Aa. S. Sudbø, D. J. Krajnovich, H. S. Kwok, Y. R. Shen, Y. T. Lee	30:379–409
Bimolecular Reactions of Virbationally Excited Molecules	M. Kneba, J. Wolfrum	31:47–79
Molecular Multiphoton Spectroscopy with Ionization Detection	P. M. Johnson, C. E. Otis	32:139–57
Dynamics of Molecules in Condensed Phases: Picosecond Holographic Grating Experiments	M. D. Fayer	33:63–87

LIQUID STATE—SIMPLE FLUIDS

Renormalized Kinetic Theory of Dense Fluids	S. Yip	30:547–77

LIQUID STATE—SOLUTIONS OF ELECTROLYTES; FUSED SALTS

Conduction in Fused Salts and Salt-Metal Solutions	N. H. Nachtrieb	31:131–56
Electrolyte Solutions at Equilibrium	H. L. Friedman	32:179–204

LIQUID STATE—STRUCTURE

NMR Relaxation Studies of Solute-Solvent Interactions	R. G. Bryant	29:167–88
Structures of Molecular Liquids	D. Chandler	29:441–71
Computer Simulations of Freezing and Supercooled Liquids	D. Frenkel, J. P. McTague	31:491–521
Simulation of Polar and Polarizable Fluids	B. J. Alder, E. L. Pollock	32:311–29

MAGNETIC RESONANCE (ELECTRON SPIN, NUCLEAR, QUADRUPOLE)

High Resolution, Solid State NMR	R. W. Vaughan	29:397–419
Recent Developments in Electron Paramagnetic Resonance: Transient Methods	S. I. Weissman	33:301–18

MISCELLANEOUS

The Avogadro Constant	R. D. Deslattes	31:435–61

MOLECULAR STRUCTURE

Spectroscopy of Molecular Ions in Noble Gas Matrices	L. Andrews	30:79–101

High Resolution Spectroscopy of Molecular Ions	R. J. Saykally, R. Claude Woods	32:403–31
Light and Radical Ions	T. A. Miller	33:257–82

PHYSICAL ORGANIC

Thermal Rearrangements	J. A. Berson	28:111–32

PHYSICAL PHENOMENA—MISCELLANEOUS

Dielectric Relaxation	W. E. Vaughan	30:103–24
Fluctuation Spectroscopy	M. B. Weissman	32:205–32
Thermodynamic Anomalies at Critical Points of Fluids	S. C. Greer, M. R. Moldover	32:233–65

POLYMERS AND MACROMOLECULES

Stiff Chain Polymer Lyotropic Liquid Crystals	W. G. Miller	29:519–35
Molecular Vibrations of Polymers	B. Fanconi	31:265–91
Dynamics of Entangled Polymer Chains	P. G. de Gennes, L. Léger	33:49–61
Polyacetylene, $(CH)_x$: The Prototype Conducting Polymer	S. Etemad, A. J. Heeger, A. G. MacDiarmid	33:443–69

PREFATORY CHAPTERS

Chemical Bonding	R. S. Mulliken	29:1–30
Molecular Spectroscopy	E. B. Wilson	30:1–27
NMR in Chemistry—An Evergreen	J. Jonas, H. S. Gutowsky	31:1–27
A History of Solution Theory	J. H. Hildebrand	32:1–23
The Way It Was	J. E. Mayer	33:1–23

QUANTUM CHEMISTRY

The Description of Chemical Bonding from Ab Initio Calculations	W. A. Goddard III, L. B. Harding	29:363–96
Singlet-Triplet Energy Separations in Some Hydrocarbon Diradicals	W. T. Borden, E. R. Davidson	30:125–53
Potential Energy Barriers in Unimolecular Rearrangements	C. E. Dykstra	32:25–52
Many-Body Perturbation Theory and Coupled Cluster Theory for Electron Correlation in Molecules	R. J. Bartlett	32:359–401
Electronic Structure Calculations Using the $X\alpha$ Method	D. A. Case	33:151–71
Complex Coordinates in the Theory of Atomic and Molecular Structure and Dynamics	W. P. Reinhardt	33:223–55

SCATTERING PHENOMENA—DYNAMICAL

Light Scattering Studies of Molecular Liquids	D. Kivelson, P. A. Madden	31:523–58

SCATTERING PHENOMENA—STRUCTURAL

Developments in Extended X-Ray Absorption Fine Structure Applied to Chemical Systems	D. R. Sandstrom, F. W. Lytle	30:215–38

SOLIDS AND ORDERED ARRAYS—STRUCTURE AND DYNAMICS

Inelastic Electron Tunneling Spectroscopy: A Probe of the Vibrational Structure of Surface Species	W. H. Weinberg	29:115–39
Coherent Energy Transfer in Solids	C. B. Harris, D. A. Zwemer	29:473–95

SPECTROSCOPY—ELECTRONIC AND PHOTOELECTRONIC

Nonlinear Spectroscopy	R. L. Swofford, A. C. Albrecht	29:421–40
Negative Ion Spectroscopy	R. R. Corderman, W. C. Lineberger	30:347–78
Optical Reflection Spectroscopy of Organic Solids	M. R. Philpott	31:97–129
Two-Photon Molecular Electronic Spectroscopy	D. M. Friedrich, W. M. McClain	31:599–77

High Pressure Studies of Molecular Luminescence	H. G. Drickamer	33:25–47
SPECTROSCOPY—INFRARED AND RAMAN		
Coherent Transient Effects in Optical Spectroscopy	R. L. Shoemaker	30:239–70
Laser Spectroscopy of Cold Gas-Phase Molecules	D. H. Levy	31:197–225
Resonance Raman Scattering: The Multimode Problem and Transform Methods	P. M. Champion, A. C. Albrecht	33:353–76
SPECTROSCOPY—MICROWAVE		
Transient Effects in Microwave Spectroscopy	R. H. Schwendeman	29:537–58
STATISTICAL MECHANICS		
Equilibrium Statistical Mechanics of Polar Fluids	M. S. Wertheim	30:471–501
Two-Body, Spherical, Atom-Atom, and Atom-Molecule Interaction Energies	G. Scoles	31:81-96
The Minimum Entropy Production Principle	E. T. Jaynes	31:579–601
Vibrational Relaxation in Liquids	D. W. Oxtoby	32:77–101
Simulation of Polar and Polarizable Fluids	B. J. Alder, E. L. Pollock	32:311–29
Polymer Collapse	C. Williams, F. Brochard, H. L. Frisch	32:433–51
SURFACES—ADSORPTION AND CATALYSIS		
Kinetic Processes on Metal Single-Crystal Surfaces	R. J. Madix, J. Benziger	29:285–306
SURFACES—STRUCTURE AND DYNAMICS		
High Resolution Electron Microscopy of Crystal Defects and Surfaces	J. M. Cowley	29:251–83
Theories of the Dynamics of Inelastic and Reactive Processes at Surfaces	J. C. Tully	31:319–43
Phase Transitions in Monomolecular Layer Films Physisorbed on Crystalline Surfaces	O. E. Vilches	31:463–90
Surface Diffusion	G. Ehrlich, K. Stolt	31:603–37
Gas-Surface Interactions Studied with Molecular Beam Techniques	M. J. Cardillo	32:331–57
Molecular Features of Metal Cluster Reactions	E. L. Muetterties, R. R. Burch, A. M. Stolzenberg	33:89–118
THERMOCHEMISTRY AND THERMODYNAMICS		
Piezoelectricity, Pyroelectricity, and Ferroelectricity in Organic Materials	R. G. Kepler	29:497–518
Hydrocarbon Bond Dissociation Energies	D. F. McMillen, D. M. Golden	33:493–532

DATE DUE

DEC 9 1982

Syringomyelia

H. J. M. BARNETT, MD, FRCP(C)
Professor and Chairman, Division of Neurology, Department of Clinical Neurological Sciences, University of Western Ontario; Chairman of Neurology, University Hospital, London, Canada.

J. B. FOSTER, MD, FRCP Lond.
Honorary Lecturer in Neurology, University of Newcastle upon Tyne; Consultant Neurologist, Newcastle University Hospital Group and the Regional Neurological Centre, Newcastle General Hospital.

P. HUDGSON, MRCP, FRACP
Principal Research Associate, Department of Neurology, University of Newcastle upon Tyne; Consultant Neurologist, Newcastle University Hospital Group.

1973

W. B. Saunders Company Ltd London · Philadelphia · Toronto

W. B. Saunders Company Ltd: 12 Dyott Street
London WC1A 1DB

West Washington Square
Philadelphia, Pa. 19105

833 Oxford Street
Toronto 18, Ontario

Syringomyelia

ISBN 0-7216-1565-1

© 1973 by W. B. Saunders Company Ltd. All rights reserved. This book is protected by copyright. No part of it may be reproduced, stored in a retrieval system, or transmitted in any form or by any means, electronic, mechanical, photocopying, recording, or otherwise, without written permission from the publisher. Library of Congress Catalog Card Number 72-95826.

Text set in 'Monophoto' Bembo 270 by Keyspools Ltd, Golborne, Lancashire

Print No: 9 8 7 6 5 4 3 2 1

Editor's Foreword

In an era of explosive medical publication, some may feel that special justification is needed for launching yet another series of specialised monographs upon the medical world. At a time when new journals and serial publications devoted to the neurosciences continue to appear almost annually, and when textbooks of neurology of varying size and scope continue to proliferate and to multiply, it is reasonable to ask what purpose can usefully be served by publishing a series of specialised monographs on neurological topics.

As Series Editor, I believe that neurologists and many of those working in the neurosciences throughout the world will accept that these volumes will more than answer this question and will serve as their own justification. The monograph has always held a special place in the medical publishing world and this is especially true in neurology. It has been and will be my purpose and that of the many authors who have agreed to participate in writing books in this series, to bring together some important recent advances in neurological knowledge arising out of clinical, investigative and fundamental research studies of a multi-disciplinary nature. This principle is underlined by the first two books to appear in this series, devoted respectively to syringomyelia on the one hand and muscle biopsy on the other. Both volumes are the products of fruitful transatlantic collaboration. The first demonstrates the importance of close collaboration and consultation between neurologists, neuroradiologists, neurosurgeons and neuropathologists in throwing light upon the many manifold clinical presentations of spinal cord cavitation, a problem of growing importance and concern and one in which recent developments have at least pointed the way to effective treatment in many cases. The second volume, shortly to appear from the pens of Professor Dubowitz and Dr Brooke, will testify to important recent advances in techniques of studying muscle biopsy samples and to the contributions which such studies have made to the diagnosis and management of neuromuscular disease.

I am happy that Dr Drake, whose neurosurgical reputation is such that he requires no introduction from myself, has agreed to write a scientific foreword to this first volume on 'Syringomyelia' by Drs Barnett, Foster and Hudgson. This first exciting volume presages the appearance of a series of other works, each similarly compact, which will appear approximately biannually, each devoted to some new and important development in clinical neurology and in the related neurosciences. I am confident that these books will fulfil a genuine need in presenting succinctly some important developments in the growing points in these disciplines. In my opinion this first volume is an admirable introduction to the series.

Newcastle upon Tyne, 1973 John N. Walton

Foreword

This book encompasses the story of syringomyelia and is unique for it is the only treatise on the subject since Schlesinger's volume in 1902 and the first comprehensive coverage of the subject in the English language. The divided authorship combines the knowledge and flavour from each side of the Atlantic.

The contributions of Foster and Hudgson from Newcastle stemmed largely from their extensive experience with this disorder from both a medical and surgical point of view. These workers had previously delineated basal arachnoiditis as another form of communicating syrinx. The senior Canadian author, because of his association with Lyndhurst Lodge in Toronto, noted and developed an early interest in the study of the post-traumatic spinal cyst. The careful follow-up arrangements at Lyndhurst Lodge of nearly 1000 paraplegics kept the Registry and their neurological supervision up to date such that when the post-traumatic form emerged, the cases were available for study and follow-up. The availability of the pathological examinations by Rewcastle, Ball and Blackwood of the post-traumatic as well as the communicating and tumour varieties has led to considerable clarification of a confusing clinical problem.

Most satisfactory is the completeness of the book. All the cystic lesions are reviewed in depth from the classic syringomyelia through that associated with arachnoiditis and tumours to the less well known post-traumatic cystic myelopathy of paraplegia. Neurologists and neurosurgeons have felt keenly the lack of a comprehensive modern review of these disorders particularly in regard to their natural life history and the results of steps taken to modify it.

Communicating and non-communicating forms of syrinx are now described. The hydrodynamic theory of Gardner caught the fancy of neurosurgeons particularly, for it provided a new concept for treatment, away from the direct approach on the syrinx which left much to be desired. Even so, it seems that foraminal occlusion may not be the whole story and in the epilogue it is suggested that other portals along the Virchow–Robin spaces may be conduits for cystic myelopathy. It appears doubtful that syringomyelia can be regarded as a single entity either from the point of view of clinical presentation, from the viewpoint of pathogenesis, or, most importantly, from the standpoint of rational therapy.

Improved radiology and isotope studies have added to the diagnostic capability and to the therapeutic challenge. These contributions now need to be related to the pathology. This volume brings our knowledge of the spinal cord cysts up to date and will become the basis for future ventures into the nature and treatment of the disorder.

London, Ontario, 1973   Charles G. Drake

Preface

The majority of the dramatic neurological disorders were described in elegant fashion by the classical neurologists and morbid anatomists of Britain, Europe and America. This 'descriptive period' of neurology came and went like a Yukon gold rush and reaped a rich harvest for the diligent, the persistent and the fortunate. Like this and other major eruptive types of human activity it left much that was unsettled, confused and, in some situations, things that were temporarily forgotten—all to be dealt with by those to follow. Much of the neurological energy of the generation after this descriptive phase was channelled into the management of epilepsy and infection as well as into the development of neurosurgical methods particularly to deal with tumours and trauma.

The present generation of neurologists is concerning itself with further enquiry into some of the many unsolved problems. Basic knowledge necessary to complete understanding of the nature of many of these disorders has been much slower in its arrival than might have been anticipated from the great ferment of the descriptive period. New thinking and new technical advances make it apparent, however, that a particular subject is due from time to time for careful scrutiny and critical reappraisal. At times the advance has consisted simply of the recognition of the fact that an entity has been erroneously accepted as part of another disorder. This was the situation, for example, with the carpal tunnel syndrome, which reposed with several other misplaced disorders within the framework of 'thoracic outlet syndromes'. Other disorders have had to be tackled afresh and put into perspective by major biochemical, pharmacological, radiological or technical neurosurgical advances. Wilson's disease, Parkinson's disease, cervical spondylosis, subarachnoid haemorrhage, all stand as examples of these respective disorders which have had to be re-examined.

Syringomyelia is not common by comparison with some of the disorders mentioned, but it has challenged neurologists and neuropathologists for well over a century—for 135 years under this name. Modern neuroradiology, isotope-scanning methods and diligent persistence on the part of neurosurgeons attempting its treatment have caused many neurologists to revive their interests in this old conundrum and to review the state of our knowledge of this condition. As happens often under such circumstances, we become aware of gaps in our knowledge, differences in opinion regarding traditional concepts, and yet an awareness that in some areas of the disorder under intensive study there are some margins coming into clearer focus.

We believe that syringomyelia in the traditional sense of a progressive disease of the spinal cord associated with dissociated sensory loss, painless ulcers, neuropathic joints, and pathological evidence of a centrally located cord cavitation, must now be regarded as a generic term for a group of disorders. The two British co-authors of this monograph have examined the two major conditions productive of syringomyelia, where there is reason to believe that some of the basis for the condition is to be found in a hydrodynamic disorder of the cerebrospinal fluid pathways. Of these two 'communicating syringomyelias', one is associated with developmental anomalies of the hindbrain and foramen magnum region, and the other with chronic arachnoiditis of the basal cisterns. The Canadian contribution considers cord cavitation where evidence of ventricular or subarachnoid communication is absent. This includes syrinx development as a sequel to severe spinal cord injury, and a much smaller group following less impressive blows to the spine. The rare development of spinal cord cavitation with arachnoiditis confined to the *spinal* meninges is reviewed. Finally, the intriguing association of syringomyelia with tumours involving the central nervous system is discussed. Taken altogether, a serious attempt has been made to use our own material and that in the literature to bring into perspective the many factors that have been alluded to in the confused and lengthy bibliography relating to syringomyelia.

Confusion of terms creeps into any condition being described by many writers over a great many years. Concepts change and ideas shift, as successive generations try to wrestle with disorders of unknown cause. In an attempt to clarify this aspect of the problem we have included an historical review relating to each aspect of the condition.

No monograph in the English language has ever been written on syringomyelia, and it is 70 years since a monograph in any language has been devoted to the subject. For this reason it is considered reasonable that the subject be covered in fair detail. It is our hope that this review will serve as a reference point for some years in the continuing study of the varieties of syringomyelia, its pathogenesis and treatment. Like all monographs that are devoted to a reasonably uncommon condition of the nervous system, it is prepared primarily for the benefit of the neurologists, neurosurgeons, neuropathologists and neuroradiologists who are called on to study and treat patients with spinal cord disease. It is obvious that some aspects of the disorder will interest orthopaedic surgeons and physicians in physical medicine. It is hoped that it will be of special interest to those charged with the care of patients with spinal cord injury.

We are including a final chapter which briefly notes some of the problems that are yet to be dealt with for a complete understanding of the varieties of syringomyelia. The incompleteness of our knowledge and understanding of the condition is exemplified by the fact that there is some difference of opinion between the Canadian and British contributors about the pathogenesis of the condition. The British contributors have come out strongly in favour of the 'idiopathic' variety, that is, the variety unassociated with spinal arachnoiditis, spinal trauma or spinal cord tumour as representing a pre-existing or a currently existing communication between the central cord and the syrinx. Evidence in the material presented in the early chapters lends some credence to this. But even here, and particularly in the later chapters, there is conflict making its complete acceptance a matter for further consideration. This problem cannot be resolved in the light of present knowledge and information, so it is left to stand as a major area of investigation for future assessment and research.

The British authors wish to record their gratitude to their neurosurgical colleagues and especially to Professor John Hankinson who has been closely associated with them in this study and has been responsible for the surgical management of the majority of the cases here presented. We also wish to thank our neurological colleagues for allowing us to examine their cases and report their findings and record our warm appreciation for the constant help given to us by our colleagues Dr G. L. Gryspeerdt and Dr A. Appleby who have been responsible for the neuroradiological investigation of our series. We would express our thanks to Dr Ennis of Dryburn Hospital who supplied us with pathological material and to the Department of Medical Photography of the University of Newcastle upon Tyne for their help in preparing the illustrations. We are grateful to the Wellcome Museum of the History of Medicine for supplying the photograph of Ollivier d'Angers and to the editors of the *Annals of the Royal College of Surgeons* and Mr E. J. Newton for permission to reproduce it. To the editor of *Brain* and Macmillan & Co., publishers, we are grateful for permission to reproduce illustrations 4.7, 5.9, 5.10, 5.13, 5.16, 7.5, and 7.7 to 7.14 and to Butterworths & Co. (Publishers) Ltd and the editor of *Modern Trends in Neurology, Volume 5* for permission to reproduce some of the surgical illustrations from Professor John Hankinson's chapter in that volume. To the editor of the *Journal of Neurological Sciences*, Professor J. N. Walton and to the Elsevier Publishing Company Ltd we are grateful for permission to reproduce Figures 4.3, 4.4, 4.5 and 4.6. Illustrations 7.1, 7.2, 7.3 and 7.4 are taken from Greenfield's *Neuropathology* and we are grateful to Professor W. Blackwood and Edward Arnold (Publishers) Ltd for permission to reproduce them. Mr B. Williams and Hospital Medicine Publications Ltd allowed us to reproduce Figures 7.23 and 8.5 from the *British Journal of Hospital Medicine* and Mr Williams and Spastics International Medical Publications, London, the publishers of *Developmental Medicine and Child Neurology* granted permission for the use of Figure 8.6. Mr Williams and the editor and publishers of *The Lancet* are thanked for Figure 8.4 a and b. Figures 8.1 and 8.2 are reproduced by kind permission of Dr W. J. Gardner and the British Medical Association, publishers of the *Journal of Neurology, Neurosurgery and Psychiatry*. Figure 8.3 is reproduced from the *Journal of Neurosurgery* by permission of the editor, the publishers and Dr L. W. Conway. The task of typing the manuscript has been shared by the secretaries of the Regional Neurological Centre and we acknowledge their assistance with gratitude.

The senior Canadian author is grateful to Dr William Geisler and Dr Megan Wynne-Jones who have provided useful statistical information on the spinal cord injuries and arranged follow-up on some of the post-traumatic paraplegic cases. We are deeply indebted to Professor William Blackwood who provided the pathological material for one of the postmortem cases in the paraplegic series, and to Dr T. J. Murray who worked in Dr Blackwood's laboratory and prepared serial sections most meticulously of this important case. Pathological material was also kindly supplied or reviewed by Dr I. Feigin and Dr John Kaufmann. We are also especially grateful to Drs C. G. Drake and John Allcock for reviewing the manuscript and offering helpful criticisms. Case records pertinent to different aspects of the disorder were kindly supplied by Drs R. R. Tasker. Wm. M.

Lougheed, T. P. Morley, C. G. Drake, R. G. Vanderlinden, Wm. S. Coxe, P. B. Allen. Neuroradiological colleagues, Drs John M. Allcock, George Wortzman and D. L. McRae kindly allowed us the use of the radiographs and one was sent by Dr A. Appleby for which we are grateful. Dr M. J. Hollenberg provided us with the scanning electron microscope photographs of the syrinx in the post-traumatic injury—Chapter 14. We are indebted to Mr Michael Donnelly of the Department of Visual Medical Education at Victoria Hospital, London, Canada, who was very painstaking about producing the illustrations and to Miss Helen Rudski for a number of the line drawings. Mrs Phyllis Doucette and Miss Doris Tanter both rendered invaluable and patient secretarial assistance which is acknowledged. Miss H. Schlemmer undertook lengthy German translations. Mr G. Sawa, as a Medical Research Council summer student did an exhaustive search of the literature and a review of the files in the Banting Institute. Mrs I. T. Borda verified the references.

London, Ontario and H. J. M. Barnett
Newcastle upon Tyne J. B. Foster
January 1973 P. Hudgson

Contents

Editor's Foreword		v
Foreword		vii
Preface		ix
Contributors		xv

Section 1: Communicating Syringomyelia

1	Historical Introduction	3
2	Traditional Concepts of Syringomyelia	11
3	The Clinical Features of Communicating Syringomyelia	16
4	Basal Arachnoiditis	30
5	The Radiology of Communicating Syringomyelia	50
6	The Surgical Treatment of Communicating Syringomyelia	64
7	The Pathology of Communicating Syringomyelia	79
8	The Pathogenesis of Communicating Syringomyelia	104

Section 2: Non-communicating Syringomyelia

9	Trauma and Syringomyelia	127
10	Syringomyelia as a Late Sequel to Traumatic Paraplegia and Quadriplegia – Clinical Features	129
11	Nature, Prognosis and Management of Post-traumatic Syringomyelia	154

12	Case Records of Authors' Series of Post-traumatic Syringomyelia	165
13	Syringomyelia Consequent on Minor to Moderate Trauma	174
14	Pathology and Pathogenesis of Progressive Cystic Myelopathy as a Late Sequel to Spinal Cord Injury	179
15	Syringomyelia Associated with Spinal Arachnoiditis	220
16	The Pathogenesis of Syringomyelic Cavitation Associated with Arachnoiditis Localised to the Spinal Canal	245
17	Syringomyelia and Tumours of the Nervous System	261
18	Epilogue	302
	Index	315

Contributors

H. J. M. Barnett, MD, FRCP(C), Professor and Chairman, Division of Neurology, Department of Clinical Neurological Sciences, University of Western Ontario; Chairman of Neurology, University Hospital, London, Ontario, Canada.

M. J. Ball, MD, FRCP(C), Assistant Professor, Neuropathology and Clinical Neurological Sciences, University of Western Ontario; Staff Neuropathologist, University Hospital, Victoria Hospital and Westminster Hospital, London, Ontario, Canada.

J. B. Foster, MD, FRCP Lond., Honorary Lecturer in Neurology, University of Newcastle upon Tyne; Consultant Neurologist, Newcastle University Hospital Group and the Regional Neurological Centre, Newcastle General Hospital.

P. Hudgson, MRCP, FRACP, Principal Research Associate, Department of Neurology, University of Newcastle upon Tyne; Consultant Neurologist, Newcastle University Hospital Group.

A. T. Jousse, BA, MD, FRCP(C), LLD(Q), Professor, Department of Rehabilitation Medicine, University of Toronto; Medical Director, Lyndhurst Lodge Hospital, Toronto; Department of Physical Medicine and Rehabilitation, Toronto General Hospital, Canada.

N. B. Rewcastle, FRCP(C), Professor and Head, Division of Neuropathology, Department of Pathology, University of Toronto; Senior Pathologist and Chief of Neuropathology Division, Toronto General Hospital; Consultant Neuropathologist, Sunnybrook Hospital and Toronto East General Hospital, Toronto, Ontario, Canada.

SECTION ONE

Communicating Syringomyelia

CHAPTER ONE

Historical Introduction

J. B. FOSTER and P. HUDGSON

Cavitation of the spinal cord was probably first described by Estienne (1546) in *La Dissection du Corps Humain*. Although it was not until 1804 that Portal recognised the clinical phenomena associated with such lesions, descriptions of pathological cavitation had been made by Brunner (1688) and Morgagni (1740) in a child and an adult. Credit for the term syringomyelia, denoting cavity formation in the cord, goes to Charles P. Ollivier d'Angers (1827) (Fig. 1.1). In a monograph published in 1824 entitled *La Moelle Épinière et ses Maladies*, Ollivier describes pathological dilatation of the central canal of the cord in continuity with the cavity of the fourth ventricle. A later two-volume work, *Traité de la Moelle Épinière et de ses Maladies*, contains the description of cord cavitation 'Syringomyélie ou cavité centrale dans la moelle' and this is cited as the first published use of the term syringomyelia by Ballantine, Ojemann and Drew (1971).

In his monograph, Ollivier uses the term to describe pathological dilatation of the central canal as a developmental anomaly. After Stilling (1859) had demonstrated that the central canal could persist in childhood and into the adult life of all vertebrates, states of pathological dilatation came to be referred to as hydromyelia and the term syringomyelia was less commonly used (Gull, 1862; Schüppel, 1865; Virchow, 1863). Pathological specimens of central cord cavities unconnected with the central canal and associated with myelitis (Hallopeau, 1870), telangiectatic tumours and gliosis were recognised and Simon (1875) suggested that the term syringomyelia be reserved for such cavities and that hydromyelia be restricted to dilatation of the central canal itself.

In the following year, Leyden offered the suggestion that the two conditions were identical and was later supported by Virchow, and Kahler and Pick (1879). Schultze had given an account of the clinical syndrome (1882), outlining the clinicopathological correlation between cysts of the cord and the clinical symptomatology, and in particular pointing to a reduced sensibility to pain and temperature. Although earlier descriptions of the symptoms of the disease had been offered by Gull (1862), Clarke (1865) and Clarke and Johnson (1868), the clinical presentation of syringomyelia was well recognised and fully described by Gowers in his *Manual of Diseases of the Nervous System* in 1886. Other clinical descriptions followed, notably those of Bäumler (1887), Hoffmann (1893) and Schlesinger (1902). Today, the clinician regards the symptom complex of syringomyelia as following a characteristic pattern and similar course and prognosis (Schaltenbrand, 1951; André, 1951; Alsen, 1957; Wells, Spillane and Bligh, 1959; Erbslöh, 1963; McIlroy and Richardson, 1965).

Fig. 1.1. Charles Prosper Ollivier d'Angers (1796–1845) from a lithograph in the Wellcome Institute of the History of Medicine. (Reproduced by permission of The Wellcome Trustees.)

Chiari in 1888 further supported the unitary hypothesis of syringomyelia and hydromyelia and confirmed that in most cases the cavities of syringomyelia connected with the central canal of the cord. A description by Langhans (1881) of a case associated with a cerebellar deformity was followed by Chiari's description in 1891 of the anomalies associated with congenital hydrocephalus, *Concerning Changes in the Cerebellum due to Hydrocephalus of the Cerebrum*.

In this original account, Chiari divided the malformations into three anatomical categories of increasing severity. In the first group (type I) he described extensions of the cerebellar tonsils into the spinal canal. He later (1896) described 14 further cases, some with hydromyelia. His second type of anomaly (type II) was associated with displacement of part of the cerebellum and fourth ventricle into the dilated vertebral canal with hydromyelia and myelocele. Other workers noted the association between these anomalies and spina bifida and subsequently Schwalbe and Gredig (1907), writing from Arnold's laboratory, suggested that the type II anomaly be called the Arnold–Chiari malformation. Russel and Donald (1935) noted the association between hydromyelia and the Chiari malformation, and Gardner and Goodall (1950) found hydromyelia in 13 of 17 adult cases of Chiari malformation.

Following this, W. J. Gardner and his colleagues in a series of papers from Cleveland (1957a, 1957b, 1958, 1965, 1967) put forward what has come to be known as the 'hydrodynamic theory of the development of syringomyelia'— that the cystic dilatation of the cord originates in embryonal life and results from overdistention of the neural tube because of a partial or complete obstruction at the foramen of Magendie; this results in a dilatation of the central canal or a ramifying diverticulum (Schüppel, 1865), originating in the central canal and in continuity with the spinal fluid pathways through the fourth ventricle. In Gardner's series of 74 cases of syringomyelia, obstruction at the foramen of Magendie was due to a Chiari malformation in 68, a Dandy–Walker malformation in three and a cyst in a further three.

Gardner's theory was reviewed by Newton (1962, 1969) Conway (1961, 1967) and Appleby et al (1968). His theory of pathogenesis has been supported historically by the description of central canal dilatation in infants with Arnold–Chiari malformations or Dandy-Walker syndromes (Schwalbe and Gredig, 1907; Ingraham and Scott, 1943; Russell, 1949; Benda, 1952) and by severe dilatation in adults with type I Chiari anomaly (Ogryzlo, 1942; Lichtenstein, 1943; Netsky, 1953; Appleby et al, 1968; Foster, Hudgson and Pearce, 1969). It has also been observed surgically that in patients with syringomyelia the foramen of Magendie is commonly occluded (Gardner and Goodall, 1950; Gardner, Karnosh and Angel, 1957b), that the cerebral ventricles are often dilated (Netsky, 1953) and that dye injected into the ventricles may be recovered from the syrinx (Conway, 1967). It has been found that the fluid in this type of cord cavity resembles cerebrospinal fluid in all respects (Gardner and Goodall, 1950; Wetzel and Davis, 1954) and, more recently, that in the experimental animal, communicating hydrocephalus has been associated with central canal dilatation as well as cavitation in the necrotic cord (McLaurin et al, 1954). Human tuberculous meningitis has led to classical syringomyelia both clinically and pathologically (Appleby et al, 1969). Surgical decompression of the posterior rim of the foramen magnum and opening of the foramen of Magendie allows significant improvement in the clinical features of syringomyelia (Chamberlain, 1939; Gustafson and Oldburg, 1940; Gardner and Goodall, 1950; Appleby et al, 1968).

Newton (1969) on the basis of this theory of pathogenesis used the term 'typical syringomyelia' to describe cord cavities containing spinal fluid and 'atypical syringomyelia' to describe other cysts of the cord containing proteinaceous fluid. In his Hunterian Lecture to the Royal College of Surgeons of England, he referred to 15 personal cases, their presentation, investigation and operative treatment. More recently, Williams (1969 and 1970) has challenged the hydrodynamic theory of causation of syringomyelia and has described 'communicating syringomyelia' as a condition in which cerebrospinal fluid pathways communicate with a cyst either at its upper or lower end in the spinal cord; the distending force in these cases is related to the movement of spinal fluid and is unrelated to the force of arterial and, hence, cerebrospinal fluid pulsation (Gardner and Angel, 1958) (see Chapter 8). Williams' theory is related to venous distension in the spinal canal and cranium and he describes as the commonest cause, obstruction at the foramen of Magendie and obliteration of the cisterna magna by a Chiari anomaly, with dural bands, cysts and arachnoiditis also responsible for the foraminal hold-up and the development of communicating syringomyelia.

The non-communicating form of hydromyelia, that is cystic dilatation of the cord not in communication with the spinal fluid pathways, has been reported following acute traumatic paraplegia (Holmes, 1915; Cossa, 1943; Freeman, 1959; Jung, 1960; Martin and Maury, 1964; Barnett et al, 1966; Rossier et al, 1968; Werner et al, 1969; Nurick, Russel and Deck, 1970; Williams and Turner, 1971). Pott's disease and tumours of the cord (Jonesco-Sisesti, 1929) have also been associated with cystic lesions of the spinal cord and spinal arachnoiditis may produce such changes (see Chapter 15). Poser in 1956 reviewed the published cases of intramedullary cord tumours associated with syringomyelia, and quoted 16·4 per cent tumours in 254 cases of syringomyelia examined at autopsy. An association between syringomyelia and haemangioblastoma of the cord was pointed out by Wyburn-Mason (1943) and Guidetti and Fortuna (1967).

Thus, the modern concept of syringomyelia suggests that the disease takes two forms, a 'non-communicating' form associated with intramedullary tumours, traumatic paraplegia or some degenerative conditions, and 'communicating' syringomyelia which may be developmentally determined and results from a persistent dilatation of the central canal under pressure, with subsequent rupture of fluid into the surrounding neural tissue and communication of the central canal or cord tissues with the cerebrospinal fluid pathways.

It has long been recognized that cases of syringomyelia may be associated with skeletal anomalies and, of these, kyphoscoliosis is perhaps the commonest. However, descriptions have been given of patients having unusually short necks, associated Sprengel's shoulder and other anomalies and, with the advent of radiology, it was found that many patients suffering from either the Chiari malformation or classical syringomyelia had associated anomalies of the cervical vertebrae and craniovertebral junction (List, 1941; Garcin and Oeconomos, 1953). Such anomalies included, notably, basilar impression or coarctation, an abnormally acute occipitovertebral angle or other anomalies of the occipital bone, atlas and axis, e.g. occipitalisation of the atlas, hypoplastic occipital condyles, odontoid separation, defects in fusion and the Klippel–Feil syndrome (1912) with short neck, low hairline and restricted neck movements due to fusion of vertebral bodies and bifid spinous processes. Wells et al (1959) measured the sagittal diameter in the lateral X-rays of the cervical spinal canal in cases of syringomyelia and found it to be wider than normal in patients whose symptoms had begun before the age of 30. However, many cases of syringomyelia have been described in which the plain X-rays of cervical vertebrae and skull are normal (Aring, 1938; Ogryzlo, 1942; Bucy and Lichtenstein, 1945; Swanson and Fincher, 1949; Gardner and Goodall, 1950; Peach, 1964; Teng and Papatheodorou, 1965; Appleby et al, 1968). The subject of developmental anomalies at the foramen magnum and the resulting syringomyelic syndromes was recently reviewed by Spillane, Pallis and Jones (1957).

The diagnosis and differentiation of cystic lesions of the cord has recently depended on the introduction of positive and negative contrast radiology, especially myelography with examination in the supine position (Baker 1963) and gas myelography in the sitting position (Greenwald et al, 1958; Gardner, 1965; Westberg, 1966; Jirout, 1966; Heinz, Schlessinger and Potts, 1966; Conway, 1967). Ellertsson (1969a and b), Ellertsson and Greitz (1969) and Greitz and Ellertsson (1969) have demonstrated by air myelocystography and radio-iodinated human serum albumin (RIHSA) myelocystography, the communication between the cerebrospinal fluid pathways and the syringomyelic cavity and have thus been able to differentiate communicating from non-communicating cysts.

Following Abbe and Coley's (1892) first decompression of the cavity, some 400 operated cases have been reported (Ballentine et al, 1971). Laminectomy was usually performed with or without myelotomy (Elsburg, 1916; Kelly, 1935; Worster-Drought, Wakeley and Shafor, 1941; Netsky, 1953; Wetzel and Davis, 1953; Frazier and Roe, 1956; Pitts and Groff, 1964). In the past 10 years, a better understanding of the pathogenesis of syringomyelia and the separation of the communicating from the non-communicating varieties has led to a more rational surgical treatment of the former condition. It is now felt that the aim of surgical treatment is to restore anatomical normality at the foramen magnum to establish adequate drainage from the ventricular system and to free the obstruction to the flow of spinal fluid (see Chapter 6).

REFERENCES

Abbe, R. & Coley, W. B. (1892) Syringomyelia, operation—exploration of cord, withdrawal of fluid, exhibition of patient. *Journal of Nervous & Mental Disease*, **19**, 512.
Alsen, V. (1957) Klinisch-statische Untersuchungen über den Status dysraphicus. *Deutsche Zeitschrift für Nervenheilkunde*, **177**, 156.
André, M. J. (1951) Études sur la syringomyélie VI Essai nosologique sur la syringomyélie et l'hydromyélie. *Acta Neurologica et Psychiatrica Belgica*, **51**, 665.
Appleby, A., Foster, J. B., Hankinson, J. & Hudgson, P. (1968) The diagnosis and management of the Chiari anomalies in adult life. *Brain*, **91**, 131.
Appleby, A., Bradley, W. G., Foster, J. B., Hankinson, J. & Hudgson, P. (1969) Syringomyelia due to chronic arachnoiditis at the foramen magnum. *Journal of Neurological Sciences*, **8**, 451.
Aring, C. D. (1938) Cerebellar syndrome in adult with malformation of cerebellum and brain-stem. *Journal of Neurology & Psychiatry*, **1**, 100.
Baker, H. L., Jr (1963) Myelographic examination of the posterior fossa with positive contrast medium. *Radiology*, **81**, 791.
Ballantine, H. T., Ojemann, R. G. & Drew, J. H. (1971) *'Syringomyelia' in Progress in Neurological Surgery*. New York: S. Karger.
Barnett, H. J. M., Botterell, H., Jousse, A. T. & Wynne-Jones, M. (1966) Progressive myelopathy as a sequel to traumatic paraplegia. *Brain*, **89**, 159.
Bäumler, A. (1887) Über Höhlenbildungen im Rückenmark. *Deutsche Archiv für Klinische Medizin*, **40**, 443.
Benda, C. E. (1952) *Developmental Disorders of Mentation and Cerebral Palsies*, p. 565. New York: Grune and Stratton.
Brunner, J. C. in Boneti, T. (1700) *Sepulchretum*, Book 1, 2nd edn., p. 396. Geneva: Cramer and Perachon.
Bucy, P. C. & Lichtenstein, B. W. (1945) Arnold–Chiari deformity in an adult without obvious cause. *Journal of Neurosurgery*, **2**, 245.

Chamberlain, W. E. (1939) Basilar impression (platybasia). *Yale Journal of Biology and Medicine*, **11**, 487.
Chiari, H. (1888) Über die Pathogenese der songenanonten Syringomyelie. *Zeitschrift für Heilkunde*, **9**, 307.
Chiari, H. (1891) Über Veränderungen des Kleinhirns infolge von Hydrocephalie des Grosshirns. *Deutsche Medizinische Wocherschrift*, **17**, 1172.
Chiari, H. (1896) Über Veränderungen des Kleinhirns des Pons und der Medulla oblongata infolge von congenitaler Hydrocephalie des Grosshirns. *Denkschrift für Akademische Wissenschaft Wien*. **63**, 71.
Clarke, J. L. (1865) On the pathology of tetanus. *Medico-chirurgical Transactions*, **48**, 255.
Conway, L. W. (1961) Radiographic studies of syringomyelia. *Transactions of the American Neurological Association*, **86**, 205.
Conway, L. W. (1967) Hydrodynamic studies in syringomyelia. *Journal of Neurosurgery*, **27**, 501.
Cossa, L. W. (1943) Syringomyélie secondaire à une blessure de la moelle dorsale supérieure. *Revue Neurologique*, **75**, 39.
Ellertsson, A. B. (1969a) Semiologic diagnosis of syringomyelia related to roentgenologic findings. *Acta Neurologica Scandinavica*, **45**, 385.
Ellertsson, A. B. (1969b) Syringomyelia and other cystic spinal cord lesions. *Acta Neurologica Scandinavica*, **45**, 403.
Ellertsson, A. B. & Greitz, T. (1969) Myelocystographic and fluorescein studies to demonstrate communication between intramedullary cysts and the cerebrospinal fluid space. *Acta Neurologica Scandinavica*, **45**, 418.
Elsberg, C. A. (1916) *Diagnosis and Treatment of Surgical Diseases of the Spinal Cord and its Membranes*. Philadelphia: Saunders.
Erbslöh, F. (1963) Dysraphie, status dysraphicus und syringomyeliekomplex. In *Differential diagnose neurologische Krankheit*, ed. Bodechtal, G., p. 710., Stuttgart: Georg Thieme.
Estienne, C. (1546) *La Dissection du Corps Humain*, Ch. 3. Paris: Simon de Colines.
Foster, J. B., Hudgson, P. & Pearce, G. W. (1969) The association of syringomyelia and congenital cervico-medullary anomalies: pathological evidence. *Brain*, **92**, 25.
Frazier, C. H. & Rowe, S. N. (1936) The surgical treatment of syringomyelia. *Annals of Surgery*, **103**, 481.
Freeman, L. W. (1959) Ascending spinal paralysis. *Journal of Neurosurgery*, **16**, 120.
Garcin, R. & Oeconomos, D. (1953) *Les Aspects Neurologiques des Malformations Congenitales de la Charnière Cranio-Rachidienne*. Paris: Masson.
Gardner, W. J. & Goodall, R. J. (1950) The surgical treatment of Arnold–Chiari malformation in adults. *Journal of Neurosurgery*, **7**, 199.
Gardner, W. J., Abdullah, A. F. & McCormack, L. J. (1957a) Varying expressions of embryonal atresia of fourth ventricle in adults. Arnold–Chiari malformation, Dandy–Walker syndrome, 'arachnoid' cyst of cerebellum and syringomyelia. *Journal of Neurosurgery*, **14**, 591.
Gardner, W. J., Karnosh, L. J. & Angel, J. (1957b) Syringomyelia: a result of embryonal atresia of the foramen of Magendie. *Transactions of the American Neurological Association*, **82**, 144.
Gardner, W. J. & Angel, J. (1958) The mechanism of syringomyelia and its surgical correction. *Clinics in Neurosurgery*, **6**, 131.
Gardner, W. J. (1965) Hydrodynamic mechanism of syringomyelia: its relationship to myelocele. *Journal of Neurology, Neurosurgery & Psychiatry*, **28**, 247.
Gardner, W. J. (1967) Myelocele: rupture of the neural tube. *Clinics in Neurosurgery*, **15**, 57.
Gowers, W. R. (1886) *A Manual of Diseases of the Nervous System*, Vol. 1, p. 433. London: Churchill.
Greenwald, C. M., Eugenio, M., Hughes, C. R. & Gardner, W. J. (1958) The importance of the air shadow of the cisterna magna in encephalographic diagnosis. *Radiology*, **71**, 695.
Greitz, T. & Ellertsson, A. B. (1969) Isotope scanning of spinal cord cysts. *Acta radiologica*, **8**, 310.
Guidetti, B. & Fortuna, A. (1967) Surgical treatment of intra-medullary hemangioblastoma of the spinal cord. Report of six cases. *Neurosurgery*, **27**, 530.
Gull, W. W. (1862) Case of progressive atrophy of the muscles of the hands: enlargement of the ventricle of the cord in the cervical region with atrophy. *Guys Hospital Reports*, 3rd series, **8**, 244.
Gustafson, W. A. & Oldberg, E. (1940) Neurologic significance of platybasia. *Archives of Neurology & Psychiatry*, **44**, 1184.
Hallopeau, F. H. (1870) Note sur un fait de sclérose diffuse de la moelle avec lacune au centre de cet organe, altération de la substance grise et atrophie musculaire. *Gazette Médicale de Paris*, **25**, 183.
Heinz, E. R., Schlesinger, E. B. & Potts, D. (1966) Radiologic signs of hydromyelia. *Radiology*, **86**, 311.

Hoffmann, J. (1893) Über chronische spinalen Muskelatrophie im Kindesalter, auf familiärer Basis. *Deutsche Zeitschrift für Nervenheilkunde*, **3**, 427.
Holmes, G. M. (1915) The pathology of acute spinal injuries (Goulstonian Lectures to the Royal College of Physicians of London). *British Medical Journal*, **ii**, 769.
Ingraham, F. D. & Scott, H. W. (1943) Spina bifida and cranium bifidum: 5 Arnold–Chiari malformation. *New England Journal of Medicine*, **229**, 108.
Jirout, J. (1966) Mobility of the spinal cord under abnormal conditions. *Neuroradiologie* (Berlin).
Johnson, Z. & Clarke, J. L. (1868) On a remarkable case of extreme muscular atrophy with extensive disease of the spinal cord. *Medico-chirurgical Transactions*, **51**, 249.
Jonesco-Sisesti, N. (1929) *Tumeurs Medullaires Associées à un Processus Syringomyélique*. Paris: Masson.
Jung, E. (1960) Syringomyelie in Kombination mit Entwicklungstoren der Nieren und mit schwerer Wirbelsaülenverletzung. *Medizinische Klinik*, **55**, 1678.
Kahler, O. & Pick, A. (1879) Beiträge zur Pathologie und pathologischen Anatomie des Centralnervensystems. *Vientaljahrschrift Praktische Heilkunde*, **142**, 70.
Kelly, R. E. (1935) Surgical treatment of syringomyelia. *Transactions of the Medical Society of London*, **58**, 141.
Klippel, M. & Feil, A. (1912) Un cas d'absence des vertèbres cervicales avec cage thoracique remontant jusqu'à la bas du crâne (cage thoracique-cervicale). *Nouvelle Iconographie de la Salpêtrière*, **25**, 223.
Langhans, T. (1881) Veber Hohlenbildung im Rückenmark infolge Blutstauung. *Archiv für Pathologische Anatomie und Physiologie*, **85**, 1.
Leyden, E. (1876) Über Hydromyelus und Syringomyelie. *Archiv für Pathologische Anatomie und Physiologie*, **68**, 1.
Lichtenstein, B. W. (1943) Cervical syringomyelia and syringomyelia-like states associated with Arnold–Chiari deformity and platybasia. *Archives of Neurology and Psychiatry*, **49**, 881.
List, C. F. (1941) Neurologic syndromes accompanying developmental anomalies of occipital bone. *Archives of Neurology and Psychiatry*, **45**, 577.
Martin, C. & Maury, M. (1964) Syndrome syringomyélique après paraplégie traumatique. *La Presse Médicale*, **72**, 2839.
McIlroy, W. C. & Richardson, J. C. (1965) Syringomyelia, a clinical review of 75 cases. *Canadian Medical Association Journal*, **93**, 731.
McLaurin, R. L., Bailey, O. T., Schurr, P. H. & Ingraham, F. D. (1954) Myelomalacia and multiple cavitations of the spinal cord secondary to adhesive arachnoiditis. *Archives of Pathology*, **57**, 138.
Morgagni, G. B. (1740) Adversaria Anatomica, book 6, Lugduni Batavorum. Animadvesio XIV, p. 18.
Morgagni, G. B. (1761) *De Sedibus et Causis Morborum*. Translated by Alexander, B. (1769). London: Millar & Cadell.
Netsky, M. G. (1953) Syringomyelia. A clinicopathologic study. *Archives of Neurology and Psychiatry*, **70**, 741.
Newton, E. J. (1962) Hydromyelia. *Journal of Neurology, Neurosurgery and Psychiatry*, **25**, 185.
Newton, E. J. (1969) Syringomyelia as a manifestation of defective fourth ventricular drainage (Hunterian lecture). *Annals of the Royal College of Surgeons of England*, **44**, 194.
Nurick, S., Russel, J. A. & Deck, H. J. F. (1970) Cystic degeneration of the spinal cord following spinal cord injury. *Brain*, **93**, 211.
Ollivier d'Angers, C. P. (1824) *De la Moelle Épinière et de ses Maladies*, p. 116. Paris: Chez Crevot.
Ollivier, d'Angers, C. P. (1827) *Traité de la Moelle Épinière et de ses Maladies*, p. 178. Paris: Chez Crevot.
Ogryzlo, M. A. (1942) Arnold–Chiari malformation. *Archives of Neurology and Psychiatry*, **48**, 30.
Peach, B. (1964 Arnold–Chiari malformation with normal spine. *Archives of Neurology*, **10**, 497.
Pitts, F. W. & Groff, R. A. (1964) Syringomyelia: current status of surgical therapy. *Surgery*, **56**, 806.
Portal, A. (1804) *Cours d'Anatomie Médicale*, Vol. 4. Paris: Baudouin.
Poser, C. M. (1956) *The Relationship between Syringomyelia and Neoplasms*. Springfield, Illinois: Thomas.
Rossier, A. B., Werner, A., Wilki, E. & Berney, J. (1968) Contribution to the study of late cervical syringomyelic syndromes after dorsal or lumbar traumatic paraplegia. *Journal of Neurology, Neurosurgery & Psychiatry*, **31**, 99.
Russell, D. S. & Donald, C. (1935) Mechanism of internal hydrocephalus in spina bifida. *Brain*, **58**, 203.
Russell, D. S. (1949) Observations on the pathology of hydrocephalus. *Medical Research Council Special Report Series No.* **265**.
Schaltenbrand, G. (1951) Syringomyelie. In *Lehrbuch der Neurologie*, p. 147. Stuttgart: Georg Thieme.
Schlesinger, H. (1902) *Die Syringomyelie*. Leipzig and Vienna: Deuticke.

Schultze, F. (1882) Ueber Spalt, Hohlen and Gliombildung im Rückenmark und in der Medulla oblongata. *Virchows Archiv*, **87**, 510.
Schüppel, O. (1865) Über Hydromyelus. *Archiv für Heilkunde*, **6**, 289.
Schwalbe, E. & Gredig, M. (1907) Über Entwicklungsstorungen des Kleinhirns, Hirsamms und Halsmarks bei Spina Bifida. Ziegler's *Beiträge zur pathologischen Anatomie und zur allegemeinen Pathologie*, **40**, 132.
Simon, T. (1875) Uber Syringomyelie und Geschwulstbildung im Rückenmark. *Archiv für Psychiatrie und Nervenkrankheiten*, **5**, 120.
Spillane, J. D., Pallis, C. & Jones, A. M. (1957) Developmental abnormalities in the region of the foramen magnum. *Brain*, **80**, 11.
Stilling, B. (1859) *Neue Untersuchungen über den Bau des Rückenmarks*, p. 13. Casel: Commissions-Verlag von Heinrich Hotop.
Swanson, H. S. & Fincher, E. F. (1949) Arnold–Chiari deformity without bony anomalies. *Journal of Neurosurgery*, **6**, 314.
Teng, P. & Papatheodorou, C. (1965) Arnold–Chiari malformation with normal spine and cranium. *Archives of Neurology*, **12**, 622.
Virchow, R. (1863) Die Bethweiligung des Rückenmark an der Spina Bifida und Hydromyelie. *Archiv für Pathologie, Anatomie und Physiologie*, **27**, 575.
Wells, C. E. C., Spillane, J. D. & Bligh, A. S. (1959) The cervical spinal cord in syringomyelia. *Brain*, **82**, 23.
Werner, A., Rossier, A., Berney, J. & Zdrojewski, B. (1969) Á propos de quatre observations de syringomélie cervicale tardive après traumatisme médullaire. *Schweizer Archiv für Neurologie und Psychiatrie*, **104**, 77.
Wetzel, N. & Davis, L. (1954) Surgical treatment of syringomyelia. *Archives of Surgery*, **68**, 970.
Westberg, G. (1966) Gas myelography and percutaneous puncture in the diagnosis of spinal cord cysts. *Acta Radiologica*, Supplement 252, 67.
Williams, B. (1969) The distending force in the production of 'communicating syringomyelia'. *Lancet*, **ii**, 189.
Williams, B. (1970) Current concepts of syringomyelia. *British Journal of Hospital Medicine*, **4**, 331.
Williams, B. & Turner, E. (1971) Communicating syringomyelia presenting immediately after trauma. *Acta Neurochirurgica*, **24**, 97.
Worster-Drought, C., Wakely, C. P. G. & Shafar, J. (1941) The surgical treatment of syringomyelia. *British Journal of Surgery*, **29**, 56.
Wyburn-Mason, R. (1943) *The Vascular Abnormalities and Tumours of the Spinal Cord and its Membranes*. London: H. Kimpton.

CHAPTER TWO

Traditional Concepts of Syringomyelia

J. B. FOSTER and P. HUDGSON

The erstwhile accepted nosological entity defined as syringomyelia has been described as a chronic, relentlessly progressive disorder, characterised by amyotrophy, dissociated anaesthesia, paraparesis with scoliosis and the development of neurogenic arthropathies and other trophic lesions. Generally considered to be more frequent in the male than the female (Brain, 1955), the disease first presents in the second or third decade, but can occur in childhood or in late life. Brain described the age of onset as varying between 10 and 60 years, usually between 25 and 40 years. Although considered by Schlesinger (1902) to be one of the commonest spinal cord disorders, the true incidence of syringomyelia has never been assessed. In a survey of neurological disease in an English city, Brewis et al (1966) found six patients to be suffering from the disease, i.e. a prevalence rate of 8·4 per 100,000.

The precise clinical phenomena of the disease depend on the situation of the destructive central cord cavitation and consequent gliosis. Monosymptomatic or unilateral forms cause difficulty in differential diagnosis but ultimately the clinical picture becomes characteristic in most cases.

Involvement of the anterior horns leads to advancing amyotrophy often first evident in or confined to the hands but later spreading to the forearm muscles and shoulder girdle.

Early damage to the first dorsal segment induces the characteristic loss of lumbrical interosseus activity with the development of the claw hand or *main en griffe*. Deep reflexes in the upper limbs are lost early in the course of the disease. Scoliosis is often an early sign and results from damage to the dorsomedial and ventrolateral spinal nuclei. Interruption of the decussating spinothalamic fibres subserving pain and thermal sensibility induces dissociated anaesthesia, an impairment and subsequent loss of pain and temperature sensation, with the retention of light touch and proprioceptive appreciation so characteristic of central cord lesions. The early limitation of the central pathology to the cervico-dorsal segments is responsible for the suspended or *cuirasse* distribution of the sensory change and, until more widespread destruction occurs, the spino-thalamic pathways are not disturbed and analgesia in the lower limbs is unusual early in the clinical course.

Destructive gliosis or compression of the lateral columns leads in most cases to early signs of spasticity in one or both legs, with ultimately an asymmetrical spastic paraplegia, absent superficial reflexes and extensor plantar responses. Sphincter disturbance is a late feature.

Pain of a deep, aching character is often felt early in the course of the disease and may become very severe. Although usually central in the neck and shoulders it may have a radicular distribution in the trunk or limbs. Cavitation at the first dorsal level may lead to an ipsilateral Horner's syndrome, and extension of the process into the medulla produces the syndrome of syringobulbia: asymmetrical weakness and wasting of the tongue, dissociated trigeminal sensory loss, palatal weakness and nystagmus.

In the upper limbs the loss of pain sensibility leads to skin thickening, callosity, scars and painless whitlows of the hands with the possibility of terminal phalangeal absorption (Morvan, 1883) or extensive subcutaneous oedema and hyperhydrosis the *main succulente* of Marinesco (1897). Neurogenic arthropathies appear in 25 per cent of cases (Meyer, Stein and Pappel, 1957), commonly within the area of analgesia, and the resulting gross deformity of the shoulder, elbow or wrist adds to the overall disfigurement.

In its later stages, the disorder is characteristic and readily recognisable but in the early case the central location of the pathology requires syringomyelia to be differentiated from other central cord lesions, notably the ependymomas and astrocytomas. When the onset is asymmetrical, sensory disturbance in a limb may mimic root compression, multiple sclerosis or even a peripheral nerve entrapment and the onset of small handmuscle wasting and weakness, with or without paraparesis where there is no dissociated anaesthesia, may suggest motor neurone disease.

The presence of dissociated anaesthesia in a limb or over the trunk reduces the diagnostic difficulty but without this the clinical pattern may resemble high cervical cord compression, especially since the classical descriptive accounts of syringomyelia fail to stress the frequency of proprioceptive abnormality in the upper limbs.

TRADITIONAL PRESENTATION

N.A., FEMALE (N.31775)

This patient presented to the Department of Neurology at the Newcastle General Hospital, at the age of 56 years, with the story that at the age of 26 she had noticed painless burns and injuries of her fingers. Thirteen years before investigation, she noticed some vague discomfort and deformity of the left shoulder joint, and during the past 10 years had developed wasting and weakness of the hands to the point where they were virtually useless, in that she could barely hold a piece of toast in her right hand and required to be fed, washed and helped to the toilet. For 20 years, she experienced progressive weakness and stiffness of the legs and at the time of presentation could walk only with help. There had been no sphincter disturbance. On examination she was obese, with a short neck and obvious scoliosis (Fig. 2.1). Neck movements were full but there was an evident gross deformity of both upper limbs, with Charcot changes in both wrists and shoulders (Fig. 2.2). The hands were atrophied and on the right she had lost the terminal phalanges of the thumb and first two fingers (Fig. 2.3). There were painless ulcers in the palm of the right hand within an area of dissociated anaesthesia which was total between the third cervical and tenth dorsal segments bilaterally (Morvan's syndrome).

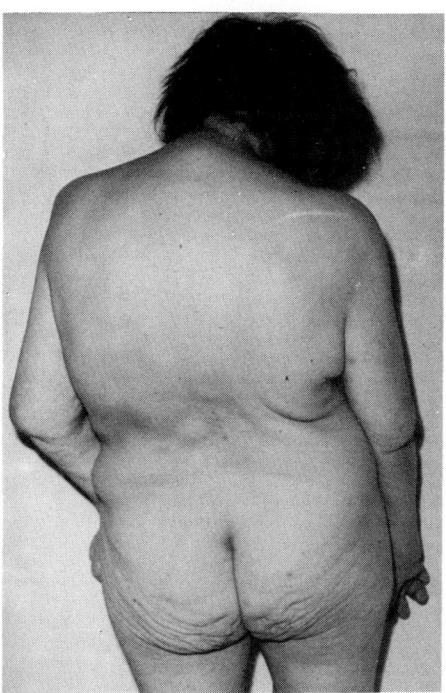

Fig. 2.1. (N.31775.) There is marked spinal deformity with evident scoliosis and short neck.

Radiological examination of the right hand showed diffuse osteoporosis and tapering of the remaining phalanges (Fig. 2.4). She showed vertical nystagmus on looking to the right and left and on looking downwards. There was reduction in light touch appreciation on the left side of the face, with loss of thermal and pain sensation over the whole of the face, apart from the right forehead. There was a brisk jaw jerk. There were no abnormalities of the lower cranial nerves, but examination of the limbs revealed a 'full-cape' sensory loss to pain and thermal appreciation to the level of D10, with gross diminution in position sense and vibration sense in both upper limbs, more marked on the right than the left. The upper limbs were very weak with absent deep tendon reflexes.

In the lower limbs, she had a spastic paraparesis with some diminution of posterior column sensation and impairment of light touch. Her gait was slightly broad based and spastic and she could walk only with help. The plantar responses were extensor and the reflexes in the lower limbs uniformly brisk. A myelogram showed lumbar arachnoiditis. Clear visualisation of the region of the foramen magnum was not possible and it was thought that she was unsuitable for surgical treatment because of her clinical state. This patient demonstrates the gradual progression and ultimate severe disability which may occur in the syringomyelic syndrome.

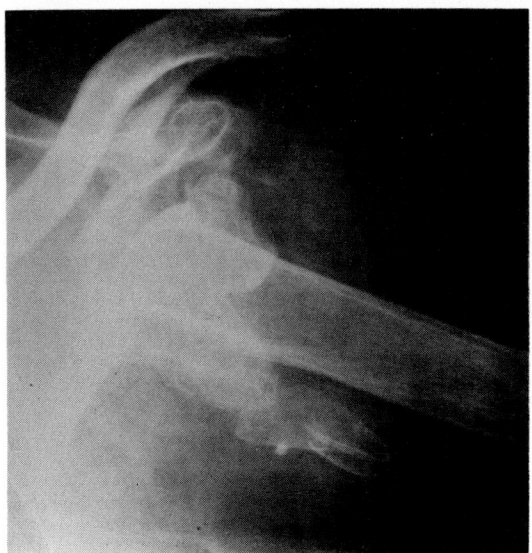

Fig. 2.2. (N.31775.) Destructive neurogenic arthropathy of the right glenohumeral joint showing complete destruction of the humeral head.

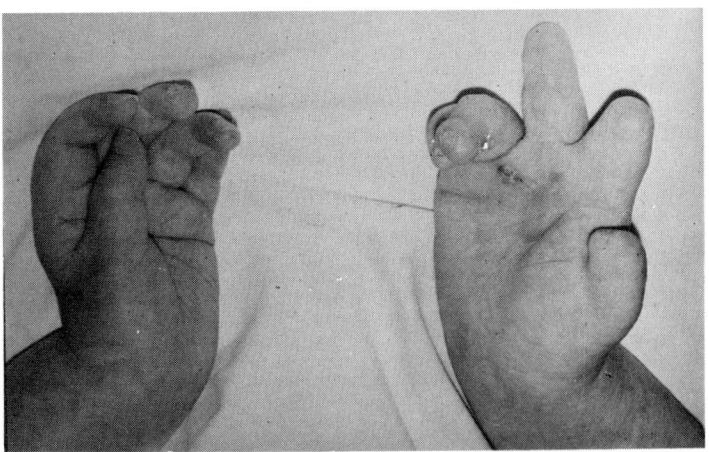

Fig. 2.3. (N.31775.) The acrodystrophic changes of Morvan's syndrome.

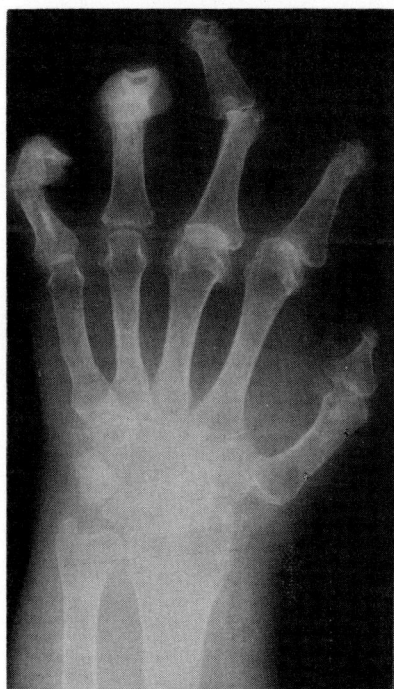

Fig. 2.4. (N.31775.) The radiological changes in the right hand showing generalised osteoporosis, dissolution of the terminal phalanges and tapering of the remaining distal phalanges.

REFERENCES

Brain, W. R. (1955) *Diseases of the Nervous System*, 5th Ed. London.
Brewis, M., Poskanzer, D. C., Rolland, C. & Miller, H. G. (1966) Neurological disease in an English city. *Acta Neurologica Scandinavica*, Supplement 24, Vol. 42.
Marinesco, G. (1897) *Main Succulente et Atrophie Musculaire dans la Syringomyélie*. MD Thesis 241, University of Paris.
Meyer, G. A., Stein, J. & Pappel, M. W. (1957) Rapid osseous changes in syringomyelia. *Radiology*, **69**, 415.
Morvan, A. M. (1883) De la parésie analgésique à panaris des extrémitiés supérieures ou paréso-analgésie des extrémitiés supérieures. *Gazette Hebdomadaire Médicine et de Chirurgie*, **35**, 580.
Schlesinger, H. (1902) *Die Syringomyelie*. Leipzig: Deuticke.

CHAPTER THREE

The Clinical Features of Communicating Syringomyelia

J. B. FOSTER and P. HUDGSON

In this chapter we describe our experience of syringomyelia in patients who have passed through the Departments of Neurology and Neurosurgery of the Regional Neurological Centre and the Royal Victoria Infirmary, in the Newcastle University Hospital Group. We report a consecutive series of 100 cases seen in these Departments since 1964, most of whom were seen initially by the medical neurologists and subsequently subjected to operation under the care of the neurosurgeons of the Centre, in particular Professor John Hankinson, Mr L. P. Lassman and Mr R. M. Kalbag. Because of Professor Hankinson's particular interest in the problem, most of the operated cases have been under his care.

The first 100 cases have shown all the manifestations of cervical cord cavitation. While some patients presented very early in the course of the disease, even before the onset of abnormal physical signs, others have been extremely disabled by an advanced degree of syringomyelic destruction of the cord and have been beyond operative help. Most of the patients studied have had symptoms of the disease for from one to six years and have been considered suitable for investigation and operative treatment. In the series there are 50 males and 50 females, i.e. there is no difference in sex prevalence. The youngest case in the group is a female aged $5\frac{1}{2}$ years at the time of presentation.

L.J., FEMALE (N.12724)

This girl, aged $5\frac{1}{2}$ years, presented with an abnormality of gait, a right-sided limp which was increasing and she seemed to 'throw' her right foot outwards. The biceps, triceps and brachioradialis reflexes were absent and on the right side the knee and ankle jerks were absent. The plantar response was unobtainable. The abdominal reflexes were absent; power and tone were diminished in the right leg, with wasting of the right calf muscle.

X-rays of the cervical and thoracic spine showed an increased interpedicular distance throughout. There was scoliosis of the lower thoracic and upper lumbar spine which was concave to the right. The appearances at myelography suggested a well-marked pathological expansion of the spinal cord extending upwards from just above the conus medullaris to the level of the third cervical vertebra. Laminectomy at D1 to 12 revealed a considerable expansion of the spinal cord; this was punctured with a very fine needle just lateral to the midline and crystal-clear fluid with the characteristics of C.S.F. was aspirated. The swelling of the cord collapsed and further laminectomy at this stage was thought to be unwise.

At follow-up, 2 years 9 months after operation, the patient's walking was found to be much improved and she was doing extremely well. At first, it was necessary for her to wear a spinal brace but this was soon discarded because her back muscles recovered power. The reflexes in the lower limbs remained unchanged.

The oldest of the cases at onset (A.R. N.33590) was a female aged 63 who presented a typical 'half-cape' dissociated anaesthesia, with wasting of the right hand and numbness of the fingers on that side. She was seen one year after the onset of symptoms but at this age it was thought unwise to undertake a myelogram and, indeed, she did not wish such an investigation to be carried out. It is possible that she could have an intramedullary tumour but the lack of clinical deterioration over a long period is very much against this.

The mean age of onset of symptoms in the group was 31 years and the mean age at investigation and/or operation was 39 years. The follow-up of cases from operation varies from three months to six years and will undoubtedly have to be prolonged before firm conclusions can be made regarding continuing improvement or the permanent arrest of deterioration following operation.

ANALYSIS OF CLINICAL FEATURES

Presenting Symptoms

Tables 3.1 and 3.2 show the presenting symptoms of the 100 patients. The commonest early symptom was stiffness of the legs, complained of by 42 of our cases. Twenty-eight presented initially with numbness of a hand or hands. A surprisingly high proportion (14 of the 100) presented with headache as their initial symptom and in seven of these the headache was induced or aggravated by coughing. In 24 patients there was pain in the neck and in 16 pain in the upper limbs. Four patients described double vision, two had drop attacks, two had vertigo and in six oscillopsia was the presenting feature. In those with oscillopsia and vertigo there were clinical features suggesting that the brain stem was involved in the cystic cavitation (Table 3.3).

Headache

Fourteen patients presented with headache. In some, the headaches were typical of raised intracranial pressure with cough headache in seven. In one patient, headache was the sole clinical feature.

E.T., FEMALE (N.28162)
For one year before admission to the unit on 26 March, 1968, this 23-year-old patient had suffered from attacks of pain, starting with short, sharp jabs at the right side of the neck which rapidly radiated up into the head and were associated with a feeling of generalised throbbing. These headaches lasted for two to three minutes and would then pass off spontaneously but the patient maintained that they were regularly induced by exertion. If she extended her head fully, then she could relieve herself of the pains. The pain could also be induced by acute flexion of the neck. There were no other symptoms and full neurological examination was entirely negative. There was no suggestion of arrested hydrocephalus nor was the neck short. She had no nystagmus and in the limbs there was no abnormality of tone, power or coordination. X-rays of the skull and cervical spine were both normal, with an antero-

COMMUNICATING SYRINGOMYELIA

Table 3.1. *Presenting symptoms (100 cases)*

Case	Headache	Oscillopsia	Numbness of hands	Pain in neck	Pain in limbs	Stiffness of legs	Vertigo	Drop attacks	Diplopia	Case	Headache	Oscillopsia	Numbness of hands	Pain in neck	Pain in limbs	Stiffness of legs	Vertigo	Drop attacks
MJ	−	−	−	+	+	+	−	−	−	RA	−	−	+	−	−	−	−	−
MT	−	−	+	+	+	−	−	−	−	CD	−	+	−	−	−	+	−	−
GS	−	−	−	−	+	−	−	−	−	MA	−	−	−	−	−	+	−	−
FD	−	−	−	−	+	−	−	−	−	LO	−	−	−	+	−	−	−	−
EW	−	−	+	−	+	−	−	−	−	SW	−	−	+	−	−	−	−	−
WH	−	−	+	−	+	−	−	−	−	WD	−	−	−	−	+	−	−	−
PG	+	−	+	−	+	−	−	−	−	MC	−	+	−	−	−	−	−	−
ES	−	−	+	+	−	+	−	−	−	RH	−	−	−	−	+	−	−	−
RJ	−	−	+	−	−	−	−	−	−	JJ	−	−	−	−	−	+	−	−
EJ	−	−	+	−	−	+	−	−	−	MG	−	−	+	−	−	−	−	−
CW	−	−	+	−	−	+	−	−	−	NA	−	−	+	−	−	+	−	−
MC	++	−	−	+	+	−	−	−	−	MB	−	−	+	−	−	−	−	−
LB	−	−	−	+	−	−	−	−	−	PR	−	−	+	+	−	−	−	−
KE	−	+	+	−	−	−	+	−	−	MM	−	−	+	−	−	−	−	−
DL	−	−	+	−	−	−	−	−	−	AR	−	−	+	+	−	−	−	−
LJ	−	−	−	−	+	−	−	−	−	GR	−	−	+	−	−	−	−	−
MC	−	−	−	−	+	−	−	−	−	JA	−	−	−	−	−	−	+	−
CH	−	−	+	+	−	−	−	−	−	EK	−	−	−	−	−	+	−	−
LT	−	−	−	−	+	−	−	−	−	RJ	−	−	−	+	+	−	−	−
NS	−	−	+	−	+	−	−	−	−	AT	+	−	−	−	−	−	−	+
JL	−	−	−	−	+	−	−	−	−	HH	−	−	−	−	−	+	−	−
EW	−	−	+	−	−	−	−	−	−	RS	−	−	+	−	−	−	−	−
JC	−	−	+	−	−	−	−	−	−	RB	−	−	+	+	−	−	−	−
KB	−	−	−	+	+	−	−	−	−	HC	−	−	−	−	−	+	−	−
WM	−	−	−	−	+	−	−	−	−	JE	−	−	−	−	−	+	−	−
RP	−	+	−	−	−	+	−	−	−	WP	−	−	−	−	−	+	−	−
BE	−	−	−	+	+	+	−	−	−	NY	+	−	−	−	−	+	−	+
JM	−	−	−	−	+	−	−	−	−	LR	−	+	−	−	−	−	−	+
MM	−	−	−	+	−	−	−	−	−	RR	−	−	−	+	−	−	−	−
MD	++	−	−	−	−	−	−	−	−	SD	−	−	−	−	−	+	−	−
RC	−	−	−	−	−	−	−	+	−	HB	−	−	−	+	−	−	−	−
FB	−	−	−	−	+	−	−	−	−	TR	−	−	−	−	−	+	−	+
PM	−	−	+	−	+	−	−	−	−	DB	−	−	−	−	−	−	−	+
HC	−	−	−	−	+	−	−	−	−	RP	+	−	−	−	−	+	−	−
MW	−	−	+	−	−	−	−	−	−	WO	−	−	−	−	−	+	−	−
JE	−	−	−	+	+	−	−	−	−	GJ	+	−	−	−	−	−	−	+
PS	−	−	−	+	−	−	−	−	−	EJ	−	−	−	+	−	−	−	−
PH	−	−	−	−	+	−	−	−	−	VN	+	+	−	−	−	−	−	−
ME	+	−	−	+	−	−	−	−	−	ER	+	−	−	−	−	−	−	−
AB	−	−	+	−	−	−	−	−	−	MG	−	−	−	−	−	+	−	−
JA	−	−	−	−	+	−	−	−	−	DO	−	−	−	+	−	−	−	−
CB	+	−	−	−	−	−	−	−	−	WI	−	−	−	−	−	+	−	−
JW	−	−	−	+	−	−	−	−	−	JJ	−	−	+	−	−	−	−	−
HF	+	−	−	−	−	−	−	−	−	RH	−	−	+	−	−	+	−	−
MJ	+	−	−	−	+	−	−	−	−	JW	−	−	+	−	−	+	−	−
ET	−	−	−	+	−	−	−	−	−	FD	−	−	−	−	−	+	−	−
JR	+	−	−	−	+	−	−	−	−	AF	−	−	−	−	−	+	+	−
CP	−	−	−	−	+	−	−	−	−	GF	−	−	−	−	−	+	−	−
FH	−	−	−	−	+	−	−	−	−	DR	−	−	+	−	−	−	−	−
JC	−	−	+	−	−	−	−	−	−	EH	−	−	−	+	−	−	−	−

posterior diameter of the canal of 18 mm at the level of C5. Air encephalography did not outline the ventricular system and there was a suspicion of cerebellar tonsillar herniation with obstruction of the subarachnoid space over the posterior surface of the spinal cord in the region of the foramen magnum. The attempted air picture was followed by myelography and in the supine position this disclosed a filling defect on the back of the spinal cord; this defect was clearly defined, smooth in outline and consistent with cerebellar tonsillar herniation.

At operation (Hankinson), the arch of atlas was removed and the spine and upper portion of the second lamina. The dura was then opened with a Y-shaped incision extending to C2 and the tonsils were seen to be hypertrophied and prolapsed to the C2 level, right more than left. There was no arachnoiditis and a free flow of fluid from the fourth ventricle was established. The tonsils were separated and under

Table 3.2. Presenting symptoms (number of cases)

Symptoms	Number of cases
Stiffness of the legs	42
Numbness of the hands	28
Pain in the neck	24
Pain in the arms	16
Headache	14
Oscillopsia	6
Diplopia	4
Vertigo	3
Drop attacks	2

Table 3.3. Brain stem features

	Oscillopsia	Diplopia	Nystagmus	Trigeminal numbness	Wasted tongue	Palatal weakness	Headache	Vertigo
K.E.		−	Phasic grade one to L. and R.	+	−	+	−	−
R.P.		−	Phasic grade one to L. and R. and vertical	−	−	−	−	+
C.D.		+	Phasic grade one to L. and vertical	−	−	−	−	−
M.C.		+	Phasic grade two to L. and R. and vertical	−	−	−	−	−
L.R.		+	Phasic grade one to L. and R. with vertical component	−	−	−	−	−
V.N.		+	Rotary to L. and R. and coarse vertical on upward gaze	−	−	−	++	−

the right one there was a depression in the posterior aspect of the medulla. The cord appeared normal, but there was a small opening into the central canal. This was occluded with a small piece of muscle and the muscles, aponeurosis and skin were repaired, leaving the dura open. Throughout the operation the patient's condition was satisfactory. Her postoperative course was normal and at follow-up she was consistently totally free from symptoms.

Stiffness of the Legs

The 42 cases with stiffness of the legs as the presenting feature showed the characteristic changes of spastic paraparesis (see below).

Numbness of the Hands

Characteristic dissociated anaesthesia was noted by 20 patients who had sustained painless burns or cuts of the hands or in whom the symptom of numbness implied an inability to differentiate between hot and cold water.

Pain in the Neck and Arms

At the onset of their disease, forty patients complained of pain, either in the neck or in one or both upper limbs. The pain was of deep character in most cases and in four the neck pain was influenced by the position of the head.

J.R., FEMALE (N.28331)

This 17-year-old girl complained that for two years before attending hospital she had suffered from occipital headache and pain in the neck. The pain was made worse by flexing the neck and was also aggravated by stooping, coughing and laughing. Often the pain, induced by laughing, would pass into the left arm. On examination, her only abnormal sign was an incomplete band of hypalgesia high on the side of her neck on the left. Her radiographs of neck and skull were normal but myelography showed gross asymmetrical ectopia of the cerebellar tonsils. At operation (Hankinson), the ectopia was confirmed and an hiatus from the fourth ventricle was seen passing down the centre of the cord. This was plugged with muscle. The decompression of the upper cervical canal region and foramen magnum has totally relieved her symptoms.

Oscillopsia and Diplopia

Six patients complained of oscillopsia; each had associated nystagmus and some, evidence of medullary involvement (Table 3.3). Four of the patients complained of diplopia. In one case, the visual disturbance was the only clinical abnormality. As will be noted, vertical nystagmus was commonly found and was often accentuated by flexion or extension of the head and neck.

L.R., MALE (R.V.I.712237)

This 34-year-old patient presented himself to his optician because of oscillopsia and occasional double vision. He was found by the optician to have nystagmus and was referred for neurological examination. The nystagmus was present in all directions of gaze, with a quick phase in the direction of gaze, but was most marked on lateral gaze to the right and the left. There was, in addition to this phasic lateral nystagmus, minimal vertical nystagmus. Full neurological examination failed to reveal any other abnormality. X-rays of skull and cervical spine were within normal limits, but a myelographic examination carried out in the supine position showed a Chiari type I deformity with descent of the cerebellar tonsils to approximately 1 cm below the foramen magnum. It was felt that the nystagmus did not itself justify decompression and therefore no operation was undertaken and the patient has been followed since he first presented in 1969. At his last review there was no evidence of further deterioration.

In one case, the oscillopsia and associated headache suggested a posterior fossa tumour.

V.N., MALE (R.V.I.659010)

This doctor, aged 30 years, presented to the neurological clinic with a history of difficulty with his vision for some 12 years, accompanied by double vision during the preceding eight years. On analysis, he was able to describe that his double vision was due to oscillopsia on conjugate gaze in all directions and that it was not a true paralytic diplopia. In addition, he had been subject to attacks of severe, throbbing, retro-orbital headache since early adolescence. The headaches were precipitated by or exacerbated by coughing, stooping and shouting and had been relieved by lying down. During the preceding nine years, he had experienced occasional attacks of tingling paraesthesiae down the outer border of the left arm into the hand. On examination, there were no abnormal neurological signs except for the coarse vertical nystagmus on fixation and on conjugate gaze upward. There was coarse rotary nystagmus on lateral gaze to either side. The combination of symptoms and signs suggested an anomaly of the hind brain at the level of the foramen magnum, and X-rays of the cervical spine showed a widening of the anteroposterior diameter of the cervical spinal canal (2·1 cm in the mid-cervical region). There was a minor degree of basilar impression. Supine myelography showed descent of the medulla and cerebellar tonsils, with an abnormally low origin of the basilar artery.

Surgical decompression was undertaken (Hankinson) when the posterior 2 cm of the foramen magnum, the arch of the atlas and the spine and lamina of C2 were removed. After opening the dura, it was evident that the tonsils were considerably prolapsed, the right 0·5 cm lower than the left, just above the upper border of C3. On separation of the tonsils, spinal fluid flowed through the fourth ventricular foramen; no orifice to the central canal could be seen. As in many such cases, the cervical cord was not sufficiently exposed to be certain that a syrinx was present but, in our opinion, this could be presumed. The dura was left widely open. Following operation, the patient made an uneventful recovery. At follow-up eight months afterwards he showed no nystagmus and was totally free from oscillopsia and headache.

Vertigo and Drop Attacks

Three patients complained of episodic vertigo, one of these, R. P., suffered from severe oscillopsia and showed nystagmus (Table 3.3).

J.A., MALE (N. 40006)
For more than a year, at the time of his presentation, this 46-year-old miner had suffered from episodes of vertigo, especially noticed when turning over in bed. He complained that turning his head to either side would cause him to feel giddy with a sense of rotation and this feeling would last for two or three seconds. Repeating the manoeuvre would cause a return of the feeling of vertigo. There was no associated tinnitus or vomiting, but the patient commented that because of a sudden episode of 'rotation', on one occasion he fell over while stooping. He complained of no diplopia or blurring of vision, nor indeed of any other symptoms, but on examination was seen to have a very short neck. He showed nystagmus, phasic grade 1 to the left and right, and in the limbs hyperreflexia with extensor plantar responses, i.e. a spastic quadriparesis. X-rays of the skull showed evidence of basilar impression and a myelogram confirmed descent of the cerebellar tonsils bilaterally. At operation (Hankinson), the removal of the occipital bone was awkward because of its very horizontal inclination and its being thick and hard. The exposed bone was removed, with the laminae of the first and second cervical vertebrae. The tonsils were found to be prolapsed on opening the dura to the level of the lower border of C1 and these were well decompressed. On separating the tonsils, a wide opening to the central canal was seen in the floor of the fourth ventricle and this was occluded with muscle. The patient made a satisfactory postoperative recovery and at follow-up 12 months later had suffered no further episodes of vertigo. His nystagmus and long-tract signs remain unchanged.

K.E., MALE (N.11329)
This 25-year-old patient was first seen in July 1964 when he gave a history of several recent attacks of vertigo, two of which were accompanied by transient loss of consciousness. For five months before admission he had also noticed a feeling of numbness down the right side of the face and in the right upper limb. On one occasion during this period he had burnt his right forefinger without feeling pain. He had also noticed impairment of thermal appreciation in the right upper limb and clumsiness in performing fine movements with the fingers of the right hand. He complained of episodic diplopia occurring for several weeks before admission on lateral gaze to the right. Physical examination showed that the head and neck were superficially normal, but flexion, extension and lateral rotation to the right produced sharp pain in the back of the neck. The patient had coarse horizontal nystagmus on lateral gaze to the right and the left with dissociated sensory loss on the right side of the face, an absent right corneal reflex and impaired sensation on the right side of the posterior pharyngeal wall. There was very slight right facial weakness of lower motor neurone type. He also had a 'half-cape' area of dissociated sensory loss from C2 down to D10 on the right side. X-rays of the cervical spine showed anomalies of the spinous processes of C2 and 3 and of the posterior arch of the atlas. X-rays of the skull were normal.
Myelography showed very slight hold-up of contrast medium posteriorly at the C1 level with the neck extended, but was otherwise normal. However, vertebral arteriography showed that the loop of the left posterior inferior cerebellar artery lay well below the foramen magnum (see Fig. 5.16). Surgical exploration was therefore carried out and showed prolapse of the cerebellar tonsils with compression of the cervical cord by the abnormal arch of the atlas.
Follow-up (2 October, 1965) showed both subjective and objective evidence of improvement in sensation on the right side of the body. The patient experienced no attacks of vertigo after the operation and head and neck movements were performed freely in all directions without pain.

Two patients had experienced sudden falls without loss of consciousness. In the first, the attacks were precipitated by head turning, while in the second they were associated with headaches.

R.C., FEMALE (N.22553)
This 35-year-old children's nurse presented with a 12-month history of numbness of the left hand and numbness of the ulnar border of the right hand, which had come on insidiously over two to three

days. In addition, she had noticed loss of pain and temperature sensation in the hand, particularly in the bath. She described episodes of unsteadiness, as though she were falling backwards, and two episodes were described in which on turning her head to the left, her legs gave way suddenly and she fell to the ground without loss of consciousness. She was immediately able to get up and continue to stand. She had two or three of these episodes. Examination of this woman revealed a 'half-cape' of dissociated anaesthesia, involving the left forequarter. The reflexes in the upper and lower limbs were unremarkable, but there was quite marked weakness of the left hand and phasic nystagmus to left and right. It was thought that she suffered from cervical cord cavitation due to a Chiari malformation, but myelography in the prone and supine position revealed no cord expansion, no tonsillar descent and no other abnormality. Operative decompression was not therefore undertaken and her condition has remained unchanged.

A.T., FEMALE (N.40499)
A 45-year-old housewife presented with a 24-month history of episodic sudden headaches. The headaches were generalised, lasting for no more than a few seconds. On several occasions associated with the headache she had suddenly fallen. Examination revealed generalised hyperreflexia with extensor plantar responses as the only abnormality. There were no features of raised intracranial pressure, i.e. no papilloedema. Because of the presentation, an attempt to outline the ventricular system with air was made by the lumbar route. This was unsuccessful. Bilateral carotid angiography revealed marked bilateral ventricular dilatation, however, and subsequent ventriculography confirmed the marked symmetrical dilatation of the lateral ventricles. The third ventricle was also dilated, as was the fourth. A lumbar route myelogram revealed an advanced Chiari malformation, with obstruction to the exit of the fourth ventricle and cerebellar tonsillar ectopia. At operation, the foramen magnum was decompressed and an evident cerebellar ectopia, more severe on the right than on the left, was seen. The cerebellar tonsils appeared to be enlarged and extended to the lower border of C1. Membranous obstruction of the foramen of Magendie was seen on retraction of the tonsils. Postoperatively, the patient made an uneventful recovery and at follow-up six months after operation was completely symptom free and leading a normal life.

Trauma

In seven patients there was a history of initiation or aggravation of the symptoms by trauma, either to the head and neck or elsewhere (Table 3.4). This phenomenon is illustrated by the following case reports:

Table 3.4. *Cases associated with trauma*

Case/Age	Type of trauma	Result
P.R./25	Football injury to R. shoulder	Dissociated anaesthesia
A.F./15	Twist of R. leg	Asymmetrical spastic paraparesis
K.B./50	Cough; manipulation of neck	Pain and dissociated anaesthesia; dysarthria and palatal weakness
R.H./21	Stretching over car seat	Pain and numbness
L.O./53	Injury to R. shoulder	Neurogenic arthropathy
G.K./53	Fall on back of head	Neck pain and stiff legs
P.S./28	Sneeze	Trigeminal numbness

P.R., MALE (N.32384)
This 25-year-old technician had previously suffered no significant illness but attended the outpatient clinic, having injured his right shoulder while playing football six weeks earlier. Within a week of this injury he noticed that he was unable to distinguish between hot and cold down the right side of the body when he was in the shower or bath and he developed shooting pains in the region of the right shoulder. He commented that in the fall, he fell horizontally, jarring the whole of the left forequarter, left side of the trunk and left lower limb. There was no weakness of the left limbs and no reflex change. The abdominal reflexes were present and the plantar responses were clearly flexor.
Straight X-rays of the cervical spine were normal, including measurement of the anteroposterior diameter. The skull showed no evidence of basilar impression or other abnormality, but there was a suggestion of atlanto-occipital fusion, although tomograms were not undertaken to confirm this. A

myelogram showed a free flow of Myodil into the cervical region, where in the prone position there was evident widening of the cord shadow at the level of C2. There was a degree of obstruction at the foramen magnum, but this was incomplete. In the supine position, the cerebellar tonsils were shown to lie at the level of the upper border of the posterior arch of C2 and a fairly high degree of obstruction was present. There was no flow into the cisterna magna or the fourth ventricle. It was concluded that the patient had tonsilar ectopia and perhaps some descent of the medulla.

Decompression was undertaken (Hankinson) and with the patient in the sitting position, a midline incision was made from the inion to the mid-cervical region and the muscles were stripped from the occipital bone and from the first two laminae. The exposed bone, including the first and second laminae, was then removed with rongeurs. The dura was opened with a Y-shaped incision, the medial limb extending to the upper border of C3. As demonstrated by supine myelography, the tonsils were prolapsed to this level and were freed; the left tonsil was prolapsed a little further than the right and seemed to be more constricted, so that the tip of the tonsil was less vascular. The cord was not seen below this level, so it was impossible to judge if there were any swelling. The tonsils were separated and a large orifice for the central canal was seen in the floor of the fourth ventricle and this was plugged with muscle. The dura was not closed, but was covered with Gelfoam. The muscles, aponeurosis and skin were repaired with multiple layers of interrupted black silk sutures. Following operation, the patient had a remarkably smooth course: he was discharged from hospital some nine days after operation and has been followed up since. He is well, is aware of his sensory disorder, but is not troubled by it, and the upper level of the spinothalamic sensory disturbance has fallen to the D3 level. He has been advised not to play soccer or rugby, but is able to participate in mountaineering.

K.B., FEMALE (N.20087)

In 1950, at the age of 31 years, this patient coughed while sitting in a chair and immediately experienced severe pain down the postaxial border of the right upper limb; the limb became weak and numb. She was found at that time to have dissociated anaesthesia in the right upper limb. Some 30 minutes after osteopathic manipulation of the neck in 1953 she developed dysarthria, palatal weakness and an abnormal gait. Investigation in 1954 revealed a 'thickening' in the spinal cord and a diagnosis of syringomyelia was made. Since that time she has burnt the right upper limb without pain and there has been deterioration of power in both the right upper and both lower limbs.

Physical examination in June 1966 revealed phasic nystagmus to the left and right, anaesthesia in the second and third divisions of the trigeminal nerve on the left, wasting of the tongue and very poor tongue movements with a left palatal palsy. There was thermal anaesthesia over the whole of the right forequarter and also the right leg. The right upper limb was globally wasted and weak and the deep tendon reflexes were absent. There was exaggeration of deep tendon reflexes in the left upper limb. The abdominal reflexes were absent and the legs showed a symmetrical spastic paraparesis with hyperreflexia and bilateral extensor plantar responses. Vibration sense was defective in both feet and joint position sense very poor in the right foot. In the right upper limb, joint position sense was defective in all the fingers and the wrist and the movements at the elbow joint were not fully appreciated. X-rays of skull were normal. X-rays of the cervical spine showed considerable narrowing of the C5 to 6 disc space with sclerosis of the adjacent disc spaces. Lumbar myelography in the supine position showed the available space for the upper cervical cord and lower brain stem in the region of the foramen magnum to be considerably narrowed. The bony canal appeared to be narrow and there was descent of the hind brain. The cerebellar tonsils were not visualised in a low position but there was undoubted hold-up of the return of Myodil from the cisterna magna into the spinal theca.

Because of a growth of coagulase-positive staphylococci from a wound of the right arm, operation was deferred and the patient has since refused surgery.

R.H., MALE (R.V.I.758079)

This patient presented at the age of 45 years, with a spastic paraparesis, proprioceptive loss in the hands and dissociated sensory loss in the upper limbs and chest. He described a sudden weakness of the hands and numbness which had come on in his 'twenties following his stretching over a car seat. He remembered pain in the neck and limbs with the development of numbness. By the time he came to investigation he had a bilateral *main succulente*. Investigation revealed no myelographic abnormality and he was not therefore subjected to operative decompression.

L.O., MALE (N.29308)

This 53-year-old patient described an injury to his right shoulder which he sustained at work.

Within a few weeks of this injury he noticed wasting of the right hand and, on examination in the clinic 12 months later, was found to have a neurogenic arthropathy involving the right shoulder joint, a dissociated sensory loss of the right upper limb and chest, and absent deep reflexes in the right upper limb. Myelography revealed no abnormality whatever and he was not therefore subjected to operative decompression.

G.K., FEMALE (N.36647)

This 53-year-old patient gave a history of trauma to the head at the age of 40 years. She fell and struck the back of her head and immediately complained of severe neck pain which persisted and was her presenting symptom. Soon after the trauma she noticed stiffness of the legs and on examination had a spastic quadriparesis, extensor plantar responses and a 'half-cape' of dissociated sensory loss on the right, with evident wasting of the right hand and dissociated anaesthesia involving that hand. Radiology of the cervical spine revealed a bifid arch of C1 and myelography confirmed a widened cervical cord with tonsillar ectopia. At operation, the angle between the occiput and the arch of the atlas was extremely acute and the tonsillar herniation was confirmed. A hiatus between the fourth ventricle and central canal was confirmed and the hole was plugged with muscle. Following operation, the patient had no complaint of occipital or neck pain and considered that the stiffness and weakness of her legs was considerably improved. Her dissociated anaesthesia and wasted hand persist.

P.S., FEMALE (N.25057)

This girl presented at the clinic at the age of 28 years and described how two years earlier she had sneezed and shortly after noticed numbness of her face. This was followed by wasting of the left hand and, on examination in the clinic, she had absent deep reflexes in the left upper limb, heightened reflexes in the lower limb but with flexor plantar responses, and a 'half-cape' of dissociated anaesthesia involving the left chest and left upper limb. At myelography her cord was widened from C3 to T9 and there was evident tonsillar descent. At operation the tonsillar descent was confirmed, a hiatus was noted between the fourth ventricle and central canal and this was plugged with muscle. She made an uneventful recovery but unfortunately the numbness of her face and left forequarter persist, although her legs, she says, are no longer stiff.

The significance of trauma in initiating symptoms in these cases cannot be questioned and must influence any theory of pathogenesis of communicating hydromyelia (see Chapter 8). Sneezing with associated venous pressure changes and venting of spinal fluid into the cranial cavity was followed by disturbance of the spinal tract and nucleus of the trigeminal complex. A fall with neck flexion was followed by the onset of long-tract signs suggestive of contusion or compression of the cord. Coughing in patient K.B. was presumed to have induced similar fluid changes resulting in pain and dissociated anaesthesia.

A.F., MALE (X.06920)

This schoolboy was seen at the age of 15 years, when he stated that a year before investigation, while playing football, he had caught his right foot in a hole in the ground and had tripped. wrenching his foot. From then on he began to limp and his mother stated that he was walking very stiffly and that gradually his walking became worse. In addition, he had complained of occasional occipital headaches and during the year developed paraesthesiae in the hands. He continued to play games at school, but his kicking power for football became greatly reduced and he began to fall. There was no history of loss of power in the upper limbs. On examination, he showed a lumbar lordosis and kyphosis and walked with a distinct limp, dragging the right leg. There were no abnormalities to be detected in the cranial nerves or upper limbs but he showed an asymmetrical spastic paraparesis, with increased reflexes, more marked on the right than the left, and extensor plantar responses. The abdominal reflexes were absent, but otherwise the neurological examination was unremarkable.

Radiological examination of the cervical spine revealed no abnormality and the dorsal spine showed a scoliosis, convex to the right. Myelography revealed a widened cord in the cervical region and evidence of arachnoiditis. The contrast medium would not enter the cisterna magna in the supine position, but in the prone position it readily ran forward on to the clivus; this is a radiological sign of basal arachnoiditis (see Chapter 4). There was considerable discussion as to whether this boy should undergo high cervical decompression or have ventricular drainage established. Ventriculography showed a slight degree of asymmetrical ventricular dilatation with dilatation of the anterior recesses of the third ventricle. The fourth ventricle was not displaced, but very little air would leave the ventricular system.

Foramen magnum decompression and posterior fossa exploration was undertaken (Hankinson). After the dura had been opened there was evident, very marked arachnoiditis, the tonsils were prolapsed some 5 mm and were firmly bound down to the medulla and to each other. Arachnoiditis was seen to pass laterally around the side of the medulla and, with considerable difficulty, vascular adhesions were divided. The fourth ventricle was entered and it was seen that there was a very deep midline cleft at the entrance to the central canal well developed at the obex. This was plugged with a small piece of muscle. Following operation, the patient made an uneventful recovery and his clinical condition has remained unchanged.

CLINICAL SIGNS

Table 3.5 illustrates the clinical signs presented by the 100 patients, with regard to general appearance, demonstration of features of syringomyelia, hydrocephalus, brain stem and cerebellar involvement, proprioceptive loss in the upper limbs, spastic paraparesis and neurogenic arthropathy. Table 3.6 shows a breakdown of the figures for the general appearance of the patient while Table 3.7 gives the numbers presenting the various features of Table 3.5.

General Appearance

The overall appearance of 65 of the 100 patients was thought to be perfectly normal in respect of head shape and size, length of neck, straightness of spine and appearance of the hands. In 13 patients, the neck was thought to be unusually short and in 22 there was a clinically recognisable scoliosis. On clinical grounds alone, five patients were thought to have a large head and in three of these, symptoms and signs suggested hydrocephalus. However, such symptoms and signs were present in a further eight cases where the head circumference was thought to be normal. In one patient (M.W., N.23535), the left upper limb was congenitally underdeveloped. One patient (C.B., N.26675), had had a meningocele repaired in early life. In three patients, there were the atrophic lesions of Morvan's disease (see Figs. 2.3 and 2.4), while three patients (J.C., N.19444; B.E., N.21652 and R.H., R.V.I. 758079) showed classical *main succulente*. In one further patient, there was persistent and distressing hyperhydrosis.

R.J., MALE (N.42447)
This 33-year-old steel dresser first noticed pain in the back of the left hand and right side of the neck twelve months before he presented himself. The pain in the hand was described as sharp and radiated up the back and outer aspect of the forearm to the elbow. The pain became continuous and was associated with the neck pain which radiated into the right shoulder and scapula. Some three weeks after the onset of pain, he noticed wasting of the small hand muscles and thinning of the muscles around the elbow. Since the onset of this wasting, the forearm has wasted considerably and one month after onset of wasting, he was unable to hold objects in the left hand and could not pick them up because of weakness of grip. The hand eventually became clawed and the fingers stiff to the point, on presentation, of the patient being unable to straighten his fingers even passively. Shortly after the onset of these symptoms, he noticed perspiration of the left hand and wrist. The perspiration was constant through the day and night and at times would be so severe that sweat would drip from the fingers to the floor. The sweating appeared to be limited to the proximal flexure crease of the left wrist. Examination showed a gross deformity due to kyphoscoliosis of the thoracic spine.

The patient showed no abnormality in his cranial nerves but in the upper limbs there was evident wasting of the muscles innervated by the C5 to D1 segments, with the deformities mentioned above. He could detect light touch in all areas except the palmar surface of the third, fourth and fifth fingers of

Table 3.5. *Clinical signs and presentation (100 cases)*

Case	Age/sex	General appearance	Syringomyelia	Hydrocephalus	Brain stem and cerebellum	Proprioceptive loss	Spastic paraparesis	Neurogenic arthropathy	Case	Age/sex	General appearance	Syringomyelia	Hydrocephalus	Brain stem and cerebellum	Proprioceptive loss	Spastic paraparesis	Neurogenic arthropathy
MJ	24/F	N	+	−	−	−	+	−	RA	57/M	N	+	−	−	−	−	−
MT	49/F	N	+	−	−	−	−	−	CD	36/M	N	−	−	+	−	−	−
GS	21/M	N	+	−	−	−	+	−	MA	36/F	N	+	−	−	−	+	−
FD	52/F	N	−	−	+	+	+	−	LO	54/M	N	+	−	−	−	−	+
FW	56/M	N	+	+	−	−	−	−	SW	47/M	N	+	−	−	−	−	+
WH	30/M	N	+	−	−	−	+	−	WD	46/M	SN	+	−	−	+	+	−
PG	33/F	LH	+	+	−	+	+	−	MC	25/F	N	−	−	−	+	−	−
ES	64/M	N	+	−	−	−	+	−	RH	58/M	N	+	−	−	−	+	+
RJ	44/M	N	+	−	−	−	−	−	JJ	63/M	SN/S	+	−	−	−	+	−
EJ	56/M	N	+	−	−	+	+	−	MG	47/M	S	+	−	−	+	+	−
CW	43/F	LH/S/SN	+	−	−	−	+	−	NA	56/F	S	+	−	+	+	+	+
MC	35/F	N	+	−	−	−	−	−	MB	34/F	N	+	−	−	−	−	−
LB	46/F	N	+	−	+	−	−	−	PR	25/M	N	+	−	−	−	−	−
KE	24/M	N	+	−	+	−	−	−	MM	32/F	N	+	−	−	−	−	−
DL	44/F	S	+	−	−	−	+	−	AR	64/F	N	+	−	−	−	−	−
LJ	5/F	N	−	−	−	−	+	−	GK	53/F	S	+	−	−	−	+	−
MC	51/F	N	+	−	−	−	+	−	JA	46/M	SN	−	−	+	−	+	−
CH	42/M	N	+	−	−	+	+	−	EK	31/F	M	+	−	−	−	−	−
LT	44/M	N	+	−	−	−	+	−	RJ	33/M	SN/S	+	−	−	−	−	−
NS	22/M	N	+	−	−	−	+	−	AT	45/F	N	−	+	−	−	+	−
JL	52/M	N	+	−	+	−	−	−	HH	48/M	S	+	−	−	−	−	−
EW	45/F	N	+	−	−	+	+	−	RS	20/M	SN	+	−	−	+	−	−
JC	56/M	L/H	+	+	−	+	+	−	RB	47/F	S	+	−	−	−	+	−
KB	50/F	S/N	+	−	+	+	+	−	HC	48/M	S/LH	+	+	−	−	+	−
WM	47/F	N	+	−	+	−	+	−	JE	36/M	SN	+	−	−	−	+	−
RP	27/M	N	−	+	+	−	−	−	WP[a]	55/M	N	+	−	−	−	+	−
BE	57/F	S	+	−	−	+	+	+	NY	47/F	S	+	−	−	−	+	−
JM	37/F	S	+	−	+	−	−	−	LR	34/M	N	−	−	+	−	−	−
MM	31/F	S	+	−	+	−	+	−	RR	37/M	N	+	−	+	−	+	−
MD	40/F	N	−	+	−	−	−	−	SD	44/F	N	−	−	−	−	+	−
RC	35/F	N	+	−	+	−	−	−	HB	60/M	S	+	−	−	−	−	−
FB	42/M	N	+	−	+	+	−	−	TR	24/M	SN	−	−	+	−	+	−
PM	42/M	N	+	−	−	−	+	−	DB	34/M	SN	−	−	+	−	+	−
HC	32/M	N	+	−	−	−	+	−	RP	26/M	N	+	+	−	+	−	−
MW	57/F	A/SN	+	−	−	−	+	−	WO	25/F	S	+	−	−	+	+	−
JE	19/F	N	+	−	−	−	−	−	GJ	24/M	N	−	+	+	−	−	−
PS	28/F	N	+	−	+	−	−	−	EJ	53/F	N	+	−	−	−	−	−
PH	20/F	N	−	−	−	−	+	−	VN	30/M	N	−	+	+	−	−	−
ME	38/F	LH	+	−	+	−	+	−	ER	48/F	N	+	+	+	−	+	−
AB	58/F	N	+	−	−	−	−	−	MG	59/F	N	+	−	−	−	−	−
JA	21/M	N	+	−	−	+	+	−	DO	20/M	N	+	−	−	−	−	−
CB	18/F	M	−	−	+	−	−	−	WI	30/M	S/SN	−	−	+	+	−	−
JW	30/F	S	−	−	−	+	−	−	JJ	50/M	S	+	−	−	−	+	+
HF	43/F	N	+	−	−	−	+	−	RH	45/M	N	+	−	−	+	+	−
MJ	53/F	N	+	−	−	−	+	−	JW	36/M	S	+	−	−	−	+	−
ET	23/F	N	−	−	−	−	−	−	FD	56/F	S	+	−	−	−	+	−
JR	17/F	N	−	−	−	−	−	−	AF	15/M	N	−	−	−	−	−	+
CP	42/F	S	−	−	−	−	+	−	GF	64/M	SN	+	−	−	−	−	−
FH	36/F	N	+	−	−	−	+	−	DR	22/M	SN	+	−	−	−	−	−
JC	26/M	SN	+	−	+	−	+	−	EH	50/F	N	+	−	−	+	+	−

[a] This man developed a complete external opthalmoplegia before death.

Key: N: Normal; SN: Short neck; S: Scoliosis; LH: Large head.

the left hand from the metacarpophalangeal joint distally. Pain was detected in all areas except the tips of the third, fourth and fifth fingers on their palmar aspect from the distal interphalangeal joints distally, and this pain loss extended on to the dorsal aspect of the second and fourth fingers. Vibration sense was intact in the hands and deep pain appreciation was intact in the hands. Position sense was lost in the fingers and at each wrist. It was noticeable that the left hand was warmer to a point just above the wrist than similar areas on the right. The patient was thought to be suffering from central cord cavita-

Table 3.6. *General appearance of the patients*

Normal	65
Short neck	13
Scoliosis	22
Large head	5
Underdeveloped limb	1
Meningocele	1
Morvan's syndrome	3
Main succulente	3
Hyperhidrosis	1

Table 3.7. *Clinical signs*

Syringomyelia	78
Spastic paraparesis	58
Hydrocephalus	11
Brain stem	29
Proprioceptive loss	20
Neurogenic arthropathy	7
Horner's syndrome	5

tion, perhaps secondary to his long-standing kyphoscoliosis, but a lumbar-route myelogram carried out with considerable difficulty showed no evidence of an abnormality at the foramen magnum and, in view of the gross distortion of the spinal canal, he was not considered suitable for operation. The hyperhydrosis of the left hand has been partially controlled with the local use of 5 per cent poldine in 80 per cent DMSO.

Comment. This is the only case in this series to show such extensive hyperhydrosis. The pathogenesis of this remains obscure. Possibly there is 'sympathetic irritation' or it may be that his tendency to palmar hyperhidrosis is bilateral but that he has a partial sympathetic block on the right so that the left hand appears to sweat more. Hyperhidrosis is much more frequent in cases of cavitation following trauma (see Chapters 10, 11 and 12).

Classical Syringomyelia

The 78 cases so designated have presented with the features generally associated with classical syringomyelia, namely an area of dissociated anaesthesia of an upper limb or forequarter, with absent deep tendon reflexes and lower motor neurone wasting and weakness of one or other upper limb. In our 78 patients, dissociated anaesthesia was present in all; in 40 it was unilateral and in a further 38, bilateral. The dissociated sensory loss showed, in all but one patient, the classical suspended sensory level of the condition as traditionally described, but in one patient (P.R., N.32384) the spinothalamic type of sensory abnormality extended from C5 down through the trunk to involve the left lower limb totally. In one patient (C.P., N.27058), there was a Brown-Séquard sensory loss involving the right lower limb with no true dissociated anaesthesia involving the shoulders or upper limbs. This man showed a scoliosis and had presented with pain in the right lower limb and analgesia of the right leg. Examination revealed spinothalamic sensory loss in the right lower limb with posterior column loss in the left lower limb and an extensor plantar response on that side, with hyperreflexia at the knee and ankle on the left. His abdominal reflexes were preserved. His X-rays of skull showed atlanto-occipital fusion and supine myelography confirmed that he had descent of the cerebellar tonsils. This patient has not as yet been subjected to operative decompression. His history is so long that we believe it very unlikely that he could have a spinal neoplasm.

Spastic Paraparesis

Fifty-eight of the patients at the time of examination showed changes interpreted as those of a spastic paraparesis. Usually symmetrical, the changes were those of corticospinal weakness involving both lower limbs, with hyperreflexia, extensor plantar responses and depression or loss of the abdominal reflexes on one or both sides. Most of the patients whose condition had existed for some years showed such a spastic paraparesis but it was particularly in those presenting with stiffness of the legs as their first symptom that the changes of upper motor neurone lesions to both legs were evident.

M.D., FEMALE (N.22319)

This 42-year-old patient presented with postural headaches of sudden onset and was found on examination to have haemorrhagic papilloedema. Her investigations included angiography, which confirmed ventricular dilatation, and exploration of the posterior fossa showed the foramen of Magendie to be obstructed by thick arachnoid membrane.

Hydrocephalus and Brain Stem Signs

Under this heading we have grouped those patients who appeared on clinical examination to have involvement of structures in the medulla and who, by their appearance, by the presence of significant headache or drop attacks were thought to have hydrocephalus. In the 11 patients thought to have hydrocephalus the head was clinically enlarged in three only, but radiological examination of their skulls showed suggestive changes (see Chapter 5). Only one patient presented with papilloedema.

Table 3.8. *Brain stem signs*

Nystagmus	27
Cerebellar incoordination	2
Wasted tongue	3
Palatal weakness	3
Trigeminal sensory disturbance	4

Cerebellar incoordination was seen in the upper limbs of two patients only, and in three there was evident wasting of the tongue and palatal weakness indicative of involvement of the dorsal nucleus of the vagus and the hypoglossal nucleus in the floor of the fourth ventricle. More significant were the cases showing nystagmus, of which there were 27, and those showing trigeminal sensory disturbance, of which there were four.

Nystagmus

Benign positional nystagmus was not seen in this group of cases. In most of the patients the nystagmus was fine, phasic grade I nystagmus to the left or right, or both. But in 10 patients there was evident phasic vertical nystagmus on upward and downward gaze. In some of those patients showing phasic nystagmus to the left or right there was a rotational element. In one patient, nystagmus was rotatory to the left, and in five there was evident vertical nystagmus with the rapid phase upwards on gaze to the left and right. No firm conclusions can be drawn from these findings and the nystagmus in these cases will be subjected to specific study.

Proprioceptive Loss

It was our initial impression throughout the first of this series of cases that loss of joint position sense in the hands was an important feature which differentiated the Chiari malformation from classical syringomyelia. However, in this series of patients, 20 have shown significant proprioceptive loss in the hands with relative sparing of the lower limbs. We now feel that this is one of the early signs of upper cervical cord damage, due either to the compression of the cord by the Chiari malformation in the foramen magnum or the extension of the cystic dilatation up to this level. The classical descriptions of syringomyelia make little of such a loss of joint position sense in the hands but several of our patients (notably F.D., N.3477) were considerably disabled by this sensory impairment and we regard it as a highly significant finding.

Neurogenic Arthropathy

On radiology, seven patients showed the changes of classical neurogenic arthropathy (Charcot joint), although these changes were not specific. In one patient (B.E., N.21652), the shoulder, elbow and wrist were involved on the right side. In four others, the right shoulder was involved (L.O., 29308; S.W., 24712; R.H., 31101 and J.J., 757353). In N.A. (31775), the left shoulder was involved. In both S.W. and N.A. there were also the changes of Morvan's syndrome in the hand on the affected side.

Horner's Syndrome

Five patients in this series showed a unilateral Horner's syndrome and in all there was significant weakness and wasting of the small hand muscles, suggestive of anterior horn cell damage in the D1 segment. In no case was a bilateral Horner's syndrome seen.

CHAPTER FOUR

Basal Arachnoiditis

J. B. FOSTER and P. HUDGSON

In 1887, and later in 1891, Joffroy and Achard first described cervical cord cavitation associated with what was then termed pachymeningitis cervicalis hypertrophica. In 1897, Schwartz presented a clinicopathological report on a patient with chronic arachnoiditis which was thought to be syphilitic. The autopsy showed extensive cavitation of the cervical cord.

Since these initial reports, occasional descriptions of syringomyelia associated with marked spinal arachnoiditis have appeared but a search for the cause has usually failed (Davison and Keschner, 1933; Lubin, 1940). Occasional cases have been described following meningitic illness and subarachnoid haemorrhage (Nelson, 1943). Most of the previous authors have considered that vascular compression, inducing ischaemic necrosis, has been responsible for the cavitation. Spinal arachnoiditis with associated cavitation will be discussed in Chapters 15 and 16.

In the chapter on pathology, we discuss cord cavitation induced in the dog by the subarachnoid injection of emulsions of talc/sodium nucleinate and fatty acids, or kaolin and Pantopaque into the cisterna magna (Camus and Roussy, 1914; McLaurin et al, 1954). These injections resulted in the development of cavitation following an induced arachnoiditis. We have examined one case at autopsy (H.C., N.22956) where tuberculous meningitis was followed by an adhesive posterior fossa arachnoiditis and subsequent central cord cavitation. In the series of patients considered here, there have been two others with whom meningitis has been followed by a similar arachnoiditis demonstrated radiologically (R.A., N.25104) and at operation (M.M., N.22171). Adhesive arachnoiditis in the posterior fossa is a recognised cause of obstructive or internal hydrocephalus and we have had the opportunity of studying three parents with hydrocephalus and syringomyelia, one of whom had a previous history of meningitis.

CASE REPORTS (HYDROCEPHALUS)

M.M., FEMALE (N.22171)

This patient, a farm worker, suffered at the age of 19 years (in 1953) from generalised headaches, nausea, vomiting, photophobia and neck stiffness, with development after four days, of diplopia and postural vertigo. A diagnosis of benign lymphocytic meningitis was made and she was treated with intramuscular penicillin for five days. Her symptoms disappeared and after three weeks she was discharged from hospital. Three weeks later, the headaches and vomiting recurred and the patient became drowsy. Lumbar puncture showed clear cerebrospinal fluid under high pressure; total protein 90 mg/100 ml, 12 lymphocytes/μl. She developed bilateral papilloedema and was transferred to the Newcastle General Hospital for further investigation. Further C.S.F. examination suggested tuberculous meningitis although direct examination and culture of the C.S.F. and urine failed to reveal tubercle bacilli or other pathogenic organisms.

Because of the chronic meningitis with obstruction to the C.S.F. pathways, and in view of her papilloedema and drowsiness, ventriculography was undertaken and showed bilateral dilatation of the lateral ventricles, the third ventricle, the aqueduct and fourth ventricle. It is doubtful if any air escaped from the fourth ventricle. A posterior fossa exploration with removal of the arch of the atlas was performed by Mr L. P. Lassman. The cerebellar hemispheres were normal in size and not bulging, and the tonsils were in normal position. Torkildsen's procedure was performed.

Subsequently, the patient was treated with antituberculous chemotherapy and cortisone. Within three months her C.S.F. had returned to normal, and her symptoms remitted. Apart from vague pains in the neck from the age of 26 she remained well until the age of 30 years when she developed a burning pain of radicular distribution under the left breast. This gradually worsened over four weeks and was made worse by contact with her clothes. Concurrently, she developed paraesthesiae, made worse by coughing, in the digits of the left hand. There was no headache, weakness or sphincter disturbance. The

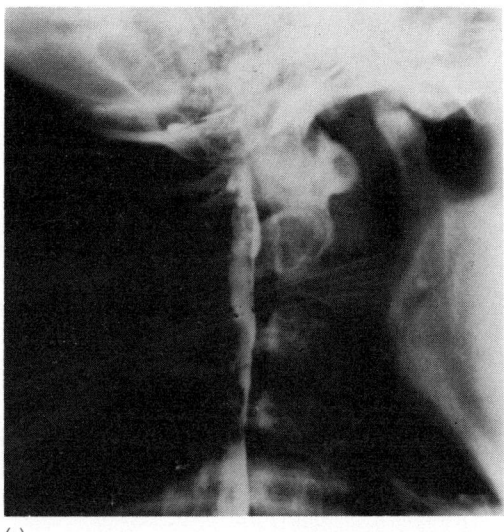
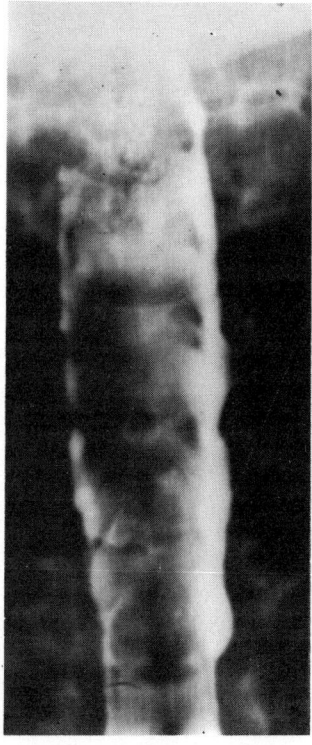

Fig. 4.1. Case M.M. (N.22171). Lateral and anteroposterior views of prone myelogram showing widening of the cord shadow and fragmentation of the contrast medium presumed to be due to arachnoiditis and dilatation of the cervical cord.

(a)

(b)

only neurological finding was a patchy but consistent mild impairment of the sensation of light touch, superficial pain and temperature over both sides of the face and neck and over the left side of the trunk to L4 but apparently sparing the left arm and the sacral dermatomes. Vibration sense was impaired in the lower part of the body up to the xiphisternum, though joint position sense was normal throughout. Because of the incapacitating nature of the pain further investigation was undertaken. The C.S.F. was under normal pressure; total protein 90 mg/100 ml (9 per cent γ-globulin), no pleocytosis. A lumbar myelogram showed the changes of arachnoiditis in the upper cervical region extending into the posterior fossa, with pathological expansion of the upper and mid-cervical regions of the spinal cord (Fig. 4.1). In the supine examination the position of the cerebellar tonsils could not be assessed due to arachnoidal adhesions preventing the flow of contrast medium into the cisterna magna (Fig. 4.2).

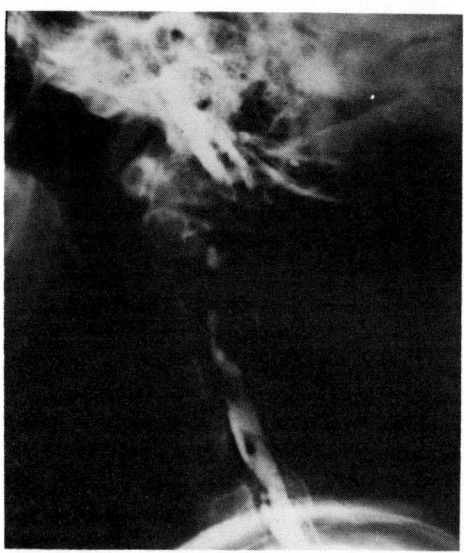

Fig. 4.2. Case M.M. (N.22171). Supine view of the same patient as in Fig. 4.1. Despite considerable head-down tilting little contrast medium would enter the upper cervical region and none passed into the cisterna magna.

It was not thought warranted to proceed further at this stage, and symptomatic relief was obtained by alcohol and phenol intercostal block. A few months later circumoral paraesthesiae developed followed after eight months by impairment of temperature sensation in the left hand, slight weakness of dorsiflexion of the left foot, and over two days, clumsiness of the left arm and leg, dysarthria and dysphagia. There was loss of joint position sense in the left hand and left foot, impairment of vibration sense bilaterally up to the level of the clavicle, and pseudoathetoid movements of the fingers of the left hand. Superficial pain and temperature sensation were impaired on both sides of the face, sparing the nose and upper lip, and on the whole of the left side of the body. The reflexes were pathologically brisk overall with a left extensor plantar response. There was slight weakness and wasting of the small muscles of the left hand, with transient weakness of the left side of the palate and tongue. There was no abnormality of external ocular movements and no papilloedema. The C.S.F. was under normal pressure with a free rise and fall; total protein 68 mg/100 ml (7 per cent γ-globulin), 72 W.B.C./μl, 79 per cent lymphocytes and 21 per cent polymorphs. Ventriculography was performed and it was seen that the Torkildsen tube was occluded. The ventricular fluid contained 8 mg/100 ml of protein, with no pleocytosis. There was moderate bilateral dilatation of the lateral and anterior part of the third ventricles, and air could not be induced to pass into the aqueduct or fourth ventricle.

It was concluded that the patient had a chronic arachnoiditis causing syringomyelia and syringobulbia due to obstruction to the outflow of C.S.F. from the fourth ventricle, with a total or partial occlusion of the aqueduct. Posterior fossa re-exploration and laminectomy of the second and third cervical vertebrae was done (Hankinson). There was a moderately bulging meningocele from the previous posterior fossa decompression and extremely thick adhesions were found around the upper cervical cord and lower cerebellum. The spinal cord was indented at the level of the arch of the atlas by tight bands of adhesions. Clear fluid was aspirated from within the dilated cervical cord. Analysis showed no cells and a total protein of 38 mg/100 ml.

Comment. Following presumed tuberculous meningitis with obstructive hydrocephalus at the age of 19 years, this patient developed the signs and symptoms of syringomyelia and syringobulbia at the age of 30. On myelography in the supine position, contrast medium failed to enter the cisterna magna but flowed forward into the prepontine cistern. At operation, there were dense adhesions around the lower cerebellum and cervical cord occluding the foramen of Magendie. It was not possible to free these adhesions, and after operation the patient was temporarily worse.

R.P., MALE (R.V.I.601798)
This man presented at the age of 26 years. He had suffered for three years from occasional attacks of rotatory vertigo precipitated by head movements which had become particularly frequent in the previous two months. At the same time, he developed pain down the left side of the neck and left arm, associated with numbness and painful dysaesthesiae. He also noted impairment of pain and temperature sensation and power in the left arm, with lesser weakness in the right arm. Examination revealed impairment of spinothalamic sensation over the left side of the face, left upper limb and trunk to the level of D6. There was also impairment of vibration sense in the left hand, though joint position sense was normal throughout. There was mild global weakness of the left arm, most evident on dorsiflexion of the wrist and in the small muscles of the hand. The deep tendon reflexes were absent in the left arm, and normal in the other limbs.

Radiography of the cervical spine and skull revealed no abnormality. C.S.F. obtained under normal pressure showed protein 31 mg/100 ml (8 per cent γ-globulin), 1 white cell/μl. The V.D.R.L. test in blood and C.S.F. was negative. Lumbar myelography was complicated by the presence of subdural contrast medium and all the abnormalities were at first thought to be due to this. An air myelogram showed the cerebellar tonsils to lie at the upper border of the atlas. No air entered the ventricles and none passed beyond the basal cistern. In retrospect, the myelographic appearances were so similar to those of the former case that they were undoubtedly significant. The patient was considered to have syringomyelia due to the Chiari type I anomaly, and a posterior fossa decompression was done (Hankinson) removing the posterior 2 cm of the foramen magnum and the arch of atlas. The tonsils lay 1 cm below the foramen magnum and were constricted at the atlanto-occipital junction. The atlanto-occipital membrane was abnormally thick, and the tonsils were bound together with adhesions. These were divided and the fourth ventricle was found to be abnormally dilated. He made a good recovery from operation and the painful dysaesthesiae disappeared, although the neurological deficit remained unchanged.

Five months after the operation, the patient developed a severe headache preceded by tinnitus and vertigo. Two days after the onset of these symptoms, he noticed diplopia on looking to the right. In addition to the previous neurological abnormalities there were now papilloedema, bilateral phasic nystagmus, diplopia on left lateral gaze and cerebellar incoordination in the left limbs. Shortly after admission to hospital, he developed respiratory failure requiring assisted respiration. A Spitz-Holter valve was inserted between the right lateral ventricle and the right internal jugular vein. Spontaneous respiration returned almost immediately and the patient made a rapid recovery to his initial state. Over the next 15 months he remained well with continuing improvement in the power and coordination in the left hand.

Comment. There was no previous history of meningitis and this patient developed the typical signs and symptoms of syringomyelia with demonstrable tonsillar ectopia. The impaired flow of contrast medium was presumed to be due to the diffuse arachnoiditis seen at operation. Following decompression he made a good recovery. However, five months after the operation he developed acute obstructive hydrocephalus with almost fatal central respiratory failure. Ventricular drainage produced a dramatic recovery.

M.D., FEMALE (N.22319)
This 42-year-old patient presented with a six-month history of headache, frontal in situation, throbbing in character and often coming on abruptly and clearing equally suddenly. Sometimes the headaches were determined by posture and on occasions she vomited. She was examined by her medical practitioner and found to have bilateral papilloedema. On examination, the bilateral papilloedema was confirmed with constricted fields and enlarged blind spots. There were flame-shaped haemorrhages at six o'clock at the lower margin of the left optic disc. Examination of her nervous system revealed no other abnormalities and X-rays of cervical spine and skull were normal.

Carotid angiography suggested ventricular dilatation, but no evidence of a space-occupying lesion in the supratentorial compartment. Ventriculography showed gross dilatation of the lateral ventricles, third ventricle and fourth ventricle. Examination of the dilated fourth ventricle, with a small amount of Myodil was undertaken and there appeared to be foraminal obstruction due to atresia, or basal arachnoiditis at the foramen of Magendie. Immediately following ventriculography, a Spitz-Holter ventriculo-atrial shunt was inserted into the right lateral ventricle and 10 days later the patient underwent posterior fossa exploration. The posterior rim of the foramen magnum and arch of C1 were removed. On opening the dura it was found that the arachnoid was extremely thick and the foramen of Magendie was obliterated by thick membrane. When this was removed, a profuse flow of C.S.F. and Myodil was seen. The arachnoid over the upper cervical cord was also abnormal. The lower part of the widely dilated fourth ventricle was inspected and a depression in the region of the obex at the site of the origin of the central canal was seen, presumably as a result of the raised pressure in the fourth ventricle. The dura was not closed. Throughout the operation, the patient's condition was satisfactory and following operation she made an uneventful recovery, and has had no further symptoms.

Comment. This patient presented with acute internal hydrocephalus, complaining of headache as her only symptom and, on examination, having papilloedema as her only physical sign. She was investigated because of the raised intracranial pressure and ventriculography revealed foraminal outlet obstruction in the posterior fossa. At operation, this was seen to be due to intense arachnoiditis, but prior to exploration, her ventricles were drained (see case N.34639, Chapter 6).

ADDITIONAL CASES OF BASAL ARACHNOIDITIS

A study of 15 further cases has allowed us to recognise a common radiological pattern in such patients although the clinical symptomatology has varied considerably from case to case. In Table 4.1, we analyse the principal features of this group of cases.

The youngest patient in the series was a schoolboy (A.F., X06920) aged 14 years at the onset of his symptoms, and whom we studied in the following year; a further patient (H.C., N.22956) who presented at the age of 32 years had suffered from symptoms since the age of 13. Our oldest patient was aged 66 years at the time of his investigation and he had suffered from symptoms for six years. In this group of 18 cases, including the three with hydrocephalus described above, there are 11 males and seven females. The clinical presentation did not serve to differentiate them from the general group of cases of cervical cord cavitation since, in nine, the features were those of classical syringomyelia. In five cases there was associated nystagmus suggesting bulbar involvement, and two presented with a spastic paraparesis. In addition to M.M. (N.22171), two further patients gave a

Table 4.1. *Cases of basal arachnoiditis*

Case	Sex	Age of onset (years)	Age at investigation (years)	Presentation	Previous mengingitis	Abnormal radiology			Myelographic findings				C.S.F. protein (mg/100 ml)	Operation
						Skull	Spine	Expanded cord	Arachnoiditis	Tonsillar prolapse				
G.S. N2677	M	18	21	Syringomyelia	–	–	–	+	+	+			33	+
J.M. N22033	F	16	37	Syringomyelia +nystagmus	–	Hydrocephalus	Wide canal	+	+	–			26	(Under consideration)
M.M. N22171	F	19	31	Syringomyelia +nystagmus	+	Hydrocephalus	Bifid arch C1	–	+	–			90	+
M.D. N22319	F	39	40	Hydrocephalus	–	–	–	Ventriculography					–	Posterior fossa exploration
F.B. N22672	M	38	42	Syringomyelia +nystagmus	–	–	–	–	+	–			53	+
P.M. N22696	M	58	42	Syringomyelia	–	–	Atlanto occipital fusion	–	+	+			72	+
H.C. N22956	M	13	32	Spastic paraparesis	+	–	–	–	Not examined				24	– (Died)
M.W. N23535	F	53	57	Syringomyelia	–	–	Wide canal	–	+	+			22	+
J.A. N26287	M	16	21	Syringomyelia	–	–	–	+	+	+			62	+
F.H. N27390	F	35	36	Syringomyelia	–	–	–	+	+	+			45	+
J.C. N28472	M	24	26	Syringomyelia +nystagmus	–	Basilar impression	Atlanto-occipital fusion	+	+	–			47	+ (And valve)
R.A. N25104	M	55	57	Syringomyelia	+	–	–	–	+	–			49	–
H.B. 552752	M	60	66	Syringomyelia	–	–	Scoliosis	–	+	–			45	(Too old)
R.P. 601798	M	23	26	Syringomyelia +hydrocephalus	–	–	–	–	+	+			31	+ (And valve)
W.O. 61061	F	22	25	Syringomyelia +nystagmus	–	–	Scoliosis	+	+	–			36	+ (And valve)
M.G. 670218	F	46	59	Syringomyelia	–	–	–	+	+	–			41	–
J.W. 491137	M	24	36	Syringomyelia	–	–	–	+	–	Prone examination			52	– (Died)
A.F. X06920	M	14	15	Spastic paraparesis	–	–	Scoliosis	+	+	–			43	+ (And valve)

BASAL ARACHNOIDITIS

history of meningitis. In three there was radiological abnormality of the skull, namely hydrocephalus and basilar impression, and in two patients the antero-posterior diameter of the spinal canal was widened, which suggested long-standing dilatation of the central canal. One patient was seen to have a bifid arch at C1 and in two others there was atlanto-occipital fusion, i.e. occipitalisation of the atlas (see Chapter 5). Scoliosis was present in the thoracic spines of three further cases.

Fifteen of the patients were examined myelographically in both the prone and supine positions. One patient (J.W.) was examined in the prone position only. In two further patients (M.D., previously described) ventriculography was undertaken and one patient (H.C.) was not examined by myelography. In eight of the patients examined the cord was seen to be expanded. In the patient whose examination was limited to the prone position, the cord was also seen to be expanded. Tonsillar prolapse or ectopia was seen on the supine myelogram of six patients and the signs which we have come to recognise as significant of arachnoiditis at the foramen magnum and posterior fossa were seen in 14 patients. Two further patients were found to have arachnoiditis at operation, and arachnoiditis was seen at post-mortem in two others (H.C. and J.W.).

The spinal fluid protein values in these 18 patients were unremarkable, despite the pathological changes in the arachnoid noted on radiography and at operation. The highest protein (M.M.) was 90 mg/100 ml and the lowest (M.W.) was 22 mg/100 ml. Only six proteins within this range were above the normal value accepted by this laboratory (45 mg/100 ml). Of the 18 patients, two came to post-mortem and were not subjected to operation (see Chapter 8). Of the 15 patients in the series to be described, nine have been explored. Of the remainder, one is under consideration for operation (J.M.), in one other the symptoms and signs are not considered severe enough to justify operation and he shows no evidence of clinical deterioration (R.A.). Case H.B. is in poor general condition and, at 66 years of age, would present too great an operative and anaesthetic risk to justify exploration. Four of the 12 cases subjected to operation have required ventriculo-atrial shunts.

FURTHER CASE REPORTS

G.S., MALE (N.2677)

In 1963, at the age of 18 years, this patient noticed numbness and loss of fine movements of the right hand. He had previously been well, though he had a mild head injury without unconsciousness eight months previously. In the right arm there was slight weakness of the small muscles of the hand, and patchy impairment of light touch, pinprick, vibration sensation, stereognosis and two-point discrimination, with normal thermal appreciation. Generalised hyperreflexia was evident with bilateral equivocal plantar responses. The C.S.F., under normal pressure showed: protein 33 mg/100 ml, no pleocytosis, Lange curve 122210000. Lumbar myelography performed only in the prone position showed a slight widening of the spinal cord in the mid-cervical region. A diagnosis of syringomyelia was made but further treatment was thought unnecessary at this stage.

Myelography a year later showed further expansion of the lower two-thirds of the cervical cord. The C.S.F. protein content was 54 mg/100 ml; no pleocytosis. There was no change in his functional state. At the age of 18 years, the patient was symptomatically unchanged though the sensory loss and diffuse wasting of all muscle groups in the right arm had become worse. Further myelography, with a supine examination, revealed expansion of the cervical cord and obstruction to the flow into the cisterna magna of contrast medium which flowed forward into the prepontine cistern (Fig. 4.3). The cerebellar tonsils were thought to lie below the foramen magnum. The C.S.F. protein was 16 mg/100 ml (21 per cent γ-globulin); no pleocytosis.

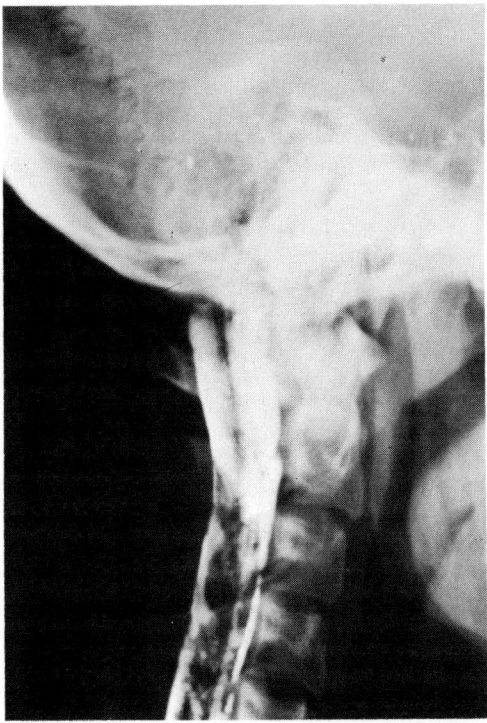

Fig. 4.3. Case G.S. (N.2677). Supine myelography showing complete failure of contrast medium to enter the cisterna magna. A few drops flow forward into the pontine cistern and illustrate the acute cranio-vertebral angle. (Reproduced with permission from *Journal of Neurological Sciences*, **8**, 451–464, 1969.)

The foramen magnum region was decompressed (Hankinson). The atlanto-occipital membrane and underlying dura were extremely thickened and there were gross arachnoidal adhesions over the whole of the cerebellum and upper cervical spinal cord. The foramen of Magendie was obliterated. Further dissection of the outlet of the fourth ventricle was impossible due to vascular adhesions. Histological examination showed connective tissue containing groups of arachnoidal cells with no tumour or inflammatory cell infiltrate. Postoperatively, the biceps and supinator reflexes on the right became sluggish, though the patient's neurological state was otherwise unaltered. In the four years since operation, there has been slight subjective improvement in the power and sensation of the right arm, although the neurological deficit remains unchanged. He is in full-time work.

Comment. There was no previous history of meningitis, but the myelographic appearances were identical to those in M.M. (N.22171), and there were gross arachnoidal adhesions around the cerebellum and upper cervical spinal cord at operation. Operation produced no objective benefit.

J.M., FEMALE (N.22033)
This 37-year-old secretary had suffered intermittent backache for many years, noticeably related to posture and involving particularly the thoracic region of her spine. Since the age of 16 years, she had

been aware of painless blisters and burns involving the right hand. She had also noticed difficulty in assessing the temperature of water with her right hand and being right-handed, she had been aware recently of clumsiness when holding a pencil. For six years, she had noticed impaired sensibility in the right foot, especially when testing bath water and had blistered her right foot in her new shoes. For six to eight years, she had tended to catch her right foot while walking. She was first seen in 1967 and at that time was unaware of the numbness of her face, which was pointed out to her after examination at that time. She showed an ill-sustained, fine, grade 1 nystagmus on lateral gaze. There was hypalgesia over all three divisions of the trigeminal distribution on the right. The right hand showed reduced sweating and she had a thermal anaesthesia and hypalgesia over the right arm and right forequarter. Vibration sense was diminished at the right wrist and the deep reflexes in the right arm were all absent. In the lower limbs she showed a similar hypalgesia involving the right leg, and joint position sense was markedly impaired in the right and left feet. The deep reflexes in the lower limbs were increased, but the plantar responses were not extensor.

X-rays of the skull showed an increase in the convolutional markings in the vault and there was thinning of the dorsum sellae, consistent with chronically raised intracranial pressure. The cervical spine was normal, except that the canal appeared wide, measuring 2·4 cm in its anteroposterio diameter at the level of C5. A lumbar-route myelogram in the prone and supine positions showed the spinal cord to be widened almost throughout its length, except in the upper cervical region. In the supine position, the contrast medium flowed freely as far as the upper border of C2 where it was held up and moved anteriorly to flow into the prepontine cistern. The obstruction was irregular and fragmented and the appearances were thought to be consistent with arachnoiditis of the posterior fossa. It was not possible to assess whether or not there was tonsillar ectopia.

Comment. This patient is currently under consideration for further treatment and it is felt that probably she would benefit by ventricular drainage rather than by an attempt at decompression in the posterior fossa. The changes she shows on myelography are characteristic of arachnoiditis and this has been sufficient to cause relative occlusion of the outlet from the ventricular system, producing mild, compensated internal hydrocephalus. There are no symptoms of chronic raised intracranial pressure and operative treatment has therefore been delayed.

F.B., MALE (N.22672)

Until four years previously, this 42-year-old engineer had been perfectly well. He then developed progressive weakness of the left hand and a tendency to veer to the left while walking. He noticed no sensory impairment, but had had recurrent painless infections of the fingers of the left hand. His neck was short with a low hairline, and the left hand had the appearance of *main succulente*. Spinothalamic sensation was impaired on the left side of the face and in a 'half-cape' distribution over the neck, left arm and upper trunk to the level of D8. Joint position and vibration sense were impaired in the left upper limb, and vibration sense was impaired at both ankles. The deep tendon reflexes were absent in both upper limbs, and there was a mild spastic paraparesis though the plantar responses were flexor.

On radiography, the skull was rather large, the dorsum sellae truncated, and the possibility of arrested hydrocephalus was suggested. The C.S.F. was clear, under normal pressure; protein 53 mg/100 ml (14 per cent γ-globulin), no pleocytosis, C.S.F. and blood W.R. negative. Lumbar myelography showed possible early widening of the cervical spinal cord, and in the supine position the contrast medium flowed forward into the prepontine cistern, but could not be induced to enter the cisterna magna (Fig. 4.4).

Foramen magnum decompression was carried out, with removal of occipital bone and the first and second cervical laminae (Hankinson). The occipital bone was extremely thick and hard. The dura over the cisterna magna was also thick and was adherent to the cerebellum. Similar, dense arachnoidal adhesions were present over the whole of the upper cervical cord. The foramen of Magendie was obliterated, and on opening the fourth ventricle, the origin of the central canal appeared to be dilated. The cerebellar tonsils were not prolapsed. Immediately after operation, the patient's neurological deficit worsened, the spastic paraparesis became so marked that walking was difficult, and the power in the left hand decreased. The sensory findings were unchanged. Over the next year, there was a slight improvement in the paraparesis and power in the left arm, though he developed new sensory symptoms with sensations of tight bands around the legs and left arm. The neurological deficit remained otherwise unchanged.

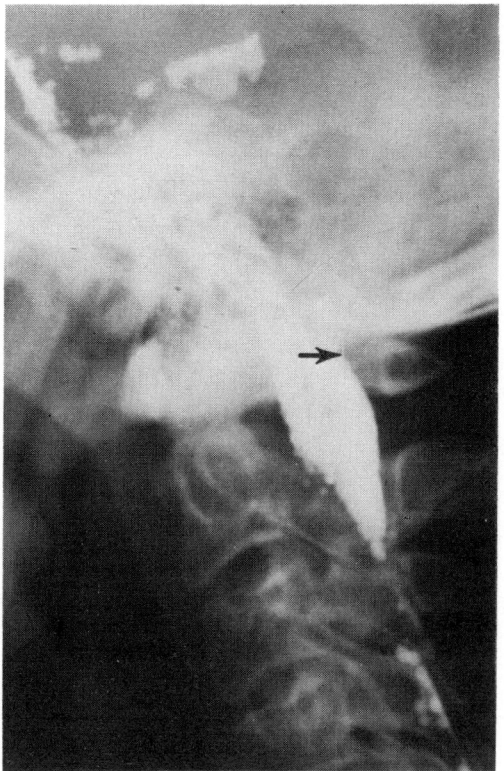

Fig. 4.4. Case F.B. (N.22672). Supine examination again showed absence of flow into the cisterna magna. The filling defect which is arrowed was erroneously thought to be caused by mild tonsillar prolapse. (Reproduced with permission from *Journal of Neurological Sciences*, **8**, 451–464, 1969.)

Comment. There was no previous history of meningitis but a supine myelogram revealed changes identical to those of cases M.M. and G.S., and there were dense arachnoidal adhesions over the whole of the cerebellum and upper cervical cord; these obliterated the foramen of Magendie. During the operation, removal of vascular adhesions to reopen the roof of the fourth ventricle produced a worsening in his clinical state immediately after the operation.

P.M., MALE (N.22696)
Four years previously this 42-year-old crane driver first became aware of paraesthesiae in the hands and legs, and unsteadiness of gait. The onset of these symptoms followed an influenza-like illness. In the six months before investigation, his right arm and both legs had become stiff and he began to drop objects from his right hand. He had a marked spastic tetraparesis more evident in the right limbs than the left, with hyperreflexia and weakness in the arms and legs of selective pyramidal distribution. The plantar responses were extensor. Vibration sense was impaired bilaterally to the level of the clavicles and there was impairment of joint position sensation and two-point discrimination in both hands, more particularly the right. Pinprick sensation was reduced in both arms, though temperature sensation was

normal. Radiography of the cervical spine demonstrated complete occipitalisation of the atlas. The C.S.F. was under normal pressure; protein 72 mg/100 ml, no pleocytosis, C.S.F. and blood W.R. negative. Myelography showed a hold-up of the flow of contrast medium at the level of the odontoid process where backward displacement reduced the sagittal diameter of the canal to 8 mm. In the supine examination, contrast medium flowed anteriorly into the prepontine cistern but could not be induced to enter the cisterna magna (Fig. 4.5). There was no definite widening of the spinal cord.

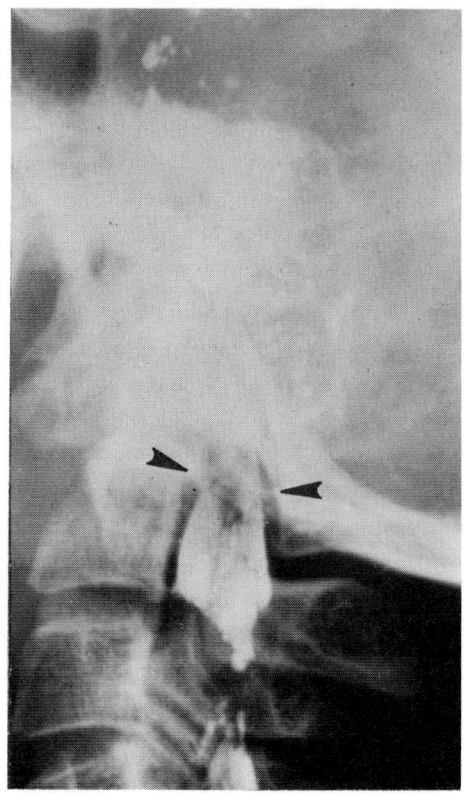

Fig. 4.5. Case P.M. (N.22696). Lateral radiograph in the supine head-down position shows once again no flow into the cisterna magna. The arrows illustrate the extreme narrowing of the theca at the foramen magnum. The occipitalisation of the axis is also discernible. (Reproduced with permission from *Journal of Neurological Sciences*, **8**, 451–464, 1969.)

It was concluded that the patient had a congenital bony narrowing of the foramen magnum associated with occipitalisation of the atlas and symptomatic syringomyelia. A foramen magnum decompression was done with the removal of part of the occipital bone and the laminae of the second and third cervical vertebrae (Hankinson). The dura was indented by bone and by a thickened atlantooccipital membrane at the foramen magnum. The cerebellar tonsils were not prolapsed but the arachnoid was extremely thick in this region and occluded the outlet to the fourth ventricle.

The flow of C.S.F. was re-established by dividing these adhesions. There was no alteration in the neurological deficit immediately after the operation but over the next 10 months there was slight improvement of gait and recovery of joint position sense in the hands. However, the patient still complained of paraesthesiae in the hands, and the neurological examination was otherwise unaltered.

Comment. With no previous history of meningitis, but with a bony congenital anomaly of the foramen magnum, this patient was shown to have supine myelographic appearances similar to those of the previous cases and arachnoidal adhesions obstructing the flow of C.S.F. from the fourth ventricle. He has shown slight improvement since decompression of the foramen magnum with reopening of the roof of the fourth ventricle.

H.C., MALE (N.22956)

In 1948, at the age of 13 years, this patient suffered from tuberculous meningitis. He was treated with streptomycin and made a fair recovery although he was left with bilateral foot drop. A year after the illness, he complained of numbness of the right hand and was found to have wasting and sensory change in the right hand. A diagnosis of syringomyelia was made. Further hospital advice was not sought for 19 years, when he was admitted with bilateral bronchopneumonia and extensive signs of syringomyelia. He had gradually deteriorated in the years after his meningitis, becoming chairbound after six years and developing urinary retention with overflow 13 years after the illness. In the two years before admission, he was aware of loss of temperature sensation over the left hand and right side of his face. He recovered from the bronchopneumonia with antibiotic therapy and was referred for neurological investigation.

At this stage he had a spastic paraplegia-in-flexion. There was marked weakness of all the upper limb muscles, particularly the small muscles of the hands, and a 'half-cape' of dissociated sensory loss spreading on to the face (though sparing the central area); this involved both upper limbs and the right side of the trunk to D8 and the whole of the left side of the trunk including the sacral dermatomes. Vibration sense was lost below the clavicles bilaterally and joint position sense was lost in the right arm and both legs. The right brachioradialis and triceps reflexes were absent and the crural reflexes were brisk, with extensor plantar responses. There was a marked scoliosis.

It was concluded that the patient had symptomatic syringomyelia from arachnoiditis due to the previous tuberculous meningitis. Radiographs of the skull and cervical spine showed flecks of calcification in the basal cisterns. The C.S.F. was under very low pressure; protein 24 mg/100 ml (14 per cent γ-globulin), no pleocytosis, blood and C.S.F. W.R. negative. It was considered that anything other than symptomatic therapy was not indicated and he was given diazepam 30 mg daily to reduce the flexor spasm in the legs. He died from a further respiratory infection.

General autopsy examination (Dr J. E. Ennis) showed bilateral bronchopneumonia and chronic pyelonephritis. The basiocciput and posterior fossa structures were removed in toto and examined after fixation. The pia-arachnoid over the surface of the cerebral hemispheres was normal, but around the inferior surface of the cerebellum, pons and medulla there were firm fibrous adhesions between the dura and pia, obliterating the subarachnoid space. The hindbrain and spinal cord were removed and fixed in continuity, after section of the brain stem. The cerebral hemispheres showed no convolutional flattening or dilatation of sulci and on section the lateral, third and fourth ventricles and aqueduct were of normal size. However, the site of the foramen of Magendie was completely obliterated by dense fibrous adhesions. On section of the cord there was a large smooth-walled cavity with a maximum diameter of 5 cm extending from the level of C2 to the conus medullaris. No macroscopic continuity of the central canal could be seen between the upper part of the syrinx at C2 and the lower portion of the fourth ventricle. However, on histological section *the central canal at C1 was found to be patent*, ependyma lined and slightly dilated (see Fig. 7.17). Many portions of the central cavity at lower levels were lined with ependyma, though the continuity of the lining was breached at many points (Fig. 7.18).

Comment. This patient developed the first signs and symptoms of syringomyelia in the year after a severe attack of tuberculous meningitis and eventually died in the final stages of syringomyelia, 19 years later. There were extensive arachnoidal adhesions around the cerebellum, brain stem and spinal cord, with apparent obliteration of the foramen of Magendie. A large central cavity extended within the spinal cord from the level of C2 to the conus medullaris and this cavity was shown histologically to be in communication with the fourth ventricle through a patent central canal at C1. Moreover, the central cavity in the remainder of the cord was partly lined with ependyma, and it was concluded that this patient had a gross hydromyelia with partial rupture of the dilated central canal into the surrounding nervous tissue. Despite the apparent obstruction to the outflow of C.S.F. from the fourth ventricle, there was no hydrocephalus.

M.W., FEMALE (N.23535)

This patient was born with micromelia of the right arm but was otherwise well until the age of 53 years when increasing weakness of the left arm developed, more proximal than distal. Concurrently, she became aware of stiffness of the legs, a tendency to trip while walking, pain in the left arm, and a sensation of tight bands around the neck. In the two months before investigation, numbness and impairment of pain and temperature sensation developed in the left hand. At this time there was gross weakness and wasting of the C5 to 6 innervated muscles bilaterally, with absent biceps and supinator reflexes, and pathologically brisk reflexes in the triceps and lower limb muscles. The plantar responses were extensor and the gait spastic. Spinothalamic sensation was impaired from C2 to D4 bilaterally, with absent vibration sense below the clavicles although joint position sense was intact. The C.S.F. was under normal pressure; protein 22 mg/100 ml, no pleocytosis, blood and C.S.F. W.R. negative. A lumbar myelogram showed no obstruction throughout the cervical canal. In the supine examination, the contrast medium flowed forwards into the prepontine cistern, but could not be induced to flow into the cisterna magna (Fig. 4.6). In view of the presumed arachnoiditis a foramen magnum decompression was considered unwise and the patient has remained symptomatically unchanged without treatment for two years.

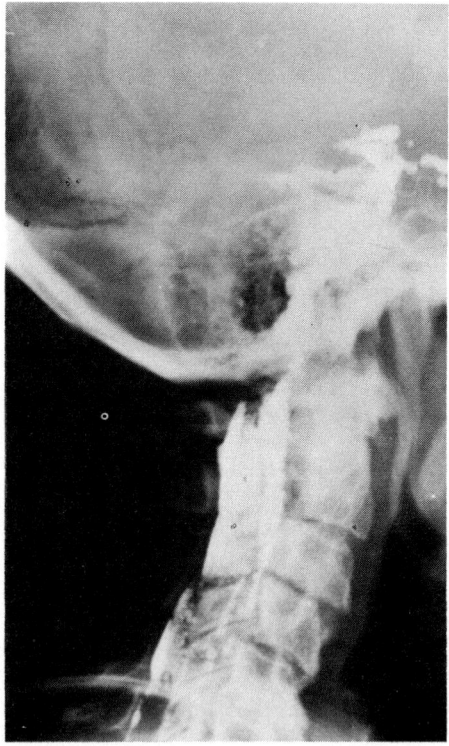

Fig. 4.6. Case M.W. (N.23535). Exactly similar appearances to N.22672 and R.V.I. 488058 with complete absence of filling of the cisterna magna. (Reproduced with permission from *Journal of Neurological Sciences*, **8**, 451–464, 1969.)

Comment. Despite the micromelia, there was no congenital bony anomaly of the foramen magnum region. Supine myelography demonstrated similar changes to those shown in other cases and in view of the probability that this patient had arachnoiditis of the foramen magnum, it was decided that there was little likelihood of benefit from decompression of the foramen magnum region.

J.A., MALE (N.26287)

This 21-year-old painter gave a seven-month history of inability to feel the temperature of the bath water with his left foot. During the seven months he had noticed that this numbness spread to involve the left leg from the thigh to the sole of his foot. For five years he had had difficulty in holding objects with his hands, especially drinking out of a glass, but these were his only symptoms.

On examination, he had lost vibration sense in each arm up to the elbows and the joint position sense was defective in the fingers of the right hand. In the lower limbs the tone was increased, particularly on the right, with a corticospinal distribution of weakness in the right lower limb. There was loss of vibration sense up to the costal margin and pain and temperature sensation were lost up to the left half of the body to the level of D4. His gait was spastic and in the lower limbs he had brisk reflexes, with bilateral extensor plantar responses. X-rays of the chest and cervical spine showed no abnormality. A lumbar-route myelogram confirmed that the cervical cord was widened and in the supine position there was hold-up of Myodil at the lower border of C1, almost certainly due to low cerebellar tonsils. However, none of the Myodil flowed into the cisterna magna and this suggested an arachnoiditis at this level.

At operation (Hankinson), it was possible to separate the dura from a thickened underlying arachnoid but it proved extremely difficult to make any progress in separating the thickened arachnoid from the tonsils or separating the tonsils from the underlying medulla. Intense localised arachnoiditis was the striking feature in this case. Following operation, the patient's physical signs were unchanged, but he thought his anaesthesia was less dense than before operation. At follow-up three years later, his neurological deficit remained unchanged.

Comment. This man presented with dissociated anaesthesia and spastic paraparesis typical of cervical cord cavitation. At operation, he showed extremely dense arachnoid adhesions at the foramen magnum and decompression of this area was extremely difficult. Following operation, he showed no significant improvement.

F.H., FEMALE (N.27390)

This 36-year-old housewife presented with a one-year history of numbness of the fingers of the right hand associated with some weakness of her grip on that side. On examination, there was evident wasting of the small muscles of the right hand, absent deep reflexes in both upper limbs and a spastic paraparesis with extensor plantar responses. There was a dissociated anaesthesia involving the C3 and 4 segments on the right side. Radiographs of the skull were normal, as were views of her cervical spine. A lumbar-route myelogram showed slight widening of the upper thoracic and lower cervical cord. There was a partial hold-up in the flow of the Myodil at the level of the odontoid process where there was thought to be mild tonsillar descent. The changes were not absolutely diagnostic of arachnoiditis, but at operation on opening the dura, there was seen to be a very widespread arachnoiditis with attachment of the dura to the arachnoid which could only be separated by sharp dissection. This was particularly noticeable over the right cerebellar hemisphere and over the tonsils. It was impossible to free the tonsils completely, but an opening was made into the fourth ventricle by separating the tonsils in the midline and there was an enlarged opening of the central canal visible in the floor of the fourth ventricle. This was plugged with a piece of muscle. It was concluded from the operative findings that this patient had a mild degree of Chiari type I malformation, but considerable arachnoiditis. Following operation, there has been no change in her clinical signs.

Comment. In this case the radiological findings were suggestive but not diagnostic of the underlying arachnoiditis seen at operation. A persisting enlarged and patent central canal was seen (see Chapter 8).

J.C., MALE (N.28472)

This man presented at the age of 26 years. For two years he had noticed increasing difficulty in appreciating pain and changes of temperature on the left side of his face, left shoulder and left upper limb and also the left side of the chest wall. He had also noticed that his eyes were wobbling when he looked to the side and he had had a mild spasmodic torticollis since the age of 13. On examination, he was a very intelligent and cooperative patient who appeared to have a short neck and low hairline. Examination of the nervous system revealed that he had rapid horizontal nystagmus on fixation and on lateral gaze in both directions, with vertical nystagmus on conjugate gaze upwards. There was diffuse wasting of the left upper limb and shoulder girdle with fasciculation around the shoulder. The deep tendon reflexes were absent in the upper limbs and the abdominal reflexes were absent, although the plantar responses were flexor. Sensory testing revealed a 'half-cape' of dissociated sensory loss over the left forequarter and left side of the face. Radiological examination of the skull showed an acute craniovertebral angle and possible concavity of the clivus. The cervical spine showed minimal basilar impression with atlanto-occipital fusion and an expanded cervical canal (anteroposterior diameter 2·2 cm). Myelography revealed gross expansion of the spinal cord from D8 up to the foramen magnum but there was no flow into the cisterna magna in the supine position. The appearances were thought to be those of arachnoiditis.

At operation (Hankinson), the dura was opened over the cerebellar hemisphere on either side and in the cervical region to the upper border of C2, but it was extremely difficult to continue this incision at the foramen magnum where the angle between the cerebellar and spinal dura was more acute than usual. There was an intense arachnoiditis so that the dura was attached to the posterior aspect of the cervical cord just caudal to the tonsils. Eventually, it was possible to open the dura in the midline but any efforts to separate it laterally at this level were found to be futile and resulted in bleeding from these vascular, but very firm adhesions. The tonsils were also adherent to one another and to the back of the medulla, although C.S.F. was seen emerging through a small foramen of Magendie. It was not possible to inspect the floor of the fourth ventricle. Following operation, the patient made a satisfactory recovery, but his sensory signs remained unchanged.

Ten months later he was readmitted to hospital because his walking was deteriorating and he complained of weakness and stiffness in the legs. On examination, the physical signs in his cranial nerves and upper limbs were unaltered, but he was developing a significant spastic paraparesis. It was thought that his arachnoiditis was probably re-forming and he was readmitted for introduction of a ventriculo-atrial shunt which later required to be revised. His shunt has continued to work satisfactorily but he has had occasional episodes of headache which have recovered spontaneously. His physical signs have regressed, with marked improvement in his walking.

Comment. In this patient, intense arachnoiditis was found at operation and it is possible that following operative decompression the adhesions re-formed and allowed further deterioration. Following the introduction of ventricular drainage through a ventriculo-atrial shunt, there has been improvement in this man's condition, with regression of his spastic paraparesis.

R.A., MALE (N.25104)

This man was first seen at the age of 57 years. He complained that 23 years prior to his being examined, while serving in the armed forces, he had suffered meningitis and had been in hospital for one month. No further details were available. He complained of episodes of giddiness and numbness of the upper limbs and on examination was found to have dissociated anaesthesia over the right shoulder and patchily in the right arm and right leg. Vibration sense was reduced at both ankles, and joint position sense was diminished in the fingers. Myelography in the prone and supine positions showed minimal changes only of arachnoiditis, but no evidence of tonsillar herniation. No operative intervention was undertaken.

Comment. This man showed minimal changes only and it was assumed that he had a slight degree of arachnoiditis consequent on his meningitis 23 years previously and that this was responsible for the 'half-cape' of dissociated anaesthesia.

H.B., MALE (R.V.I.552752)

This man was first seen in 1968. He complained of a constant aching pain in both shoulders, worse on the right than the left, with a similar pain radiating down the right arm to the fingers. In addition, he complained of steadily increasing weakness in the right arm and, more recently, he had noticed similar weakness of the left arm. He complained that all his symptoms were aggravated by movements of the head and neck. His symptoms had arisen and progressed during the past six years. He was unable to appreciate changes in temperature on the right arm and commented that he had recently burnt his fingers with matches and cigarettes without feeling the pain. He commented that his right leg felt rather numb and stiff. Examination revealed a kyphoscoliosis and neurological examination revealed gross wasting and weakness of the muscles of the right shoulder girdle, particularly the deltoid and spinati. The deep tendon reflexes were generally brisk in both upper and lower limbs and he showed an extensor left plantar response. There was no convincing dissociated anaesthesia.

Radiology of the skull and cervical spine revealed no abnormality but he was seen to have scoliosis in the mid-dorsal region. Myelographic examination in the prone and supine positions showed the characteristic pattern of arachnoiditis in the high cervical region with fragmentation of the column of dye. The dye would not enter the cisterna magna posteriorly and even in the supine position, with the head down, tended to run forward on to the clivus. At 66 years, he was not considered suitable for operative treatment and at follow-up his condition remained unchanged.

Comment. Again a classical presentation of syringomyelia, albeit with a very late onset. The myelographic changes are characteristic of arachnoiditis, but no operative confirmation of the myelographic appearances has been made.

W.O., FEMALE (R.V.I.610161)

This patient was admitted to hospital at the age of 25 years, during the previous three years having suffered some tingling in the left hand and a tendency to drop things from this hand. She had been examined at that time and there had been no abnormal findings. Four months before presentation at the clinic she complained of increasing stiffness of her legs. On examination, she was found to have a very short neck, a thoracic scoliosis to the left and it was noted that her right hand was larger than her left. She had scars from burns on all her fingers. Her eye movements were jerky, with poorly sustained horizontal nystagmus on gaze to the right. There was patchy impairment of pinprick sensation over the second and third divisions of the fifth nerve on the left side. Her gait was spastic and ataxic and she had a positive Romberg's sign. All muscles in the upper limbs were weak but in the arms there was specific weakness of the physiological extensors, and increase in tone. In the legs, she showed a spastic paraparesis with hyperreflexia and extensor plantar responses. Further examination of the upper limbs revealed slight finger/nose ataxia and she showed quite marked ataxia on knee/shin/knee testing bilaterally.

X-rays of the skull showed a small foramen magnum and a subsequent myelogram showed gross expansion of the cord from C2 to D8 with the characteristic changes of posterior fossa arachnoiditis. She was subjected to posterior fossa craniotomy (Hankinson) and localised arachnoiditis at the foramen magnum and the foramen of Magendie was seen. In addition, in the right cerebellar hemisphere, there was a large cyst, some 4 cm in diameter, containing clear fluid which appeared to be C.S.F. (subsequent analysis showed a protein of 41 mg/100 ml, a γ-globulin of 3·5 mg/100 ml or 8·5 per cent of total protein). There was no communication with the fourth ventricle and it was though that the cyst resulted from cerebellar agenesis. The upper cervical cord was considerably distended by hydromyelia and fluid was aspirated from this. Following aspiration, the cord collapsed but was seen later to re-expand with respiration. The adhesions at the foramen of Magendie were divided, but the central canal could not be seen. Following operation, the patient's mental state was grossly abnormal and for a time she remained confused, disoriented and suffered auditory and visual hallucinations. She had a persistent tachycardia and later developed a urinary infection. After some two to three weeks she improved, but because of her persisting confusion a Spitz-Holter valve was inserted. Following this her mental state improved, she was able to walk a few steps and after two week's convalescence she returned for review. Her walking had improved considerably. Her mental state had returned to normal and she remained cheerful. Her cerebellar signs were less marked than before operation, but she still showed quite evident truncal ataxia and the spastic quadriparesis was unchanged. One year later there was little change in her physical status, but mentally she was regarded as normal.

Comment. This patient at operation was found to have a cerebellar cyst responsible for some of her symptoms and also adhesive arachnoiditis sufficient to have caused internal hydrocephalus. Following operation, her mental condition required drainage of her ventricular system, presumably due to the disturbance in the C.S.F. dynamics, produced by the sudden disobliteration of the occlusion in the posterior fossa.

M.G., FEMALE (R.V.I. 488058)
This patient had no illness other than recurrent cystitis until the age of 46 years, when she first noticed loss of appreciation of temperature over the left shoulder, with pain in that region aggravated by movement and coughing. Two years later, numbness appeared in the left groin and four years after that she complained of a sensation of wetness followed by numbness down the whole of the left side of the body and rapidly developed incoordination and anaesthesia of the left leg. The crural anaesthesia involved all modalities of sensation in the left arm and trunk, and some reduction of vibration and of joint and position sense in the left arm. There was impairment of spinothalamic sensation on the left side of the face, with slight wasting of the small muscles of the left hand. Power was normal in all muscle groups. The deep tendon reflexes were absent in both legs and the left arm and diminished in the right arm. Both plantar responses were flexor. The C.S.F. was clear and under normal pressure; protein 70 mg/100 ml, 2 white cells/µl, the blood and C.S.F. W.R. were negative. It was concluded on clinical grounds that she had syringomyelia with a cavity extending into the lumbosacral region of the cord, although prone myelography at that time revealed no definite widening of the cervical cord.

Over the next seven years, there was little symptomatic deterioration although examination revealed generalised weakness and slight wasting of all muscle groups, more marked proximally in the left upper and lower limbs. All modalities of sensation had become more depressed on the whole of the left side of the body. Her gait now appeared spastic, though the plantar responses were still flexor and the deep tendon reflex pattern was unchanged. A further myelogram was performed. C.S.F. protein was 73 mg/100 ml (6 per cent γ-globulin), 1 cell/µl. In the cervical region the spinal cord was abnormally enlarged, and in the supine position contrast medium flowed forwards into the prepontine cistern but could not be induced to flow into the cisterna magna. It was concluded that the patient was suffering from syringomyelia with arachnoidal adhesions in the region of the cisterna magna, and in view of her static condition operation was not recommended.

However, 10 months later she developed increasing pain over the whole of the left side of the body, particularly in the neck and shoulder. The foramen magnum was decompressed and the arch of the atlas was also removed (Hankinson). The cerebellar tonsils were not prolapsed, but there was a marked degree of adhesive arachnoiditis binding the tonsils to the medulla, and extending down the cervical cord to involve the posterior inferior cerebellar arteries.

The foramen of Magendie appeared to be patent but further exploration was impossible owing to the nature of the vascular adhesions. Immediately after the operation there was considerable improvement in the pain, but no change in the neurological deficit.

Two months after operation the pain in the left side recurred, and by eight months after operation was as severe as before, and also involved the right groin.

Comment. Again there was no history of meningitis, but in view of our experience, the myelographic appearances were interpreted as showing arachnoiditis of the foramen magnum region. Since we had found that these cases did not do so well following operation as those with uncomplicated Chiari anomalies, it was decided initially not to offer this lady an operation. This was later undertaken because of severe pain. Operation showed extensive arachnoidal adhesions around the cerebellum and spinal cord, and relief of the pain lasted only two months.

J.W., MALE (R.V.I. 491137)
This patient, aged 36 years, gave a history of slowly progressive weakness and wasting of the small muscles of both hands for 12 years. A diagnosis of syringomyelia had been made in 1957 and he had been treated with radiotherapy to the cervical cord without result. In January 1964 he was first seen at Newcastle General Hospital where prone myelography revealed enlargement of the cervical cord. About this time, he was found to be hypertensive (blood pressure 180/120 mm Hg) and he had a haematemesis from a chronic duodenal ulcer. In February 1966, he was admitted to the Royal Victoria Infirmary for gastroenterostomy and vagotomy for his ulcer. His blood pressure was then 210/140 mm

Hg, despite treatment with methyldopa and guanethidine. He was readmitted on 31 December, 1966, in his terminal illness. He had papilloedema, was in cardiac and renal failure with a blood pressure of 250/140 mm Hg. After several attacks of 'hypertensive encephalopathy' he died on 15 January, 1967.

Inspection of the cadaver at autopsy showed gross wasting of the intrinsic muscles in both hands. Examination of the viscera showed that the heart weighed 810 g and that the left ventricular wall was grossly hypertrophied. Both lungs showed pneumonic changes, and both kidneys—although normal in size and weight—were congested; in both, the cortex and medulla contained numerous small haemorrhages.

Inspection of the brain and cord in situ disclosed no evidence of tonsillar prolapse or descent of the medulla and fourth ventricle. However, after removal of the brain the foramen of Magendie was found to be completely occluded by a thick fibrous membrane. Similar membranes extended into the lateral recesses (see Chapter 7).

Comment. This patient was investigated early in the current series, and only prone myelographic examination was undertaken. He presented with classical syringomyelia but, at post-mortem, there was evident internal hydrocephalus due to adhesive arachnoiditis of the posterior fossa and upper cervical region.

A.F., MALE (X.06920)

This schoolboy was seen at the age of 15 years, when he stated that a year before investigation, while playing football, he had caught his right foot in a hole in the ground and had tripped, wrenching his foot. From then on he began to limp and his mother stated that he was walking very stiffly and that gradually his walking became worse. In addition, he had complained of occasional occipital headaches and during the year developed paraesthesiae in the hands. He continued to play games at school, but his kicking power for football became greatly reduced and he began to fall. There was no history of loss of power in the upper limbs. On examination, he showed a lumbar lordosis and kyphosis and walked with a distinct limp, dragging the right leg. There were no abnormalities to be detected in the cranial nerves or upper limbs but he showed an asymmetrical spastic paraparesis, with increased reflexes, more marked on the right than the left, and extensor plantar responses. The abdominal reflexes were absent, but otherwise the neurological examination was unremarkable.

Radiological examination of the cervical spine revealed no abnormality and the dorsal spine showed a scoliosis, convex to the right. Myelography revealed a widened cord in the cervical region and evidence of arachnoiditis. The contrast medium would not enter the cisterna magna in the supine position, but in the prone position it readily ran forward on to the clivus; this is a radiological sign of basal arachnoiditis. There was considerable discussion as to whether this boy should undergo high cervical decompression or have ventricular drainage established. Ventriculography showed a slight degree of asymmetrical ventricular dilation with dilatation of the anterior recesses of the third ventricle. The fourth ventricle was not displaced, but very little air would leave the ventricular system.

Foramen magnum decompression and posterior fossa exploration was undertaken (Hankinson). After the dura had been opened there was evident, very marked arachnoiditis, the tonsils were prolapsed some 5 mm and were firmly bound down to the medulla and to each other. Arachnoiditis was seen to pass laterally around the side of the medulla and, with considerable difficulty, vascular adhesions were divided. The fourth ventricle was entered and it was seen that there was a very deep midline cleft at the entrance to the central canal well developed at the obex. This was plugged with a small piece of muscle. Following operation, the patient made an uneventful recovery and his clinical condition has remained unchanged.

Comment. There was no preceding meningitis in this case and the onset was associated with trauma. The radiological signs were fairly typical of arachnoiditis and the changes confirmed at operation, which was undertaken following ventriculography and the finding of minimal ventricular dilatation only. It is of interest that again the central canal was seen to be patent.

CONCLUSIONS

Despite the varying symptomatology expressed by this group of patients, there seems to be a common radiological abnormality, which may be demonstrated on supine myelography (Fig. 4.7). In the head-down position, the Myodil column is normal in the mid-dorsal region, fragments in the upper cervical region and cannot be induced to run into the cisterna magna, as though there were an obstruction at the posterior rim. Further tilting of the head will allow the contrast medium to run into the prepontine cistern. This has been a characteristic change in all the cases described and, at operation, the adhesive arachnoiditis in the posterior fossa has been confirmed. Of the 15 cases with arachnoiditis examined by supine myelography it was possible to demonstrate tonsillar ectopia in addition to this arachnoiditis in five (G.S., P.M., J.A., F.H. and R.P.). In one patient, tonsillar ectopia was not suspected on myelography, but a minimal degree of descent of the cerebellar tonsils was seen at operation (A.F., X06920). However, in the case of P.M. exploration of the foramen magnum, though confirming arachnoiditis, did not confirm the suspected tonsillar ectopia.

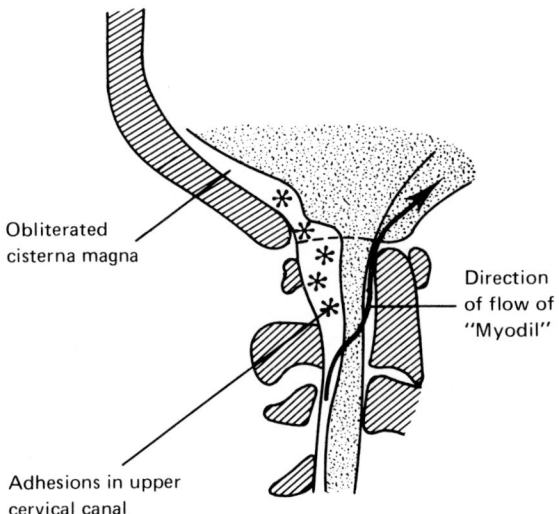

Fig. 4.7. Diagram to show the effect of adhesive arachnoiditis on the flow of the column of contrast material.

We feel that this group of patients represents a separate form of obstruction at the foramen magnum, due to a minimal amount of anomalous descent or ectopia of the cerebellar tonsils. Of the 15 patients with recognised arachnoiditis on myelography, five were found to have tonsillar ectopia at operation. The ectopic cerebellar tonsils may be responsible for the formation of the adhesive arachnoiditis, perhaps by causing mechanical obstruction at the foramen magnum. The pathogenesis of the arachnoiditis in those cases without evident cerebellar ectopia is more difficult to explain and, while meningitis may well have been the causative factor in three of our cases, in the remainder the development of this abnormality remains unexplained. It is evident, however, that for such patients to develop the hydromyelic dilatation of the cervical canal and the subsequent changes of communicating syringomyelia, the arachnoiditis must have been responsible for a persistently open central canal in communication with the fourth ventricle and hence must have arisen at or about the time of birth. It is a possibility that subarachnoid bleeding at the time of birth may have induced adhesive changes in the posterior fossa and at the foramen magnum in those cases without associated cerebellar ectopia.

Although they present a similar clinical symtomatology to that of patients with uncomplicated tonsillar ectopia, we believe that these patients require a different type of assessment and surgical approach. The surgical results attained hitherto in these cases when treated by foramen magnum decompression have been uniformly poor and no patient has been relieved of the symptoms and signs due to his cervical cord cavity. In two patients treated by shunt procedures, there was relief of the symptoms induced by the internal or obstructive hydrocephalus. It is our opinion that this adhesive arachnoiditis produces a relative obstruction in the cisterna magna and at the foramen magnum and that compensation occurs, as it may in adult aqueduct stenosis (McHugh, 1964). It may be more reasonable to relieve these patients of their symptoms by ventricular drainage rather than by attempting direct attack on the posterior fossa. While cases presenting in this way have characteristic myelographic appearances and are therefore readily recognised, we have not undertaken ventricular drainage in many of them to date. However, if there is associated hydrocephalus then this is probably the best method of approach. We have experience of 18 cases of arachnoiditis in this series of 100 cases of syringomyelia. Logue (1971), in describing his series of 35 cases of syringomyelia, in which 31 were shown to have a Chiari malformation, mentions two cases of adhesive obstruction to the fourth ventricle and one case of congenital cerebellar cyst (see case W.O. of this series). He makes no specific recommendations regarding the management of the cases of arachnoiditis, but we believe, from our experience of an almost 20 per cent incidence in a large series, that these cases should be considered separately and if they are to be treated by neurosurgical methods then ventricular drainage may prove to be the treatment of choice.

REFERENCES

Camus, J. & Roussy, G. (1914) Cavités medullaires et meningité cervicales. *Revue Neurologique*, **22**, 213.
Davison, C. & Keschner, M. (1933) Myelitic and myelopathic lesions, part 6. *Archives of Neurology & Psychiatry*, **30**, 1074.
Joffroy & Achard (1887, 1891) cited by Camus and Roussy (1914).
Logue, V. (1971) Syringomyelia: a radiodiagnostic and radiotherapeutic saga. *Clinical Radiology*, **22**, 2.
Lubin, A. W. (1940) Adhesive spinal arachnoiditis as a cause of intramedullary cavitation. *Archives of Neurology & Psychiatry*, **44**, 409.
McHugh, P. R. (1964) Occult hydrocephalus. *Quarterly Journal of Medicine*, **33**, 297.
McLaurin, R. L., Bailey, P., Schurr, P. H. & Ingraham, F. D. (1954) Myelomalacia and multiple cavitations of the spinal cord secondary to adhesive arachnoiditis. *Archives of Pathology*, **57**, 138.
Nelson, J. (1943) Intramedullary cavitation resulting from adhesive spinal arachnoiditis. *Archives of Neurology & Psychiatry*, **50**, 1.
Schwartz (1897) cited by Nelson (1943).

CHAPTER FIVE

The Radiology of Communicating Syringomyelia

J. B. FOSTER and P. HUDGSON

The initial radiological examination of our patients consisted of plain radiographs of the skull and the cervical spine. In the skull films, we paid particular attention to evidence of basilar impression and/or arrested hydrocephalus, also anomalies at the craniovertebral junction. In the cervical spine films, attention was paid to the anteroposterior diameter of the canal and the presence of anomalous vertebral bodies, bifid spinous processes or arches of the vertebrae and the Klippel–Feil deformity. The results of these initial investigations are illustrated in Table 5.1.

Cervical Spine

The cervical spines in 74 of our patients revealed no abnormality (Table 5.2). There was evident widening of the anteroposterior diameter of the canal in eight patients, the canal varying from 1·9 cm to 2·4 cm at the level of C5 in those in whom the canal was expanded (Fig. 5.1). In 16 patients there was atlanto-occipital fusion (Fig. 5.2). In one case there was an unfused, or bifid posterior arch of C1 (Fig. 5.3), and in a further case there was evident hypoplasia of the occipital condyles on one side, with maldevelopment of the lateral masses of C1 (Fig. 5.4). The angle between the occiput and the spinal axis was considered to be abnormally acute in one further patient.

In the past, it has been assumed that Arnold–Chiari and Chiari anomalies were always accompanied by abnormalities of the upper cervical spine. Isolated case reports have appeared in recent years in which no such abnormality has been found (Aring, 1938; Ogryzlo, 1942; Bucy and Lichtenstein, 1945; Gardner and Goodall, 1950; Peach, 1964; Teng and Papatheodorou, 1965). We now have a series of 100 patients, 74 of whom have normal cervical spines and yet 47 of whom, on myelography, showed ectopic cerebellar tonsils (Table 5.4). It would seem

THE RADIOLOGY OF COMMUNICATING SYRINGOMYELIA

Table 5.1. *Plain X-ray changes*

Case	Cervical spine	Skull	Case	Cervical spine	Skull
MJ	Bifid spine C2, fusion with C3	Normal	MA	Normal	Normal
MT	Normal	Normal	LO	Normal	Normal
GS	Normal	Normal	SW	Normal	Normal
FD	Normal	Normal	WD	Normal	Normal
FW	Normal	Normal	MC	Fusion of C4 and C5	Normal
WH	Normal	Normal	RH	Normal	Platybasia, no basilar impression
PG	Normal Acute cranio-vertebral angle	Hydrocephalus	JJ	Normal	Normal
			MG	Normal	Normal
			NA	Normal	Normal
ES	Normal	Normal	MB	Normal	Normal
RJ	Fusion of C6 and C7	Normal	PR	Atlanto-occipital fusion	Normal
EJ	Normal	Normal	MM	Normal	Normal
CW	Wide canal	Hydrocephalus	AR	Normal	Normal
MC	Normal	Normal	GK	Bifid arch of C1	Normal
LB	Normal	Normal	JA	Normal	Basilar impression
KE	Fusion of C2 and C3	Normal	EK	Normal	Normal
DL	Normal	Normal	RJ	Normal	Normal
LJ	Normal	Normal	AT	Normal	Normal
MC	Normal	Normal	HH	Normal	Normal
CH	Wide canal	Normal	RS	Hypoplastic R Occipital condyle	Basilar impression
LT	Normal	Normal			
NS	Normal	Normal	RB	Normal	Normal
JL	Normal	Normal	HC	Normal	Hydrocephalus
EW	Normal	Normal	JE	Atlanto-occipital fusion and fusion of C2 and C3	Basilar impression
JC	Wide canal	Hydrocephalus			
KB	Normal	Normal			
WM	Normal	Normal	WP	Normal	Normal
RP	Normal	Basilar impression	NY	Normal	Normal
BE	Normal	Normal	LR	Normal	Normal
JM	Wide canal	Hydrocephalus	RR	Fusion of C2 and C3	Normal
MM	Unfused posterior arch C1	Hydrocephalus	SD	Normal	Normal
			HB	Normal	Normal
MD	Normal	Normal	TR	Atlanto-occipital fusion and fusion of C2 and C3	Normal
RC	Normal	Normal			
FB	Normal	Normal			
PM	Complete atlanto-occipital fusion	Normal	DB	R. sided atlanto-occipital fusion	Basilar impression
HC	Normal	Normal	RP	Normal	Normal
MW	Wide canal	Normal	WO	Normal	Normal
JE	Normal	Normal	GJ	Normal	Normal
PS	Normal	Normal	EJ	Normal	Normal
PH	Normal	Normal	VN	Wide canal	Basilar impression
ME	Normal	Hydrocephalus	ER	Normal	Normal
AB	Normal	Normal	MG	Normal	Normal
JA	Normal	Normal	DO	Normal	Normal
CB	Normal	Normal	WI	Atlanto-occipital fusion and wide canal	Basilar impression
JW	Normal	Normal			
HF	Atlanto-occipital fusion	Basilar impression	JJ	Atlanto-occipital fusion	Basilar impression
MJ	Normal	Normal	RH	Normal	Normal
ET	Normal	Normal	JW	Normal	Normal
JR	Normal	Normal	FD	Normal	Basilar impression
CP	Atlanto-occipital fusion	Normal	AF	Normal	Normal
FH	Normal	Normal	GF	Acute cranio-vertebral angle	Normal
JC	Wide canal Atlanto-occipital fusion	Basilar impression			
			DR	Fusion of C2 and C3, C6 and C7	Normal
RA	Normal	Normal			
CD	Normal	Basilar impression	EH	Normal	Normal

COMMUNICATING SYRINGOMYELIA

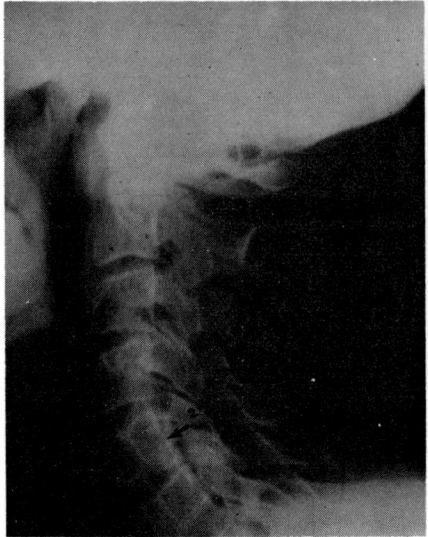 

Fig. 5.1. Increased sagittal diameter of the cervical cord (2·5 cm at C5).

Fig. 5.2. Case P.M. (N.22696). Complete occipitalisation of the atlas or atlanto-occipital fusion.

Table 5.2. *Radiographs of cervical spine*

Normal	74
Wide anteroposterior diameter	8
Atlanto-occipital fusion	
or	16
Occipitalisation of atlas	
Bifid arch of C1	1
Hypoplastic occipital condyles	1

from these figures that a normal cervical spine does not exclude or reduce the possibility of tonsillar ectopia, a Chiari malformation or communicating syringomyelia.

The Skull

In 80 of our patients radiographs of the skull were normal (Table 5.3). In 12 patients there were changes recognised as basilar impression (Figs. 5.5 and 5.6). Seven cases showed the radiological changes of arrested hydrocephalus where there was expansion of the intracranial compartment (Fig. 5.7). In one further case there was platybasia without basilar impression. Those patients thought to have hydrocephalus had been found to have a large head when examined clinically and, in the main, were those who presented with headaches, drop attacks and brain stem signs. The finding of changes suggestive of arrested hydrocephalus should determine the operative approach to these cases (see Chapter 6).

Fig. 5.3. Case M.M. (N.22171). Unfused posterior arch of C1.

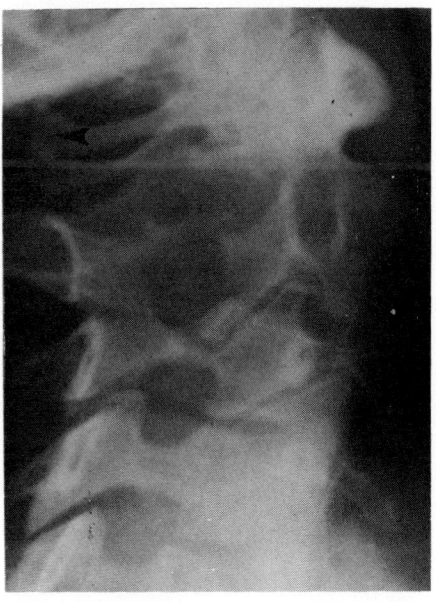

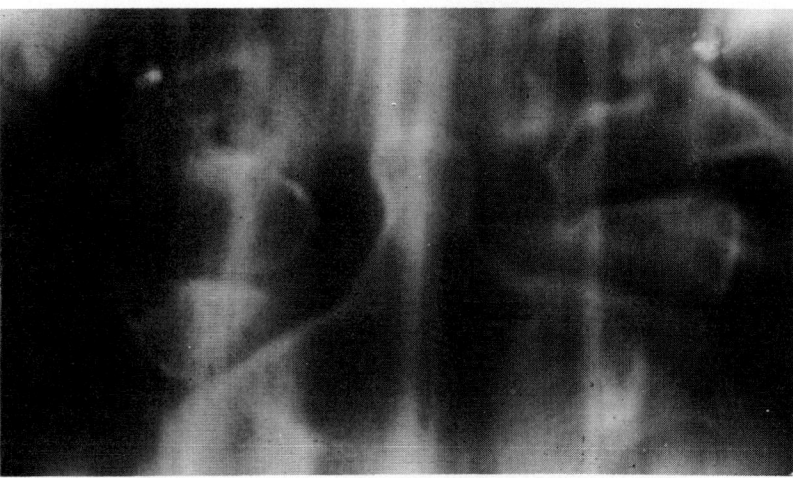

Fig. 5.4. Case R.S. (N.43177). Tomogram to show hypoplasia of an occipital condyle. Both lateral masses of C1 are hypoplastic. The odontoid process has an unusual shape.

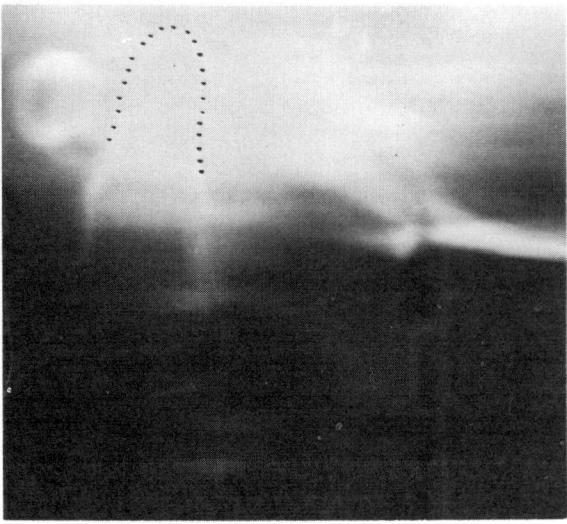

Fig. 5.5. Case R.S. (N.43177). Tomogram showing the odontoid process lying above the rim of the foramen magnum.

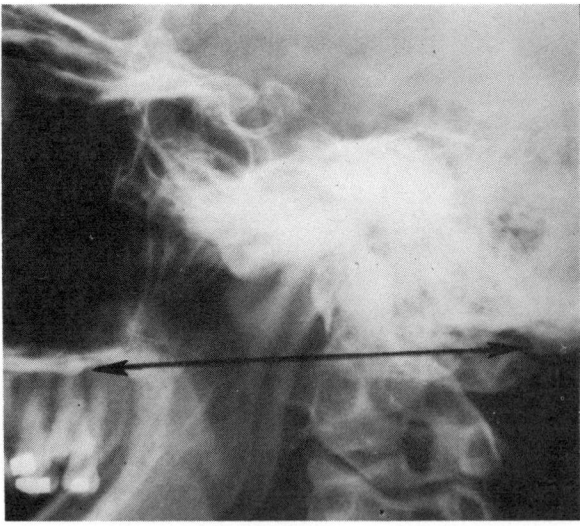

Fig. 5.6. Case H.F. (N.27001). Basilar impression. The odontoid process lies well above Chamberlain's line (arrowed).

THE RADIOLOGY OF COMMUNICATING SYRINGOMYELIA

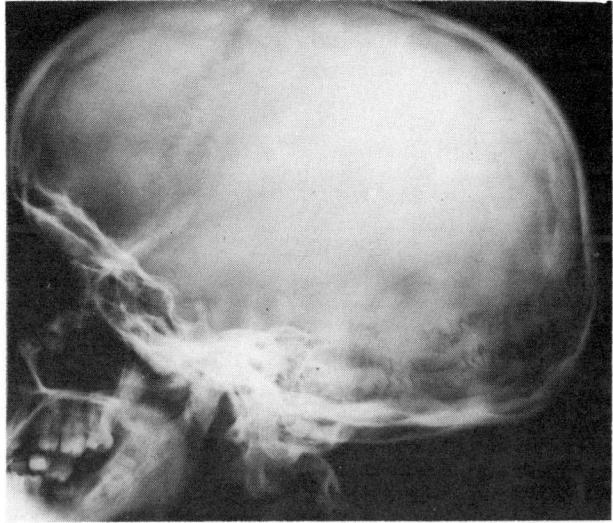

Fig. 5.7. 'Arrested hydrocephalus.' Gross disproportion between the vault and facial structures.

Table 5.3. *Radiographs of the skull*

Normal	80
Basilar impression	12
Arrested hydrocephalus	7

MYELOGRAPHY

From 1964 to 1971, during which years these cases were collected, over 4,000 myelograms were performed in the Department of Neuroradiology and, as our experience increased, so more attention was paid to the possibility of anomalies at the foramen magnum.

Myelographic examination of the spinal canal was undertaken in 92 of the 100 patients. Our technique is to introduce the positive contrast medium into the spinal canal and then to remove the lumbar puncture needle. This allows easy change from prone to supine positions. We do not believe it necessary to attempt to remove the contrast medium after completion of the examination. In most cases, the examination was carried out in both the prone and supine position, using 6 ml of iodophendylate (Myodil, Pantopaque). Spinal fluid was taken for examination on each occasion, particularly with regard to its protein content. Earlier in the series, four patients had been examined in the prone position only and in three a wide cord had been seen without cerebellar tonsillar ectopia (Fig. 5.8). In one patient examined in the prone position only, no abnormality was seen. However, he was later subjected to upper cervical decompression and tonsillar ectopia was found (case J.E.). Subsequently at post-mortem, after death from respiratory failure, the pathology was confirmed (Chapter 7). Special

56 COMMUNICATING SYRINGOMYELIA

attention has been paid to widening of the cord on myelography and this was noted in 34 patients in the series (Table 5.4 and 5.5). Lack of filling at the odontoid process as described by Malis (1958) was seen in most cases (Fig. 5.9).

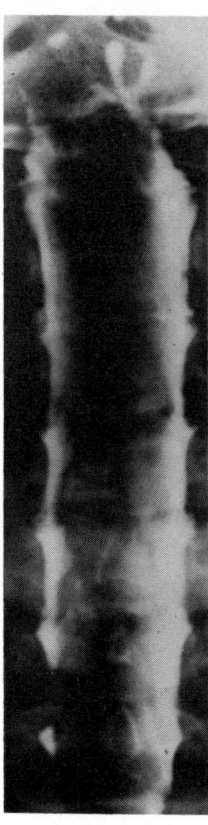

Fig. 5.8. Case F.D. (N.3477). Prone examination showing the widened cord. The dilatation passes above the level of the foramen magnum.

  The definitive abnormality in the supine myelogram has been ectopia, or descent of the cerebellar tonsils, visualised below the foramen magnum on the posterior or dorsal aspect of the upper cervical cord. In the anteroposterior projections, the descent is often seen to be asymmetrical (Figs. 5.10 and 5.11) and the appearance is very characteristic. In the lateral projection, the tonsils are readily seen on the posterior aspect of the cord outlined by the column of Myodil; the tonsils may be small and thin (Fig. 5.12) or wider and more readily visible (Figs. 5.13 and 5.14). Abnormally placed cerebellar tonsils were seen on the supine myelograms of 62 of our cases, most of whom were later subjected to operative decompression.
  The first demonstration of an anomaly by contrast myelography was in a case reported by List (1941). In the same year, Adams, Schatzki and Scoville described a further case and from that time other reports of similar examinations have been made. Baker, in 1963, suggested the importance of myelography in the supine position in examination of the foramen magnum and hind brain and we

THE RADIOLOGY OF COMMUNICATING SYRINGOMYELIA

Table 5.4. *Myelographic findings*

Case	Prone examination	Supine examination	Wide cord	Tonsillar ectopia	Arachnoiditis	Operation	Case	Prone examination	Supine examination	Wide cord	Tonsillar ectopia	Arachnoiditis	Operation
MJ	+	+	−	−	−	−	RA	+	+	−	−	+	−
MT	+	+	+	+	−	+	CD	+	+	−	+	−	+
GS	+	+	+	−	+	+	MA	+	+	−	+	−	+
FD	+	−	+	−	−	+	LO	+	+	−	−	−	−
EW	+	+	−	+	−	+	SW	+	+	−	+	−	+
WH	+	+	−	+	−	+	WD	+	+	−	+	−	+
PG	+	+	−	+	−	+	MC	+	+	+	+	−	+
ES	+	+	−	+	−	+	RH	+	+	−	−	−	−
RJ	+	+	−	−	−	−	JJ	−	−	−	−	−	−
EJ	+	+	−	−	−	−	MG	+	+	−	+	−	+
CW	+	+	+	+	−	+	NA	−	−	−	−	−	−
MC	+	−	+	−	−	−	MB	+	+	−	+	−	+
LB	+	+	−	−	−	−	PR	+	+	+	+	−	−
KE	+	−	+	+ on vert. angio.	−	+	MM	−	−	−	−	−	−
							AR	−	−	−	−	−	−
							GK	+	+	+	+	−	+
DL	+	+	−	−	−	+	JA	+	+	+	+	−	+
LJ	+	+	+	+	−	+	EK	+	+	−	+	−	−
MC	+	+	+	+	−	+	RJ	+	+	−	−	−	−
CH	+	+	−	+	−	+	AT	+	+	−	+	−	+
LT	+	+	−	−	−	+	HH	+	+	−	+	−	−
NS	+	+	+	+	−	+	RS	+	+	−	+	−	+
JL	+	+	−	+	−	+	RB	+	+	−	+	−	−
EW	+	+	−	+	−	+	HC	+	+	−	−	−	−
JC	+	+	+	+	−	+	JE	+	−	−	−	−	+
KB	+	+	−	+	−	−	WP	−	−	−	−	−	−
WM	+	+	+	−	−	−	NY	+	+	+	+	−	+
RP	+	+	+	+	−	+	LR	+	+	−	+	−	−
BE	+	+	+	+	−	−	RR	+	+	+	−	−	+
JM	+	+	+	−	+	−	SD	+	+	−	+	−	−
MM	+	+	+	−	+	−	HB	+	+	−	−	+	−
MD	+	+	−	−	+	+	TR	−	−	−	+ on vert. angio.	−	+
RC	+	+	−	−	−	−							
FB	+	+	−	−	+	+							
PM	+	+	−	+	+	+	DB	+	+	−	+	−	+
HC	+	+	−	−	+	+	RP	+	+	−	+	+	+
MW	+	+	−	−	+	−	WO	+	+	+	−	+	+
JE	+	+	+	+	−	+	GJ	+	+	−	+	−	+
PS	+	+	+	+	−	+	EJ	+	+	−	+	−	−
PH	+	+	+	+	−	+	VN	+	+	−	+	−	+
ME	+	+	−	+	−	−	ER	+	+	+	+	−	+
AB	+	+	+	+	−	−	MG	+	+	+	−	+	+
JA'	+	+	+	+	+	+	DO	+	+	+	+	−	+
CB	+	+	−	+	−	−	WI	+	+	−	+	−	−
JW	+	+	−	+	−	−	JJ	+	+	−	+	−	−
HF	+	+	−	+	−	−	RH	+	+	−	−	−	−
MJ	+	+	−	+	−	−	JW	+	+	+	−	+	−
ET	−	−	−	+ on AEG	−	+	FD	−	−	−	−	−	−
JR	+	+	−	+	−	+	AF	+	+	+	−	+	+
CP	+	+	−	+	−	−	GF	−	−	−	−	−	−
FH	+	+	+	+	+	+	DR	+	+	−	+	−	+
JC	+	+	+	−	+	+	EH	+	+	−	+	−	+

have endorsed the necessity for such an examination in this series. We have rarely undertaken gas myelography, although in one of our early cases examined by this method there was very clear visualisation of the cerebellar tonsils below the foramen magnum (case L.T., Fig. 5.15). In one further patient (E.T.), tonsillar ectopia was seen during air encephalographic examination when myelography had not been undertaken. The demonstration of a low-lying posterior inferior

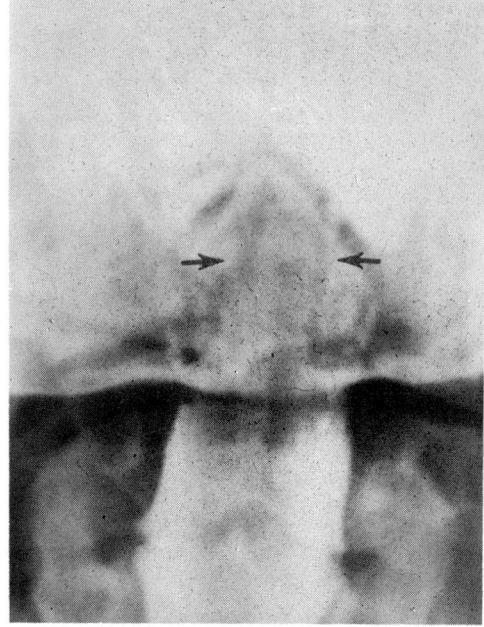

Fig. 5.9. Case K.E. (N.11329). Conventional prone myelogram illustrating the change which may be present due to descent of the hind brain. The arrows point to the odontoid peg which is clearly visualised because of the paucity of Myodil filling in this region in the prone position. Above the arrows the filling defect produced by the vertebral arteries and the basilar artery are seen. (Reproduced with permission from *Brain*, **91**, 131, 1968.)

cerebellar artery, thought to be indicative of congenital descent of the cerebellar tonsils (Occleshaw, 1970), was seen in one patient (K.E., Fig. 5.16), but patients were not examined routinely by vertebral angiography.

Table 5.5. *Myelography*

No. of cases examined	92
Prone examination only	4
Expanded cervical cord	34
Tonsillar ectopia	62
Normal	13

Ellertsson (1969a, b) and Ellertsson and Greitz (1969) have reviewed the problem of cystic lesions of the spinal cord and their radiological investigation. By use of gas myelography and tilting examination they have demonstrated in 30 cases an elongated flaccid cyst of the spinal cord. Four other cases. almost certainly suffering from syringomyelia showed no radiological evidence of a collapsing cyst. Ellertsson differentiates the fluctuating distensible cyst from non-fluctuating lesions of the spinal cord, usually gliomas. We have little experience of gas myelography and have preferred in this series to rely on the findings at positive contrast myelography. It should, however, be noted that in 13 cases with clear-cut evidence of syringomyelia (Table 5.5), contrast myelography revealed no abnormality. We cannot as yet explain the cause of this syndrome.

THE RADIOLOGY OF COMMUNICATING SYRINGOMYELIA 59

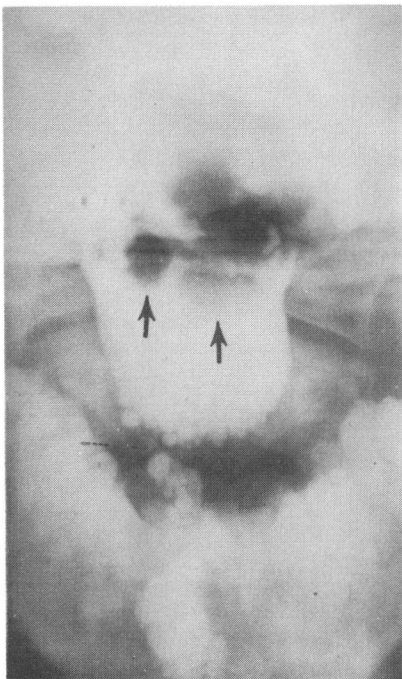

Fig. 5.10. Case N.S. (N.18401). Anteroposterior view showing asymmetrical defects caused by the cerebellar tonsils. Supine examination. (Reproduced with permission from *Brain*, **91**, 131, 1968.)

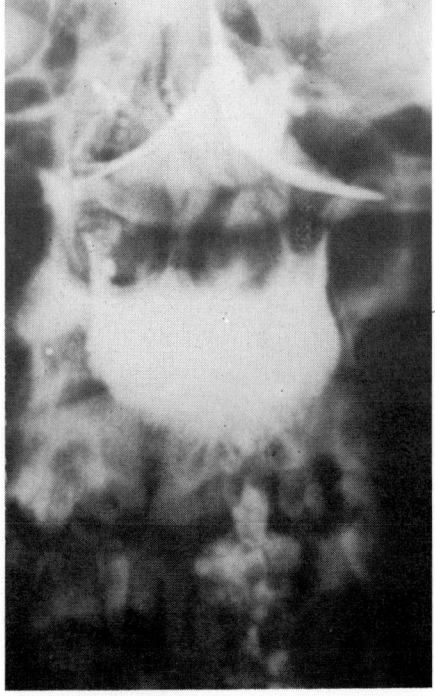

Fig. 5.11. Case C.H. (N.14793). Anteroposterior view of outline of the cerebellar tonsils, Supine myelogram.

Of the 63 cases in whom cerebellar ectopia was demonstrated by supine myelography, air encephalography or vertebral angiography, operative decompression was undertaken in 43.

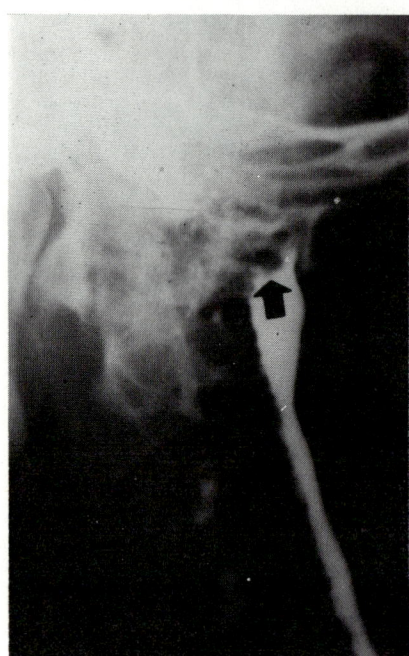

Fig. 5.12. Case E.H. (N.14895). Small tonsillar herniation (arrowed). Supine myelogram.

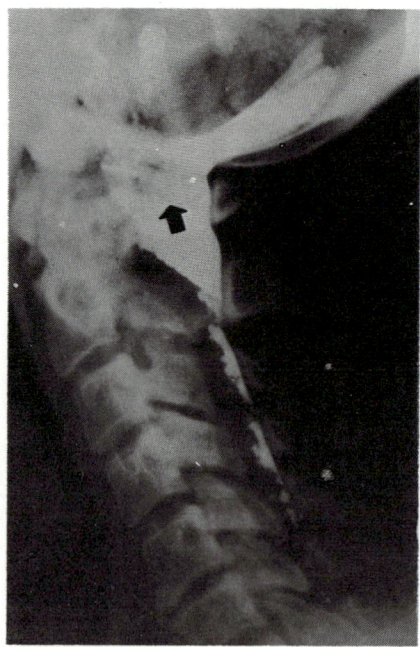

Fig. 5.13. Case N.S. (N.18401). Lateral view of supine myelogram showing large filling defect in the upper cervical region posteriorly, produced by a fairly marked degree of descent of the cerebellar tonsils. (Reproduced with permission from *Brain*, **91**, 131, 1968.)

THE RADIOLOGY OF COMMUNICATING SYRINGOMYELIA

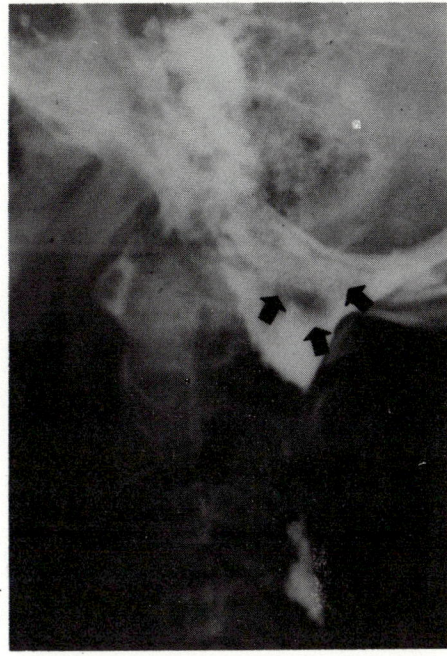

Fig. 5.14. Case N.S. (N.18401). Large area of radiolucency produced by tonsillar herniation or ectopia (arrowed). Supine myelogram.

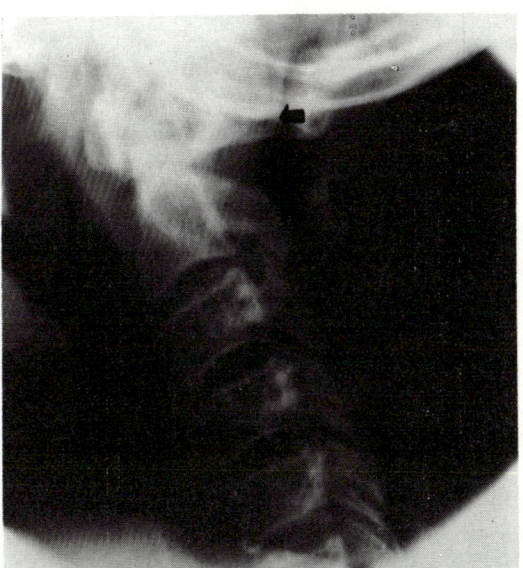

Fig. 5.15. Case L.T. (N.15611). Gas myelogram showing the outline of a 'finger' of cerebellar tonsil against the posterior aspect of the upper cervical cord (arrowed).

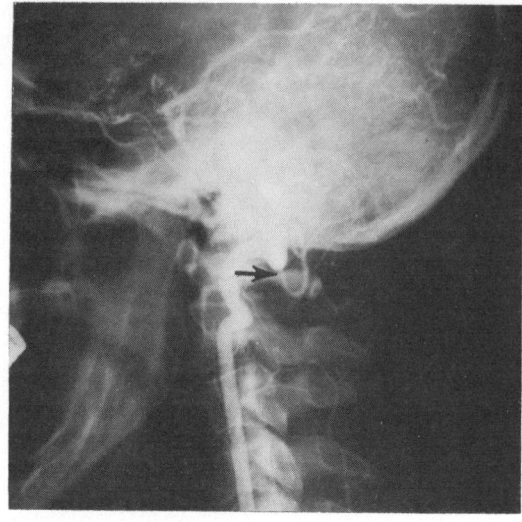

Fig. 5.16. Case K.E. (N.11329). Direct vertebral angiogram. The arrow points to the caudal loop of the posterior inferior cerebellar artery which extends down to the lower border of the first cervical vertebra. This artery was shown in the anteroposterior views to lie close to the midline of the cervical canal. When the caudal loop is low as a result of a developmental anomaly it lies far laterally in the cervical canal. Therefore, this was suggested as being brought down by the descent of the tonsil. This was confirmed at operation. (Reproduced with permission from Brain, **91**, 131, 1968.)

REFERENCES

Adams, R. D., Schatzki, R. & Scoville, W. B. (1941) The Arnold–Chiari malformation. Diagnosis, demonstration by intra-spinal lipiodol and successful surgical treatment. *New England Journal of Medicine*, **225**, 125.

Aring, C. D. (1938) Cerebellar syndrome in adults with malformation of cerebellum or brain-stem. *Journal of Neurology and Psychiatry*, **1**, 100.

Baker, H. L. Jr (1963) Myelographic examination of the posterior fossa with positive contrast medium. *Radiology*, **81**, 791.

Bucy, P. C. & Lichtenstein, B. W. (1945) Arnold–Chiari malformation in adults without obvious cause. *Journal of Neurosurgery*, **2**, 245.

Ellertsson, A. B. (1969a) Semiological diagnosis of syringomyelia related to roentgenologic findings. *Acta Neurologica Scandinavica*, **45**, 385.

Ellertsson, A. B. (1969b) Syringomyelia and other cystic spinal cord lesions. *Acta Neurologica Scandinavica*, **45**, 403.

Ellertsson, A. B. & Greitz, T. (1969) Myelocystographic and fluorescein studies to demonstrate communication between intramedullary cysts and the cerebrospinal fluid space. *Acta Neurologica Scandinavica*, **45**, 418.

Gardner, W. J. & Goodall, R. J. (1950) The surgical treatment of Arnold–Chiari malformation in adults. *Journal of Neurosurgery*, **7**, 199.

List, C. F. (1941) Neurologic syndromes accompanying developmental anomalies of occipital bone, atlas and axis. *Archives of Neurology and Psychiatry*, **45**, 577.

Malis, L. I. (1958) The myelographic examination of the foramen magnum. *Radiology*, **70**, 196.

Ogryzlo, M. A. (1942) Arnold–Chiari malformation. *Archives of Neurology and Psychiatry*, **48**, 30.

Occleshaw, J. V. (1970) The posterior inferior cerebellar arteries. Some quantitative observations in posterior cranial fossa tumours and the Arnold–Chiari malformation. *Clinical Radiology*, **21**, 1.
Peach, B. (1964) Arnold–Chiari malformation with normal spine. *Archives of Neurology*, **10**, 497.
Teng, P. & Papatheodorou, C. (1965) Arnold–Chiari malformation with normal spine and cranium. *Archives of Neurology*, **12**, 622.

CHAPTER SIX

The Surgical Treatment of Communicating Syringomyelia

J. B. FOSTER and P. HUDGSON

There are numerous references in the literature to the results of surgical treatment in this condition (Elsburg, 1916; Kelly, 1935; Worster-Drought, Wakely and Shafar, 1941; Netsky, 1953; Wetzel and Davis, 1953; Frazier and Roe, 1956; Pitts and Groff, 1964). In the majority of cases described, the syrinx has been exposed at cervical laminectomy and a variety of procedures adopted, most of which were based on the premise that much of the neurological deficit had been produced by the pressure of the fluid-filled syrinx on uninterrupted long tracts within the spinal cord, a thesis which is open to considerable doubt (see Chapter 8). The procedures performed included decompressive laminectomy alone; laminectomy with aspiration of the contents of the cyst; laminectomy followed by splitting of the median raphe and attempts at marsupialisation of the syrinx (e.g. by the insertion of polythene catheters draining into the subarachnoid space); laminectomy plus posterior fossa craniectomy and removal of the posterior rim of the foramen magnum; and more recently, insertion of a ventriculo-atrial shunt. In the first three procedures, dramatic short-term improvement is fairly common, Love and Olafson (1966) reporting 'excellent' or 'good' results in 23 of 44 patients they studied. However, long-term improvement cannot be expected in cases subjected to laminectomy alone or to laminectomy with aspiration of the syrinx, as there is no prospect of thus changing the pathophysiology of the condition. Improvement may be sustained for a reasonable period in cases where some means of drainage exists from the cyst to the spinal subarachnoid space. However, eventually the drainage tube or syringostomy often becomes blocked, probably through progressive gliosis in the wall of the syrinx, and the patient may therefore deteriorate.

The most recent comprehensive reviews have been furnished by Gardner (1965, 1967), Love and Olafson (1966), Williams (1970) and Logue (1971). A radically different approach to the operative treatment of communicating syringomyelia has been developed (Williams, 1969) and was first suggested by Gardner (1965, 1967) on the basis of his experience with the condition over two decades. Gardner's concept of the pathogenesis of syringomyelia is discussed in detail in Chapter 8. Suffice it to say at this stage that he believes that, in most cases, the condition is essentially chronic *hydromyelia* associated with impaired ventricular drainage and a patent communication between the fourth ventricle and the central canal of the cord. He and his colleagues have adduced impressive clinical, radiological and pathological evidence, the latter in the operating theatre as well as in the autopsy room, in support of this view. He believes that the neurosurgeon's attention should be switched from the syrinx itself to the cervicomedullary junction and that restoration of normal drainage of C.S.F. from the ventricles should be the principal guide to the choice of operation. This philosophy did not gain much support on either side of the Atlantic for many years but has been adopted with some enthusiasm by Conway (1967) in the United States, and in this country by Appleby et al (1968), Gordon (1969), Newton (1969), Williams (1969, 1970) and Logue (1971)—albeit with important differences in emphasis in some respects (Williams, 1970, see Chapter 8).

In Gardner's reviews (1965, 1967), he describes the operative findings and procedures in the 74 patients with syringomyelia investigated in his clinic. All these patients had developmental anomalies at the cervicomedullary junction, with effective obstruction to the outflow of C.S.F. through the foramen of Magendie and a demonstrable or inferred patent communication between the fourth ventricle and the syrinx (distended central canal). Each of these patients had a high cervical laminectomy with decompression of the neural structures impacted in the foramen magnum.

In 62 the syrinx was demonstrated at laminectomy and in 26 of these the fluid contents were aspirated and analysed. In each case the fluid was clear and watery and in 18 resembled the lumbar C.S.F. of the same patient. The protein content was similar to the lumbar C.S.F. in 10 and that of ventricular fluid in eight (less than 15 mg/100 ml). In no case was it discoloured or proteinaceous, as is the case with cysts associated with intrinsic tumours (see Chapter 10). In 52 of these patients there was *immediate* postoperative improvement in some of their symptoms, in 11 there was no change, in six there was deterioration and five died in the early postoperative period. Seventeen patients in this group had been operated on before 1950 and had been seen at intervals ranging from six to 24 years. In eight of the 17, sustained improvement was noted at their last reviews and in three others there had been improvement for 24, 13 and six years respectively, followed by recurrence. Basing his argument on these figures, Gardner (1965) contended that there was substantial evidence in favour of his views on the pathogenesis of the disease and its surgical correction.

Conway (1967) reported the operative findings and results of follow-up in 12 cases of syringomyelia with progressive symptoms ranging from two to 16 years in duration. Six of these patients had cervical laminectomies and posterior fossa craniectomies and, in each of them, maldevelopment of the cerebellar tonsils and the roof of the fourth ventricle and a cervical syrinx *in communication with the ventricle* was found. These patients were followed for periods up to nine years after operation and improvement was noted in all of them. In sharp contrast, five of the six patients who were not operated on deteriorated and one remained static. Conway's experience led him to commend Gardner's procedure (disobliteration of the exit foramina of the fourth ventricle and blocking of the communication with the central canal in the treatment of syringomyelia). He also suggested that the use of a low-pressure ventriculojugular shunt might be a worthwhile alternative, especially in advanced cases.

In the first report on the application of Gardner's operation in the United Kingdom, Appleby et al (1968) recorded the results of surgery in 15 of 17 patients who had syringomyelia or 'atypical' syringomyelia with progressive symptoms. In two of these patients (M.C., N.11193 and D.R., N.145054, see Chapter 3), a young woman and a young man with only moderately severe deficits, their abnormal signs rapidly disappeared after operation (in 1964 and 1965 respectively) and have not recurred since then. One patient (F.D.) died from myeloid leukaemia. Of the remainder, 12 felt that there had at least been arrest of the progress, and nine reported definite improvement in gait and sensation. Our observations confirmed this, and the trend has been maintained in the much larger series here reported from Newcastle.

The orthodox view of the surgical treatment of syringomyelia is summarised in the review of Love and Olafson (1966). They described 40 patients who had 44 operations of various kinds for syringomyelia, in some cases accompanied by syringobulbia. These patients all had one or other of the operations on the syrinx itself outlined in the introduction to this section. Thirty-five of these patients had 'uncomplicated' syringomyelia and five had syringomyelia associated with Arnold–Chiari malformations. Of the 35 'uncomplicated' cases, 10 were regarded as 'excellent' results (objective evidence of improvement), 12 had 'good' results (arrest in progress of neurological deficit) and eight patients had 'poor' results (deterioration continued postoperatively); five patients were lost to follow-up. Three of the five patients with Arnold–Chiari malformations were adequately followed and all three continued to deteriorate—they had suboccipital craniectomies as well as laminectomy with syringostomies. Love and Olafson (1966) suggested that their results justified the continued use of surgery in treatment of this condition. However, they cautioned that rapid deterioration or severe neurological deficit in a long-standing case militated against a successful outcome after surgery. Two problems call into question the validity of this study, namely the relatively short follow-up periods (maximum six years) and the relatively high wastage of patients (seven out of 40). More important, however, are the demonstrably poorer results, both immediate and long term, compared with patients treated by Gardner's operation or by a shunt procedure.

Hankinson (1970), in an analysis of 45 personal cases, arrived at essentially the same conclusions as Gardner (1965) and Conway (1967). In most of his cases, he found cerebellar ectopia (type I Chiari anomaly) or other developmental anomalies at the cervicomedullary junction, with effective obstruction to the drainage of C.S.F. from the ventricular system. In 26 of these he was able to demonstrate a deep V-shaped cleft in the floor of the fourth ventricle, at or about the level of the obex and in communication with the distended central canal of the cord. Hankinson dealt with all these cases using Gardner's operation and clearly believed his experience to support the hydrodynamic hypothesis for the pathogenesis of syringomyelia. Certainly, he was able to report immediate improvement in neurological deficit in 33 of his group and gradual improvement over one to two years in four others. Eight patients derived no benefit from operation. However, he and his colleagues (Appleby et al, 1969) had previously separated seven cases who had no evidence of a maldevelopment of the neuraxis but had a dense adhesive arachnoiditis in the upper cervical canal and the cisterna magna. The adhesions effectively obliterated the exit foramina of the fourth ventricle and the writers believed that the results of this apparently *acquired* lesion, as far as C.S.F. dynamics were concerned, were the same as a developmental abnormality, assuming patency of the communication with the central canal (see Chapter 4).

The outstanding difference between this group and the patients with uncomplicated developmental anomalies lay in their progress after operation. Each of these seven patients had an unusually stormy course in the immediate postoperative period and none of them derived any long-term benefit from the restoration of free drainage of C.S.F. from the fourth ventricle. All of them *deteriorated* neurologically when reviewed at intervals of up to five years and two patients deteriorated dramatically after operation, developing acute hydrocephalus with central respiratory failure and requiring ventricular drainage as a matter of urgency. The reason for their deterioration is uncertain (none of these patients was re-explored) but it is likely that the adhesions redeveloped after operation, again blocking the free flow of C.S.F. from the fourth ventricle. Because of the unsatisfactory outcome in these cases, it is suggested that others should be dealt with by insertion of a low-pressure ventriculo-atrial shunt (Conway, 1967) if a firm diagnosis of arachnoiditis can be made before operation (see Chapter 4).

This indirect approach is undoubtedly less traumatic than Gardner's operation but it is unsatisfactory inasmuch as no information can be gained at operation about the anatomical relationships of the cervicomedullary junction. In our own experience, cervical laminectomy and decompression of the foramen magnum is a safe and effective procedure in patients with demonstrated maldevelopment of the Chiari type at the cervicomedullary junction. It may, however, be positively dangerous in patients with an adhesive arachnoiditis and, in these, it may be that insertion of a ventriculo-atrial shunt will become the standard treatment.

Of the 100 patients in the present series, 49 have been subjected to operation in the cervical region, i.e. decompression of the foramen magnum and upper cervical laminectomy. Of the 62 patients shown on myelography to have tonsillar ectopia, 47 have been operated on in this way (Table 6.1). Thirteen of the 100 patients had no abnormality demonstrable on myelography and operation has

Table 6.1 *Operated cases*

	Number of patients	Operations
Total	100	59
Ectopia on myelography	62	47
Arachnoiditis on myelography	18	12
Normal myelogram	13	1 (tractotomy)
Other procedures		
Posterior fossa exploration		1
Spitz-Holter valve		4
Stereotaxic operation for pain		1
Dorsal laminectomy		1

not as yet been undertaken in these. One was submitted to a tractotomy for intractable pain. Of the 18 patients found to have radiological changes of arachnoiditis at the foramen magnum, 12 were subjected to operation. Other procedures undertaken in this group have been posterior fossa exploration alone in one, insertion of a Spitz–Holter valve in four, a stereotaxic operation for pain in one, and a dorsal laminectomy in one other (L.J.).

SURGICAL TECHNIQUE

The operations have been performed with the patient in a sitting position with controlled respiration when considered necessary. Exploration of the region of the foramen magnum has been undertaken to decompress the prolapsed tonsils adequately. In earlier cases the syrinx was exposed but this is no longer thought necessary, though in many cases the upper pole of the cavity has been seen. Hankinson (1970) quotes Gardner (1965), who stated: 'It is not only unnecessary but it may be inadvisable to carry the laminectomy low enough to expose the syrinx, for the reason that extensive laminectomy sometimes has been followed by severe flexion deformity of the cervical spine.' Hankinson states that in many of his operations the exposure and removal of the occipital bone was

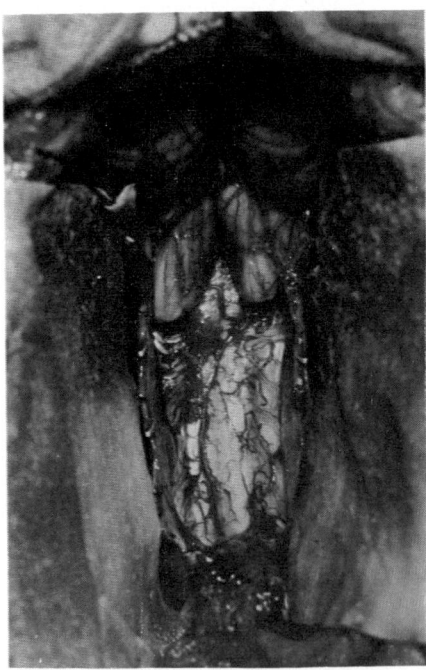

Fig. 6.1. Case M.A. (N.29275). The typical appearance of the ectopic cerebellar tonsils exposed at operation with cystic expansion of the cervical cord.

Fig. 6.2. Asymmetrical tonsillar ectopia (tonsils arrowed). (Reproduced with permission from *Modern Trends in Neurology*, 5.)

SURGICAL TREATMENT OF COMMUNICATING SYRINGOMYELIA

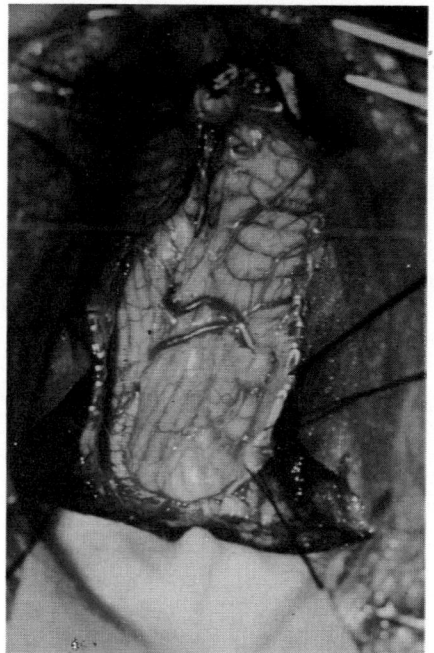
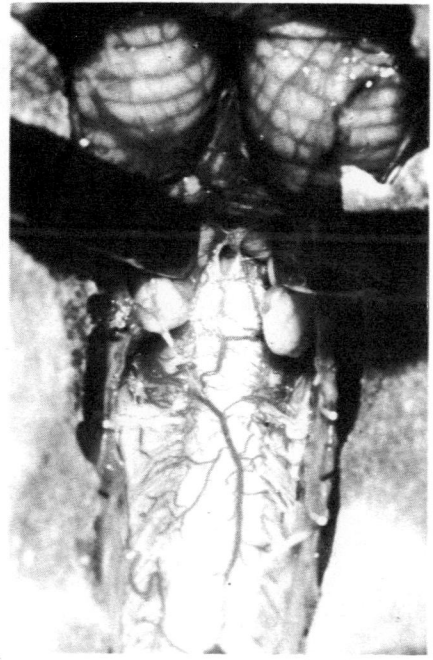

Fig. 6.3. Case D.O. (R.V.I.695114). Gross eccentric ectopia of the tonsils.

Fig. 6.4. Exposure of the origin of the central canal in the floor of the fourth ventricle. (Reproduced with permission from *Modern Trends in Neurology*, 5.)

made extremely difficult because of the acute angle between the posterior fossa and the cervical spine resulting in an upward tilt of the foramen magnum. With this abnormal configuration, there was often an unusually thick and wedge-shaped atlanto-occipital membrane, with a sharp border constricting the underlying dura.

In all cases the dura has been opened widely with a Y-shaped incision, with the lower limb extended in the midline to the upper border of the lamina of the second or third cervical vertebra. In all 47 cases, tonsillar ectopia which had been seen on supine myelography was confirmed at operation and the tonsils were found lying, to a greater or lesser extent, below the level of the foramen magnum with an asymmetrical configuration (Figs. 6.1 and 6.2). In some cases the ectopic tonsils were indented by the constricting posterior rim of the foramen magnum and atlanto-occipital membrane. In some cases also, the asymmetrical descent of the tonsils was remarkable (Fig. 6.3). The greatest degree of tonsillar ectopia was to the upper border of the lamina of the third cervical vertebra and the least amount of prolapse was some 5 mm. In these 47 cases there was no difficulty in separating the cerebellar tonsils and exposing the fourth ventricle. However, in the presence of severe arachnoiditis this presented considerable difficulty because of the tough vascular adhesions (see Fig. 6.6).

In the 47 cases without arachnoiditis the foramen of Magendie was found to be occluded by a membrane in six, and after perforation of this the fourth ventricle

70 COMMUNICATING SYRINGOMYELIA

could be entered. In all other cases the foramen appeared to be obliterated by the compressed and elongated cerebellar tonsils. In 26 cases an opening of the central canal was seen in the floor of the fourth ventricle (Fig. 6.4) and probing would allow access into the central canal of the upper cervical cord (Fig. 6.5). When this was seen, the hiatus was occluded by a small piece of muscle (Gardner, 1965). Hankinson (1970) has commented on three other abnormalities which are often seen: (1) A striking dorsal bulge at the junction of the medulla and cervical cord associated with the tonsillar herniation and a caudal position of the fourth ventricle. (2) The second cervical posterior roots take a rostral course. (3) The choroid plexus of the fourth ventricle appears more prominent than usual when it is visualised on separation of the cerebellar tonsils. The practice has been to leave the dura open to maintain the decompression, and probably as a result of this and the contamination of the cerebrospinal fluid with blood and transudate, the patients' postoperative course is rather more stormy. Often, they experience severe headache and neck stiffness with fever, and the protein content of the spinal fluid may be in the 200–300 mg/100 ml region for some weeks.

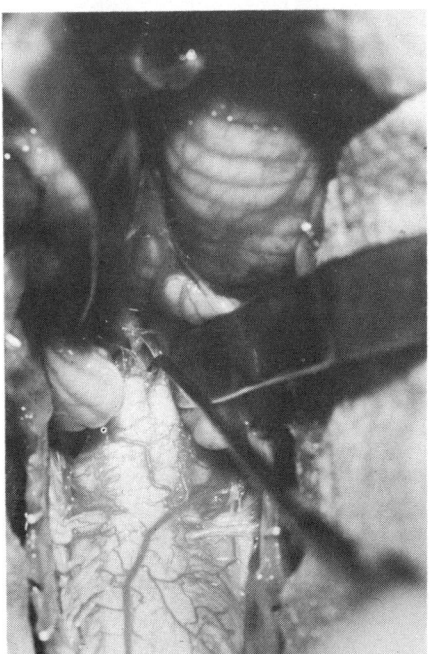

Fig. 6.5. Probe in the patent central canal.

CASES WITH COMPLETE RECOVERY

Of the 47 cases with proven cerebellar ectopia on myelography, operation was followed by complete relief of symptoms in five, and in these five patients one had an incomplete band of hypalgesia high on the side of her neck, one had had generalised hyperreflexia and extensor plantar responses, in one there had

SURGICAL TREATMENT OF COMMUNICATING SYRINGOMYELIA

been oscillopsia and on examination severe nystagmus, and in the last, a more typical syringomyelic 'half-cape' sensory abnormality with wasting of the small muscles of the ipsilateral hand. In all these patients the physical signs regressed completely and the patients have remained not only symptomatically but objectively well (Table 6.2).

Table 6.2. Results in cases with tonsillar ectopia on myelography

Number of cases	47
Cured	5
Improved	30
No change	11
Postoperative deaths	1 (M.C.)
Subsequent death	1 (Leukaemia)

J.R., FEMALE (N.28331)

This 17-year-old patient complained that for two years before attending hospital she had suffered from occipital headache and pain in the neck. The pain was made worse by flexing the neck and was also aggravated by stooping, coughing and laughing. Often the pain would pass into the left arm. On examination, her only abnormal sign was an incomplete band of hypalgesia high on the side of her neck on the left. Her radiographs of neck and skull were normal but myelography showed gross asymmetrical ectopia of the cerebellar tonsils. At operation (Hankinson), the ectopia was confirmed and an hiatus from the fourth ventricle was seen passing down the centre of the cord. This was plugged with muscle. The decompression of the upper cervical canal region and foramen magnum has totally relieved her of her symptoms.

E.T., FEMALE (N.28162)

For one year before admission to the unit on 26 March, 1968, this 23-year-old patient had suffered from attacks of pain starting with short, sharp jabs at the right side of the neck which rapidly radiated up into the head and were associated with a feeling of generalised throbbing. These headaches lasted for two or three minutes only and would then pass off spontaneously but she was of the opinion that they were regularly induced by exertion. If she extended her head fully then she could relieve herself of the pains. The pain could be induced by acute flexion of the neck. There were no other symptoms and full neurological examination was entirely negative. There was no suggestion of arrested hydrocephalus, nor was the neck short. The patient had no nystagmus and in the limbs there was no abnormality of tone, power or coordination. X-rays of the skull were normal and X-rays of the cervical spine were normal with anteroposterior diameter of the canal of 18 mm at the level of C5. Air encephalography did not outline the ventricular system and there was a suspicion of cerebellar tonsillar herniation, with obstruction of the posterior surface of the spinal cord in the region of the foramen magnum. The attempted air picture was followed by myelography and in the supine position this disclosed a filling defect on the back of the spinal cord which was well defined, smooth in outline and consistent with cerebellar tonsillar herniation.

At operation (Hankinson), the arch of the atlas was removed and the spine and upper portion of the second lamina. The dura was then opened with a Y-shaped incision extending to C2 and the tonsils were seen to be hypertrophied and prolapsed to the C2 level, right more than left. There was no arachnoiditis and a free flow of fluid from the fourth ventricle was established. The tonsils were separated and under the right one there was a depression in the posterior aspect of the medulla. The cord appeared normal but there was a small opening into the central canal. This was occluded with a small piece of muscle, and the muscles, aponeurosis and skin were repaired, leaving the dura open. Throughout the operation the patient's condition was satisfactory. Her postoperative course was normal and at follow-up she was consistently totally free from symptoms.

A.T., FEMALE (N.40499)

A 45-year-old housewife presented with a 24-month history of episodic sudden headaches. The headaches were generalised, lasting for no more than a few seconds. On several occasions associated with the headache she had suddenly fallen. Examination revealed generalised hyperreflexia with extensor plantar responses as the only abnormality. There were no features of raised intracranial pressure, i.e. no papilloedema. Because of the presentation, an attempt to outline the ventricular system with air was made by the lumbar route. This was unsuccessful. Bilateral carotid angiography revealed marked ventricular dilatation, however, and subsequent ventriculography confirmed the marked symmetrical dilatation of the lateral ventricles. The third ventricle was also dilated, as was the fourth. A lumbar-route myelogram revealed an advanced Chiari malformation, with obstruction to the exit of the fourth ventricle and cerebellar tonsillar ectopia. At operation (Hankinson), the foramen magnum was decompressed and an evident cerebellar ectopia was seen, more severe on the right than on the left. The cerebellar tonsils appeared to be enlarged and extended to the lower border of C1. Membranous obstruction of the foramen of Magendie was seen on retraction of the tonsils. Postoperatively, the patient made an uneventful recovery and at follow-up, six months after operation, was completely symptom free and leading a normal life.

V.N., MALE (R.V.I. 659010)
This doctor, aged 30 years, presented to the neurological clinic with a history of difficulty with his vision over a period of some 12 years, accompanied by double vision during the preceding eight years. On analysis, he was able to describe that his double vision was due to oscillopsia on conjugate gaze in all directions and that it was not a true paralytic diplopia. In addition, he had been subject to attacks of severe throbbing, retro-orbital headache since early adolescence. The headaches were precipitated by or exacerbated by coughing, stooping and shouting and had been relieved by lying down. During the preceding nine years he had experienced occasional attacks of tingling paraesthesiae down the outer border of the left arm into the hand. On examination, there were no abnormal neurological signs, except for coarse vertical nystagmus on fixation and on conjugate gaze upward. There was coarse rotary nystagmus on lateral gaze to either side. The combination of symptoms and signs suggested an anomaly of the hind brain at the level of the foramen magnum, and X-rays of the cervical spine showed a widening of the anteroposterior diameter of the cervical spinal canal (2·1 cm in the mid-cervical region). There was a minor degree of basilar impression. Supine myelography showed descent of the medulla and cerebellar tonsils, with an abnormally low origin of the basilar artery. Surgical decompression was undertaken (Hankinson) when the posterior 2 cm of the foramen magnum, the arch of the atlas and the spine and lamina of C2 were removed. After opening the dura, it was evident that the tonsils were considerably prolapsed, the right 0·5 cm lower than the left, just above the upper border of C3. On separation of the tonsils, spinal fluid flowed through the fourth ventricular foramen; no orifice to the central canal could be seen. The dura was left widely open. Following operation, the patient made an uneventful recovery, at follow-up eight months afterwards he showed no nystagmus and was totally free from oscillopsia and headache.

D.R., MALE (S.G.H. 145054)
This 17-year-old boy presented to the Sunderland General Hospital in February 1966 complaining of steadily progressive weakness in the left arm. He volunteered no other symptoms and had no previous history or family history of serious illness. Physical examination showed that he was 'bull necked' and had a recent cigarette burn on the left hand. Direct questioning established that this had not occasioned him any discomfort. He had global weakness of the muscles in the left upper limb with wasting of the first dorsal interosseous space and the hypothenar eminence. The deep tendon reflexes were absent in the left upper limb and diminished in the right upper limb but were normal elsewhere. The superficial abdominal reflexes were present and both plantars were flexor. Sensory testing showed that he had a 'half-cape' area of dissociated sensory loss involving the left upper limb and the left side of the trunk from C2 to D4. X-rays of the skull were normal but X-rays of the cervical spine showed fusion of the second and third cervical spinous processes. Myelography showed prolapse of the cerebellar tonsils. This finding was confirmed at operation (Kalbag) when decompression of the upper cervical canal and the foramen magnum was carried out. The patient made a satisfactory postoperative recovery and subsequent follow-up has shown that the weakness in his left arm has improved and that his 'half-cape' sensory loss has disappeared completely.

Comment. While headache and pain in the neck were features in three of these patients, all were shown myelographically to have descent or ectopia of the cerebellar tonsils. We believe that the early presentation of the females and their early decompression may well have prevented the onset of central cord cavitation and the development of the syringomyelic syndrome which D.R. was already developing at the time of his decompression. It is evident from these cases that early recognition of tonsillar ectopia may lead to a total prevention of, or recovery from, incipient central cord cavitation.

CASES WITH IMPROVEMENT AFTER OPERATION

Thirty patients of the 47 were improved by operation at initial follow-up and subsequently, with a follow-up period varying from six years to several months. Improvement occurred not only in the spasticity but in many cases there was also regression of dissociated anaesthesia and improvement in the power of an upper limb suffering from lower motor neurone weakness and wasting.

M.C., FEMALE (N.11193)

This 36-year-old patient was first seen in June 1964 when she gave a history of intermittent attacks of occipital headache over a period of five years. These attacks were originally regarded as being 'tension' in type and she was treated, without improvement, with a variety of mild analgesics and tranquillising agents. There was a gradual increase in frequency and severity of the headaches and she complained that the attacks were often precipitated by standing up suddenly and that they could usually be relieved by lying down flat. For some months before admission the headache was accompanied by nausea and vomiting. For five months she had noticed paraesthesiae in the right hand accompanied by pain in the right side of the chest radiating into the right upper limb. The latter symptom was made worse by coughing, sneezing or stooping. Previous medical history and family history were both clear. Physical examination at the time of admission showed a clear-cut 'half-cape' area of dissociated sensory loss involving the right upper limb and the right side of the trunk from C2 to D10. The deep tendon reflexes in the right upper limb were depressed but there were no other abnormal physical signs. X-rays of the skull were normal and X-rays of the cervical spine showed loss of cervical lordosis only.

Myelography showed that the spinal cord was greatly expanded from the upper cervical region down to D10 but there was no apparent abnormality in the region of the foramen magnum. Surgical exploration (Hankinson) revealed prolapse of the right cerebellar tonsil to the level of the second cervical lamina. The left tonsil lay 5 mm above this. The right tonsil was larger than the left and there was an associated hydromyelia. The upper cervical canal and the foramen magnum were decompressed and the hydromyelia was aspirated. Her headaches and chest pain were relieved immediately after operation and recent follow-up has shown that there has been virtually complete recovery of thermal appreciation and response to pinprick over the 'half-cape' area previously described.

F.D., FEMALE (N.3477)

This patient, aged 54 years, first attended in July 1964. Her history at that time was of asymmetrical weakness in the legs, left worse than right, accompanied by similarly asymmetrical stiffness in the upper limbs. She also complained of clumsiness in performing fine movements with her fingers and of disturbed sensation (a 'spongy' feeling) in the upper limbs and the trunk. The latter symptoms were aggravated by coughing, sneezing and stooping. More recently, she had developed difficulty in moving food around her mouth with her tongue and had obviously become very depressed. On examination she was short and plump although there were no superficial abnormalities in the head and neck. However, movements of the head and neck were limited in all directions by muscle spasm. She had coarse horizontal phasic grade 1 nystagmus on lateral gaze to the left and right. Both sternomastoids were weak and the right trapezius was weak and wasted. The right half of the tongue was extremely atrophied and was fasciculating. She had an asymmetrical spastic tetraparesis, more evident on the left side, with absent superficial abdominal reflexes and bilateral extensor plantar responses. Sustained clonus could be elicited at both knees and both ankles. Sensory testing revealed impaired joint position sense in all four limbs, the upper limbs being more severely affected than the lower and the left side more than the right. There was no spinothalamic sensory loss. X-rays of the skull were normal and X-rays of the cervical spine showed minimal degenerative changes only.

Prone myelography showed expansion of the cervical cord (see Fig. 5.8). The vertebral and basilar arteries were outlined and appeared to be normal. No abnormalities were defined in the region of the foramen magnum. Surgical exploration on 9 October, 1964 (Hankinson) showed that both cerebellar tonsils were prolapsed to the level of the second cervical lamina and were thin and atrophic. The patient made a satisfactory postoperative recovery and until 1 June, 1965, showed significant improvement in gait and joint position sense. She died in Darlington Memorial Hospital on 1 October, 1965, from acute myeloid leukaemia. Unfortunately, the brain and spinal cord were not examined at autopsy.

C.H., MALE (N.14793)

This patient was first seen in May 1965 at the age of 42 years. Eight years earlier, he noticed that his hands and arms were always cold and blue irrespective of the environmental temperature, and these symptoms became steadily worse up to the time of admission to hospital. The left upper limb was always more severely affected than the right. For 12 months before admission, he noticed that the left side of his face sweated more than the right and that his left leg was restless and 'jumpy' in bed at night. During the same time, he experienced increasing difficulty in performing fine movements with the fingers of his left hand although the grip on that side was normal. For several weeks before admission, he found that he could tolerate very hot objects in his left hand and that his left foot was dragging when he walked. He also complained that his left arm felt as though it was constricted by 'a tight band'. A patent ductus arteriosus had been successfully ligated at the age of 30 years, but his past history was otherwise clear. There was no relevant family history.

Physical examination failed to reveal any superficial abnormalities in the head and neck but both hands were cold and cyanosed, left more than right. Multiple old and recent burn scars were noted on the fingers of the left hand. Examination of the nervous system showed that he was extremely myopic in both eyes with visual acuity correcting to J2 bilaterally. Ophthalmoscopy revealed a large myopic crescent on the left side. The jaw jerk was abnormally brisk but the cranial nerves were otherwise normal. Fine finger movements were clumsily performed on the left side although power was normal in both upper limbs. He had a mild spastic paraparesis, the left side being more severely affected than the right. Both triceps jerks were diminished, the right radial jerk was inverted and the other upper limb jerks were absent. The deep tendon reflexes in the lower limbs were symmetrically exaggerated, the superficial abdominal reflexes were diminished and the left plantar response was extensor. There were bilateral 'half-cape' areas of dissociated sensory loss extending from C2 to D12 on the right and from C2 to D4 on the left. Vibration sense was lost below the waist and joint position sense was grossly impaired in both upper and lower limbs, left more than right. X-rays of the skull revealed no abnormality but X-rays of the cervical spine showed suspicious widening of the upper half of the cervical canal.

Myelography confirmed that the anteroposterior and lateral diameters of the cervical canal were increased and that the cord was expanded, with a posterior filling defect 1 cm below the foramen magnum. Vertebral arteriography was normal. Surgical exploration on 28 May, 1965 (Hankinson) confirmed the cystic dilatation of the cord and tonsillar prolapse to C2. The foramen magnum was decompressed and the hydromyelia was aspirated. The patient made a satisfactory postoperative recovery and recent follow-up has shown that he has made considerable progress. Joint position sense and perception of pinprick have both improved and both plantar responses are now flexor. Symptomatically, he feels that his gait has improved and that he can use his left hand with much greater facility. He is no longer troubled by facial hyperhidrosis but his hands are still blue and cold, especially during the winter.

L.T., MALE (N.15611)

This patient, aged 44 years, was first seen in July 1965 when he gave a three- to four-year history of difficulty in walking due to dragging of the left leg. At first intermittent, his limping eventually became permanent. He had also complained for two years of 'pumping' sensations in his legs, which usually came on when he was standing still or lying in bed. After admission to hospital he developed persistent sharp pain in the back of the neck and occipital region. He had no previous history of serious illness and his family history was clear. Physical examination showed that he was of short stature but his neck length and height were in proportion. The only significant abnormality on general examination was a recent unnoticed cigarette burn on the left hand. Examination of the nervous system revealed an asymmetrical spastic paraparesis. The deep tendon reflexes were absent in both upper limbs and symmetrically exaggerated in the lower limbs. The superficial abdominal reflexes were all absent and both plantar responses were abnormal. Examination of sensation showed that he had impairment of response to pinprick accompanied by loss of thermal appreciation in a patchy distribution over the trunk and both upper limbs. Light touch, vibration sense and joint position sense were all appreciated normally. Flexion and extension of the head and neck produced an acute exacerbation of the neck pain. Routine investigation, including X-rays of the skull, cervical spine, dorsal spine and lumbosacral spine failed to reveal any significant abnormality.

SURGICAL TREATMENT OF COMMUNICATING SYRINGOMYELIA

An air myelogram showed that the cerebellar tonsils were lying in the cervical canal (see Fig. 5.15). Accordingly the upper cervical spine and the posterior fossa were explored on 7 September, 1965 (Hankinson). The cerebellar tonsils were prolapsed through the foramen magnum down to the second cervical lamina. In addition, there was a complete agenesis of the left cerebellar hemisphere. The area was decompressed and the patient made a satisfactory postoperative recovery.

Recent follow-up has shown that there has been a considerable improvement in the patient's spastic gait and that pain and temperature sensation on the trunk and in the upper limbs has returned to near normal. He is at present troubled by a severe degenerative arthritis in the left knee joint. We believe this to be due to an old cartilage injury which had recently been exacerbated by his abnormal gait.

CASES WITH NO IMPROVEMENT AFTER OPERATION

In 11 patients submitted to operation there has been no change in their clinical status. Thirty-four patients, plus six cases with arachnoiditis, were not submitted to operative treatment. In 13 of these there had been no abnormality demonstrable on prone and supine myelography, although in many the presentation had been classical (see Chapter 3). This group presents a challenging problem to the dynamic concept of communicating hydromyelia, for if we are unable to demonstrate any hold-up of Myodil at the foramen magnum and no degree of cerebellar tonsillar ectopia, it is difficult to visualise the mechanism responsible for the persistent central canal dilatation and subsequent dissection. We feel that these

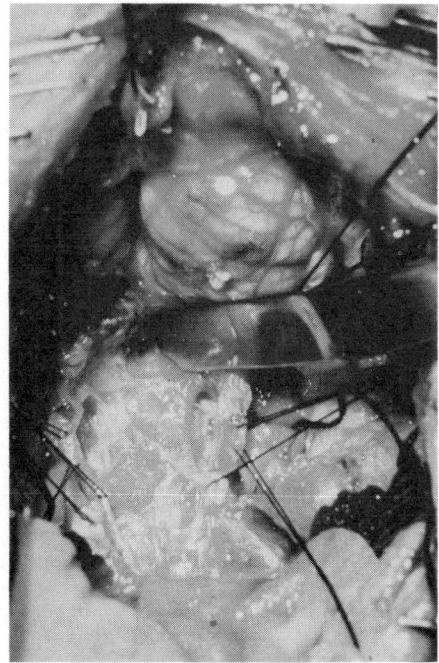

Fig. 6.6. Case A.F. (X06920). The appearances of severe adhesive arachnoiditis.

cases should possibly be subjected to exploratory operation at the foramen magnum to determine whether or not minimal degrees of atresia of the foramen of Magendie can be discovered and held responsible for their clinical presentation.

Four patients were thought to have disease which was too advanced to benefit from operation; in six cases the disease was evidently non-progressive and the patients were minimally disabled and therefore not subjected to operation; two cases await operation; four patients have refused surgery and three have died (Table 6.3). Of the deaths, one was from myeloid leukaemia and there has been one true postoperative death in the series.

Table 6.3. *Cases of syringomyelia in the series not operated upon*

Cases not submitted to operation	34 (+6 arachnoiditis)
Normal myelography	13
Clinical state too advanced	4
Non-progressive	6
Awaiting operation	2
Refused operation	4
Died	3
Other	2

M.C., MALE (N.34639)

This man, 48 years old, attended the outpatient clinic, having been well until 14 months previously when he first developed lumbar pain which tended to radiate into the left leg. The pain was made worse by lying in bed and was relieved by analgesics. He had noticed some numbness of the leg and some dragging of the feet on walking. He had suffered a scoliosis from the age of eight years but otherwise his past and current histories were unremarkable. Examination showed that he had a rather large head but there was no abnormality in the cranial nerves. There was no weakness in either upper limb but he had a spastic gait and an asymmetrical spastic paraparesis more marked on the left than the right, with bilateral extensor plantar responses and absent abdominal reflexes. There was a suggestion of some dissociated anaesthesia over the left forequarter but this was minimal.

X-rays of his skull confirmed the possibility of arrested hydrocephalus. X-rays of his cervical spine were in all respects normal. A myelogram confirmed the kyphoscoliosis but there was no convincing evidence of widening of the spinal cord except, perhaps, at the level of C1 where the Myodil column was thin. In the supine position, Myodil would not readily flow into the cisterna magna, most of the contrast medium flowing forwards into the pontine cistern and a little passing laterally into the side. The appearance suggested either tonsillar ectopia or arachnoiditis. Accordingly this man was subjected to decompression (Hankinson). After a midline incision had been made from the inion to the mid-cervical region and the muscles stripped from the bone, the exposed bone was removed. The posterior rim of the foramen magnum and atlanto-occipital membrane were deeply indenting the dura at this point. The dura was opened with a Y-shaped incision to the upper border of C2. It was noted that the tonsils were slightly prolapsed, right more than left, and there was a band of arachnoid adhesions at the level of the foramen magnum but these were fairly easily separated. The tonsils were found to be bound together by a thick membrane attached to the medulla. This was opened at the midline showing a very large fourth ventricle and there was an enormous outflow of cerebrospinal fluid. Haemostasis was established, the dura was left widely open, as is the practice, and the muscles, aponeurosis and skin were repaired. After operation, the patient failed to regain consciousness and subsequent frontal burr holes and air ventriculography showed extensive hydrocephalus with collapse of the cerebral mantle. An attempt was made to reinflate the ventricles with saline and following this he became conscious, obeyed commands and appeared to be improving. However, he again deteriorated and despite ventricular drainage being established later in the course of his illness, he died some 10 months after operation. At post-mortem, the enormous hydrocephalus, including involvement of the fourth ventricle, was confirmed. There was continuity with an extensive hydromyelia. The tonsillar ectopia noted at operation was confirmed. In the cerebral hemisphere there was a large laminated intracerebral haematoma which had almost certainly arisen at the time of cerebral collapse following decompression of the posterior fossa in the sitting position.

Comment. This case illustrates the problem of management in arrested hydrocephalus due to anomalies at the foramen magnum and foramen of Magendie. In retrospect, it was evident that this man should have undergone a ventricular drainage before any attempt at posterior fossa decompression. We feel that cases such as this may be diagnosed more accurately by the use of carotid angiography, carried out preferably in the non-dominant hemisphere, to determine the degree of ventricular dilatation. In the face of such dilatation, then a drainage procedure without posterior fossa decompression may be advisable, or even, in the non-progressive case, surgery ought to be deferred.

CONCLUSIONS

Our experience over the past few years with a consecutive series of 100 cases of syringomyelia suggests that this disease has a basis in anomalous drainage of the fourth ventricle, often associated with anomalies at the foramen magnum. The condition can be arrested by surgery if cerebellar ectopia at the foramen magnum can be demonstrated on supine myelography. We have patients whose symptoms and signs have regressed totally following operation, and of 47 cases decompressed at the foramen magnum, we have noted improvement in 30, in addition to the five cured by this procedure. We would advocate that syringomyelia no longer be regarded as a chronic degenerative condition of the spinal cord, but that an active investigational attitude be adopted so that early cases of cervical cord disorder can be recognised to be due to anomalies at the foramen magnum and subjected to early decompressive surgery. We realise that our follow-up is as yet incomplete and that in patients beyond a certain stage of disability, useful function is not likely to return. We therefore make a plea for early recognition and early treatment. Our principle is to restore, as far as possible, anatomical normality at the foramen magnum, to establish adequate drainage from the ventricular system and to relieve the obstruction to the flow of cerebrospinal fluid.

REFERENCES

Appleby, A., Foster, J. B., Hankinson, J. & Hudgson, P. (1968) The diagnosis and management of the Chiari anomalies in adult life. *Brain*, **91**, 131.
Conway, L. W. (1967) Hydrodynamic studies in syringomyelia. *Journal of Neurosurgery*, **27**, 501.
Elsburg, C. A. (1916) *Diagnosis and Treatment of Surgical Diseases of the Spinal Cord and its Membranes.* London: Saunders.
Frazier, C. H. & Roe, S. N. (1956) The surgical treatment of syringomyelia. *Annals of Surgery*, **103**, 481.
Gardner, W. J. (1965) Hydrodynamic mechanisms of syringomyelia. *Journal of Neurosurgery, Neurology & Psychiatry*, **28**, 247.
Gardner, W. J. (1967) Myelocele: rupture of the neural tube? *Clinical Neurosurgery*, **15**, 57.
Gordon, D. S. (1969) Neurological syndromes associated with craniovertebral anomalies. *Proceedings of the Royal Society of Medicine*, **62**, 725.
Hankinson, J. (1970) Syringomyelia and the Surgeon. In *Modern Trends in Neurology*, ed. Williams, D., Ch. 7, p. 127. London: Butterworths.
Kelly, R. E. (1935) Surgical treatment of syringomyelia. *Transactions of the Medical Society of London*, **58**, 141.
Logue, V. (1971) Syringomyelia: a radiodiagnostic and radiotherapeutic saga. *Clinical Radiology*, **22**, 2.
Love, J. G. & Olafson, R. A. (1966) Syringomyelia. A look at surgical therapy. *Journal of Neurosurgery*, **24**, 714.

Netsky, M. G. (1953) Syringomyelia: a clinico-pathologic study. *Archives of Neurology*, **70**, 741.
Newton, E. J. (1969) Syringomyelia as a manifestation of defective fourth ventricular drainage. (Hunterian Lecture), *Annals of the Royal College of Surgeons*, **44**, 194.
Pitts, F. W. & Groff, R. A. (1964) Syringomyelia: current status of surgical therapy. *Surgery*, **56**, 806.
Wetzel, N. & Davis, L. (1954) Surgical treatment of syringomyelia. *Archives of Surgery*, **68**, 570.
Williams, B. (1969) The distending force in the production of communicating syringomyelia. *Lancet*, **ii**, 189.
Williams, B. (1970) Current concepts of syringomyelia. *British Journal of Hospital Medicine*, **4**, 331.
Worster-Drought, C., Wakeley, C. P. & Shafar, J. (1941) The surgical treatment of syringomyelia. *British Journal of Surgery*, **29**, 56.

CHAPTER SEVEN

The Pathology of Communicating Syringomyelia

J. B. FOSTER and P. HUDGSON

J. G. Greenfield, who maintained a special interest in syringomyelia throughout his career, defined the condition as 'tubular cavitation of the spinal cord extending over many segments' (Greenfield, 1963). He did not regard cavities extending for only two or three segments as necessarily being syringomyelic in nature and he clearly considered *hydromyelia* (cystic expansion of the central canal of the cord) to be a nosologically distinct entity. He indicated that the syringomyelic spinal cord appeared swollen and tense in the cervical region when exposed at operation or autopsy and that the syrinx contained clear, yellow fluid with the consistency of serum.

He further suggested that transverse section of the cord in such cases revealed that the syrinx was largest in the cervical expansion although it was often absent (we would prefer to say 'undetectable') in the first cervical segment. Greenfield went on to describe the typically irregular appearance of the fully developed syrinx with extensions dorsally, ventrally and into the central canal. He even likened this to the cruciform shape of the central canal of the fetus although he regarded the resemblance as being fortuitous. He found that the walls of the long-standing cavities were lined by a dense feltwork of glial fibres 1–2 mm thick, often covered in places by a thin layer of collagen (Fig. 7.1). He acknowledged that parts of the cavity may be lined by ependyma but evidently considered this to be due to chance communication with the central canal, particularly in the cervical expansion.

As far as syringobulbia is concerned, Greenfield's observations were in line with those of Jonesco-Sisesti (1932), who defined three possible anatomical sites for the cavities. These are:

1. Most commonly, a slit passing ventrolaterally from the floor of the fourth

ventricle external to the hypoglossal nucleus. These slits have thin walls composed of neuroglial tissue (Fig. 7.2).
2. Almost equally commonly, an extension from the fourth ventricle along the median raphe, usually lined by ependyma (Fig. 7.2).
3. Much less commonly, a cavity between the pyramid and the inferior olive (Fig. 7.3), almost invariably unilateral.

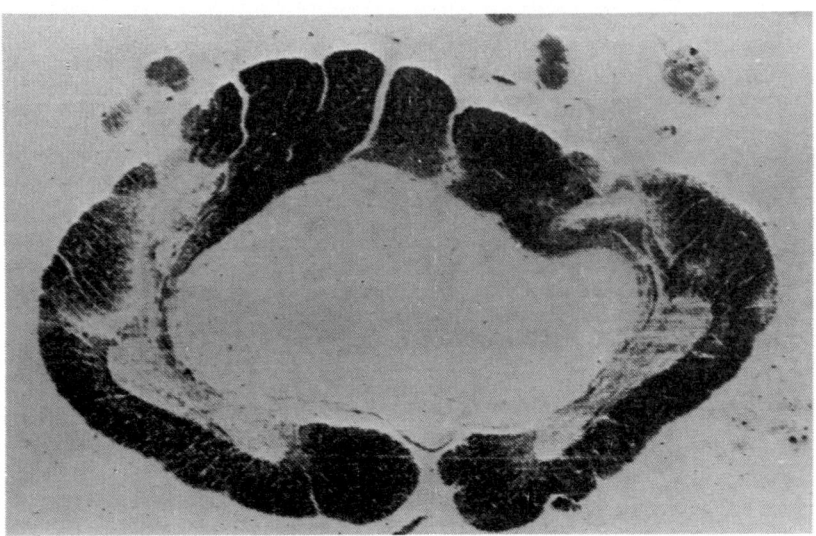

Fig. 7.1. Photomicrograph of a typical syringomyelic cavity lined by a feltwork of glial fibres and expanding the cord. Some demyelination of the lateral corticospinal tracts can also be seen. (Reproduced from *Greenfield's Neuropathology*, 2nd edition, with permission.)

Greenfield noted that secondary degeneration of long tracts occurred in both syringomyelia and syringobulbia and he recognised that syringomyelia may be secondary to trauma (see Chapters 3, 9, 10, 11 and 12), tumours (Chapter 18), chronic arachnoiditis (Chapters 4, 5, 16 and 17) and the resorption of haematomyelia. However, it is clear from his discussion of aetiology and pathogenesis that he does not regard 'secondary' syringomyelia as being of much significance. He found evidence for the syrinx being a developmental anomaly in only a minority of cases, associated with the Arnold–Chiari malformation (Fig. 7.4), and he discounted the testimony of earlier writers, e.g. Gowers and Taylor (1899), who found embryonal cells in the walls of some of the cavities they dissected. Greenfield clearly believed that in most cases syringomyelia was a primary degenerative process (an abiotrophy), sometimes associated with developmental anomalies of the axial skeleton in the cervicocranial region. However, he did accept that mechanical factors played a part in the enlargement and extension of cavities, particularly the variations in mobility between different parts of the spinal cord.

It is apparent from our discussion of the development of the cervicomedullary junction and the pathogenesis of what arguably should be called com-

THE PATHOLOGY OF COMMUNICATING SYRINGOMYELIA

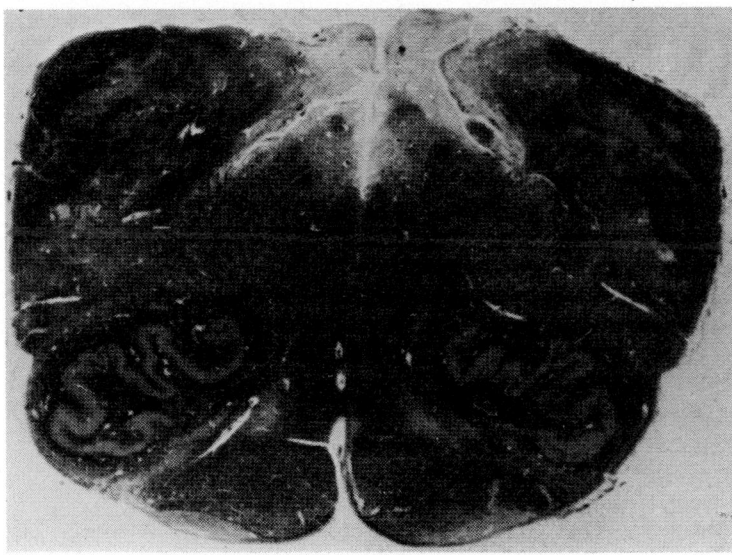

Fig. 7.2. Transverse section of the medulla from a patient with syringobulbia showing bilateral dorsolateral and dorsomedial slits, the latter in communication with the fourth ventricle. (Reproduced from *Greenfield's Neuropathology*, 2nd edition, with permission.)

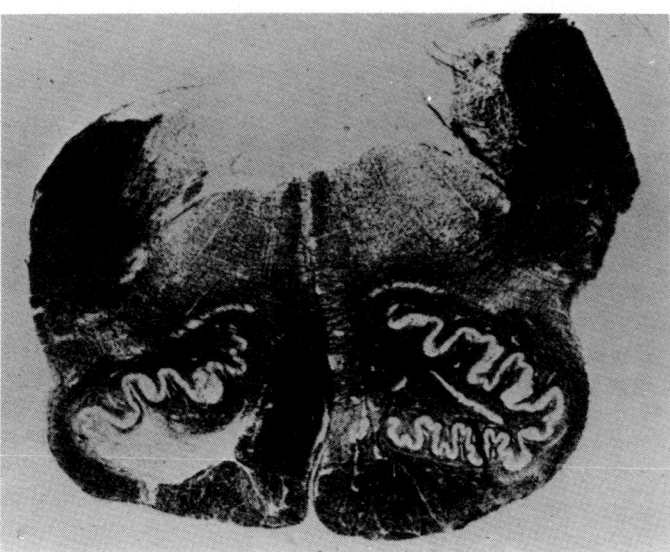

Fig. 7.3. Transverse section of the medulla from a case of syringobulbia with a cleft between the superior olivary nucleus and the pyramid. (Reproduced from *Greenfield's Neuropathology*, 2nd edition, with permission.)

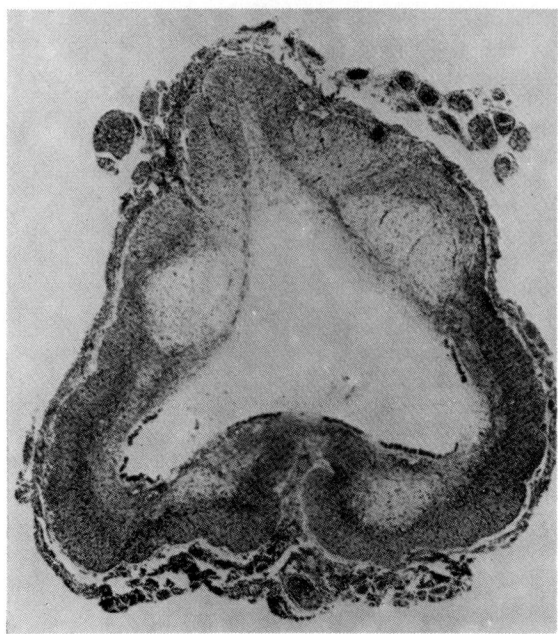

Fig. 7.4. Transverse section of a hydromyelic cord in which the distended central canal is still partially lined by ependyma from a patient with the Arnold–Chiari malformation. (Reproduced from *Greenfield's Neuropathology*, 2nd edition, with permission.)

municating syringomyelia (see Chapter 8) that we take issue with Greenfield in several important respects. We do not dispute that 'pure' syringomyelia may occur with no evidence of an underlying cause, but our clinical and pathological experience suggests that it is rare. We believe that *secondary* syringomyelia (Greenfield's terminology) is much commoner in practice and that its development in some instances is a direct consequence of disordered cerebrospinal fluid dynamics. Indeed the earlier pathological literature, including some of Greenfield's own observations (Taylor, Greenfield and Martin, 1922), at least suggested this as a possibility.

These workers described in detail the clinical and pathological findings in two patients with syringomyelia followed for many years at the National Hospital, Queen Square. In the second case it is noteworthy that they demonstrated continuity between a glial-lined syrinx in the dorsal cord and a dilated central canal in the lumbar enlargement which terminated abruptly at the third lumbar segment (see their Fig. 22). They also reported moderate ventricular dilatation with ependymal lining of some of the glial tissue in the syrinx at the second cervical segment (although the cavity apparently was not in communication with the patient central canal at this level—see their Fig. 18). In their first case, the syringomyelic cavity was lined by a thick layer of glial tissue throughout and was at no point in communication with the central canal, but they postulated that the thickness of the glial lining was due to distension of the cavity under pressure at some time. In addition, they noted that the aqueduct of Sylvius was dilated and they considered that these observations, together with the hydrocephalus and hydromyelia noted in their second case, constituted good evidence for the presence of abnormally high C.S.F. pressure in both cases. They further suggested that in the first case the pressure had either fallen or had become compensated in some way during the course of the illness. In this context, they commented on the clinical association of syringomyelia with symptoms of raised intracranial pressure and the unexpected finding of internal hydrocephalus at autopsy in some instances (see below).

Further pathological evidence for what is now called the hydrodynamic theory for the pathogenesis of syringomyelia appeared in a number of subsequent papers. The most important of these was by Lichtenstein (1943) who carried out an elegant study of the surgical and pathological findings in patients with cervical syringomyelia and a variety of developmental abnormalities in the cervical spine and base of the skull, particularly platybasia and the Arnold–Chiari malformation. In some cases, Lichtenstein believed that cord cavitation was the result of ischaemic necrosis of neural tissue associated with impaction of the cervicomedullary junction and related structures within an abnormal foramen magnum.* In most, however, he showed that the presence of a developmental anomaly interfered with the drainage of C.S.F. from the ventricles. He suggested that this led to dilatation of the central canal of the cord with subsequent formation of a syrinx and, in some cases, he found that the cavities were lined by ependyma (see his Fig. 7).

Lichtenstein clearly believed that what he called 'true' syringomyelia was a separate entity and that it was a primary degenerative process. However, he concluded that the development of cord cavitation in those cases associated with developmental anomalies of the axial skeleton and the neuraxis itself depended on ischaemia of the cervical cord or abnormal C.S.F. dynamics dilating the central canal, or both. Our own experience has in general terms corresponded with that of Lichtenstein (1943) and particularly of Gardner (1965, 1967). It will now be reviewed in detail.

SURGICAL PATHOLOGY

As mentioned in Chapter 6, of the 100 patients studied clinically, 47 with cerebellar ectopia seen on myelography have been explored and decompressive procedures carried out on the upper cervical canal and the foramen magnum in each of them (Hankinson). In most of these cases, there was simple prolapse of the cerebellar tonsils to various levels in the cervical canal (cerebellar ectopia, type I Chiari anomaly) (Figs. 6.1, 6.2 and 6.3). However, other types of Chiari anomaly were found, including one example of the fully developed Arnold–Chiari malformation, and 18 patients had chronic, adhesive arachnoiditis with no evidence of a developmental anomaly (Fig. 6.6).

---

* Lichtenstein (1943) found pathological evidence of constriction of the cervicomedullary junction in several of his cases, a point also noted by Foster, Hudgson and Pearce (1969). Constriction is produced by undue prominence of the arch of the atlas and the posterior rim of the foramen (Lichtenstein, 1943) and by a ridge of thickened dura mater (Hankinson, 1970).

In the cases with developmental abnormalities, the ectopic cerebellar tonsils tend to be flattened around the posterior surface of the cervical cord and in some cases, to overlap. It has been suggested by Appleby et al (1968) and by Hudgson and Foster (1973) that the tonsils effectively obliterate the foramen of Magendie and possibly any egress of C.S.F. through the lateral recesses of the fourth ventricle as well. Certainly, Hankinson (1970) has been unable to demonstrate any drainage of C.S.F. from the ventricle when the anomalous cervicomedullary junction is first exposed at operation. However, free drainage of the fluid occurs when the ectopic tonsils, which are usually bound down by flimsy arachnoidal adhesions, are separated and patency of the foramen of Magendie is restored.

In the patients with chronic adhesive arachnoiditis, an analogous situation exists as far as C.S.F. dynamics are concerned. Thick, white fibrous adhesions bind the arachnoid mater and the underlying neural structures to the dura lining the posterior surface of the upper cervical canal and the floor of the posterior fossa. These effectively occlude the exit foramina of the fourth ventricle and are very difficult to dissect free without damaging the underlying neural tissue or producing venous and capillary ooze which is difficult to control. Notwithstanding, it is usually possible to effect patency of the foramen of Magendie, although clinical experience (see Appleby et al, 1969; Hankinson, 1970; Hudgson and Foster, 1973, and Chapter 4) suggests that the adhesions may re-form relatively quickly in this situation. In only one of these cases was there a history of antecedent illness which may have been responsible for the development of adhesions (a female aged 30 who had had presumed tuberculous meningitis at the age of 19).

As we point out in Chapter 8, the essential requirements for the hydrodynamic theory of the pathogenesis of cord cavitation are, first, effective obstruction to the outflow of C.S.F. from the ventricular system and, second, the presence of a patent communication between the fourth ventricle and the central canal of the cord during fetal development and for some time afterwards. In fact, such a communication was found in 26 of the 47 patients undergoing operation, usually at the level of the obex of the fourth ventricle (Fig. 6.4), not at the bottom near the foramen of Magendie. However, it is not unreasonable to speculate that a patent communication may have existed in some of the other cases and may have been obliterated by the gliotic reaction accompanying expansion of the syrinx (see below and Chapter 8). In most cases subjected to surgery, the cervical cord was found to be collapsed and flattened, not expanded and tense as suggested by Greenfield (see above). The explanation is not certain, although the operations were all carried out with the patient in the sitting position and, assuming patency of the central canal throughout the length of the cord, the contents of the syrinx may simply have run down into the conus. This has been confirmed myelographically in some cases where 'shifting' expansion of the cord can be demonstrated on screening with the head up and down. In the few cases with cystic dilatation of the cervical cord, aspiration has shown that the syrinx contained clear, watery fluid with the biochemical constitution of normal C.S.F., not yellow fluid with a high protein content (Hankinson, 1970). If yellow serous fluid is found on aspiration, it is more likely to be due to syringomyelia associated with an intramedullary neoplasm than to other forms of syringomyelia, particularly communicating syringomyelia (see Chapter 10).

AUTOPSY STUDIES

Pathological Findings

Detailed autopsy studies on the central nervous system have been carried out on four cases of the syringomyelic syndrome and have been reported by Foster et al (1969) and by Appleby et al (1969). The first three of these patients had 'classical' syringomyelia although one (J.E.) was shown to have cerebellar ectopia at cervical laminectomy in 1961. The fourth developed syringomyelia some years after an attack of tuberculous meningitis treated by intrathecal streptomycin. At autopsy, two of these patients had clearly recognisable cerebellar ectopia with apparent occlusion of the foramen of Magendie and the other two had membranous adhesions occluding the foramen and the lateral recesses of the fourth ventricle without descent of the hind brain. In all four cases, the cervical expansions were collapsed in the anteroposterior diameter giving the cords a thin, ribbon-like appearance and, in the first two cases, the cervicomedullary junctions were constricted, presumably by the ectopic tonsils impacted with the cord in the congenitally undersized foramen magnum. The details of the autopsy findings in the cases are as follows:

W.P., MALE (N.3028)

This patient, aged 55 years, had clinical syringomyelia for 25 years. He was admitted to hospital on 22 February, 1965, with a history of diplopia and extreme weakness of sudden onset one month before admission. He stated that these symptoms were minimal on waking but became worse during the day and were most troublesome in the evenings. On examination he had bilateral 'half-cape' areas of dissociated sensory loss on the trunk and upper limbs, intrinsic muscle wasting in both hands, absent deep tendon reflexes in the upper limbs and a mild spastic paraparesis. In addition, he had disproportionate

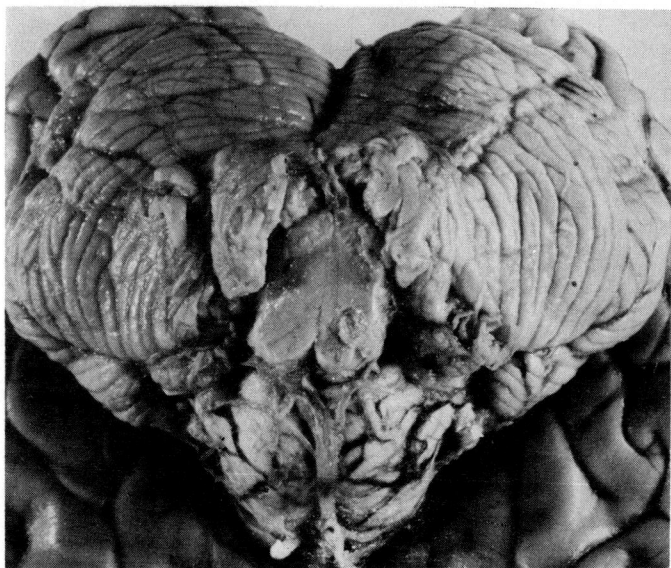

Fig. 7.5. Case W.P. Inferior view of the transected upper cervical cord, cerebellar hemispheres and brain stem showing gross ectopia of the cerebellar tonsils. The foraminal groove on the left side is deeper than that on the right. (Reproduced from *Brain*, 92, 25, 1969, with permission.)

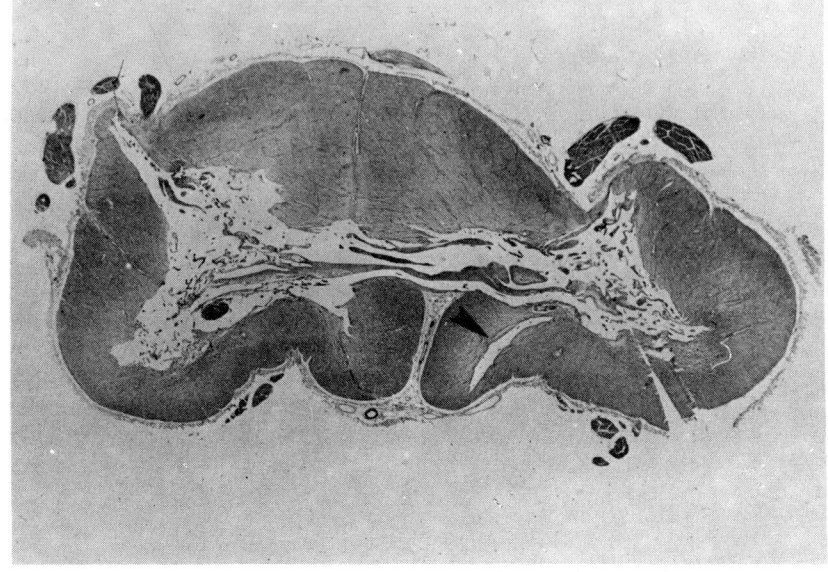

(a)

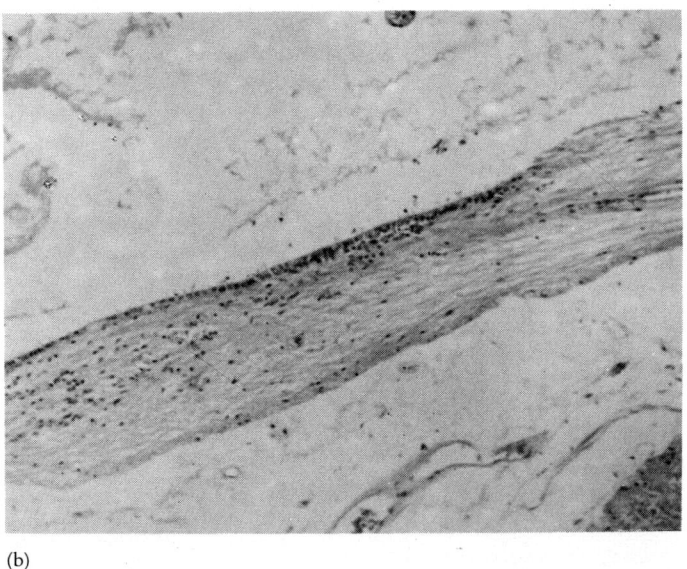

(b)

Fig. 7.6. Case W.P. (a) Transverse section of the cord at the seventh cervical segment showing a large syringomyelic cavity containing numerous glial 'septa' which give rise to the myelographic appearance of 'haustrations' when the cavity is filled with a radio–opaque dye (see Chapter 10). An apparently separate ependymal-lined cleft (arrowed) can also be seen in the anterior part of the section on the left. Haematoxylin and eosin, approx. ×6. (b) Photomicrograph of one of the 'septa' from the cavity illustrated in Fig. 7.6(a) showing that one side is lined with ependyma. Haematoxylin and eosin, ×105.

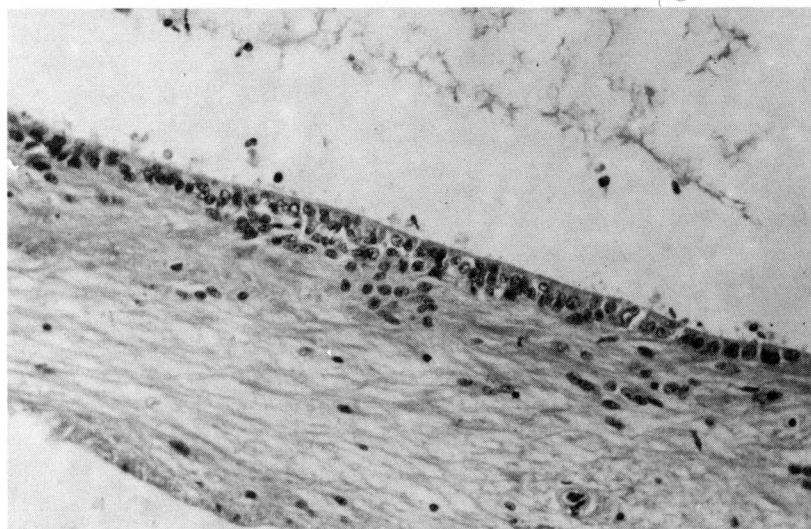

(c)

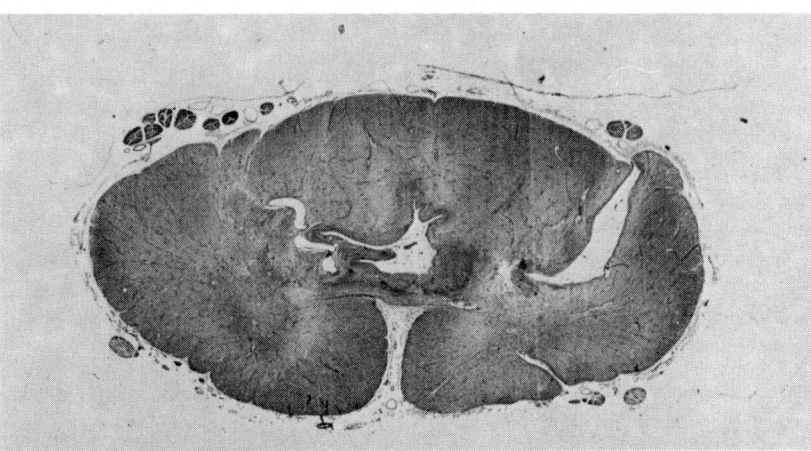

(d)

Fig. 7.6. (c) Higher power view of septum shown in Fig. 7.6(b). Haematoxylin and eosin, × 270. (d) Transverse section of the cord at the fifth dorsal segment showing a dilated central canal lined by ependyma with several apparently unconnected lateral clefts, the largest of which is completely lined by ependyma. Haematoxylin and eosin, approx. × 6.

proximal weakness in the upper limbs, external ocular muscle weakness in all directions of gaze and bilateral ptosis. A provisional diagnosis of myasthenia gravis superimposed on syringomyelia was made but the patient showed only a delayed and relatively unimpressive response to the intravenous injection of edrophonium chloride 10 mg. Further, he was only slightly improved by prostigmin parenterally or orally and not at all by pyridostigmin. Tetanic stimulation during electromyography failed to produce evidence of myasthenia or a myasthenic syndrome. In an attempt to clarify the issue the patient was submitted to an intravenous decamethonium iodide test and, after a dose of only 3·5 mg, became apnoeic, and spontaneous respiration was never resumed. He subsequently developed a severe aspiration pneumonitis and although successful external cardiac massage was carried out once after cardiac arrest, he eventually succumbed. It was presumed that he died from cardiac infarction.

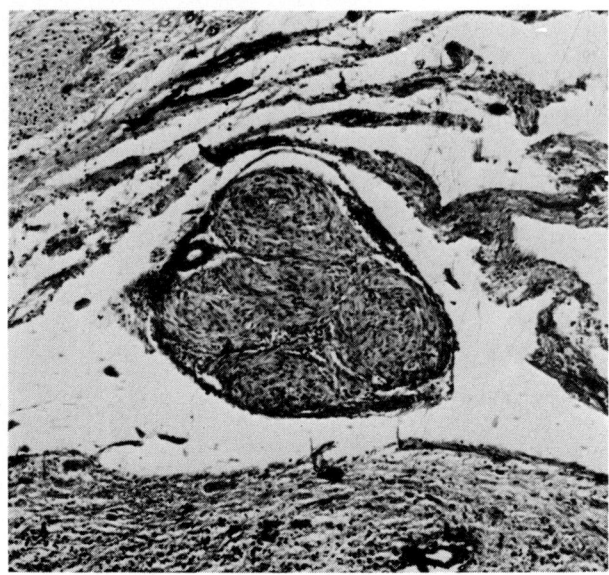

Fig. 7.7. An example of 'peripheral nerve sprouting' found within the syringomyelic cavity illustrated in Fig. 7.6(a). Haematoxylin and eosin, × 54. (Reproduced from *Brain*, **92**, 25, 1969, with permission.)

Macroscopic findings. Only slight generalised muscle wasting was apparent on inspection of the cadaver. The heart weighed 570 g, the left ventricular wall was hypertrophied (2·2 cm) and a recent full-thickness infarct was found in the posterior wall of the left ventricle. Ante-mortem thrombus was adherent to the endocardium overlying the infarcted muscle. The left lung was congested and oedematous and the lower lobe of the right lung was consolidated. Abnormal findings were otherwise confined to the central nervous system.

The cerebral hemispheres were normal on the external surfaces and on cross-section there was no ventricular dilation. Inspection of the ventral aspect of the cerebellum showed that both tonsils were elongated and were lying alongside the junction of the medulla and cervical cord which appeared to be constricted (Fig. 7.5). The spinal cord was enlarged in its lateral diameter (2·0 cm) but was thin and attenuated in its anteroposterior diameter. Random transverse sections of the cord at levels down as far as the conus showed that the central grey matter had been replaced or destroyed by a large central cavity which extended from the junction of the first and second cervical segments to the conus. This cavity had collapsed producing the flat, ribbon-like appearance of the cord.

Microscopic findings. No abnormalities were found in sections from the mid-brain, pons, medulla or cerebellum and, therefore, no clinicopathological correlates exist for this man's terminal ophthalmoplegia. The sections of the spinal cord all showed a typical syringomyelic cavity lined by a feltwork of glial fibres and collagen with a separate central canal. The Nissl stain showed substantial reduction of the anterior horn cells at all levels and the Loyez stain showed demyelination of the crossed pyramidal tracts. The picro-Mallory stain revealed a considerable increase in the amount of interstitial fibrous tissue around the small vessels and running across the wall of the cavity as fine 'septa' (Fig. 7.6a). Sections taken at all levels showed occasional patches of surviving ependyma lining the wall of the cavity or the 'septa' (Fig. 7.6b and c) and, in the mid-dorsal region, the cavity was *entirely* lined by ependyma and consisted of a dilated central canal and a separate ependyma-lined cleft (Fig. 7.6d). In a few sections, peripheral nerve-fibre sprouting of the type described by Hughes and Brownell (1963) and by Demyer (1965) was found (Fig. 7.7). Histological examination of skeletal muscle from the triceps, deltoids and sacrospinales showed patches of 'grouped' denervation atrophy. Some of the atrophic fibres contained chains of plump, vesicular nuclei, the so-called 'fetal' appearance, implying that these fibres had *never* established normal contact with their corresponding α-motoneurones (Adams, Denny-Brown and Pearson, 1962).

J.E., MALE (N.3035)

This 36-year-old patient's illness began in 1955 when he developed numbness and paraesthesiae in the right upper limb which gradually spread to involve the whole of the right side of the body. At the same time, he began to experience difficulty in performing fine movements with the fingers of the right hand and found that his gait was becoming increasingly unsteady. These symptoms were progressive. In 1959, he was admitted to hospital and was found to have an asymmetrical spastic tetraparesis, the right side being more severely affected than the left, with truncal ataxia and limb ataxia on the right. He also had gross impairment of joint position sense in both upper limbs and the right lower limb, with loss of spinothalamic sensation down the right side of the body. X-rays of the cervical spine and skull showed fusion of the bodies of C2 and C3, atlanto-occipital fusion and basilar impression. A myelogram in the prone position failed to reveal any significant abnormality. The C.S.F. protein was 72 mg per cent. At this time it was considered that surgical intervention was unlikely to help but his condition continued to deteriorate. He was readmitted to hospital in October 1961 and cervical laminectomy was carried out on 1 November, 1961 (Hankinson). At operation, both cerebellar tonsils were prolapsed and were compressing the upper cervical segments and the medulla. An effective decompression was carried out and postoperatively the patient made a rapid recovery. In particular, his walking improved and he was able to use his right hand with much greater facility. Steady improvement continued until his last recorded outpatient follow-up in March 1963. He subsequently developed severe chronic bronchitis and died in respiratory failure on 6 October, 1966. Unfortunately, a clear account of his neurological status at this time is not available.

Macroscopic findings. At autopsy the body was that of a well-nourished male in early middle life. Apart from the axial skeleton and the nervous system, the only abnormalities found were in the lungs which contained large numbers of small emphysematous bullae and foci of atelectasis. There was a

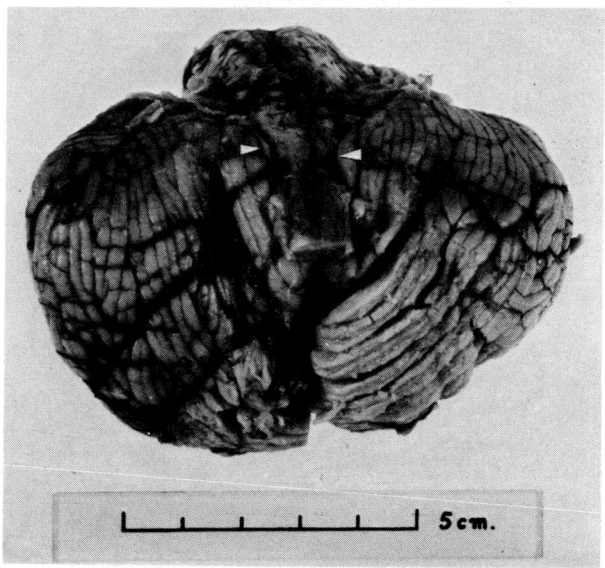

Fig. 7.8. Case J.E. Inferior view of the hind-brain and upper cervical cord showing ectopic cerebellar tonsils compressing the cervicomedullary junction (arrowed) with striking peritonsillar grooving of the undersurface of the cerebellar hemispheres, particularly on the right side. (Reproduced from *Brain*, **92**, 25, 1969, with permission.)

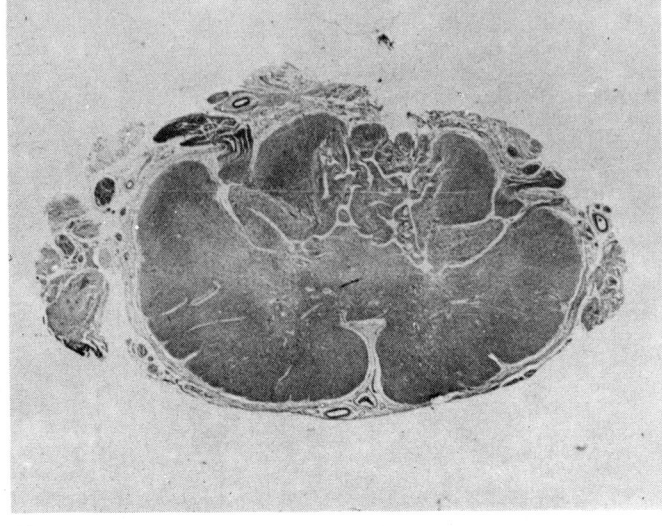

(a)

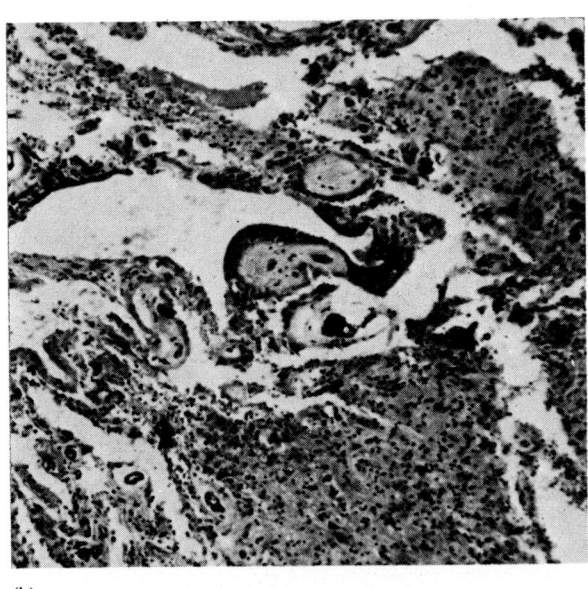

(b)

Fig. 7.9. Case J.E. (a) Low power photomicrograph of transverse section of the cord at the second cervical segment. Numerous clefts can be seen, particularly in the dorsal one-half of the cord, but there is no discrete syringomyelic cavity. Haematoxylin and eosin, approx. × 6. (b) Higher power view of the same section showing what appears to be a dilated cruciform central canal with its ependyma breached in several places and communicating with several clefts. Haematoxylin and eosin, × 63. (Reproduced from Brain, **92**, 25, 1969, with permission.)

moderate amount of mucopus in the trachea and the major bronchi but there did not appear to be any consolidation in the lungs themselves.

Examination of the upper cervical vertebrae and the base of the skull showed that the posterior rim of the foramen magnum had been removed and had been replaced by a band of dense fibrous tissue. The internal diameter of the foramen was only 1·5 cm. The cervicomedullary junction lay at the level of the foramen and the cerebellar tonsils were prolapsed down to the junction of the second and third cervical segments on either side; the spinal cord lying between the tonsils appeared to be constricted (Fig. 7.8). Examination of the cerebellum after removal confirmed the impression that the structure lying within the foramen had been constricted. The most striking feature was the very obvious evidence of tonsillar 'herniation', the impression produced on the cerebellum by the rim of the foramen being much deeper on the right than the left (Fig. 7.8). The upper two segments of the cord were compressed and attenuated and the junction of the spinal cord and medulla was extremely thin, measuring only 0·4 cm in width. Section of the brain stem and cerebellum showed that the cavity of the fourth ventricle had been completely obliterated by pressure from the cerebellum. Serial cross-sections of the spinal cord at 1 cm intervals showed central softening, with cleft, but *not* cavity formation down to the level of the fourth dorsal segment (Fig. 7.9a).

Microscopic findings. No abnormalities were found in sections from the brain stem. Sections of the spinal cord showed that the central canal appeared to be patent in the upper cervical segments, although it was not in communication with the fourth ventricle. In this region, it was surrounded by numerous fine clefts lined by glial fibres and occasional patches of ependyma. In a section taken from the junction of C1 and C2 one of these clefts could be seen communicating with the central canal (Fig. 7.9b). Similar appearances were found in all sections to the upper dorsal level.

J.W., MALE (R.V.I.491137)

This 36-year-old patient gave a history of slowly progressive weakness and wasting of the small muscles of both hands for 12 years. A diagnosis of syringomyelia had been made in 1957 and he had been treated with radiotherapy to the cervical cord without result. In January 1964 he was first seen by a neurologist when prone myelography revealed enlargement of the cervical cord. About this time he was found to be hypertensive (blood pressure 180/120 mm Hg) and he had a haematemesis from a chronic duodenal ulcer. In February 1966, he was admitted to hospital for gastroenterostomy and vagotomy for his ulcer. His blood pressure was then 210/140 mm Hg, despite treatment with methyldopa and guanethidine. He was readmitted to hospital on 31 December, 1966, in his terminal illness. He had papilloedema, was in cardiac and renal failure with a blood pressure of 250/140 mm Hg. After several attacks of 'hypertensive encephalopathy' he died on 15 January, 1967.

Macroscopic findings. Inspection of the cadaver at autopsy showed gross wasting of the intrinsic muscles in both hands. Examination of the viscera showed that the heart weighed 810 g and that the left ventricular wall was grossly hypertrophied. Both lungs showed pneumonic changes, and both kidneys, although normal in size and weight, were congested; in both, the cortex and medulla contained numerous small haemorrhages.

Inspection of the brain and cord in situ disclosed no evidence of tonsillar prolapse or descent of the medulla and fourth ventricle. However, after removal of the brain the foramen of Magendie was found to be completely occluded by a thick fibrous membrane (Fig. 7.10). Similar membranes extended into the lateral recesses. Coronal sections of the cerebral hemispheres showed marked ventricular dilatation and this was also apparent in the aqueduct and fourth ventricle when the hind brain was examined (Figs. 7.11 and 7.12). Careful examination of the floor of the fourth ventricle failed to establish the presence of a patent connection with the central canal of the cord. On external examination, the cord was thin and flattened although in life it had been distended (Fig. 7.13). Cross-section showed a thin cleft-like cavity but the centre was soft and stained a red-brown colour (Fig. 7.14).

Microscopic findings. Transverse sections of the cord showed a central cavity lined by glial fibres extending down into the mid-dorsal region. Aggregations of numerous pigment-laden macrophages were seen in the neural tissue surrounding the cavity. The pigment stained intensely with Prussian blue, suggesting that it was haemosiderin and that it was the residuum of haemorrhage into the cavity over a long period (Fig. 7.15). This phenomenon is well recognised in syringomyelia (Gowers, 1904; Perot, Feindel and Lloyd-Smith, 1966). Transverse sections at the third cervical segment showed that the central canal was patent, slightly dilated and, at one point, in communication with the syringomyelic cavity (Fig. 7.16).

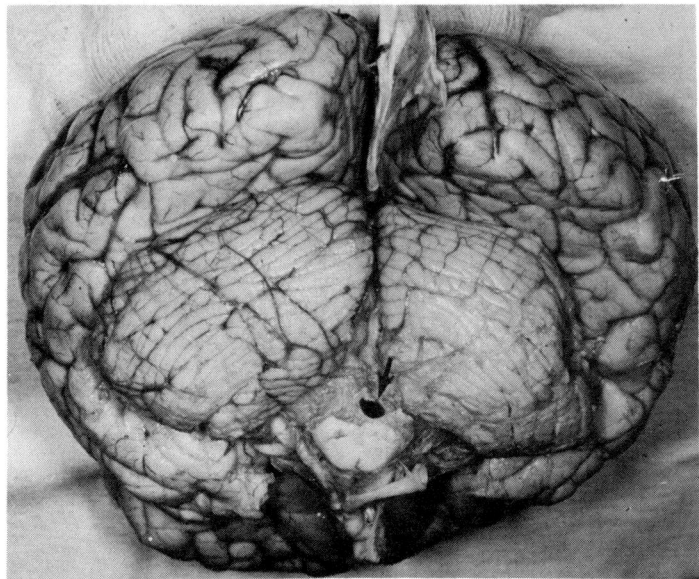

Fig. 7.10. Case J.W. Inferior view of cerebellum and occipital lobes with the neuraxis transected at the cervicomedullary junction. A dense fibrous membrane is seen obliterating the foramen of Magendie and extending out into the lateral recesses of the fourth ventricle. The arrowed defect was produced during removal of the brain from the cadaver. (Reproduced from *Brain*, **92**, 25, 1969, with permission.)

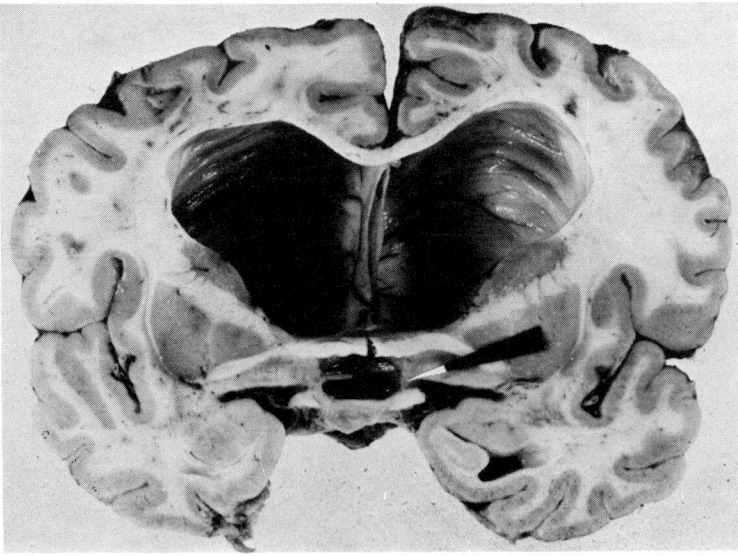

Fig. 7.11. Coronal section of the cerebral hemispheres from case J.W. showing gross dilatation of both lateral ventricles and the third ventricle (arrowed). (Reproduced from *Brain*, **92**, 25, 1969, with permission.)

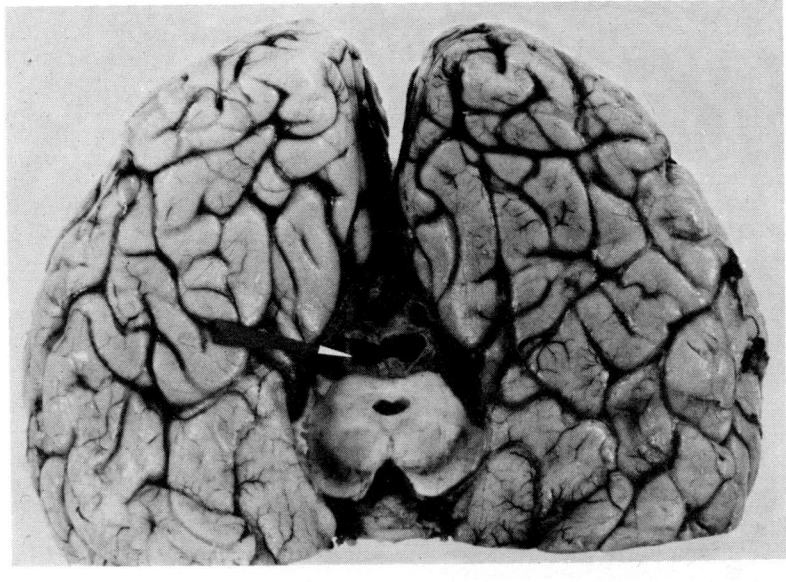

(a)

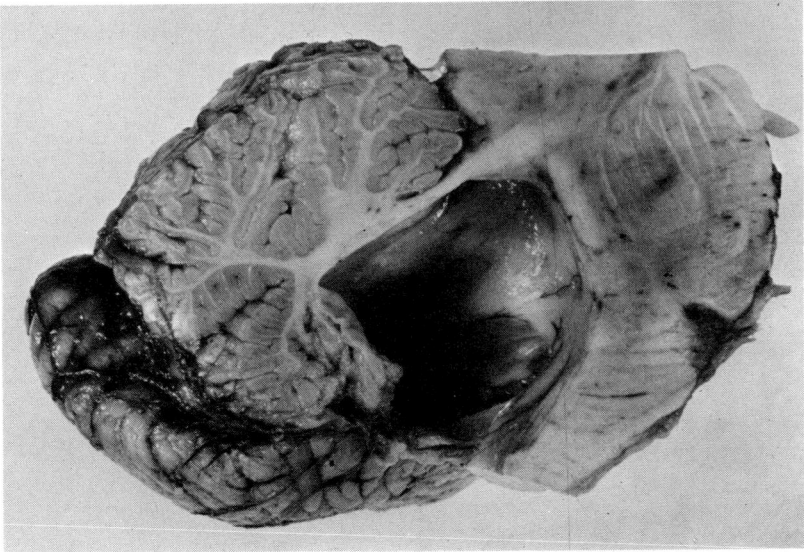

(b)

Fig. 7.12. (a) Transverse section of mid-brain from J.W. showing dilatation of the aqueduct of Sylvius. The dilated uppermost part of the fourth ventricle overlies the mid-brain at this point. (Reproduced from *Brain*, **92**, 25, 1969, with permission.) (b) Sagittal section of the hind-brain from J.W. showing gross dilatation of the fourth ventricle.

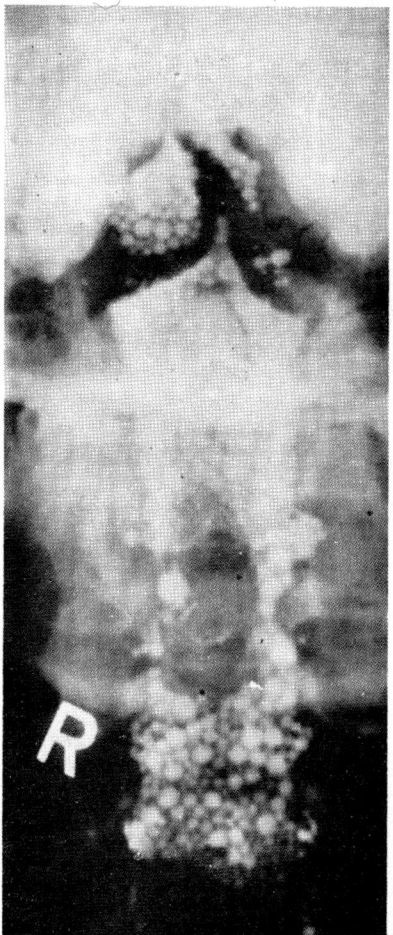

Fig. 7.13. Case J.W. Anteroposterior view of Myodil myelogram in prone position showing dilatation of the cervical cord. (Reproduced from *Brain*, **92**, 25, 1969, with permission.)

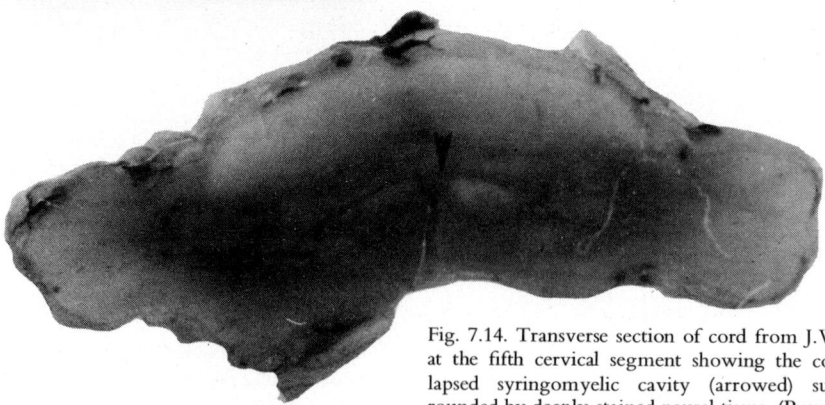

Fig. 7.14. Transverse section of cord from J.W. at the fifth cervical segment showing the collapsed syringomyelic cavity (arrowed) surrounded by deeply-stained neural tissue. (Reproduced from *Brain*, **92**, 25, 1969, with permission.)

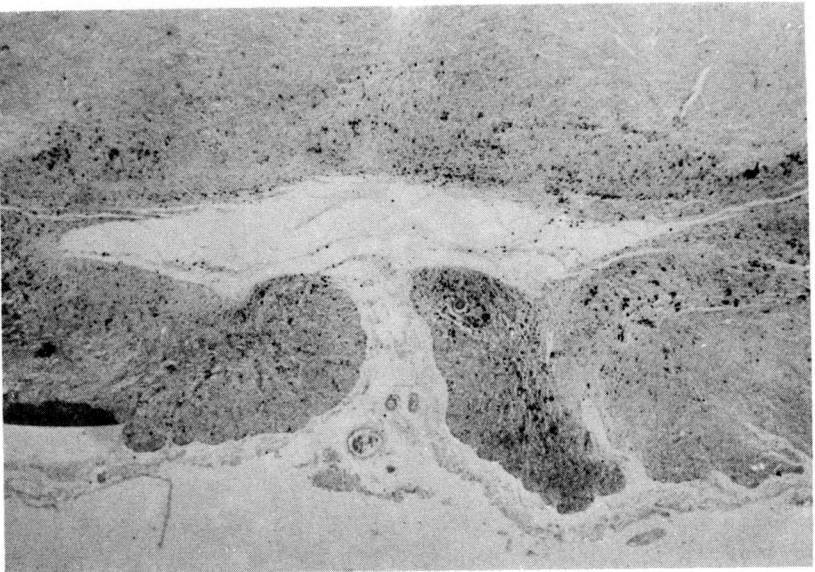

Fig. 7.15. Photomicrograph of transverse section from same area illustrated showing deposition of haemosiderin pigment at varying levels in the neural tissue surrounding the cavity (which communicates with the subarachnoid space at this level). Prussian blue, × 17.

H.C., MALE AGED 28 YEARS (N.22956)
Case report. See Chapter 4, p. 41.
Macroscopic findings. General autopsy examination (Dr J. E. Ennis) showed bilateral bronchopneumonia and chronic pyelonephritis. The basiocciput and posterior fossa structures were removed in toto and examined after fixation. The pia-arachnoid over the surface of the cerebral hemispheres was normal but around the inferior surface of the cerebellum, pons and medulla there were firm fibrous adhesions between the dura and pia, obliterating the subarachnoid space. The hind brain and spinal cord were removed and fixed in continuity, after section of the brain stem. The cerebral hemispheres showed no convolutional flattening or dilatation of sulci and on section the lateral, third and fourth ventricles and aqueduct were of normal size. However, the site of the foramen of Magendie was completely obliterated by dense fibrous adhesions. On section of the cord, there was a large smooth-walled cavity with a maximum diameter of 5 mm extending from the level of C2 to the conus medullaris. No macroscopic continuity of the central canal could be seen between the upper part of the syrinx at C2 and the lower portion of the fourth ventricle.

Microscopic findings. Serial histological sections showed that the central canal at C1 was patent, ependyma lined and slightly dilated (Fig. 7.17) and it was in continuity with the lumen of the ventricle. Interestingly, the canal retained the developmental cruciform shape referred to by Greenfield (see above). Many portions of the central cavity at lower levels were lined with ependyma, though the continuity of the lining was breached at many points and the cavity appeared in places to be a diverticulum of the central canal (Fig. 7.18).

It is acknowledged that the mere association of syringomyelia with anatomical abnormalities, developmental or otherwise, at the cervicomedullary junction does not necessarily imply a cause-and-effect relationship. It is significant, however, that such anomalies were found in each of the four patients with syringomyelia who came to autopsy in a clinical experience of 100 cases during a period of 10 years. It is important at this stage to review the significance of the

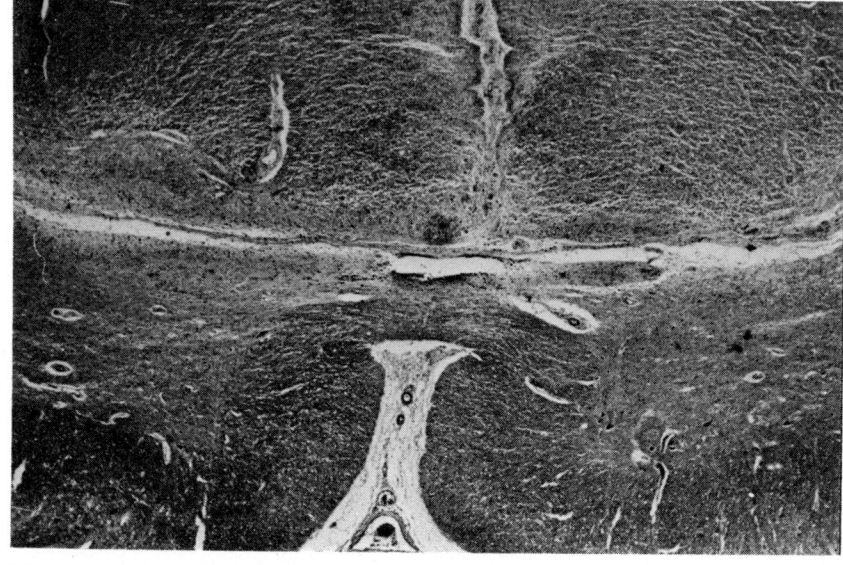

(a)

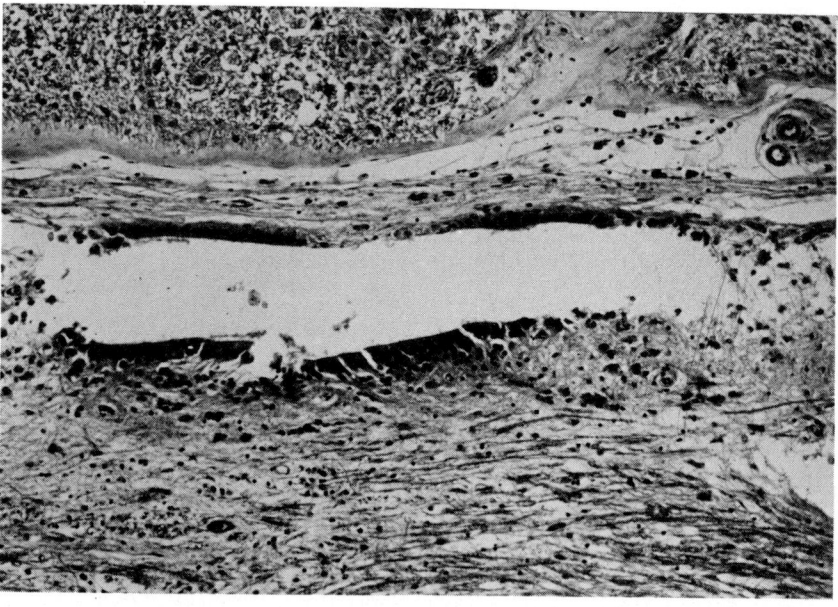

(b)

Fig. 7.16. (a) Transverse section of cord from J.W. at the third cervical segment showing a patent and slightly dilated canal with a small breach in the ependyma communicating with the syringomyelic cavity on one side. Haematoxylin and eosin. × 17. (b) Higher power view of section illustrated in Fig. 7.16(a). × 105.

THE PATHOLOGY OF COMMUNICATING SYRINGOMYELIA

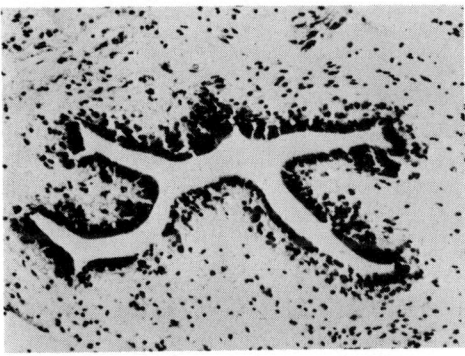

Fig. 7.17. Case H.C. Transverse section of the cord at the first cervical segment showing the patent cruciform central canal. Nissl, × 105. (Reproduced from the *Journal of the Neurological Sciences*, **8**, 451, 1969, with permission.)

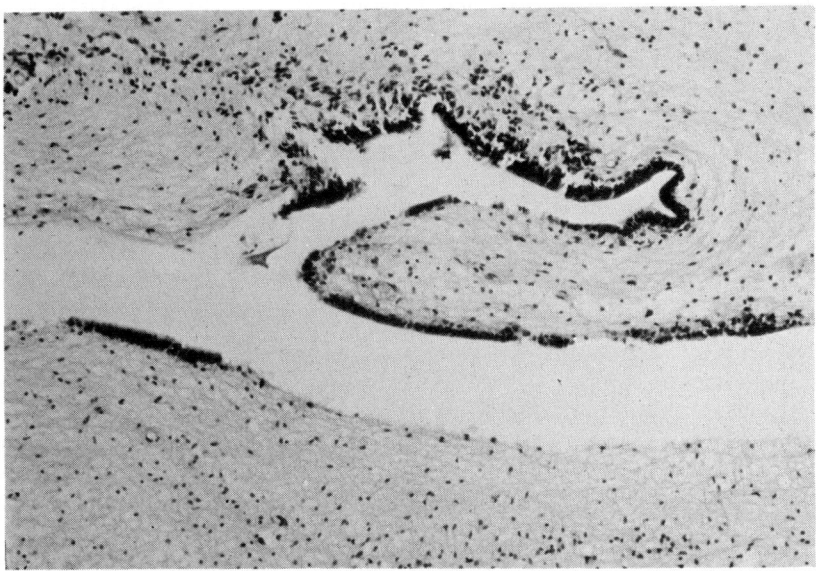

Fig. 7.18. Transverse section of the cord from H.C. at the seventh cervical segment showing the dilated central canal in communication with a syringomyelic 'diverticulum' which is partially lined by ependyma. Nissl, × 105.

findings in each of these cases in the light of current views on the pathogenesis of cord cavitation.

Summing up in reverse order, it can fairly be claimed that case H.C. (N.22956), the patient who developed adhesive arachnoiditis followed by classical syringomyelia after tuberculous meningitis, represents the model mechanism for the hydrodynamic theory of cord cavitation. This man had a patent communication between the fourth ventricle and the central canal (Fig. 7.17); when the exit foramina of the ventricle were blocked by adhesions, the central canal of the cord dilated under the influence of pulsatile C.S.F., eventually becoming a syrinx virtually completely lined by ependyma (Fig. 7.18). In this case at least, a cause-and-effect relationship between obstruction to the free egress of C.S.F. from the ventricular system, a pre-existing communication between the fourth ventricle and the central canal and gradual distension of the latter by the fluid cannot be argued. It is perhaps only surprising that this patient did not develop an obstructive hydrocephalus as well. In this particular instance we believe that the C.S.F. dynamics must have been the same as those in the experimental animals studied by McLaurin et al (1954) (see also Chapters 8 and 11).

A direct cause-and-effect relationship in the other three cases is less clear. However, each of these patients presented features in life which could reasonably be accepted as the stigmata of syringomyelia. In case W.P. the diagnosis had never seriously been questioned in life and the macroscopic appearance of the spinal cord was compatible with the conventional pathological concepts of syringomyelia. However, two neuropathological features merit further comment, the cerebellar ectopia illustrated in Fig. 7.5 and the histological appearance of the walls and glial strands within the syrinx. The walls of the cavity and the strands were composed principally of densely packed glial fibres but scattered 'islands' of ependyma were seen (Fig. 7.6b and c) and at one level the cavity appeared simply as a gigantic expansion of the lower central canal which had been cut off from its upper part (which was seen to be separate from the cavity in the upper cervical segments only) and the fourth ventricle by ongoing gliosis.

In case J.E. there was no discrete syrinx but numerous fine clefts lined by glial fibres and patches of ependymal cells were found in the cervical cord. One of these clefts communicated with the central canal (Fig. 7.9) but again no communication could be demonstrated between canal and ventricle. However, this patient also had cerebellar ectopia (Fig. 7.8) with constriction of the cervico-medullary junction. It may be that ischaemic necrosis of the cord associated with compression of arterial and venous channels in the structures impacted in the foramen magnum was the operative mechanism in this case (see Lichtenstein, 1943) although communicating syringomyelia with subsequent gliotic occlusion of the communication between central canal and ventricle cannot be ruled out.

The third of these three cases is probably the most interesting from the theoretical standpoint. This subject (J.W.) had an undoubted obstruction to the drainage of C.S.F. from the ventricles (Fig. 7.10), with what was thought to be 'typical' syringomyelia with consonant pathologic changes within the spinal cord (Fig. 7.14), slight dilatation of the canal which communicated with the cavity (Fig. 7.16), and gross internal hydrocephalus (Fig. 7.11). Despite this, he had no symptoms referable to the latter during a clinical course of over 10 years (see Chapter 3). Yet again no communication was demonstrated between central canal and fourth ventricle at autopsy. However, it is difficult to escape the conclusion that such a communication existed at some time during the patient's life, if only to assist in the compensation of his hydrocephalus. In such a case it is not unreasonable to suggest that, in the face of progressive gliosis in the cavity wall and repeated haemorrhages within it, eventual obliteration of such a communication was inevitable.

IMPLICATIONS FOR THE PATHOGENESIS OF SYRINGOMYELIA

Gardner (1965, 1967) has consistently maintained that one of the fundamental requirements for the development of syringomyelia is failure of effective perforation of the roof of the fourth ventricle during the early part of fetal life (Weed, 1917; see Chapter 8). We believe, however, that the important factor is obstruction to the outflow of C.S.F. from the ventricles and that the *nature* of this obstruction is relatively unimportant in the pathogenesis of cord cavitation* (although this may be a crucial factor in determining the outcome of treatment—see Chapter 6).

From the evidence adduced from both surgical and autopsy studies, it certainly is clear that many cases of either 'pure' or 'atypical' syringomyelia have demonstrable abnormalities, developmental or acquired, at the cervicomedullary junction.

In the operating theatre, the commonest single finding was descent of the cerebellar tonsils into the cervical canal, although a substantial minority of cases had what appeared to be chronic adhesive arachnoiditis in the cervical canal and basal cisterna (see Chapter 4). At autopsy, there were two examples of cerebellar ectopia, one of arachnoiditis and one of the Dandy-Walker syndrome. Only the last corresponds to Gardner's (1965) concept of failure of perforation of the roof of the fourth ventricle but in *each* case the exit foramina of the ventricles were occluded. It is apparent from the experience of ourselves and others (Gardner, 1965; Newton, 1969; Gordon, 1969) that the greater the number of patients with the syringomyelic syndrome who are adequately investigated, the more frequently will abnormalities of the kinds described above be found. It can also be shown that such abnormalities produce partial or complete obstruction to the outflow tract of the fourth ventricle in life.

Given the morbid anatomical situations described above, the hydrodynamic hypothesis for the development of cord cavities rests on persistent patency of the embryonal/fetal communication between the ventricular system and the central canal of the cord. We have been able to demonstrate such a communication only in a minority of cases, 26 during life and one at autopsy, although in all four autopsy cases, communication between central canal and cavity was found. It clearly is difficult, however, to accept the validity of the hypothesis in those cases in the *absence* of a patent communication. Notwithstanding, it seems reasonable to speculate that an initially persistent communication may become sealed off late in the natural history of the disease by the intense gliosis and fibrosis accompanying cavitation (see above).

The pathological evidence presented does not help to resolve the dispute between Gardner's (1965, 1967) and Williams' (1969) disparate views on the mechanisms responsible for distension of the central canal (see Chapter 8). It does, however, give strong support to the thesis that, in most cases, syringomyelia is an epiphenomenon associated with a variety of pathological processes at or near

---

* A point which was suggested by Symonds and Meadows (1937) in their description of the clinical presentation of *any* lesion at or near the foramen magnum (one of their cases had an expanded cervical cord at operation).

100 COMMUNICATING SYRINGOMYELIA

the foramen magnum. It is also at least consistent with the hydrodynamic theory of the development of cord cavitation.

SYRINGOBULBIA

Our pathological experience of syringobulbia is limited and we have been able to find Jonesco-Sisesti's clefts (1932) in only one of our cases—the patient who developed arachnoiditis after tuberculous meningitis (Fig. 7.19). In this case it seems only reasonable to suggest that this developed as a result of lateral dissection from the fourth ventricle in much the same way as syringomyelic clefts in the cord. We have also been privileged to study a case of syringobulbia examined at autopsy by Dr John Pearce, who has kindly allowed us to report the details.

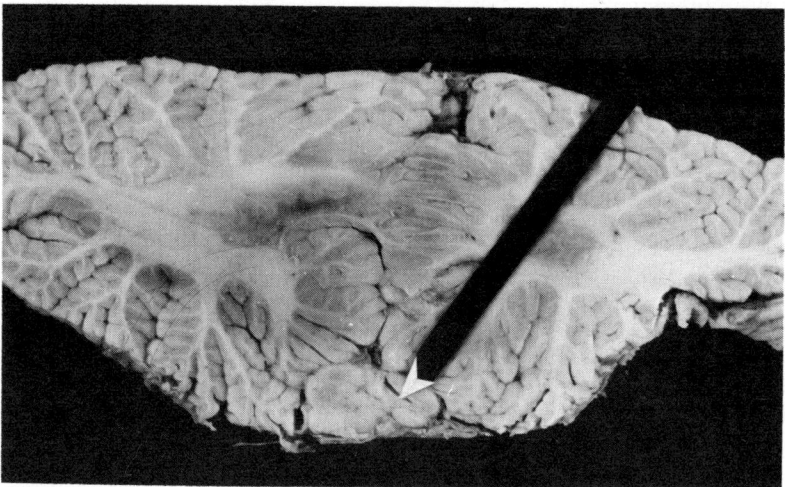

Fig. 7.19. Section of the medulla and cerebellum from H.C. showing a syringobulbic cleft in the dorsal medulla (arrowed).

The patient, a 40-year-old female, developed facial sensory disturbance, dysarthria and dysphagia at the age of 13. Her condition deteriorated steadily and she developed a typical cape of dissociated sensory loss and a spastic paraparesis. She died aged 40 with an aspiration pneumonitis and the autopsy was performed at the Massachusetts General Hospital (autopsy number 29,598). This revealed a syrinx extending from the lumbar enlargement of the cord, through the medulla, pons and mid-brain (Fig. 7.20) into the cerebral hemispheres, involving the thalamus on the same side. No communication was demonstrated between the central canal of the cord and the fourth ventricle, although the cavity in the brain stem communicated with the ventricle (Fig. 7.21), and there was obvious cerebellar ectopia (Fig. 7.22a and b) but no hydrocephalus.

This extraordinary case of syringomyelia, syringobulbia and syringocephalus again suggests the possibility of a hydrodynamic mechanism for cavitation of the cord and bulb. Interestingly, Greenfield found in his cases that most syringobulbic clefts were in communication with the ventricle, and Williams (1970) suggests that the mechanisms operating in the cord and the bulb are the same (see Chapter 8). However, his illustration indicates that he believes the bulbar cleft is an upwards extension of a diverticulum from the central canal (Fig. 7.23).

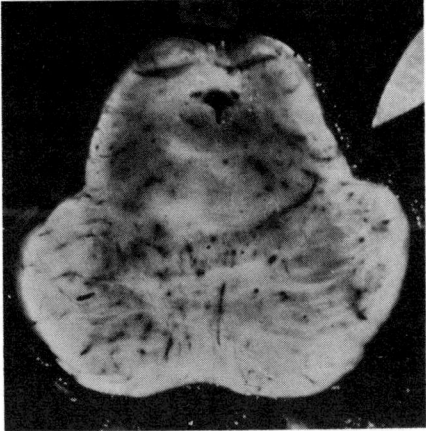
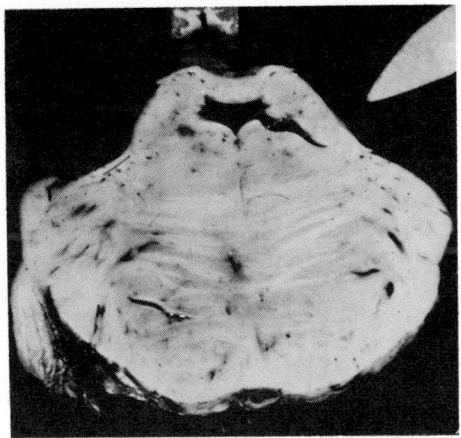

Fig. 7.20. Transverse section of mid-brain from the M.G.H. patient show a syringobulbic cleft (arrowed) in the region of the substantia nigra on the right. The optic nerves can just be seen at the bottom of the illustration. (Reproduced by permission of Dr J. M. S. Pearce.)

Fig. 7.21. Transverse section of the pons from the same patient showing two clefts, one in the left side of the pontine nucleus and the second running into the fourth ventricle of the base of the right superior cerebellar peduncle. (Reproduced by permission of Dr J. M. S. Pearce.)

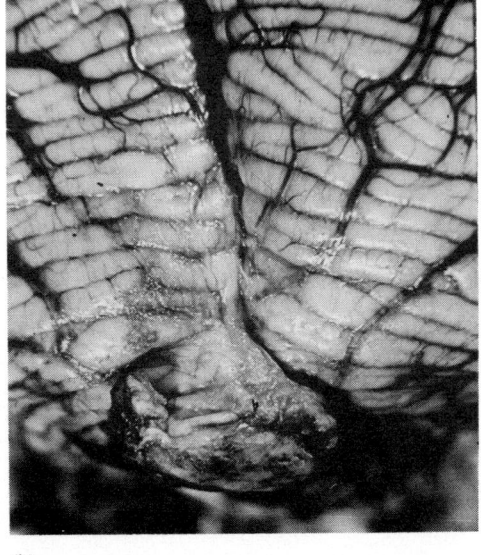

(a) (b)

Fig. 7.22. (a) Posterior aspect of the enlarged cervical cord and cervicomedullary junction showing the two ectopic cerebellar tonsils at the top. (b) Enlarged view of the cervicomedullary junction showing the splayed out left tonsil enveloping the posterior and lateral aspects of the upper cord. (Reproduced by permission of Dr J. M. S. Pearce.)

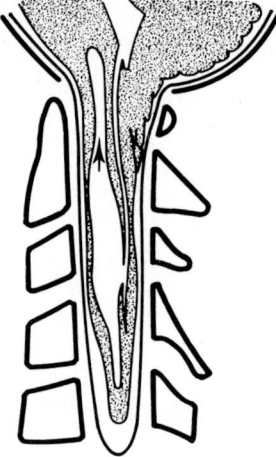

Fig. 7.23. Diagrammatic illustration of the upward extension of a 'diverticulum' from the central canal of the cord, producing a syringobulbic cleft. (Reproduced from Williams, B., *British Journal of Hospital Medicine*, **4**, 331, 1970, with permission.)

LUMBOSACRAL SYRINGOMYELIA

In the past, a small number of patients with dissociated sensory loss in the lower limbs with painless ulceration of the feet and neuropathic disruption of the subtaloid, tarsal and tarsometatarsal and interphalangeal joints have been diagnosed as 'lumbosacral syringomyelia'. Indeed it can be shown that in a minority of cases, hydromyelic cavitation in the lower dorsal cord and the conus may be associated with such a presentation, particularly in patients with spinal dysraphism (spina bifida occulta) (Lassman, James and Foster, 1968). However, most of these cases are due to a genetically determined sensory radiculoneuropathy which has been called 'acrodystrophic neuropathy' (Spillane and Wells, 1969; Banna and Foster, 1972).

REFERENCES

Adams, R. D., Denny-Brown, D. & Pearson, C. M. (1962) *Diseases of Muscle*, 2nd edition, Chapter 3. New York: Hoeber.
Appleby, A., Bradley, W. G., Foster, J. B., Hankinson, J. & Hudgson, P. (1969) Syringomyelia due to chronic arachnoiditis at the foramen magnum. *Journal of Neurological Sciences*, **8**, 451.
Appleby, A., Foster, J. B., Hankinson, J. & Hudgson, P. (1968) The diagnosis and management of the Chiari anomalies in adult life. *Brain*, **91**, 131.
Banna, M. & Foster, J. B. (1972) The radiological features of acrodystrophic neuropathy. *American Journal of Roentgenology*, **115**, 186.
Demyer, W. (1965) Aberrant peripheral nerve fibres in the medulla oblongata of man. *Journal of Neurology, Neurosurgery & Psychiatry*, **28**, 121.
Foster, J. B., Hudgson, P. & Pearce, G. W. (1969) The association of syringomyelia and congenital cervicomedullary anomalies: pathological evidence. *Brain*, **92**, 25.
Gardner, W. J. (1965) Hydrodynamic mechanism of syringomyelia: its relationship to myelocele. *Journal of Neurology, Neurosurgery & Psychiatry*, **28**, 247.
Gardner, W. J. (1967) Myelocele: rupture of the neural tube? *Clinical Neurosurgery*, **15**, 57.
Gordon, D. S. (1969) Craniovertebral and spinal anomalies: surgical management. *Proceedings of the Royal Society of Medicine*, **62**, 725.
Gowers, W. R. (1904) Lectures on diseases of the nervous system, 2nd series. *Lecture VIII Syringal haemorrhage into the spinal cord*. London: Churchill.

Gowers, W. R. & Taylor, J. (1899) *A manual of diseases of the nervous system*, 3rd edition, Vol. 1. London: Churchill.
Greenfield, J. G. (1963) Syringomyelia and syringobulbia. In *Greenfield's Neuropathology*, 2nd edition, ed. Blackwood, W., McMenemey, W. H., Meyer, A., Norman, R. M. & Russell, D. S., p. 331 (posthumous contribution). London: Arnold.
Hankinson, J. (1970) Syringomyelia and the surgeon. In *Modern Trends in Neurology*, Series 5, ed. Williams, D., p. 127. London: Butterworth.
Hudgson, P. & Foster, J. B. (1973) Syringomyelia, disease or syndrome? *Proceedings of the Australian Association of Neurologists*, **9**, 9.
Hughes, J. T. & Brownell, B. (1963) Aberrant nerve fibres within the spinal cord. *Journal of Neurology, Neurosurgery & Psychiatry*, **26**, 528.
Jonesco-Sisesti, N. (1932) *La Syringobulbie*. Paris: Masson.
Lassman, L. P., James, C. C. M. & Foster, J. B. (1968) Hydromyelia. *Journal of Neurological Sciences*, **7**, 149.
Lichtenstein, B. W. (1943) Cervical syringomyelia and syringomyelia-like states associated with Arnold–Chiari deformity and platybasia. *Archives of Neurology & Psychiatry*, **49**, 881.
McLaurin, R. L., Bailey, O. T., Schurr, P. H. & Ingraham, F. D. (1954) Myelomalacia and multiple cavitations of spinal cord secondary to adhesive arachnoiditis. *Archives of Pathology*, **57**, 134.
Newton, E. J. (1969) Syringomyelia as a manifestation of defective fourth ventricular drainage (Hunterian lecture). *Annals of the Royal College of Surgeons of England*, **44**, 194.
Perot, Ph., Feindel, W. & Lloyd-Smith, D. (1966) Hematomyelia as a complication of syringomyelia: Gowers' syringal haemorrhage. *Journal of Neurosurgery*, **25**, 447.
Spillane, J. D. & Wells, C. E. C. (1969) *Acrodystrophic Neuropathy*. London: Oxford University Press.
Symonds, C. P. & Meadows, S. P. (1937) Compression of the spinal cord in the neighbourhood of the foramen magnum. *Brain*, **60**, 52.
Taylor, J., Greenfield, J. G. & Martin, J. P. (1922) Two cases of syringomyelia and syringobulbia studied clinically over many years and examined pathologically. *Brain*, **45**, 323.
Weed, L. W. (1917) The development of the cerebrospinal spaces in pig and in man. *Carnegie Institute Contribution to Embryology*, **5**, 3.
Williams, B. (1969) The distending force in the production of 'communicating syringomyelia'. *Lancet*, **ii**, 189.
Williams, B. (1970) Current concepts of syringomyelia. *British Journal of Hospital Medicine*, **4**, 331.

CHAPTER EIGHT

The Pathogenesis of Communicating Syringomyelia

J. B. FOSTER and P. HUDGSON

Syringomyelia has traditionally been regarded as a primary degenerative disease of the spinal cord, despite a well-recognised association with developmental anomalies of the axial skeleton, at the cervicocranial junction (Spillane, Pallis and Jones, 1957). However, Gardner (1965, 1967) in recent reviews of many years' experience with the surgical treatment of syringomyelia, suggested that the disease was due in many cases to chronic *hydromyelia* associated with obstruction to the drainage of cerebrospinal fluid from the ventricles and accompanied by a patent communication between the fourth ventricle and the central canal of the cord. He further postulated that obstruction of the exit foramina of the fourth ventricle was due to maldevelopment of the hind brain in utero, and surgical relief of the obstruction would arrest the progress of cord cavitation.

It is the purpose of this chapter critically to assess the evidence of Gardner and others for what may be called the 'hydrodynamic' hypothesis for the development of syringomyelia. We will also review our own clinical, radiological and pathological observations (see Chapters 3 to 7) in relation to this hypothesis and will briefly consider the development of the C.S.F. pathways and the cervicomedullary region of the neuraxis, as maldevelopment of these structures is central to Gardner's concept of the pathogenesis of syringomyelia.

DEVELOPMENT OF THE VENTRICULAR SYSTEM AND CEREBROSPINAL FLUID PATHWAYS

The first steps in the laying down of the nervous system occur during the early stages of embryogenesis. The neural groove develops as the invagination of a thickened ectodermal plate in the midline of the dorsum of young embryos. This closes over and is separated from the parent ectoderm to form the neural tube

(Patten, 1968). The neural tube, which contains a moderately large lumen, rapidly enlarges rostrally to form the embryonal brain with five recognisable vesicles at the five- to six-week stage (Patten, 1968; see his Fig. V.17c and d). Caudally the tube enlarges more gradually and at a uniform rate throughout its length to form the spinal cord. The primary lumen of the tube is reduced in diameter caudally but dilates and is locally modified rostrally to form the ventricular system of the hemispheres and hind brain. The ependymal cells are the first of the neuroglia to differentiate from primitive spongioblasts in the neural tube, and they rapidly emigrate to line the developing ventricles and the shrinking lumen of the tube caudally. All the ependymal cells are ciliated at this stage but the majority lose their cilia later in the course of development, only a few clumps of ciliated cells persisting in adult life.

At the time of rapid expansion of the primordial brain (fifth to sixth weeks of intrauterine life in the human) the choroid plexuses begin to secrete C.S.F., the primordial ventricles distend and the fluid seeks egress from the lumen of the neural tube into the subarachnoid space. Patten (1968) takes the view that drainage normally takes place via foramina developing in the lateral recesses of the fourth ventricle (foramina of Luschka) and that the foramen of Magendie if found, is an artefact of preparation and/or examination. In this context, it is interesting that Coben (1967) has shown that the foramen of Magendie is normally absent in the adult dog, cat, rabbit and goat.

On the other hand, Weed (1917), whose classic studies in the human and the pig form the basis for much of our knowledge of the development of C.S.F. pathways, believed that the foramen of Magendie was at least of equal importance in draining fluid from the ventricles.* He suggested that the rhombic roof gradually attenuated from the sixth week onwards under pulsatile pressure from what was effectively embryonal hydrocephalus and hydromyelia (Fig. 8.1). He also claimed that frank perforation of the rhombic roof occurred during the eighth week, in the anatomical sites of the foramina of Luschka and Magendie. If this did not occur, leakage of C.S.F. from the ventricles would not be sufficient to dissect open the subarachnoid space and communicating hydrocephalus was inevitable, even if perforation occurred subsequently. Total failure of perforation would result in a Dandy–Walker syndrome and partial failure would produce one or other of the various Chiari anomalies, with hydrocephalus, *hydromyelia* (our italics) and concealment of the anomalous exit foramina within the congenital hernia. This seems to us to be at least as rational an explanation for the findings in the fetal Arnold–Chiari malformations described by Patten (1953), Barry, Patten and Stewart (1957) and Duckett (1966) as the failure of closure of the neural tube postulated by these authors.

This phenomenon also explains the association between hydromyelia (communicating syringomyelia) and congenital hydrocephalus noted by Cameron (1957). It is highly likely that it is the basis for the association between

---

* In an attempt to confirm or exclude the presence of three exit foramina from the fourth ventricle, Appleby (1971) has carried out Myodil ventriculograms on intact human fetuses of varying ages. This procedure was attended by considerable technical difficulty but in one 20-week fetus, he was able to outline with clarity the ventricular system and to show contrast medium apparently escaping from the fourth ventricle into the cisterna magna via three separate routes.

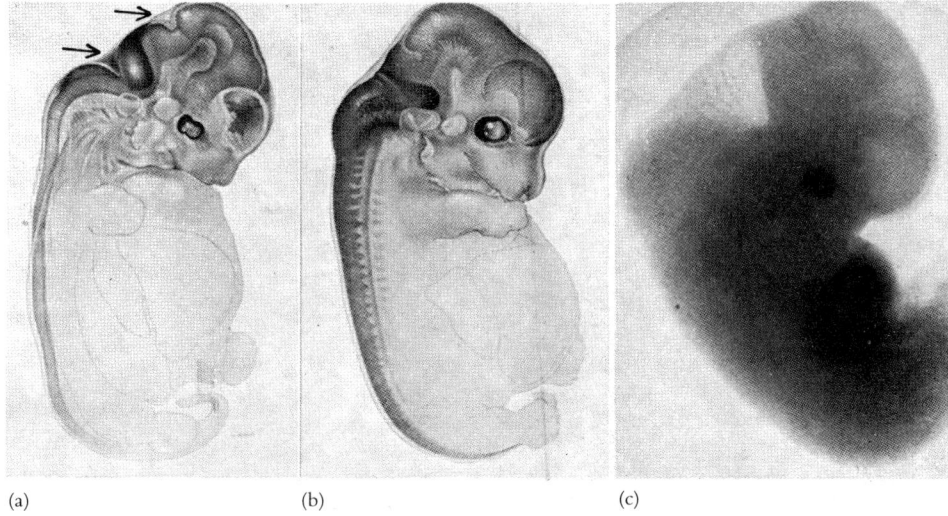

Fig. 8.1. (a) An 18 mm pig embryo the appearance of which approximates to the sixth week in the human, illustrating the physiological hydrocephalus of embryonal life. The C.S.F. spaces have been outlined by precipitated crystals of Prussian blue and the ventricular fluid is beginning to emerge through two permeable areas in the rhombic roof—the area membranacea superior and inferior (arrowed). (b) A 26 mm pig embryo corresponding to the eighth week in the human illustrating the progressive distension of the lumen of the neural tube. The subarachnoid spaces were open and Prussian blue crystals were found in the dural venous sinuses (reduced to same size as Fig. 8.1(a)). (c) A six-week human embryo photographed by transillumination while still inside the amniotic sac. The *lighter* areas in this illustration correspond with the darker areas in Fig. 8.1(a). (Slightly modified from Gardner, W.J., *Journal of Neurology, Neurosurgery and Psychiatry*, **28**, 247, 1965, and reproduced with permission.)

communicating syringomyelia and asymptomatic or arrested hydrocephalus seen in seven of our patients during life (see Chapter 5) and in the case with the Dandy-Walker syndrome diagnosed only at autopsy (see Chapter 7). Above all, it is the lynchpin of Gardner's hydrodynamic hypothesis for the development of cord cavitation.

The other popular theory for the development of the Chiari anomalies is that the hind brain structures involved are drawn into the upper cervical canal by traction due to tethering of the conus in the lower spinal canal by a lumbosacral meningomyelocele—the 'traction' theory of Penfield and Coburn (1938). It should be noted, however, that none of our adult cases of the fully developed Arnold–Chiari malformation had such a lesion, nor did they have any radiological evidence of occult spinal dysraphism. In addition, Goldstein and Kepes (1966), in an experimental study of the effects of tethering the conus in large numbers of new-born rats and opossums★ found that none of these animals developed any signs of the Arnold–Chiari malformation. Certainly, this theory does not fit the observed facts, clinical, radiological or pathological, in cases of communicating syringomyelia.

★ It should be noted that the conus ascends from the S1 vertebral body to the L3 or 4 body during the first six weeks of extrauterine life in the rat, and the opossum is born after only 13 days' gestation, 'fetal' development continuing in the marsupial pouch.

THE HYDRODYNAMIC HYPOTHESIS OF CORD CAVITATION

For over two decades Dr James Gardner of Cleveland, Ohio, has maintained that syringomyelia was nothing more or less than chronic *hydromyelia* due to impaired drainage of C.S.F. from the ventricular system. He also believes that this condition was the least severe form of a spectrum of developmental anomalies including spinal dysraphism, the Dandy-Walker syndrome (so named by Benda in 1952) and meningomyelocele. In the earliest published evidence for this thesis, Gardner and Goodall (1950) described a group of 17 patients varying in age from 14 to 54 years. In each of these patients the Arnold–Chiari malformation was found at operation but a bewildering variety of neurological diagnoses had been made before surgery. However, 12 of the patients had suspended dissociated sensory loss of the kind seen in syringomyelia and, of these, 10 had cystic expansions of the cervical cord at operation. (Another patient had an expanded cervical cord at operation *without* syringomyelic sensory loss.) On the basis of their observations on this group, Gardner and Goodall suggested that the Arnold–Chiari malformation was merely a foraminal herniation of the hind brain produced by a congenital hydrocephalus and that this anomaly was likely to be accompanied by hydromyelia. They emphasised the importance of air encephalography in diagnosing the condition and further suggested that surgical relief of the hydrocephalus was likely to be beneficial.

In a subsequent paper (Gardner, Abdullah and McCormack, 1957), it was suggested for the first time that the Arnold–Chiari malformation, the Dandy-Walker syndrome, 'arachnoid cysts' of the cerebellum and syringomyelia may all have been different pathological expressions of a single developmental anomaly, namely embryonal atresia of the fourth ventricle. In support of their contention, the writers quoted the cases described by Gardner and Goodall (1950) and they noted that the entities described above had several common pathological features. These were basilar impression, scoliosis, hydromyelia, a band of thickened meninges constricting the neural structures in the foramen magnum and obstruction of the outlets of the fourth ventricle by membranes. They also associated three others, with similar clinical and radiological findings, who were diagnosed preoperatively as adult cases of the Arnold–Chiari malformation. At exploration, one patient had a typical Dandy-Walker syndrome, another a so-called arachnoid cyst of the cerebellum and the third, classical syringomyelia. All of them had cystic expansions of the spinal cord and in the first two, injection of indigocarmine into the lateral ventricles was followed by appearance of the dye in the fluid filling the hydromyelic cavity. Because of this, Gardner et al (1957) suggested that hydromyelic cysts *always* communicated with the fourth ventricle by a patent central canal. However, they acknowledged that this communication was difficult to demonstrate because it was seldom if ever dilated. This occurred for two reasons: (1) The immediately adjacent neural structures (cervicomedullary junction, parts of the foraminal herniae) were firmly impacted in the foramen magnum. (2) The network of decussating fibres in the cord at this level offered further resistance to dilatation. Gardner et al (1957) believed that surgical disobliteration of the exit foramina of the fourth ventricle produced improvement in these patients by deflecting the ventricular fluid pulse wave away from the entrance to the hydromyelic cavity and out into the cisterna magna.

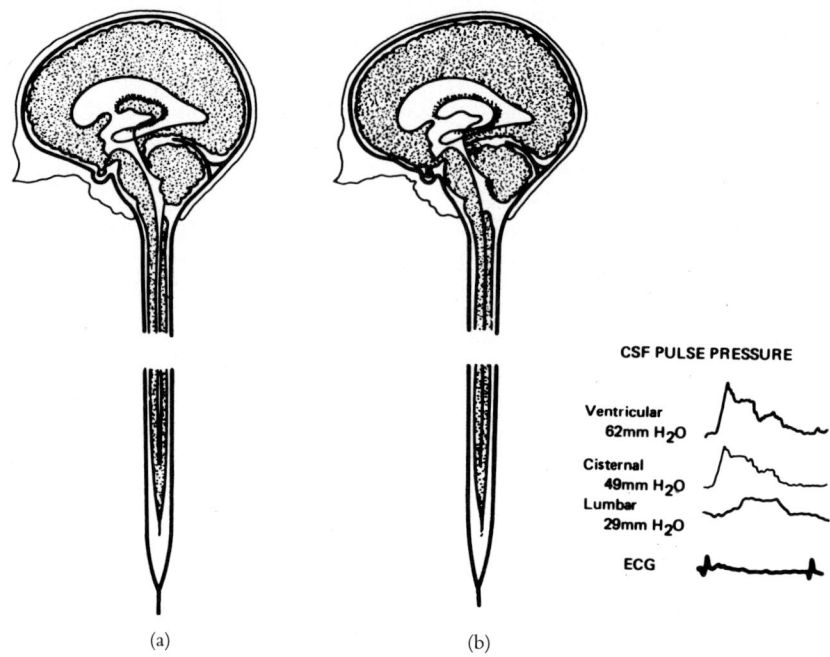

Fig. 8.2. (a) A diagram of the embryonal neural tube superimposed on the outline of the fully developed neuraxis. The outlet of the fourth ventricle is occluded by a permeable membrane which channels the pulsating ventricular fluid into the central canal of the cord. In spite of its exposed position, persistence of this membrane (the area membranacea inferior of Weed, 1917) will be found in all Dandy–Walker anomalies and in many Arnold–Chiari malformations as well.
(b) After the exit foramina of the ventricle open, the pulsating fluid escapes into the subarachnoid space surrounding the spinal cord and compressing the central canal which becomes a vestigial structure. According to Gardner (1965) the ventricular fluid pressure wave is increasingly 'damped' as it descends the spinal canal and he supported this with the pressure recordings illustrated in the Figure (from Bering, 1955).

The claims outlined above were vigorously contested at the time of presentation, for example by Kahn (1957) who claimed that Gardner and his colleagues (1957) were not describing the Arnold–Chiari malformation at all, although he acknowledged the importance of the clinical phenomena associated with the anomalies. (He suggested they collectively should be called the 'Gardner syndrome'.) However, Gardner published a series of papers defending his hypothesis (Gardner and Angel, 1958; Gardner, 1959, 1960, 1961, 1964). In three of these (Gardner, 1960, 1961, 1964), he outlined the phases through which the hydromyelic neural tube of the fetus passes in the development of meningomyelocele. If the progressive dilatation of the lumen of the tube is not relieved by perforation of the rhombic roof during the eighth week of intrauterine life, disproportionate expansion of the tube occurs where its coverings are most yielding, usually at the cephalad and caudad ends. Initially, localising bulging (syringomyelocele or syringomyelia) or splitting (diastematomyelia) occurs and, if expansion continues unrelieved, rupture (myeloschisis) in the immature caudal segments is inevitable (Gardner, 1961; see his Figs. 1 to 5). Gardner (1960) indi-

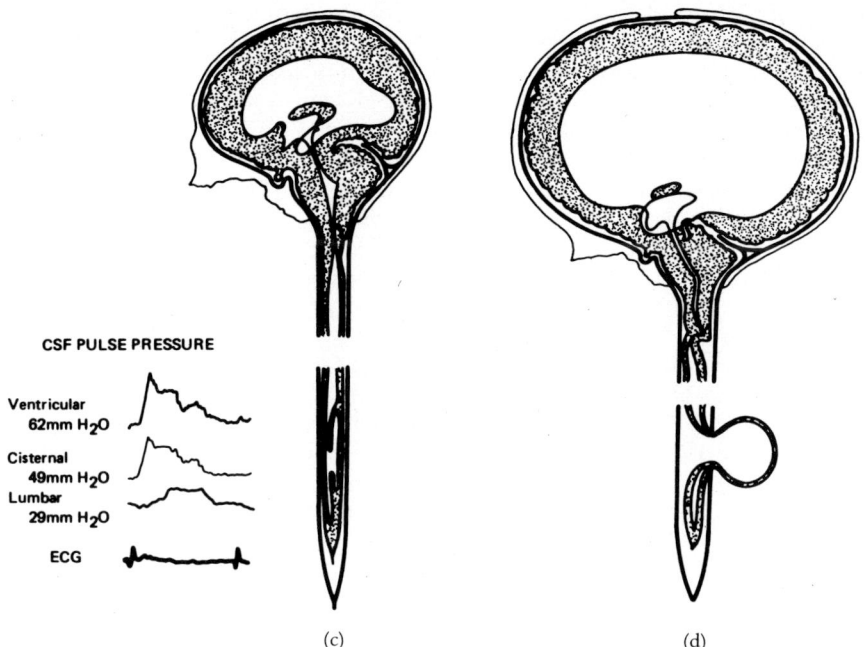

(c) A diagram of Gardner's concept of the neural tube in syringomyelia. The ventricular system is often enlarged but in most cases the intracranial pressure is normal. The foramina of the fourth ventricle are occluded by impaction of a developmental hernia of the hind-brain within the foramen magnum and the central canal of the cord at this level is patent but cannot dilate because of the impacted hernia. Below the impaction the central canal is greatly dilated and may have lost much of its ependymal lining (although this is by no means invariable—see Figs. 7.6(d), 7.16(b) and 7.18). This dilatation is maximal in the cervical canal where the amplitude of the ventricular fluid pressure wave is widest. A 'diverticulum' from the central canal may develop at any level (see lower half of illustration and Fig. 7.18) to form a true syrinx which may also be lined with ependyma (Figs. 7.6(d) and 7.18). Such a diverticulum may develop in the floor of the fourth ventricle producing syringobulbia, the downward extension of which may collapse the central canal.
(d) A diagram of the ruptured neural tube in myelocele. A much larger hind-brain hernia has telescoped the cervicomedullary junction and occluded the communication between the fourth ventricle and the dilated central canal. In this case intracranial tension is elevated, there is overt hydrocephalus and diverticula of both the dilated ventricles and the central canal may be present. (Redrawn from Gardner, W.J., *Journal of Neurology, Neurosurgery & Psychiatry*, **28**, 247, 1965 and reproduced by permission.)

cates that the level of the rupture is determined by the age of the embryo and he noted that it occurred before dissection of the spinal subarachnoid space has taken place, that is between the eighth and tenth weeks of intrauterine life (Weed, 1917).

Gardner (1960) emphasised the frequency of the association between these anomalies and hydromyelia, a point recently underlined by Lassman, James and Foster (1968), and he claims that finding myelodysplasia *without* hydromyelia at autopsy does not mean that it was produced by pre-existing hydromyelia. He suggests that, in the same way that the physiological hydromyelia in every normal fetus is adequately compensated, then relatively mild degrees of pathological hydromyelia may also develop if the rhombic roof becomes sufficiently permeable to allow some drainage of fluid early enough after the eighth week to permit the growing neural tube to enlarge into its distended lumen. Gardner (1964) also comments on the familial occurrence of cranioschisis, anencephalus, hydrocephalus, meningomyelocele, hydromyelia, diastematomyelia and other 'dysraphic' conditions and their production in the litter of pregnant rats treated with a variety of teratogens. He believes that this suggests a common 'cause' (or mechanism of production) and he states that Morgagni's (1761) hydromyelic theory is a more logical explanation for all these conditions than von Recklinghausen's later (1886) insistence on failure of closure of the neural groove. His concept of this mechanism is illustrated schematically in Fig. 8.2.

Gardner (1965, 1967) has summarised his views on the pathogenesis of syringomyelia and myelocele. At the time of writing he had investigated 74 patients with syringomyelia over 25 years and in each case found a developmental anomaly of the hind brain like that seen in the infant with a myelocele. In many of these cases the foramen of Magendie was occluded by a membrane and in all of them the fluid aspirated from the syrinx resembled normal C.S.F. A patent communication between the fourth ventricle and the syrinx was suggested by the passage of a dye or of radiographic contrast medium from the lateral ventricles into the cyst. This combination of obstructed drainage from the ventricle and access from the ventricle to the central canal constitutes the basis for the hydrodynamic or hydromyelic theory of the pathogenesis of cord cavitation.

In discussing the hydrodynamic theory, Gardner (1965) points out that every biological membrane is permeable to fluid. It is possible therefore for an 'imperforate' rhombic roof to permit the egress of C.S.F. at approximately the rate it is formed in the ventricles, presumably via ultramicroscopic 'pores'. It will not, however, permit the *pulsatile* egress of fluid which normally occurs with the rhythmical, slight accessions in C.S.F. pressure produced by arterial pulsation in the choroid plexus. It follows therefore that the oft-repeated spurts of fluid travelling through the fourth ventricle will be deflected by the membranes 'sealing' the exit foramina into the central canal of the cord, exerting what Gardner has called a 'water hammer' effect on its walls. The sustained pulse pressure of the fluid column travelling down into the central canal gradually distends it by what Gardner calls 'a process of hydrodissection comparable to that which is responsible for developing the subarachnoid space in the embryo' (Weed, 1917). It is clear from the above that while the 'sealed' exit foramina of the fourth ventricle are capable of equalising the mean *hydrostatic* pressures in the ventricular system and in the subarachnoid space (patients with syringomyelia do not as a rule develop raised intracranial pressure), they are *not* capable of equalising the different *pulse pressures*.

Conway (1967) has provided indirect but persuasive evidence for the existence of a patent communication between the fourth ventricle and the central canal in his studies on C.S.F. dynamics in 12 patients with syringomyelia during air encephalography or operation on the cervicocranial junction. During either procedure, he noted that the cervical cord rapidly collapsed, becoming a thin ribbon-like structure, whereas it appeared swollen in a positive contrast myelogram (see his Fig. 2). He suggested that the collapse was due to C.S.F. flowing from the subarachnoid space over the cerebral hemispheres into the spinal subarachnoid space (air encephalography) or the open cisterna magna (at operation). This flow

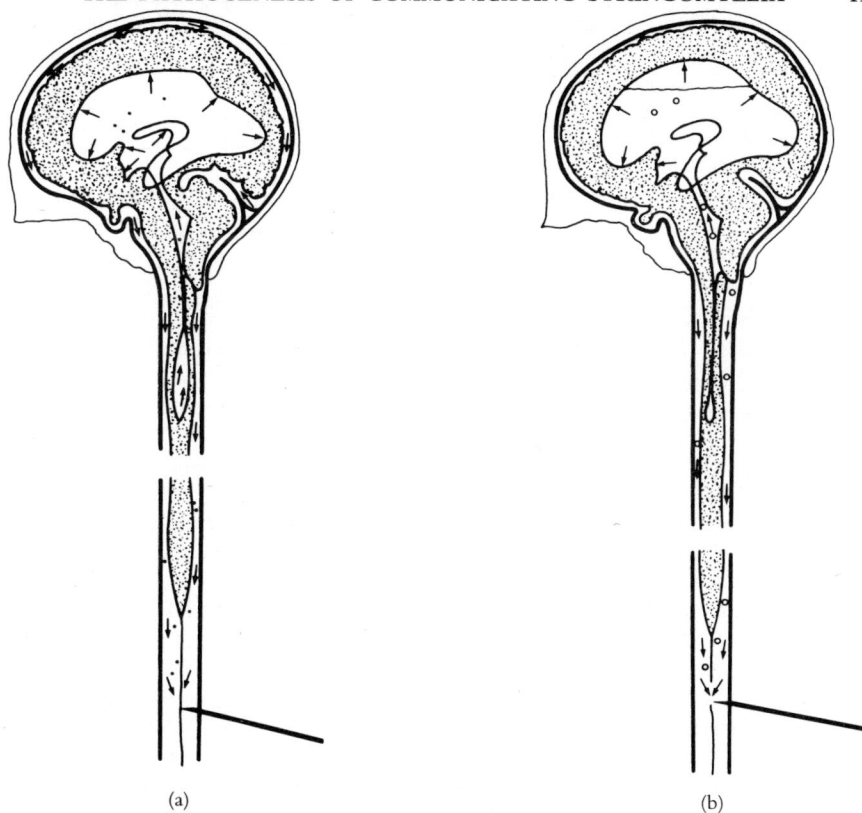

Fig. 8.3 (a) and (b) Diagrammatic illustration of the changes in cerebrospinal fluid dynamics during lumbar puncture in a patient with a cervicodorsal 'syrinx' in communication with the fourth ventricle. Drainage of fluid from the lumbar canal results in an outflow of C.S.F. from the ventricles down into the spinal canal and over the convexities of the cerebral hemispheres. The resultant relative 'vacuum' within the ventricular system is compensated by a flow of fluid from the 'syrinx' into the fourth ventricle via a communication near the obex and the syrinx and surrounding cord collapses (the 'collapsing cord sign' seen on gas myelography or at operation). (Redrawn from Conway, L. W., *Journal of Neurosurgery*, **27**, 501, 1967 and reproduced with permission.)

induced a pressure gradient between the fluid under tension in the cervical syrinx and the intracranial cavity which is equalised by fluid flowing *up* from the syrinx into the ventricles, clearly through a patent communication. Conway's proposed mechanism is illustrated diagrammatically in Fig. 8.3. Conway was able to confirm the presence of such a communication in four out of six patients he operated on, by injecting a dye into a lateral ventricle during the procedure and locating it within the syrinx a few minutes later.

Further evidence of this kind has been adduced by Ellertson and Greitz (1969) who gave injections of fluorescein to eight cases of syringomyelia and an intrathecal injection of radio-iodinated human serum albumin (RIHSA) to a ninth case. In each of these cases the injected material appeared in the fluid in the syringomyelic cyst, whereas it did not appear in the cystic fluid of other spinal cord cysts which were used as control material. Conversely, RIHSA injected into the *spinal cord cyst* in three cases of syringomyelia was shown to be concentrated in the basal cisterns two to three hours after the injection, suggesting that C.S.F. *circulates* into and out of syringomyelic cysts.

It may be relevant in this context that the mechanism envisaged by Gardner for the expansion of the central canal of the cord in syringomyelia is virtually identical to that proposed by Hakim and Adams (1965) and Adams (1966) to explain the so-called 'normal pressure' hydrocephalus syndrome. They suggest that Pascal's law for fluids (the total centrifugal force exerted by a fluid in an elastic container = pressure × area) applies to this phenomenon, and Ojemann et al (1969) believe that C.S.F. dynamics in normal pressure hydrocephalus behave 'as if' the principle operated. In practice, this means that in an enlarged ventricle (or equally, we suggest, within a central canal) the hydrostatic pressure could be normal and still exert greater force on the surrounding neural tissue than the same (or even higher) pressures acting in a smaller volume. The situation is clearly complex because the wall of the ventricles or of the central canal must have variable elasticity, and alterations in venous blood volume and the subarachnoid space are difficult to measure. The analogy is none the less interesting and the variable resistance of the wall of a distended canal may account for lateral cleft formation separate from the original syrinx. If some areas of the wall are more vulnerable than others, these may permit lateral dissection by the pulsatile fluid column producing clefts which subsequently are sequestrated from the syrinx by ongoing gliosis.

Carrying the analogy still further, Ojemann and his colleagues (1969) claimed good results in the treatment of normal pressure hydrocephalus by the insertion of ventriculo-atrial shunts; they even suggested that this measure was likely to be useful in treating some cases of syringomyelia. We have certainly found it helpful in patients with syringomyelia associated with arachnoiditis at the foramen magnum (Appleby et al, 1969; see below and Chapter 4).

At this stage, it can reasonably be argued that what should be called Gardner's triad (cord cavitation, defective ventricular drainage and a communication between cavity and fourth ventricle) (Williams, 1969a) is firmly established as a morbid anatomical substrate for the syringomyelic syndrome. However, doubt still exists as to the means whereby dilatation of the central canal of the cord is achieved. An alternative mechanism to Gardner's 'water hammer' pulse in C.S.F. has been proposed by Williams (1969a and b, 1970, 1972), who believes that intermittent sharp rises in fluid pressure associated with central *venous* pressure swings are of fundamental importance.

Introducing this hypothesis, Williams (1969b) discussed the differing physical characteristics of the intracranial cavity and the spinal canal which permit distension of the dura in the latter but not the former. He also analysed Gardner's arterial pressure wave theory in detail, pointing out that the inconsistency obvious in the association between gross hydromyelia and the modest hydrocephalus usually accompanying it★ (hydrocephalus can occasionally be gross *without*

★ It should be remembered that patients with 'pure' syringomyelia may develop symptoms of raised intracranial pressure with postural change (Appleby et al, 1968) or during the course of investigation (Newton, 1969).

necessarily causing symptoms; see Chapters 3 and 7). He also noted the similarity between Gardner's proposed mechanism and the one invoked by Hakim and Adams (1965) to explain the normal pressure hydrocephalus syndrome, suggesting that if the arterial pressure wave dilated the ventricles in some cases of impaired ventricular drainage, then it should do so in all of them. In a subsequent communication (Williams, 1969a), he even suggested that if the arterial pulse wave in C.S.F. was destructive to neural tissue, a Corrigan pulse would cause presenile dementia and few people would reach maturity with normal intelligence preserved. This seems to us to overstate the case, to put it mildly, particularly as many patients with syringomyelia *do* have significant hydrocephalus (symptomatic or otherwise) and the correlation of ventricular size with intellectual capacity is a notoriously difficult exercise anyway.

A number of other objections to Gardner's theory were raised by Williams (1969a and b). The first was that the syringomyelic cord was often collapsed at myelography or operation (see Conway, 1967) whereas it should be distended and continuously increasing in size if it is being 'pumped up' by the arterial pulse wave. He also claimed that the foramina of Luschka were patent in 'many cases' of syringomyelia (shown by inducing Myodil to enter the fourth ventricle at myelography) and suggested that this was the most convincing evidence against the arterial pulse wave mechanism. Williams later (1969a) implied that the Queckenstedt test usually produced a free rise and fall in subjects with syringomyelia whereas the arterial pulse wave mechanism would be associated with a complete block of the subarachnoid pathway. Gardner (1969), in an energetic riposte to Williams' first paper (1969b), stated that he had *never* observed a free rise and fall in C.S.F. pressure during the Queckenstedt test in this situation. While we are not in a position to comment directly from personal observations (we have never performed the test in a patient due to have a myelogram), we have never found a complete myelographic block at the foramen magnum in the patients with syringomyelia so examined. However, we should add that we have never been able to induce Myodil to enter the fourth ventricle via the lateral recesses or any other route in these patients, and Hankinson (1970) was impressed by the *absence* of any evidence of drainage from the ventricle in the cases he has explored. Nevertheless, the observation about the collapsed cord is important and it constitutes a valid objection to Gardner's postulated mechanism (although it is entirely consistent with the concept of communicating syringomyelia as discussed previously).

In his analysis of the venous pressure wave as a potential distending force in the central canal, Williams (1969b) pointed out that the venous compartment of the intracranial volume was the only one likely to increase suddenly in size. Venous drainage from the cranial contents is by way of the internal jugular veins and sudden rises in pressure arise in these channels if they are obstructed or if flow is reversed. Such a pressure rise can be produced by performing the Queckenstedt test, the Valsalva manoeuvre (forced expiration against a closed glottis) or by placing the head below the level of the right atrium. The accessions in venous pressure produced by these manoeuvres were followed by an increase in C.S.F. pressure, easily demonstrated during the Queckenstedt test. Williams (1969b) also claimed that the fluid pressure rise was accompanied by a sudden flow of C.S.F. in a *caudad* direction through the foramen magnum and this formed the basis for his proposals for the pathogenesis of cord cavitation. Gardner (1969), in his reply, stated that myelographic observations clearly showed that the *first* movement of the C.S.F. after one of these manoeuvres was in a *cephalad* direction, but Williams (1969a) insisted that, while this was true, the secondary and more forceful caudad movement was the important factor in distending the canal. (Williams (1972) has since revised his views on C.S.F. flow after venous pressure swings, without altering his basic concept of the mechanism for the development of hydromyelia.)

The term 'communicating' syringomyelia was introduced by Williams (1969a). We certainly find it an appropriate one, worthy of universal adoption because, to use Williams' own words (1969b), 'the adjective at once suggests both the pathogenesis and the treatment' of the condition. He suggests that it can be extended to cover syringomyelia developing above a traumatic paraplegia, the 'communications' occurring at the site of initial cord damage (1971). This is a departure from his earlier standpoint (Williams, 1970) when he classified post-traumatic syringomyelia as 'non-communicating' in nature. However, there is no pathological evidence in favour of this change of attitude and we still regard this condition as a non-communicating phenomenon in most cases (see Chapter 14).

The anatomical substrate for the development of cord cavitation in Williams' (1969a, 1970) hypothesis is identical with that laid down by Gardner (1965, 1968) and by Appleby et al (1968). The fundamental difference between his proposed mechanism and that of Gardner and his supporters is the insistence that distension of the central canal is an intermittent process produced by sudden rises in C.S.F. pressure, rather than a steady rhythmical one associated with the 'water hammer' of transmitted arterial pulsation in the fluid (although he does not exclude the latter as an accessory factor). Williams' concept can best be summarised by his equations:

Normal anatomy + Venous distension → Uniform pressure rise throughout subarachnoid space

Defective ventricular drainage + venous distension with a patent central canal → Communicating syringomyelia

The ways in which these changes may be brought about are illustrated in Figs. 8.4 and 8.5.

Williams (1971, 1972) has carried out detailed studies of the pressure changes occurring in the C.S.F. following sudden changes in systemic venous pressure and of the changes in the patterns of flow of the fluid associated with these. He has inserted lumbar puncture needles into the lumbar theca, the cisterna magna and the cervical syrinx itself in some patients and has recorded the pressure changes in the fluid in these situations during the Queckenstedt test and the Valsalva manoeuvre. He has been able to show that, at various times, there are sudden upswings in the fluid pressure in all three sites immediately after the rise in venous pressure produced by these manoeuvres. He believed that his earlier studies (Williams, 1971) confirmed his views on the importance of the secondary caudad wave

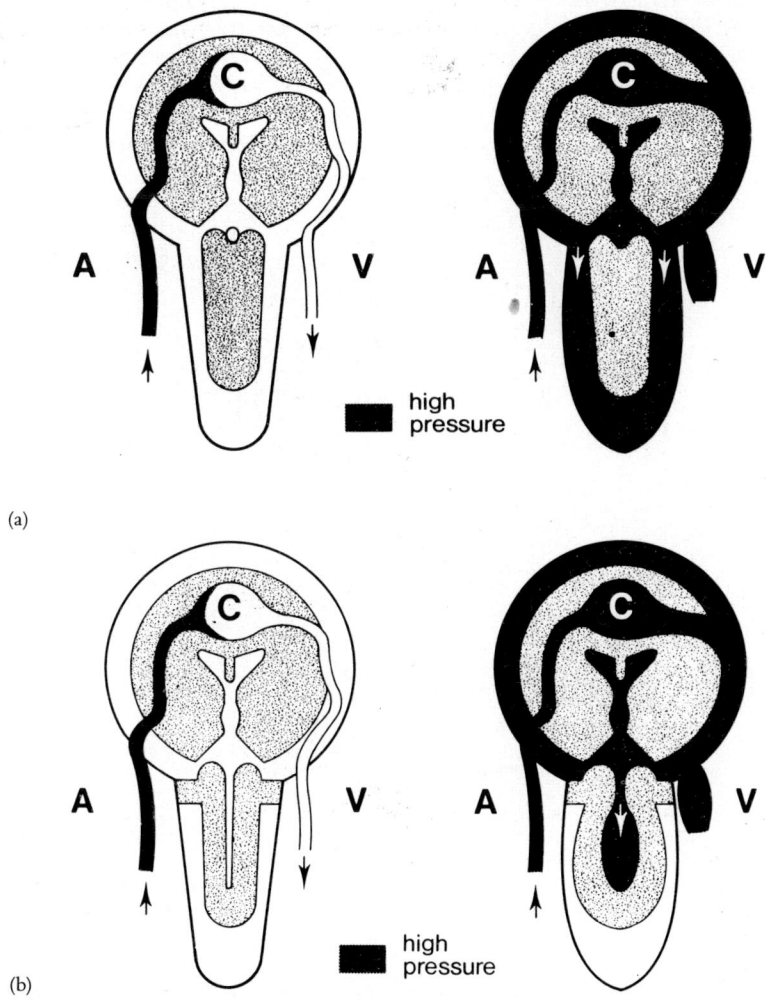

Fig. 8.4. A diagrammatic illustration of Williams' equations for the cerebrospinal fluid dynamic changes following various distension with a normal neuraxis (a) and with defective ventricular drainage and a patent central canal (b). A = arterial; V = venous; C = capillary. (Reproduced from Williams, B, *Lancet*, **ii**, 189, 1969, with permission.)

motion in pushing fluid down into the central canal of the cord, thus distending it. However, it is extremely difficult to see how fluid could pass down into the cavity in patients in whom the upper cervical cord and ectopic cerebellar tonsils are tightly impacted in the foramen and any communication between the fourth ventricle and the cavity is inevitably compressed and presumably occluded.

This difficulty may have been resolved by more recent studies (Williams,

1971, 1972) in which the cough impulse in C.S.F. (equivalent to a Valsalva manoeuvre) was studied by combined lumbar and cisternal manometry. In this study on 17 patients with spondylotic myelopathy and in two infants, one with a meningomyelocele and the other with spina bifida, Williams confirmed the previously recorded rise in C.S.F. pressure with increased venous pressure. He also showed that the lumbar pressure wave occurred earlier and was of greater

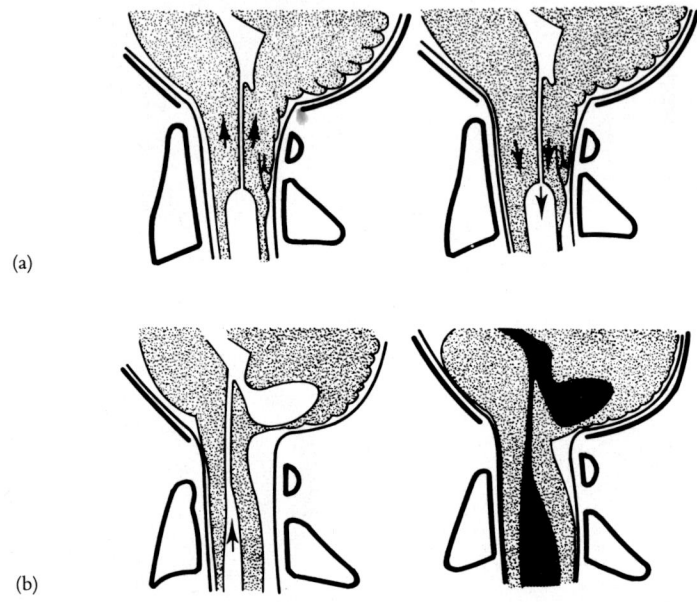

Fig. 8.5. (a) Schematic illustration of the potential ball-and-socket valve mechanism operative in syringomyelia with tonsillar ectopia. The subarachnoid pathways temporarily are wider during upward passage of C.S.F. (b) Schematic illustration of the same potential mechanism in the presence of a distended fourth ventricle with imperforate exit foramina, a cyst or a neoplasm at the level of the foramen magnum. (Reproduced from Williams, B., *British Journal of Hospital Medicine*, **4**, 331, 1970, with permission.)

Fig. 8.6. (a) Combined lumbar and cisternal pressure traces from an upright patient with advanced cervical spondylosis (the vertical scale is in millimetres of mercury above the manubrium sterni). Upward deflections of the differential trace show that lumbar pressure rises higher and earlier than cisternal pressure with coughing (the three peaks represent coughs) and it drops sooner and lower subsequently. The arterial pulse is more obvious in the cisternal trace (in line with Gardner's observations).
(b) This trace shows the lumbar and ventricular pressures in a two-week-old unanaesthetised baby after repair of a meningomyelocele. The child was crying and struggling during the examination which showed that the lumbar trace was of greater amplitude although generally lower than that in the ventricle. However, Williams (1971) emphasises the points where the ventricular pressure rises almost exactly with the lumbar pressure, indicating that the valve (i.e. foraminal hernia) may be open and equalise the pressures.
(c) Pressure traces taken under the same conditions as in Fig. 8.6(b) in an unanaesthetised 12-month-old child with spina bifida. The lumbar and ventricular pressures are much the same but the lumbar pressures change much more rapidly and, in particular, fall more quickly after crying ceases. (Reproduced from Williams, B., *Developmental Medicine and Child Neurology*, **13**, Supplement 25, 1971, with permission.)

THE PATHOGENESIS OF COMMUNICATING SYRINGOMYELIA

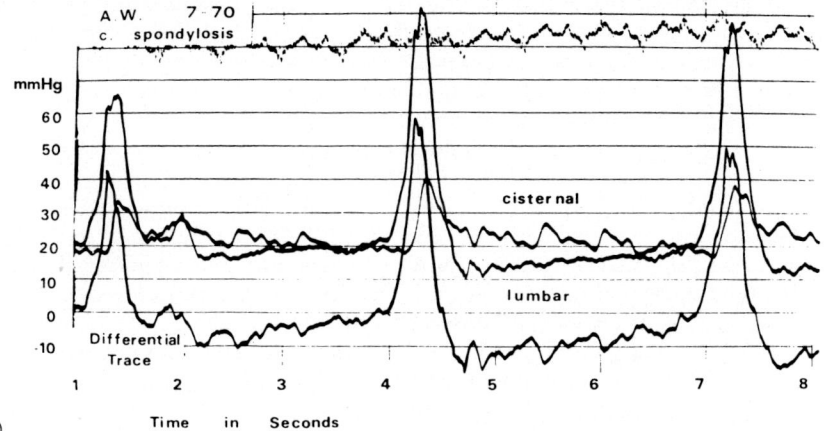

(a)

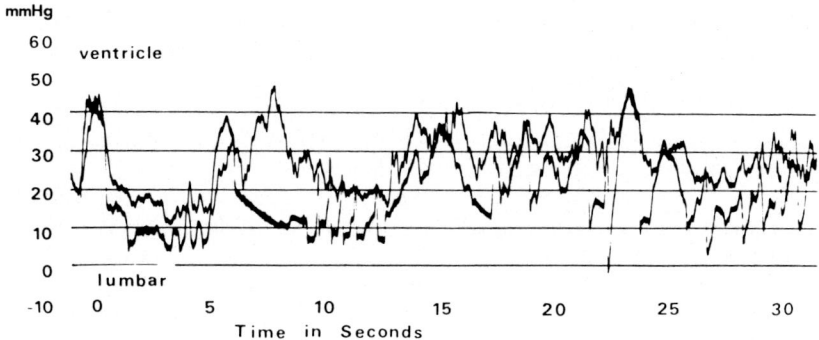

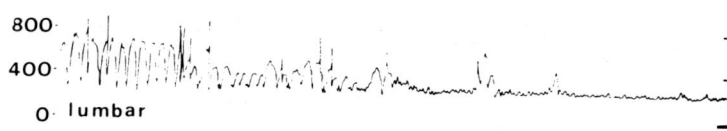

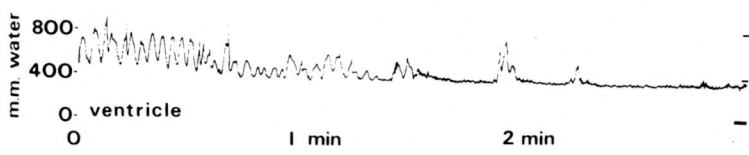

(c)

magnitude than the cisternal wave (Fig. 8.6). However, the latter fell away much more slowly than the lumbar wave until there was a crossover point when the pressures were equal, the cisternal pressure then becoming higher than the lumbar pressure. Prior to this point the flow of C.S.F. was in a cephalad direction (cf. Gardner, 1969) but afterwards it became caudad. Graphical analysis of these pressure waves showed that the cephalad impulse was more forceful than the caudad.

Reasoning from this, Williams suggests that in communicating syringomyelia associated with cerebellar ectopia, the cephalad pressure wave in the fluid temporarily dislocated the ectopic tonsils from the foramen magnum, thus permitting fluid under pressure to enter the fourth ventricle retrogradely by way of its own patent exit foramina. This fluid would then pass down into the central canal and the patent communication would then be sealed by reimpaction of the tonsils with the secondary downward movement of the fluid. This process repeated many times over with accumulation of fluid within the central canal on each occasion would gradually distend it, producing hydromyelia.

This sophistication of Williams' earlier proposals is plausible and would fit the observed facts with respect to communicating syringomyelia/hydromyelia in the presence of developmental anomalies like cerebellar ectopia. It is, however, unlikely to apply in lesions such as the Dandy-Walker syndrome where there is an imperforate roof of the fourth ventricle or chronic adhesive arachnoiditis, and retrograde flow of C.S.F. into the ventricle is impossible. Nevertheless, the venous pressure wave hypothesis is not invalidated in these situations and could still be the principal distending force in producing hydromyelia. Intraventricular pressure certainly rises sharply with upsurges in venous pressure and the pressure wave so formed may be channelled directly into the canal through the patent communication in these conditions (there is no question of the communication being compressed in either of these two situations).

Williams (1970) has also offered a reasoned explanation for the predilection of communicating syringomyelia for the cervical and lumbar enlargements of the spinal cord. The first and most obvious factor is that there is more room for the cord to expand in these regions of the spinal canal. The second is the different distensibility of grey matter compared with white matter. In the lumbar and cervical enlargements there is much more soft grey matter than white compared with the dorsal region and these are thus much more likely to distend with increased pressure in the central canal (or even to rupture in the case of myeloschisis).

As far as our personal observations are concerned, our clinical, radiological and pathological experiences (Appleby et al, 1968, 1969; Foster, Hudgson and Pearce, 1969) substantially confirm the fundamental importance of Gardner's triad (see above) in the pathogenesis of communicating syringomyelia. It is clear that the more often one investigates cases of syringomyelia, clinically or in the autopsy room, the stronger becomes the conviction that it is not a disease entity per se but the common clinical expression of a number of very different pathological processes. It is only fair at this stage to remind the reader that we were able to demonstrate a patent communication between the central canal and the ventricle in only one of our four autopsy cases. However, we have advanced what we believe are cogent reasons for believing that such a communication must have been present earlier in the natural history of the disease in each of the other three, and in all four we found communications between the cavity and a patent central canal (see Chapter 7). Gardner (1965) emphasised the importance of developmental anomalies like the Arnold–Chiari malformation in obstructing drainage of C.S.F. from the fourth ventricle. It is clear, however, that such an obstruction may be developmental or acquired, a point which was made unintentionally but with force and clarity by Symonds and Meadows (1937).

In the absence of data deriving from hydrodynamic studies such as those of Williams (1972), we cannot comment directly on the controversy existing between his views on the mechanism of cavitation and those of Gardner. It is possible, however, to draw guidelines from ancillary observations in the patients we have studied. Some patients first developed symptoms while coughing or after minor injury to the neck (see Chapter 3), a point in favour of William's hypothesis. A proportion of our patients (six of the 96 living patients and one of the four autopsy cases—*not* the man with the patent communication) had evidence of hydrocephalus (symptomatic and otherwise) or arrested hydrocephalus. This suggests that in some cases at least, a common mechanism—possibly that invoked by Hakim and Adams (1965) to explain normal pressure hydrocephalus (or indeed any other form of obstructive ventricular dilatation)—was operating to produce both hydrocephalus and hydromyelia. The 'water hammer' effect of transmitted arterial pulsation seems at least as likely a pathogenetic factor as intermittent, violent upswings in C.S.F. pressure in this situation. (The logical corollary of this observation would be to determine the incidence of syringomyelia in patients with dementia due to normal pressure hydrocephalus.) Two other important observations bearing on Williams' hypothesis were Gardner's (1969) failure to find a 'positive' Queckenstedt phenomenon in any of his patients, and the fact that Hankinson (1970) has never been able to demonstrate drainage of C.S.F. from the fourth ventricle *before* disobliteration of the exit foramina during an exploration.

Further light on the development of cord cavitation may be shed by our group of patients with chronic adhesive arachnoiditis. As indicated above, the hydrodynamic situation envisaged by Williams is simply not tenable in these patients who have complete obstruction to the exit foramina although venous pressure waves within the ventricles may play a part in distending the central canal. However, we believe that such patients are excellent 'experimental' models for the development of cord cavitation, and indeed, the pathological substrate for cavitation is virtually identical with that produced in dogs by the intrathecal injection of Pantopaque (Myodil) and kaolin (McLaurin et al, 1954).

In a series of experiments, McLaurin and his colleagues were able to produce communicating hydrocephalus by the cisternal injection of Pantopaque in 13 animals and kaolin in 26. Twenty-five of these animals developed macroscopic cavitation of the cord in association with myelomalacia and, in some cases, dilatation of the central canal. Some of these animals received injections of Indian ink into the fourth ventricle a short time before death and, in these, the dye was found in the central canal of the cord and in the cavities. These observations underline two important facts not emphasised by McLaurin et al: (1) That the fourth ventricle communicates with the central canal of the cord in adult dogs, a point also noted by Coben (1967) in his essay on absence of the foramen of

Magendie in certain mammals. (2) That the cavities in the cord communicated with the central canal.

McLaurin and collaborators inferred that the changes in the spinal cord were due to ischaemia associated with an *endarteritis obliterans* in vessels on the surface of the cord produced by the irritant effects of the kaolin and Pantopaque and they tentatively extended this thesis to the pathogenesis of syringomyelia.* However, their histological evidence for this was relatively unimpressive (see their Figs. 4 and 5) and it seems much more likely to us that the cord cavitation was a direct result of distension of the central canal (with lateral rupture at various levels into the substance of the cord) by C.S.F. under increased pressure and pulsating because of transmitted arterial *or* venous pressure waves. Indeed, we believe that the elegant laboratory model created by McLaurin et al (1954) is precisely the same in pathological and functional terms as syringomyelia associated with chronic adhesive arachnoiditis in the human. Unfortunately, it does not elucidate the thorny problem of whether arterial or venous pressure waves are of primary importance. In this context the response of patients with arachnoiditis to insertion of a ventriculo-atrial shunt may give some guidance. It is reasonable to expect that this ventricular drainage procedure would effectively reduce the *pulse* pressure of the transmitted arterial pulsation in the fluid and thus its 'water hammer' effect on either the walls of the ventricles or the spinal canal. It is perhaps less likely that the valve would effectively deal with a sudden large increase in C.S.F. pressure produced by a sudden rise in central venous pressure.

What might be called the 'ischaemic theory' for the genesis of cord cavitation (Lichtenstein, 1943; McLaurin et al, 1954) recently received a measure of support in a study by Feigin, Ogata and Budzilovich (1971). They reported the autopsy findings in 16 patients with 'syringomyelia' (perusal of the case protocols indicates that *none* of these patients appeared to conform to the generally accepted clinical criteria for making the diagnosis), and concluded that none of them had any pathological evidence for their cord cavities being due to maldevelopment of the brain stem and upper cervical cord. They presented histological evidence purporting to show that oedema associated with inflammation, circulatory insufficiency and trauma were important pathogenetic factors in cavitating the cord and that syringomyelia was a totally non-specific pathological reaction to a host of unrelated underlying conditions. However, in the absence of any convincing evidence that Feigin et al (1971) were dealing with patients who actually *had* syringomyelia, which is, after all, a well-defined clinical entity, it is unlikely that their observations have any real relevance to the problems studied by Gardner, Williams and ourselves. This is not to say that ischaemia and oedema in the cord play no part in the development of a syringomyelic cavity, but such factors, if operative, are likely to be secondary to the mechanical and hydrodynamic ones discussed previously.

The only other factor to be considered in this section is infection. Schlesinger (1902) in his monograph on syringomyelia stated that the condition could occur in both neurosyphilis and leprosy, but in all probability the only connection between cord cavitation and such infections was indirect. There is no doubt that

---

* A similar mechanism is envisaged for the development of syringomyelia in *spinal* arachnoiditis in the human (Chapter 11) as opposed to *basal* arachnoiditis (Chapter 4).

syringomyelia can develop in patients with chronic adhesive arachnoiditis and many such cases have an infective background; certainly one case following documented tuberculous meningitis was described by Appleby et al (1969). The only alternative form of 'postinfective' syringomyelia is the intriguing suggestion that cord cavitation may be a late sequel of poliomyelitis. This was first mooted in the English language literature by Wilfred Harris (1901). A further example was reported by Leri and Wilson (1904) and the most recent example was described by Carroll (1967). Carroll's patient had a severe attack of poliomyelitis at the age of nine years and sixteen years later developed what was clearly 'pure' syringomyelia. However, he was unable to suggest a cause-and-effect relationship between the two phenomena other than the rather overworked one of cord ischaemia. While it is impossible to discard out of hand the possibility of a direct relationship between the two, it is likely that the association in Carroll's case and the earlier ones reported was a chance one and that the patients had co-incident communicating syringomyelia.

In summary, it can be stated with confidence that Gardner's triad (cord cavitation, defective ventricular drainage, patent central canal) constitutes the morbid anatomical substrate for the development of most cases of communicating syringomyelia. On the evidence available, it is impossible to decide whether Gardner's or Williams' views on the mechanism of dilatation of the central canal is correct. There are a number of objections to both hypotheses and these have been discussed in detail above. Nevertheless, they are both physiologically possible and are certainly not mutually exclusive, so that the truth may well lie in varying combinations of the two. Undoubtedly, they offer a much more reasonable explanation for the pathophysiology of syringomyelia than the other possibilities discussed, for example, cord ischaemia and primary degeneration. More important, however, is the fact that together they form the only rational basis for surgical treatment of the condition.

REFERENCES

Adams, R. D. (1966) Further observations on normal pressure hydrocephalus. *Proceedings of the Royal Society of Medicine*, **59**, 1135.
Appleby, A. (1971) Unpublished observations.
Appleby, A., Bradley, W. G., Foster, J. B., Hankinson, J. & Hudgson, P. (1969) Syringomyelia due to chronic arachnoiditis at the foramen magnum. *Journal of Neurological Sciences*, **8**, 451.
Appleby, A., Foster, J. B., Hankinson, J. & Hudgson, P. (1968) The diagnosis and management of the Chiari anomalies in adult life. *Brain*, **91**, 131.
Barry, A., Patten, B. M. & Stewart, B. H. (1957) Possible factors in the development of Arnold–Chiari malformation. *Journal of Neurosurgery*, **14**, 285.
Benda, C. E. (1952) *Developmental Disorders of Mentation and Cerebral Palsies*, p. 565. New York: Grune and Stratton.
Bering, E. A. Jr (1955) Choroid plexus and arterial pulsation of cerebrospinal fluid: demonstration of the choroid plexuses as a cerebrospinal fluid pump. *Archives of Neurology & Psychiatry*, **73**, 165.
Carroll, J. D. (1967) Syringomyelia as a possible complication of poliomyelitis. *Neurology* (Minneap.), **17**, 213.
Cameron, A. H. (1957) Arnold–Chiari and other neuro-anatomical malformations associated with spina bifida. *Journal of Pathology & Bacteriology*, **73**, 195.
Coben, L. A. (1967) Absence of a foramen of Magendie in the dog, cat, rabbit and goat. *Archives of Neurology*, **16**, 524.
Conway, L. W. (1967) Hydrodynamic studies in syringomyelia. *Journal of Neurosurgery*, **27**, 501.

Duckett, S. (1966) Foetal Arnold–Chiari malformation. *Acta Neuropathologica*, **7**, 175.
Ellertson, A. B. & Greitz, T. (1969) Myelocystographic and fluorescein studies to demonstrate communication between intramedullary cysts and the cerebrospinal fluid space. *Acta Neurologica Scandinavica*, **45**, 418.
Feigin, I., Ogata, J. & Budzilovich, G. (1971) Syringomyelia: the role of edema in its pathogenesis. *Journal of Neuropathology & Experimental Neurology*, **30**, 216.
Foster, J. B., Hudgson, P. & Pearce, G. W. (1969) The association of syringomyelia and congenital cervicomedullary anomalies: pathological evidence. *Brain*, **92**, 25.
Gardner, W. J. (1959) Anatomic features common to the Arnold–Chiari and Dandy-Walker malformations suggest a common origin. *Cleveland Clinic Quarterly*, **26**, 206.
Gardner, W. J. (1960) Myelomeningocele, the result of rupture of the embryonic neural tube. *Cleveland Clinic Quarterly*, **27**, 88.
Gardner, W. J. (1961) Rupture of the neural tube. *Archives of Neurology*, **4**, 1.
Gardner, W. J. (1964) Diastematomyelia and the Klippel–Feil syndrome. *Cleveland Clinic Quarterly*, **31**, 19.
Gardner, W. J. (1965) Hydrodynamic mechanism of syringomyelia: its relationship to myelocele. *Journal of Neurology, Neurosurgery & Psychiatry*, **28**, 247.
Gardner, W. J. (1967) Myelocele: rupture of the neural tube? *Clinical Neurosurgery*, **15**, 57.
Gardner, W. J. (1969) The distending force in the production of 'communicating syringomyelia'. *Lancet*, **ii**, 541 (correspondence).
Gardner, W. J., Abdullah, A. F. & McCormack, L. J. (1957) The varying expressions of embryonal atresia of the fourth ventricle in adults. *Journal of Neurosurgery*, **14**, 591.
Gardner, W. J. & Angel, J. (1958) The cause of syringomyelia and its surgical treatment. *Cleveland Clinic Quarterly*, **25**, 4.
Gardner, W. J. & Goodall, R. J. (1950) The surgical treatment of Arnold–Chiari malformation in adults. *Journal of Neurosurgery*, **7**, 199.
Goldstein, F. & Kepes, J. J. (1966) The role of traction in the development of the Arnold–Chiari malformation. *Journal of Neuropathology & Experimental Neurology*, **25**, 654.
Hakim, S. & Adams, R. D. (1965) The special clinical problem of symtomatic hydrocephalus with normal cerebrospinal fluid pressure. *Journal of Neurological Sciences*, **2**, 307.
Hankinson, J. (1970) Syringomyelia and the surgeon. In *Modern Trends in Neurology*, Series 5, ed. Williams, D., p. 127. London: Butterworth.
Harris, W. (1901) Case of syringomyelia following infantile paralysis. *Proceedings of the Neurological Society*, **24**, 173.
Kahn, E. A. (1957) Discussion of paper by Gardner, Abdullah & McCormack cited above.
Lassman, L. P., James, C. C. M. & Foster, J. B. (1969) Hydromyelia. *Journal of Neurological Sciences*, **7**, 149.
Leri, A. & Wilson, S. A. K. (1904) Un cas de poliomyelité anterieure aigüe de l'adulte avec lesions medullaires en foyers. *Nouvelle Iconographie de la Salpêtrière*, **17**, 432.
Lichtenstein, B. W. (1943) Cervical syringomyelia and syringomyelia-like states associated with Arnold–Chiari deformity and platybasia. *Archives of Neurology & Psychiatry*, **49**, 881.
McLaurin, R. L., Bailey, O. T., Schurr, P. H. & Ingraham, F. D. (1954) Myelomalacia and multiple cavitations of spinal cord secondary to adhesive arachnoiditis. *Archives of Pathology*, **57**, 134.
Morgagni, G. B. (1761) *De Sedibus et Causis Morborum*, translated by Alexander, B. (1769). London: Millar and Cadell.
Newton, E. J. (1969) Syringomyelia as a manifestation of defective fourth ventricle drainage. (Hunterian Lecture) *Annals of the Royal College of Surgeons of England*, **44**, 194.
Ojemann, R. G., Fisher, C. M., Adams, R. D., Sweet, W. H. & New, P. F. J. (1969) Further experience with the syndrome of 'normal' pressure hydrocephalus. *Journal of Neurosurgery*, **31**, 279.
Patten, B. M. (1953) Embryological stages in the establishing of myeloschisis with spina bifida. *American Journal of Anatomy*, **93**, 365.
Patten, B. M. (1968) *Human Embryology*, 3rd edition, Ch. 13. New York and London: McGraw-Hill.
Penfield, W. & Coburn, D. F. (1938) Arnold–Chiari malformation and its operative treatment. *Archives of Neurology & Psychiatry*, **40**, 328.
Schlesinger, H. (1902) *Die Syringomyelie*. Leipzig and Vienna: Deuticke.
Spillane, J. D., Pallis, C. & Jones, A. M. (1957) Developmental abnormalities in the region of the foramen magnum. *Brain*, **80**, 11.

Symonds, C. P. & Meadows, S. P. (1937) Compression of the spinal cord in the neighbourhood of the foramen magnum. *Brain*, **60**, 52.

von Recklinghausen, F. (1886) Untersuchungen über die Spina bifida. *Archiv für Pathologische Anatomie*, **105**, 373.

Weed, L. W. (1917) The development of the cerebrospinal spaces in pig and in man. *Carnegie Institute Contributions to Embryology*, **5**, 3.

Williams, B. (1969a) The distending force in the production of 'communicating syringomyelia'. *Lancet*, **ii**, 696 (correspondence).

Williams, B. (1969b) Hypothesis: the distending force in the production of 'communicating syringomyelia'. *Lancet*, **ii**, 189.

Williams, B. (1970) Current concepts of syringomyelia. *British Journal of Hospital Medicine*, **4**, 331.

Williams, B. (1971) Further thoughts on the valvular action of the Arnold–Chiari malformation. *Developmental Medicine & Child Neurology*, **13**, Supplement 25.

Williams, B. (1972) Combined cisternal and lumbar pressure recordings in the sitting position using differential manometry. *Journal of Neurology, Neurosurgery & Psychiatry*, **35**, 142.

SECTION TWO

Non-communicating Syringomyelia

CHAPTER NINE

Trauma and Syringomyelia

H. J. M. BARNETT

HISTORICAL

Trauma has appeared in nearly every discussion of the aetiology of syringomyelia written in this century. During the nineteenth century, Strümpell (1880) reported a case of a 17-year-old boy who suffered a 'blow to the vertebral column and four years later presented with a syringomyelic syndrome'. At post-mortem the presence of spinal intramedullary cavitation was established. Bawli (1896) reported syringomyelia developing in a patient rendered paraplegic three years before by trauma; at autopsy a cavity was found in the lumbar spinal cord that was 'three centimetres long and several millimetres wide'. Without furnishing conclusive proof, Minor (1897) (1904) and Guillain (1902) both argued for spinal cord concussion as a factor in the production of syringomyelia. Minor regarded the existence of intramedullary haemorrhage above the traumatic site as the factor responsible for the later cavitation. Trauma was included among the causes of syringomyelia discussed in the monumental treatise of Schlesinger (1902), a review of the subject based on the 1,175 references available at that time.

The clinical simulation of a syringomyelic clinical picture, localised to the site of trauma, was also recognised by early writers. This was discussed in one of Charcot's late contributions (1892) and was quite extensively reviewed by Cushing (1898) in one of his earliest papers. Indeed, the preliminary report of the two cases of haematomyelia that formed the basis of Cushing's review was published in 1897, while he was still a surgical resident at Johns Hopkins, and he commented later that 'this represents therefore my first piece of writing'. At about the same time as the writings of Charcot and Cushing, Lloyd (1894) in Philadelphia discussed the simulation of syringomyelia by traumatic cervical cord disruption.

With the exception of Strümpell's and Bawli's patients, these early cases merely *simulated* syringomyelia by the constellation of signs and did not survive to develop a progressing neurological deficit. Some of them, however, survived long enough to have replaced the necrotic and haemorrhagic early lesions of the spinal cord by cavitation. Such localised cavities were known from the descriptions of von Leyden and Goldscheider (1895) and Mann (1896), but even earlier, from that of Bastian (1867). This last writer documented a case in great detail and it is summarised in Chapter 14 (Fig. 14.25).

Recognising that there is overlap between categories and that variable importance should be attached to each of the mechanisms involved, we shall discuss post traumatic syringomyelia under the following circumstances:

1. As a late sequel to traumatic paraplegia and quadriplegia (Chapters 10, 11 and 12).
2. As a late sequel to minor or moderate spinal cord injury (Chapter 13).
3. Where the trauma appears to precipitate symptoms, or to aggravate pre-existing but unrecognised syringomyelia (general discussion in Chapters 3, 10 and 17).
4. As a late sequel to spinal trauma producing adhesive arachnoiditis (Chapters 15 and 16).

REFERENCES

Bastian, H. C. (1867) On a case of concussion-lesion with extensive secondary degeneration of the spinal cord. *Proceedings of the Royal Medical and Chirurgical Society of London*, **50**, 499.

Bawli, J. (1896) *Syringomyélie et Traumatisme Médullaire*. Thése inaugurale, Konigsberg.

Charcot, J. M. (1892–3) Hospice de la Salpêtrière. *Clinicques des Maladies du Système Nerveux*, 2v. Vve. Paris: Babé et Cie. 333.

Cushing, H. (1897) Haematomyelia from gunshot wounds of the cervical spine. *Johns Hopkins Hospital Bulletin*, **8**, 195.

Cushing, H. (1898) Haematomyelia from gunshot wounds of the spine. A report of two cases with recovery following symptoms of hemilesion of the cord. *American Journal of the Medical Sciences*, **115**, 654.

Guillain, G. (1902) *La Forme Spasmodique de la Syringomyélie. La Névrite Ascendante et le Traumatisme dans l'étiologie de la Syringomyelie*. Paris: G. Steinheil.

von Leyden, A. & Goldscheider, A. (1895) *Die Erkankungen des Rückenmarks und der Medulla oblongata*, p. 85. Vienna: Alfred Holder.

Lloyd, J. H. (1894) Traumatic affections of the cervical region of the spinal cord, simulating syringomyelia. *Journal of Nervous & Mental Disease*, **21**, 345.

Mann, L. (1896) Klinische und anatomische Beiträge zur Lehre von der spinalen Hemiplegia. *Deutsche Zeitschrift für Nervenheilkunde*, **10**, 1.

Minor, L. (1897) Récherches cliniques et anatomiques sur les affections traumatiques de la moëlle suivies d'hematomyélie centrale et de formations cavitaires centrales—Congrés de Moscou.

Minor, L. (1904) Traumatische Erkrankungen des Rückenmarks. In *Handbuch der Pathologischen Antomie des Nervensystems*, ed. Flatau, E., Jacobsohn, L. & Minor, L., **2**, 1008. Berlin: S. Karger.

Schlesinger, H. (1902) *Die Syringomyelie. Eine Monographie*, 2nd edn., 255. Leipzig and Vienna: F. Deuticke.

Strümpell, A. (1880) Beitrage zur Pathologie des Rückenmarks. *Archiv für Psychiatrie und Nervenkrankheiten*, **10**, 676.

CHAPTER TEN

Syringomyelia as a Late Sequel to Traumatic Paraplegia and Quadriplegia—Clinical Features

H. J. M. BARNETT and A. T. JOUSSE

Before the introduction of antibiotics, few patients were expected to survive serious cord injury. This situation changed during World War II and their survival necessitated the setting up of spinal cord injury centres. These special units have continued to function for civilians as well as military personnel. This concentration of long-term supervised patients afflicted with spinal injury has presented the opportunity to observe the development of late sequelae from severe spinal injury. Lyndhurst Lodge in Toronto is such a centre. By 1961, a small but significant number of paraplegic individuals under surveillance at this centre had begun to manifest a progressive ascending syndrome depriving them of some motor and/or sensory function of one or both upper limbs. In 1963 eight patients with this syndrome were presented to the annual meeting of the Canadian Neurological Society. By 1964 we had fairly convincing evidence that the condition was the result of a cystic lesion extending cephalad from the site of the original trauma.

Our personal experience with this condition extends to 17 patients and we have been advised of a number of unpublished cases. The literature has been extensively reviewed. Apart from the very early cases of Strümpell (1880) and Bawli (1896) (see Chapter 9), Weisenburg (1907) reported two examples of this condition to the American Neurological Association and Strong (1919) described an isolated example. In the latter case, a conus medullaris injury in 1891 was followed in 1897 by weakness and sensory loss in one upper limb.

It would appear from the old literature cited above (five cases), from the medical literature over the past 40 years (32 cases) (Cossa, 1943; Freeman, 1959;

Finkle, 1960; Jung, 1960; Marvin, 1960; Schott et al, 1962; Martin and Maury, 1964; Riffat and Domenach, 1964; Schmitt et al, 1965; Bischof and Frowein, 1967; Klawans, 1968; Rossier et al, 1968; Werner et al, 1969; Zdrojewski, Werner and Rossier, 1969; Nurick, Russell and Deck, 1970; Feigin, Ogata and Budzilovich, 1971), from personal experience (17 cases), and from personal communications (19 cases) (Cheshire, 1967; Foster, 1972; Watson, 1972; Newman, 1972; Allen, 1972; Coxe, 1972; Tucker, 1972), that 73 cases of this condition have been recorded to date. Table 10.1 summarises the sources of all cases known to us.

Table 10.1. *Acceptable as cases of post-traumatic progressive cystic myelopathy*

Source	Number	Number proven to be 'cystic'
Medical literature pre-1920	5	3[a]
Medical literature post-1940	32	12[b]
Authors' series	17	8[c]
Personally communicated to authors	19	4
Total	73	27

[a] 2 at postmortem
[b] 3 at postmortem
[c] 1 at postmortem
21 were surgically proven

INCIDENCE

At Lyndhurst Lodge Hospital, records are available on 1387 patients with spinal cord injuries studied from 1945 to 1971. Most of these patients have been under surveillance for the duration of their paraplegia or quadriplegia and receive occasional follow-up neurological examination. Any reporting symptoms suggesting new developments have been examined in detail. Those with frank evidence of the condition have been kept under close scrutiny and re-examined at regular intervals. Of the 1387, some 864 have been rendered paraplegic from injuries at and below T4, 452 have suffered complete and 412 partial lesions; 16 (1·8 per cent of the 864) have developed syringomyelia (Table 10.2). The records reveal 523 patients with quadriplegia, 185 of them with complete, and 338 with partial lesions. One quadriplegic (case 14) has been recognised with this new progressive lesion (Table 10.2).

Table 10.2. *1387 traumatic spinal cord injuries (1945–1971)*

	Paraplegia	Quadriplegia
Total	864	523
Complete	452	185
Partial	412	338
Progressive cystic myelopathy	16 (1·8%)	1 (0·2%)

The only other series available with which to compare these figures on the expected incidence of syringomyelia in a population of patients with serious spinal cord injury is that personally communicated to us by Watson (1972). In the unit at Sheffield there are 'about 1000 cases' on the registry of spinal cord injuries, and of these, 11 have developed a syringomyelic syndrome (slightly more than 1 per cent). It has occurred in eight of their paraplegic, and three of their quadriplegic patients. Of the cases reported in the literature, only three other instances are known as quadriplegic patients (Feigin et al, 1971; Werner et al, 1969; Marvin, 1960).

Age and sex are not factors of consequence here. They merely reflect the increased tendency for young people, especially males, to suffer violent injury. One of the cases reported by Werner et al (1969) was six years old at the time of an injury which rendered her partially paraplegic from a T2 fracture dislocation. Thirty-two years elapsed before the onset of ascending myelopathy, proven at operation to be cystic. This lapse of time was the longest in the literature and double the longest time lapse in our own patients. It is not known, though, what bearing, if any, this early age of spinal injury had on the fact that there was such a long delay.

TIME OF ONSET OF PROGRESSIVE SYMPTOMS

The onset of symptoms from this complicating lesion has varied in our series from between 1 to 15 years after initial injury. There was no relationship between the latent period and the distance that the new lesion had to extend from the level of the trauma to produce symptoms referable to the cervical cord (Tables 10.3 and 10.4). In the seven patients in whom *complete* paraplegia or quadriplegia persisted after the original trauma, the average time interval from injury to new symptoms was 4·0 years, with a spread varying from 1 to 9 years (Table 10.3). The interval in the nine cases with a *partial* cord lesion averaged 9·7 years, with a spread varying from two to 15 years (Table 10.4). One case (case 17) was not aware of the progressive signs and since she was not seen for 15 years, It is not known when the advance occurred. Her case was not considered in arriving at this average.

Table 10.3. *Complete lesions*

Level of injury	Interval (years) (average 4·0)
T9	1
T11	1
T4	2
T9	5
T12	7
T8	9
C6	3

These data indicate that the more severe spinal cord injuries were associated with more rapid onset of this delayed complication. Of the cases reported in the literature with sufficient information given to be compared with our series, 14 were described as *completely* paraplegic and the time lapse from injury to new symptoms was 5·2 years. There were 14 described as *partially* paraplegic and here the time was 9·0 years. Again, it was apparent that the symptoms appeared more rapidly in the more severely injured.

Table 10.4. *Partial lesions*

Level of injury	Interval (years) (average 8·8)
T10	2
L1	2
L1	2
T11	8
T12	10
L2	13
T12	14
L1	14
T12	15

CLINICAL DATA

Initial Manifestations

Initial symptoms have presented some striking similarities. (Table 10.5.) Seven patients complained of discomfort brought on above the original spinal level by straining, coughing, sneezing or other stressful activity. Six others complained of spontaneous pain in the upper extremities and the neck. In one instance, the first symptom was an awareness of loss of temperature appreciation, and in one case (case 4) an awareness of excessive sweating from the level of the original injury at T12 upwards, to involve the upper limb, neck and face on the side of the new symptoms. One patient has been identified (case 17) on the basis of an extended area of loss of pain and temperature, but has no symptoms as yet after two years of further follow-up. In one case reported to us (Cheshire, 1967), intractable hiccup was the presenting feature, followed in a few weeks by other symptoms of a cervical intramedullary lesion.

Table 10.5. *Initial symptoms in 17 cases*

Pain on coughing, straining, sneezing	7
Spontaneous pain	7
Awareness of spinothalamic loss	1
Excessive sweating above the injury	1
Remaining asymptomatic	1

The early and later symptoms and signs clearly reflected an asymmetrical spinal cord syndrome concentrated in the central grey matter. Invariably, it began as a unilateral lesion involving fibres conducting pain, temperature and tickle and/or the reflex fibres on the same side. The picture most usually suggested an early lesion in the medial part of the posterior horn, and variable extension from this area. In some instances, the ipsilateral anterior horn cells were involved early.

Pain and Sensory Disturbance

In nine of the patients there was a history of segmental pain ascending gradually from the level of the initial injury, and proceeding to the higher cervical segments in an orderly sequence. In some, the pain was associated with, or gradually replaced by a sensation of numbness. One patient experienced an orderly progression of numbness without pain. A clear history of either ascending pain or sensory symptoms or signs of both was recorded also in 13 of the cases in the literature or personally communicated to us.

These observations appear to be of considerable significance when considering the possible pathogenesis of this condition. They draw attention repeatedly to the fact that the symptoms begin at or near the original injury and extend cephalad. Caudal extension would be masked by anaesthesia below the level of injury in complete lesions. In the partial lesions, none has noted worsening of sensation or pain below the lesion as an early symptom. In one instance, the ascent of a sensory level developed unnoticed and was found on routine neurological follow-up examination (case 17). In several instances, the area between the new and the original sensory loss was interrupted by a gap within which skin sensibility was normal (Fig. 10.1). The explanation for this, most probably, is the variable size of the cavity at different spinal cord levels and its tendency to be maximal in the cervical region (see Chapter 14).

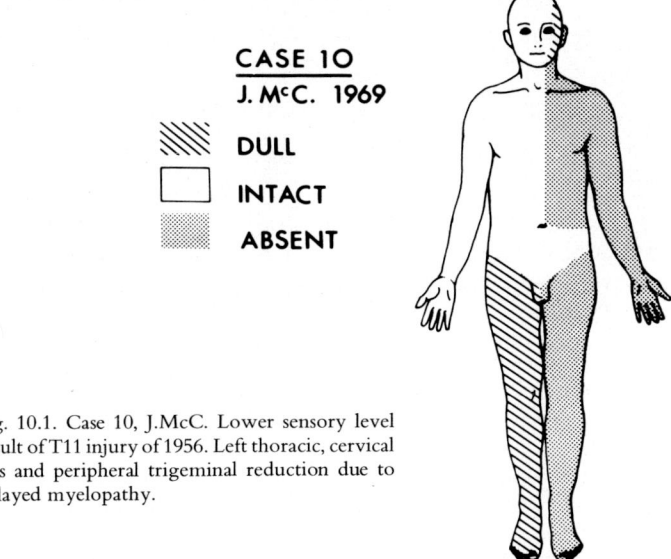

Fig. 10.1. Case 10, J.McC. Lower sensory level result of T11 injury of 1956. Left thoracic, cervical loss and peripheral trigeminal reduction due to delayed myelopathy.

Deep pain was lost in most patients in whom the superficial pain appreciation was eliminated. In six instances with this deep pain loss, neuropathic joints (two elbows and four shoulders) were observed (Fig. 10.2). These represent considerable handicaps to patients already paraplegic and very dependent on their arms. Coughing, straining or sneezing resulted in quite violent pain in some patients (e.g. case 10) and might be followed by a number of hours of distressing discomfort. In one instance (case 12), a violent sneeze abruptly brought on severe pain in the upper cervical dermatomes. As the pain eased off over a few hours the patient became aware of a new sensory loss in the upper limbs. One of Marvin's cases (1960) precipitously became aware of the new and eventually progressive disability following a very violent sneeze. The sneeze produced immediate pain in his shoulder and arm, followed by numbness in this extremity. Similarly, a case with a paraplegia from T12 fracture reported by Nurick et al (1970) 'while sneezing developed a searing pain below the costal margin lasting 5–10 minutes. Whenever he sneezed, coughed or blew his nose it occurred. After a month he became aware that the site of his pain seemed to be rising up the anterior chest wall and that the skin below the site of the pain had become numb.'

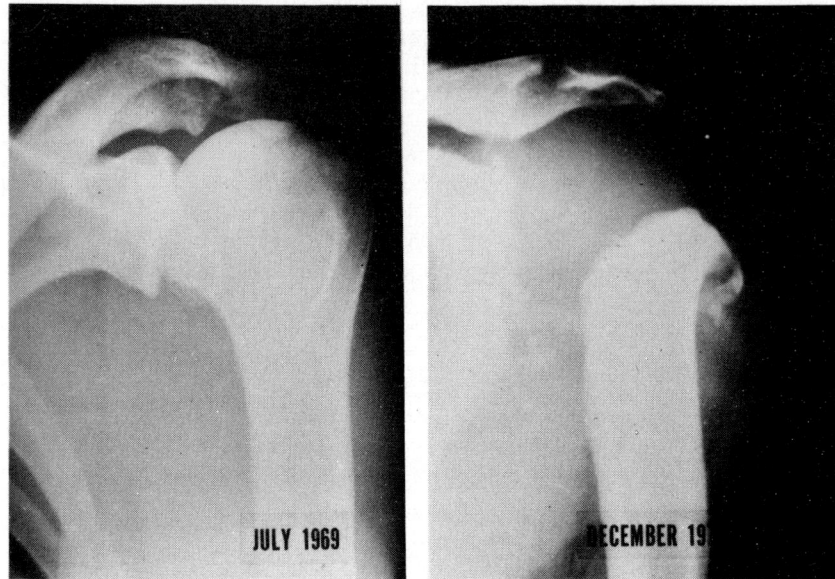

Fig. 10.2. Case 7, R.S. Neuropathic shoulder lesion developing between 1969 and 1971. Original injury at T11–12, 1955.

Foster's (1972) case with a partial paraplegia from a T12–L1 fracture dislocation, 18 months after the injury developed pain in the right side of the trunk on sneezing. Months later an 'electric current sensation' in the right arm was experienced after sneezing. At about the same time as the arm symptoms developed, a similar feeling developed in the right side of the neck *when the head was turned to the left side*. These observations are of great interest in considering the factors responsible for the progression of the developing cavities which will be discussed in detail in Chapter 14.

Motor Involvement

New and progressive motor involvement in the patients originally affected with a *complete* spinal lesion was confined, obviously, to the upper limbs. In all instances, the motor worsening in the upper limbs was indicative of a lower neurone lesion (Fig. 10.3). In four of the 17 cases, there have been no signs or symptoms of new motor involvement to date. At the other extreme, were two patients (cases 2 and 9) who were originally partially paraplegic from lesions at L1 and T12, respectively, and who died in states of quadriplegic helplessness with spastic paralysis of legs and flaccid, amyotrophic and areflexic paralysis of the upper limbs. Upper limb involvement in the group of 17 cases is summarised in Table 10.6. The only patient in the series with an original quadriplegic weakness (case 14) experienced worsening in both affected upper limbs.

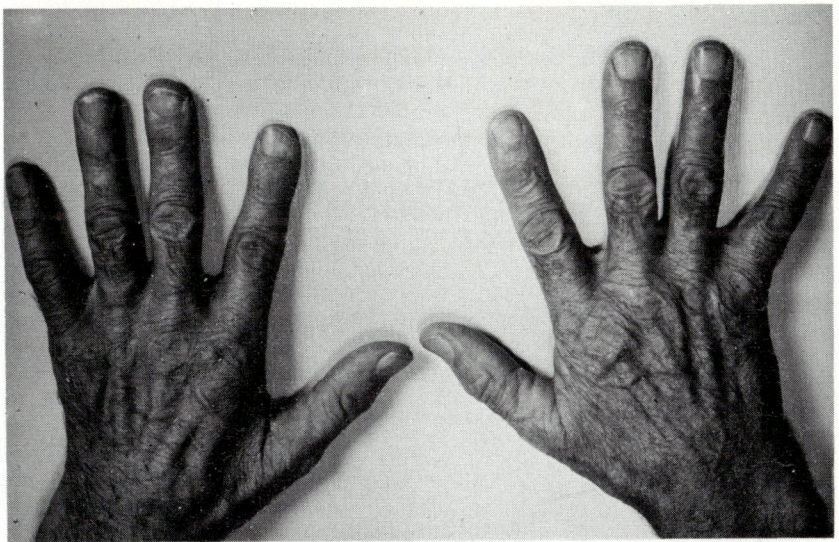

Fig. 10.3. Case 15, P.B. 1969. Asymmetrical hand wasting. Original injury at T12–L1, 1953.

Table 10.6. *Progression to produce upper limb motor weakness (17 cases)*

None	4
Mild to moderate unilateral	6
Severe unilateral	1
Severe bilateral	4
Quadriplegic helplessness	2

There was a *partial* paraplegia in nine cases, but the motor function of the legs was at risk in only eight of them. This was so because one case was designated 'partial' only by virtue of sensory sparing in the lumbar segments (case 10), but his legs were functionless. In the other eight cases lower limb motor function declined, and disappeared altogether in five, during the progress of the extending lesion. These patients ceased to be able to get about with canes, crutches, and braces, and became, in effect, totally paraplegic, either because of the leg weakness or because associated hand and arm weakness prevented use of auxiliary aids with the upper limbs, or because a combination of both factors existed.

Bilateral Extension of Symptoms and Signs Above the Original Lesion

Without exception, at the beginning, our cases *developed the progressive lesion unilaterally*. Eventually, but again asymmetrically, the opposite side became involved in 11 of the 17 cases. In eight of them, the bilateral involvement was both motor (Fig. 10.4 and 10.5) and sensory. In one, bilateral motor involvement was only a temporary phenomenon during a curious, delayed and temporary postoperative exacerbation (case 13). Twice there was merely reflex evidence of contralateral involvement. On one occasion, the only bilateral event was the pathological sweating above the original lesion (case 6). Of the reported cases, there was striking confirmation of the tendency for this complication to begin, and often to persist, as a unilateral extension of spinal cord abnormality. Sufficient information is available in 35 cases to judge this point and 23 of them were still unilateral at the time they were recorded or reported to us.

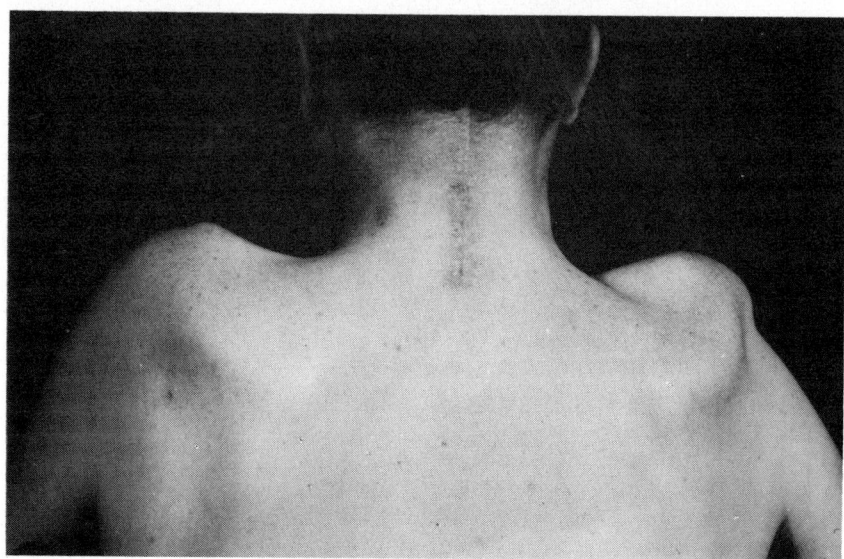

Fig. 10.4. Case 8, R.W. 1967. Bilateral shoulder girdle wasting. Original injury L1, 1949.

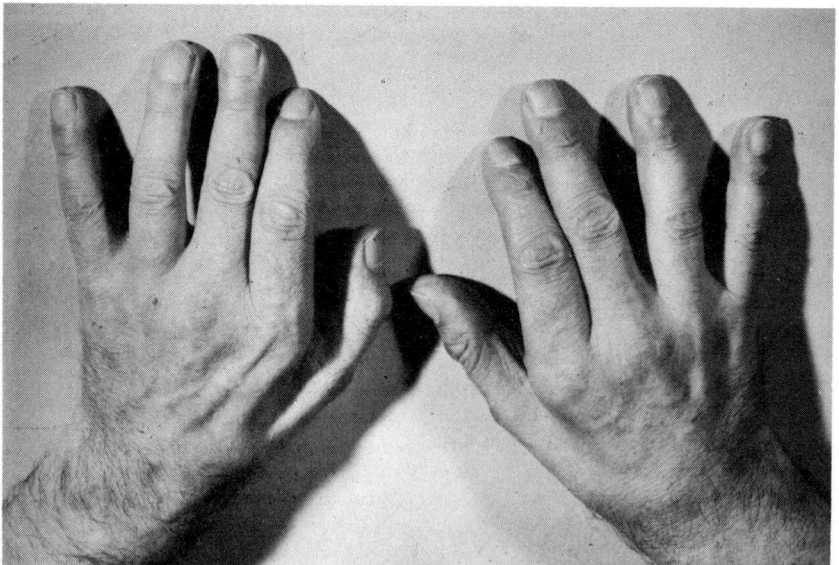

Fig. 10.5. Case 8, R.W. 1967. Bilateral hand wasting. Original injury L1, 1949.

Rapid Progression of Signs and Symptoms

Three of our cases recounted an abrupt worsening of pain and extension of disability. One (case 12) became aware of more pain and extended numbness after a violent sneeze. Another (case 14) had a minor neck blow with immediate extension of pain and evidence of more neurological signs which persisted until surgical drainage of the syrinx was carried out. The third case (case 13) experienced a sudden spontaneous flare-up of severe pain with marked sensory and motor worsening in the seventh week after surgical drainage. He had enjoyed decided benefit from the surgery and after this period of worsening went on again to a rather abrupt spontaneous return to his pre-exacerbation condition.

The sudden worsening and even the abrupt presentation of a syringomyelic syndrome has been recognised for a long time. Gowers (1904) described four clinical cases in whom this occurred and substantiated his description with pathological observation in one of his four cases. Wilson (1954) refers to the condition as Gowers' syringal haemorrhage. Perot, Feindel and Lloyd-Smith (1966) have described an instance of this syndrome, with known cord dilatation for six years, abrupt worsening and then successful operative aspiration of the clot in the cavity. Traumatic incidents have precipitated the development or progression of clinical signs and symptoms in syringomyelia of communicating type (see Chapter 3, also Williams and Turner, 1971); they have been noted in the onset of symptoms of cavitation with tumour (see case 6, Chapter 17), and have been recorded in this variety complicating serious spinal injury. Trauma or other stress (e.g. violent straining) may precipitate haemorrhage from vessels within the lumen of the cavity. We have evidence that thin-walled blood vessels may traverse these cavities in glial bridges (see Fig. 14.6). It is also evident, though, that stress and trauma will produce exacerbation of symptoms by a haemodynamic effect. This matter will be discussed further in considering the extension of these cavities (Chapter 14).

Sensory Impairment in Trigeminal Nerve Territory

Eight of the 17 patients have been aware of, or have been found to have, diminished sensibility to pain over dermatomes classically regarded as trigeminal in location. In all patients, the loss of sensibility in the face began on the side of the early upper limb and neck involvement; in all but one it was unilateral. In one case (case 8), as the opposite cervical spinal cord function became impaired for conduction of pain and temperature, the second side of the face showed similar early loss of sensation (Fig. 10.6). By contrast with the cervical dermatomes it was never a complete loss, but tended to involve all three divisions and always showed a tendency to spare at least some of the central portions of the face (Fig. 10.6, case 7). Temperature reduction tended to be less remarkable and extensive than pain appreciation. Two of the cases (cases 8 and 12) have shown a decided improvement in facial sensory disturbance after surgical drainage of their spinal cord cysts.

Further pathological studies will be needed before a definitive answer can be given as to the origin of the trigeminal sensory loss. Three mechanisms are possible:

1. The cyst may extend into the medulla. As noted below, there are three cases

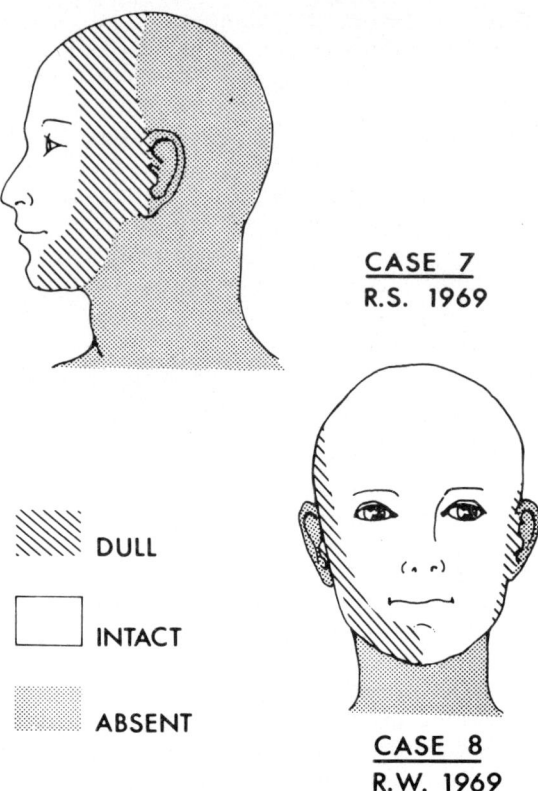

Fig. 10.6. Cases 7 and 8. Two cases with involvement of trigeminal nerve territory. Pain reduced more extensively than temperature, sparing central portion of face.

in the literature in whom the brain stem was involved, but trigeminal loss was described in none of them. In Klawans' case (1968) (see Fig. 14.2), it would have been expected, but he reports (Klawans, 1972) that the case was not re-examined neurologically in the few months before death so that this is not on record. Our own case with possible brain stem extension has yet to show trigeminal disturbance.

2. The classical trigeminal territory may be overlapped, particularly peripherally, by sensory supply arising from the uppermost cervical nerves. Zander (1897), in detailed dissections, found terminal filaments of the upper cervical nerves extending to all parts of the face, save for its central portion. Sherrington (1893) demonstrated the extent of overlap between segmental sensory nerve territories. In later experiments with macaques (1898), he was satisfied that there was a good deal of overlap of the trigeminal territory by upper cervical segments. Indeed, the illustrations from this classic paper depicting the sensory deficit when the upper cervical roots were severed are very similar to the area of diminished sensibility in our present group of spinal lesions. Lewy (1938) considered that there was only

a small central area of the face that was 'the pure trigeminal field'. Carmichael (1933) illustrated considerable variability, particularly in the second and third divisions of the trigeminal territory, where anaesthesia may be found after surgical removal of the Gasserian ganglion. Hunter and Mayfield (1949) and Mayfield (1955) have demonstrated considerable reduction of pain sensibility overlapping into the face and forehead when the second cervical sensory root was cut for relief of neck pain. Their patients experienced reduction of peripheral pain sensibility with sparing of the central parts of the face.

3. The extension of the spinal tract of the fifth nerve into the cervical spinal cord may be sufficiently extensive to allow a purely spinal lesion to deprive all three trigeminal divisions of normal appreciation. Déjerine (1914) evolved the so-called 'onion skin' pattern of innervation of the face. His concept was that the peripheral portions of facial pain appreciation terminated as low as the fourth cervical segment, whereas impulses from the central portion of the face terminated in the highest cervical and lowest medullary areas. McKenzie (1955) performed tractotomies at the spinal cord/medullary junction, 3–5 mm below the obex, and reported retention of pain sensation confined to a small area around the lips. Crosby, Humphrey and Lauer (1962) believe that most fibres mediating facial pain appreciation terminate within the spinal cord. The decided tendency in our cases to spare the central face area is in keeping with these observations and we do not believe that involvement of the medulla oblongata need be postulated.

Brain Stem Extension

Only one of our cases (case 3) has had evidence suggesting that the brain stem was involved. Even here, we have rather incomplete evidence. It was suggested by a period when hemi-atrophy of the tongue was apparent. Coincident with a subsidence of some of the cervical dermatome pain and temperature impairment (Fig. 10.7), this sign disappeared and raised the possibility that spontaneous drainage of the intramedullary syrinx had occurred. Three cases with very good evidence of brain stem extension are known to us. Klawans (1968) has described the condition in his autopsied case, and Rossier et al (1968) have referred to a case. Watson (1972) has described a patient with classical features of this condition, with sensory loss up to the top of the C2 dermatome who has developed hiccup, hoarseness and dysphagia indicative of an extension into the medulla. Facial sensation in this case has not been disturbed as yet.

Sweating

A remarkable pattern of sudorific disturbance has presented itself in this group of patients. The only comparable disturbance is that described by Foster and Hudgson in Chapter 3. Abnormal piloerection following a similar distribution was observed within the territory of the excess sweating in two of the patients. Eight of the group of 17 have complained of sweating segmentally involving the upper half of the body, and always above the level of the original injury. As well as the trunk and axilla, the excessive sweating in most instances involved the ipsilateral upper limb, neck, head and face. It was of sufficient severity to be the source of spontaneous comment for all those affected, and in three was bad enough to require frequent daily changes of clothing. Seven cases exhibited this excessive

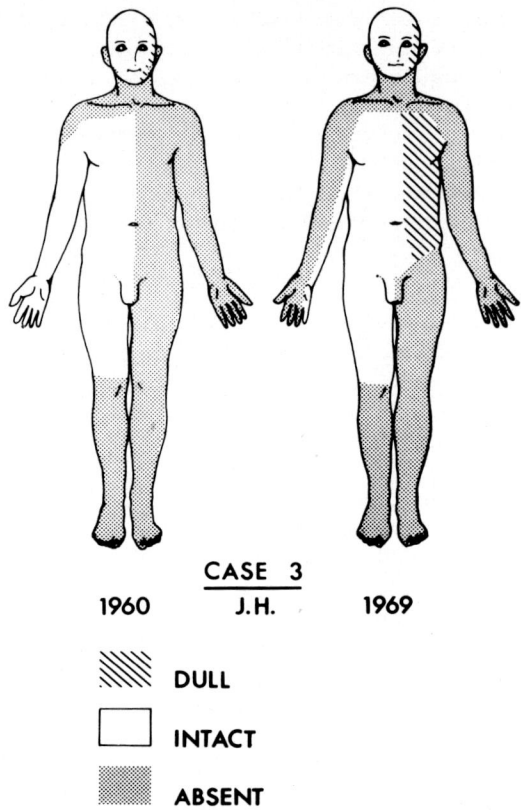

Fig. 10.7. Case 3, J.H. Note the improvement on the left side between 1960 and 1969. Pain and temperature loss increasing on right side. Original injury L1, 1951.

sweating on the side, and in the area of, the early and major new sensory symptoms. Once it was on the side opposite the major new development, but in this patient (case 6) a bilateral lesion did in time develop. Furthermore, he went on to develop gustatory sweating in the forehead on the same side as the main ascending lesion. In some patients, several factors were capable of evoking the excessive sweating. Six noted it accompanying distension of a viscus, six with exercise, seven with extra heat in the room, three with anxiety and excitement, and two with gustatory stimuli.

Enquiry was made of the patients who eventually had this peculiar excess of sweating of their perspiration pattern between the time that they had their original injury and the development of their new progressive symptoms. In this pre-complication, post-traumatic period six of the eight experienced only normal sweating above the injury level. Two of the patients noted some excess immediately after their injury; both had complete lesions, one at T11 (case 13) and one at T8 (case 11). The former remarked only on a slight increase 'from the waist up' with excitement, exercise and extra heat. The latter experienced a mild nocturnal tendency distributed both above and below the level of the injury, with neck, back, groins, legs and feet involved. This minor excess was inconsequential and not lateralised, compared with the later remarkable development that ensued in these two plus the six other patients in this series with excessive upper limb and head sweating.

The curious distribution of this unusual sudorific disturbance is, of course, associated with a unique neurological deficit. As well as a major traumatic transverse myelopathy, these patients were afflicted with a lesion in the dorsal horn extending above the initial traumatic level. This unusual situation, therefore, combines the expected sweating abnormalities of the paraplegic, with the additional disturbance which might develop from a superimposed dorsal horn lesion at a higher level. In a previous study of sweating in paraplegics, Kendrick et al (1953) concluded that reflex sweating occurred in those dermatomes from which sensory nerves entered the brain stem or spinal cord above the level of the lesion, and which retained normal sensation, but for which the sympathetic supply arose below the level of the lesion. Conversely, excessive sweating did not occur in the dermatomes for which the autonomic supply came from above the lesion, as is the case in a thoracic injury. In reflex response to a distended bladder, excess sweating extended over the head, neck and arms in patients with a lesion in the cervical or first two thoracic spinal cord segments. If the lesion was in the next three or four thoracic segments, the sweating occurred as a belt around the body just above the level of the sensory loss. Patients with lesions below T7 were not afflicted with this excess sweating. With the two exceptions mentioned, the present cases conformed to this general pattern after their injury and before the extending lesion developed.

The opinion of Kendrick et al (1953) was that sensory impulses from a dermatome exert an inhibitory effect on reflex sweating in the same dermatome. The addition of an extending sensory lesion in the upper thoracic and cervical levels might be considered to remove the inhibitory effect dermatome by dermatome, and allow the excess sweating experienced by these patients. It developed even though the area of excess sweating in these cases received sympathetic supply from above the original lesion. This inhibition alternatively might be effected by the inhibitory activity of the fine medullated efferent fibres in the posterior roots carrying the antidromic impulses described by Bayliss (1901). These have been considered to carry parasympathetic impulses causing vasodilatation and inhibition of sweat and pilomotor activity. The lesion in this syndrome would be located ideally, anatomically, to cause interruption of these fibres, to remove their inhibitory effect and allow excess sweating.

Above the level of the original injury, the descending sympathetic pathways in the spinal cord, as well as the effector neurones, must be intact in patients exhibiting this excessive sweating. The former appear to descend (Foerster, 1936; List and Peet, 1939; Hyndman and Wolkin, 1941) just anterior to, but close to, the pyramidal tract or, alternatively more widely through the anterolateral column (Johnson, Roth and Craig, 1952). The effector neurones are in the intermediolateral cell column and the sparing of these areas is further evidence of the rather circumscribed location of the pathological process in this paraplegic complication. Of interest also, in this regard, is case 3, in whom the excess sweating eventually disappeared spontaneously. Concomitant with its disappearance, a Horner's syndrome developed, the patient's effector neurones having become involved with the cystic expansion. No other patient in the series developed a Horner's syndrome, and no other developed excess sweating only to lose it spontaneously.

New Symptoms Related to the Original Injury Site and Below

No case in our series lost function at or below the level of his original injury as an initial symptom of the development of the progressive lesion. All who did lose some lower limb or sphincter function in the evolution of the new condition had already become aware of symptoms referable to the spinal cord above the site of the original trauma. In reviewing the cases in the literature, two cases (Feigin, Ogata and Budzilovich, 1971; Marvin, 1960) have been exceptions to this. In Feigin's case, at post-mortem, the extension was entirely *below* the level of the C4 to 5 fracture dislocation (Fig. 14.1). Marvin described a clinical history in which the significant and progressive new disability was referable to the site of the original injury, and did not extend above this level after a lapse of several more years.

Myelography

INCIDENCE OF MYELOGRAPHIC DEFECTS

In 14 of the 17 cases, oil myelography was carried out at least once, and in 10, a widened cord in the cervical region was demonstrated. Two of the 17 had dilatation below the second thoracic segment to T4 and T5, respectively. In this condition, the lowest level of cord dilatation known to us as demonstrated by oil myelography is illustrated in Fig. 10.8. Crush fractures at L1 and T7 were associated with a complete block at the former level and cisternal myelography demonstrated cord dilatation beginning just above T7, continuing to C2 but maximal at C4 to 5 levels. In two cases, myelography was still normal three and seven years, respectively, after the new syndrome developed. In one case, myelography was normal in the sixth year, and in the seventh year had become abnormal. In the two others with normal contrast studies, the new symptoms had been present for only six months and one year, respectively, when their myelograms were carried out.

RADIOLOGICAL EVIDENCE OF ARACHNOIDITIS

At the level of the original trauma there was always considerable evidence of abnormality in the spinal subarachnoid space. Usually, it was of such severity that radiologists described it as being 'in keeping with an arachnoiditis' (Figs. 10.9 and 10.10). In some, this was so severe that myelography to demonstrate the cervical and thoracic region had to be performed by cisternal injection. The 'arachnoiditis' never extended more than three segments above the level of the original trauma. The extent of the 'arachnoiditis' was not related to the type of initial treatment received at the time of the original trauma; it was present in those who had been treated by laminectomy, by laminectomy with fusion, and in those who had been given no initial surgical treatment.

INTRAMEDULLARY MYELOGRAPHY

Regular oil contrast myelography, even when indicative of dilatation in the cervical region, fails to give a true picture of the extent of the cavitation below

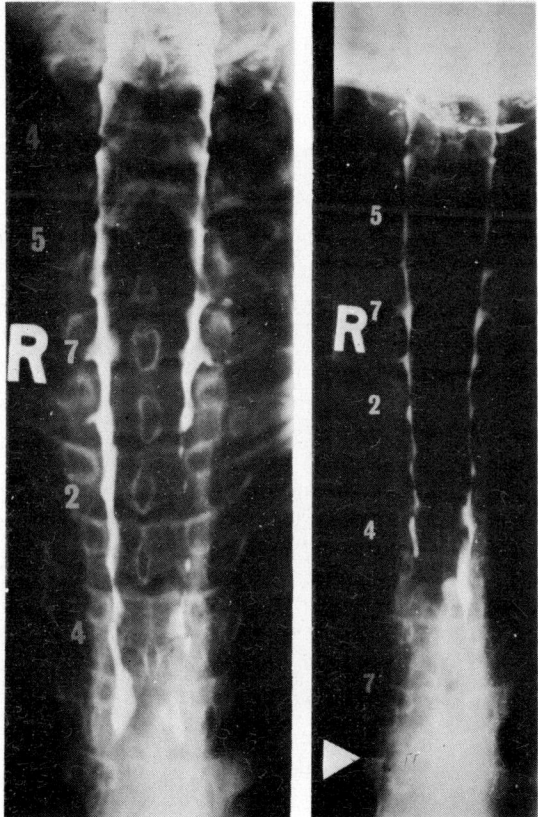

Fig. 10.8. Patient paraplegic from T7 (arrow) and L1 fracturing (1967). Developed upper limb signs 1969. Cord dilatation 9 months after new symptoms. Maximal C4–5, but visible C3 to T6. (By courtesy of Dr A. Appleby.)

the cervical area. To supplement this routine procedure, intramedullary myelography was employed twice, once in conjunction with air in the subarachnoid space. In one instance (case 8), the dilatation of the canal was visualised by regular cisternal myelography extending from the third cervical to the second thoracic level (Fig. 10.11). Contrast material was injected into the cyst with the spinal cord exposed at operation. It was then determined that the cavity extended up to the cervicomedullary junction (Fig. 10.12). It went down to T11, almost as far as the original injury site (L1) (Fig. 10.13). In the other instance (case 14), the spinal injury had been a fracture dislocation at C5 to 6. The investigation was first carried out with a routine lumbar oil contrast injection. A widened cervical cord was demonstrated. Fifty ml of air was later injected into the lumbar subarachnoid space and through a percutaneous puncture at the C6 to 7 level a needle was put into the spinal cord cavity using fluoroscopic control. A cavity was detected within the cord from C2 (Fig. 10.14) to T12 (Fig. 10.15). It was of interest that the contrast material at the time of the injection went only to T7 but that one month later it had passed down to the T12 level. Despite this lapse of four weeks, the cavity was still filled with contrast material, apparently the same quantity as at the time of injection. None could be seen, despite careful scrutiny, in the ventricular system.

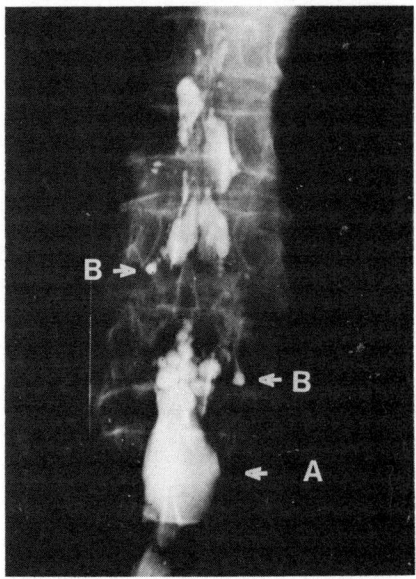 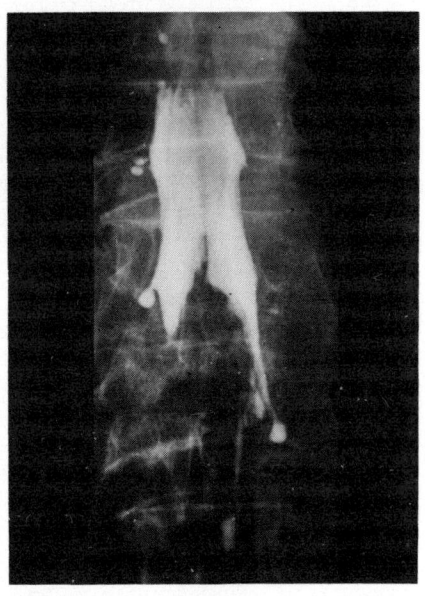

Fig. 10.9. Case 15, P.B. 1969. Fracture at T12 (arrow A), arachnoiditis extending cephalad by two vertebrae. Note 'candle-guttering' (arrows B). Original injury, 1953.

Fig. 10.10. Case 15, P.B. 1969. Close up of level of injury with excellent 'candle-guttering' of arachnoiditis.

Segmentation of Cavities

The appearance of these cavities is curiously segmented. These segmental indentations, reminiscent of the X-ray appearance of haustra in the bowel, are irregularly spaced. The opportunity to observe them at post-mortem in two more usual 'syringohydromyelia' cases (Ball, 1971) has indicated that astroglial proliferation with blood vessels and some collagen within the laterally indenting folds is the anatomical substrate to this segmentation (see Chapter 14). The post-mortem X-ray studies carried out in one of these cases by Dr J. Allcock (1971) are illustrated (Fig. 10.16). Their post-mortem appearance is demonstrated in Fig. 14.36. This appearance was observed in the operative case transected by Dr T. P. Morley (case 11). He noted that the inner lining of the cavity had a 'chamois-leather appearance'. This segmental radiographic appearance has been recorded by others employing iodoventriculography (Tjaden et al, 1969). The significance of these glial ridges in the pathogenesis of such cavities is discussed in Chapter 14.

Correlation of Myelographic Defect with Clinical Signs

The extent and severity of the clinical signs could not be accurately cor-

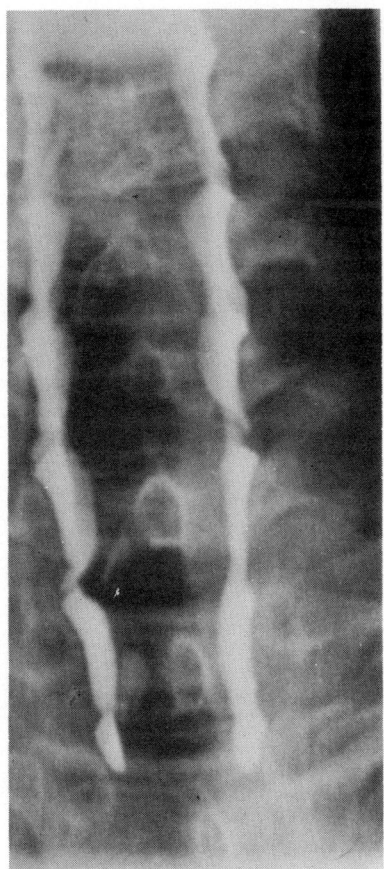

Fig. 10.11. Case 8, R.W. Widened cervical cord shadow. This widening extended from C3 to T2. Original injury L1, 1949.

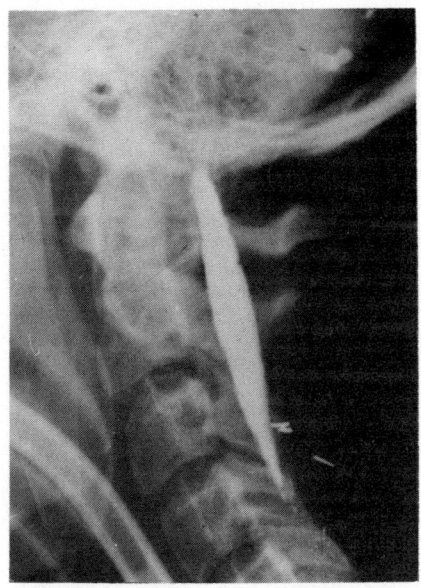

Fig. 10.12. Case 8, R.W. Contrast material injected into cystic lesion within the spinal cord at cervical laminectomy, and extending to highest cervical level. (Intracranial contrast medium from previous lumbar myelogram.)

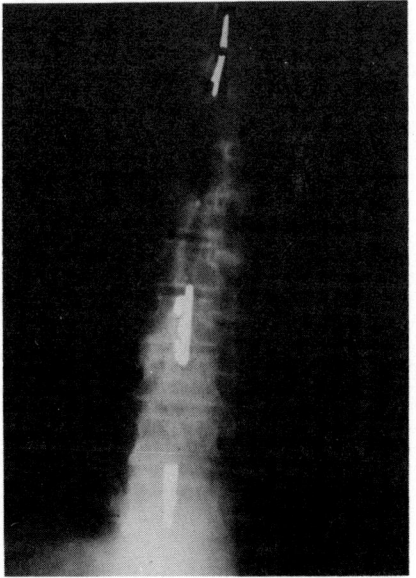

Fig. 10.13. Case 8, R.W. Intramedullary myodil extending down in cystic cavity to T11 vertebral level (highlighted to improve reproduction).

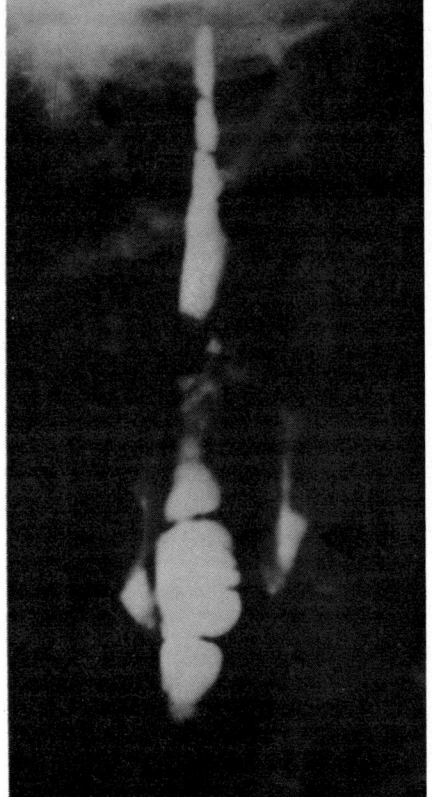 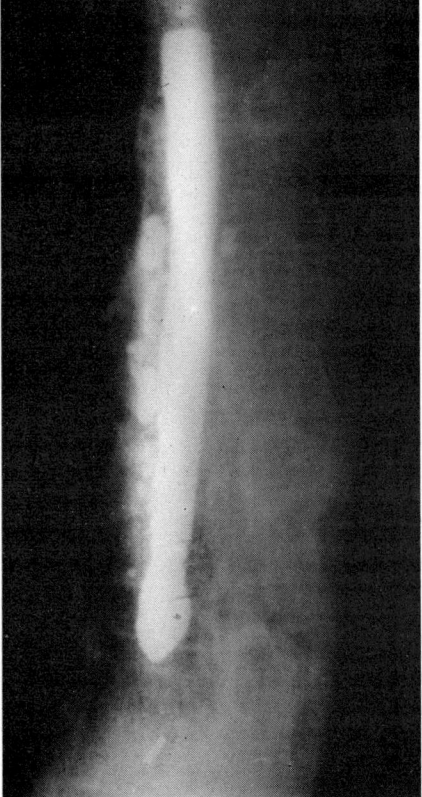

Fig. 10.14. Case 14, D.L. Irregular loculated appearance of cavity outlined by percutaneous injection of Myodil into intramedullary cavitation. Some contrast material in subarachnoid space outlines cord expansion (arrows). Cavity extends above and below the level of injury. Original major fracture at C5, 1965.

Fig. 10.15. Case 14, D.L. Extension of intramedullary cavitation down to T12 level. Injection was made by percutaneous needle inserted into cervical cystic dilatation.

related with the degree of the dilatation of the cervical cord. The most striking example of this negative correlation was seen in case 15 in whom bilateral upper limb wasting and weakness was apparent seven years after the new syndrome developed. A myelogram did not indicate spinal cord enlargement and yet a cyst was identified and drained in the cervical region at laminectomy. This was the only case in our series whose cord was definitely proved to be the site of a cyst and who still had a normal oil myelogram at the time of operative confirmation of the existence of the cavity. Several times operation was postponed until the cord enlargement was demonstrable and this may be an unwise procrastination in some instances.

GAS MYELOGRAPHY

It has been demonstrated by Conway (1961) and further elaborated by

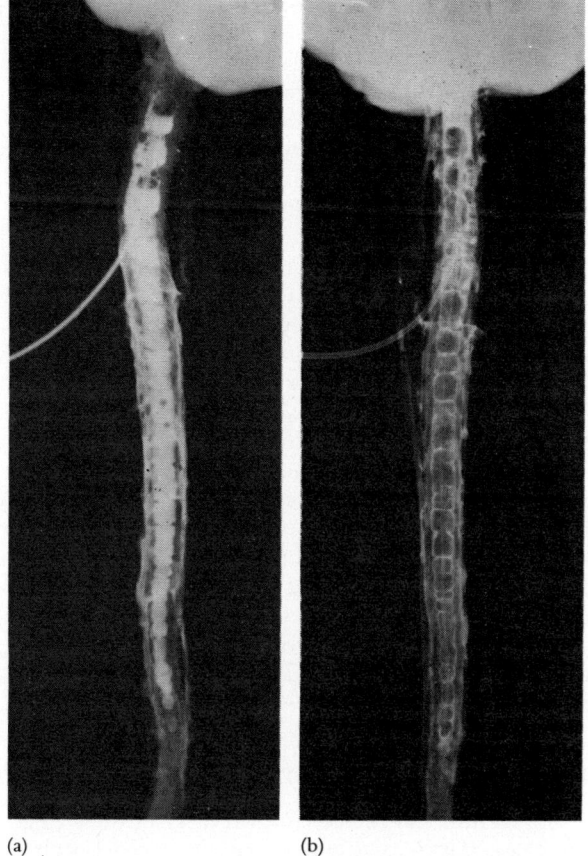

(a) (b)

Fig. 10.16. A case of non-traumatic communicating syringohydromyelia with post-mortem injection (a) of the removed spinal cord to show the irregularly segmented nature of the cavitation due to lateral glial ridges (see Fig. 14.36, Chapter 14). The contrast material, when partially removed, (b) delineates the segments more clearly.

Scandinavian neuroradiologists (Westberg, 1966) that a mobile column of fluid exists within some syringomyelic cavities. This column of fluid will expand the cord in the 5° head-down tilt of the radiographic table and the size of the cord shadow will diminish visibly in the 15° feet-down position. Gas myelography with lateral tomography, have been used to make these observations. Until recently, the facilities for this technique were not available to the authors so that it was used in only one of these 17 cases. However, it has been reported, by Nurick et al (1970), in similar cases of postparaplegic cord cavitation and this changing cord size demonstrated.

Air myelogram and lateral tomography techniques must be pursued, in conjunction with pathological data, in these post-traumatic cases to determine whether or not a 'flaccid' cyst is to be considered as unequivocal evidence of direct subarachnoid or ventricular communication, and whether or not a 'non-flaccid' cyst denotes a predictable absence of such communication (Ellertsson and Greitz, 1969).

COMMUNICATION WITH VENTRICLE OR WITH SUBARACHNOID SPACE

Pathological studies in this post-traumatic group have so far failed to demonstrate any communication between the cavity and either the subarachnoid space or the ventricle (see Chapter 14). However, there is one other radiological observation (Allan, 1972) and two surgical observations (Werner et al, 1969; Tucker 1972) which indicate that such a communication may develop. In the case of Werner et al, an air myelogram demonstrated a cord enlargement extending from C3 to 4 down to the level of original fracturing at T9 to 10. When laminectomy at C7 to T3 was carried out, the cystic dilatation of the underlying cord yielded 30 ml of clear fluid; protein content 28·6 mg/100 ml. When jugular compression was carried out there was a 'reflux of fluid into the intramedullary cavity which thereby re-expanded'. Werner et al took this as evidence of communication of the cavity with the fourth ventricle. Tucker (1972) has also seen evidence, at operation, highly suggestive of a communication of a post-traumatic cavitation with the fourth ventricle. Neither of these cases had any bulbar signs or symptoms.

Allen (1972) has studied a patient with a long-standing paraplegia from trauma at T4. Seven years later, an ascent of symptoms into the arms led to myelography in the lumbar region. The contrast medium entered the spinal cord at the level of the trauma and ascended within the cord to the cervical region. In this regard, this case is unique and it does not follow that this is the pathogenesis of the usual post-traumatic cavitation. The evidence, in fact, indicates that most cases are not like this. However, in any given case this mechanism should be sought.

CEREBROSPINAL FLUID AND INTRACAVITY FLUID

The cerebrospinal fluid has been examined in 11 of these cases. The protein content of the fluid was not elevated significantly in seven cases but four of these were from cisternal samples only, obtained at the time of cisternal myelography. Once the protein was minimally elevated at 56 mg/100 ml (lumbar), twice moderately elevated at 102 and 115 mg/100 ml, respectively (both lumbar), and once (case 6 with a fracture dislocation at T4) initial normal readings in 1964 and 1967 were followed in 1968 by a level of 1500 mg/100 ml (lumbar). The very high protein content presumably does nothing more than reflect the degree of block in the subarachnoid space.

Four samples of intracavity fluid, obtained at operation, have been available for analysis. Twice there was normal protein content in clear fluid (38 and 32 mg/100 mls), and twice an elevated protein in faintly yellow fluid (224 and 200 mg/100 ml). The lack of correlation between the cerebrospinal and cystic fluids in these four cases is illustrated in Table 10.7.

In case 14, spinal fluid was sampled simultaneously from the cervical subarachnoid space and from the intramedullary cavity by percutaneous needle. The spinal fluid contained seven white blood cells, two red blood cells, 135 mg protein, 55 mg sugar and 125 mEq/l of chloride. At the same time, the cavity fluid contained 13 white blood cells, 1140 red blood cells, 200 mg, protein, 40 mg sugar and 125 mEq/l of chloride. There was sufficient fluid from the cavity in one case (case 15) to examine it electrophoretically and it contained 17 mg/100 ml of albumin and 15 mg/100 ml of globulin (A/G ratio 1·14). Calcium and magnesium were measured in the same case (case 15) both in the C.S.F. and in the cavity fluid. An atomic absorption method was used and in the C.S.F. (cisternal) the calcium content was 1·7 mEq/l and the magnesium content 1·7 mEq/l. In the cavity, the readings were calcium 1·25 mEq/l and magnesium 2·1 mEq/l.

Table 10.7. *Protein content of C.S.F. and intracystic fluid (authors' cases)*

Case No.	C.S.F. protein (mg/100 ml)	Cavity fluid protein (mg/100 ml)
6	1500 (L)	38
12	24 (C)	224
14	82 (L)	
	135 (C) post-myelogram	200
15	22 (C)	32

L = Lumbar
C = Cisternal

Four other communications are known to us in which intracavity fluid protein analysis have been reported in this post-traumatic syndrome. In Freeman's (1959) case, the protein content of 'cloudy white fluid' was 130 mg/100 ml; no simultaneous reading was given for the subarachnoid fluid. Werner et al (1969) made spinal fluid and cavity fluid observations on three of their four post-traumatic syringomyelic cases. Their observations on the three cases are summarised in Table 10.8 and indicate normal or slightly elevated readings of intracavity protein in fluid described as clear. As mentioned above, one of these cases was judged to have communicated with the fourth ventricle. The levels in the case of Nurick et al (1970) were 55 mg/100 ml in the cavity and 20 mg/100 ml in the cisternal C.S.F. In Foster's (1972) case, the cisternal protein content was 14 mg/100 ml, while that in the cord cavity was 71 mg/100 ml. (Table 10.8).

Table 10.8. *Protein content of C.S.F. and intracystic fluid (reported cases)*

Author	C.S.F. protein (mg/100 ml)	Cavity fluid protein (mg/100 ml)
Freeman	—	130
Werner et al		
Case 1	98 (L)	77
	28 (C)	
Case 3	116 (L)	28
	28 (C)	
Case 4	29 (L)	37
Nurick et al	20 (C)	55
Foster	14 (C)	71

L = Lumbar
C = Cisternal

Gardner (1965) studying communicating syringomyelia, examined the protein content of lumbar and syrinx fluid in 18 cases and found ranges between 15 and 59 mg/100 ml in the subarachnoid space fluid and between 16 and 90 mg/100 ml in the syrinx fluid. In 10 cases, the protein content in the syrinx was below the lumbar C.S.F. reading and 'more in keeping with cisternal fluid levels or intraventricular levels'. Love and Olafson (1966) had levels similar to Gardner's for the subarachnoid and syrinx protein content. They found 17 of 27 patients had lumbar C.S.F. levels above 45 mg/100 ml, four of which were over 750 mg. The syrinx protein content ranged from 10 to 180 mg/100 ml, and the latter in a patient with lumbar subarachnoid space reading of 900 mg/100 ml. With this exception, they recorded a syrinx protein content at least 20 mg/100 ml below the lumbar C.S.F. value. Gardner contrasted the findings in his group who had 'syringohydromyelia' with readings of syrinx protein levels in cases of spinal cord tumour. Here, he described a protein content of 2500–5000 mg/100 ml. Gardner used these figures as part of his argument favouring the hypothesis that syringomyelia is generally the result of failure of the fourth ventricle to be properly drained of C.S.F. and of the consequent dilatation of the central canal connecting the syrinx and/or dilated central canal to the fourth ventricle. He argued that the normal or near-normal protein content of fluid in a syrinx indicated that it was of transcellular origin as opposed to the intracellular (interstitial) origin of the fluid from tumour cases.

There would appear to be good reason to distinguish between the content of protein in tumour cases and the ordinary case of communicating syringomyelia. In most of the former, there is no communication between the cavity and the subarachnoid space or ventricle. In some at least of the latter, there is increasing evidence of just such a communication. The assumption that this demonstrable communication is the basis for the equality between syrinx fluid protein and cerebrospinal fluid protein, however, is open to some dispute for the following reasons:

1. The group of cases presented here is known to have normal or slightly elevated protein levels and yet most have no communication with the C.S.F. circulation.
2. Few detailed analyses are on record but at times the communication between the central canal and its diverticulum, the syrinx, is reduced to an opening of 1 to 2 mm or less (Ball 1971, Rewcastle 1972). It is difficult to accept that such a narrow orifice would be adequate to maintain circulation in a long intramedullary syringal cavitation (Ball and Roach, 1972). Furthermore, evidence is presented in Chapter 7 that this communication, if it exists, may not persist beyond the development period.
3. The elevation of C.S.F. protein in a case of mechanical block in the subarachnoid space is generally regarded as a result of the obstruction interfering with the circulation of the fluid. Other factors may be responsible as well, among them the effect of the compression on the blood vessels, including the capillaries. It is probably inaccurate to accept the converse and to presume a good circulation exists if the cavity has a protein content comparable to ventricular or subarachnoid space fluid. Some cases of syringohydromyelia with minute connecting channels must have very little circulation and in the present group of cases most of them cannot be claimed to have any, at least as far as the ventricles and the subarachnoid space are concerned.
4. Observations were made by Nassar, Correll and Housepian (1968) on a series of cases with syrinxes confined to the conus medullaris with 'clear colourless

fluid'. No actual values were presented but they stated that the 'protein content of the cystic fluid was not consistent with tumour'.

5. In syringomyelia accompanying a series of cases of spinal arachnoiditis (Chapter 15), the fluid from the cavities was always clear and watery and in the two with protein content assessed, it was within the range of normal cerebrospinal fluid.

The question of the origin and resorption of this intracystic fluid will be further discussed in Chapter 14.

Summary of Clinical Data

1. Approximately 2 per cent of patients who survive with traumatic paraplegia and a smaller number with traumatic quadriplegia will develop progressive signs of ascending myelopathy.
2. The delayed disability comes on after an average lapse of four years in patients who have had a complete spinal lesion and after 8·8 years in those with a partial spinal lesion.
3. The time of onset is independent of the level of initial trauma.
4. The early symptoms and signs frequently exhibit an ascending picture, beginning at or just above the original traumatic level and eventually reaching the uppermost cervical cord and occasionally the medulla.
5. The signs reflect a central lesion, which invariably begins unilaterally.
6. Quadriplegic helplessness may be the final stage in a patient initially partially paraplegic.
7. Myelography eventually indicates a syrinx.
8. The fluid in the cavities resembles C.S.F.

REFERENCES

Allcock, J. M. (1971) Personal communication.
Allen, P. B. (1972) Personal communication.
Ball, M. J. (1971) Personal communication.
Ball, M. J. & Roach, M. R. (1972) Paper presented to Canadian Neurological Congress, Banff, Canada.
Bawli, J. (1896) *Syringomyélie et Traumatisme Médullaire*. Thése inaugurale, Konigsberg.
Bayliss, W. M. (1901) On the origin from the spinal cord of the vasodilator fibres of the hind-limb, and on the nature of these fibres. *Journal of Physiology*, **26**, 172.
Bischof, W. & Frowein, R. A. (1967) Zur Spätmyelomalazie nach Kompressionsfraktur. *Zentrablatt für Neurochirurgie*, **28**, 61.
Carmichael, E. A. (1933) Some observations on the fifth and seventh cranial nerves. *Brain*, **56**, 109.
Cheshire, D. J. E. (1967) Personal communication.
Conway, L. W. (1961) Radiographic studies of syringomyelia. The hydrodynamics of the syrinx in relation to therapy. *Transactions of the American Neurological Association*, **86**, 205.
Cossa, M. P. (1943) Syringomyelie secondaire à une blessure de la moëlle dorsale supérieure. *Revue Neurologique*, **75**, 39.
Coxe, W. S. (1972) Personal communication.
Crosby, E. C., Humphrey, T. & Lauer, E. W. (1962) *Correlative Anatomy of the Nervous System*, p. 173. New York: MacMillan.
Déjerine, J. J. (1914) *Sémiologie des Affections du Systeme Nerveux*. Paris: Masson.
Ellertsson, A. B. & Greitz, T. (1969) Myelocystographic and fluorescein studies to demonstrate communication between intramedullary cysts and the cerebrospinal fluid space. *Acta Neurologica Scandinavica*, **45**, 418.

Feigin, I., Ogata, J. & Budzilovich, G. (1971) Syringomyelia: the role of edema in its pathogenesis. *Journal of Neuropathology & Experimental Neurology*, **30**, 216.
Feigin, I. (1972) Personal communication.
Finkle, J. R. (1960) Lesions ascending from spinal cord injuries. *Proceedings of the 9th Annual Clinical Spinal Cord Injury Conference*, Long Beach, California, p. 45.
Foerster, O. (1936) Symptomatologie der Erkrankungen des Rückenmarks und seiner Wurzeln. II. Die sympathische Seitenhornkette. *Handbuch für Neurologie*, **5**, 32.
Foster, J. B. (1972) Personal communication.
Freeman, L. W. (1959) Ascending spinal paralysis. Case presentation. *Journal of Neurosurgery*, **16**, 120.
Gardner, W. J. (1965) Hydrodynamic mechanism of syringomyelia: its relationship to myelocele. *Journal of Neurology, Neurosurgery & Psychiatry*, **28**, 247.
Gowers, W. R. (1904) *Lectures on diseases of the nervous system*, 2nd series. Lecture VIII, Syringal hemorrhage into the spinal cord. London: J. & A. Churchill.
Hunter, C. R. & Mayfield, F. H. (1949) Role of the upper cervical roots in the production of pain in the head. *American Journal of Surgery (New Series)*, **78**, 743.
Hyndman, O. R. & Wolkin, J. (1941) Sweat mechanism in man. *Archives of Neurology & Psychiatry*, **45**, 446.
Johnson, D. A., Roth, G. M. & Craig, W. McK. (1952) Autonomic pathways in the spinal cord. *Journal of Neurosurgery*, **9**, 599.
Jung, E. (1960) Syringomyelie in Kombination mit Entwicklungsstorung der Nieren und mit schwerer Wirbelsaulenvertetzung. *Medizinische Klinik Part II*, **55**, 1678.
Kendrick, W. W., Scott, J. W., Jousse, A. T. & Botterell, E. H. (1953) Reflex sweating and hypertension in traumatic transverse myelitis. *D.V.A. Treatment Services Bulletin*, **8**, 437.
Klawans, H. L. (1968) Delayed traumatic syringomyelia. *Diseases of the Nervous System*, **29**, 525.
Klawans, H. L. (1972) Personal communication.
Lewy, F. H. (1938) The role of cervical nerves in facial sensations and the quantitative disturbance of sensitivity in major trigeminal neuralgia. *American Journal of the Medical Sciences*, **196**, 564.
List, C. F. & Peet, M. M. (1939) Sweat secretion in man. V. Disturbances of sweat secretion with lesions of the pons, medulla and cervical portion of cord. *Archives of Neurology & Psychiatry*, **42**, 1098.
Love, J. G. & Olafson, R. A. (1966) Syringomyelia: a look at surgical therapy. *Journal of Neurosurgery*, **24**, 714.
Martin, C. & Maury, M. (1964) Syndrome syringomyélique après paraplégie traumatique. *La Presse Medicale*, **48**, 2839.
Marvin, S. L. (1960) Arachnoiditis. In *Proceedings of 9th Annual Clinical Spinal Cord Injury Conference*, Long Beach, California, p. 50.
Mayfield, F. H. (1955) The role of the cervical roots as a source of head pain. *Clinical Neurosurgery*, **3**, 83.
McKenzie, K. G. (1955) Trigeminal tractotomy. *Clinical Neurosurgery*, Vol. 2, pp. 50–70. Baltimore: Williams & Wilkins.
Nassar, S. I., Correll, J. W. & Housepian, E. M. (1968) Intramedullary cystic lesions of the conus medullaris. *Journal of Neurology, Neurosurgery & Psychiatry*, **31**, 106.
Newman, M. J. (1972) Personal communication.
Nurick, S., Russell, J. A. & Deck, M. D. (1970) Cystic degeneration of the spinal cord following spinal cord injury. *Brain*, **93**, 211.
Perot, P., Feindel, W. & Lloyd-Smith, D. (1966) Hematomyelia as a complication of syringomyelia: Gowers' syringal hemorrhage. Case report. *Journal of Neurosurgery*, **25**, 447.
Rewcastle, N. B. (1972) Personal communication.
Riffat, G. & Domenach, J. (1964) Syndrome syringomyélique succédant à un traumatisme médullaire ancien. *Le Journal de Médécine de Lyon*, **212**, 1043.
Rossier, A. B., Werner, A., Wildi, E. & Berney, J. (1968) Contribution to the study of late cervical syringomyelic syndromes after dorsal or lumbar traumatic paraplegia. *Journal of Neurology, Neurosurgery & Psychiatry*, **31**, 99.
Schmitt, J., Laxenaire, M., Barrucand, D. & Duprez, A. (1965) Paraplégie traumatique—syndrome syringomyélique unilatéral sus-jacent d'apparition retardée. *Annales Medicales de Nancy*, **28**, 760.
Schott, B., Trillet, M., Vauterin, C. & Koshbin. (1962) Syndromes syringomyéliques tardifs suslésionnels après traumatisme médullaire. *Revue Neurologique*, **106**, 751.
Sherrington, C. S. (1893) IX. Experiments in examination of the peripheral distribution of the fibres of the posterior roots of some spinal nerves. *Philosophical Transactions of the Royal Society*, **184B**, 641.

Sherrington, C. S. (1898) II. Experiments in examination of the peripheral distribution of the fibres of the posterior roots of some spinal nerves. *Philosophical Transactions of the Royal Society*, **190B**, 45.

Strong, O. S. (1919) A case of sacral cord injury and a subsequent unilateral syringomyelia. *Neurological Bulletin (Columbia University, New York)*, **2**, 277.

Strümpell, A. (1880) Beiträge zur Pathologie des Rückenmarks. *Archive für Psychiatrie und Nervenkrankheiten*, **10**, 676.

Tjaden, R. J., Ethier, R., Vezina, J. L. & Melancon, D. (1969) Iodoventriculography in hydromyelia. *Journal of the Canadian Association of Radiologists*, **20**, 265.

Tucker, H. H. (1972) Personal communication.

Watson, N. (1972) Personal communication.

Weisenburg, T. H. (1907) Sensory and motor disturbances in parts above the distribution involved by definite organic lesions of the spinal cord. *Journal of Nervous & Mental Diseases*, **34**, 434.

Werner, A., Rossier, A., Berney, J. & Zdojewski, B. (1969) A propos de quatre observations de syringomyélie cervicale tardive aprés traumatisme medullaire. *Schweizer Archiv für Neurologie, Neurochirurgie und Psychiatrie*, **104**, 77.

Westberg, G. (1966) Gas myelography and percutaneous puncture in the diagnosis of spinal cord cysts. *Acta Radiologica*, Supplement, **252**, 5.

Williams, B. & Turner, E. (1971) Communicating syringomyelia presenting immediately after trauma. A case description and some theoretical concepts. *Acta Neurochirurica*, **24**, 97.

Wilson, S. A. K. (1954) *Neurology*, 2nd edn., London: Butterworth.

Zander, R. (1897) Beiträge zur Kenntnis der Hautnerven des Kopfes. *Anatomische Hefte*, **9**, 1.

Zdrojewski, B., Werner, A. & Rossier, A. (1969) Syndrome syringomyélique cervical tardif aprés paraplégie traumatique dorsale. *Neuro-Chirurgie, Paris*, **15**, No. 2, 153.

CHAPTER ELEVEN

Nature, Prognosis and Management of Post-traumatic Syringomyelia

H. J. M. BARNETT and A. T. JOUSSE

NATURE OF THE LESION IN PROGRESSIVE MYELOPATHY

When paraplegic patients began to be observed with a delayed extension of their initial disability, the nature of the process was uncertain. Cord cavitation was suspected and by continued observation there is now unequivocal evidence for the existence of a cervical cord syrinx in eight of the 17 cases of our series. In the 56 other acceptable examples of this condition reported in the literature or personally communicated to the authors, 27 have been shown by post-mortem (6) or surgical exploration (21) to have spinal cord cavitation (see Table 10.1). The data accumulated as to the cystic nature of the extending lesion in the authors' series may be summarised as follows:

Cystic Cavitation from Traumatic Level to Cervical Region

The first definite evidence in this group of cases of cavitation extending from approximately the level of injury (L1) to the cervical region was in case 8. Contrast medium was introduced at operation into the cystic cavity in the cervical cord and was followed upwards to the level of C1 and downwards as far as T11 (see Figs. 10.12 and 10.13). Equally unequivocal evidence of an intramedullary dilatation extending up from the level of the original trauma to the top of the cervical cord is available from the detailed post-mortem study carried out in case 2. A syrinx extended from the level of trauma at the spinal cord segment L5 up to, but not above, the uppermost extent of the cervical cord. It was not connected at any level with the central canal. A smaller cyst extended from L5 to L3 (see Chapter 14 for pathological details, also Fig. 14.1).

Extension of the cavity from the original site of trauma up to the cervical region has been confirmed in two instances by the passage of a small catheter within the spinal cord. In one case (case 6), laminectomy was done at T4 and, as the patient was paraplegic below this level, the traumatised segment of the cord was totally excised. Contrast medium could not be seen for more than a few millimetres above the area of operative incision into the cyst (Fig. 11.1). However, a small catheter could be passed without difficulty up to the cervical level within the substance of the expanded cord. In the other case (case 11), two cavities were identified at the site of original trauma. One of them was small, but the other large enough to admit a small catheter which passed easily from the operative site (T7) up to the lower cervical region. No contrast medium injection was attempted in this case.

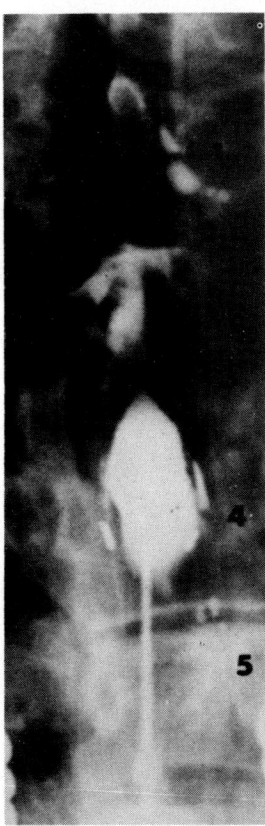

Fig. 11.1. Case 6, G.S. Old fracture-dislocation T4. Laminectomy at T4 with contrast medium injected into cyst expanding cord at level of original trauma. Contrast material would not pass beyond this local level. A catheter was inserted and passed to cervical level.

Cervical Cystic Cavitation but Connection to Traumatic Level Unproven

The one instance in which this situation obtained (case 12), had a rather stepwise development of his upper limb signs. A lumbar myelogram indicated a block at L1 (the original injury level), and a cisternal myelogram revealed cord

156 NON-COMMUNICATING SYRINGOMYELIA

dilatation (Fig. 11.2) in the cervical region. This was drained by a Silastic tube inserted at a C6 to 7 laminectomy (Dr Wm. Lougheed). Myodil would not advance up or down from the site and, although a catheter was slipped up the cord at the time of the surgical procedure, no attempt was made to pass it down to the lower cord. We are left not knowing whether this cervical cavitation communicates with the lower levels of the cord at the traumatic site. All indications from the other cases suggest that it does.

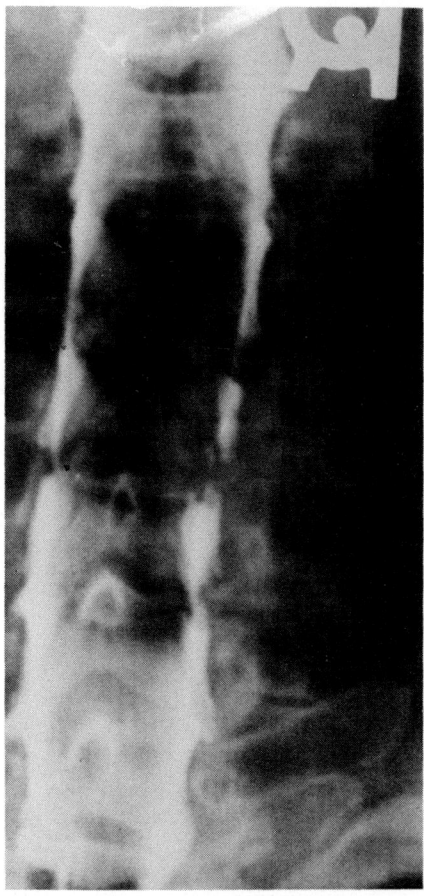

Fig. 11.2. Case 12, B.J. Original injury L1. Cisternal myelogram with widened cervical cord shadow. Widening visualised from C5 to T3, and cavity drained at C6, C7 laminectomy.

Cystic Cavitation at Traumatic Level and Cervical Cord Myelographically Expanded

In one instance at operation a cyst was identified at the site of the original trauma, and efforts to catheterise or to follow injected contrast medium beyond the local segment were unsuccessful. In this instance (case 13), we had unequivocal myelographic evidence of a cord expansion in the cervical region. Possibly the cyst was double, and the one identified the smaller of the two, or possibly the cavity was segmented (as in case 14, Fig. 11.13), and the narrow portions held up the catheter or contrast medium. Our post-mortem material indicates that a continuous cavity may be small in the lower part of the cord and larger in the high thoracic and cervical area. Another alternative might be that the cervical cavity was not connected to the cavity at the traumatic level. The latter suggestion is given some plausibility by the study of the acute spinal injury cases. An interrupted lesion of haemorrhagic necrosis may expand above and/or below a traumatic disruption of the cord but may taper off to a very minute lesion, only to expand beyond this to a larger area of necrotic tissue (see Fig. 14.17). If these necrotic lesions are the precursors to the cavitation, as we believe them to be, a theoretical possibility exists that there may be discontinuity between two separate cavities, both a sequel to the original trauma. An important consideration in the case in question, however, must be that the drainage of the cyst at the site of initial trauma produced definite improvement in the upper limb signs and symptoms—indirect, but quite convincing, evidence of the extension of the cavitation to the cervical segment.

Cystic Cavitation at Traumatic Level, Cervical Cord not Myelographically Expanded

In one instance (case 15), a cord cavitation was drained at the site of the original trauma (T12). A preoperative myelogram failed to demonstrate a cervical cord expansion and no attempt at catheterization was carried out. However, the patient's upper limb signs stopped progressing postoperatively. It was reasonable, therefore, to presume that these upper limb changes resulted from a cavitation communicating with that drained in the lowest cervical region.

Cystic Cavitation Extending Both Up and Down from the Level of Trauma

The extension of a cyst, both upwards and downwards from the level of the original trauma, was demonstrated by percutaneous injection, under fluoroscopic control, of 2 ml of contrast medium into a cyst within the substance of the spinal cord (case 14, Figs. 10.14 and 10.15). The patient, rendered quadriplegic by a fracture dislocation at C5 and 6, had an extension upwards of symptoms to the uppermost cervical level. The contrast medium, Ethiodan, was visible up to the highest cervical level, and down as low as T11. The cavity was irregularly interrupted by many indentations in its upper expanded area, giving it a segmented appearance.

Unproven but Suspected of Cavitation

In one case the cystic nature of the progressive lesion was not proven, even at operation. The operative intervention in this case (case 4) preceded arousal of our suspicion of the existence of this syndrome; the patient was submitted to surgery (1961) with a suspicion of an intramedullary tumour. Six other cases remain unproven, since mild clinical symptoms and normal myelograms have not justified surgical exploration (cases 1, 5, 7, 10, 16 and 17). One has declined surgery, even though he has an expanded cervical cord and a serious progressive upper limb disability (case 3). Finally, one (case 9) has died without myelogram, surgical exploration or autopsy. He had reached a stage of quadriplegic helplessness in the years before this complication of traumatic cord injury was recognised.

In summary then, we have established by various methods in eight of the 17 cases (Table 11.1) that the spreading signs and symptoms are due to a cyst extending from the site of original trauma up to the cervical region. Confirmation in the remaining cases has not yet been achieved. There would seem to be little doubt that the pathogenesis and pathological process is identical in all such cases.

Table 11.1. *Proof of cystic nature of cervical lesion (17 cases)*

Intramedullary myelography	2
Intramedullary catheter from site of trauma	2
Cyst drainage at site of trauma with symptomatic improvement	2
Cyst drainage in cervical region	1
Postmortem	1
Not yet proven to be cystic	9

PROGNOSIS IN PROGRESSIVE CYSTIC MYELOPATHY

Sufficient time has elapsed since recognition of this complication of spinal injury for us to assess the course followed over many years. Our experience is divided between patients for whom no operation has been done to alleviate their new symptoms (nine cases), and those for whom some surgical procedure for the progressive lesion has been carried out (eight cases) (Table 11.2).

Table 11.2. *Prognosis of progressive post-traumatic cystic myelopathy*

I	Tolerating minor or moderate abnormality 7, 7, 9 and 9 years duration	4
	Unaware of disability	1
	Tolerating serious disability 19 years duration	1
II	Dead—direct result 11 and 16 years duration	2
	—indirect result (uraemia) 8 years (moderate disability)	1
	—unrelated cause 9 years duration—mild	1
III	Operated upon—for severity of disability	7
IV	Operated upon—for diagnostic reasons	1
	(Durations refer to delayed myelopathy symptoms)	

The Clinical Course in the Cases Treated without Operation
TOLERATING DISABILITY

Six patients have tolerated the disability and have not been operated on. In four instances (cases 5, 7, 10 and 16) they have been aware of symptoms of the condition for nine, nine, seven and seven years, respectively, but none of them has progressed beyond a moderate disability. One is troubled by distressing pain (case 10) and one by a Charcot's shoulder joint lesion (case 7, Fig. 10.2). Myelograms in three of these patients carried out 12 months, six months and three years, respectively, after the onset of the new symptoms, were all thought normal. One patient (case 17) is unaware of the new ascending lesion. Another (case 3) has a serious disability including bilateral upper limb wasting and weakness and a neuropathic arthropathy of his left elbow joint. He can still function usefully and is reluctant to accept surgical treatment at the present time. Slow progression of his disability suggests that he will probably require surgical therapy in the foreseeable future. Nevertheless, his 19-year history is an indication of the fact that the condition may progress in an indolent manner.

Natural Remission

Two of the patients have insisted on a definite spontaneous reduction of their new symptoms. One (case 5) has had a decided increase in power of his hand. One (case 3) was noted in 1960 to have a wasted and fasciculating tongue on the side of his new cervical and thoracic sensory affliction. This was no longer apparent on re-evaluation (1970). In this same patient, an area of his thorax and abdomen had become less densely hypoaesthetic between the 1960 and the 1969 examinations (see Fig. 10.7). In view of the finding in the second case of Werner et al (1969) and the surgical observation of Tucker (1972), of a possible connection developing between the ascending cyst and the fourth ventricle, it is interesting to speculate on the possibility of such an extension occurring in our case, with spontaneous evacuation of the cyst through a connection to the fourth ventricle. Unlike syringomyelia of the kind associated with lesions at the base of the brain and foramen magnum region, a normal outflow of C.S.F. through the foramina of the fourth ventricle would be expected in these post-traumatic cases. Effective spontaneous drainage of the spinal cystic lesion might be accepted as theoretically possible. Once the new communication had been established, the pulsatile force acting on the ventricular fluid would simply drive the cerebrospinal fluid out through the patent and normal openings rather than distending the cavity communicating with the ventricles, as is believed to occur in communicating syringomyelia, or the 'syringohydromyelic syndrome' of Lichtenstein (1943), Gardner (1965) and of Appleby et al (1969).

Death

Four patients are dead. Two of them (cases 2 and 9) died almost totally quadriplegic and helpless after 11 and 16 years disability, respectively. One patient (case 1) died of an incidental cause and had been carrying a minor affliction from the complication of his paraplegia for nine years. One (case 15) died of uraemia from urinary tract infection.

SURGICAL THERAPY

Eight of the authors' patients have been operated on. The operations were carried out by Drs E. H. Botterell, T. P. Morley, Wm. Lougheed and H. W. K. Barr.

Indications for Surgery

One of the eight was operated on before we recognised the nature of this condition. A cervical laminectomy was carried out (1961) and the cord was not considered to be abnormal (case 4). The patient's disability was minor and had been progressing for only three months preoperatively. For eight years his lesion was reasonably stationary. But now, after 11 years, he has progressed to a moderate amount of additional disability, including a neuropathic shoulder joint, involvement of both upper limbs and a patchy sensory loss in the trigeminal territory. He will probably require further surgical therapy.

The other seven patients were all operated on because they were developing moderately severe, or severe disabilities (cases 6, 8, 11, 12, 13, 14 and 15). Follow-up extended from one to six years. The two situations demanding surgery were, firstly, the development of a serious upper limb weakness which in three instances was bilateral, and, secondly, varying amounts of severe pain. Pain afflicted all patients in the operative series. The surgical procedures varied and the primary consideration in deciding on the operative procedure was the completeness or otherwise of the original spinal cord injury.

Type of Operative Procedure
PARTIAL SPINAL INJURY

In the three cases in our series in whom the initial injury left them with a partial lesion, cyst drainage was carried out in the cervical enlargement (two cases) or at the site of original trauma (one case) (Table 11.3). In one case, cervical drainage was carried out twice (Fig. 11.3a), and on the second occasion a Silastic tube inserted into the cyst (Fig. 11.3b). In the second case, a Silastic catheter was inserted into the cervical cavitation at the initial procedure, and after five years the patient retains his improvement and has not shown neurological deterioration. In the third case (case 15), the operation was carried out at the site of original trauma and a drain was inserted at this level into the spinal cord cyst and carried down to the lumbar subarachnoid space. Progression was arrested for the remaining five months of the patient's life and his pain was considerably diminished. He died of uraemia. If he had survived longer it might have evolved that his improvement was temporary and that an insertion of the catheter *above* the arachnoiditis would have been preferable in the long run.

Table 11.3. Surgical therapy—'partial' paraplegia group

Case	Level of injury	Level of procedure	Type of procedure	Length of follow-up	Results
8	L1	Procedure 1 C2–5	Simple drainage	3 years	Initial arrest of progression, then decline
		Procedure 2 C2–5	Catheter inserted	5 years	Strength and pain improved, no further decline
12	L1	C6–T1	Catheter inserted	5 years	Power and some sensory recovery maintained
15	T12–L1	T9–11	Catheter inserted	5 months	Pain improved, no worsening; died of uremia

COMPLETE SPINAL INJURY

Four of the cases submitted to operation had complete loss of cord function below the level of initial injury (Table 11.4). In three, transection and reaction of a segment of the cord was carried out (Fig. 11.3c). One patient specifically requested that his cord not be transected so that only a partial section was done. A Silastic tube was inserted into the proximal cut end of the cord in two of the three cases in which complete transection was done (Fig. 11.3d). These procedures were

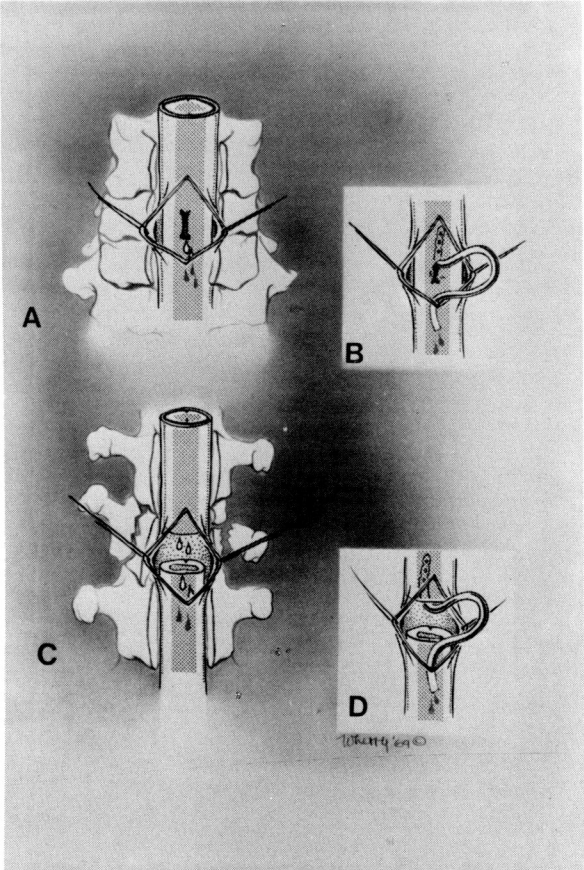

Fig. 11.3. Operative approaches. A—simple drainage in cervical enlargement. B—Catheter drainage in cervical enlargement or in general area of original trauma. C—Transection of cord at site of original trauma. D—Transection of cord with intra cystic catheter.

carried out at the level of the old injury with the exception of the one case in whom the original trauma was in the cervical region, and then the resection was carried out below the level of the old injury (case 14). In the cases operated on at the site of the original injury, there was a very significant amount of meningeal reaction which required dissection. After such dissection, two cases were then submitted to cord transection and resection of the portion of the grossly damaged part of the cord. In one case (case 14), there was half-inch retraction of both cut ends of the transected cord. This phenomenon of retraction was not observed in the other case transected at the damaged site (case 11) nor was it observed in the case (case 6) in whom transaction was below the site of original trauma. Using the same technique of freeing the adherent cord and then sectioning it, a similar retraction of cord tissue was observed in a patient afflicted with cavitation related to non-traumatic arachnoiditis and will be further discussed in Chapter 15.

NON-COMMUNICATING SYRINGOMYELIA

Table 11.4. *Surgical therapy—'complete' paraplegia group*

Case	Level of injury	Level of procedure	Type of procedure	Length of follow-up	Results
6	T4–5	Procedure 1 T1–5	Cord transection	2 years	Initial improvement, trauma caused decline
		Procedure 2 T1–5	Cord transection Catheter insertion	2 years	Sustained improvement (pain and weakness)
11	T6–7	T5–8	Cord transection Catheter insertion	5 years	Marked, sustained improvement (weakness and pain)
13	T11	T9–10	Partial transection Catheter insertion	4 years	Considerable, sustained improvement (weakness and pain) after brief delayed worsening at 7 weeks
14	C5–6 [a]	T3–4	Cord transection (cut-end retraction)	5 years	Power and sensation improved, pain gone; sustained

[a] Quadriplegic.

Surgical Results and Recommendations

Experience has taught caution in anticipating results from surgery in any form of syringomyelia. Nevertheless, a satisfactory result from operation was obtained in every instance in this post-traumatic group. A complete or reasonable relief of pain was effected, progression of the neurological deficit was arrested and a moderate or substantial return of power and sensation was witnessed in the newly affected upper limb or limbs. This recovery or alleviation of progression has been preserved for a satisfactory period of time ranging from two to six years. In one patient, as mentioned, it was temporary on the first occasion, but was more satisfactorily achieved on the second occasion. In one case (case 13), a curious and abrupt exacerbation occurred in the convalescent period after an initially successful relief of pain and some reduction of motor and sensory abnormalities. A little more than seven weeks after a drain had been put into his spinal cord cavity at laminectomy, he was precipitously readmitted because of sudden worsening. A repeat myelogram did not indicate anything other than the same appearance as before the operation—a dilated cord shadow between T1 and C3. He was watched for a few days and slowly regained function. It seemed possible that effective drainage was established, became suddenly obstructed and then rather quickly was re-established—possibly aided by the change in dynamics effected by the repeat myelography. Alternatively, the patient may have had bleeding into the syrinx which again drained away effectively and spontaneously. He has retained his subsequent improvement for three years.

The indications for surgery appear to be the development of distressing symptoms (pain), a neuropathic joint, or significant motor disability coming as a progressing sequel to an underlying spinal cord injury. Our experience with these cases, as well as with those who have syringomyelia with arachnoiditis, leads us to conclude that the presence of a normal spinal cord size in the myelogram should not by itself lead to withholding of operation. Cord transection is not possible in the partially paraplegic individual but instead a tube inserted into the cavity at, or just below, the level of original trauma may be the surgical treatment of choice (Fig. 11.3b). The insertion of the distal end of the catheter into normal subarachnoid space will usually demand that this be above the site of the original injury. Operation at the site of cervical enlargement without insertion of a drain seems to be the least desirable (Fig. 11.3a). Cord transection at, or below, the level of injury and insertion of a small drainage tube into the cavity in the cephalad stump of the transected cord appears to be a reasonable method of treatment in the totally paraplegic individual (Figs. 11.3c and 11.3d).

Operative procedures, such as plugging of the obex, designed to deal with a communicating type of syringomyelia should not be carried out as the initial procedure in this type of cavitation. In one of the two cases reported by Nurick, Russell and Deck (1970), an initial procedure at C7 to T1 identified a cavity containing only 55 mg/100 ml of protein. This low protein content was interpreted apparently as indicative of ventricular communication, as was the fact that jugular compression caused fluid to well up into the lower cavity. Accordingly, a second procedure was carried out and the opening of the central canal at the obex, 2 mm in diameter, was plugged with muscle. It is our opinion that this level of protein is by itself not an indication of such communication. Nevertheless, some patients may have this communication as described above and both a spinal and a posterior fossa procedure may be considered justified, as they were in this case and in the case reported to us by Tucker (1972). Iodoventriculography or intramedullary myelography might be considered prior to the posterior fossa approach and our intention in these cases is to give serious consideration to this procedure as a preliminary investigation in most, if not all, of them.

Finally, the unique experience of Allen (1972) has to be recalled in deciding on surgery for these cases. As mentioned in Chapter 10, he was able to demonstrate a direct communication with the spinal subarachnoid space and the syrinx at the level of original injury.* He demonstrated the appearance which had been the subject of a hypothetical sketch produced by Williams (1970). In any similar instance that can be identified, it is obvious that a procedure at, and just above the level of original injury, would be indicated. In the case of complete paraplegia, cord transection and drain insertion would be ideal. In a partial lesion, an intracavity drain brought out to the unobstructed subarachnoid space would likely be effective.

Summary of Nature, Prognosis and Management
1. Unequivocal evidence exists that the ascending myelopathy occurring as a sequel to traumatic spinal injury is due to progressive cavitation within the central spinal cord.
2. The progression of this cavitation may be indolent or much more rapidly and seriously progressive.
3. Serious disability may develop in the upper limbs before the cervical myelogram is unequivocally abnormal.
4. Consideration of a surgical approach to the condition is directed by several factors. It may be the development and continuation of spontaneous pain, or pain on straining, in the upper limbs or neck. It may be the development of a

---

* Both air and positive contrast injected in the lumbar region entered the cord at the level of injury, at the site of complete block and ascended within its substance. Both were identifiable within the third ventricle.

neuropathic joint in the upper limbs. It may be the development of significant weakness in one or both upper limbs and/or a worsening of the strength in the lower limbs in cases of partial injury.
5. The surgical approach will be determined by the original disability.
6. If the original lesion, excepting cervical injuries, led to a *complete permanent disruption* of spinal cord function, the best treatment for the new lesion is to drain the cavitation at, or just above, the level of original injury. This will be best accomplished by complete transection of the cord at this level and by the insertion and fixation of a Silastic tube into the cavity in the proximal stump of the resected spinal cord.
7. If the original injury led to a *partial disruption* of spinal cord function, the approach most beneficial will likely be at, and just above, the original injury site with the insertion of a Silastic tube into the cavity and its fixation into the normal subarachnoid space above the injury level.
8. Surgical results have been satisfactory in the seven cases in whom these methods were employed under the surveillance of the authors.

REFERENCES

Allen, P. (1972) Personal communication, paper in preparation.
Appleby, A., Bradley, W. G., Foster, J. B., Hankinson, J. & Hudgson, P. (1969) Syringomyelia due to chronic arachnoiditis at the foramen magnum. *Journal of Neurological Sciences*, **8**, 451.
Gardner, W. J. (1965) Hydrodynamic mechanism of syringomyelia: its relationship to myelocele. *Journal of Neurology, Neurosurgery & Psychiatry*, **28**, 247.
Lichtenstein, B. W. (1943) Cervical syringomyelia and syringomyelia-like states associated with Arnold–Chiari deformity and platyblasia. *Archives of Neurology & Psychiatry*, **49**, 881.
Nurick, S., Russell, J. A. & Deck, M. D. (1970) Cystic degeneration of the spinal cord following spinal cord injury. *Brain*, **93**, 211.
Tucker, H. H. (1972) Personal communication.
Werner, A., Rossier, A., Berney, J. & Zdrojewski, B. (1969) A propos de quatre observations de syringomyélie cervicale tardive après traumatisme medullaire. *Schweizer Archiv für Neurologie, Neurochirurgie und Psychiatrie*, **104**, 77.
Williams, B. (1970) The distending force in the production of communicating syringomyelia. *Lancet*, **ii**, 41.

CHAPTER TWELVE

Case Records of Authors' Series of Post-traumatic Syringomyelia

H. J. M. BARNETT and A. T. JOUSSE

The previous two chapters discuss our experience with the post-traumatic cystic myelopathy complicating serious spinal cord injury. An extensive review of the literature and personal communication with interested individuals in various centres in Canada, the United States, Britain and Australia has determined that, to date, 73 cases have been recognised with this condition. Since many of these cases were incompletely documented and were followed often for brief periods, it seemed worthwhile to summarise the data of our 17 cases, accumulated and followed over a 10-year period. Some details of cases 1 to 8 have already been published (Barnett et al, 1966) so that they are given in shorter summary, but the details of their longer follow-up is recorded.

Summary of Original Eight Cases with Follow-up Data

CASE 1, M.B.
At the age of 44 years, in 1951, this man suffered *complete* paraplegia from a T9 fracture. Unilateral upper limb pain, followed by numbness and areflexia, but not accompanied by weakness, developed in 1952. With no major progression this condition existed until death in 1962 from unrelated causes.

CASE 2, M.B.
In 1941, a woman, aged 43 years, sustained a fracture dislocation of L1 to 2 with *partial* paraplegia. Eventually she walked with braces and crutches. Examination 12 years later showed sensory and motor abnormalities of a lumbar cord lesion but no neurological abnormalities above this level. In 1955, 14 years after the trauma, paraesthesiae developed in the left hand, with reduced pain sensation in the left arm. Gradually the left arm, and later the right arm, became weak and the weakness in her legs became more marked so that she was no longer able to get around on crutches. In 1960, five years after the onset of her arm symptoms, she had a neuropathic joint in the left shoulder. Myelogram in Toronto showed an expanded cervical cord.

In 1963 and 1965, she was admitted to the National Hospital, Queen Square, London, England, under the care of Dr M. J. McArdle. On the later examination there was marked weakness of both legs

166   NON-COMMUNICATING SYRINGOMYELIA

with complete sensory loss below the L4 level, and weakness of both arms, left more than right. Vibration and position sense were absent on the left side and in the right leg, but normal in the right arm. Pain and temperature sensation were reduced over dermatomes C3 to T6 on the left, and over C5 on the right (Fig. 12.1). The patient became bedridden and virtually helpless and died in Barnet Hospital, Hertfordshire, in December 1966. Post-mortem was carried out and is reported in Chapter 14.

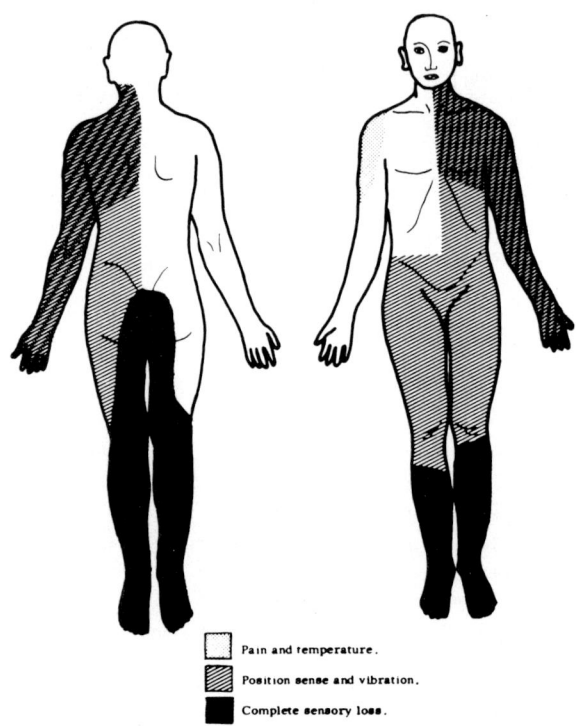

Fig. 12.1. Case 2, M.B. Gross sensory loss in left upper thoracic and cervical segments. Slight loss over right C5 dermatome. Original injury at L1–2, 1941.

CASE 3, J.H.

In 1951, at the age of 31 years, this man developed *partial* paraplegia, with an L1 fracture. Unilateral neck and arm pain developed in 1953. Bilateral limb pain, wasting, weakness, loss of pain and temperature sensation and peripheral facial numbness continued to progress from 1960. A neuropathic left elbow joint lesion developed in 1963. Curiously, by 1969, the patient reported spontaneous improvement in sensation over the left thoracic segments, T2 to 12 (see Fig. 10.7). An earlier tendency to profuse left trunk and upper limb sweating was no longer present. A partial left Horner's syndrome had now developed. An earlier wasting and fasciculation of the left half of the tongue was not evident. By December 1971, he was still working but with further deterioration in the power of both arms. Buttons and writing have been progressively difficult to manage.

CASE 4, R.M.

At the age of 44 years, in 1955, this man suffered *complete* paraplegia resulting from a T9 to 10 fracture dislocation. In 1961, he developed unilateral lack of awareness of temperature appreciation, followed by weakness and clumsiness of his right hand. The whole of the involved side began to perspire more freely. Cervical laminectomy in 1961 revealed no recognisable cord lesion. By 1969, he had further right upper limb pain and temperature and deep reflex loss with a neuropathic shoulder joint. By December 1971, the patient was aware of early left shoulder weakness. He is the least physically active individual in this series.

Case 5, D.P.

In 1949, a 0·22 calibre bullet wound, involving an L2 injury rendered this man, then aged 16 years, *completely* paraplegic save for preserved sexual potency. In 1962, pain gradually ascended unilaterally from abdomen to neck with ipsilateral excessive perspiration on the right side of the head and right upper limb. Minimal right hand weakness developed. In 1963, pain and temperature appreciation was unilaterally lost below C5 on the right and impaired from C4 up to, and including, the trigeminal territory. Deep pain was absent in this areflexic limb. Myelogram in 1963, and again in 1970, was normal in the cervical region. By 1969, the patient had minimal evidence of contralateral left sensory involvement but was convinced that his right hand and arm had regained their original strength. In December 1971, he judged himself to be weaker in his arms but on re-examination, the strength, bulk and reflex function was unchanged. His pain and temperature loss now extended up to C2 on the right and up to T2 on the left.

Case 6, G.S.

In 1959, at the age of 22 years, this man was rendered *completely* paraplegic from a T4 to 5 fracture dislocation. In 1961, he developed pain at the old fracture site along with unilateral left upper limb pain and sensory changes with ipsilateral reflex diminution. By 1965, constant burning pain in the left hand was requiring codeine. Severe pain in the left shoulder was precipitated by trunk extensor spasms. By 1967, additional pain developed in the contralateral arm and shoulder, along with minimal weakness of his left hand. He could not grip the wheel of his chair tightly enough to go up a ramp. He was distressed by excessive spontaneous right upper trunk, upper limb and neck sweating of about one year's duration (on the side *opposite* to his major signs and symptoms). There was sweating on the forehead of the left side, aggravated by eating. Myelograms in 1964 and 1967 were normal. In January 1968, a third myelogram indicated slight but definite enlargement of the cervical cord behind the fourth, fifth, sixth and possibly the seventh cervical vertebrae.

On 18 January, 1968, laminectomy from T1 to 5 was carried out (Wm. Lougheed). Adherent arachnoid was dissected away, during which procedure a large cyst inside the cord at the T3 to 4 level was opened. The fluid within the cavity was clear and watery, but was mixed with subarachnoid fluid before a sample was obtained. The spinal cord, with its contained cyst, were transected at T4. At this level, the cyst extended almost the full diameter of the spinal cord leaving only 1 to 2 mm of cord tissue surrounding it. Contrast medium injected into the cyst (see Fig. 11.1) could not be induced to run cephalad, and subsequent pictures indicated that the contrast medium had drained from the cord. A small catheter could be passed within the cyst and made to pass up into the cervical region. By July 1969, the patient had lost all but a minimum of pain. Excessive perspiration stopped. It had been requiring two or three daily changes of his shirt. His grip had become normal although he was still unable to adduct the fifth finger.

In April 1970, he fell from a chair and suffered a fractured femur. Pain and weakness returned to his left hand accompanied by further wasting. In August 1970, he was readmitted and further laminectomy carried out at the site of the earlier procedure at T1 to 5. The exposed spinal cord was yellow and atrophic. A small cyst in the cord was aspirated. The fluid was clear with a protein content of 38 mg/100 ml. The cord was transected at the site of the old injury, a Teflon tube inserted into the cyst and the fluid allowed to drain into the subarachnoid space. The catheter was sewn into position and left there. In December 1971, the patient had very substantially recovered strength in the left hand without a further sensory decline. Left posterior head and posterior neck pain was episodic and severe, but the upper limb pain was less troublesome. Left forehead sweating was still apparent with eating and his right neck, upper limbs and upper trunk sweating was now associated with right lower trunk and right lower limb sweating.

Case 7, R.S.

In 1955, at the age of 24 years, this man suffered a T12 to L1 fracture dislocation produced *complete* paraplegia with sensory level at T11. By 1962, he developed left chest pain on sneezing, followed by left upper limb and occipital pain. Cisternal myelogram within six months of these new symptoms was judged to be normal in the cervical region, with an irregular block at T10 and complete block at T12.

The painful area was succeeded by numbness and, in 1963, a razor felt less sharp on the left (see Fig. 10.6); the involved limb was anaesthetic to deep and superficial pain, to temperature and was areflexic but strong. In July 1969, the patient's left shoulder was slightly swollen. Minor changes were visible in the X-ray. By January 1970, a more marked painful swelling developed in this shoulder with radiological features of an advanced neuropathic joint lesion. A repeat cisternal myelogram was still normal in the cervical region. By December 1971, the patient had decreased function in the left arm at the elbow and shoulder and had lost all deep reflexes in the right upper limb. A neuropathic left shoulder joint dissolution made shoulder power assessment difficult (see Fig. 10.2). Trigeminal sensory loss was encroaching further into the central face area.

CASE 8, R.W.

At the age of 25 years, in 1949, this man was rendered *partially* paraplegic from an L1 compression fracture. In 1951, he developed pain progressing from the chest into the upper limb and neck on one side, followed by numbness. In 1953, bilateral cervical sensory loss and unilateral facial sensory disturbance was detected, but a myelogram was normal in the cervical region. From 1958 to 1963, gradual bilateral upper limb weakness and wasting developed (see Figs. 10.4 and 10.5). By 1963, he was aware of loss of proper appreciation of feeling over both sides of the face on shaving and was found to have bilateral diminished appreciation of pain, temperature and tickle, sparing the central part of his face and involving second and third divisions more than the first (see Fig. 10.6). Myelography disclosed a complete hold-up of contrast material opposite L1, but cisternal myelogram revealed a widening in the cervical region which extended from C3 at the top to T2 below, (see Fig. 10.11). In 1964, laminectomy in the cervical region revealed a cyst, and simple drainage was carried out from C2 to 5 (Dr T. P. Morley). Cessation of progression lasted several months.

Further progression of upper limb weakness and aching in arms led to a second exploration in the cervical region in 1967. A bluish cystic lesion was apparent between the posterior columns, and a fine needle removed clear fluid; 1·5 ml of Myodil injected into the cyst extended to the uppermost cervical subarachnoid space (see Fig. 10.12) and down to T11 (see Fig. 10.13) within the spinal cord. A Silastic tube was inserted into the cyst and fixed with two small clips to the surrounding arachnoid. Pain diminished and the patient felt improvement in facial sensation when shaving. After five years, in February 1972, he claims his condition is stationary. He is regularly employed and has not had any further deterioration since his second laminectomy in 1967. The worst wasting and weakness is in his shoulders, so that abduction, forward flexion and external rotation are almost absent bilaterally. There is still detectable loss of pain sensation over the periphery of all three trigeminal divisions.

Details of Further Cases

CASE 9, W.R.

In 1924, at the age of 19 years, this man suffered a T12 fracture dislocation with *partial* paraplegia. He was able to walk with a spastic gait. In 1945, he became aware of diminished appreciation of temperature in his right arm, and subsequently in his left. Within five years he developed weakness of both arms, and by 1960, was unable to feed himself, although he could walk with diminished facility. He had bilateral sensory loss to pain and temperature in his cervical and upper thoracic segments, no reflexes in his upper limbs, and only a flicker of extension in the left shoulder and left thumb and in the right trapezius and right supinator. All other upper limb muscles were totally paralysed, flaccid and wasted. He died in 1961 from chronic pyelonephritis and uraemia. Post-mortem was not carried out.

CASE 10, J.McC.

At the age of 22 years, in 1956, this man was rendered *partially*, but severely, paraplegic due to a T11 to 12 fracture dislocation. Complete loss of sensation was observed below T12 on the left, and below S1 on the right, with marked reduction on the right from T12 to S1. He had totally paralysed legs, with normal abdominal reflexes, brisk knee and ankle jerks, more active tone and reflex activity on the left than the right, and extensor plantar responses. Neither myelography nor laminectomy was pursued in the immediate post-traumatic period.

In 1964, he experienced pain in his left chest, shoulder and down into his left arm to the elbow, brought on by coughing and sneezing. After two or three months, pain became a minor spontaneous experience in the left shoulder, and he had a sensation of pins and needles in the arm down to the mid-forearm. In August 1965, examination disclosed left-sided pain and temperature loss up to the C2-trigeminal junction and down to the mid-abdominal area (see Fig. 10.1), loss of left-sided abdominal reflexes, decreased left upper limb reflexes, but no strength reduction. By 1966, the patient was periodically aware of loss of power in the two medial fingers of the left hand and at the left elbow, wrist and fingers there was minimal, but detectably reduced strength. Over the first and second divisions of the left trigeminal territory there was diminished, but not absent, pain and temperature appreciation. Lumbar myelogram indicated slight delay at T11 but no major obstruction, and the cervical area was judged normal. The C.S.F. protein content was 56 mg/100 ml.

Beginning in 1967 and continuing to the last follow-up in July 1969, repeated bouts of coughing precipitated hours of pain in his left upper, anterior chest and posterior shoulder, the top of the shoulder, and down to the left elbow. Minimal strength reduction and reflex loss were detected in the left upper limb. Over the peripheral area supplied by the three trigeminal divisions, diminished pain and temperature awareness were detected (see Fig. 10.1). Excessive sweating frequently enveloped his left face, head, neck, upper limbs and trunk down to the umbilicus. Excitement and bowel distention triggered this remarkable phenomenon. Heat or exercise were not necessary to precipitate it. Repeat myelogram continued to be normal in the cervical area, approximately three years after the onset of the new symptoms. He depends on narcotics for the control of pain and its severity would justify operative treatment. He is not a good subject for surgery but should further weakness develop it will be mandatory.

CASE 11, G.S.

In 1957, at the age of 47 years, this woman was rendered *completely* paraplegic with fracture dislocation at T8. A laminectomy was performed, and within a year she returned to teaching from a wheelchair despite a complete motor and sensory paraplegia below T6 to 7. Early in 1966, she became aware of some aching pain in the right scapula, right arm and hand. In September 1966, weakness and, within a few weeks, numbness was noted in this right hand. Right arm and hand weakness slowly progressed and small muscle wasting developed. Writing and blackboard writing became increasingly difficult and her livelihood was threatened. The area of numbness gradually ascended the arm into the shoulder and then into the neck and up to the angle of the jaw.

Myelography was performed in December 1966 at the Winnipeg General Hospital, and a moderate enlargement of the upper thoracic and cervical cord was detected. In July 1967, the patient was admitted to the Toronto General Hospital. She had wasting of the small muscles of her right hand, considerable weakness of grip, moderate weakness in right wrist flexion and extension but, above this and on the left side, strength was normal. Sensory examination indicated that pinprick sensation was absent over the entire right upper extremity, except for slight preservation in the palm of the hand, and the distal portion of the dorsum of the fingers. It was absent up to the lowest occipital level but was present to a minimal, but detectable amount, over the chest and dermatomes T2 to 6. There was loss of tickle and temperature appreciation from the level of the old deficit up to, and including, the angle of the jaw and the lowest occipital area on the right. Sensation was entirely intact on the left side above the level of the old lesion. Deep reflexes were normal in the left, but absent in the right, upper limb.

On 17 July, 1967, Dr T. P. Morley operated at the site of the previous laminectomy and the site of her old injury. At the level of the lesion (T8) and extending over a distance of 4 cm, there was 'quite severe arachnoid inflammatory reaction attaching the cord on all its aspects to the dura. Above this the cord was completely free, but obviously filled with a large cavity of pulsating fluid'. The surgeon was content that these pulsations were not coming from the surrounding arachnoid space. The adherent, injured cord was mobilised completely so that it could be elevated, and it was then cut across. The cephalad portion was lifted from its bed, and 1 in of the cord resected. The cord was not under longitudinal tension and did not retract, but it was noted that there was no possibility of mobility of this adherent cord. No visible communication was present between the subarachnoid space and this cord cavity. Two cavities were visible in the proximal cut end of the cord. One was small and not communicating with the larger cavity. Both were yellowish-grey, and semi-translucent, 'tending to have a wash-leather appearance'. A catheter was passed up the larger cavity until it stopped at the lower cervical region. A Silastic tube, into which had been cut some side holes, was passed up for 2 in into this cavity. The protruding end of the tube was cut down the middle and each half clipped to the cut tip of the cord, to allow central cavity drainage.

Postoperatively, and indeed while still in the recovery room, the patient noted that her hand was stronger. Within 10 days she could use her right hand to wind her watch, which had not been possible for several months. She was free of all pain. Two months later bulk was returning to the small muscles of the hand and gave very useful power in this hand. The only significant weakness was impairment of abduction of the little finger. Her sensory loss at its upper level was less abrupt, and the level was down to the base of the neck, instead of the angle of the jaw. Almost five years later, she maintains neurological improvement, has minimal pain, no significant weakness of her hand and arm, no loss of pain and temperature sensation above T2 and a return of biceps and brachioradialis reflexes. She has returned to teaching and can cope with writing and blackboard writing.

CASE 12, B.J.

At the age of 22 years, in 1949, this man suffered a fracture dislocation of L1 with *partial* paraplegia. Power in his lower limbs was confined to his inner hamstrings. Superficial sensation was absent below the knees, with diminished proprioception in the feet. At the knee, reflexes were absent, at the ankle, hyperactive. A laminectomy was done at L1 and L2 with fusion from T11 to L4. He became ambulatory with braces and crutches. Early in 1959, he experienced aching pain in his lower abdomen, which lasted a few months and disappeared spontaneously. In October 1959, a slight decrease in pain and temperature appreciation in the right hand and right trunk was detected, with normal upper limb reflexes, normal left but absent right abdominal reflexes. He complained that he perspired excessively over the right head, neck, upper limb and trunk.

In 1965, he had a sudden violent sneeze, and developed acute, very severe pain behind his right ear, extending up over the occiput and down towards the neck. Over a few hours this eased off, but he was aware, as it left, of a gradual numbness involving the fingertips of his right hand. After that, he noted that he burned himself more readily in this extremity and his handwriting deteriorated. In the next six months, he was aware that a sneeze would produce a pain in the back of his neck, over both shoulders and up into his right face and temple. In early 1966, over a three-week period, he had a rapid progression of numbness involving the whole arm, the right neck, occiput and face, the right thorax and later down to below the umbilicus. He noted more weakness of the right arm. He could not write legibly and was no longer able to walk.

In September 1966, there was weakness in the intrinsic muscles of the right hand. Pinprick sensation was lost from C2 to T10 on the right, but normal on the left. There was hypoaesthesia in all divisions of the right trigeminal nerve. This involved the forehead down to the hairline, the preauricular region as far forward as the middle of the zygoma, and an area of the chin three-quarters of an inch over the bony prominence of the jaw. Position and vibration sense were now absent at the foot and ankle level. Right upper limb reflexes were absent while those on the left were normal. All abdominal and lower limb reflexes were absent including plantar responses. A lumbar myelogram revealed a complete block at L1. Cisternal myelogram indicated a widening of the cord shadow from C5 to T3 (see Fig. 11.2). C.S.F. from this cisternal procedure was normal, with protein 24 mg/100 ml and sugar of 60 mg/100 ml.

On 23 December, 1966, laminectomy (Dr Wm. Lougheed) at C6, C7 and T1 disclosed a dilated translucent cord from which a few ml of slightly xanthochromic fluid were removed by aspiration. A small incision was made into the cavity, and a Silastic tube inserted. Through the tube a few drops of Myodil were inserted, and would not advance although the tube could be moved up freely. The tube was left in the cavity and held in position with a silver clip. The only analysis performed on the fluid obtained from the cavity was a protein estimation, of 224 mg/100 ml. Within a few days, the patient remarked that the right side of his face was no longer the site of an abnormal sensation. Two months later, his hand strength had returned to normal, but sensation and reflexes were as before. Slight improvement in pain and temperature appreciation in the first and third trigeminal territories were noted. By December 1971, he continued to enjoy the initial improvement and no further decline had developed. The only complaint that he had of his right hand was weakness in adduction of the right index and middle fingers. Sensory loss now extended only to the C3–4 junction or two inches in from the tip of the shoulder. His left abdominal reflexes had returned and his ankle jerks had come back to normal, but his clonus and extensor plantar responses had not returned.

CASE 13, R.E.L.

At the age of 20 years, in September 1966, this man suffered a compression fracture at T11 with resultant permanent *complete* motor and sensory paraplegia. He learned to stand with braces and move with crutches, but for the most part used a wheelchair. In August 1967, some 11 months after his accident, he became aware of pain in his right axilla, resulting from straining, coughing and sneezing. A few weeks later this discomfort extended into the lateral aspect of his arm, later into the right side of his neck, and within two months had extended into the right occipital area. With the onset of this discomfort he noted progressive numbness extending from the right abdominal area, to the anterior chest and up to the axilla. Numbness then spread down the arm and by the end of the tenth week reached the angle of the jaw and the occiput.

When he was admitted to Sunnybrook Hospital, Toronto, in December 1967, he had intact motor power in his right upper limb. Within three weeks he had slight, but definite reduction in his grip, and weakness in the interossei of this hand. Pain, temperature and tickle sensation were absent on the right side up to the C2 trigeminal junction, and down to the level of his pre-existing sensory loss. All deep reflexes in this limb were absent, while those on the left were normal. There was very excessive perspiration on the right, from the level of his original injury up to the axilla. On 8 January, 1968, a myelogram demonstrated marked, but incomplete, obstruction at T11 and a spindle-shaped dilatation of the cord between T1 and C3. The spinal fluid obtained at this time revealed 52 mg/100 ml protein, 102 mg/100 ml sugar, and 121 mEq/l of chloride.

On 22 January, 1968, a T9 and 10 laminectomy was carried out (Dr Wm. Lougheed). Dense arachnoid adhesions were detected at this site, and when removed, a small spinal cord cyst was opened in the midline posteriorly at T10 and a rectangular incision effected to promote drainage. Catheter and contrast substance could be passed cephalad for only $\frac{1}{2}$ to $\frac{3}{4}$ cm in this cavity. The patient had expressly asked that his cord not be subjected to complete transection, so any attempt to find a further possible cavitation in the cord was not pursued.

Within six weeks his grip was back to normal. He was no longer aware of discomfort with coughing. Over the back of the forearm, the outer aspect of the arm and the region of the occiput, there was a return of appreciation of pinprick sensation; temperature appreciation returned from C2 to 6 inclusive. On 14 March, 1968, the patient experienced a sudden right upper limb pain, a decrease of his right had grip and, on 21 March, of his left hand grip. On 22 March, wrist extensor weakness was detected bilaterally, and he had now lost the left biceps and brachioradialis reflexes. A repeat myelogram was carried out, and an identical picture to the preoperative state identified. Gradually, over the next few days, his left, and then his right hand weakness disappeared and his left reflexes returned. Almost three years later, in December 1971, he retained his recovery with no further deterioration, had perfectly normal strength in his upper limbs bilaterally, normal left arm deep reflexes, and the right triceps reflex was obtainable for the first time. Sensory examination indicated that he had retained the recovery noted in the weeks after his operation. No further deterioration occurred in three years of follow-up.

CASE 14, D.L.

In August 1965, at the age of 17 years, this man became quadriplegic from a fracture dislocation of C5 on 6. With caliper traction, sensory function settled out at a C8 level, and although at first *totally* paralysed at and below the wrists, this eventually recovered to moderate strength in flexion and extension of the left wrist, and flexion of the right wrist. By November 1965, the upper limb reflexes were 2+ on the right, and 3+ on the left. In April 1969, three days after a minor blow on the back of his head, he became aware of posterior neck pain with spasms in his neck muscles. Left wrist power and bilateral elbow extension had become enfeebled and there was a decline in right wrist flexion, left elbow flexion and bilateral shoulder extension. Sensation was lost to pain and temperature over the entire left upper limb to the level of the C2 dermatome, and on the right side to the C5 to 6 level. Light touch was intact down to C6. The left deep reflexes were absent, and the right recorded as 3+.

On 8 April, 1969, a myelogram indicated a complete block at C5. The cord shadow was widened from the mid-thoracic area up to C5. On 17 April, 50 ml of air were injected into the lumbar subarachnoid space and then, at C6 to 7, a needle was inserted into the spinal cord under fluoroscopic control, using an image intensifier. A cyst was encountered. Four ml of Myodil were injected into this cavity after removal of 5 ml of clear fluid. A cavity with irregular contour (see Fig. 10.14), widest at C7 but tapering to a thin tail at C2 and extending down to T7, was visualised. A severe incidental systemic infective illness prevented further X-rays or investigation until almost a month later. Then 23 days later, on 10 May, a repeat X-ray indicated that the contrast medium had now extended down to the level of the upper border of the eleventh thoracic vertebral body (see Fig. 10.15). C.S.F. obtained from the lumbar region on 8 April contained no cells, 82 mg/100 ml protein, 52 mg/100 ml sugar, 122 mEq/l of chloride. From the cervical region, on 17 April, the C.S.F. contained 7 W.B.C. and 2 R.B.C./μl, 135 mg/100 ml of protein, 55 mg/100 ml sugar, and 125 mEq/l of chloride. On the same occasion, fluid from the intramedullary cyst contained 13 W.B.C./μl, 1140 R.B.C./μl, 200 mg/100 ml protein, 49 mg/100 ml sugar, and 125 mEq/l of chloride.

In July 1968, a T3 to 4 laminectomy was carried out bilaterally (Dr Wm. Lougheed). The spinal cord was ballooned out, occupied most of the intradural space and bulged into the dural opening. Clear fluid was aspirated. The cavity, when opened, had glistening walls. A small segment of the spinal cord was excised, and a retraction of the cord, both superiorly and inferiorly, was noted for about half an inch in both directions. Within two weeks, the patient had recovered some of the right biceps and wrist flexor strength, had considerable increase in his left shoulder and elbow strength, and on the right side there was a return of feeling down to the ring finger, but no change in pain or temperature appreciation on the left. By July 1969, he was free of neck pain and his motor power had returned to the precomplication level. He had not regained his deep left reflexes. Sensory improvement had been retained on the right, and on the left he now had a sensory level at C4 to 5 instead of at the highest cervical level. In December 1971, his condition was stationary at this level of recovery.

CASE 15, P.B.

In 1953, at the age of 52 years, this man suffered a fracture dislocation of T12 on L1. Three months of total paraplegia were succeeded by *partial* disability, by gradual return of sufficient function allowing for active life with crutches. In 1963, aching developed in the fingers of his right hand and in 1965 right hand weakness developed and slowly progressed. In 1967, diminished sensibility developed in the ulnar two fingers on the left hand and gradually spread to involve the whole hand. In 1969, the power in his legs began to decline. In early 1970, he was aware of pain on coughing and straining that extended from the upper anterior chest to the top of his left shoulder and the outer aspect of the arm to the elbow. It disappeared in one month. In July 1970, he was hospitalised and had a spastic weakness of his lower limbs but was still capable of standing and crutch walking. His lower limb deep reflexes and plantar responses were absent. There was asymmetrical but widespread wasting and weakness of both upper limbs with greatest wasting in the right deltoid, right spinati and left intrinsic hand muscles (see Fig. 10.3). Upper limb reflexes were absent. Pain and temperature appreciation were normal in the right upper limb, decreased or absent on the left from C2 to T10, sparing segments T2 to 6. As from the time of his original injury, he had a patchy loss of sensation in right and left lumbosacral dermatome areas.

A swollen left shoulder joint was the site of painless destructive change of the joint surfaces interpreted in an X-ray as compatible with an early neuropathic joint. Lumbar myelography encountered a block at L1 so that an Ethiodan cisternal myelogram in June 1970 suggested the cervical spinal cord to be within normal limits. At the T2 level there was partial hold-up of the contrast and it then trickled slowly down to T12 in irregular channels; at T12 there was almost a complete hold-up (Figs. 10.9 and 10.10), but some contrast trickled past and the lumbar subarachnoid space was normal. Cisternal fluid had protein of 22 mg/100 ml, calcium 6·9 mg/100 ml and magnesium 1·7 mEq/l.

On 26 October, 1970, laminectomy was done at T9 to L1 (Dr H. W. K. Barr). The dura was opened at T12 to L1 and the conus medullaris was found to be adherent to the dura. The cord was distended by a bluish cyst from which 5 ml of clear fluid was aspirated. The protein content was 32 mg/100 ml, calcium 5·0 mg/100 ml and magnesium 2·1 mEq/l (Ca and Mg levels were obtained by atomic absorption). The protein was electrophoretically examined and contained 17 mg/100 ml of albumin and 15 mg/100 ml of globulin, with an A.G. ration of 1·14. The cyst was opened by incising through the posterior columns. It was large enough to contain the surgeon's index finger and was opened in a vertical midline direction for a distance of 2½ cm. Its walls were sutured on each side to the dura and a Pudenz tube was inserted into the cavity for 5 cm in a cephalad direction and placed at the bottom into the lumbar subarachnoid space over the cauda equina. The dura was not closed. Five months later, the patient had not deteriorated further neurologically and had less upper limb pain. In April 1971, he died of uraemia and post-mortem was not obtained.

CASE 16, A.M.

In March 1966, at the age of 24 years, this man was rendered *partially* paraplegic with a compression fracture of T10 and 11. After several months of rehabilitation he could walk with crutches but had moderately spastic weakness of hips, knees, ankles and toes, with more involvement on the left than the right. Lower abdominal reflexes were absent, upper abdominal reflexes were intact. Sensory loss was confined to the right with a level of sensation to pain and temperature at the inguinal ligament, but sparing most of the lower four sacral segments. Position and vibration sense and light touch were preserved. In July 1968, the patient began to complain of right lower flank pain aggravated by sitting and walking. By January 1969, this area was the site of hypalgesia. Pinprick sensation on the right was now absent from T9 and only the right side of the scrotum was spared in the previously intact sacral segments. Vibration appreciation was lost at the ankles.

By September 1969, he had developed pain in the right arm aggravated by straining at stool, and reported an awareness of numbness in this arm. There was no weakness but he had lost all deep reflexes in this upper limb. Sensory loss to pinprick was now apparent on the right side up to C5. By June 1970, he continued to have pain throughout the right arm but it now extended into the occiput, particularly behind the right ear. Pain appreciation was lost up to C3 on the right. By November 1971, his gait was more spastic and his bladder function showed further impairment. Manual compression was becoming necessary to ensure bladder emptying. By this date, he was no longer aware of arm or head pain.

By December 1971, he had lost his right upper limb deep reflexes, but retained normal upper limb strength and bulk. On sensory testing, he had diminished pain and temperature appreciation up to C4 on the right, with pain sensation spared on a slight area over the thumb, index and middle fingers and the radial part of the palm. This sensory loss extended down to involve right sacral segments that were previously spared, with relative normality confined only to the third sacral dermatome on this right side. Tickle appreciation was absent in the right axilla and groin. Normal sensation to touch and position existed everywhere. Vibration was absent on the right at and below the anterior superior spine. On the left side no sensory loss existed and the left upper limb reflexes were normal. The only remaining abdominal reflex was in the left upper quadrant. The evidence in this case suggests an extension upward in the right dorsal horn central spinal cord area to the mid-cervical region with downward extension on the same side to the conus medullaris. The patient is being kept under surveillance.

CASE 17, D.M.P.

In 1952, at the age of 18 years, this woman suffered a fracture dislocation of L1 on 2, rendering her *partially* paraplegic, with flaccidity and sensory loss below the first lumbar dermatome. Initial treatment consisted of a laminectomy with reduction of the fracture dislocation. She recovered no power below the knees; useful but diminished strength returned for hip and knee movement. Initially, superficial sensation was largely absent in all lumbar and sacral dermatomes but eventually the upper lumbar segments became normal. The only certain evidence of an involvement of the conus medullaris, after the injury, that could be pointed to with certainty, was the absence of lower abdominal reflexes. She was not seen from 1954 to 1969 and then came under review because of a recurrent patellar dislocation. Examination then, and again in July 1970, indicated an asymptomatic ascent of pain and temperature loss to T5 level on the left, with absent left abdominal reflexes. She is being kept under surveillance. Involvement has not extended to the cervical segment so that her upper limbs remain normal to motor, sensory and reflex testing.

CHAPTER THIRTEEN

Syringomyelia Consequent on Minor to Moderate Trauma

H. J. M. BARNETT

Disorders of uncertain and baffling origin involving the nervous system have been attributed too often, on insufficient evidence, to syphilis, arteriosclerosis and trauma. The first of these is losing its role in this regard because of its great decline in incidence. The role of the other two is being scrutinised within any given disorder with justifiable scepticism. Rarely should trauma be invoked to explain a neurological disorder unless the traumatic event is significant. Furthermore, a recognised pathological process should be sought or known before the traumatic event is accepted confidently as a causal factor. In addition, the general rule applies that the maximum damage to nervous tissue is closely related to the traumatic event, and apart from accumulating blood clots, a delayed or late lesion requires particularly careful analysis before its traumatic basis is allowable.

A review of the old literature in respect to the aetiology of syringomyelia and cord cavitation disclosed that these constraints were not followed rigidly. Newer knowledge of the recognisable varieties of spinal cavitation makes it imperative though, that older and traditional concepts be reviewed with care. It is established now without equivocation that the individual with a severely injured spinal cord and resultant paraplegia or quadriplegia may fall victim to a variety of post-traumatic syringomyelia with characteristic features. These cases are discussed in Chapters 10, 11, 12 and 14. On equally good pathological evidence, presented in Chapters 15 and 16, there is a variety of spinal cord cavitation related to trauma when it is productive of arachnoiditis. This is plainly a rare situation and other factors such as myelography and infection, possibly contributory to the arachnoiditis, may play a role in its development in a given post-traumatic case.

The third post-traumatic variety forming the substance of this chapter is that which follows spinal cord contusion of lesser severity than is necessary to produce paraplegia. It may be associated with very transient weakness or sensory disturbance and the trauma may or may not be associated with vertebral fracturing. It is a rare condition and cases in this category may overlap with those of nearly equal rarity arising as a sequel to traumatically induced arachnoiditis.

As discussed in Chapters 3, 10, 15 and 17, trauma may unmask symptoms which are due to a pre-existing syrinx. This syrinx may be of the so-called communicating variety or may be related to an arachnoiditis (e.g. case 7, chapter 15) or tumour (e.g. case 6, Chapter 17). It does not qualify for a designation of post-traumatic syringomyelia, as the trauma is merely a precipitating and not a causal factor. Further discussion of this problem is to be found in the papers of McIlroy and Richardson (1964) and of Williams and Turner (1971).

HISTORICAL RECOGNITION

The historical recognition of minor to moderate spinal trauma producing cavitation of the cord cannot be traced accurately because inadequate history, radiological information and surgical or post-mortem correlation with the clinical picture is so frequently lacking. It may be that this was the condition mentioned in Chapter 9, where Strümpell (1880) published a report of a boy who developed a syringomyelic syndrome four years after a 'blow to the vertebral column'. Lloyd (1894) read a paper on 'Traumatic affections of the cervical region of the spinal cord simulating syringomyelia' to the Philadelphia Neurological Society. His two patients each had two spinal injuries and both completely recovered from the first incident. Then in each case, a second blow resulted in a clinical picture simulating syringomyelia. They were *not* of *delayed* or *progressive* nature and, although quoted in the literature as post-traumatic syringomyelia, this designation is inaccurate. They were afflicted with immediate post-traumatic central spinal cord syndromes.

The delayed development of a spinal cord dysfunction after trauma, however, has intrigued many authors. Lhermitte (1932) reviewed the problem extensively and quoted particularly from Marburg and Foerster. The problem, as he reviewed it, was one of deciding on the importance of vascular disturbance or else of arachnoiditis in its production. He described a post-traumatic case who went on to develop typical syringomyelia and who had been studied by Pierre Marie. At post-mortem, he had neither arachnoiditis nor vascular disorder but hydrocephalus and hydromyelia (what Schlesinger (1902) described as 'syringomélie-hydrocéphalique'). Lhermitte's opinion was that a syringomyelic syndrome after spinal or cerebrospinal trauma would result from cavitation either as a sequel to ischaemic softening of the cord or as a hydromyelic complication of hydrocephalus.

ACCEPTABLE EXAMPLES

Despite these voluminous early writings, cavitation of delayed onset, but produced by mild to moderate spinal cord injury remains a rarity. Proven cases are very scarce. Three examples are sufficiently well documented to be reviewed here and the condition deserves to be kept under consideration:

CASE 1, MALE (studied by Dr Frank Turnbull, Vancouver, who has kindly supplied this information).

In 1947, at the age of 37 years, this patient was struck on his back by a falling log. He suffered a severe compression fracture of T12 and L1 and was in bed for about two months. He does not recall the events of the first week clearly but at least after that there was no paraplegia or hand weakness. When first seen in 1967, he was complaining of incipient but progressive weakness of his right hand and had a periodic burning feeling in both feet. For three years he had been troubled with pains from the left loin around to the groin. He had a partial right Horner's syndrome. There was gross wasting of the small muscles of the hand with some fasciculation and slight wasting of the forearm. There was no leg weakness. Sensation to touch, vibration and joint position was normal. There was also dissociated loss of pinprick and cold sensation over the left foot and left lower leg. Superficial abdominal reflexes were depressed on the right side. The right knee jerk was more active than the left. Ankle jerks were both present. Both plantar reflexes were indefinite. X-ray showed gross crushing of T12 and L1, with ridges of bone between T12, L1 and L2. A myelogram with injection of contrast by cisternal puncture indicated a block at T11.

A laminectomy was done from T10 to L1. There seemed to have been a rotation of the twelfth thoracic vertebra, so that the right lamina which had broken off was left as a large chunk of bone inside the canal. At the point of fracturing, the dural sac was rotated so that the root of T12 was projecting posteriorly. At this level, the cord was swollen and pale and not pulsating with a very obvious cystic appearance. It was more prominent on the right side. A fine needle was introduced into the sac and about 3 ml of clear fluid removed. This caused the sac to collapse like a slack balloon and after this it pulsated. The foot of the table was tilted down to try to fill up the slack sac, but this manoeuvre was only partly successful. An attempt was made to free the roots of T12 and L1 on the left from dense adhesions. The operative appearance indicated that there had been considerable damage to the surface of the cord even though the patient had not been rendered paraplegic. A piece of No. 15 Intracath about 7 cm long was introduced into the cavity above and then threaded into the subarachnoid space below and anchored to the dura. The patient left hospital with left leg weakness as well as his old sensory loss and upper limb wasting.

He returned to hospital two years later for reinvestigation. Neurological findings in his hand had not changed. He showed some atrophy of the left buttock and gross weakness of movements of the right foot at the ankle, particularly dorsiflexion. The weakness of his left leg had improved since he left hospital in 1968. The dissociated anaesthesia on the right side remained as before. It was now a little more extensive on the left side extending up to T12. There was sparing of the L4 segment on both sides. Tendon reflexes were as before. On follow-up, the impression was that there had been some progression of his lesion in the thoracolumbar region.

Summary. A T12 to L1 fracture dislocation without early detected spinal cord injury was followed 22 years later by evidence of a syringomyelic syndrome, including hand wasting. Florid evidence of old trauma at the original site was disclosed by the laminectomy and a cystic expansion of the cord was confirmed. Arachnoiditis was present at the traumatic site but did not appear to be the most prominent factor.

CASE 2, MALE (reported from Cologne by Bischof and Frowein (1967)).

In 1960, at the age of 58 years, this patient suffered a compression fracture of T10 without positive neurological findings. Four years later, he began to experience slowly progressive difficulty in walking. By 1965, he had posterior leg pain that was associated with spastic legs, bilateral Babinski responses and dissociated sensory loss in the legs. In particular, pain and temperature sensation was lost below L1. He had urinary retention. Myelogram revealed a complete block at T10. At operation, there was a clear bulging intramedullary cyst in the area of the gibbus. The cyst was drained, the ventral bony projection removed and the dura left open. These authors attributed the cavitation to vascular causes—central ischaemic myelomalacia succeeded by cavity formation. They did not note significant arachnoiditis.

Summary. A T10 compression fracture was followed four years later by a lower limb picture compatible with an intramedullary lesion in the lower cord which proved at laminectomy to be an intramedullary cavitation. Arachnoiditis was not described.

CASE 3, MALE (case 1, Chapter 15).

This patient was injured in 1947, at the age of 18 years, and was told at that time that he had fractured two cervical vertebrae. He had an immediate paresis of his right upper limb but during 10 weeks of traction and immobilisation, it cleared completely to become normal. He was able to join the Air Force, passing a medical examination in 1952. New symptoms developed in 1956 and at first involved only the opposite (left) upper limb with weakness, numbness and pain. Evidence of a high cervical lesion progressed and his spinal cord was explored at C6 to 7, the level of an obstructive myelographic defect. This operation indicated the presence of an arachnoiditis.

By 1970, he had a Horner's syndrome, spastic legs and bilateral sensory disturbance indicating a progression of central spinal cord function. No old fracture could be seen in his plain X-rays and the myelogram picture suggested an irregular defect from C5 to T1 in keeping with arachnoiditis. Laminectomy revealed a gross thickening and adherence of the meninges as well as a significant arachnoid cyst within this chronic leptomeningitis. A cavitation within the spinal cord was confirmed and contained clear fluid which had a protein content of 49 mg/100 ml.

Summary. A C6 to 7 injury, probably without fracturing, was followed nine years later by a clinical picture compatible with syringomyelia. This progressed and a cord cavitation was proved at laminectomy. A severe arachnoiditis and a coexistent arachnoid cyst was encountered. The intracystic fluid had a protein content resembling C.S.F.

Comment. It may seem arbitrary to delineate these cases from those which arise as a delayed sequel in patients who have had severe trauma and are thereby rendered paraplegic or quadriplegic. However, they have certain features that are clinically distinctive. In the first place they are, to date, decidedly rare. Those following more severe injuries are by no means common but are turning up, when looked for, in about 2 per cent of paraplegic populations. Secondly, they have had no major persistent sequelae from their injuries until the cord cavitation begins to produce symptoms. With continued observation, it is possible that there will be more cases resulting from rapidly reversible or initially unrecognised spinal trauma as well as those which are a sequel to severe spinal injury. It is likely that certain pathogenic mechnisms are shared by both varieties. These mechanisms may include all or some of the following:

1. The initial trauma produces a central area of softening. This central cord necrosis, as reviewed extensively by Schneider, Cherry and Pantek (1954), is more often recognised in the cervical region than elsewhere and may follow hyperextension injuries without detectable fracture. Dorothy Russell (1932) discussed its occurrence in the lumbar cord in a patient who remained without urinary control after a fracture dislocation of the lumbar spine. Ten years later, 'bilateral cystic softening of the grey matter was present from the fifth lumbar segment downwards', most conspicuous at the third sacral segment.
2. As discussed in Chapter 14, vascular factors—including arterial, venous and capillary effects—may contribute to the production of this central softening.
3. The central softening may be aggravated because this central part of the cord is in the marginal zone of blood supply (Turnbull, Brieg and Hassler, 1966) and may even be damaged further by norepinephrine release at the time of initial injury (Osterholm and Mathews, 1972).
4. It is possible, but not probable, that the necrotic cores of central myelomalacic and pulped tissue described by Holmes (1915) and produced experimentally by McVeigh (1923) have occurred in these cases. The pulping of central tissue might be expected to occur only with more florid and persisting evidence of its existence, but smaller amounts might be asymptomatic.
5. Haemorrhage of localised nature can occur in a cord even without fracture of the overlying bone. Bennett (1860) described a central pea-sized haemorrhage in

the centre of the cervical cord of a woman kicked in the neck by her spouse as she lay in a drunken stupor. A haemorrhage of this size would be expected to be as conspicuous as central cord pulping, but a smaller amount might contribute to the central damage and not be disabling at the time of initial trauma.

6. Arachnoiditis, secondary to the trauma, may further interfere with cord blood supply and mobility. This will add to the initial insult to central spinal cord structures.

Summary of Data Regarding Minor Spinal Trauma Producing Subsequent Cavitation:

1. Spinal trauma of a severity insufficient to be immediately productive of a permanent spinal cord dysfunction in a few cases will be followed by syringomyelia.
2. These cases are distinguished from those in whom trauma precipitates syringomyelic symptoms; the latter cases are followed immediately by the syringal symptoms, whereas in the cases discussed here there is an interval between the trauma and the development of symptoms.
3. Spinal fracture may or may not be associated with the injury.
4. Arachnoiditis may or may not be present to a significant degree.

REFERENCES

Bennett, J. H. (1860) *Clinical Lectures on the Principles and Practice of Medicine*, p. 392. New York: S. S. & W. Wood.

Bischof, W. & Frowein, R. A. (1967) Zur, Spätmyelomalazie nach Kompressionsfraktur. *Zentrablatt für Neurochirurgie*, **28**, 61.

Holmes, G. M. (1915) The pathology of acute spinal injuries. (Goulstonian Lectures to the Royal College of Physicians of London) *British Medical Journal*, **ii**, 769.

Lhermitte, J. (1932) Étude de la Commotion de la Moelle. *Revue Neurologique*, **1**, 210.

Lloyd, J. H. (1894) Traumatic affections of the cervical region of the spinal cord, simulating syringomyelia. *Journal of Nervous & Mental Diseases*, **21**, 345.

McIlroy, W. J. & Richardson, J. C. (1965) Syringomyelia: a clinical review of 75 cases. *Canadian Medical Association Journal*, **93**, 731.

McVeigh, J. F. (1923) Experimental cord crushes, with special reference to the mechanical factors involved and subsequent changes in the areas of the cord affected. *Archives of Surgery*, **7**, 573.

Osterholm, J. L. & Mathews, G. J. (1972) Altered norepinephrine metabolism following experimental spinal cord injury. Part 1: relationship to hemorrhagic necrosis and post-wounding neurological deficits. *Journal of Neurosurgery*, **36**, 386.

Russell, D. S. (1932) Capillary haemangioma of spinal cord associated with syringomyelia. *Journal of Pathology & Bacteriology*, **35**, 103.

Schlesinger, H. (1902) *Die Syringomyelie. Eine Monographie.* 2nd Edn., p. 255. Leipzig and Vienna: F. Deuticke.

Schneider, R. C., Cherry, G. R. & Pantek, H. (1954) Syndrome of acute central cervical spinal cord injury with special reference to the mechanisms involved in hyperextension injuries of cervical spine. *Journal of Neurosurgery*, **11**, 546.

Strümpell, A. (1880) Beiträge zur Pathologie des Rückenmarks. *Archiv für Psychiatrie und Nervenkrankheiten*, **10**, 676.

Turnbull, I. M., Breig, A. & Hassler, O. (1966) Blood supply of cervical spinal cord in man. A microangiographic cadaver study. *Journal of Neurosurgery*, **24**, 951.

Williams, B. & Turner, E. (1971) Communicating syringomyelia presenting immediately after trauma. A case description and some theoretical concepts. *Acta Neurochirurgica*, **24**, 97.

CHAPTER FOURTEEN

Pathology and Pathogenesis of Progressive Cystic Myelopathy as a Late Sequel to Spinal Cord Injury

H. J. M. BARNETT, A. T. JOUSSE and M. J. BALL

A characteristic clinical syndrome complicating serious spinal cord injury, developing many months or, more often, many years after the original trauma has been described (Barnett et al, 1966) and is reviewed in detail in Chapters 10, 11 and 12. The clinical picture is such as to indicate a lesion of central grey matter beginning on one side, and in some cases becoming bilateral. It involves predominantly pain and temperature appreciation, a loss of lower motor neurone function and of the deep reflexes. When severe enough, the lesion converts a paraplegic into a quadriplegic patient who may reach total helplessness, and the outcome may be fatal. Clinical evidence has been accumulated which indicates that the basis of this syndrome is a syrinx which arises within the spinal cord at, or near, the level of trauma, and ascends to the cervical region and, on rare occasions, into the medulla.

In this chapter, post-mortem material from a patient who died with this syndrome will be described, as will the material from one who died after developing cystic cavitation extending above and below the traumatic site four months after his injury. Pathological material from two cases in which the spinal cords were surgically transected will be reviewed, as will the three other post-mortem cases to be found in the literature of this condition. All this material will form the basis for a discussion of pathogenesis.

POST-MORTEM MATERIAL

Jung's Case

The first published post-mortem of this condition in modern times was probably that of Jung (1960). He recorded the occurrence of a congenital abnormality of the kidney in a patient who died from syringomyelia. It was reported in support of the argument favouring a congenital origin of syringomyelia. However, the cervical spinal lesion had become symptomatic five-and-a-half years after the patient became paraplegic from a T12 injury. The spinal cord was described as 'hollow with sac-like cavities of asymmetrical size extending almost throughout the cord, but not up to the level of the fourth ventricle' (Fig. 14.1a). It was not clear if it went above C1, but it was clearly stated that it did not open into the medulla.

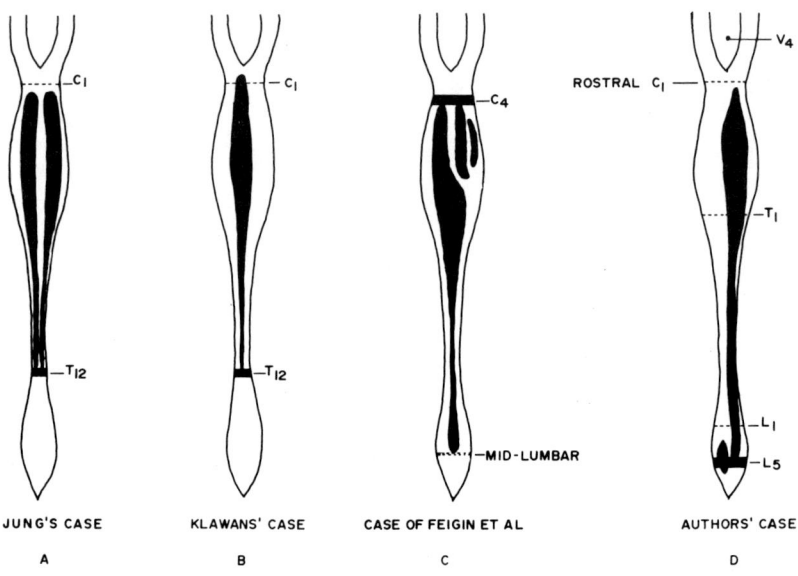

Fig. 14.1. Extent of cavitation in the four cases proven at post-mortem. Original levels of trauma at T12, T12, C4 and L5 respectively.

Klawans' Case

Klawans (1968) described the second reported case, with post-mortem proof of the extension of a central cavity from the level of a spinal cord injury at T12 up to the top of the cervical cord and on into the medulla oblongata (Fig. 14.1b). Left upper limb symptoms developed 12 years after an accident had rendered the patient partially paraplegic. His walking, as well as the strength and feeling in his left upper limb, deteriorated during the last two years of his life. The syrinx was most extensive in the cervical region and clearly separated from the central canal at all levels, including the level of the medulla (Fig. 14.2). The cavity in the cervical region was lined, in some places at least, with collagen.

Case of Feigin et al

Feigin, Ogata and Budzilovich (1971) have provided the third reported post-mortem example. A 66-year-old woman was rendered quadriplegic from a C4 to 5 lesion 16 years before death. No history of worsening or extension was reported. The spinal cord at the level of injury was described as flattened in its anteroposterior diameter and with this there was fibrotic thickening of the dura and leptomeninges. 'From the level of injury down to the mid-lumbar level there was a cystic cavitation within the spinal cord most prominent in the mid-thoracic segments.' The cavitation immediately below the level of trauma consisted of several channels but coalesced to one cavity at the lower cervical region (Fig. 14.1c). It was 'lined by gliotic tissue but not by connective tissue or ependymal cells'. The spinal cord above the

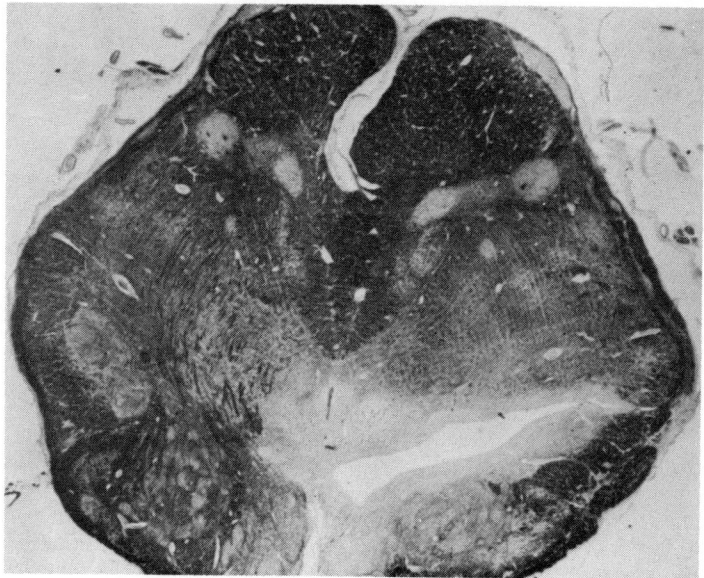

Fig. 14.2. Trauma at T12 14 years previously. Cavitation proven at post-mortem from injury site to medulla. Cavity in medulla separate from central canal. (Reproduced, by permission, from Klawans, H. L., *Diseases of the Nervous System*.)

level of trauma did not contain any cavity and merely revealed degeneration of ascending tracts. There was no connection between the cavity and the subarachnoid space and no connection with the fourth ventricle.

Authors' Case

This patient (case 2, Chapter 12) had sustained an L1 to 2 fracture dislocation with partial paraplegia 14 years before her upper limb symptoms developed and 25 years before she died in quadriplegic helplessness. The spinal cord and brain were sent to Professor William Blackwood, National Hospital, Queen Square, and we are indebted to him and to Dr T. J. Murray for a detailed analysis of the material.

The spinal cord had a large left-sided cavity extending from the middle of the scar up to C1 and a small cavity in segments L4 and L5 on the right (Fig. 14.1d). There was a fibrous scarred area at L5 with no preservation of cord structure and no extension of the major cavity caudal to this scar (Fig. 14.3). A cavity was found beginning on the left within this scar and, slightly lower, a similar cavity was found on the right side. At the rostral end of L5 the cavities were equal in size, but the cavity on the right narrowed and disappeared in the L4 segment above (Fig. 14.4). The cavity on the left enlarged and at L3 extended across the midline to the right side of the cord. The cavity remained bilateral, but was principally on the left in the thoracic area (Fig. 14.5) and narrowed slightly at the T1 level.

At C7 a duplication of the cavity was noted; this was due to a small bridge of glial tissue containing blood vessels (Fig. 14.6). At C6, the cavity was widest in relation to the width of the cord and involved both posterior horns, extending close to the pial surface in the dorsal root entry zone and extending across the midline posterior to the region of the central canal (Fig. 14.7). At C4, the cavity began to narrow and was again mostly on the left side. Serial sectioning through the C1 segment showed that the cavity disappeared in the glial scar a short distance below the rostral end of the C1 segment (Fig. 14.8).

Throughout the cord, the cavity remained posterior to the central canal region and did not connect with it at any level. The central canal area consisted of a solid mass of ependyma-like cells but was not patent in the cord or lower medulla caudal to the inferior olives. Above this, a small but patent central canal could be seen. The detail of the rostral and caudal limits of the cavities was examined in serial sections. The small cavity on the right was closed at both ends, with only a glial scar seen at either

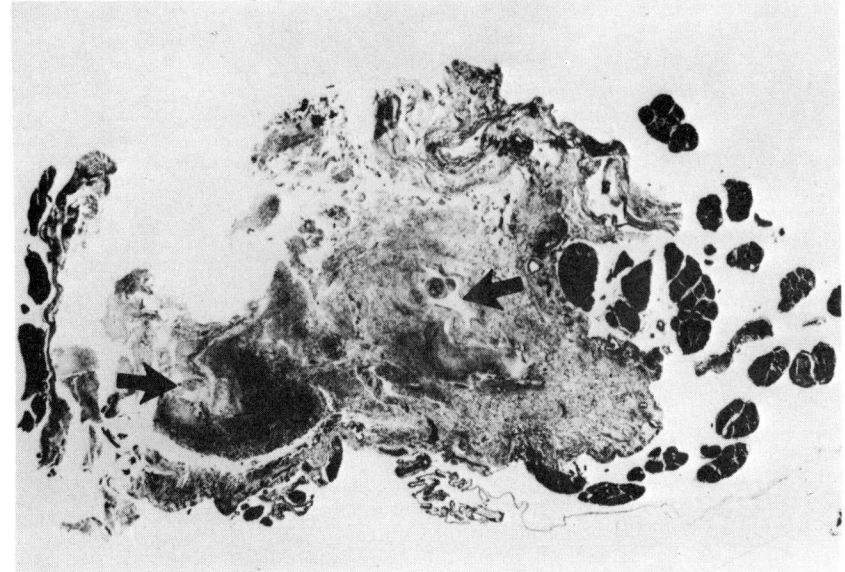

Fig. 14.3. Case 2. Area of primary injury (L5) with evidence of two cavities in dorsal horns (arrows).

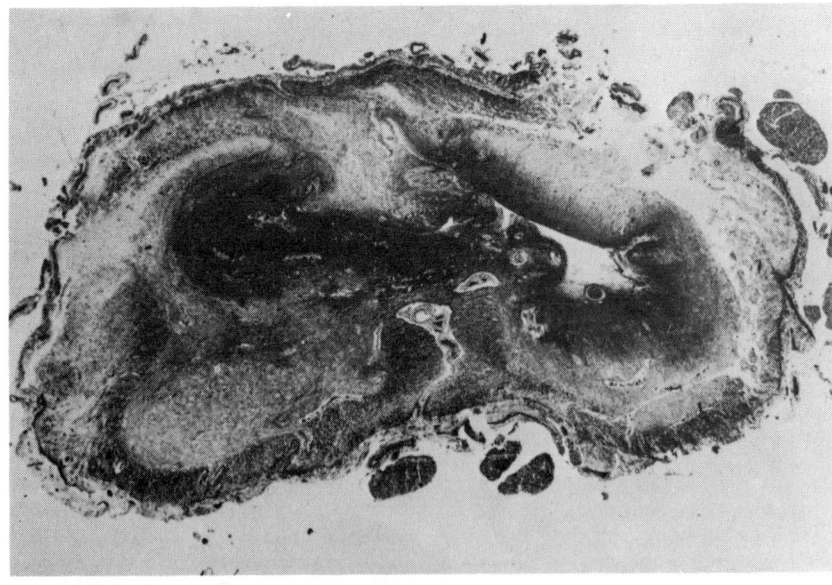

Fig. 14.4. Case 2. L4–5 area: right cavity (left of picture), left cavity (right of picture) expanding.

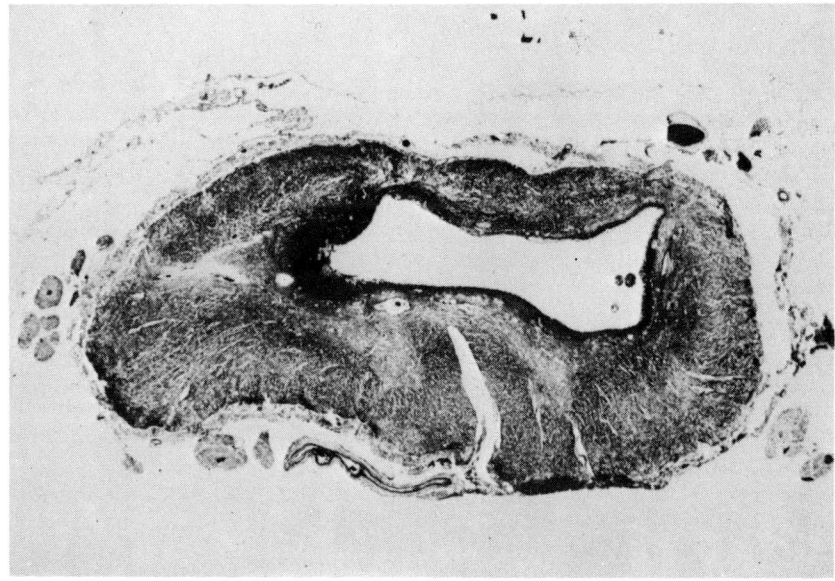

Fig. 14.5. Case 2. T11, a single cavity, more to the left than the right of the midline.

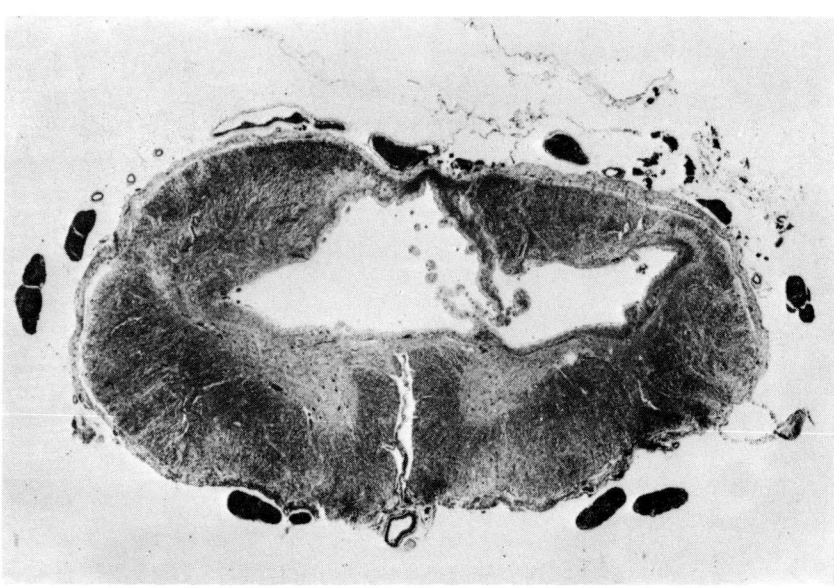

Fig. 14.6. Case 2. C7, expansion of cavity in cervical region with bridge of glial fibres and vascular tissue.

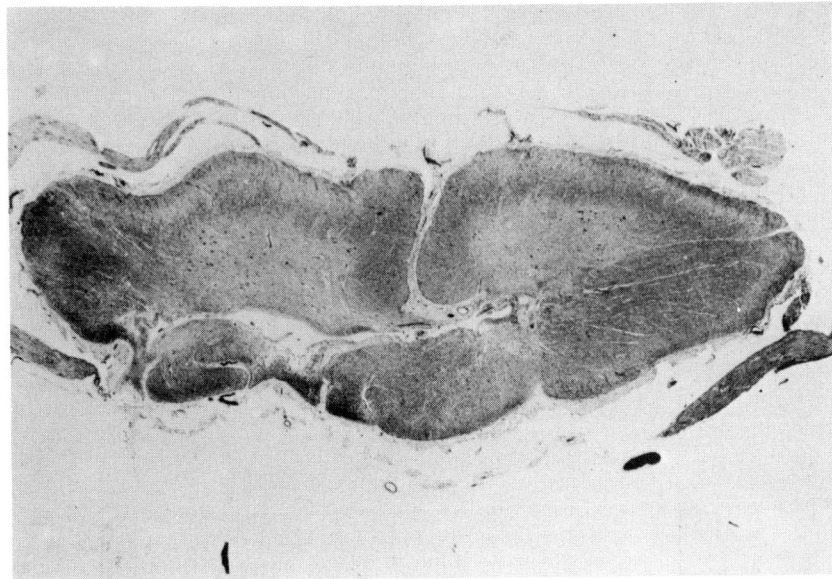

Fig. 14.7. Case 2. C6, single cavity extending well out into left posterior horn and dorsal to central canal into right posterior horn.

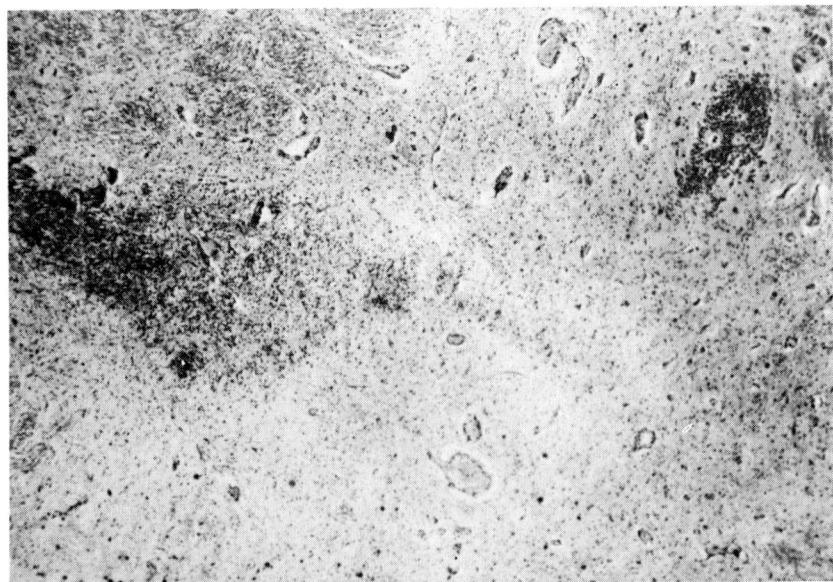

Fig. 14.8. Case 2. Rostral C1, cavity not present, glial scar in its place.

end. The caudal end of the left cavity was found in the L5 scar where it seemed to divide into many irregular interstices and channels which in the sections appeared at first glance to connect with the subarachnoid space in this region. This apparent connection, however, was considered to be an artefact due to removal of the scarred cord from the canal at necropsy. At its rostral end, in C1, the left cavity ended in a glial scar (Fig. 14.8).

The wall of the large cavity was essentially similar at all levels. The glial scar and compressed tissues making up the cavity wall were asymmetrical, being thicker in the anterior and right lateral walls (Fig. 14.9a), but there was great variability in thickness at different levels. The inner lining was mostly thin collagen, but in some areas the wall consisted of a loose meshwork of astrocytic fibrils (Fig. 14.9b). A prominent feature throughout many levels of the cord, but especially the thoracic, was the presence of a definite endothelial-like lining which in many areas lay on the collagenous tissue but in other areas seemed to cover the mat of glial fibres abutting on the cavity (Fig. 14.9c). Many of the trabeculae which traversed the cavity contained blood vessels (Fig. 14.6) in a mat of glial fibres admixed with collagen, and often had a covering of endothelial-like cells. Another feature of the wall of the cavity was that occasionally collections of thin-walled blood vessels lay in the wall very close to the lumen of the syrinx. The central canal was always outside the wall.

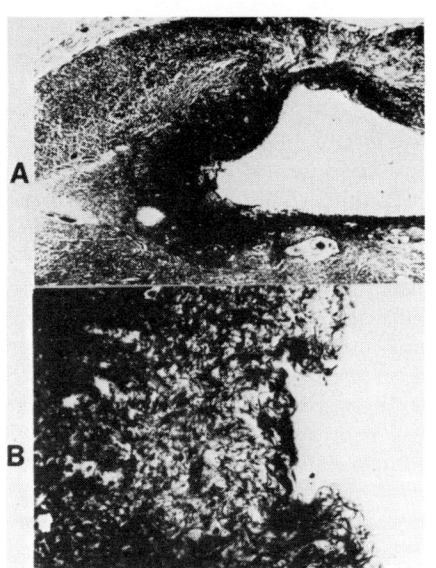
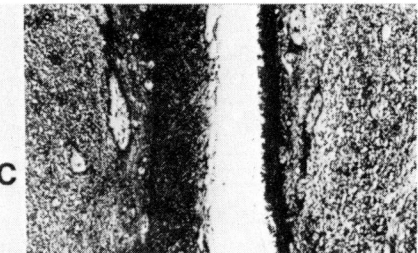

Fig. 14.9. Case 2. (a) Irregular thickness of lining of cavity, greatest laterally and anteriorly. Largely made up of collagen.
(b) High power of area marked with arrow from (a). This portion of wall consists of loose meshwork of astrocytic fibres.
(c) Endothelial-like lining overlying collagen and glial tissue.

Summary of Post-mortem Material

The four cases are strikingly similar in that the initial injury to the spinal cord was severe, the new lesions extended from the site of original trauma and in each case cystic cavitation was located in the central spinal cord substance. In three of the four cases, there was more than one channel to the cavitation or more than one cavity. In all but one of them, furthest from the site of the injury, only a single cavity persisted. None of the cases had communication between the subarachnoid space and the cavitation in the cord, none were dilatations of the central canal and none communicated with the fourth ventricle. One extended up to the level of the medulla, but was clearly described as distinct from the central canal and the fourth ventricle. Three of the four had extension *upwards* from the level of initial injury. One of the four (Feigin et al, 1971) had *downward* extension and, in that this was the only injury originally in the cervical spinal cord in the post-mortem series, it may be of significance in considering what factors cause the extension of an existing intraspinal cavitation.

PATHOLOGICAL MATERIAL FROM SURGICAL PROCEDURES

1. In 1957, a 57-year-old woman (case 11, Chapter 12) was rendered paraplegic at T8. Nine years later, in 1966, she became aware of pain with wasting and weakness in the right arm and hand. In 1967, the cord was explored at T8 and since she was totally paraplegic below this level, the cord was surgically transected. Cavitation was identified in the cord, and 1 in. of the cord was resected to the top level of the area of the surrounding arachnoidal adhesions. When the wall of the cavity was exposed, the lining had an infolded appearance and resembled a 'chamois wash-leather'.

The section of spinal cord was serially blocked and many sections cut. There was a large central cavity (Fig. 14.10) which, in a large portion of the specimen, was completely lined by dense fibrous tissue probably of meningeal origin. This fibrous tissue was covered on its innermost side by a definite layer of flattened plate-like cells resembling endothelial cells. The origin of such cells is obscure. Metaplasia from ependymal cells is one possibility. A mesothelial origin from perivascular structures is also possible. Towards one end of the specimen, residual central nervous system tissue was apparent but its architecture had been largely replaced by a dense mat of glial fibres in which there were blood vessels of varying size. Where this gliosed cord tissue abutted upon the cavity these glial fibres were in direct contact with the lumen, but more often there appeared to be a thin layer or collar of collagen separating the glial fibres from the lumen; on the luminal side of the collagen there was again a single layer of somewhat flattened plate-like cells. This surrounding collar of connective tissue appeared to have originated from the meningeal region. It varied in thickness and contained nerve roots showing moderate degrees of demyelination and fibrosis, together with focal areas of peripheral nerve regeneration. Scattered clumps of haemosiderin pigment were present both within the fibrous tissue collar and within the residual gliosed cord tissue.

2. This man (case 14, Chapter 12) was the victim of a C5 to 6 fracture dislocation, with complete paralysis of his lower extremities and trunk. He retained only some wrist function. Nearly three years later, he developed an increasing weakness in his left upper limb and posterior neck pain. With the percutaneous technique, an intramedullary injection of Myodil was performed and a cavity demonstrated, extending from C2 to T11 (see Figs. 10.14 and 10.15).

In July 1968, a T3 to 4 laminectomy was carried out, and a segment of the spinal cord 1·5 cm long was excised. The specimen was inadvertently sectioned before fixing but reconstruction was attempted and an artist's sketch made of the location of the cyst (Fig. 14.11). A large part of one-half of the spinal cord was replaced completely by a cyst-like structure. The cavity involved the medial part of the posterior horn, the anterior part of the dorsal column and the lateral portion

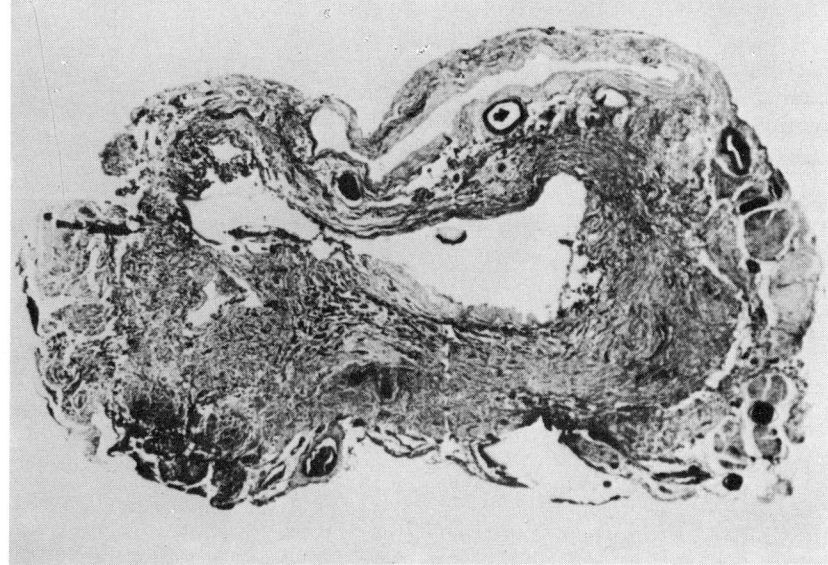

Fig. 14.10. Case 11. Surgical specimen from transected cord at traumatic site. Cavity lined by mixture of collagen and glial tissue.

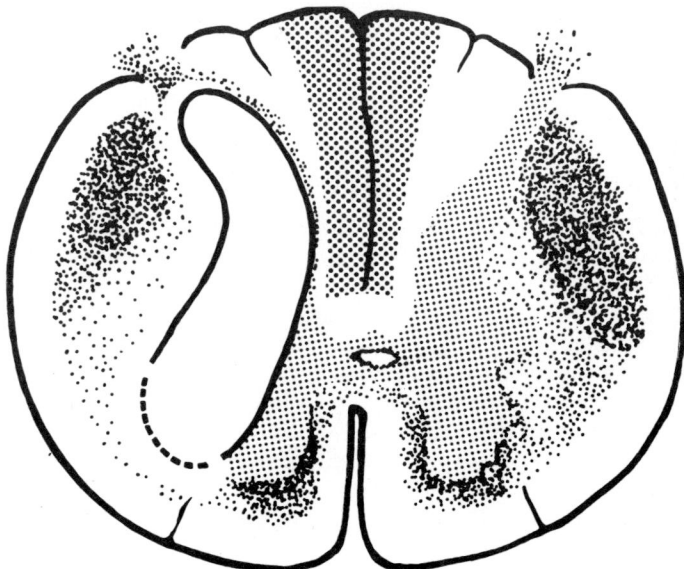

Fig. 14.11. Case 14. Sketch to show portion of cord and cavitation at site of transection (T3–4), original injury being at C5–6.

of the anterior horn of the grey matter. It extended laterally as a narrow slit-like cavity along the course of the dorsal horn on the same side. The lining of this cyst was composed of a very loosely packed thin layer of glial fibres (Fig. 14.12). Scattered among these were seen occasional reactive astrocytes with prominent eosinophilic cytoplasm. At no point was any connective tissue forming a lining seen, but a few areas with flattened endothelial-like cells were evident.

Fig. 14.12. Case 14. Lining of cavitation consisting of loosely packed glial fibres and no ependymal cells (P.T.A.H. × 100).

PATHOGENESIS

The Early Origins of the Lesions Later to Become Cystic
ACUTE MYELOMALACIC CORES

The cystic nature of the lesion responsible for this post-traumatic myelopathy of delayed onset is now established. It remains to examine the origin and the progression of the lesion.

Reference was made in an earlier communication (Barnett et al, 1966) to the central cores of necrotic tissue described by Holmes (1915) (Fig. 14.13), and which appeared at the site of spinal trauma, extending up and down the spinal cord for as many as four or five segments. This necrotic tissue tended to be in the dorsal columns and dorsal horns. Cushing (1898) remarked on lesions in the cord akin to those later described by Holmes. In an early corroboration of Holmes' observations, McVeigh (1923) produced experimental lesions with strikingly similar cores of myelomalacic tissue. The development and the extension of this pathological process was in the ventral portion of the posterior columns and the contiguous dorsomedial portions of the dorsal horns (Figs. 14.14 and 14.15).

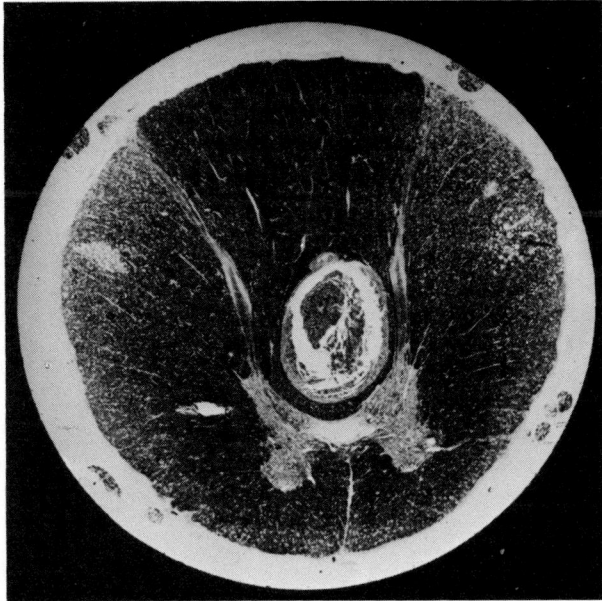

Fig. 14.13. Acute myelomalacic core in acute spinal cord injury (from Holmes' original material by permission of Prof. Wm. Blackwood).

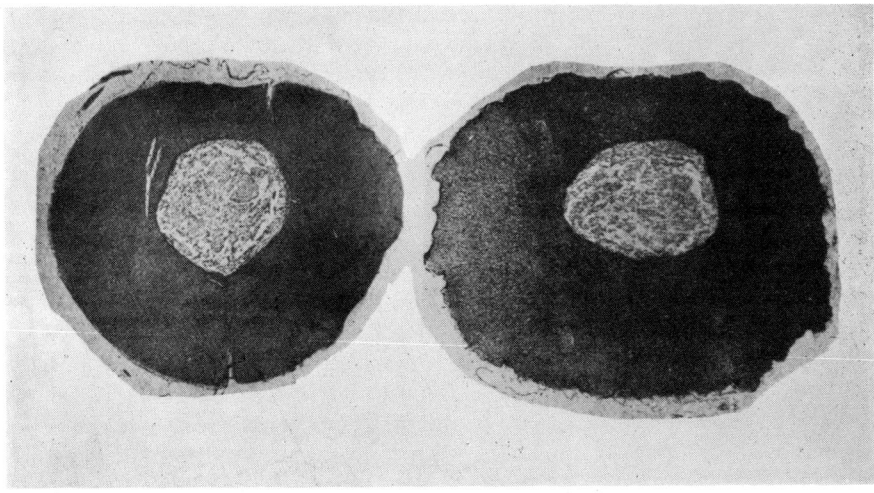

Fig. 14.14. Experimental spinal cord injury and myelomalacic cores. (Reproduced from McVeigh (1923), by permission of *Archives of Surgery*.)

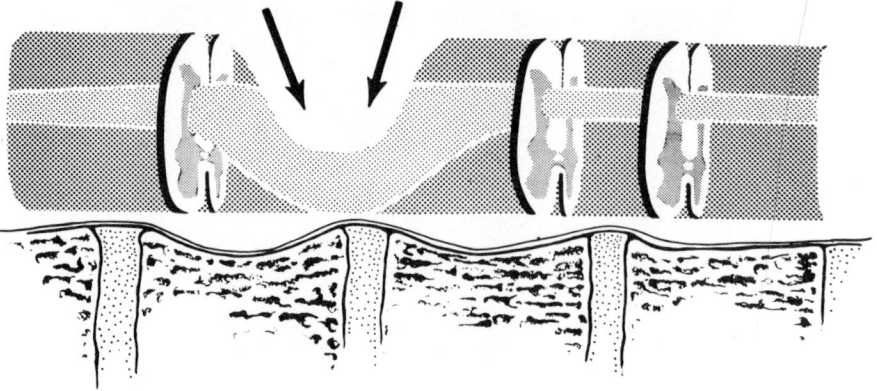

Fig. 14.15. Artist's sketch of McVeigh's experiments producing digital pressure and central myelomalacic cores. Pulped necrotic tissue (lighter stippling) located centrally in dorsal horns. (Adapted from McVeigh (1923), by permission of *Archives of Surgery*.)

Rewcastle and Hugenholtz (1969) in a series of unpublished experiments carried out in rabbits demonstrated that digitally compressing the cord will squeeze pulped tissue cephalad and caudad from this compressed area and will do so in many cases with a minimum of haemorrhage. Their observations confirmed those of McVeigh, that an intact pia is sufficiently strong to contain the cord tissue against the compressive force applied at right angles to the length of the cord and that the displaced tissue moves up and down the cord in the same central sites as were determined by McVeigh and observed by Holmes in gunshot wounds in World War I. The lesions in all these circumstances might be unilateral or bilateral and were similar to those seen in acute civilian spinal injuries as illustrated in a number of recent publications (Davidson, 1943; Barnett et al, 1966 (Fig. 14.16); Hughes, 1966; Blackwood, Dodds and Sommerville, 1970). These were reports concerning the result of fractures and fracture dislocations, while Schneider, Cherry and Pantek (1954) have illustrated the same type of lesions in acute forced hyperextension injuries.

CENTRAL HAEMORRHAGE

The amount of haemorrhage in this type of injury may be, as in Fig. 14.16, an insignificant aspect of the cord lesion or, at other times, may be prominent enough to justify the use of the term 'traumatic haematomyelia'. In this regard, the earliest recorded case reflects a good deal on human frailty while providing evidence of the fact that trauma may initiate intramedullary haemorrhage. Bennett (1860) described a clot of pea size in the upper cervical cord (just below the medulla) of a woman who died four days after a blow to the back of the neck. The blow, which rendered her quadriplegic, had been administered to her while she was drunk, by the boot of her husband. No bony damage or change in the surface of the cord was visible at the post-mortem. We have observed two recent cases who succumbed shortly after injury, which are instructive in regard to the problem of significant early intramedullary haemorrhage:

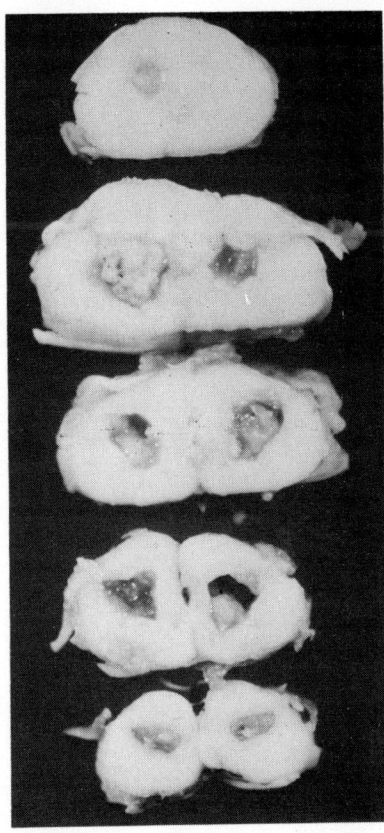

Fig. 14.16. Bilateral myelomalacic cores in cervical segments C4 to C2. Patient quadriplegic from C5–6 fracture dislocation seven days previously.

S.A.F., MALE

A 63-year-old man, in a motor vehicle accident in February 1970 suffered a fracture dislocation of C4 to 5 with immediate quadriplegia. All limbs were completely paralysed and the sensory level was at C4. Death from bronchopneumonia and massive gastric haemorrhage occurred 14 days after the injury.

Pathological Findings. At post-mortem a purplish-red swollen area on the cord, affecting mainly the fourth and fifth cervical segments, corresponded to the fracture. Transverse sections showed that a large, fresh haematoma had extended rostrally for 1·5 cm to the cervicomedullary junction and caudally 5·5 cm to the T1 segment (Fig. 14.17a). This acute haemorrhage, which had no connection with the central canal, was situated in both posterior columns in its cephalad portion, and confined to the right posterior horn and adjacent posterior column in its caudal portion (Fig. 14.18).

T.M., MALE

This patient, aged 44 years, in July 1971, suffered a fracture dislocation of C3 on C4. He was quadriplegic, with a sensory level at C3. On the fifth day after his injury he suffered a cardiac arrest and died.

Pathological Findings. At post-mortem, the fracture dislocation was shown to have inflicted the primary trauma on the third cervical cord segment. Above and below this primary trauma was a fresh haemorrhage affecting mainly the right dorsal column and extending from C2 to T1 (Fig. 14.17b). A 4 mm length of the mid-cervical cord was shown on gross inspection to be almost spared, and the largest portion of the haematoma was at C6 to 8, some 5 to 6 cm below the site of disruptive trauma. Microscopy revealed a minute track of blood joining these two regions (Fig. 14.19).

NON-COMMUNICATING SYRINGOMYELIA

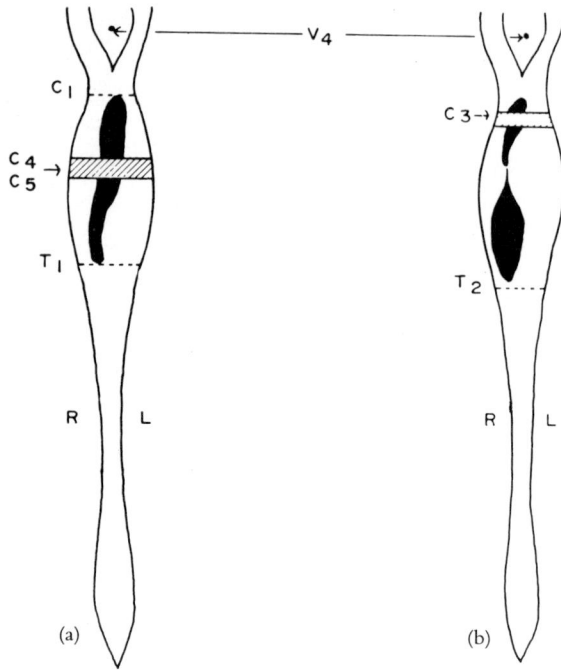

Fig. 14.17. (a) Case S.A.F. C4–5 injury 14 days before death, haemorrhagic lesion extending from C1 to T1. (b) Case T.M. C3 fracture, haemorrhagic lesion C2–T1 with mid-cervical area approaching normal and wide expansion of haemorrhage below.

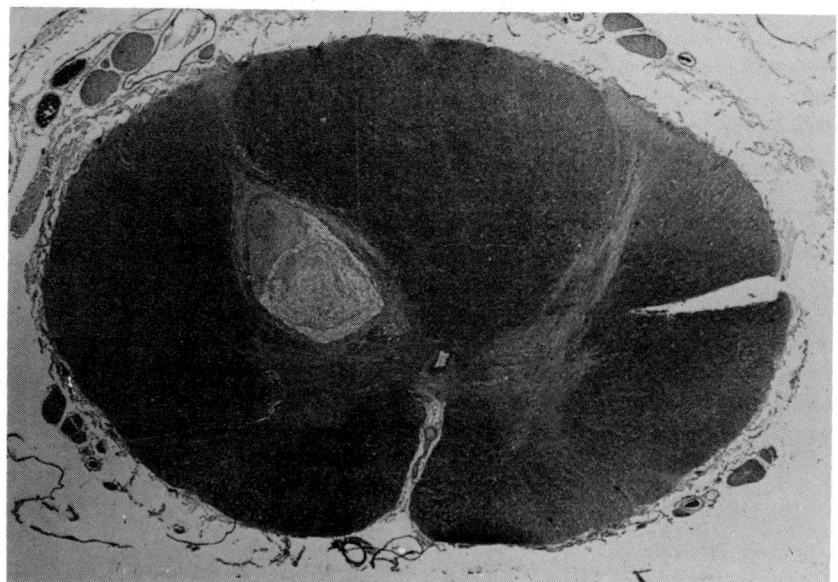

Fig. 14.18. Case S.A.F. Trauma at C4–5, 14 days before death. Haemorrhagic lesion (at C7) in dorsal horn extending caudad (see also Fig. 14.17(a)).

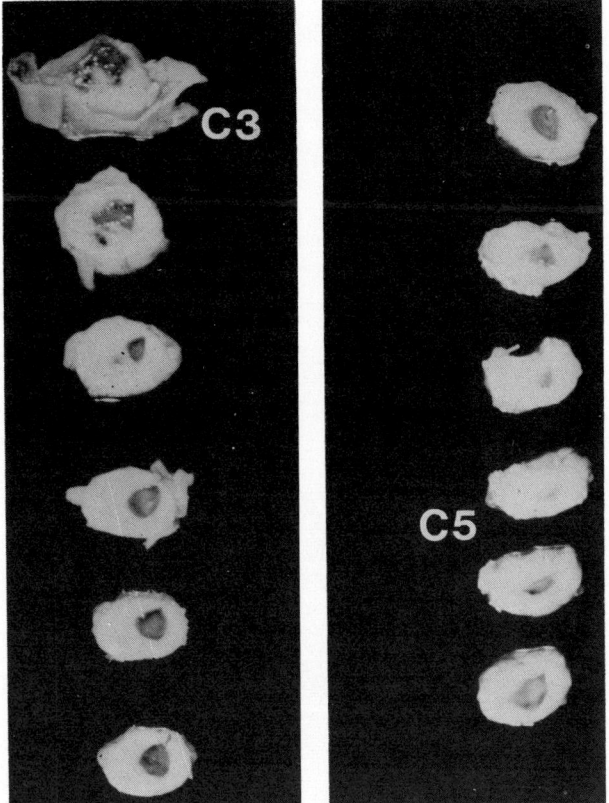

Fig. 14.19. Haemorrhagic necrotic pulp beginning at site of trauma (C3), diminishing at C5 and then expanding below. (T.M.—as in Fig. 14.17(b).)

These two cases, therefore, are indicative of the fact that significant haemorrhagic lesions within the central cord may occur above and below the site of the trauma. Also, and more significantly, they demonstrate that the haemorrhagic lesion may be greater in size at a site removed from the maximally traumatised segment. The lower level of the lesion cannot have developed simply as a direct extension of the haemorrhage from the site of trauma, since an intervening segment was almost free of any evidence of haemorrhage.

In a series of experiments, Allen (1911, 1914) produced spinal cord injuries in dogs. Grey matter haemorrhages were demonstrated and the possibility suggested that the serous fluid and haemorrhage, which at times were under significant pressure, might secondarily damage the adjacent tissues.

Wagner, Dohrmann and Bucy (1971) have recently added to Allen's observations. They produced transient traumatic paraplegia in rhesus monkeys and illustrated small haemorrhages in the pericentral regions and in the dorsal horns of the grey matter. They detected early changes in the central, compared with the peripheral vessels of the spinal cords subjected to controlled but moderate trauma (Fig. 14.20). Krogh's work (1945) suggested that these early central lesions might be partly of venous origin since the veins were considered more susceptible than the arterioles to trauma. Fairholm and Turnbull (1971) studied the microvasculature in experimental spinal cord trauma and have delineated a central area of capillary damage (Fig. 14.21), with pericapillary haemorrhage and necrosis (Fig. 14.22) in the posterocentral portion of the cord. This necrotic zone is surrounded by a second area in which the intimate vasculature is intact, but despite this, neuronal and axonal damage is severe. This necrosis can be detected histologically within two hours of the trauma, more rapidly, therefore, than the time required to detect changes from ischaemia by the same light-microscope technique.

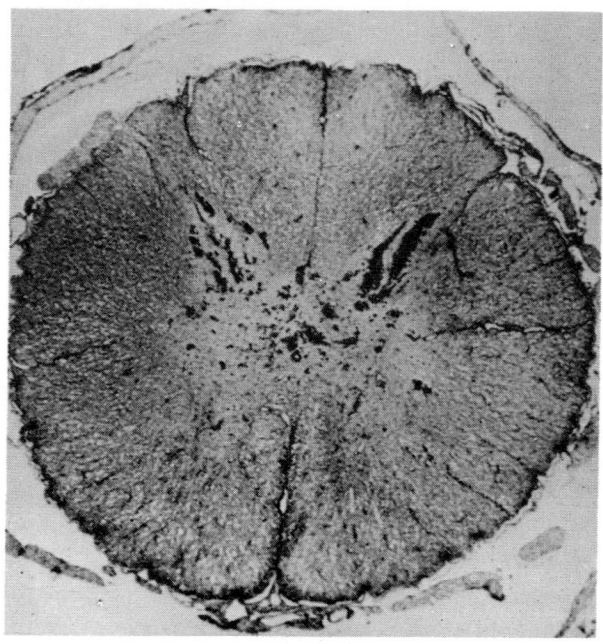

Fig. 14.20. Experimental trauma to spinal cord of Rhesus monkey, producing small haemorrhages in grey matter. (Reproduced by permission of Wagner et al, 1971, and *Journal of Neurosurgery*.)

Recent work by Osterholm and Mathews (1972) suggests that chemical, as well as mechanical (pressure) and vascular factors may mediate and cause extension of some of this cord necrosis. Central haemorrhagic necrosis in their experiments did not appear for the first hour after a blow of controlled force (500 g/cm) was applied. During this hour, there was a heavy local build-up of norepinephrine. They demonstrated that there was a correlation between the norepinephrine content of the injured tissue and the presence of central grey haemorrhage. They claimed that they were able to inject minute quantities of norepinephrine atraumatically into the grey matter and within two hours visualise large central haemorrhages resembling the lesions of severe spinal injury. They postulated

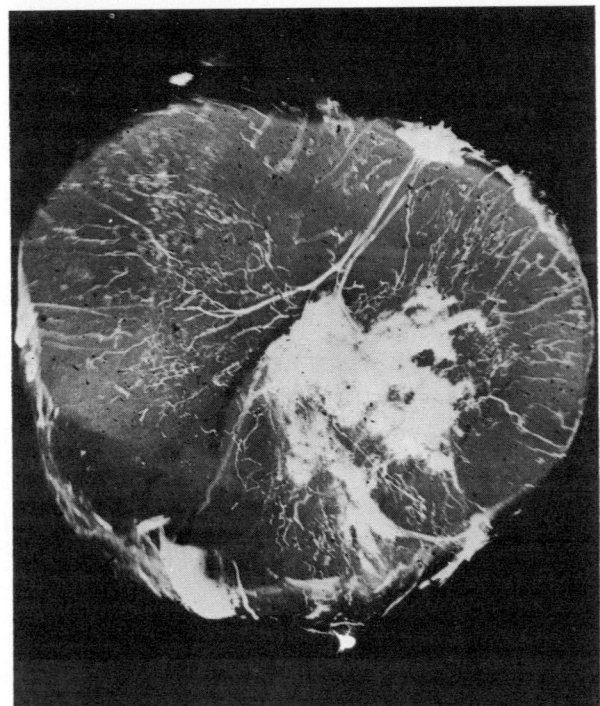

Fig. 14.21. Experimental cord trauma with microvascular injection indicating capillary damage. (Reproduced by permission of Fairholm and Turnbull, 1971, and *Journal of Neurosurgery*.)

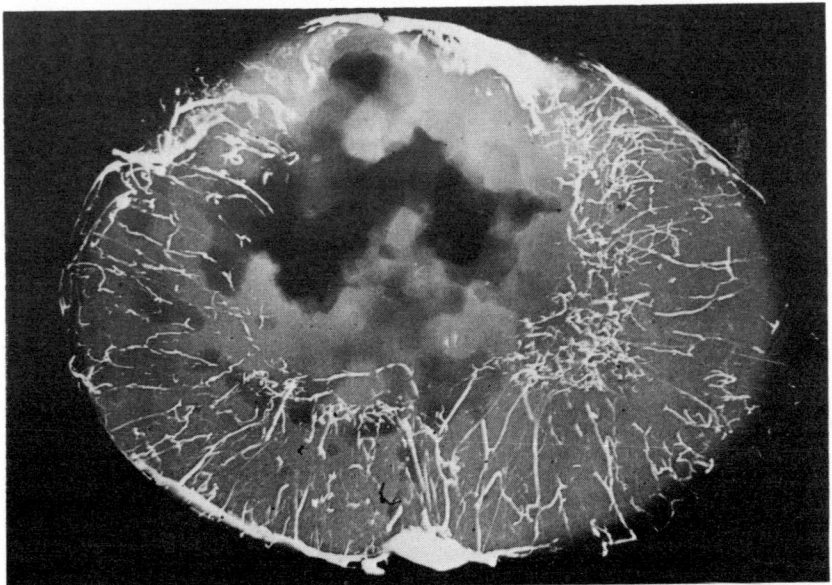

Fig. 14.22. Experimental cord trauma with microvascular injection at a later stage than Fig. 14.21 indicating posterocentral necrosis. (Reproduced by permission of Fairholm and Turnbull, 1971, and *Journal of Neurosurgery*.)

that the tissue norepinephrine induces intense vasospasm severe enough to impede cord perfusion.

Unlike the fatal human cases and the experiments of McVeigh (1923) and of Allen (1911, 1914), these later experimental injuries were less violent and reflect the more subtle vascular and other tissue changes resulting from less devastating blows. Nevertheless, from all this evidence the conclusion seems justified that variable amounts of haemorrhage and/or disintegrating nervous tissue are to be expected at the site of trauma in the spinal cord and that it will be particularly prominent in the ventral part of the dorsal columns and the medial parts of the dorsal horns. It will extend cranially and caudally from this site for variable distances. Resorption of this myelomalacic tissue, or of this haemorrhage, or both, may result in cavity formation.

The Role of Ischaemia in the Pathogenesis of this Early Lesion

Trauma producing infarction of the spinal cord has been discussed since the days of the classical writings of the descriptive neurologists. Collier (1916), writing about gunshot wounds of the spinal cord, described a condition of 'total necrosis of the distal segment' of the cord in a case afflicted with complete severance of the cord at T3. The patient survived for three months with a total motor, sensory and reflex loss below the level of injury and, at post-mortem, the cord below the injury was completely necrotic, shrunken and 'resembled a strip of greenish-yellow, wet wash-leather'. There was no sign of haemorrhage or infection. Collier referred to a similar case recorded by Beevor; in this instance a C6 birth injury was survived for six weeks and the cord appeared necrotic below C6 at post-mortem. Without anatomical verification, Collier attributed the lesions to severance of the anterior and posterior spinal arteries and distal spread of thrombosis. Cavity formation was not described, and although the patients had the same signs as in syringomyelia, the lesions were not progressive and did not extend cephalad.

Tauber and Langworthy (1935) reviewed the aetiology of syringomyelia and laid great emphasis on ischaemic infarction in its production. They equated myelomalacia with cavity formation and referred to a number of writings to support their claims (Spiller, 1906, 1909; Martin, 1925; Chung, 1926; Grinker and Guy, 1927; Ornsteen, 1931). All of these were well-documented reports of spinal cord infarction, often they were in patients with syphilis, some were associated with struggling and effort, most had signs compatible with a syrinx (or any other centrally located spinal lesion) but *none truly involved cavitation*.

It is not reasonable to ascribe to ischaemia a direct role in the development of a progressive cervical cord lesion many years after a lumbar or lower thoracic cord trauma. However, ischaemia at the original site of injury may be a factor in determining the nature and location of the haemorrhagic necrotic cores of tissue which appear to be the precursors of the later cystic cavitation beginning at, and near the traumatic site. Davison (1943) demonstrated thrombosis of meningeal arteries in relation to spinal cord injury with associated haemorrhagic infarction of the underlying traumatised cord. What part was played in the production of the cord damage by the vascular occlusion is the question at issue. In a review of circulatory disturbances in paraplegia, Wolman (1965) stated that traumatic occlusion of major spinal arteries is a rare event, but that the smaller pial arteries were more commonly found to be occluded. In 95 cases dying with spinal cord injury, only three had evidence of major arterial occlusion.

Woodard and Freeman (1956) produced cavitation in the dog by tying off paired anterior and posterior nerve roots extradurally over six successive thoracic segments. The radicular arteries were obliterated and in 12 of their 16 animals cavitation occurred in the midline grey matter dorsal to the central canal. The larger cavities extended from this central point along the dorsal horns and often involved the adjacent white matter of the posterior or lateral funiculi. They concluded from their histological studies that the milder degrees of cavitation were simply a passive separation of tissue comparable to a *hydrocephalus ex-vacuo*, but that the more severe lesions produced cavities that were true myelomalacic ischaemic lesions. It was their opinion that these lesions did not represent ischaemia in any particular vascular territory but that they reflected a 'general reduction in circulation', and they speculated on the importance of impaired venous drainage brought about by cord swelling.

The work of Hassler (1966) and of Turnbull, Breig and Hassler (1966) on the pattern of intimate vasculature of the spinal cord would indicate that the vulnerable 'marginal zone' area of spinal cord blood supply (Fig. 14.23) coincides with the territory of occurrence of many of these experimental ischaemic necroses and many of these local cavities. (This marginal zone, as illustrated in Fig. 14.23, is to be compared with the sites of lesions as in Figs. 14.13, 14.14 and 14.29.) It also corresponds in a remarkable way to the territory of infarction in the case of cord ischaemia described by Garland, Greenberg and Harriman (1966) and in some, but not all, the cases documented by Hughes and Brownell (1966). In the latter

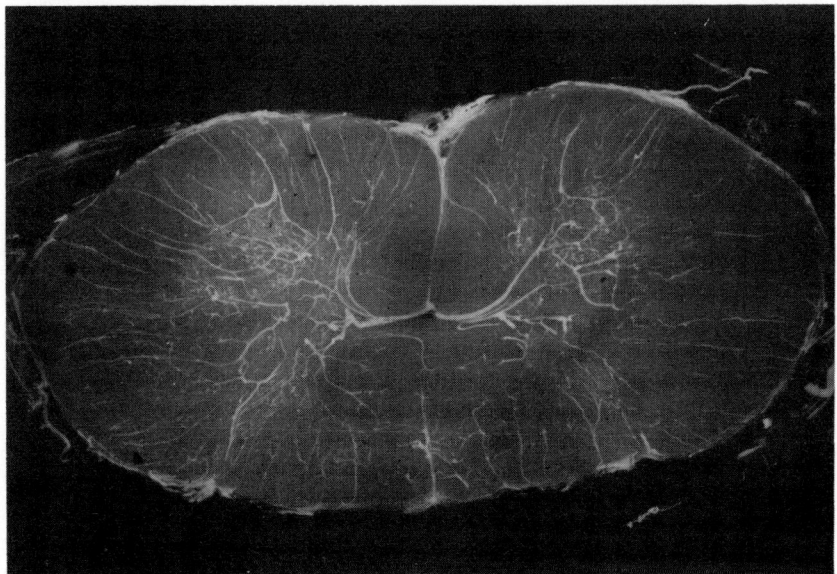

Fig. 14.23. Microvascular injection of spinal cord indicating central zone of marginal blood supply. (Photograph loaned by Dr Ian Turnbull.)

report, some of the specimens were the site of many patchy areas of infarction in grey and white matter as opposed to the centrally located marginal zone infarctions of the former. Zulch (1962), reviewing the pathological lesions of spinal infarction, felt that the evidence indicates two areas prone to marginal-zone infarction, one in the posterior horn, the other in the ventral part of the dorsal column. The marginal zone as interpreted by Zulch would embrace the site of involvement of other lesions described in Holmes' original (1915) paper, as well as that illustrated here (Fig. 14.13), and would contain the lesions in Fig. 14.28 and 14.29.

Schneider et al (1970) have illustrated cystic cavitation in the central cervical cord at C2 in a hyperextension football injury in which they judged the vertebral artery to have been mechanically compromised. The patient died within five weeks of the injury, a well-defined cystic lesion replacing the necrotic tissue (Fig. 14.24). Mechanical factors, alone or in combination with the ischaemia, could not be excluded. In one of his cases, Jellinger (1967) has written on progressive arteriosclerotic vascular myelopathy, with cystic cavitation illustrated in the dorsal horn. Wilson et al (1969), in a recent report of experimental cord ischaemia and a review of ischaemic cervical myelopathy noted the predilection for the central grey and inner part of the white matter to be involved in these processes. Cavitation was illustrated in the 'cleft between grey and white matter'.

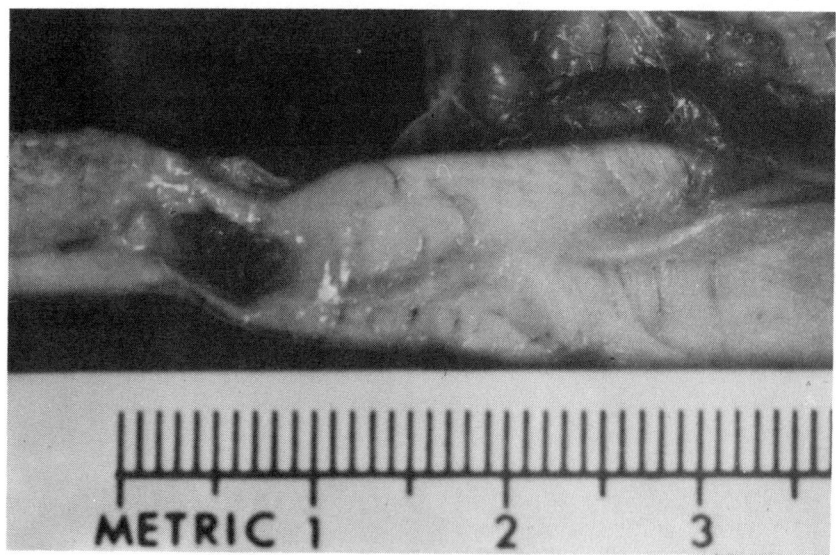

Fig. 14.24. Cavitation at C2 in spinal cord of patient afflicted with hyperextension injury of 5 weeks duration. (Reproduced by permission of Schneider et al, 1970, and *Journal of Neurosurgery*.)

It must be noted that the extent of cavitation developing in all these un-equivocally ischaemic lesions has been small—never as large at the site of maximum damage as the post-traumatic condition described in this paper. Furthermore, and of considerable significance, *there is no report available of a known arterial, ischaemic lesion becoming cystic and then extending up or down the cord*. Our pathological material is too limited for sweeping conclusion, but no significant lesions of the spinal vessels were ever detectable in the early and later cases recorded here. Localised ischaemic cavitation, however, does have a predilection to involve the same part of the cord as these post-traumatic cavitations. If survival is prolonged, reports may yet appear of extensive ischaemic cavitation. For the time being though, it seems reasonable to conclude that central infarction may play a contributory part in determining the early lesion at the site of trauma, but that the extensive later lesions cannot be accepted as basically ischaemic.

Cystic Lesions at the Site of Trauma

It is recognised that severe local trauma inflicted on the spinal cord may result in a local cystic lesion replacing the necrotic and haemorrhagic tissue. Bastian (1867) published a detailed post-mortem examination of a 26-year-old man who 'was sleeping on the top of an unfinished hayrick, 25 feet in height, and whilst asleep, rolled off, falling on his back. He found himself unable to move, and was conveyed to the Barnet Union.' Quadriplegia led to his death *six months later*. 'A transverse section through his hardened cord, through the upper part of the cervical enlargement, showed a large rupture extending obliquely from before, backwards across the grey matter of the right side' (Fig. 14.25). The 'rupture' is clearly a cavitation throughout the dorsal horn on one side. At one level, it extended as well into the ventral horn. Other early writers on spinal injuries (Lloyd, 1894; von Leyden and Goldscheider, 1895; Mann, 1896) recorded late cavitations developing at the site of spinal injury and ascribed them to the resorption of a traumatic haematomyelia. The condition was said to be exclusively confined to the cervical area.

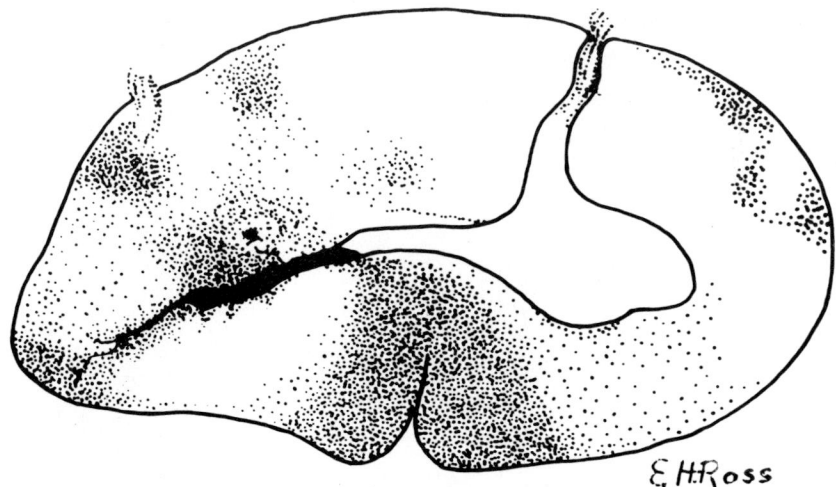

Fig. 14.25. Cavitation at site of trauma 6 months after injury. (Adapted from Bastian's publication of 1867.)

Freeman and Wright (1953) illustrated experimental spinal cord concussion and contusion in dogs which they kept alive for many weeks and months. They described several instances in which cystic lesions were demonstrable at the site of trauma. The cysts were bulging under tension and collapsed when excised. In their opinion, these cysts were less likely to form at the site of trauma if the injury was followed immediately by a midline longitudinal incision of the pia.

Davison (1943) studied post-traumatic cavities in association with pulp-like cylindrical necrotic cores and noted their occurrence mostly in the dorsal horns or dorsal columns near the grey matter. The necrotic lesions were thought to be explained on the basis of vascular constriction or thrombosis, as well as rupture of the walls of small blood vessels within the cord, brought about by invaginated blood and pulp displacing normal tissue within the cord. Haemorrhage, when it occurred, was variable in amount and was considered to derive from this indirect mechanism of rupture of the blood vessel walls as well as directly to the trauma. All this necrotic and haemorrhagic tissue was judged liable to liquefy and resorb, and was considered to be the forerunner of a cavity that might be of considerable size and extent. Cavitation was detected in cases surviving longer than six weeks.

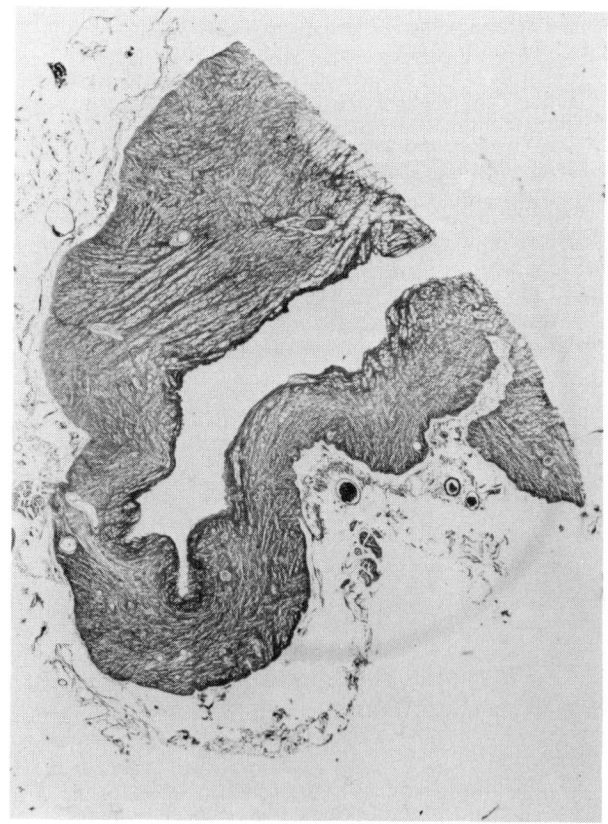

Fig. 14.26. Half-section of cross section of cord in patient paraplegic T9–10 for 4 years. This cystic dilatation was taken from cord at T5 level. (Reproduced from *Brain*, by permission, after Barnett et al, 1966.)

Wolman (1965) carried out a series of post-mortem analyses of 95 patients dying with both recent and long-standing spinal cord injuries. In the early cases, haemorrhage was a frequent finding. It was confined at times to the damaged segment, but more frequently it extended upwards and downwards in a tapering, elongated tube involving as many as a dozen segments. It was more prevalent in the posterior than the anterior horns, and more in the grey than the white matter. More than half the cases survived more than a month. Older haemorrhagic lesions demonstrated an increasing tendency for cavity formation when the haemorrhage was being absorbed. Twice, much larger cavities were detected that extended over many segments. These were described as the end-result of haematomyelia.

In our initial report on this condition (Barnett et al, 1966), reference was made to a patient rendered paraplegic in 1948 by a fracture dislocation of T9 to 10. The patient had died in 1952 and two half-sections of his spinal cord were available— one at T9 to 10 showing a traumatic disruption of the cord, the other at T5 to 6 with a cavity extending through much of the available remnant of cord tissue (Fig. 14.26). The cystic lesion was lined by a mixture of glial and connective tissue elements, not by ependyma. The material in this case was incomplete, the patient had developed no upper limb symptoms but cavitation extended well above the level of original trauma.

We have had the opportunity to study in detail the spinal cord of a patient with a four-month survival after fracture dislocation. Cavitation was found similar in appearance and location to the cavitations in the earlier clinicopathological descriptions, similar to those produced in the experiments referred to, and similar to those seen in recent post-mortem studies on patients with delayed syringomyelia after serious spinal cord injury. The importance of this case is that it represents an intermediate stage between the acute type of lesion and the post-traumatic progressive myelopathy of delayed onset.

A.G., MALE
On 26 June, 1969, a 54-year-old man was admitted to Victoria Hospital, London, Ontario, having suffered a fracture dislocation of C6 on C7. The dislocation was anterior and practically the depth of the vertebral body. He had complete sensory loss below the C6 dermatome bilaterally, feeble power of elbow flexion bilaterally, but complete paralysis in both triceps and in the muscles below the elbow on either side. An interbody C6 to 7 fusion was carried out 10 days after the injury. Due to sepsis at the operative site, coupled with pulmonary complications, he died on 28 October, 1969, four months after injury. He had made no detectable recovery from his quadriplegia.

Pathological Findings. The entire spinal cord, with attached medulla, was removed and the location of the lesions is sketched in Fig. 14.27. A large, irregularly shaped syrinx extended in two directions from the C7 segment of main trauma: rostrally to about C3, and caudally as far as the caudal portion of T1. It was situated primarily within the left dorsal horn and posterior column (Fig. 14.28) and at all levels caused considerable displacement of the adjacent grey matter (particularly the left posterior horn and posterior commissure), the posterior median septum and the surrounding white matter of both dorsal columns. The cavity appeared to taper at both ends, with no extension into any glial scar. In one area it crossed the midline dorsal to the central canal (Fig. 14.29). A second, small syrinx was situated in the right dorsal column, involving the C4 and C5 cord segments. Its rostral end was abrupt, but its caudal end extended into a dense linear glial scar within the ventral half of the right posterior column. An irregular slit-like cavity could be discerned in the depths of this scar (Fig. 14.30). *It did not extend down to the level of the trauma.*

The central canal had a tiny lumen at all levels. Both the syrinxes were situated dorsal to the central canal but neither communicated with it, nor did they connect with one another. The walls of both syrinxes were composed of a mixture of dense astrocytic fibres, large numbers of reactive astrocytes (many of which still had abundant eosinophilic cytoplasm), and small amounts of collagenous fibres, especially around vessels (Fig. 14.31). The astroglial fibres were generally arranged haphazardly, although sometimes they ran in a radial manner towards the lumen of the syrinx, or occasionally, parallel to its circumference. Except for the rare focus in which a few very attenuated endothelial-like cells could be found covering the glia (Fig. 14.32), the cavities appeared to have no true lining. Certainly no ependymal-type epithelium was seen. Trabeculae crossing the cavities were not a feature of this case. The absence of any obvious epithelial lining to the syrinx was confirmed with the scanning electronmicroscope (Fig. 14.33), which clearly showed the dense matting of interdigitating glial fibres forming the syrinx wall.

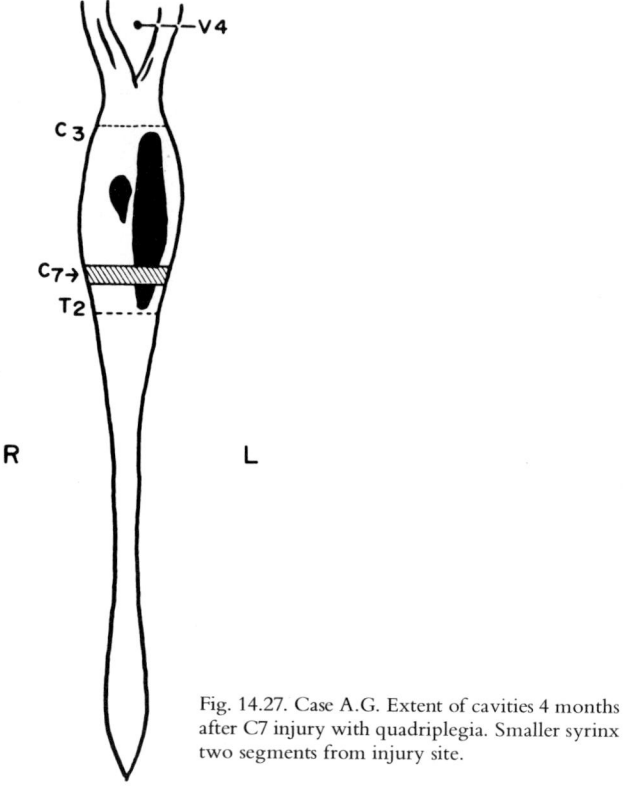

Fig. 14.27. Case A.G. Extent of cavities 4 months after C7 injury with quadriplegia. Smaller syrinx two segments from injury site.

The usual degree of architectural distortion was present at the traumatised zone and mild gliosis extended rostrally in all funiculi to the level of the inferior olives, and caudally for a short distance as well. A mild degree of leptomeningeal fibrosis was noted, involving most cervical segments. No vascular occlusions were detected in any sections; nor was there evidence of gross communication between the spinal subarachnoid space and the damaged central portion of the cord, despite serial microscopic sections of the site of major trauma.

Recognition of this early cavity formation localised to the segments close to the traumatic site has implications for the consideration of the pathogenesis of the late extending lesion. It may also have some significance in the management of the paraplegic patient before the ascending symptoms develop. This consideration was brought to our attention by the following case described to us by Drs Wm. Lougheed and Wm. Geisler:

Fig. 14.28. Case A.G. Magnified photograph of C6 cervical segment with a large syrinx affecting the left posterior column and horn. Fracture was at C7.

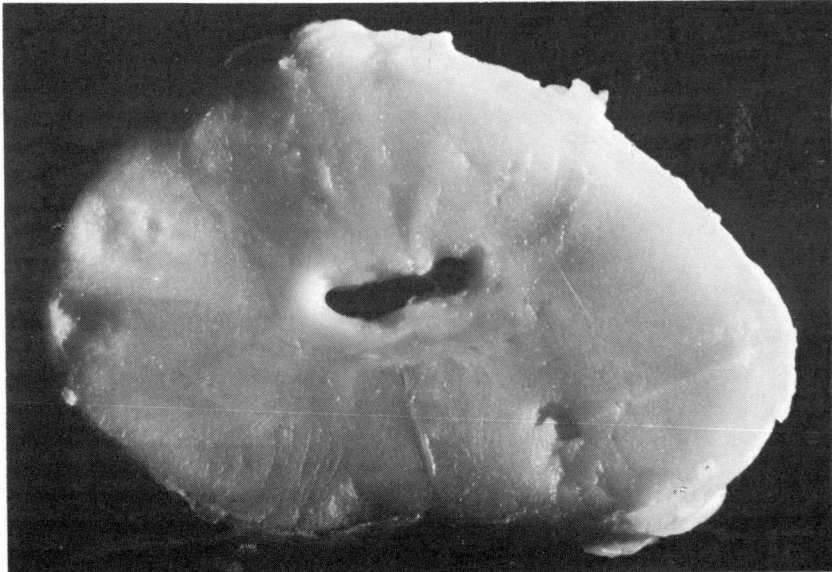

Fig. 14.29. Case A.G. Magnified photograph of C6 cervical segment with a large syrinx crossing the midline centrally and dorsal to central canal. Fracture was at C7.

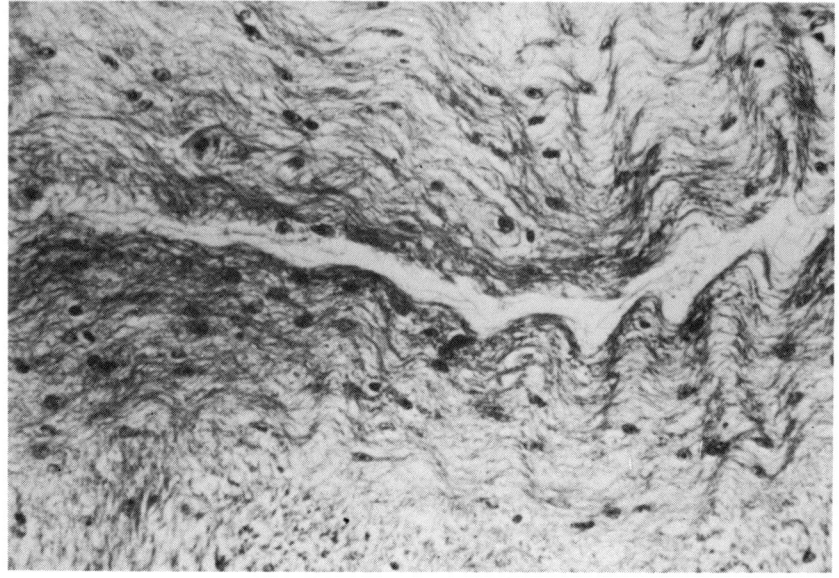

Fig. 14.30. Case A.G. Photomicrograph of the caudal termination of the smaller, right-sided syrinx. Note the slit-shaped cavity within this dense, glial scar. (× 400.)

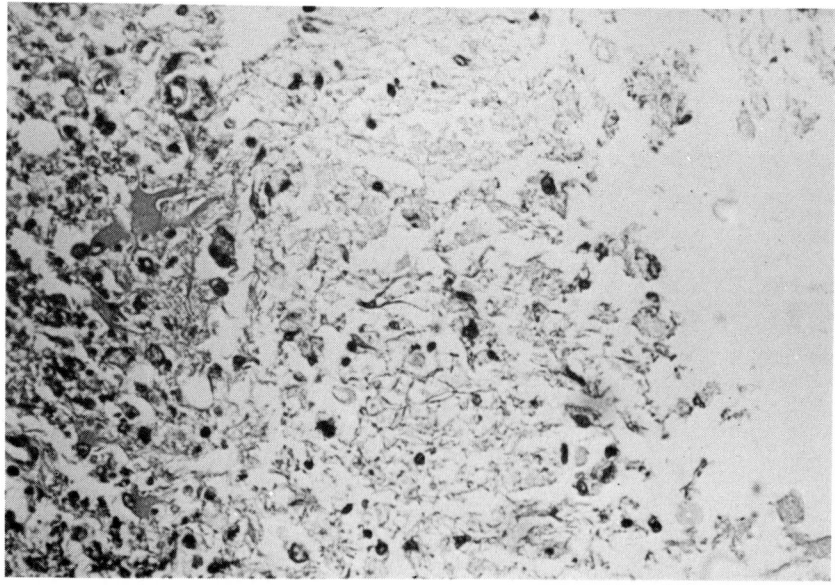

Fig. 14.31. Case A.G. Cavity lined by loosely packed layer of glial fibres. Occasional reactive astrocyte with prominent eosinophilic cytoplasm.

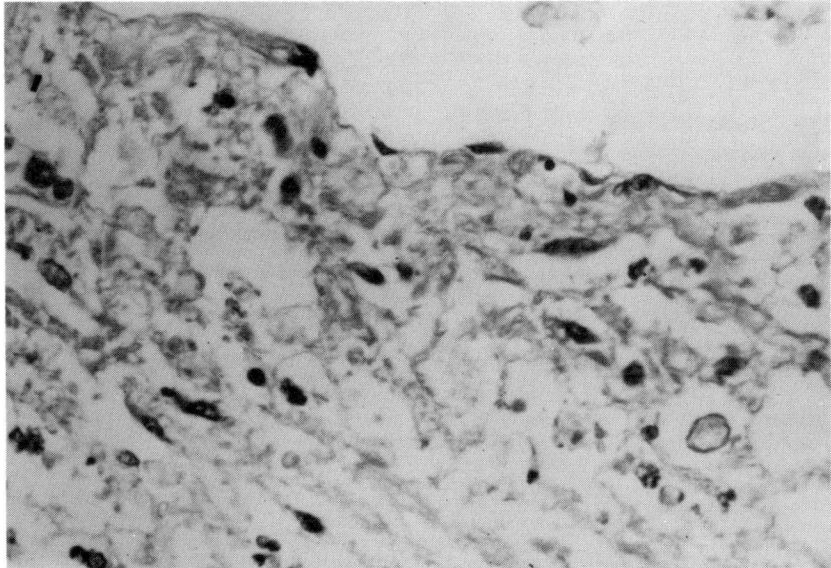

Fig. 14.32. Case A.G. Lining of cavity largely of loose astrocytic fibrils, but a few flattened endothelial-like cells.

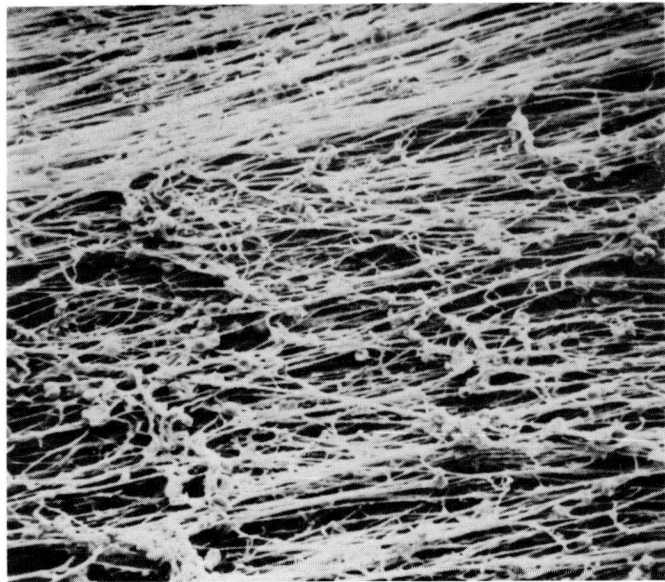

Fig. 14.33. Case A.G. Scanning electron-micrograph of the inner aspect of the syrinx wall. Occasional red blood cells can be seen within the dense glial mesh. No epithelium is present. (× 500).

M.Z., MALE

On 9 April, 1971, this patient, then aged 37 years, was rendered paraplegic from a fracture dislocation of T6 and T7 as a result of a motor accident. His motor and sensory deficit at and below T7 was complete and a fairly marked kyphosis was present at the site of the bony lesion. No initial operative procedure was carried out. Within two months of the accident he began to experience pain of girdle type, beginning in the back at the level of the deformity and radiating just above his lower costal margin bilaterally. At first it was of spontaneous origin, but within a few weeks was clearly precipitated and aggravated by trunk movement. It was most severe anteriorly and to the level of the inferior posterior axillary areas. The pain consistently was referred to a narrow band located just above the area of complete sensory loss. The sensory loss was to all modalities and became abruptly normal so that the painful area was not in an area of reduced sensibility. In December 1971, the pain was very distressing and sensory nerve root blocks at T5, 6 and 7 were carried out without relief.

In March 1972, his pain was intractable so that a T4 to 7 laminectomy was carried out. The cord was enveloped in adhesions and was angulated at the site of the gibbus. It was freed and was completely transected at T5. When this was done, two significant observations were made. Firstly, *there was a syrinx occupying the centre of the cord*, surrounded by a rim of remaining spinal cord. Secondly, the *cord retracted cephalad three-quarters of an inch*, demonstrating that it had been tethered at this site. Posterior rhizotomy at T5, 6 and 7 was done bilaterally. Postoperatively the patient was free of all pain in his chest wall and his sensory loss had extended upwards by two dermatomes. Despite the persistence of this sensory deprivation, some chest wall pain returned with the eventual addition of burning dysaesthesiae in his insensitive legs.

This problem remains unresolved, but it does illustrate the development of a syrinx at the site of injury. It appears to be an example of a localised post-traumatic cavitation contributing to a painful syndrome in a paraplegic. It vividly illustrates the lack of normal mobility that may be a sequel to the reaction in the tissues surrounding a segmental cord injury. The possibility that such tethering contributes to the extension of an intramedullary spinal cavity will be discussed later in this chapter.

Comments on Dorsal Column and Dorsal Horn Predilection for Early and Late Lesions

Clinical or experimental, traumatic, necrotic lesions and post-traumatic cavitations favour the same areas of the dorsal horn and anterior dorsal column. Cushing's comment (1898) in this regard was that the grey matter gave less firm support to its blood vessels than did the white matter. The greater amount of traumatic haemorrhage in the cervical cord was ascribed to the relatively more extensive grey matter in the cervical region than elsewhere. Cushing (1898) referred to some injections into the spinal cord of cadavers, made by von Leyden and Goldscheider (1895), in which it was found that the grey matter, chiefly in the posterior horn, was favourable to the spread of injected substance and presumably also to the extension of spontaneous haemorrhage.

Davison (1943) explained the location of the cavities in the ventral part of the posterior column, and the dorsal horn, on the basis of comparatively 'poor blood supply to this area'. He related the extension up and down the cord to a 'stasis of serum or cerebrospinal fluid which, in trying to escape, tracks up and down the cord in places of least resistance, thus leading to the formation of a cavity'. Wolman (1965), on the other hand, felt that the 'richer blood supply and less rigid consistency of the tissue' in the grey matter was responsible for the location of the lesions.

The recent studies of the intimate vasculature of the spinal cord (Hassler, 1966; Turnbull et al, 1966) are germane to this problem. In the first place, they indicate that the anterior portion of the dorsal column and the medial portion of the posterior horn are least supported by an arterial vascular framework. Injections of the spinal veins (Gillilan, 1958, 1971) demonstrate a similar absence of a supporting venous framework by comparison with surrounding tissue. The possibility exists that the supporting architectural stroma of the penetrating arteries and veins is a protection against the extension of cavities into the anterior, lateral and more peripheral parts of the cord. The absence of such supporting stroma may be a contributing factor in the localisation in the areas of predilection—plainly the areas least supported (Fig. 14.23).

Additionally, these injection experiments reveal that the favoured sites for the post-traumatic cores of necrotic tissue and of the subsequent extending cavities are the areas of poorest collateral blood supply, the so-called 'border zone areas'. As discussed, this may suggest that arterial ischaemia plays a significant role in the initial necrotic haemorrhagic lesions. What role it may play in the later extension cephalad and caudad is more difficult to determine. Possibly the cyst extension at its upper and lower margins can occur more readily because of ischaemic necrosis produced mechanically in the zone of marginal blood supply.

Venous infarction may be an additional factor determining the site as well as the nature of the early lesion. Hughes' (1971) recent pathological studies have clearly identified the central posterior horns and ventral part of the dorsal columns as the major sites of haemorrhage and haemorrhagic infarction in cases of thrombosis of spinal veins (Fig. 14.34). Turnbull et al (1966) demonstrated that the veins originate in these critical central parts of the grey matter. The condition of the spinal veins in acute cases where haemorrhage is present is deserving of more careful scrutiny.

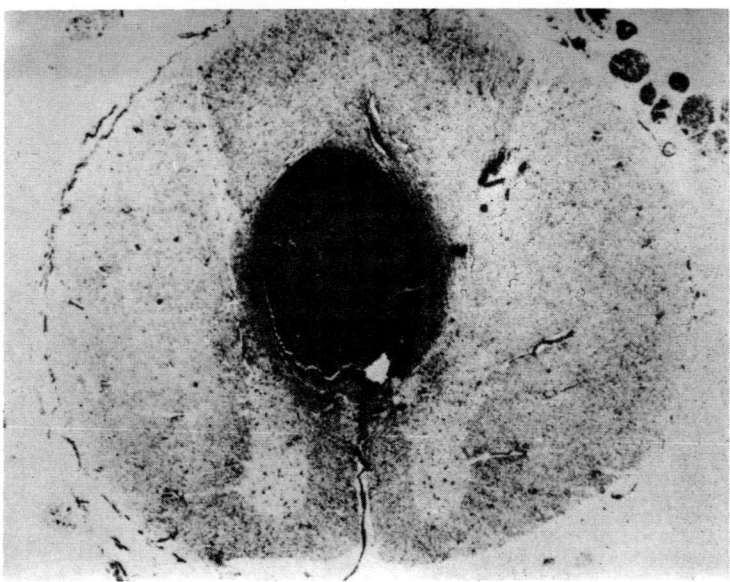

Fig. 14.34. Central haemorrhage and haemorrhagic infarction in a case of spinal vein thrombosis. (Reproduced by permission of Dr Trevor Hughes, 1971, and *Neurology*.)

Finally, the capillary circulation in the spinal cord is known to be much richer in grey than in white matter. Wolman (1965) spoke of the richness of the blood supply of the central part of the cord. Craigie's (1972) work in rabbits, and the work of Fazio in humans (1938) clearly identifies the fact that the central grey matter is much more heavily vascularised with capillaries than is any part of the white matter. This is typical of grey matter, with the capillary supply in direct proportion to the number of nerve cell bodies. The dorsal horn area is even more richly supplied than the rest of the grey matter (Fig. 14.35). This area of greatest capillary concentration is to be compared with the early lesions illustrated in Figs. 14.16, 14.18, 14.19 and 14.20.

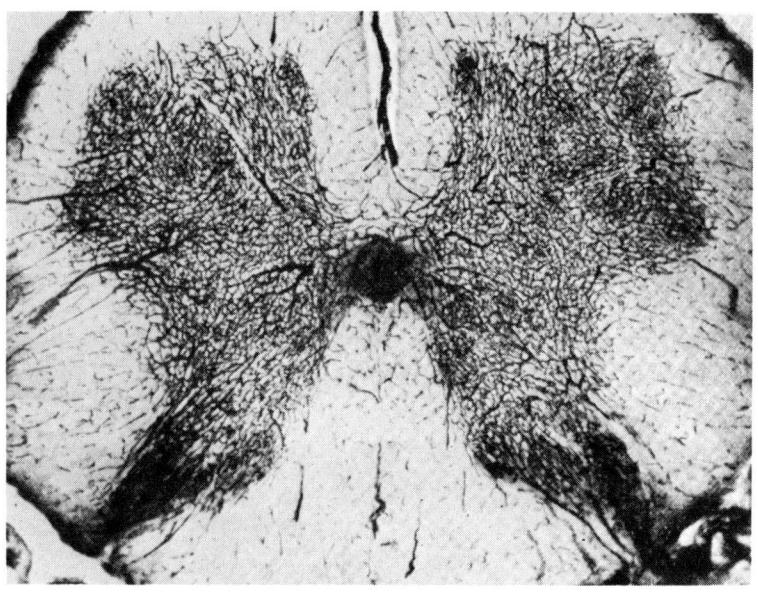

Fig. 14.35. Microvascular studies indicating capillary prominence in grey matter and dorsal horn, with particular prominence in posterior grey matter. (Reproduced by permission from Fazio, C., 1938, and *Rivista di Patologia nervosa e mentale*.)

A combination of vascular and mechanical factors most likely contribute to the predilection for traumatic haemorrhage and/or necrosis to favour the dorsal horns and central portion of the posterior columns. It is of interest that this is the site of extending cavitation whether one considers cavitation with arachnoiditis or cavitation associated with tumour. The information available to explain why this central dorsal area is the site of major involvement may be summed up as follows:

1. The capillary supply of the grey matter, particularly of the dorsal horns, is greater than that of the periphery of the cord.
2. Blows to the cord lead to early haemorrhage in this area of rich capillary supply.
3. The release of norepinephrine in the area of early damage may lead to a spread of necrosis, probably by inducing vasospasm of the intrinsic spinal cord arterial

supply. Confirmation of this work is needed but it is, none the less, intriguing.
4. This central dorsal area is the watershed area of arterial blood supply.
5. Infarction from venous obstruction, as judged by the area of haemorrhagic lesions with venous thrombosis, is maximal in this central dorsal portion of the cord.
6. The intrinsic veins and arteries of the cord, as judged by studies of the microvasculature, give a less significant supporting framework in this central area than they do to the rest of the white and grey matter of the spinal cord.
7. The dense fibre tracts in the lateral and posterior columns prevent disrupted necrotic tissue and/or blood from extending either laterally or posteriorly. The anterior commissure presumably constrains its anterior spread. The result is an extension up and/or down in the central dorsal area.

Cavitation Remote from the Trauma

The evidence so far presented indicates that the cavitation in these cases communicates with lesions at the site of trauma. Davison (1943) describes one instance of injury to the third lumbar vertebra in which the lumbar and lower thoracic regions of the cord were said to be intact. Pulp-like necrotic areas were present in the grey matter and in the ventral part of the posterior columns in the lower cervical and upper thoracic regions. He quotes three earlier German writers on spinal cord injuries who describe a blow at one point of the vertebral column associated with necrotic lesions confined to other distant areas of the cord.

Schneider and Crosby (1959) reported a patient with a fracture of T12 and a spinal cord lesion whose sensory level was at T7. In the myelogram, the cord at T7 was locally swollen and this swelling was sufficient to cause a block to the passage of the contrast material. They related this swelling to ischaemic necrosis due to interference with the arterial circulation at the level of injury. In their review, Henson and Parsons (1967) consider the level of the thoracolumbar vertebral junction to be the critical level for the most important of the lower radicular arteries (arteria radicularis magna—the artery of Adamkiewicz) and its damage from a T12 fracture could produce a major infarction remote from the primary traumatic site. In a further publication, Schneider and Schemm (1961) reported on a central cord necrosis six segments *below* a cervical injury, and again this was attributed to a radicular vessel supplying a part of the cord remote from the point of entry of the vessel into the spinal canal.

We have described and illustrated above a case with central cord haemorrhage distant from the trauma and with an area nearly normal between the upper and lower areas of the haemorrhagic lesion (Figs. 14.17b and 14.19). The presence of this intervening, nearly normal segment negates the possibility that the total haemorrhage originated at the site of injury and forced its way through the cord under pressure. It also raises the possibility that cavitation may begin and extend from an area of spinal cord remote from the initial site of trauma. To date no case of later progressive cystic myelopathy, distinctly removed from the original site, has been examined pathologically. The nearest approach to this might be case A.G. (Figs. 14.28, 14.29 and 14.30) also described above, where there was a syrinx above and below the level of trauma but with a second cavity involving the mid-cervical cord only, starting two full segments above the original C7 level of injury.

The Source of Fluid in the Cavity

The source of fluid in these cavities is of interest in the proper management and appropriate surgical approach to these disabling lesions. It is of even greater significance in the more fundamental problem of the origin of a fluid resembling C.S.F. from a cavity within the spinal cord itself that rarely has an identifiable communication with the subarachnoid or ventricular spaces. Our data on the constituents of the fluid is incomplete and only a small amount of further information is available from other recorded cases. Yet, we have noted in Chapter 11 that the fluid may be of low or even normal protein content and may be considerably lower in this regard than in the C.S.F. below the obstruction in the subarachnoid space. For example, in case 6 the content of the protein in the low lumbar space was 1500 mg/100 ml while that in the fluid from the cavity was 38 mg/100 ml (see Table 10.7).

One possible source for this cavity fluid is through a *sinus at the level of spinal cord injury* connecting the subarachnoid space with the central cord substance. Williams (1970) has published an illustration, supporting his thesis on 'communicating' syringomyelia in which this mechanism of cavity formation and extension in these post-traumatic cases is argued. There is only one observation to date to support this hypothesis, and it was in a myelogram reported to us by Allen (1972) and described briefly in Chapter 10. This phenomenon was not observed in any of our own myelographic studies, nor in any of our patients operated on at the original traumatic site. It was not visualised in our own, or the other three autopsy cases in the literature. In case 2, a communication appeared to have been present but from serial sections it was considered artefactual. An attempt was made to find a sinus at the site of injury between the subarachnoid space and the earlier cavitation in the patient who died four months after injury and had a well-defined cavity (Figs. 14.27, 14.28 and 14.29). Again, this search by serial sectioning of the cord was unsuccessful. It would be likely that the reaction to the injury and the constrictive nature of the arachnoidal adhesions about the injured site as a rule would obliterate such a sinus rather than encourage it to remain patent. Allen's case seems to be an outstanding exception to this situation.

The discrepancy between the lumbar fluid content and the intracystic fluid in case 6 would argue further against this 'sinus' concept. The 1500 mg/100 ml of protein in the lumbar space cannot be dismissed as an accidental sampling of fluid from a small location of concentrated and proteinaceous liquid in an area of arachnoiditis. In this particular case, the injured site was T4 and the fluid was obtained at the time of initiating *lumbar* myelography. The myelogram indicated a lumbar and lower thoracic subarachnoid space free of obliterative adhesions. Quite obviously, the fluid in the syrinx in this instance, with its normal protein content, did not come from the subarachnoid space at and below the site of injury where the protein content was 1·5 g/100 ml.

Secondly, fluid might enter the cavity by *connection between the fourth ventricle and the central canal*. In our case at post-mortem, the cyst was closed at the rostral end and the central canal patent only within the medulla. There was no evidence of any connection with the central canal at any level, despite the fact that the cord was serially sectioned. Neither did any connection exist with the fourth ventricle. The three reported cases were also devoid of any connection between the ventricle and the spinal cavity. This was true even in Klawans' case (1968) where there was some extension of the spinal syrinx into the medulla oblongata. Although this mechanism has been cited as a major factor in many non-traumatic syringomyelic cavitations, it does not seem to be the usual situation in this group of patients. As mentioned in Chapter 12, there is some clinical evidence from two surgical cases that post-traumatic syrinxes may establish communication with the fourth ventricle (Tucker, 1972; Nurick, Russell and Deck, 1970). This remains to be confirmed, but even then, the evidence suggests that if such a communication ever truly exists it develops as a sequel to the ascending lesion and not as the initiating defect.

The third possibility, to explain the slow enlargement of an apparently closed cavity, would be the *production of fluid by the cells lining the cavity wall*. For the most part, glial cells have been demonstrated to surround the cavity. These astrocytes may be in no way unusual, but the possibility must be considered that they are derived from precursors of the ependymal cells that are known embryologically to form the posterior median septum (Chambers and Liu, 1972). This may make it possible for them to secrete a fluid resembling C.S.F. On the other hand, if there is production of C.S.F. within the spinal cord, it may be by cells other than the ependymal epithelium of the central canal. In the cat, experiments have recently been conducted which demonstrate that in this animal, at any rate, the ependymal cells do not have a function in producing C.S.F. (Becker, Wilson and Watson, 1972).

If the glial cells which line these cavities are able to produce a fluid comparable to C.S.F. they must also be capable of an absorptive mechanism. The fluid increases slowly in amount and yet retains a clear consistency and a fairly low protein content after many years of symptoms indicative of the presence of an intramedullary lesion. The protein content of this post-traumatic cavity fluid is distinctly lower than that encountered in cavities associated with tumours of the brain and spinal cord. Gardner, Collis and Lewis (1963) postulate that the fluid in cysts surrounding brain tumours is the result of diffusion of plasma from the capillaries of the tumour. The absence of a lymphatic drainage system is said to be the reason for its accumulation in such tumour cases. The blood brain-barrier mechanism is said to be abolished in them because the capillaries of the tumour acquire the permeability characteristics of extracerebral capillaries because of the presence of mesodermal elements in the tumour, ordinarily foreign to brain. It would appear as if there is a comparable situation in the cord with respect to intramedullary tumours. However, with the post-traumatic cavities the blood-brain barrier mechanism appears to persist; the fluid apparently is transcellular and not interstitial. The flow back and forth through glial linings (as opposed to fibrous tissue linings) is apparently effected in the normal way for C.S.F. production and resorption.

Irregularly placed glial ridges identical to those noted in the more usual type of syringohydromyelic case (Fig. 14.36), which are the anatomical substrate of the lateral indentations visualised radiographically (Fig. 10.14), have been seen in these post-traumatic cavitations. The rather prominent vessels present within these glial ridges may also be of considerable importance, not only in the production of fluid but also in its continuous movement and turnover.

212 NON-COMMUNICATING SYRINGOMYELIA

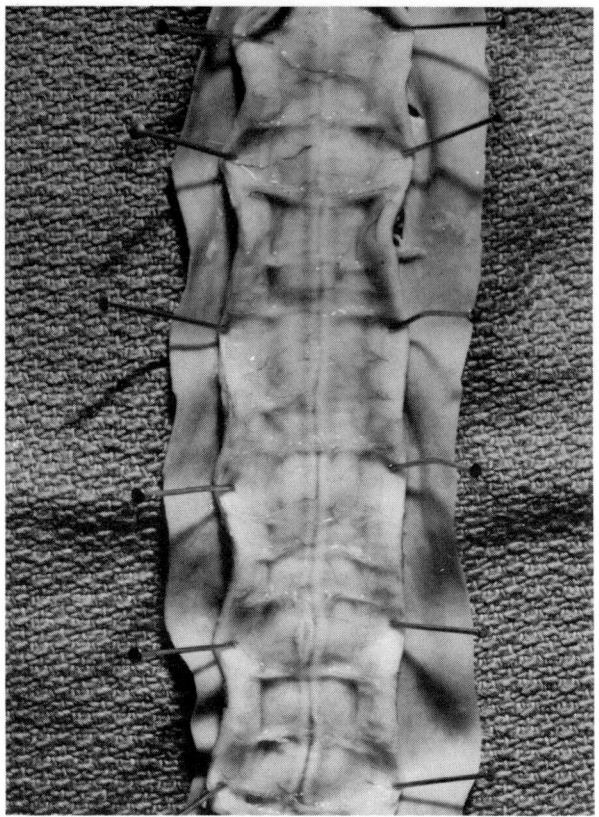

Fig. 14.36. Irregularly spaced glial bands across lumen of syrinx from case of non-traumatic syringomyelia. These glial bands do not follow a segmental pattern. (Cf. Fig. 10.14, Chapter 10.)

The Extension of the Cavity

One of the impressive aspects of the syrinx in this type of cavitation, and for that matter in any other variety of spinal cavity formation, is the tendency to gradual and occasionally sudden increase in size. This contrasts with the usual intracerebral cystic lesion so common within areas of previous infarction. There probably are many reasons why there is the difference in the tendency for cavity extension in the intracranial and intraspinal locations. They share one common characteristic which makes it difficult for them to handle extracellular fluid accumulation—both lack lymphatic drainage systems. Evidence is available though, that major differences in the two areas exist in respect to the pulsations within the subarachnoid space in the spinal canal and those of the intracranial cisternal areas. The effect of Valsalva manoeuvres within the spinal and the cerebral C.S.F. pathways is not the same. Everyday accustomed movements change the shape and tensions exerted on the spinal cord much more than they influence the brain.

Du Boulay (1966) has been a pioneer in studying pulsations in the C.S.F.

pathway. He has observed a striking difference between the amplitude of pulsatile movement of oily contrast medium above and below the foramen magnum. The excursion of the contrast medium in the upper cervical canal is about $\frac{1}{2}$ in. whereas that in the lower part of the medullary cistern is 1/16 in. Du Boulay (1972) has observed at myelography on numerous occasions that, with coughing or other Valsalva-like manoeuvres, the epidural veins become greatly distended and that from 2 to 8 ml of spinal fluid is forced out of the spinal into the cerebral subarachnoid space under great pressure. This head of venous pressure leads to marked narrowing of the spinal subarachnoid space and 'squeezing' of the cord readily visible at fluoroscopy. The squeezing process starts in the lower part of the canal but is maximal in the cervical region. It is probable that the same venous pressure increase exerts a similar effect on the intracystic fluid. Direct observations on this have not yet been made. The clinical histories clearly indicate an effect from Valsalva manoeuvres, but do not distinguish between an effect on the subarachnoid space and on the intracystic fluid.

Paraplegic and quadriplegic patients have to substitute vigorous muscular activity in their shoulders and, if possible, their upper limbs and neck muscles to compensate for their disabilities and to remain mobile. Using wheelchairs, getting in and out of beds and cars, and other activities which involve moving the body with the arms alone, produce repeated Valsalva manoeuvres which raise the spinal fluid pressure. Coughing and sneezing have the same effect and were frequently noted by the patients in this series to be the immediate precursors of earliest painful and other sensory symptoms. This rise in pressure in the C.S.F. around the cord would be transmitted to the side walls of the cyst and cause it to extend. By the time the cyst has become large, thin-walled and near the surface, these pressure changes would be even more noticeable. The absence of valves in this venous system probably contributes to the prolongation of the effect of a Valsalva manoeuvre. As mentioned above, the raised venous pressure may be transmitted directly to the intracystic fluid with a similar tendency to produce extension.

The tendency to extend below the injury site might be discouraged by strong fibrous scar tissue blocking the lower end. On the other hand, in the highest traumatic lesion reported developing this syndrome, the case of Feigin et al (1971) with a C4 injury, the extension of the cavitation took place in a *downwards* direction only, reaching to the mid-lumbar region and not going into the cervical cord above the site of traumatic disruption. In this case, raised intraspinal pressure of an intermittent but recurrent nature would have been less effective in transmitting stress to the uppermost cervical segment because of the arachnoid block at C4. If this hypothesis is correct in this instance, it may also explain why this complication of spinal trauma was found in 1·8 per cent of paraplegic patients injured at or below T4, while the incidence dropped in the quadriplegic cases to 0·2 per cent.

It would seem likely, then, that increased intraspinal and probably intrasyringal pressure will be transmitted to the spinal cord and that the gradual effect on the developed cavity will be to cause it to extend into the part of the cord offering least resistance—to move cephalad and/or caudad, whichever offers the least resistance. As the syrinx gradually extends, it will cause the tissues to be separated from their intimate blood supply. It would be expected that the leading edge of the cavitation might be the site of gliosis because of pressure and ischaemia, and this was noted in our post-mortem studies (Figs. 14.8 and 14.30) where the tapering ends of the syrinxes terminated in a very dense glial scar. Such a leading edge can well be thought of as dissecting along favourable longitudinal tissue planes.

The irregular positioning of the glial ridges (see Figs. 10.14 and 10.17) and their lateral location within the cavities (Fig. 14.36) suggest that they may represent the remnants of previous 'leading edges' of the syrinx. A temporary dam to further extension of the lesion may have been caused by this glial scar, then later it may have been ruptured by excessive forces leading to increased intramedullary pressure. In this way, the ridges may be considered to represent episodic extension of the cavities and might be likened to the growth rings in the cross-section of a tree trunk.

The Site of Maximal Dilatation

In this delayed clinical syndrome, symptoms are most marked in the area subserved by the cervical cord. In fact, the cord below the cervical region and down to the level of original trauma may manifest little, if any, overt dysfunction. The cavity in the cervical cord has been larger than in the thoracic and lumbar cord in all autopsy cases except the high cervical injury of Feigin et al (1971). The extra stresses and tensions that can be transmitted to the cervical cord are obviously important in determining this particular cervical distention of extending cavities.

The observations of Breig (1960) have indicated how much these tensions in the spinal cord can be altered with neck movements (Fig. 14.37). The neck and the underlying cervical cord is capable of the greatest excursion. Intrinsic movements here have a greater range than the rest of the spinal axis and much more than any of the intracranial central nervous tissue. A marker placed on the surface of the cervical spinal cord will move as much as the distance of one spinal segment between the maximum laxity of the cord in neck extension and its greatest elongation in neck flexion (Fig. 14.38). What additional influence the preponderance of grey matter over white matter might have in altering the susceptibility to cavity formation in the cervical region is a matter for speculation.

The Role of the Arachnoiditis in Pathogenesis

Leptomeningeal thickening and adhesions due to spinal cord trauma are common in the traumatic paraplegic, but extend only a few segments above and below the level of trauma. This post-traumatic arachnoiditis is not an ascending lesion, although it may be progressively constricting. Nevertheless, it may be associated with cystic cavitation of the cord under various circumstances, and the subject is reviewed in Chapter 15. It cannot be dismissed as a possible contributing factor in the condition under discussion.

The arachnoiditis might play a part by compressing the blood supply with resulting cystic degeneration of the underlying structures secondary to ischaemia. We have no histological evidence, though, of vascular occlusion or stenosis in our autopsy cases nor in the surgical material we have reported. This constrictive effect might operate only at capillary level. If vascular factors are significant they probably apply for the most part to the early lesions and a constrictive vasculopathy is not likely to be of great consequence in the later development of this sequel to injury.

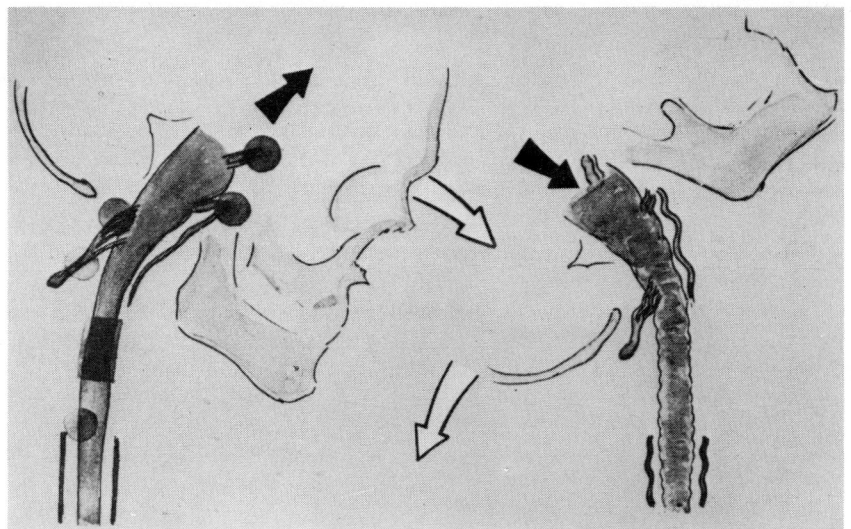

Fig. 14.37. Flexibility of spinal cord as demonstrated by laxity in hyperextension and taut appearance in flexion. (Reproduced by permission of A. Breig, 1960, and Almquist & Wiksell, Stockholm.)

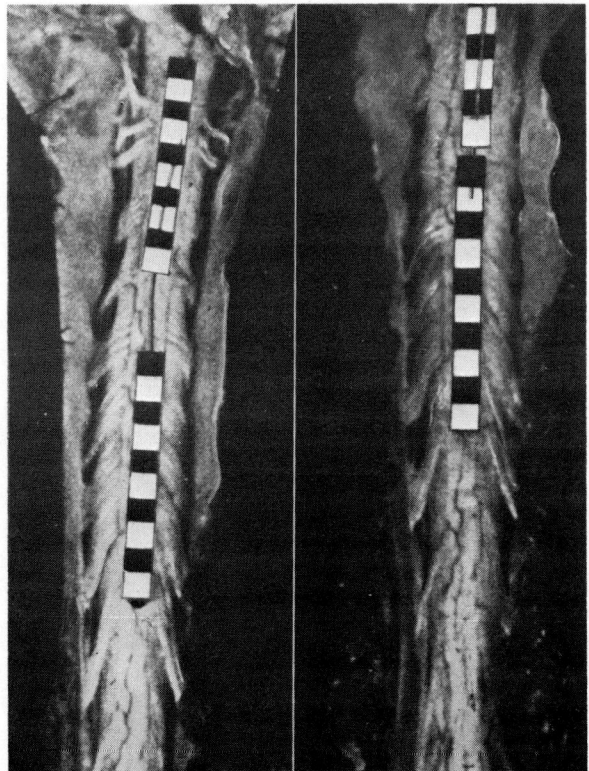

Fig. 14.38. Mobility and flexibility of cervical spinal cord. Markers placed on cadaver's cord in position of neck extension (right). Markers separate the width of one cord segment in full neck flexion (left). (Reproduced by permission of A. Breig, 1960, and Almquist & Wiksell, Stockholm.)

The possibility exists that by fixing and tethering the cord, arachnoidal adhesions play some role in determining the extension of the cavitation. The tethering has been shown to be so severe in some instances that freeing the adhesions and then transecting the cord in completely paraplegic individuals has been followed immediately by a retraction of the transected cord for a distance of as much as 3/4 in. It is reasonable to presume that the flexibility of the cord in normal neck movements (Fig. 14.37 and 14.38) is also an adaptation of importance in other conditions known to produce intraspinal pressure change (e.g. Valsalva manoeuvres). Since we have good evidence that the arachnoidal adhesions produce an unequivocal tethering of the cord, and even retraction when this is released (case M.Z., this chapter; case 14, Chapter 12; case 4, Chapter 15), it is highly probable that the tethered cord is deprived of the mobility which will allow it to adapt to movement and to coughing, sneezing and straining. Our case histories contain many references to the aggravation of symptoms, of both temporary and persisting nature, that will follow these stresses. It is difficult, then, to deny the significance of the arachnoidal adhesions, for some of these cases and the cord tethering as a potential factor in cavity extension. The extent of the arachnoiditis at the site of the injury may affect the speed of progression of the cavitation. This is speculative and we have no substantiating evidence for it.

The Role of Compression on Syrinx Production

Chronic extramedullary spinal cord compression has also been associated with cystic cavitation of the cord. It is an unusually rare situation, but a progressing cystic intramedullary degeneration has been described with it. It is reviewed in Chapter 17, on tumours. Since a number of the cases in this series have been treated by early laminectomy and bony decompression, this type of compression cannot be considered an important factor. If it has a role it may be as part of the mechanism operating in association with arachnoiditis.

Summary of Data on Pathology and Pathogenesis

1. The delayed myelopathy as a sequel to spinal cord injury is due to a syrinx which extends upwards from the level of the injury and in high cervical lesions may extend downwards.
2. In early death from spinal cord injury, necrotic and haemorrhagic lesions are located in the dorsal horns and the ventral parts of the dorsal columns. In this same area, cavities are located after a survival of a few months. These cavities then extend to produce the long lesions, centrally and asymmetrically located, which produce the clinical picture of delayed traumatic cystic myelopathy.
3. The cavities have clear fluid with a protein content that may be normal or only slightly elevated, but they rarely have any communication with the subarachnoid space, and equally rarely, with the fourth ventricle.
4. The cavities are lined for the most part with a glial network but there is some collagen in the lining, particularly in the vicinity of the original trauma. The suspicion exists that these glial structures may produce and resorb a fluid resembling C.S.F.
5. The extension of the cavities is regarded as a result of transmission of venous back-pressure—brought about by coughing, straining and sneezing—to the spinal fluid and thence to the spinal cord. Once a cavity has become large enough

to distend the cord, it will be more susceptible to these pressures and extend more readily.

6. The tethering of the cord at the site of trauma by dense arachnoidal adhesions has been clearly established and, by freeing the cord of these adhesions and transecting it, considerably retraction of the cut ends of the cord has been demonstrated. Presumably this restores its flexibility and hence its adaptability to the extremes of movement and intraspinal pressure changes. The existence of this arachnoiditis and tethering appears to be an important factor in the extension within the cord of these postmyelomalacic cavities.

REFERENCES

Allen, A. R. (1911) Surgery of experimental lesion of spinal cord equivalent to crush injury of fracture dislocation of spinal column—a preliminary report. *Journal of the American Medical Association*, **57**, 878.
Allen, A. R. (1914) Remarks on the histopathological changes in the spinal cord due to impact—an experimental study. *Journal of Nervous & Mental Diseases*, **41**, 141.
Allen, P. B. (1972) Personal communication, paper in preparation.
Barnett, H. J. M., Botterell, E. H., Jousse, A. T. & Wynn-Jones, M. (1966) Progressive myelopathy as a sequel to traumatic paraplegia. *Brain*, **89**, 159.
Bastian, H. C. (1867) On a case of concussion-lesion with extensive secondary degeneration of the spinal cord. *Proceedings of the Royal Medical and Chirurgical Society of London*, **50**, 499.
Becker, D. P. Wilson, J. A. & Watson, G. W. (1972) The spinal cord central canal: response to experimental hydrocephalus and canal occlusion. *Journal of Neurosurgery*, **36**, 416.
Bennett, J. H. (1860) *Clinical Lectures on the Principles and Practice of Medicine*, p. 392. New York: S. S. & W. Wood.
Blackwood, W., Dodds, T. C. & Sommerville, J. C. (1970) *Atlas of Neuropathology*, p. 147. Edinburgh: E. & S. Livingstone Ltd.
Breig, A. (1960) *Biomechanics of the central nervous system; some basic normal and pathologic phenomena*. Stockholm: Almqvist & Wiksell.
Chambers, W. & Liu, Chan-Nao (1972) In *The Spinal Cord*, ed. Austin, G., 2nd edn., p. 111. Springfield: C. C. Thomas.
Chung, Mon-Fah (1926) Thrombosis of the spinal vessels in sudden syphilitic paraplegia. *Archives of Neurology & Psychiatry*, **16**, 761.
Collier, J. (1916) An address on gunshot wounds and injuries of the spinal cord. *Lancet*, **i**, 711.
Craigie, E. H. (1972) In *The Spinal Cord*, ed. Austin, G., 2nd edn., p. 111. Springfield: C. C. Thomas.
Cushing, H. (1898) Haematomyelia from gunshot wounds of the spine. A report of two cases with recovery following symptoms of hemilesion of the cord. *American Journal of the Medical Sciences*, **115**, 654.
Davison, C. (1943) Trauma of the central nervous system. Pathology of the spinal cord as a result of trauma. *Proceedings: Association for Research in Nervous & Mental Diseases*, **24**, 151.
Du Boulay, G. H. (1966) Pulsatile movements in the CSF pathways. *British Journal of Radiology*, **39**, 255.
Du Boulay, G. H. (1972) Personal communication.
Fairholm, D. J. & Turnbull, I. M. (1971) Microangiographic study of experimental spinal cord injuries. *Journal of Neurosurgery*, **35**, 277.
Fazio, C. (1938) L'angioarchitettonica del midollo spinale umano e i suoi rapporti con la cito-mieloarchitettonica. *Rivista di Patologia Nervosa e Mentale*, **52**, 252.
Feigin, I., Ogata, J. & Budzilovich, G. (1971) Syringomyelia: the role of edema in its pathogenesis. *Journal of Neuropathology & Experimental Neurology*, **30**, 216.
Freeman, L. W. & Wright, T. W. (1953) Experimental observations of concussion and contusion of the spinal cord. *Annals of Surgery*, **137**, 433.
Garland, H., Greenberg, J. & Harriman, D. G. F. (1966) Infarction of the spinal cord. *Brain*, **89**, 645.
Gardner, W. J., Collis, J. S. & Lewis, L. A. (1963) Cystic brain tumours and the blood-brain barrier. Comparison of protein fractions in cyst fluids and sera. *Archives of Neurology*, **8**, 291.

Gillilan, L. A. (1958) The arterial blood supply of the human spinal cord. *Journal of Comparative Neurology*, **110**, 75.

Gillilan, L. A. (1971) *Cerebral Vascular Diseases, Seventh Conference*. New York: Grune & Stratton.

Grinker, R. R. & Guy, C. C. (1927) Sprain of cervical spine causing thrombosis of anterior spinal artery. *Journal of the American Medical Association*, **88**, 1140.

Hassler, O. (1966) Blood supply to human spinal cord. A microangiographic study. *Archives of Neurology*, **15**, 302.

Henson, R. A. & Parsons, M. (1967) Ischaemic lesions of the spinal cord: an illustrated review. *Quarterly Journal of Medicine*, **36**, 205.

Holmes, G. M. (1915) The pathology of acute spinal injuries. *British Medical Journal*, **ii**, 769.

Hughes, J. T. (1966) *Pathology of the Spinal Cord*. London: Lloyd-Luke.

Hughes, J. T. (1971) Venous infarction of the spinal cord. *Neurology*, **21**, 794.

Hughes, J. T. & Brownell, B. (1966) Spinal cord ischemia due to arteriosclerosis. *Archives of Neurology*, **15**, 189.

Jellinger, K. (1967) Spinal cord arteriosclerosis and progressive vascular myelopathy. *Journal of Neurology, Neurosurgery & Psychiatry*, **30**, 195.

Jung, E. (1960) Syringomyelie in Kombination mit Entwicklungsstorung der Nieren und mit schwerer Wirbelsaulenverletzung. *Medizinische Klinik*, **55**, 1678.

Klawans, H. L. (1968) Delayed traumatic syringomyelia. *Diseases of the Nervous System*, **29**, 525.

Krogh, E. (1945) Studies on the blood supply to certain regions in the lumbar part of the spinal cord. *Acta Physiologica Scandinavica*, **10**, 271.

Lloyd, J. H. (1894) Traumatic affections of the cervical region of the spinal cord, simulating syringomyelia. *Journal of Nervous & Mental Diseases*, **21**, 345.

Mann, L. (1896) Klinische und anatomische Beiträge Lehre von der spinalen Hemiplegie. *Deutsche Zeitschrift für Nervenheilkunde*, **10**, 1.

Martin, J. P. (1925) Amyotrophic meningo-myelitis (spinal progressive muscular atrophy of syphilitic origin). *Brain*, **48**, 153.

McVeigh, J. F. (1923) Experimental cord crushes, with special reference to the mechanical factors involved and subsequent changes in the areas of the cord affected. *Archives of Surgery*, **7**, 573.

Nurick, S., Russell, J. A. & Deck, M. D. (1970) Cystic degeneration of the spinal cord following spinal cord injury. *Brain*, **93**, 211.

Ornsteen, A. M. (1931) Thrombosis of the anterior spinal artery. *American Journal of the Medical Sciences*, **181**, 654.

Osterholm, J. L. & Mathews, G. J. (1972) Altered norepinephrine metabolism following experimental spinal cord injury. Part I: relationship to hemorrhagic necrosis and post-wounding neurological deficits. *Journal of Neurosurgery*, **36**, 386.

Rewcastle, N. B. & Hugenholtz, H. (1969) Personal communication.

Schneider, R. C., Cherry, G. R. & Pantek, H. (1954) Syndrome of acute central cervical spinal cord injury with special reference to the mechanisms involved in hyperextension injuries of cervical spine. *Journal of Neurosurgery*, **11**, 546.

Schneider, R. C. & Crosby, E. C. (1959) Vascular insufficiency of brain stem and spinal cord in spinal trauma. *Neurology*, **9**, 643.

Schneider, R. C., Gosch, H. H., Norrell, H., Jerva, M., Combs, L. W. & Smith, R. A. (1970) Vascular insufficiency and differential distortion of brain and cord caused by cervico-medullary football injuries. *Journal of Neurosurgery*, **33**, 363.

Schneider, R. C. & Schemm, G. W. (1961) Vertebral artery insufficiency in acute and chronic spinal trauma. *Journal of Neurosurgery*, **18**, 348.

Spiller, W. G. (1906) Syringomyelia. *British Medical Journal*, **ii**, 1017.

Spiller, W. G. (1909) Thrombosis of the cervical anterior median spinal artery; syphilitic acute anterior poliomyelitis. *Journal of Nervous & Mental Diseases*, **36**, 601.

Tauber, E. S. & Langworthy, O. R. (1935) A study of syringomyelia and the formation of cavities in the spinal cord. *Journal of Nervous & Mental Diseases*, **81**, 245.

Turnbull, I. M., Breig, A. & Hassler, O. (1966) Blood supply of cervical spinal cord in man. A microangiographic cadaver study. *Journal of Neurosurgery*, **24**, 951.

Tucker, H. H. (1972) Personal communication.

von Leyden, A. & Goldscheider, A. (1895) *Die Erkankungen des Rückenmarks und der Medulla oblongata*, p. 85. Vienna: Alfred Holder.

Wagner, F. C., Dohrmann, G. J. & Bucy, P. C. (1971) Histopathology of transitory traumatic paraplegia in the monkey. *Journal of Neurosurgery*, **35**, 272.
Williams, B. (1970) The distending force in the production of communicating syringomyelia. *Lancet*, **ii**, 41.
Wilson, C. B., Bertan, V., Norrell, H. A. & Hukada, S. (1969) Experimental cervical myelopathy. II. Acute ischemic myelopathy. *Archives of Neurology*, **21**, 571.
Wolman, L. (1965) The disturbance of circulation in traumatic paraplegia in acute and late stages: a pathological study. *Paraplegia*, **2**, 213.
Woodard, J. S. & Freeman, L. W. (1956) Ischemia of the spinal cord—an experimental study. *Journal of Neurosurgery*, **13**, 63.
Zulch, K. J. (1962) Reflexions sur la physiopathologie des troubles vasculaires medullaires. *Revue Neurologique*, **106**, 632.

CHAPTER FIFTEEN

Syringomyelia Associated with Spinal Arachnoiditis

H. J. M. BARNETT

Spinal cord cavitation as a consequence of lesions interfering with drainage of cerebrospinal fluid from the fourth ventricle and leading to the development of hydromyelia has gained widespread acceptance. Lichtenstein (1943) was the first modern writer to suggest that syringomyelia was a diverticulum from a hydromyelia and backed up his thesis with a pathological specimen. Schüppel (1865) had speculated in this direction much earlier, without morphological proof. Harris (1886) had also offered the suggestion to explain the association of tumour and syringomyelia. The most common abnormality found in surgical series has been a Chiari malformation (Gardner, 1965; Foster, Hudgson and Pearce, 1969; Logue, 1971). Equally convincing evidence has accumulated that syringohydromyelia (or as we prefer to call it, 'communicating' syringomyelia) on occasions, may be a sequel to inflammatory and postinflammatory leptomeningitis ('arachnoiditis') involving the basal cisterns and obstructing the foramen of Magendie (Appleby et al, 1969). It is essential now to reassess the role of *spinal* forms of arachnoiditis in the pathogenesis of spinal cord cavitation since 'arachnoiditis' and 'chronic meningitis' have appeared prominently in most early writings on the aetiology of syringomyelia.

A complete review of the whole subject of arachnoiditis is not within the scope of this chapter. It appears that it was more common in the era of the classical descriptive neurologists. Charcot and Joffroy (1869), and Joffroy (1873), first wrote about it and over the ensuing decades several varieties came to be identified. They were generally distinguished by a circumscribed or diffuse distribution and by the presence or absence in the arachnoid of loculated fluid (e.g. meningitis serosa spinalis—Mendel and Adler, 1908). This terminology is not based on aetiology as a rule, nor does it take into account the fact that the arachnoid is seldom involved alone. The three enveloping membranes are usually actively involved and adherent—a 'pachymeningitis with a leptomeningitis' rather than an arachnoiditis, frequently would be an acceptable designation for the condition under discussion but despite the inaccuracies implicit in the term, arachnoiditis will be freely used here.

Reviews of arachnoiditis have appeared sporadically, the most comprehensive being those of Horsley (1909), Elkington (1936), Hinds Howell (1936), French (1946), Winkelman, Gotten and Scheibert (1953), Lombardi, Passerini and Migliavacca (1962), Weiss, Sweeney and Dreyfuss (1962). The more usual aetiological factors have been syphilis (pachymeningitis cervicalis hypertrophica), tuberculosis (both Pott's disease and healed tuberculous meningitis), healed pyogenic meningitis, trauma, non-traumatic haemorrhage into the subarachnoid space, reaction to spinal anaesthetic agents, reaction to radio-opaque material (including thorium dioxide as well as more modern agents that are not radioactive), reaction to detergents in syringes used to introduce anaesthetic or radio-opaque substances and finally, a large and unidentified group. Elkington (1936), for example, could recognise no determining factor in 18 of 43 cases. Our present concern is to determine when this disorder, with these varying causes, may provoke cavitation in the underlying spinal cord.

EARLY WRITINGS

The earliest report of spinal arachnoiditis producing spinal cord cavitation was that of Vulpian (1861), followed shortly by Charcot and Joffroy (1869), Joffroy (1873), Joffroy and Achard (1887) and Rosenblath (1893); these reports involved the condition described either as 'chronic meningitis' or 'hypertrophic cervical pachymeningitis'. Joffroy and Achard (1887) attributed the cavitation to 'venous stasis and arterial thrombosis'.

A 33-year-old woman came to post-mortem in Guy's Hospital, London, England, in 1882, with signs of a progressive cervical spinal cord lesion (Taylor, 1884). An adaptation of the woodcut illustration of 1884 (Fig. 15.1) clearly demonstrates a syrinx in the spinal cord from the upper cervical to the lumbar region, associated with a gross pachymeningitis and a gumma contained in the meninges of the lower dorsal region. The same chronic meningitis bound the medulla and pons to the cerebellum and the ventricles were dilated. The unfortunate part of this old record is this statement: 'The cavity is found throughout the whole extent of the cord which has been kept, but the medulla oblongata was not preserved and I cannot say how the cavity began at its upper extremity.'

Schwarz's (1897) case was described as syphilitic meningomyelitis. A 31-year-old woman died in a state of complete paraplegia after a progressive decline over two-and-a-half years. 'Intense meningitis' over the lower cervical cord and to a lesser degree the entire dorsal section of the cord was recorded at post-mortem. An additional finding was a meningeal cyst at the second dorsal level compressing the cord anteriorly. Cystic cavitation extended throughout the cord from mid-cervical to lower dorsal sections involving 'almost the entire grey substance in the dorsal segment of the cord, both anterior horns, and a great portion of the posterior columns'. The changes were attributed to venous and lymphatic obstruction. The description does not say that the cranial cavity was inspected. No description is given of upper cervical cord involvement with cavita-

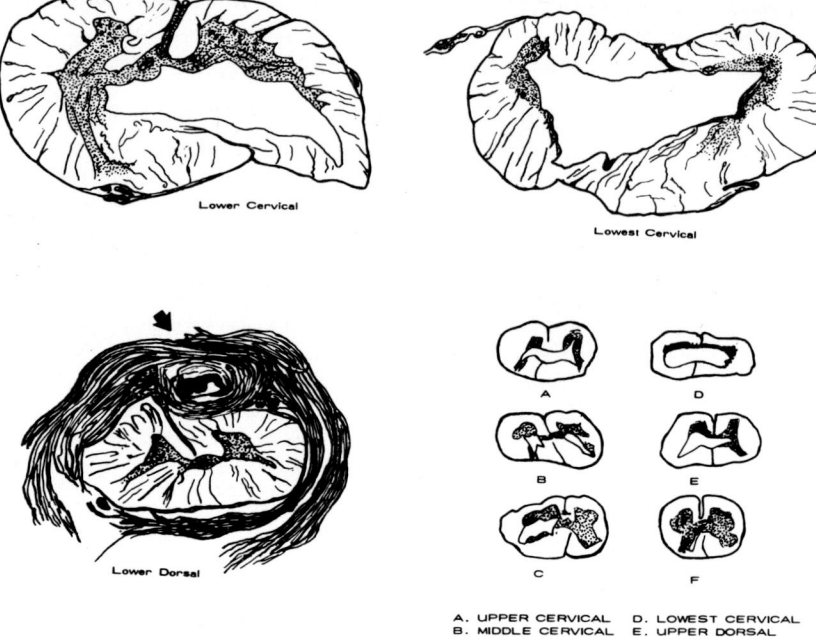

Fig. 15.1. Extensive cavitation extending from upper cervical to lumbar cord associated with syphilitic pachymeningitis and gumma (arrow). (Adapted from Taylor, 1884.)

tion or arachnoiditis so that it may be reasonable to presume that it was examined and found intact. If this be accepted, then it is a good early description of the condition under examination.

Phillipe and Oberthür (1900), in a series of 16 cases of syringomyelia from La Salpêtriére, recorded three that were associated with the condition described by Charcot and Joffroy (1869) as 'pachymeningitis cervicalis hypertrophica'. A singularly rapid evolution of an otherwise typical syringomyelic syndrome had been observed clinically in the three cases. In one of their cases they described cavitation in the lower dorsal and lumbar region. The upper dorsal region was the site of the pachymeningitis and instead of cavitation in the underlying spinal cord, there was a condition of 'tissus spongieux'. They do not describe the cervical cord, but the suggestion was that it was normal. They felt that they were describing a sufficiently particular condition as to call it 'syringomyélie pachyméningitique'. Schlesinger (1902) in his exhaustive review of the pathogenesis of syringomyelia ascribed one form to 'pachymeningitis and leptomeningitis'. Alquier and Touchard (1909) described cavitation of the cervical cord and of the medulla oblongata, independent of the central canal, in a case of syphilis with pachymeningitis and they illustrated a cord encircled by a greatly thickened pachymeningitis. They attributed the cavity to a breakdown of myelomalacia.

Many of the earliest writings are in the German and French literature but Rhein (1908) presented from Philadelphia an admirable review of the syringomyelic syndrome with 'pachymeningitis' or 'chronic meningitis'. Most cases which he reviewed to that date were syphilitic patients. Many had basal cisternal involvement, and in some, vascular involvement was prominent. In others, no comment was available about the blood vessels or the arachnoid at the base. His own case, from the description and the illustrations, was probably not a syringohydromyelia of the communicating variety although it did extend up to the medulla. In the same year (1908), Holmes and Kennedy illustrated a remarkable picture of syphilitic pachymeningitis with an underlying syringomyelia (Fig. 15.2) and syringobulbia. It was significant, however, that they stated clearly that he had sufficient 'occlusion of the foramen of Magendie by meningitis to produce a hydrocephalic ventricular dilatation'. Unfortunately, they too made no comment on the condition of the central canal in the lower medulla or of the important area of the cervicomedullary junction.

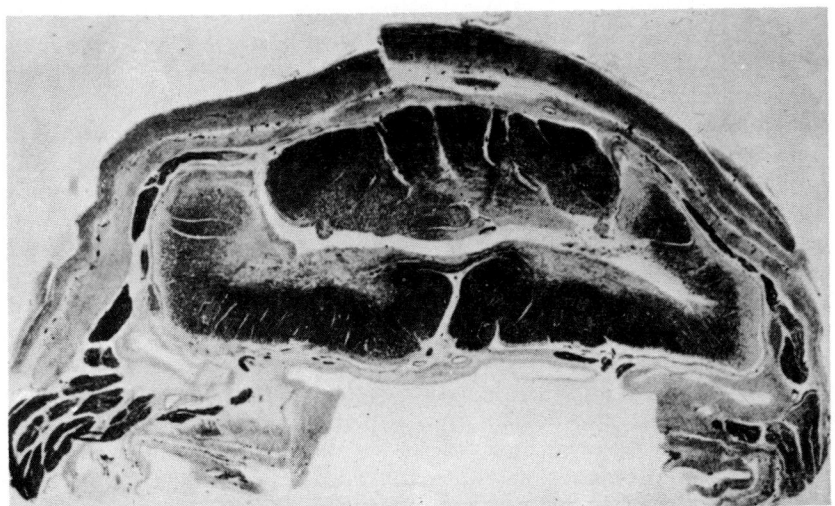

Fig. 15.2. Syphilitic pachymeningitis associated with cervical cavitation. (Reproduced from Holmes and Kennedy (1908) by permission of the Royal Society of Medicine.)

Martin (1925), describing a case with syphilitic amyotrophic meningomyelitis, commented that 'glial overgrowth within the degenerated cord may possibly break down and lead to the formation of syringomyeloid cavities'. In Martin's report, as in that of Holmes and Kennedy, the chronic leptomeningitis extended up to the region about the medulla.

Frequently, the earlier reports were not supported by radiological, post-mortem or surgical confirmation of a cystic and expanded spinal cord. Dissociated sensory loss and other clinical features of a central spinal lesion were often accepted as adequate diagnostic evidence of syringomyelia. It is clear that a constrictive arachnoiditis alone can present with a clinical picture similar to classical syringomyelia and yet no cyst be present in the underlying spinal cord (Stookey, 1927; Elkington, 1936; Nielson, 1953).

RECENT REPORTED EXAMPLES OF SPINAL CORD CAVITATION WITH SPINAL ARACHNOIDITIS

The literature has been reviewed in an attempt to identify proven examples of syringomyelia causally related to chronic adhesive leptomeningitis or pachymeningitis (loosely accepted as arachnoiditis) where reasonable certainty exists that the meningeal reaction did not extend above the foramen magnum. Taking this critical approach to the subject, there are few examples that can be accepted.

A curious case reported by Harbitz and Lossius (1929) is frequently referred to as an example of spinal cord cavitation in association with 'cystic arachnoiditis'. There are too many unsettled aspects of the case, and it typifies the difficulty of assigning an aetiological role to a particular case. There was an 'endothelioma' arising in the posterior fossa and wedged between the medulla and the cerebellar hemispheres but extending in the spinal canal down to the fourth cervical level. There was a syrinx in the cord at C7, but its upper and lower limits were not described. Pachymeningitis was present from C6 down to T8. At the lowest level, there was cyst formation within the adherent arachnoid. The pachymeningitis may have contributed to the syrinx but the foramen magnum tumour cannot be overlooked, especially as no description was given of the spinal cord sections above C6.

As a further instance of this difficulty of ascertaining that the condition causing the cavitation is purely a spinal arachnoiditis, Feigin et al (1971) reported eight cases of syringomyelia with 'fibrosing arachnoiditis'. In three of them a posterior fossa congenital abnormality existed, another had a posterior fossa inflammatory lesion, and in a fifth, the brain was 'not available for examination'. The other three are acceptable.

Tables 15.1 and 15.2 and Fig. 15.3, summarise these three cases plus the other four 'acceptable' cases from the literature of the past 40 years. They total only seven and all but one were studied at post-mortem. Vogel's case (1946) was studied by myelography and at operation only. Two other reported cases clearly associated with trauma have been omitted and are discussed in Chapter 13. The extent of the arachnoiditis was not impressive in these two. It is realised that this may be an artificial separation but it has been made for convenience of discussion. For reasons mentioned later, two cases (Thomas and Hauser, 1901; Shaw, 1972) with arachnoiditis related to Pott's disease are also omitted here and discussed in Chapter 16. If we added the traumatic and the Pott's disease cases to the reported series it would still only total 11 cases.

The clinical data from the seven cases are summarised in Table 15.1. Three of the seven had no known antecedent affliction to be invoked as causative. Two had preceding inflammation, one quite clearly a pyogenic meningitis. Two had preceding evidence of subarachnoid bleeding and are therefore analogous to the cases of bleeding in the intracranial subarachnoid space developing obliterative arachnoiditis and leading to sequelae such as communicating hydrocephalus. This appears to be further evidence that bleeding in the subarachnoid space may have more significance for the future than once was presumed to be so.

The interval from the incident to which the arachnoiditis was ascribed and the onset of symptoms related to the latter is available in three of the cases. In the pyogenic meningitis case the interval was only four months. In the two cases

Table 15.1. *Reported cases of spinal arachnoiditis and cord cavitation—clinical data*

Author	Case No.	Age and sex	Cause of arachnoiditis	Interval to symptoms	Duration of arachnoiditis	Symptoms and signs	Signs due to syrinx?
Davison and Keschner (1933)	1	50 female	Unknown	Not stated	4 months	Paraparesis Atrophy of hands Dissociated sensory loss	Probably
Mackay (1937)	2	24 male	Pyogenic meningitis	4 months	3 months	Paraplegia Wasting upper limbs	Not definite
Nelson (1943)	3	50 male	Subarachnoid haemorrhage[a]	3 years	18 months	Flaccid paraparesis and sacral sensory loss	?
Vogel (1946)	4	43 male	Unknown	Not stated	9 years	Spastic paraparesis Dissociated sensory loss to C4 on one side, T3 on the other	Upper limbs
Feigin, Ogata and Budzilovich (1971)	5	69 male	Unknown	Not stated	15 years	Pain, thoracic roots (15 years) Paraplegia and thoracic sensory level (1 year)	Probably the later symptoms
	6	37 female	Subarachnoid haemorrhage[b]	'Immediate'	5 years	'Transverse myelitis at thoracic levels'	Probably combined
	7	43 male	'Inflammatory' (?)	Uncertain	25 years	'Paraplegia below T4'	?

[a] Thrombocytopenic Purpura
[b] Polyarteritis Nodosa

Table 15.2. *Reported cases of spinal arachnoiditis and cord cavitation—pathological data*

Author	Location of arachnoiditis	Cysts in arachnoid?	Dura adherent	PATHOLOGY Location of cavitation in cord	Condition of blood vessels
Davison and Keschner (1933)	Cervical to lumbar	Over lower third of cord-bulged with 30 ml fluid	Yes	Mid-medulla to lumbar	'Extensively thickened with narrow lumens'
Mackay (1937)	Cervical to lumbar	Numerous, small in cervical and lumbar segment	?	'Extensive'	'Normal vessels were few' Thickened with occasional occlusion and hyalinisation
Nelson (1943)	Mid-thoracic to cauda equina	Small, insignificant	Yes	Lower dorsal to lumbar enlargement	Appeared normal in sub-arachnoid space Within cord had 'sclerosis'
Vogel (1946)	T3—very localised	No	No	Upper dorsal—? cervical	Considered normal
Feigin, Ogata and Budzilovich (1971)	Mid-cervical to upper lumbar	No	Yes	Lower cervical to mid-thoracic	Severe periadventitial fibrosis, especially about veins No luminal obstruction
	Lower thoracic	No	Yes	Lower thoracic to lumbar	No arterial changes seen
	Upper thoracic cord only	No	Yes	Upper thoracic to caudal segments	No arterial changes seen

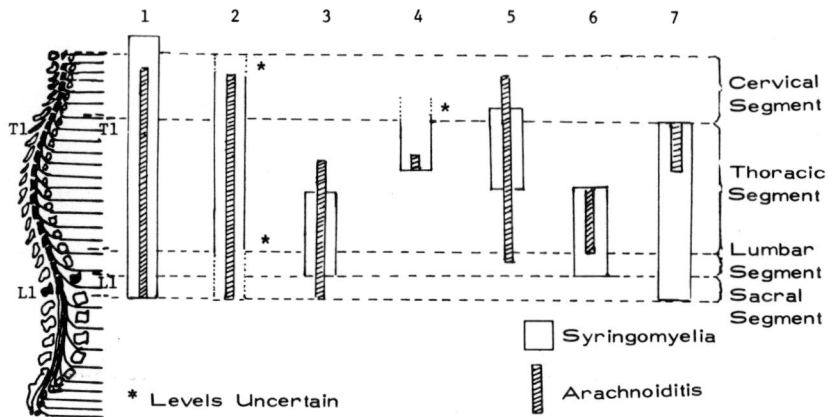

Fig. 15.3. Schematic representation of the seven cases reported with syrinx underlying arachnoiditis. (See Table 15.1 for sources of cases.)

subsequent to intraspinal haemorrhage, symptoms of arachnoiditis developed rapidly as a sequel in one and after three years in the other. Some of the contemporary cases appear to follow the seriously progressive pattern described by Phillipe and Oberthür (1900). The progression of the lesion was either over a few months, as in the cases of Mackay (1939), Nelson (1943) and Davison and Keschner (1933), or over a great many years, as in Vogel's case (1946) and those of Feigin, Ogata and Budzilovich (1971). All ultimately developed major neurological disability. The signs were usually those of progressive cord compression, with or without evidence of an intramedullary type of dissociated sensory loss. Occasionally, it was possible to decide that some of the signs were due to the cystic cavitation above the level of the arachnoiditis. This was best seen in Vogel's (1946) case in whom a lesion of circumscribed nature at T3 was accompanied nevertheless by upper limb signs.

The pathological data from these seven cases is summarised in Table 15.2 and Fig. 15.3. The location of the lesion in the arachnoid was extensive in three of the seven (Davison and Keschner, 1933; Mackay, 1939; Nelson, 1943), and would qualify for the old descriptive phrase of 'arachnoiditis adhesiva diffusa spinalis'. It was circumscribed and local in the other four (Vogel, 1946; Feigin et al, 1971), and could be designated 'arachnoiditis adhesiva circumscripta spinalis'. Once there was an associated cystic lesion of significant size *within the arachnoid* and twice there were smaller cystic lesions in this location. The dura and the leptomeninges were intimately involved in the process in all but one of these reported cases where sufficient description was available.

The cavity in the cord always involved the area beneath the thickened meninges and it is likely that it had its origin in this location. Twice the cavity was very extensive and involved most of the spinal cord (Mackay, 1939; Davison and Keschner, 1933). In the latter case it extended into the medulla. In the other instances it was less extensive and in the four instances where the arachnoiditis spared the uppermost thoracic and the cervical cord, so too did the cord cavitation.

Dr Irwin Feigin has kindly sent us pathological material from his three published cases of spinal arachnoiditis with underlying syringomyelia. They all demonstrate unequivocally thickened and adherent dural and leptomeningeal coverings with cavitation having a predilection for the grey matter and central cord. None of them had thick walls to the cavities so characteristic of the tumour cases, and none had Rosenthal fibres in the walls. There were no essential differences between the cavity location and cavity lining in any of them, despite varying aetiology (posthaemorrhagic, Fig. 15.4; postinflammatory(?), Fig. 15.5; idiopathic, Fig. 15.6).

The condition of the blood vessels within the arachnoid was the subject of comment by all the above authors. In Vogel's case (1946), it was judged that vascular changes 'did not contribute to the intramedullary cavitation since the arachnoid lesion was so localised'. This comment might be open to dispute, especially if one is considering the microvasculature. In three reports, the vessels were thought to be sufficiently abnormal as to be a major factor in producing the underlying lesion in the spinal cord. Feigin et al (1971), in their publication, did not give any description of the blood vessels in their cases. However, in a later communication, Feigin (1972) reports as follows:

'The blood vessels seen in the affected meningeal tissues show no major morphological changes. There is a severe periadventitial fibrosis, particularly about the veins, but nothing is recognized which clearly blocks the lumen. Nonetheless, it seems logical to assume that some vessels, particularly veins, were

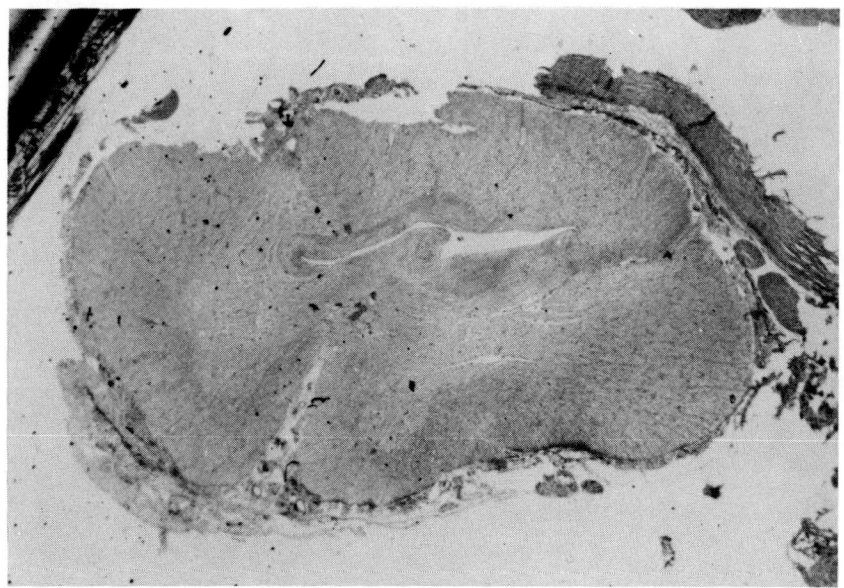

Fig. 15.4. Arachnoiditis and central cavitation in patient afflicted with polyarteritis nodosa who had experienced subarachnoid bleeding. (Loaned by kindness of Dr I. Feigin.)

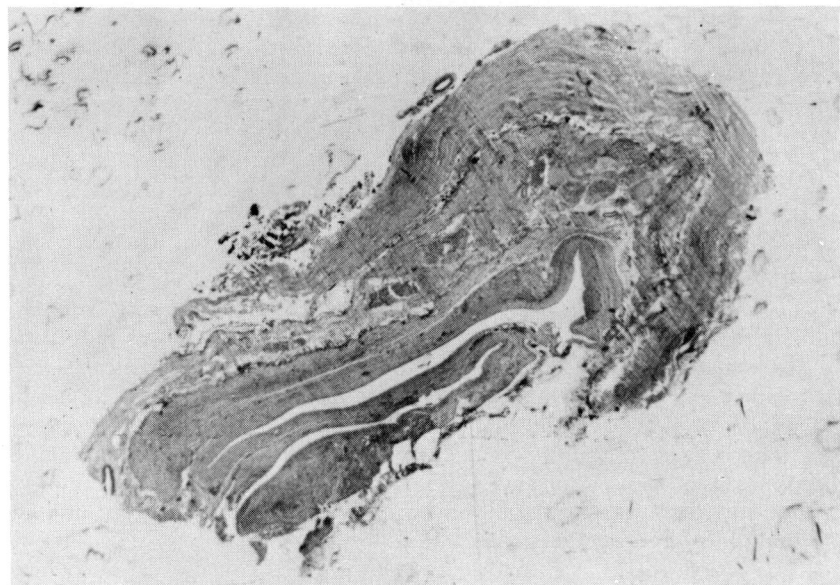

Fig. 15.5. Extensive multilocular cavitation in postinflammatory arachnoiditis. Cord remnants distorted by removal of thickened arachnoid ventrally. Gross dorsal thickening prominent. (Loaned by kindness of Dr I. Feigin.)

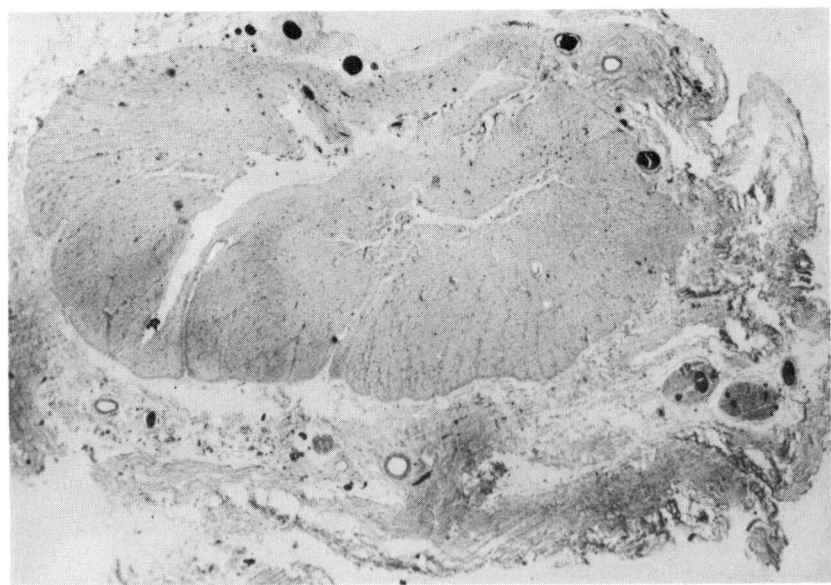

Fig. 15.6. Idiopathic arachnoiditis producing underlying syrinxes. Cavitation predominantly central and dorsal, the latter extending into the anterior horn. (Loaned by kindness of Dr I. Feigin.)

compressed in passing through the dense fibrous connective tissues that were formed. Some vessels might have been completely obliterated, and be no longer recognizable as such.'

AUTHOR'S EXPERIENCE WITH SPINAL ARACHNOIDITIS AND SYRINGOMYELIA

My own experience involves six cases. A seventh is reported also, as the details were kindly supplied by Dr W. S. Coxe, of St. Louis. All were afflicted with a chronic and severe neurological spinal disability. All had a myelographic appearance compatible with 'chronic adhesive arachnoiditis'. In all instances, the arachnoiditis and the underlying cystic cavitation of the spinal cord have been proved at operation. The details were collected over the past decade during which the author was known by colleagues to be interested in spinal cord cavitations and during which time Dr A. T. Jousse, director of Lyndhurst Lodge Hospital for Spinal Injuries, was alert to unusual spinal lesions in traumatic and non-traumatic cases. Although it remains a rare phenomenon, it is doubtful if it is as rare as the seven previously recorded cases, plus the seven of the author, would suggest.

CASE 1, A.B., MALE

At the age of 18 years, in 1947, and as a result of a car accident, this patient suffered 'an immediate right arm paresis', associated with what he understood was a 'fracture of two cervical vertebrae'. Immobilisation and traction was employed for 10 weeks and he was normal at the time of hospital discharge. In 1952, he passed an Air Force medical examination without difficulty. In 1956, he became aware of numbness over the fifth finger of the left hand (opposite to his original right-sided traumatic weakness). Gradually, the entire left hand and arm became numb and weak and pain developed in his neck. In 1959 and 1960, Drs R. T. Ross and D. Parkinson, in Winnipeg, studied his problem. He now had evidence of a central spinal cord lesion in the cervical region as manifest by a left Horner's syndrome, a sensory decrease to pain and temperature over the second and third divisions of the fifth cranial nerve, and from C2 to T3, sparing the C7 and 8 dermatomes. There was some weakness of his left pectoralis, triceps, wrist extensors and of the dorsal interossei. The latter were wasted and some fasciculation was present. The only sign in the legs was a briskness of the left knee jerk.

In 1960, a myelogram revealed an obstruction at C6 to 7. At laminectomy, the cord was judged to be thin and atrophic and the myelographic obstruction was plainly the result of an adherent and fibrosed arachnoid. No cyst was encountered. The density of the adhesions was greatest on the left side. C.S.F. protein at the time of myelography was 20 mg/100 ml.

Between 1960 and 1970, the patient's left hand became slowly weaker to the point of approaching uselessness. Weakness developed in his left leg by 1965 and slowly progressed. By 1967, his right little finger became affected with the same numbness that had begun on the left side in 1956. Examination in December 1970 indicated more marked left upper limb wasting and weakness, with considerable right-sided weakness as well. Both upper limbs were areflexic. Legs were hyperreflexic and spastic, with weakness on the left more than on the right. Sensation to pain and temperature was lost in the areas depicted in the chart (Fig. 15.7), with light touch spared everywhere except in the left hand. Position sense was impaired in both lower limbs at the toes and he had absent vibration appreciation below the sternum.

Cervical spine X-rays demonstrated hypertrophic lipping at C5, but nothing was seen to indicate an old fracture. Myelography, on 26 January, 1971, with 10 ml of Myodil, was normal in the lumbar and lower thoracic areas. In the upper thoracic space the contrast medium broke up more than usual at this level, flow was inconstant and a partial block was indicated at C7. From T1 to C5 many irregular indentations in the contrast column and a few small diverticula were noted ('candle-guttering') in the upper thoracic area on the left. The cord size was judged to be normal or small, not enlarged. A separate density was seen over the C7 to T1 area. The total appearance was interpreted as indicative of arachnoiditis (Fig. 15.8).

In March 1971, Dr Ross Fleming performed a laminectomy at C5, 6 and 7. As soon as the dura was opened, significant arachnoidal adhesions were encountered, thicker and denser on the right than on the left (Fig. 15.9). An arachnoid cyst was present on the right dorsolaterally displacing the cord to the left

(Fig. 15.10). This was drained and opened widely. A needle was inserted into the cord and faintly yellowish fluid aspirated. Through a midline myelotomy incision the ventricular end of a Pudenz shunt was inserted (Fig. 15.11) and passed downwards for 3 cm. It was sewn into place by a ligature through the thickened arachnoid. Fluid was injected into the tube and reaspirated causing the cord to collapse visibly. The fluid from the cyst in the cord was clear and contained 49 mg/100 ml of protein.

When reviewed in September 1971, the patient's walking was a little better but his hands had not improved. Sensation was a little better in that his highest cervical segments on the left were no longer insensitive to pinprick.

Summary. A post-traumatic arachnoiditis extended from C5 to upper thoracic levels. After many years, hand wasting and dissociated sensory loss slowly extended to the top of the cervical segments and an intramedullary cervical cavity with protein content of 49 mg/100 ml was drained. An arachnoid cyst was present within the adherent arachnoid as well.

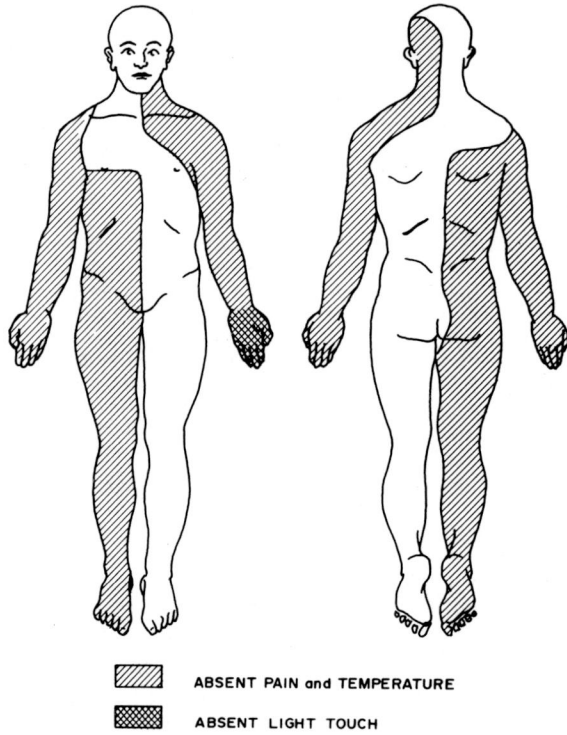

ABSENT PAIN and TEMPERATURE

ABSENT LIGHT TOUCH

Fig. 15.7. Case 1, A.B. Area of sensory change the result of post-traumatic arachnoiditis and underlying syrinx. The meningeal lesion extended from C7 to T2, the syrinx from C1 to C8.

CASE 2, L.S., FEMALE

In 1942, at the age of 14 years, this patient was desperately ill, with unconsciousness lasting one week because of a severe pyogenic meningitis. At first all four limbs were paralysed, but her arms rapidly returned to normal and she learned to walk with a spastic gait, using crutches and later canes. In 1950, flexor spasms were troublesome and a surgeon performed a partial posterior rhizotomy from the lower lumbar region to T8 or 10 as judged by the incision. The exact details of the procedure are unknown but it rendered her paralysed and insensitive in the lower extremities. Between 1950 and 1958 the level of her numbness moved up to the upper thoracic regions. In 1958, she became aware of right hand weakness. In 1963, numbness was noted along the ulnar border of the right forearm and hand. In 1965, weakness was apparent in the left hand.

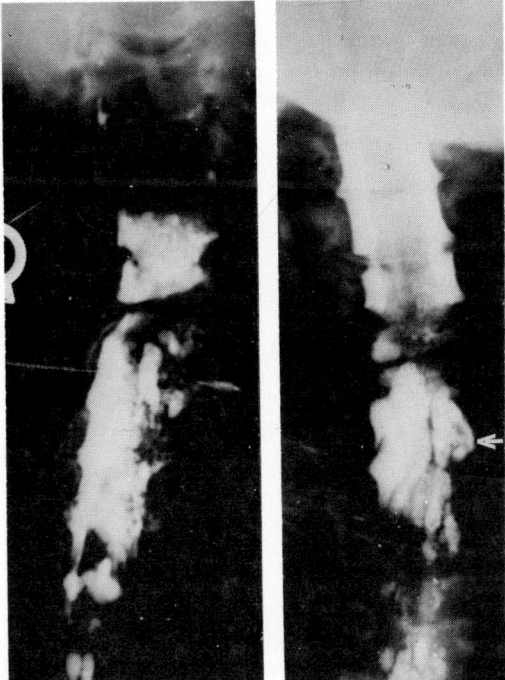

Fig. 15.8. Case 1, A.B. Myelogram 1971, with partial block at C7 and irregular indentation C5 to T1, due to arachnoiditis. Note double-density in right picture at C8 to T1 level, due to arachnoid cyst (arrow).

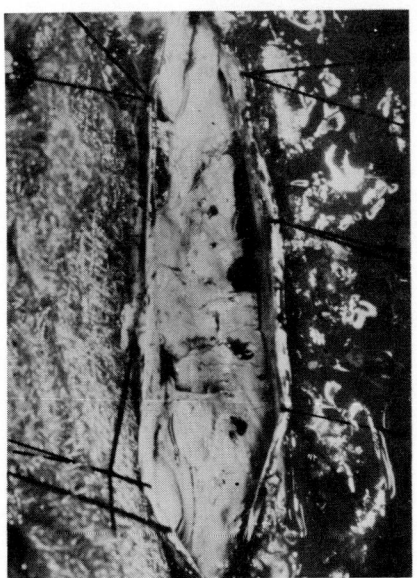

Fig. 15.9. Case 1, A.B. Dense arachnoid adhesions.

Fig. 15.10. Case 1, A.B. Cystic dilatation *within* arachnoiditis (right) pushing cord to left.

232 NON-COMMUNICATING SYRINGOMYELIA

Fig. 15.11. Case 1, A.B. After removal of the arachnoid cyst and the insertion of Pudenz catheter.

She was admitted to Sunnybrook Hospital, Toronto, in October 1968. In the upper limbs there was wasting of the small hand muscles with mild weakness, the left worse than the right. The left triceps reflex was absent, but the other upper limb reflexes were brisker than normal. Sensory loss in the upper limbs was confined to decreased pain and temperature loss along the inner aspects of arms, forearms and hands. Her lower limbs were insensitive, paralysed and her reflexes were pathologically brisk with extensor plantar responses. The total loss of feeling extended up to T4 and then lessened gradually to the area described for the upper limbs.

Lumbar myelogram demonstrated a complete block at T4 so a cisternal myelogram was done. Dr D. McRae reported 'numerous adhesions in the upper thoracic subarachnoid space deforming from T2 downwards and completely obstructing this space at the T4 level' (Fig. 15.12). The spinal cord in the mid and lower thoracic regions appeared to be normal in size and shape. On 11 November, 1968 Dr Wm. Lougheed performed a laminectomy at T3, 4 and 5. The dura was adherent to a thickened arachnoid and as the thickened arachnoid was being dissected from the cord the thin wall of the cyst within the cord was opened and clear fluid appearing like ordinary C.S.F. 'gushed forth'. Into this cystic lesion in the cord a catheter was passed cephalad for 5 cm and down the cord for 10 cm. An attempt to leave oily contrast substance within the cavity for later radiography was unsuccessful.

A Pudenz catheter was inserted into the cavity (Fig. 15.13), carried cephalad subcutaneously and its other end inserted into the cisterna magna. This was not successful; the patient's hands became weaker and when the tube was removed six weeks later it was plugged at its lower end with yellowish-white tissue. Three years later, she has continued to have significant hand weakness probably slightly more marked each year, but has no further loss of sensation. A further drainage attempt is being considered.

Summary. As a result of pyogenic meningitis an extensive arachnoiditis extended up to T4. After 16 years, weakness and dissociated sensory loss gradually involved the upper limbs. Surgical exploration indicated a lengthy cavity within the cord, extending above the arachnoiditis as well as caudally. The fluid within the cavity was clear.

CASE 3, T.T., FEMALE

In 1952, at the age of 26 years, this woman fell down a flight of stairs and developed low back pain with some tendency for clumsiness in the right leg. In 1956, her right foot was dragging when tired but not until about 1960 was it very noticeable. Myelogram in 1962 indicated a block in the lower thoracic region and an appearance suggesting an arachnoiditis (Fig. 15.14). In 1963, Dr T. P. Morley performed a laminectomy at T5 to 9, with dissection of some arachnoidal adhesions. This gave a

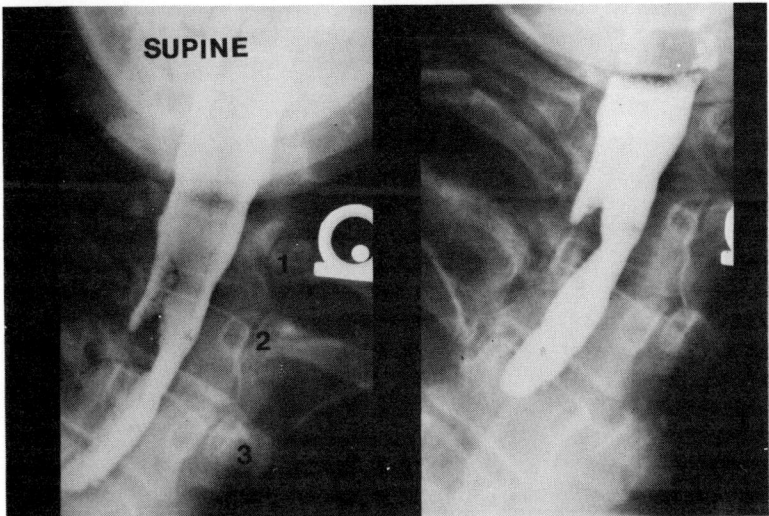

Fig. 15.12. Case 2. Irregularity of contrast medium T1 to T4, in cisternal myelogram. Contrast was held up at T4. Previous pyogenic meningitis led to progressively constricting arachnoiditis.

Fig. 15.13. Case 2. Laminectomy at T3–5 with Pudenz catheter into cord cavitation beneath arachnoiditis.

temporary improvement, but by 1964 the patient's right leg was spastic, her left leg to a lesser degree and a sensory level to all superficial modalities extended to T3 on the right and T5 on the left. Position and vibration sense were diminished in the lower extremities.

Spasticity increased so that in 1967 a Bischof's myelotomy was performed. In the lower thoracic cord, extending up to the T1 level, an intramedullary cystic cavity was discovered and drainage effected by Silastic tubing placed within it. From 1967 until 1972 she has not deteriorated. The sensory level has not ascended. The spasticity of her legs has been tolerable since the myelotomy and cyst drainage.

234 NON-COMMUNICATING SYRINGOMYELIA

Summary. Following a minor traumatic incident an insidiously progressive paraparesis came on over a ten-year period. A sensory and motor level developed in the upper thoracic region and at laminectomy clear fluid was drained from an intramedullary cavity present throughout the entire length of the thoracic spinal cord.

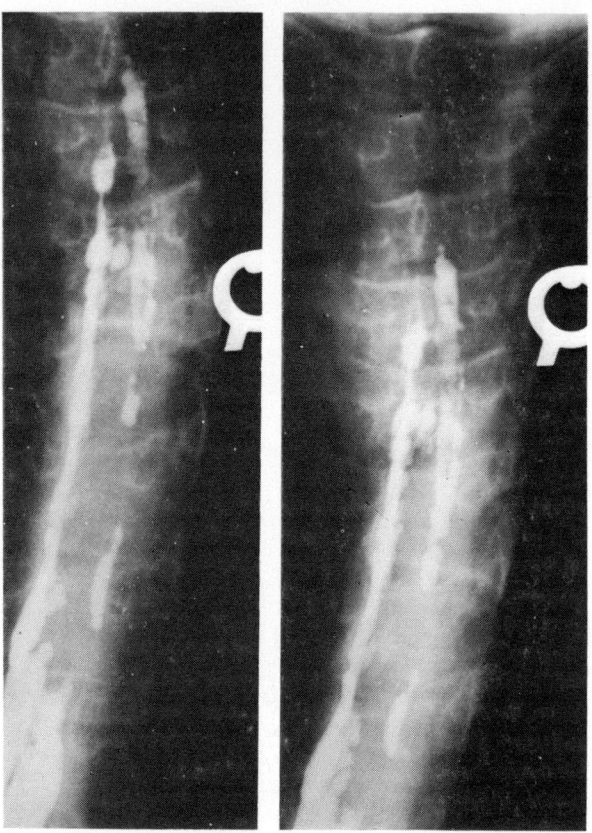

Fig. 15.14. Case 3. Thoracic myelogram with irregularity of oil column but no evidence of cord dilatation. Cavitation encountered at laminectomy. Previous history of small (? significant) trauma.

CASE 4, C.E., MALE

In April 1947, at the age of 39 years, this patient was given a spinal anaesthetic for an abdominal operation. On the day following this he was aware of severe backache which lasted for one-and-a-half months. When he got out of bed his back was still stiff and he was aware that his feet and legs were clumsy. In a few weeks his stiff legs became the site of numbness and of muscle spasms. Within six months this weakness and numbness in his lower extremities was so bad that he had to stop work. In December 1947, a lumbar laminectomy by the late Dr K. G. McKenzie revealed an adhesive arachnoidal thickening, with adhesive fibrosis binding the dura to the arachnoid and the arachnoid to the filum terminale and cauda equina region. The patient was not improved by this procedure. Slowly over the next 17 years his legs became weaker and he developed complete sensory loss up to about T2, with complete motor paraplegia in his lower limbs. Sphincter difficulties progressed with the other symptoms. Eventually, in 1961 he required a suprapubic cystotomy and was totally paraplegic, with sensory loss complete up to T4.

In 1964, an *upward extension* of his neurological disability became apparent. He began to have paraesthesiae in the left hand and arm and fine movements were interfered with in his left hand. He had sensory loss to pain and temperature in his left upper limb in the seventh and eighth cervical dermatomes and this became continuous with the loss in the thoracic dermatomes in the trunk. The left triceps reflex was absent. By 1968, he was complaining of pain down the left arm with coughing, sneezing or using the arm. He had wasting of the left hand muscles with weakness of grip and of wrist, elbow and shoulder. Pinprick and temperature sensation were lost below C7 on the left and diminished up to the cervical-trigeminal dermatome margins. He was completely paraplegic. Myelogram was done in May 1968. The appearance in the cervical region indicated minor changes compatible with a modest degree of cervical spondylosis. The spinal cord in the cervical region did not appear enlarged. The opaque substance, injected cisternally, encountered a complete obstruction at the T5 level.

In June 1968, a laminectomy at T2, 3 and 4 was carried out by Dr T. P. Morley. The dura and arachnoid were thickened and adherent and were freed by blunt dissection. The cord was transected and segments T2, 3 and 4 were resected at this level. As this was done *the upper part of the cord retracted.* This retraction was sufficiently significant for the surgeon to comment that 'with the freeing of adhesions from the superior stump, the cord retracted almost out of sight. The distance between the two cut ends after resection of the segment was much more than double the length of the section resected—more like three or four times'. The spinal cord was smaller than normal but in its centre there was a cystic cavity (Fig. 15.15). It was lined by a combination of glial and connective tissue and was not connected to the central canal. No Rosenthal fibres were present.

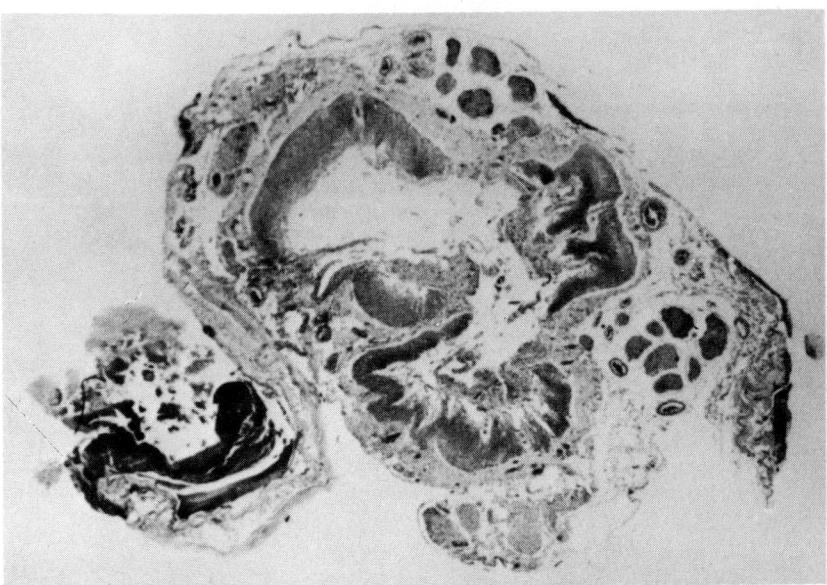

Fig. 15.15. Case 4. Cross section of transected cord with cavitation and arachnoiditis at T3 level. Cavity lined by glial and connective tissue. (Gomar, C. T., stain.)

At follow-up in March 1972, the patient had maintained improvement of the strength in his left hand. He was able to use the hand to pull himself up by grasping the bar over his bed, using strength in muscles incapable of such performance before the operative procedure described. Sensory reduction above C7 was no longer detectable. Dr Jousse, who conducted the follow-up examination, commented: 'It is apparent, beyond a shadow of a doubt, that the operative procedure has arrested the progression and brought about a remission'.

Summary. A spinal anaesthetic precipitated an early paraparesis which progressed to total paraplegia at a T4 level after 17 years. After this, the patient's upper limbs became affected with weakness and dissociated sensory loss. A resection of his upper thoracic spinal cord encountered a cord with central cavitation and surrounding arachnoiditis. The normal cord mobility was interfered with by the adhesions. As a result of this procedure he achieved and maintained significant improvement.

CASE 5, C.LaC., FEMALE

At the age of 15 years, in 1940, this woman was afflicted with tuberculous osteomyelitis affecting T2, 3 and 4 with a profound paraparesis. After prolonged immobilisation her spinous processes were fused from C7 to T6 and gradually she made a recovery which she judged to be complete. In 1962 she became weak again, rather abruptly in her legs, and by 1965 had deteriorated to the point of a seriously spastic gait and had a cold sensation in her lower limbs. Examination indicated spastic lower limbs with knee and ankle clonus, pathologically brisk reflexes and bilateral Babinski responses. Weakness of moderate degree involved chiefly the flexors in the lower extremities. Sensory loss included diminished pain and temperature below T4, and yet it was only completely lost between T6 on the right and L3 on the left. Position sense was intact but vibration appreciation was decreased at the right and absent at the left ankle. The upper limbs were entirely normal.

Plain films of the thoracic spine revealed a kyphotic angulation at T2, with collapse of T2 and 3 and an immobile spine from the previous fusion. A myelogram demonstrated a slight hold-up at T2, but not a block. The cord was considered expanded for at least three segments above and below the gibbus. At laminectomy on 19 May, 1965 (Dr T. P. Morley) numerous arachnoidal adhesions were encountered at T2, but the cord was not judged to be fixed by them. The dura was not densely adherent to the underlying membranes. The cord was exposed from C7 to T4 and was cleared of arachnoidal adhesions. The posterior aspect of the cord was tissue-paper thin (Fig. 15.16) and 'with each inspiration could be seen to ripple as waves on water might'.

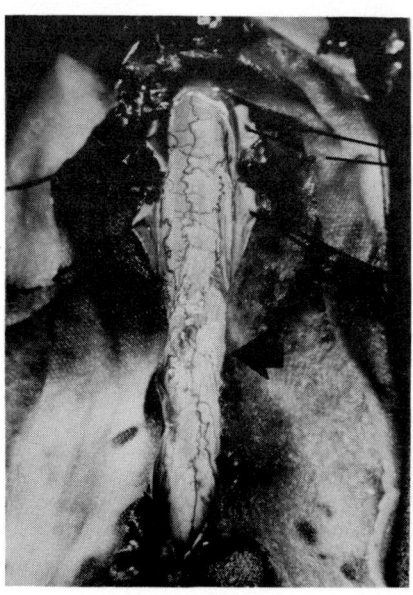

Fig. 15.16. Case 5. Laminectomy at T2–4 exposing dilated cord in a case of old Pott's disease. Arrow indicates site of moderate arachnoiditis at level of long-standing gibbus deformity.

A needle was passed into the syrinx and a few drops of clear fluid obtained. No analysis was conducted on this fluid. An opening was then made into the cord by a small incision and the remainder of the fluid drained off. A small catheter determined that the syrinx extended down into the thoracic cord beyond the limits of the exposure and extended cephalad in the cervical cord beyond the exposure, to approximately C7 level. Within three weeks the patient was walking much better but was still spastic and had similar sensory findings.

Summary. A paraparesis from upper thoracic Pott's disease was followed 22 years later by the rather abrupt onset of a return of spinal cord dysfunction at and below the old level of disease. Laminectomy disclosed a syrinx beneath the old gibbus and the cavity extended at least to the low cervical and midthoracic levels.

CASE 6, S.H., MALE

In 1953, at the age of nineteen years, this patient was troubled by numbness, tingling and pain in a band around his lower chest, aggravated by spine flexion and a mild leg weakness. An oil myelogram was done and a T6 disc protrusion was identified. A T5 to 7 laminectomy was carried out and a T6 to 7 intervertebral disc removed via a left extradural approach. A large bony spur, located in the midline just inferior to the disc was not removed. The dentate ligaments were divided and dural openings left at the operative site.

By 1956, the patient was aware of increasing leg stiffness and early flexor spasms and had pathological lower limb reflexes and absent abdominal reflexes. By 1967, his gait was spastic with spontaneous left ankle clonus. The motor findings were the same in quality, but much worse. He had absent pain, temperature and tickle appreciation bilaterally from T6 to 12, with impaired position and vibration sense in both lower limbs. A myelogram (Fig. 15.17) was judged to be 'compatible with an arachnoiditis' predominent at T6. The C.S.F. contained 42 mg/100 ml of protein.

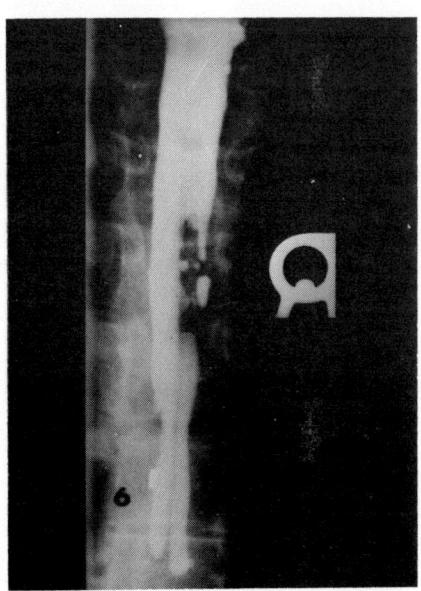

Fig. 15.17. Case 6. Myelogram of very circumscribed arachnoiditis at and above site of old thoracic (T5–7) discectomy. Cavitation at and above was proven by surgical exploration. No dilatation of cord was seen by myelography.

In 1967, he declined laminectomy but by 1969 was more disabled and aware of minimal early tiring of his arms. A slight but definite increase in upper limb reflexes was all that was found in the upper limbs. The lower limbs were more spastic and the sensory loss to pain and temperature below T6 included all but the sacral segments. A repeat myelogram in 1969 was unchanged and on 25 April, 1969 he was re-explored at T5 to 7 (Dr T. P. Morley). A fibrous membrane was encountered when the dura was opened and there was one loculus of C.S.F. within this adherent membrane. Adhesions of varying density bound the cord to this thickened covering and the cord was considered to be of normal size, while above the lesion it was enlarged, fluid-containing and 'with every respiration it blew up like a balloon'. A needle in the sac within the cord drew forth clear fluid resembling C.S.F. and its protein content was 48 mg/100 ml.

Simply freeing the cord of all adhesions restored it to normal size so that it was decided not to leave in any type of permanent drain. He has remained virtually unchanged in his legs in the three years since this drainage and his arms are not at all abnormal. In retrospect, Dr G. Wortzman, neuroradiologist, reviewed the myelogram and was unable, even then, to state that the spinal cord above the lesion was enlarged.

Summary. A T6 to 7 discectomy was followed, after a three-year lapse, by a slowly progressive spastic paraparesis with sensory level up to T6. After 13 years of symptoms an adherent arachnoiditis, with one loculated cyst within it, surrounded a spinal cord that was distended with clear fluid with a protein content of 48 mg/100 ml. Freeing the arachnoidal adhesions and simple drainage have effected remission.

CASE 7, M.E.B., FEMALE (Details supplied by Dr W. S. Coxe, Barnes Hospital, St. Louis, Missouri.)

By the age of 28 years, in 1941, this woman had suffered two years of low back pain and sciatica. Myelography and lumbar discectomy were performed. Persistent symptoms led, in 1943, to a T4 anterolateral cordotomy on the right side. Right leg weakness, left lower limb analgesia and impaired bladder function persisted for some months but remitted and she was free of pain. In 1956, she slipped on a rug and fell, striking the upper thoracic spine against the edge of an open door. For 20 minutes she was totally paraplegic, then began to improve slowly, and after many days recovered to a spastic gait, worse on the right, requiring crutches. Dissociated sensory loss was present on the left at and below the cordotomy level at T4, but on the right went up to T1.

Progressive spastic leg weakness led to myelography in 1959 and there was a block at T4, some irregularity below this, and a normal lower thoracic and lumbar subarachnoid space. Further decline led, in October 1961, to a laminectomy at T4 and 5 disclosing an arachnoid densely adherent to the overlying dura. The surgeon (Dr W. S. Coxe) stated that 'the cord was angulated upward as the dura was retracted'. The dura and arachnoid were gradually separated, an intramedullary cystic area was encountered and a large amount of fluid drained from it which was 'apparently clear spinal fluid'. The cavity had a glistening white lining. A dissection of the surrounding adhesions was carried out dorsally and laterally in an attempt to allow free flow of fluid from the cavity into a subarachnoid space that would be open above and below, surrounding the normal area of the cord. Gelfoam was used in an attempt to hold open the dorsal myelotomy. The patient's spasticity improved and a temporary worsening of posterior column function gradually cleared.

Within a few months, weakness, wasting and sensory loss to pain and temperature over T1, C8 and 7 became apparent. In April 1962, re-exploration of the cord was carried out at C7 to T5. There were dense adhesions at T3 to T5, but none above this. The cord 'appeared very saccular. It was distended with clear fluid. When the fluid was removed, the cord was extremely thin.' The cavity was probed with a catheter and appeared to extend into the mid-cervical region. A right-angle plastic catheter was left in the cord, brought out into the subarachnoid space, and held in position by a suture to the thickened overlying arachnoid. During the next few months the patient's hands became stronger, with their power approaching normal, and she regained normal sensation in the upper limbs.

When re-examined in 1963 and 1964 she had no return of upper limb symptoms or signs; sensory loss was only as high as T3 to 4 but she had excessive distress from flexor spasms. Since 1964 she has been lost to follow-up.

Summary. An anterolateral cordotomy at T4, for lower lumbar discogenic pain was followed, 13 years later, by a blow to the upper thoracic region. This was immediately succeeded by the abrupt onset of a spinal lesion appropriate to the cordotomy level. Within five years upper limb signs indicated extension to the cervical region, of an intramedullary cavity. Drainage by an indwelling catheter relieved the signs of the syrinx but the deficit at the level of the local lesion, ascribed to arachnoiditis, has progressed.

Analysis of Author's Cases

The cases were all proven by myelography to have arachnoiditis and demonstrated at operation to have underlying spinal cord cavitation. Only one of the cases, case 5, the patient with previous Pott's disease was known preoperatively to have an expanded cord with cavitation. Myelograms in all the others failed to indicate this and, even in retrospect, a careful review of the picture did not allow one to confirm cord dilatation. The cases were divided between male and female and were from the fourth to the seventh decade at the time of diagnosis.

The aetiology, as shown in Table 15.3, was reasonably certain in most cases. One followed a very severe pyogenic meningitis, one was associated with a long-standing deformity due to Pott's disease, one followed myelography and discectomy, one occurred as a sequel to anterolateral cordotomy (done at T4), one followed a spinal anaesthetic and two were probably a sequel to trauma. In one of the latter cases the trauma was minor but in the other it was sufficiently significant to cause transient monoparesis. The offending episode preceded by a very variable period of time the onset of symptoms indicative of abnormal spinal cord function from arachnoiditis. Twice it was almost immediate (pyogenic meningitis and spinal anaesthesia) while in others there was a delay of many years (Table 15.3).

All the cases went on to develop serious spinal cord dysfunction but many years of chronically progressive neurological disability occurred in most before operative identification of the combination of arachnoiditis and intramedullary cavitation was recognised. Four times the upper limbs became involved and a spastic paraparesis due to the original arachnoiditis was coupled with an upper limb dysfunction due to the ascent of the spinal cavitation. Despite the upper limb involvement in four of the seven cases, there was an arachnoiditis above the thoracic region in only one case and here it was known to extend as high as C5 (Table 15.4). In this case, sensory loss went even higher and was present up to the trigeminal-second cervical dermatome junction. In the other six cases, the upper limit of the arachnoid lesion, myelographically and/or surgically recognised was at or below T2 (Fig. 15.18).

In the seven cases from the literature, all had thoracic involvement and in three the only involvement was in the thoracic region. One of these was confined to the lower and the other two to upper thoracic segments. The other four had thoracic involvement as part of a more widespread lesion. Altogether, only two of the cases did not have some involvement of the upper thoracic segments (one of the cases of Feigin et al, 1971, and the case of Nelson, 1943). Thoracic predilection was remarked by Lombardi et al (1962) in a discussion of arachnoiditis without cavitation—68 per cent of 41 cases were confined to this area.

The symptoms and signs in all cases were probably the result of a combination of cord and root damage due to the arachnoiditis with the addition of syringomyelic cavitation. In one instance (case 1), a large secondary cyst of the arachnoid was obviously a part of the picture as it was under tension at operation and was distorting the normal position of the spinal cord (Fig. 15.10). It was recognisable in the myelogram as a separate density shadow (Fig. 15.8). A smaller cyst of doubtful significance was present in one other case (case 6).

In the reported cases (Table 15.2) arachnoid cysts occurred in three of the seven, and in one (Davison and Keschner, 1933) there was a large and clinically significant tense cavity. Thus, they occurred in five of 14 cases with syringomyelia from arachnoiditis. This feature of arachnoiditis containing cystic loculations was recognised by Schwarz (1897) and by Spiller, Musser and Martin (1903) and was termed 'meningitis serosa spinalis' by Mendel and Adler (1908). Elkington (1936), in his extensive review of arachnoiditis, reported that arachnoid cysts were found in 44 per cent of those subjected to operation. In managing arachnoiditis recognition of these cystic lesions is important. Horsley (1909) noted that they added to the confusion existing between arachnoiditis and tumour, and Bliss (1909) was

Table 15.3. *Author's cases of spinal arachnoiditis and cavitation*

Case	Age (at diagnosis) and sex	Aetiology of arachnoiditis	Interval to symptoms	Duration (preoperative) of arachnoiditis	Signs	Signs due to syrinx
Case 1, A.B.	42 male	Trauma	9 years	4 years (1st op.) 15 years (2nd op.)	Spastic paraparesis, wasted areflexic arms Sensory loss to C2 Horner's syndrome	Yes—upper limbs at least
Case 2, L.S.	40 female	Pyogenic meningitis	Immediate and delayed 16 years	26 years	Spastic paraparesis, wasted, areflexic arms Sensory loss to C7–8	Yes—upper limbs
Case 3, T.T.	41 female	? Trauma	4 years	15 years	Spastic paraparesis	No
Case 4, C.E.	60 male	Spinal anaesthetic	Immediate	9 months (1st op.) 21 years (2nd op.)	Complete paraplegia, wasting and areflexia, one upper limb C7 sensory level	Yes—upper limb
Case 5, C.LaC	40 female	Pott's disease	17 years	3 years	Spastic paraparesis Sensory level T4	Most of signs
Case 6, S.H.	35 male	Thoracic discectomy (? myelography)	3 years	11 years	Spastic paraparesis Sensory level T6	Not certain
Case 7, M.E.B.	46 female	Thoracic (T4) spinothalamic tractotomy	13 years	5 years	Spastic paraparesis and T4 sensory level Ascent to wasting, weakness hands Sensory level to C7	Yes—upper limbs

SYRINGOMYELIA ASSOCIATED WITH SPINAL ARACHNOIDITIS

Table 15.4. *Author's cases of spinal arachnoiditis and cavitation*

Case	Location of arachnoiditis	Apparent known extent of cavity	'Cyst' in arachnoid	Nature of cystic fluid in cord
Case 1, A.B.	C5 to upper thoracic	Whole cervical	Yes	Faintly yellow—49 mg/100 ml of protein
Case 2, L.S.	T4 and below	Low cervical to mid-thoracic	No	Clear
Case 3, T.T.	T5 and below	Entire thoracic	No	Clear fluid
Case 4, C.E.	T5 and below	Upper thoracic (? more extensive)	No	—
Case 5, C.LaC.	T2 to 4	Lower cervical to mid-thoracic	No	Clear
Case 6, S.H.	T6 to 7	Probably lower cervical to mid-thoracic	Yes	Clear—48 mg/100 ml of protein
Case 7, M.E.B.	T3 to T5	Mid-cervical to T5	No	Clear fluid

the first to remark on the tension within such arachnoid cysts. Whether it adds to the likelihood of an associated syrinx is unknown but it needs to be kept in mind that the symptoms in a given case of arachnoiditis can be due to one, or all, of three factors: (1) The arachnoiditis itself. (2) The presence of a cystic loculation under pressure within the arachnoid. (3) An underlying syrinx.

The fluid in the cavities was of interest. It was always described as 'clear' or 'clear and watery' except in one instance when it was 'faintly yellow'. Twice, protein levels were assessed and both were within the range of normal for C.S.F. (49 and 48 mg/100 ml). In view of the fact that these cavitations, at post-mortem, as well as those in the cases reported in the literature, had no connection with the fourth ventricle or the spinal subarachnoid space the possibility must be considered that this fluid is being made and absorbed within the syrinx (see Chapters 10 and 14).

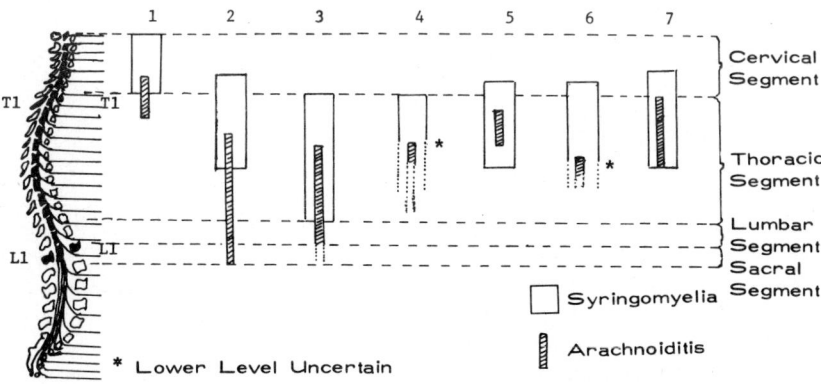

Fig. 15.18. Schematic representation of the author's seven cases of syrinx underlying arachnoiditis.

NON-COMMUNICATING SYRINGOMYELIA
PROGNOSIS AND SURGICAL MANAGEMENT

Cases—both from the literature and my own experience—who have developed a syrinx in association with an arachnoiditis have all been relentlessly progressive in so far as neurological disability is concerned. The pace of progression varied, but the eventual disturbance of function has been most serious in them all. It is apparent, therefore, that a particularly serious form of arachnoiditis is present when underlying cavitation develops.

Treatment of constrictive arachnoiditis, either medical or surgical, is usually very unrewarding. The process generally extends over several segments at least and may be extremely diffuse. I have attempted intrathecal and oral steroid therapy without convincing benefit. The cases reported here have not had lasting benefit in respect to the arachnoiditis itself from procedures designed to cut and free extensive adhesions. They re-form and the constrictive and tethering process is renewed.

However, a study of our own cases, including the record of the case following thoracic cordotomy provided by Dr W. S. Coxe, convinces us that there are situations in which serious progression of a neurological disability in the presence of an arachnoiditis may reach the stage when it will demand an exploratory laminectomy. The following situations are those which warrant this approach, in an attempt to improve or arrest a neurological disability, or to prevent its ascent:

1. An associated intra-arachnoid cyst may coexist, and may be large enough and under sufficient pressure to compress and distort the spinal cord. Its preoperative recognition may not be possible, but a separate density shadow in the myelogram may reflect its presence.

2. Constrictive arachnoiditis is not an extending lesion—rather it is a process that worsens at the site of its origin and, of course, this may be localised or diffuse. An ascending neurological disability, particularly likely to be detectable if upper limbs that are originally spared become abnormal, is good evidence, therefore, that an associated syrinx is extending up the cord. In this situation, it is probably best that an exploratory laminectomy be performed *just at and above the upper limit of the arachnoiditis*. If a syrinx coexists beneath the arachnoiditis a plastic (Silastic) drain should be introduced into it and fixed in the subarachnoid space, preferably above the adhesive process. The best long-term results in our series of patients partially paraplegic from the combination of factors (arachnoiditis, arachnoid cyst and syrinx) were obtained when this procedure was carried out (cases 1, 3, 5 and 7). Where it was not done (case 6) or could not be left in for technical reasons (case 2), less satisfactory results were obtained. Progression of the disability due to the ascending lesion continues in at least one of these patients (case 2).

3. Some cases of arachnoiditis eventually become totally paraplegic as a result of the inexorable progression of the constrictive process. Attempts to excise the arachnoidal adhesions probably will not restore spinal cord function. The spinal cord completely deprived of its function under these circumstances will be fixed and tethered by these adhesions and deprived of its normal flexibility. This flexibility allows it to adjust to neck and back movement (see Chapter 14) and to the altered intraspinal pressures resulting from straining, coughing and sneezing. For this reason, exploratory laminectomy may be indicated in the completely

paraplegic person with a view to dealing with the factors already mentioned, plus an attempt to restore flexibility to the part of the cord above the tethering. In these cases, the proximal portion of the cord may be normal or it may contain an upwardly extending syrinx. For example, in case 4 a total paraplegia developed and the arachnoidal constrictive lesion was at and below T5. The upper limb involvement was evidently due to a syrinx even though it was not demonstrable myelographically.

Basing his decision on an experience gained from a post-traumatic cystic myelopathy in a completely paraplegic patient (case 11, Chapter 12), Dr T. P. Morley reasoned that the cord could be severed and the tethering negated. At laminectomy, and in confirmation of this hypothesis, when the cord was severed it retracted 'almost out of sight of the operative field'. With this procedure the upper portion of the cord was presumed to return to its original state of flexibility. At the same time, a very effective drainage of the syrinx (Fig. 15.15) was possible from the proximal stump of the severed cord. To obviate the possible closure of the cavity by adhesions developing over the stump, there is an advantage in fixing a catheter into the cavity in the severed proximal end of the cord (see Fig. 11.3), with the other end fixed in the normal subarachnoid space.

REFERENCES

Alquier, L. & Touchard, P. (1909) Syphilis probable du néuraxe à forme anormale, méningite scléreuse cerebro-spinale, petites lésions bulbaires en foyer, cavités médullaires, syringoméliformes. *Encéphale*, **4**, 404.
Appleby, A., Bradley, W. G., Foster, J. B., Hankinson, J. & Hudgson, P. (1969) Syringomyelia due to chronic arachnoiditis at the foramen magnum. *Journal of the Neurological Sciences*, **8**, 451.
Bliss, M. A. (1909) Cysts within the spinal canal. *Journal of the American Medical Association*, **52**, 885.
Charcot, J. M. & Joffroy, A. (1869) Deux cas d'atrophie musculaire progressive avec lesions de la substance grise et des faisceaux anterolateraux de la moëlle épinière. *Archives de Physiologie*, **2**, 354.
Davison, C. & Keschner, M. (1933) Myelitic and myelopathic lesions. VI. Cases with marked circulatory interference and a picture of syringomyelia. *Archives of Neurology & Psychiatry*, **30**, 1074.
Elkington, J. St. C. (1936) Meningitis serosa circumscripta spinalis (spinal arachnoiditis). *Brain*, **59**, 181.
Feigen, I. (1972) Personal communication.
Feigin, I., Ogata, J. & Budzilovich, G. (1971) Syringomyelia: the role of edema in its pathogenesis. *Journal of Neuropathology & Experimental Neurology*, **30**, 216.
Foster, J. B., Hudgson, P. & Pearce, G. W. (1969) The association of syringomyelia and congenital cervico-medullary anomalies: pathological evidence. *Brain*, **92**, 25.
French, J. D. (1946) Clinical manifestations of lumbar spinal arachnoiditis. A report of 13 cases. *Surgery*, **20**, 718.
Gardner, W. J. (1965) Hydrodynamic mechanism of syringomyelia: its relationship to myelocele. *Journal of Neurology, Neurosurgery & Psychiatry*, **28**, 247.
Harris, T. (1886) On a case of multiple spinal and cerebral tumours (sarcomata), with a contribution to the pathology of syringomyelia. *Brain*, **8**, 447.
Harbitz, F. & Lossius, I. (1929) Extramedullary tumour arachnitis fibrosa cystica et ossificans. Gliosis of the medulla. *Acta Psychiatrica et Neurologica*, **4**, 51.
Hinds Howell, C. M. (1936/37) Arachnoiditis. *Proceedings of the Royal Society of Medicine*, **30**, Part 1, 33.
Holmes, G. & Kennedy, R. F. (1908) Two anomalous cases of syringomyelia. *Proceedings of the Royal Society of Medicine*, **2**, 1.
Horsley, Sir V. (1909) A clinical lecture on chronic spinal meningitis. Its differential diagnosis and surgical treatment. *British Medical Journal*, **i**, 513.
Joffroy, A. (1873) *De la pachyméningité cervicale hypertrophique*. Thesis, Paris.
Joffroy, A. & Achard, C. (1887) De la myélite cavitaire. *Archives Physiologie Normale Pathologie*, **10** 435.

Lichtenstein, B. W. (1943) Cervical syringomyelia and syringomyelia-like states associated with Arnold–Chiari deformity and platybasia. *Archives of Neurology & Psychiatry*, **49**, 881.
Logue, V. (1971) Syringomyelia: a radio-diagnostic and radiotherapeutic saga. *Clinical Radiology*, **22**, 2.
Lombardi, G., Passerini, A. & Migliavacca, F. (1962) Spinal arachnoiditis. *British Journal of Radiology*, **35**, 413.
Martin, J. P. (1925) Amyotrophic meningo-myelitis (spinal progressive muscular atrophy of syphilitic origin). *Brain*, **48**, 153.
Mackay, R. P. (1939) Chronic adhesive spinal arachnoiditis. A clinical and pathologic study. *Journal of the American Medical Association*, **112**, 802.
Mendel, K. & Adler, S. (1908) Zur Kenntnis der Meningitis serosa spinalis. *Berliner Klinische Wochenschrift*, **45**, 1596.
Nelson, J. (1943) Intramedullary cavitation resulting from adhesive spinal arachnoiditis. *Archives of Neurology & Psychiatry*, **50**, 1.
Nielsen, J. M. (1953) Progressive course of arachnoiditis simulating syringomyelia. *Bulletin of the Los Angeles Neurological Society*, **18**, 127.
Philippe, Cl. & Oberthür, J. (1900) Classification des cavités pathologiques intra-médullaires. *Revue Neurologique*, **8**, 171.
Rhein, J. H. W. (1908) Syringomyelia with syringobulbia. *Journal of Medical Research*, **18**, 127.
Rosenblath, W. (1892–93) Zur Casuistik der Syringomyelie und Pachymeningitis cervicalis hypertrophica. *Deutsche Archiv für Klinische Medizin*, **51**, 210.
Schlesinger, H. (1902) *Die Syringomyelie. Eine Monographie*, 2nd Edn., p. 255. Leipzig and Vienna: F. Deuticke.
Schüppel, O. (1865) Uber Hydromyelus. *Archiv für Heilkunde*, **6**, 289.
Schwartz, E. (1897) Präparate von einem Falle syphilitischer myelomeningitis mit Höhlenbildung im Rückenmark und besonderen degenerativen Veränderungen der Neuroglia. *Wiener Klinische Wochenschrift*, **10**, 177.
Shaw, D. A. (1972) Personal communication.
Spiller, W. G., Musser, J. H. & Martin, E. (1903–4) A case of intradural spinal cyst, with operation and recovery. With a brief report of 11 cases of tumour of the spinal cord or spinal column. *University of Pennsylvania Medical Bulletin*, **16**, 27 & 56.
Stookey, B. (1927) Adhesive spinal arachnoiditis simulating spinal cord tumor. *Archives of Neurology & Psychiatry*, **17**, 151.
Taylor, F. (1884) Syphilitic meningitis and gumma of the spinal dura mater, with tubular cavity in the spinal cord (syringomyelia). *Transactions of the Pathological Society of London*, **25**, 36.
Thomas, A. & Hauser, G. (1901) Cavités médullaires et mal de Pott. *Revue Neurologique*, **9**, 117.
Weiss, R. M., Sweeney, L. & Dreyfuss, M. (1962) Circumscribed adhesive spinal arachnoiditis. *Journal of Neurosurgery*, **19**, 435.
Winkelman, N. W., Gotten, N. & Scheibert, D. (1953) Localized adhesive spinal arachnoiditis. A study of 25 cases with reference to etiology. *Transactions of the American Neurological Association*, **15**.
Vogel, P. J. (1946) Circumscribed spinal arachnoiditis with cavitation of the spinal cord. Report of case. *Bulletin of the Los Angeles Neurological Society*, **11**, 58.
Vulpian (1861) Quoted by Rhein, J. H. W. (see above).

CHAPTER SIXTEEN

The Pathogenesis of Syringomyelic Cavitation Associated with Arachnoiditis Localised to the Spinal Canal

H. J. M. BARNETT

Syrinx formation in the spinal cord secondary to pathological lesions that are confined within the spinal axis is of paramount importance in considering the aetiology of all types of syringomyelia. For this reason, it is important to review the experimental and clinical observations that may relate to the pathogenesis of this variety of syringomyelia.

EXPERIMENTAL PRODUCTION OF ARACHNOIDITIS

Camus and Roussy (1914) produced severe arachnoiditis in dogs by the intracisternal injection of a suspension in water of fatty acid, sodium nucleinate and talc. In the animals that died or were sacrificed early, necrotic lesions were found in the grey matter and the adjacent posterior columns. In those animals which survived for several months, cavities were present in these same areas. The arachnoiditis was maximal in the upper cervical region and here it constituted a 'veritable fibrous ring' about the spinal cord (Fig. 16.1). It stopped in the lower cervical area. The meninges, at the base, were involved up to the level of the peduncles. Although cavitation was most prominent in the cervical cord (Fig. 16.2), Camus and Roussy reported on one dog who survived nearly six months with a cavity found in the medulla at the level of the inferior olives adjacent to, but not communicating with, the fourth ventricle. They do not describe any connection between this cavity and the larger cervical cystic lesions. There may have been one but in the light of the developments of recent years, this important point requires further clarification.

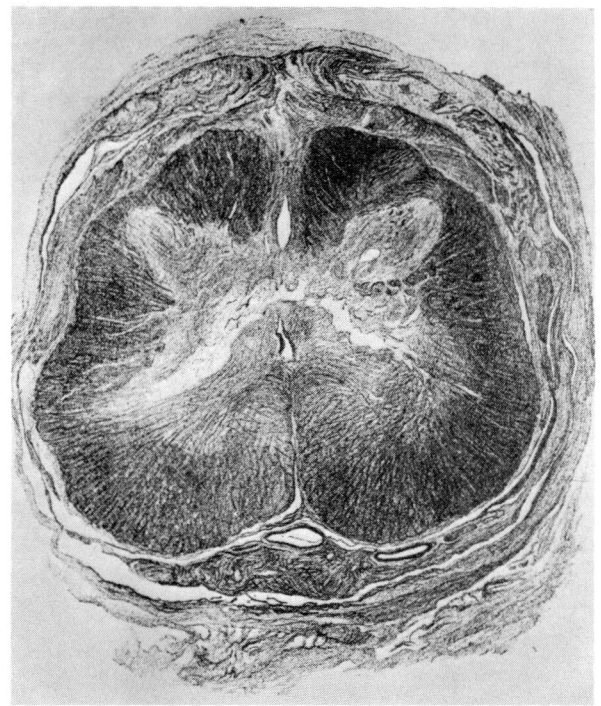

Fig. 16.1. Experimental arachnoiditis encircling cervical spinal cord with necrosis and early cavitation centrally. (Reproduced from Camus and Roussy, 1914, by permission of Masson et Cie, Paris.)

Fig. 16.2. Central cavitation in cervical cord of dog from experiments of Camus and Roussy. (Reproduced by permission of Masson et Cie, Paris.)

They judged that they were producing lesions of an ischaemic nature, and that softened areas progressed to intramedullary cavitation. The addition of an inflammatory element could not be excluded. They likened their experimental lesions to the cavities already described in the condition of 'hypertrophic cervical pachymengitis'.

McLaurin et al (1954) followed up this early experimental work by the production, again in dogs, of a chronic adhesive arachnoiditis. They injected kaolin into the cisterna magna in 42 animals and ethyl iodophenylundecylate (Pantopaque) in 13 animals. The animals were sacrificed between nine days and four-and-a-half months later. The leptomeninges at the base of the brain and in the posterior fossa were adherent sufficiently to obstruct the foramina of Magendie and Luschka and lead to hydrocephalus. All the animals exhibited 'arachnoiditis' below the foramen mangnum, surrounding the cervical and thoracic portions of the spinal cord. In more than half of them, some degree of macroscopic cavitation was visible. In all animals injected with kaolin, a ring of dense collagenous tissue was found, extending from the pia to the dura and blending with the fibrous structure of the latter. The vessels of the subarachnoid space were embedded in the fibrous tissue and appeared to be constricted by it. Thrombosis was not observed in the constricted vessels, nor was haemorrhage visible in the areas of the cord within their supply.

The early cord lesions consisted of necrosis and softening in the posterior half of the cord, centrally, at the junction of the grey and white matter (Fig. 16.3). These lesions were 'patchy and irregular without macroscopic continuity between successive degenerative areas' (Fig. 16.4). In animals surviving longer, McLaurin found cystic cavitation in the areas where necrosis alone was seen in the earlier lesions. The cavities in the longer survivors became continuous and were largest in the cervical region. They were connected with the ventricles since Pantopaque

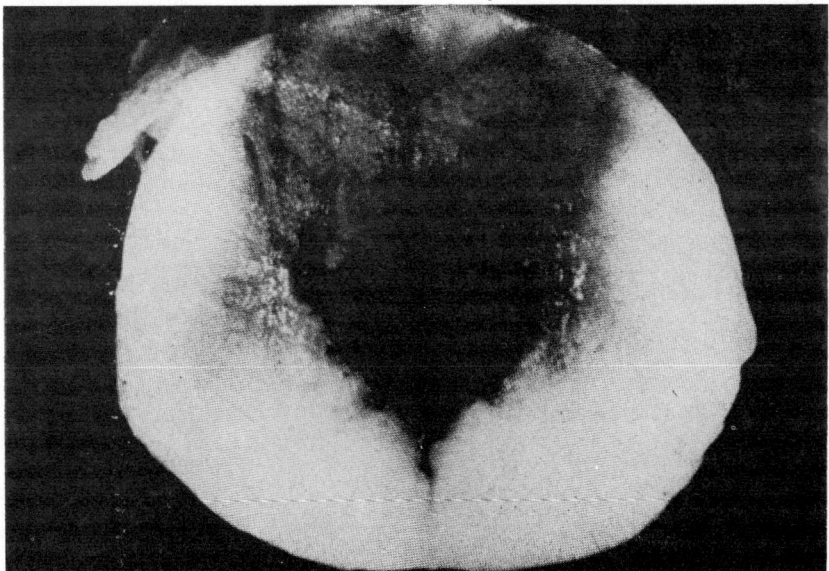

Fig. 16.3. Central necrosis and early cavitation produced experimentally by cisternal Kaolin injection. (Reproduced from McLaurin et al, 1954, by permission of the authors and the *Archives of Pathology*.)

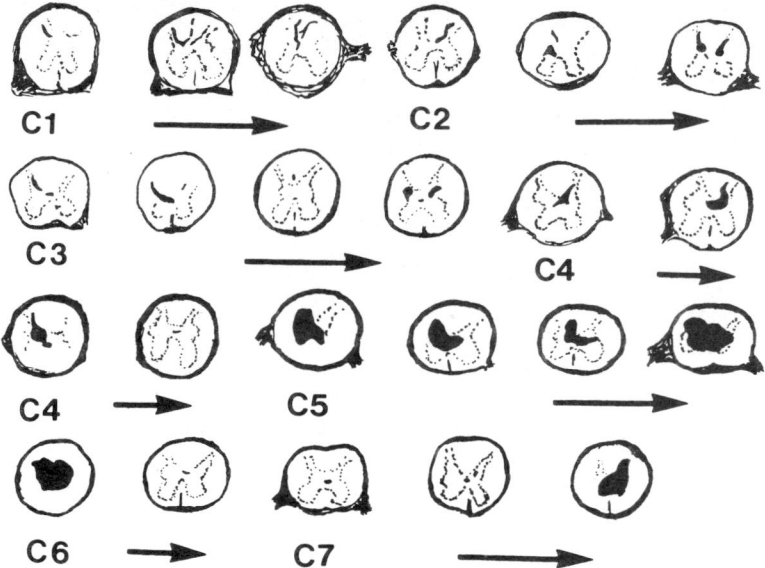

Fig. 16.4. Sequential cervical segments with necrotic lesions, produced by experimental arachnoiditis. Note the lack of continuity of early lesions. (Adapted, from McLaurin et al, 1954, by permission of the authors and the *Archives of Pathology*.)

injected into the fourth ventricle in early cystic cases outlined both the central canal and numerous irregular projections from it. Indian ink injected into the ventricle was found post-mortem in phagocytes surrounding the vessels in, or near, the lining of the spinal cord cavities.

A study of the late lesions alone in McLaurin's material might lead one to think that the arachnoiditis at the base of the brain could be the mechanism of production of the cord cavitation and that 'communicating syringomyelia' had been produced. This is in conflict, though, with the observations in the early lesions where discontinuity was quite apparent. Later development of a continuous cavitation communicating with the fourth ventricle may have led to the *extension* of such a cavity but the published evidence suggests that it did not *initiate* the lesion.

This experimental work on arachnoiditis has importance in any consideration of the overall problem of the pathogenesis of a syrinx with arachnoidal lesions. An experimental model, however, is needed that presents an opportunity to study the effect on the spinal cord of an arachnoiditis or pachymeningitis confined, unequivocally, to the spinal axis and which has no cranial extension to the basal cisterns.

EXPERIMENTAL PRODUCTION OF SPINAL CORD ISCHAEMIA

Both of the series of experimental observations referred to arouse interest in the part played by ischaemia in the production of cord cavitation by extramedullary constrictive and/or compressive lesions. Ischaemia is well known to be associated with centrally located necrotic lesions and the clinicopathological evidence respecting cavitation as a sequel to central necrosis is reviewed in detail in Chapter 14. From the experimental viewpoint, it has been studied by several groups of investigators. Tauber and Langworthy (1935), reviewing the aetiology of syringomyelia, were impressed by the significant number of cases who had a preceding history of trauma and thereafter developed progressive syringomyelic signs and symptoms. They postulated that trauma produced infarction, and therefore tried several experimental methods to produce ischaemia and, hopefully, subsequent cavitation.

First they sectioned the posterior roots bilaterally in the lumbar region. This did not result in sufficient ischaemia to produce cavitation. Then they tied silk ligatures around the dorsal or the ventral halves of the thoracic spinal cords of cats and kept them alive for several months. With dorsal ligatures, only an area of damage close to the ligature site resulted. In those with ventral ligatures, a large cavity extended for several segments in the ventral portions of the posterior columns. In the group with dorsal ligations, they considered that they were interfering with a non-continuous arterial trunk but that in the ventral ligations they were depriving the thoracic cord of an important blood supply from the anterior spinal artery. Finally, they did an unspecified number of paraffin injection experiments and, in one instance, produced mechanical occlusion of the anterior spinal artery with a secondary central cavity. It was a small lesion and the experiment was inconclusive.

Expanding on the attempt to produce spinal cord cavitation by ischaemia, Woodard and Freeman (1956) and most recently Wilson et al (1969) have reported on experiments carried out in dogs. In 42 dogs, Woodard and Freeman sectioned six adjacent pairs of spinal roots. They postulated that this would interrupt the blood vessels passing along the roots and as it was done extradurally, the lesion produced would not be confused by incidental operative trauma to the cord. Twenty-seven animals survived the procedure and were sacrificed in periods ranging from one to four weeks. The 16 spinal cords were examined in detail and of these, 12 had central cavitation. The cavities were small, and indeed, with one exception, they were microscopic. They occurred, however, in the grey matter, centrally and always dorsal to the central canal.

Wilson et al (1969) carried out a similar experiment, but varied the site and pattern of radicular extradural arterial ligation. A few of their animals surviving between 30 and 43 days were found to have small cystic lesions in the central grey matter and adjacent white matter. It is of interest that these experiments designed to interfere with spinal cord circulation without provoking arachnoidal reaction, did not produce cavitations as large or extensive as those in the experimental arachnoiditis studies.

CLINICAL OBSERVATIONS RELATED TO THE PATHOGENESIS OF CORD CAVITATION WITH SPINAL ARACHNOIDITIS

If ischaemia due to strangulation of blood supply within constrictive arachnoiditis is to be considered as a major factor in the development of the cord damage leading to cavity formation, it is important to review the association of cavity formation with any type of recognised arterial lesions affecting the spinal cord blood supply. It soon becomes apparent that, just as in the experimental production of ischaemia to the cord, clinical conditions associated with ischaemia rarely occur with extensive cavity formation. In most reported instances, the amount of cavitation has been small and localised to the two or three segments of most marked infarction (Zülch, 1962; Hughes and Brownell, 1966; Jellinger, 1967).

In the localised cavitation underlying cervical spondylosis described by Mair and Druckman (1953), obliterative changes in the anterior spinal arteries appropriate to the underlying central ischaemia and cavity formation were demonstrated. Interference with venous drainage and to the microcirculation (Turnbull, 1971) may have been important as well, because of the compressive element in the condition. Again, the cavitation was small and unimportant. The case reported by Osetowska-Wiechowska (1956) stands as unusual in this regard. The medullary arteries were the site of extensive inflammatory and thrombotic obliterative disease and major cavitation extended from lumbar to upper thoracic segments of the cord. It seems possible that the extensive involvement of the blood vessels in this case determined that a greater degree of cavitation developed than was noted in the usual circumstance of cord ischaemia.

It is not easy to determine the extent or importance of a vascular or other component to the pathogenesis of the more extensive cavitation encountered in our cases of arachnoiditis. The nature of the lesion producing the arachnoiditis must be considered in any attempt to weigh the importance of aetiological factors. In the *postmeningitis cases*, for example, there is reason to believe that arteritis followed by obliterative arteriopathy is an important factor in early cord infarction and that this softened area is succeeded by cavitation. Brooks, Fletcher and Wilson (1954) have illustrated early cavitation occurring in a case of tuberculous meningitis (Fig. 16.5) and have indicated the importance of the arterial lesion in the myelomalacia preceding cavity formation.

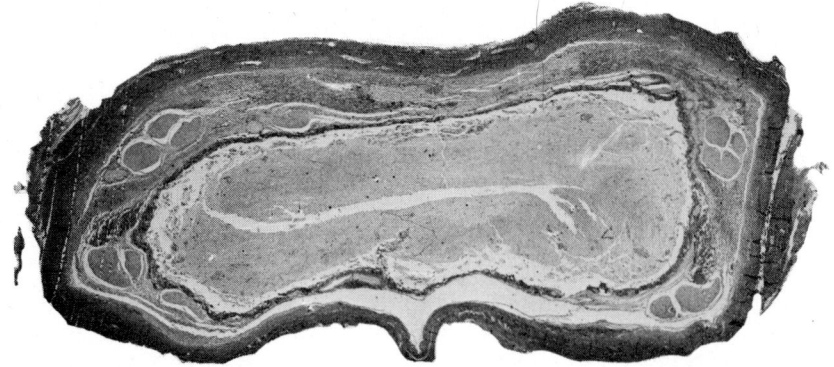

Fig. 16.5. Tuberculous meningitis with extensive meningeal thickening and cavitation in underlying spinal cord. (Reproduced from Brooks et al, 1954, by permission of the authors and the *Quarterly Journal of Medicine*.)

THE PATHOGENESIS OF SYRINGOMYELIC CAVITATION

The eminent pathologist, Oscar Klotz (1913) reported on a case of interest in this regard; the patient, a woman, had multiple systemic abscesses and died as a result of sepsis. She had become abruptly paraplegic several weeks before she died. This complete paraplegia had come on over a period of *two to three minutes* and never changed. At post-mortem, osteomyelitis and some extradural pus were found at T7, 8 and 9. The dura was decidedly thickened, but within the dura there was neither pus, exudate nor adhesions. The underlying cord was grossly necrotic and thin. From the necrotic area, a single cavity passed rostrally for three segments to T5 and multiple small cavities (Fig. 16.6) passed caudally to the T10 and 11 levels. No association existed between these cavities and the central canal. The dural blood vessels were inflamed and one recently thrombosed vein was detected.

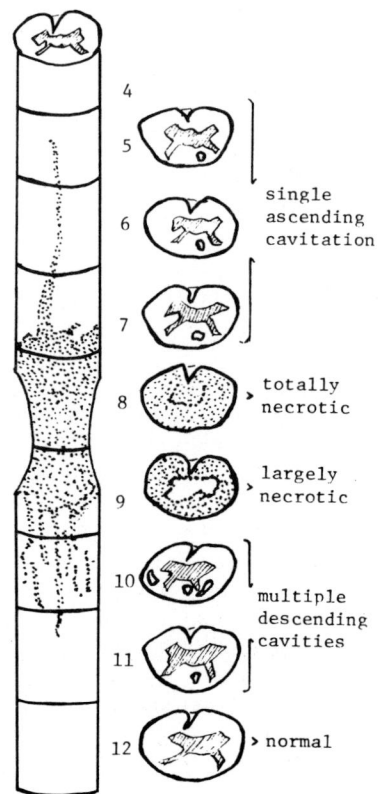

Fig. 16.6. Artist's adaptation of cord necrosis and multiple cavity formation from extradural abscess and ischaemic change. (Adapted from Klotz, 1913, by permission of the *American Journal of Medical Science*.)

It is improbable, though not impossible, that an abscess of the cord or a pre-suppurative myelitis may contribute to the production of the postmeningitic cavitation. A syrinx as a late sequel to spinal meningitis is a rare phenomenon nowadays and abscess of the cord is equally rare. The combination has not been recognised, although Jelsma (1972) has recently described something approaching this. An upper and lower respiratory illness was succeeded five weeks later by signs of meningitis and a spinal lesion. At operation, dense arachnoiditis and an abscess of the mid-cervical spinal cord was drained and the organism grown proved to be *Corynebacterium diphtheriae gravis*. The fluid in the cavity was under pressure, was orange in colour, and contained 4·4 g of protein and 150 W.B.C./ µl. Within seven months the patient's legs became more spastic and a myelographic irregularity was visualised from T4 to L1. A constrictive arachnoiditis was found at a T4 to L1 laminectomy and at T6 a cavitation of the cord was again identified, filled with a 'semicongealed yellow fluid' from which no organisms grew. This was drained, the adhesions removed as much as possible, and within a few months he learned to walk without aid. Eighteen months later he fell from a bicycle and within 48 hours became abruptly worse; Jelsma postulated that flexibility of the cord had been lost and the patient was thereby vulnerable to a torsion effect. This is a unique case in that it is a cord cavity associated with arachnoiditis as a sequel to meningitis and an accompanying spinal cord abscess. The high protein in the cavity betrayed the fact that it was an exudate, not a transudate or a fluid resembling C.S.F.

In the *postmyelogram* and *postspinal anaesthetic* varieties, signs of spinal cord dysfunction may come on abruptly. In some of these cases, infection or foreign irritants (e.g. detergent solutions) introduced with the oil and/or the anaesthetic agent (Joseph and Denson, 1958; Winkelman et al, 1953) appear to be the primary factors in the immediate neurological sequelae and causally related to initiation of the delayed reaction. The iodinised oil used for myelography (Hurteau, Baird and Sinclair, 1954; Howland and Curry, 1966) appears in others to be the important agent in the immediate reaction and is likely to be so in the delayed reaction also. Whether an allergic vasculitis is a part of the clinical picture in the latter group is not proven but is a possibility. The later progression appears to be determined by arachnoidal reaction, while a more immediate event is needed to explain the initial symptoms. No doubt the mixture of blood and oil is a potent combination in producing and aggravating the extent of the ultimate arachnoiditis (Howland and Curry, 1966) but the fact remains that the onset of early symptoms is coincident with the procedure in some of the cases. The experiments of McLaurin et al (1954) suggest that vascular factors may contribute to the early, as well as the later lesions. One of the irritants used in his experiments was Pantopaque, more highly concentrated than in the strength employed clinically.

In the *post-traumatic* variety of arachnoiditis there may not be an important initial ischaemic factor. Bleeding into the subarachnoid space may be sufficient to set in motion the train of events beginning with arachnoiditis and ending with underlying cord cavitation. However, where the trauma is more severe there may have been an initial central myelomalacic pulping of the variety described by Holmes (1915) and discussed in detail elsewhere (Chapter 14). Arterial, venous and even capillary disturbances are likely to be involved to some extent in this lesion.

Discectomy is regarded by some as a cause of subsequent arachnoiditis (Smolik and Nash, 1951). This point is not to be argued in detail here. However, trauma producing the disc protrusion, myelography to delineate it, as well as the trauma of the procedure may act as a concatenation of factors initiating the arachnoiditis. Operative occlusion of spinal arterial supply and venous drainage may have been a contributory factor in the example (case 6) quoted, in whom cord cavitation became linked with arachnoiditis to produce a progressing neurological disability as a late sequel to thoracic discectomy.

In the case recorded with arachnoiditis and cord cavitation as a sequel to *thoracic anterolateral cordotomy* (case 7), the pathogenesis of the arachnoiditis was never established but operative haemorrhage, with or without minor local infection and/or other local irritants, was probably important. Operative hematomyelia, along with associated overlying arachnoiditis, constricting the cord and its blood supply appear to have combined to give rise to the expanding intramedullary cavitation.

Regardless of the role of vascular change in the primary insult that initiates the meningeal reaction and may also damage the cord coincidentally, there is obviously a potential for interference with circulation through the dense meningeal constrictive fibrosis of the later stages. The vascular lesions in constrictive meningitis of this chronic variety are well described in experimental and clinical studies on arachnoiditis itself. The obliteration of venous drainage may be important, even when arterial lesions are not too obvious, in determining underlying cord necrosis and late cavitation (Feigin, 1972).

POTT'S DISEASE AND SYRINGOMYELIA

Pott's disease presents a situation of peculiar interest. Paraplegia may develop in the acute stage of the disease and myelography may indicate an extradural, obstructing mass. On other occasions, the subarachnoid space at myelography or post-mortem may be normal. Myelitis of vascular origin, and possibly with added toxic factors, was the explanation frequently advanced to explain the cause of the spinal cord lesion in patients with a normal myelogram. Evidence favouring this hypothesis was reported as early as 1888 by Kroger (quoted by Seddon, 1946). Vascular changes were also referred to by Spiller (1898). Gordon (1904) reported on a case paraplegic for three months with Pott's disease, studied at post-mortem. The ninth and tenth thoracic vertebrae were carious, but not compressing the spinal cord. From T9 to L5 there were two small cavities near the central canal and the overlying meninges were mildly thickened with arteritis of the vessels on the periphery of the cord, 'the majority filled with thrombi'. The same process extended into the intramedullary arterioles. Even the vertebral artery itself may be at risk as judged by a case of acute cervical caries reported by Seddon (1935). Vasculitis of this artery was followed by thrombosis and fatal brain stem infarction. These acute cases were recognised to be associated at times with central softening and even local cavity formation (Butler, 1935; Gordon, 1904).

Cord Cavitation as a Late Sequel

As well as these observations in acute active Pott's disease, spinal lesions were recognised also in cases afflicted with a major deformity from Pott's disease which was 'healed' and 'inactive' for a number of years. The coexistence of a pachymeningitis in relation to these late spinal lesions of Pott's disease was recognised and described by a number of early writers, among them Charcot and Joffroy (1869), Lloyd (1894), Oppenheim (1907), and von Bruns (1908). Accompanying cord cavitation was not recorded by them. The question always arose as to whether or not the lesion was a mechanical one due to the prolonged, sharp angulation of the spinal cord around the major deformity thereby 'indenting' the theca and cord (Alexander, 1946), or a flare-up of the granuloma. Both factors probably contributed although Seddon (1935) and Butler (1935) favoured the latter as most usual.

The question that our case (case 5) poses, though, is how frequently is Pott's disease followed by a syrinx and if it is, do vascular, constrictive or simply straightforward compressive factors operate. A review of the literature on Pott's disease establishes that associated cord cavitation is very rare. It is not mentioned in recent extensive reviews of Pott's disease although it is listed by Wilson and Bruce (1954) and Elliott (1971) in their neurological textbooks among the causes of syringomyelia. Blackwood et al (1971) make no mention of it. Neither does Butler (1935) in his description of a series of 96 cases of late-onset paraplegia. Kocen and Parsons (1970) in a recent review of the neurological complications of tuberculosis do not make reference to cord cavitation.

Thomas and Hauser (1901) described the case of a young woman who had a spinal deformity from the age of seven years, with a draining lesion adjacent to the vertebral canal. Deformity developed, centred on T12, between the ages of seven and 12. At the age of 23 years, she became afflicted with the rapid onset of a total paralysis of her legs and incomplete loss of superficial sensation in them. During the next eight years, her lower limbs completely lost sensibility to pain and temperature and her ankle reflexes disappeared. She died of pulmonary tuberculosis and, at autopsy, the ninth to the twelfth segments of the thoracic cord were the site of cavitation (Fig. 16.7). This cavitation involved the posterior horn and ventral part of the posterior column largely on one side. There was an extradural caseation at this site and a thickening of the dura. However, Thomas and Hauser specifically noted that the dura was not adherent to the underlying leptomeninges. These leptomeninges were greatly thickened but independently of the dural thickening. The meningeal reaction was 'intense', particularly anteriorly. They noted that the vessels were 'relatively few in number' but 'those that persisted were not obliterated'. They did not consider that they could implicate a vascular lesion in the production of the cavity formation. They speculated that it might be a tuberculous process of the cord itself—a conclusion that would not be acceptable on the evidence presented.

Alquier and Lhermitte (1906) described another single case from the La Salpêtriére in a patient with sacral Pott's disease. From the second lumbar to the first sacral segment there was a dilatation of the central canal. It was asymptomatic. They refer to three contemporary case reports of Pott's disease complicated by a cavitation in which the cavitation appeared to originate below the site of compression and extended up the cord for three segments in one case, for an unspecified distance in a second, and for three segments above the twelfth dorsal compression in a third. The usual location of a cavity dorsal to the central canal was described in two of these cases.

In the modern literature, there is a similar paucity of reported cases of syringomyelia associated with Pott's disease. Martin and Maury (1964) described a patient who developed Pott's disease in the eleventh thoracic region at the age of 11 years, with a persistent gibbus and became paraplegic as an immediate sequel to an osteotomy at the age of 16. This slowly recovered, but 10 years later the right upper limb became insensitive to pain and temperature and the deep reflexes disappeared in this limb. The sensory level extended up to C6 and yet five years

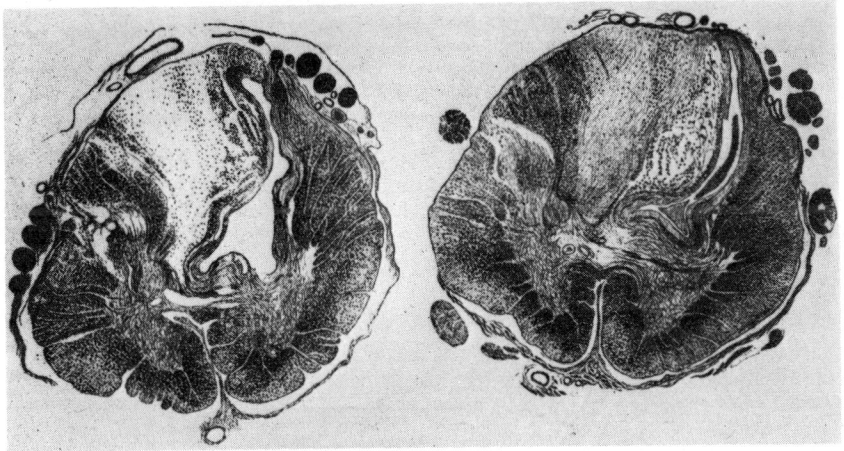

Fig. 16.7. Thomas and Hauser's (1901) illustration of cord cavitation at T9–12, the result of Pott's disease and deformity of 24 years' duration. (Reproduced by permission of Masson et Cie, Paris.)

later weakness had not supervened. A cavitation in the cord was never proven but is almost certainly the explanation of the upper limb symptoms.

Feigin, Ogata and Budzilovich (1971) made brief reference to one case, in whom the cord was surgically transected. Their description is as follows:

'This was a 52-year-old woman with a spastic paraplegia of long duration attributed to spinal cord injury due to Pott's disease of the spine with kyphoscoliosis. A segment of lower thoracic spinal cord 4·5 cm long and 1·3 cm in diameter was removed surgically. Microscopically, a cystic cavitation 2 mm in diameter was present in the central tissue on one side, and an area of marked rarefaction was symmetrically placed on the other side of the spinal cord. The cavitation contained red cells, a small amount of debris and a few macrophages. Its walls were composed of a dense layer of glial fibres.'

In this instance, the size of the cavity was unimpressive. However, had the process persisted undisturbed, an extension of this initial process may well have occurred.

Because of its great rarity, the following case reported to us by Dr D. A. Shaw (1972), Newcastle, is also worthy of record:

In 1940, at the age of 17 years, this man developed a progressive spastic paraparesis over a three-month period. This proved to be due to Pott's disease involving the seventh thoracic vertebral body. He deteriorated despite conservative therapy. In September 1941, he was submitted to a dorsal laminectomy and the operative note stated that a dense white dura was encountered with a sclerotic non-pulsatile cord. There was no block, however, and biopsy showed only fat and fibrous tissue. Possibly there was slight improvement but it was not definite. In December 1942, a myelogram now showed a block at D4 to 6. In January 1943, he had a second laminectomy at the D3 to 6 level and some arachnoidal adhesions were found, but there was no sign of any compression. He was immobilised in a plaster shell, regained control of bladder and bowel and became able to walk again. By May 1944, he was again walking with sticks, and he was able to earn his living as a telephonist.

His condition remained relatively static and in 1959, on a follow-up examination, it was recorded that he had a spastic paraparesis and 80 mg/100 ml of protein in his spinal fluid, but no arm involvement was described. He discharged himself from hospital at that time but was readmitted in February 1960, and for the first time it was discovered that he had a sensory loss in his right arm. A neurologist in Manchester diagnosed syringomyelia. He was having some troublesome flexor spasms in his legs at this time and he had some intrathecal phenol injections with partial relief. Over the next few years the right forequarter sensory loss increased and the arm became weak. Loss of sphincter control had returned and by November 1966 he was no longer able to perform manual evacuation of his rectum because of the weakness in his right hand.

Examination in November 1966 then showed impaired pain and temperature sensation in the right trigeminal distribution as the only cranial nerve abnormality. He had gross weakness and spasticity in the legs and bilateral extensor plantar responses and a sensory level at D4 on both sides. Above this level, however, there was also a 'half-cape' loss of pain and temperature sensation on the right side. There was considerable distal weakness in the right upper limb and all the right arm reflexes were absent, whereas those on the left were all present.

The picture was that of syringomyelia and it seemed reasonable to conclude that *he had an ascending syrinx associated with his long-standing dorsal cord lesion.* Because of the patient's wishes, myelography and definitive therapy were not attempted. It stands, therefore, as an unproven but likely example of this rare type of syringomyelic cavitation.

Table 16.1 summarises briefly some of the data on the five reported cases and of the present author's single case of Pott's disease complicated later by a progressive neurological disability due to underlying cord cavitation. As mentioned, Alquier and Lhermitte refer briefly to three additional cases known to them, but give no further details. The level of the cavitation was closely related to the old tuberculous lesion in most of the cases but in four of the six, the cavitation extended cephalad to a varying degree and in Martin and Maury's (1964) case and that reported here by the author from Shaw (1972), the signs were frankly related to the cervical cord although the infective process had been below this. Figure 16.8 schematically indicates the relationship between the syrinx and the previous tuberculous infection of bone and extradural tissues.

Table 16.1. *Pott's disease with delayed syringomyelia*

Author	Case number	Level of infection	Arachnoiditis	Level of cavity	Interval to cord lesion
Thomas and Hauser (1901)	1	T12	T9–12	T9–12	16 years
Alquier and Lhermitte (1906)	2	Lumbo-sacral	?	L2–S1	?
Martin and Maury (1964)	3	T11	?	Cervical	15 years
Feigin et al (1971)	4	Lower thoracic	?	Thoracic	Many years
Shaw (1972)	5	T4–6	Some adhesions Not compressive	Cervical	15 years
Author's case	6	T2–4	T2–4	Lower cervical to upper thoracic	22 years

The rarity of spinal cavitation with long-standing Pott's disease is of special interest when one considers that it is a condition with a recognised association of vascular myelomalacia coupled with a lesion having compressive capability. It reinforces the fact referred to in Chapter 17, that compression from an extradural tumour along the spinal axis is rarely associated with cord cavitation.

THE PATHOGENESIS OF SYRINGOMYELIC CAVITATION

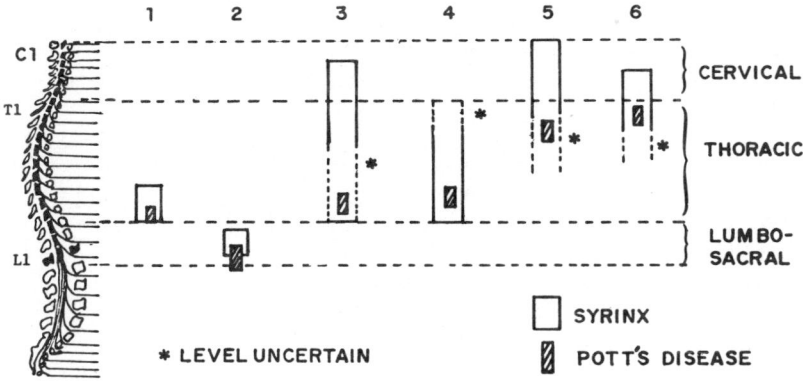

Fig. 16.8. Schematic representation of all known cases of syringomyelia underlying longstanding Pott's disease. (See Table 16.1 for source of cases.)

SUMMARY OF INFORMATION RESPECTING SYRINGOMYELIC CAVITATION COMPLICATING SPINAL ARACHNOIDITIS

The association between chronic spinal leptomeningitis or spinal pachymeningitis and a cavitation of the underlying cord is a rare phenomenon. The following would appear to summarise the essentials of the condition, as it has been detailed in the previous two chapters:

1. The condition was recognised and more commonly discussed in the decades prior to the adequate prevention of neurosyphilis.
2. Basal intracranial arachnoiditis must be excluded before it is reasonable to accept spinal arachnoiditis as a factor in the causation of spinal cavitation. Otherwise, obstruction of the outflow from the fourth ventricle may be provoking the development of a syrinx by the mechanism of the communicating syringomyelia syndrome (syringohydromyelia).
3. In the more modern literature, there are only seven acceptable examples of this condition reported with adequate pathological substantiation. Two others related to trauma and two others as a late sequel to Pott's disease are also known.
4. The author is adding seven further examples.
5. In the author's series the primary condition of spinal arachnoiditis was produced once by pyogenic meningitis, once definitely and once more doubtfully as a sequel to trauma, once after spinal anaesthesia, once following thoracic discectomy (and ? myelography), once after thoracic tractotomy and once as a late sequel to Pott's disease.
6. The cavitation in the spinal cord accounted for an extension of the signs and symptoms above the level of the arachnoiditis in one-half of the author's

cases. These upper limb signs occurred despite the fact that only in one case did the arachnoiditis extend above the thoracic level.
7. This complication occurred in cases of arachnoiditis that were either of indolent or rapid progression, but in all cases the neurological disability caused by the arachnoiditis was a severe one.
8. Arachnoid cysts, loculations within the arachnoiditis, were noted in addition to the spinal cord cavitation, in some of those recorded in the literature as well as in one of the author's series. The cord may be compromised further by these extramedullary cysts.
9. The fluid content of the spinal syrinx has been clear, watery and where measured, the protein content has been that of cerebrospinal fluid.
10. No communication has ever been demonstrated to exist between the syrinx and either the ventricle or the subarachnoid space.
11. Experimental observations suggest that an important factor in the production of the spinal syrinx is ischaemia of the central area of the spinal cord followed by subsequent cavity formation.
12. Late paraplegia from Pott's disease has been discussed for many decades and chronic pachymeningitis is an expected finding in many such cases. Despite this, the complication of an underlying spinal cord cavitation is very rare. Only two old and four modern examples, including one of the author's cases, and one communicated to the author, have been identified.
13. Myelography failed to visualise the syrinx in most cases in this series as the spinal canal was so distorted by the arachnoiditis.
14. Exploratory laminectomy is to be considered in cases of arachnoiditis where the disability is progressing lest an expensile cystic lesion be present within the arachnoiditis acting as a mass lesion.
15. Laminectomy is also to be contemplated in seriously progressive neurological disability in case an underlying syrinx in the cord is adding to the disability.
16. Laminectomy is obligatory if the neurological disability extends above the upper limits of the arachnoiditis. This is usually identified because the usual site of the arachnoiditis is in the thoracic region and the upper limb involvement is incompatible with this localisation (a cervical syrinx is to be expected under these circumstances).
17. Evidence is available that a tethered cord, which has lost its flexibility is more susceptible to trauma. It may even be one of the factors productive of the cord damage resulting in cavity formation.
18. Where a syrinx is present and the spinal lesion is a complete one, a resection of a section of the spinal cord is the treatment of choice. If carried out above the arachnoiditis it will restore flexibility to the cord and also be an effective method of maintaining cavity drainage.
19. Where a syrinx is present and the spinal lesion is a partial one, a plastic catheter inserted into the spinal cord and brought out into normal subarachnoid space is probably the best treatment programme. It is likely to leave the cord still deprived of its flexibility but the restoration of cord mobility is not attainable with certainty since the cord cannot be resected.
20. Surgical treatment for the arachnoiditis alone is unrewarding. The constrictive process continues and the disability progresses despite attempts to 'free the adhesions'.

REFERENCES

Alexander, G. L. (1946) Discussion on spinal caries with paraplegia. *Proceedings of the Royal Society of Medicine*, **39**, 730.
Alquier, L. & Lhermitte, J. (1906) Mal de Pott et syringomyélie. *Revue Neurologique*, **14**, 1141.
Blackwood, W., McMenemey, W. H., Meyer, A., Norman, R. M. & Russel, D. S. (1971) *Greenfield's Neuropathology*, 2nd. edn. London: Edward Arnold.
Brooks, W. W. D., Fletcher, A. P. & Wilson, R. R. (1954) Spinal cord complications of tuberculous meningitis. A clinical and pathological study. *Quarterly Journal of Medicine*, **23**, 275.
Butler, R. W. (1935) Paraplegia in Pott's disease with special reference to pathology and aetiology. *British Journal of Surgery*, **22**, 738.
Camus, J. & Roussy, G. (1914) Cavités médullaires et méningites cervicales. Étude expérimentale. *Revue Neurologique*, **I**, 213.
Charcot, J. M. & Joffroy, A. (1869) *Archives de Physiologie*, **2**, 354.
Elliott, F. A. (1971) *Clinical Neurology*. Philadelphia: W. B. Saunders Company.
Feigin, I., Ogata, J. & Budzilovich, G. (1971) Syringomyelia: the role of edema in its pathogenesis. *Journal of Neuropathology & Experimental Neurology*, **30**, 216.
Feigin, I. (1972) Personal communication.
Gordon, A. (1904) A microscopical study of the spinal cord, not compressed by displaced vertebrae, in a case of Pott's disease; areas of necrosis in the cerebellum, one superior cerebellar peduncle and cord; Reynaud's bodies in one sciatic nerve. *Journal of Nervous & Mental Diseases*, **31**, 526.
Holmes, G. M. (1915) The pathology of acute spinal injuries. *British Medical Journal*, **ii**, 769.
Howland, W. J. & Curry, J. L. (1966) Experimental studies of pantopaque arachnoiditis. I. Animal studies. *Radiology*, **87**, 253.
Hughes, J. T. & Brownell, B. (1966) Spinal cord ischemia due to arteriosclerosis. *Archives of Neurology*, **15**, 189.
Hurteau, E. F., Baird, W. C. & Sinclair, E. (1954) Arachnoiditis following the use of iodized oil. *Journal of Bone Joint Surgery*, **36A**, 393.
Jellinger, K. (1967) Spinal cord arteriosclerosis and progressive vascular myelopathy. *Journal of Neurology, Neurosurgery & Psychiatry*, **30**, 195.
Jelsma, F. *Cervical Intramedullary Cyst—due to Corynbacterium Diphtheriae Gravis* (in press).
Joseph, S. I. & Denson, J. S. (1958) Spinal anesthesia, arachnoiditis, and paraplegia. *Journal of the American Medical Association*, **168**, 1330.
Klotz, O. (1913) Syringomyelia; with autopsy findings in two cases. *American Journal of the Medical Sciences*, **146**, 681.
Kocen, R. S. & Parsons, M. (1970) Neurological complications of tuberculosis: some unusual manifestations. *Quarterly Journal of Medicine*, **39**, 17.
Lloyd, J. H. (1894) Traumatic affections of the cervical region of the spinal cord, simulating syringomyelia. *Journal of Nervous & Mental Diseases*, **21**, 345.
Mair, W. G. P. & Druckman, R. (1953) The pathology of spinal cord lesions and their relation to the clinical features in protrusion of cervical intervertebral discs. A report of four cases. *Brain*, **76**, 70.
Martin, C. & Maury, M. (1964) Syndrome syringomyélique après paraplégie traumatique. *La Presse Medicale*, **72**, 2839.
McLaurin, R. L., Bailey, O. T., Schurr, P. H. & Ingraham, F. D. (1954) Myelomalacia and multiple cavitations of spinal cord secondary to adhesive arachnoiditis; an experimental study. *Archives of Pathology*, **57**, 138.
Oppenheim, H. (1907) *Beiträge zur Diagnostik und Therapie der Geschwülste im Berich des zentralen Nervensystems*. Berlin: S. Karger.
Osetowska-Wiechowska, E. (1956) Sur une forme spéciale de pseudosyringomyélie. Étude anatomo-clinique. *Revue Neurologique*, **95**, 310.
Seddon, H. J. (1935) Pott's paraplegia: prognosis and treatment. *British Journal of Surgery*, **22**, 769.
Seddon, H. J. (1946) Discussion on spinal caries with paraplegia. *Proceedings of the Royal Society of Medicine*, **29**, 723.
Shaw, D. A. (1972) Personal communication.
Smolik, E. A. & Nash, F. P. (1951) Lumbar spinal arachnoiditis: a complication of the intervertebral disc operation. *Annals of Surgery*, **133**, 490.
Spiller, W. G. (1898) A microscopical study of the spinal cord in two cases of Pott's disease. *Johns Hopkins Hospital Bulletin*, **9**, 125.

Tauber, E. S. & Langworthy, O. R. (1935) A study of syringomyelia and the formation of cavities in the spinal cord. *Journal of Nervous & Mental Diseqses*, **81**, 245.

Thomas, A. & Hauser, G. (1901) Cavités médullaires et mal de Pott. *Revue Neurologique*, **9**, 117.

Turnbull, I. M. (1971) Microvasculature of the human spinal cord. *Journal of Neurosurgery*, **35**, 141.

von Bruns, L. (1908) Zur Frage der idiopathischen Form der 'Meningitis spinalis serosa circumscripta'. *Berliner Klinische Wochenschrift*, **45** (Part 2), 1753.

Wilson, C. B., Bertan, V., Norrell, H. A. & Hukuda, A. (1969) Experimental cervical myelopathy. II. Acute ischemic myelopathy. *Archives of Neurology*, **21**, 571.

Wilson, S. A. K. & Bruce, A. N. (1954) *Neurology*, 2nd edn., Vol. 2, p. 1189. London: Butterworth.

Winkelman, N. W., Gotten, N. & Scheibert, D. (1953) Localized adhesive spinal arachnoiditis. A study of 25 cases with reference to etiology. *Transactions of the American Neurological Association, 78th Annual Meeting*, **15**.

Woodard, J. S. & Freeman, L. W. (1956) Ischemia of the spinal cord. An experimental study. *Journal of Neurosurgery*, **13**, 63.

Zülch, K. J. (1962) Réflexions sur la physiopathologie des troubles vasculaires médullaires. *Revue Neurologique*, **106**, 632.

CHAPTER SEVENTEEN

Syringomyelia and Tumours of the Nervous System

H. J. M. BARNETT and N. B. REWCASTLE

DEFINITION OF TERMS

Some writers would object to the use of the term syringomyelia in connection with the cavitation associated with spinal cord tumours. To justify its use in the present context a brief note is indicated regarding the origin of the word. The term syringomyelia is a combination of the Greek 'syrinx' referring to a 'tube' or 'pipe' and the Greek 'myelós' referring to the 'marrow', and was first used by Ollivier in 1827. Since this first use of the word and early descriptions of the condition there has been a search for its aetiology with a plethora of literature and confusion of terms. Some authorities accept the term syringomyelia within restricted limits. If it begins early in life, has associated bony or soft tissue anomalies and pathologically is distinct from the central canal, it is generally accepted as a congenital defect and designated 'true' or 'primary' syringomyelia. To some, this is the only justifiable use of the term syringomyelia. The acquired cavitations associated with pachymeningitis or with spinal cord tumour or following trauma are designated as 'secondary' or even 'pseudosyringomyelia'. Most writers have distinguished 'hydromyelia' from 'syringomyelia', with the former term restricted to a dilatation of an ependymal lined central canal. It is of interest that Ollivier in introducing the term syringomyelia was of the opinion that he was describing a condition resulting from persistence of a central canal.

Changing concepts of the pathogenesis of syringomyelia further elaborated elsewhere (see Chapter 8) have led to the introduction of the terms 'communicating' and 'non-communicating' syringomyelia and to the use of a compromise term 'syringohydromyelia'. It appears that there are a variety of conditions that could reasonably be given the generic name of syringomyelia and that the following definition would be acceptable: *Syringomyelia is a cavitation in the spinal cord which has a wall largely composed of glial tissue.*

This definition is flexible enough to allow the cavity in some instances to have some contact with the central canal, and thus some lining by ependymal cells. It also allows us to recognise some instances where the cavity may be partly lined by collagen, or associated with and even partly lined by tumour cells, yet, at the same time, enables us to use an acceptable term in a semantic sense. The association of syringomyelia with tumours of the nervous system will be discussed under the following headings:

1. Intramedullary Spinal Cord Tumours
2. Extramedullary Spinal Cord Tumours
3. Intracranial Tumours
   (a) Infratentorial
   (b) Supratentorial

For the purpose of this discussion, tumours will not include abscesses and granulomas, but will include neoplastic growths, developmental cysts (e.g. dermoids) and certain lesions of uncertain nature such as histiocytosis X.

INTRAMEDULLARY SPINAL CORD TUMOURS

Historical Recognition

The association of spinal cord cavitation, syringomyelia in its broadest sense, with spinal cord tumours is well established. Simon (1875) appears to have been the first to comment on it and suggested that syringomyelia might be the result of softening developing in spinal cord gliomas. Langhans (1881), in reporting four cases of syringomyelia, considered three of them aetiologically related to co-existing tumours of the posterior fossa. Baumler (1887) reported a very large series (96 cases) of autopsy proven syringomyelia, of which there were 17 with tumours. Dimitroff, in 1897, contributed another 84 autopsy proven cases with 12 tumours. Schlesinger was another of the early writers to gather a large series of cases of syringomyelia and, in his monograph of 1902, listed tumour among the aetiological factors. In 1898, he reported two intramedullary tumours in a series of 20 autopsied cases of syringomyelia.

The first report in the English literature of cavitation accompanying tumour, and incidentally, the first report of multiple tumours associated with syringomyelia, was from Thomas Harris (1886). The patient had a two-and-a-half-year illness with progressive development of a picture compatible with syringomyelia and syringobulbia. At necropsy there were 'sarcomatous tumours' in the cauda equina, the dorsal cord, the pons, lungs and pericardium. A cavitation, said to have been of the central canal, existed from the cervicomedullary junction down to the upper dorsal region; from here down to the mid-dorsal region there were two cavities, one separated from the central canal. Below this, one cavity extended to the lower cord region. No cavitation or dilatation of central canal was seen in the lower medulla. Harris regarded the second cavity as a diverticulum from the central canal, and stated that he judged it to be due to a 'damming up of the cerebrospinal fluid in the central canal of the cord whereby great pressure was exerted, diverticula formed, and as a consequence of that pressure, a sclerosis ensued'. As well as being the first report of multiple spinal tumours with cavitation, this also appears to be one of the first suggestions of a syrinx forming as a diverticulum from a hydromyelia (for an earlier reference by Schüppel (1865) of this concept in non-tumorous circumstances see Chapter 1).

Bullard (1899) reported an early case and emphasised that the syringomyelic condition was not as prominent as was the infiltrative glioma of the cord. He referred to the latter as 'gliomatosis'. The difficulty in distinguishing between excess 'gliosis' and 'glial tumour' then existed. It still confronts us.

Oppenheim (1911) thought that most cases of syringomyelia were of congenital origin due to the inclusion of cell rests along the dorsal raphe in the region of the central canal or the dorsal septum. He thought that these cells might proliferate, causing syringomyelia, or change character producing a spinal cord tumour with cavitation. The opinion was expressed that this proliferation might be in response to trauma or a spontaneous phenomenon. Thus, the curious association of spinal cord tumour and cavitation in the cord has been recognised for a century, and has aroused speculation that both conditions may be an expression of one primary process—heterotopic tissue which may proliferate and may then degenerate centrally, producing a cavity.

Incidence of Syringomyelia with Tumour

It is difficult to give an estimate of the incidence of syringomyelic cavitation in cases of spinal cord tumour. An early Mayo Clinic series of nine spinal cord tumours examined at necropsy (Kernohan, Woltman and Adson, 1931) was reported to include five with syringomyelia. Russell (1932), in England, reported three examples of intramedullary cord tumours from the London Hospital records and two of them (both ependymomas) were associated with syringomyelia. Two early major reviews of this association were carried out. Jonesco-Sisesti collected the 19 cases reported to 1929 and Dimitri and Aranovich the 24 cases to 1940. Poser (1956) reviewed all cases of syringomyelia that had come to post-mortem at the Presbyterian Hospital in New York and of 18 cases, six were associated with tumour. In a comprehensive review of all cases in the literature shown to have syringomyelia proven at autopsy, he determined that 234 cases were associated with tumours of the central nervous system. These 234 cases included 48 that did not involve the spinal cord or brain stem (they were extramedullary or supratentorial) so that there were 186 cases of spinal cord cavitation including his own six cases associated with intramedullary neoplasm.

From a different perspective and a study of 10 review series containing 245 cases examined at autopsy with a clinical diagnosis of syringomyelia, Poser (1956) noted that there were 40 cases who had intramedullary tumours—an incidence of 16·4 per cent (Table 17.1). In 10 other series comprising 209 patients coming to post-mortem because of intramedullary tumour, Poser determined that 65, or 31 per cent, of these spinal tumours were associated with syringomyelia (Table 17.2). Reviewing the Mayo Clinic's experience, Slooff, Kernohan and MacCarthy (1964) reported on 33 cases of intramedullary tumour of the spinal cord at post-mortem. Of these 33 cases, 19 (57·6 per cent) were associated with a syrinx.

An illustration of the difficulty of assessing the true incidence of the association between spinal cord tumour and syrinx is provided by the report from the Armed Forces Institute of Pathology by Ferry, Hardman and Earle (1969). They reviewed all the post-mortem material in the institute coded for 'syringomyelia' and found that of 38 cases so diagnosed, the astonishing number of 29 had

Table 17.1. *The incidence of intramedullary tumours found in autopsied cases of syringomyelia*

Author	Number of cases in series	Number of cases with tumour
Baumler	96	17
Dimitroff	84	12
Schlesinger	18	2
Margulis	7	0
Wangel	3	1
Jonesco-Sisesti	3	2
Van Dam and Van der Zwan	5	0
Grossman	3	0
Netsky	8	0
Presbyterian Hospital	18	6
TOTAL	245	40

Incidence: 16·4%
Reproduced by permission of C. M. Poser and Charles C. Thomas, Publisher. The Presbyterian Hospital (Columbia University) data is Poser's own material.

Table 17.2. *The incidence of syringomyelia found in autopsied cases of intramedullary tumour*

Author	Number of cases in series	Number of cases with syrinx
Schlesinger	10	0
Schlesinger	13	0
Kernohan and Sayre	16	10
Foerster and Bailey	70	34
Hamby	18	6
Wolf	12	4
Shenkin and Alpers	27	7
Russe	3	2
Wyburn-Mason	14	5
Wyburn-Mason	26	7
TOTAL	209	65

Incidence 31%
Reproduced by permission of C. M. Poser and Charles C. Thomas, Publisher.

associated intramedullary neoplasms. As they comment, it is probable that they receive material that is weighted in favour of the unusual. Furthermore, the early onset of a syringomyelia syndrome might well prevent a young man or woman from entering the armed forces.

At the University of Toronto, the division of neuropathology at the Banting Institute is associated with a large general hospital dealing with active disease. From these records we determined that 16 patients had come to post-mortem over a 40-year period with a diagnosis of intramedullary spinal cord tumour (including one with histiocytosis X and one with sarcoidosis), and that of these, four had an associated syrinx or major cavitation within the tumour to such a degree that the secondary diagnosis of syringomyelia was entered in the records. In one instance (case 3, see Figure 17.4), the syrinx was so lined by tumour, and there was so much breakdown to cavity formation of a necrotic tumour as to be questionably included within the category of a 'syrinx with tumour'. Our present custom would not have categorised it thus, but it is included since it appeared in the records in this way and also since it allows us to make certain comparisons with the more generally acceptable syrinx and tumour association. Two cases had metastatic spinal cord tumour and neither of them had cavitation. In this 40-year period, 75 cases with extramedullary tumour were examined and none of them had an underlying syrinx except for the two which eventually invaded the spinal cord. A total of 12 cases came to post-mortem with a primary diagnosis of syringomyelia and one of them had a spinal cord tumour. Accordingly, our material consists of 17 intramedullary tumours associated on five occasions with a syrinx (Table 17.3). Finally, one patient who died because of a parieto-occipital glioblastoma had a small syrinx isolated in the spinal cord which was judged by his symptoms to have been of several years' duration.

Table 17.3 *Intramedullary tumours (autopsy) at the Banting Institute 1932–1971*

Type of tumour	Number	Number with syrinx
Astrocytoma	6	2
Ependymoma	3	–
Malignant schwannoma	1	1
Central neurofibroma	1	–
Haemangioblastoma	1	1
Teratoma	1	–
Histiocytosis X	1	1
Sarcoidosis	1	–
Metastatic carcinoma	2	0
TOTAL	17	5

A summary of the reported autopsy incidence of syringomyelia in patients dying with a clinical diagnosis of intramedullary tumour is given in Table 17.4. It includes the massive review of cases available up to 1956 by Poser, the Mayo Clinic's experience (Slooff et al, 1964) and the review carried out at the Banting Institute by the authors. The incidence of syringomyelia in the cases with intramedullary tumour varied from 25 to 57·6 per cent. Much of the material from the Mayo series appears to have been previously reported and therefore included by Poser. The dilution of Poser's material by this duplication is not significant by comparison with the value of a large series from a single clinic and pathology unit, so that both series are quoted.

Table 17.4. *Syringomyelia with intramedullary tumours (autopsy)*[a]

	Number with tumour	Number with syrinx	%
10 series (Poser)	209	65	31
Mayo series	33	19	57·6
Banting Institute (1932–1971)	16	4	25

[a] Where clinical diagnosis was spinal cord tumour

Similarly, the reported and available incidence of intramedullary tumour at autopsy in patients dying with a clinical diagnosis of syringomyelia is given in

Table 17.5, and here there was a range of incidence from 16·4 to 75 per cent. The weighting of each available series is so great and depends so much on the interest and emphasis of the individual doctors and institutions as to make these figures no more than a rough guide to the extent of this problem.

Table 17.5. *Intramedullary tumour with syringomyelia (autopsy)[a]*

	Cases with syrinx	Number with tumour	%
10 series (Poser)	245	40	16·4
Armed Forces Institute of Pathology	44	38	75
Banting Institute	12	1[b]	8·3

[a] Where clinical diagnosis was syringomyelia.
[b] One other died with parieto-occipital glioblastoma multiforme.

Type of Tumour Associated with Syringomyelia
THE COMMON TUMOUR TYPES—GLIAL TUMOURS

The histological features of the tumours associated with cavitation do not differ from those of spinal cord tumours without syrinxes. This matter was the subject of particular scrutiny in our small series and in the larger series reported by Slooff et al (1964). In neither series did particular or special features characterise the tumours associated with syringomyelia. The histological types of intramedullary tumour that have been associated with syringomyelic cavitation at autopsy can be seen from Table 17.6. Tumours designated as ependymoma, astrocytoma, glioma and glioblastoma were commonest as one would expect from studies on overall incidence of spinal cord tumour. Of Poser's compilation, the 'gliomatous tumours' (glioblastoma, 'glioma' and astrocytoma) formed 32·7 per cent of the cases with spinal tumour and cavitation. The single intramedullary tumours alone formed 44 per cent of this series. The whole series contained 50·9 per cent of cases which he stated would qualify as 'congenital'-type tumours. Poser felt, therefore, that he could arrive at a figure of 83·6 per cent of cases in which the syringomyelic cavity may be considered either another instance of the well-recognised multiple occurrence of congenital anomalies, or related glial metaplasia. The series of Ferry et al (1969) and Slooff et al (1964) confirmed the high incidence of tumours of gliomatous origin, and the specially high relationship to congenital tumour and hereditary tumour-forming syndromes, specifically von Recklinghausen's and von Hippel–Lindau's diseases.

As noted above, Slooff et al (1964) have reported one of the largest series of autopsy examined spinal cord tumours from one clinical laboratory. This makes the incidence of syringomyelia and the variety of tumours in their collection of special value and interest (Table 17.7). The commonest tumour (ependymoma) had a little more than a 50 per cent incidence of syrinx. Astrocytoma had a slightly reduced tendency. In all tumour types in their collection there were some examples with cavitation. The highest incidence was in those forming part of the two phakomatoses, von Hippel–Lindau's disease and von Recklinghausen's disease.

In the authors' smaller series (Table 17.3) of 17 tumours proven at autopsy

Table 17.6. Histological types of intramedullary neoplasms with syringomyelic cavity

Tumour type	Poser's[a] compilation	Slooff et al[a] (Mayo)	Ferry et al (AFIP)	Banting series
Ependymoma	31	9	9	—
Glioblastoma	30	—	—	—
Astrocytoma	26	3	15	2
Glioma	25	—	2	—
Blood vessel tumour	24	4	2	1
Central neurinoma	16	—	—	1
Unclassified tumour	5	—	—	—
Teratoma	4	1	—	—
Ganglioglioma	2	—	1	—
Epidermoid	2	—	—	—
Neuroepithelioma	2	—	—	—
Medulloblastoma	2	—	—	—
Histiocytosis X	—	—	—	1
Dermoid	1	—	—	—
Melanoma	1	—	—	—
Sarcoma	1	—	—	—
Oligodendroglioma	—	1	—	—
Spongioblastoma	—	1	—	—
TOTAL	172	19	29	5

[a] Modified to include the histological type of spinal cord tumour in the cases of von Recklinghausen's disease.

Table 17.7. Intramedullary tumours with syringomyelia (autopsy)

Type	Number	Syrinx
Ependymoma[a]	17	9
Astrocytoma	8	3
Haemangioblastoma	4	4
Oligodendroglioma	1	1
Spongioblastoma	2	1
Teratoma	1	1
TOTAL	33	19

Adapted by permission of Slooff et al (1964) and the W. B. Saunders Company.
[a] 4 cases had these as part of von Recklinghausen's disease, 3 of which had cord cavitation.

(including one sarcoidosis of the spinal cord), none of the three ependymomas accompanied a syrinx but there were two examples of astrocytoma and one each of haemangioendothelioma, malignant Schwannoma and histiocytosis X with associated cavitation.

The Phakomatoses

The high incidence of syrinx with vascular tumours (mostly as part of von Hippel–Lindau's disease) and with von Recklinghausen's neurofibromatosis has been put forward as indicating a special link between cavity formation and these phakomatoses. Seventeen cases in Poser's review had syringomyelia with von Recklinghausen's disease and nine had von Hippel–Lindau's disease. Rodriguez and Berthrong (1966) reviewed the literature on von Recklinghausen's disease from 1822 to 1966 and noted that if the disorder showed multiple intraspinal or intracranial neurinomata or meningiomata, there was close to a 25 per cent chance that there would also be syringomyelia (12:49 cases). This association has been used by some, including Lichtenstein (1949), to promote the argument for the neoplastic nature of syringomyelia as a derivative of degeneration in heterotopic tissue within the spinal cord.

Kernohan and Parker (1932) reported in detail one remarkable case of neurofibromatosis from whose intradural space, at T7 to 8, a meningioma had been removed nine years before death. After a lapse of six years, he displayed new spinal cord dysfunction and three intramedullary tumours (two astrocytomas and one ependymoma) were present. Above and below each tumour, there was cavitation unrelated to the central canal. Each cavity was surrounded by a dense zone of glial tissue and those parts of the cord related to the cavities, between the frank tumour masses, were free of any evidence of neoplasia. They noted that a central gliosis continued up and down the cord beyond both the upper and lower cavities and that this glial tissue represented a 'presyringomyelia stage'. This gliosis extended over several cord segments and involved the anterior two-thirds of the posterior columns. They commented on the lack of blood vessels in the centre of the gliosis and judged the gliosis to be thicker in its depths. In places it was rarefied centrally. Worster-Drought, Dickson and McMenemey (1937) described another example of multiple intramedullary tumours in a case of neurofibromatosis, and illustrated most strikingly the severely marked central gliosis within which the cavitation occurred.

The tumours in the central nervous system in cases of von Recklinghausen's disease associated with syringomyelia are of many types. In the four cases of von Recklinghausen's disease in the collection of Slooff et al (1964), there were four spinal intramedullary ependymomas in one case, three in two others and one in a fourth. All but one of these four cases had associated spinal cavitation. In the 13 cases collected by Poser, where spinal cord intramedullary tumours were present with von Recklinghausen's disease, there were seven central neurinomas, four astrocytomas, three ependymomas, two examples of glioblastoma multiforme and one unclassified glioma for a total of 17 tumours.

There is considerable interest in a report (Fracassi, Ruiz, and Garcia, 1935) of a patient who had a spinal extramedullary *neurofibroma*, a spinal intramedullary *haemangioblastoma*, and who was afflicted as well with a syrinx in the cord related to the latter tumour. He had the cutaneous lesions of von Recklinghausen's disease. In addition to this example of the concomitant occurrence of spinal neurofibroma and haemangioblastoma, there is also a significant correlation between these two major forms of phakomatosis and phaeochromocytoma. Poser (1956) has suggested that syringomyelia and phaeochromocytoma both form links between neurofibromatosis and haemangioblastomatosis.

Blood vessel tumours of the spinal cord are rare, but in collected series, after ependymomas and gliomas, they are the next commonest tumour associated with syringomyelic cavitation. There is lack of unanimity as to the classification of blood vessel tumours, but the commonest type reported has been designated haemangioblastoma. Our custom has been to describe this as a haemangio-endothelioma, and in this review both terms are used and are taken to be synonymous. Approximately half the reported cases of haemangioblastoma of the cord are associated with other features of von Hippel-Lindau's disease and most of these have syringomyelia. Of the first 11 cases of spinal haemangioblastoma in the literature, 10 had an associated syrinx (Kinney and Fitzgerald, 1947). Melmon and Rosen (1964) in a review of von Hippel–Lindau's disease gave a figure of 80 per cent for the association of syrinx and spinal haemangioblastoma. On the other hand, the incidence of syringomyelia reported in examples of isolated spinal haemangioblastoma is closer to 50 per cent.

Brain, Greenfield and Northfield (1943), commenting on a post-mortem case of retinal and spinal haemangioblastomata, felt that the syrinx associated with the spinal tumour was a 'true' syringomyelia in that it was 'only fortuitously connected to the central canal'. They judged that 'the syringomyelia resulted from a separate developmental abnormality and was not caused, as the cerebellar cysts appear to be, by transudation of plasma into the nervous tissue from the tumour'. This speculative comment can neither be corroborated nor yet completely denied in the light of present knowledge. However, those who would support the concept of a common aetiology producing growth and/or degenerative process in heterotopic tissue as the basis for both central nervous tissue tumour and spinal cavitation might be disturbed to contemplate that in all but one possible case reported to date with syringomyelia developing in von Recklinghausen's disease or in von Hippel–Lindau's disease there has been tumour either within the spinal cord or in the posterior fossa and/or the foramen magnum region.

The exception is not sufficiently convincing to be unequivocally acceptable. It concerns a report by D'Antona, quoted by Ebbers (1941), of a patient exhibiting cutaneous features of von Recklinghausen's disease and dying with the clinical picture of an intramedullary cervical spinal cord lesion. At post-mortem, 'instead of tumour, from C2 to T5 there was a cavity which superiorly was filled with a paste-like mass and inferiorly was lined with greyish-red layers'. Inadequate microscopic description has been recorded to make it certain that this was not a growth. If the phakomatoses give rise to tumours and syringomyelia due to their heterotopic qualities per se, one would expect to find genuine examples of cord cavitation without cord, brain stem or posterior fossa tumour. No thoroughly convincing examples are known to the authors from their own material or from those in the literature.

Because of the multiplicity of intramedullary neoplasms in von Hippel–Lindau's and von Recklinghausen's diseases, it is difficult to know if the apparent increased incidence of cavity formation is related to the increased number of tumours, or if it is evidence of a true tendency of these phakomatoses to develop syringomyelia.

Miscellaneous Tumours

Metastatic Tumour. Secondary cancer has been discussed in this context on few occasions. An isolated instance of this condition was that of a secondary (renal) carcinoma in the cervical cord (Weitzner, 1969). This carcinoma tissue existed as a tumour nodule located at C3 to C4 within a cavity extending from C2 to 7 dorsal to the central canal. Since the cavity extended beyond the tumour, it was accepted as a syrinx and not merely as a cystic degeneration in the centre of a tumour. Describing the genesis of cavitation in intramedullary tumour, Russell

(1932) used as an example a case of lung carcinoma with a secondary deposit in the first and second lumbar spinal segments without vertebral or meningeal involvement. She described 'haemorrhagic softening extending up to the tenth, and anaemic softening up to the sixth thoracic segment. The remainder of the lumbosacral cord below the growth was also the site of haemorrhagic softening'. The lesions described were judged to be the precursors of cavitation. Metastatic carcinoma in the spinal cord is rare, and when it does occur the cases do not go on to cavitation, probably because of the brevity of this illness. The files at the Banting Institute contain records of only two examples of metastatic intramedullary spinal tumours. One was an incidental finding in a patient who died from breast carcinoma, the other was a deposit of metastatic melanoma to the cervical spinal cord resulting in six weeks of clinical symptoms. Neither had associated cavitation.

Eosinophilic Pituitary Tumour—Acromegaly. A few reports of the association of acromegaly and syringomyelia have appeared (MacBride, 1925; Atkinson, 1932; Worster-Drought and Shafar, 1940). Joffroy-Gillo et al (1967) in reporting a further isolated example of this association, stated that it was 'too frequent to be imputable to mere chance'. This might be disputed as there would appear to be less than a dozen reasonably well-documented cases in the entire literature. However, Joffroy-Gillo et al referred to the recognised association between acromegaly and neurofibromatosis and suggested a link between these three involving some common teratogenic pathogenesis.

Histiocytosis X. Histiocytosis X may or may not be a neoplasm in the light of the contemporary understanding of the disease and use of the term. Without entering into this controversy, which is beyond the scope of this discussion, it is noted that a brief mention of spinal cord cavitation associated with one case of this condition is on record from the Banting Institute (Ezrin, Chaikoff and Hoffman, 1963). The case was reported because of panhypopituitarism. The following is a review of the salient manifestations of the spinal involvement:

CASE A.S., MALE

In 1948, at the age of 42 years, this patient developed diabetes insipidus which was controlled with Pitressin tannate in oil injections. In 1952 sudden low back pain developed, with early urinary difficulty. By 1953, weak legs and a T6 sensory level led to the myelographic demonstration of a T5 intradural tumour, located anterior to the cord. The patient became totally paraplegic with a T4 level and this proved to be permanent. Further evidence of a suprasellar lesion (bitemporal field defect) and serious panhypopituitarism developed and by 1956 he had right hand weakness, right arm numbness and weakness. Within a year both hands and arms were the site of wasting, great weakness, absent triceps reflexes and a pain, temperature and tickle sensory level developed from C5 to T1 (inclusive) on the right and in a gauntlet fashion in the left hand and lower forearm. The sensory level from the old lesion persisted at T4, with complete sparing of the thoracic segments T2 and 3 on the right. By 1957, he had completely lost motor power in the fingers of both hands and in the right wrist and triceps and the new dissociated sensory level extended from T1 on both sides to C4 on the right and C6 on the left. He was treated with radiotherapy and steroids and his spinal function remained stationary until his death in 1960.

The details of his post-mortem have been reported elsewhere by Ezrin et al (1968). The important pathological aspects of his spinal lesion are as follows. The cord at the T5 to 6 level, the site of the original compression, was severely flattened and what cord remained contained several slit-like, intercommunicating cavities which were continuous with the cord cavitation above and below this level (Fig. 17.1). Caudally, the syrinx became large, down to the conus, the cord remaining as a rim around it (Fig. 17.2b). Above T5 to 6, the cyst became unilateral, replacing the anterior and posterior horns (Fig. 17.2a) and progressively narrowed as it extended rostrally, until only a tiny cavity remained at the C6 level where a second, large intradural tumour was causing severe cord compression and invasion. Associated with this compressive lesion was an extensive rarefaction of the central areas of the cord with loss of nerve cells and myelinated axons, extending up at least as far as C3. Histologically, the wall of the syrinx was thin and contained no Rosenthal fibres (Fig. 17.11b).

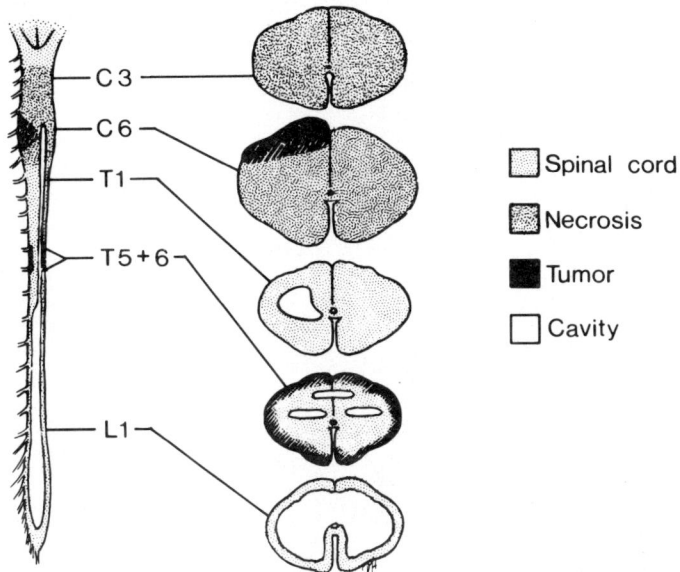

Fig. 17.1. Diagrammatic representation of the extensive cord cavitation present in case 1, a patient with histiocytosis X and cord compression and infiltration at T5–6 and subsequently at C6.

Distinctive Pathological Features of Syrinx Occurring with Tumour

The availability in the literature of large numbers of reported cases of syringomyelia, spinal tumour, and cases combining syringomyelia with spinal tumour make it possible to compare the anatomical, pathological and clinical features of these various single or combined situations. Our concern is to seek similarities and/or differences between the tumours and the syrinxes when they are in combination and when they occur alone.

LOCATION OF THE TUMOURS ASSOCIATED WITH SYRINGOMYELIA

In published series of tumours without reference to syringomyelia, the lower thoracic and lumbosacral regions have been the site of predilection for the intramedullary neoplasms. Kernohan and Sayre (1952) for example, noted that 50 per cent of their gliomas were in the lumbosacral segment, and Elsberg (1941), reporting 20 cases of ependymoma, found that 'most arose in the lower part of the

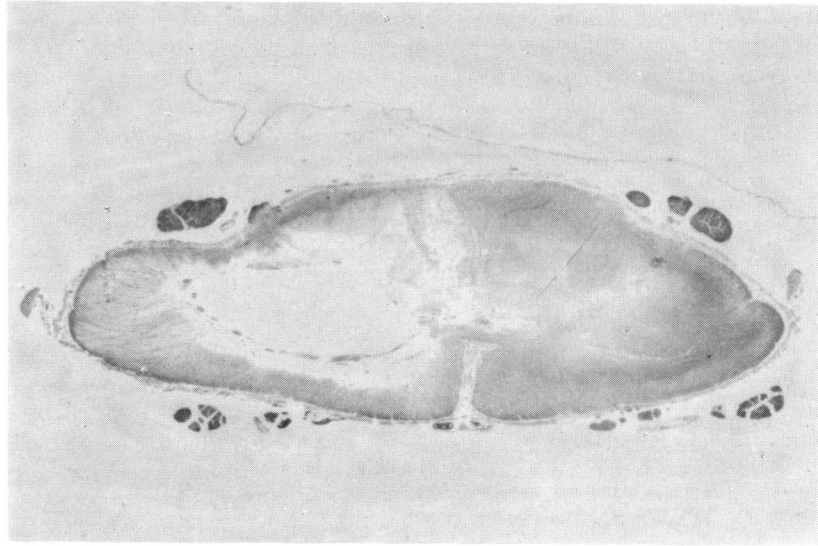

(a)

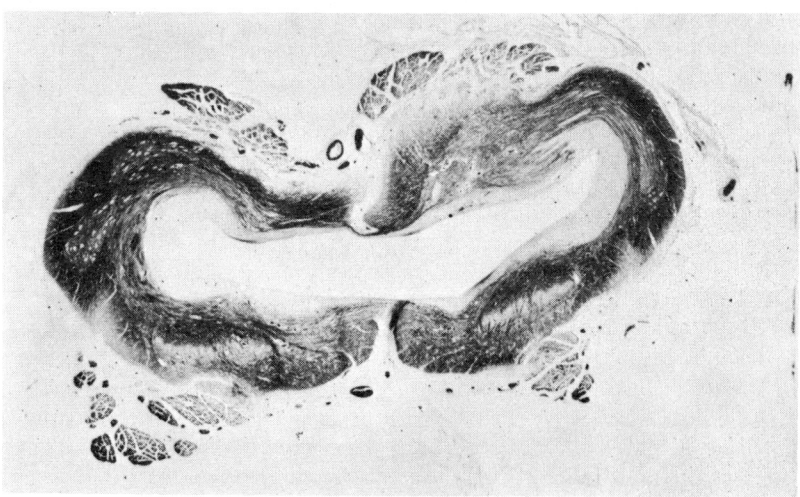

(b)

Fig. 17.2. Spinal cord on each side of the T5–6 compression and infiltration by intradural histiocytosis X. (a) Thoracic cord rostral to compression with unilateral cavitation; (b) huge cavity in lumbo-sacral region.

cord'. In Adson's series of 112 intramedullary spinal tumours (1938), approximately 70 per cent were thoracic, thoracolumbar or lumbosacral and 30 per cent were cervical or cervicothoracic.

The material in our review of primary intramedullary tumours coming to post-mortem at the Banting Institute (Table 17.8) did not confirm the predisposition of the ependymomas to favour the lower part of cord (two of the three were above the low-thoracic region). If one included, in addition, those which arose from the filum terminale this reversed the ratio. The neurofibromas, the teratomas and the histiocytosis X were below the mid-thoracic level. The von Hippel–Lindau's case had a haemangioendothelioma at each level, cervical, thoracic and lumbar. All six of the intramedullary astrocytomas had involvement of the cervical cord, with two of them extending as well to the thoracic cord and two up into the medulla.

Table 17.8. *Type and location of primary intramedullary spinal cord tumours (autopsy)*

Tumour	Location of tumour
1. Astrocytoma	Cervical, medulla
2. ,,	Cervical, upper thoracic, medulla
3. ,,	Cervical
4. ,,	Cervical
5. ,, [a]	Cervical
6. ,, [a]	Lower cervical, upper thoracic
7. Ependymoma	Cervical
8. ,,	Cervical
9. ,,	Thoracic
10. Malignant schwannoma[a]	Mid and lower thoracic
11. Central neurinoma	Lumbar
12. Haemangioblastoma[a]	Cervical, thoracic, lumbar
13. Teratoma	Lower thoracic, lumbar
14. Histiocytosis X[a]	Mid-thoracic
15. Sarcoidosis	Mid-thoracic

[a] With syrinx.

Despite the fact that our small series does not lend it support, there appears to be a trend to a thoracolumbar predilection for spinal cord tumours. The reverse is found in those which have cavitation. Ferry et al (1969) noted that the 'tumour and the syrinx were most frequently located in the lower cervical and upper thoracic spinal cord'. In Poser's review, where tumour and syrinx coexisted, the tumour location figures for the cervical and cervicothoracic region approximated one-half (47.5 per cent) and the other half (52.5 per cent) were thoracic and below. Figure 17.3 demonstrates that in the Mayo series the tumour with syrinx also favoured the upper part of the cord. In our own smaller series, the tumour always involved the cervical or upper thoracic part of the cord (Fig. 17.4). The large series of primary spinal cord tumours, disregarding cavitation, presented by Adson (1938) and by Slooff et al (1964) are compared with Poser's series for tumour with syringomyelia and are summarised in Table 17.9.

The higher proportion of cervical and cervicothoracic tumours developing significant cavitation might be explained either by the occurrence of a peculiar variety of tumour in this area liable to produce cavitation, or by the peculiar liability of any tumour in this anatomical area to cause breakdown of adjacent tissue with extending cavitation. The lack of recognisable histological peculiarity of the tumours with cavitation from those without a syrinx has been discussed already and is not a factor of apparent importance. A more probable explanation for the predilection to cavitation with cervical and cervicothoracic tumours may be sought in the extra mobility and other mechanical factors demonstrated in the cord in the cervicothoracic region (Breig, 1960). A cervical cord containing a tumour mass, with possibly some surrounding meningeal reaction, may be more susceptible to mechanical stresses than is normal cord tissue. Oedema, ischaemic softening and even haemorrhage related to tumours may be spread more readily in these more mobile parts of the cord. For further discussion see Chapter 14 and Figure 14.38, as well as later discussion in this chapter.

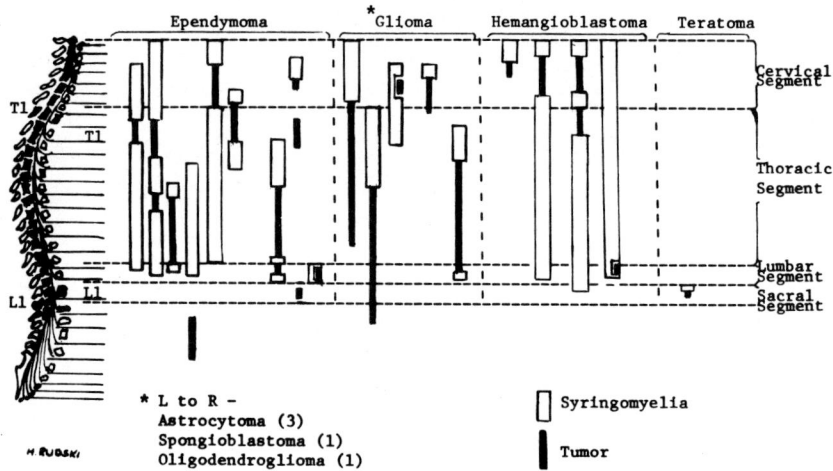

Fig. 17.3. Diagrammatic representation of relative extent of syringomyelia and tumour as modified from Slooff et al, 1964. (Reproduced by permission of authors and W. B. Saunders Company Ltd.)

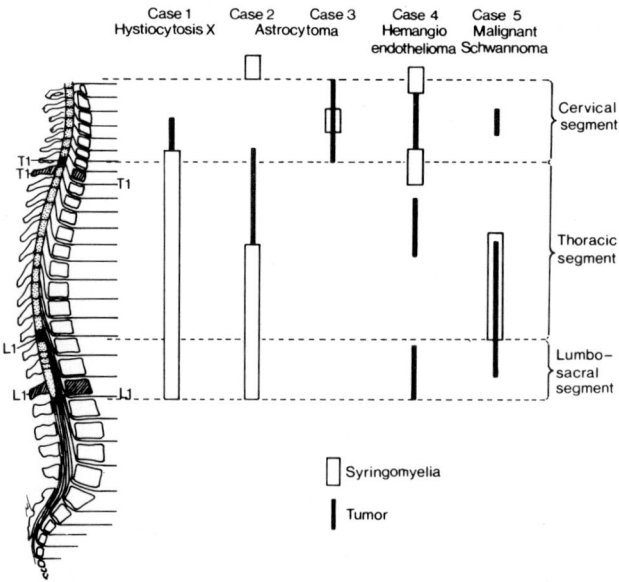

Fig. 17.4. Diagrammatic representation of relative extent of syringomyelia and tumour in the authors' series.

SYRINGOMYELIA AND TUMOURS OF THE NERVOUS SYSTEM

Table 17.9. *Anatomical location of all intramedullary tumours compared with intramedullary tumours with syringomyelia*

	Intramedullary tumours				Combined with syringomyelia Poser's compilation	
	Adson, 1938		Slooff et al, 1964[a]			
	Cases	%	Cases	%	Cases	%
Cervical and cervico-thoracic	33	29·5	60	33·3	47	47·5
Thoracic to lumbar	79	70·5	200	66·7	52	52·5
TOTAL	112	100	260	100	99	100

[a] Ependymomas and gliomas only.

The Structure of the Cavity Wall

There are several features of the lining of the cavity in tumour cases that are unlike ordinary syringomyelia. Firstly, there is the obvious fact that a nubbin of tumour tissue may lie within the cavity (Fig. 17.5). Secondly, the cavity may be lined in part by tumour tissue (Fig. 17.6). A mural nodule or tumour partly lining the cavity was found in half the cases reported by Ferry et al (1969). This intimate location of tumour to cavity was found in only four of the 19 cases reviewed in detail by Slooff et al (1964) (Fig. 17.3). It existed in two of our five cases.

Thirdly, the presence of extra thickness to the glial wall of the cavity is a feature of cavitation with tumour. In some instances this thickening may be such

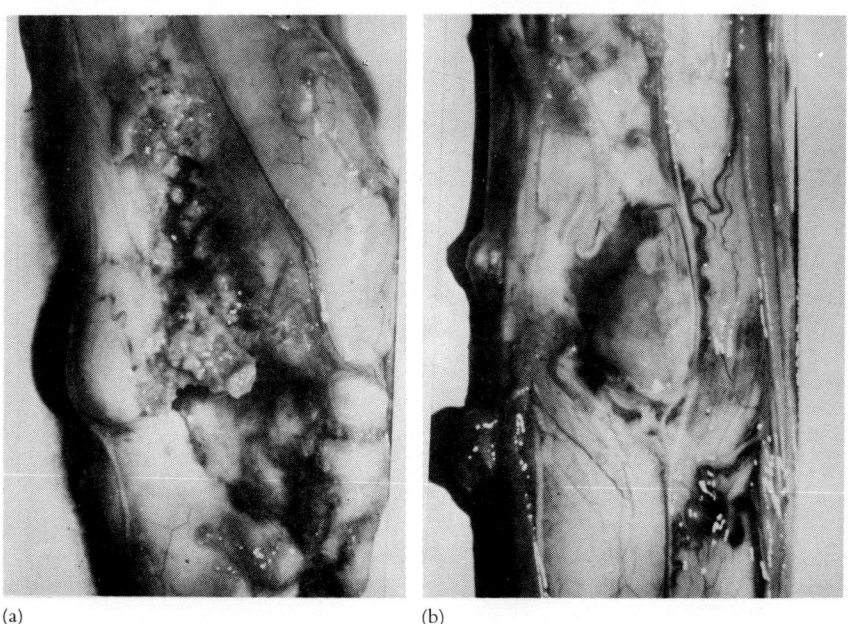

(a)　　　　　　　　(b)

Fig. 17.5. Gross photographs show a cystic haemangioendothelioma (a) as seen with cyst wall intact, (b) tumour visible following opening of cyst. (Reproduced by permission of Dr Y.-M. Rho and *Canadian Medical Association Journal*.)

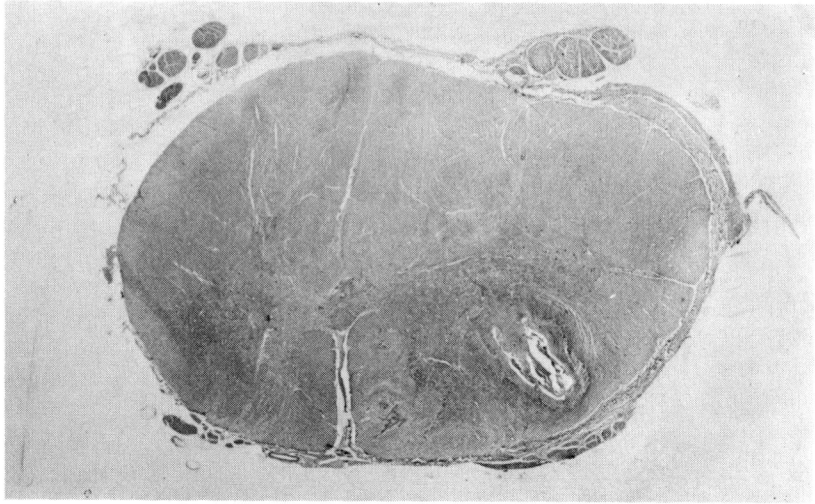

Fig. 17.6. An example of a primary cervical astrocytoma with foci of anaplasia resulting in cystic cavitation. Note also the aberrant (anterolateral) location of the cavitation.

that it is difficult to distinguish between excessive gliosis and an early well-differentiated astrocytoma. A striking example of this thickness, from our own material, is illustrated in Fig. 17.7. This patient died at the age of 26 years, having

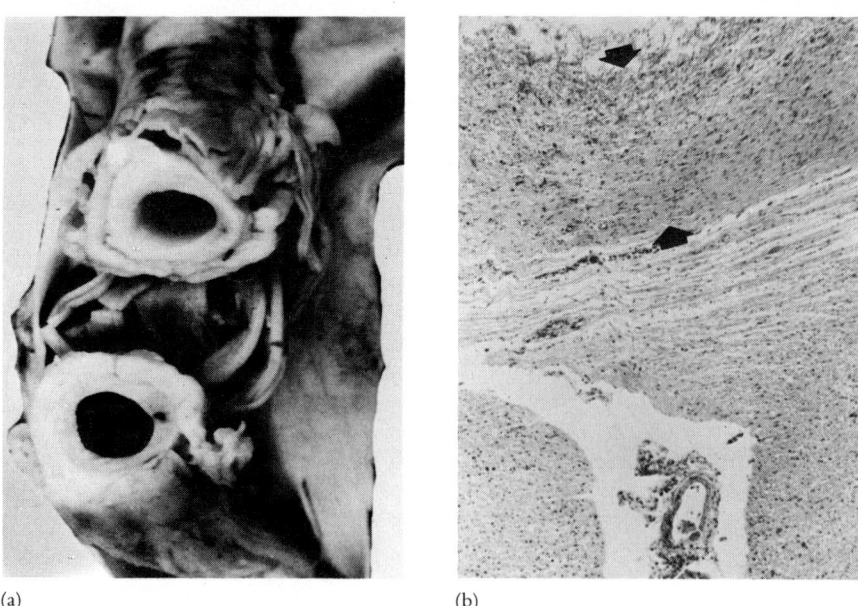

(a) (b)

Fig. 17.7. (a) Below a thoracic astrocytoma is seen the large cavity in the cord surrounded by a thick (2·0 mm) collar of glia. (b) At low magnification (× 34) glial collar (between arrows) is distinct from the adjacent cord tissue. The small dark structures represent Rosenthal fibres.

had a progressive spinal lesion for four years, with a tumour extending from the lower to the upper thoracic levels and with a large cavitation occupying the lowest thoracic and the lumbosacral part of the spinal cord. The tumour was an *astrocytoma* and was not radiated. *Ependymoma*, as well, may be accompanied by excess glial tissue in the cavity wall as was illustrated by Ferry et al (1969) (Fig. 17.8). In this instance, a tumour extended from the lower medulla to a point 8 cm further down in the cervical cord. The syrinx extended from the lower cervical to the upper thoracic cord (Earle, 1972). This patient had an eight-year history from early symptoms to death and had been radiated. It is doubtful if radiation contributes to the wall thickness. Many times it is recognised when radiation has not been given, and was noted before radiotherapy was used. Radiation, indeed, has been commonly administered to non-tumourous syringomyelias without apparently producing excessive syrinx wall gliosis.

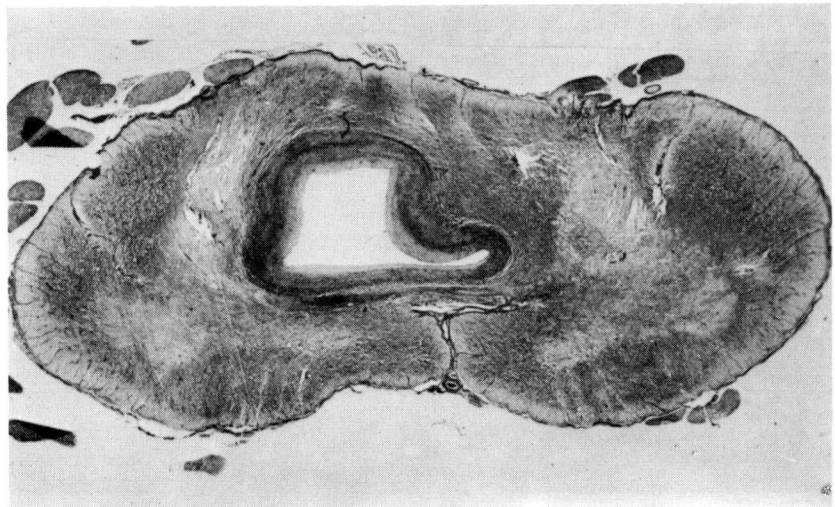

Fig. 17.8. This cervico-thoracic syrinx with thick glial lining extended caudal from a medullo-cervical ependymoma. (Reproduced by permission of Ferry et al, 1969, and the *Medical Annals of the District of Columbia*.)

The case of von Recklinghausen's disease, described by Kernohan and Parker and already referred to, had such gross glial excess about the cavities as to raise the question of whether or not it was neoplastic. In this instance as mentioned, two of the multiple intramedullary tumours were ependymomas and one was an astrocytoma. Similarly, in the case described by Worster-Drought et al (1937) (Fig. 17.9), the glial excess is so striking in the central region of the cord as to raise serious doubts about the possible neoplastic nature of this thick collar surrounding the central cavity. Thick glial excess in cavity walls is not a peculiarity confined to tumours of glial and ependymal origin. It was also noted in our case with von Hippel–Lindau's disease (Fig. 17.10). Old blood pigment lined this thick-walled cavity and haemorrhage may have contributed to the particular thickness of the wall in this instance.

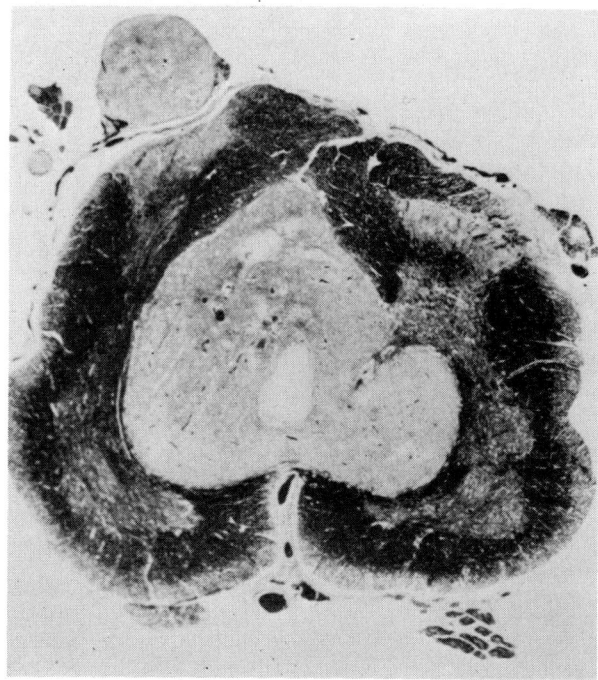

Fig. 17.9. Excessive glial proliferation about a small cystic cavity in the cord may be so gross as to raise the question of neoplasm. (Reproduced from Worster-Drought et al, 1937, by permission from *Brain*.)

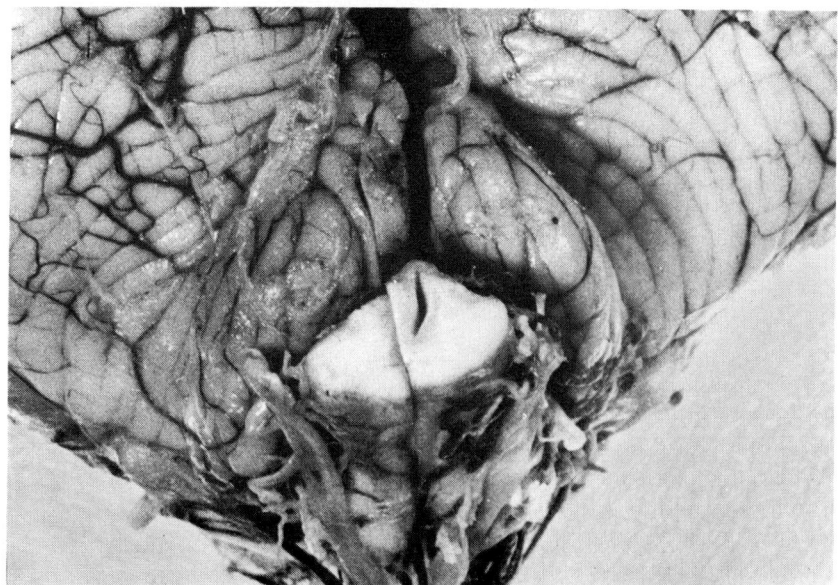

Fig. 17.10. The lower medulla exhibits a slit-like cavity extending rostrally from a cervical haemangio-endothelioma.

The final distinctive feature of the cavitation relative to cord tumours is Rosenthal fibres. In 1898, Rosenthal described glistening objects in the glial tissue surrounding a spinal cord cavity in one of his spinal cord tumour cases. In his opinion they represented degenerating glial fibres. Rosenthal fibres may occur in some non-tumour syringomyelia cases but tend to be a particular feature of the syrinx wall in tumour cases. In the Armed Forces Institute of Pathology series, they were found in 17 of 38 cases and occurred in 75 per cent of the cases with astrocytomas, 63 per cent of those with ependymomas, but also in one case with a non-glial tumour—a haemangioblastoma. Slooff et al (1964) reported that 'they were commonly found in the wall of the cavity in cases in which syringomyelia was associated with intramedullary tumour, whereas they were rare in cases of syringomyelia not accompanied by a neoplasm'. The appearance of a typical Rosenthal fibre is seen (Fig. 17.11a) in the wall of a lumbar syrinx where a thoracic astrocytoma (case 2) was associated with a cavity extending down to the lumbosacral cord level. In our four other cases they were found with the malignant Schwannoma, with the second astrocytoma, but not with the haemangioendothelioma or with the syrinx with histiocytosis X (Fig. 17.11b). Rosenthal fibres are known to occur in association with gliomas of the cerebrum, cerebellum, brain stem and optic nerve, apart from syringomyelia (Russell and Rubinstein, 1971).

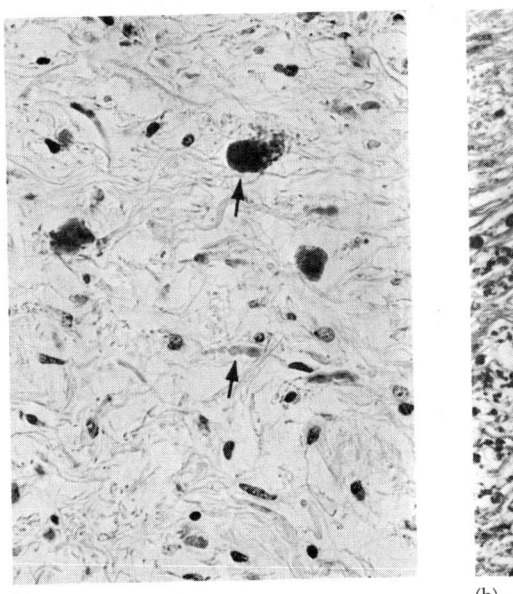 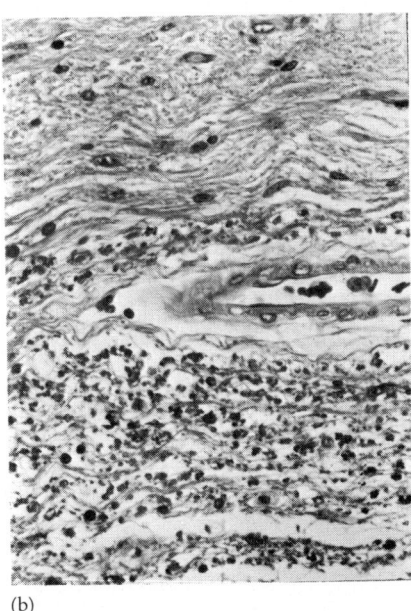

(a)   (b)

Fig. 17.11. (a) Rosenthal fibres of varying sizes (arrows) in the wall of the cavity associated with a spinal astrocytoma (× 240). (b) Note the absence of Rosenthal fibres from the gliotic wall of the syrinx due to histiocytosis X secondary to cord compression and infiltration (× 240).

The Location and Extent of the Cavitation

Location in the vertical plane. The location of the syrinx in non-tumourous cases of syringomyelia is discussed in Chapter 7. The site of predilection is the cervical spinal cord. From the recorded series with tumours, most of the cavities (75 to 80 per cent) involve the cervical region. This leaves 20 to 25 per cent that are present within the thoracic and/or lumbosacral cord alone. These figures are approximate since it is not always possible from reported descriptions to be sure of the exact upper limit of a cavity. In our own five cases, the cavity location in the vertical plane is shown in Fig. 17.4. Twice the cavity extended all the way to the bottom of the cord (cases 1 and 2). Once it was confined to the mid-cervical cord, and in this case it would be more accurately described as cavitation in a tumour rather than a syrinx (case 3). In two of our cases more than one cavity was present (cases 2 and 4) and the medulla was the site of the second cavity in both, so that two examples of syringobulbia are noted in this series (Figs. 17.10 and 17.12).

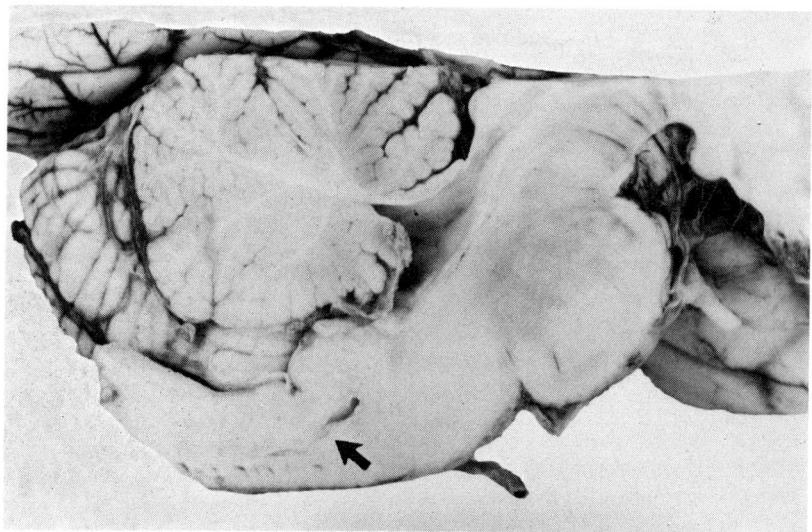

Fig. 17.12. Sagittal section of brain stem shows the isolated medullary cavitation in case 2, a patient with a thoracic cord astrocytoma and extensive cord cavitation caudal to the latter.

Syringobulbia coextensive with syringomyelia is reported in the literature in 16 cases in whom the tumour was a primary intramedullary spinal cord tumour. Of this 16, three had von Hippel–Lindau's disease and one other a vascular tumour. Where tumours coexisted in the cranium they were not included in this figure, so that it can be seen that a syrinx developing in relation to a spinal cord tumour alone may extend up to the medulla. None was found extending beyond the medulla in this survey.

The cavity tends to extend rostral to the tumour. In none of the cases in the series of Slooff et al was the tumour rostral to the uppermost part of the cavity (Fig. 17.3). Of their 19 cases, five had no cavity caudal to the tumour, and the other 14 had cavity both caudal and rostral (including three in whom the tumour was presenting within the cavity as a nodule of tissue). In multiple tumours, as for example in those reported by Wolf and Wilens (1934) as well as in those in the larger collected series, there was a tendency for tumours and cavities to alternate through the spinal axis. In one of our cases (case 2) with an astrocytoma at T2 to 8 the major cavity descended from T8 to the end of the conus medullaris. This cavitation did not extend rostral to the tumour. After a portion of normal thoracic and normal cervical cord, cavity was found in the medulla (Fig. 17.12).

Some part of the cavity was related to the tumour in most cases, as one would have expected. However, in 21 instances (13·6 per cent) in Poser's cases, this affinity was not present. It is to be noted, though, that he included intracranial and even supratentorial tumours in his review. Slooff et al (1964) had only one case out of 19 in whom the tumour was not in contact with the cavitation. The only case in our small series which showed this discontinuity was case 2 (Fig. 17.4) where a second cavity existed in the medulla. Blood vessel tumours at times have a very obvious relationship to the cyst wall. Rho (1969) described five cases of von Hippel–Lindau's disease, two with spinal cord haemangioblastomas and of these, one had two tumour nodules within the walls of the cavity. The cavity extended the entire length of the cord and was expanded at the two levels where the tumour nodules were located (lower cervical and lower thoracic cord).

Syringomyelic cavitation can be extensive (see Figs. 7.20–7.22). Spiller (1906) reported a case that extended from the conus medullaris up to the level of the internal capsule. Such involvement of the entire spinal cord is unusual in non-tumourous cavitation. In the seven cases of syringomyelia without tumours of the spinal cord or posterior fossa examined at post-mortem in the Banting Institute, the extent of the cavitation is shown in Fig. 17.13; only two of the seven extended as far as the lumbar enlargement. Nevertheless, there were three instances where the cervical cord was not involved. By contrast, in Poser's collection associated with tumour there was a considerable tendency to extensive cavitation and 24·3 per cent of his cases extended throughout the length of the cord. In six out of 38 cases in one series of autopsy proven syringomyelia (Ferry et al, 1969), the cavity extended from the medulla oblongata to the caudal end of the spinal cord. Five of these six were in cases with tumour.

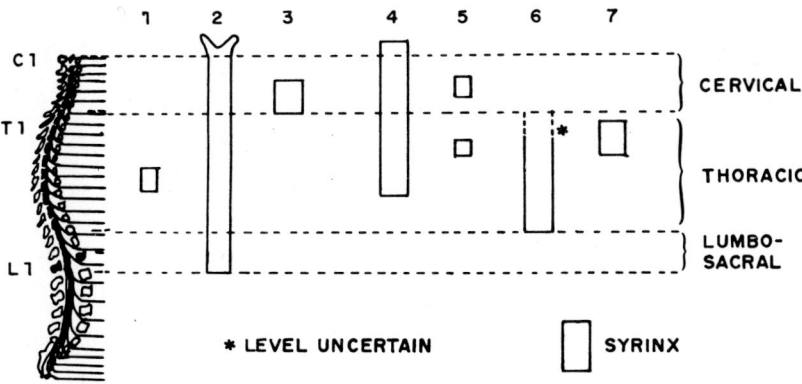

Fig. 17.13. Extent of cavitation in syringomyelia (non-tumourous) from Banting Institute files, 1931–1972.

Location in the sagittal plane. The cavities were located in the posterior half of the spinal cord favouring the dorsal horns, the posterior commissure and the ventral part of the posterior columns. In this way they followed the same pattern of involvement encountered in cord cavitation from any cause, whether it is 'communicating' syringomyelia, the type found in association with spinal arachnoiditis or that following trauma. The question of central and posterolateral predilection for any cord cavitation is reviewed in detail in Chapter 14. Among the factors discussed that may lead to the favouring of this location was the possible role of the supporting vascular stromal architecture preventing easy extension where the intramedullary arteries and veins are most prominent and allowing it to occur more readily where they are less abundant (see Fig. 14.23). Another factor of importance may be the reduced resistance to stress offered by grey matter in comparison with white matter (Ommaya, 1968). The capillary supply of grey matter exceeds that of white matter, as is the case in all areas of the C.N.S. where there is a relative concentration of neurones to cell processes (Fazio, 1938; Craigie, 1972; Bertram, 1972). The watershed in arterial supply (Turnbull, 1971) coincides with this central area predisposed to cavitation and the extending tumour and coexisting syrinx may exert maximal damage to cord tissue within the centre of such an area of marginal blood supply, with the result that the extending cavity follows the least resistant path through tissue rendered ischaemic.

The exception to this central cavitation is that which is confined within necrotic tumour. Here, an aberrant location of the cystic degeneration may be seen. This was so in our case 2, in whom a malignant astrocytoma was accompanied by considerable tissue breakdown. The necrotic tumour within the anterior horn and anterolateral segment of the cord was the site of cavitation (Fig. 17.6). This is an unusual location for cord cavitation except under this circumstance.

The Content of the Cavity

The subject of protein content in other forms of syringomyelia is reviewed elsewhere in this monograph and need not be repeated here. Except for abscess, no other form of spinal cord cavity has the high protein content found in those with spinal cord tumour. Gardner (1965) has given figures ranging from 2·5 g/100 ml to 5·0 g/100 ml. The fluid varies from light to dark yellow in colour. Even when the tumour was attached to the cauda equina, as in the cases of Gooding (1972) and the present authors, there was a protein content of 4·6 g/100 ml and 2·2 g/100 ml, respectively. The protein content undoubtedly varies with the nature of the tumour and its rate of growth. Occasionally, the fluid contains a high content of refractile cholesterol crystals.

In case 6 of our series, an aspiration was carried out of the symptomatic syrinx at T1 to T3; the protein content of the intracystic fluid was 188 mg/100 ml. At the same time, the lumbar C.S.F. fluid contained 130 mg/100 ml of protein. Three years later, when the patient's condition deteriorated and operation was carried out at T12, the cavity adjacent to the tumour contained the usual high protein content associated with cavities adjacent to tumours—1200 mg/100 ml. In the literature, the reported very high protein content in cavities associated with tumours is nearly always from fluid obtained near the tumour site and at a time when the tumour has made its presence sufficiently evident to demand a direct surgical attack. This present case would suggest that in the early stages, or in the part of the syrinx remote from the tumour, the protein may be closer to that of the C.S.F.

From a surgical viewpoint, it is important to appreciate that a cavity may have a tumour associated with it even though the fluid contains elements more usually associated with the C.S.F. and non-tumourous cavities. Aspiration of a cavity at operation will often pose the question of the origin of the fluid, whether from within a syrinx or within a cavitated tumour. It should be part of every operative aspiration of an intraspinal cavitation to have the fluid analysed at once. Decisions respecting management may be altered by this analysis. A high protein content will make a tumour a reasonable certainty. Such a feature characterised a case brought to our attention by Dr R. G. Vanderlinden, of Toronto. In this instance, mucinoid material was aspirated from an intraspinal cavitation and it is of sufficient rarity to warrant the following brief summary:

A woman born in 1933 developed a scoliosis by the age of 11 years and was afflicted with a paraparesis by the age of 23. A cystic lesion in the C8 to T1 cord segment was aspirated and mucinoid material sucked out (Dr W. Keith). Respiratory epithelium was lining the cavity and a diagnosis of teratoma was established. The cavity has extended during her period of surveillance from lowest cervical to upper lumbar levels. Four times over the decade of observation the re-forming cavity has been aspirated and relieved of its content of thick, pale-yellow mucinous material. Pain into the upper limb heralded the most recent extension and its aspiration by suction (Fig. 17.14) required a higher cervical (C5 to 6) laminectomy to remove the thick content of the cavity and to relieve pain. The nature of the extension is presently unknown. It may be extension of the cystic teratoma itself or the extrusion of mucin beyond the tumour into a syringal cavity.

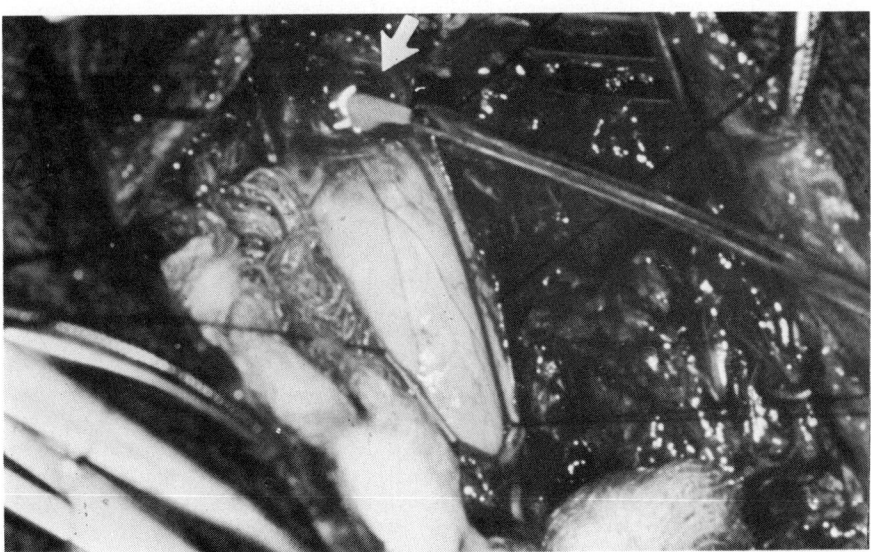

Fig. 17.14. Mucin (arrow) being sucked from cavity associated with recurrent and extensive teratoma of the spinal cord. (Loaned by Dr R. G. Vanderlinden.)

The Pathogenesis of Syrinx with Intramedullary Tumours

It is not possible to give a definitive explanation of the mechanism of production of cavities associated with intramedullary tumours. Cyst formation in conjunction with tumour occurs in cerebral, cerebellar, brain stem and spinal cord locations. It may form as a sequel to degeneration or necrosis within a neoplasm, or may occur as a major cavity with a disproportionately small mural nodule of tumour tissue, the latter situation being encountered most regularly with astrocytoma and haemangioendothelioma. However, in the spinal cord, the cavitation may not necessarily be directly contiguous to the neoplasm.

Kernohan and Parker (1932) regarded central gliosis, in association with intramedullary tumour, as a presyringomyelic lesion with the denser gliosis breaking down to cavity formation. Poser (1956) concluded that this glial excess was due to faulty differentiation based on incomplete closure of the dorsal raphe and inclusion of glial and mesodermal elements in the spinal cord, so that neoplasm and syrinx were both the result of one basic developmental abnormality. Russell (1932) considered that the spread of a tumour interfered with the blood supply to the spinal cord and produced an ischaemic area of softening which went on to cavity formation. Liber and Lisa (1937) considered that tumour produced cavitation by interference with tissue fluid egress from the cord, because of the blockage of the perivascular spaces from adventitial thickening associated with arachnoiditis overlying the tumour. Certainly some tumours present the phenomenon of an overlying arachnoiditis (Kernohan and Parker, 1932) but it is not always seen even when cavitation is present. In a recent study of the various factors including tumour associated with syrinx formation, Feigin, Ogata and Budzilovich (1971) concluded that spread of oedema initiated by these factors, including intramedullary neoplasms, played a major role in syrinx formation. Where cavitation is directly contiguous to the neoplasm, then the recent observation of Lumsden (1971) may be relevant. He demonstrated the development of microcystic space formation in tissue culture preparations of astrocytic gliomas which he thought most likely due to spontaneous autolytic liquefaction, a phenomenon that may prove to be a useful area for further investigation in this regard.

The extension of an established cavity in the spinal cord has already been discussed in Chapter 14 where the pathogenesis of the post-traumatic syrinx was considered. A cavity developing with tumour may have other unknown factors contributing to its extension but most of the comments related to the post-traumatic variety are probably germane here and need not be repeated.

Clinical Manifestations of the Syrinx

Slooff et al (1964) state that 'syringomyelia in the presence of an intramedullary tumour of the spinal cord remains silent'. This statement is in conflict with the fact that in the 38 cases carrying a clinical diagnosis of syringomyelia who were examined in the Armed Forces Institute of Pathology, tumour was an unexpectedly common finding. Furthermore, the breakdown of reported cases in Table 17.4 suggests that a diagnosis of syringomyelia will be made in a proportion of cases accompanied by tumour. Of 245 collected post-mortem cases of syringomyelia, 16·4 per cent had a tumour. In our 12 cases coming to autopsy with a clinical diagnosis of syringomyelia, a tumour was found in one of them as an unexpected finding.

The following illustrative histories from surgically proven cases in the authors' experience are included to emphasise some important clinical points. They have been selected because they have all had significant cavitation with

SYRINGOMYELIA AND TUMOURS OF THE NERVOUS SYSTEM

tumour and are representative of the diagnostic and treatment problems presented by this combination:

CASE 6, S.M., MALE

In 1957, at the age of 35 years, this patient became aware of low back pain with bilateral groin radiation. In 1965, he fell striking his back and almost at once became conscious of clumsiness of his legs with numbness after prolonged sitting. Additionally, his right hand grip became diminished and pain made worse by coughing and straining radiated into his right upper limb along its inner and ulnar border down to the fifth finger. In March 1967 examination disclosed spastic legs, the left worse than the right, and extremely brisk lower limb reflexes with Babinski signs. Vibration sense was absent in the legs. His right biceps power was diminished as were the muscles of his right hypothenar eminence. His right upper limb was areflexic, with pain and temperature sensation decreased from C3 to T10 but absent over C8 and T1 (Fig. 17.15).

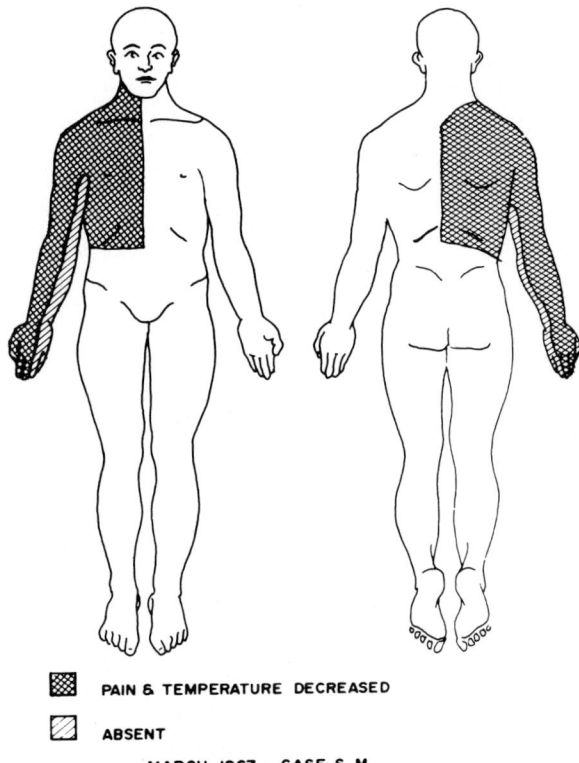

PAIN & TEMPERATURE DECREASED

ABSENT

MARCH 1967 - CASE S.M

Fig. 17.15. Case 6, S.M. Sensory loss from glioma at T12 with cavitation up to cervical level.

Myelography in March 1967 revealed a widening of the cord shadow in the T9 to 11 region. An ependymoma or glioma were suggested by the neuroradiologist and because of its location syringomyelia was thought to be unlikely. C.S.F. protein was 130 mg/100 ml. Laminectomy was carried out in March 1967 (Dr R. R. Tasker), at T1 to 3, and a cystic dilatation of the cord was seen at this level, presenting to the right of the midline of the cord posteriorly, judged to be about the lateral aspect of the dorsal column. The fluid was aspirated and was yellow with 188 mg/100 ml protein, 70 mg/100 ml of sugar and 129 mEq/l of chloride. The cyst was incised and no tumour seen when its inner lining was inspected. A Silastic T-tube was inserted into the opening of the cavity, passed up and down within the cavity and the free end of the tube was inserted up into the subarachnoid space to the mid-cervical level.

Hand strength recovered, there was temporary improvement in leg function and the patient's sensory level regressed to a band from T2 to 10 of diminished pain and temperature. Within 18 months he was experiencing further gait disturbance as well as back and leg pain. This worsened and by 1970 he was more spastic in his legs, had marked sensory impairment to light touch, pain and temperature at and below T10 bilaterally, with sparing of the L4 and 5 and S1 and 5 segments. Slight hypoaesthesia to pain and temperature was present over C6 to T3 on the right only (Fig. 17.16). His arms were strong, but lacked deep reflexes on the right.

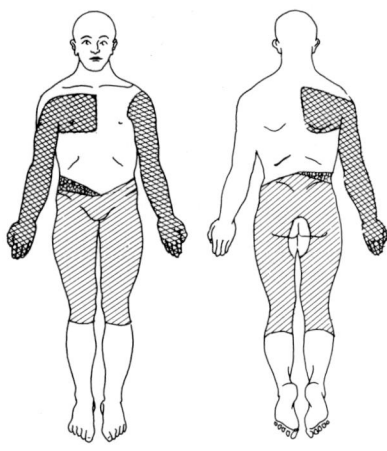

■ HYPERAESTHETIC
▨ DECREASED PAIN & TEMPERATURE
▧ ABSENT PAIN, TEMPERATURE & LIGHT TOUCH
AUGUST 1970 - CASE S.M

Fig. 17.16. Case 6, S.M. Sensory loss after cyst drainage—cervical signs have improved but low thoracic tumour signs more manifest.

Myelogram (Fig. 17.17) now indicated a cord dilatation at T9 to 10, similar to the 1967 picture but with considerable obstruction from a 'tight swelling at the T11 to 12 level'. Laminectomy at T10 to L1 in August 1970 exposed a cystic cord dilatation from which 5 ml of yellowish fluid were aspirated. The cavity had a smooth yellowish lining. The lowermost bulge of this swelling did not collapse with the aspiration and when incised a fleshy tumour the size of a grape was encountered, centred at T12. It merged with the cord but a small biopsy was taken and diagnosed as a glioma. Radiation was given. Spinal fluid at the time of myelogram contained 122 mg/100 ml of protein, 66 mg/100 ml of sugar, and the fluid from the cystic cavity contained 1200 mg/100 ml of protein.

Summary. A gliomatous tumour at T12 presented with a progressive unilateral cervical spinal cord syndrome. The presentation was obviously due to the cavitation which had extended well above the tumour. This upper limb neurological condition was remedied by cyst drainage and tube insertion and improvement persisted. Leg symptoms eventually worsened due to the progress of the T12 glioma—proven three years later by biopsy. The syrinx extended from T12 to the uppermost cervical cord. Trauma precipitated the symptoms of the syrinx. The fluid at the initial aspiration from the upper part of the spinal cord had a protein content close to that of the C.S.F. The fluid from the second procedure at the level of the tumour (T12) contained the excessive protein content generally described in cavities related to tumours.

CASE 7, F.B.L., MALE

At the age of 30 years, in 1962, this patient was struggling beneath a vehicle in which he was attempting to replace the transmission. Suddenly he developed sharp neck pain radiating to the top of

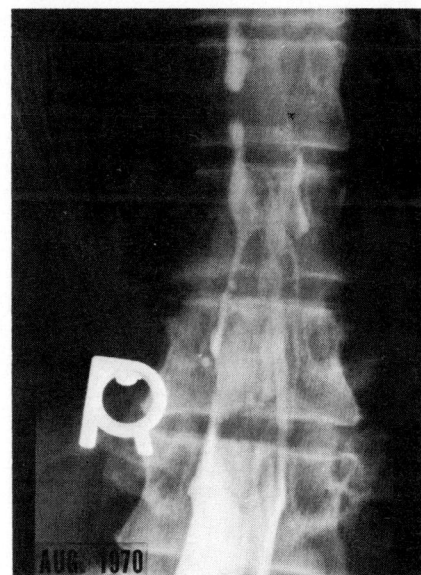

Fig. 17.17. Case 6, S.M. Myelography illustrating glioma at T12. Enlargement up to T10 only despite extensive thoracic and cervical cavitation.

his shoulder. Later that day coughing resulted in sharp discomfort in the right hand. For nine or 10 days thereafter he was unduly sensitive to cold over the right side of the neck, right upper limb and right trunk. As this subsided this entire right-sided area became insensitive to hot and cold and his razor became dull on the right side of his neck and chin margins. All this persisted but did not progress.

In mid-1970, he developed progressive leg weakness, progressive difficulty in initiating urination, defaecation, and diminished sexual potency. A tightness and numbness developed in his perineum. A low back ache was mildly troublesome for four months before admission. In November 1971, he was hospitalised and examination revealed normal muscle function in his upper limbs but deep reflexes in the right upper limb were absent. Sensory loss to pain, temperature and tickle was nearly complete from C2 to T8 and over T8 to 9 there was a band of hyperaesthesia (Fig. 17.18). His lower extremities were of normal tone but with weakness in the dorsiflexors of the ankles, right more than left, and of the right hip flexors. All superficial and deep reflexes were absent in the legs and he had superficial sensory loss over the sacral segments and slightly in the right lumbar segments. Vibration was absent at and below the knees. His anal sphincter was patulous.

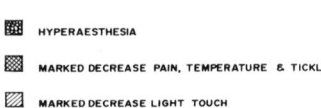

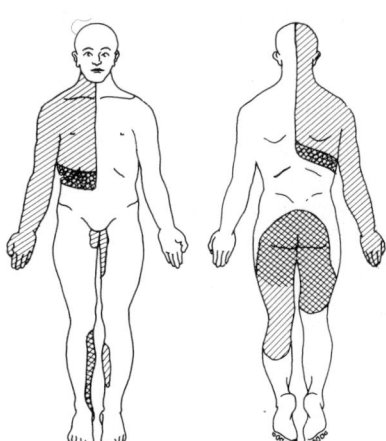

Fig. 17.18. Case 7. Sensory loss due to cauda equina tumour and cervical cavitation.

A clinical diagnosis of arteriovenous malformation of the cord was made. Lumbar myelogram was unsatisfactory with an apparent block at L1 to 2. Cisternal myelogram indicated a lower spinal cord dilatation centred on T10 (Fig. 17.19). The reasons for these findings were evident at a laminectomy carried out (Dr C. G. Drake) from T9 to L2 (Fig. 17.20). By separating the roots, it was possible to remove a tumour 2 in long and 1 in wide. Over the lower part of the cord and overlying the tumour were very prominent blood vessels. It was not absolutely certain that the lowest part of the cavity was in direct contact with the tumour. This appeared probable, although the residual firmness in the terminal part of the sac may have been reactive gliosis. The pathological diagnosis was haemangioendothelioma. No other evidence of von Hippel–Lindau's disease was present. Communication between the lowest cystic lesion and the long lesion extending from C2 to T8 was not established, but it would seem most improbable that they were separate lesions. Fluid from the lumbar subarachnoid space contained 156 mg/100 ml of protein and from the cavity in the conus contained 2150 mg/100 ml of protein.

After six months follow-up, the patient was free of pain, had recovered much of his lower limb weakness and had a slight decrease in lumbar sensory loss. His right upper limb signs remain as they were preoperatively.

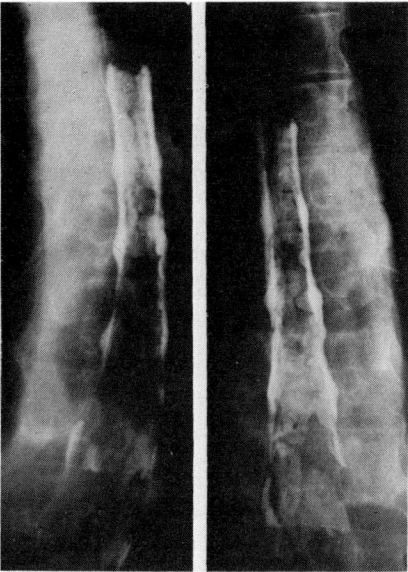 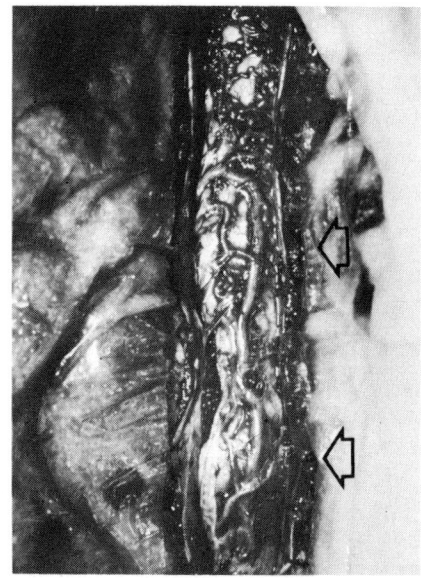

Fig. 17.19. Case 7. Myelogram showing dilation of cord shadow at T12–L1 level due to syrinx, and tumour at cauda equina level. Vascular markings prominent.

Fig. 17.20. Case 7. Operative photograph with lower arrow pointing to haemangioendothelioma at the level of cauda equina and upper arrow pointing to cystic dilation of lower cord.

Summary. A cauda equina tumour possibly involving the conus medullaris was associated with an obvious syrinx in the lowest part of the spinal cord. The presenting symptoms were of cervical syringomyelia. The upper limb signs came on abruptly eight years before the signs of the tumour developed in the legs, and were a sequel of violent straining. As it was a haemangioendothelioma, other manifestations of von Hippel–Lindau's disease may declare themselves eventually.

CASE 8, J.N., MALE

At the age of 24 years, in 1966, this patient began to experience pain across the top of both shoulders and within six months was found to have a wasted weak right hand and a sensory level at T4 on the right. Myelogram indicated a widened cord at T1. Laminectomy (Dr W. W. Blue) from C7 to T2 disclosed a cyst with fluid (yellow) of presumed high protein content distending the lower cervical, upper thoracic cord. An adjacent tumour was suspected, sought, but not found. The cord cavitation was drained and progression was halted.

Two years later, the patient developed pain in the left anterior thigh, radiating to the medial calf. His upper limb findings were indicative of a lesion in the cervical cord at C7 level and below. The right hand muscles were wasted and weak and the right wrist flexors and the triceps were weak. Both triceps reflexes were absent. Pain and temperature sensation was decreased from C4 and even more diminished as one descended the trunk on the left side. In the lower limbs there was no weakness but the left knee jerk was absent. He had loss of pain and temperature in the left lower limb up to the mid-abdominal region and decrease from here to C4. Abdominal reflexes were lost.

In August 1968, he was seen at the Montreal Neurological Institute. There was pedicular erosion and widening of the interpedicular distance from C6 to T2 and a myelogram indicated an intramedullary expansion from C4 to T12. A laminectomy (Dr Garretson) at C4 to 6 was carried out and a sub-total removal of an intramedullary astrocytoma done. The tumour extended from C6 to T1. Radiotherapy was given. A few weeks of back pain and increasing right leg weakness led to further investigation and removal of a T12 intradural meningioma in June 1971. His cervical myelogram on this occasion was improved and the dilatation was largely confined to the C7 to T1 area.

When reviewed in April 1972, his right hand was wasted and weak, with weakness of wrist flexion as well. Deep reflexes were lost in the arms. His legs were minimally spastic. Pain and temperature appreciation was lost from C4 to S5 on the left side. Von Recklinghausen's disease was suggested by the presence of many pigmented moles over his skin.

Summary. The patient had two tumours, one an astrocytoma in the cervical cord, the other an intradural meningioma at the T12 level. From C4 to T12 he had a syrinx which must have contributed to his symptoms (upper limbs) since its drainage produced a two-year subsidence of these symptoms.

Signs and symptoms were confusing due to the concomitant development of two spinal tumours, one intramedullary and one extramedullary, along with an extensive syrinx.

Case 9, S.H., male

Three years and again two years before admission in 1953, this patient, then aged 32 years, suffered two periods, each of several weeks, of acute neck and shoulder- blade pain. In the few months before admission, pain returned with radiation into the left arm associated with numbness and weakness in the arm. Eventually his legs and bladder were affected and he was admitted. A weak, wasted areflexic left upper limb, sparing the hand, with a slight weakness of the right biceps muscle was accompanied by loss of pain and temperature sensation up to C5 in the left arm, but not extending over the trunk. His legs were minimally involved with slightly increased tone and an absent ankle reflex.

Myelogram revealed a cord dilatation from C3 to T12 and, since there was a block at C3, it was considered that he might have an Arnold–Chiari malformation. Laminectomy in 1953 (Dr E. H. Botterell) disclosed an ependymoma at C3 to 7 and it was possible to enucleate it. Cystic dilatation of the cord was recognised above and below the tumour. In 1968, the patient had pain in the right leg as well as the left neck and shoulder. Radiation by cobalt over a 10-day period led to subsidence of the pain. In 1972, he was working every day with a stationary neurological disability.

Summary. A cervical cord ependymoma had the unexpected association of a syrinx dilating almost the entire spinal cord. The clinical signs and symptoms were almost exclusively those of the cervical tumour, and the syrinx was unsuspected until myelography was done. The cavitation appears to have been largely, if not entirely, *below* the tumour.

Case 10, T.M., male

In 1956, at the age of 37 years, and 13 years before tumour diagnosis was made, this patient's right hand became weak and the fingers numb. The condition evolved over a 10-year period, with atrophy and weakness in the right upper limb and a spastic weakness of the right leg. By 1966, in a New York hospital he was found to have decreased pain and temperature appreciation from C3 to S5 on the right and from T3 to 6 on the left. His legs were asymmetrically spastic with bilateral Babinski responses. Myelography determined that his cord was dilated in the cervical region. No therapy was given but in 1968 progressive weakness, including respiratory weakness, led to a course of radiotherapy. This was repeated a year later.

By 1969, his signs were more advanced with sensory level to C2 on the left and C4 on the right, the left extending to the sacral but the right only to mid-thoracic levels. Light touch was extensively lost as well by this time. Vital capacity was down to 900 ml. Myelogram was repeated and an upper cervical cord dilatation, judged to be 'compatible with syringomyelia' was demonstrated. This was particularly felt to be a 'flaccid cyst' as the size of the cord increased in the head-down position and noticeably diminished in the position of elevation.

At combined craniotomy and laminectomy, from occiput to C3 (Dr T. P. Morley), a cavity distended the visible parts of the cord and extended up to the level of the fourth ventricle. The cyst was filled with clear deep-golden fluid and the lining of the cyst was yellow and wrinkled. A biopsy of this cyst (judged by Dr Morley to be tumour lined) showed it to be an astrocytoma, giant-cell type. Drainage of the cyst was then carried out, with a Silastic tube inserted into it and then down into the cervical subarachnoid space.

Summary. A 10-year history of 'syringomyelia' in the cervical region, and two courses of radiotherapy for advancing symptoms, proved to be due to a cervical tumour with a cavity. The cord size fluctuated at myelography in the fashion most typical of a 'flaccid cyst' or 'communicating syringohydromyelia'.

CASE 11, A.F., MALE

Six years before the diagnosis of tumour was established, this patient developed numbness in both hands and arms. After five years of this he became unsteady in his gait and his right hand became weak. They were no longer sensitive to pain. His only weakness, when examined preoperatively, was in the right elbow extension. Sensation was decreased, including light touch, over the third and fourth cervical dermatomes bilaterally. Myelogram indicated a normal cervical cord in 1961 but a distended one from C3 to 5 by 1962. At laminectomy (Dr C. G. Drake), from C1 to 6, a cord enlargement from C2 to 6 was determined. A tumour was biopsied and proved to be an ependymoma. A cavity was encountered at the lower end of this tumour which contained old liquid blood which was removed by irrigation when the cord 'literally collapsed like an empty sac'. Radiotherapy was given.

Summary. An elderly man developed dissociated sensory loss in both upper limbs. A cervical cord ependymoma had a lower cervical syrinx associated with it. Despite a normal myelogram one year before, the tumour diagnosis was highly likely due to his 70 years. His 'haematomyelia' had occurred without known symptoms.

Table 17.10 summarises the data regarding histological type; tumour location; syrinx location as far as was determined by physical signs, myelography and laminectomy; and finally, an indication of which lesion, syrinx or tumour gave the initial clinical presentation.

Table 17.10. *Authors' clinical cases of syrinx with tumour*

Case	Tumour	Tumour location	Syrinx location	Primary presentation
6	Glioma	T12	Cervico-thoracic	Syrinx
7	Haemangio-endothelioma	Conus medullaris to filum	Cervical to conus medullaris	Syrinx
8	Ependymoma	Cervical	C4 to T12	Syrinx and tumour
9	Ependymoma	C3 to C7	C3 to T12	Tumour
10	Astrocytoma, giant-cell type	C1 to C3	Whole cervical	Syrinx
11	Ependymoma	C2 to C6	Lower cervical	Tumour

EXTRAMEDULLARY TUMOURS AND SYRINGOMYELIA

Introduction

This presents an intriguing situation. It is known that cavitation in the cord may be related at times to interference with the blood supply and venous drainage of the cord. This situation occurs with cavitation developing in conjunction with experimental arachnoiditis, it is probably a factor in syrinx development in the clinical condition of chronic pachymeningitis, and there is evidence of its occurrence in experimental and clinical states producing a spinal cavity related to arterial occlusion without reaction in the meninges. The question that arises then is whether or not extramedullary tumours or non-neoplastic expanding and compressing lesions behave in a comparable fashion to these other extramedullary conditions and result in 'secondary' syringomyelia. The answer is simple—they do so exceedingly rarely.

Compression by Intraspinal Extramedullary Tumour

In the Banting Institute in 40 years, 75 cases with cord compression due to extramedullary tumours were examined at post-mortem. None had cord cavitation though it must be realised that most of these were associated with malignant disease and therefore ran a short clinical course. Nevertheless, in addition, there were two cases in which an extramedullary lesion had initially caused symptoms, had been removed, recurred and then infiltrated the cord. One of these was the case of histiocytosis X with extensive syringomyelia referred to and illustrated (Figs. 17.1, 17.2) above. The other was the example of the malignant Schwannoma which recurred after removal from its extramedullary location, surrounded the cord like a collar and then invaded the parenchyma of the cord. The cavitation was lined by tumour cells and was more in the nature of a necrotic degeneration in the centre of a malignant growth (Fig. 17.6).

Ferry et al (1969) at the Armed Forces Institute of Pathology reviewed their cases of syringomyelia and although they had 38 cases of syringomyelia and 29 of them had an intramedullary tumour, none was noted to have a tumour that was located outside the cord. In Poser's (1956) review of 234 cases of syringomyelia associated with tumours involving the central nervous system, he accepted only nine instances of extramedullary spinal cord tumour apparently related to an underlying syrinx. Some of these nine might be questioned. Thus a case reported by Elliott (1884) had two neurofibromata associated with a spina bifida and an underlying cord cavitation. However, only the segment of the spinal cord at the level of the spina bifida was available for post-mortem examination and on the basis of this incomplete scrutiny of the spinal axis other abnormalities capable of causing the associated syrinx could not be excluded. Similarly, Harbitz and Lossius (1929) are quoted sporadically in the literature as reporting such a case. Close scrutiny of the case reveals that the cavity was not related to the extramedullary compression which was at the foramen magnum but was associated with a wedge of 'gliomatosis with central necrosis' within the substance of the lower cervical cord. Because of the difficulty already mentioned, of distinguishing excess glial growth from a slowly growing astrocytoma it is not possible to deny that this represents an intramedullary tumour. Muscatello (1894) reported on a case with a cervical syrinx coexisting with an extramedullary thoracic dermoid cyst and an extramedullary lumbar tumour which he designated a 'cholesteatoma', but also with a thoracic spina bifida, with diastematomyelia and reduplication of the spinal cord. What role to assign to the various potential factors in such a case is difficult to decide.

If one is careful to exclude cases that are incompletely examined or have other aetiological factors that could equally likely be causing the cavitation, it becomes evident that there are a very few cases in this category. The seven acceptable cases that have come to post-mortem analysis are contained in Table 17.11. The absence of any case since 1935 possibly reflects neurosurgical diligence and diagnostic skill but even before this only a total of six cases are on record—none in the English or American literature. The tumours were various: two of the seven were meningiomas, one a glioma, one a glioblastoma, one a lipoma, one was a myeloma and one was not classified. Five of the seven were in the extramedullary thoracic and two in the extramedullary cervical canal. All but one produced a cavitation in the cervical and thoracic cord; one extended as well into the medulla.

Table 17.11. *Extramedullary tumour and spinal cord syrinx*

Author	Type of tumour	Location of tumour	Location of syrinx
Alexandroff and Minor (1889)	Meningioma	Cervical	Cervical–Thoracic
Schaffer and Preisz (1892)	Unclassified	Thoracic	Cervical–Thoracic
Leupold (1919)	Glioma	Thoracic	Thoracic
Antoni (1920)	Meningioma	Thoracic	Cervical–Thoracic
Bychovskaja (1926)	Glioblastoma	Cervical	Bulbar–Cervical–Thoracic
Bielschowsky and Valentin (1927)	Lipoma	Thoraco-Lumbar	Cervical–Thoracic
Tauber and Langworthy (1935)	Myeloma	Thoracic	Cervical–Thoracic

The one which the syrinx was confined to the thoracic region is of particular interest (Leupold, 1919). It caused a patient to be afflicted with progressive paraplegia for four years. When he died of an incidental cause, a mass interpreted as an extramedullary glioma was compressing the cord at the T9 to 11 segment (Fig. 17.21). It was sharply demarcated from the compressed underlying lower dorsal spinal cord. Many cavities, with glial lining, were located within segment T9 to 11.

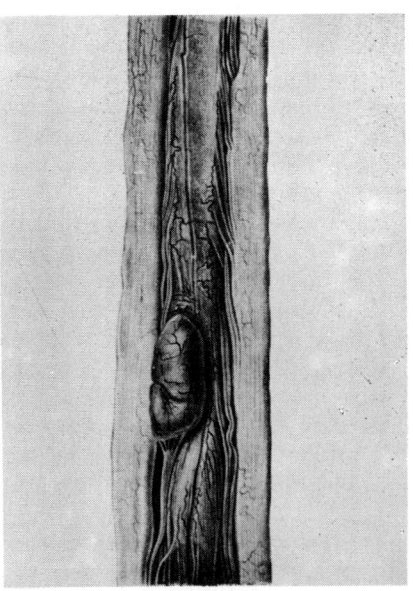

Fig. 17.21. Extramedullary glioma compressing spinal cord and producing cavitation. (Reproduced from Leupold, 1919, by permission of *Beiträge zur Pathologischen Anatomie und zur Allgemeinen Pathologie*.)

Cauda Equina and Filum Terminale Tumours

Cauda equina tumours with cavitation of the cord are just as rare as other extramedullary intraspinal growths causing a syrinx. Seven cases are known to us, five proven at autopsy (Bernstein and Horwitt, 1913; Wolf and Wilens, 1934; Voss, 1938; Poser, 1956; Slooff et al, 1964). Of those proven at post-mortem, three were ependymomas, one a teratoma and one a haemangioblastoma (Table 17.12). The remarkable aspect of the haemangioblastoma was the fact that there were two other separate tumours in the spinal axis. One involved the uppermost cervical cord with a cavity in the medulla above it and with a cavity below it down to T2. The second was in the cord at T10 and the cavity extended from here to L1. The third was attached to the cauda equina and a cavity involved the lumbo-sacral enlargement. The most dramatic case of all the cauda equina group at post-mortem was that described by Slooff et al (1964). An ependymoma of the cauda equina was accompanied by a syrinx that began in the mid-lumbar segment of the spinal cord and extended as high as T4. There was a distinct separation between the upper end of the tumour which was below the conus medullaris and the lower pole of the cavity.

Table 17.12. Cauda equina tumour and spinal cord syrinx

Author	Type of tumour	Location of syrinx
Bernstein and Horwitt, 1913—autopsy	Ependymoma	Cervical, thoracic, lumbo-sacral
Wolf and Wilens, 1934—autopsy	Haemangioblastoma	Lumbar[a]
Voss, 1938—autopsy	Teratoma	Thoracic, lumbo-sacral
Poser, 1956—autopsy	Ependymoma	Thoracic, lumbo-sacral
Sloof, Kernohan and MacCarthy, 1964—autopsy	Ependymoma	Thoracic, lumbar
Author's case 2 (F.B.L.) 1971—surgical	Haemangioendothelioma	Cervical, thoracic, lumbo-sacral
Gooding, 1972—surgical	Neurofibroma	Thoracic, lumbo-sacral

[a] 2 other tumours and 2 other related cavities higher up.

In Gooding's case (1972) the neurofibroma attached to the filum terminale was also completely separate from the conus medullaris, with no communication between the tumour and the cavity which extended from T5 to the conus. A cyst surrounding the tumour contained yellow fluid and the fluid in the spinal cavity contained 4·6 g/100 ml of protein. The possibility of a second tumour in the spinal cavity cannot be completely disregarded. The same possibility exists in the authors' case (described above, case 7) which was attached to the cauda equina. This proved to be a haemangioendothelioma. As described elsewhere, this tumour *presented* with cervical syringomyelic symptoms. It is probable but not proven that the uppermost part of the tumour was touching the lowest part of the conus.

Compression by Non-tumorous Lesions

Extreme angulation of the vertebral column from idiopathic scoliosis, old Pott's disease and Paget's disease are all capable of causing cord compression. The adaptability of the spinal cord to such extremes of deformity as may occur in these cases is the most remarkable feature of many of them. In a classic paper on scoliosis with paraplegia, McKenzie and Dewar (1949) illustrated five cases with gross deformities and accompanying cord compression. None of their cases had the clinical picture of syringomyelia. Curtis et al (1969) reported on 22 cases with a combination of kyphoscoliosis, paraplegia and neurofibromatosis from their own and the reported cases. Excluding the cases with accompanying intraspinal tumour, none had syringomyelia; nor did the cases with kyphoscoliotic paraplegia as a late sequel to Pott's disease, reported by Love and Erb (1949) or the single case described by Cole and Keim (1971). Several cases from these reports, in particular that of Love and Erb, had truly remarkable degrees of angulation, but nevertheless no recognised cavitation in the compromised cord.

On rare but proven occasions, Pott's disease is accompanied by a syrinx. This has been seen with and without severe angular deformity and with or without extensive pachymeningitis. This matter is reviewed in Chapter 15 and will not be discussed here. Suffice it to say that it has been recognised early in the post-paraplegic state and as a delayed phenomenon after a lapse of many years. Paget's disease and other conditions associated with osteoporosis can cause equally striking angulations but no reports of syringomyelia have been encountered by the authors nor discovered in the literature.

Cervical spondylosis is the commonest cause of benign chronic extramedullary cord compression. Although Mair and Druckman (1953) described and illustrated cavitation in one of their four cases of spondylotic cord compression examined at autopsy, it was no more than a small focal area of cavity formation. They regarded the cavitation as secondary to the effect of the compression on the blood vessels supplying the cervical cord. Wilkinson (1971) illustrated cavitation in the grey matter of 17 patients, examined at post-mortem, who had severe cervical spondylosis and secondary myelopathy. The size of the cavity in one illustrative case was considerably larger than in Mair and Druckman's case. No proven case is on record, however, in which a syringomyelic cavitation began at the site of bony compression from spondylosis and then extended within the cord.

Experimental Cord Compression

Experiments designed to promote spinal cord compression fail to mimic exactly the circumstances presented by a tumour. The procedure needed to insert the artificial mass must interfere to a certain extent with the blood supply to the cord and frequently will produce an associated arachnoiditis. Despite these shortcomings, they are of interest in this context.

Lhermitte and Boveri (1912) in an effort to study the effects of experimental cord compression inserted a 'tige laminaire' (a stick which swells when it absorbs fluid, thereby producing gradual cord compression) into the spinal canal of four dogs. Two dogs died within two days and softening had occurred in grey matter one to two segments above the compression. The third dog lived six weeks without paralysis and the cord was only mildly oedematous. The fourth, who survived eight days, had a cavity develop confined to the area of the cord above the compression. The cord was softened in its posterior central area from T7 (the compressed area) to T3 and the cavity was present in the necrotic tissue at T4 to 5.

In a series of experiments using inflatable balloons, Tarlov and his co-workers have produced both gradual and abrupt spinal cord compression (Tarlov, Klinger and Vitale, 1953; Tarlov and Klinger, 1954; Tarlov, 1954; Gelfan and Tarlov, 1955; Tarlov, 1972). Their lesions were associated with local cavitation at times extending a few segments above and below the compressive site. Most recently, Tator (1972) has compressed the spinal cords of monkeys by an inflatable extradural Silastic cuff. He was performing acute experiments and inflated the cuff rapidly. The early changes observed consisted of venous dilatation, venous stasis and subpial petechial haemorrhage. With mild injuries degenerative changes were seen, whereas in the more severe lesions after 12 weeks gross local cavitation was apparent.

In spite of these experiments, it is rare to encounter a syrinx underlying a chronic spinal cord compression. Furthermore, no experiments have been carried out, of which we are aware, which have the chronicity of the usual benign lesion, tumourous or non-tumourous, compressing the cord.

INTRACRANIAL TUMOURS

Infratentorial Tumours

The association of syringomyelia and syringobulbia with developmental abnormalities and also with chronic meningitis in the posterior fossa and foramen magnum region is now well known and fully described elsewhere in this monograph. These developmental and acquired lesions are both associated with 'communicating' syringomyelia or 'syringohydromyelia' and the relationship to foraminal obstruction is described elsewhere in this monograph. If this pathogenic mechanism is accepted in some instances, then it follows that some posterior fossa tumours should be associated with spinal and medullary cavitation. Such cases are, in fact, on record. Lhermitte and Boveri (1912), as noted above, reported a bony tumour arising from the basiocciput with massive cord cavitation extending from C1 to T10. Kozary et al (1969) described a case in which a meningioma arising in the peritorcular region was believed to be blocking the fourth ventricle by remote pressure and causing a syringomyelia, acceptable as such by clinical and radiographic examination. Removal of the tumour was followed by a regression of the syringomyelic signs and a distinct reduction in the width of the cord shadow to repeat myelography.

Both these cases were reported because of the suspicion of an association between the tumour and the syringomyelia. Lhermitte and Boveri (1912) regarded the tumour as interfering with the blood supply of the upper cord and thereby causing softening and cavitation. Kozary et al (1969) following the work of Gardner (1965), Appleby et al (1969) and Foster, Hudgson and Pearce (1969), were disposed to consider that the tumour was interfering with fourth ventricle outflow and thereby causing the syndrome of 'communicating' syringomyelia.

Tauber and Langworthy (1935) claimed that 'the occurrence of a neoplasm in the brain stem is known not infrequently to cause cavity formation in the cervical portion of the cord'. They said it did so by causing tonsillar herniation and 'decreasing the lumen of the anterior spinal artery'. They said that it was 'reasonable to assume a vascular basis' inasmuch as the brain stem tumour they were reporting had no connection at all with the cord cavity. The relationship to brain stem tumours is known but it is rare.

Eighteen cases of tumour in the posterior fossa described in association with spinal cord and bulbar syrinx formation in Poser's (1956) review were located in the region of the fourth ventricle six times, the cerebellum on five occasions, four times in the medulla and three times in the 'brain stem' not specified as to exact location. All would have been in a position to cause fourth ventricle outflow interference or blood vessel disturbance but the exact mechanism was not sought in the light of contemporary thinking about the syringomyelic syndrome. Future cases in this area demand much more careful and detailed clinical and post-mortem study.

It is probable that these posterior fossa tumours act in much the same way as Chiari malformations and like arachnoiditis at the base act to interfere with the cerebrospinal fluid outflow from the fourth ventricle. Posterior fossa arachnoid cyst may simulate posterior fossa tumour and it too may be associated, as one would expect, with the occasional recurrence of a syringomyelia. Such a lesion was noted in Logue's (1971) series of cases operated on for syringomyelia. The authors are familiar with another similar example.

Supratentorial Tumours

There are so few tumours in this location which are related to spinal cord cavitation that they must be regarded as occurring by chance. The pituitary tumours have already been mentioned. There was one third ventricular, one lateral ventricular and two cerebral hemisphere cases in Poser's (1956) review—an insignificant number out of 235 collected cases. The association, when it occurs, as in one of our cases from the Banting Institute files, is more likely to confuse diagnosis than it is to shed light on the aetiology of syringomyelia. Kaelber (1952) recorded this difficulty in one case with syringomyelia as an incidental occurrence

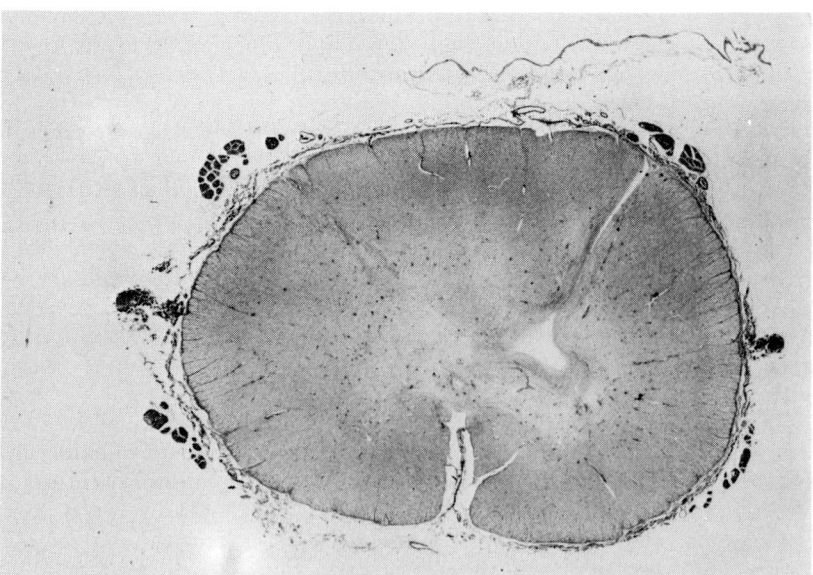

Fig. 17.22. The thoraco-cervical junction contains a thin-walled unilateral syrinx at C3–T2 that resulted in a seven-year history of numbness and decreased pain and temperature awareness in the right upper limb. The patient died with supratentorial glioblastoma multiforme.

in conjunction with a parasagittal meningioma. In our own patient there was a seven-year history of numbness with decreased pain and temperature awareness in the right upper limb. He was found to have reduced pain and temperature appreciation from C2 to 7 on the right side and no right upper limb deep reflexes. Two years before he died he developed gustatory hallucinations and eventually papilloedema was found. He succumbed to a parieto-occipital glioblastoma multiforme. The tiny cavity (Fig. 17.22) that extended from C3 to T2 was confined to the posterior horn region on the one side. It was in no way connected with the collapsed central canal nor did it extend rostrally into the upper two cervical segments. In our opinion, cases such as this are of more significance in understanding the pathogenesis of non-tumorous syringomyelia than in the pathogenesis of syrinx associated with tumour.

SUMMARY OF DATA ON SYRINGOMYELIA WITH TUMOURS OF THE NERVOUS SYSTEM

1. Intramedullary spinal cord tumours have an incidence of syringomyelia that varies between 25 to 57·6 per cent in reported series.
2. Autopsy studies on patients dying with a clinical diagnosis of syringomyelia reveal an incidence of associated spinal cord tumour that is between 8·3 and 16·4 per cent.
3. There is a high incidence of tumours of glial origin in the spinal cord associated with syringomyelia.
4. The tumours with the highest overall syrinx rate are those in the spinal cord as a manifestation of von Hippel–Lindau's or von Recklinghausen's disease.
5. The cavitations tend to be more extensive than in syringomyelia generally, tend to be associated with cervical and upper thoracic tumours, may be multiple and may be in the medulla alone or in combination with the spinal cord.
6. The syrinxes are in the same sagittal plane of the cord as in syringomyelia from other causes.
7. The lining of the syrinxes is most often glial tissue, but there may be intruding tumour nodules or partial lining by tumour cells.
8. The cavity walls tend to be thicker than in other forms of syringomyelia and have a considerably higher incidence of Rosenthal fibres.
9. Extramedullary spinal cord tumours are rarely associated with underlying cavitation, although seven examples of this combination are recorded and seven examples of cauda equina tumours have been described with a syringomyelic cord lesion.
10. A few examples are recorded of posterior fossa tumours and cysts with associated syringomyelia. They are probably analogous to the 'communicating' syringomyelia related to the Chiari anomalies and posterior fossa arachnoiditis.
11. The association of syringomyelia with supratentorial tumours appears to be a chance one and yet may confuse the occasional clinical situation in which it develops.
12. The syringomyelic syndrome may be the presenting feature of a tumour.
13. The syrinx may be an unexpected finding in a tumour case.

14. The tumour may be very indolent and the syringomyelia disabling and of long duration.
15. The syringomyelic symptoms may come on very abruptly, just as they may in non-tumourous syringomyelia (McIlroy and Richardson, 1964). This abrupt onset may be related to minor trauma or excessive straining.
16. The clinical picture in a given case may be confusing, as it may be due to the combination of symptoms and signs from a circumscribed tumour with those from a long syrinx.
17. Symptoms indicative of very lengthy cord dysfunction or of cord dysfunction at more than one level should raise the suspicion that there is a tumour and a syrinx coexisting.
18. A syrinx presenting in the thoracic or lumbar region should raise serious concern about the coexistence of a tumour.
19. The protein content of the fluid in the cyst is high, but remote from the tumour the level of protein may be less striking.
20. Determination of the protein content in the cavity should be performed at operation in all cases as it may influence the immediate and long-term management of a given case, and may disclose an unsuspected tumour.
21. Trauma may precipitate symptoms of the tumour and syrinx.
22. Haemorrhage may occur into a cavity with tumour just as it may into non-tumorous syringomyelic cavitation.

REFERENCES

Adson, A. (1938) Intraspinal tumors—surgical considerations. *International Abstracts of Surgery*, **67**, 225.
Alexandroff, L. & Minor, L. (1889) Intervention chirurgicale dans deux cas de lésion de la moëlle épinière et du cerveau chez des enfants. *Archives de Neurologie (Paris)*, **3**, 481.
Antoni, N. (1920) *Über Rückenmarkstumoren und Neurofibrome*. Munich: Bergmann.
Appleby, A., Bradley, W. G., Foster, J. B., Hankinson, J. & Hudgson, P. (1969) Syringomyelia due to chronic arachnoiditis at the foramen magnum. *Journal of the Neurological Sciences*, **8**, 451.
Atkinson, F. R. B. (1932) *Acromegaly*. London: John Bale & Co.
Baumler, A. (1887) *Über Höhlenbildungen im Rückenmark*. Leipzig: J. B. Hirschfeld.
Bernstein, E. P. & Horwitt, S. (1913) Syringomyelia with pathological findings. *Medical Record (New York)*, **84**, 698.
Bertram, E. G. (1972) Personal communication.
Bielschowsky, M. & Valentin, B. (1927) Über ein Lipom am Rückenmark mit Hydro-Syringomyelie und anderen Missbildungen. *Zeitschrift für Psychologie und Neurologie*, **34**, 225.
Brain, W. R., Greenfield, J. G. & Northfield, D. W. C. (1943) A case of atypical Lindau's disease. *Journal of Neurology & Psychiatry*, **6**, 32.
Breig, A. (1960) *Biomechanics of the Central Nervous System; Some Basic Normal and Pathologic Phenomena*. Stockholm: Almqvist & Wiksell.
Bullard, W. N. (1899) Syringomyelia and glioma of the spinal cord. *Medical & Surgical Reports of the Boston City Hospital*, **10**, 197.
Bychovskaja, G. (1926) Ein Fall von Kombination eines Rückenmarkstumors mit Syringomyelie. *Zentralblatt für die Gesamte Neurologie und Psychiatrie*, **43**, 556.
Cole, J. R. & Keim, H. A. (1971) Impending paraplegia secondary to paralytic scoliosis. A case report. *Journal of Bone and Joint Surgery*, **53A**, 591.
Craigie, E. H. (1972) In *The Spinal Cord*, ed. Austin, 2nd edn., p. 111. Springfield, Illinois: C. C. Thomas.
Curtis, B. H., Fisher, R. L., Butterfield, W. L. & Saunders, F. P. (1969) Neurofibromatosis with paraplegia. A report of 8 cases. *Journal of Bone and Joint Surgery*, **51A**, 843.

Dimitri, V. & Aranovich, J. (1940) Ependiomoma de evolución seudomielitica, siringomielia intra y extrablastomatosa. *Revista Neurológica de Buenos Aires*, **5**, 95.
Dimitroff, S. (1897) Über Syringomyelie. *Archiv für Psychiatrie*, **29**, 299.
Earle, K. M. (1972) Personal communication.
Ebbers, H. (1941) Über das gleichzeitige Vorkommen von Syringomyelie mit Recklinghausenschekrankheit und Hirntumor. *Archiv für Psychiatrie*, **113**, 605.
Elliott, G. R. (1884) Neurofibroma complicating spina bifida. *Medical Records (New York)*, **25**, 194.
Elsberg, C. A. (1941) *Surgical Diseases of the Spinal Cord, Membranes and Nerve Roots*. New York: Hoeber.
Ezrin, C., Chaikoff, R. & Hoffman, H. (1963) Panhypopituitarism caused by Hand-Schüller-Christian disease. *Canadian Medical Association Journal*, **89**, 1290.
Fazio, C. (1938) L'angioarchitettonica del midollo spinale umano e i suoi rapporti con la cito-mieloarchitettonica. *Rivista di Patologia Nervosa e Mentale*, **52**, 252.
Feigin, I., Ogata, J. & Budzilovich, G. (1971) Syringomyelia: the role of edema in its pathogenesis. *Journal of Neuropathology & Experimental Neurology*, **30**, 216.
Ferry, D. J., Hardman, J. M. & Earle, K. M. (1969) Syringomyelia and intramedullary neoplasms. *Medical Annals of the District of Columbia*, **38**, 363.
Foster, J. B., Hudgson, P. & Pearce, G. W. (1969) The association of syringomyelia and congenital cervico-medullary anomalies: pathological evidence. *Brain*, **92**, 25.
Fracassi, T., Ruiz, F. R. & Garcia, D. E. (1935) Angiomatosis medular; siringomielia y otras formaciones cavitarias coexistentes. *Revista Argentina de Neurología y Psiquiatría*, **1**, 4.
Gardner, W. J. (1965) Hydrodynamic mechanism of syringomyelia: its relationship to myelocele. *Journal of Neurology, Neurosurgery & Psychiatry*, **28**, 247.
Gelfan, S. & Tarlov, I. M. (1956) Physiology of spinal cord, nerve root and peripheral nerve compression. *American Journal of Physiology*, **185**, 217.
Gooding, M. R. (1972) Syringomyelia in association with a neurofibroma of the filum terminale (in press).
Harbitz, F. & Lossius, I. (1929) Extramedullary tumour arachnitis fibrosa cystica et ossificans. Gliosis of the medulla. *Acta Psychiatrica et Neurologica*, **4**, 51.
Harris, T. (1886) On a case of multiple spinal and cerebral tumours (sarcomata), with a contribution to the pathology of syringomyelia. *Brain*, **8**, 447.
Joffroy-Gillo, L., Retif, J., Stenuit, J. & Brihaye, J. (1967) A propos d'une association d'hydromyélie et d'acromégalie chez le même patient. *Acta Neurologica et Psychiatrica Belgica*, **67**, 548.
Jonesco-Sisesti, N. (1929) *Tumeurs Médullaires Associées à un Processus Syringomyélique*. Paris: Masson.
Kaelber, W. W. (1952) Dissimilar lesions of the nervous system: brain tumor and syringomyelia. *Journal of Neuropathology & Experimental Neurology*, **11**, 79.
Kernohan, J. W., Woltman, H. W. & Adson, A. W. (1931) Intramedullary tumors of the spinal cord. A review of 51 cases with an attempt at histologic classification. *Archives of Neurology & Psychiatry*, **25**, 679.
Kernohan, J. W. & Parker, H. L. (1932) A case of Recklinghausen's disease with observations of the associated formation of tumors. *Journal of Nervous & Mental Diseases*, **76**, 313.
Kernohan, J. W. & Sayre, G. P. (1952) Atlas of Tumor Pathology. Tumors of the central nervous system. Section X—Fascicles 35 and 37. Armed Forces Institute of Pathology, Washington, D.C.
Kinney, T. D. & Fitzgerald, P. J. (1947) Lindau–von Hippel disease with hemangioblastoma of the spinal cord and syringomyelia. *Archives of Pathology*, **43**, 439.
Kosary, I. Z., Braham, J., Shaked, I. & Tadmor, R. (1969) Cervical syringomyelia associated with occipital meningioma. *Neurology*, **19**, 1127.
Langhans, T. (1881) Über Höhlenbildung im Rückenmark infolge von Blutstauung. *Virchows Archiv für Pathologische, Anatomie und Physiologie*, **85**, 1.
Leupold, E. (1919) Ein Beitrag zur Kenntnis der Syringomyelie. *Beiträge zur Pathologischen, Anatomie und zur allgemeinen Pathologie*, **65**, 370.
Lhermitte, J. & Boveri, P. (1912) Sur un cas de cavité médullaire consécutive à une compression bulbaire chez l'homme et étude expérimentale des cavités spinales produites par la compression. *Revue Neurologique*, **20**, 385.
Liber, A. F. & Lisa, J. R. (1937) Rosenthal fibres in non-neoplastic syringomyelia: A note on the pathogenesis of syringomyelia. *Journal of Nervous & Mental Diseases*, **86**, 549.
Lichtenstein, B. W. (1949) Neurofibromatosis (von Recklinghausen's disease of the nervous system). Analysis of the total pathologic picture. *Archives of Neurology & Psychiatry*, **62**, 822.

Logue, V. (1971) Syringomyelia: a radiodiagnostic and radiotherapeutic saga. *Clinical Radiology*, **22**, 2.

Love, J. G. & Erb, H. R. (1949) Transplantation of the spinal cord for paraplegia secondary to Pott's disease of the spinal column. *Archives of Surgery*, **59**, 409.

Lumsden, C. E. (1971) The study of tissue culture of tumors of the nervous system. In *Pathology of Tumours of the Nervous System*. 3rd edn, ed. Russell, D. S. & Rubinstein, L. J. London: Edward Arnold.

MacBride, H. J. (1925) Syringomyelia in association with acromegaly. *Journal of Neurology & Psychopathology*, **6**, 114.

Mair, W. G. P. & Druckman, R. (1953) The pathology of spinal cord lesions and their relation to the clinical features in protrusion of cervical intervertebral discs (a report of four cases). *Brain*, **76**, 70.

McIlroy, W. J. & Richardson, J. C. (1965) Syringomyelia: a clinical review of 75 cases. *Canadian Medical Association Journal*, **93**, 731.

McKenzie, K. G. & Dewar, F. P. (1949) Scoliosis with paraplegia. *Journal of Bone & Joint Surgery*, **31B**, 162.

Melmon, K. L. & Rosen, S. W. (1964) Lindau's disease. Review of the literature and study of a large kindred. *American Journal of Medicine*, **36**, 595.

Muscatello, G. (1894) Über die angeborenen Spalten des Schädels und der Wirbelsäule. *Archiv für Klinische Chirurgie*, **47**, 257.

Ommaya, A. K. (1968) Mechanical properties of tissues of the nervous system. *Journal of Biomechanics*, **1**, 127.

Ollivier, C. P. (1827) *De la Moëlle Épinière et de ses Maladies*. 2nd edn., Vol. 1, p. 178. Paris: Crevot.

Oppenheim, H. (1911) *Textbook of Nervous Diseases for Physicians and Students*, 5th edn. Edinburgh: Otto Schultze.

Poser, C. M. (1956) *The Relationship Between Syringomyelia and Neoplasm*. Springfield, Illinois: C. C. Thomas.

Rho, Yong-Myun. (1969) Von Hippel–Lindau's disease. A report of five cases. *Canadian Medical Association Journal*, **101**, 135.

Rodriguez, H. A. & Berthrong, M. (1966) Multiple primary intracranial tumors in von Recklinghausen's neurofibromatosis. *Archives of Neurology*, **14**, 467.

Rosenthal, W. (1898) Über eine eigenhümliche mit Syringomyelie complicirte Geschwulst des Rückenmarks. *Beiträge für Pathologische Anatomie*, **23**, 112.

Russell, D. S. (1932) Capillary haemangioma of spinal cord associated with syringomyelia. *Journal of Pathology & Bacteriology*, **35**, 103.

Russell, D. S. & Rubinstein, L. J. (1971) *Pathology of Tumours of the Nervous System*, 3rd edn, p. 122. London: Edward Arnold.

Schaffer, K. & Preisz, H. (1892) Über Hydromyelus und Syringomyelie. *Archiv für Psychiatrie*, **22**, 1.

Schüppel, O. (1865) Über Hydromyelus. *Archiv für Heilkunde*, **6**, 289.

Schlesinger, H. (1898) *Beiträge zur Klinik der Rückenmarks—und Wirbeltumoren*. Jena: Fischer.

Schlesinger, H. (1902) *Die Syringomyelie. Eine Monographie*, 2nd Edn., p. 255. Leipzig and Vienna. F. Deuticke.

Simon, T. (1875) Beiträge zur Pathologie und pathologischen Anatomie des Zentralnervensystems. *Archiv für Psychiatrie*, **5**, 108.

Slooff, J. L., Kernohan, J. W. & MacCarthy, C. S. (1964) *Primary Intramedullary Tumors of the Spinal Cord and Filum Terminale*. Philadelphia: W. B. Saunders Company.

Spiller, W. G. (1906) Syringomyelia. *British Medical Journal*, **ii**, 1017.

Tarlov, I. M. (1954) Spinal cord compression studies. III. Time limits for recovery after gradual compression in dogs. *Archives of Neurology & Psychiatry*, **71**, 588.

Tarlov, I. M. & Klinger, H. (1954) Spinal cord compression studies. II. Time limits for recovery after acute compression in dogs. *Archives of Neurology & Psychiatry*, **71**, 271.

Tarlov, I. M., Klinger, H. & Vitale, S. (1953) Spinal cord compression studies: I. Experimental techniques to produce acute and gradual compression. *Archives of Neurology & Psychiatry*, **70**, 813.

Tarlov, I. M. (1972) Acute spinal cord compression paralysis. *Journal of Neurosurgery*, **36**, 10.

Tator, C. H. (1972) Studies of acute spinal cord injury produced by circumferential compression of primate spinal cord. *Paper presented to 7th Canadian Congress of Neurological Sciences*, Banff, June 1972.

Tauber, E. S. & Langworthy, O. R. (1935) A study of syringomyelia and the formation of cavities in the spinal cord. *Journal of Nervous & Mental Diseases*, **81**, 245.

Turnbull, I. M. (1971) Microvasculature of the human spinal cord. *Journal of Neurosurgery*, **35**, 141.
Voss, W. (1938) Über Syringomyelie und Teratombildung am Rückenmark. *Gesämte Neurologie und Psychiatrie*, **163**, 289.
Weitzner, S. (1969) Coexistent intramedullary metastasis and syringomyelia of cervical spinal cord. Report of a case. *Neurology*, **19**, 674.
Wilkinson, M. (1971) *Cervical Spondylosis—Its Early Diagnosis and Treatment*. London: Heinemann.
Wolf, A. & Wilens, S. L. (1934) Multiple hemangioblastomas of the spinal cord with syringomyelia. A case of Lindau's disease. *American Journal of Pathology*, **10**, 545.
Worster-Drought, C., Dickson, W. E. C. & McMenemey, W. H. (1937) Multiple meningeal and perineural tumours with analogous changes in glia and ependyma (neurofibroblastomatosis), with report of 2 cases. *Brain*, **60**, 85.
Worster-Drought, C. & Shafar, J. (1940) The association of acromegaly and syringomyelia. *Clinical Journal*, **69**, 281.

CHAPTER EIGHTEEN

The Epilogue

H. J. M. BARNETT

This monograph has attempted to review the clinical and pathological features of spinal cord cavitation. The descriptions have been given under the general term 'syringomyelia' since long usage of the term indicates that it will persist to embrace these related conditions. It can be regarded as a generically useful word referring to several nosologically distinct forms of extending spinal cord cavitation.

What might be specified as the 'classical variety' has been reviewed under the heading of Communicating Syringomyelia. It is commonly associated with recognisable developmental abnormalities at the foramen magnum, most often a Chiari anomaly. A subgroup of this type is found with acquired basal lesions, predominant among which is basal cisternal arachnoiditis. Secondly, there is a most distinct variety which arises as a sequel to serious spinal cord injury with the cavitation usually extending cephalad from the injury. It is possible that the mechanism of its production is shared by the occasional case which develops after a much less significant spinal injury. Thirdly, there is a variety of syringomyelia identified with spinal cord tumours, an association known since the earliest descriptions of cord cavitation. The fourth variety is the one that has been recognised most infrequently and develops in conjunction with arachnoiditis of a constrictive nature that is confined to the spinal canal.

Some of these varieties are less clearly distinct pathogenetically than are others. Overlap and a concatenation of aetiological mechanisms are undoubtedly present in the production of cavitation in some of the cases. This appears to be particularly true in the cases linked with spinal arachnoiditis and those associated with minor to moderate trauma. Some, but not all, of the latter have significant arachnoiditis. Conversely some, but not all, of the cases secondary to spinal arachnoiditis have this arachnoidal reaction as a sequel to trauma.

Although this monograph has attempted to present as complete a review of the subject of extending spinal cord cavitation as the authors' collective experience and the world's literature will allow, it is apparent that there are significant gaps in our knowledge and understanding of the condition. All cases of syringomyelia which come under clinical, radiological and, most particularly, pathological observation should be carefully scrutinised lest it be possible to obtain answers to some of these persisting perplexities. Some of the outstanding problems for further study are mentioned below.

OUTSTANDING PROBLEMS

The Importance of Vascular Factors

Extramedullary spinal cord lesions are common but rarely are associated with underlying cavitation. Spinal arachnoiditis, by comparison, is a rare condition but with a disproportionate number of recognised underlying cavitations. If vascular factors play a part in both instances, as is often claimed, it remains to be explained why there is this discrepancy. The answer may lie in a consideration of the total length of the constrictive lesion in the arachnoiditis cases compared with the tumorous ones. The experimental production of chronic long-survival spinal cord compression, as well as the development of an animal model with arachnoiditis confined to the spinal meninges, may help clarify this matter. It remains a striking phenomenon that any cavitation associated with experimental or naturally-developing spinal ischaemic lesions has not been described, as yet, extending beyond a very local site.

Syringomyelia as Evidence of Faulty Closure of the Neuraxis

Is the association with tumour evidence of heterotopic tissue overgrowth and central cavitation within such abnormal tissue? This has been frequently suggested and strongly argued by Poser (1956). The question is discussed in the tumour chapter but remains unanswered. It is noteworthy that no case with multiple tumours, as evidence of one of the phakomatoses, has yet been described where the syrinx was associated with no more than excessive gliosis. There has always been a spinal intramedullary tumour, or a tumour in a position to obstruct the foramina of exit of the fourth ventricle, or a mass in a position to interfere with the rostral flow of cerebrospinal fluid from the spinal subarachnoid space during periods of raised intraspinal pressure.

The Significance of Flaccid and Non-flaccid Cavitations

Can the radiological demonstration of a flaccid cyst be taken unequivocally to represent evidence of a communicating syrinx? Evidence is presented in the authors' series that there are exceptions and that a mobile fluid column within a cavity may be associated with a tumour. Westberg (1966) and Ellertsson and Greitz (1969), in early descriptions of this useful radiographic manoeuvre, have not given sufficient pathological data to allow acceptance as yet of the concept of the non-flaccid cavity being pathognomonic of tumour nor of the flaccid variety being invariably related to a communicating syrinx. Further observations correlating the radiological and pathological data are required.

The Source of the Cavity Fluid

This matter has been discussed already in some detail and unanswered questions persist in all varieties, more in some than in others. In the variety of syrinx developing with a *tumour*, the fluid has a high protein content, and the origin of the proteinaceous fluid is either from tumour cells, adjacent damaged tissues, or possibly, since it has the features of a plasma transudate, from abnormally permeable blood vessels associated with the tumour. The fluid within the cavities in the *post-traumatic* variety and in the form associated with *spinal arachnoiditis* is of most uncertain origin and demanding of further study. It is intriguing that the fluid in these two types resembles the fluid from the subarachnoid space and yet for the most part no macroscopic communication to this space, nor to the fourth ventricle, can be demonstrated by detailed pathological and radiological investigation. It would seem that either it has to come from the subarachnoid space through perivascular or other microscopic interstitial channels, or else that it is produced by the cells of the glial lining and is in equilibrium with the intracellular fluid. Detailed simultaneous analysis of the fluid in the subarachnoid space and in the cavity to determine any disproportion between the ionic content in the two fluids is needed. No radioisotope or post-mortem injection studies are available in these varieties and they too are needed.

The early part of this monograph presents data interpreted by Foster and Hudgson as substantiating Gardner's hypothesis (1965) that in the *classical variety* of syringomyelia the fluid content of the syrinx derives from the choroid plexuses and passes directly from the fourth ventricle to the central canal and thence to the connected syrinx. Indeed, in some instances, there is unequivocal evidence from iodo-ventriculography (Debrun et al, 1964; Heinz, Schlesinger and Potts, 1966; Tjaden et al, 1969) or from myeloendogram (Vitek, 1928; Logue, 1971) of a major communication, radiologically detectable, and confirming this connection. Radioisotope methods have also demonstrated a communication between the syrinx in the spinal cord and the fourth ventricle. The serious question that arises, however, is whether or not the identification of an occasional case by the positive contrast method and the demonstration of a connecting channel by the isotope method, proves that most of the fluid in most cases of so-called communicating syringomyelia is produced by the choroid plexus and that its circulation is adequate to retain its identity as cerebrospinal fluid. Even more important is the question of whether in fact the circulation of this fluid in the manner described is a major factor in the initial development of, as well as the perpetuation of, a syrinx. Certain observations are unexplained if this hypothesis of the origin of the intracavity fluid and its circulation in classical syringomyelia are to be accepted.

The pathological literature, prior to the resurgence of interest in surgical therapy of syringomyelia, contains incomplete data to identify the size and frequency of connection between the ventricular system and the cavities of hydromyelia and/or syringomyelia. Schlesinger (1902) reported 10 cases of syrinx-formation in the medulla but gave no description of any channels connecting to the ventricle. It might be argued that his senses were not alerted to the possibility, but it has to be acknowledged that his descriptions were exhaustively detailed. Spiller (1906) also provided a very extensive report of a syrinx extending from the termination of the spinal cord all the way up to the internal capsule, and neither described nor illustrated any communication to the central canal in the medulla, to the fourth ventricle or to the aqueduct.

Discontinuous lumbar and thoracic hydromyelic dilations in conjunction with diastematomyelia, with spina bifida cystica and not uncommonly as incidental unexpected post-mortem findings (Greenfield, 1963) were clearly described. Greenfield, who took a special interest in syringomyelia, stated that the syrinx was 'often absent in the first cervical segment'. Netsky (1953) is much quoted as a recent authority on the pathology of syringomyelia. His cases are of no value to the solution of the present problem as essential parts of the medulla were either not available or not described in any of his eight cases. On the positive side, there is the one isolated case described by Lichtenstein (1943) in which a connection between a syrinx and a hydromyelic dilatation of the central canal descending from the obex was illustrated. No pictures of the connecting channel were presented and the opening apparently was a small one.

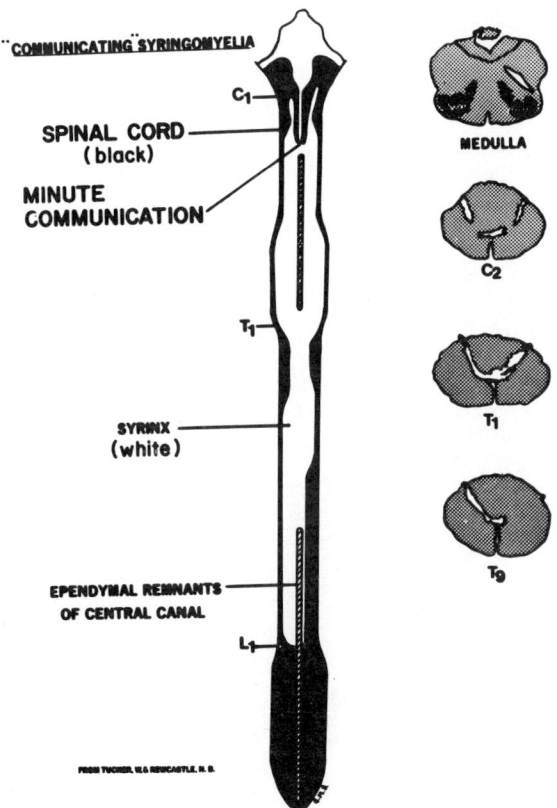

Fig. 18.1 Patent central canal connecting to the lower syrinx through a minute communication. Cavities of syringo-bulbia are connected to the syrinx but not to the fourth ventricle or directly to the central canal. (Reproduced by permission of Drs W. Tucker and N. B. Rewcastle.)

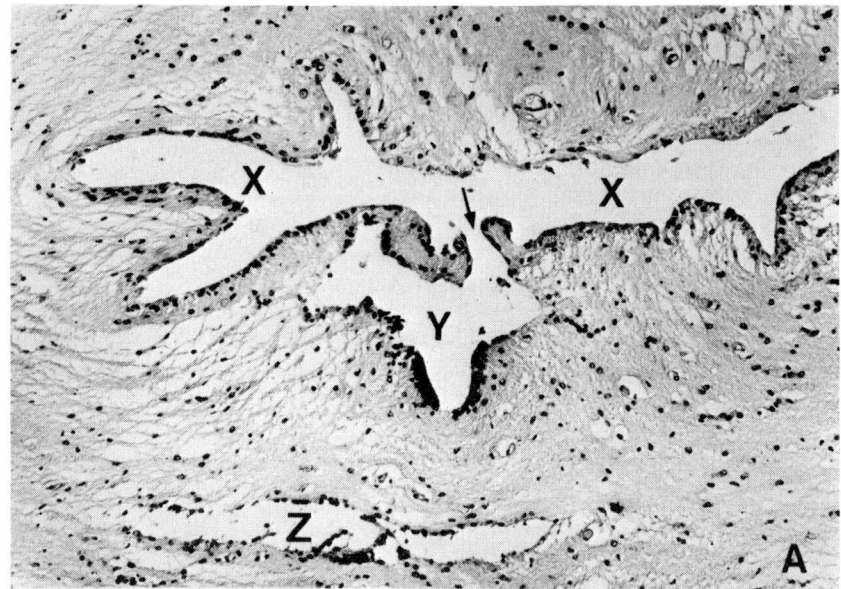

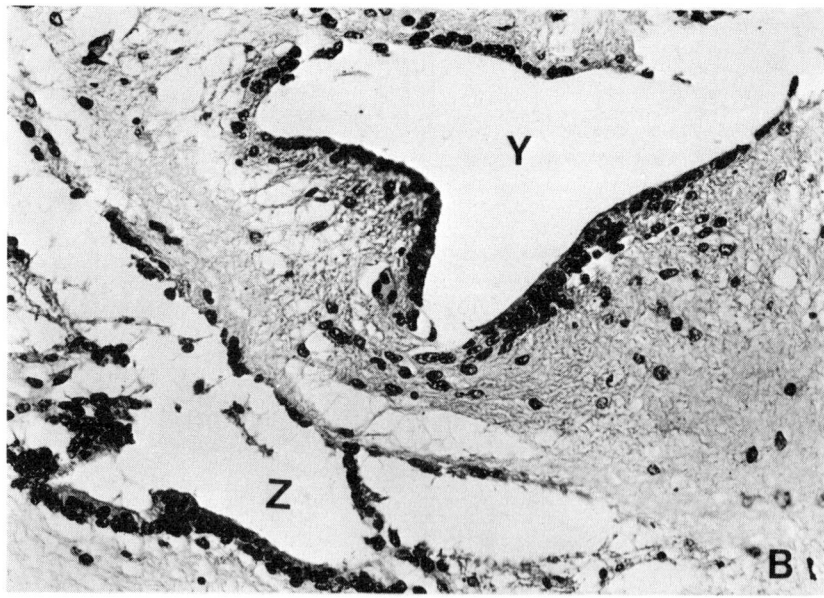

Fig. 18.2. Case of syringo–hydromyelia at the cervico–medullary junction. (a) Forked termination of the central canal (X) connects to a diverticulum through a narrow (40 μm) channel (arrow). (b) Diverticulum (Y) ependymal-lined, approximates the upper syrinx (Z) and one section (50 μm) caudad, communicates through a second minute (40 μm) channel (sections loaned by M. J. Ball. Same case as Fig. 14.36).

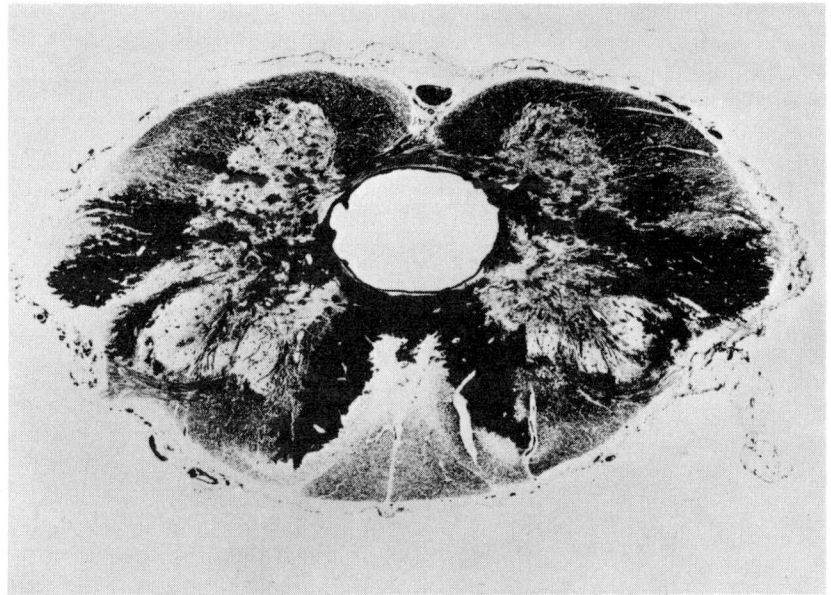

Fig. 18.3. Transverse section of the lower cervical cord after India ink injection into the lumen of a syrinx. (Reproduced, with permission, from M. J. Ball & A. D. Dayan, 1972, *Lancet*, **ii**, 799.)

Pathological material examined since the newer concepts were being discussed is of more value and interest but once more there are unanswered questions. In some of the pathological data presented by Foster and Hudgson (Chapter 7) a complete connection between the fourth ventricle, the central canal and the syrinx could not be demonstrated. They explain the absence of a communication as a late stage of an evolving process, speculate that it was present in early life and that it became closed. It is perplexing that no glial scarring whatsoever remained to mark the site of this previous communicating channel. Newton (1969) examined two of his 15 surgically-treated cases of communicating syringomyelia at post-mortem, removing the brain-stem and spinal cord in one piece. In one of these cases a bulbar syrinx extended upwards from the main cavity and lay very close to the floor of the fourth ventricle at the obex but no communication was observed with the ventricle. In the second case, no obvious communication between the fourth ventricle and the syrinx was recognised in the gross examination. Serial sectioning was said to show a bulbar syrinx connecting the fourth ventricle to the spinal syrinx. It must be assumed to have been of microscopic size, although this was not stated. A short length of central canal persisted in the medulla, *occluded* at its upper end, and *assumed* by Newton to communicate at its lower end with the spinal syrinx.

In the six cases of syringomyelia, wherein adequate material was available for complete pathological examination at the Banting Institute in Toronto (reviewed in Chapter 17, Fig. 17.13), a minimal connection between the syrinx and the rest of the cerebrospinal fluid pathway existed in two cases. The size of the persistent central canal, its connection with the syrinx and its relation to the remainder of the spinal cavitation is graphically illustrated in one of these six cases (Fig. 18.1) carefully dissected and sectioned by Tucker and Rewcastle (1973). In four others the syrinx was confined entirely below the upper cervical cord. Ball (1973) has made similar observations by serially sectioning the medulla and spinal cord in three recent autopsies on cases believed clinically to be classical, communicating syringomyelia. In one, the patent central canal that extended from the fourth ventricle to the C1-level was connected to the major syrinx by a tortuous channel which at one point was a mere 40 μm in diameter (Fig. 18.2). In the other two, no connecting channel at all was ever found. Hill and Newman (1969) reported on four patients with syringomyelia and anomalies at the foramen magnum. One came to post-mortem after an operation on the posterior fossa had revealed cerebellar tonsils extending down to C1 and bound to the underlying medulla. At the operation, a constricting band across the lower end of the fourth ventricle was identified and cut, and a dural graft and muscle pledgets were inserted into the 'large opening in the floor of the fourth ventricle, thought to be the central canal'. At post-mortem the cleft beginning in the floor of the

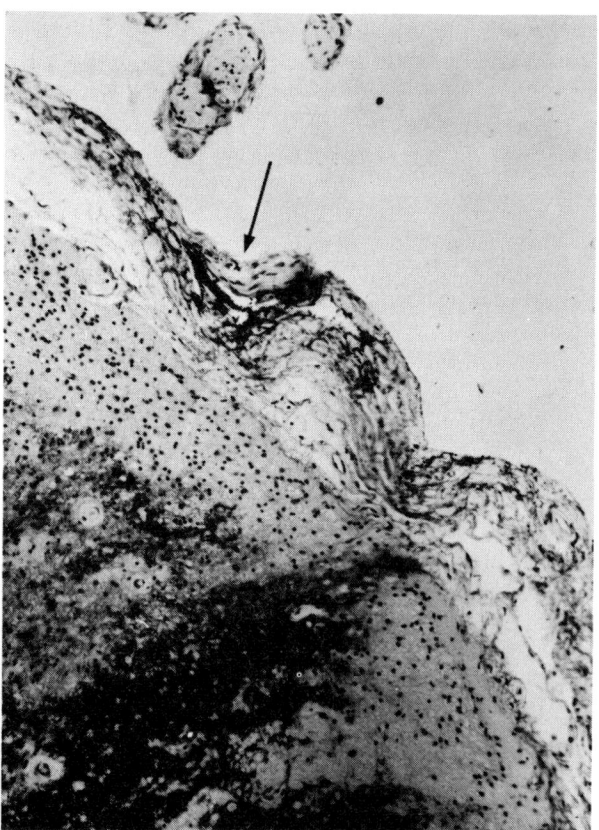

Fig. 18.4. India ink lying in arachnoid overlying cervical cord, having passed through perivascular spaces from syrinx. (Loaned and reproduced by permission of M. J. Ball.)

fourth ventricle was identified and followed 'as a pinpoint canal down through the medulla to the upper cervical cord, where it opened into a large syrinx'.

None of the strong proponents of the 'Gardner hypothesis' have presented convincing substantiating pathological data. The absence of pathological substantiation is conspicuous in the largest surgical series of Gardner (1965), Hankinson (1970), Ballantine, Ojemann and Drew (1971), and Logue (1971).

Recent observations: Ball and Dayan (1972) and Ball (1973) have made some isolated observations suggesting that the fluid in a syrinx may come from the spinal subarachnoid space and reach the syrinx through Virchow–Robin spaces. These perivascular spaces are reported to be larger than normal in cords affected by syringomyelia. India ink injection was made into a syringomyelic cavity at a post-mortem investigation. This entailed the gentle injection of 5 ml of the marking solution into the cavity after removing the eighth cervical lamina, opening the dura and identifying the syrinx-distended spinal cord. The particulate matter was encouraged to pass cephalad in the syrinx by placing the cadaver in a head-down position for two hours. None of the material ever entered the fourth ventricle, and no communication with it could be histologically identified in the serial sections. The material however, was seen tracking through the perivascular spaces (Fig. 18.3) and some passed into the overlying subarachnoid space (Fig. 18.4). Ball has suggested that the inability of cerebrospinal fluid to

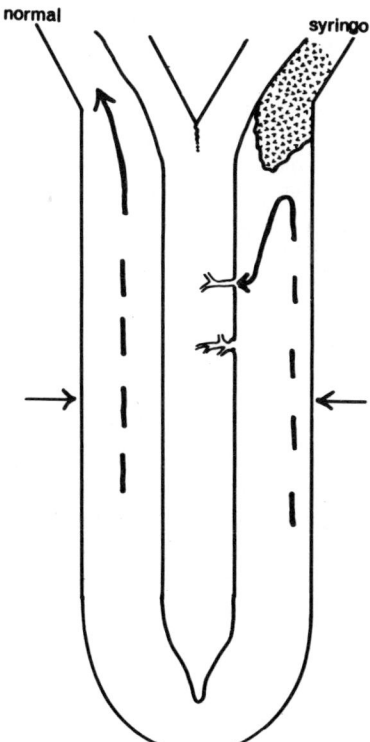

Fig. 18.5. Subarachnoid obstruction at the craniocervical junction in syringomyelia, causing redirection of the C.S.F. pressure wave into the spinal cord along perivascular channels. (Reproduced, with permission, from M. J. Ball & A. D. Dayan, 1972, *Lancet*, **ii**, 799.)

pass up into the cisterna magna, due to obstruction by the cerebellar tonsils or by chronic basal arachnoiditis may determine that the Virchow–Robin spaces will carry fluid into the substance of the cord during periods of increased intraspinal pressure (Fig. 18.5).

In summary, there may be three possible sources for the syrinx fluid in the non-tumourous cases. Firstly, it may reach the syrinx from the fourth ventricle (the Gardner variety of communicating hydromyelia and syringomyelia). Secondly, it may enter from the spinal subarachnoid space through innumerable enlarged Virchow–Robin spaces as postulated by Ball and Dayan. Thirdly, it may be formed and resorbed by the glial cells lining the cavities. Future observations are required to determine whether one, two or even all of these possible sources are utilised in the varieties of syringomyelia.

The Extension of Cavities in the Varieties of Syringomyelia

The speed and ultimate extent of cavitation are variable. It is apparent from the series presented here that any variety of syringomyelia may progress in a malignant fashion, or at the other extreme, there may be a most indolent progression. On occasions, no apparent progression at all may occur after a certain size and extent have been reached. The reasons for this variability await further investigation.

Patients with the classical as well as the other varieties may describe clearly the extension of symptoms in situations where the intraspinal pressure would be raised. A number of patients in our own and other series have had an intermittent awareness of this progression while straining, sneezing or otherwise performing a function that would increase the intra-abdominal pressure. One 'acute syringomyelia' is on record where both the acute symptoms of a syrinx and of an obstructive hydrocephalus developed abruptly as a sequel to an operation for a Dandy–Walker syndrome (Newton, 1969). Failure to achieve lasting decompression of obstructed cerebrospinal fluid pathways occasioned the hydrocephalus and it was presumed that there was a direct caudal communication through a patent central canal. Pathological proof of this situation was not available. Possibly an asymptomatic syrinx or hydromyelia existed already.

The glial-vascular bands which pass across the cavities may be the source of bleeding in some instances. Reference to this phenomenon of so-called Gowers' syringal haemorrhage is discussed in earlier chapters. It was a possible mechanism in the case described by Williams and Turner (1971) with onset of signs immediately after a fall from a tree. The fluid aspirated from the cavity was bloodstained. The patient had a Chiari malformation and an alternative explanation to syringal haemorrhage was favoured by these authors and used in their argument about the formation and extension of cavitation in the spinal cord. Noting du Boulay's (1966) observations on normal subjects, that elevated intraspinal pressure resulted in a spurt of spinal subarachnoid dye passing cephalad into the 'reservoir' of the cisterna magna, they speculated that the tonsils in the Chiari malformation were elevated by just such a jet of fluid. The resultant rise in intracranial pressure, producing re-impaction of the tonsils in the foramen magnum, was judged to result in transmission of the pressure wave against the intracranial structures and descent of intraventricular fluid within the supposedly patent central canal and down into the syrinx.

To this hypothesis, an expansion of the Gardner hypothesis reviewed in Chapter 8, as a general explanation of the hydrodynamics involved in syrinx extension, serious objections must be kept in mind. One is the fact that major ventricular dilatation is not a usual and certainly not a constant accompaniment of syringomyelia related to foramen magnum lesions. Logue (1971) has described 70 cases of Chiari malformation with syringomyelia in 31 of them. Of these, only two had hydrocephalus. Four of Newton's 15 cases had enlarged ventricles. Furthermore, Newton (1969), operating on 12 cases of Chiari malformation with syringomyelia, described displaced cerebellar tonsils in 5 of the 12 cases, 'matted together and bound to the medulla and upper spinal cord by adhesions'. This was hardly the sort of condition that would allow the tonsils to act as a ball-valve. Furthermore, in the sub-variety of communicating syringomyelia with a basal arachnoiditis as the major lesion, the tonsils are adherent and could never act as a ball-valve. Another objection, already registered to the Gardner hypothesis and applicable equally to this extension thereof, is the paucity of supporting pathological evidence.

The cessation of progression claimed in operations designed to obliterate an enlarged central canal at the obex of the fourth ventricle has been taken as evidence that the extension of the cavity is dependent on the water-hammer pulsations of the fourth ventricular fluid denied its customary exit through the foramina of Magendie and Luschka and therefore transmitted down the central canal. This may be wholly or partially true in certain instances. Logue is quoted (Ball and Dayan, 1972) however, as stating that the surgical outcome was no better for the cases in his operative series of 50 patients where the patent canal at the obex was plugged than in those where this was not done and they were merely subjected to posterior-fossa decompression and reconstruction.

Gardner's hydrodynamic theory of extending cavitation is difficult to accept where nothing more than a microscopic communication exists between the central cord and the syrinx. Ball and Roach (1972) have made calculations in one post-mortem case which they studied. The diameter of the narrow channel connecting the syrinx to the hydromyelic dilatation of the cervical cord was as small as $40 \mu m$ (Fig. 18.2). They calculated that a pulse-pressure wave of the order of only 4×10^{-5} mmHg would be operating through the tiny channel. This force they judged inadequate to produce a syrinx.*

Chiari malformations or basal arachnoiditis may be major factors in the pathogenesis of so-called communicating syringomyelia. The possibility exists that they operate mechanically at times to prevent pressure increases in the spinal subarachnoid space being effectively dissipated by fluid escaping into the basal cisterns. In some of these cases the spinal Virchow–Robin spaces may be of more consequence than was previously recognised. The good results from surgery may be achieved because the cervico-occipital operative approach restores some communication between the spinal and the cranial subarachnoid spaces to cushion the physical effects of a Valsalva-induced raised intraspinal pressure.

If an experimental model of the extending cavity with spinal arachnoiditis can be developed, observations on the pressures under varying conditions within

---

* The Newcastle authors wish to point out that the alternative route suggested by Ball and Dyan (1972), i.e. the Virchow–Robin spaces, is not materially larger than this and would refer the reader to the operative findings described in Chapter 6.

A Fifth Variety of Syringomyelia?

the cavity, as well as within the spinal subarachnoid space, will be of interest. In the meantime, further studies at post-mortem comparable to those started by Ball and Dayan are needed as they may help to clarify the mechanism of formation and extension as well as the direction of extension of intraspinal cavities.

The case for a variety of syringomyelia to be regarded as communicating has been presented. Much work needs to be done before its limits are clearly delineated. On the other hand, an accumulation of pathological data, summarised above, makes it unreasonable to accept all cases of syringomyelia that are not associated with tumour, spinal arachnoiditis or following on serious spinal injury as examples of this communicating variety.

The question that follows, therefore, is: is there not a fifth variety of syringomyelia which is unassociated with any of the aetiological factors described in this monograph, either of the communicating or the non-communicating varieties, which arises for reasons unknown and which produces cavities which do not communicate with the spinal subarachnoid space in a macroscopic way nor with the fourth ventricle in a detectable fashion? If there is such a distinct variety, its comparative natural history is unknown and, of course, even to establish its existence requires more study. Pathological data available at the present time would favour its existence. It is clear that rational therapy designed for one of the other varieties might be unsuitable for the 'fifth variety'.

PROPOSED CLASSIFICATION OF THE VARIETIES OF SYRINGOMYELIA

In the light of what is known to-date, and having in mind that much further information has still to be collected before the final word can be written about the varieties of syringomyelia, the following working classification is proposed:
1. Communicating syringomyelia (syringo-hydromyelia).
   (a) With developmental anomalies at foramen magnum and in the posterior fossa.
   (b) Associated with acquired abnormalities at the base (basal arachnoiditis, posterior fossa tumours and cysts).
2. Syringomyelia as a late sequel to trauma.
   (a) Serious spinal cord injury.
   (b) Mild to moderate spinal cord injury.
3. Syringomyelia as a sequel to arachnoiditis confined to the spinal canal.
4. Syringomyelia associated with spinal cord tumours.
   (a) Intramedullary.
   (b) Extramedullary.
5. Idiopathic syringomyelia.
   This, the fifth variety, appears to be unrelated to any of the above pathogenic factors, is usually confined to the spinal cord but may extend into the brain stem and yet does not communicate through detectable channels to the subarachnoid space or fourth ventricle.

REFERENCES

Ball, M. J. (1973) Personal communication.
Ball, M. J. & Dayan, A.D. (1972) Pathogenesis of syringomyelia. *Lancet,* **ii,** 799.
Ball, M. J. & Roach, M. (1972) Syringomyelia—is Gardner correct? Presented to 7th Canadian Congress of Neurological Sciences, Banff, June 1972.
Ballantine, H. T., Ojemann, R. G. & Drew, J. H. (1971) Syringohydromyelia. *Progress in Neurological Surgery,* **4,** 227.
du Boulay, G. H. (1966) Pulsatile movements in the CSF pathways. *British Journal of Radiology,* **39,** 255.
Debrun, G., Doyon, D., Lefevre, J. & Lepintre, J. (1964) Opacification fortuité du canal épendymaire lors d'une iodoventriculographie. *Presse Médicale,* **72,** 239.
Ellertsson, A. B. & Greitz, T. (1969) Myelocystographic and fluorescein studies to demonstrate communication between intramedullary cysts and the cerebrospinal fluid space. *Acta Neurologica Scandinavica,* **45,** 418.
Gardner, W. J. (1965) Hydrodynamic mechanism of syringomyelia: its relationship to myelocele. *Journal of Neurology, Neurosurgery & Psychiatry,* **28,** 247.
Greenfield, J. G. (1963) Syringomyelia and syringobulbia. In *Neuropathology:,* p. 306. London; Arnold.
Hankinson, J. (1970) Syringomyelia and the surgeon. In *Modern Trends in Neurology,* **5,** ed. Williams, D., p. 127. London: Butterworth.
Heinz, E. R., Schlesinger, E. B. & Potts, D. (1966) Radiologic signs of hydromyelia. *Radiology,* **86,** 311.
Hill, N. & Newman, M. (1969) Secondary syringomyelia. Presented at the Canadian Neurological Congress, June 1969.
Lichtenstein, B.W. (1943) Cervical syringomyelia and syringobulbia-like states associated with Arnold–Chiari malformations and platybasia. *Archives of Neurology & Psychiatry,* **49,** 881.
Logue, V. (1971) 14th Crookshank Lecture. Syringomyelia: a radiodiagnostic and radiotherapeutic saga. *Clinical Radiology,* **22,** 2.
Newton, E. J. (1969) Syringomyelia as a manifestation of defective fourth ventricular drainage. *Annals of the Royal College of Surgeons of England,* **44,** 194.
Netsky, M. G. (1953) Syringomyelia. A clinicopathologic study. *Archives of Neurology & Psychiatry,* **70,** 741.
Poser, C. M. (1956) The relationship between syringomyelia and neoplasm. Springfield, Illinois: Thomas.
Schlesinger, H. (1902) *Die Syringomyelie. Eine monographie.* 2 Aufl. p. 255. Leipzig and Vienna: Deuticke.
Spiller, W. G. (1906) Syringomyelia. *British Medical Journal,* **ii,** 1017.
Tjaden, R. J., Ethier, R., Vezina, J. L. & Melancon, D. (1969) Iodoventriculography in hydromyelia. *Journal of the Canadian Association of Radiology,* **20,** 265.
Tucker, W. & Rewcastle, N. B. (1973) Personal communication. *a*
Vitek, J. (1929) La fonction dorsale thérapeutique et diagnostique des cavitées syringomyéliques. *Bruxelles Medical,* **9,** 311.
Westberg, G. (1966) Gas myelography and percutaneous puncture in the diagnosis of spinal cord cysts. *Acta Radiologica,* Supplementum, 252.
Williams, B. & Turner, E. (1971) Communicating syringomyelia presenting immediately after trauma. *Acta Neurochirurgica,* **24,** 97.

Index

Abscess of cord, 251
Acrodystrophic neuropathy, 102
Acromegaly and syringomyelia, 270
Amyotrophic meningomyelitis, syphilitic, 223
Amyotrophy, 11
Aqueduct stenosis, 49
Arachnoiditis, 6, 142, 178
  experimental production of, 119, 245
  fibrosing, 224
  following cordotomy, 253
    discectomy, 252
    myelography, 252
    spinal anaesthesia, 252
    trauma, 252
  pathogenesis of cord cavitation and, 214
  post-traumatic, 252
  spinal, 220–244
  thorium dioxide and, 221
  tuberculous meningitis and, 33, 41, 97, 221
Arachnoiditis adhesiva circumscripta spinalis, 226
Arachnoiditis, basal, 30–49, 302
  radiological features of, 48
  spinal fluid protein in, 36
  treatment of, 49
  tuberculous meningitis and, 41
Arachnoid cyst, of posterior fossa and syringomyelia, 296
Arteria radicularis magna (Adamkiewicz), 209
Astrocytoma, 12, 265, 277
Atlanto-occipital fusion, 6, 50, 52, 86

Basilar coarctation, 6
Basilar impression, 6, 52, 54, 86
Brain-stem signs, in syringomyelia, 139
Brown-Séquard syndrome, 27

Carcinoma, metastatic, 265, **266**
Cavitation, flaccid, 302
Cerebellar incoordination, 28
Cerebrospinal fluid, in post-traumatic syringomyelia, 148
Cervical spondylosis and non-communicating syringomyelia, 294

Chiari malformation, 4, 49, 50, 72, 89, 119, 220, 302
Communicating syringomyelia, 1–123
Concussion of cord, 127
Cord compression, experimental, **294**
Cordotomy, arachnoiditis following, 253
Craniectomy, suboccipital, 66
Cuirasse, sensory loss, 12
Cyst, arachnoid, of cerebellum, 45, 107
Cystic myelopathy, post-traumatic, pathogenesis of, 179–219

Dandy-Walker syndrome, 99, 105, 106, 107
Dermoid, and syringomyelia, 267
Diastematomyelia, 108, 291, 305
Diplopia, 20
Discectomy, arachnoiditis following, 252
Dissociated anaesthesia, 11
Drainage, of cyst, 160
Drop attacks, 21
Dysraphism, spinal, 102, 106, 107

Ectopia, cerebellar (see Chiari malformation), 48, 50, 55, **56**, 66, 72, 89, 99
Endarteritis obliterans, 120
Ependymoma, 12, 265, **266**, 267, 277, 293
Epidermoid, and syringomyelia, 267

Fluid, in cavity, source of, **210**
Ganglioglioma, 267
Gardner's hypothesis, 304, 309
  triad, 112, 118
Glioblastoma, 267, 292
Glioma of cord, 262, **266**, 292
Gliomatosis of cord, 263
Gliosis, 3, 98
Gowers' syringal haemorrhage, 90, 310

Haemangioblastoma, 6, 265, 268, 293
Haematomyelia, 80
Haemorrhage, intramedullary, 127, **190**, 201, 207, 209

Haustra, in cavities, 144, 211, 214
Headache, 17
von Hippel-Lindau disease, 266, 269
Histiocytosis X, 262, 264, 267, **270**, 291
Horner's syndrome, 12, **29**, 141, 166, 177, 229
Hydrocephalus, **28**, 105
    arrested, 52, 55, 71, 76, 106
    ex vacuo, 197
    normal pressure, 112, 113
Hydrodynamic theory of pathogenesis, 107–120
Hydromyelia, 3, 65, 79, 104, 107
Hyperhidrosis, 27, **139**, 141, 169
Hypertensive encephalopathy, 91

Infarction, venous, 207
Intracavity fluid, **148**, 210, 241, 282
    source of, **304**
Iodo-ventriculography, 304
Ischaemia, in the pathogenesis of myelomalacia, **196**
    of the cord, experimental production of, **248**
Ischaemic theory of cord cavitation, 120

Klippel-Feil syndrome, 6
Kyphoscoliosis, 6

Laminectomy, cervical, 64
Leprosy, and syringomyelia, 120
Leukaemia, myeloid, 75
Lipoma, and non-communicating syringomyelia, 292

Main en griffe, 11
Main succulente, 12
Marsupialisation, of syrinx, 64
Medulloblastoma, and syringomyelia, 267
Melanoma, and syringomyelia, 267
Meningioma, 268, 292, 295, 297
    posterior fossa, 295
Metaplasia, of ependymal cells, 186
Metastatic tumours, **269**
Micromelia, 42
Morvan's syndrome, 12, 13, 14, 15
Multiple sclerosis, 12
Myelitis, 3
Myelocele, 5, 106, 107
Myelocystography, air, 7
Myelography, 55–61, 292
    arachnoiditis following, **252**
    experimental, 119
    gas, 57, 146
    intramedullary, 143
    after paraplegia, 142
Myeloma, 292
Myelomalacia cores, acute, 188, 207
Myeloschisis, 108
Myelotomy, 7

Neuroepithelioma, 267, 268, 291
Neurinoma, central, 267
Neurogenic arthropathy, 11, 12, **29**, 158, 160, 167, 168
Non-communicating syringomyelia, 125–301
Nystagmus, 12, **28**, 34, 71, 72

Occipital condyles, hypoplastic, 6, 50, 53
Occipitalisation of atlas, 6, 50
Oligodendroglioma, 267
Oscillopsia, 20, 71
Osteomyelitis, tuberculous, 236

Pachymeningitis cervicalis hypertrophica, 30, 221, 222, 247
Paget's disease, and cord compression, 293
Pain, in syringomyelia, 19, 132
Palate, weakness of, 28
Papilloedema, 34
Paraplegia, traumatic, syringomyelia as late sequel of, 129–153
    clinical data, 134–148
    incidence of, 130
Pascal's law, 110
Phaeochromocytoma, 286
Phakomatoses, and syringomyelia, **267**
Piloerection, 139
Platybasia, 52
Poliomyelitis, and syringomyelia, 120
Post-traumatic syringomyelia, case records, 165–171
Pott's disease, 221, 236, 238, **253–257**, 293
Proprioceptive loss, 28
Protein content, of intracyst fluid, 148, **304**
    of C.S.F., 148, 210, 241
Pseudo-syringomyelia, 261
Pudenz shunt, 230, 232

Radiculoneuropathy, sensory, 102
Radio-iodinated human serum albumin (RIHSA), 111
von Recklinghausen's disease, 268, 277, 289
Rosenthal fibres, 227, 279

Sagittal diameter of cervical canal, 6, 50
Sarcoidosis, 264
Sarcoma, 267
Schwannoma, malignant, 265
Scoliosis, and cord compression, 293
Sensory loss, trigeminal, 28
Silastic tubing, 156, 160, 169, 170
Sneezing, and onset of symptoms, 22, 133, 137, 167, 168, 213, 216
Spasmodic torticollis, 44
Spinal anaesthesia, and arachnoiditis, 252
Spina bifida, 5
Spitz-Holter valve, 67

Spondylosis, cervical, and syringomyelia, 294
Spondioblastoma, and syringomyelia, 267
Spongioblastoma, 267
Sprengel's shoulder, 6
Stereotaxic operation, 67
Stiffness of legs, 19
Streptomycin, intrathecal, 85
Surgical treatment, 64–78, 159–164
Sweating, in syringomyelia, 27, 139, 141, 169
Syphilis, 5, 41, 84, 85, 97, 120, 174
   arachnoiditis and, 221, 222
Syringobulbia, 12, 79, 81, **100**, 223, 295
   and cord tumours, 280
Syringocephalus, 100
Syringohydromyelia, 159, 295
Syringomyelia
   acromegaly and, 270
   astrocytoma and, 12, 265, 267, 277
   atypical, 5
   autopsy studies, 85–98
   basal arachnoiditis and, **30–49**
   central neurinoma and, 267
   cervical spondylosis, 294
   clinical features of, 16–29
   clinical signs of, 25–29
      appearance, 25
      brain-stem signs, 28
      classical syringomyelia, 27
      Horner's syndrome, 12, **29**, 141, 166, 177, 229
      neurogenic arthropathy, 11, 12, **29**, 158, 160, 167, 168
      nystagmus, 12, **28**, 34, 71, 72
      proprioceptive loss, 28
      spastic paraparesis, 28
   differential diagnosis of, 12
   dermoid tumour and, 267
   ependymoma and, 12, 265, **266**, 267, 277, 293
   epidermoid and, 267
   ganglioglioma and, 267
   glioblastoma and, 267, 292
   haemangioblastoma and, 6, 265, 268, 293
   histiocytosis X and, 262, 264, 267, **270**, 291
   hydrocephalus and, 28, 105
   hydrodynamic theory of, development of, 5, 84, 97, 104, **107–121**
   incidence of, 11
   intracranial tumours and, 295–297
   leprosy and, 120
   lumbosacral, 102
   origin, 114
   medulloblastoma and, 267
   melanoma and, 267
   meningioma and, 268, 292, 295, 297
   metastatic tumours and, **269**
   myeloma and, 292
   neuroepithelioma and, 267, 268, 291
   oligodendroglioma and, 267

   pathogenesis of, 104–123
   pathology of, 79–103
   phakomatoses and, 267
   poliomyelitis and, 121
   post-traumatic, 154–164
   radiology of, 50–63
      angiography, 62
      cervical spine in, 50
      myelography in, 55
         gas, 57, 61, 74
      skull in, 52
   sarcoma and, 267
   spinal arachnoiditis and, **220–244**
      pathogenesis, 245–260
      prognosis, **242**
   spongioblastoma and, 267
   surgical treatment of, 64–78, 159–164
      laminectomy, 64
      syringotomy, 64
      technique, 68
   symptoms of, 17–22
      age of onset, 17
      diplopia, 20
      drop attacks, 21
      headache, 17
      numbness of hands, 19
      oscillopsia, 20
      pain in neck, 19
      presenting symptoms, 17
      stiffness of legs, 19
      vertigo, 21
   syphilis and, 5, 41, 84, 85, 97, 120, 174
   teratoma and, 265, 267
   tumours of the nervous system and, **261–301**
   traditional concepts of, 11–15
   trauma and, **127–128**
   trauma, minor to moderate, and, **174–178**
   traumatic paraplegia and, **129–153**
   tuberculous meningitis and, 5, 41
   von Hippel-Lindau disease and, 266, 267, 269
Syringostomy, 64

Teratoma, 265, 267
Thorium dioxide, and arachnoiditis, 221
Tige laminaire, 294
Tongue, hemiatrophy of, 139
   wasted, 28
Tractotomy for pain, 68
Transection of cord, 161, 162
Trauma and syringomyelia, 127–128
   experimental, to the cord, 193, 194
   minor to moderate, **174–178**
Trigeminal sensory loss, 137, 168
Tuberculosis, as cause of arachnoiditis, 221, 250
Tuberculous meningitis, 5, 33, 41, 84, 85
Tumours, intramedullary and syringomyelia, 262–290

Tumours *(continued)*
   contents of cavity in, 282
   incidence of, 263
   location of, 272–282
   pathogenesis of, 283
   extramedullary and syringomyelia, 290–295
   intracranial and non-communicating syringomyelia, 295–297
   of the cauda equina, **293**

Tumours *(continued)*
   telangiectatic, 3

Valsalva manoeuvre, 212, 216
Ventriculo-atrial shunt, 44, 64, 66
Ventricular system, development of, **104–106**
Vertebral angiogram, 62
Vertigo, 21
Virchow-Robin spaces, in the pathogenesis of syringomyelia, **310**